Geometry of Vector Sheaves

Mathematics and Its Applications



Volume 439

# Geometry of Vector Sheaves

## An Axiomatic Approach to Differential Geometry

### Volume I: Vector Sheaves. General Theory

*by*

Anastasios Mallios

*Department of Mathematics,*
*University of Athens,*
*Athens, Greece*

KLUWER ACADEMIC PUBLISHERS
DORDRECHT / BOSTON / LONDON

A C.I.P. Catalogue record for this book is available from the Library of Congress.

ISBN 0-7923-5004-9 (Vol. I)
ISBN 0-7923-5005-7 (Vol. II)
Set ISBN 0-7923-5006-5

---

Published by Kluwer Academic Publishers,
P.O. Box 17, 3300 AA Dordrecht, The Netherlands.

Sold and distributed in North, Central and South America
by Kluwer Academic Publishers,
101 Philip Drive, Norwell, MA 02061, U.S.A.

In all other countries, sold and distributed
by Kluwer Academic Publishers,
P.O. Box 322, 3300 AH Dordrecht, The Netherlands.

*Printed on acid-free paper*

## Contents



# Part One. Vector Sheaves. General Theory

## CHAPTER I. Sheaf Theory

## CHAPTER II. Sheaves and Presheaves with Algebraic Structure



## CHAPTER III. Sheaf Cohomology

## CHAPTER IV. Linear and Multilinear Algebra of Vector Sheaves



## CHAPTER V. Vector Sheaves and Cohomology

## **General Preface** (to both Volumes I, II)

Our aim by this study is to exhibit and also exploit a quite general technique, yet simple, by its very nature, which enables one to formulate, at least, a substantial amount (if not all!) of the fundamental notions of the classical differential geometry of $\mathcal{C}^\infty$-manifolds and obtain too several standard results thereof. More interestingly, this can be achieved (we might say here, surprisingly enough!), by no use at all (!) of any concept of tangent vectors or of differential forms, in the classical sense of these words; indeed, no calculus is employed altogether! This is, generally speaking, our program, through the present two-volume work, whose secondary title to the adopted main one might be, of course, *An Introduction to* (an) *Abstract* (treatise of) *Differential Geometry*. Such a perspective, even if subconciously evident, more or less, to the experts, has never been, however, quite explicit, to the extent and generality, at least, which is undertaken by this discussion, so that the popularization of this point of view might be of some interest, much more because, as we shall see, there do exist considerable possibilities for applications.

Treatments aiming at the exposition of several aspects of differential geometry, outside the classical framework of (finite-dimensional) $\mathcal{C}^\infty$-manifolds, range, of course, from the standard theory of infinite-dimensional $\mathcal{C}^\infty$-manifolds, with a vast already literature thereon, to the more sophisticated aspects of "differential geometry" on topological spaces, *more general than* $\mathcal{C}^\infty$*-manifolds.* However, this has always been done, through a suitable formulation of the basic notions of the standard case, as for instance, vector fields, differential forms and the like.

By contrast, our treatment is quite general and, so to speak, *axiomatic.* So the basic tools here are *sheaves of modules,* with respect to an appropriate *sheaf of* **C***-algebras* (alias, **C***-algebra sheaf*), our *"domain of coefficients",* by analogy with the classical case. Yet, the sheaves involved are mostly over an arbitrary *topological space* $X$, that eventually is decreed to be *paracompact* (Hausdorff). The latter requirement for $X$, base space of the sheaves considered, is only entered into the discussion, to profit from it *cohomologically,* just as occasion serves; so this is, for instance, the case with the counterpart here of *Weil's integrality theorem,* pertaining, as we know, to the cohomology class, determined by the curvature form of a connection of a given line bundle (sheaf).

Thus, *sheaf cohomology* is the second basic constituent of our armament, in order to carry out a rather ambitious(!) program, as is namely, the transfer, to the present abstract setting, of several fundamental results, of a global and/or local nature, of differential geometry of manifolds. Indeed, it is, at least, interesting (and also encouraging!, for an undertaking, like the one at hand) to realize how a considerable amount of the classical theory can be carried over to the present formalism, especially in what concerns an important differential geometric aspect, as is the theory of connections and its consequences, as e.g. curvature, characteristic classes (à la Chern-Weil) and so on.

Our study is, as already mentioned, quite general and essentially of an *algebraic* (topological) *nature,* based, as actually does, on *sheaf theory.* So it contains, as a very special case, to the extent, of course, that is accomplished herewith, the classical aspect of finite (and, partly, of infinite) dimensional differential geometry of $\mathcal{C}^\infty$-manifolds, as well as, several other generalizations thereon, as alluded to at the beginning of this Preface. The latter cases are also discussed, as particular non-standard examples, at the pertinent places in the sequel (see thus Volume II, Part Three).

In this context, another item worth pointing out here is the crucial rôle played, in a number of places, by the corresponding to our case *de Rham complex* of the standard theory. This refers, in particular, to the exactness of that complex (here deemed to be in force), in the classical case being, of course, a consequence of the famous *Poincaré Lemma* (viz. that every closed form is exact). Yet, in the aforementioned (non-standard) examples exactness of the previous complex is still valid, as a consequence again of the same lemma, which thus is proved to hold true in those extended cases, as well. On the other hand, in our abstract setting the situation can be remedied by demanding, for instance, the section algebras of the $\mathbf{C}$-algebra sheaf $\mathcal{A}$, our *"domain of coefficients",* to carry further a topological algebraic structure, being, thus, in effect, appropriate *topological algebras.* So our *"sheaf of coefficients"* becomes now, precisely speaking, a *topological* $\mathbf{C}$*-algebra sheaf* on $X$.

The insert at this place of *topological algebra theory* is both crucial and decisive. Of course, this can be construed as revealing, within the present general framework, the oblique, nontheless, significant rôle, played in the classical case by the important (certainly, *non-normed!*) topological algebra

$\mathcal{C}^{\infty}(X)$ of (say, $\mathbf{C}$-valued) $\mathcal{C}^{\infty}$-functions on a given $\mathcal{C}^{\infty}$-manifold $X$. (We refer the reader to A. Mallios [2] for a general and detailed account on topological algebra theory, as well as, for the particular algebra at issue). Thus, by employing a suitable *topological algebra sheaf* $\mathcal{A}$, one is able to get an (exact) *resolution of* (the *constant sheaf*) $\mathbf{C}$, of the form

$$(*) \qquad 0 \longrightarrow \mathbf{C} \xrightarrow{\varepsilon} \mathcal{A} \xrightarrow{\partial} \Omega^1 \xrightarrow{d^1} \Omega^2 \xrightarrow{d^2} \cdots,$$

where the $\Omega^i$'s stand for appropriately defined $\mathcal{A}$*-modules* (modules of *"differential forms"*) and the $d^i$'s, with $d^0 \equiv \partial$, for suitable also ($\mathbf{C}$-linear) *morphisms* of the $\mathcal{A}$-modules concerned. Furthermore, a similar aspect (in fact, more feasible than $(*)$), which is still worth remarking here, is the exploitation of a *(short exact) exponential sheaf sequence* of the form

$$(**) \qquad 0 \longrightarrow \mathbf{Z} \longrightarrow \mathcal{A} \xrightarrow{e} \mathcal{A}^{\cdot} \longrightarrow 1,$$

where $\mathcal{A}^{\cdot}$ denotes the *sheaf of units* of $\mathcal{A}$, while $e$ represents the standard *exponential* (sheaf) *morphism* of the classical theory. Here again an exact sequence, like $(**)$, is further realizable in some particular cases of *topological algebra sheaves,* as before. Indeed, it was the relevant work of S.A. Selesnick [1; 2], in the context of *Banach(-Šilov) algebra theory,* aiming also at applications to harmonic analysis, which, by its intense geometrical flavour, kindled actually our first thoughts about this subject matter. (See thus the first announcement thereon in A. Mallios [3: (i) – (iv)]). Moreover, the same exponential sequence $(**)$ substitutes in many instances for the more stringent exact sequence $(*)$, more so, when the $\mathcal{A}$-modules involved are, in particular, the analogues, in our sheaf-theoretic approach, of the usual line bundles.

We mentioned already that the prominent features throughout this discussion are certain sheaves of $\mathcal{A}$-modules, in fact, *vector sheaves;* that is, locally free sheaves of $\mathcal{A}$-modules of finite rank, over an arbitrary, in general, topological space $X$, while $\mathcal{A}$ is our "sheaf of coefficients", as above. This type of $\mathcal{A}$-modules substitute here for the familiar sheaves (of germs) of sections (continuous or otherwise) of *vector bundles* (of finite rank) in the classical theory. If, in particular, the vector sheaf, under consideration, has rank 1, we then speak of a *line sheaf* on $X$. (See also the Preface to Volume I, concerning this particular terminology).

Within the same context, we further remark that the exploitation of *"structures more general than manifolds"*, pertaining to concrete problems, as for example, calculation of the cohomology of *quotient spaces,* or even of classifying spaces (à la Milnor) and of semi-simplicial manifolds (see e.g. M.A. Mostow [1] and J.W. Smith [1]), indicates a further potential concrete application of the point of view of this treatise. Thus, all the previous structures are indeed a very particular instance of those considered herein, so that the present technique exhibits a possible further extension, within a similar setting (in fact, much more general), not only of the topology, as is actually the case with the above structures, but also of the *geometry* (e.g. connections and curvature) *of* such *"generalized spaces".*

Finally (but certainly, not least!), another fact that supports the framework and terminology adopted by this study, is the ever growing usage of a *sheaf-theoretic language,* compiled with that one of the standard differential geometry, in what actually concerns the *"geometrization of physical thought",* as expressed, for instance, by the nowdays *"gauge theories".* (See, for example, Yu.I. Manin [1] and/or S.A. Selesnick [3], or even R.S. Ward – R.O. Wells, Jr. [1]). All the preceding, along with an extremely positive reaction towards this language of several audiences, both at the University of Athens and elsewhere, has been a strong further inducement to the undertaking of the present more systematic account of the subject.

To make the exposition sufficiently self-contained, we prefix a pertinent introduction to *sheaf theory,* a theme in itself already classical, including the necessary for the present purpose material from *sheaf cohomology.* Yet, we adequately expose the indispensable elements for this study of what we may call *linear and multilinear algebra of vector sheaves,* or even of *sheaf modules,* in general, a subject for which a systematic account is still lacking. *Topological algebra sheaves* are also discussed, to the extent that we need it here, namely, in Part Two of this work. On the other hand, the individual Prefaces to each one of the two Volumes of this treatise contain more details and comments on the material of the corresponding Volume.

**Preface** (to Volume I)

In this first Volume of our two-volume work we give, as the title suggests, the necessary preliminary material for the study of the *Geometry of Vector Sheaves,* which, strictly speaking, is further undertaken in the second Volume.

Thus, to make the whole treatment as self-contained as possible, we consider here all the necessary elements from *Sheaf Theory,* in general (Chapter I), specializing in Chapter II to the pertinent notions of *Sheaves* (and *Presheaves*) with *Algebraic Structures,* that is, of the types of sheaves we exclusively consider throughout the sequel. The term *vector sheaf* (respectively, *line sheaf*) I actually borrowed from Serge Lang [2], who applied it, within another context; thus, I preferred that term, instead, namely, of the more standard one of a *locally free $\mathcal{A}$-module of finite rank* ($n \in \mathbf{N}$, with $n > 1$, or of *rank 1,* respectively). Yet, I considered Lang's terminology more geometrical, than the classical term for these sheaves, due also to the applications that this type of sheaves has throughout our whole study. (The adopted terminology, however, does not seem to be entirely devoid of objections!). The *Linear and Multilinear Algebra of Vector Sheaves* is examined in Chapter IV, having essentially to do with the action and corresponding interrelations of the ("linear algebra") *functors* $\mathrm{Hom}_{\mathcal{A}}$ (and $\mathcal{H}om_{\mathcal{A}}$), along with $\bigotimes_{\mathcal{A}}$, as well as, with the *exterior (power) functor* $\wedge_{\mathcal{A}} \equiv \wedge$, the resulting conclusions being still of concrete use in the sequel.

On the other hand, a basic tool for the whole treatise is *Sheaf Cohomology,* as a *derived functor cohomology* over an arbitrary topological space $X$, with respect to a given *$\mathcal{A}$-module* (i.e., sheaf of $\mathcal{A}$-modules), or to a *sheaf of* (not necessarily abelian) *groups.* Thus, following the same spirit, as before, of keeping, namely, the whole treatment self-contained, as far as possible, we present the pertinent material from *Sheaf Cohomology* in Chapter III of this Volume. The topological spaces considered, base spaces of the sheaves involved, are eventually *paracompact* (Hausdorff), so that *Čech cohomology* is then applied; we further present this important particular case too, to the extend that is necessary for the ensuing discussion. Furthermore, having in mind an application given in Volume II (viz. a cohomological classification of *line sheaves with $\mathcal{A}$-connections*), we also examine in the same Chapter the necessary material from *hypercohomology* and, in particular, the *Čech* counterpart of it. The motivation to this item was a recent treatment by

J.-L. Brylinski [1], pertaining to the classical *Kostant-Weil Theorem,* which also we present in Volume II, in full generality. (We use, instead, the term *Weil's Integrality Theorem,* that is also of common usage).

The indispensable for our discussion *cohomological classification of vector sheaves,* in general, is given in the ensuing final Chapter V of this Volume. The corresponding particular case of *line sheaves (Picard group)* is examined here too, along with the notion of the *determinant line sheaf,* that is canonically associated with a given vector sheaf. Furthermore, the functors $\mathcal{E}xt_{\mathcal{A}}$ (and $\mathcal{H}om_{\mathcal{A}}$), already introduced in Chapter III, are also examined here, in conjunction with their use in Volume II for the study of the important geometrical notion of an *$\mathcal{A}$-connection* and its *existence.*

Yet, it is worth mentioning at this point the consideration (in Chapter IV too) of the corresponding notion to our case of *vector sheaves with inner-products.* (The situation thus serves as another example of the motto that *"any good function can be conceived as section of an appropriate sheaf"*). So *Riemannian* and *hermitian vector sheaves*, together with the pertinent concepts of (special) *linear group sheaves* (corresponding, of course, to the analogous ones of the *classical groups*), are also treated in Chapter IV. This, as a further application, within the present abstract framework, of the respective classical notions in finite-dimensional geometry (*linear or not*).

ANASTASIOS MALLIOS

Athens
June, 1997

## Acknowledgements

It is a special pleasure to record here my indebtedness to all those, who, in one way or another, contributed to giving me the indispensable impetus and countenance to bring this project to an end. First, to *S. A. Selesnick* for his continuing interest in my work, in particular, in what concerns its relation with topological algebra theory; thus, his own earlier work in harmonic analysis on compact groups, written, within the context of Banach algebra theory, with a strong geometrical flavour, was, in effect, my initial inspiration for the abstract framework that has finally been adopted by the present study. Second, to *E. E. Rosinger,* who, during my visits to the University of Pretoria (South Africa), gave me great encouragement to finish this project, by his extremely enthusiastic reaction to early versions of this material. The final boost, and while the manuscript was almost ready, came from *S. Negrepontis,* who, as Head of our Department in Athens, at that time, offered me kindly the first financial support for converting the manuscript into LaTeX form. Further financial help to finishing the project has been supplied by the University of Athens, thanks, especially, to the kind willingness of *Y. Dracopoulos,* Vice-Rector of the University of Athens. My special thanks are due here to *Dr. M. Gikas* for his excellent rendering of a quite difficult text in LaTeX form, as well as, for an extremely friendly and helpful cooperation. My singular thanks are also due to *my colleagues* in the *Math. Department* of the *University of Athens*, who allowed me an extended leave of absence, very helpful, indeed, to accomplish this undertaking.

Yet, a quite decisive influence to attempt the present venture, in spite of appearing circumspections, as to its applicability, due mainly to the entirely abstract (sheaf-theoretic) language employed, was the very positive final facing of the subject from *various colleagues in Greece and abroad.* For all those I want to keep my very best and thankful memories. For a similar interest I wished, especially, to mention and thank from this place too, *Prof. Ky Fan,* who, as member of the Editorial Board of JMAA, was very kind, as well as, enthusiastic in handling my earlier papers on the subject, connected, in particular, with topological algebra theory, while his response, at that time, was among the first strong impulses for me to continue working on the topic. For their kind help, for the LaTeX writing, during the final corrections of the camera-ready copies of this two-volume work, I want also to thank my (cousins and) colleagues, at the Math. Dept. in Athens, *M. Papatriantafillou* and *E. Vassiliou.* (For any remaining errors, typing or otherwise, it is me, of course, who is to be blamed, not including, certainly, LaTeX itself!).

Last, but not least, I wish to express my sincere appreciation to *Prof. M. Hazewinkel* for accepting this book to his Series, as well as, to *Dr. P. Roos* for his extremely kind interest, enthusiasm and editorial endeavour in bringing this work to the light.

# Contents of Volume II

# PART ONE

## VECTOR SHEAVES. GENERAL THEORY

# CHAPTER I

# Sheaf Theory

"... the notion of a sheaf has proved to be an effective conceptual tool in describing a considerable variety of sorts of spaces; the elementary notion of space thus acquires (and requires) a sophisticated conceptual development."

S. Mac Lane in "Mathematics Form and Function" (Springer-Verlag, New York, 1986) p. 255.

In this Chapter we collect together those elements from *Sheaf Theory* which are directly relevant to the sequel. Of course, there is hardly something that cannot be found in the standard texts on the subject. However, our aim here is to select only the necessary material for the sequel, so that by omitting everything else, we seek to give a better picture of the rest, while the whole becomes thus more self-contained. In fact, this is a pervasive aspect that we look after throughout our exposition. Standard definitions and basic results come thus to the foreground, where for the results, we are dealing with, we mostly supply complete proofs.

# 1. Preliminaries. Definitions

We start with commenting upon the next notion, a fundamental one for all that follows. Thus, we have.

**Definition 1.1.** By a **sheaf** (of sets) one means a triple

(1.1) $$(\mathcal{S}, \pi, X),$$

where $\mathcal{S}, X$ are topological spaces and

(1.2) $$\pi : \mathcal{S} \longrightarrow X$$

a (surjective) *local homeomorphism.*

Now, by the last term in the previous definition one means that $\pi$ in (1.2) has, *by definition*, the property that:

(1.3) *for every element $z \in \mathcal{S}$, there exists an open neighborhood $V$ of $z$ in $\mathcal{S}$ such that $\pi(V)$ is an open neighborhood of $\pi(z)$ in $X$ and the restriction of $\pi$ to $V$, viz. the map*

(1.3.1) $$\pi|_V : V \longrightarrow (\pi|_V)(V) = \pi(V),$$

*a homeomorphism.*

As we shal presently see, by considering *open neighborhoods in (1.3),* instead of arbitrary ones, *is not, in effect, any restriction* whatsoever (see (1.3′) below). Yet, the surjectivity of the map $\pi$ in (1.2) is not necessary; however, this will usually be our assumption in the sequel, unless otherwise specified. (So cf., for instance, the case of a *subsheaf* below). On the other hand, instead of (1.1), we shall usually refer just to $\mathcal{S}$, by simply saying that $\mathcal{S}$ *is a sheaf over* $X$. Moreover, we call $\mathcal{S}$ the *sheaf space,* although now one actually looks at the "sheaf" as a map, viz. as a triple like (1.1); in that case too, $X$ is called the *base space* of the sheaf concerned, while the map $\pi$ (in the case it is, in particular, onto) the *projection* of (the sheaf space) $\mathcal{S}$ on $X$.

**Scholium 1.1.**– Before we proceed further, let us settle our previous claim, referring to the *"open neighborhoods in (1.3)"*. So one proves that;

(1.3′) *condition (1.3) is equivalent with that in which, instead of open neighborhoods, one considers arbitrary ones, so that (1.3.1) be a homeomorphism.*

Of course, *(1.3) implies (1.3′);* to prove the converse we also apply the following.

**Remark.**– *Suppose that* $\pi : \mathcal{S} \longrightarrow X$ *is a local homeomorphism in the* (seemingly(!) wider) *sense of (1.3′). Then, the map* $\pi$ *is continuous.*

Indeed, let $z \in \mathcal{S}$ and $U \subseteq X$, open with $\pi(z) \in U$. Now, there exists $V' \subseteq \mathcal{S}$, open such that $z \in V' \subseteq V$, where $V$ is a neighborhood of $z$ realizing $\pi$ as a local homeomorphism (in the sense of (1.3′)). So $\pi(V') \cap U$ is an open subset of $\pi(V')$ (in the relative topology from $X$), such that *the set* $\pi^{-1}(\pi(V') \cap U) \subseteq V'$ *is* an open subset of $V'$ (according to the assumption for $\pi$), hence, by hypothesis for $V'$, an *open* set *in* $\mathcal{S}$, as well. Thus, $z \in \pi^{-1}(\pi(V') \cap U)$ such that

$$\pi\left(\pi^{-1}(\pi(V') \cap U)\right) = \pi(V') \cap U \subseteq U,$$

which proves our claim in the above Remark.

Now, as a result of the preceding, one concludes that;

*the map* $\pi$ (as in the previous Remark) *is a local homeomorphism in the "strict" sense* (i.e., one applies *open neighborhoods in (1.3.1)*).

For, suppose that $z \in \mathcal{S}$ and $V$ is an (arbitrary!) neighborhood of $z$ mapped by $\pi$ homeomorphically onto a neighborhood $\pi(V)$ (not necessarily open!) of $\pi(z) = x$ in $X$. Thus, there exists open $U \subseteq X$, with $x = \pi(z) \in U \subseteq \pi(V)$, so that, by the preceding, $\pi^{-1}(U)$ *is an open set in* $\mathcal{S}$, with $z \in \pi^{-1}(U) \subseteq V$, while the map

$$\pi|_{\pi^{-1}(U)} : \pi^{-1}(U) \longrightarrow U$$

*is* still *a homeomorphism.*

So, as a consequence of the above Scholium 1.1, we are free in the sequel to apply either one of (1.3) or (1.3′), by referring to a given local homeomorphism (*sheaf!*); however, *for convenience, we do employ (1.3).*

Thus, an easy implication of (1.3) is the fact that;

(1.4) *a given local homeomorphism, as in (1.3), is a continuous and open map, as well.*

(We note that the proof of (1.4) is here more direct than the corresponding claim in the Remark of the previous Scholium 1.1; on the other hand, it will be an exercise to *prove directly the openness of $\pi$ under the "wider" assumption (1.3′)*).

Now, as a result of (1.4), one gets the following useful information, pertaining to the topology of (the sheaf space) $\mathcal{S}$, as in (1.1). That is, one has.

**Lemma 1.1.** *Let $\mathcal{S}$ be a sheaf over $X$. Then, the open subsets of $\mathcal{S}$ to which the restriction of $\pi$ is a homeomorphism, constitute a basis of the topology of $\mathcal{S}$.*

**Proof.** If $W \subseteq \mathcal{S}$ is open and $z \in W$, let $V$ be an open neighborhood of $z$ in $\mathcal{S}$, for which the map (1.3.1) is a homeomorphism. So $W \cap V$ is an open set in $\mathcal{S}$, as well, with $z \in W \cap V \subseteq W$, while the map

$$\pi|_{W\cap V} : W \cap V \longrightarrow \pi(W \cap V)$$

is still a homeomorphism, which thus proves the assertion. ∎

On the other hand, given an element $x \in X$ (we assume, of course, that $x \in \operatorname{im}\pi$, as in (1.2)), the set

$$\mathcal{S}_x := \pi^{-1}(\{x\}) \equiv \pi^{-1}(x) \tag{1.5}$$

is called the *fiber of* (the sheaf) $\mathcal{S}$ *over* (the point) $x \in X$, or even the *stalk of $\mathcal{S}$ at $x$*. Of course, the latter notion concerns, in fact, the *fiber of the map $\pi$ at $x$*, so that one obtains a *partition of $\mathcal{S}$* in its fibers; i.e., one has the following *direct sum* (viz. disjoint union) *of sets*

$$\mathcal{S} = \sum_{x\in X} \pi^{-1}(x) \equiv \sum_{x\in X} \mathcal{S}_x. \tag{1.6}$$

Moreover, according to our hypothesis for $\pi$ (cf. (1.3)), one obtains that;

(1.7) *for every $x \in X$, the corresponding stalk $\mathcal{S}_x$ of the given sheaf $\mathcal{S}$ over $X$ is a discrete space, when endowed with the relative topology from $\mathcal{S}$.*

(*Hint:* Given $z \in \mathcal{S}_x$, one applies an open $V$, as in (1.3.1), so that one gets $V \cap \mathcal{S}_x = \{z\}$). Moreover, by the continuity of $\pi$ (cf. (1.4)), one concludes that $\mathcal{S}_x$ *is a closed subspace of* $\mathcal{S}$, in the case, at least, that $X$ is a $T_1$-*space*!).

Now, by a *subsheaf* of a given sheaf $\mathcal{S}$ over $X$, one means, by definition, any open subset of $\mathcal{S}$; thus, if $A \subseteq \mathcal{S}$ is open, then the triple

$$(A, \pi|_A, X) \tag{1.8}$$

is a sheaf over $X$, in the sense of Definition 1.1. Of course, in this case the "projection" $\pi|_A : A \longrightarrow X$ may not be an "onto" map. However, *it is* a local homeomorphism. Yet, by considering here open subsets of $\mathcal{S}$ only, is not a restriction: thus (cf. (1.3));

(1.8′) a subspace of a given sheaf $\mathcal{S} \equiv (\mathcal{S}, \pi, X)$, say $A \subseteq \mathcal{S}$, is a *subsheaf* of $\mathcal{S}$ (viz. (1.8) is a sheaf on $X$), *if, and only if, $A$ is an open subset of* $\mathcal{S}$.

In particular, *for any open set $U \subseteq X$, the open set $\pi^{-1}(U) \subseteq \mathcal{S}$* (see (1.4)) *defines a subsheaf of* $\mathcal{S}$. Thus, by restriction to $U$, *the triple*

$$\left(\pi^{-1}(U), \pi|_{\pi^{-1}(U)}, U\right) \tag{1.9}$$

*defines a sheaf over* $U$; we call it the *restriction of* $\mathcal{S}$ *to* $U$, denoted by

$$\mathcal{S}|_U := \pi^{-1}(U), \tag{1.10}$$

while the full notation is, of course, (1.9).

We come now to another fundamental notion for what is to follow, namely, to that of a *"morphism of sheaves"*. So we have the next.

**Definition 1.2.** Let $(\mathcal{E}, \pi, X)$ and $(\mathcal{F}, \rho, X)$ be two given sheaves over a topological space $X$. Then, a **morphism of $\mathcal{E}$ in $\mathcal{F}$** is a continuous map $\phi : \mathcal{E} \longrightarrow \mathcal{F}$ which is also "fiber preserving" (or even "stalk preserving"), in the sense that the following diagram is commutative

(1.11)

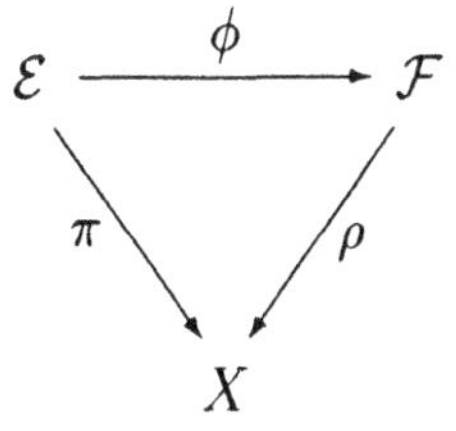

One also speaks of a **sheaf morphism** $\phi : \mathcal{E} \longrightarrow \mathcal{F}$.

So, according to (1.11), one has by definition the relation

$$\pi = \rho \circ \phi, \tag{1.12}$$

or, equivalently, the relation;

$$\phi(\mathcal{E}_x) \subseteq \mathcal{F}_x, \tag{1.13}$$

for every $x \in X$, that also explains the terminology adopted by the previous Definition 1.1 (viz. $\phi$ is a *"fiber preserving"* (continuous) map). Thus, one obtains via the given map $\phi$ a *"fiber-wise"* (or even *"stalk-wise"*) map

$$\phi_x := \phi|_{\mathcal{E}_x} : \mathcal{E}_x \longrightarrow \mathcal{F}_x, \tag{1.14}$$

for every point $x \in X$. *Equivalently,* one can consider a sheaf morphism, in the previous sense, as a family of maps,

$$(\phi_x)_{x \in X}, \tag{1.15}$$

with $\phi_x : \mathcal{E}_x \longrightarrow \mathcal{F}_x$, $x \in X$, such that the map $\phi : \mathcal{E} \longrightarrow \mathcal{F}$, defined by

$$\phi(z) := \phi_x(z), \text{ where } x = \pi(z), \ z \in \mathcal{E}, \tag{1.16}$$

(see also (1.6)) be continuous. (But, see also the next Section 2).

In fact, *any sheaf morphism* $\phi : \mathcal{E} \longrightarrow \mathcal{F}$, as above, *displays* $\mathcal{E}$ *as a sheaf over* $\mathcal{F}$; that is, one has the following.

**Lemma 1.2.** *Let* $\phi : \mathcal{E} \longrightarrow \mathcal{F}$ *be a morphism of sheaves over a topological space* $X$. *Then,* $\phi$ *is a local homeomorphism.*

**Proof.** If $z \in \mathcal{E}$, let $V, W$ be open neighborhoods of $z$ in $\mathcal{E}$ and $\phi(z)$ in $\mathcal{F}$, respectively, to which the restrictions of (the projections) $\pi$ and $\rho$, respectively, are homeomorphisms (cf. (1.3)), while, *by the continuity of* $\phi$, *we may also assume that* $\phi(V) \subseteq W$. So the set $U = \pi(V) \cap \rho(W)$ is an open subset of $X$ such that $\pi(z) = \rho(\phi(z)) \in U$. Moreover, it is not hard to see (cf. also (1.12)) that the continuous map

$$\phi : \pi^{-1}(U) \longrightarrow \rho^{-1}(U), \tag{1.17}$$

between the open sets in $\mathcal{E}$ and $\mathcal{F}$, respectively, as indicated, is still a bijection with a continuous inverse, thus, a *homeomorphism,* such that $z \in \pi^{-1}(U)$ and $\phi(z) \in \rho^{-1}(U)$, and this completes the proof. ∎

Now, given two sheaf morphisms, say $\phi : \mathcal{E} \longrightarrow \mathcal{F}$ and $\psi : \mathcal{F} \longrightarrow \mathcal{G}$, as above, *the composite map* $\psi \circ \phi : \mathcal{E} \longrightarrow \mathcal{G}$ *is* still *a morphism of* the *sheaves* concerned. Moreover, the *identity morphism* (or even *identity automorphism*), $1_{\mathcal{E}} : \mathcal{E} \longrightarrow \mathcal{E}$ is defined, of course, for any given sheaf $\mathcal{E}$ over $X$, having the obvious properties with respect to other sheaf morphisms "from" and "into" $\mathcal{E}$ ("left" and "right" identity). Yet, the usual "associative law" (for composition of morphisms) is readily verified.

Therefore (see, for instance, S. Lang [1: p.26, §7]), one may speak of the *category of sheaves over* $X$ (with $X$ being an arbitrary topological space), which we denote by

$$\mathcal{S}h_X. \tag{1.18}$$

In this context, one can also consider the notion of a *sheaf isomorphism,* this being thus an isomorphism in the category $\mathcal{S}h_X$.

So we say that a given *sheaf morphism*

$$\phi : \mathcal{E} \longrightarrow \mathcal{F} \tag{1.19}$$

(see (1.11)) is an *isomorphism,* whenever $\phi$ has a two-sided inverse in $\mathcal{S}h_X$; thus, *equivalently,* it suffices to provide a continuous map $\psi : \mathcal{F} \longrightarrow \mathcal{E}$ such that

$$\phi \circ \psi = 1_{\mathcal{F}} \text{ and } \psi \circ \phi = 1_{\mathcal{E}} \tag{1.20}$$

(such a $\psi$ is then uniquely defined); now, the same $\psi$ is also "fiber-preserving", i.e., one has $\pi \circ \psi = \rho$, as easily follows from the hypothesis for $\phi$ and (1.20). Therefore, one concludes that,

(1.21) $\phi$ *is an isomorphism if, and only if, this holds true fiber-wise.* Analogously (see (1.14), (1.16)), one infers that $\phi$ *is one-to-one,* respectively, *onto, if, and only if, this is valid fiber-wise.*

# 2. Sections of a sheaf

We comment next on another key-notion of Sheaf Theory; in fact, it is about an equivalent (see also Section 3 below) and usually a more convenient substitute, so to speak, of the same concept of a sheaf. Now, since we are going to consider throughout, almost exclusively, "continuous sections", we presume in the relevant definition continuity from the outset. So we have.

**Definition 2.1.** Let $(\mathcal{S}, \pi, X)$ be a given sheaf and $U$ an arbitrary subset of $X$. Then, by a **section of $\mathcal{S}$ over** $U$ one means a continuous map, say, $s : U \longrightarrow \mathcal{S}$, such that the following diagram be commutative

(2.1)

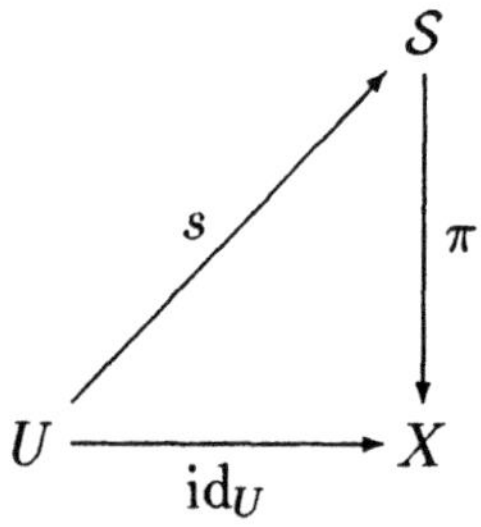


That is, one has (by definition) the relation

$$\pi \circ s = \mathrm{id}_U, \tag{2.2}$$

where $\mathrm{id}_U$ stands for the identity map of $U$.

Of course, one can consider instead as a *"section of the map $\pi$"* (over a set $U \subseteq X$) a function $s$, as above, but not necessarily continuous, making (2.1) commutative. Yet, we refer in the sequel to a (continuous) *local section* of a given sheaf $\mathcal{S}$ over a subset $U \subseteq X$, whenever one has a map $s$, as in (2.1), with $U \subsetneqq X$. We denote the set of all such sections of $\mathcal{S}$ by (see also Section 9 below)

$$\mathcal{S}(U) \equiv \Gamma(U, \mathcal{S}). \tag{2.3}$$

In particular, for $U = X$, the elements of the repective set

$$\mathcal{S}(X) \equiv \Gamma(X, \mathcal{S}) \tag{2.4}$$

are called (continuous) *global sections* of $\mathcal{S}$. In this context, it is an immediate consequence of (2.2) that *any section* (whatsoever, of the map $\pi$) *is a 1-1 map.*

On the other hand, a given section yields actually an *open map.* Precisely, one has the following.

**Lemma 2.1.** *Let $(\mathcal{S}, \pi, X)$ be a sheaf, $U$ open in $X$ and $s$ a local section* *of $\mathcal{S}$ over $U$, i.e., $s \in \Gamma(U, \mathcal{S})$. Then, the set $s(U)$ is open in $\mathcal{S}$, so that*

$$s : U \longrightarrow s(U) \tag{2.5}$$

*yields a homeomorphism.*

**Proof.** Let $z \in s(U)$ and $V$ an open neighborhood of $z$ in $\mathcal{S}$, as given by (1.3). So, *by the continuity of* $s$, there exists open $A \subseteq U$, with $x \in A \subseteq U$ and $z = s(x)$, such that $s(A) \subseteq V$. Yet, $\pi(s(A)) = A \subseteq \pi(V)$ is an open subset of $\pi(V)$, so that, by hypothesis for $\pi$, $s(A)$ is open in $V$, hence in $\mathcal{S}$, as well, such that $z = s(x) \in s(A) \subseteq s(U)$; this proves the first part of our claim, the rest being now straightforward by the preceding. ∎

In fact, a given sheaf has a plethora of local sections; indeed, it can be determined by them (cf., for instance, the next Section 3, as well as, Sections 7, 9 in the sequel). Thus, we first remark that;

(2.6) *any local homeomorphism provided by the map $\pi$, say,*

$$\pi|_V : V \longrightarrow \pi(V) \tag{2.6.1}$$

(cf. (1.3)), *yields through the map*

$$(\pi|_V)^{-1} : \pi(V) \longrightarrow V \subseteq \mathcal{S} \tag{2.6.2}$$

*a* (continuous *local*) *section of $\mathcal{S}$ over the open set $\pi(V) \subseteq X$.*

The same section (2.6.2), denoted now by $s$, is still defined by

$$s : \pi(V) \longrightarrow V : \; x = \pi(z) \longmapsto z \in V \cap \mathcal{S}_x, \tag{2.7}$$

where, in fact, $\{z\} = V \cap \mathcal{S}_x$, by hypothesis for $V$ (see also (1.7)).

Based on (2.2), we further conclude that

(2.8) *(2.6.2) is actually the more general form that (continuous) sections of $\mathcal{S}$ are displayed, when suitably restricted to open subsets of $X$.*

Precisely, for any $s \in \Gamma(U, \mathcal{S})$, with $U$ an open subset of $X$, one gets that $s$ *defines a local inverse of $\pi$, when further suitably restricted;* i.e., one has

$$s = \left(\pi|_{s(U)\cap V}\right)^{-1} = \pi^{-1}|_{U \cap \pi(V)} \tag{2.9}$$

(see also Lemma 2.1), where $V$ is defined by (1.3) such that, for some $x \in U$,

$s(x) \in V$. (In this context, cf. also (2.13) below). So one concludes that: *For any given sheaf, the notions projection and (continuous) section determine locally one another.*

Now, as another information from (2.9), one obtains the following basic fact;

(2.10) *given a sheaf* $(\mathcal{S}, \pi, X)$, *for every element* $z \in \mathcal{S}$, *there exists an open neighborhood* $U$ *of* $\pi(z) = x$ *in* $X$ *and a* (continuous local) *section* $s \in \Gamma(U, \mathcal{S})$ *such that*

(2.10.1) $$z = s(x).$$

(*Hint:* Take $U = \pi(V)$ and $s = (\pi|_V)^{-1} := \pi^{-1}|_{\pi(V)}$, with $V$ associated with $x$, as in (1.3)). Applying here a terminology that will be justified later on (cf. Section 4 in the sequel), we call the element $s(x) = z \in \mathcal{S}$, as in (2.10.1) (see also (2.6)) the *germ of the section* $s$ *at* $x$, which is also denoted by

(2.11) $$s_x \equiv s(x).$$

So (2.10) can now be restated in the following convenient form:

(2.12) *Every element of a given sheaf can be displayed as the germ at the projection of this element of a suitably defined (continuous local) section of the sheaf.*

As a consequence (cf. also (1.6)), we can state, for the present, that *any sheaf is* (made out of) *the germs of its sections,* a claim that will become more precise later on (see Section 3). Moreover, as we shall presently see (cf. (2.15) below), *the topology of the sheaf space is still determined by the (local) sections of the sheaf,* a result that should be compared with Lemma 1.1, in conjunction with the previous comments pertaining to (2.9).

Yet, we also have the following basic information, concerning the (local) behaviour of sections of a given sheaf. That is,

(2.13) *whenever two (local) sections of a given sheaf coincide* (viz. take the same value) *at some point (of the base space), then this happens "locally"; i.e., there exists an open neighborhood of that point on which the two given sections agree.*

Indeed, let $s, t$ be two (continuous local) sections of a given sheaf $(\mathcal{S}, \pi, X)$

such that $s(x) = t(x) \equiv z \in \mathcal{S}$, for some point $x \in X$. So let $V$ be an open neighborhood of $z$ in $\mathcal{S}$, provided by (1.3). Hence, by the continuity of $s, t$, there exist open neighborhoods $U_1 \subseteq \mathrm{Dom}(s)$ (domain of definition of the map $s$) and $U_2 \subseteq \mathrm{Dom}(t)$ of $x$, respectively, with $s(U_1) \cup t(U_2) \subseteq V$. Thus, $U \equiv U_1 \cap U_2$ is now an open neighborhood of $x$ in $X$ on which, due to the behaviour of $\pi$ on $V$, the two given sections $s, t$ coincide. ∎

We come now to the promised result about the topology of a sheaf space. But, first we still remark the following;

(2.14) *given a sheaf $(\mathcal{S}, \pi, X)$, any open subset of $\mathcal{S}$ on which the restriction of $\pi$ is a homeomorphism, is of the form $s(U)$, for some suitably chosen $s \in \Gamma(U, \mathcal{S})$* (cf. also (2.3)).

Indeed, if $A$ is an open subset of $\mathcal{S}$, as in (2.14), then we first note that $\pi(A)$ *is an open subset of $X$* too (viz. $(A, \pi|_A, \pi(A))$ satisfies the conditions in (1.3.1)). So taking now $U \equiv \pi(A)$ and $s := (\pi|_A)^{-1} = \pi^{-1}|_{\pi(A)}$, one concludes by (2.6.2) that $s \in \Gamma(U, \mathcal{S})$ with $s(U) = A$, that proves our claim. ∎

Thus, the following alternative of Lemma 1.1 is now an easy consequence of the preceding (see also (2.14)):

(2.15) *Given a sheaf $(\mathcal{S}, \pi, X)$, the sets $s(U) \subseteq \mathcal{S}$, where $U$ runs over the open subsets of $X$ and $s \in \Gamma(U, \mathcal{S})$, constitute a basis of the topology of $\mathcal{S}$. In particular those sets, on which the restriction of $\pi$ is a homeomorphism.*

As another application of the foregoing, we further obtain the following supplementary information, pertaining to a sheaf morphism (cf. Definition 1.2). So we have.

**Proposition 2.1.** *Let $(\mathcal{E}, \pi, X)$ and $(\mathcal{F}, \rho, X)$ be two given sheaves, while let also $\phi : \mathcal{E} \longrightarrow \mathcal{F}$ be a continuous map. Then, the three following assertions are equivalent:*

*1) The map $\phi : \mathcal{E} \longrightarrow \mathcal{F}$ is a sheaf morphism.*

*2) For any open $U \subseteq X$ and $s \in \Gamma(U, \mathcal{E})$ one gets $\phi^*(s) \equiv \phi \circ s \in \Gamma(U, \mathcal{F})$;* we express this by saying that $\phi$ *is section preserving. Equivalently, one has*

(2.16) $$\phi^*(\Gamma(U, \mathcal{E})) \subseteq \Gamma(U, \mathcal{F}),$$

*for any open $U \subseteq X$.*

*3) For every element $z \in \mathcal{E}$, there exists an open set $U$ in $X$ and $s \in \Gamma(U, \mathcal{E})$ such that $z \in s(U)$ and $\phi \circ s \in \Gamma(U, \mathcal{F})$.*

**Proof.** If 1) is valid and $s \in \Gamma(U, \mathcal{E})$, then the *continuous map* $\phi^*(s) \equiv \phi \circ s : U \longrightarrow \mathcal{F}$ satisfies the relation

$$\rho((\phi \circ s)(x)) = \rho(\phi(s(x))) = x, \tag{2.17}$$

for any $x \in U$, due to the hypothesis for $\phi$ and $s$, so that $\phi \circ s \in \Gamma(U, \mathcal{F})$, i.e., *1) implies 2)*. Conversely, (2.16) implies (1.13), as easily follows from (2.10) and (2.17), so that *2) implies 1)*, as well. Thus, putting it otherwise, we have proved so far that;

(2.18) *a continuous map of sheaves is fiber preserving* (hence, a *sheaf morphism*) *if, and only if, it is section preserving.*

(Here we still note, in passing, that a *section preserving map* of sheaves, *not necessarily continuous, is fiber preserving,* as well. But, see also (9.22) in the sequel.) Now, suppose that 2) is true, and let $z \in \mathcal{E}$. Then (cf. (2.9)), $z = s(x) \in s(U)$, for some $s \in \Gamma(U, \mathcal{E})$, so that by hypothesis $\phi \circ s \in \Gamma(U, \mathcal{F})$, that is, *2) implies 3)*. Conversely, assuming 3), one easily proves (1.13) by a similar argument to that in (2.16), so *3) implies 1)*, as well, and this completes the proof. ∎

# 3. A sheaf is its sections. Defining families of sections

We give below a first formal justification of our claim in the title of this section, while this same aspect has been supported already by the comments after (2.12). We shall also return to this point of view later on (see e.g. Sections 8, 9 and 13 in the sequel) still in another formal way.

Thus, suppose we are given a (disjoint) *family of sets*, say,

$$(S_x)_{x \in X}, \tag{3.1}$$

*parametrized by the points of a topological space* $X$. Moreover, assume that we have a *family of local functions* on $X$, say,

$$\mathfrak{S} = \{s\}; \tag{3.2}$$

that is, we suppose that *the generic element of* $\mathfrak{S}$ *is* a function $s : U \longrightarrow \mathcal{S}$ or, equivalently, *a triple*

$$(U, s, \mathcal{S}), \tag{3.3}$$

such that $U \equiv \mathrm{Dom}\,(s)$ *is an open subset of* $X$ and $\mathcal{S}$ is given by (the following *direct sum of sets*)

$$\mathcal{S} := \bigcup_{x \in X} S_x = \sum_{x \in X} S_x. \tag{3.4}$$

Finally, we assume that *the family* $\mathfrak{S}$, as in (3.2), *satisfies the following three conditions:*

*i) For every* $x \in \mathrm{Dom}\,(s) \equiv U$ (cf. (3.3)), *with* $s \in \mathfrak{S}$, *one has*

$$s(x) \in S_x; \tag{3.5}$$

*equivalently, any* $s \in \mathfrak{S}$ *is given by*

$$s \in \prod_{x \in U} S_x, \tag{3.5$'$}$$

*where* $U \equiv \mathrm{Dom}\,(S)$ (in the notation of (3.3)) *is an open subset of* $X$.

*ii) Each set* $S_x$, $x \in X$, *is covered by the images of the functions in* $\mathfrak{S}$; *i.e., one has*

$$S_x \subseteq \bigcup_{s \in \mathfrak{S}} \mathrm{im}\,(s) \tag{3.6}$$

*or, equivalently, we assume that*

$$\mathcal{S} = \bigcup_{s \in \mathfrak{S}} s(U). \tag{3.6$'$}$$

*iii) For any two elements* $s, t$ *of* $\mathfrak{S}$, *with* $U \equiv \mathrm{Dom}\,(s)$ *and* $V \equiv \mathrm{Dom}\,(t)$ (cf. (3.3)), *and for every* $z \in s(U) \cap t(V)$, *there exists an open set* $W$ *in* $X$, *with*

$$x \in W \subseteq U \cap V \text{ (where } z = s(x) = t(x)) \textit{ and } s = t|_W. \tag{3.7}$$

(In this context, we note that if $z \in s(U) \cap t(V)$, as above, then $z = s(x) =$

$t(x')$, with $x \in U$ and $x' \in V$, so that $x = \pi(s(x)) = \pi(t(x')) = x'$, which explains the first relation in (3.7)).

Now, one further defines the following *(projection) map* (see also (3.4))

$$\pi : \mathcal{S} \longrightarrow X \tag{3.8}$$

such that

$$\pi(S_x) := \{x\}, \; x \in X. \tag{3.9}$$

Therefore, one obtains that (see also (1.5))

$$\mathcal{S}_x := \pi^{-1}(x) = S_x, \; x \in X. \tag{3.10}$$

The conditions set forth for the previous families (3.1) and (3.2) represent, in effect, within the above context, similar ones satisfied by a given sheaf on $X$ and its sections. So (3.5) along with (3.9) imply that *any* $s \in \mathfrak{S}$ *is a section of the map* $\pi$ (viz. (2.2) is satisfied; of course, no continuity of $s$ is established as yet. But, see the next Theorem 3.1). On the other hand, *(3.6′) corresponds to (2.12),* while *iii) stands for (2.13).* Thus, our next task is to exhibit the above triple $(\mathcal{S}, \pi, X)$ as a sheaf over $X$. Indeed, one has the following.

**Theorem 3.1.** *Consider the triple* $(\mathcal{S}, \pi, X)$*, as defined above, and assume that the previous conditions for the family* $\mathfrak{S}$ (cf. (3.2) and i) – iii), as before) *are satisfied. Then, the family*

$$\mathcal{B} = \left\{ s(V) : s \in \mathfrak{S} \text{ and } V \text{ open in } X \text{ with } V \subseteq \mathrm{Dom}\,(s) \right\} \tag{3.11}$$

*is a basis for the topology in* $\mathcal{S}$*, with respect to which* $(\mathcal{S}, \pi, X)$ *is a sheaf on* $X$*, viz. the map* $\pi : \mathcal{S} \longrightarrow X$ (cf. (3.9)) *is a local homeomorphism.*

**Proof.** We first remark that, by virtue of (3.6′) and cond. iii), it is easy to see that $\mathcal{B}$ *can be taken as a basis of a topology in* $\mathcal{S}$. So we prove now that *the map* $\pi : \mathcal{S} \longrightarrow X$, as given by (3.9), *yields a local homeomorphism* (cf. (1.3)), when $\mathcal{S}$ is topologized by means of (3.11): Indeed, if $z \in \mathcal{S}$, then by (3.6′) $z \in s(U)$, for some $s \in \mathfrak{S}$; on the other hand, as already remarked, any $s \in \mathfrak{S}$, being a section of $\pi$, is a 1-1 map, so that one obtains for the triple $(U, s, \mathcal{S})$, under consideration (see also (3.3)), the map

$$\pi|_{s(U)} = s^{-1} : s(U) \longrightarrow U, \tag{3.12}$$

that is, a *bijection* through the restriction of $\pi$, as indicated. Now, in view of (3.11) and our hypothesis for the topology of $\mathcal{S}$, *$s(U)$ is an open set in $\mathcal{S}$, with $z \in s(U)$*, which is thus *bijectively mapped, via $\pi$, onto the open set $U$ in $X$, such that $\pi(z) \equiv x \in U$*. We further prove that (3.12) *is a homeomorphism:*

Indeed, we first observe that $s : U \longrightarrow \mathcal{S}$ *is an open map,* as immediately follows by (3.11) and the very definition of the topology of $\mathcal{S}$. *The same map is continuous,* as well; for, if $A \subseteq \mathcal{S}$ is open and $x \in s^{-1}(A) \subseteq U$, $s(x) \in t(V)$, for some $t \in \mathfrak{S}$, with $V$ an open subset of Dom $(t)$, such that $s(x) \in t(V) \subseteq A$ (see also (3.11)). Hence (cf. cond. iii)), there exists an open $W \subseteq U \cap V$, with $x \in W$ and $s(W) = t(W) \subseteq t(V) \subseteq A$, so that $x \in W \subseteq s^{-1}(A)$, which thus proves our claim, pertaining to the map $s$, and this terminates the proof. ∎

Now, it is a consequence of the last part of the previous proof that;

(3.13) *every element* $(U, s, \mathcal{S}) \in \mathfrak{S}$ (cf. (3.3)) *yields a continuous and open map $s : U \longrightarrow \mathcal{S}$, when $\mathcal{S}$ is topologized by taking as a basis of its topology the family (3.11).*

In fact, something more, pertaining to the continuity of the maps $s$ ($s \in \mathfrak{S}$) is true. That is, one has.

**Lemma 3.1.** *Suppose we are given the families (3.1) and (3.2) such that cond. i), ii) and iii), as given above, be satisfied. Moreover, let $(\mathcal{S}, \pi, X)$ be the triple defined by (3.4) and (3.9). Then, the two following assertions are equivalent:*

*1) The family $\mathcal{B}$, as given by (3.11), is a basis of a topology in $\mathcal{S}$.*

*2) The set $\mathcal{S}$* (cf. (3.4)) *is endowed with the finest topology making the maps $s \in \mathfrak{S}$ continuous..*

**Proof.** If 1) is valid, to establish prop. 2) we prove, *equivalently,* that;

(3.14) *a set $A \subseteq \mathcal{S}$ is open* (in the topology defined by $\mathcal{B}$) *if, and only if, $s^{-1}(A) \subseteq U$ is open, for every $s \in \mathfrak{S}$.*

Thus, according to (3.13), the "only if" part of (3.14) is already established. So assume that $A \subseteq \mathcal{S}$, satisfies the condition in the "if" part of (3.14), and let $z \in A$, so that (cf. (3.6′)) $z = s(x) \in s(U)$, for some $(U, s, \mathcal{S})$ in $\mathfrak{S}$. Now, by hypothesis, $s^{-1}(A) \subseteq U$ is an open set, while by (3.13) $s : U \longrightarrow \mathcal{S}$ is an open map, so that $s(s^{-1}(A)) = s(U) \cap A$ is an open set in $\mathcal{S}$, hence $A$ is open, which proves (3.14), so *1) implies 2)*. Conversely, if 2) is valid then, by hypothesis, each $s \in \mathfrak{S}$ yields a *continuous* (local) *map* $s : U \longrightarrow \mathcal{S}$, while *the same map is open,* as well; indeed, suppose that $N \subseteq U$ is open and let $x \in t^{-1}(s(N))$, for some $(V, t, \mathcal{S})$ in $\mathfrak{S}$ (see (3.3)). Thus, $t(x) = s(x) \in s(N)$, with $x \in N \cap V$, so that (cf. cond. iii)), there exists open $W$ with $x \in W \subseteq U \cap V$ and $s = t|_W$. Therefore, $x \in W \cap N \subseteq t^{-1}(s(N))$, that is $s(N)$ is open in $\mathcal{S}$, according to our hypothesis for the topology of $\mathcal{S}$, which thus proves our claim about the openness of $s$. Thus, assuming as a topology of $\mathcal{S}$ that one defined by prop. 2), we have proved so far that;

(3.15) *the family $\mathcal{B}$,* as defined by (3.11), *consists of open subsets of $\mathcal{S}$.*

Now, let $A \subseteq \mathcal{S}$ be open and $z \in A$, so that (cf. (3.6′) $z = s(x) \in s(U)$, for some $(U, s, \mathcal{S})$ in $\mathfrak{S}$. But, by hypothesis, $s^{-1}(A) \subseteq U$ is open, hence $s(s^{-1}(A)) \in \mathcal{B}$, while $z \in s(s^{-1}(A)) \subseteq A$; so $\mathcal{B}$ *is a basis of the topology of* $\mathcal{S}$, as the latter is defined by prop. 2), hence *2) implies 1),* as well, and the proof is complete. ∎

Thus, as a consequence of the preceding, one concludes that;

(3.16) *under either one of the two equivalent conditions of Lemma 3.1, the map* $\pi$ (cf. (3.8), (3.9)) *is a local homeomorphism, that is, the triple* $(\mathcal{S}, \pi, X)$ *is a sheaf.*

Now, as already remarked, *any sheaf* $(\mathcal{S}, \pi, X)$ *fulfils the above cond. i) - iii),* by taking as $\mathfrak{S}$ the set of all continuous (local) sections of $\mathcal{S}$ over the open subsets of (the base space) $X$. So, according to the preceding, *a sheaf is* revealed as *determined by a suitable set of "local data"*, like, for instance, the family $\mathfrak{S}$ in Theorem 3.1. Indeed, the point of view here is that;

(3.17) *a sheaf is completely determined by a suitable family of local* (continuous) *sections, like $\mathfrak{S}$ in Theorem 3.1,* thus, not necessarily by all of them. Such a family $\mathfrak{S}$ is called a *defining family of sections* of the sheaf concerned.

We shall apply the previous technique of definition of a sheaf in the sequel (cf., for instance, Section 4.(b) below), while we shall also encounter a similar aspect in the case of a *"presheaf"* (of sets) on a given topological space (cf. Sections 5, 9, 10 in the sequel).

# 4. Examples. Germs and sheaves

Examples of sheaves will be supplied, of course, in abundance throughout our discussion. However, to begin with we give here some of the most basic ones, that we shall also encounter several times in the sequel. So we first have the following:

**4.(a). The sheaf of germs of locally constant functions. The constant sheaf.**– The simplest example of a sheaf is that of a "constant sheaf".

So by a *constant sheaf* one means a triple (cf. Definition 1.1)

$$(X \times M, \pi, X), \tag{4.1}$$

where $X \times M$ denotes the cartesian product of a topological space $X$ with a given set $M$, the latter being endowed with the discrete topology. Moreover,

$$\pi := \mathrm{pr}_1 : X \times M \longrightarrow X \tag{4.2}$$

stands for the (canonical) projection of $X \times M$ onto the first factor.

Now, it is clear that *the map* $\pi$, as given by (4.2), *is a local homeomorphism.* Yet, the *fiber of a constant sheaf,* as above, over an arbitrary point $x \in X$, is given by (see also (1.5))

$$(X \times M)_x := \pi^{-1}(x) = \{x\} \times M \cong M, \tag{4.3}$$

the last (*canonical*) bijection being also a *homeomorphism;* thus, *(4.1) is a sheaf,* with $\pi$ given by (4.2), *if, and only if, $M$ has the discrete topology.*

Thus, by the preceding *a constant sheaf is completely determined by* (the discrete space) $M$, the latter being, in effect, (canonically) identified with any one of its fibers (cf. (4.3)). Therefore, one simply refers to (4.1) as *"the constant sheaf $M$"*, as well. Yet, a "bold face" $\mathbf{M}$ is also used, i.e., we set

$$\mathbf{M} = (X \times M, \pi, X), \tag{4.4}$$

while we also apply for (4.1) the notation

$$M_X. \tag{4.4$'$}$$

(In this context, see also Sections 10, 11 in the sequel). We next apply the argument of the preceding Section 3 (viz. looking at a sheaf as an object defined by "local data"), in order to *exhibit a* given *constant sheaf* $M_X$, as above, *as a sheaf of germs of* (suitably defined) *local functions:*

So consider any (continuous) local section of $M_X$ over an open set $U \subseteq X$, say,

$$s : U \longrightarrow X \times M \tag{4.5}$$

(cf. (4.1)), such that one has (see also (2.11) and (4.3))

$$s_x \equiv s(x) = (x, m) \in (X \times M)_x = \{x\} \times M, \tag{4.6}$$

with $x \in U$. But, for any $m \in M$ and open $U \subseteq X$, one obtains the following *continuous "constant section" of $M_X$ over $U$, associated with $m \in M$,*

$$\widetilde{m} : U \longrightarrow X \times M : x \longmapsto \widetilde{m}(x) \equiv \widetilde{m}_x := (x, m). \tag{4.7}$$

(Thus, the applied terminology here can be justified by (4.3)). So, in view of (4.6) and (4.7), one can apply (2.13), pertaining to the given point $m \in M$, so that one obtains an open neighborhood $V$ of $x$, with $x \in V \subseteq U$ and $s(x) = \widetilde{m}(x)$, such that

$$s = \widetilde{m}|_V. \tag{4.8}$$

So one concludes that;

(4.9) *any* (continuous) *local section of a constant sheaf is locally a "constant function"* (section). Thus, (see also (1.6) and (2.12)), *a constant sheaf is entirely composed of germs of "locally constant functions"* (sections).

In fact, as we shall see later on (cf. Theorem 11.1), *the above is the only way that local sections of $M_X$ arrive.* This also justifies the terminology adopted in the heading of this subsection. (But see also Scholium 4.1 below).

**4.(b). The sheaf of germs of continuous functions.**– The functions under consideration herewith will be defined, in general, on arbitrary open subsets of a given *topological space* $X$ and will be, by assumption, **C**-*valued,* where **C** stands for the (field of) *complexes.* (Of course, *any topological space $Y$ could be considered instead,* as well). So the sheaf in the title will be denoted henceforth by

$$\mathcal{C}_X^{\mathbf{C}} \equiv \mathcal{C}_X. \tag{4.10}$$

Now, by a *local function of* $X$, we shall mean in the sequel a **C**-*valued continuous map,* defined *on an open subset of* $X$. Thus, such an object is given by a triple, say,

$$(U, f, \mathbf{C}), \tag{4.11}$$

where $U \equiv \mathrm{Dom}(f)$ is an open subset of $X$ and the map $f : U \longrightarrow \mathbf{C}$ continuous. (The term *locally defined function of* $X$ is also of use concerning (4.11)). The set of all such triples (functions), whose common domain (of definition) is the open set $U \subseteq X$, is denoted by

$$\mathcal{C}(U, \mathbf{C}) \equiv \mathcal{C}_{\mathbf{C}}(U) \equiv \mathcal{C}(U). \tag{4.12}$$

Furthermore, we denote by

$$\mathcal{C}_{\mathrm{loc}}(X) := \bigcup_{U \in \tau_X} \mathcal{C}(U) = \sum_{U \in \tau_X} \mathcal{C}(U) \tag{4.13}$$

(in place of the more consistent notation $\mathcal{C}^{\mathbf{C}}_{\mathrm{loc}}(X)$) the *set of all local functions of* $X$, where $\tau_X$ denotes the topology of $X$.

In particular, given an *arbitrary point* $x \in X$, consider the set

$$\mathcal{C}_{\mathrm{loc}}(X)_x := \bigcup_{U \in \mathcal{V}(x)} \mathcal{C}(U) = \sum_{U \in \mathcal{V}(x)} \mathcal{C}(U), \tag{4.14}$$

obtained from (4.13) by restricting oneself to the set $\mathcal{V}(x) \subseteq \tau_X$ of open neighborhoods of $x$ in $X$. We next consider on the set (4.14) the following relation:

Given two elements $(U, f, \mathbf{C})$ and $(V, g, \mathbf{C})$ of (4.14) (for simplicity we also refer to the latter elements just as $f, g$, the rest being understood), we say that $f, g$ *have the same germ at* $x$, and we write

$$f \underset{x}{\sim} g, \tag{4.15}$$

whenever there exists an *open* $W \subseteq X$ such that

$$x \in W \subseteq U \cap V \ \textit{and}\ f|_W = g|_W. \tag{4.15'}$$

It is easy to see that *(4.15) defines an equivalence relation in the set (4.14).* The resulting quotient set is denoted by

$$\mathcal{C}_{X,x} := \mathcal{C}_{\mathrm{loc}}(X)/\underset{x}{\sim}, \tag{4.16}$$

while a generic element of the latter is denoted by

$$[f]_x \equiv f_x, \tag{4.17}$$

where $(U, f, \mathbf{C}) \equiv f \in \mathcal{C}_{\mathrm{loc}}(X)_x$ (cf. (4.14)). We call (4.17) the *germ of $f$ at* $x$, a terminology that will be justified, in connection with (2.11), presently below (see thus (4.23) in the sequel).

So we have actually assembled so far almost all (barring, namely, $\mathfrak{S}$ as in (3.2)) the necessary ingredients to define the sought sheaf $\mathcal{C}_X$ (cf. (4.10)), by applying the technique of Section 3 in the preceding:

Thus, we have a *family of (disjoint) sets,* parametrized by the points of the given topological space $X$, namely,

$$(\mathcal{C}_{X,x})_{x \in X}, \tag{4.18}$$

so that the *sheaf space* of the sheaf in question is defined by

$$\mathcal{C}_X := \bigcup_{x \in X} \mathcal{C}_{X,x} = \sum_x \mathcal{C}_{X,x} \tag{4.19}$$

(see also (3.4)). Yet, one further defines a *projection map*

$$\pi : \mathcal{C}_X \longrightarrow X, \tag{4.20}$$

given by the relation (cf. also (3.9))

$$\pi(\mathcal{C}_{X,x}) := \{x\}, \ x \in X. \tag{4.21}$$

So to proceed further, following the pattern of Section 3, we have to invent the analogue here of the family $\mathfrak{S}$, as in (3.2), that should also satisfy conds. i) – iii) of Section 3, provided we are going to topologize (4.19) through the latter family. Indeed, we undertake this task right away.

Thus, for any *open* $U \subseteq X$, one gets the following map (cf. also (2.3), (3.3) and (4.11))

$$\mathcal{C}(U) \longrightarrow \mathcal{C}_X(U) \equiv \Gamma(U, \mathcal{C}_X) : (U, f, \mathbf{C}) \equiv f \longmapsto \tilde{f} \equiv (U, \tilde{f}, \mathcal{C}_X) \tag{4.22}$$

such that one sets (cf. (4.17))

$$\tilde{f}(x) := [f]_x, \ x \in U. \tag{4.23}$$

In this context, we applied before, in anticipation, the notation for "continuous sections", concerning the range of (4.22), a fact that is finally justified by just taking the various $\tilde{f}$'s, as above, as the sought elements of $\mathfrak{S}$; so we set (see also (4.13))

$$\mathfrak{S} := \left\{\tilde{f} : f \in \mathcal{C}_{\mathrm{loc}}(X)\right\}. \tag{4.24}$$

Now, based on the very definitions (cf. e.g. (4.15′), in conjunction with cond. iii)), it is easily seen that;

(4.25) *the family* $\mathfrak{S}$, as given by (4.24), *satisfies conds. i) – iii) of Section 3.* Therefore (Theorem 3.1), *the triple*

(4.25.1) $$(\mathcal{C}_X, \pi, X),$$

as given by (4.19) and (4.21), *yields a sheaf on* $X$. We call it the *sheaf of germs of* (**C**-*valued) continuous functions on* $X$.

In particular, based on the same Theorem 3.1, we conclude that;

*the family*

(4.26) $$\mathcal{B} = \left\{ \tilde{f}(V) : f \in \mathcal{C}_{\mathrm{loc}}(X) \text{ and } \mathrm{Dom}\,(f) \supseteq V, \text{ open} \right\}$$

(see also (3.11) and (4.24)) *constitutes a basis of the topology of* $\mathcal{C}_X$.

The preceding exhibit an elementary way of defining the sheaf $\mathcal{C}_X$ by applying the argument of Section 3; a more sophisticated approach to the same object will be given in Section 4 below in terms of the notion of a *"presheaf"* (of sets, leading, among other things, to the *bijectiveness of (4.22)*. However, we do prove presently below, *directly,* as an application of the foregoing, that *the latter map is one-to-one,* this having also an interest per se.

So consider the following map

(4.27) $$\alpha : \mathcal{C}_X \longrightarrow \mathbf{C},$$

defined by the relation (cf. also (4.15), (4.17) and (4.23))

(4.28) $$\alpha([f]_x) \equiv \alpha\left(\tilde{f}(x)\right) = \left(\alpha \circ \tilde{f}\right)(x) := f(x), \; x \in X,$$

so that *one obtains*

(4.29) $$\alpha \circ \tilde{f} = f,$$

*for every* $(U, f, \mathbf{C}) \equiv f \in \mathcal{C}(U) \subseteq \mathcal{C}_{\mathrm{loc}}(X)$ (see (4.13)). Thus, as a result of (4.29) (cf. also (4.23)), one concludes that;

(4.30) *(continuous local) sections of the sheaf $\mathcal{C}_X$, which are of the form $\tilde{f}$, for some $f \in \mathcal{C}_{\text{loc}}(X)$, uniquely determine (and are uniquely determined, as well, by) $f$.*

Furthermore, based on Lemma 3.1 and (4.24), we deduce that

(4.31) *the map $\alpha : \mathcal{C}_X \longrightarrow \mathbf{C}$,* as given by (4.28), *is continuous.*

Thus, in view of (4.31), one infers that *any local (continuous) section of $\mathcal{C}_X$ gives rise to an element of $\mathcal{C}_{\text{loc}}(X)$* (cf. (4.13)). However, as already mentioned, we shall prove later on (thus see, for instance, (11.10) and Proposition 11.1) that *any* such *section of $\mathcal{C}_X$ is actually of the form exhibited by (4.30).* Hence, as a result of the preceding, we can now restate (4.22) in the form

(4.32) $$\mathcal{C}(U) \subsetneq \Gamma(U, \mathcal{C}_X) \equiv \mathcal{C}_X(U),$$

for every open $U \subseteq X$. (Here the symbol "$\subsetneq$" stands for the above *injective map* (4.22), as yielded by (4.30)). So one can assert, along with S. Mac Lane [1: p.357] that;

(4.33) *"any good class of functions can be represented simply as continuous cross sections."*

We are still going to verify the above aphorism presently below by considering two more important examples, apart from $\mathcal{C}_X$, of *"functional sheaves"*. Yet, the same situation, even more generally, will be appeared by considering *"sectional representations"* of suitable *topological algebras* (cf. Chapter 11), that will be of importance in Part Two of this work.

**Scholium 4.1.**– Suppose we have a *constant sheaf* $\mathbf{M}$ (see (4.1) or even (4.4)). Thus, according to (4.9), its elements are germs of locally constant sections of $\mathbf{M}$ (suitably interpreted, via (4.3), as locally constant functions on the base space $X$ of $\mathbf{M}$). Now, based on the preceding, we can look at $\mathbf{M}$ as the *sheaf of germs of $M$-valued continuous functions on $X$*, when $M$ carries the *discrete topology* (see also (4.1)), or, what amounts to the same, as the *sheaf of germs of* ($M$-valued) *locally constant functions on $X$*, denoted by

(4.34) $$\mathcal{C}_X^M$$

(for the notation applied, cf. also the comments before (4.10)). Namely, one employs here the above technique for $\mathcal{C}_X$ by considering instead, for any open $U \subseteq X$, the set

$$\mathcal{C}(U, M) \tag{4.35}$$

of all *M-valued continuous functions on U, with M carrying the discrete topology.* Thus, based on the very definitions (see also (4.4′)), one verifies that

$$\mathcal{C}(U, M) = \Gamma(U, M_X) \equiv \mathbf{M}(U), \tag{4.36}$$

within a *bijection, for any open* $U \subseteq X$, which thus still justifies (4.9). On the other hand, we also note, in anticipation, that the same relation (4.36) yields

$$\mathcal{C}_X^M = M_X, \tag{4.37}$$

within a *sheaf isomorphism* (see also (1.20)); in this context, cf. Theorem 9.1, (11.17) and Theorem 11.1, in conjunction with Proposition 11.1 in the sequel.

As a result, we improve on (4.9) by realizing that;

(4.38) *the germ at $x \in X$ of a (local) section of* **M** *is completely determined by the germ at $x$ of a locally constant function of $X$, say, $f$, such that $x \in$* Dom $(f)$.

**4.(c). Sheaves of germs of $\mathcal{C}^\infty$-functions and of holomorphic ones.**– Let $X$ be a (finite-dimensional) $\mathcal{C}^\infty$*-manifold.* Then, applying a similar argument to that one, as before, for the sheaf $\mathcal{C}_X$ (cf. (4.25)), one defines the *sheaf of germs of* (**C**-valued) $\mathcal{C}^\infty$*-functions on X*, denoted by

$$\mathcal{C}_X^\infty \tag{4.39}$$

(an abbreviation, in fact, of ${}^{\mathbf{C}}\mathcal{C}_X^\infty$; of course, **R**-*valued* ($\mathcal{C}^\infty$-) functions, with **R** standing for the *reals,* or even, more generally, $\mathcal{C}^\infty$*-functions on X taking values in another* given $\mathcal{C}^\infty$*-manifold Y*, could be considered, as well, as the case may be).

Similarly, if $X$ is a (finite-dimensional) *complex* (analytic) *manifold,* one defines the *sheaf of germs of holomorphic functions on X,* denoted by

$$\mathcal{O}_X. \tag{4.40}$$

So in both the previous two cases one defines the corresponding sheaves by considering, instead of (4.12), the sets $\mathcal{C}^\infty(U)$ or $\mathcal{O}(U)$ of $\mathcal{C}^\infty$-functions or of holomorphic ones, respectively (appropriately valued), on an open set $U \subseteq X$. (In this context, cf., for instance, H. Grauert - R. Remmert [1: p.2, ftn.], concerning the history of the symbol "$\mathcal{O}$", as applied above).

## 5. Presheaves. Basic definitions

We discuss below another important aspect of the theory of sheaves, being, to a certain extend (cf. Section 11 in the sequel), an equivalent notion to that of a sheaf. But, we first consider the general concept of a *"presheaf of sets"*, so that we start with giving the following *1st definition* of this notion.

**Definition 5.1.** Let $X$ be a given topological space. Then, by a **presheaf (of sets)** on $X$, one means an assignment (correspondence), say, $S$, which, with every open subset $U$ of $X$, associates a set $S(U)$, in such a way that the following conditions are satisfied:

1) For any open sets $U, V$ of $X$ with $V \subseteq U$, there exists a *("restriction")* map

$$\rho_V^U : S(U) \longrightarrow S(V). \tag{5.1}$$

2) These (restriction) maps obey further the relations;

$$\rho_U^U = \mathrm{id}_{S(U)}, \tag{5.2}$$

for every open $U \subseteq X$, as well as,

$$\rho_W^U = \rho_W^V \circ \rho_V^U, \tag{5.3}$$

for any open sets $U, V, W$ in $X$, with $W \subseteq V \subseteq U$.

On the other hand, we do have another useful (although more technical) description of the previous fundamental notion, as given by Definition 5.1, based on the concept of an *"inductive system"* (of sets). Thus, we come next to the rudiments, concerning this *2nd definition:*

So suppose that we are given a *topological space* $X$, whose topology (viz. the set of open subsets of $X$) is denoted by $\tau$; thus, we can further consider

$\mathcal{T}$ as an *(upwards) directed set,* by setting for any two members $U, V$ of $\tau$ the relation

$$V \succeq U \text{ if, and only if (by definition), } V \subseteq U. \tag{5.4}$$

Using the above relation, we can give now the next, *equivalent,* formulation of Definition 5.1. That is, one has.

**Definition 5.2.** Let $X$ be a given topological space and $\tau$ the set of its open subsets (viz. the topology of $X$), considered as a directed set, according to (5.4). So by a **presheaf on** $X$, one means any *inductive system of sets, with respect to the (directed) index set* $\tau$, say

$$\left(S(U), \rho_V^U\right). \tag{5.5}$$

Thus, according to the very definition of an inductive system of sets (see, for instance, N. Bourbaki [1: p.88, $n^o$ 5]), one is given, along with a family of sets, indexed by $\tau$, say,

$$(S(U))_{U\in\tau}, \tag{5.6}$$

a family of maps (cf. also (5.4))

$$\left(\rho_V^U\right)_{U\preceq V}, \tag{5.7}$$

as well (the *"restriction maps"* of Definition 5.1), such that the relations (5.2) and (5.3) be satisfied. Therefore, the *equivalence of the previous two definitions,* as claimed.

The preceding very convenient, indeed, interpretation of the notion of a presheaf (of sets), as given by Definition 5.2, will be often applied throughout the sequel. We thus get here a more sophisticated, however useful, formulation of the said notion, in terms of "inductive systems", in comparison with a rather "naïve", so to speak, point of view of the same, as exhibited by our first Definition 5.1.

Furthermore, we do have another *third interpretation* of the above notion of a presheaf, still useful under several circumstances, by employing *categorical language.* (We take here for granted the relevant notions from *category theory;* however, cf. also Appendix A at the end of this treatise):

Thus, one can consider the *topology* $\tau$ of a given topological space $X$ as a *(small) category,* whose *objects* are the elements of $\tau$ (viz. the open subsets of $X$), while its *morphisms* are defined as follows; for any $U, V$ in $\tau$, with $V \subseteq U$, one sets

$$\text{Hom}\,(V, U) := \left\{ i_U^V \right\}. \tag{5.8}$$

That is, one defines here the respective *"set of morphisms"*, Hom $(V, U)$, to be the *singleton* consisting of the unique (canonical) inclusion map $V \overset{\subseteq}{\rightarrow} U$, denoted by $i_U^V$. Otherwise, for any other pair $(U, V)$ of elements of $\tau$, one sets

$$\text{Hom}\,(V, U) := \emptyset. \tag{5.9}$$

Thus, we come now to the following *3rd definition* of the previous notion. Namely,

(5.10) given a topological space $X$, by a **presheaf (of sets) on** $X$, one means any *contravariant functor* of the *category of open subsets of* $X$ into the *category of sets.*

It is a routine matter to check that *(5.10) is, in effect, the interpretation of Definition 5.2 in the language of category theory.*

Now, in this context, we also comment, for later use, on the following notion: Namely, suppose that $A$ is an *open subspace* of a given topological space $X$, and let $S$ be a *presheaf (of sets) on* $X$.

Thus, we call *restriction of* $S$ *to* $A$ (or even a *subpresheaf of* $S$ *on* $A$), denoting it by

$$S|_A, \tag{5.11}$$

the presheaf (of sets) on the (topological space) $A$, which is given by the relation

$$(S|_A)(U) := S(U), \tag{5.12}$$

*for any open* $U \subseteq A$. The corresponding restriction maps of $S|_A$ (see (5.1)) are, of course, defined as those of the given presheaf $S$ on $X$, "restricted" now to the open subsets of $A$. As a result, one concludes that;

(5.13) *the restriction of a given presheaf* (of sets) *to any open subset of its base space,* as defined by (5.12), *is still a presheaf* (of sets) *on that subset.*

# 6. Morphisms of presheaves

The notion referred to in the title of this Section is, of course, easily understood if one looks at a given presheaf on a topological space $X$ through (5.10) or, equivalently, in the sense of Definition 5.2 (see below).

Thus, by first employing categorical language, a *morphism* between two given presheaves on $X$, as above, is (by definition) any *natural transformation of* the respective (contravariant) *functors,* defining the given presheaves (cf. (5.10)).

Now, equivalently, applying Definition 5.2, one can look at a morphism of two given presheaves, as a *morphism of* the respective *inductive systems* (of sets), defining the presheaves in question. (In this context, see, for instance, N. Bourbaki [1: p.92, Corollaire 1]).

Specifically, suppose we are given two presheaves (of sets) on a topological space $X$, say,

(6.1) $$S \equiv \left(S(U), \rho_V^U\right) \text{ and } E \equiv \left(E(U), \lambda_V^U\right)$$

(cf. (5.5) - (5.7)). Now, by a *morphism of $S$ in $E$,* say, $\phi$, denoted by

(6.2) $$\phi : S \longrightarrow E,$$

one means a family of maps, indexed by $\tau$ (the topology of $X$), say,

(6.3) $$\phi \equiv (\phi_U)_{U \in \tau},$$

in such a manner that

(6.4) $$\phi_U : S(U) \longrightarrow E(U),\ U \in \tau,$$

while, for any $U, V$ in $\tau$, with $V \succeq U$ (cf. (5.4)), the following diagram (of maps) commutes

(6.5) $$\begin{array}{ccc} S(U) & \xrightarrow{\phi_U} & E(U) \\ {\scriptstyle \rho_V^U}\downarrow & & \downarrow{\scriptstyle \lambda_V^U} \\ S(V) & \xrightarrow[\phi_V]{} & E(V) \end{array}$$

That is, one has (by definition), concerning the given family of maps (6.4), the relation

$$\lambda_V^U \circ \phi_U = \phi_V \circ \rho_V^U, \tag{6.6}$$

for any open sets $U, V$ in $X$, with $V \subseteq U$.

So based on (6.3) and (6.4), *one obtains that;*

$$\phi \equiv (\phi_U) \in \prod_{U \in \tau} \mathrm{Hom}\,(S(U), E(U)). \tag{6.7}$$

We have applied in (6.7) categorical notation concerning the set of all maps between the two sets $S(U), E(U)$ $(U \in \tau)$, the latter being considered as objects of the *category of sets,* in accordance with (5.10). So one further considers the *set of morphisms of $S$ in $E$,* denoted by

$$\mathrm{Hom}\,(S, E), \tag{6.8}$$

as that subset of the cartesian product set in (6.7), determined by the relation (6.6). Yet, one can still define *"composition of morphisms"* in the obvious way, obeying the appropriate associativity relations.

Therefore, given an arbitrary topological space $X$, one considers the *category of presheaves (of sets) on $X$*, denoted by

$$\mathcal{PS}h_X; \tag{6.9}$$

its *morphisms are given by (6.3)* (cf. also (6.4)), in *such* a manner *that (6.6) be satisfied.* In this context, we also note, in anticipation (see thus Section 13, in particular, (13.20), in the sequel), that the category of sheaves on $X$, $\mathcal{S}h_X$ (cf. (1.18)), is in fact a *full subcategory* of (6.9), being identified with its image in the latter, via an appropriate *equivalence (of functors);* this is, of course, of a particular significance, concerning morphisms of sheaves in that sense (cf. e.g. (13.19)).

On the other hand, based again on the categorical interpretation of a presheaf on $X$, as in (5.10), on sets the following definition, pertaining to standard particular morphisms in the category (6.9).

**Definition 6.1.** Given a *morphism* $\phi : S \longrightarrow E$ in the category (6.9), viz. a family of maps $\phi \equiv (\phi_U)_{U \in \mathcal{T}}$, as in (6.3) (satisfying thus (6.6)), we shall say that $\phi$ is an **injection,** respectively, a **surjection,** whenever each one of the individual maps $\phi_U : S(U) \longrightarrow E(U)$, with $U \in \tau$ (cf. (6.4)), is of the pertinent type, respectively.

So one concludes, in particular, that;

(6.10) a morphism $f : S \longrightarrow E$ of presheaves on $X$ (cf. (6.7)) is an *isomorphism* if, and only if, each one of the individual maps $\phi_U$, $U \in \tau$, is a bijection.

# 7. Stalks of a presheaf. Sheafification, or sheaf generated by a presheaf

Consider a presheaf (of sets), say,

$$S \equiv \left(S(U), \rho_V^U\right), \tag{7.1}$$

on a given topological space $X$ (cf. Definition 5.2), as well as, an arbitrary point $x \in X$.

Now, by letting (the index) $U$ in (7.1) vary over the *set of open neighborhoods of* $x$, say $\mathcal{V}(x)$, one gets through (7.1) and the *restriction of (5.4) to* $\mathcal{V}(x) \subseteq \tau$, still an *inductive system of sets.* In this context, it suffices to consider a *fundamental system of open neighborhoods of* $x$, say, $\mathcal{B}(x)$ (in fact, a *cofinal subset* of $\mathcal{V}(x)$, with respect to (5.4), suitably restricted). Indeed, we make full use of these two possibilities in the sequel. Thus, we can further consider the corresponding *inductive limit set,* say,

$$S_x \equiv \varinjlim_{U \in \mathcal{B}(x)} S(U) \cong \varinjlim_{U \in \mathcal{B}(x)} S(U), \tag{7.2}$$

where the second relation stands for a *bijection* (of sets; cf. also e.g. N. Bourbaki [1: p.89, and p.95, Proposition 8]).

We call the above set $S_x$ the *stalk* (or even the *fiber*) of the given presheaf $S$ on $X$, as in (7.1), *at the point* $x \in X$.

Our next objective is to collect together the stalks of $S$ at the various points of $X$, so that to build up a sheaf on $X$, say, $\mathcal{S}$, having the same stalks as the given presheaf $S$, while the topology of $\mathcal{S}$ will also depend on $S$. Thus, by considering the family $(S_x)_{x\in X}$ as *disjoint,* we further set

$$\mathcal{S} := \bigcup_{x\in X} S_x = \sum_{x\in X} S_x, \tag{7.3}$$

the latter being thus the corresponding union (*direct sum*) of the family in question. Yet, one has the obvious *projection* map

$$\pi : \mathcal{S} \longrightarrow X, \tag{7.4}$$

such that

$$\pi(S_x) := \{x\},\ x \in X. \tag{7.5}$$

Therefore, *one concludes that* (cf. also (1.5))

$$\mathcal{S}_x \equiv \pi^{-1}(x) = S_x,\ x \in X, \tag{7.6}$$

as already claimed, while we further prove that *the triple*

$$(\mathcal{S}, \pi, X), \tag{7.7}$$

as defined by (7.3) and (7.5), *can actually be made into a sheaf, by exhibiting thus* $\pi$ *as a local homeomorphism.* The idea here is to apply to the previous family $(S_x)_{x\in X}$ the argument of Section 3 in the preceding; so, by analogy with (3.2), we next define, through the given presheaf $S$, a suitable family of local maps (sections) $\mathfrak{S}$ :

Namely, for any open set $U \subseteq X$, one defines a *(canonical) map*

$$\rho_U : S(U) \longrightarrow \mathcal{S}(U, \pi) \tag{7.8}$$

(the range of $\rho_U$ is the set of *(local) sections of* $\pi$, we cannot refer to their continuity, as yet(!)), such that, *for any* $s \in S(U)$ *and* $x \in U$, one sets (see also (7.6));

$$\rho_U(s) \equiv \tilde{s} : x \longmapsto [\rho_U(s)](x) \equiv \tilde{s}(x) := \rho_U^x(s) \equiv [s]_x \in S_x = \mathcal{S}_x. \tag{7.9}$$

Here we still denote by

$$\rho_U^x : S(U) \longrightarrow S_x \equiv \varinjlim_{U\in\mathcal{V}(x)} S(U) \tag{7.10}$$

the corresponding (canonical) map, as defined, by (7.2), for any $x \in X$ and $U$ varying over the open neighborhoods of $x$. So, as follows from (7.10) and

(7.9), *the map (7.8) is well-defined,* as claimed.

**(Terminological) Scholium 7.1.**- The above map $\rho_U$, with $U \in \tau$, justifies the terminology, usually applied to the elements of the set $S(U)$, as, namely, *"sections"* of the given presheaf $S$ (cf. (7.1)) over the open set $U \subseteq X$. It concerns, in effect, the elements

$$\rho_U(s) \equiv \tilde{s} \in S(U, \pi), \tag{7.11}$$

as defined by (7.9), for any $s \in S(U)$, with $U$ open in $X$, the latter being in fact *sections of* $\pi$. (They will also become *continuous,* presently below, after we define a topology on the sheaf space $\mathcal{S}$; the latter topology will actually be the strongest one making these "sections" continuous. Cf. (7.12) below). However, several times (therefore, the previous abuse of language), the map (7.8) happens to be an injection (*"monopresheaves"*, cf. Section 11 in the sequel), or even a *bijection* (*"complete presheaves"*; ibid.).

Now, we next consider the following family (of *local $\mathcal{S}$-valued maps of $X$*, see also (7.9), or even (7.11))

$$\mathfrak{S} = \{\tilde{s} : s \in S(U), \textit{ with } U \textit{ open in } X\}, \tag{7.12}$$

which is proved to be a *defining family of sections of $\mathcal{S}$*, in the sense of (3.17); that is, $\mathfrak{S}$ *satisfies conds. i) – iii) of Section 3* (viz. (3.5), (3.6) and (3.7)). As a result (cf. Theorem 3.1), *the triple* $(\mathcal{S}, \pi, X)$ (see (7.7)) *will be a sheaf on* $X$.

So we come now to the *proof of our claim about* $\mathfrak{S}$*:* As already remarked (cf. (7.11)), each member of $\mathfrak{S}$ defines, in view of (7.9), a map

$$\tilde{s} : U \equiv \mathrm{Dom}\,(\tilde{s}) \longrightarrow \mathcal{S}, \tag{7.13}$$

such that one obtains

$$\pi \circ \tilde{s} = \mathrm{id}_U. \tag{7.14}$$

Thus, *$\tilde{s}$ is a local section of $\mathcal{S}$ over $U$*, for any $s \in S(U)$ and $U$ open in $X$, which will also be *continuous,* at the very end, as a result of the definition of the topology of $\mathcal{S}$. Now, for every $x \in U \equiv \mathrm{Dom}\,(\tilde{s})$, one obtains (see also (7.9))

$$\tilde{s}(x) := \rho_U^x(s) \in S_x = \mathcal{S}_x, \tag{7.15}$$

where $s \in S(U)$, so that $\mathfrak{S}$ *fulfils (3.5)*. Yet, if $z \in \mathcal{S}_x = S_x$ (cf. (7.6)), then, by the very definition of (7.2), there exists an open $U \in \mathcal{V}(x)$ and $s \in S(U)$ such that (cf. (7.9))

$$z = \rho_U^x(s) \equiv \tilde{s}(x), \tag{7.16}$$

that is, $\mathfrak{S}$ *satisfies (3.6)*, as well. Finally, suppose that

$$z \in \tilde{s}(U) \cap \tilde{t}(V), \tag{7.17}$$

for some $s \in S(U)$ and $t \in S(V)$, so that $\pi(z) \equiv x \in U \cap V$. Therefore (cf. (7.15)), one has

$$z = \rho_U^x(s) = \rho_V^x(t), \tag{7.18}$$

so that, by the very definition of (7.2), *there exists* an *open* $W \subseteq X$, *with* $x \in W \subseteq U \cap V$, such that

$$\rho_W^U(s) = \rho_W^V(t) \in S(W). \tag{7.19}$$

(In this respect, cf. also, for instance, A. Mallios [2: p.110, (1.13) and p.111, (1.14)]). Therefore (see also (7.9), as well as, (8.14) in the sequel), one obtains, for every $x \in W \subseteq U \cap V$,

$$\begin{aligned}(\tilde{s}|_W)(x) &= \tilde{s}(x) \equiv \rho_U^x(s) = \rho_W^x\left(\rho_W^U(s)\right) \equiv \widetilde{\rho_W^U(s)}(x) = \widetilde{\rho_W^V(t)}(x) \\ &\equiv \rho_W^x\left(\rho_W^V(t)\right) = \rho_V^x(t) \equiv \tilde{t}(x) = \left(\tilde{t}|_W\right)(x),\end{aligned} \tag{7.20}$$

that is,

$$\tilde{s} = \tilde{t}|_W, \tag{7.21}$$

so that $\mathfrak{S}$ *satisfies (3.7)* too, and this actually proves our claim, concerning the family $\mathfrak{S}$, as defined by (7.12). ∎ Now, one defines the topology of $\mathcal{S}$, according to Theorem 3.1 (cf. also Lemma 3.1), so that, as already promised, one can actually improve on the range of the map (7.8). That is, *one has the map* (see also (2.3))

$$\rho_U : S(U) \longrightarrow \Gamma(U, \mathcal{S}) \equiv \mathcal{S}(U), \tag{7.22}$$

*for every open* $U \subseteq X$, given by (7.9). (In this context, cf. also (11.6) in the sequel).

The sheaf $(\mathcal{S}, \pi, X)$, thus obtained from the given presheaf $S$, is called the *sheafification of* $S$, or even the *sheaf generated by the presheaf* $S$.

So by summing up the previous discussion, we conclude that;

(7.23) *with any given presheaf (of sets)* $S \equiv (S(U), \rho_V^U)$ *on a topological space* $X$, *there is associated (in a canonical way,* see also the next Section) *a sheaf on* $X$, $(\mathcal{S}, \pi, X)$, *viz. the sheafification of* $S$. *By this association, the individual members of* $S$ *and the corresponding (local) sections of* $\mathcal{S}$ (over the same subset of $X$) *are related by the canonical maps (7.8)* (cf. also (7.22)).

The previous conclusion may be viewed, indeed, as a prelude to the existence of an important functor, viz. the *"sheafification functor"*, between the categories $\mathcal{PSh}_X$ and $\mathcal{Sh}_X$, as they were defined in the preceding (cf. (6.9) and (1.18), respectively). We examine this functor presently below.

# 8. The sheafification functor

Suppose we are given an arbitrary *topological space,* and let $\mathcal{PSh}_X$ and $\mathcal{Sh}_X$ be the *category of presheaves* (of sets) *on* $X$ and the *category of sheaves* (of sets) *on* $X$, respectively (cf. (6.9) and (1.18)). So let us further consider the assignement

$$\mathbf{S} : \mathcal{PSh}_X \longrightarrow \mathcal{Sh}_X \tag{8.1}$$

between the two previous categories, in such a manner that, with a given presheaf (of sets) on $X$, say, $S \equiv (S(U), \rho_V^U)$ (cf. (7.1)), one associates the following sheaf (of sets) on $X$

$$\mathbf{S}(S) := (\mathcal{S}, \pi, X), \tag{8.2}$$

as the latter was defined from $S$ in the preceding Section 7 (cf. (7.7)); in other words, *with any given presheaf on* $X$, *one associates the sheaf on* $X$, *generated by the given presheaf,* according to (7.23). Thus, our next objective is now to show that;

(8.3) the correspondence

(8.3.1) $$\mathbf{S} : \mathcal{PS}h_X \longrightarrow \mathcal{S}h_X,$$

as given by (8.2), provides, in effect, a *(covariant) functor* between the respective categories. We call it the *sheafification functor.*

Indeed, what we actually have to prove is the *"functorial behaviour" of (8.3.1) relative to the morphisms* of the said categories: Namely, if $S$ and $E$ are two objects of $\mathcal{PS}h_X$, suppose that

(8.4) $$\phi \equiv (\phi_U) \in \mathrm{Hom}_{\mathcal{PS}h_X}(S, E) \equiv \mathrm{Hom}\,(S, E)$$

is a given *morphism of $S$ in $E$* in the category $\mathcal{PS}h_X$ (cf. (6.7), (6.8)). So one has to define a corresponding *morphism,* say,

(8.5) $$\mathbf{S}(\phi) \equiv \tilde{\phi} \in \mathrm{Hom}_{\mathcal{S}h_X}(\mathbf{S}(S), \mathbf{S}(E)) \equiv \mathrm{Hom}\,(\mathcal{S}, \mathcal{E}),$$

of the sheafification of $S$, $\mathbf{S}(S) \equiv \mathcal{S}$, into that of $E$, $\mathbf{S}(E) \equiv \mathcal{E}$ (cf. also (8.2)), *in the category $\mathcal{S}h_X$.* In other words (cf. Definition 1.2), we have to prove that the sought morphism $\tilde{\phi}$ is a *fiber preserving,* as well as, a *continuous map* of $\mathcal{S}$ into $\mathcal{E}$:

So, in this respect, we first remark that the commutative diagram (6.5) guarantees already, *for every* $x \in X$, the *existence of a (canonical) map* (cf. also (7.6))

(8.6) $$\phi_x : \mathcal{S}_x = S_x \longrightarrow E_x = \mathcal{E}_x$$

between the respective fibers of $\mathcal{S}$ and $\mathcal{E}$ (viz. the corresponding *inductive limit sets,* as in (7.2); cf. also, for instance, N. Bourbaki [1: p.92, Corollaire 1]). Thus, one has (see also ibid., p.93)

(8.7) $$\phi_x := \varinjlim_{U \in \mathcal{V}(x)} \phi_U, \; x \in X,$$

such that, *for any $U$ open* in $X$, *with* $x \in U$, (viz. $U \in \mathcal{V}(x)$), *one obtains*

(8.8) $$\phi_x \circ \rho_U^x = \lambda_U^x \circ \phi_U$$

(cf. also (6.1) and (7.10)). So the family of maps

(8.9) $$(\phi_x)_{x \in X},$$

as given by (8.7), yields further a map (cf. also (7.3))

(8.10) $$\widetilde{\phi} : \mathcal{S} = \sum_x S_x \longrightarrow \mathcal{E} = \sum_x E_x,$$

according to the relation;

(8.11) $$\widetilde{\phi}|_{S_x} \equiv \widetilde{\phi}_x := \phi_x, \; x \in X.$$

So, as follows from the last relation, *the following diagram is commutative*

(8.12) $$\begin{array}{ccc} \mathcal{S} & \xrightarrow{\widetilde{\phi}} & \mathcal{E} \\ & {\scriptstyle \pi} \searrow \quad \swarrow {\scriptstyle \kappa} & \\ & X & \end{array}$$

that is, one has

(8.13) $$\pi = \kappa \circ \widetilde{\phi},$$

where we set $\mathbf{S}(E) \equiv (\mathcal{E}, \kappa, X)$ (see also (7.7)). Therefore, the map $\widetilde{\phi}$ *is fiber preserving.* Yet, its continuity is provided by the following Lemma 8.1; however, it is instructive to point out, first, the ensuing remark, which will be employed by the proof of that lemma. So we have.

**Remark 8.1.**– The following *extension of (7.9)* will be applied presently below: Namely, *for any open sets $U, V$ in $X$ and $x \in U \subseteq V$, one first obtains* (cf. also (6.1))

(8.14) $$\rho_U^x = \rho_V^x \circ \rho_V^U.$$

(see e.g. N. Bourbaki [1: p.89; (22)], or even A. Mallios [2: p.110; (1.11)]). So as a result, *one concludes,* in view of (7.9), *that*

(8.15) $$\widetilde{s}(x) := \rho_U^x(s) = \rho_V^x\left(\rho_V^U(s)\right) =: \widetilde{\rho_V^U(s)}(x),$$

*for any* $x \in V \equiv \mathrm{Dom}\left(\widetilde{\rho_V^U(s)}\right) \subseteq U \equiv \mathrm{Dom}\,(\widetilde{s})$.

Thus, we come next to the proof of the following.

**Lemma 8.1.** *Let $\widetilde{\phi} : \mathcal{S} \longrightarrow \mathcal{E}$ be the map (8.10), as defined by (8.11) and satisfying thus (8.13). Then, $\widetilde{\phi}$ is continuous, so that it also yields a morphism of $\mathcal{S}$ into $\mathcal{E}$ in the category $\mathcal{S}h_X$* (cf. (1.18)).

**Proof.** Let $z \in \mathcal{S} = \sum_x \mathcal{S}_x$, so that (cf. (7.16)) $z = \widetilde{s}(x)$, for some $s \in S(U)$ with $\pi(z) = x$ and $\mathrm{Dom}(\widetilde{s}) \equiv U \in \mathcal{V}(x)$. Therefore, if (see also (8.13))

$$\phi_x(\widetilde{s}(x)) \equiv \widetilde{t}(x) \in E(x) = \mathcal{E}_x, \tag{8.16}$$

with $t \in E(V)$ and $V \equiv \mathrm{Dom}(\widetilde{t}) \in \mathcal{V}(x)$, let $N$ be an open neighborhood of $\widetilde{t}(x)$ in $\mathcal{E}$. Thus (cf. (3.11)), there exists an open $A \subseteq X$, with $x \in A \subseteq V$ and such that

$$\widetilde{t}(x) \in \widetilde{t}(A) \subseteq N. \tag{8.17}$$

Now, based on (8.16), (7.15) and (8.8), one has

$$\widetilde{t}(x) = \phi_x(\widetilde{s}(x)) = \phi_x\left(\rho_U^x(s)\right) = \lambda_U^x(\phi_U(s)) \equiv \widetilde{\phi_U(s)}(x), \tag{8.18}$$

where, by hypothesis (cf. (6.4)), $\phi_U(s) \in E(U)$ (*hence,* $\mathrm{Dom}\left(\widetilde{\phi_U(s)}\right) = \mathrm{Dom}(\widetilde{s}) \equiv U$), with $x \in U \cap V$. So (cf. also (2.12) or even (3.7)), there exists $W \in \mathcal{V}(x)$, with $W \subseteq U \cap A$, such that

$$\widetilde{t} = \widetilde{\phi_U(s)}|_W. \tag{8.19}$$

Now, *we claim that*

$$\widetilde{\phi}(\widetilde{s}(W)) \subseteq \widetilde{t}(W), \tag{8.20}$$

*which,* in fact (cf. also (8.17)), *supplies the continuity of $\widetilde{\phi}$ at $z = \widetilde{s}(x)$*: Indeed, *for every $x' \in W$ ($\subseteq U \cap A \subseteq U \cap V$), one obtains* (see also (6.6), (7.9), (8.8), (8.11) and (8.19), along with the preceding Remark 8.1);

$$\begin{aligned} \widetilde{\phi}(\widetilde{s}(x')) &= \phi_{x'}(\widetilde{s}(x')) = \phi_{x'}\left(\rho_U^{x'}(s)\right) = \phi_{x'}\left(\rho_W^{x'}\left(\rho_W^U(s)\right)\right) \\ &= \lambda_W^{x'}\left(\phi_W\left(\rho_W^U(s)\right)\right) = \lambda_W^{x'}\left(\lambda_W^U(\phi_U(s))\right) = \widetilde{\lambda_W^U(\phi_U(s))}(x') \\ &= \widetilde{\phi_U(s)}(x') = \widetilde{t}(x') \in \widetilde{t}(W), \end{aligned} \tag{8.21}$$

which proves our claim. ∎

Now, it is easy to prove further the rest of the conditions in order **S**, as given by (8.2), to be a *covariant functor;* namely, composition of morphisms, along with the corresponding associativity relations (in this regard, cf., for instance, N. Bourbaki [1: p.93, Corollaire 2]), as well as, the analogous identity morphisms with their properties.

By terminating this discussion, we still note, in anticipation (cf. thus Section 13 below), that *the* previously defined *functor* **S**, when suitably restricted to a subcategory of $\mathcal{PSh}_X$, *has a "left adjoint"* (the so-called *section functor,* see the next Section 9) by which *one can identify the category of sheaves over* $X$, $\mathcal{Sh}_X$, *with the source of the previously restricted sheafification functor.* So, according to this, very convenient, indeed, interpretation (this, in effect, was also the initial point of view in sheaf theory!), *sheaves* are displayed as those *objects in* $\mathcal{PSh}_X$, *for which the canonical maps (7.22) are bijections* (cf. Theorem 13.1 in the sequel). We refer thus to Section 13 for details, but see also the following Section 9.

# 9. Presheaf of sections of a sheaf. The section functor

The presheaf we refer to in the title of this section supplies, in fact, a fundamental example of this notion. (Further important examples will be considered in the next Section 10). In particular, presheaves arising from sheaves, in the way we describe it below, are further characterized, amongst all the presheaves on a given topological space, in Section 11 (cf. thus Theorem 11.2). Yet, the same type of presheaves provide, as already said, a basic, in effect, an *equivalent way of defining sheaves.*

Thus, suppose that we are given a *sheaf*

$$(\mathcal{S}, \pi, X) \tag{9.1}$$

on a topological space $X$. (As usual, we denote the topology of $X$ by $\tau$; cf. also Definition 1.1).

Now, one obtains a *presheaf (of sets) on* $X$, through the following *functor* (we defer the proof of the "functoriality" of the next correspondence for another place; cf. thus Section 13 in the sequel):

$$\Gamma(\cdot, \mathcal{S}) \equiv \Gamma_{\mathcal{S}} : \tau \longrightarrow \text{Set} \tag{9.2}$$

(the target of (9.2) stands for the *category of sets,* while its source is *the* (small) *category of the open subsets of* the given topological space $X$), such that

$$\Gamma(\cdot, \mathcal{S})(U) \equiv \Gamma_{\mathcal{S}}(U) := \Gamma(U, \mathcal{S}) \equiv \mathcal{S}(U), \tag{9.3}$$

for any open set $U \subseteq X$ (see also (2.3)), as well as, (5.10)). So, for any open sets $U, V$ in $X$, with $V \subseteq U$ (viz. $V \succeq U$, in the sense of (5.4)), one defines the corresponding *restriction map*

$$\sigma_V^U : \Gamma(U, \mathcal{S}) \longrightarrow \Gamma(V, \mathcal{S}), \tag{9.4}$$

by the relation;

$$\sigma_V^U(s) := s|_V, \quad s \in \Gamma(U, \mathcal{S}). \tag{9.5}$$

Here by the second member of (9.5) we really mean the *restriction to* $V$ of the given local map (section) $s : U \longrightarrow \mathcal{S}$ (see Definition 2.1); thus, we find at this point a first justification of the terminology applied to the maps (5.1) in the preceding. (See also Examples 10.(b) below).

So it is now readily seen that *the family*

$$\left(\Gamma(U, \mathcal{S}), \sigma_V^U\right), \tag{9.6}$$

as given by (9.3) and (9.5), with $U$ ranging over $\tau$, *yields a presheaf on* $X$; that is, an *inductive system of sets,* with respect to the (directed) index set $\tau$ (cf. Definition 5.2). We call it the *presheaf of sections of* $\mathcal{S}$, denoted by

$$\Gamma(\mathcal{S}) \equiv \Gamma_{\mathcal{S}} \equiv \left(\Gamma(U, \mathcal{S}), \sigma_V^U\right). \tag{9.7}$$

Now, since (9.6) defines an inductive system of sets, one can consider the corresponding *inductive limit set,* at any point $x \in X$, by letting the open set $U$, as above, varying over the open neighborhoods of $x$ (cf. e.g. (7.2)); thus, we set

$$\Gamma(\mathcal{S})_x \equiv (\Gamma_{\mathcal{S}})_x := \varinjlim_{U \ni x} \Gamma(U, \mathcal{S}) \tag{9.8}$$

(for simplicity we set above "$U \ni x$", in place of the more accurate notion, as we did in the preceding, "$U \in \mathcal{V}(x)$"; but, we do employ the same notation as in (9.8) occasionally in the sequel, as well). So the relation of the last set with the fiber (stalk) at $x \in X$ of the given sheaf $\mathcal{S}$ is now the subject matter of the next.

**Lemma 9.1.** *Let $(\mathcal{S}, \pi, X)$ be a given sheaf on a topological space $X$ and $\Gamma_{\mathcal{S}} \equiv (\Gamma(U, \mathcal{S}), \sigma_V^U)$ the corresponding presheaf of sections of $\mathcal{S}$* (cf. (9.6)). *Then, for every point $x \in X$, one obtains* (see also (9.8))

$$\mathcal{S}_x = (\Gamma_{\mathcal{S}})_x, \tag{9.9}$$

*within a bijection.*

**Proof.** We first remark that, for every $U \in \mathcal{V}(x)$ (viz. *open neighborhood of* $x$), one obtains a map

$$\delta_x^U : \Gamma(U, \mathcal{S}) \longrightarrow \mathcal{S}_x \subseteq \mathcal{S}, \tag{9.10}$$

according to the relation,

$$\delta_x^U(s) := s(x), \ s \in \Gamma(U, \mathcal{S}) \tag{9.11}$$

(*"evaluation map"* at $x \in U$). Now, it is easy to prove that the family

$$\left(\delta_x^U\right)_{U \in \mathcal{V}(x)}, \tag{9.12}$$

as given by (9.11), provides an *inductive system of maps* (viz. one has the relation $\delta_x^V \circ \sigma_V^U = \delta_x^U$, for any $U, V$ in $\mathcal{V}(x)$, with $V \subseteq U$; cf. (9.6), as well as, (5.4)). Therefore, there exists a corresponding (uniquely defined) *inductive limit map*

$$\delta_x \equiv \varinjlim_{U \ni x} \delta_x^U : (\Gamma_{\mathcal{S}})_x \equiv \varinjlim_{U \ni x} \Gamma(U, \mathcal{S}) \longrightarrow \mathcal{S}_x, \tag{9.13}$$

such that

$$\delta_x \circ \sigma_U^x = \delta_x^U, \tag{9.13$'$}$$

where $\sigma_U^x : \Gamma(U, \mathcal{S}) \longrightarrow (\Gamma_{\mathcal{S}})_x$, $U \in \mathcal{V}(x)$, stands for the corresponding (canonical) map, as provided by (9.8) (see also e.g. (7.10). In this context, cf., for instance, N. Bourbaki [1: p.91, Proposition 6]). We show now that *the map $\delta_x$ is the sought bijection in (9.9):* Indeed, we *first* observe that *$\delta_x$ is an onto map;* for, if $z \in \mathcal{S}_x \subseteq \mathcal{S}$, then (cf. also (2.10) and (9.11)), $z = s(x) = \delta_x^U(s)$, for some $s \in \Gamma(U, \mathcal{S})$. Therefore (cf. (9.13$'$)), $z = \delta_x^U(s) = \delta_x(\sigma_U^x(s))$, where $\sigma_U^x(s) \in (\Gamma_{\mathcal{S}})_x$, which proves our claim.

Now, we further prove that $\delta_x$ *is one-to-one;* equivalently (cf. the last citation), we prove that

(9.14) for any $U \in \mathcal{V}(x)$, if $s, t \in \Gamma(U, \mathcal{S})$, with $\delta_x^U(s) = \delta_x^U(t)$, then there exists $V \in \mathcal{V}(x)$, with $V \subseteq U$, such that $\sigma_V^U(s) = \sigma_V^U(t)$. That is (cf. also (9.5)), one has

$$s = t|_V. \tag{9.14'}$$

Thus, given $U \in \mathcal{V}(x)$, let $s, t \in \Gamma(U, \mathcal{S})$, with $s(x) = t(x)$ (cf. (9.11)). Hence (see (2.13)), there exists $V \in \mathcal{V}(x)$, with $V \subseteq U$, satisfying (9.14′), which proves our claim, and this also terminates the proof. ∎

As a consequence of the previous lemma, one still obtains the following useful.

**Corollary 9.1.** *Let* $S \equiv (S(U), \rho_V^U)$ *be a given presheaf (of sets) on a topological space* $X$, $\mathbf{S}(S) \equiv (\mathcal{S}, \pi, X)$ *its sheafification* (cf. (7.7) or even (8.2)) *and* $\Gamma_{\mathcal{S}} \equiv (\Gamma(U, \mathcal{S}), \sigma_V^U)$ *the presheaf of sections of* $\mathcal{S}$ (cf. (9.7)). *Then, for every point* $x \in X$, *one obtains*

$$S_x = (\Gamma_{\mathcal{S}})_x, \tag{9.15}$$

*within a bijection.*

**Proof.** By definition (cf. (7.6)), one has $S_x = \mathcal{S}_x$, $x \in X$, where $\mathcal{S} \equiv \mathbf{S}(S)$; hence, in view of (9.9), *one* actually *obtains,* within a bijection for the second of the next relations,

$$S_x = \mathcal{S}_x = (\Gamma_{\mathcal{S}})_x, \tag{9.16}$$

*for every* $x \in X$, which thus proves our assertion. ∎

In this context, it is worth pointing out here the above very convenient, indeed, behaviour of the stalks of the presheaves $S$ and $\Gamma_{\mathcal{S}}$, with $\mathcal{S} \equiv \mathbf{S}(S)$, as exhibited by (9.15), by contrast with what happens with the corresponding "sections" of $S$ (see also Scholium 7.1) and those (no inverted commas here!) of $\mathcal{S}$; see thus (7.22), along with Proposition 11.1 in the sequel.

Now, as another consequence of Lemma 9.1, we further obtain the following basic result, an application of which will be given in Section 11 below. So we have.

**Theorem 9.1.** *Let* $(\mathcal{S}, \pi, X)$ *be a given sheaf on a topological space* $X$, $\Gamma_{\mathcal{S}}$ *its presheaf of sections* (cf. (9.7)) *and* $\mathbf{S}(\Gamma_{\mathcal{S}})$ *the corresponding sheafification of* $\Gamma_{\mathcal{S}}$ (cf. (7.7)). *Then, one has*

$$\mathcal{S} = \mathbf{S}(\Gamma_{\mathcal{S}}) \equiv \mathbf{S}(\Gamma(\mathcal{S})), \tag{9.17}$$

*within an isomorphism of the sheaves concerned* (see also (1.20)).

**Proof.** As an application of Lemma 9.1, *one obtains, for every point* $x \in X$, *a bijection* (see also (7.6))

$$\phi_x : \mathcal{S}_x \longrightarrow (\Gamma_{\mathcal{S}})_x \equiv \Gamma(\mathcal{S})_x = \mathbf{S}(\Gamma(\mathcal{S}))_x. \tag{9.18}$$

(We note for later use, cf. (9.20) below, that we actually put here $\phi_x \equiv \delta_x^{-1}$, according to (9.13)). So one next defines the following *map of sheaves* (viz. fiber preserving)

$$\phi : \mathcal{S} \longrightarrow \mathbf{S}(\Gamma_{\mathcal{S}}), \tag{9.19}$$

such that , for every $x \in X$, one sets (see also (9.9))

$$\phi(\mathcal{S}_x) := \phi_x(\mathcal{S}_x) \cong (\Gamma_{\mathcal{S}})_x = \mathbf{S}(\Gamma(\mathcal{S}))_x. \tag{9.20}$$

In a similar way, one defines a fiber preserving map $\psi : \mathbf{S}(\Gamma_{\mathcal{S}}) \longrightarrow \mathcal{S}$, as *inverse of* $\phi$ (cf. for instance, (1.20)), by taking into account that $\phi_x$, $x \in X$, is a bijection. That is, $\phi$ *is a bijection,* as well.

Now, we prove further that *the map* $\phi$, as above, *is section preserving:* Thus, (see e.g. (2.16)), if $s \in \Gamma(U, \mathcal{S})$, we show that $\phi \circ s \in \mathbf{S}(\Gamma(\mathcal{S}))(U)$; for, if $x \in U$, then (cf. also (9.20), (9.11) and (9.13′)),

$$\begin{aligned}(\phi \circ s)(x) = \phi(s(x)) = \phi_x(s(x)) = \phi\left(\delta_x^U(s)\right) &= \phi_x\left(\left(\phi_x^{-1} \circ \sigma_U^x\right)(s)\right) \\ &= \sigma_U^x(s) \equiv \tilde{s}(x),\end{aligned} \tag{9.21}$$

where $\tilde{s} \in \Gamma(U, \mathbf{S}(\Gamma_{\mathcal{S}}))$ see also (7.9), (7.12)), which thus proves our claim. Of course, *this provides already the continuity of* $\phi$, according to the definition of the topology of $\mathcal{S}$ (cf. (2.15), or even Lemma 3.1), that is, $\phi$ *is a sheaf morphism* (cf. Definition 1.2). So $\phi$ being thus a local homeomorphism (Lemma 1.2), *its inverse* $\phi^{-1}$ (viz. $\psi$, as above) *is* a continuous map too (cf. (1.4)), hence *a sheaf morphism, as well,* so that (1.20) is satisfied, and this terminates the proof. ∎

By the last part of the previous proof, we have also established, incidentally, a very useful fact, indeed, yet, as we shall see, concerning later applications, as well, which also supplements Proposition 2.1 (cf. (2.18)). So it is worth pointing out it formally; that is, one concludes that

(9.22) *a map between sheaves is a sheaf morphism if, and only if, it is section preserving.* In particular (that was, in fact, the case above), *if it is also a bijection, then it is a sheaf isomorphism.*

(Indeed, (9.22) strengthens the *"if"* part of (2.18), in that *no continuity* of the map in question is required, *if* it is *section preserving*(!), due to the very definition of the topology of a sheaf).

Before ending this Section, we still give for later use another application of the preceding, related with the notion of a *subpresheaf* (cf. (5.11)). So one has the next.

**Lemma 9.2.** *Let $S \equiv (S(U), \rho^U_V)$ be a given presheaf (of sets) on a topological space $X$. Then, the sheafification of the restriction of $S$ to an open subset of $X$ is (isomorphic with) the restriction to the same subset of the sheafification of $S$; that is, for every open set $A \subseteq X$, one has the relation*

$$\mathbf{S}(S|_A) = \mathbf{S}(S)|_A, \tag{9.23}$$

*within an isomorphism of the sheaves concerned.*

**Proof.** We first remark that *the two sheaves in (9.23) have the same* (within a bijection) *defining family of sections;* see (3.17) for the terminology applied, as well as, (1.9) (and/or (1.10)), (5.12) and (7.12). In addition, *they also have the same* (within a bijection too) *sheaf space.* That is, one obtains

$$\mathbf{S}(S)_A = \pi^{-1}(A) = \sum_{x\in A} S_x \cong \mathbf{S}(S|_A). \tag{9.24}$$

See (1.10), (7.3) and (7.6); yet, the last isomorphism (of sets) in (9.24) is a consequence of the following one, valid *for every* $x \in A$,

$$S_x := \varinjlim_{U \ni x} S(U) \cong \varinjlim_{A \supseteq U \ni x} S(U) =: (S|_A)_x. \tag{9.25}$$

(See also (5.12); yet, the set $\{U \subseteq X, \text{ open} : x \in U \subseteq A\}$ is a *cofinal subset* of $\mathcal{V}(x)$, cf. also (5.4), hence, the bijection in (9.25)). Now, the preceding along with (3.17) provide the proof of our assertion. ∎

We shall see a first application of (9.23) in Section 11 below (cf. (11.39), as well as, Corollary 11.1), while some more ones will appear in Chapter IV in the sequel.

# 10. Further examples of presheaves

In this Section we consider two more examples of presheaves (of sets), apart from the one supplied by any given sheaf, through the "section functor" $\Gamma$, as described in the preceding Section. These are indeed amongst the more important examples of presheaves, while some more will appear, of course, throughout the sequel. Now, the *constant presheaf,* which we treat next, supplements the relevant picture we already had in Section 4.(a).

**10.(a). The constant presheaf.**– As is the case with a constant sheaf, the presheaf we consider below is the simplest one.

So let $X$ be any given *topological space* and $M$ an arbitrary *set;* thus, by looking as usual at the topology $\tau$ of $X$ as a (small) category (cf. (5.8)), one defines a *contravariant functor* (cf. also (9.2))

$$\mathbf{M} : \tau \longrightarrow \text{Set}, \tag{10.1}$$

by setting, for every open set $U \subseteq X$,

$$\mathbf{M}(U) := M. \tag{10.2}$$

Moreover, for any $V \subseteq U$ in $\tau$ (see also (5.4)) one defines, in view of (10.2), the respective *"restriction map"* (see (5.1)) by

$$\rho_V^U := \mathrm{id}_M \; (: \mathbf{M}(U) \longrightarrow \mathbf{M}(V)), \tag{10.3}$$

that is, the identity map on $M$.

Now, it is easy to verify our claim, concerning (10.2); in other words (cf. (9.2)), in view of (10.2) and (10.3), one obtains the following *inductive system of sets,* with respect to the (index) set $\tau$;

$$\mathbf{M} \equiv \left(\mathbf{M}(U) = M, \rho_V^U = \mathrm{id}_M\right) \equiv (M, \mathrm{id}_M). \tag{10.4}$$

We call it the *constant presheaf* $\mathbf{M}$ (or even simply $M$, viz. no "boldface em", if no confusion occurs. Yet, the use here of the same (boldface) $\mathbf{M}$, as for the constant sheaf, with fiber $M$ (cf. (4.4)), is made only for simplicity of the notation, while it is also justified by (10.8) below.

On the other hand, applying the technique of Section 7, one gets for the fiber at $x \in X$ of the presheaf (10.4)

$$\mathbf{M}_x := \varinjlim_{U \ni x} \mathbf{M}(U) \cong M, \tag{10.5}$$

within a *bijection* for the second relation (cf. also (7.2). In this respect see also, for instance, N. Bourbaki [1: p.90, Exemple 2]). Thus, one obtains as the corresponding *sheaf space,* say, $\mathcal{M}$, of $\mathbf{S}(\mathbf{M})$, the *sheafification of* $\mathbf{M}$, the set (cf. (7.3))

$$\mathcal{M} := \bigcup_{x \in X} M \; (\equiv \{x\} \times M) = X \times M. \tag{10.6}$$

Now, based on (7.12) and (4.7), one obtains, for any open $U \subseteq X$, the *open set* in $X \times M$

$$\widetilde{m}(U) = U \times \{m\}, \tag{10.7}$$

so that $X \times M$ carries the *cartesian product topology,* with the factor $M$ being endowed with the *discrete topology.* So one concludes that

$$\mathbf{S}(\mathbf{M}) = X \times M \equiv M_X \tag{10.8}$$

(see also (4.4′). Yet, cf. (11.21) in the sequel for another point of view, concerning (10.8)).

As a final remark, at this point, about the constant presheaf (10.4), we still note that the latter can be viewed as a *"presheaf of (continuous) functions"* in the point of view of the next Subsection 10.(b); that is, one can consider the elements of $M = \mathbf{M}(U)$, $U$ open in $X$ (cf. (10.2)), as *continuous functions on* $U$ *with values in the discrete space* $M$ (see also (4.7)). But, *the former do not exhaust all the latter functions!* See (11.15) in the sequel and the comments following it. So, thus viewed, $\mathbf{M}$ *is a monopresheaf,* in the sense of Remark 11.1, i) below; however, cf. also (11.14).

**10.(b). Presheaves of functions.**- We look next at the presheaf obtained by the *(local) continuous functions* on a given topological space $X$, as another fundamental example of a presheaf on $X$. The functions involved here are $\mathbf{C}$*-valued,* but *any other topological space, in place of the complexes* $\mathbf{C}$, *could be taken, as well.*

So consider the following family of sets, along with the corresponding restriction maps,

$$C \equiv \left(C(U), \rho_V^U\right) \tag{10.9}$$

such that one has, by definition,

$$C(U) := \mathcal{C}(U, \mathbf{C}) \equiv \mathcal{C}(U), \tag{10.10}$$

for any *open set* $U \subseteq X$ (cf. also (4.12)), while we still define

$$\rho_V^U : \mathcal{C}(U) \longrightarrow \mathcal{C}(V), \tag{10.11}$$

for any open subsets $U, V$ of $X$, *with* $V \subseteq U$ (cf. (5.4)), as the *(canonical) restriction map;* i.e., one sets, *for every* $f \in \mathcal{C}(U)$,

$$\rho_V^U(f) := f|_V. \tag{10.12}$$

So one finds here, through the above classical example, a full justification of the relevant terminology, that has been applied, thus, *by extension* of the previous situation, to the maps (5.1) in the preceding. (But, see also Remark 11.1, i) below).

Now, it is readily seen that the previous family (10.9) yields an inductive system (of sets), with respect to $\tau$, that is (cf. Definition 5.2), it provides a presheaf (of sets) on $X$. We call it the *presheaf of* (**C**-*valued) continuous functions on* $X$.

In this context, we still remark, in anticipation (see Section 11 below), that *the above presheaf can,* in effect, *be identified* (in a sense, however, that will be made precise later on (ibid.)) *with the sheaf* $\mathcal{C}_X$, considered by (4.10).

Now, continuing our examples of *"functional presheaves"*, as above, we can further consider the

(10.13) *presheaf of* (**C**-valued) $\mathcal{C}^\infty$*-functions on a given* $\mathcal{C}^\infty$*-manifold* $X$, viz. the family (inductive system of sets)

$$C^\infty \equiv \left(\mathcal{C}^\infty(U), \rho_V^U\right). \tag{10.13.1}$$

So, by analogy with (10.10), we put here $C^\infty(U) := \mathcal{C}^\infty(U)$, i.e., the set of **C**-valued $\mathcal{C}^\infty$-functions on the open set $U \subseteq X$, while the (restriction!) maps $\rho_V^U$, with $V \subseteq U$ open subsets of $X$, have a meaning analogous to that one in (10.12).

Yet, in a similar way, we finally consider the

(10.14) *presheaf of holomorphic functions on a complex* (analytic) *manifold* $X$, i.e.,

(10.14.1) $$O \equiv \left(\mathcal{O}(U), \rho_V^U\right),$$

where we set $O(U) := \mathcal{O}(U)$ for the holomorphic functions on the open $U \subseteq X$, along with the maps $\rho_V^U$ analogously to (10.12).

Yet, in both the previous two examples, as with (10.9), one can consider, in place of **C**, *any* other *smooth* (viz. $\mathcal{C}^\infty$-), *or complex* (analytic) *manifold, respectively, as range of the functions involved.* Finally, in a similar manner, as with the case of (10.9), *the last two presheaves can be identified,* in effect (cf. Section 11), *with the sheaves* $\mathcal{C}^\infty{}_X$ *and* $\mathcal{O}_X$, that were considered by (4.39) and (4.40), *respectively.* However, we shall return to all the above important examples of presheaves right in the next Section.

## 11. Complete presheaves. Examples

We examine below an important subcategory of the category of presheaves (of sets) on a given topological space $X$. In fact, it is standard to show that this subcategory of $\mathcal{PSh}_X$ (cf. (6.9)) provides essentially an equivalent manner, to that employed so far, of describing sheaves on $X$. Yet, as already mentioned, it is actually the way we consider next, in which sheaves were initially introduced (*Leray,* 1946; but see also H. Grauert – R. Remmert [2: p.VII, Introduction]). So we start with the following fundamental.

**Definition 11.1.** Let

(11.1) $$S \equiv \left(S(U), \rho_V^U\right)$$

be a presheaf (of sets) on a topological space $X$ (cf. Definition 5.1). Then, we say that $S$ is a **complete presheaf** (of sets) **on** $X$, if the two following conditions are satisfied:

(S1) If $U$ is an open subset of $X$ and $\mathcal{U} = (U_\alpha)_{\alpha \in I}$ an open covering of $U$, let $s, t$ be two elements ("sections") of $S(U)$ such that

(11.2) $$\rho_{U_\alpha}^U(s) \equiv s_\alpha = t_\alpha \equiv \rho_{U_\alpha}^U(t),$$

for every $\alpha \in I$. Then, $s = t$ (the converse is certainly true).

(S2) Let $U$ and $\mathcal{U} = (U_\alpha)_{\alpha \in I}$ be as in (S1); moreover, let $(s_\alpha) \in \prod_\alpha S(U_\alpha)$ such that, for any $U_{\alpha\beta} \equiv U_\alpha \cap U_\beta \neq \emptyset$ in $\mathcal{U}$, one has

$$\rho^{U_\alpha}_{U_{\alpha\beta}}(s_\alpha) \equiv s_\alpha|_{U_{\alpha\beta}} = s_\beta|_{U_{\alpha\beta}} \equiv \rho^{U_\beta}_{U_{\alpha\beta}}(s_\beta). \tag{11.3}$$

Then, there exists an element $s \in S(U)$ such that

$$\rho^{U}_{U_\alpha}(s) \equiv s|_{U_\alpha} = s_\alpha, \tag{11.4}$$

for every $\alpha \in I$.

**Remarks 11.1.**– i) It is customary to call a presheaf on $X$ for which (S1) is valid a *monopresheaf* (of sets) *on* $X$.

ii) Concerning a complete presheaf, it is easily seen, by virtue of (S1), that *the element* $s \in S(U)$, whose existence is assured by (S2), *is,* in fact, *uniquely defined.*

We further prove that *the presheaf of sections of a given sheaf is a complete presheaf;* as we shall see, this is, in effect (cf. Theorem 11.2 below), the only way that such presheaves arise. But, first we shall need the following observation, pertaining to the maps (7.22):

So let $S \equiv \left(S(U), \rho^U_V\right)$ be a given *presheaf on* $X$, $\mathcal{S} \equiv \mathbf{S}(S)$ the corresponding *sheafification of* $S$ and $\Gamma_{\mathcal{S}} \equiv \Gamma(\mathcal{S}) = (\Gamma(U, \mathcal{S}), \sigma^U_V)$ the associated *presheaf of sections of* $\mathcal{S}$ (see (7.7) and (9.7)). Thus, according to (7.22) (but, see also (7.9)), one obtains a family of maps

$$\rho_U : S(U) \longrightarrow \Gamma(U, \mathcal{S}), \tag{11.5}$$

with $U$ ranging over the open subsets of $X$ (viz. the index set (topology of $X$) $\tau$; cf. (5.4)). Now, our claim here is that;

(11.6) *the map*

(11.6.1) $$\rho \equiv (\rho_U)_{U \in \tau} : S \longrightarrow \Gamma_{\mathcal{S}} \equiv \Gamma(\mathbf{S}(S)),$$

as defined by (11.5) (viz., in fact, by (7.9)), *is a morphism of the presheaves concerned* (see also (6.3)).

So what one has to prove here is that, for any $U, V$ in $\tau$, with $V \succeq U$ (viz. $V \subseteq U$, cf. (5.4)), *the following diagram is commutative* (cf. also (6.5));

$$
\begin{array}{ccc}
S(U) & \xrightarrow{\quad \rho_U \quad} & \Gamma(U, \mathcal{S}) \\
\rho_V^U \Big\downarrow & & \Big\downarrow \sigma_V^U \\
S(V) & \xrightarrow{\quad \rho_V \quad} & \Gamma(V, \mathcal{S})
\end{array}
\tag{11.7}
$$

Thus, for any $s \in S(U)$ and $x \in V \subseteq U$, one obtains (cf. (7.9), (8.15) and (9.5))

$$
\begin{aligned}
&\sigma_V^U(\rho_U(s))(x) = (\rho_U(s)|_V)(x) \equiv (\tilde{s}|_V)(x) = \tilde{s}(x) = \rho_U^x(s) \\
&= \rho_V^x\left(\rho_V^U(s)\right) \equiv \widetilde{\rho_V^U(s)}(x) \equiv \rho_V\left(\rho_V^U(s)\right)(x),
\end{aligned}
\tag{11.8}
$$

that is, *one has*

$$
\rho_V \circ \rho_V^U = \sigma_V^U \circ \rho_U,
\tag{11.9}
$$

*for any* $V \subseteq U$ *in* $\mathcal{T}$, as desired.

So we come next to the promised result, concerning the presheaf of sections of a sheaf. That is, we have.

**Lemma 11.1.** *Let* $\mathcal{S}$ *be a given sheaf on a topological space* $X$ *and* $\Gamma_{\mathcal{S}} \equiv (\Gamma(U, \mathcal{S}), \sigma_V^U)$ *the corresponding presheaf of sections of* $\mathcal{S}$. *Then,* $\Gamma_{\mathcal{S}}$ *is a complete presheaf.*

**Proof.** We have to prove that $\Gamma_{\mathcal{S}}$ *satisfies conds. (S1) and (S2)* of Definition 11.1; now, based on (9.5) and (11.2), it is easy to prove that $\Gamma_{\mathcal{S}}$ satisfies (S1). Now, if (11.3) is valid for a given family of elements of $\Gamma(U, \mathcal{S})$, say, $(s_\alpha)$, then by a standard argument one defines a *common extension* of the given $s_\alpha$'s, say, $s : U \longrightarrow \mathcal{S}$; yet, the latter, by its very definition, is a section of $\mathcal{S}$ (i.e., of the map $\pi$ in (1.2), not yet continuous, a fact that has to be proved!), satisfying (11.4). Finally, the continuity of $s$ is still an easy consequence of its definition (cf. also (1.3)), so that $\Gamma_{\mathcal{S}}$ satisfies (S2), as well, and this completes the proof. ∎

Now, as a corollary of the previous proof, one realizes, in general, that

(11.10) *any presheaf* (of sets) *on a topological space* $X$, *whose terms* (sets, $S(U)$) *are composed of (local) functions* (defined on the various open sets $U \subseteq X$, with $\rho_V^U$ *the usual restriction of functions*), *the latter functions being also "locally characterized", provides, in fact, a complete presheaf.*

So one concludes that.

(11.11) *The following "functional presheaves"* (viz. *presheaves of functions*) *are, in effect, complete presheaves:*
i) The *presheaf of* (**C**-*valued) continuous functions* on a given topological space $X$ (cf. (10.9)).
ii) The *presheaf of* (**C**-*valued)* $\mathcal{C}^\infty$-*functions* on a $\mathcal{C}^\infty$-manifold $X$ (see (10.13)).
iii) The *presheaf of holomorphic functions* on a given complex (analytic) manifold $X$ (cf. (10.14)).

Of course, concerning the *range of the functions considered above,* analogous remarks to those at the beginning of Subsection 10.(b) and at the end of Section 10 are here too in force. In this context, we further remark that;

(11.12) *the sheafification of the complete presheaves considered by (11.11) are the corresponding "functional sheaves", as they were derived in Subsections 4.(b) and 4.(c). That is, one has*

(11.12.1) $\mathbf{S}(C) = \mathcal{C}_X,\ \ \mathbf{S}(C^\infty) = \mathcal{C}^\infty{}_X$ *and* $\mathbf{S}(O) = \mathcal{O}_X$.

The assertion is straightforward by the very definitions; so cf. Subsections 4.(b), 4.(c), as well as, Section 7. (We have thus here, through, namely, the notion of a (complete) presheaf, another (more sophisticated!) way of looking at the relevant constructions of Section 4 in the preceding). As a result of (11.12) (see thus Proposition 11.1 in the sequel), one still obtains, for instance, the relation (cf. (9.7) and (10.9))

(11.13) $$C = \Gamma(\mathcal{C}_X),$$

within an *isomorphism of* the *presheaves* concerned (see also (6.10)). Analogous relations to (11.13) for the rest two (complete) presheaves $C^\infty$ and $O$, as above, hold certainly true.

**Scholium 11.1.**– Concerning the comments in (11.10), it has still to be remarked here that the requirement there of being the relevant functions *"locally characterized"* is quite essential: So by considering, for instance, the *presheaf on* $\mathbf{C}$, whose (local) "sections" are the *bounded* (non "local property"!) *holomorphic functions,* one obtains a *monopresheaf on* $\mathbf{C}$, *which is not a complete presheaf (Liouville's Theorem;* see e.g. O. Forster [1: p.44, Exerc. 6.1]). In this context, it is readily seen, however, that, in general,

(11.14) *"functional presheaves"* (viz. presheaves of functions), *whose "restriction morphisms" are the usual restriction of functions, are always monopresheaves.*

Now, within the same context, as already seen in Subsection 10.(a), *the sheafification of a constant presheaf* $\mathbf{M}$, derived from a given *set* $M$, *is* (but, cf. also (10.5)) *the constant sheaf* $M_X$, associated with $M$, when the latter set carries the discrete topology (cf. (10.8)). This resembles, of course, the construction involved in (11.12.1). But, at this point, we still note that;

(11.15) *a constant presheaf* $\mathbf{M}$, corresponding to a given set $M$, as in (10.4), *being always a monopresheaf* (cf. (11.14)), *is not, in general, a complete presheaf.*

Namely, if the set $M$ contains, for instance, more than one point, while the open set $U \subseteq X$ is not connected (see also (10.2)), the inclusion (cf. (4.7) and (4.35))

(11.16)
$$\mathbf{M}(U) \equiv M \overset{\subseteq}{\rightarrow} \mathcal{C}(U, M)$$

may, of course, *not* be *an onto map!*

On the other hand, by analogy with the previous Example (11.11; i)) (see also (11.10), as well as, the comments following (11.11)), one concludes that

(11.17)
$$\mathcal{C}_X^M \equiv \left(\mathcal{C}(U, M), \rho_V^U\right),$$

with $\rho_V^U$ the usual restriction of functions (cf. e.g. (10.11)), *is a complete presheaf.* Now, the corresponding sheafification of (11.17), i.e., the sheaf

(11.18)
$$\mathbf{S}\left(\mathcal{C}_X^M\right),$$

is called the *sheaf of germs of* $M$*-valued locally constant functions on* $X$, a terminology justified, in view of the (discrete) topology of $M$. The same sheaf is also occasionally denoted simply by $M$. Yet, by considering the *sheafification of the constant presheaf* $\mathbf{M}$ (see (10.4)), one obtains for the respective *stalks*

(11.19) $$\mathbf{M}_x = \mathbf{S}(\mathbf{M})_x = (M_X)_x \cong M, \quad x \in X,$$

(cf. also (7.6) (10.5) and (10.8)). Moreover, based on (4.38), one has

(11.20) $$C_X^M(U) \equiv \mathcal{C}(U, M) \cong \Gamma(U, M_X) \equiv M_X(U),$$

within a *bijection,* as indicated, *for every open set* $U \subseteq X$, so that one obtains (cf. Section 3) the following *sheaf isomorphisms;*

(11.21) $$\mathbf{S}(\mathbf{M}) = M_X = \mathbf{S}\big(C_X^M\big).$$

Thus, one can now recapitulate the preceding into the following.

**Theorem 11.1.** *Let $X$ be a topological space and $M$ any given set. Then, the following sets coincide, within (canonical) bijections:*

i) The set of *local sections of the constant sheaf* $M_X$ over an open set $U \subseteq X$,

ii) that of *$M$-valued continuous functions on $U$,* when $M$ carries the discrete topology, and

iii) that of *$M$-valued locally constant functions on $U$*, with $M$ as in ii). ∎

We give next the already promised important *characterization of complete presheaves,* in terms of the canonical maps $\rho_U$, as they were defined by (7.22) (cf. also (7.9)). So one has the following. (But, see also (11.37) below).

**Proposition 11.1.** *Let $S \equiv \big(S(U), \rho_V^U\big)$ be a given presheaf (of sets) on a topological space $X$ and $\mathcal{S} \equiv \mathbf{S}(S)$ the corresponding sheafification of $S$. Then, the following two assertions are equivalent:*

*1) $S$ is a complete presheaf* (cf. Definition 11.1).

*2) For any open set $U \subseteq X$, the respective (canonical) map* (cf. (7.22))

(11.22) $$\rho_U : S(U) \longrightarrow \Gamma(U, \mathcal{S}) \equiv \mathcal{S}(U)$$

*is a bijection.*

**Proof.** Let $U$ open in $X$, and $s, t \in S(U)$ such that (cf. (7.9)

(11.23) $$\tilde{s} \equiv \rho_U(s) = \rho_U(t) \equiv \tilde{t};$$

therefore, for any point $x \in U$, one obtains (still consult (7.9))

$$[s]_x \equiv \tilde{s}(x) := \rho_U^x(s) = \rho_U^x(t) =: \tilde{t}(x) \equiv [t]_x, \tag{11.24}$$

so that there exists an open neighborhood of $x$, say, $V_x \subseteq U$, such that

$$\rho_{V_x}^U(s) \equiv s|_{V_x} = t|_{V_x} \equiv \rho_{V_x}^U(t), \quad x \in U. \tag{11.25}$$

(That is, (11.24) entails that $s \underset{x}{\sim} t$, in the analogous sense, for instance, to (4.15), when considering $\tilde{s}(x) = \tilde{t}(x)$ in $\mathcal{S}_x$; cf. (7.2) and (7.10). See also e.g. N. Bourbaki [1: p.89] or even A. Mallios [2: p.109; (1.6)]). In other words, the given *"sections"* (cf. (Terminilogical) Scholium 7.1, concerning the "inverted commas" here) $s, t$ *of* $S$ *over* $U$, namely, the elements $s, t$ in $S(U)$, as above, satisfy (11.2), relative to the *open covering* $\mathcal{U} = (V_x)_{x\in U}$ *of* $U$. Consequently, *if (S1) is valid,* one concludes that $s = t$, that is, *the map (11.22) is injective.*

Now, suppose that $t \in \Gamma(U, \mathcal{S})$; then (cf. also (7.6), (7.10)), one obtains, for any $x \in U$,

$$t(x) = \rho_V^x(s) \in \mathcal{S}_x = S_x := \varinjlim_{V\in\mathcal{V}(x)} S(V), \tag{11.26}$$

for some $s \in S(V)$ with $V \in \mathcal{V}(x)$. Therefore, setting $V \equiv V_x$ and $s \equiv s_x$, one obtains an open covering $\mathcal{U} = (V_x)_{x\in U}$ of $U$, along with an element $(s_x) \in \prod_{x\in U} S(V_x)$, associated with $\mathcal{U}$; moreover, based on (11.26) and (2.13), *we may assume* here that

$$t = \tilde{s}_x|_{V_x}, \textit{ such that } (\text{cf. } (7.9))\ \tilde{s}_x := \rho_{V_x}(s_x), \ x \in U. \tag{11.27}$$

So we further claim that *the previous data satisfy the analogous relation to (11.3):* That is, one has to prove that, for any $x, y$ in $U$, with $W \equiv V_x \cap V_y \neq \emptyset$, one has

$$\rho_W^{V_x}(s_x) \equiv s_x|_W = s_y|_W \equiv \rho_W^{V_y}(s_y). \tag{11.28}$$

Thus, by further *assuming that (S1) is valid* (hence, as already proved, *(11.22) is one-to-one),* it suffices to prove (11.28) for the corresponding *sections of* $\mathcal{S}$, i.e.,

$$\tilde{s}_x = \tilde{s}_y|_W. \tag{11.28'}$$

But, *this is already clear, by virtue of (11.27).* Therefore, *if (S2) is true, then there exists an element* $s \in S(U)$ such that one has

$$\rho^U_{V_x}(s) \equiv s|_{V_x} = s_x, \ x \in U. \tag{11.29}$$

Furthermore, by considering the corresponding to $s \in S(U)$ section of $\mathcal{S}$, i.e., the element $\tilde{s} := \rho_U(s) \in \Gamma(U, \mathcal{S})$, one obtains (cf. also (7.9), (8.14), (11.27) and (11.29))

$$\tilde{s}(x) := \rho^x_U(s) = \rho^x_{V_x}\left(\rho^U_{V_x}(s)\right) \equiv \rho^x_{V_x}(s_x) \equiv \tilde{s}_x(x) = t(x); \tag{11.30}$$

that is, one has

$$\tilde{s} := \rho_U(s) = t, \tag{11.31}$$

which thus proves that *(11.22) is a surjective map* (if it is already one-to-one), so that one finally concludes that *1) implies 2)*.

Conversely, *assume that (11.22) is a bijection;* thus, taking $s, t$ in $S(U)$, for some *open* set $U \subseteq X$, for which *(11.2) holds true,* one obtains (cf. also (8.14))

$$\tilde{s}(x) := \rho^x_U(s) = \rho^x_{U_\alpha}\left(\rho^U_{U_\alpha}(s)\right) = \rho^x_{U_\alpha}\left(\rho^U_{U_\alpha}(t)\right) = \rho^x_U(t) =: \tilde{t}(x), \tag{11.32}$$

*for any* $x \in U = \bigcup_\alpha U_\alpha$, so that $x \in U_\alpha$, for some $\alpha \in I$; hence, $\tilde{s} = \tilde{t}$, so that, by hypothesis for (11.22), one has $s = t$, as well, that is, *(S1) is satisfied.* On the other hand, let $(s_\alpha)$ be a given family of "sections" of $S$, satisfying (11.3), and let

$$(\tilde{s}) \in \prod_{\alpha \in I} \Gamma(U_\alpha, \mathcal{S}) \tag{11.33}$$

be the corresponding family of sections of $\mathcal{S}$, as given by (11.22). Thus, by (11.3), (11.7), one gets for (11.33) the analogous relation to (11.3). Hence (Lemma 11.1, viz. *(S2) is valid for* $\Gamma(\mathcal{S})$*)*, there exists a section $t \in \Gamma(U, \mathcal{S})$ such that

$$t|_{U_\alpha} = \tilde{s}_\alpha, \ \alpha \in I. \tag{11.34}$$

But, by hypothesis for (11.22), one also has $t = \tilde{s} := \rho_U(s)$, for some $s \in S(U)$. Therefore, in view of (11.7) and (11.34), one obtains

$$\begin{aligned} \widetilde{s|_{U_\alpha}} \equiv \widetilde{\rho^U_{U_\alpha}(s)} &:= \rho_{U_\alpha}\left(\rho^U_{U_\alpha}(s)\right) = \sigma^U_{U_\alpha}(\rho_U(s)) \equiv \sigma^U_{U_\alpha}(\tilde{s}) \\ &= \sigma^U_{U_\alpha}(t) \equiv t|_{U_\alpha} = \tilde{s}_\alpha \equiv \rho_{U_\alpha}(s_\alpha), \end{aligned} \tag{11.35}$$

for any $\alpha \in I$, so that, by virtue of the hypothesis for the maps (11.22), one obtains

(11.36) $$\rho^U_{U_\alpha}(s) \equiv s|_{U_\alpha} = s_\alpha,\ \alpha \in I.$$

Thus, *(S2) is also satisfied,* so that *2) implies 1), as well,* and this completes the proof. ∎

As a byproduct of the previous proof, we still point out for steady use below that:

(11.36′) *a given presheaf (of sets) $S$ on a topological space $X$ is a monopresheaf if, and only if, the maps (11.22) are injective.*

Now, as already observed (cf. (11.6)), *the maps (11.22) define a morphism of* the corresponding *presheaves* (viz. (11.7) is satisfied). Consequently, one can have the following quite convenient *restatement of Proposition 11.1* (see also (6.10)):

(11.37) *A given presheaf (of sets) $S$ on a topological space $X$ is complete if, and only if, it is isomorphic with the presheaf of sections of its sheafification; that is, if, and only if, one has*

(11.37.1) $$S = \Gamma(\mathbf{S}(S)),$$

within an *isomorphism of* the *presheaves* involved.

Employing categorical language, (11.37.1) provides, as we shall see later (cf. Section 13), a *natural isomorphism of the functors* $\Gamma$ *and* $\mathbf{S}$*;* so it supplies half of the *"adjoint equivalence"* between the categories $\mathcal{S}h_X$ (cf. (1.18)) and that of the *complete presheaves* on $X$ (see (11.63) below, as well as, Section 13 in the sequel). Indeed, one can actually consider in (11.37.1) *any other sheaf* $\mathcal{E}$ *on* $X$*, in place of* $\mathbf{S}(S)$; but, on this matter, see Theorem 11.2 below.

Now, as an application of Proposition 11.1, in conjunction with Lemma 9.2, one obtains that;

(11.38) *the restriction of a complete presheaf to an open subset of the base space is still a complete presheaf on that set.*

Indeed, if $A$ is an open subset of $X$, base space of the presheaf considered, say, $S$, by employing the notation of (9.23), one obtains, for every open set $U \subseteq A$ (cf. also (2.3) and (5.12)),

$$(11.39)\qquad \begin{aligned}(S|_A)(U) &:= S(U) \cong \Gamma(U, \mathbf{S}(S)) \equiv \mathbf{S}(S)(U) \\ &\cong (\mathbf{S}(S)|_A)(U) = \mathbf{S}(S|_A)(U),\end{aligned}$$

which, in view of Proposition 11.1, proves the assertion. ∎

**Remark.**– Concerning the second *bijection*, "≅", in (11.39), we implicitly made use here of the (obvious!) fact that;

$$(11.40)\qquad (\mathcal{S}|_A)(U) = \mathcal{S}(U),$$

within a *bijection* of the respective sets, true, *for any sheaf* $\mathcal{S}$ on a topological space $X$, *and any open* sets $U, A$ in $X$, *with* $U \subseteq A$. (See also (1.10)).

As a result of the foregoing we now have the following.

**Corollary 11.1.** *Let* $\mathcal{S}$ *be a given sheaf on a topological space* $X$ *and* $\Gamma(\mathcal{S})$ (cf. (9.7)) *the corresponding (complete,* see Lemma 11.1*) presheaf of sections of* $\mathcal{S}$. *Moreover, let* $A$ *be any open subset of* $X$. *Then, one obtains the following isomorphism of the complete presheaves concerned* (cf. also (1.10)),

$$(11.41)\qquad \Gamma(\mathcal{S}|_A) = \Gamma(\mathcal{S})|_A.$$

**Proof.** First, based on (9.23) and (9.17), we get the following *sheaf isomorphisms;*

$$(11.42)\qquad \mathbf{S}(\Gamma(\mathcal{S})|_A) = \mathbf{S}(\Gamma(\mathcal{S}))|_A = \mathcal{S}|_A.$$

Therefore, since *a functor,* like $\Gamma$ (cf. Section 13), *preserves isomorphisms,* one obtains, by virtue of (11.37.1) and (11.42), the following *isomorphisms of presheaves*

$$(11.43)\qquad \Gamma(\mathcal{S}|_A) = \Gamma(\mathbf{S}(\Gamma(\mathcal{S})|_A)) = \Gamma(\mathcal{S})|_A,$$

which thus proves our claim. ∎

We come next to the promised variant of Proposition 11.1 (cf. the comments following (11.37.1)). So we have.

**Theorem 11.2.** *Let $X$ be a topological space and $S$ a given presheaf (of sets) on $X$.* (cf. Definition 5.1). *Then, the three following assertions are equivalent:*

*1) $S$ is a complete presheaf on $X$* (cf. Definition 11.1).

*2) $S$ is isomorphic with the (complete,* cf. Lemma 11.1*) presheaf of sections of its sheafification $\mathbf{S}(S)$.*

*3) $S$ is isomorphic with the (complete) presheaf of sections of a sheaf on $X$.*

**Proof.** Of course, the first two assertions are equivalent, due to Proposition 11.1, which, in turn, proves also (through the sheaf $\mathbf{S}(S)$) that *1) implies 3)*. Now, if $S \cong \Gamma(\mathcal{E})$, for some sheaf $\mathcal{E}$ on $X$, then by taking sheafifications in the last relation, one further obtains (within *sheaf isomorphisms*): $\mathbf{S}(S) = \mathbf{S}(\Gamma(\mathcal{E})) = \mathcal{E}$ (see also (9.17)). Therefore, applying the *functor* $\Gamma$ on the last relations and based on (11.37.1), for the sheaf $\mathcal{E}$, as well as, on the comments after (11.37.1), one finally obtains (yet, within *isomorphisms of presheaves*); $\Gamma(\mathcal{E}) = \Gamma(\mathbf{S}(S)) = S$, that is, (11.37.1) for $S$, so that *3) implies 2), hence, 1), as well,* and this completes the proof. ∎

As a byproduct of the previous proof, we still observe that;

(11.44) *isomorphisms of presheaves preserve "completeness".*

In this context, we further remark that the above *characterization of complete presheaves, through prop. 3) of Theorem 11.2,* a seemingly stronger condition than that provided by (11.37.1), is also taken, occasionally, as a *definition of a sheaf* on $X$. (See, for instance, J. R. Strooker [1: p.205, Definition 4.5.1]; but, in this concern, cf. also Theorem 13.1 in the sequel).

We terminate the discussion in this Section with one more remark, pertaining to complete presheaves; namely,

(11.45) given a *presheaf* $S$, the complete presheaf $\Gamma_{\mathcal{S}}$, with $\mathcal{S} \equiv \mathbf{S}(S)$ (cf. Lemma 11.1), is the *"smallest" complete presheaf containing* $S$.

So $\Gamma_{\mathcal{S}} \equiv \Gamma(\mathbf{S}(S))$ (cf. e.g. (11.6)) is a kind of *"complete presheaf envelope"*(!) for $S$. Precisely speaking, we thus have the following. (But, see also (11.53) below).

**Theorem 11.3.** *Let* $S \equiv \left(S(U), \rho_V^U\right)$ *be a given presheaf (of sets) on a topological space* $X$, $\mathcal{S} \equiv \mathbf{S}(S)$ *the corresponding sheafification of* $S$ *and* $\Gamma_{\mathcal{S}} \equiv \Gamma(\mathcal{S}) \equiv \left(\Gamma(U, \mathcal{S}), \sigma_V^U\right)$ *the presheaf of sections of* $\mathcal{S}$. *Finally, let* $E \equiv \left(E(U), \lambda_V^U\right)$ *be a complete presheaf on* $X$ *and* $\phi : S \longrightarrow E$ *a presheaf morphism. Then, there exists a unique morphism of (complete) presheaves* $\psi : \Gamma_{\mathcal{S}} \longrightarrow E$ *which factorizes* $\phi$. *That is, one has the following commutative diagram* (cf. also (11.6));

(11.46)
$$\begin{array}{ccc} S & \xrightarrow{\ \phi\ } & E \\ & {\scriptstyle\rho}\searrow \quad \nearrow{\scriptstyle\psi} & \\ & \Gamma_{\mathcal{S}} & \end{array}$$

*or, equivalently, the relation*

(11.47)
$$\phi = \psi \circ \rho,$$

*where the presheaf morphism* $\rho : S \longrightarrow \Gamma_{\mathcal{S}}$ *is given by (11.6).*

**Proof.** First, by applying the sheafification functor $\mathbf{S}$ (cf. (8.3)) to the given morphism $\phi : S \longrightarrow E$, one finds a *sheaf morphism* (see (8.5))

(11.48)
$$\mathbf{S}(\phi) \equiv \tilde{\phi} : \mathbf{S}(S) \equiv \mathcal{S} \longrightarrow \mathcal{E} \equiv \mathbf{S}(E).$$

Hence, by further employing the section functor (see also Section 13 in the sequel) to the morphism (11.48), one obtains the following *presheaf morphism* (cf. (13.4) below)

(11.49)
$$\Gamma(\mathbf{S}(\phi)) \equiv \Gamma(\tilde{\phi}) \equiv \overline{\phi} : \Gamma(\mathbf{S}(S)) \equiv \Gamma_{\mathcal{S}} \longrightarrow \Gamma(\mathbf{S}(E)) \equiv \Gamma_{\mathcal{E}} \cong E,$$

the last *isomorphism* in the range of $\overline{\phi}$ being due to the hypothesis for $E$ (cf. (11.37.1)). So *one defines*

(11.50)
$$\psi := \lambda^{-1} \circ \overline{\phi} \equiv \lambda^{-1} \circ \Gamma\mathbf{S}(\phi),$$

where $\lambda : E \longrightarrow \Gamma(\mathbf{S}(E))$ stands for the (presheaf) *isomorphism,* as above. We prove now that $\psi$ *satisfies (11.47) and* (as such) *is unique:* Indeed, one obtains

(11.51)
$$\lambda \circ \phi = \overline{\phi} \circ \rho,$$

*which,* in conjunction with (11.50), *yields (11.47).* In this context, we still note that (11.51) follows immediately from

$$\lambda_U^x(\phi_U(s)) = \phi_x(\rho_U^x(s)), \tag{11.52}$$

valid, for any open $U \subseteq X$, $s \in S(U)$ and $x \in U$, as one can easily verify by the very definitions. (In this concern, see also (12.49) and (12.52) in the sequel, concerning the map $\phi_x : \mathcal{S}_x \longrightarrow \mathcal{E}_x$, $x \in X$, as well as, (8.8) in the preceding).

It remains to prove the *uniqueness of* $\psi$; thus, if $\psi' : \Gamma_{\mathcal{S}} \longrightarrow E$ is another (presheaf) morphism, satisfying (11.47), then one obtains $\lambda \circ \psi = \lambda \circ \psi'$, *from which,* of course, *follows our claim.* Yet, the last relation is still easily verified, by the very definitions (cf. also (7.6), (7.10)), and this actually terminates the proof. (Concerning the uniqueness of the map $\psi$, see also the following Lemma 11.2, along with Scholium 11.2). ∎

(11.53) **Universal problem of complete presheaves.**– The above Theorem 11.3 supplies, in effect, a *solution to the* following *universal problem;* given a presheaf $S$ on a topological space $X$, one has to find a complete presheaf, along a morphism of $S$ into the latter, which factorizes (uniquely) any other similar morphism of $S$. Thus, *the pair* $(\Gamma(\mathbf{S}(S)), \rho)$, as in (11.46), *constitutes,* for the given presheaf $S$, *the solution to the universal problem, under consideration.*

As already said at the end of the preceding proof of Theorem 11.3, the next result provides also another proof of the uniqueness of the map $\psi$ in (11.47); see thus the next Scholium 11.2. So we first have the following.

**Lemma 11.2.** *Let* $S \equiv (S(U), \rho_V^U)$ *and* $E \equiv (E(U), \lambda_V^U)$ *be two given presheaves (of sets) on a topological space* $X$*, such that* $E$ *is a monopresheaf* (viz. it satisfies (S1); cf. Remark 11.1, i)). *Moreover, let* $\phi, \psi$ *be two presheaf morphisms between* $S$ *and* $E$*, such that one has* (see also (8.6, (8.7))

$$\phi_x = \psi_x, \tag{11.54}$$

*for every point* $x \in X$*. Then, one also obtains* $\phi = \psi$.

**Proof.** Based on (11.54) and (8.8), one first obtains

$$\lambda_U^x \circ \phi_U = \phi_x \circ \rho_U^x = \psi_x \circ \rho_U^x = \lambda_U^x \circ \psi_U, \tag{11.55}$$

for any open $U \subseteq X$ and $x \in U$. Therefore, for any $s \in S(U)$, one has (see also (7.9))

$$\widetilde{\phi_U(s)}(x) := \lambda_U^x(\phi_U(s)) = \lambda_U^x(\psi_U(s)) \equiv \widetilde{\psi_U(s)}(x), \tag{11.56}$$

valid in the fiber $\mathcal{E}_x$ $(= E_x$, cf. (7.6)), with $\mathcal{E} \equiv \mathbf{S}(E)$. Hence (see (2.13)), there exists an open $V \subseteq X$, with $x \in V \subseteq U$, such that (see also (9.5))

$$\widetilde{\phi_U(s)}|_V =: \mu_V^U\left(\widetilde{\phi_U(s)}\right) = \mu_V^U\left(\widetilde{\psi_U(s)}\right) \equiv \widetilde{\phi_U(s)}|_V, \tag{11.57}$$

relative to the (complete) presheaf $\Gamma_{\mathcal{E}} \equiv (\Gamma(U, \mathcal{E}), \mu_V^U)$, with $\mathcal{E}$ as before. But, one still obtains (cf. also (11.9))

$$\begin{aligned} \mu_V^U\left(\widetilde{\phi_U(s)}\right) &= \mu_V^U\left(\lambda_U(\phi_U(s))\right) = \lambda_V\left(\lambda_V^U(\phi_U(s))\right) \\ &= \lambda_V\left(\lambda_V^U(\psi_U(s))\right) = \mu_V^U\left(\widetilde{\psi_U(s)}\right), \end{aligned} \tag{11.58}$$

so that, in view of the *property (S1) for E* (see (11.2)), one has

$$\lambda_V^U(\phi_U(s)) \equiv \phi_U(s)|_V = \psi_U(s)|_V \equiv \lambda_V^U(\psi_U(s)). \tag{11.59}$$

So (11.59) is valid for an open covering of $U$, consisting of the various $V$, as above, hence, *by virtue* again *of (S1), we* finally *obtain, for any* $s \in S(U)$ *and* $U$ *open in* $X$,

$$\phi_U(s) = \psi_U(s), \tag{11.60}$$

that is, $\phi_U = \psi_U$, *for any open* $U \subseteq X$, which thus proves the assertion (cf. also (6.3)). ∎

**Scholium 11.2.**– The previous lemma can be applied, as already mentioned, to get *another proof* (more transparent!) *of the uniqueness of the map* $\psi$ *in (11.47):* So, within the context of Theorem 11.3, the conditions of Lemma 11.2 are fulfilled for two morphisms $\psi$ and $\psi'$ satisfying the relation

$$\psi \circ \rho = \psi' \circ \rho \; (= \phi), \tag{11.61}$$

as in (11.47). Namely, since the presheaf $E$ in the same theorem is *complete,* it is a fortiori a *monopresheaf* (see Remark 11.1, i)). Moreover, by virtue of (11.61), *one has* $\psi_x = \psi'_x$, $x \in X$: For, if $z \in (\Gamma_S)_x = \mathcal{S}_x = S_x := \varinjlim_{U \in \mathcal{V}(x)} S(U)$ (see also (7.6), (7.2) and Lemma 9.1), then $z = \rho_x(\rho_U^x(s))$, for some open $U \subseteq X$, with $x \in U$ and $s \in S(U)$, while we also set $\rho_x := \delta_x^{-1}$ (cf. (9.3)).

Now, applying on (11.61) the *"passage to the limit" (covariant) functor,* for $x \in X$, one gets $(\psi \circ \rho)_x = \psi_x \circ \rho_x = \psi'_x \circ \rho_x$, $x \in X$ (cf. (8.7), as well as, e.g. N. Boubaki [1: p.93, Corollaire 2]). Therefore, one obtains

$$\psi_x(z) = \psi_x\Big(\rho_x\big(\rho_U^x(s)\big)\Big) = \psi'_x\Big(\rho_x\big(\rho_U^x(s)\big)\Big) = \psi'_x(z), \tag{11.62}$$

which proves our claim, hence (Lemma 11.1), $\psi = \psi'$, as desired. ∎

Now, given a topological space $X$, we denote by

$$\mathcal{CoPSh}_X \tag{11.63}$$

the *category of complete presheaves on* $X$. The corresponding *morphisms of (11.63)* are the *same as in* $\mathcal{PSh}_X$, the category of presheaves on $X$ (cf. (6.9)). So one has that

(11.64) $\mathcal{CoPSh}_X$ is a *full subcategory* of $\mathcal{PSh}_X$.

More on (11.63) and its relation with $\mathcal{Sh}_X$, the category of sheaves on $X$ (cf. (1.18)), will be given in the sequel. (Thus, see, for instance, Theorem 13.1 below).

## 12. Morphisms of presheaves (contn'd)

We consider below certain useful criteria in order given morphisms of presheaves to be *injective* and/or *surjective*. As we shall see in the next Section 13 (cf., for instance, (13.19)), such criteria provide, in turn, in a canonical way, corresponding ones for the sheaves, which are generated by the presheaves concerned.

Thus, we first have the following.

**Lemma 12.1.** *Let $S \equiv (S(U), \rho_V^U)$, $E \equiv (E(U), \lambda_V^U)$ be presheaves (of sets) on a topological space $X$ and $\phi : S \longrightarrow E$ a given morphism of the presheaves concerned. Now, consider the following assertions:*

*1) $\phi$ is injective* (cf. Definition 6.1).

*2) For any point $x \in X$ and open set $U \subseteq X$, with $x \in U$, there exists an open set $V \subseteq X$, with $x \in V \subseteq U$, such that the map* (see also (6.4))

$$\phi_V : S(V) \longrightarrow E(V) \tag{12.1}$$

*be one-to-one. Equivalently,*

(12.2) *every point $x \in X$ has a fundamental system (basis) of open neighborhoods, such that the corresponding maps ("components" of $\phi$), like (12.1), be one-to-one.*

*3) For every point $x \in X$, the respective map of stalks* (see (8.7))

$$\phi_x : S_x \longrightarrow E_x \tag{12.3}$$

*is one-to-one.*

*Then, one has the following implications:*

(12.4) *1) $\Longrightarrow$ 2) and 3).*

*In particular, if $E$ is a monopresheaf, then 2) $\Longrightarrow$ 3), as well, while if $S$ is a monopresheaf, then 3) $\Longrightarrow$ 1) too.* (So *in the latter case 1) and 3) are equivalent*). *Therefore, if both of the given presheaves are, in particular, monopresheaves, then all the above three assertions are equivalent.*

**Proof.** Of course, as immediately follows from Definition 6.1, *cond. 1) implies 2)*. (In fact, something *much more is actually true, than cond. 2), if 1) is valid*). On the other hand, if each one of the *maps $\phi_U$* in Definition 6.1 (cf. (6.4)) is *injective*, then *the same holds true for the maps $\phi_x$*, $x \in X$, in (12.3); thus, see, for instance, N. Bourbaki [1: p.93, Proposition 7]. That is, *1) implies 3),* as well.

We prove now that *2) implies 3),* under the further assumption that *$E$ is a monopresheaf:* So consider a point $x \in X$ and two elements $z, z'$ in $S_x$, with $\phi_x(z) = \phi_x(z')$. So (cf. (7.2), (7.10) or even (7.16)) one has

$$z = \rho_U^x(s) \text{ and } z' = \rho_U^x(s'), \tag{12.5}$$

*for some open $U \subseteq X$, with $x \in U$ and $s, s'$ in $S(U)$.* (We note here, in passing, that a similar relation to (12.5) is actually true, *for any finite number of elements of $S_x$;* see, for instance, N. Bourbaki [1: p.90, Lemme 1] or even A. Mallios [2: p.111, Lemma 1.1]). Therefore, in view of our hypothesis for $z$ and $z'$, one obtains (cf. also (8.8)),

$$\phi_x(z) = \phi_x\big(\rho_U^x(s)\big) = \lambda_U^x\big(\phi_U(s)\big) \equiv \widetilde{\phi_U(s)}(x) = \phi_x(z') = \widetilde{\phi_U(s')}(x); \tag{12.6}$$

hence (cf. (2.13)), there exists an open $V \subseteq X$, with $x \in V \subseteq U$, such that

$$\widetilde{\phi_U(s)}|_V = \widetilde{\phi_U(s')}|_V, \tag{12.7}$$

so that one still obtains (cf. also (7.9), (9.5), as well as, (11.57), (11.58));

$$\begin{aligned} \widetilde{\phi_U(s)}|_V &= \lambda_U(\phi_U(s))|_V \equiv \mu_V^U(\lambda_U(\phi_U(s))) = \lambda_V\left(\lambda_V^U(\phi_U(s))\right) \\ &= \lambda_V\left(\lambda_V^U(\phi_U(s'))\right) \equiv \widetilde{\phi_U(s')}|_V. \end{aligned} \tag{12.8}$$

Thus, *in view of the hypothesis for E*, one gets (cf. (6.6))

$$\lambda_V^U(\phi_U(s)) = \phi_V\left(\rho_V^U(s)\right) = \phi_V\left(\rho_V^U(s')\right) = \lambda_V^U(\phi_U(s')). \tag{12.9}$$

Now, by further *assuming that (12.2) holds,* we can consider $V$ in (12.9) as such for which *$\phi_V$ is one-to-one*. Therefore, by virtue of (12.9), one obtains $\rho_V^U(s) = \rho_V^U(s')$, so that one has (cf. also (12.5)

$$z = \rho_U^x(s) = \rho_V^x\left(\rho_V^U(s)\right) = \rho_V^x\left(\rho_V^U(s')\right) = \rho_U^x(s') = z', \tag{12.10}$$

which thus proves our assertion.

Finally, suppose that $S$ is a *monopresheaf;* so let $s, s'$ be elements of $S(U)$, for some open $U \subseteq X$, with $\phi_U(s) = \phi_U(s')$. Thus, one obtains, for every $x \in U$ (cf. also (8.8)),

$$\phi_x\left(\rho_U^x(s)\right) = \lambda_U^x\left(\phi_U(s)\right) = \lambda_U^x\left(\phi_U(s')\right) = \phi_x\left(\rho_U^x(s')\right), \tag{12.11}$$

so that, *if cond. 3) is valid,* one gets (see (7.9))

$$\rho_U^x(s) \equiv \rho_U(s)(x) = \rho_U(s')(x), \; x \in U, \tag{12.12}$$

that is, $\rho_U(s) = \rho_U(s')$; hence, *in view of our hypothesis for $S$*, $s = s'$. So *cond. 3) implies 1),* as well, under our supplementary hypothesis for $S$, as above, hence, in that case too, *the two conditions are* thus *equivalent,* and this essentially terminates the proof. ∎

**(Terminological) Scholium 12.1.**– A *morphism of presheaves* $\phi : S \longrightarrow E$ satisfying (12.2), will be called *B-injective* ("*B*" stands here for *basis*). Analogous terminology will be employed in the sequel by considering *B-surjective* (presheaf) *morphisms* (cf. Lemma 12.2) or even *B-isomorphisms* (of presheaves; cf. Theorem 12.1 and Corollary 12.1 below). This occasionally *weaker condition than 1)* (see thus Definition 6.1, along with the last part of Lemma 12.1 or Theorem 12.1) acquires a special interest in that *it also implies cond. 3)* (eventually, under further assumptions; cf. e.g. Lemma 12.1). The latter implication still appears in the case of *B-surjective morphisms* (yet, no supplementary assumptions are needed now, cf. Lemma 12.2 below) or *B-isomorphisms* (see Theorem 12.1; in fact, all the relevant assertions here are equivalent). Of course, *cond. 3) is always a consequence of cond. 1).* Now, this particular weakening of *cond. 1),* proper or not, is proved to be of use in later applications; thus, see, for instance, Chapters II, III in the sequel.

We come now to the following companion of Lemma 12.1. So we have.

**Lemma 12.2.** *Suppose we are given two presheaves $S, E$ on a topological space $X$, along with a presheaf morphism $\phi : S \longrightarrow E$. Now, consider the following assertions:*

*1) $\phi$ is surjective* (cf. Definition 6.1).

*2) $\phi$ is B-surjective* (cf. Scholium 12.1; i.e., a condition here, like (12.2), is fulfilled, with the respective components of $\phi$ (see (12.1)) being surjective).

*3) For every point $x \in X$, the corresponding map of stalks* (cf. (8.7))

$$\phi_x : S_x \longrightarrow E_x \tag{12.13}$$

*is surjective.*

*Then, one has the following implications:*

$$1 \Longrightarrow 2) \text{ and } 3), \text{ while } 2) \Longrightarrow 3). \tag{12.14}$$

**Warning!** (We employed at this point the classic *"tournant dangereux"* of N. Bourbaki to point out an eventual peculiarity in (12.14), with $\phi$ being surjective). Thus, by contrast with Lemma 12.1, *cond. 3)* of the previous lemma, *does not, in general, imply 1), even if $S$ and $E$ are complete presheaves!* (See thus Example 12.1 below).

**Proof of Lemma 12.2.** By Definition 6.1, *cond. 1) implies 2)* (and, of course, much more), while *1) implies 3), as well,* as follows e.g. from N. Bourbaki [1: p.93, Proposition 7]; cf. also (8.7). Now, let $x \in X$, and $z \in E_x = \varinjlim_{U \in \mathcal{V}(x)} E(U)$ (cf. (7.2)), so that one has

$$z = \lambda_U^x(t), \tag{12.15}$$

for some *open* $U \subseteq X$, with $x \in U$ and $t \in E(U)$. Thus, assuming cond. 2), there exists open $V \subseteq X$, with $x \in V \subseteq U$, such that the corresponding map $\phi_V : S(V) \longrightarrow E(V)$ be *surjective.* So one obtains (cf. (12.15), (8.14) and (8.8))

$$z = \lambda_U^x(t) = \lambda_V^x\left(\lambda_V^U(t)\right) = \lambda_V^x(\phi_V(s)) = \phi_x\left(\rho_V^x(s)\right) \in \operatorname{im}\phi_x, \tag{12.16}$$

where, by hypothesis for $\phi_V$, $\lambda_V^U(t) = \phi_V(s)$, for some $s \in S(V)$, while we still have (cf. (7.9)) that $\rho_V^x(s) \in S_x$; thus, *cond. 2) implies 3),* as well. ∎

We consider next the aforementioned *counterexample to 3) $\Longrightarrow$ 1) in Lemma 12.2.* (See also R. Godement [1: p.115]).

**Example 12.1.**– Let $\mathcal{O}_{\mathbf{C}} \equiv \mathcal{O}$ be the *sheaf of (germs of* $\mathbf{C}$*-valued) holomorphic functions on* $\mathbf{C}$*;* now, apart from the point of view of Section 4 (cf. (4.40), for $X = \mathbf{C}$), one can consider $\mathcal{O}$ as the *sheaf generated by the complete presheaf* $O \equiv (\mathcal{O}(U), \rho_V^U)$ (see (10.14.1), yet for $X = \mathbf{C}$). Thus (cf. Proposition 11.1), one has

$$\mathcal{O}(U) = \Gamma(U, \mathcal{O}) \equiv \mathcal{O}_{\mathbf{C}}(U), \tag{12.17}$$

within a (canonical) *bijection,* for any open $U \subseteq \mathbf{C}$ (see (11.20)).

On the other hand, we denote by

$$\mathcal{O}^{\cdot}_{\mathbf{C}} \equiv \mathcal{O}^{\cdot} \tag{12.18}$$

the *sheaf on* $\mathbf{C}$, generated by the *presheaf of nowhere vanishing holomorphic functions on* $\mathbf{C}$; so, by definition, one considers the maps

$$U \longmapsto \mathcal{O}(U)^{\cdot}, \ \textit{for any open } U \subseteq \mathbf{C}, \tag{12.19}$$

where $\mathcal{O}(U)^{\cdot}$ stands for the *set of non-vanishing* ($\mathbf{C}$*-valued) holomorphic functions on* $U$*,* while one also defines, in the obvious way, the corresponding to (12.19) restriction maps (see e.g. (10.12)). Now, it is not hard to see that (12.19) *yields a complete presheaf.* (A fuller treatment of a much more general situation than (12.18) will be given in later chapters of this book; cf. e.g. Chapters II, IV). Thus, one obtains a (canonical) *bijection,* concerning the last of the following relations (cf. (12.18), (12.19) and also (9.3));

$$\Gamma(U, \mathcal{O}^{\cdot}_{\mathbf{C}}) \equiv \mathcal{O}^{\cdot}_{\mathbf{C}}(U) \equiv \mathcal{O}^{\cdot}(U) = \mathcal{O}(U)^{\cdot}, \tag{12.20}$$

*for any open set* $U \subseteq \mathbf{C}$.

Furthemore, we can define a *morphism of presheaves* (see also (9.7))

$$\phi : \Gamma(\mathcal{O}_{\mathbf{C}}) \longrightarrow \Gamma(\mathcal{O}^{\cdot}_{\mathbf{C}}); \tag{12.21}$$

that is, *for any open* $U \subseteq \mathbf{C}$, one has (cf. (6.4), (12.17) and (12.20))

$$\phi_U : \Gamma(\mathcal{O}_{\mathbf{C}})(U) = \mathcal{O}(U) \longrightarrow \Gamma(\mathcal{O}^{\cdot}_{U})(U) = \mathcal{O}(U)^{\cdot}, \tag{12.22}$$

such that one sets;

$$\phi_U(f) := \exp 2\pi i f, \quad f \in \mathcal{O}(U). \tag{12.23}$$

Due to the last relation, we also employ, henceforth, the notation

$$\phi_U \equiv \exp_U, \; \textit{for any open } U \subseteq \mathbf{C}, \tag{12.24}$$

defined thus by (12.23), and call (12.21) the *exponential* (presheaf) *morphism* (cf. also the (terminological) Scholium 12.2 below).

Now, *every non-vanishing holomorphic function is locally of the form* $\exp 2\pi i f$, or yet any such function *has locally a logarithm* (see e.g. B. P. Palka [1: p.176, Theorem 4.1]). So, as a result, one easily concludes, that;

(12.25) *for every point* $x \in \mathbf{C}$, *the* corresponding *fiber map* of $\phi$ (cf. (8.6))

$$\phi_x : \mathcal{O}_x \longrightarrow \mathcal{O}^{\cdot}_x \tag{12.25.1}$$

*is surjective.*

(Concerning (12.25.1), cf. also (7.6) along with Lemma 9.1). Namely, if $z \in \mathcal{O}^{\cdot}_x$, then there exists $f \in \mathcal{O}(U)^{\cdot}$, with $z = \lambda^x_U(f) = \lambda^x_V(\lambda^U_V(f)) = \lambda^x_V(\phi_V(h)) \equiv \lambda^x_V(\exp_V(h)) = \phi_x(\rho^x_V(h))$, where of course $\rho^x_V(h) \in \mathcal{O}_x$, which thus proves our claim about (12.25.1).

However, *if the open set* $U \subseteq \mathbf{C}$ *is not simply-connected, the map (12.22),* as given by (12.23), *may be not surjective:* Thus, the identity map, for instance, does not have a single-valued logarithm on $\mathbf{C}^{\cdot} := \mathbf{C} - \{0\}$; cf. e.g. O. Forster [1: p.121, Remark 15.7]. (In this context, our hypothesis of being $U$, as above, *not simply-connected,* is, of course, characteristic of the situation, as exhibited by the previous counter-example; thus, see again, for instance, B. P. Palka [1: p.196, Theorem 6.2]).

**(Terminological) Scholium 12.2.**- The (complete) presheaf morphism (12.21), as defined by (12.23), can be further extended to a corresponding *sheaf morphism* (cf. (8.10)), which we still denote by exp, viz.

$$\exp : \mathcal{O}_{\mathbf{C}} \longrightarrow \mathcal{O}^{\cdot}_{\mathbf{C}}. \tag{12.26}$$

Thus, if $\mathbf{f}_x \equiv [f]_x \in \mathcal{O}_x$ denotes *the germ at* $x \in \mathbf{C}$ *of a* ($\mathbf{C}$-valued) *holomorphic function,* say, $f \in \mathcal{O}(U)$, with $x \in U$, one has (cf. (9.16), (8.8) and (8.11));

$$\begin{aligned}\exp(\mathbf{f}_x) &:= \phi_x(\mathbf{f}_x) = \phi_x\big(\rho^x_U(f)\big) = \lambda^x_U(\phi_U(f)) \\ &= \lambda^x_U(\exp 2\pi i f) \equiv [\exp 2\pi i f]_x.\end{aligned} \tag{12.27}$$

We are going to employ often in the sequel the above important sheaf morphism, even for any *complex* (analytic) *manifold* $X$; yet, in Part Two of this treatise in a much more general, quite abstract, setting. Now, motivated by the previous classical context, we call (12.26), along with any of its forthcoming generalizations, the *exponential sheaf morphism,* denoted thus, deliberately, by exp (or even, occasionally, by $\exp 2\pi i$).

In this context, we still obtain, however, the following useful result. So we have.

**Lemma 12.3.** *Let* $S, E$ *be two given presheaves on a topological space* $X$, *along with a presheaf morphism* $\phi : S \longrightarrow E$ (see also Lemma 12.1 for the notation applied). *Moreover, assume that these data satisfy further the following conditions:*

*i)* $S$ *satisfies (S2) and* $E$ *(S1)* (cf. Definition 11.1).

*ii)* $\phi$ *is injective.*

*iii) For every point* $x \in X$, *the respective map of stalks,* $\phi_x : S_x \longrightarrow E_x$, *is surjective.* (Hence, in view of ii) and of Lemma 12.1, *each one of these maps is, in fact, a bijection*).

*Then, the given morphism* $\phi : S \longrightarrow E$ *is surjective, hence* (by virtue of ii), cf. also (6.10)), *an isomorphism, as well.*

**Proof.** If $U$ is any open set in $X$, let $t \in E(U)$, so that if $x \in U$, one obtains

$$\lambda_U^x(t) \equiv \tilde{t}(x) \in E_x; \tag{12.28}$$

(cf. (7.9), (7.10), as well as, Lemma 12.1 for the notation applied hereafter). Therefore, in view of *cond. iii)* of our hypothesis, there exists $z \in S_x$, such that

$$\lambda_U^x(t) \equiv \tilde{t}(x) = \phi_x(z). \tag{12.29}$$

Hence (cf. (7.2)), there exist open $V \subseteq X$, with $x \in V \subseteq U$, and $s \in S(V)$ such that $z = \rho_V^x(s)$; so (cf. also (8.8) and (8.14)), one obtains

$$\begin{aligned}\lambda_U^x(t) &= \lambda_V^x\left(\lambda_V^U(t)\right) \equiv \widetilde{\lambda_V^U(t)}(x) = \phi_x(z) = \phi_x\left(\rho_V^x(s)\right)\\ &= \lambda_V^x(\phi_V(s)) \equiv \widetilde{\phi_V(s)}(x).\end{aligned} \tag{12.30}$$

So, by considering (12.30) within $\mathcal{E}_x = E_x$ cf. e.g. (9.16)), with $\mathcal{E} \equiv \mathbf{S}(E)$, the shefification of $E$ (see (8.2)), one finds, for every $x \in U$, an open $U_x \subseteq X$, with $x \in U_x \subseteq V \subseteq U$, such that (cf. (2.12); but, see also (11.57) for the notation applied, as well as, (7.9) and (11.9))

$$\begin{aligned}\widetilde{\lambda_V^U(t)}|_{U_x} &= \mu_{U_x}^V\left(\widetilde{\lambda_V^U(t)}\right) = \mu_{U_x}^V\left(\lambda_V\left(\lambda_V^U(t)\right)\right) = \lambda_{U_x}\left(\lambda_{U_x}^V\left(\lambda_V^U(t)\right)\right)\\ &= \widetilde{\phi_V(s)}|_{U_x} = \mu_{U_x}^V\left(\widetilde{\phi_V(s)}\right) = \mu_{U_x}^V\left(\lambda_V\left(\phi_V(s)\right)\right) = \lambda_{U_x}\left(\lambda_{U_x}^V\left(\phi_V(s)\right)\right),\end{aligned} \tag{12.31}$$

so that, in view of the *hypothesis for* $E$ (cf. also the proof of Proposition 11.1, (5.3) and (6.6)), one has

$$\lambda_{U_x}^V\left(\lambda_V^U(t)\right) = \lambda_{U_x}^U(t) = \lambda_{U_x}^V(\phi_V(s)) = \phi_{U_x}\left(\rho_{U_x}^V(s)\right). \tag{12.32}$$

To use right below, we also remark here that an analogous argument to that employed for (10.32), still provides (more generally!) the relation

$$\phi_W\left(\rho_W^V(s)\right) = \lambda_W^U(t), \tag{12.32'}$$

*for any open* $W \subseteq X$, *with* $x \in W \subseteq U_x (\subseteq V \subseteq U)$. Thus, by further setting

$$s^{(x)} := \rho_{U_x}^V(s) \in S(U_x),\ x \in U, \tag{12.33}$$

one obtains an *open covering* $(U_x)_{x\in U}$ *of* $U$, along with *a family of "sections"*

$$\left(s^{(x)}\right) \in \prod_{x\in U} S(U_x), \tag{12.33'}$$

*satisfying (11.3):* Namely, for any $x, y$ in $U$, with $U_{xy} \equiv U_x \cap U_y \neq \emptyset$, one has (see also (12.32′), (12.33) and (5.3))

$$\begin{aligned}(12.34)\qquad &\phi_{U_{x,y}}\left(\rho^{U_x}_{U_{x,y}}(s^{(x)})\right) = \phi_{U_{x,y}}\left(\rho^{U_x}_{U_{x,y}}\left(\rho^{V}_{U_x}(s)\right)\right) = \phi_{U_{x,y}}\left(\rho^{V}_{U_{x,y}}(s)\right)\\ &= \lambda^{U}_{U_{x,y}}(t) = \phi_{U_{x,y}}\left(\rho^{U_y}_{U_{x,y}}\left(\rho^{V}_{U_y}(s)\right)\right) = \phi_{U_{x,y}}\left(\rho^{U_y}_{U_{x,y}}(s^{(y)})\right).\end{aligned}$$

Therefore, since, by hypothesis, $\phi$ *is injective,* one obtains

$$(12.35)\qquad \rho^{U_x}_{U_{x,y}}\left(s^{(x)}\right) = \rho^{U_y}_{U_{x,y}}\left(s^{(y)}\right),$$

as claimed. Accordingly, by *hypothesis for* $S$, *there exists* $s' \in S(U)$, such that (cf. e.g. (11.4))

$$(12.36)\qquad s'|_{U_x} \equiv \rho^{U}_{U_x}(s') = s^{(x)},\ x \in U.$$

Therefore, one obtains (cf. also (6.6), (12.33) and (12.32′))

$$(12.37)\ \lambda^{U}_{U_x}(\phi(s')) = \phi_{U_x}\left(\rho^{U}_{U_x}(s')\right) = \phi_{U_x}\left(s^{(x)}\right) = \phi_{U_x}\left(\rho^{V}_{U_x}(s)\right) = \lambda^{U}_{U_x}(t),\ x \in U,$$

so that, since, by hypothesis, $E$ *satisfies (11.2),* one finally has $t = \phi_U(s')$, as desired. ∎

As a consequence of the preceding, we now have the following.

**Theorem 12.1.** *Let* $S \equiv \left(S(U), \rho^U_V\right)$ *be a complete presheaf and* $E \equiv \left(E(U), \lambda^U_V\right)$ *a monopresheaf (of sets) on a topological space* $X$, *as well as, a presheaf morphism* $f : S \longrightarrow E$. *Then, the following assertions are equivalent:*

*i)* $f$ *is an isomorphism.*

*ii)* $f$ *is a* $B$*-isomorphism* (cf. (Terminological) Scholium 12.1).

*iii) Each one of the fiber maps* $\phi_x : S_x \longrightarrow E_x$, $x \in X$ (cf. (8.7)), *is a bijection.*

**Proof.** First, looking at $\phi$ as an *injective* morphism, we conclude from Lemma 12.1, based on the hypothesis for $S$ and $E$, that all three assertions are equivalent. Now, given this, the rest is a consequence of our hypothesis for $S$ and $E$, along with all the above three Lemmas 12.1 – 12.3. ∎

We end our discussion in this section by still commenting a bit on the case of a *presheaf,* exhibited as the *image of a* given *presheaf morphism:*

Thus, suppose that

(12.38) $$\phi \equiv (\phi_U) : S \equiv \left(S(U), \rho_V^U\right) \longrightarrow E \equiv \left(E(U), \lambda_V^U\right)$$

is a *morphism of presheaves (of sets)* on a topological space $X$ (cf. also (6.3)).

First, it is readily seen that the family

(12.39) $$\operatorname{im}\phi \equiv \left(\phi_U(S(U)) \equiv \operatorname{im}\phi_U, \lambda_V^U|_{\operatorname{im}\phi_U}\right)_{U \subseteq X,\, open}$$

defines a *presheaf (of sets) on* $X$; in fact, as immediately follows from (6.6), (12.39) is an *inductive system of subsets of* $E \equiv \left(E(U), \lambda_V^U\right)$ (see also e.g. N. Bourbaki [1: p.94]). So one usually considers (ibid.) the following *identification* (of sets, viz. *bijection*)

(12.40) $$(\operatorname{im}\phi)_x := \varinjlim_{U \ni x} \phi_U(S(U)) \overset{\subseteq}{\underset{j_x}{\rightarrow}} E_x, \; x \in X,$$

where $j_x := \varinjlim_{U \ni x} j_U$, with $j_U : \phi_U(S(U)) \longrightarrow E(U)$, $U$ open in $X$, the (canonical) *injection* map. Indeed, one actually obtains (see also (7.2), (8.7), as well as, N. Bourbaki [1: p.94, Corollaire, (i)])

(12.41) $$(\operatorname{im}\phi)_x := \left(\varinjlim_{U \ni x} \phi_U\right)\left(\varinjlim_{U \ni x} S(U)\right) \equiv \phi_x(S_x) \subseteq E_x, \; x \in X.$$

On the other hand, let us consider next the *sheafifications* of the presheaves involved so far; so if

(12.42) $$\mathbf{S}(\phi) \equiv \tilde{\phi} : \mathcal{S} \equiv \mathbf{S}(S) \longrightarrow \mathbf{S}(E) \equiv \mathcal{E}$$

is the *sheaf map,* as indicated, associated with the given morphism $\phi$ (cf. (8.11)), one concludes that;

(12.43) $\tilde{\phi}(\mathcal{S})$ *is the sheafification of* $\operatorname{im}\phi \equiv \phi(S)$ (cf. (12.39)); precisely, one has

(12.43.1) $$\mathbf{S}(\operatorname{im}\phi) \equiv \mathbf{S}(\phi(S)) = \tilde{\phi}(\mathcal{S}) = \tilde{\phi}(\mathbf{S}(S)),$$

within a *sheaf isomorphism.*

First, $\tilde{\phi}(\mathcal{S})$ *is a sheaf* (in fact, a *subsheaf of* $\mathcal{E}$), given that $\tilde{\phi}$, being a local homeomorphism (cf. Lemma 1.2), is an open map (cf. (1.4)), so that $\tilde{\phi}(\mathcal{S})$ *is an open subset of* $\mathcal{E}$ (cf. also (1.8) and the comments following it). Now, in view of (12.41) (see also (7.6)), one obtains

$$\mathbf{S}(\operatorname{im}\phi)_x = (\operatorname{im}\phi)_x = \phi_x(S_x) = \tilde{\phi}_x(\mathcal{S}_x) = \tilde{\phi}(\mathcal{S})_x,\ x \in X, \tag{12.44}$$

so that we do have, first, the *same sheaf spaces* of the two sheaves in question. Thus, we next apply the technique of Section 3 to show that the latter have *equal defining families of sections,* as well (cf. also (3.17)): So, for any *open* $U \subseteq X$ and $\bar{s} \in \Gamma(U, \tilde{\phi}(\mathcal{S}))$, one obtains, for any $x \in U$ (see also (12.44)), $\bar{s}(x) \in \tilde{\phi}(\mathcal{S})_x = \phi_x(S_x) = \phi_x(\mathcal{S}_x)$, hence,

$$\bar{s}(x) = \phi_x(\rho^x_U(s)) = \lambda^x_U(\phi_U(s)) \equiv \widetilde{\phi_U(s)}(x) \tag{12.45}$$

(cf. also (8.8)), where (12.45) is actually considered within $E_x = \mathcal{E}_x = (\Gamma_{\mathcal{E}})_x$ (cf. (9.16)). Therefore (cf. (2.12)), there exists *open* $V \subseteq X$, *with* $x \in V \subseteq U$, such that

$$\bar{s} = \widetilde{\phi_U(s)}|_V. \tag{12.46}$$

As a result, one concludes that;

(12.47) *sections of* $\tilde{\phi}(\mathcal{S}) \equiv \tilde{\phi}(\mathbf{S}(S))$, *which are locally of the form (12.46), constitute a defining family of sections of* $\tilde{\phi}(\mathcal{S})$.

On the other hand, sections (of $\mathcal{E}$) of the form $\widetilde{\phi_U(s)}$, with $s \in S(U)$ and $U$ open in $X$, provide, of course, a *similar family for* $\mathbf{S}(\operatorname{im}\phi)$ (see (7.12)), yet, *those, in particular,* which are *defined by (12.46),* as easily follows from the very definitions. Now, this yields, in fact, our assertion, by virtue of Theorem 3.1 and Lemma 3.1. ∎

Within the same context, we finally comment on the following *relations pertaining to a* given *sheaf morphism,* which will also be of use later on.

So assume that we are given a *sheaf morphism* (cf. Definition 1.2)

$$\phi : \mathcal{S} \longrightarrow \mathcal{E} \tag{12.48}$$

between the given sheaves (of sets) $\mathcal{S}$ and $\mathcal{E}$ on a topological space $X$. Now, based on Proposition 2.1 (cf. (2.16)), one defines, through (12.48), a *morphism of* (complete) *presheaves*

$$\bar{\phi} \equiv (\bar{\phi}_U) : \Gamma(\mathcal{S}) \longrightarrow \Gamma(\mathcal{E}) \tag{12.49}$$

(see also (9.7)), such that

$$\bar{\phi}_U(s) := \phi \circ s, \tag{12.50}$$

with $s \in \Gamma(U, \mathcal{S}) \equiv \mathcal{S}(U)$ and $U$ *open* in $X$. (As we shall see in the next Section 13, (12.49) is but the application on $\phi$ of the *section functor;* cf. (13.1)). Therefore, applying to (12.49) the technique of Section 8 (cf. (8.7)), one obtains, for every point $x \in X$, the corresponding *fiber maps* (see also (9.8))

$$\overline{\phi}_x : \Gamma(\mathcal{S})_x \longrightarrow \Gamma(\mathcal{E})_x, \ x \in X. \tag{12.51}$$

On the other hand, the given morphism $\phi$ induces, by restriction, a corresponding *fiber map* (cf. (1.14))

$$\phi_x := \phi|_{\mathcal{S}_x} : \mathcal{S}_x \longrightarrow \mathcal{E}_x, \ x \in X. \tag{12.52}$$

Now, by applying a (convenient) abuse of notation, one has

$$\overline{\phi}_x = \phi_x, \ x \in X. \tag{12.53}$$

Namely, *one has, in effect,*

$$\overline{\phi}_x = \varepsilon_x^{-1} \circ \phi_x \circ \delta_x, \ x \in X, \tag{12.54}$$

where we set

$$\delta_x : \Gamma(\mathcal{S})_x \longrightarrow \mathcal{S}_x \ \ and \ \ \varepsilon_x : \Gamma(\mathcal{E})_x \longrightarrow \mathcal{E}_x, \ x \in X, \tag{12.55}$$

for the (canonical) *bijections,* as indicated, defined by Lemma 9.1 (cf. also (9.13)).

# 13. The section functor and its (left) adjoint: The sheafification functor (contn'd)

We prove in this Section that the two categories $\mathcal{Sh}_X$ and $\mathcal{CoPSh}_X$, as defined in the preceding (cf. (1.18) and (11.63), respectively), are, in effect, equivalent (see thus Theorem 13.1 below). So, as often happens in practice, one identifies, occasionally, a sheaf with a complete presheaf, the transfer from each concept to the other being achieved, as we shall presently see, through the two functors $\mathbf{S}$ and $\Gamma$. The first one, viz. the *sheafification functor,* has been defined already in Section 8.

So we begin with next defining the *section functor* $\Gamma$, something that has also repeatedly been used so far in the foregoing, however, implicitly (viz. not as a categorical notion, as it really is!).

Thus, given a *sheaf* (of sets) $\mathcal{S} \equiv (\mathcal{S}, \pi, X)$ on a topological space $X$, one considers the corresponding *complete presheaf of sections of* $\mathcal{S}$ (cf. Lemma 11.1) as the value of the functor we are looking for, at the given object $\mathcal{S}$ of the category $\mathcal{S}h_X$. (So we claim that the already familiar procedure, cf. Section 9, is, in fact, a *functorial correspondence!*). That is, one has, by definition, the following rule (assignment) between the categories concerned;

$$\Gamma : \mathcal{S}h_X \longrightarrow \mathcal{C}o\mathcal{PS}h_X : \mathcal{S} \longmapsto \Gamma(\mathcal{S}) := \Gamma_{\mathcal{S}} \equiv \left(\Gamma(U, \mathcal{S}), \sigma_V^U\right) \tag{13.1}$$

(cf. also (9.7), as well as, (9.2) and (9.3)). Of course, by an obvious *abuse of notation,* we used here the same symbol for the *"functor* $\Gamma$*"* and the notation, already employed in the preceding, $\Gamma(\mathcal{S}) \equiv \Gamma_{\mathcal{S}}$ (cf. (9.7)), which, however, is thus finally justified, as well.

Now, to proceed further, one has first to verify that, with any given *sheaf morphism* (cf. Definition 1.2), say,

$$\phi : \mathcal{S} \longrightarrow \mathcal{E} \tag{13.2}$$

in $\mathcal{S}h_X$, there is associated, via $\Gamma$, a *morphism of complete presheaves*

$$\Gamma(\phi) \equiv \overline{\phi} : \Gamma(\mathcal{S}) \longrightarrow \Gamma(\mathcal{E}). \tag{13.3}$$

However, such a map has already been defined by (12.49) (cf. also (12.50)). Yet, by virtue of (11.64), the morphism (13.3) is *the same, either as a morphism of* $\mathcal{C}o\mathcal{PS}h_X$ *or of* $\mathcal{PS}h_X$.

Now, it is easy to prove further the usual compatibility conditions of functors, with respect to given morphisms (cf. e.g. S. Lang [1: p.29]). So one concludes that;

(13.4) (13.1) defines a *covariant functor* between the categories concerned. We call it the *section functor*.

To proceed we next employ the following basic notion of *category theory.* Namely,

(13.5) two given categories are *equivalent* if, and only if, there is a *functor* between them, which is *fully faithful* and *essentially surjective.*

The above categorical jargon will become more accessible in a moment, through its concrete application in the proof of the next Theorem 13.1. Yet, concerning (13.5), see, for instance, S. Mac Lane [1: p.91, Theorem 1], or even N. Jacobson [1: p.27, Proposition 1.3].

So we come next to our main objective in this Section, namely, the following.

**Theorem 13.1.** *Given a topological space $X$, the corresponding section functor*

$$\Gamma : \mathcal{S}h_X \longrightarrow \mathcal{C}o\mathcal{PS}h_X, \tag{13.6}$$

as given by (13.1), *provides an equivalence of the respective categories, viz. it is fully faithful and essentially surjective.*

**Proof.** We first prove that the functor $\Gamma$ *is fully faithful:* Namely, by definition, we show that, *for any two* given *sheaves* $\mathcal{S}, \mathcal{E}$ *on* $X$, *one has* the relation;

$$\mathrm{Hom}_{\mathcal{S}h_X}(\mathcal{S}, \mathcal{E}) = \mathrm{Hom}_{\mathcal{C}o\mathcal{PS}h_X}(\Gamma(\mathcal{S}), \Gamma(\mathcal{E})) \tag{13.7}$$

*within a bijection.* Thus, suppose that $\phi, \psi$ are two given sheaf morphisms between $\mathcal{S}$ and $\mathcal{E}$, viz. two elements of the first set in (13.7), such that (cf. (13.3)) $\Gamma(\phi) \equiv \overline{\phi} = \overline{\psi} \equiv \Gamma(\psi)$. Hence, *for any* $z \in \mathcal{S}$, one obtains (see also (2.10.1), (12.49) and (12.50))

$$\phi(z) = \phi(s(x)) = (\phi \circ s)(x) = \overline{\phi}_U(s)(x) = \overline{\psi}_U(s)(x) = \psi(z), \tag{13.8}$$

that is, $\phi = \psi$. So the functor $\Gamma$ establishes (through (13.3) a *one-to-one correspondence* between the two sets in (13.7); thus, by definition, $\Gamma$ is a *faithful functor.* Now, we prove further that *the same map is onto,* as well, or in categorical language $\Gamma$ is a *full functor* too: But, this will require a bit more work, through an attractive(!), as we shall see, "diagram chessing".

Thus, suppose that

$$\psi \equiv (\psi_U) : \Gamma(\mathcal{S}) \longrightarrow \Gamma(\mathcal{E}) \tag{13.9}$$

is a given *morphism in* $\mathcal{C}o\mathcal{PS}h_X$, as indicated (cf. also Lemma 11.1, (6.3), as well as, (11.64)). So our objective is to show that *there exists a sheaf morphism* $\phi : \mathcal{S} \longrightarrow \mathcal{E}$, *such that*

$$\Gamma(\phi) \equiv \overline{\phi} = \psi. \tag{13.10}$$

Now, applying a similar argument to that in Section 8, one first obtains, through the given morphism $\psi$ as in (13.9), a family of *fiber maps*

$$(\psi_x)_{x\in X}, \tag{13.11}$$

where one sets (see (8.7) and (13.1));

$$\psi_x := \varinjlim_{U\ni x} \psi_U : (\Gamma_{\mathcal{S}})_x \longrightarrow (\Gamma_{\mathcal{E}})_x,\ x \in X. \tag{13.12}$$

In turn, the above family (13.11) provides a *fiber preserving map*

$$\phi : \mathcal{S} \longrightarrow \mathcal{E}, \tag{13.13}$$

such that one sets (cf. also (12.54), along with diagram (13.16) below)

$$\phi|_{\mathcal{S}_x} \equiv \phi_x := \varepsilon_x \circ \psi_x \circ \delta_x^{-1},\ x \in X. \tag{13.14}$$

(Thus, here too, the sought map $\phi$, as in (13.13), is given through the corresponding family of the individual stalk maps, each one of them being given by (13.14); yet, the same relation proves, of course, that *$\phi$ is fiber preserving*. Cf. also (13.16)). So our main task is yet to prove that *the* previous fiber preserving *map $\phi$*, as given by (13.14), *is section preserving, as well;* this, in turn, would imply, according to (9.22), that $\phi$ should be a *sheaf morphism*. Thus, we show that, *for any open* set $U \subseteq X$, *and* $s \in \Gamma(U, \mathcal{S})$, one obtains (cf. (13.9))

$$\phi \circ s = \psi_U(s) \in \Gamma(U, \mathcal{E}), \tag{13.15}$$

which, of course, yields the assertion. Now, (13.15) is just a consequence of the following *commutative diagram* (cf. also (8.8), (9.13′) and (13.14));

(13.16)

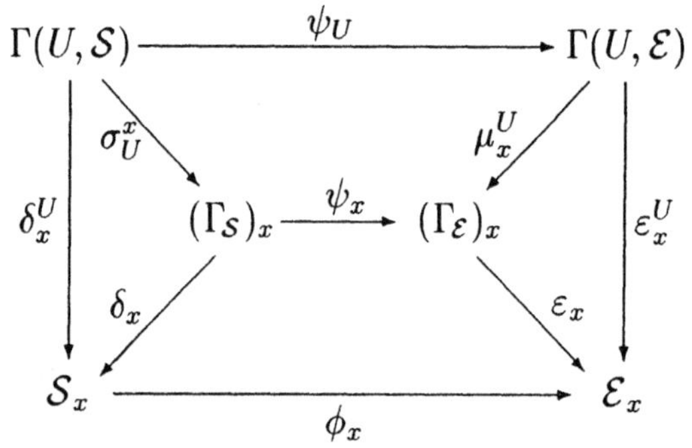


So, based on the previous commutative diagram, one now obtains, *for any*

$x \in U$,

$$(13.17) \quad \begin{aligned}(\phi \circ s)(x) &= \phi(s(x)) = \phi_x(s(x)) = \phi_x\left(\delta_x^U(s)\right) = \phi_x\left(\delta_x\left(\sigma_U^x(s)\right)\right) \\ &= \varepsilon_x(\psi_x(\sigma_U^x(s))) = \varepsilon_x(\mu_x^U(\psi_U(s))) = \varepsilon_x^U(\psi_U(s)) = (\psi_U(s))(x),\end{aligned}$$

which thus proves (13.15), as desired. Yet, the same relation (13.15) *proves (13.10), as well,* where $\overline{\phi}$ is given by (12.50). So the preceding proved that $\Gamma$ is also a full functor, which, along with the first part of the proof, entails that *the functor* $\Gamma$ *is fully faithful* (viz. full and faithful, as already proved).

Finally, it still remains to prove (cf. (13.5)) that $\Gamma$ is also *essentially surjective.* So, by definition, we must show that *for any object* $S$ *of* $\mathcal{CoPSh}_X$, *there exists an object* $\mathcal{S}$ *of* $\mathcal{Sh}_X$ *such that* $\Gamma(\mathcal{S}) = S$, *within an isomorphism.* However, in ordinary language(!) this is just prop. 3) of Theorem 11.2, so that, in practice (cf. also (11.37.1)), it suffices to take, for any $S$, as above, $\mathcal{S} := \mathbf{S}(S)$, viz. its sheafification, which also terminates the proof. ▮

We also express the preceding by setting, for convenience,

$$(13.18) \qquad \mathcal{Sh}_X \underset{\Gamma}{\cong} \mathcal{CoPSh}_X.$$

Thus, based also on (11.64), one obtains the following

$$(13.19) \quad \mathrm{Hom}_{\mathcal{Sh}_X}(\mathcal{S}, \mathcal{E}) \cong \mathrm{Hom}_{\mathcal{CoPSh}_X}(\Gamma(\mathcal{S}), \Gamma(\mathcal{E})) = \mathrm{Hom}_{\mathcal{PSh}_X}(\Gamma(\mathcal{S}), \Gamma(\mathcal{E})),$$

*for any sheaves* $\mathcal{S}, \mathcal{E}$ *on* $X$ (the first relation denoting a *bijection* of the sets concerned).

As we see, the last two relations (13.18) and (13.19) justify also the usual practice of identifying objects and morphisms of either one of the two categories appeared in (13.18) with those of the other. We also apply it in the sequel, as the case may be. Yet, the identifications in question are performed through the section functor $\Gamma$, as above, or even through its *left adjoint,* the sheafification functor (cf. (8.3.1)), in the sense that *one has*

$$(13.20) \qquad \mathbf{S}\,\Gamma = 1_{\mathcal{Sh}_X},$$

indeed, an equivalent form of Theorem 9.1. In categorical language the last relation indicates, in fact, a *natural isomorphism of functors.* (For the applied terminology, cf. S. Mac Lane [1:p.16], or even N. Jacobson [1: p.23, Definition 1.3]). Furthermore, we still have the next *natural isomorphism of* the same *functors,* viz.

(13.21) $$\Gamma\mathbf{S} = 1_{\mathcal{C}o\mathcal{P}\mathcal{S}h_X};$$

this, of course, is but an *equivalent version of (11.37.1)*. So by further applying here categorical language, we see that the above two last relations prove that;

(13.22) the functors $\Gamma$ and $\mathbf{S}$ establish an *adjoint equivalence* of the categories $\mathcal{S}h_X$ and $\mathcal{C}o\mathcal{P}\mathcal{S}h_X$, $\mathbf{S}$ being thus the *left adjoint,* as well as, the *right adjoint* of $\Gamma$ (and "dually" for $\Gamma$, according to the same relations).

Thus, *(13.20) along with (13.21) provide,* in fact, *another equivalent way of getting the isomorphism* (viz. the *category equivalence*) *(13.18),* as established in the preceding by Theorem 13.1. (In this context, see, for instance, S. Mac Lane [1: p.91, Theorem 1] or even J. R. Strooker [1: p.34, Theorem 1.7.11]; yet, cf. B. Pareigis [1: p.55, Corollary 5]). The above categorical language will be applied in the sequel too, in particular, in Chapter III, pertaining to *Sheaf Cohomology.*

We terminate our discussion in the present Section with the following useful application of (13.19). That is, we have.

**Lemma 13.1.** *Let $\mathcal{S}$,$\mathcal{E}$ be two given sheaves (of sets) on a topological space $X$ and $A$ an open subset of $X$. Moreover, let $\mathcal{S}|_A$ and $\mathcal{E}|_A$ be the sheaves on $A$, defined by restriction on it of the given sheaves $\mathcal{S}$ and $\mathcal{E}$, respectively* (cf. (1.10)). *Then, one obtains the following bijection* (see also (5.11));

(13.23) $$\mathrm{Hom}_{\mathcal{S}h_A}(\mathcal{S}|_A, \mathcal{E}|_A) = \mathrm{Hom}_{\mathcal{P}\mathcal{S}h_A}(\Gamma(\mathcal{S})|_A, \Gamma(\mathcal{E})|_A).$$

**Proof.** Indeed, by applying (13.19), as well as, Corollary 11.1, one has the following bijections;

(13.24) $$\begin{aligned}\mathrm{Hom}_{\mathcal{S}h_A}(\mathcal{S}|_A, \mathcal{E}|_A) &= \mathrm{Hom}_{\mathcal{P}\mathcal{S}h_A}(\Gamma(\mathcal{S}|_A), \Gamma(\mathcal{E}|_A)) \\ &= \mathrm{Hom}_{\mathcal{P}\mathcal{S}h_A}(\Gamma(\mathcal{S})|_A, \Gamma(\mathcal{E})|_A),\end{aligned}$$

which thus proves the assertion. ∎

# 14. Change of base space

In this final Section of the present Chapter we examine the variation that may result, concerning basic concepts of the preceding, from letting a continuous map act on the base space of the (pre)sheaves involved. Thus, we start with the following notion.

**14.(a). The direct image** (or **push-out**) **functor.**– We have, in effect, two such functors corresponding to each one of the two categories $\mathcal{PSh}_X$ and $\mathcal{Sh}_X$: Thus, suppose that we are given a *continuous map*

$$f : X \longrightarrow Y \tag{14.1}$$

between two topological spaces, as indicated. Now, we first consider the following *functor*

$$\widetilde{f}_* : \mathcal{PSh}_X \longrightarrow \mathcal{PSh}_Y; \tag{14.2}$$

namely, for any object $S$ of $\mathcal{PSh}_X$, viz. a *presheaf* (of sets) *on* $X$, one sets

$$\widetilde{f}_*(S)(V) := S(f^{-1}(V)), \tag{14.3}$$

with $V$ *varying over the open subsets of* $Y$. So it is a routine matter to verify that (14.2), as given by (14.3), *yields indeed a covariant functor* between the categories concerned. In particular, to any *morphism in* $\mathcal{PSh}_X$, say, $\phi : S \longrightarrow S'$, there corresponds a *morphism in* $\mathcal{PSh}_Y$, viz.

$$\widetilde{f}_*(\phi) : \widetilde{f}_*(S) \longrightarrow \widetilde{f}_*(S'), \tag{14.4}$$

in such a manner that one sets;

$$\widetilde{f}_*(\phi)_V := \phi_{f^{-1}(V)}, \tag{14.5}$$

*for any open* $V \subseteq Y$ (cf. also (6.3), as well as, (14.3)).

We call (14.2) the *direct image* (or even *push-out*) *presheaf functor,* associated with the given continuous map $f$. Yet, the presheaf on $Y$, $\widetilde{f}_*(S)$, as defined by (14.3), is called the *(direct) image presheaf* (or even *push-out*) *of* $S$, via $f$.

In this context, one can check further, based on the very definitions, that;

(14.6) *the (direct) image presheaf functor,* as above, *preserves both the conditions (S1) and (S2)* (cf. Definition 11.1). Hence, *in particular, it maps complete presheaves to similar ones.*

We come next to the *functor*

$$f_* : \mathcal{S}h_X \longrightarrow \mathcal{S}h_Y, \tag{14.7}$$

which is associated *with any given* $f$, *as in (14.1)*. So one has, by definition,

$$f_* := \mathbf{S}_Y \circ \tilde{f}_* \circ \Gamma_X, \tag{14.8}$$

where, of course, $\tilde{f}_*$ is given by (14.3), while $\Gamma_X, \mathbf{S}_Y$ stand for the section and the sheafification functors, respectively, on the pertinent categories $\mathcal{S}h_X$ and $(\mathcal{C}o\mathcal{PS}h_Y \overset{\subsetneq}{\rightarrow})\ \mathcal{PS}h_Y$ (hence, the attached indices to the respective functors). The meaning of (14.8) is certainly clear in perspective with Theorem 13.1 and (14.6). Thus, by its very definition $f_*$ *is* indeed *a covariant functor* between the categories concerned, being, namely, a composition of such functors.

Furthermore, *for any* given *sheaf* $\mathcal{S}$ *on* $X$, one obtains the following (canonical) *bijection* (cf. also (2.3))

$$f_*(\mathcal{S})(V) = \mathcal{S}(f^{-1}(V)), \tag{14.9}$$

*for every open set* $V \subseteq Y$. Indeed, based on (14.8), (14.6), (11.37.1) and (14.3), one has

$$\begin{aligned} f_*(\mathcal{S})(V) &\equiv \left(\mathbf{S}_Y \tilde{f}_* \Gamma_X\right)(\mathcal{S})(V) = \mathbf{S}\left(\tilde{f}_*(\Gamma(\mathcal{S}))\right)(V) \cong \tilde{f}_*(\Gamma(\mathcal{S}))(V) \\ &= \Gamma(\mathcal{S})(f^{-1}(V)) \cong \mathcal{S}(f^{-1}(V)), \end{aligned} \tag{14.10}$$

(see also (9.3)), as desired.

As an application of (14.9), *one* also *obtains* the following useful relation;

$$(g \circ f)_* = g_* \circ f_*, \tag{14.11}$$

*for any* given *continuous maps* $f : X \longrightarrow Y$ *and* $g : Y \longrightarrow Z$. In fact, (14.11) stands for a *natural isomorphism of functors,* as indicated (cf. also Theorem 9.1; yet, cf. e.g. (13.20), concerning the applied terminology).

One further application of (14.9) is also the existence of the following (canonical) *fiber map;*

$$\tilde{f}_x : f_*(\mathcal{S})_{f(x)} \longrightarrow \mathcal{S}_x, \quad x \in X, \tag{14.12}$$

with $f$ *as in (14.1)* and $\mathcal{S}$ *any sheaf on* $X$: That is, for any $z \in f_*(\mathcal{S})_{f(x)}$, one gets (cf. (2.10.1)) $z = t(f(x))$, for some $t \in f_*(\mathcal{S})(V)$, with $f(x) \in V$ and $V$ open in $Y$. Hence, by virtue of (14.9), one also obtains a *unique* $s \in \mathcal{S}(f^{-1}(V)) = f_*(\mathcal{S})(V)$; therefore, one now defines;

$$\tilde{f}_x(z) := s(x) \in \mathcal{S}_x, \tag{14.13}$$

which thus yields (14.12). Yet, one easily checks here that the various $t$'s, that eventually represent $z \in f_*(\mathcal{S})_{f(x)}$, as before, yield *equivalent sections* $s$ (cf. e.g. (4.15)), as well, due again to (14.9) and (2.13), which, in turn, provide thus the same *germ* in $\mathcal{S}_x$, as in (14.13). ∎

**14.(b). Inverse image** (or **pull-back**) **functor.**– Given a *continuous map* $f$, as in (14.1), and a *sheaf* $\mathcal{E} \equiv (\mathcal{E}, \pi, Y)$ *on* $Y$, the range of $f$, we next define the *inverse image sheaf* (or *pull-back*) *of* $\mathcal{E}$, denoted by

$$f^*(\mathcal{E}) \equiv f^{-1}(\mathcal{E}), \tag{14.14}$$

as that *sheaf on* $X$, domain of definition of $f$, which is given by the relation;

$$f^*(\mathcal{E}) := \{(x, z) \in X \times \mathcal{E} : f(x) = \pi(z)\}. \tag{14.15}$$

Of course, we should explain here our (abused) terminology, concerning the last relation; so *firstly* (14.15) stands actually for the *sheaf space* of the sought sheaf $f^*(\mathcal{E})$, which thus is further considered as a *topological space* endowed *with the relative topology from* $X \times \mathcal{E}$, being, by its definition, a subset of the latter space. *Secondly,* one defines the corresponding *projection map*

$$f^*(\mathcal{E}) \longrightarrow X, \tag{14.16}$$

by the relation

$$(x, z) \longmapsto x, \ (x, z) \in f^*(\mathcal{E}); \tag{14.17}$$

viz. as the restriction to $f^*(\mathcal{E}) \subseteq X \times \mathcal{E}$ of the first (canonical) projection map of $X \times \mathcal{E}$ onto $X$. Thus, *thirdly,* it remains to prove that *the map (14.16),* as given by (14.17), *yields a local homeomorphism:* Indeed, if $(x, z) \in f^*(\mathcal{E}) \subseteq X \times \mathcal{E}$, let $z = t(f(x))$, for some $t \in \mathcal{E}(V)$ and $V$ open in $Y$, with $f(x) \in V$ (cf. (2.10.1)). Now, *the map*

$$(f^{-1}(V) \times t(V)) \cap f^*(\mathcal{E}) \longrightarrow f^{-1}(V), \tag{14.18}$$

a restriction, in fact, of (14.16), as indicated, *yields a homeomorphism,* its domain (of definition) being, of course, an *open neighborhood of* $(x, z)$ *in* $f^*(\mathcal{E})$. In fact, the assertion follows straightforwardly by the relation

$$(f^{-1}(V) \times t(V)) \cap f^*(\mathcal{E}) = \left\{(x, t(f(x)) : x \in f^{-1}(V)\right\}, \tag{14.19}$$

whose second member maps now (under (14.16)) homeomorphically onto

$f^{-1}(V)$, as immediately follows from the very definitions. ∎

We have thus completely established our claim that $f^*(\mathcal{E})$, *along with the* (projection) *map (14.16), constitutes a sheaf* (of sets) *on* $X$. Yet, we are going to give another useful interpretation of the same sheaf presently below. However, we first comment on a certain *particular,* in fact, very convenient, *form of the* (local) *sections of* $f^*(\mathcal{E})$ (see e.g. (14.21) or even (14.22), as well as (14.28) below):

So, for any *open set* $V \subseteq Y$, one has a (canonical) *"map of sections"* (or even an *adjunction map*), as follows;

$$f_V^* : \mathcal{E}(V) \longrightarrow f^*(\mathcal{E})(f^{-1}(V)), \tag{14.20}$$

such that one sets, *for any* given $t \in \mathcal{E}(V)$ (cf. also (4.15)),

$$f_V^*(t)(f^{-1}(V)) := \left\{(x, t(f(x))) : x \in f^{-1}(V)\right\} \subseteq f^*(\mathcal{E}). \tag{14.21}$$

Therefore, based on (14.19), *one obtains*

$$f_V^*(t)(f^{-1}(V)) = (f^{-1}(V) \times t(V)) \cap f^*(\mathcal{E}). \tag{14.22}$$

In particular, *one has*

$$(f_V^*(t))(x) = (x, t(f(x))) \in f^*(\mathcal{E})_x, \tag{14.23}$$

*for any* $x \in f^{-1}(V) \equiv \operatorname{Dom}(f_V^*(t))$, *with* $t \in \mathcal{E}(V)$.

Thus, based on the preceding, one further obtains, *for any point* $x \in X$, a corresponding (canonical) *fiber map*

$$f_x^* : \mathcal{E}_{f(x)} \longrightarrow f^*(\mathcal{E})_x, \tag{14.24}$$

such that one sets, in view of (14.23),

$$f_x^*(z) := (x, t(f(x))) \equiv (f_V^*(t))(x), \tag{14.25}$$

where $z \in \mathcal{E}_{f(x)}$, such that (cf. (2.10.1)) $z = t(f(x))$, for some $t \in \mathcal{E}(V)$ and $f(x) \in V$, with $V$ open in $Y$. (In this context, it is further easily seen, according to the very definitions, that *(14.25) is well-defined,* viz. independently of the particular $t \in \mathcal{E}(V)$ involved, representing the given *germ* $z \in \mathcal{E}_{f(x)}$; see also (2.13)).

Indeed, one easily proves, by virtue of (14.23), that;

$$\mathcal{E}_{f(x)} = f^*(\mathcal{E})_x, \; x \in X, \tag{14.26}$$

*within a* (canonical) *bijection, established by (14.25).*

Our next objective about the sheaf $f^*(\mathcal{E})$, as above, is to show now that;

(14.27) the set (cf. also (14.21))

$$(14.27.1) \qquad \mathfrak{S} = \{f_V^*(t) : t \in \mathcal{E}(V), \text{ with } V \textit{ open in } Y\}$$

provides a *defining family of sections of* $f^*(\mathcal{E})$ (cf. (3.17)), so that its members yield a *basis of the topology of* $f^*(\mathcal{E})$ (see Theorem 3.1).

Indeed, the assertion about (14.27.1) follows easily from (14.22) and the fact that (14.18) supplies a homeomorphism. (See also the relevant terminology in Section 3). So, as a result of (14.27), one still concludes, yet within the perspective of Section 3 in the preceding, that, in practice,

(14.28) *the sheaf $f^*(\mathcal{E})$ is actually determined by its sections over those open subsets of $X$, which are of the form $f^{-1}(V)$, with $V$ open in $Y$.*

Thus, as an application of (14.27) — and for that matter, of (14.28), as well — we come now to the aforementioned *variation of the definition of* $f^*(\mathcal{E})$*:*

Namely, for any *open* $U \subseteq X$, let us consider the set

$$(14.29) \qquad \mathcal{E}_f(U) = \{t : U \longrightarrow \mathcal{E}, \text{ cont.} : \pi \circ t = f\} \subseteq \mathcal{C}(U, \mathcal{E}),$$

viz. all *continuous maps* $t : U \longrightarrow \mathcal{E}$, obeying the indicated condition; we deliberately call the members of (14.29) continuous (local) *sections of* $\mathcal{E}$ *along* $f$ (over $U$, a terminology reminiscent of analogous situations, for instance, in differential geometry; viz. e.g. vector fields, or connections, along (differentiable) maps).

Now, based on the very definitions, it is not hard to see that *the correspondence*

$$(14.30) \qquad U \longmapsto \mathcal{E}_f(U), \ U \textit{ open in } X,$$

along with the obviously defined *restriction maps,* provides a *complete presheaf* on $X$. We call it the *presheaf of (continuous) sections of* $\mathcal{E}$ *along* $f$ and denote by

$$(14.31) \qquad \mathcal{E}_f.$$

To make thus the résumé, so far, we have that;

(14.32) *for any continuous map* $f : X \longrightarrow Y$ *and any sheaf* $\mathcal{E}$ *on* $Y$, $\mathcal{E}_f$ *is a complete presheaf on* $X$.

In this context, we will remark that by taking, in particular, $f = \mathrm{id}_X : X \longrightarrow X$, viz. the *identity map* on $X$, one obtains, *for any* $\mathcal{E}$ *in* $\mathcal{S}h_X$, the *(complete) presheaf* $\mathcal{E}_{\mathrm{id}_X} = \Gamma(\mathcal{E})$ (!); see also (9.7), and Lemma 11.1.

So the claim now is that one can actually recapture our previous sheaf on $X$, $f^*(\mathcal{E})$, just by taking the sheafification of $\mathcal{E}_f$; yet, in the language of Theorem 13.1, $\mathcal{E}_f$ *is, in effect,* $f^*(\mathcal{E})$. Precisely, *one* thus *obtains*

$$\mathbf{S}(\mathcal{E}_f) = f^*(\mathcal{E}), \tag{14.33}$$

within a *sheaf isomorphism.*

Now, *to prove* (14.33), we shall need some further auxiliary results, which we presently exhibit. So we first have the following *bijection*

$$\mathcal{E}_f(f^{-1}(V) = f^*(\mathcal{E})(f^{-1}(V)), \tag{14.34}$$

*for any open* $V \subseteq Y$. That is, to any $t \in \mathcal{E}_f(f^{-1}(V))$ (cf. (14.29)) one associates the section $s \in f^*(\mathcal{E})(f^{-1}(V))$, such that $s(x) := (x, t(x))$, $x \in f^{-1}(V)$ (cf. also (14.15)). That is, *we set*

$$s := (\mathrm{id}_{f^{-1}(V)}, t), \tag{14.35}$$

*for any* $t \in \mathcal{E}_f(f^{-1}(V))$. The correspondence is, of course, *one-to-one;* yet, for any $s \in f^*(\mathcal{E})(f^{-1}(V))$, *one sets* $t := \mathrm{pr}_2 \circ s$ where $\mathrm{pr}_2 : X \times \mathcal{E} \longrightarrow \mathcal{E}$ stands for the second (canonical) *projection map,* as indicated, which proves that the previous correspondence is an *onto* map, as well.

Furthermore, we can supplement (14.26) by the following *bijection;*

$$\mathcal{E}_{f(x)} = (\mathcal{E}_f)_x, \; x \in X. \tag{14.36}$$

Namely, to any $z \in \mathcal{E}_{f(x)}$, with $z = t(f(x))$ (cf. (14.25)), one associates the *germ*

$$\rho^x_{f^{-1}(V)}(\hat{t}) \in (\mathcal{E}_f)_x = \varinjlim_{U \ni x} \mathcal{E}_f(U), \tag{14.37}$$

where $\hat{t} \in \mathcal{E}_f(f^{-1}(V))$, such that $\hat{t}(x) := t(f(x))$, $x \in f^{-1}(V)$ (viz. $\hat{t} := t \circ f$; cf. also (7.10) for the notation applied in (14.37)). Now, it is not hard to see

(see also (14.26) and (14.35)) that the above correspondence is, in fact, a *bijection.*

Now, based on (14.34) and on (14.36), one further concludes that

(14.38) (14.27.1) (see also (14.21) and (14.34)) *yields a defining family of sections of the sheaf* $\mathbf{S}(\mathcal{E}_f)$, *as well;*

So, not only of the sheaf $f^*(\mathcal{E})$, as already observed (cf. (14.27)). In this context, we still note that, $\mathcal{E}_f$ being a complete presheaf (cf. (14.32)), the sections of $\mathbf{S}(\mathcal{E}_f)$ *are,* within *bijections* (cf. Proposition 11.1), *those of* $\mathcal{E}_f$ (see also Scholium 7.1), hence the usefulness of (14.34). Thus, this establishes, of course, (14.33), according to Theorem 3.1. ∎

In this context, we further note the following useful definition, concerning the *inverse image,* or *pull-back,* of a given presheaf $E$ on $Y$, via a *continuous map* $f : X \longrightarrow Y$: That is, one sets;

(14.39)
$$f^*(E)(U) := \varinjlim_{V \supseteq f(U)} E(V),$$

for every *open* $U \subseteq X$, while $V$ in the second member of (14.39) runs over the *open neighborhoods of* $f(U) \subseteq Y$.

Finally, to prove the *functional character of (14.14),* when $\mathcal{E}$ is varying in $\mathcal{S}h_Y$, as already claimed, one has to define for any given *morphism* in $\mathcal{S}h_Y$, say,

(14.40)
$$\phi : \mathcal{E} \longrightarrow \mathcal{E}',$$

a corresponding *morphism* in $\mathcal{S}h_X$, viz.

(14.41)
$$f^*(\phi) : f^*(\mathcal{E}) \longrightarrow f^*(\mathcal{E}').$$

Indeed, one sets *point-wise*

(14.42)
$$f^*(\phi)(x,z) := (x, (\phi \circ t)(f(x))),$$

*for any* $(x,z) \in f^*(\mathcal{E})_x \subseteq f^*(\mathcal{E})$, where one has $z = t(f(x)) \in \mathcal{E}_{f(x)}$, as before (see thus (14.26)); yet, we can further check, as we did before e.g. for (14.25), that (14.42) *is well-defined,* concerning the section $t \in \mathcal{E}(V)$ involved. Now, as a consequence of (14.41), along with (14.21), *one* further *obtains*

(14.43)
$$f^*(\phi)(f_V^*(t)) = f_V^*(\phi \circ t),$$

*for any* $t \in \mathcal{E}(V)$ and $V$ *open in* $Y$. This proves, in particular, that;

(14.44) *the map (14.41) is section preserving, with respect to the defining family of sections (14.27.1)* for the sheaf $f^*(\mathcal{E})$. Hence, *the same map yields a sheaf morphism* of the sheaves concerned, according to (9.22) and Lemma 3.1.

(In particular, the lesson here from (14.44) is, therefore, that, *to get a sheaf morphism, it suffices to check (9.22) just for a defining family of sections,* an immediate consequence, for that matter, of Lemma 3.1).

Therefore, one concludes that;

(14.45) *the correspondence*

(14.45.1) $$f^* : \mathcal{S}h_Y \longrightarrow \mathcal{S}h_X,$$

as given by (14.15), *yields, for any continuous map* $f : X \longrightarrow Y$, *a covariant functor* of the categories concerned.

In this context, for any given *continuous functions*

$$f : X \longrightarrow Y \text{ and } g : Y \longrightarrow Z,$$

*one* further *obtains;*

(14.46) $$(g \circ f)^* = f^* \circ g^*,$$

as easily follows, by the very definitions.

The previous two functors $f_*$ and $f^*$ will also be considered in the sequel (cf. the next Chapters II – IV). In this regard, we further note that, *in general, one obtains*

(14.47) $$f_* \circ f^* \neq 1_{\mathcal{S}h_Y} \text{ and } f^* \circ f_* \neq 1_{\mathcal{S}h_X},$$

concerning an *arbitrary continuous map* $f : X \longrightarrow Y$, relative to the *natural isomorphisms of functors,* as indicated. In this context, see also, for instance, J. R. Strooker [1: p.213, Example].

Within the same context, we finally remark that one can also define, *for any continuous map* $f$, as before, a corresponding *inverse image* (or *pull back*) *functor for presheaves,* in terms of more involved (categorical) considerations as, for example, *"Kan extensions"* and the associated terminology (see, for instance, J. R. Strooker [1: p. 210, Proposition 4.6.3]). However, cf. also (14.39) in the preceding.

# CHAPTER II

# Sheaves and Presheaves with Algebraic Structure

"... The methods of sheaf theory are algebraic."

H. Grauert – R. Remmert in "Coherent Analytic Sheaves" (Springer - Verlag, Berlin, 1984). p.VII.

"... find a purely algebraic theory for the description of reality."

A. Einstein in "The meaning of Relativity" (Princeton Univ. Press, Princeton, NJ, 1984). p.166.

The sheaves, which are considered next, have their stalks enriched with an algebraic structure, a fact, providing a similar structure for the various sets of sections of the particular sheaf concerned. Moreover, the algebraic operations, thus involved, are *"continuous"* in an appropriate sense. Similarly, the constituent sets of the presheaves considered are also endowed with analogous algebraic structure, while the standard sheafification procedure transfers, in turn, the algebraic structure in question to (the individual stalks of) the respective sheaves.

Now, the sort of algebraic structure on the (pre)sheaves involved, which we are going to use in the sequel, is essentially that of a *module over an* appropriate **C**-*algebra* (commutative and with, or without, an identity element). However, we begin with first giving the corresponding definition of a *sheaf of abelian groups;* the latter constitutes, in fact, the classical notion of a *sheaf of coefficients,* when looking, in particular, at the associated (complete) presheaf of sections.

# 1. Preliminaries. Basic definitions

As already noted throughout the previous chapter, what we have named heretofore sheaf (or presheaf) on a given topological space $X$ was, in fact, a (pre)sheaf of *sets* on $X$, in the sense that the individual stalks, or to put it differently (but, in effect, *equivalently,* see below), the corresponding sets of sections of the (pre)sheaf involved, were just *sets, without any particular algebraic structure.* This, by contrast with what we are going to apply in the sequel. So, *algebraically,* an additional structure will be considered *in the stalks of the sheaf* under discussion, while, *topologically,* the same structure is further transferred to the whole sheaf space, by means of an appropriate notion of *"continuity of the algebraic operations"* thus resulted. The above amounts, in a sense, to a *"fiber-wise manner"* of introducing an extra structure in the initial definition of a sheaf, the latter being thus, taken just as a local homeomorphism.

We start with first endowing a given sheaf with a particular algebraic structure; the analogous situation for presheaves is also dealt with in the sequel. (Cf. Subsection 1.(b) below). So we consider first:

**1.(a). Sheaves with algebraic structure. $\mathcal{A}$-modules.**– We begin with the following basic.

**Definition 1.1.** Given a sheaf (of sets) $\mathcal{S} \equiv (\mathcal{S}, \pi, X)$ on a topological space $X$ (cf. Chapt. I, Definition 1.1), we say that $\mathcal{S}$ is a **sheaf of groups** (or even a **group sheaf**) on $X$, whenever the two following conditions are satisfied:

i) Each stalk of $\mathcal{S}$, $\mathcal{S}_x = \pi^{-1}(x)$, $x \in X$, (cf. I; (1.5)), is a group.

ii) *"Addition is continuous"* (in a sense that is presently explained below).

In particular, a **sheaf of abelian groups** on a topological space $X$, is a sheaf of grouprs on $X$, having each one of its stalks an abelian group.

First, we comment on the phrase *"addition is continuous"* of the previous definition: Thus, given a sheaf (of sets) $\mathcal{S} \equiv (\mathcal{S}, \pi, X)$ on a topological space $X$, let us consider the set

$$\mathcal{S} \circ \mathcal{S} := \{(z, z') \in \mathcal{S} \times \mathcal{S} : \pi(z) = \pi(z')\}; \tag{1.1}$$

in other words, one takes by (1.1) the *"fiber-wise product"* (or just *"fiber product"*) of $\mathcal{S}$ with itself, denoted also by $\mathcal{S} \times_X \mathcal{S}$.

Now, if each one of the stalks of $\mathcal{S}$ is a *group*, one transfers this structure *"fiber-wise"* to the whole sheaf space $\mathcal{S}$, in the sense that one has a map

$$\mathcal{S} \circ \mathcal{S} \longrightarrow \mathcal{S} : (z, z') \longmapsto z + z' \in \mathcal{S}_x \subseteq \mathcal{S}, \tag{1.2}$$

where, in view of (1.1), one has $\pi(z) = \pi(z') \equiv x \in X$, viz. both $z$ and $z'$ are elements of the same stalk (group) $\mathcal{S}_x$, so that *(1.2) is well-defined.* So cond. ii) of Definition 1.1 means now, *by definition,* that *the map (1.2) is continuous,* when $\mathcal{S} \circ \mathcal{S}$ carries the relative topology from $\mathcal{S} \times \mathcal{S}$.

As we shall also see in the sequel, the above constitute, in effect, the *sheaf-theoretic aspect of continuity of an algebraic operation,* already defined in the stalks of a given sheaf (of sets).

So, as a first consequence of the preceding, we come next to the definition of the so-called *zero-section* of any given *sheaf of groups.* Namely, we have.

**Lemma 1.1.** *Let $\mathcal{S} \equiv (\mathcal{S}, \pi, X)$ be a given sheaf of groups on a topological space $X$. Then, for any open set $U \subseteq X$, one has the following section of $\mathcal{S}$ over $U$;*

$$0 : U \longrightarrow \mathcal{S} : x \longmapsto 0(x) := 0_x \in \mathcal{S}_x \subseteq \mathcal{S}, \tag{1.3}$$

*where $0_x$, $x \in X$, stands for the zero (neutral) element of the group (stalk) $\mathcal{S}_x$, $x \in X$.*

**Proof.** Of course, (1.3) is, by its very definition, a *section of* $\pi$, in the set-theoretic sense, viz. I; (2.2) is fulfilled. So we further prove the *continuity of (1.3):* Thus, given $x \in U$, let $V$ be an *open neighborhood of* $0_x$ *in* $\mathcal{S}$. In this context, it is first useful to remark that;

(1.4) based in I; (2.14), (2.15), one may assume that the particular neighborhood $V$, as above, is a *domain of homeomorphicity of* $\pi$, in the sense of I; (1.3.1). (This will also be applied in the sequel, *without further notice*).

Now, given that *"addition is continuous"* in $\mathcal{S}$ (Definition 1.1, ii)) and $0_x + 0_x = 0_x$ in $\mathcal{S}_x$ ($x \in X$), one concludes that there exist *open neighborhoods* $V_1, V_2$ *of* $0_x$ *in* $\mathcal{S}$ such that $V_1 + V_2 \subseteq V$. (*"Addition"*, as in the last relation, is always meant, of course, according to (1.1) and (1.2), *fiber-wise*!). Now, setting

$$W = V \cap V_1 \cap V_2 \subseteq V, \tag{1.5}$$

thus, *still an open neighborhood of* $0_x$ *in* $\mathcal{S}$, one further proves that

$$W = 0(\pi(W)); \tag{1.6}$$

that is,

(1.7) *the open set* $W \subseteq \mathcal{S}$, *as given by (1.5), consists of zeroes only, while it is also mapped, through* $\pi$, *homeomorphically onto the open set* $\pi(W) \subseteq X$.

Indeed, let $z \in W$, such that $\pi(z) = x' \in \pi(W)$; hence, $z \in \mathcal{S}_{x'}$, so that $z + z \in \mathcal{S}_{x'}$. Thus, $\pi(z+z) = x' = \pi(z)$; but, in view of (1.4) and (1.5), $\pi|_{W \subseteq V}$ is a homeomorphism, so that $z+z = z$, hence $z = 0_{x'} = 0(x') \in \mathcal{S}_{x'}$ (cf. also (1.3)), *which proves (1.6)*. Of course, *(1.7) proves already the continuity of (1.3).* ∎

As already said, we call (1.3) the *zero-section of* $\mathcal{S}$ *over* $U$. Thus, (1.3) being also an *open map* (cf. Chapt. I, Lemma 2.1), *its image in* $\mathcal{S}$ *is* an open subset of $\mathcal{S}$, hence, *a subsheaf of* $\mathcal{S}$. In particular, for $U = X$,

(1.8) the respective *global zero-section* of $\mathcal{S}$, defines through its image, the so-called *set of zeros of* $\mathcal{S}$, an open set in $\mathcal{S}$, therefore, a *subsheaf of* $\mathcal{S}$; we call it the *zero subsheaf of* $\mathcal{S}$, as well, being also *homeomorphic to* $X$.

The above sheaf is certainly *isomorphic*(!) to the so-called *zero sheaf on* $X$, viz., by definition, the *constant sheaf on* $X$, defined by the *singleton* $\{0\}$. (See also, for instance, Chapt. I; Subsection 4.(a)). The latter sheaf is denoted in the sequel by $0_X$ (cf. e.g. I; (4.4′)), or usually *just by*

$$0. \tag{1.9}$$

In this respect, cf. also the next Definition 1.2, as well as, Example 1.1 below.

So, before we proceed further, and by reason of the previous comments, we first give the following.

**Definition 1.2.** Given two sheaves of groups $\mathcal{S}$ and $\mathcal{S}'$ on a topological space $X$ (Definition 1.1), a **morphism of $\mathcal{S}$ in $\mathcal{S}'$** (or even **morphism of** the **group sheaves** concerned), is a morphism, say $\phi : \mathcal{S} \longrightarrow \mathcal{S}'$ of the respective sheaves (Definition I, 1.2) such that the corresponding fiber maps (cf. I; (1.14))

$$\phi_x : \mathcal{S}_x \longrightarrow \mathcal{S}'_x, \ x \in X, \tag{1.10}$$

are *group morphisms* of the groups (stalks) concerned.

Thus, on the basis of the two above Definitions 1.1 and 1.2, one can further consider, for any given topological space $X$,

$$\mathcal{S}h\mathcal{G}r_X, \tag{1.11}$$

viz. the *category of sheaves of groups on* $X$.

In this context, it is clear, of course, by the very definitions, that the isomorphism connecting the two sheaves in (1.8) and (1.9) is still valid in the sense of the previous Definition 1.2.

We come now to some further basic consequences of Definition 1.1. So one has the next.

**Lemma 1.2.** *Let $\mathcal{S} \equiv (\mathcal{S}, \pi, X)$ be a sheaf of groups on a given topological space $X$. Then, the operation of "taking opposites" and that one of subtraction in $\mathcal{S}$* (considered, of course, *fiber-wise*), *viz. the maps*

$$\mathcal{S} \longrightarrow \mathcal{S} : z \longmapsto -z \ \text{ and } \ \mathcal{S} \circ \mathcal{S} \longrightarrow \mathcal{S} : (z, z') \longmapsto z - z', \tag{1.12}$$

*are continuous.*

**Proof.** Let $z \in \mathcal{S}_x \subseteq \mathcal{S}$, so that $-z \in \mathcal{S}_x$, *as well,* and let $V$ be an *open neighborhood of* $-z$ *in* $\mathcal{S}$. But, $z + (-z) = 0_x \in \mathcal{S}_x$, so that let $W$ be the *open neighborhood of* $0_x$ *in* $\mathcal{S}$, *as in (1.6).* Thus, since *"addition in $\mathcal{S}$ is continuous"*, there exist open neighborhoods of $z$ and $-z$, say, $V_1$ (cf. also (1.4)) and $V_2$, respectively, such that $V_1 + V_2 \subseteq W$. Now, setting

$$V' \equiv V \cap V_2, \tag{1.13}$$

thus, still an *open neighborhood of* $-z$ *in* $\mathcal{S}$, one obtains

$$V_1 + V' \subseteq V_1 + V_2 \subseteq W. \tag{1.14}$$

Furthermore, setting $U \equiv \pi(V_1) \cap \pi(V')$, thus an *open subset of* $\pi(V_1)$, *with* $x \in U$, let

(1.15) $$V_1' = (\pi|_{V_1})^{-1}(U) \subseteq V_1$$

viz. an *open neighborhood of* $z$ *in* $\mathcal{S}$. So *we assert that;*

(1.16) $$-V_1' \subseteq V' \subseteq V,$$

in the sense that, *for any* $z' \in V_1' \cap \mathcal{S}_{x'}$, *one has* $-z' \in V' \cap \mathcal{S}_{x'}$: For, by hypothesis and (1.15), $\pi(z') \in U \equiv \pi(V_1) \cap \pi(V')$, so that $\pi(z') = \pi(z'')$, for some $z'' \in V'$; hence (cf. also (1.14) and (1.15)), $z' + z'' \in V_1' + V' \subseteq V_1 + V' \subseteq W$, thus (cf. (1.6)) $z' + z'' = 0_{x'}$, that is, $-z' = z'' \in V'$, as claimed. Now, (1.16) proves already the *continuity of the first of the maps in (1.12).*

On the other hand, based on the above part of the proof, one further concludes that *the map*

(1.17) $$(z, z') \longmapsto (z, -z') : \mathcal{S} \circ \mathcal{S} \longrightarrow \mathcal{S} \circ \mathcal{S},$$

*is continuous.* (Of, course, $\mathcal{S} \circ \mathcal{S}$ carries always, by definition (cf. (1.1)), the *relative topology from* $\mathcal{S} \times \mathcal{S}$.) Therefore, *"subtraction in* $\mathcal{S}$*" is continuous,* as well, being thus the composition of the continuous maps (1.17) and (1.2), which also completes the proof. ∎

**Example 1.1.** We discuss below a particular instance of a *sheaf of groups,* which also appears several times, in various forms, throughout the sequel. Namely, it refers to the *constant sheaf of groups* on a topological space $X$, which is *associated with a given group* $G$.

Thus, following Chapt. I, Subsection 4.(a), one considers the *constant sheaf*

(1.18) $$G_X \equiv (X \times G, \pi \equiv \mathrm{pr}_1, X)$$

(see e.g. I, (4.4′)), where the given group $G$ *carries the discrete topology.* Of course, every fiber of (1.18) is (a set) *isomorphic to* $G$ (cf. I, (4.3)), hence, *a group,* where the *"group operation in* $G_X$*"* is thus defined (*fiber-wise*) by the relation;

(1.19) $$(x, \alpha) + (x, \beta) := (x, \alpha + \beta),$$

for any $\alpha, \beta$ in $G$, and every $x \in X$. Yet, it is straightforward, by the very definition of the (discrete) topology in $G$, as well as, that of $X \times G$, that (cf. also (1.1));

(1.20) *"addition in $G_X$", viz. the map $G_X \circ G_X \longrightarrow G_X$, as given by (1.19), is continuous.*

So (Definition 1.1), (1.18) is really a *sheaf of groups* on $X$. In the sequel we shall also denote it usually, *just by*

$$G, \tag{1.21}$$

saying that the given (discrete) group $G$ is considered as the *constant sheaf* $G$ on $X$. In this context, the constant sheaf (1.9) in the preceding corresponds thus to the particular case of the *trivial group* $\{0\}$.

On the other hand, several examples of sheaves, endowed with a particular algebraic structure, will also be considered below in abundance, especially, concerning *sheaves of modules* and/or *sheaves of algebras* (see below for the corresponding definitions). Thus, we do not insist further on that point here.

So we come now to the definition of the next *sheaf-theoretic algebraic structure,* which will also be of use (however, implicitly) in the sequel. Namely, one has.

**Definition 1.3.** A **sheaf of rings** on a topological space $X$, is a sheaf of abelian groups on $X$, say, $\mathcal{R} \equiv (\mathcal{R}, \pi, X)$, such that the two following conditions are fulfilled:

i) Each stalk of $\mathcal{R}$, $\mathcal{R}_x = \pi^{-1}(x)$, $x \in X$, is a ring.

ii) The *"multiplication in $\mathcal{R}$"* is continuous; viz. the map $\mathcal{R} \circ \mathcal{R} \longrightarrow \mathcal{R}$: $(z, z') \longmapsto z \cdot z' \in \mathcal{R}_x \subseteq \mathcal{R}$, such that $\pi(z) = \pi(z') = x \in X$ (see also (1.1)), is continuous.

In particular, one speaks of a **sheaf of commutative rings** on $X$, in the case the various stalks $\mathcal{R}_x$, $x \in X$, of a sheaf of rings $\mathcal{R}$ on $X$, as before, are commutative rings.

As a ramification of the previous definition, that will also occur in the sequel, we further consider the case that the various stalks $\mathcal{R}_X$, $x \in X$, of a given sheaf of rings $\mathcal{R}$, as above, have *identity elements,* denoted henceforth by $1_x$, $x \in X$. Precisely, one has the next.

**Definition** 1.3$'$. A **sheaf of rings with an identity element** on a topological space $X$, is a sheaf of rings on $X$, say, $\mathcal{R} \equiv (\mathcal{R}, \pi, X)$ (cf. Definition 1.3), such that the stalks (rings) $\mathcal{R}_x$, $x \in X$, of $\mathcal{R}$ have *identity elements* $1_x$,

$x \in X$, in such a manner that the corresponding *identity section* (of $\pi$), viz. the map

$$(1.22) \qquad \mathbf{1} : X \longrightarrow \mathcal{R} : x \longmapsto \mathbf{1}(x) := 1_x \in \mathcal{R}_x,$$

is *continuous.* (Hence, it supplies, by definition, a *global continuous section* of $\mathcal{R}$; cf. also Definition I, 2.1).

On the other hand, a *morphism of sheaf of rings,* say, $\phi : \mathcal{R} \longrightarrow \mathcal{R}'$, is a morphism of the given *sheaves* (of rings) $\mathcal{R}$ and $\mathcal{R}'$ (when thus the latter are considered just as sheaves (of sets), cf. Definition I, 1.2), such that the individual fiber maps (cf. also I; (1.14))

$$(1.23) \qquad \phi_x : \mathcal{R}_x \longrightarrow \mathcal{R}'_x,\ x \in X,$$

are *ring morphisms* of the corresponding rings (Definition 1.3, i)).

Now, by analogy with Example 1.1, we further note for later use that, for any given *ring* $R$ and *topological space* $X$, one can consider the corresponding *constant sheaf of rings* on $X$, denoted, by a convenient(!) abuse of notation, yet by

$$(1.24) \qquad R;$$

viz. the *constant sheaf* $R_X \equiv (X \times R, \pi \equiv \mathrm{pr}_1, X)$, where the given ring $R$ carries the *discrete topology* (see also I, (4.1), (4.2)). So the *"multiplication in $R$"* (viz. in $R_X$) and, of course, the addition too, are defined analogously to (1.19), while one can easily prove, by the very definition of the topology in $X \times R$, that *both the* above *two "operations in $R$" are continuous,* so that (1.24), as defined is, indeed, a *sheaf of rings.* (Definition 1.3). Yet, if the given ring $R$ has an *identity element,* then it is also straightforward to see that the above *constant sheaf of rings* $R$, is such *with an identity element* too, in the sense of Definition 1.3′, viz. the respective map (1.22) is continuous.

We come next to a notion that will mostly appear in the sequel, namely, that of a *sheaf of* ($\mathbf{C}$-)*algebras* on a given topological space $X$. So one has the following.

**Definition 1.4.** Given a topological space $X$, by a **sheaf of C-algebras,** or even a **C-algebra sheaf,** on $X$, one means a triplet

$$(1.25) \qquad \mathcal{A} \equiv (\mathcal{A}, \tau, X),$$

such that the following conditions hold true:

i) $\mathcal{A}$ is a sheaf of rings on $X$ (see Definition 1.3).

ii) Each stalk of $\mathcal{A}$, $\mathcal{A}_x = \tau^{-1}(x)$, $x \in X$, is, in particular, a **C**-*algebra* (viz. a linear associative algebra over the complexes **C**).

iii) The *"scalar multiplication in $\mathcal{A}$"*, viz. the map

$$\mathbf{C} \times \mathcal{A} \longrightarrow \mathcal{A} : (\lambda, a) \longmapsto \lambda \cdot a \in \mathcal{A}_x \subseteq \mathcal{A}, \tag{1.26}$$

with $\tau(a) = x \in X$, is *continuous,* where **C** carries the discrete topology.

**Scholium 1.1.**- Concerning the previous map (1.26), one can look at it, *equivalently,* as follows: Namely, consider first the *constant sheaf*

$$\mathbf{C} \tag{1.27}$$

(cf. also, for instance, (1.21) or even (1.24)), viz. the *sheaf* $\mathbf{C}_X \equiv (X \times \mathbf{C}, \pi \equiv \mathrm{pr}_1, X)$, where **C** is viewd as a topological space in the *discrete topology* (cf. I; (4.1)). Yet, let us also consider the following map;

$$\mathbf{C}_X \circ \mathcal{A} \longrightarrow \mathcal{A} : (\lambda_x, a) \longmapsto \lambda_x \cdot a := \lambda \cdot a \in \mathcal{A}_x \subseteq \mathcal{A}, \tag{1.28}$$

such that $\lambda_x \equiv (x, \lambda) \in X \times \mathbf{C}$, $\pi(\lambda_x) \equiv \mathrm{pr}_1(x, \lambda) = x = \tau(a)$ (that also explains the notation in the domain of (1.28). See also (1.1) and (1.25)). So one now concludes that

(1.29) *the continuity of (1.26) is equivalent with that of the map (1.28).*

Indeed, one remarks that *the map*

$$\nu : \mathbf{C} \times \mathcal{A} \longrightarrow \mathbf{C}_X \circ \mathcal{A} : (\lambda, a) \longmapsto \nu(\lambda, a) := (\lambda_x, a), \tag{1.30}$$

such that $\tau(a) = x = \pi(\lambda_x) \equiv \mathrm{pr}_1(x, \lambda)$, *is a homeomorphism,* when **C** *carries the discrete topology:* Namely, first, (1.30) *is,* of course, *a bijection.* Second, one further obtains

$$\nu(\{\lambda\} \times V) = (U \times \{\lambda\}) \times V, \tag{1.31}$$

such that $U = \tau(V)$, with $V$ an open neighborhood of $a \in \mathcal{A}$, satisfying I; (1.3.1). So the last relation proves now our claim for (1.30), according to Chapt. I, Lemma 1.1. In turn, *this also yields (1.29).* ∎

Now, within the same context, we stil note that *the constant sheaf* $\mathbf{C}_X$ (≡

**C**), as above, *is* certainly *a* **C**-*algebra sheaf* on $X$, in the sense of Definition 1.4. Thus, what actually (1.29) says is that;

(1.32) *any* **C**-*algebra sheaf* on a given topological space $X$, say, $\mathcal{E} \equiv (\mathcal{E}, \pi, X)$, *is, in fact,* (algebra sheaf *isomorphic to*) *an* $\mathcal{A}$-*algebra* on $X$, *with* $\mathcal{A} = \mathbf{C}_X$.

We have employed above certain as yet undefined terms, as e.g. *algebra sheaf isomorphism* and $\mathcal{A}$-*algebra;* in this context, see, however, Definition 6.2 in the sequel. Yet, the meaning of the term *algebra sheaf isomorphism* should be already clear by the preceding; thus, this should be referred to an isomorphism of the relevant sheaves (cf., for instance, Chapt. I; (1.20)), preserving also (*fiber-wise*) the algebra structures of the sheaves concerned. In this regard, see also the following Remark 1.1, as well as, Section 2 below (cf. e.g. (2.1), (2.2)).

**Remark 1.1.**– Concerning the previous two maps (1.26) and (1.28), we further remark that the relevant argument is still valid if one takes, *instead of* **C**, more generally, *any* **C**-*algebra* **A** (with an identity element). Of course, one has to assume here that *the stalks* $\mathcal{A}_x$, $x \in X$, of the given **C**-algebra sheaf $\mathcal{A}$ on $X$ (cf. Definition 1.4, ii)) *are* **A**-*algebras* too. So we are led, in fact, to the notion of a *sheaf of* **A**-*algebras* on a given topological space $X$, or else an **A**-*algebra sheaf* on $X$. (But, see also (1.40) below for the term **A**-*sheaf* on $X$).

Indeed, the preceding constitute a special case of the notion of an $\mathcal{A}$-*algebra* on a topological space $X$, as applied just before (cf. (1.32)); so in the latter case one actually considers as *"domain of coefficients"*, more generally, any **C**-*algebra sheaf* $\mathcal{A}$ on $X$, instead of a particular **C**-*algebra* **A**, as above. Thus, the (algebra) *sheaf,* say, $\mathcal{E} \equiv (\mathcal{E}, \pi, X)$, under consideration (see also e.g. (1.32), is, by definition, such that the individual stalks $\mathcal{E}_x$, $x \in X$, are $\mathcal{A}_x$-*algebras.* However, we shall not actually have the opportunity to employ here systematically the previous more general type of *algebra sheaves,* a notion, otherwise, quite useful and natural, as well, in some other context. (See, for example, A. Mallios [5], as well as, A. Kyriazis [1], [2], [3]). But, cf. also (6.29) below and Chapter IV in the sequel.

On the other hand, a *unital* **C**-*algebra sheaf,* or else a *sheaf of* **C**-*algebras with an identity element,* on a given topological space $X$, is, of course, a sheaf on $X$ in the sense of the previous Definition 1.4, say, $\mathcal{A} \equiv (\mathcal{A}, \tau, X)$, such that the **C**-*algebras* (stalks of $\mathcal{A}$) $\mathcal{A}_x$, $x \in X$, are, in addition, *unital* ones (viz. they have *identity elements* $1_x$, $x \in X$) in such a manner that;

(1.33) *the corresponding map (1.22) is continuous.*

That is, one has a combination of the relevant concepts in Definitions 1.3′ and 1.4.

In this context, by analogy with the classical situation (in algebra), one obtains a *canonical sheaf injection*

$$\mathbf{C}_X \equiv \mathbf{C} \underset{\varepsilon_X}{\hookrightarrow} \mathcal{A} \tag{1.34}$$

(we also denote it, occasionally, *just by* $\varepsilon$, see e.g. Part Two of this treatise), in such a manner that one sets

$$\varepsilon_X(\lambda) := \lambda \cdot \mathbf{1}_{\mathcal{A}}, \tag{1.34′}$$

for any *(local) constant section* $\lambda$ of $\mathbf{C} \equiv \mathbf{C}_X$, while $\mathbf{1}_{\mathcal{A}} \equiv \mathbf{1}$ stands, of course, for the *identity* (global continuous, by (1.33)) *section of* $\mathcal{A}$ (cf. also (1.22)). Thus, it is clear, in view of (1.34′), that $\varepsilon_X$ is fiber preserving and 1-1, yet, by the same relation, *section preserving.* (See also Chapt. I; (4.7), as well as, (9.22)).

In fact, (1.34) provides a **C**-*linear (iso)morphism* of the sheaves concerned (the latter being, in effect, **C**-*vector space sheaves* on $X$, as we shall presently see in the sequel), in the sense that the corresponding fiber maps (cf. also I; (1.14))

$$\varepsilon_{X,x} : (\mathbf{C}_X)_x \equiv \mathbf{C}_{X,x} \cong \mathbf{C} \longrightarrow \mathcal{A}_x, \ x \in X, \tag{1.35}$$

are, in particular, *(one-to-one)* **C**-*linear maps* (viz. **C**-*linear injections*). But, more on this we shall see presently below; cf. also the next Section 2.

The notion that has been established by the above Definition 1.4 gives rise to the following commentary, anticipating thus the application of a fundamental concept, which we shall encounter in the sequel.

**Scholium 1.2.**– Based on Definition 1.4 and by further specializing a standard concept (of Complex Analysis and of Algebraic Geometry, as well), namely, that of the so-called *"ringed space"* (see, for instance, H. Grauert – R. Remmert [2: p.1]), consider a pair, as follows;

$$(X, \mathcal{A}). \tag{1.36}$$

So this consists of a *topological space* $X$ and a *sheaf of* **C**-*algebras* $\mathcal{A}$ on $X$; in particular, we also assume that $\mathcal{A}$ is *unital* and *commutative*. Thus, we further suppose that the individual stalks of $\mathcal{A}$, $\mathcal{A}_x$, $x \in X$, are *unital commutative* **C**-*algebras,* whilst (1.33) is still satisfied, of course, by definition. (In this regard, cf. also Subsection 1.(c) below).

Thus, we call a pair $(X, \mathcal{A})$, as above, a **C**-*algebraized space,* while $X$ is still called the *base space* and $\mathcal{A}$ the *structure sheaf* of the **C**-algebraized space considered.

Indeed, the previous concept is significant for the sequel, being, as it is, at the basis of our considerations in Part Two of this book; so we still refer to that part of our treatise for applications, as well as, further relevant comments pertaining, in particular, to the connection of (1.36) with other classical (geometric) concepts.

Now, the notion which is given by the next definition, is a fundamental one for all that follows. In fact, the relevant objects, that are defined, are the `sine qua non` in all the remainder part of our discussion. So we have.

**Definition 1.5.** Let $X$ be a given topological space and $\mathcal{A} \equiv (\mathcal{A}, \tau, X)$ a unital **C**-algebra sheaf on $X$. Now, a **sheaf of $\mathcal{A}$-modules** (or even an **$\mathcal{A}$-module**) on $X$, is a sheaf on $X$, $\mathcal{E} \equiv (\mathcal{E}, \pi, X)$, such that the following conditions are fulfilled:

i) $\mathcal{E}$ is a sheaf of abelian groups on $X$ (cf. Definition 1.1).

ii) For every point $x \in X$, the corresponding stalk $\mathcal{E}_x$ of $\mathcal{E}$ is a (*left*) $\mathcal{A}_x$-module (in fact, due to the hypothesis for $\mathcal{A}$, a *unitary* one).

iii) The "*exterior module multiplication in* $\mathcal{E}$", viz. the map

$$\mathcal{A} \circ \mathcal{E} \longrightarrow \mathcal{E} : (a, z) \longmapsto a \cdot z \in \mathcal{E}_x \subseteq \mathcal{E}, \tag{1.37}$$

with $\tau(a) = \pi(z) = x \in X$, is *continuous.* (We have employed above the already familiar notation, as in (1.1), i.e., $\mathcal{A} \circ \mathcal{E} = \{(a, z) \in \mathcal{A} \times \mathcal{E} : \tau(a) = \pi(z)\}$).

As a first remark on the previous Definition 1.5, one has, of course, that *each stalk of a sheaf of $\mathcal{A}$-modules is,* in particular (due to the hypothesis for $\mathcal{A}$), *a* **C**-*vector space.* Therefore, since (cf. also (1.34)) $\mathbf{C}_X \circ \mathcal{E} \overset{\subseteq}{\rightarrow} \mathcal{A} \circ \mathcal{E}$, as a *topological subspace,* and in view of the continuity of the map (1.37), one concludes, yet for later use too, that;

(1.38) *every $\mathcal{A}$-module* on a given topological space $X$ *yields,* in particular, *a sheaf of* $\mathbf{C}$*-vector spaces* (or else a $\mathbf{C}$*-vector space sheaf*) on $X$.

Now, the following special case of the previous notion *of an $\mathcal{A}$-module* will also be of use in the sequel. So consider a $\mathbf{C}$*-algebra* $\mathbf{A}$ (with an identity element), along with the corresponding *constant sheaf* on $X$,

$$\mathbf{A}_X \equiv (X \times \mathbf{A}, \mathrm{pr}_1, X), \tag{1.39}$$

where $\mathbf{A}$ carries the *discrete topology.* It is also customary to denote (1.39) *just by*

$$\mathbf{A}, \tag{1.39'}$$

if no confusion arises. Now, it is not hard to see that $\mathbf{A}$, viz. $\mathbf{A}_X$, *is a* (unital) $\mathbf{C}$*-algebra sheaf* on $X$ (see Definition 1.4, along with (1.24) and (1.33)). So one defines;

(1.40) a *sheaf of* $\mathbf{A}$*-modules,* or even an $\mathbf{A}$*-sheaf,* on $X$ is just an $\mathcal{A}$-module, say, $\mathcal{E} \equiv (\mathcal{E}, \pi, X)$, on $X$, with $\mathcal{A} = \mathbf{A}_X$, viz. the constant sheaf (1.39).

Of course, it is *equivalent* to say that $\mathcal{E}$, as before, is a sheaf of abelian groups on $X$ (cf. Definition 1.1), whose stalks are, in particular, (left) $\mathbf{A}$*-modules,* in such a manner that *the map*

$$\mathbf{A} \times \mathcal{E} \longrightarrow \mathcal{E} : (a, z) \longmapsto a \cdot z \in \mathcal{E}_x \subseteq \mathcal{E}, \tag{1.41}$$

with $\pi(z) = x \in X$, *is continuous.* Namely, by analogy with (1.30), one can consider, instead, as domain of the last map the space (see also (1.39))

$$\mathbf{A}_X \circ \mathcal{E} = \Big\{(a_x, z) \in \mathbf{A}_X \times \mathcal{E} : \mathrm{pr}_1(a_x \equiv (x, a)) = \pi(z)\Big\}, \tag{1.42}$$

being, in fact, *homeomorphic to* $\mathbf{A} \times \mathcal{E}$ (cf. also (1.28), (1.29) and (1.37)).

On the other hand, *morphisms* between $\mathcal{A}$-modules will be considered in the next Section 2. Furthermore, we are going to examine presently below (cf. Subsection 1.(c) in the sequel) a very convenient alternative of the previous way of defining an algebraic structure in a given sheaf (of sets). In fact, this will be a consequence of the definition of algebraic structure in a given presheaf (of sets), to which case we thus turn right in the next Subsection.

**1.(b) Presheaves with algebraic structure. *A*-presheaves.**– By definition, a *presheaf of sets* on a given topological space $X$ is an *inductive system of sets,* having as (directed) index set the open subsets of $X$ (cf. Chapt. I, Definition 5.2). So according to standard reasoning, one can consider the individual sets of the system as carrying, instead, some particular algebraic structure. Of course, *the corresponding "restriction maps"* (cf. I; (5.1)) *should then be morphisms between the algebraic structures involved.*

Thus, one can speak of a *presheaf of (abelian) groups* or *rings* or **C** (or even **A**)-*algebras* on a given topological space $X$, yet of any other algebraic structure, whatsoever. However, one can also employ, with much profit, indeed(!), the corresponding *categorical language,* already used for presheaves (cf. Chapt. I; (5.10)). So one sets the following definition;

(1.43) a *presheaf "of type, a given algebraic structure"* (e.g. that one of groups, rings, etc.) on a topological space $X$, is a *contravariant functor* of the (small) category of the open subsets of $X$ (cf. I; (5.8), (5.9)) into that particular *category,* whose objects are (*all* the) *sets endowed with the type of algebraic structure, under consideration.*

So, based on the very definitions, it is a routine matter to check that (1.43) provides, in fact, an *equivalent formulation* of the concept of a presheaf (of sets) endowed with some particular algebraic structure, as given, previously, in terms of inductive systems.

In this context, it is still important to note, for what follows, that;

(1.44) *the algebraic structure* that appears in the individual sets of a given presheaf of sets equipped with that structure, *is passed over to the inductive limit set* (*stalk* of the presheaf considered), which is *associated with every point of the* topological *(base) space,* parametrizing the given presheaf.

So for convenience of reference, consider, in particular, the case (fundamental, anyhow) of a *presheaf of groups,* say,

$$S \equiv \left(S(U), \rho_V^U\right), \tag{1.45}$$

on a given *topological space* $X$. Thus, by definition, each one of the sets $S(U)$, with $U$ open in $X$, is a *group,* while for any open sets $U, V$ in $X$, with $V \subseteq U$, the corresponding "restriction map"

$$\rho_V^U : S(U) \longrightarrow S(V) \tag{1.46}$$

a *group morphism,* in such a manner that the relations I; (5.2) and (5.3) are fulfilled. (In this regard, cf. also e.g. N. Bourbaki [2: Chap. 1, §10, no 3; p.114 ff]). So the corresponding *stalk of* $S$ *at* $x \in X$ (see I; (7.2)),

$$S_x = \varinjlim_{U \ni x} S(U), \tag{1.47}$$

*is a group* (in particular, an *abelian* one, if this happens for the individual sets (groups) $S(U)$, $U$ open in $X$). Yet, concerning the last (inductive) *limit set,* one can apply, of course, any fundamental system of open neighborhoods of the point $x \in X$ considered, instead of all such neighborhoods (see also Chapt. I; (7.2)).

Now, as already remarked (cf. (1.44)), one obtains an *analogous conclusion for the stalks* $S_x$, $x \in X$, as above, if one considers, instead, *any other type of algebraic structure,* whatsoever, on the various sets $S(U)$, as before. (See thus N. Bourbaki [2], as cited above, or even e.g. A. Mallios [2: p.112, Definition 1.1] for the case of **C**-*algebras*).

Furthermore, we also look at a basic notion for the sequel as, namely, exhibited by the next.

**Definition 1.6.** Suppose we are given a topological space $X$, as well as, a presheaf of **C**-algebras $A \equiv (A(U), \tau_V^U)$ and a presheaf of abelian groups $E \equiv (E(U), \rho_V^U)$, both on $X$, such that the following data are satisfied:

i) $E(U)$ is a (left)-$A(U)$-module, for every open set $U \subseteq X$.

ii) For any open sets $U, V$ in $X$, with $U \subseteq V$, one has

$$\rho_V^U(a \cdot s) = \tau_V^U(a) \cdot \rho_V^U(s), \tag{1.48}$$

for any $a \in A(U)$ and $s \in E(U)$.

Thus, under the previous circumstances, $E$ is called a **presheaf of modules** (more precisely, of $A(U)$**-modules**) on $X$. Yet, it is customary to say, simply, that $E$ is an $A$**-presheaf** on $X$.

So given an $A(U)$-*module* $E$, as before, the respective *stalk of* $E$ *at* a point $x \in X$ (see I; (7.2)),

(1.49) $$E_x = \varinjlim_{U \ni x} E(U),$$

is a (left) $A_x$-*module,* where

(1.50) $$A_x = \varinjlim_{U \ni x} A(U)$$

stands for the corresponding *stalk at* $x \in X$ of the given presheaf of **C**-algebras $A$ on $X$, (1.50) being thus a **C**-*algebra* too. Yet, our assertion concerning (1.49), is based, in principle, on (1.44) and, in particular, on (1.48), where the last relation assures the passing of the *"module multiplication"* to the respective limits. (In this regard, cf. also N. Bourbaki [2: Chap. 1, §10, no 4, p.117, and Chap. 2, p.90] and [1: p.91, Proposition 6, and p.97, Proposition 10], as well as, A. Mallios [2], as cited above).

In this context, one further obtains that;

(1.51) if $A$ is a *presheaf of (unital)* **C**-*algebras* on $X$, then *each one of the above* $A(U)$, resp. $A_x$, *-modules is,* in fact, *a* **C**-*vector space.* Therefore, in that case too, *a given presheaf of* $A(U)$-*modules* on $X$ *is,* in particular, *a presheaf of* **C**-*vector spaces* on $X$.

On the other hand, by analogy with the concept of an **A**-sheaf (cf. (1.40)), one can also consider (for later use too) an

(1.52) **A**-*presheaf* on a given topological space $X$.

Namely, given a **C**-*algebra* **A**, one further considers the corresponding *constant presheaf* on $X$, associated with (the set, underlying the given algebra) **A**; cf. Chapt. I, Subsection 10.(a). Now, it is obvious that the latter is, in effect, a **C**-*algebra presheaf* on X (see I; (10.2), (10.3), along with (1.43) above). Therefore, one further concludes that;

(1.52′) a *presheaf of* **A**-*modules,* or even an **A**-*presheaf,* on $X$ (see also (1.43)) can be viewed as an $A$-*presheaf on* $X$ too, in the sense of Definition 1.6, where $A$ denotes the above *constant presheaf* on $X$, *associated with* the given **C**-algebra **A**.

Yet, as a variation of (1.51), within the above context, one obtains that;

(1.53) if the given **C**-algebra **A**, as above, is *unital,* then *an* **A**-*presheaf* on $X$ *provides,* in particular, *a presheaf of* **C**-*vector spaces* on $X$.

Now, as one might expect, we verify next (however, in the particular cases, of course, that will concern us in the sequel) that;

(1.54) *the sheafification of a given presheaf endowed with an algebraic structure, yields, in fact, a sheaf with a similar structure.* Put in another way, *the sheafification functor* (cf. Chapt. I; §§8,13) *preserves algebraic structures.*

First, by definition of the sheafification procedure for a given presheaf of sets (cf. Chapt. I; Section 7) and according to (1.44), one immediately concludes that;

(1.55) *the stalks of the sheafification considered inherit the algebraic structure of the individual sets of the given presheaf.*

Thus, to verify (1.54), we have still to show the *"continuity of the algebraic operations"*, appeared after the application of the sheafification procedure. As an example (anyhow, fundamental), we consider the case of a presheaf of groups. So we have.

**Lemma 1.3.** *The sheafification of a given presheaf of (abelian) groups,* $S \equiv (S(U), \rho_V^U)$, *on a topological space* $X$ *is a sheaf of (abelian) groups on* $X$.

**Proof.** Let $\mathcal{S} \equiv \mathbf{S}(S)$ be the sheafification of $S$ (see Chapt. I; (7.23)). Now, in view of (1.55), one has only to prove the *continuity of the map (1.2):* So let $z, z'$ be elements of $\mathcal{S}$, with $\pi(z) = \pi(z') = x \in X$, so that $z, z'$, as well as, $z + z'$ are elements of $\mathcal{S}_x$. Moreover, let $N$ be an *open neighborhood of* $z + z'$ *in* $\mathcal{S}$. Now, since $z + z' \in \mathcal{S}_x = \varinjlim_{U \ni x} S(U)$ (cf. I; (7.6)), one obtains (see I; (7.16))

$$z + z' = \rho_U^x(r) \equiv \tilde{r}(x), \tag{1.56}$$

where $r \in S(U)$, with $U$ an open neighborhood of $x$ in $X$. Yet, *we may suppose* that (see also I; (7.22))

$$\tilde{r}(U) \subseteq N. \tag{1.57}$$

Furthermore, since $z, z' \in \mathcal{S}_x$, applying a similar argument, as above, one obtains;

$$z = \rho_W^x(s) \equiv \tilde{s}(x) \text{ and } z' = \rho_W^x(t) \equiv \tilde{t}(x), \tag{1.58}$$

where $W$ is an *open* set in $X$, *with* $x \in W \subseteq U$, and $s, t \in S(W)$, *hence* $s + t \in S(W)$, as well. But, then one has (see also (1.56), (1.58), as well as, I; (7.9))

$$(\widetilde{s+t})(x) = \rho_W^x(s+t) = \rho_W^x(s) + \rho_W^x(t) = \tilde{s}(x) + \tilde{t}(x) = z + z' = \tilde{r}(x). \tag{1.59}$$

(In this concern, we remark that we have also applied in (1.59) the fact that *the canonical map* $\rho_W^x : S(W) \longrightarrow S_x$, by analogy with I; (7.9), *becomes a group morphism,* a consequence, indeed, of the very definitions; so see e.g. N. Bourbaki [2: Chap. 1; p.114, Proposition 1], or even A. Mallios [2: p.111; (1.17), and p.112, the relevant comments thereon]. The last Ref. is actually concerned with inductive systems of **C**-algebras. The above remarks supplement the respective commentary on (1.47) and are, of course, in the spirit of (1.44)).

Thus, based now on (1.59) and on I; (2.13), we conclude the existence of an *open* set $V \subseteq X$, *with* $x \in V \subseteq W \subseteq U$, such that

$$\widetilde{s+t} = \tilde{r}|_V. \tag{1.60}$$

The proof is based further on the following.

**Remark 1.2.**– To proceed, we shall need first the following comments on the *canonical map*

$$\rho_U : S(U) \longrightarrow S(U, \pi), \tag{1.61}$$

with $U$ *open* in $X$ (cf. Chapt. I; (7.5), (7.11)). Namely, although the range of (1.61) is the set $\Gamma(U, \mathcal{S}) \subseteq S(U, \pi)$ (see I; (7.22)), we deliberately consider as such the set $S(U, \pi)$ of *all sections of* $\pi : \mathcal{S} \longrightarrow X$ *over* $U$. Thus, based on the *group structure* of $S_x = \mathcal{S}_x$ (cf. (1.47) and I; (7.6)), one can further define in $S(U, \pi)$, *point-wise, a similar algebraic structure;* that is, given $\alpha, \beta$ in $S(U, \pi)$, one defines $(\alpha + \beta)(x) :=$

$\alpha(x) + \beta(x)$, $x \in U$, so that $\alpha + \beta \in S(U, \pi)$, as well, and $S(U, \pi)$ becomes thus a *group*. Now, for any $s, t$ in $S(U)$, one sets;

$$\rho_U(s+t)(x) := \rho_U(s)(x) + \rho_U(t)(x), \; x \in U, \tag{1.62}$$

so that *one obtains*

$$\rho_U(s+t) = \rho_U(s) + \rho_U(t), \tag{1.62'}$$

*with* $s + t \in S(U)$, that is, *the map (1.61) becomes a group morphism,* a fact that we shall use right below (cf. (1.63)). In this context, we still note, in anticipation, that $\Gamma(U, \mathcal{S})$ *inherits any* eventual *algebraic structure of* $\mathcal{S}$, yet, *in a characteristic way* (!). See thus the next Subsection 1.(c).

**End of the proof of Lemma 1.3.** Now, in view of the hypothesis for the maps (1.46), the previous relation (1.62′), as well as, of Chapt. I; (7.9), (7.22), (9.7) and (11.9), *one obtains from (1.60);*

$$\begin{aligned} \tilde{r}|_V &= \widetilde{s+t}|_V = \sigma_V^W(\rho_W(s+t)) = \rho_V\left(\rho_V^W(s+t)\right) \\ &= \rho_V\left(\rho_V^W(s) + \rho_V^W(t)\right) = \rho_V\left(\rho_V^W(s)\right) + \rho_V\left(\rho_V^W(t)\right) \\ &= \sigma_V^W(\rho_W(s)) + \sigma_V^W(\rho_W(t)) \equiv \tilde{s}|_V + \tilde{t}|_V. \end{aligned} \tag{1.63}$$

That is, one has (see also (1.57))

$$\tilde{s}(V) + \tilde{t}(V) = \tilde{r}(V) \subseteq \tilde{r}(U) \equiv N; \tag{1.64}$$

this, of course, proves now the continuity of (1.2), in view of the hypothesis for $N$ and (1.58), and this terminates the proof of Lemma 1.3. ∎

As already remarked, the above proof gives also a *general idea of verifying (1.54)*. So, within the same context, we further note, yet, for later use, that similar comments to those following (1.59) lead to a corresponding remark for the *canonical map* $\rho_U^x : E(U) \longrightarrow E_x$ in the particular case that $E(U)$ *is an* $A(U)$*-module* (cf. Definition 1.6). Thus, by virtue of (1.48), one still has;

$$\rho_U^x(a \cdot s) = \tau_U^x(a) \cdot \rho_U^x(s), \tag{1.65}$$

for any $a \in A(U)$ and $s \in E(U)$, where $\tau_U^x : A(U) \longrightarrow A_x$ stands for the corresponding *canonical map* referred to the presheaf $A$ (see also I; (7.10)).

So the next result, which we shall also use in the sequel, is now a consequence of an analogous procedure to that one applied in the proof of the preceding lemma. Namely, one concludes in particular, that;

(1.66) *the sheafification of a presheaf of $A(U)$-modules,* or yet of an *$A$-presheaf,* on a topological space $X$ *is an $\mathcal{A}$-module* on $X$ (see Definition 1.5).

Yet, more formally, one has, in effect, the following.

**Proposition 1.1.** *Let $A \equiv (A(U), \tau_V^U)$ be a presheaf of (unital) $\mathbf{C}$-algebras on a topological space $X$ and $E \equiv (E(U), \rho_V^U)$ a presheaf of $A(U)$-modules on $X$* (cf. Definition 1.6). *Moreover, let $\mathcal{A} \equiv \mathbf{S}(A)$ and $\mathcal{E} \equiv \mathbf{S}(E)$ be the corresponding sheafifications of $A$ and $E$* (see Chapt. I, Section 7, and also (8.2)). *Then, $\mathcal{E}$ is an $\mathcal{A}$-module on $X$.* ∎

Now, as already promised, we consider right in the next Subsection the continuity of eventual algebraic operations in a given sheaf, through the (continuous) sections of the sheaf in question. So we have the following.

**1.(c). Section-wise definition of algebraic structure in a given sheaf** (of sets).– The ensuing discussion is concerned with a *"functional way"*, of defining an algebraic structure in a given sheaf of sets; thus, one looks at the *"closedness"* of the individual sets of local (continuous) sections of the given sheaf, with respect to the algebraic operations involved. Precisely, one has the following:

(1.67) *A given sheaf of sets $\mathcal{S}$ on a topological space $X$ is endowed with some particular algebraic structure if, and only if, this is the case for its (complete) presheaf of sections $\Gamma(\mathcal{S})$ on $X$.*

Now, as a (basic) instance of the above assertion, we presently consider below the particular case of a *sheaf of groups* on $X$ (cf. Definition 1.1). Of course, *a similar argument can be supplied for any other algebraic structure.* (In fact, the sheaves we are going to deal with in the sequel are mostly *particular cases of group sheaves;* however, we hope that the preceding are adequately developed so as to leave the verification of the last remark as an easy exercise for the reader!).

Thus, suppose that $\mathcal{S} \equiv (\mathcal{S}, \pi, X)$ is a *group sheaf* on a given topological space $X$ and let

(1.68) $$\Gamma(\mathcal{S}) \equiv \left(\Gamma(U, \mathcal{S}), \sigma_V^U\right)$$

be the (complete) *presheaf of sections of* $\mathcal{S}$ (cf. I; (9.6)). Now, our first claim here is that;

(1.69) *for any open set $U \subseteq X$, $\Gamma(U, \mathcal{S})$ is a group.*

Indeed, *the assertion follows from the continuity of the map (1.2)* (*to which* actually *it is equivalent,* as it will become clear from the sequel; cf. thus (1.75)). Namely, given $s, t \in \Gamma(U, \mathcal{S})$ and $x \in U$, let $V$ be an *open neighborhood of* $(s+t)(x) = s(x) + t(x) \in \mathcal{S}_x$ in $\mathcal{S}$. Hence, *by the continuity of (1.2),* there exist open neighborhoods, say, $V_1$ and $V_2$ in $\mathcal{S}$ of $s(x)$ and $t(x)$, respectively, such that $V_1 + V_2 \subseteq V$. Thus, due to the continuity of $s$ and $t$, *there exists open $U' \subseteq X$, with $x \in U' \subseteq U$,* such that $s(U') \subseteq V_1$ and $t(U') \subseteq V_2$; hence, one finally obtains

(1.70) $$(s+t)(U') \subseteq s(U') + t(U') \subseteq V_1 + V_2 \subseteq V,$$

which, of course, proves the continuity of $s + t$ at the given point $x \in U$, otherwise arbitrary, so that $s + t \in \Gamma(U, \mathcal{S})$, as desired. Now, this along with Lemma 1.2 prove (1.69). ∎

Furthermore, one proves that;

(1.71) *for any open $U, V$ in $X$, with $V \subseteq U$,* the (restriction) map

(1.71.1) $$\sigma_V^U : \Gamma(U, \mathcal{S}) \longrightarrow \Gamma(V, \mathcal{S})$$

(cf. I; (9.4)) *is a group morphism.*

Namely, for any $s, t$ in $\Gamma(U, \mathcal{S})$, and every $x \in V \subseteq U$, one obtains;

(1.72) $$\begin{aligned} \sigma_V^U(s+t)(x) &\equiv (s+t)|_V(x) = (s+t)(x) = s(x) + t(x) \\ &= (s|_V)(x) + (t|_V)(x) \equiv \sigma_V^U(s)(x) + \sigma_V^U(t)(x), \end{aligned}$$

which thus proves our claim for (1.71.1). ∎ Now, *(1.69) along with (1.71) provide the "only if" part of (1.67)* (group case, of course).

Conversely, *suppose that $\Gamma(\mathcal{S})$ is a presheaf of groups on $X$.* Thus, *we* shall *prove that* the map (1.2) *is continuous;* that is, let $z, z' \in \mathcal{S}$, such that $\pi(z) = \pi(z') = x \in X$, so that $(z, z') \in \mathcal{S} \circ \mathcal{S}$ (see (1.1)). Now, in view of

I; (9.8), (9.9) and (1.44) above, *the stalks* $\mathcal{S}_x$, $x \in X$, of $\mathcal{S}$ *are groups,* so that $z + z' \in \mathcal{S}_x$, as well. Thus, if $N$ is an *open neighborhood of* $z + z'$ *in* $\mathcal{S}$, let $r \in \Gamma(U, \mathcal{S})$, with $x \in U$, and $z + z' = r(x) \in r(U) \subseteq N$ (see I; (2.15)). Moreover, there exists open $U' \subseteq X$, with $x \in U' \subseteq U$, as well as, $s \in \Gamma(U', \mathcal{S})$ and $t \in \Gamma(U', \mathcal{S})$ such that $z = s(x)$ and $z' = t(x)$. Hence, $z + z' = s(x) + t(x) = (s + t)(x) = r(x)$, so that (cf. I; (2.13)) one has

$$s + t = r|_V, \tag{1.73}$$

for some open $V \subseteq X$, with $x \in V \subseteq U \cap U'$. Therefore, one finally obtains

$$s(V) + t(V) = r(V) \subseteq r(U) \subseteq N, \tag{1.74}$$

which, of course (cf. also I; Lemma 2.1), proves our claim about the map (1.2). ∎ The above provide thus the *"if" part of (1.67),* the latter assertion being now fully established. Yet, it is still instructive to state (1.67) in the following form:

(1.75) *Given a sheaf (of sets)* $\mathcal{S}$ *on a topological space* $X$, *the continuity of any "algebraic operation in* $\mathcal{S}$*" is tantamount* (but, see also the ensuing comments) *to have the sets of sections of* $\mathcal{S}$ *"closed", with respect to the operation concerned.*

In this context, we still note, however, that, as follows from the latter part of the preceding, the *"sufficiency" of (1.75)* presupposes the *stalks* $\mathcal{S}_x$, $x \in X$, of $\mathcal{S}$ to be already *endowed with the algebraic structure in question,* a fact that was guaranteed above by considering $\Gamma(\mathcal{S})$ as a presheaf equipped with that structure, viz., in particular, by (1.67).

So by still taking as fundamental example the case of a *group sheaf* (that is, *analogous arguments are valid for any other algebraic structure on sheaves,* we consider below), we further conclude the following *equivalent form of (1.67):*

(1.76) Let $\mathcal{S}$ be a *sheaf* (of sets) on a given topological space $X$. Then, $\mathcal{S}$ is a *sheaf of groups* on $X$ *if, and only if,* the following holds true:

(1.76.1) *For every point* $x \in X$ *and any open* $U \subseteq X$, *the corresponding sets* $\mathcal{S}_x$ *and* $\Gamma(U, \mathcal{S})$ *are groups.*

Now, as already mentioned, the above condition (1.67) (or even equivalently, (1.76)) constitutes, in effect, a very convenient way of defining *sheaves with algebraic structure,* in particular, when one has to employ, for instance, a *topological algebraic structure;* yet, this particular case will also be considered in the sequel, where we deal, for example, with *topological algebra sheaves* (so see Chapter XI, Vol. II). Thus, as we shall see, the latter type of structure is customarily imposed on the various sets of sections of the sheaf involved.

## 2. $\mathcal{A}$-morphisms. Exact sequences

Suppose we have a *sheaf of* **C**-*algebras* (with identity elements) on a *topological space* $X$, say, $\mathcal{A} \equiv (\mathcal{A}, \tau, X)$ (cf. Section 1 above for the terminology applied). Moreover, let $\mathcal{E} \equiv (\mathcal{E}, \pi, X)$ and $\mathcal{F} \equiv (\mathcal{F}, \rho, X)$ be two (left) $\mathcal{A}$-*modules* on $X$.

Thus, we say that a given map

$$\phi : \mathcal{E} \longrightarrow \mathcal{F} \tag{2.1}$$

is a *morphism of* $\mathcal{A}$-*modules,* or just an $\mathcal{A}$-*morphism,* if $\phi$ is a morphism of the sheaves concerned (cf. Chapt. I, Definition 1.2) such that, *for every point* $x \in X$, the corresponding fiber-map (cf. I; (1.14))

$$\phi_x := \phi|_{\mathcal{E}_x} : \mathcal{E}_x \longrightarrow \mathcal{F}_x \tag{2.2}$$

is a *morphism of* (left) $\mathcal{A}_x$-*modules* (or else, as we also say, an $\mathcal{A}_x$-*morphism*).

Now, we first remark for later use that, by virtue of (1.38) and our hypothesis for $\mathcal{A}$, one obtains that;

(2.3) each $\mathcal{A}$-morphism, as above, is, in particular, a **C**-*linear morphism of the* **C**-*vector space sheaves* involved; thus, such a map $\phi$, as in (2.1), is a sheaf morphism such that each one of the maps $\phi_x$, $x \in X$, is, in particular a **C**-*linear map* of the **C**-vector spaces concerned.

By the preceding we have thus singled out the *category of* (left) $\mathcal{A}$-*modules* on a given topological space $X$, denoted by

$$\mathcal{A}\text{-}\mathcal{M}od_X . \tag{2.4}$$

In fact, our main concern in the sequel will be a certain full subcategory of (2.4). (So cf. Section 4 below). We come next to examine several basic properties of the above category (2.4) that we shall need in the sequel. Indeed, these are actually particular instances of the fact that (2.4) is an *abelian category* (see, for example, R. Godement [1: p.127]); but, we shall not exploit it to that extend, at least formally. So we further consider those notions connected with that of $\mathcal{A}$-morphism, that are directly relevant to the formulation of the concept of an *exact sequence,* the latter being, in effect, in conjunction also with certain important functors, a crucial item in this, as well as, in other Sections of our treatise. (So see e.g. Section 6 below).

Thus, given an $\mathcal{A}$-*morphism* $\phi : \mathcal{E} \longrightarrow \mathcal{F}$, as in (2.1), one defines its *kernel,* denoted by $\ker\phi$, through the following relation;

$$\ker\phi := \{z \in \mathcal{E} : \phi(z) = 0\}. \tag{2.5}$$

Therefore, since (cf. Chapt. I; (1.12))

$$\pi(z) = \rho(\phi(z)) = x \in X,\ z \in \mathcal{E}, \tag{2.6}$$

one obtains, by (2.5),

$$\phi(z) = \phi_x(z) = 0 \equiv 0_x \in \mathcal{F}_x,\ z \in \ker\phi \tag{2.7}$$

(see also (1.3)). So, by considering the *set of zeroes of* $\mathcal{F}$, denoted by $0_X^{\mathcal{F}}$, in fact, an *open subset of* $\mathcal{F}$, viz. a *subsheaf of* $\mathcal{F}$ (cf. (1.8)), one obtains, by the preceding, that;

$$\ker\phi = \phi^{-1}\left(0_X^{\mathcal{F}}\right) \subseteq \mathcal{E}, \tag{2.8}$$

that is, an *open subset of* $\mathcal{E}$, hence, itself a *subsheaf of* $\mathcal{E}$ (see I; (1.8)). Thus, by applying the notation of (I; (1.8)), one has the following *sheaf on* $X$;

$$\ker\phi \equiv (\ker\phi, \pi|_{\ker\phi}, X)\,. \tag{2.9}$$

Indeed, (2.9) yields a *sub-$\mathcal{A}$-module of* $\mathcal{E}$ in the sense that

(2.10) $\ker\phi$ *is an $\mathcal{A}$-module on $X$, as well as, a subsheaf of* $\mathcal{E}$; viz. an *open subset of* $\mathcal{E}$, being also *"closed" with respect to the operations of the $\mathcal{A}$-module* $\mathcal{E}$.

So by (2.6) one has that every *stalk of* $\ker\phi$ is given by

$$(\ker\phi)_x = \ker(\phi_x) \subseteq \mathcal{E}_x,\ x \in X, \tag{2.11}$$

that is, an *$\mathcal{A}_x$-submodule of $\mathcal{E}_x$* (cf. also (2.2)). Moreover, the operations of *addition* and *module multiplication* in $\ker\phi$ *are continuous,* being obviously the restrictions to $\ker\phi$ of the same in $\mathcal{E}$. (The same assertion, in view of (2.11) and the hypothesis for $\mathcal{E}$, follows quite easily too from a mere *application of (1.76)* for the particular case at hand).

Now, an analogous situation appears when one considers the *image of an $\mathcal{A}$-morphism;* i.e., one sets

$$\operatorname{im}\phi := \phi(\mathcal{E}) \subseteq \mathcal{F}, \tag{2.12}$$

thus, an *open subset of $\mathcal{F}$*, hence still a *subsheaf of $\mathcal{F}$* (see Chapt. I; (1.4) and Lemma 1.2). Furthermore, similarly to the case of $\ker\phi$, one obtains that *the sheaf* on $X$

$$\operatorname{im}\phi \equiv (\operatorname{im}\phi, \rho|_{\operatorname{im}\phi}, X) \tag{2.12'}$$

*is,* in fact, *a sub-$\mathcal{A}$-module of $\mathcal{F}$*, whose stalks are given by

$$(\operatorname{im}\phi)_x = \operatorname{im}(\phi_x) \subseteq \mathcal{F}_x,\ x \in X \tag{2.13}$$

(see also (2.2)). Analogous comments to those for $\ker\phi$, are here in force too, concerning the *continuity of the algebraic operations in* $\operatorname{im}\phi$.

In this context, by extending standard terminology, a given *$\mathcal{A}$-morphism* $\phi : \mathcal{E} \longrightarrow \mathcal{F}$, as above, is said to be a *monomorphism* (precisely, *$\mathcal{A}$-monomorphism*), whenever $\phi$ is *one-to-one,* viz. one has (see also (2.11))

$$\ker\phi = 0, \tag{2.14}$$

where in the second member of (2.14) one means the *zero sheaf,* in fact, the *zero $\mathcal{A}$-module on $X$* (cf. (1.9)). Yet, $\phi$ is called an *epimorphism* (after the Greek word $\epsilon\pi\iota$ (epi)=on), if it is an *onto* map; so one has (cf. also (2.13))

$$\operatorname{im}\phi = \mathcal{F}. \tag{2.15}$$

Finally, $\phi$ is called an *$\mathcal{A}$-isomorphism,* or just an *isomorphism* (we look thus at $\phi$ as a morphism of the category (2.4)), if it is both a *mono-,* as well as, an *epi*-morphism.

**Remark 2.1.**- All the preceding notions can be formulated, of course, in the case of a *sheaf of groups* on $X$. Anyhow, *a group sheaf on $X$ is but a* **Z***-module on $X$*, with respect to the *constant sheaf of integers* **Z** *on $X$* (cf. also (1.24)). Yet, the same remarks are in force here for any other type of algebraic structure on a given sheaf on $X$, specializing that of a group sheaf on $X$.

We come next to the main concept of this Section, being, as already said, an important ingredient for all the ensuing discussion. So one has the following.

**Definition 2.1.** A given sequence of $\mathcal{A}$-modules on a topological space $X$ and $\mathcal{A}$-morphisms between them (we call it too an **$\mathcal{A}$-sequence** on $X$),

$$\cdots \longrightarrow \mathcal{E}_{i-1} \xrightarrow{\phi_{i-1}} \mathcal{E}_i \xrightarrow{\phi_i} \mathcal{E}_{i+1} \longrightarrow \cdots, \tag{2.16}$$

is said to be **exact at $\mathcal{E}_i$**, if one has

$$\ker \phi_i = \operatorname{im} \phi_{i-1}. \tag{2.17}$$

In particular, one speaks of an **exact sequence of $\mathcal{A}$-modules** (or, simply, of an **exact $\mathcal{A}$-sequence**) on $X$, if (2.16) is exact at any $\mathcal{E}_i$ (cf. also the following Remark).

Yet, an exact sequence of $\mathcal{A}$-modules of the form

$$0 \longrightarrow \mathcal{F} \xrightarrow{\psi} \mathcal{S} \xrightarrow{\phi} \mathcal{E} \longrightarrow 0 \tag{2.18}$$

(where 0 stands for the *zero $\mathcal{A}$-module* on $X$), is called a **short exact sequence of $\mathcal{A}$-modules,** or just a **short exact $\mathcal{A}$-sequence,** on $X$.

> **Remark.**- The indices in (2.16) are varied, in principle, over the *integers* $\mathbf{Z}$; however, we shall not consider that case below, restricting thus ourselves to the *positive integers* $\mathbf{Z}_+ \equiv \mathbf{N} \cup \{0\}$ only (unless othewise indicated).

Concerning (2.16), see also Chapt. III; Lemma 1.1 in the sequel. On the other hand, *(2.18) implies,* by the very definitions (see (2.17)), *that;*

(2.18′) i) $\psi$ is a *monomorphism,* ii) $\phi$ is an *epimorphism* and iii) $\ker \phi = \operatorname{im} \psi$.

Yet, with any given *$\mathcal{A}$-morphism* $\phi : \mathcal{E} \longrightarrow \mathcal{F}$, as in (2.1), there is associated the following *short exact $\mathcal{A}$-sequence;*

$$0 \longrightarrow \ker \phi \xrightarrow{\iota} \mathcal{E} \xrightarrow{\phi} \operatorname{im} \phi \longrightarrow 0. \tag{2.19}$$

Here $\iota$ stands for the restriction to $\ker\phi \subseteq \mathcal{E}$ (cf. (2.8)) of the *identity* $\mathcal{A}$*-endomorphism of* $\mathcal{E}$, still denoted by $\iota \equiv \mathrm{id}_{\mathcal{E}} : \mathcal{E} \longrightarrow \mathcal{E}$ $(: z \longmapsto \iota(z) := z)$.

Now, we shall need in a moment to look at the analogous situation to the preceding, when considering $A$*-presheaves* (cf. Definition 1.6). So we come to our next objective, as exhibited by the following Subsection.

**2.(a). Morphisms of $A$-presheaves.**– Suppose that

$$E \equiv \left(E(U), \rho_V^U\right) \text{ and } F \equiv \left(F(U), \lambda_V^U\right) \tag{2.20}$$

are two given $A$*-presheaves* on a *topological space* $X$, with $A \equiv (A(U), \tau_V^U)$ a given *presheaf of* $\mathbf{C}$*-algebras* on $X$ (cf. Subsection 1.(b) above). Moreover, let

$$\phi \equiv (\phi_U) : E \longrightarrow F \tag{2.21}$$

be a *morphism* of the given *presheaves* (cf. Chapt. I; (6.3)).

Now, we shall say that a given presheaf morphism $\phi$, as before, is, in particular, a *morphism of $A$-presheaves,* or simply an $A$*-morphism,* whenever, for any open $U \subseteq X$, the corresponding map

$$\phi_U : E(U) \longrightarrow F(U) \tag{2.22}$$

is, in addition, an $A(U)$*-morphism* of the $A(U)$-modules concerned (see also Definition 1.6).

Thus, we can further consider the *category of $A$-presheaves* on $X$, along with the corresponding morphisms, the latter being thus the $A$*-morphisms,* as before; this category is denoted in the sequel by

$$A\text{-}\mathcal{PSh}_X, \tag{2.23}$$

a notion that will also be of use presently below.

In this context, we first observe that;

(2.24) *any $A$-morphism* in (2.23) *is extended to an $\mathcal{A}$-morphism* in (2.4), *by taking* the corresponding *sheafifications* of the presheaves involved.

First, it is a consequence of (1.66) that the sheafifications concerned yield $\mathcal{A}$-modules (see also Proposition 1.1). Yet, if $\phi : E \longrightarrow F$ is a given $A$-morphism in (2.23), then (cf. I: (8.5)), $\widetilde{\phi} \equiv \mathbf{S}(\phi) : \mathcal{E} \equiv \mathbf{S}(E) \longrightarrow \mathbf{S}(F) \equiv \mathcal{F}$

*is a sheaf morphism* (see also Chapt. I; Lemma 8.1). Furthermore, we also remark that (cf. also Chapt. I; (8.6), (8.7) and (8.11));

*the fiber maps*

(2.25) (2.25.1) $\tilde{\phi}_x := \phi_x := \varinjlim_{U \ni x} \phi_U : E_x = \mathcal{E}_x \longrightarrow F_x = \mathcal{F}_x, \; x \in X,$

*are* $\mathcal{A}_x (= A_x)$*-morphisms.*

Indeed, this follows from our hypothesis for $\phi$ and (1.44), along with the analogous comments to those following (1.59) in the preceding, adapted to the present case. So this, in view of (2.2), provides now our assertion in (2.24). ∎

On the other hand, by analogy with the $\mathcal{A}$-modules (2.9) and (2.12′), one defines, *for any* given *A-morphism* $\phi$, as in (2.21), the two following *A-presheaves*

$$\ker\phi \subseteq E \;\; and \;\; \operatorname{im}\phi \subseteq F. \tag{2.26}$$

The "inclusion" relations above are only, simply to indicate that the said $A$-presheaves are, in fact, *sub-A-presheaves* of $E$ and $F$, respectively; indeed, one has (see also (I; (6.6)), along with (2.20), (2.21) above)

$$\ker\phi \equiv \left(\ker\phi_U, \, {}^{\cdot}\rho_V^U\right), \tag{2.27}$$

as well as,

$$\operatorname{im}\phi \equiv \left(\operatorname{im}\phi_U, \, {}^{\cdot}\lambda_V^U\right). \tag{2.28}$$

Yet, the "dots" in the previous relations indicate the corresponding *restrictions* of the maps involved to $\ker\phi_U \subseteq E(U)$ and $\operatorname{im}\phi_U \subseteq F(U)$, respectively, for any open sets $U, V$ in $X$, with $V \subseteq U$. Namely, by employing standard terminology (cf. e.g. N. Bourbaki [1: p.94, Corollaire]), one obtains through (2.27), resp. (2.28), an *inductive system of sub-$A(U)$-modules of $E(U)$, resp. of $F(U)$*, with $U$ open in $X$. This also explains the term applied above for the presheaves in question.

Now, concerning the *A-presheaf (2.28),* by further using I; (12.43), one still obtains the relation (cf. also (2.12′))

$$\mathbf{S}(\operatorname{im}\phi) = \operatorname{im}\mathbf{S}(\phi) \equiv \operatorname{im}\tilde{\phi}, \tag{2.29}$$

within an *$\mathcal{A}$-isomorphism* of the $\mathcal{A}$-modules concerned.

We next look at a useful application of the previous concepts by considering the corresponding notion for the case at hand of an exact sequence, by analogy with (2.16). This will also be used right below.

Thus, given a *sequence of A-presheaves* on a topological space $X$ and *A-morphisms* (cf. also the Remark following Definition 2.1)

$$\cdots \longrightarrow E_{i-1} \xrightarrow{\phi_{i-1}} E_i \xrightarrow{\phi_i} E_{i+1} \longrightarrow \cdots, \tag{2.30}$$

we say that (2.30) is an *exact sequence of A-presheaves,* or just an *exact A-sequence,* if it is exact at any $E_i$; viz. whenever one has

$$\ker \phi_i = \operatorname{im} \phi_{i-1}, \quad \textit{for all } i. \tag{2.31}$$

Furthermore, analogously with (2.18) one considers a *short exact sequence of A-presheaves* on $X$, or just a *short exact A-sequence,*

$$0 \longrightarrow F \xrightarrow{\psi} S \xrightarrow{\phi} E \longrightarrow 0 \tag{2.32}$$

(with 0 denoting the (constant) zero $A$-presheaf on $X$; cf. also Chapt. I; Subsection 10.(a)). So one obtains, by definition,

$$\ker \phi = \operatorname{im} \psi, \tag{2.33}$$

while, in addition, $\psi$ *is injective* and $\phi$ *surjective* (cf. Chapt. I; Definition 6.1).

Indeed, our interest here lies actually in the following fact. That is,

(2.34) *exactness of (2.30) is transferred to the stalks* (relative to the various points of the base space $X$) *of the A-presheaves involved.*

The last claim follows immediately from the fact that *inductive limits preserve exactness* (of sequences; see, for instance, N. Bourbaki [1: p.92, Corollaire 1 and p.93, Proposition 7], [2: Chap. 2; p.91, Proposition 3]). So by considering, in particular, (2.32) one obtains, by virtue of (2.34) (cf. also I; (7.2)), the following *short exact sequence of $A_x$-modules, for every point* $x \in X$;

$$0 \longrightarrow F_x \xrightarrow{\psi_x} S_x \xrightarrow{\phi_x} E_x \longrightarrow 0. \tag{2.35}$$

Thus, by taking now the *sheafification of (2.32),* one further obtains, in view of (2.35), the following *short exact $\mathcal{A}$-sequence,* in the terminology of (2.18);

(2.36) $$0 \longrightarrow \mathcal{F} \xrightarrow{\psi} \mathcal{S} \xrightarrow{\phi} \mathcal{E} \longrightarrow 0.$$

For simplicity of the notation, we also set above $\widetilde{\psi} = \psi$ and $\widetilde{\phi} = \phi$ (see also Chapt. I, Lemma 8.1). Now, as already mentioned, we apply the foregoing presently below, in the ensuing Subsection.

**2.(b). Quotients of $\mathcal{A}$-modules.**- We come next to another important notion for the sequel, as this is expressed by the sheaves referred to in the title of this Subsection.

Thus, suppose that

(2.37) $$\mathcal{E} \equiv (\mathcal{E}, \pi, X) \text{ and } \mathcal{F} \equiv (\mathcal{F}, \rho, X)$$

are two given $\mathcal{A}$-*modules* (cf. Definition 1.5 above) on a topological space $X$ such that $\mathcal{E}$ is a *sub-$\mathcal{A}$-module of* $\mathcal{F}$; viz. $\mathcal{E}$ is a subsheaf of $\mathcal{F}$ such that, for every $x \in X$, $\mathcal{E}_x$ is a sub-$\mathcal{A}_x$-module of $\mathcal{F}_x$ (see e.g. (2.10), (2.11) and the ensuing comments there). So we can next define the following *set*

(2.38) $$\mathcal{F}/\mathcal{E} := \sum_{x \in X} \mathcal{F}_x/\mathcal{E}_x,$$

along with the natural *projection map*

(2.39) $$\sigma : \mathcal{F}/\mathcal{E} \longrightarrow X,$$

such that one has

(2.40) $$\sigma(\mathcal{F}_x/\mathcal{E}_x) := \{x\} \equiv x, \ x \in X.$$

Therefore, one gets, by definition,

(2.41) $$(\mathcal{F}/\mathcal{E})_x = \sigma^{-1}(x) = \mathcal{F}_x/\mathcal{E}_x, \ x \in X.$$

So *the triple* thus defined

(2.42) $$(\mathcal{F}/\mathcal{E}, \sigma, X)$$

can be made into an $\mathcal{A}$-*module* on $X$, called the *quotient $\mathcal{A}$-module of $\mathcal{F}$ by $\mathcal{E}$:*

Namely, let us take the *complete presheaves of sections of $\mathcal{E}$ and $\mathcal{F}$*, respectively, viz.

(2.43) $$\Gamma(\mathcal{E}) \equiv \left(\Gamma(U, \mathcal{E}), \sigma_V^U\right) \text{ and } \Gamma(\mathcal{F}) \equiv \left(\Gamma(U, \mathcal{F}), \omega_V^U\right).$$

So $\Gamma(\mathcal{E})$ *is a sub-$\Gamma(\mathcal{A})$-module of* $\Gamma(\mathcal{F})$, where

$$\Gamma(\mathcal{A}) \equiv \left(\Gamma(U,\mathcal{A}), \kappa_V^U\right) \tag{2.44}$$

is the *complete presheaf of sections of* $\mathcal{A}$, that is, a (complete) *presheaf of* **C**-*algebras on* $X$ (see also (1.67) above). Hence, one can further consider the following *presheaf on* $X$;

$$U \longmapsto S(U) := \Gamma(U,\mathcal{F})/\Gamma(U,\mathcal{E}), \tag{2.45}$$

*with* $U$ *open in* $X$, having the obvious *restriction maps* for any open $U, V$ in $X$, with $V \subseteq U$. Thus, (2.45) *is a* $\Gamma(\mathcal{A})$-*presheaf on* $X$, in fact, as easily seen (cf. also (2.47) below), a *monopresheaf,* which we still denote, for convenience, by

$$S \equiv \Gamma(\mathcal{F})/\Gamma(\mathcal{E}). \tag{2.46}$$

Now, let us further look at the *sheafification* of $S$. That is, we set

$$\mathcal{S} := \mathbf{S}(S) \equiv \mathbf{S}(\Gamma(\mathcal{F})/\Gamma(\mathcal{E})). \tag{2.47}$$

Therefore, one has the following *short exact* $\Gamma(\mathcal{A})$-*sequence,* in the sense of (2.32);

$$0 \longrightarrow \Gamma(\mathcal{E}) \xrightarrow{\iota} \Gamma(\mathcal{F}) \xrightarrow{\phi} \Gamma(\mathcal{F})/\Gamma(\mathcal{E}) \longrightarrow 0, \tag{2.48}$$

where $\phi$ stands for the *(canonical) quotient* $\Gamma(A)$-*morphism,* as defined by (2.45). Thus, by further applying (2.34), we first obtain the following *short exact sequence of* $\mathcal{A}_x$-*modules* (or even *short exact* $\mathcal{A}_x$-*sequence,* see e.g. N. Bourbaki [2: Chap. 2; p.8, Définition 5]); cf. also Chapt. I; (8.6), (8.11),

$$0 \longrightarrow \mathcal{E}_x \xrightarrow{\iota_x} \mathcal{F}_x \xrightarrow{\phi_x} \mathcal{S}_x \longrightarrow 0, \tag{2.49}$$

*for any* $x \in X$, such that *one has*

$$\mathcal{S}_x = S_x = \mathcal{F}_x/\mathcal{E}_x, \ x \in X, \tag{2.50}$$

*within an* $\mathcal{A}_x$-*isomorphism* (see also I; (9.16) in the preceding, as well as, N. Bourbaki [2: Chap. 2; p.8, Proposition 3, d) and p.9, Remarque 2)]). Consequently, using now a similar reasoning to that one in (2.36), one obtains the following *short exact* $\mathcal{A}$-*sequence* on $X$ (cf. Definition 2.1)

$$0 \longrightarrow \mathcal{E} \xrightarrow{\iota} \mathcal{F} \xrightarrow{\phi} \mathcal{S} \longrightarrow 0. \tag{2.51}$$

As in (2.36), we still set here, for convenience, $\tilde{\iota} = \iota$ and $\tilde{\phi} = \phi$. (Of course, according to Subsection 2.(a) above, (2.51) *is* but *the sheafification of (2.48)*; cf. also Chapt. I, Theorem 9.1). Yet, concerning the $\mathcal{A}$-*module* $\mathcal{S}$ on $X$ (see (2.47)), one concludes the following;

(2.52) *the topology of* $\mathcal{S}$, as in (2.51), *is the final one, induced on it by* $\phi$. This also ensures the *uniqueness of* $\mathcal{S}$ *as* (*the* quotient) $\mathcal{A}$-*module* on $X$.

The assertion follows immediately from the fact that $\phi$ *is a sheaf epimorphism* (hence, a *continuous open* and *onto* map; cf. Chapt. I; Lemma 1.2). Now, we collect together the preceding into the following result. (See also, for instance, S. Iitaka [1: p.36, Proposition 1.14]).

**Lemma 2.1.** *Let* $\mathcal{F}$ *be an* $\mathcal{A}$-*module on a topological space* $X$ *and* $\mathcal{E}$ *a sub-*$\mathcal{A}$-*module of* $\mathcal{F}$. *Then, there exists a uniquely defined* $\mathcal{A}$-*module* $\mathcal{S}$ *on* $X$, *whose stalks are given by (2.41)* (cf. also (2.50)), *yielding further the short exact* $\mathcal{A}$-*sequence (2.51). Yet, more generally, for every short exact* $\mathcal{A}$-*sequence*

$$0 \longrightarrow \mathcal{E} \stackrel{\iota}{\longrightarrow} \mathcal{F} \stackrel{\phi}{\longrightarrow} \mathcal{S} \longrightarrow 0$$

*one has* (canonically, viz. within an $\mathcal{A}$-*isomorphism*)

$$\mathcal{S} = \mathcal{F}/\mathcal{E}. \blacksquare$$

As an application of the preceding, we also define for later use the following concepts: Thus, given an $\mathcal{A}$-*morphism*

$$\phi : \mathcal{E} \longrightarrow \mathcal{F}, \tag{2.53}$$

as in (2.1), we can look at the following $\mathcal{A}$-*modules;*

$$\operatorname{coker}\phi := \mathcal{F}/\operatorname{im}\phi \tag{2.54}$$

(see (2.12) and (2.42)). Yet, one sets

$$\operatorname{coim}\phi := \mathcal{E}/\ker\phi. \tag{2.55}$$

So one now obtains the following *exact* $\mathcal{A}$-*sequence*

$$0 \longrightarrow \ker\phi \stackrel{\iota}{\longrightarrow} \mathcal{E} \stackrel{\phi}{\longrightarrow} \mathcal{F} \stackrel{\eta}{\longrightarrow} \operatorname{coker}\phi \longrightarrow 0, \tag{2.56}$$

with $\eta$ being the *canonical sheaf epimorphism,* supplied by (2.54). (See also N. Bourbaki [2: Chap. 2; p.8, Proposition 3,e)]).

**2.(c). The (left) exactness of the section functor.**- Consider the following *short exact $\mathcal{A}$-sequence* on a given topological space $X$ (cf. (2.18));

$$0 \longrightarrow \mathcal{F} \xrightarrow{\psi} \mathcal{S} \xrightarrow{\phi} \mathcal{E} \longrightarrow 0. \tag{2.57}$$

Thus, by applying the *section functor* to (2.57), one obtains the following *exact sequence of (complete) presheaves,* in fact, an *exact sequence of* $\Gamma(\mathcal{A})$-*presheaves*

$$0 \longrightarrow \Gamma(\mathcal{F}) \xrightarrow{\overline{\psi}} \Gamma(\mathcal{S}) \xrightarrow{\overline{\phi}} \Gamma(\mathcal{E}). \tag{2.58}$$

In this context, cf. (1.67) above, as well as, Chap. I; (13.1), (13.3) and (12.49), (12.50).

Indeed, concerning (2.58), the point here is that starting from a short exact $\mathcal{A}$-sequence, as in (2.57), the resulting *exact* $\Gamma(\mathcal{A})$-*sequence* has its last morphism *not, in general, surjective.* We express this fact by saying that;

(2.59) *the section functor*

(2.59.1) $$\Gamma : \mathcal{A}\text{-}\mathcal{M}od_X \longrightarrow \Gamma(\mathcal{A})\text{-}\mathcal{C}o\mathcal{PS}h_X$$

*is left exact.*

For the notation applied in (2.59.1), see also (2.4), (2.23) above, as well as, Chapt. I; (13.6). Now, the *exactness of (2.58)* follows easily from that one of (2.57), according to the very definitions. So cf. I; (12.50) and (12.52), along with the proof of Proposition 11.1, in particular, of the surjectivity of the map I; (11.22). On the other hand, concerning our claim that $\overline{\phi}$ *is not always surjective,* see Chapt. I, Warning after Lemma 12.2, as well as, Example 12.1.

**2.(d). Change of base space. The direct, resp. inverse, image functors for $\mathcal{A}$-modules and $A$-presheaves.**- Suppose we have a *continuous map*

$$h : X \longrightarrow Y. \tag{2.60}$$

Thus, by applying the argument of Chapt. I; Section 14, one obtains the analogous functors to (14.2) and (14.7). So one has the following *direct image functors,* associated with $h$. Namely, first, one gets (see also (2.23))

$$\tilde{h}_* : A\text{-}\mathcal{P}Sh_X \longrightarrow \tilde{h}_*(A)\text{-}\mathcal{P}Sh_Y; \tag{2.61}$$

here $\tilde{h}_*(A)$ denotes the *presheaf of* $\mathbf{C}$*-algebras on* $Y$*, direct image* through $h$ of the given one $A \equiv (A(U), \tau_V^U)$ on $X$ (cf. also Definition 1.6 and I; (14.3)). Second, one further obtains

$$h_* : \mathcal{A}\text{-}\mathcal{M}od_X \longrightarrow h_*(\mathcal{A})\text{-}\mathcal{M}od_Y. \tag{2.62}$$

Similarly, $h_*(\mathcal{A})$ stands here for the $\mathbf{C}$*-algebra sheaf on* $Y$, obtained as the *direct image*, via $h$, of the given such sheaf $\mathcal{A}$ on $X$. In this context, we still remark that *(2.62) is well defined;* namely, cf. I; (14.8), along with (2.59.1), (2.61) and (1.66) in the preceding. Furthermore, due to the same relation I; (14.8) and to (2.59) above (viz. the *insertion of the left exact section functor),* one concludes that;

(2.63) *the direct image functor* $h_*$ (cf. (2.62)) *is left exact.*

The last term has been already applied in (2.59) for the section functor too. Now, specializing to the particular functor $h_*$, as above, we formally explain right below the term in question, anticipating thus its use in similar cases later on; viz. when considering analogous functors between the appropriate categories, for which, namely, the term *"exact sequence"* has a meaning. So by the last term in (2.63) one actually means that;

(2.64) given any *short exact* $\mathcal{A}$*-sequence in* $\mathcal{A}\text{-}\mathcal{M}od_X$ (cf. (2.4) and (2.57)), the application on it of the functor $h_*$, under consideration, provides an *exact* $h_*(\mathcal{A})$*-sequence on* $Y$, whose *last morphism* (to the right) *is not onto* (in general).

The analogous case for the section functor $\Gamma$ has been already observed in the preceding, according to (2.59); see also (2.57) and (2.58).

On the other hand, by analogy with I; (14.44) (cf. also I; (14.15)), we further define the *inverse image functor,* associated with any given continuous map $h : X \longrightarrow Y$, as above. That is, one has

$$h^* : \mathcal{A}\text{-}\mathcal{M}od_Y \longrightarrow h^*(\mathcal{A})\text{-}\mathcal{M}od_X. \tag{2.65}$$

Thus, we assume that $\mathcal{A} \equiv (\mathcal{A}, \tau, Y)$ is a given $\mathbf{C}$*-algebra sheaf on* $Y$, so that

$$h^*(\mathcal{A}) = \left\{(x, z) \in X \times \mathcal{A} : h(x) = \tau(z)\right\} \tag{2.66}$$

is a **C**-*algebra sheaf on* $X$: The assertion can easily be derived from an application of the associated with the given $h$, *complete presheaf of sections of* $\mathcal{A}$ *along* $h$, $\mathcal{A}_h$ (cf. I; (14.31)). The latter is indeed a *(complete)* **C**-*algebra presheaf on* $X$, as one easily verifies by the very definitions and I; (14.36). Therefore (see (1.54)), the respective *sheafification of* $\mathcal{A}_h$, $\mathbf{S}(\mathcal{A}_h)$ *is a* **C**-*algebra sheaf on* $X$. But, now one has (cf. I; (14.33));

(2.67) $$h^*(\mathcal{A}) = \mathbf{S}(\mathcal{A}_h),$$

within a *sheaf isomorphism,* which thus proves our claim about (2.66), in view also of Chapt. I; (14.26), (14.36) and (7.6).

Furthermore, as a result of I; (14.26), one also concludes that:

(2.68) *the inverse image functor* $h^*$, associated with any given continuous map $h : X \longrightarrow Y$ (cf. (2.65)), *is exact.*

By complementing our previous comments in (2.64), we further note here, by still anticipating an analogous use in the sequel, that the last term in (2.68) about the functor $h^*$, means that

(2.69) for any *short exact* $\mathcal{A}$-*sequence in* $\mathcal{A}$-$\mathcal{M}od_Y$, the corresponding $h^*(\mathcal{A})$-*sequence on* $X$, resulted from the former after an application on it of the functor $h^*$, *is exact* too (cf. also Definition 2.1).

Now, concerning the *inverse image of A-presheaves on* $Y$, via a continuous map $h$, as above, see the analogous comments at the end of Chapt. I, Subsection 14.(b); thus, cf. for instance, J. R. Strooker [1: p.209, Definitions 4.6.2].

# 3. Direct (or Whitney) sum of $\mathcal{A}$-modules. Free $\mathcal{A}$-modules

We deal here with a special class of $\mathcal{A}$-modules, which are of a particular significance for the sequel. Indeed, the $\mathcal{A}$-modules we consider below are at the base of the notion of *locally free $\mathcal{A}$-modules* (see the next Section 4), which, as already said, are going to play a prominent rôle in the ensuing discussion. So the present Section is to be considered, in fact, as a preparatory to the next one. We thus begin with the following basic notion for the sequel.

Namely, given two *$\mathcal{A}$-modules* $\mathcal{E} \equiv (\mathcal{E}, \pi, X)$ and $\mathcal{F} \equiv (\mathcal{F}, \rho, X)$ on a *topological space* $X$, consider the set

(3.1) $$\mathcal{E} \oplus \mathcal{F} := \{(z, z') \in \mathcal{E} \times \mathcal{F} : \pi(z) = \rho(z')\}.$$

This is further endowed with the *relative topology from* $\mathcal{E} \times \mathcal{F}$, as well as, with the obvious *projection map* (cf. (3.1))

(3.2) $$\sigma : \mathcal{E} \oplus \mathcal{F} \longrightarrow X : (z, z') \longmapsto \sigma(z, z') := \pi(z)\ (= \rho(z')).$$

Now, we first claim that;

(3.3) *the triple*

(3.3.1) $$(\mathcal{E} \oplus \mathcal{F}, \sigma, X),$$

as given by the preceding, *is a sheaf on* $X$.

Namely, for any $(z, z') \in \mathcal{E} \oplus \mathcal{F}$, if $V, V'$ are the neighborhoods of $z \in \mathcal{E}$ and $z' \in \mathcal{F}$, respectively, as given by I; (1.3), one readily verifies that

(3.4) $$\pi^{-1}(U) \times \rho^{-1}(U) \xrightarrow{\sigma} U,$$

such that $U = \pi(V) \cap \rho(V')$ (cf. (3.1)), yields a *local homeomorphism* for $\sigma$, as well. ∎ Yet, concerning the stalks of (3.3.1), one easily obtains, by (3.1), that

(3.5) $$(\mathcal{E} \oplus \mathcal{F})_x = \mathcal{E}_x \times \mathcal{F}_x \cong \mathcal{E}_x \oplus \mathcal{F}_x, \ x \in X.$$

Consequently,

(3.6) *(3.3.1) yields,* in effect, *an $\mathcal{A}$-module on* $X$.

Indeed, in view of the last relation, $(\mathcal{E} \oplus \mathcal{F})_x$ *is an $\mathcal{A}_x$-module,* for any $x \in X$, while one further concludes the *continuity of the algebraic operations in* $\mathcal{E} \oplus \mathcal{F}$, being, in fact, the restrictions to the latter space of those in $\mathcal{E} \times \mathcal{F}$ (defined, of course, fiber-wise; but, cf. also (1.76)). Thus, we come to the following.

**Definition 3.1.** Given two $\mathcal{A}$-modules $\mathcal{E}$ and $\mathcal{F}$ on a topological space

$X$, the $\mathcal{A}$-module $\mathcal{E} \oplus \mathcal{F}$ on $X$, as given by (3.3) is called the **direct** (or **Whitney) sum** of $\mathcal{E}$ and $\mathcal{F}$. (Yet, $\mathcal{E} \oplus \mathcal{F}$ is but the *fiber product* of $\mathcal{E}$ and $\mathcal{F}$ over $X$, viewed as an $\mathcal{A}$-module on $X$).

Similarly, one defines the *direct sum of any finite number of $\mathcal{A}$-modules on $X$*. In particular, for any given *$\mathcal{A}$-module* $\mathcal{E}$ on $X$, one considers, for any *natural number* $n \geqq 1$, the *$\mathcal{A}$-module on $X$*

$$\mathcal{E}^n := \mathcal{E} \underbrace{\oplus \cdots \oplus}_{n-\text{times}} \mathcal{E}. \tag{3.7}$$

We call it the *n-fold* (*direct* or) *Whitney sum* of $\mathcal{E}$ with itself.

On the other hand, the preceding notions can also be defined *through* an application of the *sheafification* procedure, yet *for any family of $\mathcal{A}$-modules:* Namely, suppose that we are given a family

$$(\mathcal{E}_i)_{i \in I} \tag{3.8}$$

of *$\mathcal{A}$-modules* on $X$. So one can further consider the next two *$\mathcal{A}(U)$-presheaves* on $X$; first,

$$U \longmapsto \prod_{i \in I} \Gamma(U, \mathcal{E}_i), \tag{3.9}$$

and second

$$U \longmapsto \bigoplus_{i \in I} \Gamma(U, \mathcal{E}_i), \tag{3.10}$$

where, by definition, *$U$ varies over the open subsets of $X$*. Now, it is easy to see that;

(3.11) *the previous two $\mathcal{A}(U)$-presheaves* on $X$, as given by (3.9) and (3.10), *are,* in fact, *complete* (cf. I; Definition 11.1).

The *sheaves generated by the previous* two *presheaves* are called the *cartesian* (or *direct*) *product* and *direct* (or *Whitney*) *sum,* respectively, of the given family of $\mathcal{A}$-modules on $X$, being, in fact, as readily seen, *$\mathcal{A}$-modules on $X$*. Yet, they are denoted by

$$\prod_{i \in I} \mathcal{E}_i \quad \textit{and} \quad \bigoplus_{i \in I} \mathcal{E}_i, \quad \textit{respectively}. \tag{3.12}$$

Thus, as an application of (3.11) and of Chapt. I; Proposition 11.1, one has the following useful formulas, pertaining to the respective sections of the previous two $\mathcal{A}$-modules; viz., *for any open set* $U \subseteq X$, one obtains

$$\Gamma\left(U, \prod_{i\in I} \mathcal{E}_i\right) = \prod_i \Gamma(U, \mathcal{E}_i), \tag{3.13}$$

as well as,

$$\Gamma\left(U, \bigoplus_{i\in I} \mathcal{E}_i\right) = \bigoplus_i \Gamma(U, \mathcal{E}_i), \tag{3.14}$$

within $\mathcal{A}(U)$*-isomorphisms.*

Furthermore, since *inductive limits commute with direct sums* (cf. e.g. R. Kultze [1: p.14, Satz 1.15]), one gets for the *stalks* of a *direct sum* of $\mathcal{A}$-modules, as above, the relation

$$\left(\bigoplus_{i\in I} \mathcal{E}_i\right)_x = \bigoplus_i (\mathcal{E}_i)_x, \quad x \in X, \tag{3.15}$$

within an $\mathcal{A}_x$*-isomorphism:* Indeed (cf. also Chapt. I; (9.8) and (9.16), as well as, (3.13) above), one obtains;

$$\begin{aligned}\left(\bigoplus_{i\in I} \mathcal{E}_i\right)_x &= \varinjlim_{U\ni x} \Gamma\left(U, \bigoplus_i \mathcal{E}_i\right) = \varinjlim_{U\ni x} \bigoplus_i \Gamma(U, \mathcal{E}_i) \\ &= \bigoplus_i \varinjlim_{U\ni x} \Gamma(U, \mathcal{E}_i) = \bigoplus_i (\mathcal{E}_i)_x,\end{aligned} \tag{3.16}$$

*with* $x \in X$, all the previous relations being valid, of course, within $\mathcal{A}_x$*-isomorphisms,* which thus proves our claim. ∎ However, in case the *index set* $I$ in (3.8) is *infinite,* one has, *in general,*

$$\left(\prod_{i\in I} \mathcal{E}_i\right)_x \subsetneqq \prod_i (\mathcal{E}_i)_x, \quad x \in X; \tag{3.17}$$

that is, the indicated (canonical) *injection is not always surjective.* See, for instance, H. Grauert – R. Remmert [1: p.4, Warning].

On the other hand, if $I$ is *finite,* it is clear (cf. (3.9) and (3.10)) that *the* above *two notions coincide,* within an $\mathcal{A}$*-isomorphism;* hence, *the corresponding relation to (3.15) is,* in the latter case, *in force for the direct product too* (up to $\mathcal{A}_x$*-isomorphisms,* $x \in X$). Yet, in the same case, Definition 3.1 yields an equivalent notion to that one obtained, as above, through an application of (3.10); of course, this, within an $\mathcal{A}$-*isomorphism* of the $\mathcal{A}$-modules involved.

Thus, based on the preceding, we come next to another crucial concept for the sequel, as this is expressed by the following.

**Definition 3.2.** Let $\mathcal{E}$ be an $\mathcal{A}$-module on a topological space $X$. Then, $\mathcal{E}$ is called a **free $\mathcal{A}$-module of rank** $n$ (a natural number), whenever one has

$$\mathcal{E} = \mathcal{A}^n, \tag{3.18}$$

within an $\mathcal{A}$-*isomorphism* (cf. also (3.7)).

The previous condition (3.18) *is* certainly *equivalent with a similar one, pertaining to the sections, or* even to the *stalks,* of the $\mathcal{A}$-modules involved. Thus, one has (cf. also (3.13))

$$\Gamma(U, \mathcal{E}) = \Gamma(U, \mathcal{A}^n) = \Gamma(U, \mathcal{A})^n, \tag{3.19}$$

within $\mathcal{A}(U)$-*isomorphisms,* for any open set $U \subseteq X$; moreover, one obtains

$$\mathcal{E}_x = (\mathcal{A}^n)_x = (\mathcal{A}_x)^n, \; x \in X, \tag{3.20}$$

within $\mathcal{A}_x$-*isomorphisms* (see e.g. (3.15)). So *each one* of the latter relations is *an equivalent formulation of (3.18)* (where in the last one we assume, of course, the existence of an $\mathcal{A}$-morphism $\mathcal{E} \longrightarrow \mathcal{A}^n$). In this respect, see also Chapt. I; Theorems 12.1 and 13.1.

Now, using the *identity section of* $\mathcal{A}$ (cf. (1.22)), one obtains the so-called *canonical basis (of sections) of the free* $\mathcal{A}$-*module* $\mathcal{A}^n$ ($n \in \mathbf{N}$); that is, we set

$$\varepsilon_i := \delta_{ij} \in \mathcal{A}^n(X) \cong \mathcal{A}(X)^n, \; 1 \leqq i, j \leqq n, \tag{3.21}$$

such that one has

$$\delta_{ij} := \begin{cases} 1 \in \Gamma(X, \mathcal{A}), & \textit{for } i = j \\ 0 \in \Gamma(X, \mathcal{A}), & \textit{for } i \neq j. \end{cases} \tag{3.22}$$

So one gets, by definition (cf. also (3.20)),

$$\varepsilon_i(x) = (\delta_{ij}(x))_{1 \leqq j \leqq n} \in \mathcal{A}_x^n, \; 1 \leqq i \leqq n, \tag{3.23}$$

*for any* $x \in X$, where *we set,* by (3.22),

$$\delta_{ij}(x) = 1_x \in \mathcal{A}_x, \; \textit{if} \;\; i = j, \;\; \textit{and} \;\; \delta_{ij}(x) = 0_x \in \mathcal{A}_x, \; \textit{if} \;\; i \neq j, \tag{3.24}$$

*for every* $x \in X$, *and* $1 \leqq i, j \leqq n$. Yet, we call the latter sections, as given by (3.21), the *Kronecker sections of* $\mathcal{A}^n$.

Consequently, by virtue of (3.19) (cf. also (3.20)), one gets, *for any open* set $U \subseteq X$, a corresponding *canonical basis of the* $\Gamma(U, \mathcal{A})$*-module* $\Gamma(U, \mathcal{A}^n) \cong \Gamma(U, \mathcal{A})^n$. (An analogous remark is certainly true, concerning the $\mathcal{A}_x$*-module* $\mathcal{A}_x^n \cong (\mathcal{A}_x)^n$, $x \in X$; (cf. (3.20) and (3.23)).

Therefore, by considering any *free* $\mathcal{A}$*-module* $\mathcal{E}$ *of rank* $n$ on $X$, one concludes that;

(3.25) *for every open* set $U \subseteq X$ *and for any* $s \in \Gamma(U, \mathcal{E})$, *one has*

(3.25.1) $$s = \sum_{i=1}^{n} \alpha_i e_i,$$

*where* $\alpha_i \in \Gamma(U, \mathcal{A})$ *and* $e_i = \phi(\varepsilon_i)$, $1 \leqq i \leqq n$, with $\phi \equiv \psi^{-1}$ denoting the (canonical) $\mathcal{A}$*-isomorphism (3.18).*

(Concerning the classical theory, see, for instance, N. Bourbaki [2: Chap. 2, p.24ff]). Now, more on this and on related results will be found in Chapt. IV in the sequel. But, see also the following Section 4.

## 4. Locally free $\mathcal{A}$-modules. Vector sheaves

The $\mathcal{A}$-modules in the title of this Section will mainly be considered in Chapt. IV below, while they also constitute the principal ingredients in the second part of this treatise. For convenience, however, being also a natural important generalization of the sheaves in the previous Section, we present herewith the most basic features of these objects. So we start with the following.

**Definition 4.1.** Let $\mathcal{E} \equiv (\mathcal{E}, \pi, X)$ be a given $\mathcal{A}$-module on a topological space $X$ (cf. Definition 1.5). Then, $\mathcal{E}$ is called a **locally free $\mathcal{A}$-module of finite rank** on $X$, whenever there exists *a natural number* $n$, such that the following condition holds true;

(4.1) *for every point $x \in X$, there exists an open neighborhood $U$ of $x$ in $X$, such that one has* (cf. also I; (1.10))

(4.1.1) $$\mathcal{E}|_U = \mathcal{A}^n|_U,$$

*within an $\mathcal{A}|_U$-isomorphism* of the $\mathcal{A}|_U$-modules involved.

We also call a given $\mathcal{A}$-module $\mathcal{E}$ on $X$, for which (4.1) is valid, a **locally free $\mathcal{A}$-module of rank** $n$ on $X$. Yet, we write

(4.2) $$n = \mathrm{rk}_{\mathcal{A}}(\mathcal{E}) \equiv \mathrm{rk}\, \mathcal{E}.$$

Now, concerning the terminology applied in (4.1.1), we still remark that, *for any given $\mathcal{E}$, as in the previous definition,* one concludes that;

(4.3) *for every open $U \subseteq X$, $\mathcal{E}|_U$ is an $\mathcal{A}|_U$-module on $U$, where $\mathcal{A}|_U$ is a (unital) $\mathbf{C}$-algebra sheaf on $U$.*

This is straightforward from I; (11.40) and (1.67); in particular, one gets

(4.4) $$(\mathcal{E}|_U)(V) = \mathcal{E}(V),$$

within an $(\mathcal{A}|_U)(V) = \mathcal{A}(V)$*-isomorphism, for any open $V \subseteq U$*. Yet, we note for later use the relation

(4.5) $$\mathcal{E}|_U = \mathcal{A}^n|_U = (\mathcal{A}|_U)^n, \ n \in \mathbf{N},$$

valid within $\mathcal{A}|_U$*-isomorphisms,* for *any open $U \subseteq X$*. On the other hand, another useful corollary of (4.1) is that:

(4.6) *for any locally free $\mathcal{A}$-module $\mathcal{E}$ of rank $n$ on a topological space $X$, and for every point $x \in X$, there exists a basis of open neighborhoods of $x$, say, $\mathcal{B}(x)$, such that one has*

(4.6.1) $$\mathcal{E}|_U = \mathcal{A}^n|_U = (\mathcal{A}|_U)^n,$$

*within $\mathcal{A}|_U$-isomorphisms,* (see also (4.5)), *for any $U \in \mathcal{B}(x)$.*

Thus, one obtains such a basis $\mathcal{B}(x)$ by taking any fundamental system of open neighborhoods of $x$ and then its intersection with an open $U \subseteq X$, as in (4.4.1). Indeed, *(4.6) is equivalent to (4.1), for any given $\mathcal{A}$-module $\mathcal{E}$ on $X$.*

Now, another consequence of (4.1) (yet, an *equivalent formulation* of it) is obtained by remarking that *(4.1) entails the existence of an open covering of* $X$, say, $\mathcal{U} = (U_\alpha)_{\alpha \in I}$, *such that one has;*

$$\mathcal{E}|_{U_\alpha} = \mathcal{A}^n|_{U_\alpha}, \tag{4.7}$$

within an $\mathcal{A}|_{U_\alpha}$*-isomorphism, for any* $\alpha \in I$. Consequently, one gathers that:

(4.8) the previous conditions (4.1), (4.6) and (4.7) *are all equivalent formulations,* in order a given $\mathcal{A}$*-module* $\mathcal{E}$ on a topological space $X$ to be locally free (of finite rank).

The motivation to the following terminology comes from the considerations in the previous Section 3, in particular, from (3.21), (3.22) and (3.25). But, see also Part Two in the sequel. So we have the next.

**Definition 4.2.** Given a locally free $\mathcal{A}$-module $\mathcal{E}$ of rank $n$ (with $n \in \mathbf{N}$) on a topological space $X$, an open covering of $X$, satisfying (4.7), is called a **coordinatizing open covering** (or even a **local choice of basis**) **for** $\mathcal{E}$. Yet, we call such a covering of $X$, simply, a **local frame of** $\mathcal{E}$. Moreover, any open set $U$ in $X$ for which (4.1.1) holds (within an $\mathcal{A}|_U$-isomorphism), is called a **local gauge** of $\mathcal{E}$.

In this context, it is still useful to point out at this place, the special rôle that play, in the set of all *"proper"* coverings of $X$, those associated with any given $\mathcal{E}$, as before. Thus, we first recall that, although *one may not speak* of the *set of all* (open) *coverings of* $X$ (due to logical problems), we can consider, however, the *set of all proper (open) coverings of* $X$; so an element of the latter set is, by definition, a set of open subsets of $X$, whose union is $X$ (the former set being thus merely a subset of the *"2nd power set of* $X$*"*). Yet, a covering of the latter type can further be viewed as *a family* of open sets, *indexed by itself* (viz. by the sets themselves. In this context, see also e.g. C. H. Dowker [1: p.60f], or even F. Hirzebruch [1: p.17]).

On the other hand, we can further look at the previous set as a *partially (pre)ordered set*, with respect to the *refinement* relation between (proper) coverings, being, in effect, an (upwards) *directed set* (see also Chapt. III; (4.9) in the sequel). So we come now, as already promised, to the following fact which will also be of use below. That is, one concludes that:

(4.9) *given a locally free $\mathcal{A}$-module $\mathcal{E}$* (of finite rank) *on a topological space $X$, the (proper) local frames of $\mathcal{E}$* (cf. Definition 4.2) *constitute a cofinal subset of the set of all proper open coverings of $X$.*

Namely, for any given (proper) open covering of $X$, its intersection with a local frame of $\mathcal{E}$, provides a new covering, consisting of the pair-wise non-void intersections (associated with any given point of $X$), being also *finer than the given one and yet a local frame of $\mathcal{E}$ (the restriction of an $\mathcal{A}$-isomorphism to an open subset is still an isomorphism,* in the appropriate sense). ∎

Now, throughout the rest of our discussion we shall employ the following terminology.

**Definition 4.3.** By a **vector sheaf** on a given topological space $X$, we mean a locally free $\mathcal{A}$-module of finite rank over $X$ (cf. Definition 4.1). In particular, a locally free $\mathcal{A}$-module of rank one, is named a **line sheaf** on $X$.

The above type of sheaves will mainly be considered in later chapters of this treatise (see, in particular Chapter IV, and especially Part Two in the sequel). Now, the previous nomenclature is, of course, reminiscent of the classical case of *vector* (and *line*) *bundles,* whose sheaves (of germs) of sections are, respectively, of the above type (cf. Chapter V below). The terminology is actually due, to my knowledge, to S. Lang [2: p.1], who applied it within another context.

Henceforth we shall denote the corresponding categories of such a type of $\mathcal{A}$-modules on a given topological space $X$ by

$$\mathcal{V}ect\mathcal{S}h_X \tag{4.10}$$

and

$$\mathcal{LS}h_X, \tag{4.11}$$

respectively, the morphisms being, of course, $\mathcal{A}$-*morphisms* (cf. also (2.4)). So one has the following inclusion (functor)

$$\mathcal{V}ect\mathcal{S}h_X \hookrightarrow \mathcal{A}\text{-}\mathcal{M}od_X, \tag{4.12}$$

this pertaining, in fact, to a *full subcategory. Warning!* In this regard, denoting by

(4.13) $$\mathbf{C}\text{-}\mathcal{V}ect\mathcal{S}h_X$$

the *category of* **C**-*vector space sheaves on* $X$, with morphisms the **C**-*linear morphisms* (cf. (2.3)), we note that $\mathcal{A}$-$\mathcal{M}od_X$ *is not a subcategory of* **C**-$\mathcal{V}ect_X$. (So different $\mathcal{A}$-morphisms might coincide as **C**-linear morphisms).

Finally, it is an immediate consequence of (2.68) (cf. also (2.69)) and the very definitions (cf. e.g. (4.6)), that:

(4.14) *the pull-back of any vector sheaf is a vector sheaf too, of the same* (finite) *rank as the initial one.*

Yet, we are going to apply in the sequel the following notation

(4.15) $$\Phi^n_{\mathcal{A}}(X),$$

which stands for the *set of isomorphism classes* (modulo, namely, $\mathcal{A}$-*isomorphisms*) *of vector sheaves of rank* $n$ *on* $X$. In this context, we still mention, in anticipation, that a *cohomological description* of the latter set, with applications in Part Two, will also be supplied in the sequel (cf. Chapter V).

# 5. Tensor products of $\mathcal{A}$-modules

For convenience, we assume below, as well as, throughout the rest of our discussion, that whenever one considers tensor product modules, the corresponding *"domain of coefficients"* is always a *commutative* **C**-*algebra with an identity element.* So the tensor product modules concerned, which are always taken (unless otherwise stated) with respect to the coefficient (commutative) algebra (even if this is not explicitly indicated) are, in particular, **C**-*vector spaces.* (Concerning the classical theory, we refer, for instance, to N. Bourbaki [2]).

Thus, suppose that

(5.1) $$\mathcal{A} \equiv (\mathcal{A}, \tau, X)$$

is a given (fixed throughout the sequel) *sheaf of commutative* **C**-*algebras with an identity element* on a topological space $X$ (cf. Definition 1.4 and (1.33)). Moreover, let

(5.2) $$\Gamma(\mathcal{A}) \equiv \left(\Gamma(U, \mathcal{A}), \tau_V^U\right)$$

be the corresponding (complete) *presheaf of sections of* $\mathcal{A}$.

Finally, let

(5.3) $$\mathcal{E} \equiv (\mathcal{E}, \pi, X) \ \ and \ \ \mathcal{F} \equiv (\mathcal{F}, \rho, X)$$

be two given $\mathcal{A}$-modules on $X$, along with the corresponding (complete) *presheaves of sections*

(5.4) $$\Gamma(\mathcal{E}) \equiv \left(\Gamma(U, \mathcal{E}), \rho_V^U\right) \ \ and \ \ \Gamma(\mathcal{F}) \equiv \left(\Gamma(U, \mathcal{F}), \sigma_V^U\right),$$

respectively. Now, it is easy to see that

(5.5) the correspondence

(5.5.1) $$U \longmapsto \Gamma(U, \mathcal{E}) \otimes_{\Gamma(U, \mathcal{A})} \Gamma(U, \mathcal{F}),$$

where $U$ runs over the open sets in $X$, along with the obvious restriction morphisms (tensor product of those in (5.4)), provide a *presheaf of* $\Gamma(\mathcal{A})$*-modules* on $X$. We denote it by

(5.5.2) $$\Gamma(\mathcal{E}) \otimes_{\Gamma(\mathcal{A})} \Gamma(\mathcal{F}).$$

*Warning!* The previous presheaf *is not always complete.* Cf., for instance, S. Iitaka [1: p.37, Example 1.17] or yet B. R. Tennison [1: p.106, Example 5.8].

Thus, we now set the following.

**Definition 5.1.** Given two $\mathcal{A}$-modules $\mathcal{E}$ and $\mathcal{F}$ on a topological space $X$, we call **tensor product of $\mathcal{E}$ and $\mathcal{F}$ (over $\mathcal{A}$)** and denote it by

(5.6) $$\mathcal{E} \otimes_{\mathcal{A}} \mathcal{F},$$

the following sheaf on $X$ (see also I; (7.7), (8.2))

(5.7) $$\mathcal{E} \otimes_{\mathcal{A}} \mathcal{F} := \mathbf{S}\left(\Gamma(\mathcal{E}) \otimes_{\Gamma(\mathcal{A})} \Gamma(\mathcal{F})\right);$$

viz. the *sheafification of (5.5.2)*.

In fact, based on the preceding, it is readily seen that;

(5.8) $$\mathcal{E} \otimes_{\mathcal{A}} \mathcal{F} \ \textit{is an } \mathcal{A}\textit{-module on } X.$$

See Proposition 1.1, as well as, Chapt. I; Theorem 9.1.

On the other hand, by contrast with the above warning (concerning the local sections of (5.7)), by referring to the *stalks* at the various points of $X$ of the $\mathcal{A}$-module $\mathcal{E} \otimes_{\mathcal{A}} \mathcal{F}$, as defined by (5.7), one obtains the next relation (5.9), due to the fact that *inductive limits commute with tensor products.* (See e.g. N. Bourbaki [2: Chap. 2; p.93, Proposition 7] or even R. Kultze [1: p.16, Satz 1.19]). That is, we have.

(5.9)
$$\left(\mathcal{E} \otimes_{\mathcal{A}} \mathcal{F}\right)_x = \mathcal{E}_x \otimes_{\mathcal{A}_x} \mathcal{F}_x,$$

within an $\mathcal{A}_x$*-isomorphism, for any* $x \in X$: Indeed, one has (cf. (5.5.1), as well as, I; (9.8) and (9.16));

(5.10)
$$\begin{aligned}\left(\mathcal{E} \otimes_{\mathcal{A}} \mathcal{F}\right)_x &= \varinjlim_{U \ni x} \left(\Gamma(\mathcal{E}) \otimes_{\Gamma(\mathcal{A})} \Gamma(\mathcal{F})\right)(U) = \varinjlim_{U \ni x} \left(\Gamma(U, \mathcal{E}) \otimes_{\Gamma(U,\mathcal{A})} \Gamma(U, \mathcal{F})\right) \\ &= \varinjlim_{U \ni x} \Gamma(U, \mathcal{E}) \otimes_{\varinjlim_{U \ni x} \Gamma(U,\mathcal{A})} \varinjlim_{U \ni x} \Gamma(U, \mathcal{F}) = \mathcal{E}_x \otimes_{\mathcal{A}_x} \mathcal{F}_x,\end{aligned}$$

which proves our claim. ▮ Now, as a consequence of (5.9), one further obtains that (see also (5.8));

(5.11) for any $\mathcal{A}$*-module* $\mathcal{E}$ on $X$, *the functor* (cf. also (5.14) below)

(5.11.1)
$$\otimes_{\mathcal{A}} \mathcal{E} : \mathcal{A}\text{-}\mathcal{M}od_X \longrightarrow \mathcal{A}\text{-}\mathcal{M}od_X$$

*is* covariant and *right exact.*

That is, *for any exact* $\mathcal{A}$*-sequence,*

(5.12)
$$\mathcal{S}' \longrightarrow \mathcal{S} \longrightarrow \mathcal{S}'' \longrightarrow 0,$$

*the* $\mathcal{A}$*-sequence*

(5.13)
$$\mathcal{S}' \otimes_{\mathcal{A}} \mathcal{E} \longrightarrow \mathcal{S} \otimes_{\mathcal{A}} \mathcal{E} \longrightarrow \mathcal{S}'' \otimes_{\mathcal{A}} \mathcal{E} \longrightarrow 0$$

*is also exact.* ▮

Furthermore, as readily follows from the very definitions and the corresponding properties of the classical theory of modules, one also obtains the following $\mathcal{A}$*-isomorphisms* of the $\mathcal{A}$*-modules* concerned:

(5.14)
$$\mathcal{E} \otimes_{\mathcal{A}} \mathcal{F} = \mathcal{F} \otimes_{\mathcal{A}} \mathcal{E} \quad (\textit{commutativity}).$$

(5.15)
$$\mathcal{A} \otimes_{\mathcal{A}} \mathcal{E} = \mathcal{E} \otimes_{\mathcal{A}} \mathcal{A} = \mathcal{E}.$$

$$\mathcal{E} \otimes_{\mathcal{A}} (\mathcal{F} \otimes_{\mathcal{A}} \mathcal{S}) = (\mathcal{E} \otimes_{\mathcal{A}} \mathcal{F}) \otimes_{\mathcal{A}} \mathcal{S} \quad (associativity). \tag{5.16}$$

$$\mathcal{E} \otimes_{\mathcal{A}} (\mathcal{F} \oplus \mathcal{S}) = (\mathcal{E} \otimes_{\mathcal{A}} \mathcal{F}) \oplus (\mathcal{E} \otimes_{\mathcal{A}} \mathcal{S}) \quad (distributivity). \tag{5.17}$$

(Concerning the last relation, cf. also (3.6)). Of course, *(5.17) holds for any finite Whitney sum* of $\mathcal{A}$-modules (see also (3.12)). In particular, as follows from (5.17) (yet, cf. (3.7), (5.15)), one has the following *$\mathcal{A}$-isomorphisms,* for any given *$\mathcal{A}$-module* $\mathcal{E}$ on $X$;

$$\mathcal{A}^n \otimes_{\mathcal{A}} \mathcal{E} = \mathcal{E} \otimes_{\mathcal{A}} \mathcal{A}^n = \mathcal{E}^n, \; n \in \mathbf{N}. \tag{5.18}$$

We apply the foregoing right below in the following result, that also supplements (5.11) above. (See also R. Kultze [1: p.48, Satz 8.3]). So one has.

**Theorem 5.1.** *For every vector sheaf $\mathcal{E}$ on a topological space $X$, the corresponding tensor product functor $\otimes_{\mathcal{A}}\mathcal{E}$* (cf. (5.11.1)) *is exact. In other words, given any short exact $\mathcal{A}$-sequence* (see (2.18))

$$0 \longrightarrow \mathcal{S}' \xrightarrow{\psi} \mathcal{S} \xrightarrow{\phi} \mathcal{S}'' \longrightarrow 0, \tag{5.19}$$

*tensoring with a vector sheaf $\mathcal{E}$ on $X$, yields an exact $\mathcal{A}$-sequence too,*

$$0 \longrightarrow \mathcal{S}' \otimes_{\mathcal{A}} \mathcal{E} \longrightarrow \mathcal{S} \otimes_{\mathcal{A}} \mathcal{E} \longrightarrow \mathcal{S}'' \otimes_{\mathcal{A}} \mathcal{E} \longrightarrow 0. \tag{5.20}$$

**Proof.** In view of (5.11), it suffices to prove the *exactness of (5.20) at $\mathcal{S}' \otimes_{\mathcal{A}} \mathcal{E}$;* viz. that *the $\mathcal{A}$-morphism* (cf. (5.19))

$$\psi \otimes \mathrm{id}_{\mathcal{E}} : \mathcal{S}' \otimes_{\mathcal{A}} \mathcal{E} \longrightarrow \mathcal{S} \otimes_{\mathcal{A}} \mathcal{E} \tag{5.21}$$

*is injective* (of course, we set $\mathrm{id}_{\mathcal{E}} \equiv 1_{\mathcal{E}}$ for the *identity $\mathcal{A}$-morphism* (*automorphism of $\mathcal{E}$* in itself): In turn, our inquiry is reduced to the *injectivity of the $\mathcal{A}_x$-morphism*

$$(\psi \otimes \mathrm{id}_{\mathcal{E}})_x = \psi_x \otimes (\mathrm{id}_{\mathcal{E}})_x : \mathcal{S}'_x \otimes_{\mathcal{A}_x} \mathcal{E}_x \longrightarrow \mathcal{S}_x \otimes_{\mathcal{A}_x} \mathcal{E}_x, \tag{5.22}$$

*for any point $x \in X$* (cf. also (5.9) and I; (1.14)). But, due to the hypothesis for $\mathcal{E}$, if $n = \mathrm{rk}_{\mathcal{A}}(\mathcal{E})$, one obtains (see (4.7) and (5.18))

$$\mathcal{S}'_x \otimes_{\mathcal{A}_x} \mathcal{E}_x = \mathcal{S}'_x \otimes_{\mathcal{A}_x} \mathcal{A}^n_x = (\mathcal{S}'_x)^n. \tag{5.23}$$

Therefore, by the exactness of (5.19), *the $\mathcal{A}_x$-morphism*

$$\psi^n_x : (\mathcal{S}'_x)^n \longrightarrow \mathcal{S}^n_x \tag{5.24}$$

*is one-to-one* (see also (3.15), as well as, e.g. N. Bourbaki [2: Chap. 2; p.14, Corollaire 1]), and this terminates the proof. ∎

Finally, we still note for later use the following. (See e.g. Chapt. IV; (6.12)).

**Lemma 5.1.** *Let $\mathcal{E}$,$\mathcal{F}$ be given $\mathcal{A}$-modules on a topological space $X$, and $U$ any open subset of $X$. Then, one has*

$$(\mathcal{E} \otimes_{\mathcal{A}} \mathcal{F})|_U = \mathcal{E}|_U \otimes_{\mathcal{A}|_U} \mathcal{F}|_U, \tag{5.25}$$

*within an $\mathcal{A}|_U$-isomorphism.*

**Proof.** Indeed, based on (5.5) and (5.7), as well as, on Chapt. I; (5.12), (5.13), (9.23) and (11.41), one has:

$$\begin{aligned}(\mathcal{E} \otimes_{\mathcal{A}} \mathcal{F})|_U &= \mathbf{S}(\Gamma(\mathcal{E}) \otimes_{\Gamma(\mathcal{A})} \Gamma(\mathcal{F}))|_U = \mathbf{S}((\Gamma(\mathcal{E}) \otimes_{\Gamma(\mathcal{A})} \Gamma(\mathcal{F}))|_U) \\ &= \mathbf{S}\big(\Gamma(\mathcal{E})|_U \otimes_{\Gamma(\mathcal{A})|_U} \Gamma(\mathcal{F})|_U)\big) = \mathbf{S}\big(\Gamma(\mathcal{E}|_U) \otimes_{\Gamma(\mathcal{A}|_U)} \Gamma(\mathcal{F})|_U)\big) \\ &= \mathcal{E}|_U \otimes_{\mathcal{A}|_U} \mathcal{F}|_U,\end{aligned} \tag{5.26}$$

which thus proves (5.25). ∎

Now, as a straightforward application of (5.25), one concludes that;

(5.27) *the tensor product $\mathcal{E} \otimes_{\mathcal{A}} \mathcal{F}$ of two vector sheaves $\mathcal{E}$ and $\mathcal{F}$ on a topological space $X$, with $\operatorname{rk} \mathcal{E} = n$ and $\operatorname{rk} \mathcal{F} = m$, is a vector sheaf on $X$, as well, with $\operatorname{rk}(\mathcal{E} \otimes_{\mathcal{A}} \mathcal{F}) = n \cdot m$.* (An analogous statement is valid, *for any finite number of factors*).

In particular, *the tensor product of two* (or even *finite many*) *line sheaves on $X$ is* again *a line sheaf on $X$.*

Indeed, by using, for instance, a similar argument to that we applied for the justification of (4.9), one can find, first, an open covering of $X$, common for both sheaves, satisfying further (4.7) for $n$ and $m$, respectively. (An analogous argument applies, of course, in the case of *any finite number of factors*). Thus, (5.27) is now an immediate consequence of (5.25) and (5.18). ∎

**Scholium 5.1.**– An analogous argument to that employed in (5.5.1) can be used, as well, to yield, more generally, the notion of the *tensor product of any two A-presheaves* on a given topological space $X$ (viz. not necessarily of the type considered by (5.5.1)); hence, of *any finite number* of such, as well (see also Definition 1.6). However, we shall not pursue further this notion here, confining ourselves, instead, to deal with it at the pertinent places of our discussion, whenever the latter notion appears.

## 6. The functor $\mathcal{H}om_{\mathcal{A}}$

The functor we consider in this section is, in effect, the most important among those we employ in our discussion, in what, in particular, concerns the applications we deal with in Part Two. It is actually a *bifunctor,* viz. a functor of two variables, both from $\mathcal{A}$-$\mathcal{M}od_X$ (cf. (2.4)), i.e., from the *category of $\mathcal{A}$-modules on a topological space $X$*, having the same category as range (target), as well. In this context, the *sheaf of coefficients $\mathcal{A}$* of the $\mathcal{A}$-modules involved is, as usual, a *unital* and *commutative* **C**-*algebra sheaf* on $X$ (see (5.1)).

Thus, to start with, suppose that we are given two *$\mathcal{A}$-modules* $\mathcal{E}$ and $\mathcal{F}$ on $X$, as in (5.3). Now, for any *open set* $U \subseteq X$,

(6.1) let

(6.1.1) $$\mathrm{Hom}_{\mathcal{A}|_U}(\mathcal{E}|_U, \mathcal{F}|_U)$$

be the *set of $\mathcal{A}|_U$-morphisms of the $\mathcal{A}|_U$-modules concerned.* Indeed,

(6.1.2) the set (6.1.1) *is an $\mathcal{A}(U)$-module;* hence, in particular, a **C**-*vector space.*

That is, for any $\alpha \in \mathcal{A}(U)$ and $\phi \equiv (\phi_V) \in \mathrm{Hom}_{\mathcal{A}|_U}(\mathcal{E}|_U, \mathcal{F}|_U)$, where $V$ runs over the open subsets of $U$ (see Chapt. I; (6.3) and (13.23)), one sets

(6.2) $$(\alpha \cdot \phi)_V(t)(x) := \alpha(x) \cdot \phi_V(t)(x),$$

for any open $V \subseteq U$ and $t \in (\mathcal{E}|_U)(V) = \mathcal{E}(V)$ (cf. I; (11.40)), with $x \in V$. So one may write (6.2) in the form;

(6.2′) $$(\alpha \cdot \phi)_V := (\alpha|_V) \cdot \phi_V \equiv \alpha \cdot \phi_V,$$

*for every open* $V \subseteq U$, which thus provides the $\mathcal{A}(U)$*-module structure of (6.1.1)*. Therefore, according to the very definitions, one obtains

$$\mathrm{Hom}_{\mathcal{A}|_U}(\mathcal{E}|_U, \mathcal{F}|_U) \subseteq \prod_{U \supseteq V, \text{open}} \mathrm{Hom}_{\mathcal{A}(V)}(\mathcal{E}(V), \mathcal{F}(V)) \tag{6.3}$$

(see also Chapt. I; (6.7), in conjunction with (13.23)).

On the other hand, it is readily verified that;

(6.4) given the $\mathcal{A}$-modules $\mathcal{E}$ and $\mathcal{F}$ on $X$, as before, the correspondence

$$U \longmapsto \mathrm{Hom}_{\mathcal{A}|_U}(\mathcal{E}|_U, \mathcal{F}|_U), \tag{6.4.1}$$

where $U$ runs over the open subsets of $X$, along with the obvious *restriction maps* yield a *complete presheaf of* $\mathcal{A}$*-modules* on $X$ (cf. also (6.1.2)). So (6.4.1) defines a *complete* $\Gamma(\mathcal{A})$*-presheaf* on $X$ (cf. Definition 1.6 and (6.1.2)), with $\Gamma(\mathcal{A})$ given by (5.2).

*Warning!* Concerning the correspondence (6.4.1), we still note that we could not use, instead, as target of the said map, for instance, the $\mathcal{A}(U)$*-module*

$$\mathrm{Hom}_{\mathcal{A}(U)}(\mathcal{E}(U), \mathcal{F}(U)), \tag{6.5}$$

since then *one wouldn't have e.g. restriction maps*(!).

So we come now to the following basic notion.

**Definition 6.1.** Let $\mathcal{E}$ and $\mathcal{F}$ be two given $\mathcal{A}$-modules on a topological space $X$. Then, we denote by

$$\mathcal{H}om_{\mathcal{A}}(\mathcal{E}, \mathcal{F}) \tag{6.6}$$

the sheaf on $X$ generated by the (complete) presheaf defined by (6.4.1). We call it the **sheaf of germs of $\mathcal{A}$-morphisms of $\mathcal{E}$ in $\mathcal{F}$.**

Indeed (cf. Proposition 1.1, along with Chapt. I; Theorem 9.1), one concludes that:

(6.7) for any given $\mathcal{A}$-modules $\mathcal{E}$,$\mathcal{F}$ on $X$, *the sheaf* $\mathcal{H}om_{\mathcal{A}}(\mathcal{E}, \mathcal{F})$, as above, *is* still *an* $\mathcal{A}$*-module* on $X$.

(But, see also Lemma 6.1 below). Thus, based now on Chapt. I; Proposition 11.1, one obtains (see also (6.3)), for every *open set* $U \subseteq X$,

$$\mathcal{H}om_{\mathcal{A}}(\mathcal{E},\mathcal{F})(U) = \mathrm{Hom}_{\mathcal{A}|_U}(\mathcal{E}|_U,\mathcal{F}|_U), \tag{6.8}$$

within an $\mathcal{A}(U)$*-isomorphism* of the $\mathcal{A}(U)$-modules concerned (cf. (6.1.2) and (6.7)). In particular, one has

$$\mathcal{H}om_{\mathcal{A}}(\mathcal{E},\mathcal{F})(X) = \mathrm{Hom}_{\mathcal{A}}(\mathcal{E},\mathcal{F}), \tag{6.9}$$

within an $\mathcal{A}(X)$*-isomorphism* of the $\mathcal{A}(X)$-modules considered.

In this regard, we further note that, *in general, one has;*

$$\mathcal{H}om_{\mathcal{A}}(\mathcal{E},\mathcal{F})_x \neq \mathcal{H}om_{\mathcal{A}_x}(\mathcal{E}_x,\mathcal{F}_x), \tag{6.10}$$

*with* $x \in X$. See e.g. R. Godement [1: p.115], or even G. E. Bredon [1: p.15]; cf. also S. Iitaka [1: p.39, Remark and Example 1.18].

Furthermore, for any given $\mathcal{A}$*-module* $\mathcal{E}$ on $X$, *the functor* (cf. also (6.7))

$$\mathcal{H}om_{\mathcal{A}}(\mathcal{E},\cdot) : \mathcal{A}\text{-}\mathcal{M}od_X \longrightarrow \mathcal{A}\text{-}\mathcal{M}od_X \tag{6.11}$$

such that

$$\mathcal{H}om_{\mathcal{A}}(\mathcal{E},\cdot)(\mathcal{F}) := \mathcal{H}om_{\mathcal{A}}(\mathcal{E},\mathcal{F}), \tag{6.12}$$

for any $\mathcal{A}$-module $\mathcal{F}$ on $X$, *is covariant.* Analogously, one defines *the functor*

$$\mathcal{H}om_{\mathcal{A}}(\cdot\,,\mathcal{F}) : \mathcal{A}\text{-}\mathcal{M}od_X \longrightarrow \mathcal{A}\text{-}\mathcal{M}od_X \tag{6.13}$$

with

$$\mathcal{H}om_{\mathcal{A}}(\cdot\,,\mathcal{F})(\mathcal{E}) := \mathcal{H}om_{\mathcal{A}}(\mathcal{E},\mathcal{F}), \tag{6.14}$$

for any $\mathcal{A}$-module $\mathcal{E}$ on $X$, which *is* thus *contravariant.* Thus, as we say,

(6.15) *the (bi)functor* $\mathcal{H}om_{\mathcal{A}} \equiv \mathcal{H}om_{\mathcal{A}}(\cdot\,,\cdot)$ *is contravariant in the first variable and covariant in the second* variable.

Of course, if $\alpha : \mathcal{F} \longrightarrow \mathcal{F}'$ is a given $\mathcal{A}$*-morphism,* one defines

$$\mathcal{H}om_{\mathcal{A}}(\mathcal{E},\cdot)(\alpha) : \mathcal{H}om_{\mathcal{A}}(\mathcal{E},\mathcal{F}) \longrightarrow \mathcal{H}om_{\mathcal{A}}(\mathcal{E},\mathcal{F}'), \tag{6.16}$$

by an obvious *composition of the respective morphisms.* An analogous remark holds true for the (contravariant) functor (6.13) (of course, *contravariantly!*). On the other hand, one concludes that:

(6.17) *the* (covariant) *functor* $\mathcal{H}om_{\mathcal{A}}(\mathcal{E}, \cdot)$ (see (6.11), (6.12)) *is left exact;* that is, *for any exact* $\mathcal{A}$*-sequence*

$$(6.17.1) \qquad 0 \longrightarrow \mathcal{S}' \longrightarrow \mathcal{S} \longrightarrow \mathcal{S}''$$

*the* $\mathcal{A}$*-sequence*

$$(6.17.2)\; 0 \longrightarrow \mathcal{H}om_{\mathcal{A}}(\mathcal{E}, \mathcal{S}') \longrightarrow \mathcal{H}om_{\mathcal{A}}(\mathcal{E}, \mathcal{S}) \longrightarrow \mathcal{H}om_{\mathcal{A}}(\mathcal{E}, \mathcal{S}'')$$

*is exact.*

In this context, see e.g. R. Kultze [1: p.49, Satz 9.1]. ∎ *Warning!* By contrast with (6.17.1), when considering the (contravariant) functor $\mathcal{H}om_{\mathcal{A}}(\cdot, \mathcal{F})$, the exact $\mathcal{A}$-sequence (6.17.1) *does not yield, in general, an exact* $\mathcal{A}$*-sequence of the form*

$$(6.17') \qquad 0 \longrightarrow \mathcal{H}om_{\mathcal{A}}(\mathcal{S}'', \mathcal{F}) \longrightarrow \mathcal{H}om_{\mathcal{A}}(\mathcal{S}, \mathcal{F}) \longrightarrow \mathcal{H}om_{\mathcal{A}}(\mathcal{S}', \mathcal{F}).$$

However, *the following sequence of* $\mathcal{A}_x$*-modules is exact, for any* $x \in X$;

$$(6.17'') \quad 0 \longrightarrow \mathrm{Hom}_{\mathcal{A}_x}(\mathcal{S}''_x, \mathcal{F}_x) \longrightarrow \mathrm{Hom}_{\mathcal{A}_x}(\mathcal{S}_x, \mathcal{F}_x) \longrightarrow \mathrm{Hom}_{\mathcal{A}_x}(\mathcal{S}'_x, \mathcal{F}_x).$$

(See e.g. H. Grauert – R. Remmert [2: p.232]; yet, cf. N. Jacobson [1: p.99, Theorem 3.1′]).

Furthermore, one obtains the following $\mathcal{A}$*-isomorphism*

$$(6.18) \qquad \mathcal{H}om_{\mathcal{A}}(\mathcal{A}, \mathcal{E}) = \mathcal{E},$$

*for any* $\mathcal{A}$*-module* $\mathcal{E}$: Indeed, we prove that, *for any open* $U \subseteq X$, *one has* (cf. also (6.8))

$$(6.18') \qquad \mathcal{H}om_{\mathcal{A}}(\mathcal{A}, \mathcal{E})(U) = \mathrm{Hom}_{\mathcal{A}|_U}(\mathcal{A}|_U, \mathcal{E}|_U) = \mathcal{E}(U),$$

within an $\mathcal{A}(U)$*-isomorphism;* that is, for any $\phi \in \mathrm{Hom}_{\mathcal{A}|_U}(\mathcal{A}|_U, \mathcal{E}|_U)$, one defines a section $s_\phi \in \mathcal{E}(U)$, such that (see also (1.22))

$$(6.19) \qquad s_\phi := \phi \circ (1|_U),$$

so that one has (see also I; (1.14))

$$(6.19') \qquad s_\phi(x) = \phi(1_x) \equiv \phi_x(1_x), \; x \in U,$$

which proves our claim for $s_\phi$. Yet, the same correspondence yields further the desired *bijection* between the above $\mathcal{A}(U)$-modules, *being* also *an* $\mathcal{A}(U)$*-morphism,* due to the hypothesis for $\phi$ (cf. (6.2)). ∎

On the other hand, one has the following $\mathcal{A}$-*isomorphisms* of the $\mathcal{A}$-*modules* involved:

$$\mathcal{H}om_{\mathcal{A}}(\mathcal{E}, \mathcal{F}_1 \oplus \mathcal{F}_2) = \mathcal{H}om_{\mathcal{A}}(\mathcal{E}, \mathcal{F}_1) \oplus \mathcal{H}om_{\mathcal{A}}(\mathcal{E}, \mathcal{F}_2) \tag{6.20}$$

and

$$\mathcal{H}om_{\mathcal{A}}(\mathcal{E}_1 \oplus \mathcal{E}_2, \mathcal{F}) = \mathcal{H}om_{\mathcal{A}}(\mathcal{E}_1, \mathcal{F}) \oplus \mathcal{H}om_{\mathcal{A}}(\mathcal{E}_2, \mathcal{F}). \tag{6.21}$$

Of course analogous relations hold true, *for any finite number of factors* of the respective Whitney sums, as before: The previous assertions follow immediately by the very definitions and the corresponding formulae of the classical theory of modules (cf. e.g. N. Bourbaki [2: Chap. 2; p.12 and p.13, Corollaire 1]).

In particular, one gets the following $\mathcal{A}$-*isomorphism* (see also (6.18), (6.21) and (3.7));

$$\mathcal{H}om_{\mathcal{A}}(\mathcal{A}^n, \mathcal{E}) = \mathcal{E}^n, \; n \in \mathbf{N}, \tag{6.22}$$

for any given $\mathcal{A}$-*module* $\mathcal{E}$on $X$.

Moreover, we shall presently employ the following fact; namely, one concludes the next $\mathcal{A}|_U$-*isomorphism* of the $\mathcal{A}|_U$-modules involved, *for any open set* $U \subseteq X$ (see also (6.7))

$$\mathcal{H}om_{\mathcal{A}}(\mathcal{E}, \mathcal{F})|_U = \mathcal{H}om_{\mathcal{A}|_U}(\mathcal{E}|_U, \mathcal{F}|_U). \tag{6.23}$$

Indeed, for every open set $V \subseteq U$, one obtains (cf. (6.8) and I; (11.40));

$$\begin{aligned} (\mathcal{H}om_{\mathcal{A}}(\mathcal{E}, \mathcal{F})|_U)(V) &= \mathcal{H}om_{\mathcal{A}}(\mathcal{E}, \mathcal{F})(V) = \mathrm{Hom}_{\mathcal{A}|_V}(\mathcal{E}|_V, \mathcal{F}|_V) \\ &= \mathcal{H}om_{\mathcal{A}|_U}(\mathcal{E}|_U, \mathcal{F}|_U)(V), \end{aligned} \tag{6.24}$$

which proves (6.23). Yet, if, in particular, $\mathcal{E}$ is a *vector sheaf* of *rank* $n$ and $U$ a *local gauge* of it, one has the following $\mathcal{A}_U$-*isomorphisms;*

$$\mathcal{H}om_{\mathcal{A}}(\mathcal{E}, \mathcal{F})|_U = \mathcal{F}^n|_U = (\mathcal{F}|_U)^n. \; ▮ \tag{6.24'}$$

We come now to the following useful result for the sequel, supplementing (6.7). That is, one has the next.

**Lemma 6.1.** *Let* $\mathcal{E}$, $\mathcal{F}$ *be two given vector sheaves on a topological space* $X$. *Then, the* $\mathcal{A}$-*module*

$$\mathcal{H}om_{\mathcal{A}}(\mathcal{E}, \mathcal{F}) \tag{6.25}$$

*is a vector sheaf on* $X$ *too, such that*

$$\operatorname{rk}(\mathcal{H}om_{\mathcal{A}}(\mathcal{E},\mathcal{F})) = (\operatorname{rk}\mathcal{E})\cdot(\operatorname{rk}\mathcal{F}). \tag{6.26}$$

**Proof.** Based on the hypothesis for $\mathcal{E}$, $\mathcal{F}$, one may consider an open covering $\mathcal{U} = (U_\alpha)_{\alpha\in I}$ of $X$, satisfying (4.7), with $n = \operatorname{rk}\mathcal{E}$ and $m = \operatorname{rk}\mathcal{F}$, respectively (viz. a *common local frame of* $\mathcal{E}$, $\mathcal{F}$; cf. Definition 4.2). So one has (see also (6.23) and (6.22));

$$\begin{aligned}\mathcal{H}om_{\mathcal{A}}(\mathcal{E},\mathcal{F})|_{U_\alpha} &= \mathcal{H}om_{\mathcal{A}|_{U_\alpha}}(\mathcal{E}|_{U_\alpha},\mathcal{F}|_{U_\alpha}) = \mathcal{H}om_{\mathcal{A}|_{U_\alpha}}(\mathcal{A}^n|_{U_\alpha},\mathcal{A}^m|_{U_\alpha})\\ &= \mathcal{H}om_{\mathcal{A}}(\mathcal{A}^n,\mathcal{A}^m)|_{U_\alpha} = \mathcal{A}^{n\cdot m}|_{U_\alpha},\ \alpha\in I,\end{aligned} \tag{6.27}$$

within $\mathcal{A}|_{U_\alpha}$*-isomorphisms* ($\alpha \in I$), which proves the assertion for (6.25), and with it (6.26), as well. ∎

Now, terminating this Section and with it the present Chapter, we still introduce the following terminology, which will also be of use later on. So, given any $\mathcal{A}$*-module* $\mathcal{E}$ on $X$, one sets;

$$\mathcal{E}nd_{\mathcal{A}}\mathcal{E} \equiv \mathcal{E}nd\,\mathcal{E} := \mathcal{H}om_{\mathcal{A}}(\mathcal{E},\mathcal{E}), \tag{6.28}$$

viz. the *sheaf of germs of* $\mathcal{A}$*-endomorphisms of* $\mathcal{E}$ (that is, $\mathcal{A}$*-morphisms of* $\mathcal{E}$ *into itself;* cf. Definition 6.1). Thus (cf. (6.7)), another $\mathcal{A}$*-module* on $X$.

On the other hand, taking in the previous $\mathcal{A}$-module the usual *composition of morphisms* (in fact, $\mathcal{A}$-endomorphisms of $\mathcal{E}$), as *multiplication,* one concludes that:

(6.29) $\mathcal{E}nd\,\mathcal{E}$ is an $\mathcal{A}$*-algebra (sheaf)* on $X$.

Indeed, the only thing that we have to check here (cf. also the next Definition 6.2), is the *continuity of the* (ring) *multiplication* in $\mathcal{E}nd\,\mathcal{E}$; however, based on (1.67) (or yet (1.75)), it suffices to prove that, *for any given* $\phi,\psi \in (\mathcal{E}nd\,\mathcal{E})(U)$, one obtains that $\psi\circ\phi \in (\mathcal{E}nd\,\mathcal{E})(U)$, as well, *for every open* $U \subseteq X$. In fact, this is immediate by the very definitions (cf. Chapt. I, Definition 1.2, as well as, (6.8) above, and even (6.3), for $\mathcal{E} = \mathcal{F}$).

In this context, we simply give here, for convenience, the formal definition of the notion employed in (6.29), which will further be used in the ensuing discussion, as well. (So see Chapter IV). Thus, we have the next.

**Definition 6.2.** Given a **C**-algebra sheaf $\mathcal{A}$ on a topological space $X$, an **$\mathcal{A}$-algebra sheaf** (or simply an **$\mathcal{A}$-algebra**) on $X$, is an $\mathcal{A}$-module $\mathcal{E}$ on $X$, which is also a sheaf of rings, such that the stalks of $\mathcal{E}$, $\mathcal{E}_x$ ($x \in X$), are $\mathcal{A}_x$-algebras.

Indeed, the above sort of sheaves are going to be considered in the sequel, in particular, under the form of *matrix algebra sheaves* (see thus Chapter IV); this latter type of algebra sheaves will often be employed in Part Two, of our discussion, where we also refer for this sort of applications.

Finally, specializing (6.9), for $\mathcal{E} = \mathcal{F}$, one sets, in particular, for any *$\mathcal{A}$-module* $\mathcal{E}$ on $X$,

$$\text{End}\,\mathcal{E} := (\mathcal{E}nd\,\mathcal{E})(X) = \text{Hom}_{\mathcal{A}}(\mathcal{E}, \mathcal{E}), \tag{6.30}$$

where the last relation is valid, of course, within an *$\mathcal{A}(X)$-isomorphism.* In fact, one infers that:

(6.31) for every $\mathcal{A}$-module $\mathcal{E}$ on $X$, $\text{End}\,\mathcal{E}$ *is an $\mathcal{A}(X)$-algebra;* hence, in particular, a $\mathbf{C}$-*algebra.*

Yet, we consider the important particular case of (6.28), where $\mathcal{E} = \mathcal{L}$, a *line sheaf* on $X$, that will often be applied in the sequel. So one has the following.

**Lemma 6.2.** *Let $\mathcal{L}$ be a line sheaf on a given topological space $X$. Then, one obtains* (see (6.28), for $\mathcal{E} = \mathcal{L}$);

$$\mathcal{E}nd\,\mathcal{L} = \mathcal{A}, \tag{6.32}$$

*within an $\mathcal{A}$-isomorphism. In particular, one has*

$$\text{End}\,\mathcal{L} = \mathcal{A}(X), \tag{6.33}$$

*within an $\mathcal{A}(X)$-isomorphism.*

**Proof.** Based on I; (13.19), it suffices to manufacture first a *section preserving map,* say,

$$\phi : \Gamma(\mathcal{A}) \longrightarrow \Gamma(\mathcal{E}nd\,\mathcal{L}). \tag{6.34}$$

In this context, we still note that the bijectiveness of the last map is, by I; (13.19), tantamount to that one of the analogous map between $\mathcal{A}$ and $\mathcal{E}nd\,\mathcal{L}$, according to the very functorial properties of $\Gamma$, in fact, of its left adjoint, the sheafification functor $\mathbf{S}$; thus, cf., for instance, I; (13.20) and (13.21).

So, *for any open set* $U \subseteq X$, one defines (see also (6.3), (6.8) and I; (6.7))

$$\begin{aligned} &\phi_U : \mathcal{A}(U) \longrightarrow (\mathcal{E}nd\,\mathcal{L})(U) : \alpha \longmapsto \phi_U(\alpha) \\ &= (\phi_U(\alpha)|_V)_{V \subseteq U,\,\text{open}} \in (\mathcal{E}nd\,\mathcal{L})(U) \\ &= \mathrm{Hom}_{\mathcal{A}|_U}(\mathcal{L}|_U, \mathcal{L}|_U) \subseteq \prod_{V \subseteq U,\,\text{open}} \mathrm{Hom}_{\mathcal{A}(V)}(\mathcal{L}(V), \mathcal{L}(V)), \end{aligned} \tag{6.35}$$

such that

$$(\phi_U(\alpha)|_V)(s) := (\alpha|_V) \cdot s = \alpha \cdot s, \tag{6.36}$$

for any $s \in (\mathcal{L}|_U)(V)$ and for every open $V \subseteq U$. Of course, one has

$$(\alpha|_V) \cdot s \in \mathcal{L}(V), \tag{6.37}$$

*for any* $\alpha \in \mathcal{A}(U)$, with $V$ *open in* $U$, *and* $s \in \mathcal{L}(V)$, $\mathcal{L}$ being an $\mathcal{A}$-module (cf. also (1.67) and (1.75)). Thus, (6.36) *is well defined.*

Now, we further prove that (6.34) *is a bijection:* So, since $\Gamma(\mathcal{A})$ and $\Gamma(\mathcal{E}nd\,\mathcal{L})$ are *complete presheaves* (cf. Chapt. I, Lemma 11.1), the assertion is reduced to the proof that $\phi$ *is a* $B$*-isomorphism* (see Chapt. I; Theorem 12.1, (12.2), as well as, Scholium 12.1); thus, we further apply the fact that $\mathcal{L}$ is, by hypothesis, a *line sheaf* on $X$. Namely, we prove that *the map* $\phi_U$ *in (6.35) is a bijection, when* $U$ *runs over a local frame of* $\mathcal{L}$. Therefore (cf. also Definition 4.2 and (4.7), for $n = 1$, as well as, (6.18)), one obtains

$$\begin{aligned} \phi_U &\in \mathrm{Hom}_{\mathcal{A}(U)}(\mathcal{A}(U), (\mathcal{E}nd\,\mathcal{L})(U)) \\ &= \mathrm{Hom}_{\mathcal{A}(U)}(\mathcal{A}(U), \mathrm{Hom}_{\mathcal{A}|_U}(\mathcal{L}|_U, \mathcal{L}|_U)) \\ &= \mathrm{Hom}_{\mathcal{A}(U)}(\mathcal{A}(U), \mathrm{Hom}_{\mathcal{A}|_U}(\mathcal{A}|_U, \mathcal{A}|_U)) \qquad (6.38) \\ &= \mathrm{Hom}_{\mathcal{A}(U)}(\mathcal{A}(U), \mathcal{H}om_{\mathcal{A}}(\mathcal{A}, \mathcal{A})(U)) \\ &= \mathrm{Hom}_{\mathcal{A}(U)}(\mathcal{A}(U), \mathcal{A}(U)) \cong \mathcal{A}(U). \end{aligned}$$

So, as follows from (6.37), $\phi_U$ is, in effect, the (left) *regular representation* of the *unital* (commutative) **C**-*algebra* $\mathcal{A}(U)$, yielding thus a *bijection* with $\mathcal{A}(U)$ (cf. also e.g. (6.19)), which terminates the proof, (6.33) being, of course, an immediate consequence of (6.32) (yet, cf. (6.30), as well as, Chapt. I; (13.19) and (6.10)). ∎ Of course, (6.32) is still an immediate consequence of (6.26). (see also the proof of Lemma 6.1)

Now, by anticipating our discussion in Part Two, we also note at this place that an important notion connected with (6.30) is the *group of units* of the latter algebra (cf. (6.31)), resp., the *sheaf of units* of (6.28) (see also (6.29)). This, in particular, when $\mathcal{E}$ is a *vector sheaf* on $X$ (cf. also Lemma 6.1 above); thus, one is then naturally led to employ (locally) the notion of what we call, by analogy with the classical case, a *general linear group sheaf* on $X$, with respect to the given (**C**-algebra sheaf) $\mathcal{A}$, as above. However, all these and related material are deferred for Chapter IV.

**6.(a). The functor** $\mathrm{Hom}_{\mathcal{A}}$.– In the sequel we shall also need to consider the functor in the title of this Subsection, being naturally associated with the previous functor $\mathcal{H}om_{\mathcal{A}}$ (cf. (6.15)). It is also a *bifunctor,* as the latter one, that is (cf. (2.4) and (1.67)),

$$\mathrm{Hom}_{\mathcal{A}} : \mathcal{A}\text{-}\mathcal{M}od_X \times \mathcal{A}\text{-}\mathcal{M}od_X \longrightarrow \mathcal{A}(X)\text{-}\mathcal{M}od, \qquad (6.39)$$

where the range of (6.39) is the *category of* $\mathcal{A}(X)$*-modules,* along with the corresponding $\mathcal{A}(X)$*-morphisms.* Thus, denoting by

$$\Gamma_X : \mathcal{A}\text{-}\mathcal{M}od_X \longrightarrow \mathcal{A}(X)\text{-}\mathcal{M}od \qquad (6.40)$$

the *global section functor,* associated with the section functor $\Gamma$ (see (1.67) above and Chapt. I; (13.1)), one sets;

$$\Gamma_X(\mathcal{E}) := \mathcal{E}(X) \equiv \Gamma(X, \mathcal{E}), \qquad (6.41)$$

for any $\mathcal{A}$*-module* $\mathcal{E}$ on $X$.

So (6.39) is now given by the relation

$$\mathrm{Hom}_{\mathcal{A}} := \Gamma_X \circ \mathcal{H}om_{\mathcal{A}}, \tag{6.42}$$

such that one has

$$\mathrm{Hom}_{\mathcal{A}}(\mathcal{E},\mathcal{F}) = \Gamma_X(\mathcal{H}om_{\mathcal{A}}(\mathcal{E},\mathcal{F})) = \mathcal{H}om_{\mathcal{A}}(\mathcal{E},\mathcal{F})(X) = \mathrm{Hom}_{\mathcal{A}}(\mathcal{E},\mathcal{F}), \tag{6.43}$$

for any two $\mathcal{A}$-modules $\mathcal{E}$, $\mathcal{F}$, the last relation in (6.43) being valid up to an $\mathcal{A}(X)$*-isomorphism* (see (6.9)). The *(bi)functor* (6.39) will also appear several times in the sequel.

Now, by analogy with the (partial) functors (6.11) and (6.13), one defines the *covariant functor*

$$\mathrm{Hom}_{\mathcal{A}}(\mathcal{E},\cdot) : \mathcal{A}\text{-}\mathcal{M}od_X \longrightarrow \mathcal{A}(X)\text{-}\mathcal{M}od, \tag{6.44}$$

for any given $\mathcal{A}$-module $\mathcal{E}$ on $X$, *according to (6.43)*. (A first application of (6.43) is given below, Chapt. III; (1.49)). Similarly, one defines, by (6.43), the *contravariant functor*

$$\mathrm{Hom}_{\mathcal{A}}(\cdot,\mathcal{F}) : \mathcal{A}\text{-}\mathcal{M}od_X \longrightarrow \mathcal{A}(X)\text{-}\mathcal{M}od, \tag{6.45}$$

for every $\mathcal{A}$-module $\mathcal{F}$ on $X$.

In this regard, it is easy to prove, based on the very definitions, that;

(6.46) *the functors (6.44) and (6.45),* as above, *are both left exact.*

But, see also below, Chapt. III; (1.51).

On the other hand, as an application of (6.42) (or even (6.43)), we still have the following $\mathcal{A}(X)$*-isomorphisms* of the $\mathcal{A}(X)$-modules involved;

$$\mathrm{Hom}_{\mathcal{A}}(\mathcal{A}^n,\mathcal{E}) = \mathcal{E}^n(X) = \mathcal{E}(X)^n, \tag{6.47}$$

for any given $\mathcal{A}$*-module* $\mathcal{E}$ on $X$, and $n \in \mathbf{N}$ (cf. also (6.22) and (3.14)). In particular, *for* $n = 1$, we obtain

$$\mathrm{Hom}_{\mathcal{A}}(\mathcal{A},\mathcal{E}) = \mathcal{E}(X), \tag{6.48}$$

up to an $\mathcal{A}(X)$-*isomorphism.* (Indeed, one has here a (natural) *isomorphism of* the *functors* involved, viz. of $\mathrm{Hom}_{\mathcal{A}}(\mathcal{A}, \cdot)$ and $\Gamma(X, \cdot) \equiv \Gamma_X(\cdot)$ between the obvious categories; see e.g. (6.40)).

# CHAPTER III

# Sheaf Cohomology

" To set up a cohomology theory requires a lot of work, but ... it is well worth the effort."

R. Hartshorne in "*Algebraic Geometry*" (Springer - Verlag, 1977) p.57.

"... the structure underlying an intrinsic approach to physics is "essentially" de Rham-cohomology."

C.v.Westenholz in "*Differential Forms in Mathematical Physics*" (North - Holland, 1981). p.321.

In this Chapter we present the rudiments of *sheaf cohomology theory,* as the latter will be employed in the ensuing discussion. So, although the theory can be formulated, at least, within an arbitrary *abelian category* (cf., for instance, A. Grothendieck [2], R. Godement [1], B. Iversen [1]), for convenience, we confine ourselves here, in particular, to the *category of $\mathcal{A}$-modules* on an arbitrary *topological space* $X$ (cf. II; (2.4)), with $\mathcal{A}$ being, as usual, a given $\mathbf{C}$-*algebra sheaf* on $X$ (see Chapt. II; Definitions 1.4, 1.5). This will be actually the case in the sequel, while, in effect, locally free $\mathcal{A}$-modules (of finite rank, viz. *vector sheaves,* see Chapt. II; Definition 4.3) will steadily be employed throughout the ensuing chapters. So we start with the most basic definitions and related fundamental results.

# 1. Preliminaries. Fundamental concepts

As already said, we work in the sequel within the *category of $\mathcal{A}$-modules,* and related *$\mathcal{A}$-morphisms,* on an arbitrary *topological space* $X$,

$$\mathcal{A}\text{-}\mathcal{M}od_X \tag{1.1}$$

(cf. II; (2.4)). Here $\mathcal{A}$ stands for a *sheaf of* $\mathbf{C}$*-algebras* on $X$, which are further assumed to be *commutative* and *unital;* hence, the corresponding *stalks* of $\mathcal{A}$ at the various points of $X$, $\mathcal{A}_x$ $(x \in X)$, have, as $\mathbf{C}$-algebras, the said properties (see Chapt. II, Definitions 1.3′, 1.4 along with (1.55)).

We start with the following basic notion for all that follows: Namely, by an (open) *$\mathcal{A}$-complex* (or even a *differential $\mathcal{A}$-sequence*) on $X$, one means a sequence of $\mathcal{A}$-modules and $\mathcal{A}$-morphisms (cf., for instance, Chapt. II; Definition 2.1)

$$\mathcal{E}^{\cdot} \equiv (\mathcal{E}^{\cdot}, d) := \{(\mathcal{E}^n, d^n)\}_{n\in\mathbf{Z}}\ , \tag{1.2}$$

that is,

$$\cdots \longrightarrow \mathcal{E}^{n-1} \xrightarrow{d^{n-1}} \mathcal{E}^n \xrightarrow{d^n} \mathcal{E}^{n+1} \longrightarrow \cdots \tag{1.3}$$

such that one has

$$d^n \circ d^{n-1} = 0,\ n \in \mathbf{Z}. \tag{1.4}$$

Yet, by referring to the map (in fact, *sequence of $\mathcal{A}$-morphisms*)

$$d \equiv (d^n)_{n\in\mathbf{Z}} \tag{1.5}$$

in (1.2), we also write instead of (1.4), for short,

$$d^2 \equiv d \circ d = 0. \tag{1.6}$$

(*Warning!* The last "$d^2$" should not be confused with the corresponding $\mathcal{A}$-morphism $d^2 : \mathcal{E}^2 \longrightarrow \mathcal{E}^3$ in the $\mathcal{A}$-sequence (1.3), as above).

So, concerning (1.4), one assumes, *equivalently,* that;

$$\operatorname{im} d^{n-1} \subseteq \ker d^n,\ n \in \mathbf{Z} \tag{1.7}$$

(see also Chapt. II; (2.5) and (2.12)).

The preceding *$\mathcal{A}$-morphisms* $d^n$ $(n \in \mathbf{Z})$ are also called *differentials,* or even *coboundary maps (operators)* of the $\mathcal{A}$-complex concerned. (The motivation of the terminology applied stems, of course, from Differential Analysis and/or Geometry; cf., for instance, Chapter VI in the sequel). Furthermore, the term *cochain complex of $\mathcal{A}$-modules,* or just *cochain $\mathcal{A}$-complex,* is still employed for an $\mathcal{A}$-complex, as before. Yet, by referring to the map $d$ in (1.2) (cf. (1.5)), we also speak of the *differential* or *coboundary map* (or even *coboundary operator*) of the given $\mathcal{A}$-complex $\mathcal{E}^{\cdot}$.

Now, in view of (1.7), the preceding $\mathcal{A}$-sequence (1.3) *is not, in general, exact* at every stage (cf. Chapt. II; Definition 2.1). So, *for every $n \in \mathbf{Z}$,* one can consider the respective *quotient $\mathcal{A}$-module* (see also Chapt. II, Lemma 2.1)

$$\ker d^n / \operatorname{im} d^{n-1}, \tag{1.8}$$

measuring thus the deviation of being the given $\mathcal{A}$-sequence (1.3) exact. For reasons that will become soon apparent, we call (1.8) the *n-th* (or even *n-dimensional*) *cohomology $\mathcal{A}$-module* of the given $\mathcal{A}$-complex (1.2), so that we set;

$$h^n(\mathcal{E}^{\cdot}) := \ker d^n / \operatorname{im} d^{n-1}, \; n \in \mathbf{Z}. \tag{1.9}$$

On the other hand, for convenience, we still apply to the preceding notions the following terminology, by again anticipating its motivation (cf. the above comments for the given $\mathcal{A}$-morphisms, *"differentials"*, $d^n$, $n \in \mathbf{Z}$).

Thus, *for any $n \in \mathbf{Z}$,* one sets;

$$\ker d^n \equiv Z^n(\mathcal{E}^{\cdot}) \subseteq \mathcal{E}^n \; (\textit{sub-}\mathcal{A}\textit{-module of } \mathcal{E}^n), \tag{1.10}$$

as well as,

$$\operatorname{im} d^{n-1} \equiv B^n(\mathcal{E}^{\cdot}) \subseteq \mathcal{E}^n \; (\textit{sub-}\mathcal{A}\textit{-module of } \mathcal{E}^n) \tag{1.11}$$

(see also (1.3)), calling the elements of (1.10) the $n$- (or even *$n$-dimensional*) *cocycles of $\mathcal{E}^{\cdot}$* and $Z^n(\mathcal{E}^{\cdot})$ the *$\mathcal{A}$-module of $n$-cocycles of the* (cochain) *$\mathcal{A}$-complex $\mathcal{E}^{\cdot}$*. Moreover, the elements of (1.11) are called the $n$- (or *$n$-dimensional*) *coboundaries of $\mathcal{E}^{\cdot}$* and $B^n(\mathcal{E}^{\cdot})$ the *$\mathcal{A}$-module of $n$-coboundaries of $\mathcal{E}^{\cdot}$*. Thus, the quotient $\mathcal{A}$-module (1.9) can now be put in the form;

$$h^n(\mathcal{E}^{\cdot}) := Z^n(\mathcal{E}^{\cdot}) / B^n(\mathcal{E}^{\cdot}), \; n \in \mathbf{Z}. \tag{1.12}$$

The elements of the last $\mathcal{A}$-module are also called the *$n$-dimensional cohomology classes of* (the *$\mathcal{A}$-complex*) $\mathcal{E}^{\cdot}$.

So the *cohomology of the $\mathcal{A}$-complex $\mathcal{E}^{\cdot}$*, is the following *graded*
(1.13) *$\mathcal{A}(X)$-module,* viz. the $\mathcal{A}(X)$-sequence,

$$h^*(\mathcal{E}^{\cdot}) := \{h^n(\mathcal{E}^{\cdot})\}_{n\in\mathbf{Z}} \ ;$$

here the individual terms of the previous sequence are given by (1.9) (cf. also (1.12)), while *the* same *sequence displays,* as already said, *the obstruction to the exactness of (1.3).*

Indeed, as we shall realize by the ensuing discussion, *cohomology of a space with coefficients in some sheaf* (of $\mathcal{A}$-modules or, more generally, of (abelian) groups) means the *cohomology of a certain particular $\mathcal{A}$-complex,* (canonically) associated with the initial sheaf; in effect, this is the content of a rough formulation of the so-called *abstract de Rham's theorem* (see also Section 3 below). So we continue to establish the necessary terminology for the sequel.

Now, it is instructive to consider first at this point the *category of $\mathcal{A}$-complexes* on a given topological space $X$, so we have next to define its morphisms. Thus, first, by a *morphism of $\mathcal{A}$-complexes of degree $p$,* with $p \in \mathbf{Z}_+$ (non-negative integers),

$$f : (\mathcal{E}^{\cdot}, d) \longrightarrow (\mathcal{F}^{\cdot}, d') \tag{1.14}$$

one means a sequence of $\mathcal{A}$-morphisms, that is,

$$f \equiv (f^n)_{n\in\mathbf{Z}} \tag{1.15}$$

with (see also (1.2))

$$f^n : \mathcal{E}^n \longrightarrow \mathcal{F}^{n+p}, \ n \in \mathbf{Z}, \tag{1.16}$$

such that the following diagram is *commutative*

$$\begin{array}{ccc} \mathcal{E}^{n-1} & \xrightarrow{\ f^{n-1}\ } & \mathcal{F}^{n-1+p} \\ {\scriptstyle d^{n-1}}\downarrow & & \downarrow{\scriptstyle d'^{n-1+p}} \\ \mathcal{E}^{n} & \xrightarrow{\ f^{n}\ } & \mathcal{F}^{n+p} \end{array} \tag{1.17}$$

That is, *equivalently,* one has

$$d'^{n-1+p} \circ f^{n-1} = f^n \circ d^{n-1}, \ n \in \mathbf{Z}. \tag{1.18}$$

Indeed, the morphisms of the category, under consideration, will be, by definition, *morphisms of (cochain) $\mathcal{A}$-complexes of degree 0,* commuting thus, in view of (1.18), with the corresponding differentials. So for such a morphism one gets the following relation, putting $p = 0$ in (1.18),

$$d'^{n-1} \circ f^{n-1} = f^n \circ d^{n-1}, \ n \in \mathbf{Z}. \tag{1.19}$$

*Equivalently,* one has the following *commutative diagram* (yet, a consequence of (1.17), for $p = 0$);

$$\begin{array}{ccc} \mathcal{E}^{n-1} & \xrightarrow{f^{n-1}} & \mathcal{F}^{n-1} \\ {\scriptstyle d^{n-1}}\big\downarrow & & \big\downarrow{\scriptstyle d'^{n-1}} \\ \mathcal{E}^{n} & \xrightarrow{f^{n}} & \mathcal{F}^{n} \end{array} \tag{1.19'}$$

Furthermore, by applying a similar notation, for instance, to (1.6), *we express (1.19) by simply writing* (see also (1.5) and (1.15))

$$d' \circ f = f \circ d. \tag{1.19''}$$

Now, given a *morphism* $f = (f^n)$ *of cochain $\mathcal{A}$-complexes,* as in (1.14) (cf. also (1.15)), it is clear from (1.19) (or even (1.19″)) that;

(1.20) $f$ *preserves cocycles and coboundaries.*

In other words, *one has the relations* (cf. (1.10), (1.11))

$$\begin{aligned} f^n(Z^n(\mathcal{E}^\cdot)) &\subseteq Z^n(\mathcal{F}^\cdot) \\ f^n(B^n(\mathcal{E}^\cdot)) &\subseteq B^n(\mathcal{F}^\cdot), \end{aligned} \tag{1.21}$$

*for every* $n \in \mathbf{Z}$. Hence, $f$ induces a *morphism* of the respective *cohomology $\mathcal{A}$-modules,* denoted by

(1.22) $$f^* \equiv (f^{*,n})_{n\in\mathbf{Z}},$$

such that (cf. (1.9))

(1.23) $$f^{*,n} : h^n(\mathcal{E}^\cdot) \longrightarrow h^n(\mathcal{F}^\cdot) : [z] \longmapsto f^{*,n}([z]) := [f^n(z)],$$

*with* $z \in Z^n(\mathcal{E}^\cdot)$.

Thus, applying the notation of (1.13), one gets a *morphism (of degree* 0*) of* the corresponding *graded* $\mathcal{A}$*-modules,* viz.

(1.24) $$f^* \equiv h^*(f) : h^*(\mathcal{E}^\cdot) \longrightarrow h^*(\mathcal{F}^\cdot),$$

given by (1.23). In fact (based on (1.13) and (1.24)), one proves that;

(1.25) $h^*$ *is a* (covariant) *functor from the category of* $\mathcal{A}$*-complexes and their morphisms,* as before, *to the category of graded* $\mathcal{A}$*-modules* (cf. (1.13)) *and their morphisms of such of degree* 0 (cf. (1.22), (1.23)).

Another notion that will be of use presently below is that of a *homotopy,* connecting two given morphisms, say, $f \equiv (f^n)$ and $g \equiv (g^n)$ (see (1.15)), between two $\mathcal{A}$-complexes $\mathcal{E}^\cdot$ and $\mathcal{F}^\cdot$, as above. So, again, one means by this a sequence of $\mathcal{A}$*-morphisms,*

(1.26) $$\alpha \equiv (\alpha^n)_{n\in\mathbf{Z}},$$

each one *of degree* $-1$ (cf. also e.g. (1.16)), that is, one has

(1.27) $$\alpha^n : \mathcal{E}^n \longrightarrow \mathcal{F}^{n-1},\ n \in \mathbf{Z},$$

such that (see also (1.14))

(1.28) $$f^n - g^n = \alpha^{n+1} \circ d^n + d'^{n-1} \circ \alpha^n,\ n \in \mathbf{Z}.$$

(The algebraic operations appeared in (1.28) are justified by the fact that $\mathrm{Hom}_{\mathcal{A}}(\mathcal{E}^n, \mathcal{F}^n)$ *is an* $\mathcal{A}(X)$*-module;* cf. Chapt. I; (9.22) and Chapt. II; (6.1.2) and (6.9)). In this context, we also say that the previous two morphisms $f$ *and* $g$ *are homotopic,* or even that the morphism (cf. the last comments on (1.28)) $f-g$ *is nullhomotopic.* Yet, by analogy with (1.19″), we *simply write,* instead of (1.28),

(1.28′) $$f - g = \alpha \circ d + d' \circ \alpha.$$

Now, as in the classical case, the significance of the above terminology lies in the fact that;

(1.29) *given two homotopic morphisms* $f, g$ between the cochain $\mathcal{A}$-complexes $\mathcal{E}^{\cdot}$ and $\mathcal{F}^{\cdot}$, as before, *the induced maps in cohomology* (cf. (1.24)) *are equal; viz. one has*

(1.29.1) $$f^* = g^*.$$

Indeed, the assertion follows immediately from (1.28) and (1.23). ∎ (Cohomology is always a *"homotopy invariant"*, anyway!)

Thus, two given cochain $\mathcal{A}$-complexes $\mathcal{E}^{\cdot}$ and $\mathcal{F}^{\cdot}$ are said to be *homotopy equivalent,* or even we talk of a *homotopy equivalence* between them, whenever we are given two morphisms $f : \mathcal{E}^{\cdot} \longrightarrow \mathcal{F}^{\cdot}$ and $g : \mathcal{F}^{\cdot} \longrightarrow \mathcal{E}^{\cdot}$ (cf. (1.14)), such that $g \circ f$ and $f \circ g$ are homotopic (see (1.28)) to the identity morphisms of $\mathcal{E}^{\cdot}$ and $\mathcal{F}^{\cdot}$, respectively. In that case too, the two given $\mathcal{A}$-complexes $\mathcal{E}^{\cdot}$ *and* $\mathcal{F}^{\cdot}$ *entail,* of course, *the same cohomology* (cf. (1.13) and (1.29.1)).

We come now to an important feature of the *cohomology of cochain* $\mathcal{A}$*-complexes* (cf. (1.13) and Theorem 1.1 below), whose consequences, under various forms, will appear in the sequel. But, first we need some more terminology.

Thus, by a *short exact sequence of* $\mathcal{A}$*-complexes,*

(1.30) $$0 \longrightarrow \mathcal{F}^{\cdot} \xrightarrow{i} \mathcal{S}^{\cdot} \xrightarrow{f} \mathcal{E}^{\cdot} \longrightarrow 0,$$

one means a finite sequence of $\mathcal{A}$-complexes, as indicated, such that the appeared *morphisms* $i$ *and* $f$ in (1.30) *of* the $\mathcal{A}$*-complexes* concerned (cf. (1.14)) satisfy the relation

(1.30′) $$\ker f = \operatorname{im}(i),$$

while we still assume that $i$ *is injective* and $f$ *onto* (exact sequence). Yet, the latter terms mean, by definition, that one has the following *short exact* $\mathcal{A}$*-sequence* (cf. also (1.15) and II; (2.18))

(1.30″) $$0 \longrightarrow \mathcal{F}^n \xrightarrow{i^n} \mathcal{S}^n \xrightarrow{f^n} \mathcal{E}^n \longrightarrow 0,$$

*for every* $n \in \mathbf{Z}$.

Now, before we proceed further, it is also useful to give at this point the following lemma, which will presently be applied below, supplementing thus our previous considerations in Chapt. II; Section 2. So we have.

**Lemma 1.1.** *A given $\mathcal{A}$-sequence* (see also Chapt. II; Definition 2.1)

$$\mathcal{F} \xrightarrow{\phi} \mathcal{S} \xrightarrow{\psi} \mathcal{E} \tag{1.31}$$

*is exact, viz. one has*

$$\ker \psi = \operatorname{im} \phi \tag{1.32}$$

(see Chapt. II; (2.8), (2.12)), *if, and only if, this happens fiber-wise. That is, if, and only if, the following sequence of $\mathcal{A}_x$-modules,*

$$\mathcal{F}_x \xrightarrow{\phi_x} \mathcal{S}_x \xrightarrow{\psi_x} \mathcal{E}_x\,, \tag{1.33}$$

*is exact, for every $x \in X$, or, equivalently, in case one has*

$$\ker \psi_x = \operatorname{im} \phi_x\,, \tag{1.34}$$

*for every $x \in X$.*

**Proof.** Indeed, we first observe that both $\ker \psi$ *and* $\operatorname{im} \phi$ *are,* by definition (cf. Chapt. II; (2.8) and (2.12)) *subsheaves* (in fact, *sub-$\mathcal{A}$-modules*) *of* $\mathcal{S}$; thus, the condition is, of course, *necessary,* due to Chapt. II; (2.11), (2.13). Conversely, *(1.34)* obviously *implies (1.32),* and this completes the proof. ∎

So we come now to the following.

**Theorem 1.1** *(of the long exact cohomology sequence). Suppose we are given a short exact sequence of $\mathcal{A}$-complexes* (see (1.30)),

$$0 \longrightarrow \mathcal{F}^{\cdot} \xrightarrow{i} \mathcal{S}^{\cdot} \xrightarrow{f} \mathcal{E}^{\cdot} \longrightarrow 0. \tag{1.35}$$

*Then, one obtains the following long exact sequence of cohomology* (cf. (1.13));

$$\cdots \longrightarrow h^n(\mathcal{F}^{\cdot}) \xrightarrow{h^n(i)} h^n(\mathcal{S}^{\cdot}) \xrightarrow{h^n(f)} h^n(\mathcal{E}^{\cdot}) \xrightarrow{\delta} h^{n+1}(\mathcal{F}^{\cdot}) \longrightarrow \cdots, \tag{1.36}$$

where we set $h^n(i) \equiv i^{*,n}$ and $h^n(f) \equiv f^{*,n}$ (cf. (1.22), (1.23)), while $\delta \equiv \delta^n$ stands for an $\mathcal{A}$-*morphism* of the $\mathcal{A}$-modules concerned (see (1.9)), that has to be defined (see (1.46) below).

**Proof.** First, we remark that the maps, in fact, $\mathcal{A}$-*morphisms*

$$i^{*,n} \equiv h^n(i) \;\; and \;\; f^{*,n} \equiv h^n(f), \;\; n \in \mathbf{Z}, \tag{1.37}$$

are already given by (1.23) (see also (1.25)). So we next define the map (in effect, a *sequence of maps*)

$$\delta^n \equiv \delta : h^n(\mathcal{E}^{\cdot}) \longrightarrow h^{n+1}(\mathcal{F}^{\cdot}),\ n \in \mathbf{Z}, \tag{1.38}$$

*fiber-wise*, as follows:

Let (see also (1.9))

$$\xi \equiv [z] \in h^n(\mathcal{E}^{\cdot}) \equiv \ker d^n / \operatorname{im} d^{n-1}, \tag{1.39}$$

such that $z \in Z^n(\mathcal{E}^{\cdot}) \equiv \ker d^n \subseteq \mathcal{E}^n$ (cf. (1.10)), be any *representative of* $\xi$;

(1.40) here in this proof, as well as, *throughout the sequel,* when appropriate, we eventually employ, *for notational convenience,* the *same symbol "d" for* the various *"differentials"* of a given, or even of different $\mathcal{A}$-*complexes* involved, the respective diversification being obvious from the context.

(Same practice, for instance, for the map (1.38) above). Now, if $\pi(z) = x \in X$, $\mathcal{E}^n$ being a sheaf on $X$, having *projection map* $\pi_{\mathcal{E}^n} \equiv \pi$ (analogous notational practice here, as with the differentials above), one obtains by the hypothesis for (1.35) and (1.30″) (cf. also Lemma 1.1), an element $z' \in \mathcal{S}^n_x$, with $f^n(z') \equiv f^n_x(z') = z \in \mathcal{E}^n_x$ (see also Chapt. I; (1.14), (1.15) and Chapt. II; (2.1)). Thus (cf. (1.40)),

$$d^n(z')\ (\equiv d^n_x(z')) \in d^n\mathcal{S}^n \subseteq \mathcal{S}^{n+1}, \tag{1.41}$$

such that (cf. also (1.19))

$$f^{n+1}(d^n z') = d^n(f^n(z')) = d^n z = 0, \tag{1.42}$$

due to the hypothesis for $z \in Z^n(\mathcal{E}^{\cdot})$ (cf. (1.39)). Therefore,

$$d^n(z') \in \ker f^{n+1} = \operatorname{im}(i^{n+1}) \cong \mathcal{F}^{n+1}, \tag{1.43}$$

$i^n$, $n \in \mathbf{Z}$, being, by hypothesis (cf. (1.30″)), an $\mathcal{A}$-*isomorphism* (into); that is, there exists, by (1.43), an element

$$z'' \in \mathcal{F}^{n+1}_x \subseteq \mathcal{F}^{n+1}, \tag{1.44}$$

such that

$$d^n(z') = i^{n+1}_x(z'') \equiv i^{n+1}(z'') \approx z'' \in \mathcal{F}^{n+1}, \tag{1.45}$$

where the last relation is justified by that one in (1.43). So one now defines;

$$(1.46)\qquad \delta([z]) := [dz'] \in h^{n+1}(\mathcal{F}^{\cdot}),$$

where we set $dz' \equiv d^n(z') \approx z'' \in \mathcal{F}^{n+1}$, in view of (1.45), while from the same relation one concludes, of course, that $z'' \approx dz' \in \ker d^{n+1} \equiv Z^{n+1}(\mathcal{F}^{\cdot})$, due to (1.4) (cf. also (1.40)). Furthermore, it is clear, by the very definitions (see (1.5) and (1.38)), that the previously defined map $\delta$ *is,* in fact, *an $\mathcal{A}$-morphism* of the $\mathcal{A}$-modules concerned.

So it remains only to prove the *exactness of (1.36):* That is, *we claim that one has;*

$$(1.47)\quad \ker h^n(f) = \operatorname{im} h^n(i), \quad \ker h^n(\delta) = \operatorname{im} h^n(f) \quad \textit{and} \quad \operatorname{im}\delta = \ker h^{n+1}(i),$$

*for every* $n \in \mathbf{Z}$. Now, similarly to (1.46), one proves the last three relations *fiber-wise,* based first on the hypothesis for (1.35), viz., equivalently, on that for (1.30″), by applying the above Lemma 1.1. Thus, the proof goes further analogously with the classical case in Homological Algebra, taking also into account (1.25), (1.19) and (1.23). (In this regard, cf., for instance, J. J. Rotman [1: p.172, Theorem 6.3]). ∎

The above map $\delta$ ($\equiv \delta^n$), as given by (1.46), is called the *connecting $\mathcal{A}$-morphism,* or even *Bockstein $\mathcal{A}$-morphism,* associated with the given short exact sequence of cochain $\mathcal{A}$-complexes (1.35).

Now, as already said, the *"sheaf cohomology"* of a space is actually that one of a particular complex, associated with the given *"sheaf* (in effect, *$\mathcal{A}$-module) of coefficients";* in fact, it is computed through an appropriate *(acyclic) resolution* of the sheaf, under consideration, in the category $\mathcal{A}$-$\mathcal{M}od_X$ (cf. (1.1)). Thus, we continue by commenting on the relevant terminology.

So an important notion, which we are going to employ in the sequel, is that of an *"injective"* object of the category $\mathcal{A}$-$\mathcal{M}od_X$. Namely, one has the following.

**Definition 1.1.** A given $\mathcal{A}$-module $\mathcal{E}$ on $X$ is said to be **injective,** whenever for any exact $\mathcal{A}$-sequence (cf. Chapt. II; Definition 2.1)

$$(1.48)\qquad 0 \longrightarrow \mathcal{S} \xrightarrow{\ i\ } \mathcal{F},$$

the corresponding sequence of $\mathcal{A}(X)$-modules

$$(1.49)\qquad \operatorname{Hom}_{\mathcal{A}}(\mathcal{F},\mathcal{E}) \xrightarrow{\ i^*\ } \operatorname{Hom}_{\mathcal{A}}(\mathcal{S},\mathcal{E}) \longrightarrow 0$$

is exact, as well.

In other words, $\mathcal{E}$ is an *injective $\mathcal{A}$-module,* if for every $\mathcal{A}$-morphism $i : \mathcal{S} \longrightarrow \mathcal{F}$, with $\ker(i) = 0$ (cf. (1.48)), and every $\mathcal{A}$-morphism $\phi : \mathcal{S} \longrightarrow \mathcal{E}$, there exists an $\mathcal{A}$-morphism $\psi : \mathcal{F} \longrightarrow \mathcal{E}$, "extending $\phi$", viz. for which one has $\phi = \psi \circ i \ (\equiv i^*(\psi)$; cf. (1.49), as well as, the next *commutative diagram*

(1.50)

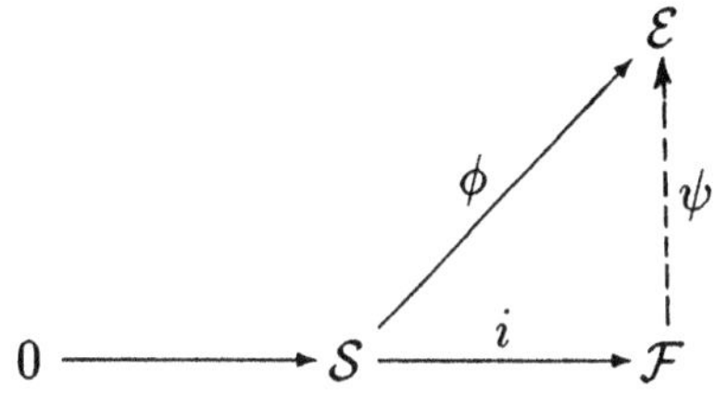


Yet, the same definition, as above, guarantees that a given $\mathcal{A}$-module $\mathcal{E}$ is *injective,* whenever (and only then!) *the* contravariant left exact *functor* (cf. Chapt. II; (6.45), (6.46))

$$\mathrm{Hom}_{\mathcal{A}}(\cdot, \mathcal{E}) : \mathcal{A}\text{-}\mathcal{M}od_X \longrightarrow \mathcal{A}(X)\text{-}\mathcal{M}od, \tag{1.51}$$

*is exact.* That is, for any short exact $\mathcal{A}$-sequence

$$0 \longrightarrow \mathcal{S}'' \longrightarrow \mathcal{S} \longrightarrow \mathcal{S}' \longrightarrow 0, \tag{1.52}$$

the following $\mathcal{A}(X)$-sequence (viz. its image through (1.51)),

$$0 \longrightarrow \mathrm{Hom}_{\mathcal{A}}(\mathcal{S}', \mathcal{E}) \longrightarrow \mathrm{Hom}_{\mathcal{A}}(\mathcal{S}, \mathcal{E}) \longrightarrow \mathrm{Hom}_{\mathcal{A}}(\mathcal{S}'', \mathcal{E}) \longrightarrow 0, \tag{1.53}$$

is also exact.

Now, injective $\mathcal{A}$-modules will be the constituents of certain particular $\mathcal{A}$-complexes that play an important rôle in the definition of (sheaf) cohomology theory. So we first have the following notion.

**Definition 1.2.** Given an $\mathcal{A}$-module $\mathcal{E}$, by an **$\mathcal{A}$-resolution** of $\mathcal{E}$, or simply a **resolution** of $\mathcal{E}$, one means a (cochain) $\mathcal{A}$-complex (of positive degree, cf. (1.2) for $n \in \mathbf{Z}_+$)

$$\mathcal{F}^\cdot : 0 \longrightarrow \mathcal{F}^0 \xrightarrow{d^0} \mathcal{F}^1 \xrightarrow{d^1} \mathcal{F}^2 \longrightarrow \cdots, \tag{1.54}$$

such that the following (*augmented* by $\mathcal{E}$) $\mathcal{A}$-sequence

$$\tilde{\mathcal{F}}^\cdot : 0 \longrightarrow \mathcal{E} \xrightarrow{i} \mathcal{F}^0 \xrightarrow{d^0} \mathcal{F}^1 \xrightarrow{d^1} \mathcal{F}^2 \longrightarrow \cdots \tag{1.55}$$

be exact, hence, an *exact $\mathcal{A}$-complex.* We also employ for (1.55) the notati

(1.55′) $$0 \longrightarrow \mathcal{E} \xrightarrow{i} \mathcal{F}^{\cdot}, \text{ or even } \mathcal{E} \xrightarrow{i} \mathcal{F}^{\cdot}.$$

In particular, we say that the $\mathcal{A}$-complex $\mathcal{F}^{\cdot}$, as in (1.54), defines **injective $\mathcal{A}$-resolution** of the given $\mathcal{A}$-module $\mathcal{E}$, in case each one of $\mathcal{A}$-modules $\mathcal{F}^n$ in (1.54), $n = 0, 1, \ldots$, is injective. (So one may call then an **injective $\mathcal{A}$-complex** of positive degree).

Now, referring to the *cohomology of a* given *$\mathcal{A}$-resolution* $\mathcal{F}^{\cdot}$ of $\mathcal{E}$, as (1.54) (see also (1.9) and (1.13)), one obtains, in view of (1.55);

(1.56) $$h^0(\tilde{\mathcal{F}}^{\cdot}) = \mathcal{E},$$

within an *$\mathcal{A}$-isomorphism,* as well as,

(1.57) $$h^n(\tilde{\mathcal{F}}^{\cdot}) = 0, \; n > 0.$$

On the other hand, a given $\mathcal{A}$-complex $\mathcal{F}^{\cdot}$ (cf. (1.2)) is said to be *acyc* in case it has a *trivial cohomology in positive degrees;* viz. one has (see a (1.9))

(1.58) $$h^n(\mathcal{F}^{\cdot}) = 0, \; \textit{for } n > 0.$$

Thus, by applying the latter terminology to the situation described the previous Definition 1.2, one easily concludes that;

(1.59) *an $\mathcal{A}$-complex $\mathcal{F}^{\cdot}$ is a resolution of a given $\mathcal{A}$-module $\mathcal{E}$, if, a only if, the augmented $\mathcal{A}$-complex $\mathcal{E} \longrightarrow \mathcal{F}^{\cdot}$* (cf. (1.55), (1.55 *is acyclic.*

In this context, the idea further is to *compute sheaf cohomology by plying* (certain particular *acyclic*) *resolutions* from the category $\mathcal{A}$-$\mathcal{M}od$ in effect, *injective ones* (see also Section 3 below). But, of course, one first to be sure for the existence of such resolutions; indeed, there are su in abundance! (So cf. Theorem 1.2 below). In this regard, one first obta the following.

**Lemma 1.2.** *Let $(X, \mathcal{A})$ be a* **C**-*algebraized space* (cf. Chapt. II; (1.36 *Then the category of $\mathcal{A}$-modules on $X$, $\mathcal{A}$-$\mathcal{M}od_X$* (see (1.1)), *contains eno injectives; viz., for every given $\mathcal{A}$-module $\mathcal{E}$ on $X$, there exists an inject $\mathcal{A}$-module $\mathcal{F}$* (cf. Definition 1.1) *and an $\mathcal{A}$-monomorphism, as follows;*

(1.60) $$0 \longrightarrow \mathcal{E} \xrightarrow{i} \mathcal{F}$$

(so $\ker(i) = 0$). *Thus, in other words, every $\mathcal{A}$-module on $X$ can be embedded in an injective object of $\mathcal{A}$-$\mathcal{M}od_X$.*

**Proof.** The proof is essentially based on the following classical fact:

(1.61) *for any unital* **C**-*algebra* **A**, *its algebraic dual (***C***-vector space) is an injective (left)* **A**-*module.*

See, for instance, P. J. Hilton - U. Stammbach [1: p.35, Theorem 8.2]. So one proves further that *every (left)* **A**-*module can be embedded in an injective* **A**-*module* (cf. also e.g. J. J. Rotman [1: p.71, Theorem 3.27], or even H. Cartan - S. Eilenberg [1: p.9, Theorem 3.3]).

Thus, taking an *$\mathcal{A}$-module $\mathcal{E}$*, one applies the preceding *fiber-wise,* that is, for any point $x \in X$, $\mathcal{E}_x$, the stalk of $\mathcal{E}$ at $x$, *is an $\mathcal{A}_x$-module* (with $\mathcal{A}_x$ being, by hypothesis for $\mathcal{A}$, a *unital commutative* **C**-*algebra;* cf. Chapt. II; Definition 1.4 and Scholium 1.2). Hence, by what has been said above, *$\mathcal{E}_x$ is embedded in an injective $\mathcal{A}_x$-module,* say, $F_x$ ($x \in X$).

Now, by means of the so-called *"skyscraper sheaf"* (in fact *$\mathcal{A}$-module*) on $X$ defined by (the constant sheaf, $\mathcal{A}_x$-module, on $\{x\} \subseteq X$) $F_x$, viz. by the *image sheaf* $(i_x)_*(F_x)$, with $i_x : \{x\} \longrightarrow X$, the (canonical) continuous injection (cf. Chapt. I; (14.8), (14.9) and Chapt. II; Subsection 2.(d)), one further defines an *injective $\mathcal{A}$-module $\mathcal{F}$* on $X$, satisfying the requirements. Indeed, based on the relevant material in Chapter II in the preceding, the proof goes further exactly as, for instance, in R. Hartshorne [1: p.207, Proposition 2.2], where we thus refer for the rest of the details. (Cf. also e.g. J. -L. Brylinski [1: p.8, Lemma 1.1.13]). ∎

On the other hand, based on the previous Lemma 1.2, one now obtains the following important consequence.

**Theorem 1.2.** *Let $(X, \mathcal{A})$ be a* **C**-*algebraized space. Then, for any given $\mathcal{A}$-module $\mathcal{E}$ on $X$, there always exists an injective resolution of $\mathcal{E}$*

(1.62) $$\mathcal{E} \longrightarrow \mathcal{F}^{\cdot}$$

(see also (1.55′) for the notation applied, as well as, (1.64) below).

**Proof.** By Lemma 1.2 one finds first an *exact $\mathcal{A}$-sequence* (see Chapt. II; Definition 2.1)

$$0 \longrightarrow \mathcal{E} \xrightarrow{\ i\ } \mathcal{F}^0, \tag{1.63}$$

with $\mathcal{F}^0$ an *injective $\mathcal{A}$-module.* Now, one further proceeds by *induction on $n$, $n \in \mathbf{N}$,* so that if the *injective $\mathcal{A}$-module $\mathcal{F}^n$* is defined, one next considers the *quotient $\mathcal{A}$-module $\mathcal{F}^n/\mathrm{im}\, d^{n-1}$* (see also the next diagram (1.64) and Chapt. II; Subsection 2.(b)). Furthermore, based on Lemma 1.2 again, the latter $\mathcal{A}$-module can be imbedded in an *injective $\mathcal{A}$-module $\mathcal{F}^{n+1}$*, this process providing thus the desired *injective resolution $\mathcal{F}^{\cdot}$* (cf. also (1.54)). The previous procedure is depicted by the following diagram (see also Chapt. II; (2.54))

$$\begin{array}{ccccccccc}
0 \longrightarrow \mathcal{E} & \xrightarrow{\ i\ } & \mathcal{F}^0 & \xrightarrow{\ d^0\ } & \mathcal{F}^1 & \xrightarrow{\ d^1\ } & \mathcal{F}^2 & \xrightarrow{\ d^2\ } & \cdots \\
& & \mu \searrow & \nearrow \nu & & \searrow & \nearrow & & \\
\mathrm{coker}(i) \equiv \mathcal{F}^0/i(\mathcal{E}) \cong & & & \mathcal{F}^0/\mathcal{E} & & \mathcal{F}^1/\mathrm{im}\, d^0 \equiv \mathrm{coker} d^0 & & & \cdots \\
& & \nearrow & & \searrow \nearrow & & \searrow & & \\
& & 0 & & 0 & & 0 \cdots, & &
\end{array} \tag{1.64}$$

where the *differentials* $d^n$, $n \geqq 0$, involved are, by definition, the indicated composites of the *$\mathcal{A}$-morphisms* concerned (e.g., one has $d^0 = \nu \circ \mu$, as shown). ∎

As already said, the above will be applied presently below (see e.g. Section 3 in the sequel). Furthermore, apart from the previous injective resolution of a given $\mathcal{A}$-module $\mathcal{E}$ on $X$, the so-called (canonical) *Godement resolution,* or even *flabby resolution,* of $\mathcal{E}$ will also be considered (see (3.27), (3.34) below).

## 2. Derived functors

The notion which we deal with in the sequel, has as its natural content that one of an *abelian category.* So in this regard, one easily proves, for instance, that (see (1.1))

(2.1) *$\mathcal{A}$-$\mathcal{M}od_X$ is an abelian category,*

a fact which is actually patterned on an analogous conclusion for the *category* $\mathbf{A}$-Mod of all (left) $\mathbf{A}$-modules, with $\mathbf{A}$ any given (*unital* commutative) $\mathbf{C}$-*algebra.* In particular, taking $\mathbf{A} = \mathbf{C}$, one gets a similar result, which will also be applied below, for the *category of* all **C**-*vector spaces,* denoted by

$$\mathbf{C}\text{-}\mathcal{V}ect, \tag{2.2}$$

having, of course, as *morphisms* the **C**-*linear maps.* (See, for example, S. Mac Lane [1: p.187ff and p. 27ff], or even S. Lang [1: p.116ff and p.141ff]. However, see also Chapt. II; (2.4), (1.67) or (1.75), (3.12), (6.39), as well as, (2.9) and (2.54)).

On the other hand, since the *bifunctor* $\mathrm{Hom}_{\mathcal{A}}(\cdot\,,\cdot)$, defining the respective set of morphisms of the two given objects in $\mathcal{A}\text{-}\mathcal{M}od_X$, takes values from the *category of* $\mathcal{A}(X)$-*modules,* $\mathcal{A}(X)\text{-}\mathcal{M}od$ (cf. Chapt. II; (6.39)), the objects of the latter being, in particular, **C**-*vector spaces* (see II; (1.34), (1.35)), we still speak below of **C**-*linear functors,* instead just of *additive functors;* thus, saying in the sequel that

$$T : \mathcal{A}\text{-}\mathcal{M}od_X \longrightarrow \mathbf{C}\text{-}\mathcal{V}ect \tag{2.3}$$

is a **C**-*linear functor,* we mean that, for any given $\mathcal{A}$-modules $\mathcal{E}$ and $\mathcal{F}$, the map

$$\mathrm{Hom}_{\mathcal{A}}(\mathcal{E},\mathcal{F}) \longrightarrow \mathrm{Hom}_{\mathbf{C}}(T(\mathcal{E}),T(\mathcal{F})) \equiv L(T(\mathcal{E}),T(\mathcal{F})) \tag{2.4}$$

is **C**-*linear.* The range of (2.4) is, of course, the set (indeed, **C**-*vector space*) of the **C**-linear maps of the **C**-vector spaces concerned; in other words, the *set of morphisms* in the category $\mathbf{C}\text{-}\mathcal{V}ect$, associated with the two given objects of the latter, as indicated (cf. (2.2)).

So we come now to the definition of the following fundamental concept for the ensuing discussion. Namely, we have.

**Definition 2.1.** Suppose that

$$T : \mathcal{A}\text{-}\mathcal{M}od_X \longrightarrow \mathbf{C}\text{-}\mathcal{V}ect \tag{2.5}$$

is a given covariant left exact (cf. e.g. II; (6.17)) **C**-linear functor between the abelian categories concerned. Then, the **right derived functors** of $T$ is a sequence of functors

$$(R^nT)_{n\geqq 0} \ \text{(i.e., n} \in \mathbf{Z}_+), \tag{2.6}$$

between the same (abelian) categories as (2.5), such that, for any $\mathcal{A}$-module $\mathcal{E}$, one sets (see e.g. (1.9), as well as, Scholium 2.1 below)

(2.7) $$(R^nT)(\mathcal{E}) := h^n(T(\mathcal{S}^{\cdot})),$$

where

(2.8) $$\mathcal{S}^{\cdot} : 0 \longrightarrow \mathcal{S}^0 \longrightarrow \mathcal{S}^1 \longrightarrow \mathcal{S}^2 \longrightarrow \cdots$$

is any *injective $\mathcal{A}$-resolution of $\mathcal{E}$*(cf. Definition 1.2 and Theorem 1.2).

Thus, according to the above definition, *(2.7) is the $n$-th cohomology (**C**-vector) space* of the following *complex*

(2.9) $$T(\mathcal{S}^{\cdot}) : 0 \longrightarrow T(\mathcal{S}^0) \longrightarrow T(\mathcal{S}^1) \longrightarrow T(\mathcal{S}^2) \longrightarrow \cdots,$$

*image of $\mathcal{S}^{\cdot}$ in the category* **C**-$\mathcal{V}ect$, through the given functor $T$, as in (2.5). Indeed, (2.7) is justified, in view of the following.

**Scholium 2.1.**- Concerning the above relation (2.7), one actually defines the *right derived functors,* for any given covariant left exact *additive functor* between two abelian categories, the first of which (domain of definition of the given functor) has *enough injectives* (cf. (2.5) and Lemma 1.2). By the term "additive functor", we mean that the functor in question yields a group morphism between the respective "Hom-*sets*", the latter being, by definition, *abelian groups* (see e.g. R. Hartshorne [1: p.203f]).

Therefore, **C**-$\mathcal{V}ect$ being an *abelian category,* the second member of (2.7) *is well-defined,* given that one has, by definition (see also (1.9), as well as (2.9));

(2.10) $$h^n(T(\mathcal{S}^{\cdot})) := \ker T(d^n)/\operatorname{im} T(d^{n-1}),\ n \geqq 0,$$

that is, a **C**-*vector space,* thus, an object of **C**-$\mathcal{V}ect$. Yet, according to general patterns in Homological Algebra, (2.10) yields, by definition, the *cohomology of the complex $T(\mathcal{S}^{\cdot})$*, as given by (2.9), (in the category **C**-$\mathcal{V}ect$; cf. e.g. R. Godement [1: p.22ff]).

Furthermore, the same relation (2.10) *is well-defined, independently of the injective resolution of $\mathcal{E}$*, under consideration (in fact, this within an *isomorphism* of the respective (cohomology) functors; see also (3.5) in the sequel). Indeed, this is assured by the following theorem; for convenience, here too, we restrict ourselves to the same categories, as above, although *the analogous conclusion is valid for two abelian categories,* with the same restriction, as before, for *the first one* (viz. the domain of definition of the functor involved is an *abelian category with enough injectives*). See, for instance, Theorem 3.1 in the next Section 3, as well as, R. Hartshorne [1: p.204, Theorem 1.1A]. Thus, we come now to the following basic result. For its proof, within the context of *abelian categories,* we refer to A. Grothendieck [2]. But, see also S. Lang [1: p.142, Theorem 6.1].

**Theorem 2.1.** *Suppose we are given the abelian categories* $\mathcal{A}$-$\mathcal{M}od_X$ *and* $\mathbf{C}$-$\mathcal{V}ect$, *as above* (cf. (1.1), (2.1) and (2.2)), *and let*

$$T : \mathcal{A}\text{-}\mathcal{M}od_X \longrightarrow \mathbf{C}\text{-}\mathcal{V}ect \tag{2.11}$$

*be a covariant left exact* $\mathbf{C}$-*linear functor of the categories concerned* (see (2.3)). *Then, the following assertions hold true:*

*1) For every integer* $n \geqq 0$, *the n-th right derived functor of* $T$,

$$R^nT : \mathcal{A}\text{-}\mathcal{M}od_X \longrightarrow \mathbf{C}\text{-}\mathcal{V}ect, \tag{2.12}$$

*as given by (2.7)* (see also (2.10)), *is a* $\mathbf{C}$-*linear functor. Furthermore, (2.7) is independent of the injective* $\mathcal{A}$-*resolution of the* $\mathcal{A}$-*module* $\mathcal{E}$ *concerned, within a unique (natural) isomorphism of functors* (cf. also Scholium 2.2 below).

*2) One has the following natural isomorphism of functors;*

$$R^0T \approx T. \tag{2.13}$$

*3) For any short exact* $\mathcal{A}$-*sequence,*

$$0 \longrightarrow \mathcal{F} \longrightarrow \mathcal{S} \longrightarrow \mathcal{E} \longrightarrow 0, \tag{2.14}$$

*and for every integer* $n \geqq 0$, *there exists a (natural)* $\mathbf{C}$-*linear morphism* (called *connecting* ($\mathbf{C}$-linear) *morphism,* or even *Bockstein* $\mathbf{C}$-*linear operator*)

$$\delta^n : (R^nT)(\mathcal{E}) \longrightarrow (R^{n+1}T)(\mathcal{F}), \tag{2.15}$$

*such that one obtains the following long exact sequence in* $\mathbf{C}$-$\mathcal{V}ect$ (viz. *of* $\mathbf{C}$-*vector spaces*)

$$\cdots \longrightarrow (R^nT)(\mathcal{F}) \longrightarrow (R^nT)(\mathcal{S}) \longrightarrow (R^nT)(\mathcal{E}) \xrightarrow{\delta^n} (R^{n+1}T)(\mathcal{F}) \longrightarrow \cdots. \tag{2.16}$$

*4)* For any $\mathcal{A}$-*morphism of short exact* $\mathcal{A}$-*sequences,* viz., for any given *commutative "*$\mathcal{A}$-*diagram"* (all arrows are $\mathcal{A}$-*morphisms*),

$$
\begin{array}{ccccccccc}
0 & \longrightarrow & \mathcal{F} & \longrightarrow & \mathcal{S} & \longrightarrow & \mathcal{E} & \longrightarrow & 0 \\
 & & \downarrow & & \downarrow & & \downarrow & & \\
0 & \longrightarrow & \mathcal{F}' & \longrightarrow & \mathcal{S}' & \longrightarrow & \mathcal{E}' & \longrightarrow & 0,
\end{array} \tag{2.17}
$$

*one gets the following commutative diagram in* **C**-$\mathcal{V}ect$

$$
\begin{array}{ccc}
(R^nT)(\mathcal{E}) & \xrightarrow{\delta^n} & (R^{n+1}T)(\mathcal{F}) \\
\downarrow & & \downarrow \\
(R^nT)(\mathcal{E}') & \xrightarrow{\delta^n} & (R^{n+1}T)(\mathcal{F}'),
\end{array} \tag{2.18}
$$

*for every integer* $n \geqq 0$.

*5) For every injective* $\mathcal{A}$*-module* $\mathcal{E}$ (cf. Definition 1.1), *and any integer* $n \geqq 1$, *one has*

$$(R^nT)(\mathcal{E}) = 0. \quad \blacksquare \tag{2.19}$$

The following terminology will be of use right below: Thus, given a functor $T$, as in (2.11), an $\mathcal{A}$-module $\mathcal{E}$ is said to be $T$*-acyclic,* in case one has (see also (2.10));

$$(R^nT)(\mathcal{E}) = 0, \; n > 0. \tag{2.20}$$

Yet, a given $\mathcal{A}$-resolution of $\mathcal{E}$, $\mathcal{E} \longrightarrow \mathcal{S}^{\cdot}$ (cf. (1.55′)), is called a $T$*-acyclic resolution* of $\mathcal{E}$, if each one of its members is $T$-acyclic (cf. also (1.54)). So, as an obvious consequence of (2.19), one concludes that:

(2.21) given a functor $T$, as in (2.11), *every injective* $\mathcal{A}$*-module is* $T$*-acyclic;* in particular, *every injective* $\mathcal{A}$*-resolution of a given* $\mathcal{A}$*-module* $\mathcal{E}$ (cf. also Theorem 1.2) *is* $T$*-acyclic.*

The previous terminology is, in fact, an extension of a more motivated (concrete) one, as this becomes clear by the discussion in the next Section.

**Scholium 2.2.**– Concerning the assertion in prop. 1) of the previous

Theorem 2.1, one first obtains that:

(2.22) *any two injective resolutions of a given $\mathcal{A}$-module $\mathcal{E}$ are homotopy equivalent.* Moreover, *this equivalence is uniquely defined, up to homotopy.* (Cf. Section 1 above for the terminology applied).

Namely, any two injective resolutions of a given $\mathcal{A}$-module $\mathcal{E}$ are (canonically) connected by a *morphism of $\mathcal{A}$-complexes* (cf. Section 1, corresponding to these resolutions). Moreover, *any two such morphisms are homotopic.* (*"Comparison Theorem"* (for cohomology); see, for instance, S. Lang [1: p.139, Lemma 5.2], or even J. J. Rotman [1: p.179, Theorem 6.9, Remark]). Now, the image of the previous data through the (covariant) functor $T$, as in (2.11), yields, due to the covariance of $T$, *homotopic complexes within the category* $\mathbf{C}$-$\mathcal{V}ect$(cf. also (1.28)). Therefore (cf. e.g. (1.29)), one obtains the *same cohomology, within* $\mathbf{C}$-$\mathcal{V}ect$, as an application, by definition (cf. (2.7), (2.10)), of the right derived functors of $T$. Yet, *this isomorphism in cohomology,* in fact, *(natural) isomorphism* of the respective functors (cf. (2.12)), *is* actually *unique.* (See also e.g. J. J. Rotman [1: p.184, Theorem 6.14], as well as, S. Lang [1: p.142, comments preceding Theorem 6.1]. In this respect, see also J. -L. Brylinski [1: p.11, Corollary 1.1.16]).

On the other hand, based on the very definitions, it is easy to verify (2.13).

Now, having in mind later applications (cf. Part Two in the sequel), one can consider, instead of (2.11), an *$\mathcal{A}(X)$-linear functor*

$$T : \mathcal{A}\text{-}\mathcal{M}od_X \longrightarrow \mathcal{A}(X)\text{-}\mathcal{M}od \tag{2.23}$$

(see also e.g. Chapt. II; (6.39)), where again $\mathcal{A}(X)$-$\mathcal{M}od$ *is an abelian category.* This is actually the case in the next Section 3, by considering the *global section functor* $\Gamma_X$ (cf. Chapt. II; (6.40)). Thus, for any two given $\mathcal{A}$-modules $\mathcal{E}$,$\mathcal{F}$, the resulting map by (2.23),

$$\mathrm{Hom}_{\mathcal{A}}(\mathcal{E},\mathcal{F}) \longrightarrow \mathrm{Hom}_{\mathcal{A}(X)}(T(\mathcal{E}),T(\mathcal{F})) \tag{2.24}$$

is an *$\mathcal{A}(X)$-morphism.* Notice that *an $\mathcal{A}(X)$-linear functor is,* of course, *a* $\mathbf{C}$*-linear functor,* but, in general, *not conversely!* (See also, for instance, the relevant comments, following II; (4.13)).

In that case *the corresponding cohomology spaces* as, for example, in (1.9), *will be $\mathcal{A}(X)$-modules,* due to the hypothesis for the functor (2.23), as this results in (2.24) (cf. thus, for instance, (3.5) in the sequel).

## 3. Derived functor cohomology

We give below the most salient features of the theory in the title of the present Section, the same being, in fact, a rough description of *sheaf cohomology* (cf. Definition 3.1 below), while we also provide some elementary indications, how this theory works in certain concrete cases. Now, although the cohomology theory that we shall often apply will be *Čech cohomology* (with coefficients in a (pre)sheaf, cf. the next Section 4), the former theory will be a convenient one, as well, in several instances (see, for instance, Sections 9, 10 below). Furthermore, the two theories in question are indeed equivalent, in some important instances, in view, for example, of the classical *Leray's Theorem* (cf. Theorem 4.1 in the sequel). Thus, we start with the following basic.

**Definition 3.1.** Consider the abelian categories $\mathcal{A}$-$\mathcal{M}od_X$ and $\mathcal{A}(X)$-$\mathcal{M}od$, as in (2.23), and let

$$\Gamma_X \equiv \Gamma(X, \cdot) : \mathcal{A}\text{-}\mathcal{M}od_X \longrightarrow \mathcal{A}(X)\text{-}\mathcal{M}od; \tag{3.1}$$

be the global section functor (cf. Chapt. II; (6.40)), being a covariant left exact $\mathcal{A}(X)$-linear functor between the (abelian) categories concerned (see also Chapt. II; (2.59)). Now, the $n$-th derived functor of $\Gamma_X$ (cf. Definition 2.1 along with (2.23) above) is called the **$n$-th cohomology functor,** denoted by

$$H^n(X, \cdot) : \mathcal{A}\text{-}\mathcal{M}od_X \longrightarrow \mathcal{A}(X)\text{-}\mathcal{M}od; \tag{3.2}$$

that is, one has, by definition (cf. also (2.6))

$$H^n(X, \cdot) := R^n\Gamma_X \equiv R^n\Gamma(X, \cdot), \tag{3.3}$$

for every integer $n \geqq 0$.

**Scholium 3.1.**– In the above Definition 3.1 we have considered as range of the functors involved the *abelian category* $\mathcal{A}(X)$-$\mathcal{M}od$, instead just of the (abelian) category $\mathbf{C}$-$\mathcal{V}ect$, that was the case in the previous Section 2. As already said (cf. the comments at the end of the preceding Section 2), this is essentially suggested by applications, which we shall see in Part Two of this treatise. So the functors, which are employed there, will be either **C**-*linear* or, in particular, $\mathcal{A}(X)$-*linear.* We have thus treated these two cases in the previous and the present Section respectively, while the argument applied holds, as already noticed, *for any abelian category,* in general (see also the remarks following Scholium 2.1 in the preceding). However, for convenience, we restricted ourselves to the above two cases, which will actually be encountered in the sequel. In particular, the analogous conclusion to the previous Theorem 2.1 is still valid; thus, cf. Theorem 3.1 below.

Now, before we proceed further, we first comment on certain immediate consequences of the above Definition 3.1, which also clarify it.

Thus, suppose we are given an $\mathcal{A}$-*module* $\mathcal{E}$ on $X$ and let

$$\mathcal{S}^{\cdot}: 0 \longrightarrow \mathcal{S}^0 \xrightarrow{d^0} \mathcal{S}^1 \xrightarrow{d^1} \cdots \longrightarrow \mathcal{S}^{n-1} \xrightarrow{d^{n-1}} \mathcal{S}^n \xrightarrow{d^n} \cdots \tag{3.4}$$

be an *injective resolution of* $\mathcal{E}$ (cf. Theorem 1.2). Then, by (2.7) and (3.3), one obtains (see also (2.10));

$$\begin{aligned}[H^n(X,\cdot)](\mathcal{E}) &\equiv H^n(X,\mathcal{E}) := (R^n\Gamma_X)(\mathcal{E}) \equiv [R^n\Gamma(X,\cdot)](\mathcal{E}) \\ &\equiv R^n\Gamma(X,\mathcal{E}) := h^n(\Gamma_X(\mathcal{S}^{\cdot})) \equiv h^n(\Gamma(X,\mathcal{S}^{\cdot})) \\ &:= \ker\Gamma_X(d^n)/\operatorname{im}\Gamma_X(d^{n-1}),\end{aligned} \tag{3.5}$$

*for every integer* $n \geq 0$.

Now, by hypothesis for (3.4) (cf. Definition 1.2), one obtains the following *exact* $\mathcal{A}$-*sequence*

$$0 \longrightarrow \mathcal{E} \xrightarrow{i} \mathcal{S}^0 \xrightarrow{d^0} \mathcal{S}^1 \xrightarrow{d^1} \cdots \longrightarrow \mathcal{S}^{n-1} \xrightarrow{d^{n-1}} \mathcal{S}^n \xrightarrow{d^n} \cdots. \tag{3.7}$$

Thus, by applying the left exact functor $\Gamma_X \equiv \Gamma(X,\cdot)$ (see Chapt. II; (2.59)) to (3.4), one obtains the following $\mathcal{A}(X)$-*complex,*

$$\begin{aligned}&0 \longrightarrow \Gamma(X,\mathcal{S}^0) \xrightarrow{\Gamma_X(d^0)} \Gamma(X,\mathcal{S}^1) \xrightarrow{\Gamma_X(d^1)} \Gamma(X,\mathcal{S}^2) \longrightarrow \cdots \\ &\cdots \xrightarrow{\Gamma_X(d^{n-1})} \Gamma(X,\mathcal{S}^n) \xrightarrow{\Gamma_X(d^n)} \Gamma(X,\mathcal{S}^{n+1}) \longrightarrow \cdots,\end{aligned} \tag{3.8}$$

giving, by definition (cf. (3.5)), the (*derived functor,* or even *sheaf*) *coho-*

*mology* of the space $X$, with *coefficients in the sheaf* (in fact, $\mathcal{A}$-*module*) $\mathcal{E}$ involved.

On the othe hand, applying the same functor $\Gamma_X$ to the exact $\mathcal{A}$-sequence (3.7), one obtains, in view of the *(left) exactness of* $\Gamma_X$, the following *exact* $\mathcal{A}(X)$-*sequence;*

$$0 \longrightarrow \Gamma(X,\mathcal{E}) \xrightarrow{\Gamma_X(i)} \Gamma(X,\mathcal{S}^0) \xrightarrow{\Gamma_X(d^0)} \Gamma(X,\mathcal{S}^1). \tag{3.9}$$

(In this respect, cf. also e.g. S. Lang [1: p.141]). Therefore, one has, in particular (see also (3.5), (3.7));

$$\begin{aligned} H^0(X,\mathcal{E}) &= (R^0\Gamma_X)(\mathcal{E}) = h^0(\Gamma_X(\mathcal{S}^{\cdot})) = \ker \Gamma_X(d^0) \\ &= \operatorname{im}\Gamma_X(i) \cong \Gamma(X,\mathcal{E}), \end{aligned} \tag{3.10}$$

up to an $\mathcal{A}(X)$-*isomorphism* of the $\mathcal{A}(X)$-*modules* involved. In fact, one has

$$H^0(X,\cdot) = \Gamma(X,\cdot) \equiv \Gamma_X, \tag{3.11}$$

within a *natural isomorphism of functors* (cf. also the next Theorem 3.1). Indeed, (3.11) is *one of the characteristic properties* of the sequence *of cohomology functors,* as the latter is given by (3.2) and (3.3) (see also Section 8 in the sequel). Thus, one has the following fundamental result, a *characterization of the "derived functor cohomology"* (cf. (3.24.2) below); it is, in fact, a consequence of Theorem 2.1 (see also Scholium 3.1) and the previous Definition 3.1. Namely, one obtains the following *"Fundamental Theorem of sheaf cohomology":*

**Theorem 3.1.** *Let $(X,\mathcal{A})$ be a* **C**-*algebraized space and $\mathcal{A}$-$\mathcal{M}od_X$ the abelian category (with enough injectives) of $\mathcal{A}$-modules on $X$. Then, the sequence of cohomology functors*

$$\{H^n(X,\cdot)\}_{n\in\mathbf{Z}_+}, \tag{3.12}$$

*as defined by (3.3), satisfies the following conditions:*

*1) For any given $\mathcal{A}$-module $\mathcal{E}$ on $X$, the $\mathcal{A}(X)$-sequence*

$$\{H^n(X,\mathcal{E})\}_{n\geqq 0} \tag{3.13}$$

(see (3.5)) *is independent of the injective $\mathcal{A}$-resolution of $\mathcal{E}$ considered* (in (3.4); cf. Scholium 2.2).

*2) There is the following natural isomorphism of functors;*

$$H^0(X, \cdot) \approx \Gamma(X, \cdot). \tag{3.14}$$

*3) For every short exact $\mathcal{A}$-sequence*

$$0 \longrightarrow \mathcal{F} \longrightarrow \mathcal{S} \longrightarrow \mathcal{E} \longrightarrow 0 \tag{3.15}$$

*and, for every integer $n \geqq 0$, there are associated $\mathcal{A}(X)$-linear morphisms* (the so-called *connecting ($\mathcal{A}(X)$-linear) morphisms,* or yet *Bockstein operators*)

$$\delta^n : H^n(X, \mathcal{E}) \longrightarrow H^{n+1}(X, \mathcal{F}), \tag{3.16}$$

*such that one has the following long exact $\mathcal{A}(X)$-sequence (in cohomology);* cf. also (3.10));

$$\begin{aligned} &0 \longrightarrow \Gamma(X, \mathcal{F}) \longrightarrow \Gamma(X, \mathcal{S}) \longrightarrow \Gamma(X, \mathcal{E}) \xrightarrow{\delta^0} H^1(X, \mathcal{F}) \\ &\longrightarrow H^1(X, \mathcal{S}) \longrightarrow H^1(X, \mathcal{E}) \xrightarrow{\delta^1} H^2(X, \mathcal{F}) \longrightarrow \cdots \\ &\cdots \longrightarrow H^n(X, \mathcal{E}) \xrightarrow{\delta^n} H^{n+1}(X, \mathcal{F}) \longrightarrow \cdots . \end{aligned} \tag{3.17}$$

*4) For every commutative diagram of $\mathcal{A}$-modules* (or, as we say, for any *morphism of short exact $\mathcal{A}$-sequences*)

$$\begin{array}{ccccccccc} 0 & \longrightarrow & \mathcal{F} & \longrightarrow & \mathcal{S} & \longrightarrow & \mathcal{E} & \longrightarrow & 0 \\ & & \downarrow & & \downarrow & & \downarrow & & \\ 0 & \longrightarrow & \mathcal{F}' & \longrightarrow & \mathcal{S}' & \longrightarrow & \mathcal{E}' & \longrightarrow & 0, \end{array} \tag{3.18}$$

*one has the following commutative diagram of $\mathcal{A}(X)$-modules* (cf. also (3.16));

$$\begin{array}{ccc} H^n(X, \mathcal{E}) & \xrightarrow{\delta^n} & H^{n+1}(X, \mathcal{F}) \\ \downarrow & & \downarrow \\ H^n(X, \mathcal{E}') & \xrightarrow{\delta^n} & H^{n+1}(X, \mathcal{F}'), \end{array} \tag{3.19}$$

*for all integers $n \geqq 0$.* (In (3.19) we used, by an obvious abuse of notation, the same symbol for the two different, in general, *$n$-th connecting $\mathcal{A}(X)$-linear* *morphisms).*

*5) For every injective $\mathcal{A}$-module $\mathcal{E}$, one has*

$$H^n(X, \mathcal{E}) = 0, \tag{3.20}$$

*for every integer $n > 0$.* ∎

The following easy consequences of the previous theorem will be of use later on.

Thus, based on the *naturality of the isomorphism (3.14)* (see also, for instance, S. Mac Lane [1: p.16]), if $\phi : \mathcal{E} \longrightarrow \mathcal{F}$ is any given $\mathcal{A}$-*morphism* between the $\mathcal{A}$-modules involved, one gets the following *commutative diagram;*

$$\begin{array}{ccc} H^0(X, \mathcal{E}) & \xrightarrow{\quad\cong\quad} & \Gamma(X, \mathcal{E}) \\ \big\downarrow & & \big\downarrow \\ H^0(X, \mathcal{F}) & \xrightarrow{\quad\cong\quad} & \Gamma(X, \mathcal{F}') \,. \end{array} \tag{3.21}$$

Furthermore, to any *commutative triangle in $\mathcal{A}$-$\mathcal{M}od_X$,*

$$\begin{array}{ccc} \mathcal{E} & \xrightarrow{\quad\phi\quad} & \mathcal{F} \\ & {\scriptstyle \psi\circ\phi=\tau}\searrow \quad \swarrow{\scriptstyle \psi} & \\ & \mathcal{S} & \end{array} \tag{3.22}$$

there corresponds a *commutative diagram in cohomology;* viz., the following diagram commutes, *for all integers $n \geq 0$,*

$$\begin{array}{ccc} H^n(X, \mathcal{E}) & \longrightarrow & H^n(X, \mathcal{F}) \\ & \searrow \quad \swarrow & \\ & H^n(X, \mathcal{S}) \,. & \end{array} \tag{3.23}$$

Our claim is an immediate consequence of (3.22) and the functor transformation of a composition of morphisms, in view of (3.3). ∎

Thus, according to Definition 3.1, we have defined (the) *sheaf cohomology* of a topological space $X$, *with coefficients in a given* $\mathcal{A}$*-module* $\mathcal{E}$, as the *derived functor cohomology of the global section functor* $\Gamma_X$ between suitably chosen, for our purposes!, (abelian) categories (cf. (3.1), or even (2.5)), the first being that one to which belongs the *"coefficient sheaf"* of the cohomology concerned. Yet, the same cohomology (procedure, cf. also Section 8 below for the terminology applied) has been actually defined by employing *any* injective resolution of the coefficient sheaf ($\mathcal{A}$-module; cf. (2.22)), such a resolution being, in effect, $\Gamma_X$*-acyclic* (see (2.21) and (3.20)). Indeed, one can use, instead, just *any* $\Gamma_X$*-acyclic resolution of the coefficient sheaf.* Namely one has the following fundamental fact:

(3.24) *consider the global section functor* $\Gamma_X$*, as in (3.1), and let*

(3.24.1) $$0 \longrightarrow \mathcal{E} \longrightarrow \mathcal{F}^{\cdot}$$

*be a* $\Gamma_X$*-acyclic resolution of a given* $\mathcal{A}$*-module* $\mathcal{E}$ *on* $X$ (cf. also (1.55′)). *Then, one obtains*

(3.24.2) $$(R^n\Gamma_X)(\mathcal{E}) = H^n(X, \mathcal{E}) \cong h^n(\Gamma_X(\mathcal{F}^{\cdot})),$$

*within a natural isomorphism of functors* (see also (3.5) and (2.10)).

The proof of (3.24) can be based on the theory of *spectral sequences;* so cf., for example, R. Godement [1: p.178, Théorème 4.6.1], or even R. S. Ward - R. O. Wells, Jr. [1: p.230, Example 3.6.5]. (However, see also G. E. Bredon [1: p.34, Theorem 4.1], or even L. Kaup - B. Kaup [1: p.216, Theorem 50.4]).

On the other hand, (3.24) stated in another, less technical manner, means that;

(3.25) *the (sheaf) cohomology of a topological space* $X$*,* relative to the global section functor (see also (3.38) below), *having coefficients from a given* $\mathcal{A}$*-module* $\mathcal{E}$ *is* (up to a natural isomorphism of functors) *the cohomology of an acyclic* (with respect to the above functor) *resolution of* $\mathcal{E}$*.*

The above is, in effect, the content of the so-called *abstract de Rham theorem* (cf. the previously cited Refs.).

In particular, a $\Gamma_X$*-acyclic* resolution of a given $\mathcal{A}$-module $\mathcal{E}$ is the so-called *canonical* (or even *flabby*) *resolution* of $\mathcal{E}$ (cf. thus Lemma 3.1 in the sequel).

In this context, we first recall that a given *sheaf* (of sets) on a topological space $X$ is said to be *flabby,* whenever for any open set $U \subseteq X$, the canonical restriction map (cf. Chapt. I; (9.4))

$$\Gamma(X,\mathcal{S}) \longrightarrow \Gamma(U,\mathcal{S}) \tag{3.26}$$

is surjective; i.e., every continuous section of $\mathcal{S}$ over any open $U \subseteq X$ can be extended to the whole of $X$. Equivalently, one can assume, instead, the corresponding relation to (3.26), for any pair of open sets $U, V$ in $X$, with $V \subseteq U$. Of course, one can consider the previous notion, in particular, *for any $\mathcal{A}$-module $\mathcal{E}$* on $X$.

Thus, following R. Godement [1: p.167, §4.3], one proves that (see also (3.34) below);

(3.27) *every $\mathcal{A}$-module $\mathcal{E}$ on $X$ admits a flabby resolution in $\mathcal{A}$-$\mathcal{M}od_X$.*

We briefly recall the corresponding construction (ibid.): Thus, one first defines

$$\mathcal{C}^0(X;\mathcal{E}) \equiv \mathcal{C}^0(\mathcal{E}), \tag{3.28}$$

as the *sheaf of germs of all sections* (continuous or not) *of $\mathcal{E}$ over $X$;* equivalently (3.28) is *the sheaf on $X$ generated by the presheaf*

$$U \longmapsto S(U;\mathcal{E}) := \left\{s : U \longrightarrow \mathcal{E} : \pi \circ s = \mathrm{id}_X\right\} \equiv \prod_{x\in U} \mathcal{E}_x \tag{3.29}$$

*(all sections of $\mathcal{E}$ over $U$).* It is easy to see that *(3.29) is a complete presheaf on $X$* (cf. Chapt. I; Definition 11.1). Furthermore, one easily verifies that;

(3.30) for any open $U \subseteq X$, $S(U;\mathcal{E})$ is an $\mathcal{A}(U)$*-module;* hence (Chapt. II; Proposition 1.1), $\mathcal{C}^0(\mathcal{E})$ *is an $\mathcal{A}$-module on $X$.* Yet, one has the following *exact $\mathcal{A}$-sequence*

$$0 \longrightarrow \mathcal{E} \stackrel{i}{\longrightarrow} \mathcal{C}^0(\mathcal{E}) \tag{3.30.1}$$

(cf. also (3.29) and the ensuing comments there).

In this regard, one also obtains (cf. Chapt. I; (14.7) and (14.44))

$$\mathcal{C}^0(\mathcal{E}) = f_* f^* \mathcal{E}, \tag{3.31}$$

where we set $f : X_d \longrightarrow X$ for the natural (identity) map, with $X_d$ denoting the set $X$ with the *discrete topology.* (Thus, one has, through (3.31), another proof that $\mathcal{C}^0(\mathcal{E})$ *is an* $\mathcal{A}$*-module, whenever* $\mathcal{E}$ *is* such. Cf. Chapt. II; (2.62), (2.65)). Yet, one also speaks of $\mathcal{C}^0(\mathcal{E})$, as the *"sheaf of discontinuous sections of* $\mathcal{E}$*"*.

Now, one further defines (see also (3.30.1), as well as, Chapt. II; Lemma 2.1):

$$\mathcal{C}'(\mathcal{E}) := \mathcal{C}^0(\mathcal{E})/\mathcal{E}, \tag{3.32}$$

thus, another $\mathcal{A}$*-module* on $X$. So one now sets, *inductively* (cf. also (3.28));

$$\begin{array}{rcl} \mathcal{C}^1(\mathcal{E}) & := & \mathcal{C}^0(\mathcal{C}'(\mathcal{E})) \\ \mathcal{C}^2(\mathcal{E}) & := & \mathcal{C}^0(\mathcal{C}'(\mathcal{C}^1(\mathcal{E}))) \\ & & \cdots \\ \mathcal{C}^n(\mathcal{E}) & := & \mathcal{C}^0(\mathcal{C}'(\mathcal{C}^{n-1}(\mathcal{E}))) \\ & & \cdots \end{array} \tag{3.33}$$

It is easy to see that the above $\mathcal{A}$*-modules* $\mathcal{C}^n(\mathcal{E})$, $n \geqq 0$, *are flabby.* Thus, one obtains the following *flabby* $\mathcal{A}$*-resolution of* $\mathcal{E}$*;*

$$0 \longrightarrow \mathcal{E} \xrightarrow{i} \mathcal{C}^0(\mathcal{E}) \xrightarrow{d} \mathcal{C}^1(\mathcal{E}) \xrightarrow{d} \cdots \longrightarrow \mathcal{C}^n(\mathcal{E}) \xrightarrow{d} \cdots, \tag{3.34}$$

where the $\mathcal{A}$*-morphisms* $d$ involved are defined by composition of the obvious injections and quotient maps, according to the preceding definitions in (3.33); see thus R. Godement [1: p.168].

In this context, one now obtains the following.

**Lemma 3.1.** *Let* $(X, \mathcal{A})$ *be a given* **C***-algebraized space. Then, every flabby* $\mathcal{A}$*-module* $\mathcal{E}$ *on* $X$ *is* $\Gamma_X$*-acyclic; i.e., one has*

$$H^n(X, \mathcal{E}) = 0, \; n > 0. \tag{3.35}$$

**Proof.** The proof is based on the fact that;

(3.36) *any injective* $\mathcal{A}$*-module is flabby.*

However, for this and the same proof of Lemma 3.1, see, for instance, R.

Hartshorne [1: p.208, Proposition 2.5]. ∎

Therefore, according to (3.24) (cf. also (3.25)), we conclude that:

(3.37) *one can define sheaf cohomology* (with respect to the global section functor) *of a given topological space, by using the canonical flabby resolution of the coefficient sheaf ($\mathcal{A}$-module)* involved.

So, in this context, see e.g. R. Godement [1: p.173, §4.4], or even G. E. Bredon [1: p.28, Definition 2.1]. Moreover, cf. L. Kaup - B. Kaup [1: p.216, Theorem 50.4], as well as, Section 8 in the sequel of our discussion.

Now, concerning (3.24), one concludes, *more generally,* the following (a result, of course, of theoretical significance, for that matter! Indeed, see also Section 9 below):

(3.38) the *(sheaf) cohomology of a topological space $X$,* with respect to a given *covariant left exact $\mathcal{A}(X)$-linear functor*

(3.38.1) $$T : \mathcal{A}\text{-}\mathcal{M}od_X \longrightarrow \mathcal{A}(X)\text{-}\mathcal{M}od,$$

having *coefficients from an $\mathcal{A}$-module $\mathcal{E}$, is* (up to a natural isomorphism of functors) *the cohomology of any $T$-acyclic* resolution of $\mathcal{E}$; i.e., if $0 \longrightarrow \mathcal{E} \longrightarrow \mathcal{S}^{\cdot}$ is such a resolution of $\mathcal{E}$ (see (1.55′) and (2.20)), then *one has*

(3.38.2) $$(R^nT)(\mathcal{E}) = h^n(T(\mathcal{S}^{\cdot})),$$

*within a natural isomorphism of functors.*

In this respect, cf. also (2.10), as well as, R. Hartshorne [1: p.205, Proposition 1.2A]. (Here again one can give an analogous proof, as for (3.24), in terms of *spectral sequences*). ∎

**Scholium 3.2.**– In view of (3.38.2), we have above abused terminology (cf. thus Definition 3.1) by referring to *"sheaf cohomology"* of a topological space $X$ ( *"with respect to a given functor"* as in (3.38.1)), as the *derived functor cohomology* (cf. Definition 2.1, or even (2.10)) of the functor in question. So, strictly speaking (Definition 3.1), the relevant terminology is actually referred to the (*derived functor cohomology* of the) *global section functor* $\Gamma_X$.

On the other hand, the *abstract de Rham's theorem,* as given by (3.25) (or even its *further abstraction*(!) by (3.38)), provides also an indication (in fact, a weak one) of the *"uniqueness of (the) cohomology"* of a given space, a fact that we shall discuss again, in detail, in Section 8 below. However, before that, we further consider right in the next Section 4 another technique, *more convenient to computations,* namely, *Čech theory,* for defining the (sheaf) cohomology of a given topological space $X$. The latter theory will also be the sort of cohomology, which will often be employed throughout the rest of our discussion.

# 4. Čech cohomology

We present in this and the following two Sections the basic facts about Čech cohomology of a topological space, with coefficients in a given (pre)sheaf on that space, as these will be employed in later chapters of this treatise. On the other hand, as we shall see later on (cf. Sections 6, 8 below) the present cohomology theory coincides with the derived functor cohomology, we already considered in the previous Section 3, when we restrict ourselves to the important special case of (Hausdorff) *paracompact spaces.* Indeed, it is actually the latter case, that we resort to, when we have to consider *"long exact sequences in cohomology"* of the type exhibited by Theorems 2.1 and 3.1 in the preceding. Yet, this specialization to the aforementioned class of topological spaces (an important class for that matter), is, in effect, a major drawback of the general (arbitrary topological spaces) Čech cohomology theory, the same being, otherwise, as already said, *very convenient to computations;* on the other hand, this latter feature of Čech cohomology makes it, however, superior, in several respects, to the previously examined (abstract) derived functor cohomology (cf. Section 3 above).

So we begin with establishing the necessary preliminary material for the sequel:

**4.(a). The directed set of (proper) open coverings of a given topological space.**– Suppose we are given an arbitrary *topological space* $X$, and let

$$\mathcal{U} = \{U_\alpha\}_{\alpha \in I} \text{ and } \mathcal{V} = \{V_\beta\}_{\beta \in J} \tag{4.1}$$

be two *open coverings* of $X$. Now, we shall say that $\mathcal{V}$ is *finer* than $\mathcal{U}$, or even that $\mathcal{V}$ is a *refinement* of $\mathcal{U}$, and we write

(4.2) $$\mathcal{V} \succeq \mathcal{U} \text{ (or even } \mathcal{U} \preceq \mathcal{V}),$$

whenever there exists a *refining map*

(4.3) $$\mu : J \longrightarrow I,$$

such that one has;

(4.4) $$V_\beta \subseteq U_{\mu(\beta)}, \text{ with } \beta \in J.$$

On the other hand, as already noticed (cf. Chapt. II; Section 4), it is standard that *there are "logical" difficulties by considering "the set of all open coverings of a given topological space"* $X$. (In fact, *this is not a set,* in the "technical" sense of the word). Now, to avoid this inconvenience, one considers, instead, *proper open coverings* of $X$: So, according to the last term, one looks at a given open covering of $X$ (indexed or not) by indexing each one of the members of the given family by itself. Thus, the covering in question becomes simply a subset of the power set of $X$, $\mathcal{P}(X)$, so that one can speak of *the set of all proper (open) coverings of* $X$. It is now clear that

(4.5) *given any open covering of a topological space* $X$, *there always exists a proper* (open) *refinement* (cf. (4.2)) *of the initial covering.*

Indeed, it suffices to take the members of the given family (covering $X$) indexed now by themselves, as above, which thus constitute the desired open refinement of the initial covering. (In this respect, cf. also, for instance, C. H. Dowker [1: p.60, Definition], or even F. Hirzebruch [1: p.17]). ∎

Furthermore, it is easily seen that (4.2), as given by (4.4), supplies the set, under consideration, with a *partial preorder;* viz. the latter set endowed, through (4.2), with a relation, which is *reflexive* and *transitive.* Moreover, the same relation *"directs"* the set in question, in the sense that;

(4.6) *for any two proper coverings* of a given topological space $X$, say, $\mathcal{U}$ and V, *there exists a common refinement of both,* say $\mathcal{W}$.

Namely, taking $\mathcal{U}$ and V, as in (4.1), one considers the open covering $\mathcal{U} \cap \mathcal{V}$, whose members are all possible non-void open sets in $X$ of the form $U_\alpha \cap V_\beta$, for some $(\alpha, \beta) \in I \times J$. So one defines

(4.7) $$K \equiv K_{\mathcal{U}\cap\mathcal{V}} = \{(\alpha, \beta) \in I \times J : W_{(\alpha,\beta)} \equiv U_\alpha \cap V_\beta \neq \emptyset\},$$

as the *index set of* $\mathcal{U} \cap \mathcal{V}$, along with its two projections on $I$ and $J$, respectively, as the corresponding *refining maps* (cf. (4.3)), getting thus $\mathcal{U} \cap \mathcal{V}$ as a common refinement of $\mathcal{U}$ and $\mathcal{V}$. Now, an application of (4.6) to the latter covering and the transitivity of (4.2) provide the desired proper open covering $\mathcal{W}$ of $X$, while we still set, by an eventual abuse of notation,

$$\mathcal{W} = \mathcal{U} \cap \mathcal{V}. \tag{4.8}$$

We thus have proved by the preceding the following useful fact.

(4.9) *the set of proper open coverings of a topological space $X$ becomes an upwards directed set, under the relation* $\succeq$ (cf. (4.2)), *as this is defined by (4.4).*

(In this context, we recall that a set $S$ equipped with a partial (pre)order "$\succeq$", is called an *(upwards) directed set,* whenever, for any two elements $x, y$ in $S$, there exists a third one $z \in S$, with $z \succeq x$ and $z \succeq y$).

**4.(b) Čech cochain complex.**– Consider an *open covering*

$$\mathcal{U} = \{U_\alpha\}_{\alpha \in I} \tag{4.10}$$

of a given *topological space* $X$, and let $\mathcal{S}$ be an $\mathcal{A}$-*module* on $X$. Now, *for every integer* $n \geqq 0$, one defines

$$C^n(\mathcal{U}, \mathcal{S}) := \prod_{(\alpha_0, \ldots, \alpha_n)} \mathcal{S}(U_{\alpha_0 \ldots \alpha_n}), \tag{4.11}$$

where we set

$$U_{\alpha_0 \ldots \alpha_n} := U_{\alpha_0} \cap U_{\alpha_1} \cap \cdots \cap U_{\alpha_n}, \tag{4.12}$$

for any $n+1$-*tuple of open sets* $U_{\alpha_0}, \ldots, U_{\alpha_n}$ *in* $\mathcal{U}$. In this context, one sets, of course,

$$\mathcal{S}(\emptyset) \equiv \Gamma(\emptyset, \mathcal{S}) := \{0\}, \tag{4.13}$$

viz. the *trivial* $\mathcal{A}(X)$-*module* (being also, analogously, the *trivial* $\mathcal{A}(U)$-*module, for any open* $U \subseteq X$). Thus, by hypothesis for $\mathcal{S}$ (cf. e.g. Chapt. II; (1.67)), it is clear from (4.11) that

(4.14) *the set* $C^n(\mathcal{U}, \mathcal{S})$ *is an* $\mathcal{A}(X)$-*module* (hence, in particular, a $\mathbf{C}$-*vector space;* (cf. also Remark 4.1 in the sequel). Its elements are called the *n-cochains of* $\mathcal{U}$ *with coefficients in* $\mathcal{S}$.

Thus, one further defines the following *n-th coboundary operator,* in effect (cf. (4.14)), an $\mathcal{A}(X)$-*morphism,*

$$\delta^n : C^n(\mathcal{U}, \mathcal{S}) \longrightarrow C^{n+1}(\mathcal{U}, \mathcal{S}),\ n \in \mathbf{Z}_+, \tag{4.15}$$

by the relation (cf. Chapt. I; (9.4), (9.5)),

$$\delta^n(f)_{\alpha_0 \cdots \alpha_{n+1}} := \sum_{i=0}^{n+1} (-1)^i \sigma_{U_{\alpha_0 \cdots \alpha_{n+1}}}^{U_{\alpha_0 \cdots \hat{\alpha}_i \cdots \alpha_{n+1}}} (f_{\alpha_0 \alpha_1 \cdots \hat{\alpha}_i \cdots \alpha_{n+1}}), \tag{4.16}$$

for every *n-cochain* $f \equiv (f_{\alpha_0 \alpha_1 \cdots \alpha_n}) \in C^n(\mathcal{U}, \mathcal{S})$. (The symbol "^" in the last relation indicates omission of the corresponding index). In this respect, it is a routine matter to check that (cf. e.g. R. C. Gunning - H. Rossi [1: p.187, Lemma 1]);

$$\delta^{n+1} \circ \delta^n = 0, \ \textit{for any integer}\ n \geqq 0. \tag{4.18}$$

We also write for short, instead,

$$\delta \circ \delta \equiv \delta^2 = 0. \tag{4.18$'$}$$

We have thus arrived at the following *(cochain) complex of* $\mathcal{A}(X)$*-modules;*

$$\check{C}^{\cdot}(\mathcal{U}, \mathcal{S}) := \left(C^n(\mathcal{U}, \mathcal{S}),\ \delta \equiv \{\delta^n\}\right)_{n \in \mathbf{Z}_+}. \tag{4.19}$$

We call it the *Čech cochain complex of* $\mathcal{U}$ *with coefficients in* $\mathcal{S}$.

Thus, we come next to the following fundamental.

**Definition 4.1.** Suppose we have a **C**-algebraized space $(X, \mathcal{A})$ and let $\mathcal{U}$ be an open covering of $X$. Moreover, let $\mathcal{S}$ be an $\mathcal{A}$-module on $X$. Then, one defines the **$n$-th Čech cohomology $\mathcal{A}(X)$-module of $\mathcal{U}$, with coefficients in $\mathcal{S}$,** by the relation (cf. also (4.23) below)

$$\check{H}^n(\mathcal{U}, \mathcal{S}) := h^n(\check{C}^{\cdot}(\mathcal{U}, \mathcal{S})),\ n \in \mathbf{Z}_+; \tag{4.20}$$

viz. (cf. also (1.9)), as the *n-th dimensional cohomology* $\mathcal{A}(X)$*-module of the* corresponding (to the previous data $\{(X, \mathcal{A}); \mathcal{U}, \mathcal{S}\}$), according to (4.19), *Čech cochain complex* $\check{C}^{\cdot}(\mathcal{U}, \mathcal{S})$. Yet, we call (4.20) the **$n$-th (dimensional) Čech cohomology $\mathcal{A}(X)$-module of the** given open **covering $\mathcal{U}$** of $X$, **with coefficients in** the $\mathcal{A}$-module $\mathcal{S}$.

Thus, applying a similar notation to (1.12), one first defines (cf. also (4.15));

$$\check{Z}^n(\mathcal{U}, \mathcal{S}) := \ker \delta^n, \ n \in \mathbf{Z}_+. \tag{4.21}$$

We call (4.21) the $\mathcal{A}(X)$**-module of the** $n$**-th (dimensional Čech) cocycles of** $\mathcal{U}$**, with coefficients in** $\mathcal{S}$ (or, by an abuse of language, those **of** $\mathcal{S}$**, with respect to** $\mathcal{U}$). Furthermore, the elements of the $\mathcal{A}(X)$*-module*

$$\check{B}^n(\mathcal{U}, \mathcal{S}) := \operatorname{im} \delta^{n-1}, \ n \in \mathbf{Z}_+, \tag{4.22}$$

are called the $n$**-th (dimensional Čech) coboundaries of** $\mathcal{U}$**, with coefficients in** $\mathcal{S}$ (or even **of** $\mathcal{S}$**, with respect to** $\mathcal{U}$). So, according to standard notation in Homological Algebra, (4.20) is, *by definition,* given through the following relation, being justified by (4.18) (since, then $\check{B}^n(\mathcal{U}, \mathcal{S}) \subseteq \check{Z}^n(\mathcal{U}, \mathcal{S})$, $n \in \mathbf{Z}_+$);

$$\check{H}^n(\mathcal{U}, \mathcal{S}) := h^n(\check{C}^{\cdot}(\mathcal{U}, \mathcal{S})) := \ker \delta^n / \operatorname{im} \delta^{n-1} \equiv \check{Z}^n(\mathcal{U}, \mathcal{S}) / \check{B}^n(\mathcal{U}, \mathcal{S}). \tag{4.23}$$

In this context, we also say that *two* given *n-th (dimensional) cocycles* are **cohomologous,** if their difference (cf. (4.21)) is an $n$-th (dimensional) coboundary; namely, whenever *they define the same cohomology class in* $\check{H}^n(\mathcal{U}, \mathcal{S})$.

> **Remark 4.1.**– According to the hypothesis for the *structure sheaf* $\mathcal{A}$ (cf. Chapt. II; Scholim 1.2), one has the relation
>
> $$\mathbf{C} \underset{\varepsilon}{\overset{\subsetneq}{\rightarrow}} \mathcal{A}, \tag{4.24}$$
>
> where $\mathbf{C}$ stands for the corresponding *constant sheaf of complexes* on $X$ and $\varepsilon \equiv \varepsilon_X$ for the respective *canonical sheaf injection* (see Chapt. II; (1.27) and (1.34)). Thus, *all* the previously considered $\mathcal{A}(X)$*-modules are, in particular,* $\mathbf{C}$*-vector spaces,* a fact which will be of use later on, the respective morphisms being, in particular, $\mathbf{C}$*-linear morphisms.*

Now, as a first application of the preceding, one obtains the next result, reminiscent of Theorem 3.1; (3.14) (see also Theorem 2.1; (2.13)). In effect, as we shall see in the sequel (cf. Section 7 below), it is just a particular instance of the fact that *"Čech cohomology"*, as this will be defined below, is, indeed, (when restricted to paracompact spaces) a *"cohomology theory"* (cf. Section 8). So we now have.

**Lemma 4.1.** *Let $(X, \mathcal{A})$ be a* **C**-*algebraized space, $\mathcal{S}$ an $\mathcal{A}$-module on $X$ and $\mathcal{U} = \{U_\alpha\}_{\alpha \in I}$ an open covering of $X$. Then, one has* (cf. (4.20))

$$\check{H}^0(\mathcal{U}, \mathcal{S}) = \Gamma(X, \mathcal{S}) \equiv \mathcal{S}(X), \tag{4.25}$$

*within an $\mathcal{A}(X)$-isomorphism.*

**Proof.** First, one sets, *by definition;*

$$\delta^p = 0, \text{ for every integer } p < 0. \tag{4.26}$$

Thus, in view of (4.22), one has

$$\check{B}^0(\mathcal{U}, \mathcal{S}) := \operatorname{im} \delta^{-1} = 0. \tag{4.27}$$

Therefore (cf. (4.23)), one obtains (see also (4.11))

$$\check{H}^0(\mathcal{U}, \mathcal{S}) = \check{Z}^0(\mathcal{U}, \mathcal{S}) \subseteq C^0(\mathcal{U}, \mathcal{S}) = \prod_{\alpha \in I} \mathcal{S}(U_\alpha). \tag{4.28}$$

So, let $f \equiv (f_\alpha)$, with $f_\alpha \in \mathcal{S}(U_\alpha)$, $\alpha \in I$, be a 0-cochain of $\mathcal{S}$ being, in particular, a 0-*cocycle;* That is, suppose that (cf. also (4.16))

$$\delta^0(f) = \delta^0((f_\alpha)) \equiv (f_{\alpha\beta}) = (f_\beta - f_\alpha) = 0, \tag{4.29}$$

for any $\alpha, \beta$ in $I$, with $U_{\alpha\beta} \equiv U_\alpha \cap U_\beta \neq \emptyset$. Hence, one actually obtains that

$$f_\alpha = f_\beta|_{U_\alpha \cap U_\beta}, \tag{4.30}$$

for every pair $(\alpha, \beta) \in I^2$, as before. Now, given that $\Gamma(\mathcal{S})$, the *presheaf of sections of $\mathcal{S}$, is complete* (cf. Chapt. I; Lemma 11.1), one concludes the existence of a *global section $g \in \mathcal{S}(X) \equiv \Gamma(X, \mathcal{S})$*, extending the given $f_\alpha$'s (see Chapt. I; Definition 11.1). Furthermore, *the correspondence*

$$\check{H}^0(\mathcal{U}, \mathcal{S}) = \check{Z}^0(\mathcal{U}, \mathcal{S}) \ni f \equiv (f_\alpha) \longmapsto g \in \Gamma(X, \mathcal{S}), \tag{4.31}$$

as previously established, *is,* in effect, *a bijection* of the respective sets, as easily proved (cf. also Chapt. I; cond. (S1) in Definition 11.1), *preserving the $\mathcal{A}(X)$-module structures* of the same sets, as well (same properties of $\Gamma(\mathcal{S})$). ∎

**4.(c). Passage to the limit.**– We examine next the variation of the Čech cochain complex $\check{C}^\cdot(\mathcal{U}, \mathcal{S})$, as defined by (4.19) above, with the open covering $\mathcal{U}$ of $X$ under consideration. So we continue to work within a given **C**-*algebraized space* $(X, \mathcal{A})$, and let $\mathcal{S}$ be a fixed $\mathcal{A}$-*module* on $X$.

Now, suppose further that $\mathcal{U}$ and $\mathcal{V}$ are two given open coverings of $X$, such that $\mathcal{V}$ *is finer that* $\mathcal{U}$; viz. we have (cf. (4.2))

(4.32) $$\mathcal{V} \succeq \mathcal{U},$$

so that, adopting the notation of (4.1), assume that

(4.33) $$\mu : J \longrightarrow I$$

is a *refining map,* yielding (4.32) (cf. (4.3), (4.4)). Thus, *for every integer* $n \geqq 0$, one gets a *map of the* corresponding $\mathcal{A}(X)$*-modules of* $n$*-cochains* (see (4.14))

(4.34) $$\mu^n \equiv \mu : \check{C}^n(\mathcal{U}, \mathcal{S}) \longrightarrow \check{C}^n(\mathcal{V}, \mathcal{S})$$

(we still employ here, for simplicity, an obvious abuse of notation), given by the relation

(4.34′) $$\mu(f)_{\beta_0\beta_1\cdots\beta_n} := f_{\mu(\beta_0)\mu(\beta_1)\cdots\mu(\beta_n)}|V_{\beta_0\beta_1\cdots\beta_n},$$

for any $f \equiv (f_{\alpha_0\alpha_1\cdots\alpha_n}) \in \check{C}^n(\mathcal{U}, \mathcal{S})$ (see also (4.11) and (4.4)).

Now, based on the very definitions, one easily proves that:

(4.35) *for every integer* $n \geqq 0$, *the map* $\mu$, as given by (4.34′) (and associated with the given refining map (4.33)), *is an* $\mathcal{A}(X)$*-morphism, commuting with the* $\delta$'s (cf. (4.15), or even (4.18′)); *viz. one has;*

(4.35.1) $$\mu \circ \delta = \delta \circ \mu.$$

*Precisely speaking, one thus obtains;*

(4.35.2) $$\mu \circ \delta^n = \delta^n \circ \mu, \; n \in \mathbf{Z}_+.$$

Thus, as consequence of (4.35) (cf. (4.35.1), along with (4.21) and (4.22)), *one obtains, for every integer* $n \geqq 0$;

(4.36) $$\mu(\check{Z}^n(\mathcal{U}, \mathcal{S}) \subseteq \check{Z}^n(\mathcal{V}, \mathcal{S}),$$

*as well as,*

(4.37) $$\mu(\check{B}^n(\mathcal{U}, \mathcal{S})) \subseteq \check{B}^n(\mathcal{V}, \mathcal{S}).$$

Therefore, one finally gets *a map in (n-th dimensional,* $n \in \mathbf{Z}_+$*) cohomolo-*

*gy,* in fact, an $\mathcal{A}(X)$*-morphism* (notice the employed, here again, abuse of notation),

$$\mu^{*,n} \equiv \mu^* : \check{H}^n(\mathcal{U}, \mathcal{S}) \longrightarrow \check{H}^n(\mathcal{V}, \mathcal{S}),\ n \in \mathbf{Z}_+, \tag{4.38}$$

such that *one has*

$$\mu^*([f]) = [\mu(f)], \tag{4.39}$$

*for any* $[f] \in \check{H}^n(\mathcal{U}, \mathcal{S})$, *with* $f \in \check{Z}^n(\mathcal{U}, f)$ (cf. (4.36)).

Now, the important fact at this point is that:

(4.40) *(4.38) does not actually depend on the particular refining map* $\mu$, *under consideration* (cf. (4.33)); hence, *only on the relation* $\mathcal{U} \preceq \mathcal{V}$ (cf. (4.32)) *between the two given open coverings* $\mathcal{U}$, $\mathcal{V}$ *of* $X$.

More precisely speaking, one obtains the following result. (However, see also (6.11) below).

**Lemma 4.2.** *Let* $(X, \mathcal{A})$ *be a* $\mathbf{C}$*-algebraized space and* $\mathcal{S}$ *an* $\mathcal{A}$*-module on* $X$. *Moreover, let* $\mathcal{U} = (U_\alpha)_{\alpha \in I}$ *and* $\mathcal{V} = (V_\beta)_{\beta \in J}$ *be two given open coverings of* $X$ *such that* $\mathcal{V} \succeq \mathcal{U}$ (cf. (4.2)), *and let* $\mu, \nu : J \longrightarrow I$ *be two refining maps* (cf (4.3)). *Then, concerning the corresponding cochain maps* (cf. (4.34), (4.34′)), *one obtains, for every integer* $n \geqq 0$,

$$(\mu^n \equiv)\ \mu = \nu\ (\equiv \nu^n). \tag{4.41}$$

*Therefore, the analogous relation to (4.41) holds true for the respective maps in cohomology* (cf. (4.38)); *viz. one has*

$$\mu^*\ (\equiv \mu^{*,n}) = (\nu^{*,n} \equiv)\ \nu^*\ \ (n \in \mathbf{Z}_+). \tag{4.42}$$

**Proof.** First, due to the $\mathcal{A}(X)$*-isomorphism* (4.25), one obtains, *for* $n = 0$, the corresponding relation to (4.42), since in that case each one of the maps $\mu^*$, as in (4.38) (with $\mu$ given by (4.3), satisfying (4.4)), is essentially the *identity map on* $\Gamma(X, \mathcal{S})$; namely, denoting by

$$\tau_{\mathcal{U}} : \check{H}^0(\mathcal{U}, \mathcal{S}) = \check{Z}^0(\mathcal{U}, \mathcal{S}) \longrightarrow \Gamma(X, \mathcal{S}) \tag{4.43}$$

the $\mathcal{A}(X)$*-isomorphism,* established by Lemma 4.1, *for any given open covering* $\mathcal{U}$ *of* $X$, one obtains, concerning a given refining map $\mu$ (cf. also (4.34) and (4.38)),

$$\tau_{\mathcal{V}} \circ \mu \circ \tau_{\mathcal{U}}^{-1} = 1, \tag{4.44}$$

where we set $1 \equiv \mathrm{id}_{\Gamma(X,\mathcal{S})}$, which thus proves our claim, for $n = 0$.

On the other hand, *for any integer* $n \geqq 1$, one first defines the following auxiliary map, yet (by its very definition, cf. (4.45′)) an $\mathcal{A}(X)$*-morphism,*

$$\lambda : \check{C}^n(\mathcal{U}, \mathcal{S}) \longrightarrow \check{C}^{n-1}(\mathcal{V}, \mathcal{S}), \tag{4.45}$$

in such a manner that

$$\lambda(f)_{\beta_0\beta_1\cdots\beta_n} := \sum_{i=0}^{n-1} f_{\mu(\beta_0)\cdots\mu(\beta_i)\nu(\beta_i)\cdots\nu(\beta_{n-1})}|_{V_0\cap\cdots\cap V_{n-1}}, \tag{4.45′}$$

for any $f \equiv (f_{\alpha_0\alpha_1\cdots\alpha_n}) \in \check{C}^n(\mathcal{U}, \mathcal{S})$. Now, a routine computation yields that

$$\mu - \nu = \lambda \circ \delta + \delta \circ \lambda, \tag{4.46}$$

on the respective "dimension level" (viz. *for any integer* $n \geqq 1$; see also (4.15), as well as, (4.18′) for the applied notation). Thus, for any given $n$*-cocycle* $f \in \check{Z}^n(\mathcal{U}, \mathcal{S})$, so that $\delta(f) = 0$, one obtains, in view of (4.46),

$$\mu(f) - \nu(f) = \delta(\lambda(f)) \in B^n(\mathcal{V}, \mathcal{S}) \tag{4.47}$$

(cf. also (4.22), (4.45)); viz. *the elements* $\mu(f), \nu(f)$ in $\check{Z}^n(\mathcal{V}, \mathcal{S})$ (see (4.36)) *are cohomologous* (cf. the comments following (4.23)). Hence (cf. also (4.39)), *one has*

$$\mu^*([f]) = [\mu(f)] = [\nu(f)] = \nu^*([f]), \tag{4.48}$$

*for any* $f \in \check{Z}^n(\mathcal{U}, \mathcal{S})$, which thus terminates the proof. ∎

The preceding enables thus one to consider, for any pair $(\mathcal{U}, \mathcal{V})$ of open coverings of $X$, *with* $\mathcal{V} \succeq \mathcal{U}$ (see (4.2)), a *uniquely defined map,* in fact, *an* $\mathcal{A}(X)$*-morphism,* say,

$$\tau_{\mathcal{V}\mathcal{U}} : \check{H}^n(\mathcal{U}, \mathcal{S}) \longrightarrow \check{H}^n(\mathcal{V}, \mathcal{S}), \tag{4.49}$$

*for every* $n \in \mathbf{Z}_+$ (hence, in effect, $\tau_{\mathcal{V}\mathcal{U}} \equiv \tau_{\mathcal{V}\mathcal{U}}(n, \mathcal{S})$), *independently of the refining map* $\mu$ (cf. (4.1), (4.3)), that was actually employed to yield (4.49) (see (4.38)). Therefore, one can also write, for simplicity, unambiguously, just $\mu^* \equiv \mu^{*,n}$, as in (4.38), in place of $\tau_{\mathcal{V}\mathcal{U}}$, as the case may be (cf. thus Lemma (4.3) in the sequel).

On the other hand, by considering the *(upwards) directed set of all proper open coverings of* $X$ (cf. (4.9)), one easily proves, based on the very definitions and Lemma 4.2, the relations (cf. also e.g. Chapt. I; (5.2), (5.3)):

$$\tau_{\mathcal{U}\mathcal{U}} = \mathrm{id} \tag{4.50}$$

and

$$\tau_{\mathcal{W}\mathcal{U}} = \tau_{\mathcal{W}\mathcal{V}} \circ \tau_{\mathcal{V}\mathcal{U}}, \tag{4.51}$$

for any *(proper) open coverings* $\mathcal{U}, \mathcal{V}, \mathcal{W}$ *of* $X$, *with* $\mathcal{W} \succeq \mathcal{V} \succeq \mathcal{U}$ (cf. (4.2)). These relations are valid, of course, *for any integer* $n \geqq 0$ (cf. (4.49)). In other words, we thus conclude the following;

(4.52) *for every integer* $n \geqq 0$, one obtains an *inductive system of* $\mathcal{A}(X)$*-modules* (cf. also (4.20), or even (4.23))

$$(\check{H}^n(\mathcal{U}, \mathcal{S}), \tau_{\mathcal{V}\mathcal{U}}), \tag{4.52.1}$$

*with respect to the above directed index set* (see (4.9)).

So we come now to the following fundamental concept, which thus supplements our previous one, as given by Definition 4.1. Namely, one sets the next.

**Definition 4.2.** Let $(X, \mathcal{A})$ be a **C**-algebraized space and $\mathcal{S}$ an $\mathcal{A}$-module on $X$. Then, for every integer $n \geqq 0$, the $n$**-th (dimensional) Čech cohomology** $\mathcal{A}(X)$**-module of** $X$ **with coefficients in** (the given $\mathcal{A}$-module) $\mathcal{S}$, denoted by $\check{H}^n(X, \mathcal{S})$, is the (inductive) *limit* $\mathcal{A}(X)$*-module* of (4.52); viz.

$$\check{H}^n(X, \mathcal{S}) := \varinjlim_{\mathcal{U}} \check{H}^n(\mathcal{U}, \mathcal{S}). \tag{4.53}$$

Yet, one defines the **Čech cohomology of** $X$ **with coefficients in** $\mathcal{S}$ by

$$\check{H}^*(X, \mathcal{S}) \equiv \left\{\check{H}^n(X, \mathcal{S})\right\}_{n \in \mathbf{Z}_+}. \tag{4.53'}$$

Concerning the terminology, which we applied in (4.53), see also, for instance, S. Lang [1: p.107, Theorem 9.1]. To see the preceding at work, we first note, for instance, an easy consequence of (4.53) in conjunction with Lemma 4.1. Namely, one has

$$\check{H}^0(X, \mathcal{S}) = \check{H}^0(\mathcal{U}, \mathcal{S}) = \Gamma(X, \mathcal{S}), \tag{4.54}$$

within $\mathcal{A}(X)$-*isomorphisms* of the $\mathcal{A}(X)$-modules concerned. (In this respect, see also, for instance, N. Bourbaki [1: p.90, Example 2)]).

Indeed, something more is true, concerning (4.54); thus, one has (cf. also Definition 3.1)

$$H^0(X,\mathcal{S}) = \check{H}^0(X,\mathcal{S}) = \check{H}^0(\mathcal{U},\mathcal{S}) = \check{Z}^0(\mathcal{U},\mathcal{S}) = \Gamma(X,\mathcal{S}), \tag{4.55}$$

within $\mathcal{A}(X)$-*isomorphisms* of the respective $\mathcal{A}(X)$-modules; the assertion follows from (3.10) (see also (3.14), (4.28) and (4.54)). ∎ See also Section 8 in the sequel, concerning a similar result to (4.55) in dimension 1, pertaining to the first two terms of the last relation.

On the other hand, the above map (4.49) *is*, in fact, *one-to-one in dimension 1.* Namely, one has the following useful result.

**Lemma 4.3.** *Suppose we are given a* **C**-*algebraized space* $(X,\mathcal{A})$ *and let* $\mathcal{S}$ *be an* $\mathcal{A}$-*module on* $X$. *Moreover, let* $\mathcal{U}$ *and* $\mathcal{V}$ *be two open coverings of* $X$, as in (4.1), *with* $\mathcal{V} \succeq \mathcal{U}$ (cf. (4.2)), *and let* $\mu$ *be any refining map* (cf. (4.3)). *Then, the* $\mathcal{A}(X)$-*morphism*

$$\mu^* \, (\equiv \mu^{*,1} \equiv \tau_{\mathcal{V}\mathcal{U}}) : \check{H}^1(\mathcal{U},\mathcal{S}) \longrightarrow \check{H}^1(\mathcal{V},\mathcal{S}) \tag{4.56}$$

(see also (4.38) and (4.49), for $n=1$) *is one-to-one.*

**Proof.** Namely, assume that (cf. also (4.39))

$$\mu^*([f]) = [\mu(f)] = 0, \tag{4.57}$$

for some $[f] \in \check{H}^1(\mathcal{U},\mathcal{S})$, with $f \in \check{Z}^1(\mathcal{U},\mathcal{S})$. Thus (cf. (4.23)), $\mu(f) \in B^n(\mathcal{U},\mathcal{S})$, so that (cf. (4.15), (4.22)) there exists $g \equiv (g_\beta) \in C^0(\mathcal{V},\mathcal{S})$ (cf. also (4.1) and the hypothesis for $\mathcal{U},\mathcal{V}$) such that one has;

$$\mu(f) = \delta(g) \equiv \delta^0((g_\beta)) = (g_{\beta_1} - g_{\beta_0}) \tag{4.58}$$

(cf. (4.16)). Therefore, based on (4.34′) and (4.58), one obtains

$$\mu(f)_{\beta_0\beta_1} := f_{\mu(\beta_0)\mu(\beta_1)} = g_{\beta_1} - g_{\beta_0}, \tag{4.59}$$

as elements of (the $\mathcal{A}(X)$-module) $\mathcal{S}(V_{\beta_0} \cap V_{\beta_1})$, for any pair $(\beta_0,\beta_1) \in J^2 \equiv J \times J$ (cf. also (4.13)).

On the other hand, one gets an *open covering of an open set* $U_\alpha \in \mathcal{U}$, say $\mathcal{U}^{(\alpha)}$, by setting (cf. (4.1));

$$\mathcal{U}^{(\alpha)} = \{U_\alpha \cap V_\beta\}_{\beta\in J}. \tag{4.60}$$

Furthermore, one gets the following 0-*cocycle of* $\mathcal{U}^{(\alpha)}$

$$h_\alpha \equiv (h_\beta^{(\alpha)}) \in \check{Z}^0(\mathcal{U}^{(\alpha)}, \mathcal{S}) \subseteq \check{C}^0(\mathcal{U}^{(\alpha)}, \mathcal{S}), \tag{4.61}$$

such that

$$h_\beta^{(\alpha)} := g_\beta + f_{\mu(\beta)\alpha} \in \mathcal{S}(U_\alpha \cap V_\beta), \; \beta \in J. \tag{4.61$'$}$$

Now, concerning our claim for $h_\alpha$ (cf. (4.61)), by considering the open set

$$U_\alpha \cap V_{\beta_0} \cap V_{\beta_1} \subseteq U_\alpha \cap U_{\mu(\beta_0)} \cap U_{\mu(\beta_1)} \tag{4.62}$$

(see also (4.4)), as well as, the given $f \in \check{Z}^1(\mathcal{U}, \mathcal{S})$, one obtains (see also (4.16));

$$\delta(f) \equiv (\delta(f)_{\alpha_0\alpha_1\alpha_2}) = 0. \tag{4.63}$$

Therefore, one has (cf. also (4.16) and (4.59))

$$\begin{aligned} \delta(f)_{\mu(\beta_0)\mu(\beta_1)\alpha} &= f_{\mu(\beta_1)\alpha} - f_{\mu(\beta_0)\alpha} + f_{\mu(\beta_0)\mu(\beta_1)} \\ &= f_{\mu(\beta_1)\alpha} - f_{\mu(\beta_0)\alpha} + g_{\beta_1} - g_{\beta_0} = 0. \end{aligned} \tag{4.64}$$

Consequently, in view of (4.61$'$), one obtains

$$\delta(h_\alpha) \equiv \delta\left(\left(h_\beta^{(\alpha)}\right)\right) = \left(h_{\beta_1}^{(\alpha)} - h_{\beta_0}^{(\alpha)}\right) = 0. \tag{4.65}$$

Therefore, one concludes that (cf. (4.25), (4.60))

$$h_\alpha \equiv \left(h_\beta^{(\alpha)}\right)_{\beta \in J} \in \check{Z}^0(\mathcal{U}^{(\alpha)}, \mathcal{S}) = \check{H}^0(\mathcal{U}^{(\alpha)}, \mathcal{S}) \cong \Gamma(U_\alpha, \mathcal{S}) \equiv \mathcal{S}(U_\alpha), \tag{4.66}$$

which thus justifies (4.61). Now setting

$$h \equiv (h_\alpha)_{\alpha \in I} \in \check{C}^0(\mathcal{U}, \mathcal{S}), \tag{4.67}$$

one further obtains, by (4.61$'$) and the hypothesis for $f$ (cf. also (4.16) and (4.72) below),

$$\begin{aligned} \delta(h)_{\alpha_0\alpha_1} &\equiv \delta((h_\alpha))_{\alpha_0\alpha_1} = h_{\alpha_1} - h_{\alpha_0} \\ &= g_\beta + f_{\mu(\beta)\alpha_1} - g_\beta - f_{\mu(\beta)\alpha_0} = f_{\mu(\beta)\alpha_1} - f_{\mu(\beta)\alpha_0} \\ &= f_{\alpha_0\mu(\beta)} + f_{\mu(\beta)\alpha_1} = f_{\alpha_0\alpha_1}, \end{aligned} \tag{4.68}$$

for any $\alpha_0$, $\alpha_1$ in $I$, so that one has (cf. (4.67) and (4.22))

$$f = \delta(h) \in \check{B}^1(\mathcal{U}, \mathcal{S}); \tag{4.69}$$

hence, $[f] = 0$, which thus terminates the proof. ∎

**Remark 4.2.**- Suppose we have the context of the above Lemma 4.3; moreover, consider the 1-*cocycle*

$$g \equiv (g_{\alpha\beta}) \in \check{Z}^1(\mathcal{U}, \mathcal{S}), \tag{4.70}$$

with $\mathcal{U} = \{U_\alpha\}_{\alpha \in I}$ a given open covering of $X$. Then, *one has*

$$g_{\alpha\alpha} = 0, \quad \textit{for any } \alpha \in I. \tag{4.71}$$

Indeed, by taking the triple $(\alpha, \alpha, \beta) \in I^3$, one has, in view of (4.70) and (4.16); $(\delta g)_{\alpha\alpha\beta} = g_{\alpha\beta} - g_{\alpha\beta} + g_{\alpha\alpha} = 0$, which proves (4.71).

Yet, *every* 1-*cocycle is "alternating"*; viz. *one has*

$$g_{\beta\alpha} = -g_{\alpha\beta}, \quad \textit{for any } (\alpha, \beta) \in I^2. \tag{4.72}$$

For, by considering the triple $(\alpha, \beta, \alpha) \in I^3$, one gets, in view of (4.70), (4.16) and (4.71); $(\delta g)_{\alpha\beta\alpha} = g_{\beta\alpha} - g_{\alpha\alpha} + g_{\alpha\beta} = g_{\beta\alpha} + g_{\alpha\beta} = 0$, i.e., (4.72). ∎

Now, an immediate consequence of Lemma 4.3 is the following.

**Corollary 4.1.** *Let $(X, \mathcal{A})$ be a* **C**-*algebraized space and $\mathcal{S}$ a given $\mathcal{A}$-module on $X$. Then, one has $\check{H}^1(X, \mathcal{S}) = 0$ if, and only if, $\check{H}^1(\mathcal{U}, \mathcal{S}) = 0$, for any open covering $\mathcal{U}$ of $X$ (in the set (4.9) or, equivalently, in a cofinal subset of (4.9)).*

**Proof.** According to (4.53) (see also (4.52)), *one has*

$$\check{H}^1(X, \mathcal{S}) := \varinjlim_{\mathcal{U}} \check{H}^1(\mathcal{U}, \mathcal{S}) = \bigcup_{\mathcal{U}} \check{H}^1(\mathcal{U}, \mathcal{S}). \tag{4.73}$$

(In this respect, cf. also e.g. N. Bourbaki [1: p.92, Remarque 1)]). This proves the *"if"* part of our claim, the rest being obvious, by the very hypothesis. (But see also N. Bourbaki [1: p.90, Example 2)]). ∎

**4.(d). Leray's Theorem.**- Concerning the previous Lemma 4.3, one can obtain an analogous conclusion, *for any integer* $n \geq 1$ (in fact, one gets an *identification* of the $\mathcal{A}(X)$-modules concerned, modulo an $\mathcal{A}(X)$-isomorphism, see Corollary 4.2 below), as a consequence of the classical *Leray's* *theorem.* The latter theorem is, however, formulated for *paracompact (Hausdorff) spaces* (cf., for instance, R. C. Gunning - H. Rossi [1: p.189, Theorem 4]).

On the other hand, following the standard argument, we can also get an *abstract form of that classical result* (cf. Theorem 4.1 below), by imposing, instead, on the part of the open coverings and the $\mathcal{A}$-modules involved, appropriate restrictions, illuminating also, for that matter, the rôle played in the classical technique by various fundamental concepts, already considered in the foregoing.

Thus, given an *arbitrary* **C**-algebraized space, as before (we do not thus consider any restriction on the topological space $X$), an *open covering* $\mathcal{U} \equiv \{U_\alpha\}_{\alpha\in I}$ of $X$ and an *$\mathcal{A}$-module* $\mathcal{S}$ on $X$, we shall say that $\mathcal{U}$ is *acyclic with respect to* $\mathcal{S}$, whenever one has

$$\check{H}^n(U_{\alpha_0\alpha_1\cdots\alpha_p}, \mathcal{S}) = 0, \tag{4.74}$$

*for every integer* $n \geqq 1$ *and for any finite intersection* (of elements) *in* $\mathcal{U}$; viz. *for every* $U_{\alpha_0\alpha_1\cdots\alpha_p} \equiv U_{\alpha_0} \cap U_{\alpha_1} \cap \cdots \cap U_{\alpha_p}$, *with* $U_{\alpha_p} \in \mathcal{U}$, *for all* $p \in \mathbf{Z}_+$. (Note that a so-called *"good covering"* is such, no matter what $\mathcal{S}$ we consider; see, for instance, R. Bott - L. W. Tu [1: p.147], as well as, Section 8 below).

Now, another important ingredient of the classical argument is the use of a *long exact cohomology sequence, associated with any given short exact sequence;* we can assure this for Čech cohomology theory in the case of paracompact spaces (cf. the next Section 5, Theorem 5.2), while *we do have this in* the case of (abstract) *sheaf cohomology* (viz. of a *derived functor cohomology,* as already seen; of course, a significant advantage, for that matter, of the latter theory, against Čech cohomology, on an arbitrary topological space, as already said. Thus, see e.g. Theorem 3.1 in the preceding). So we can now state the following.

**Theorem 4.1.** *(Abstract Leray's theorem).* Suppose we have a **C**-*algebraized space* $(X, \mathcal{A})$, $\mathcal{S}$ an $\mathcal{A}$-*module* on $X$ and $\mathcal{U}$ an *open covering of* $X$, *acyclic with respect to* $\mathcal{S}$ (cf. (4.74)). Moreover, let

$$0 \longrightarrow \mathcal{S} \longrightarrow \mathcal{S}^0 \longrightarrow \mathcal{S}^1 \longrightarrow \cdots \tag{4.75}$$

be a *(Čech) strongly acyclic $\mathcal{A}$-resolution of $\mathcal{S}$, with respect to $\mathcal{U}$;* we assume that

(4.76) $$\check{H}^n(X, \mathcal{S}^i) = \check{H}^n(\mathcal{U}, \mathcal{S}^i) = 0, \textit{ for all } n \in \mathbf{N} \textit{ and } i \in \mathbf{Z}_+.$$

(Cf. also Definitions 1.2 and 4.1 in the preceding). Finally, suppose that *to every short exact $\mathcal{A}$-sequence, there corresponds a long exact (Čech) cohomology sequence* (cf., for instance, (3.15) and (3.17) above). Then, *one obtains*

(4.77) $$\check{H}^n(X, \mathcal{S}) = \check{H}^n(\mathcal{U}, \mathcal{S}), \; n \in \mathbf{Z}_+,$$

*within an $\mathcal{A}(X)$-isomorphism,* of the $\mathcal{A}(X)$-modules concerned.

**Proof.** This can be patterned, within the above (abstract) context, along the classical steps in the case of paracompact (Hausdorff) spaces, while, to this effect, the crucial items have been just axiomatized (thus, abstracted), as before; so see, for instance, R. C. Gunning - H. Rossi [1: p.189, Theorem 4]. ∎

As already said, the assumptions of the previous theorem are all satisfied in the case of a paracompact (Hausdorff) space, by taking, for instance, a *"good covering"* $\mathcal{U}$ of $X$ (in fact a *"Leray covering"* suffices, viz. such a one that (4.74) holds true) and a *finite resolution* of the given $\mathcal{A}$-module $\mathcal{S}$ (cf. also the next Section 5).

An immediate consequence of the previous theorem is now the following.

**Corollary 4.2.** Suppose we have *the context of the previous Theorem 4.1,* while now $\mathcal{U}$, $\mathcal{V}$ are two given *open coverings of $X$, both of which satisfy the corresponding conditions, with respect to the $\mathcal{A}$-module $\mathcal{S}$* under consideration, with $\mathcal{V}$ *being* also *a refinement of $\mathcal{U}$.* Moreover, let $\mu$ be *any refining map* (cf. (4.3)). *Then,* the corresponding $\mathcal{A}(X)$-morphism (cf. also (4.38))

(4.78) $$\mu^* : \check{H}^n(\mathcal{U}, \mathcal{S}) \longrightarrow \check{H}^n(\mathcal{V}, \mathcal{S})$$

*is an $\mathcal{A}(X)$-isomorphism.*

**Proof.** First, note that *for $n = 0$*, this already happens, *in general,* due to (4.54). Now, *for $n \geq 1$*, one considers a *resolution of $\mathcal{S}$, as in (4.75)* above (so that, by hypothesis, *(4.76) is valid for both $\mathcal{U}$,$\mathcal{V}$*), applying further the corresponding classical argument; so cf., for instance, R. C. Gunning - H. Rossi [1: p.191, Corollary 5]. ∎

# 5. Lifting of cocycles

Suppose we are found within the *context of* the previous *Section 4,* as it concerns the basic objects we are dealt with. So, based on (4.11), one can consider, for any *open covering* $\mathcal{U}$ of $X$, the following *functor* (correspondence)

(5.1) $$\check{C}^n_{\mathcal{U}} : \mathcal{A}\text{-}\mathcal{M}od_X \longrightarrow \mathcal{A}(X)\text{-}\mathcal{M}od,\ n \in \mathbf{Z}_+,$$

(see also (1.1) and Chapt. II; (6.39) for the notation applied), such that one sets

(5.2) $$\check{C}^n_{\mathcal{U}}(\mathcal{S}) := \check{C}^n(\mathcal{U}, \mathcal{S}),$$

for any $\mathcal{A}$*-module* $\mathcal{S}$ on $X$. So one easily proves, by the very definitions that:

(5.3) for any given fixed open covering $\mathcal{U}$ of $X$, and any integer $n \geqq 0$, *the* mapping (5.1), as given by (5.2), yields a *left exact covariant functor* of the categories concerned.

Now, *the above functors are not, in general, exact.* Namely, if

(5.4) $$0 \longrightarrow \mathcal{F} \xrightarrow{\phi} \mathcal{S} \xrightarrow{\psi} \mathcal{E} \longrightarrow 0$$

is a *short exact* $\mathcal{A}$*-sequence* on $X$ (cf. Chapt. II; (2.18)), one gets (in general, only!) the following *exact sequence of* $\mathcal{A}(X)$*-modules*

(5.5) $$0 \longrightarrow \check{C}^n(\mathcal{U}, \mathcal{F}) \xrightarrow{\phi} \check{C}^n(\mathcal{U}, \mathcal{S}) \xrightarrow{\psi} \check{C}^n(\mathcal{U}, \mathcal{E}),$$

with $\mathcal{U}$ *a given open covering of* $X$ *and* $n \in \mathbf{Z}_+$. (By an obvious *abuse of notation,* we have used here, *for simplicity,* the same symbols for the given $\mathcal{A}$-morphisms $\phi, \psi$ in (5.4) and their images, through the functor (5.1), in (5.5)). In this context, note also that *the above inconvenience does not arise,* as we shall see in the next Section 6, in the case of *Čech cohomology with coefficients in a given presheaf* (of modules). (Yet, concerning our assertion about (5.5), cf. e.g. E. H. Spanier [1: p.325, Example 8]).

So our objective in the ensuing discussion is to look nearer at the above *inconvenience, arising from being the* $\mathcal{A}(X)$*-morphism* $\psi$ *in (5.5) not, in general, surjective,* as well as, to see thus the way this can be circumvented at the expense, however, of the generality of the topological space $X$ considered.

Thus, setting (cf. also (5.5), as well as, (4.11) and (4.14))

$$(5.6) \qquad \check{C}^n_0(\mathcal{U},\mathcal{E}) := \operatorname{im}\psi \equiv \psi(\check{C}^n(\mathcal{U},\mathcal{S}) \subseteq \check{C}^n(\mathcal{U},\mathcal{E}),$$

one gets the following *short exact sequence of* $\mathcal{A}(X)$*-modules* (or even *short exact* $\mathcal{A}(X)$*-sequence*)

$$(5.7) \qquad 0 \longrightarrow \check{C}^n(\mathcal{U},\mathcal{F}) \xrightarrow{\phi} \check{C}^n(\mathcal{U},\mathcal{S}) \xrightarrow{\psi} \operatorname{im}\psi \equiv \check{C}^n_0(\mathcal{U},\mathcal{E}) \longrightarrow 0;$$

hence, the obviously justified terminology, calling, namely, (5.6) the $\mathcal{A}(X)$-*module of liftable* $n$*-cochains* (of $\mathcal{U}$ with coefficients in $\mathcal{E}$), $n \in \mathbf{Z}_+$.

Now, it is readily seen, by the very definitions, that *the coboundary operators* $\delta$ (in fact, each $\delta^n$, $n \geqq 0$, cf. (4.15)) *commute with* $\mathcal{A}(X)$*-morphisms* (cf., for instance, (4.35.2) above); so one obtains, in particular,

$$(5.8) \qquad \delta^n \circ \psi = \psi \circ \delta^n, \ n \in \mathbf{Z}_+.$$

(An analogous relation to (5.8) holds also, of course, for the $\mathcal{A}(X)$-morphism $\phi$ in (5.7)). Yet, we also write, instead of (5.8), briefly,

$$(5.8') \qquad \delta \circ \psi = \psi \circ \delta.$$

In other words, *coboundary operators preserve liftable cochains;* thus, precisely speaking, one obtains the following $\mathcal{A}(X)$-morphism, for every $n \in \mathbf{Z}_+$,

$$(5.9) \qquad \check{C}^n_0(\mathcal{U},\mathcal{E}) \xrightarrow{\delta(\equiv\delta^n)} \check{C}^{n+1}_0(\mathcal{U},\mathcal{E})$$

(See also (5.6)). So one sets, in particular,

$$(5.10) \ \check{Z}^n_0(\mathcal{U},\mathcal{E}) := \check{Z}^n(\mathcal{U},\mathcal{E}) \cap \check{C}^n_0(\mathcal{U},\mathcal{E}) \equiv \check{Z}^n(\mathcal{U},\mathcal{E}) \cap \psi(\check{C}^n(\mathcal{U},\mathcal{S})) \subseteq \check{C}^n(\mathcal{U},\mathcal{E})$$

(*liftable* $n$*-cocycles* of $\mathcal{U}$ with coefficients in $\mathcal{E}$). Similarly, one defines;

$$(5.11) \ \check{B}^n_0(\mathcal{U},\mathcal{E}) := \check{B}^n(\mathcal{U},\mathcal{E}) \cap \check{C}^n_0(\mathcal{U},\mathcal{E}) \equiv \check{B}^n(\mathcal{U},\mathcal{E}) \cap \psi(\check{C}^n(\mathcal{U},\mathcal{S})) \subseteq \check{C}^n(\mathcal{U},\mathcal{E})$$

(*liftable* $n$*-coboundaries* of $\mathcal{U}$ with coefficients in $\mathcal{E}$). Now, it is an immediate consequence of the above two last relations, along with (4.21) and (4.22), that

$$(5.12) \qquad \check{B}^n_0(\mathcal{U},\mathcal{E}) \subseteq \check{Z}^n_0(\mathcal{U},\mathcal{E}), \ n \in \mathbf{Z}_+,$$

as $\mathcal{A}(X)$*-modules;* hence, one further considers the following *Čech liftable cohomology* $\mathcal{A}(X)$*-modules;* that is, one sets, with $\mathcal{U}$ and $\mathcal{E}$, as above:

$$(5.13) \qquad \check{H}^n_0(\mathcal{U},\mathcal{E}) := \check{Z}^n_0(\mathcal{U},\mathcal{E}) / \check{B}^n_0(\mathcal{U},\mathcal{E}), \ n \in \mathbf{Z}_+.$$

Thus, our next objective is to show (see Theorem 5.1 below) that;

(5.14) *by employing Čech liftable cohomology $\mathcal{A}(X)$-modules, one can associate with every short exact $\mathcal{A}$-sequence a long exact cohomology sequence.*

So, based on the preceding (cf. (5.7), (5.8′)), one obtains, *for any given short exact $\mathcal{A}$-sequence, as in (5.4), the following commutative diagram* of $\mathcal{A}(X)$-modules, *whose rows are exact;*

$$(5.15)\qquad \begin{array}{ccccccccc} 0 & \longrightarrow & \check{C}^{n-1}(\mathcal{U},\mathcal{F}) & \xrightarrow{\phi} & \check{C}^{n-1}(\mathcal{U},\mathcal{S}) & \xrightarrow{\psi} & \check{C}_0^{n-1}(\mathcal{U},\mathcal{E}) & \longrightarrow & 0 \\ & & \downarrow{\scriptstyle\delta} & & \downarrow{\scriptstyle\delta} & & \downarrow{\scriptstyle\delta} & & \\ 0 & \longrightarrow & \check{C}^{n}(\mathcal{U},\mathcal{F}) & \xrightarrow{\phi} & \check{C}^{n}(\mathcal{U},\mathcal{S}) & \xrightarrow{\psi} & \check{C}_0^{n}(\mathcal{U},\mathcal{E}) & \longrightarrow & 0 \\ & & \downarrow{\scriptstyle\delta} & & \downarrow{\scriptstyle\delta} & & \downarrow{\scriptstyle\delta} & & \\ 0 & \longrightarrow & \check{C}^{n+1}(\mathcal{U},\mathcal{F}) & \xrightarrow{\phi} & \check{C}^{n+1}(\mathcal{U},\mathcal{S}) & \xrightarrow{\psi} & \check{C}_0^{n+1}(\mathcal{U},\mathcal{E}) & \longrightarrow & 0 \end{array}$$

(We have employed above an obvious *abuse of notation,* concerning the maps $\delta$ and $\phi, \psi$; cf. thus (5.9) and (5.5) in the preceding).

Now, our first conclusion from the previous commutative diagram is the following technical lemma, however, basic for the sequel. (In this respect, see also, for example, L. Hörmander [1: p.179f]). So we have.

**Lemma 5.1.** *Let $(X, \mathcal{A})$ be a* **C***-algebraized space, $\mathcal{U}$ an open covering of $X$ and*

$$(5.16)\qquad 0 \longrightarrow \mathcal{F} \xrightarrow{\phi} \mathcal{S} \xrightarrow{\psi} \mathcal{E} \longrightarrow 0$$

*a given short exact $\mathcal{A}$-sequence. Moreover, let $n$ be any positive (viz. non-negative) integer, so $n \in \mathbf{Z}_+$. Then, relative to the given open covering $\mathcal{U}$ of $X$, one concludes the following:*

*i) For every liftable $n$-cocycle of* ($\mathcal{U}$ with coefficients in) $\mathcal{E}$, $g \in \check{Z}_0^n(\mathcal{U},\mathcal{E})$ (cf. (5.10)), *there exists an $n$-cochain of $\mathcal{S}$, $h \in \check{C}^n(\mathcal{U},\mathcal{S})$, as well as, an $n+1$-cocycle of $\mathcal{F}$, $f \in \check{Z}^{n+1}(\mathcal{U},\mathcal{F})$, such that one has;*

$$(5.17)\qquad g = \psi(h) \text{ and } \delta(h) = \phi(f).$$

*ii) In particular, the above* (multi-valued) *correspondence*

(5.18) $$\check{Z}^n_0(\mathcal{U},\mathcal{E}) \longrightarrow \check{Z}^{n+1}(\mathcal{U},\mathcal{F}),$$

*as given by i), is "boundary preserving"; viz. if* $g \in \check{B}^n_0(\mathcal{U},\mathcal{E}) \subseteq \check{Z}^n_0(\mathcal{U},\mathcal{E})$ (see (5.12)), *then, there exists* $h \in \check{C}^n(\mathcal{U},\mathcal{S})$ *and* $f \in \check{Z}^{n+1}(\mathcal{U},\mathcal{F})$, *as in i), such that, in particular, one obtains* (see also (5.17))

(5.19) $$\delta(h) = \phi(f) \in \phi(\check{B}^{n+1}(\mathcal{U},\mathcal{F})).$$

*Therefore, one concludes, in effect, for the above* $n+1$*-cocycle* $f$ *that,* $f \in \check{B}^{n+1}(\mathcal{U},\mathcal{F}) \subseteq \check{Z}^{n+1}(\mathcal{U},\mathcal{F})$.

*iii) The previous correspondence (5.18) yields, in fact, in cohomology the following* $\mathcal{A}(X)$*-morphism* (see also (5.13));

(5.20) $$\delta^{*,n} \equiv \delta^* : \check{H}^n_0(\mathcal{U},\mathcal{E}) \longrightarrow \check{H}^{n+1}(\mathcal{U},\mathcal{F}),$$

*such that one has* (cf. (5.17))

(5.21) $$\delta^*([g]) := [\phi^{-1}(\delta(h))] = [f],$$

*for any* $g \in \check{Z}^n_0(\mathcal{U},\mathcal{E})$.

**Proof.** Suppose that $g \in \check{Z}^n_0(\mathcal{U},\mathcal{E}) = \check{Z}^n(\mathcal{U},\mathcal{E}) \cap \psi(\check{C}^n(\mathcal{U},\mathcal{S}))$, that is, a *liftable* $n$*-cocycle of* $\mathcal{E}$ (precisely speaking, of $\mathcal{U}$ with coefficients in $\mathcal{E}$; cf. (5.10)). Therefore, there exists an $n$*-cochain*, say $h \in \check{C}^n(\mathcal{U},\mathcal{S})$, such that

(5.22) $$g = \psi(h) \in \check{Z}^n(\mathcal{U},\mathcal{E}) \cap \psi(\check{C}^n(\mathcal{U},\mathcal{S})) \equiv \check{Z}^n_0(\mathcal{U},\mathcal{E}).$$

Hence, due to the hypothesis for $g$ and (5.8′), one further obtains

(5.23) $$\delta(g) = \delta(\psi(h)) = \psi(\delta(h)) = 0,$$

that is (see also (5.15)), $\delta(h) \in \ker\psi = \operatorname{im}\phi \equiv \phi(\check{C}^{n+1}(\mathcal{U},\mathcal{F}))$; namely, one has

(5.24) $$\delta(h) = \phi(f) \in \phi(\check{C}^{n+1}(\mathcal{U},\mathcal{F})),$$

for some $f \in \check{C}^{n+1}(\mathcal{U},\mathcal{F})$. But, of course, $\delta(\delta(h)) = 0$, so that

(5.25) $$\delta(\delta(h)) = \delta(\phi(f)) = \phi(\delta(f)) = 0,$$

that is, $\delta(f) = \ker\phi = 0$ (since $\phi$ is 1-1, cf. (5.15)); hence, $f \in \check{Z}^{n+1}(\mathcal{U},\mathcal{F})$. Therefore, *one finally obtains;*

(5.26) $$\delta(h) = \phi(f) \in \phi(\check{Z}^{n+1}(\mathcal{U},\mathcal{F})) \underset{\phi}{\cong} \check{Z}^{n+1}(\mathcal{U},\mathcal{F}),$$

or even

$$\delta(h) = \phi(f) \approx f, \tag{5.26'}$$

*with* $f \in \check{Z}^{n+1}(\mathcal{U}, \mathcal{F})$, which thus, along with (5.22), *proves prop. i).*

On the other hand, suppose, in particular, that (see also (5.11), (5.12))

$$g \in \check{B}^n_0(\mathcal{U}, \mathcal{E}) = \delta(\psi(\check{C}^{n-1}(\mathcal{U}, \mathcal{S})) \subseteq \check{Z}^n_0(\mathcal{U}, \mathcal{E}) \subseteq \psi(\check{C}^n(\mathcal{U}, \mathcal{S})) \tag{5.27}$$

*(liftable n-coboundary of* $\mathcal{E}$*);* that is, one has (cf. also (5.8′) and (5.22))

$$g = \delta(\psi(h')) = \psi(\delta(h')) = \psi(h), \tag{5.28}$$

with $h' \in \check{C}^{n-1}(\mathcal{U}, \mathcal{S})$ and $h \in \check{C}^n(\mathcal{U}, \mathcal{S})$. Thus, in view of (5.28), one obtains

$$\delta(h') - h \in \ker \psi = \operatorname{im} \phi \equiv \phi(\check{C}^n(\mathcal{U}, \mathcal{F})), \tag{5.29}$$

that is, *one has* $\delta(h') - h = \phi(f')$, *for some* $f' \in \check{C}^n(\mathcal{U}, \mathcal{F})$. But, one also obtains

$$\delta(\delta(h') - h) = \delta(h) = \delta(\phi(f')) = \phi(\delta(f')), \tag{5.30}$$

so that, due to (5.19) (or even (5.26′)), one has

$$\delta(h) = \phi(f) = \phi(\delta(f')), \tag{5.31}$$

that is, given that $\phi$ is one-to-one,

$$f = \delta(f') \in \delta(\check{C}^n(\mathcal{U}, \mathcal{F})) \equiv \check{B}^{n+1}(\mathcal{U}, \mathcal{F}), \tag{5.32}$$

*which thus proves (5.19).*

Now, the preceding show that *the cohomology class in* $\check{H}^{n+1}(\mathcal{U}, \mathcal{F})$ *of an* $f = \phi^{-1}(\delta(h)) \in \check{Z}^{n+1}(\mathcal{U}, \mathcal{F})$*, as in (5.26′), depends only on that of* $g \in \check{Z}^n_0(\mathcal{U}, \mathcal{E})$ *in* $\check{H}^n_0(\mathcal{U}, \mathcal{E})$. In other words, *the map (5.20), as given by (5.21),* in view of (5.26′), *is well-defined.* Moreover, it is an easy consequence of the very definitions, that *the same map is,* in fact, *an* $\mathcal{A}(X)$*-morphism* of the $\mathcal{A}(X)$-modules concerned, and this terminates the proof of the lemma. ∎

As a consequence of the previous lemma, we come next to the following basic fact of *(liftable)* Čech cohomology, a prelude, in effect, of our next main result (Theorem 5.2 in the sequel). So we now have.

**Theorem 5.1.** *Let* $(X, \mathcal{A})$ *be a given* **C***-algebraized space,* $\mathcal{U}$ *an open covering of* $X$*, and*

$$0 \longrightarrow \mathcal{F} \xrightarrow{\phi} \mathcal{S} \xrightarrow{\psi} \mathcal{E} \longrightarrow 0 \tag{5.33}$$

*a short exact $\mathcal{A}$-sequence. Then, one gets the following long exact $\mathcal{A}(X)$-sequence in (liftable) Čech cohomology;*

$$(5.34)\qquad \begin{aligned} 0 \longrightarrow \check{H}^0(\mathcal{U},\mathcal{F}) \xrightarrow{\phi^*} \check{H}^0(\mathcal{U},\mathcal{S}) \xrightarrow{\psi^*} \check{H}^0_0(\mathcal{U},\mathcal{E}) \xrightarrow{\delta^*} \check{H}^1(\mathcal{U},\mathcal{F}) \\ \xrightarrow{\phi^*} \check{H}^1(\mathcal{U},\mathcal{S}) \xrightarrow{\psi^*} \check{H}^1_0(\mathcal{U},\mathcal{E}) \xrightarrow{\delta^*} \check{H}^2(\mathcal{U},\mathcal{F}) \longrightarrow \cdots \\ \cdots \xrightarrow{\phi^*} \check{H}^n(\mathcal{U},\mathcal{S}) \xrightarrow{\psi^*} \check{H}^n_0(\mathcal{U},\mathcal{E}) \xrightarrow{\delta^*} \check{H}^{n+1}(\mathcal{U},\mathcal{F}) \longrightarrow \cdots . \end{aligned}$$

(As in (5.19), we have employed above, for simplicity, an obvious *abuse of notation,* concerning the various maps $\phi^*, \psi^*$).

**Proof.** First, the relation $\ker\psi = \operatorname{im}\phi$ that one has, by virtue of (5.33), is obviously transferred to cohomology (according to the very definitions), so that one concludes the *exactness of (5.34) at any term of the form $\check{H}^n(\mathcal{U},\mathcal{S})$, for all integers $n \geqq 0$.* (In this context, cf. thus (4.25), along with the left exactness of $\Gamma_X \equiv \Gamma$ (see Chapt. II; (2.59)), *for the case $n = 0$.* Yet, cf. (5.5) and the argument applied in the proof of prop. i), Lemma 5.1, *for $n \geqq 1$*).

On the other hand, by considering the *range of $\delta^*$ in $\check{H}^n(\mathcal{U},\mathcal{F})$* (with $\delta^*$ having then, of course, as domain of definition the $\mathcal{A}(X)$-module $\check{H}^{n-1}_0(\mathcal{U},\mathcal{E})$, cf. (5.20)), one infers, in view of (5.21), that the former consists exactly of those cohomology classes, whose representative cocycles are of the form;

$$(5.35)\qquad \phi^{-1}(\delta(h)) \in \phi^{-1}(\check{B}^n(\mathcal{U},\mathcal{S})) \cap \check{Z}^n(\mathcal{U},\mathcal{F}),$$

such that (cf.(5.22)) $h \in \check{C}^{n-1}(\mathcal{U},\mathcal{S})$, hence $\delta(h) \in \delta(\check{C}^{n-1}(\mathcal{U},\mathcal{S})) \equiv \check{B}^n(\mathcal{U},\mathcal{S})$ (see (4.22)). Thus, one obtains (cf. also e.g. (4.39), or even (5.8))

$$(5.36)\qquad \phi^*([\phi^{-1}(\delta(h))]) = [\delta(h)] = 0 \in \check{H}^n(\mathcal{U},\mathcal{S});$$

that is, one has

$$(5.37)\qquad \operatorname{im}\delta^* = \ker\phi^*,$$

which thus proves the *exactness of (5.34) at each term of the form $\check{H}^n(\mathcal{U},\mathcal{F})$, for all integers $n \geqq 1$.* (For $n = 0$, exactness of (5.34) at $\check{H}^0(\mathcal{U},\mathcal{F})$ is, in effect, a consequence of the *left exactness of* $\Gamma$; see the beginning of the proof).

We finally show *exactness of (5.34) at $\check{H}^n_0(\mathcal{U},\mathcal{E})$, for all $n \in \mathbf{Z}_+$:* Thus, suppose that $[g] \in \check{H}^n_0(\mathcal{U},\mathcal{E})$ such that (cf. also e.g. (4.39))

$$(5.38)\qquad [g] = \psi^*([h]) = [\psi(h)],$$

for some $h \in \check{Z}^n(\mathcal{U}, \mathcal{S})$. But, then $\delta(h) = 0$, so that (cf. (5.21)) one has

$$\delta^*([g]) = \left[\phi^{-1}(\delta(h))\right] = 0, \tag{5.39}$$

*that is,* $\operatorname{im}\psi^* \subseteq \ker\delta^*$. *Conversely,* consider a cohomology class $[g] \in \check{H}^n_0(\mathcal{U}, \mathcal{E})$, such that $\delta^*([g]) = 0$. Hence, one has, by (5.21), $\phi^{-1}(\delta(h)) = \delta(f')$, for some $f' \in \check{C}^n(\mathcal{U}, \mathcal{F})$, such that $g = \psi(h)$, with $h \in \check{C}^n(\mathcal{U}, \mathcal{S})$. Therefore, one has

$$\delta(h) = \phi(\delta(f')) = \delta(\phi(f')), \tag{5.40}$$

so that $\delta(h - \phi(f')) = 0$; that is, one finally has that

$$h - \phi(f') \in \check{Z}^n(\mathcal{U}, \mathcal{S}), \tag{5.41}$$

such that (see also (5.7))

$$\begin{aligned} \psi^*([h - \phi(f')]) &= [\psi(h - \phi(f'))] = [\psi(h) - \psi(\phi(f'))] \\ &= [\psi(h) - 0] = [\psi(h)] = [g]. \end{aligned} \tag{5.42}$$

So $[g] \in \operatorname{im}\psi^*$, that was to be proved, and this terminates the proof. ∎

In this context, we also note that the above proof is essentially the adaptation, within the present context, of a classical argument, pertaining to the proof of the so-called *"zig-zag lemma"* (this sort of proof being also called *"diagramm-chasing"*; cf. the *commutative diagram (5.15)* above. See also, for example, J. R. Munkres [1: p.136, Lemma 24.1]). Yet, notice that an analogous technique has already been applied to the proof of Theorem 1.1 in the preceding.

Now, our next objective is to get *a similar exact sequence to (5.34), without resorting,* however, *to "liftable cocycles".* Indeed, as we shall see, this can be done (cf. Theorem 5.2 below) by restricting the class of the topological spaces involved. But, to this end, we shall need first a bit more terminology, which thus we explain presently below.

**5.(a). Passage to the limit (of liftable cocycles).**– Suppose we are within the context of Subsection 4.(c) above. So let $\mathcal{U}$, $\mathcal{V}$ be two *open coverings* of the *topological space* $X$ considered, *with* $\mathcal{V} \succeq \mathcal{U}$ and let $\mu : J \longrightarrow I$ be a *refining map* (cf. (4.32), (4.33)). Furthermore, let

$$\mu : \check{C}^n(\mathcal{U}, \mathcal{E}) \longrightarrow \check{C}^n(\mathcal{V}, \mathcal{E}) \tag{5.43}$$

be the corresponding to (the given refining map) $\mu$ *$\mathcal{A}(X)$-morphism of $n$-cochains,* as indiacated, with $\mathcal{E}$ being a given *$\mathcal{A}$-module* on $X$ (notice the *abuse of notation* for the map $\mu$ in (5.43); but, see also (4.34), (4.34′)).

On the other hand, suppose that we are given another *$\mathcal{A}$-module* $\mathcal{S}$ on $X$, along with an *$\mathcal{A}$-morphism*

$$\psi : \mathcal{S} \longrightarrow \mathcal{E}. \tag{5.44}$$

Moreover, let

$$\check{C}^n_0(\mathcal{U}, \mathcal{E}) := \psi(\check{C}^n(\mathcal{U}, \mathcal{S})),\ n \in \mathbf{Z}_+, \tag{5.45}$$

be the associated with the given $\mathcal{A}(X)$-morphism $\psi$, as above, *liftable $n$-cochains of* ($\mathcal{U}$ with coefficients in) $\mathcal{E}$ (cf. also (5.6)). Thus, based on the very definitions (see (4.34′)), it is easy to prove that *the $\mathcal{A}(X)$-morphism $\mu$ in (5.43) preserves liftable cochains; that is, one has*

$$\mu : \check{C}^n_0(\mathcal{U}, \mathcal{E}) \longrightarrow \check{C}^n_0(\mathcal{V}, \mathcal{E}),\ n \in \mathbf{Z}_+. \tag{5.46}$$

Consequently (cf. also (4.37) and (4.38)), one defines the following *$\mathcal{A}(X)$-morphism* in cohomology;

$$\mu^* : \check{H}^n_0(\mathcal{U}, \mathcal{E}) \longrightarrow \check{H}_0(\mathcal{V}, \mathcal{E}),\ n \in \mathbf{Z}_+. \tag{5.47}$$

Yet, by analogy with (4.40) (cf. Lemma 4.2), one proves that;

(5.48) *(5.47) does not depend on the particular refining map* $\mu$ (see (4.33)), *but only on the relation* $\mathcal{V} \succeq \mathcal{U}$ (cf. (4.32)).

Therefore, setting

$$\mu^* \equiv \tilde{\tau}_{\mathcal{V}\mathcal{U}} : \check{H}_0(\mathcal{U}, \mathcal{E}) \longrightarrow \check{H}^n_0(\mathcal{V}, \mathcal{E}), \tag{5.49}$$

one easily verifies (as in (4.52)) that;

(5.50) *the family*

$$\left\{\check{H}^n_0(\mathcal{U}, \mathcal{E}), \tilde{\tau}_{\mathcal{V}\mathcal{U}}\right\} \tag{5.50.1}$$

*defines,* for any integer $n \geqq 0$, *an inductive system of $\mathcal{A}(X)$-modules,* with respect to the *directed index set of proper open coverings of* $X$ (see (4.9)).

Accordingly, one defines;

(5.51) $$\check{H}^n_0(X,\mathcal{E}) := \varinjlim_{\mathcal{U}} \check{H}^n_0(\mathcal{U},\mathcal{E}),\ n \in \mathbf{Z}_+,$$

That is, an $\mathcal{A}(X)$*-module,* by its very definition, called the *$n$-th dimensional Čech liftable cohomology* ($\mathcal{A}(X)$-module) *of $X$ with coefficients in the given $\mathcal{A}$-module $\mathcal{E}$.*

Now, as we shall presently see, the above limit module can be identified with the usual one, as in (4.53), for a suitable topological space $X$. So this will be our next objective in the ensuing subsection.

**5.(b). Long exact sequences in Čech cohomology (contn'd).**– As already said, we seek below conditions ensuring the *existence of a long exact cohomology sequence,* in the case of *Čech cohomology* (cf. Definition 4.2), which will be *associated with any given short exact sequence,* within the category under consideration, namely, that one of $\mathcal{A}$-modules on a given (fixed) topological space $X$, $\mathcal{A}\text{-}\mathcal{M}od_X$ (cf. (1.1)).

So, by restricting ourselves to the category of *paracompact (Hausdorff) topological spaces,* concerning the (common) *base space* of the $\mathcal{A}$-modules involved, we realize the following crucial fact; namely,

(5.52) by arguing, for instance, within the *context of* the previous *Lemma 5.1, every given $n$-cocycle in $\check{Z}^n(\mathcal{U},\mathcal{E})$ is liftable* (cf. (5.10)), *modulo an eventual refinement of the open covering $\mathcal{U}$* considered.

In effect, something more is true. Namely, more precisely, one arrives at the following basic conclusion, as given by the next.

**Lemma 5.2.** *Let $(X,\mathcal{A})$ be a given* **C***-algebraized space, with $X$ being a paracompact (Hausdorff) space, and let $\mathcal{U} = \{U_\alpha\}_{\alpha\in I}$ be an open covering of $X$. Moreover, let*

(5.53) $$\mathcal{S} \xrightarrow{\psi} \mathcal{E} \longrightarrow 0$$

*be an exact $\mathcal{A}$-sequence on $X$* (cf. also Chapt. II; Definition 2.1), *viz. an $\mathcal{A}$-epimorphism of the $\mathcal{A}$-modules concerned. Then, every $n$-cochain $g \in \check{C}^n(\mathcal{U},\mathcal{E})$, $n \in \mathbf{Z}_+$, is refinement-liftable; in other words,* *for any* $g \in \check{C}^n(\mathcal{U}, \mathcal{E})$, *there exists a refinement* $\mathcal{V} = \{V_\beta\}_{\beta \in J}$ *of* $\mathcal{U}$, *with refining map* $\mu : J \longrightarrow I$ (cf. (4.2), (4.3)), *such that* (cf. (4.34), (4.34′))

(5.54) (5.54.1) $$\mu(g) \in \check{C}^n_0(\mathcal{V}, \mathcal{E}) \equiv \psi(\check{C}^n(\mathcal{V}, \mathcal{S})).$$

*That is, there exists an n-cochain* $h \in \check{C}^n(\mathcal{V}, \mathcal{S})$, *such that*

(5.54.2) $$\mu(g) = \psi(h).$$

In this context, we still recall, for convenience, that a Hausdorff (topological ) space $X$ is said to be *paracompact,* whenever every (open) covering of $X$ has a *locally finite refinement* $\mathcal{V}$ (that is, every point $x \in X$, has an (open) neighborhood intersecting only finite-many members of $\mathcal{V}$). Furthermore, it is standard (J. Dieudonné) that *every paracompact space is normal* (see e.g. J. Dugundji [1: p.163, Theorem 2.2]). Thus, by the normality of (a given paracompact space) $X$, one finds, for every locally finite open covering of $X$, say $\mathcal{V} = \{V_\beta\}_{\beta \in J}$, as above, a *"shrinking"* of it; that is, a *"precise"* refinement of $\mathcal{V}$, say $\mathcal{W} = \{W_\beta\}_{\beta \in J}$, such that

(5.55) $$\overline{W}_\beta \subseteq V_\beta, \textit{ for every } \beta \in J.$$

(Cf. J. Dugundji [1: p.161] and N. Bourbaki [3: Chap. IX; p.48, Corollaire 1]).

Now, concerning the proof of the previous Lemma 5.2, which we give below, see also, for instance, J. -P. Serre [1: p.218, Lemme 2], as well as, the relevant treatments by R. C. Gunning [1: p.34, proof of Theorem 1], or even F. Hirzebruch [1: p.31, Lemma 2.9.2], F. W. Warner [1: p. 204] and S. G. Krantz [1: p.230, Lemma 6.2.20]. Yet, cf. L. Hörmander [1: p.180, Lemma 7.3.6], or even K. Kodaira [1: p.128, Lemma 3.3]. In particular, for the conditions (5.75) in the sequel, we adapt within the present context, an analogous detailed and elegant argument given by R. Kultze in [1: p.154, Hilfssatz 33.5]. But, see also C. Andreian Cazacu [1: p.147, Satz 1]. So we come next to the

**Proof of Lemma 5.2.** Since, by hypothesis, $X$ is paracompact, *one may assume that* the given covering of $X$, $\mathcal{U} = \{U_\alpha\}_{\alpha \in I}$, *is locally finite* (cf. the preceding comments). Moreover, one can further consider (the same comments, as above) a *shrinking of* $\mathcal{U}$, say $\mathcal{W} = \{W_\alpha\}_{\alpha \in I}$, so that (cf. (5.55)) one has;

(5.56) $$\overline{W}_{\alpha} \subseteq U_{\alpha}, \text{ for every } \alpha \in I,$$

(hence, a (*precise,* viz. with the same index set, as $\mathcal{U}$) *locally finite refinement of* $\mathcal{U}$, as well).

On the other hand, according to Lemma 1.1, *exactness of (5.53) is equivalently expressed fiber-wise;* thus, *the given n-cochain* $g \in \check{C}^n(\mathcal{U}, \mathcal{E})$, consisting, by its very definition (cf. (4.11)), of local sections of $\mathcal{E}$, *can be lifted locally, via* $\psi$, *to a section of* $\mathcal{S}$:

Namely, suppose that (yet, cf. (4.11))

(5.57) $$x \in U_{\alpha_0} \cap U_{\alpha_1} \cap \cdots \cap U_{\alpha_n} \equiv U_{\alpha_0\alpha_1\cdots\alpha_n},$$

with $U_{\alpha_i} \in \mathcal{U}, 0 \leqq i \leqq n$. Thus, given that (cf. (5.53) and the above comments)

(5.58) $$g(x) \in \mathcal{E}_x = \psi_x(\mathcal{S}_x)$$

(see also Chapt. I; (1.14) for the notation applied), one concludes that

(5.59) $$g(x) = \psi_x(h(x)) = (\psi \circ h)(x),$$

for some *local section* $h \in \mathcal{S}(U)$, with $U$ an open neighborhood of $x$ (cf. Chapt. I; (2.10), as well as, Proposition 2.1); therefore (see Chapt. I; (2.13)), the local sections $g$ *and* $\psi \circ h$ of $\mathcal{E}$ *coincide locally* (viz. on a suitable neighborhood of $x$). Thus, one finally obtains

(5.60) $$g = \psi \circ h \equiv \psi(h)|_{U_x} \in \mathcal{E}(U_x),$$

such that (cf. also (5.57))

(5.61) $$U_x \equiv U^{(x)}_{\alpha_0\alpha_1\cdots\alpha_n} \subseteq U_{\alpha_0\alpha_1\cdots\alpha_n}.$$

In this context, if

(5.62) $$\Gamma(\mathcal{E}) \equiv \left(\Gamma(U, \mathcal{E}), \sigma^U_V\right)$$

stands, of course, for the *(complete) presheaf of sections of* $\mathcal{E}$ (see Chapt. I; (9.6)), we also set in (5.60), as usual, for convenience,

(5.63) $$g|_{U_x} \equiv \sigma^{U_{\alpha_0\alpha_1\cdots\alpha_n}}_{U_x}(g) \in \mathcal{E}(U_x),$$

as well as,

(5.64) $$\psi(h) \equiv \psi \circ h = \psi_{U_x}(h) \in \mathcal{E}(U_x)$$

(cf. Chapt. I; (6.7) and (13.7). Yet, *for convenience(!)*, we employed in (5.60) an obvious, ***abuse of notation,*** writing $\psi(h)$, instead of, e.g., $\psi^*(h) \equiv \psi \circ h$, for the local section $h$ of $\mathcal{S}$ under consideration; cf. also (5.54.2), or even (5.73.3) below).

Accordingly, *given an element $x \in X$, as in (5.57),* one considers the two following *open neighborhoods of $x$:* Firstly,

$$V_x^{(1)} := \bigcap_{\alpha \in I_x} U_\alpha, \tag{5.65}$$

where we set

$$I_x = \{\alpha \in I : x \in U_\alpha \in \mathcal{U}\}, \tag{5.65'}$$

hence, by hypothesis for $\mathcal{U}$, a *finite subset of $I$*, which also justifies our claim for $V_x^{(1)}$. Secondly, we set (see also (5.61))

$$V_x^{(2)} := \bigcap_{(\alpha_0,\ldots,\alpha_n)} U^{(x)}_{\alpha_0\alpha_1\cdots\alpha_n}, \tag{5.66}$$

where *the index set is a finite subset of $I^{n+1}$*, in view again of the hypothesis for $\mathcal{U}$, along with (5.61), whence our claim for $V_x^{(2)}$ too.

Now, consider the *shrinking of $\mathcal{U}$*, as given by (5.56), viz. the *(precise) locally finite refinement of $\mathcal{U}$*,

$$\overline{\mathcal{W}} \equiv \left\{\overline{W}_\alpha\right\}_{\alpha \in I}. \tag{5.67}$$

Thus, the latter property of $\overline{\mathcal{W}}$, along with (5.56) and (5.65′), assure already that *the set*

$$V_x^{(3)} := \complement\left(\bigcup_{\alpha \notin I_x} \overline{W}_\alpha\right) = \bigcup_{\alpha \notin I_x} \complement\overline{W}_\alpha \tag{5.68}$$

*is an open neighborhood of $x$*, as well. Furthermore, the same set enjoys the condition (cf. also (5.65′));

$$V_x^{(3)} \cap W_{\alpha'} \neq \emptyset \text{ implies that } \alpha' \in I_x \text{ (viz. that } x \in U_{\alpha'}). \tag{5.69}$$

Indeed, our claim follows immediately from (5.68), since then one has

$$W_{\alpha'} \cap V_x^{(3)} := W_{\alpha'} \cap \left(\bigcup_{\alpha \notin I_x} \complement\overline{W}_\alpha\right) = \bigcup_{\alpha \notin I_x} W_{\alpha'} \cap \complement\overline{W}_\alpha \neq \emptyset, \tag{5.70}$$

that is, $\emptyset \neq W_{\alpha'} \cap \complement \overline{W}_\alpha \subseteq W_{\alpha'} \cap \complement W_\alpha$, *for every* $\alpha \notin I_x$, which proves the assertion.

Moreover, setting $\widetilde{I}_x = \{\alpha \in I : x \in W_\alpha\}$, we further define the set

(5.71) $$V_x^{(4)} := \bigcap_{\alpha \in \widetilde{I}_x} W_\alpha,$$

so, according to the hypothesis for $\mathcal{W}$ (cf. (5.56)), another *open neighborhood of* $x$. Thus, we finally obtain the following *open neighborhood of* $x$;

(5.72) $$V_x := \bigcap_{i=1}^{4} V_x^{(i)}.$$

On the other hand, we also consider *the map*

(5.73) $$\mu : X \longrightarrow I, \textit{ such that } x \in W_{\mu(x)},\ x \in X.$$

(Due to the hypothesis for $\mathcal{W}$, one can really define a *map*, as above). So we come now to the conclusion that *the family*

(5.74) $$\mathcal{V} = \{V_x\}_{x \in X},$$

as given by (5.72), *satisfies the following conditions:*

(5.75)

*1) If* $x \in U_\alpha$, *then* $V_x \subseteq U_\alpha$.

*2)* $V_x \subseteq W_{\mu(x)}$, *for every* $x \in X$.

*3) If* $V_x \cap W_\alpha \neq \emptyset$, *then* $V_x \subseteq U_\alpha$.

*4) For every open* $V \subseteq X$, *with*

(5.75.1) $$V \subseteq V_x \subseteq U_{\alpha_0 \ldots \alpha_n},$$

*there exists a local section* $h \in \mathcal{S}(V)$, *such that one has* (cf. also (5.64))

(5.75.2) $$\sigma_V^{U_{\alpha_0 \cdots \alpha_n}}(g) \equiv g|_V = \psi(h) \equiv \psi_V(h) \in \mathcal{E}(V);$$

*yet, we simply write*

(5.75.3) $$g = \psi(h)|_V \in \mathcal{E}(V).$$

**Proof of (5.75).** Indeed, if $x \in U_\alpha$, then (cf. (5.65′)) $\alpha \in I_x$, hence (see (5.65) and (5.72)), one obtains

$$V_x \subseteq V_x^{(1)} = \bigcap_{\alpha \in I_x} U_\alpha \subseteq U_\alpha, \tag{5.76}$$

*which proves cond. 1).* Now, based on (5.73), $x \in W_{\mu(x)}$, with $x \in X$, so that (cf. (5.71)) $\mu(x) \in \widetilde{I}_x$; therefore, one has

$$V_x \subseteq V_x^{(4)} = \bigcap_{\alpha \in \widetilde{I}_x} W_\alpha \subseteq W_{\mu(x)}, \tag{5.77}$$

*for any* $x \in X$, that is, *cond. 2) is fulfilled.* On the other hand, if $V_x \cap W_\alpha \neq \emptyset$, then a fortiori (cf. (5.72)) $V_x^{(3)} \cap W_\alpha \neq \emptyset$, so that, in view of (5.69), $\alpha \in I_x$; viz. $x \in U_\alpha$, hence, by cond. 1), $V_x \subseteq U_\alpha$, which *proves cond. 3).* Finally, if $x \in V_x \subseteq U_{\alpha_0\alpha_1\cdots\alpha_n}$, then there exists an open neighborhood of $x$, as in (5.61), such that one has (cf. also (5.66) and (5.72))

$$V_x \subseteq U^{(x)}_{\alpha_0\alpha_1\cdots\alpha_n} \subseteq U_{\alpha_0\alpha_1\cdots\alpha_n}. \tag{5.78}$$

Therefore, based now on (5.60) and (5.61), (5.63) and (5.64), as well as, on Chapt. I; (5.3) and (5.62) above, one obtains for the open $V \subseteq X$, as in (5.75.1);

$$\begin{aligned} g|_V &\equiv \sigma_V^{U_{\alpha_0\cdots\alpha_n}}(g) = \left(\sigma_V^{U^{(x)}_{\alpha_0\cdots\alpha_n}} \circ \sigma^{U_{\alpha_0\cdots\alpha_n}}_{U^{(x)}_{\alpha_0\cdots\alpha_n}}\right)(g) \\ &= \sigma_V^{U^{(x)}_{\alpha_0\cdots\alpha_n}}\left(\sigma^{U_{\alpha_0\cdots\alpha_n}}_{U^{(x)}_{\alpha_0\cdots\alpha_n}}(g)\right) = \sigma_V^{U^{(x)}_{\alpha_0\cdots\alpha_n}}\left(\psi_{U^{(x)}_{\alpha_0\cdots\alpha_n}}(h)\right) \\ &= \psi_V\left(\tau_V^{U^{(x)}_{\alpha_0\cdots\alpha_n}}(h)\right) \equiv \psi_V(h|_V) \equiv \psi(h)|_V \end{aligned} \tag{5.79}$$

*(sheaf morphisms commute with the restriction morphisms;* cf. Chapt. I; (13.7), (6.7) and (6.5). Yet, we have employed above the notation $\Gamma(\mathcal{S}) \equiv (\Gamma(U,\mathcal{S}), \tau_V^U)$, cf. e.g. Chapt. I; (6.9), concerning the respective presheaf of sections of $\mathcal{S}$). So *one actually has*

$$g = \psi(h)|_V \in \mathcal{E}(V), \tag{5.80}$$

*with* $g \in \mathcal{S}(V)$. (See also the comments, following (5.64) above, for the notation applied to the last relation, as well as, to that one in (5.79)). Thus, *(5.80) proves* already *(5.75.3),* that is, *cond. 4),* and the proof of (5.75) is complete. ∎

Now, according to *cond. 2)* in (5.75), *the family* $\mathcal{V} = \{V_x\}_{x\in X}$ in (5.74) *yields a refinement* of $\mathcal{W}$, hence (cf. (5.56)), *of the* given *open covering* $\mathcal{U}$, as well, *with* $\mu$, as given by (5.73), *as a refining map.* So *it remains only to prove (5.54.2):*

Indeed, in view of (4.34′), one first obtains

$$\mu(g)_{x_0\cdots x_n} = \sigma_{V_{x_0\cdots x_n}}^{U_{\mu(x_0)\cdots\mu(x_n)}}(g) \equiv g|_{V_{x_0\cdots x_n}}, \tag{5.81}$$

where the map $\mu$ is given by (5.73). In this context, if $V_{x_0\cdots x_n} = \emptyset$, then, of course, our claim is trivially true. Thus, suppose that (cf. (5.74))

$$V_{x_0 x_1\cdots x_n} \equiv V_{x_0} \cap V_{x_1} \cap \cdots \cap V_{x_n} \neq \emptyset, \tag{5.82}$$

so that one has

$$V_{x_0} \cap V_{x_i} \neq \emptyset,\ 0 \leqq i \leqq n. \tag{5.83}$$

Therefore (cf. (5.75); *cond. 2)*), one obtains

$$V_{x_0} \cap W\mu(x_i) \neq \emptyset,\ 0 \leqq i \leqq n, \tag{5.84}$$

so that, in view of (5.75); *cond. 3)*, one concludes that $V_{x_0} \subseteq U_{\mu(x_i)}$, $i = 0, 1, \ldots, n$. Thus, one finally obtains

$$V_{x_0\cdots x_n} \subseteq V_{x_0} \subseteq U_{\mu(x_0)\cdots\mu(x_n)}. \tag{5.85}$$

Consequently, by applying now (5.75); *cond. 4)*, with $V \equiv V_{x_0\cdots x_n}$ (cf. (5.75.1)), *one concludes the assertion,* due to (5.79) (or even (5.80)) and (5.81), that is, *one* thus *obtains (5.75.3),* and this terminates the proof of the lemma. ∎

As a result of the previous lemma, we come to our first basic conclusion; namely, one realizes that:

(5.86) *in paracompact (Hausdorff) spaces one can work with "liftable (Čech) cohomology", in place of the usual Čech (sheaf) cohomology.*

As an important profit thereon, one then obtains, for the latter theory too (see thus Theorem 5.1 above), a *long exact sequence in cohomology* (cf. Theorem 5.3 in the sequel). So, more precisely, one first obtains the following (formal version of (5.86)).

**Corollary 5.1.** *Let $(X, \mathcal{A})$ be a* **C***-algebraized space, with the base space $X$ being paracompact (Hausdorff). Moreover, let*

$$\mathcal{S} \xrightarrow{\psi} \mathcal{E} \longrightarrow 0 \tag{5.87}$$

*be an epimorphism of the $\mathcal{A}$-modules concerned* (or else an *exact $\mathcal{A}$-sequence* on $X$; see also (5.53)). *Then, one obtains*

$$\check{H}^n_0(X, \mathcal{E}) = \check{H}^n(X, \mathcal{E}), \tag{5.88}$$

*for all integers $n \geqq 0$, within an $\mathcal{A}(X)$-isomorphism of the $\mathcal{A}(X)$-modules involved.*

Before we come to the proof of Corollary 5.1, we shall need the following auxiliary comments: Thus, the cohomology $\mathcal{A}(X)$-modules appeared in (5.88) are, by definition (cf. (5.51) and (4.53)), *inductive limit $\mathcal{A}(X)$-modules,* with respect to the *directed set of proper open coverings of $X$* (with *refinement of coverings* as the corresponding *preorder relation;* (cf. (4.9)). However, in the case of a *paracompact (Hausdorff) space,* the locally finite members of the above set constitute, by the very definition of a paracompact space (cf. e.g. the comments following the statement of Lemma 5.2 in the preceding), a *cofinal subset* of the set in question. Hence, it is equivalent (within, of course, an appropriate isomorphism, as the case may be) to consider the corresponding limits, with respect to the latter (sub)set (in this context, cf. e.g. N. Bourbaki [1: p.95, Proposition 8]). Therefore, one concludes that;

(5.89) by dealing with the *Čech cohomology ($\mathcal{A}(X)$)-modules* (usual or liftable ones, see (4.53) and (5.51), respectively) of a *paracompact (Hausdorff) space $X$, it suffices to consider simply the locally finite* (proper) *open coverings* of the space.

Indeed, we are going to employ the previous principle right below in the proof of Corollary 5.1. However, the following remarks directly connected with the meaning of (4.53) and of (5.51), for that matter, will also be of use in the same proof:

So, based on the very definition of an *inductive limit* (see, for instance, N. Bourbaki [1: p.89]), one can consider e.g. $\check{H}^n(X, \mathcal{E})$ (cf. (4.53), while analogous remarks hold for $\check{H}^n_0(X, \mathcal{E})$, cf. (5.51)), as the *disjoint union* of the various $\check{H}^n(\mathcal{U}, \mathcal{E})$, where $\mathcal{U}$ *varies as in (4.9);* in particular, *for a paracompact space $X$, $\mathcal{U}$ is varying according to (5.89).* Yet, *we* also *identify any two classes,* in the previous union, which are associated, for instance, with two open coverings of $X$, say $\mathcal{U}$, V, in case there exists a common refinement, say $\mathcal{W} \succeq \mathcal{U}, \mathcal{V}$ (cf. (4.2)), such that the corresponding classes (via any refining map (see Lemma 4.2 and (4.49)) have *the same image in* $\check{H}^n(\mathcal{W}, \mathcal{E})$. Accordingly, one has, concerning (4.53);

$$\check{H}^n(X, \mathcal{E}) = \varinjlim_{\mathcal{U}} \check{H}^n(\mathcal{U}, \mathcal{E}) = \sum_{\mathcal{U}} \check{H}^n(\mathcal{U}, \mathcal{E})/\!\sim, \; n \in \mathbf{Z}_+, \tag{5.90}$$

where "$\sim$" denotes the *equivalence relation* in the previous disjoint union, that we alluded to above. Of course, an *analogous relation to (5.90)* holds true *for* $\check{H}^n_0(X, \mathcal{E})$. (In this context, still notice the *simplified form of (5.90) in the particular case that* $n = 1$ (see thus (4.73)), as a result of Lemma 4.3). So one concludes that:

(5.91) *with every* $[g] \in \check{H}^n(X, \mathcal{E})$, *there is associated an open covering* (in particular, *for* $X$ *paracompact* (Hausdorff), *a locally finite open covering* (see (5.89)) *of* $X$, *say* $\mathcal{U} = \{U_\alpha\}_{\alpha \in I}$, *such that*

$$g \in \check{Z}^n(\mathcal{U}, \mathcal{E}) \tag{5.91.1}$$

(we call this the *representative* $n$*-cocycle* of the given class $[g] \in \check{H}^n(X, \mathcal{E})$), while we also take the above equivalence relation into account.

*Similar comments hold true,* of course, *for the* $\mathcal{A}(X)$*-module* $\check{H}^n_0(X, \mathcal{E})$, $n \in \mathbf{Z}_+$.

Now, in view of the relations (5.10) and (5.11), one first obtains a map, in effect, an $\mathcal{A}(X)$*-morphism* of the $\mathcal{A}(X)$-modules involved,

$$\lambda^n_{\mathcal{U}} : \check{H}^n_0(\mathcal{U}, \mathcal{E}) \longrightarrow \check{H}^n(X, \mathcal{E}), \; n \in \mathbf{Z}_+, \tag{5.92}$$

being thus *associated with every open covering* $\mathcal{U}$ *of* $X$. So one finally gets an $\mathcal{A}(X)$*-morphism*

$$\lambda^n := \varinjlim_{\mathcal{U}} \lambda^n_{\mathcal{U}} : \check{H}^n_0(X, \mathcal{E}) \longrightarrow \check{H}^n(X, \mathcal{E}), \; n \in \mathbf{Z}_+, \tag{5.93}$$

Concerning the notation employed in (5.93), see (5.51) along, for instance, with N. Bourbaki [1: p.95, Remarque 2)]. Yet, taking into account the very definitions of the maps (in fact, $\mathcal{A}(X)$*-morphisms*) involved, the next *commutative diagram,* thus resulted, clarifies the definition of the maps $\lambda^n$, $n \in \mathbf{Z}_+$, in (5.93).

$$
\begin{array}{ccc}
\check{Z}^n_0(\mathcal{U},\mathcal{E}) & \xrightarrow{\;\subset\;} & \check{Z}^n(\mathcal{U},\mathcal{E}) \\
\downarrow & \nearrow & \downarrow \\
\check{H}^n_0(\mathcal{U},\mathcal{E}) \equiv \check{Z}^n_0(\mathcal{U},\mathcal{E})/\check{B}^n_0(\mathcal{U},\mathcal{E}) & \dashrightarrow & \check{Z}^n(\mathcal{U},\mathcal{E})/\check{B}^n(\mathcal{U},\mathcal{E}) \equiv \check{H}^n(\mathcal{U},\mathcal{E}) \\
\downarrow & \searrow \lambda^n_{\mathcal{U}} & \downarrow \\
\check{H}^n_0(X,\mathcal{E}) & \xrightarrow{\;\lambda^n\;} & \check{H}^n(X,\mathcal{E})
\end{array} \tag{5.94}
$$

Thus, our claim in (5.88) is that, *for any* $n \in \mathbf{Z}_+$, *the* above *map* $\lambda^n$ *is,* in fact, *an* $(\mathcal{A}(X)$*-)isomorphism (onto).* So we come next to the

**Proof of Corollary 5.1.** First, *for* $n = 0$, based on the very definition of the 0-*dimensional* (Čech) *cohomology* (see (5.13) and (4.54)), as well as, on (5.10) and (5.11)), one concludes that (cf. also (5.93));

(5.95) the 0-*dimendional map* $(\mathcal{A}(X))$*-morphism)*

(5.95.1) $\lambda \equiv \lambda^0 : \check{H}^0_0(X,\mathcal{E}) \longrightarrow \check{H}^0(X,\mathcal{E}) \cong \Gamma(X,\mathcal{E})$

is merely the *inclusion map.* Therefore, 1-1, *for every topological space* $X$ (not necessarily paracompact).

On the other hand, *if* $n \geqq 1$, let $[g] \in \check{H}^n(X,\mathcal{E})$, such that

$$\lambda^n([g]) = 0 \in \check{H}^n(X,\mathcal{E}). \tag{5.96}$$

Thus, by hypothesis, along with (5.51) and (5.91), there exists an open covering $\mathcal{U}$ of $X$, such that (cf. also (5.10)) $g \in \check{Z}^n_0(\mathcal{U},\mathcal{E}) \subseteq \check{Z}^n(\mathcal{U},\mathcal{E})$, so that (cf. (5.94))

$$\lambda^n([g]) = [g] = 0 \in \check{H}^n(X,\mathcal{E}). \tag{5.97}$$

Therefore (see (5.91.1) and (4.22)), $g \in \check{B}^n(\mathcal{U},\mathcal{E}) = \delta(\check{C}^{n-1}(\mathcal{U},\mathcal{E}))$, *that is,* $g = \delta(g')$, *for some* $g' \in \check{C}^{n-1}(\mathcal{U},\mathcal{E})$. But, then (cf. Lemma 5.2), *one has*

$$\mu(g') = \psi(h), \tag{5.98}$$

*for some* $h \in \check{C}^{n-1}(\mathcal{V},\mathcal{S})$, with $\mathcal{V}$ *a* (*locally finite,* cf. (5.89)) *refinement of*

$\mathcal{U}$ (and $\mu$ a *refining map,* see (5.54.2)). Thus, one obtains, by (5.98) and the hypothesis for $g$ (cf. also (4.35.1) and (5.8′));

$$\begin{aligned}\mu(g) = \mu(\delta(g')) = \delta(\mu(g')) = \delta(\psi(h)) = \psi(\delta(h)) \in \\ \psi(\delta(\check{C}^{n-1}(\mathcal{V},\mathcal{S}))) \subseteq \psi(\check{C}^n(\mathcal{V},\mathcal{S})).\end{aligned} \tag{5.99}$$

In other words (cf. also (4.34′) and (5.11)), one has

$$\mu(g) \in \check{B}^n(\mathcal{V},\mathcal{E}) \cap \psi(\check{C}^n(\mathcal{V},\mathcal{S})) = \check{B}^n_0(\mathcal{V},\mathcal{E}), \tag{5.100}$$

*so that* (cf. (5.89) too) $[g] = 0 \in \check{H}^n_0(X,\mathcal{E})$, which thus proves that $\lambda^n$ *is one-to-one* (being, by its very definition, an $\mathcal{A}(X)$-*morphism,* as well).

Yet, we finally prove that *the same map is onto:* So let $[g] \in \check{H}^n(X,\mathcal{E})$; hence (cf. (5.91.1)), $g \in \check{Z}^n(\mathcal{U},\mathcal{E})$, for some (locally finite, see (5.89)) open covering of $X$. Therefore (Lemma 5.2), one gets $\mu(g) = \psi(h)$, for some $h \in \check{C}^n(\mathcal{V},\mathcal{S})$ (see (5.54.2)), so that (cf. (4.35.1) and (5.8′)) one has;

$$\delta(\mu(g)) = \mu(\delta(g)) = \delta(\psi(h)) = \psi(\delta(h)) = 0, \tag{5.101}$$

due to the hypothesis for $g$. That is (see also (5.10)), one obtains

$$\mu(g) \in \check{Z}^n(\mathcal{V},\mathcal{E}) \cap \psi(\check{C}^n(\mathcal{V},\mathcal{S})) = \check{Z}^n_0(\mathcal{V},\mathcal{E}), \tag{5.102}$$

so that *the given* $[g] \in \check{H}^n(X,\mathcal{E})$ *lies in* $\mathrm{im}(\lambda^n)$, which thus proves our assertion, and this terminates the proof of the lemma. ∎

Now, before we come to our main conclusion of the present Section, namely, the next Theorem 5.3, we give an improved version of Theorem 5.1 above, which will also be of use for that matter. So it is standard that:

(5.103) *inductive* (or else *direct*) *limits preserve exactness* (of sequences); said otherwise, *the inductive limit functor is exact.*

In particular, here, the functor in question has, of course, as *domain of definition* the category of inductive systems of (cohomology) $\mathcal{A}(X)$-modules (see (4.52.1) and (5.50.1)), indexed by a *common* directed set, viz. that one of all (proper) open coverings of the topological space $X$ considered (see (4.9), or for $X$ *paracompact,* that of locally finite ones, (cf. (5.89)), while the *range of the same functor* is $\mathcal{A}(X)$-$\mathcal{M}od$, i.e., the category of $\mathcal{A}(X)$-modules. Yet, in this context, cf., for example, A. Solian [1: p.273, Corollary] and/or S. Eilenberg - N. Steenrod [1: p.225, Theorem 5.4], or even S. Lang [1: p.144, Ex. 2]; the crucial point in the proof of the previous argument can also be seen by consulting, for instance, N. Bourbaki [1: p. 93, Proposition 7]. (See also R. M. Switzer [1: p.119, Proposition 7.50], or even N. Bourbaki [2: Chap. 2; p.91, Proposition 3]).

So, based now on (5.103), we come to the following version of Theorem 5.1; namely, each one of the terms in (5.34) is now replaced by the corresponding inductive system, therefore, the analogous exact sequence, by passing to the limit, according to (5.103). Thus, one obtains the next.

**Theorem 5.2.** *Let* $(X, \mathcal{A})$ *be a given* **C***-algebraized space, and*

$$0 \longrightarrow \mathcal{F} \xrightarrow{\phi} \mathcal{S} \xrightarrow{\psi} \mathcal{E} \longrightarrow 0 \tag{5.104}$$

*a short exact* $\mathcal{A}$*-sequence. Then, one gets the following long exact* $\mathcal{A}(X)$*-sequence of (liftable) Čech cohomology;*

$$\begin{aligned} &0 \longrightarrow \check{H}^0(X,\mathcal{F}) \xrightarrow{\phi^*} \check{H}^0(X,\mathcal{S}) \xrightarrow{\psi^*} \check{H}^0_0(X,\mathcal{E}) \xrightarrow{\delta^*} \check{H}^1(X,\mathcal{F}) \\ &\longrightarrow \cdots \xrightarrow{\psi^*} \check{H}^1_0(X,\mathcal{E}) \xrightarrow{\delta^*} \check{H}^2(X,\mathcal{F}) \longrightarrow \cdots \xrightarrow{\psi^*} \check{H}^n_0(X,\mathcal{E}) \\ &\xrightarrow{\delta^*} \check{H}^{n+1}(X,\mathcal{F}) \longrightarrow \cdots . \blacksquare \end{aligned} \tag{5.105}$$

Analogously to (5.34), we have employed above, for simplicity, an obvious *abuse of notation,* concerning the maps $\phi^*, \psi^*$ and $\delta$. This too will be the case presently below in (5.107). Now, as an immediate consequence of (5.105), in conjunction with Corollary 5.1, we come to the next fundamental result, that we have already referred to in the preceding. Namely, one has.

**Theorem 5.3.** *(Long exact sequence in Čech cohomology). Suppose that* $(X, \mathcal{A})$ *is a given* **C***-algebraized space, with* $X$ *a paracompact (Hausdorff) space, and let*

$$0 \longrightarrow \mathcal{F} \xrightarrow{\phi} \mathcal{S} \xrightarrow{\psi} \mathcal{E} \longrightarrow 0 \tag{5.106}$$

*be a short exact* $\mathcal{A}$*-sequence. Then, one obtains the following long exact* $\mathcal{A}(X)$*-sequence in Čech cohomology;*

$$\begin{aligned} &0 \longrightarrow \check{H}^0(X,\mathcal{F}) \xrightarrow{\phi^*} \check{H}^0(X,\mathcal{S}) \xrightarrow{\psi^*} \check{H}^0(X,\mathcal{E}) \xrightarrow{\delta^*} \check{H}^1(X,\mathcal{F}) \\ &\longrightarrow \check{H}^1(X,\mathcal{S}) \xrightarrow{\psi^*} \check{H}^1(X,\mathcal{E}) \xrightarrow{\delta^*} \check{H}^2(X,\mathcal{F}) \xrightarrow{\phi^*} \cdots \longrightarrow \\ &\check{H}^n(X,\mathcal{E}) \xrightarrow{\delta^*} \check{H}^{n+1}(X,\mathcal{F}) \longrightarrow \cdots . \blacksquare \end{aligned} \tag{5.107}$$

Applications of the previous exact sequence will be found in Part Two of this treatise. Of course, we have already obtained a similar exact sequence by (3.17) in the preceding, *for any topological space* $X$ (so, not necessarily paracompact, as was the case with (5.107)), by employing (*abstract,* viz. a "derived functor") *sheaf cohomology;* however, as we shall see, in a while (cf. Section 8 below), *all cohomology theories agree "actually" on paracompact spaces* (cf. Theorem 8.1). So this, of course, will be the case too (see thus Theorem 6.1 below) for a *useful variant of Čech cohomology,* which we presently examine by the next Section (see Theorem 6.1 in the sequel).

# 6. Čech cohomology with coefficients in a presheaf

The technique of Čech cohomology, as exposed in the preceding is actually applicable when the *"domain of coefficients"* of the various cohomology modules, as before (cf. e.g. (4.52.1), or even (4.53)), is, more generally, an arbitrary *presheaf* (of suitable *modules* in our case and, in general, of *abelian groups*) on a given topological space $X$. Indeed, as we shall presently see, the former is a particular case of what follows (cf. thus (6.8) below), when, namely, one looks at a sheaf, equivalently, "as" (cf. Chapt. I; Theorem 13.1) the (complete) presheaf of its (continuous) sections.

Thus, as we did so far, suppose that we are given a **C**-*algebraized space* $(X, \mathcal{A})$, along with a *presheaf of* $\mathcal{A}(U)$*-modules* on $X$

(6.1) $$E \equiv \left(E(U), \rho_V^U\right),$$

such that

(6.2) $$A(U) := \Gamma(U, \mathcal{A}) \equiv \mathcal{A}(U), \textit{ with } U \textit{ open in } X.$$

(Of course, as usual, we set $A(\emptyset) = E(\emptyset) = (0)$, the *trivial module*). Namely, we assume that the *presheaf of coefficients of* $E$ (cf. Chapt. II; Definition 1.6) is

(6.3) $$\Gamma(\mathcal{A}) \equiv \left(\Gamma(U, \mathcal{A}), \tau_V^U\right) \equiv \left(\mathcal{A}(U), \tau_V^U\right),$$

viz. the *(complete) presheaf of sections of* $\mathcal{A}$ (with $\mathcal{A}$ being, as told, the *structure sheaf* of our standard framework $(X, \mathcal{A})$).

Of course, as already said, *any presheaf of modules* (or even of *abelian groups*) *would be enough for what follows.* However, we have chosen, in purpose, the previous context, since then *the* corresponding *sheafification of* $E$, $\mathbf{S}(E) \equiv \mathcal{E}$, *will provide an* $\mathcal{A}$*-module on* $X$ (see Chapt II; Proposition 1.1), *which,* among other things, *will, equivalently, substitute* $E$ (with appropriate $X$ too, see thus Theorem 6.1 below) *for the domain of coefficients of the cohomology modules* concerned.

Thus, with any given *open covering*

$$\mathcal{U} = \{U_\alpha\}_{\alpha \in I} \tag{6.4}$$

of $X$, one can associate, by analogy with Subsection 4.(b) in the preceding, a *Čech cochain complex,* where now, however, *the cochains* involved *take values from the* given *presheaf* $E$, as above. So one defines, for instance, *for any* given $n \in \mathbf{Z}_+$ (cf. also (4.12) for the following notation),

$$\check{C}^n(\mathcal{U}, E) := \prod_{(\alpha_0, \ldots, \alpha_n)} E(U_{\alpha_0 \cdots \alpha_n}), \tag{6.5}$$

as the $\mathcal{A}(X)$*-module of* $n$*-cochains of* $\mathcal{U}$ *with coefficients in the* given *presheaf* $E$. Thus, following Subsection 4.(b), one can repeat *verbatim* the notions of *cocycles, coboundaries* and *coboundary operators,* which correspond to the present context, taking, therefore, the respective *Čech cochain complex of* $\mathcal{U}$ *with coefficients in* $E$. Namely, one has (see also (4.19));

$$\check{C}^{\cdot}(\mathcal{U}, E) := (\check{C}^n(\mathcal{U}, E), \delta \equiv \{\delta^n\})_{n \in \mathbf{Z}_+}. \tag{6.6}$$

Yet, by analogy with (4.23), one further defines, *for any* $n \in \mathbf{Z}_+$, the respective $n$*-dimensional Čech cohomology* $\mathcal{A}(X)$*-module of* $\mathcal{U}$ *with coefficients in* $E$,

$$\check{H}^n(\mathcal{U}, E) := \check{Z}^n(\mathcal{U}, E) / \check{B}^n(\mathcal{U}, E); \tag{6.7}$$

that is, the corresponding *Čech cohomology of the (cochain) complex (6.6),* or even the *Čech cohomology of* $\mathcal{U}$ *with coefficients in the* $\Gamma(\mathcal{A})$*-presheaf* $E$, denoted by

$$h^*(\check{C}^{\cdot}(\mathcal{U}, E)) := \{h^n(\check{C}^{\cdot}(\mathcal{U}, E))\}_{n \in \mathbf{Z}_+} \equiv \{\check{H}^n(\mathcal{U}, E)\}_{n \in \mathbf{Z}_+}. \tag{6.8}$$

Accordingly, comparing the previous argument with that one in the above

Subsection 4.(b), one comes now to the conclusion that;

(6.9) *the* (usual) *Čech cohomology with coefficients in a given* $\mathcal{A}$-*module* $\mathcal{E}$ (cf. (4.20), (4.53)), *is, in effect that cohomology* (procedure), *as above* (cf. (6.8)), *applied, in particular, to the corresponding (complete) presheaf of sections of* $\mathcal{E}$, $\Gamma(\mathcal{E})$ $(\equiv \{\Gamma(\mathcal{U}, \mathcal{E}), \sigma_V^U\})$.

So, what amounts to the same thing, in the case of the (usual) Čech complex (4.19), one thus considers a particular *complete presheaf,* namely, *that one* (the only case, any way(!), cf. Chapt. I; Theorem 11.2) *deriving from the sections of a sheaf.*

Now, following (4.52), one further defines the corresponding to (6.8) *inductive system*

$$\left(\check{H}^n(\mathcal{U}, E), \tau_{\mathcal{V},\mathcal{U}}\right). \tag{6.10}$$

For simplicity, we employed here the same notation for the maps $\tau_{\mathcal{V}\mathcal{U}}$ (with $\mathcal{V} \succeq \mathcal{U}$, see (4.2)), as in (4.49), while it is still quite clear that;

(6.11) *Lemma (4.2) is actually true, within the present more general context.*

So one defines the *n-th dimensional Čech cohomology* $\mathcal{A}(X)$-*module of* $X$ *with coefficients in the presheaf* (of $\mathcal{A}(U)$-*modules*, cf. (6.2)) $E$, by the relation;

$$\check{H}^n(X, E) := \varinjlim_{\mathcal{U}} \check{H}^n(\mathcal{U}, E), \; n \in \mathbf{Z}_+. \tag{6.12}$$

Namely, one thus arrives at the *Čech cohomology of* $X$, *with coefficients in the* $\Gamma(\mathcal{A})$-*presheaf* $E$, as above (see also (6.3)), denoted by (the following *sequence of* $\mathcal{A}(X)$-*modules*);

$$\check{H}^*(X, E) \equiv \left\{\check{H}^n(X, E)\right\}_{n \in \mathbf{Z}_+}. \tag{6.13}$$

Now, an important fact here is, of course, the *existence of a long exact cohomology sequence,* within the above context, by contrast, for instance, with what happens in the case of (4.53)(!), the latter being remedied, however, through (5.34) (or, in particular, in the case of a paracompact space $X$; but, see also Theorem 6.1 in the sequel). So we come next to this matter by the following Subsection.

**6.(a). Long exact (Čech) cohomology sequence for presheaves.**– Assume that we have two *presheaves of* $\mathcal{A}(U)$*-modules* on $X$ (cf (6.2))

$$S \equiv \left(S(U), \rho_V^U\right) \text{ and } E \equiv \left(E(U), \lambda_V^U\right), \tag{6.14}$$

along with a *morphism of* $S$ *in* $E$, say, $\phi \equiv (\phi_U : S \longrightarrow E)$, such that

$$\phi_U : S(U) \longrightarrow E(U),\ U \textit{ open} \text{ in } X, \tag{6.15}$$

being an $\mathcal{A}(U)$*-morphism* of the $\mathcal{A}(U)$-modules concerned, satisfies Chapt. I; (6.6). Now, it is easy to see that, given an *open covering* $\mathcal{U}$ of $X$ (cf. (6.4)), one obtains, *for every* $n \in \mathbf{Z}_+$, a corresponding *map of* $n$*-cochains,*

$$\phi_{\mathcal{U}}^{*,n} : \check{C}^n(\mathcal{U}, S) \longrightarrow \check{C}^n(\mathcal{U}, E), \tag{6.16}$$

being, in effect, as one easily proves, an $\mathcal{A}(X)$*-morphism* of the $\mathcal{A}(X)$-modules involved (cf. (6.5). Indeed, one proves this for each one of the maps (6.15)). Moreover, each one of *the maps (6.16) commutes with the coboundary operators,* so that one finally has an $\mathcal{A}(X)$*-morphism*

$$\phi_{\mathcal{U}}^* : \check{H}^n(\mathcal{U}, S) \longrightarrow \check{H}^n(\mathcal{U}, E) \tag{6.17}$$

of the $\mathcal{A}(X)$-modules concerned (cf. also (6.7)). In fact, the various maps $\phi_{\mathcal{U}}^*$, with $\mathcal{U}$ varying as in (4.9), constitute a *morphism of inductive systems,* which, by (6.10), correspond to the given $S, E$ and $\mathcal{U}$, as above. Therefore (cf. also, for example, N. Bourbaki [1: p.92, Corollaire 1]), for any two (proper) open coverings $\mathcal{U}$,$\mathcal{V}$ of $X$, with $\mathcal{V} \succeq \mathcal{U}$ (see (4.2)), one has the following *commutative diagram*

$$\begin{array}{ccc} \check{H}^n(\mathcal{U}, S) & \xrightarrow{\phi_{\mathcal{U}}^*} & \check{H}^n(\mathcal{U}, E) \\ \downarrow{\scriptstyle \tau_{\mathcal{V},\mathcal{U}}} & & \downarrow{\scriptstyle \tau_{\mathcal{V},\mathcal{U}}} \\ \check{H}^n(\mathcal{V}, S) & \xrightarrow{\phi_{\mathcal{U}}^*} & \check{H}^n(\mathcal{V}, E), \end{array} \tag{6.18}$$

so that one finally obtains a *limit map* (cf. also (6.12))

$$\phi^* = \varinjlim_{\mathcal{U}} \phi_{\mathcal{U}}^* : \check{H}^n(X, S) \longrightarrow \check{H}^n(X, E); \tag{6.19}$$

in fact, the latter map is an $\mathcal{A}(X)$*-morphism* of the $\mathcal{A}(X)$-modules concerned.

Thus, suppose now that we are given the following *short exact sequence of presheaves of $\mathcal{A}(X)$-modules*

$$0 \longrightarrow F \xrightarrow{\phi} S \xrightarrow{\psi} E \longrightarrow 0; \tag{6.20}$$

that is, a (short) *exact sequence of* the respective *inductive systems,* provided by the given presheaves, where we also set, along with (6.14),

$$F \equiv \left(F(U), \mu_V^U\right). \tag{6.21}$$

In this regard, cf. Chapt. I; Definition 5.2, as well as, Chapt. II; (2.32). Therefore (see Chapt. I; Definition 6.1), one obtains, *for every open $U \subseteq X$, the following exact sequence of $\mathcal{A}(U)$-modules* (cf. also e.g. (6.15));

$$0 \longrightarrow F(U) \xrightarrow{\phi_U} S(U) \xrightarrow{\psi_U} E(U) \longrightarrow 0. \tag{6.22}$$

So, according to a standard argument (cf. (6.5), as well as, for instance, N. Bourbaki [2: Chap. 2; p.10, Proposition 5 and p.11, Corollaire]), one gets a corresponding to (6.22) *(short) exact sequence of cochain $\mathcal{A}(X)$-modules*

$$0 \longrightarrow \check{C}^n(\mathcal{U}, F) \xrightarrow{\phi_{\mathcal{U}}^{*,n}} \check{C}^n(\mathcal{U}, S) \xrightarrow{\psi_{\mathcal{U}}^{*,n}} \check{C}^n(\mathcal{U}, E) \longrightarrow 0, \tag{6.23}$$

*for every* $n \in \mathbf{Z}_+$ (cf. also (6.16)). In this context, we should also note that *the $\mathcal{A}(X)$-morphisms in (6.23) commute with the respective coboundary operators* (cf. e.g. (4.15), as well as, (5.7) and (5.8), or even (5.8′)); see also (5.15) and (1.19″)). So one finally obtains the following *short exact sequence of* (the corresponding, cf. (6.6)) *Čech cochain complexes*

$$0 \longrightarrow \check{C}^{\cdot}(\mathcal{U}, F) \xrightarrow{\phi_{\mathcal{U}}^{*}} \check{C}^{\cdot}(\mathcal{U}, S) \xrightarrow{\psi_{\mathcal{U}}^{*}} \check{C}^{\cdot}(\mathcal{U}, E) \longrightarrow 0. \tag{6.24}$$

(See also, for instance, (1.30), (1.30′) and (1.30″) for the analogous situation to (6.24) in the case of complexes of $\mathcal{A}$-modules). Thus, based on standard Homological Algebra (in particular, *theory of cochain complexes,* cf. e.g. Theorem 1.1 in the foregoing), one now obtains the following *long exact sequence in Čech cohomology for presheaves:*

$$\begin{aligned} &0 \longrightarrow \check{H}^0(\mathcal{U}, F) \xrightarrow{\phi^*} \check{H}^0(\mathcal{U}, S) \xrightarrow{\psi^*} \check{H}^0(\mathcal{U}, E) \xrightarrow{\delta^*} \check{H}^1(\mathcal{U}, F) \\ &\xrightarrow{\phi^*} \check{H}^1(\mathcal{U}, S) \xrightarrow{\psi^*} \check{H}^1(\mathcal{U}, E) \xrightarrow{\delta^*} \check{H}^2(\mathcal{U}, F) \xrightarrow{\phi^*} \cdots \xrightarrow{\delta^*} \check{H}^n(\mathcal{U}, F) \\ &\xrightarrow{\phi^*} \check{H}^n(\mathcal{U}, S) \xrightarrow{\psi^*} \check{H}^n(\mathcal{U}, E) \xrightarrow{\delta^*} \check{H}^{n+1}(\mathcal{U}, F) \xrightarrow{\phi^*} \cdots. \end{aligned} \tag{6.25}$$

(In this respect, cf., for example, R. Godement [1: p.22, §2.2], or even S. Lang [1: p.123, Theorem 2.1]; yet, see, for instance, J. J. Rotman [1: p.172, Theorem 6.3]). So the proof of (6.25) is analogous to that of Theorem 1.1 in the preceding, where we only consider different complexes; indeed, due to (6.8), one actually sets, *by definition* (cf. also (1.9)),

$$\check{H}^n(\mathcal{U}, E) = h^n(\check{C}^\cdot(\mathcal{U}, E)),\ n \in \mathbf{Z}_+, \tag{6.26}$$

so that (6.25) is the analogous exact sequence to (1.36). ∎

Our main conclusion in this Subsection is a direct consequence of (6.25) and the fact that *exact sequences are preserved by direct limits* (see thus (5.103) above; yet, cf. N. Bourbaki [2: Chap. 2; p.91, Proposition 3] or even R. M. Switzer [1: p.119, Proposition 7.50]). Consequently, in view of (6.25) and (6.12), one now obtains the following basic result.

**Theorem 6.1.** *Let $(X, \mathcal{A})$ be a given* **C***-algebraized space and*

$$0 \longrightarrow F \xrightarrow{\phi} S \xrightarrow{\psi} E \longrightarrow 0 \tag{6.27}$$

*a short exact sequence of $\Gamma(\mathcal{A})$-presheaves on $X$. Then, one obtains the following long exact sequence in Čech cohomology* (with coefficients in a given presheaf):

$$\begin{aligned} &0 \longrightarrow \check{H}^0(X, F) \xrightarrow{\phi^*} \check{H}^0(X, S) \xrightarrow{\psi^*} \check{H}^0(X, E) \xrightarrow{\delta^*} \check{H}^1(X, F) \\ &\xrightarrow{\phi^*} \cdots \xrightarrow{\delta^*} \check{H}^n(X, F) \xrightarrow{\phi^*} \check{H}^n(X, S) \xrightarrow{\psi^*} \check{H}^n(X, E) \xrightarrow{\delta^*} \check{H}^{n+1}(X, F) \\ &\xrightarrow{\phi^*} \cdots . \end{aligned} \tag{6.28}$$ ∎

(Yet, by an obvious abuse of notation, we used above, for simplicity, the same symbols with (6.27) for the maps involved). Now, the following easy consequence of the above theorem will be of use presently below. That is, one has.

**Corollary 6.1.** *Assume that we have a* **C***-algebraized space $(X, \mathcal{A})$, along with the following exact sequence of $\Gamma(\mathcal{A})$-presheaves on $X$*

$$0 \longrightarrow F \longrightarrow S \xrightarrow{h} G \longrightarrow E \longrightarrow 0, \tag{6.29}$$

*such that*

(6.30) $$\check{H}^n(X,F)=\check{H}^n(X,E)=0,\ n\in\mathbf{Z}_+.$$

*Then, one obtains*

(6.31) $$\check{H}^n(X,S)=\check{H}^n(X,G),\ n\in\mathbf{Z}_+,$$

*within an* $\mathcal{A}(X)$*-isomorphism* (cf. e.g. (6.19)).

**Proof.** Taking the *image* $\Gamma(\mathcal{A})$*-presheaf of* $h$, $\operatorname{im} h$ (cf. Chap. II; (2.28)), one obtains, in view of (6.29), the two following *short exact sequences of* $\Gamma(\mathcal{A})$*-presheaves;* namely,

(6.32) $$0\longrightarrow F\longrightarrow S\longrightarrow \operatorname{im} h\longrightarrow 0$$

and

(6.33) $$0\longrightarrow \operatorname{im} h\longrightarrow G\longrightarrow E\longrightarrow 0.$$

Therefore, taking (6.28) into account, for each one of the above two exact sequences, along with (6.30), *one finds*

(6.34) $$\check{H}^n(X,S)=\check{H}^n(X,\operatorname{im} h)=\check{H}^n(X,G),\ n\in\mathbf{Z}_+,$$

within $\mathcal{A}(X)$*-isomorphisms,* and this proves (6.31). ∎

We come next to our main result of the present Section, specializing to paracompact spaces $X$. We treat this in the following Subsection.

**6.(b). Čech cohomology for presheaves on paracompact spaces.**– Our aim by the ensuing discussion is to show the following fundamental fact. Namely,

(6.35) *the Čech cohomology* $\mathcal{A}(X)$*-modules of a paracompact (Hausdorff) space* $X$ (base space of a given $\mathbf{C}$*-algebraized space* $(X,\mathcal{A})$*), with coefficients in a* $\Gamma(\mathcal{A})$*-presheaf* $E$*, depend only on the sheafification of* $E$*,* $\mathbf{S}(E)\equiv\mathcal{E}$*; i.e., on the* $\mathcal{A}$*-module* $\mathcal{E}$ *on* $X$*, generated by the given presheaf* $E$*.*

The proof of the above assertion is based, in effect, on the following variant of our previous Lemma 5.2. So we have the next.

**Lemma 6.1.** *Let* $(X,\mathcal{A})$ *be a* $\mathbf{C}$*-algebraized space, with* $X$ *a paracompact (Hausdorff) space and* $E$ *a* $\Gamma(\mathcal{A})$*-presheaf on* $X$*, whose corresponding sheafification is the zero* $\mathcal{A}$*-module on* $X$*; i.e., one has* (cf. Chapt. I; (7.23)

and Chapt. II; (1.66))

$$\mathcal{E} \equiv \mathbf{S}(E) = 0. \tag{6.36}$$

*Moreover, let* $\mathcal{U} = \{U_\alpha\}_{\alpha\in I}$ *be an open covering of* $X$*, along with an* $n$*-cochain*

$$g \in \check{C}^n(\mathcal{U}, E). \tag{6.37}$$

*Then, there exists a refinement* $\mathcal{V} = \{V_\beta\}_{\beta\in J}$ *of* $\mathcal{U}$*, with refining map, say,* $\mu : J \longrightarrow I$*, such that*

$$\mu(g) = 0 \in \check{C}^n(\mathcal{V}, E). \tag{6.38}$$

**Proof.** This can be patterned after that one of Lemma 5.2: Namely, if

$$E \equiv \left(E(U), \rho_V^U\right) \tag{6.39}$$

stands for a given $\Gamma(\mathcal{A})$-*presheaf* on $X$, then, based on the hypothesis for $X$, consider a point $x \in X$, as in (5.57); thus, setting for simplicity, $U \equiv U_{\alpha_0\cdots\alpha_n}$, one obtains, in view of (6.36),

$$\rho_U(g)(x) = \rho_U^x(g) \equiv \tilde{g}(x) = 0 \in E_x = \mathcal{E}_x = (0) \tag{6.40}$$

(see also Chapt. I; (7.9), (7.10), (7.22) and (7.6)). Therefore (cf. Chapt. I; (7.12)), there exists an open $V \subseteq X$, with $x \in V \subseteq U$, such that

$$\widetilde{\rho_V^U(g)} \equiv \tilde{g}|_V = 0|_V. \tag{6.41}$$

Here we can further assume, for instance, that $V = U_x$, as in (5.61).

Now, the argument proceeds, hereafter, by analogy with that one in the proof of Lemma 5.2, taking $h = 0$. (In this context, it is still easy to see that *the* $\mathcal{A}(X)$*-morphism, associated with the refining map* $\mu$*, commutes with the corresponding ones defined by the "restriction maps" of* $E$ *and* $\Gamma(\mathcal{E})$*;* see also (4.35), as well as, (6.9) above). ∎

As an immediate consequence of the previous Lemma 6.1, one obtains the following.

**Corollary 6.2.** *Suppose we have a* **C**-*algebraized space* $(X, \mathcal{A})$*, with* $X$ *a paracompact (Hausdorff) space, as well as, a* $\Gamma(\mathcal{A})$*-presheaf* $E$ *on* $X$*, having trivial sheafification; viz. such that* $\mathcal{E} \equiv \mathbf{S}(E) = 0$*. Then, one obtains* (cf. also (6.13))

$$\check{H}^*(X, E) = 0; \tag{6.42}$$

*that is, one has*

$$\check{H}^n(X, E) = 0, \tag{6.43}$$

*for every* $n \in \mathbf{Z}_+$.

**Proof.** The assertion follows straightforwardly from (6.38) and the very definition (6.12). (In this regard, see also, for instance, the comments following (5.89) in the preceding, as well as, (5.90) and (5.91)). ∎

So we come next to our main conclusion in this Section, justifying thus (6.35). Namely, one has the following basic result.

**Theorem 6.2.** *Let* $(X, \mathcal{A})$ *be a* $\mathbf{C}$*-algebraized space, with* $X$ *a paracompact (Hausdorff) space, and* $E$ *a given* $\Gamma(\mathcal{A})$*-presheaf on* $X$, *with* $\mathcal{E} \equiv \mathbf{S}(E)$ *its corresponding sheafification, thus an* $\mathcal{A}$*-module on* $X$ (cf. Chapt. II; Proposition 1.1). *Then, one obtains*

$$\check{H}^*(X, E) = \check{H}^*(X, \mathbf{S}(E)) \equiv \check{H}^*(X, \mathcal{E}), \tag{6.44}$$

*within an isomorphism of the cohomology* (theories) *concerned, viz.* $\mathcal{A}(X)$*-isomorphisms of the respective cohomology* $\mathcal{A}(X)$*-modules* (cf. (4.53′) and (6.13)).

**Proof.** Suppose that

$$E \xrightarrow{\rho} \Gamma(\mathbf{S}(E)) \equiv \Gamma(\mathcal{E}) \tag{6.45}$$

is the *canonical* $\mathcal{A}(X)$*-morphism* of the $\Gamma(\mathcal{A})$-presheaves concerned (cf.Chapt. I; (11.6)). Therefore, one obtains the following *exact sequence of* $\Gamma(\mathcal{A})$*-presheaves;*

$$0 \longrightarrow \ker \rho \longrightarrow E \xrightarrow{\rho} \Gamma(E) \longrightarrow \operatorname{coker} \rho \longrightarrow 0. \tag{6.46}$$

For the first non-zero term in (6.46) see Chapt. II; (2.27), while for the $\Gamma(\mathcal{A})$-*presheaf* $\operatorname{coker} \rho := \Gamma(\mathcal{E})/\operatorname{im} \rho$, cf. Chapt. II; (2.28) and e.g. (2.45), as well as, N. Bourbaki [1: p.94, Corollaire].

On the other hand, from the exactness of (6.46) and based on the very definitions, one obtains, *for every open set* $U \subseteq X$, a corresponding *exact sequence of* $\mathcal{A}(X)$*-modules* (cf., for example, (6.20) and (6.22)), so that, by further applying (5.103), one gets the *analogous exact sequence to (6.46), for every stalk* of the respective presheaves, hence, in turn, for their sheafifications, in view of Lemma 1.1 and Chapt. I; (7.6). Thus, one finally obtains the following *exact $\mathcal{A}$-sequence* (see also Chapt. II; Proposition 1.1);

$$0 \longrightarrow \mathbf{S}(\ker \rho) \longrightarrow \mathbf{S}(E) \xrightarrow{\rho} \mathbf{S}(\Gamma(E)) \longrightarrow \mathbf{S}(\operatorname{coker} \rho) \longrightarrow 0. \tag{6.47}$$

But (cf. Chapt. I; Theorem 9.1), one has

$$\mathbf{S}(E) \equiv \mathcal{E} = \mathbf{S}(\Gamma(\mathcal{E})), \tag{6.48}$$

up to an *$\mathcal{A}$-isomorphism* of the $\mathcal{A}$-modules concerned, so that *the map $\rho$ in (6.47) is,* in effect, *the identity $\mathcal{A}$-isomorphism* (or even *$\mathcal{A}$-automorphism*) *of $\mathcal{E}$,* viz. one obtains

$$\rho = \mathrm{id}_{\mathcal{E}}. \tag{6.49}$$

So, in view of the exactness of (6.47) and the last relation, one has, of course;

$$\mathbf{S}(\ker \rho) = \mathbf{S}(\mathrm{coler}\, \rho) = 0. \tag{6.50}$$

Consequently (cf. Corollary 6.2), one obtains

$$\check{H}^*(X, \ker \rho) = \check{H}^*(X, \operatorname{coker} \rho) = 0 \tag{6.51}$$

hence, in view of Corollary 6.1, one gets

$$\check{H}^*(X, E) = \check{H}^*(X, \Gamma(\mathcal{E})), \tag{6.52}$$

within an *isomorphism* of the cohomologies concerned; but, due to (6.9), this is actually (6.44), and this terminates the proof. ∎

In particular, we still remark the following (cf. also F. Hirzebruch [1: p.32, Remarks 1,2]). Namely,

(6.53) *in 0-dimensional Čech cohomology Lemma 6.1 and Theorem 6.2 hold true with $X$ an arbitrary topological space; however, concerning Theorem 6.2, this, under the further assumption that the given $\Gamma(\mathcal{A})$-presheaf $E$ is, in particular, a monopresheaf* (viz. the map (6.45) is 1-1. Cf. also Chapt. I; Remark 11.1, i) and (11.36′)).

Indeed, given a *0-cochain*

$$g \equiv (g_\alpha) \in \check{C}^0(\mathcal{U}, E) = \prod_{\alpha \in I} E(U_\alpha) \tag{6.54}$$

(cf. (4.11)), one gets an analogous relation to (6.41), for any given $x \in$

$U_\alpha$, while one has $x \in V \equiv V_x \subseteq U_\alpha$; therefore, one obtains a *refinement* $\mathcal{V} \equiv \{V_x\}_{x\in X}$ of the given open covering of $X$, $\mathcal{U} \equiv \{U_\alpha\}_{\alpha\in I}$, with *refining map,* $\mu : X \longrightarrow I$, such that $V_x \subseteq U_{\mu(x)}$, as above, which thus proves our claim, concerning Lemma 6.1. So, as a result of the previous argument, one concludes, in particular, that:

(6.55) *Corollary 6.2 holds true in 0-dimensional cohomology, with $X$ an arbitrary topological space; viz.*

(6.55.1) $$\mathcal{E} \equiv \mathbf{S}(E) = 0, \text{ yields } \check{H}^0(X, E) = 0.$$

So *assume* now that the map $\rho$ *in* (6.45) is *one-to-one;* hence (cf. (6.46)), one obtains the following *short exact sequence in* $\Gamma(\mathcal{A})$*-presheaves*

(6.56) $$0 \longrightarrow E \xrightarrow{\rho} \Gamma(E) \longrightarrow \operatorname{coker}\rho \longrightarrow 0.$$

Therefore, taking the corresponding *long exact* (Čech) *cohomology sequence* (cf. Theorem 6.1), one obtains, in view of (6.55) (see also (4.55)),

(6.57) $$\check{H}^0(X, E) = \check{H}^0(X, \Gamma(\mathcal{E})) = \Gamma(X, \mathcal{E}),$$

up to $\mathcal{A}(X)$*-isomorphisms,* as desired, terminating thus the proof of (6.53). ▮

Yet, one easily proves that *(6.57) is valid, for any* $\Gamma(\mathcal{A})$*-presheaf* $E$*, with* $\mathcal{E} \equiv \mathbf{S}(E)$. (See e.g. E. H. Spanier [1: p.327, Example 13]).

# 7. Cohomology of double complexes. Hypercohomology

As the title of this Section indicates, we consider in the sequel *sheaf* and/or *Čech cohomology* as these are applied to certain *double complexes,* which we are going to employ in Part Two of this treatise. So we start with the necessary terminology. In this context, cf. also, for intance, J. -L. Brylinski [1: p. 21 ff], P. Griffiths - J. Harris [1: p.445f], or yet H. Cartan - S. Eilenberg [1: p. 362ff].

As usual so far, we restrict our attention to $\mathcal{A}$-*modules,* with respect to a given fixed **C**-*algebraized space* $(X, \mathcal{A})$, so that the objects of our complexes will belong to the (abelian) category

$$\mathcal{A}\text{-}\mathcal{M}od_X \tag{7.1}$$

(cf. (1.1)). Of course,

(7.1′) *any abelian category could be employed instead,* for much of what follows, a generality, however, which we are not going to consider in the sequel.

(In this regard, see e.g. R. Godement [1: p.32]).

Thus, by a *double complex of* $\mathcal{A}$-*modules,* or *double* $\mathcal{A}$-*complex,* or just an $\mathcal{A}$-*bicomplex,* one means a double sequence of $\mathcal{A}$-modules on $X$,

$$\mathcal{E}^{\cdot\cdot} \equiv \{\mathcal{E}^{n,m}\}_{(n,m)\in \mathbf{Z}\times\mathbf{Z}}, \tag{7.2}$$

along with *two* sort of *differentials;* viz. one has

i) the *horizontal differential*

$$\delta^{n,m} \equiv \delta : \mathcal{E}^{n,m} \longrightarrow \mathcal{E}^{n+1,m}, \ (n,m) \in \mathbf{Z}\times\mathbf{Z} \equiv \mathbf{Z}^2, \tag{7.3}$$

i.e., a *differential of bidegree* (1,0), and

ii) the *vertical differential*

$$d^{n,m} \equiv d : \mathcal{E}^{n,m} \longrightarrow \mathcal{E}^{n,m+1}, \ (n,m) \in \mathbf{Z}^2, \tag{7.4}$$

i.e., a *differential of bidegree* (0,1),

such that *the following holds true* (see also (1.6))

$$\text{i) } \delta^2\,(\equiv \delta\delta) = d^2\,(\equiv dd) = 0, \quad \textit{and} \quad \text{ii) } \delta d = d\delta. \tag{7.5}$$

Usually, one gets a picture of a given $\mathcal{A}$-*bicomplex*

$$(\mathcal{E}^{\cdot\cdot}, \delta, d), \tag{7.6}$$

as above, by depicting it on the "(n,m)-*plane*" as a family of $\mathcal{A}$-modules, as follows,

(7.7)
$$
\begin{array}{ccccccc}
 & & \vdots & & \vdots & & \\
 & & \uparrow d & & \uparrow d & & \\
\cdots & \xrightarrow{\delta} & \mathcal{E}^{n,m} & \xrightarrow{\delta} & \mathcal{E}^{n+1,m} & \xrightarrow{\delta} & \cdots \\
 & & \uparrow d & & \uparrow d & & \\
\cdots & \xrightarrow{\delta} & \mathcal{E}^{n,m-1} & \xrightarrow{\delta} & \mathcal{E}^{n+1,m-1} & \xrightarrow{\delta} & \cdots \\
 & & \uparrow d & & \uparrow d & & \\
 & & \vdots & & \vdots & &
\end{array}
$$

Hence, the meaning of (7.5), ii); that is, each one of the *"rectangles"* in (7.7) is *commutative,* so that *one has*

(7.8) $$\delta^{n,m} \circ d^{n,m-1} = d^{n+1,m-1} \circ \delta^{n,m-1},$$

*for any* $(n, m) \in \mathbf{Z}^2 \equiv \mathbf{Z} \times \mathbf{Z}$.

On the other hand, from a given double $\mathcal{A}$-complex, as in (7.6), one gets two distinct ordinary *(single)* $\mathcal{A}$*-complexes* (see (1.2)); namely, the *horizontal* (or *row*) $\mathcal{A}$*-complex of* $\mathcal{E}^{\cdot\cdot}$, *for each* $m \in \mathbf{Z}$,

(7.9) $$(\mathcal{E}^{\cdot,m}, \delta) \equiv \left\{(\mathcal{E}^{n,m}, \delta^{n,m} \equiv \delta)\right\}_{n\in\mathbf{Z}}$$

(cf. any row of (7.7)), and the *vertical* (or *column*) $\mathcal{A}$*-complex of* $\mathcal{E}^{\cdot\cdot}$, *for each* $n \in \mathbf{Z}$,

(7.10) $$(\mathcal{E}^{n,\cdot}, d) \equiv \left\{(\mathcal{E}^{n,m}, d^{n,m} \equiv d)\right\}_{m\in\mathbf{Z}}$$

(see any column in (7.7). In this context, an element $z \in \mathcal{E}^{n,m}$ is said to be of *first degree* $n$ and of *second degree* $m$, while its *total degree* is $n+m$.

Now, given a double $\mathcal{A}$-complex $\mathcal{E}^{\cdot\cdot}$, as in (7.2), one constructs an ordinary $\mathcal{A}$-complex $\mathcal{E}^{\cdot}$, called the *total* $\mathcal{A}$*-complex* (or simply, *total complex*) of $\mathcal{E}^{\cdot\cdot}$ and denoted by

(7.11) $$\mathrm{tot}\,(\mathcal{E}^{\cdot\cdot}) \equiv \mathcal{E}^{\cdot},$$

as follows;

(7.12) $$\mathcal{E}^{\cdot} \equiv (\mathcal{E}^{\cdot}, D) \equiv \left\{(\mathcal{E}^{p}, D^{p} \equiv D)\right\}_{p\in\mathbf{Z}},$$

such that (see also Chapt. II; (3.12))

$$\mathcal{E}^p := \bigoplus_{n+m=p} \mathcal{E}^{n,m}, \; p \in \mathbf{Z}, \tag{7.13}$$

while we set, as the *total differential* of $\mathcal{E}^\cdot$,

$$D := \delta + (-1)^n d, \text{ with } bidegree\, (n,m). \tag{7.14}$$

The *sign* here is necessary, so that $\mathcal{E}^\cdot$ be a complex, that is, (1.6) be satisfied by $D$. (*Sign Lemma;* cf., for instance, J. J. Rotman [1: p.320, Lemma 11.15]). In this regard, (7.14) means, in effect, the following map

$$D^p : \mathcal{E}^p \longrightarrow \mathcal{E}^{p+1}, \; p \in \mathbf{Z}, \tag{7.15}$$

such that

$$D^p := \sum_{n+m=p} \delta^{n,m} + (-1)^n d^{n,m}, \; p \in \mathbf{Z}. \tag{7.16}$$

Indeed, one proves that *(7.12) is an $\mathcal{A}$-complex,* in the sense of (1.2) (viz. (1.6) is satisfied for the total differential $D$; cf. also e.g. J. J. Rotman [1: p.320, Lemma 11.14]).

So, according to the preceding, *each one of the $\mathcal{A}$-complexes (7.9) and (7.10) has* a corresponding *cohomology,* viz. the analogous to (1.13) *graded $\mathcal{A}(X)$-module;* thus i) *for every $m \in \mathbf{Z}$*, one defines the *horizontal cohomology of $\mathcal{E}^{\cdot\cdot}$*, by

$$h_H^m(\mathcal{E}^{\cdot\cdot}) \equiv h_H^*(\mathcal{E}^{\cdot,m}) := \left\{ h_H^p(\mathcal{E}^{\cdot,m}) \right\}_{p\in\mathbf{Z}}, \tag{7.17}$$

and ii) *for every $n \in \mathbf{Z}$*, the *vertical cohomology of $\mathcal{E}^{\cdot\cdot}$*,

$$h_V^n(\mathcal{E}^{\cdot\cdot}) \equiv h_V^*(\mathcal{E}^{n,\cdot}) := \left\{ h_V^p(\mathcal{E}^{n,\cdot}) \right\}_{p\in\mathbf{Z}}. \tag{7.18}$$

Therefore, one defines the *horizontal cohomology of $\mathcal{E}^{\cdot\cdot}$*, according to (7.17), by the following *bigraded $\mathcal{A}(X)$-module;*

$$h_H^*(\mathcal{E}^{\cdot\cdot}) := \left\{ h_H^m(\mathcal{E}^{\cdot\cdot}) \right\}_{m\in\mathbf{Z}}, \tag{7.19}$$

while, through (7.18), one gets the *vertical cohomology of $\mathcal{E}^{\cdot\cdot}$*,

$$h_V^*(\mathcal{E}^{\cdot\cdot}) := \left\{ h_V^n(\mathcal{E}^{\cdot\cdot}) \right\}_{n\in\mathbf{Z}}. \tag{7.20}$$

(In this regard, cf. also R. Godement [1: p.32], concerning (7.1′)).

In turn, *one looks at (7.19) as a double $\mathcal{A}$-complex, having* (first differential) $\delta = 0$ *and* as (second differential) $d$ that one *obtained from* the given $d$ *of* $\mathcal{E}^{\cdot\cdot}$ (cf. (7.7)); an *analogous situation holds* obviously *true for (7.20).* Therefore, one can further consider the *cohomologies;*

$$h_V^*(h_H^*(\mathcal{E}^{\cdot\cdot})) \text{ and } h_H^*(h_V^*(\mathcal{E}^{\cdot\cdot})), \tag{7.21}$$

respectively. The latter are actually expressed in terms of *spectral sequences,* which are, by a standard argument, canonically *associated with any* given *$\mathcal{A}$-bicomplex* $\mathcal{E}^{\cdot\cdot}$, as before. (See e.g. the Refs. cited above for details). So we come next to the following.

**Definition 7.1.** Given an $\mathcal{A}$-bicomplex $\mathcal{E}^{\cdot\cdot}$ on $X$, as above (see (7.6)), one defines the **cohomology of** $\mathcal{E}^{\cdot\cdot}$, $h^*(\mathcal{E}^{\cdot\cdot})$, as that one of its corresponding total $\mathcal{A}$-complex, tot $(\mathcal{E}^{\cdot\cdot}) \equiv \mathcal{E}^{\cdot}$ (cf. (7.12) and (1.13)). Namely, we set;

$$h^*(\mathcal{E}^{\cdot\cdot}) := h^*(\mathrm{tot}\,(\mathcal{E}^{\cdot\cdot})) \equiv h^*(\mathcal{E}^{\cdot}). \tag{7.22}$$

We shall apply the previous concept right below, by defining the *hypercohomology of* $X$, with respect to a given $\mathcal{A}$-complex $\mathcal{E}^{\cdot}$ on $X$. However, we need first the following supplementary terminology: Namely, as already seen (cf. Theorem 1.2), *every $\mathcal{A}$-module $\mathcal{E}$ on $X$ admits an injective resolution, $\mathcal{E} \longrightarrow \mathcal{F}^{\cdot}$.* So, based on this, one proves an *analogous* result *for any $\mathcal{A}$-complex* $\mathcal{E}^{\cdot}$ on $X$ (while, for simplicity, one takes $n \in \mathbf{Z}_+$, concerning the degrees of the individual $\mathcal{A}$-modules of $\mathcal{E}^{\cdot}$; see thus, for instance, S. Lang [1: p.170, Lemma 9.5]). So one gets the following (cf. also J. -L. Brylinski [1: p.20, Proposition 1.2.6]):

(7.23) *for every $\mathcal{A}$-complex $\mathcal{E}^{\cdot}$ on $X$, of positive degree* (viz. $\mathcal{E}^n = 0$, *for any integer $n < 0$), there exists an $\mathcal{A}$-bicomplex $(\mathcal{F}^{\cdot\cdot}, \delta, d)$, such that*

$$\mathcal{F}^{\cdot\cdot} \equiv \{\mathcal{F}^{n,m}\}_{n,m \in \mathbf{Z}_+ \times \mathbf{Z}} \tag{7.23.1}$$

($\mathcal{F}^{\cdot\cdot}$ is of *positive first degree,* i.e., $\mathcal{F}^{n,m} = 0$, for $n < 0$), having further suitable properties (ibid., and/or S. Lang [1: p.169]). We call it a *(fully) injective $\mathcal{A}$-resolution* (or, simply, an *injective resolution*) of $\mathcal{E}^{\cdot}$.

Now, given an *$\mathcal{A}$-complex* $\mathcal{E}^{\cdot}$ on $X$, as above, we further consider the following $\mathcal{A}(X)$-*bicomplex,* associated with $\mathcal{F}^{\cdot\cdot}$,

$$\Gamma_X(\mathcal{F}^{\cdot\cdot}) := \{\Gamma(X, \mathcal{F}^{n,m})\}_{(n,m)\in\mathbf{Z}^2}, \tag{7.24}$$

still of positive first degree, by hypothesis for $\mathcal{F}^{\cdot\cdot}$, and with differentials $\Gamma_X(\delta), \Gamma_X(d)$. So we come next to the following.

**Definition 7.2.** Let $(X, \mathcal{A})$ be a given **C**-algebraized space and $\mathcal{E}^{\cdot}$ an $\mathcal{A}$-complex on $X$. Then, one defines the **hypercohomology of $X$, with respect to $\mathcal{E}^{\cdot}$,** by the relation (see also (7.24) and (7.22));

$$h^*(X, \mathcal{E}^{\cdot}) := h^*(\Gamma_X(\mathcal{F}^{\cdot\cdot})). \tag{7.25}$$

In this context, it should also be remarked that *the second member of (7.25) is,* in effect, *independent of the injective resolution of $\mathcal{E}^{\cdot}$ considered* (see J. -L. Brylinski [1: p.21, Definition and Proposition 1.2.7]). So (7.25) is, in fact, the following *graded $\mathcal{A}(X)$-module;*

$$h^*(X, \mathcal{E}^{\cdot}) \equiv \{h^p(X, \mathcal{E}^{\cdot})\}_{p\in\mathbf{Z}}, \tag{7.26}$$

*such that* (see (7.22) and (1.13))

$$h^p(X, \mathcal{E}^{\cdot}) := h^p(\Gamma_X(\mathcal{F}^{\cdot\cdot})) := h^p(\text{tot}\,(\Gamma_X(\mathcal{F}^{\cdot\cdot}))),\ p \in \mathbf{Z}. \tag{7.27}$$

See also (7.12), (7.13) and (1.9). Again, computations referred to (7.27), can be brought under control, through suitable spectral sequences, naturally connected with $\mathcal{E}^{\cdot}$ and $\mathcal{F}^{\cdot\cdot}$. (See, for instance, J. -L. Brylinski [1: p.23, Proposition 1.2.12], or even S. Lang [1: p.168f]).

Now, *every $\mathcal{A}$-module $\mathcal{E}$ on $X$ can be considered as an 1-term $\mathcal{A}$-complex $\mathcal{E}^{\cdot}$ on $X$*, or even as an *$\mathcal{A}$-complex $\mathcal{E}^{\cdot}$, concentrated in degree 0;* thus, the 0-th term of $\mathcal{E}^{\cdot}$ is the given $\mathcal{A}$-module $\mathcal{E}$, while all other terms are 0, the corresponding differential (cf. (1.3)) being also, of course, the trivial one. Therefore, one concludes that;

(7.28) *hypercohomology extends (the usual) sheaf cohomology* (see (1.13) and (3.13)).

On the other hand, as happens with the usual Čech (sheaf) cohomology (cf. Section 4 above), the counterpart of the preceding in the case of *Čech cohomology* is *more amenable* (concrete!) and, in fact, the one which we are going to employ in the sequel (Part Two of the book); so we discuss it presently below, by the following Subsection.

**7.(a). Čech hypercohomology.**– We always consider, as usual, a $\mathbf{C}$-*algebraized space* $(X, \mathcal{A})$, while suppose that

$$\mathcal{U} = \{U_\alpha\}_{\alpha \in I} \tag{7.29}$$

is an *open covering* of $X$. Moreover, assume that we are given an $\mathcal{A}$-*complex* $(\mathcal{E}^\cdot, d)$ on $X$ (cf. (1.2)).

Thus, one now defines the following *Čech cochain $\mathcal{A}(X)$-bicomplex of $\mathcal{U}$, with respect to $\mathcal{E}^\cdot$* (see also (7.6));

$$\{\check{C}^\cdot(\mathcal{U}, \mathcal{E}^\cdot), \delta, d\} := \left(\{\check{C}^n(\mathcal{U}, \mathcal{E}^m)\}_{(n,m)\in \mathbf{Z}^2}, \delta, d\right). \tag{7.30}$$

Here $\delta$ is the corresponding *Bockstein operator* (cf. (4.15), (4.16), as well as, (7.31) below) and $d$ the given *differential of $\mathcal{E}^\cdot$*; thus, *the relations (7.5) are* obviously *satisfied,* for the case at hand (see also (4.18′), (1.6) and (5.8′)). The preceding are usually depicted on the $(n, m)$-*plane* through the following diagram, having, by what has been said, the respective *"rectangles" commutative* (see also (7.7) and (7.8));

$$\begin{array}{ccccccc}
 & & \vdots & & \vdots & & \\
 & & \uparrow d & & \uparrow d & & \\
\cdots & \xrightarrow{\delta} & \check{C}^n(\mathcal{U},\mathcal{E}^m) & \xrightarrow{\delta} & \check{C}^{n+1}(\mathcal{U},\mathcal{E}^m) & \xrightarrow{\delta} & \cdots \\
 & & \uparrow d & & \uparrow d & & \\
\cdots & \xrightarrow{\delta} & \check{C}^n(\mathcal{U},\mathcal{E}^{m-1}) & \xrightarrow{\delta} & \check{C}^{n+1}(\mathcal{U},\mathcal{E}^{m-1}) & \xrightarrow{\delta} & \cdots \\
 & & \uparrow d & & \uparrow d & & \\
 & & \vdots & & \vdots & &
\end{array} \tag{7.31}$$

Now, suppose, for simplicity, that the given $\mathcal{A}$-complex $\mathcal{E}^\cdot$ is of positive degree, viz. *let $\mathcal{E}^n = 0$, for $n < 0$.* Moreover, let

$$\{\mathrm{tot}\,(\check{C}^\cdot(\mathcal{U}, \mathcal{E}^\cdot)), D\} \tag{7.32}$$

be the corresponding *total $\mathcal{A}$-complex of (7.30)* (see (7.12), as well as, (7.14) and (7.16)).

So one now defines the *Čech hypercohomology of $\mathcal{U}$, with respect to the $\mathcal{A}$-complex* $\mathcal{E}^{\cdot}$, by the relation (cf. also (7.22)),

$$h^{*}(\check{C}^{\cdot}(\mathcal{U},\mathcal{E}^{\cdot})) := h^{*}(\operatorname{tot}(\check{C}^{\cdot}(\mathcal{U},\mathcal{E}^{\cdot}))). \tag{7.33}$$

Yet, for the same notion, as above, we shall also use in the sequel the notation;

$$\check{H}^{*}(\mathcal{U},\mathcal{E}^{\cdot}) \equiv \{\check{H}^{n}(\mathcal{U},\mathcal{E}^{\cdot})\}_{n\in\mathbf{Z}_{+}}, \tag{7.33$'$}$$

such that one has, by the very definitions (cf. also (7.22) and (1.13)),

$$\check{H}^{n}(\mathcal{U},\mathcal{E}^{\cdot}) := h^{n}(\check{C}^{\cdot}(\mathcal{U},\mathcal{E}^{\cdot})) := h^{n}(\operatorname{tot}(\check{C}^{\cdot}(\mathcal{U},\mathcal{E}^{\cdot}))),\ n \in \mathbf{Z}_{+}. \tag{7.34}$$

On the other hand, by analogy with (4.53), one finally defines the *Čech hypercohomology of $X$, with respect to a* given *$\mathcal{A}$-complex $\mathcal{E}^{\cdot}$ on $X$* (of *positive degree*), by the relation,

$$\check{H}^{*}(X,\mathcal{E}^{\cdot}) \equiv \{\check{H}^{n}(X,\mathcal{E}^{\cdot})\}_{n\in\mathbf{Z}_{+}}, \tag{7.35}$$

such that one sets (cf. (7.34));

$$\check{H}^{n}(X,\mathcal{E}^{\cdot}) := \varinjlim_{\mathcal{U}} \check{H}^{n}(\mathcal{U},\mathcal{E}^{\cdot}). \tag{7.36}$$

Concerning the last relation, one easily proves, of course, that;

(7.36$'$) the family $\{\check{H}^{n}(\mathcal{U},\mathcal{E}^{\cdot})\}$, with $\mathcal{U}$ varying as in (4.9), yields, for any $n \in \mathbf{Z}_{+}$, an *inductive system of $\mathcal{A}(X)$-modules* (see also (7.34)).

In this respect, see also e.g. P. Griffiths - J. Harris [1: p.446]. Therefore, *analogous remarks to those following (5.89)* in the preceding (cf. e.g. (5.90)) *are still valid for the second member of (7.36)*, something that will presently be applied below in Example 7.1. On the other hand, *spectral sequences* are here again the *appropriate device* to look at (7.36). As an indication thereon, we present the following useful lemma, connected with the notion of a *morphism of $\mathcal{A}$-complexes* (cf. (1.15) in the preceding) and whose proof is an immediate application of the theory of spectral sequences. (Anyway, we are not going to employ the latter technique in the sequel; but, see also the following Example 7.1 for an application of (Čech) hypercohomology).

So a given *morphism of $\mathcal{A}$-complexes* (cf. (1.14) – (1.17), (1.19$'$))

$$f : \mathcal{E}^{\cdot} \longrightarrow \mathcal{F}^{\cdot} \tag{7.37}$$

is said to be a *quasi-isomorphism,* in case the corresponding morphism in cohomology (see (1.24))

$$h^*(f) \equiv f^* : h^*(\mathcal{E}^{\cdot}) \longrightarrow h^*(\mathcal{F}^{\cdot}) \tag{7.38}$$

is an *isomorphism* of the *graded $\mathcal{A}$-modules* concerned; cf. (1.13), (1.25), as well as, (1.23)).

Thus, one now obtains the following.

**Lemma 7.1.** *Suppose that $(X, \mathcal{A})$ is a* **C**-*algebraized space, $\mathcal{E}^{\cdot}$ and $\mathcal{F}^{\cdot}$ $\mathcal{A}$-complexes on $X$, of positive degree* (cf. (7.23)), *and $f : \mathcal{E}^{\cdot} \longrightarrow \mathcal{F}^{\cdot}$ a given quasi-isomorphism of the $\mathcal{A}$-complexes concerned. Then, one has*

$$\check{H}^*(X, \mathcal{E}^{\cdot}) = \check{H}^*(X, \mathcal{F}^{\cdot}), \tag{7.39}$$

*within an isomorphism of the respective graded $\mathcal{A}(X)$-modules* (cf. (7.36), (7.36′)), *provides by $f^*$* (see also (7.38)).

*Furthermore, for any $\mathcal{A}$-complex $\mathcal{E}$ on $X$ and for every resolution $\mathcal{F}^{\cdot}$ of $\mathcal{E}$* (cf. (1.54)), *one has* (see also (4.53′))

$$\check{H}^*(X, \mathcal{E}) = \check{H}^*(X, \mathcal{F}^{\cdot}), \tag{7.40}$$

*within an isomorphism of the graded $\mathcal{A}(X)$-modules concerned.*

*In particular, if $\mathcal{F}^{\cdot}$, as above, is acyclic* (viz. $\check{H}^n(X, \mathcal{F}^m) = 0$, $n > 0$, $m \in \mathbf{Z}$), *then one obtains* (cf. also (3.5), (3.8))

$$\check{H}^*(X, \mathcal{E}) = \check{H}^*(X, \mathcal{F}^{\cdot}) = h^*(\Gamma_X(\mathcal{F}^{\cdot})), \tag{7.41}$$

*within isomorphisms of the respective graded $\mathcal{A}(X)$-modules.*

**Proof.** Cf. P. Griffiths - J. Harris [1: p.447, Lemma]. Yet, see J. -L. Brylinski [1: p.24, Proposition 1.2.13 and p.22, Corollary 1.2.11]; here all the previous relations are given for *sheaf hypercohomology,* in general, not necessarily Čech, as we did above, for simplicity's sake. ∎

In this context, we still remark, in passing, that the above Lemma 7.1 can be employed to give, for instance, another proof of *de Rham's theorem* (see e.g. P. Griffiths - H. Harris [1: p.447], as well as, the next Section 8 below). Indeed, (7.41) supplies another aspect of *abstract de Rham's theorem,* in the sense e.g. of (3.24.2) in the foregoing.

Now, the following Example provides another instance of the preceding, which will also be of use in the sequel.

**Example 7.1.**– We consider below a special $\mathcal{A}$-complex, by abstracting a certain concrete one, we shall encounter in the ensuing discussion (see Part Two).

So suppose, as usual, that we are given a **C**-*algebraized space* $(X, \mathcal{A})$, as well as, the 2-*term* $\mathcal{A}$-*complex* on $X$,

$$\mathcal{E}^\cdot : \mathcal{E}^0 \xrightarrow{d^0 \equiv d} \mathcal{E}^1. \tag{7.42}$$

Furthermore, consider the 1-*dimensional Čech hypercohomology of* $X$, *with respect to* $\mathcal{E}^\cdot$, viz. the following $\mathcal{A}(X)$-*module* (cf. (7.36), (7.42));

$$\check{H}^1(X, \mathcal{E}^\cdot) \equiv \check{H}^1(X, \mathcal{E}^0 \xrightarrow{d} \mathcal{E}^1). \tag{7.43}$$

In particular, let us look at an element *(cohomology class)* of (7.43); thus, let

$$[z] \in \check{H}^1(X, \mathcal{E}^\cdot). \tag{7.44}$$

Therefore (cf. (7.36′), as well as, the comments following it), there exists an *open covering* $\mathcal{U}$ of $X$ such that one has (cf. also (7.34) and (1.9))

$$\begin{aligned} [z] \in \check{H}^1(\mathcal{U}, \mathcal{E}^\cdot) \equiv h^1(\check{C}^\cdot(\mathcal{U}, \mathcal{E}^\cdot)) &:= h^1(\operatorname{tot}(\check{C}^\cdot(\mathcal{U}, \mathcal{E}^\cdot))) \\ &\equiv \ker D^1 / \operatorname{im} D^0, \end{aligned} \tag{7.45}$$

where

$$(\mathcal{S}^\cdot, D) \equiv (\operatorname{tot}(\check{C}^\cdot(\mathcal{U}, \mathcal{E}^\cdot)), D), \tag{7.46}$$

stands for the corresponding *total* $\mathcal{A}$-*complex* of the $\mathcal{A}$-bicomplex $\check{C}^\cdot(\mathcal{U}, \mathcal{E}^\cdot)$ (cf. (7.12), (7.14)). Thus, suppose that a *representative of* $[z]$ in (7.44) *is the element*

$$z \in \ker D^1 \equiv \ker(\mathcal{S}^1 \xrightarrow{D^1} \mathcal{S}^2) \tag{7.47}$$

(cf. (7.45)), where we set, in view of (7.46) (cf. also (7.13))

$$\mathcal{S}^1 \equiv \operatorname{tot}(\check{C}^\cdot(\mathcal{U}, \mathcal{E}^\cdot))^1 := \check{C}^1(\mathcal{U}, \mathcal{E}^0) \oplus \check{C}^0(\mathcal{U}, \mathcal{E}^1); \tag{7.48}$$

analogously for $\mathcal{S}^2$, according to (7.13). Consequently, one obtains

$$z = (g, \theta) \in \mathcal{S}^1, \quad \textit{such that } D^1(z) = 0. \tag{7.49}$$

Furthermore, based on (7.48), one has

$$g \equiv (g_{\alpha\beta}) \in \check{C}^1(\mathcal{U}, \mathcal{E}^0) \text{ and } \theta \equiv (\theta_\alpha) \in \check{C}^0(\mathcal{U}, \mathcal{E}^1), \tag{7.50}$$

such that, due to (7.49) and (7.16), one obtains

$$D^1(z) \equiv D^1(g,\theta) = (\delta^1 - d^1)(g,\theta) = \delta^1(g) - d^1(\theta) = 0. \tag{7.51}$$

That is, *the element* $z \equiv (g,\theta) \in \mathcal{S}^1$, being a representative of $[z] \in \check{H}^1(\mathcal{U}, \mathcal{E}^\cdot) \hookrightarrow \check{H}^1(X, \mathcal{E}^\cdot)$, *satisfies the relation;*

$$\delta^1(g) = d^1(\theta). \tag{7.52}$$

Thus, by drawing the corresponding picture on the $(n,m)$-*plane* of the case at hand, one gets the following diagram (*commutative* in the *"rectangle"* concerned; cf. also e.g. (5.8′));

$$\begin{array}{ccccc} 0 & & 0 & & \\ \uparrow{\scriptstyle d} & & \uparrow{\scriptstyle d} & & \\ \theta \in \check{C}^0(\mathcal{U},\mathcal{E}^1) & \xrightarrow{\ \delta\ } & \check{C}^1(\mathcal{U},\mathcal{E}^1) & \xrightarrow{\ \delta\ } & \cdots \\ \uparrow{\scriptstyle d^0 \equiv d} & & \uparrow{\scriptstyle d^0 \equiv d} & & \\ \check{C}^0(\mathcal{U},\mathcal{E}^0) & \xrightarrow{\ \delta\ } & \underbrace{\check{C}^1(\mathcal{U},\mathcal{E}^0)}_{\substack{\cup\!\!\!\mid \\ g}} & \xrightarrow{\ \delta\ } & \cdots \end{array} \tag{7.53}$$

Namely, (7.52) *yields*

$$dg = \delta\theta, \tag{7.54}$$

a very useful relation, indeed, when in the particular case considered in the sequel (see Part Two), acquiring, as we shall see, an appropriate *geometrical interpretation,* through a particular 2-term $\mathcal{A}$-complex, as above.

Of course, *conversely,* one concludes that;

(7.55) *any pair* $z \equiv (g,\theta) \in \mathcal{S}^1$, as in (7.50) (cf. also (7.48)), *satisfying the relation* $dg = \delta\theta$ (see (7.54)), *determines a cohomology class* $[z] \in \check{H}^1(X, \mathcal{E}^\cdot)$, with respect to the given 2-term $\mathcal{A}$-complex $\mathcal{E}^\cdot$, as in (7.42).

The above, along with (7.54), characterize, in fact, the *geometrical objects* alluded to above (indeed, *line sheaves admitting an $\mathcal{A}$-connection,* when in the appropriate context, in effect, $\mathcal{A}$-complex, considered; we thus refer to Part Two of this treatise).

# 8. Cohomology theories in action. (Interrelations)

The cohomology theories we have considered so far, and which will also be employed in subsequent parts of this discussion, are actually the *sheaf cohomology,* or even (indicating thus the key-idea involved) the *derived functor cohomology,* the various *Čech cohomology techniques* (usual, viz. having sheaf coefficients, liftable and presheaf), as well as, *hypercohomology,* sheaf and/or Čech. So, in this context, we have already seen that;

(8.1) *in the case of a paracompact (Hausdorff) topological space $X$, all Čech cohomology techniques,* as considered hitherto, viz. that one with *coefficients in a sheaf* (in fact, an *$\mathcal{A}$-module*), *liftable* Čech cohomology and Čech cohomology with *coefficients in a presheaf* (in effect, that one *of $\mathcal{A}(X)$-modules*), *are actually the same,* up to appropriate isomorphisms, as the case may be (indeed, *within $\mathcal{A}(X)$-isomorphisms*).

Here, as usual, $(X, \mathcal{A})$ stands for a given $\mathbf{C}$-*algebraized space,* having $X$, as above.

The previous fundamental conclusion (but see also (8.8) in the sequel) is, indeed, a consequence of our discussion in Sections 5 and 6 in the preceding; so cf. Corollary 5.1, or even (5.86), along with Theorem 6.2. On the other hand, another important fact thereof, which will be applied later on, as well (cf. Part Two), is the

(8.2) *existence of a long exact cohomology sequence,* associated with any given *short exact $\mathcal{A}$-sequence,* for *Čech cohomology* on a *paracompact (Hausdorff)* topological space.

See Theorem 5.3 and 6.1 in the foregoing.

Indeed, as we shall presently see, the previous conclusion (8.1) *is actually valid for all cohomology theories* (*not* including *hypercohomology,* however!) on a given *paracompact* space (see Theorem 8.1 below), a fact that will be, in effect, our main concern in the sequel. Yet, concerning hypercohomology, see also (7.41).

However, to obviate the terminology applied, we have first to comment a bit on the hitherto repeatedly employed term:

(8.3) *"cohomology theory", with* respect to a given fixed *topological space* $X$*;* thus, by applying standard terminology, one means by this a *sequence* (usually indexed by the non-negative integers $\mathbf{Z}_+$) *of covariant functors* from a specific category, in our case from $\mathcal{A}$-$\mathcal{M}od_X$ (cf. (1.1)), to that one of abelian groups (*"abelian* (sheaf) *cohomology"*, but see also Section 11 in the sequel), here, in particular, to the category of $\mathcal{A}(X)$-modules, $\mathcal{A}(X)$-$\mathcal{M}od$ (cf. Chapt. II; (6.39)), with $(X, \mathcal{A})$ being a given (fixed) $\mathbf{C}$-*algebraized space.* Yet, *the same functors* have to *satisfy* certain particular *axioms,* the classical *Eilenberg-Steenrod axioms* for a cohomology theory, save the so-called *"dimension"* axiom *("generalized cohomology theory").* Furthermore, the existence of the *"coboundary operators"* (cf. (1.38), (2.15) or (3.16)) is still axiomatized, the latter satisfying suitable conditions, as well. (See e.g. (3.17), (3.19) in the preceding).

In this context, the same Theorem 3.1, for instance (cf. also Theorem 2.1), provides, in effect, an indication of the sort of "axioms" that the *functors and* $\delta$'s connected (see e.g. (3.12), (3.16)) have to fulfill, in order to define a *cohomology theory,* as before; so see, for instance, S. Eilenberg- N. Steenrod [1: p.13, §3c.], or even E. H. Spanier [1: p.240]. Yet, cf. L. Kaup - B. Kaup [1: p.215, Definition 50.1].

On the other hand, according to what we have already practiced in the foregoing, by a *cohomology theory,* one means the procedure of *associating with a given* particular (cochain, in fact, *differential cochain*) *complex,* in our case, an $\mathcal{A}$*-complex* (cf. (1.2)), *a sequence of* $\mathcal{A}$*-modules* (cf. (1.13)); this is actually the *point of view of Homological Algebra* (see also e.g. Theorem 1.1). In particular, by specializing to the context of (8.3) (viz. to that one of Theorem 3.1), starting with a suitably defined *(cochain)* $\mathcal{A}(X)$*-complex*

(8.4) (derived functor cohomology, cf. (3.5), (3.8)), one finds a *sequence of* $\mathcal{A}(X)$*-modules* (cf. (3.13) and (3.5)), viz. a *graded* $\mathcal{A}(X)$*-module.* In both cases *the sequence,* thus obtained, *is,* roughly speaking, *the cohomology of the complex* under consideration; yet, an important ingredient here is also the *existence of* the *coboundary operators* (*"$\delta$-functors"*, cf. (1.46) and (3.16)), along with the *exactness of (1.36), (3.17),* and the *commutativity of (3.19)* (*"naturality"* of $\delta$'s).

(*Terminological warning.* We obviously abused language just before by identifying, namely, a *function* (in effect, a *functor*), *cohomology theory* (see thus (8.5)), with *its value, cohomology* (sequence), at a given *point,* the *complex* in question). Yet, within the same context, as in (8.4), we further note, for use right below (cf. (8.6)), that an *analogous conclusion to (3.19) for* $\mathcal{A}$*-complexes is* still *in force* (see, for example, J. J. Rotman [1: p.173, Theorem 6.4]; *"naturality" of* $\delta$, cf. (1.46), *for* $\mathcal{A}$*-complexes*).

Now, as already hinted at, just above, the preceding (say, still, the same Theorem 3.1) realize the situation one has by considering a so-called *"$\delta$-functor"*, by which one actually means a very *concise and axiomatic formulation* (à la Grothendieck) *of* the picture one gets via *Theorem 3.1.* So, by employing that language, one concludes that;

(8.5) a *(sheaf) cohomology theory,* as above, is simply a *cohomological* $\delta$*-functor over the global section functor* $\Gamma_X$ *between the categories,* in our case, $\mathcal{A}$-$\mathcal{M}od_X$ *and* $\mathcal{A}(X)$-$\mathcal{M}od$ (cf. e.g. (3.1)).

Namely, a functor between the aforementioned categories, yielding actually a sequence of such, like (3.12), and satisfying, as alluded to already, the conditions that these functors along with $\delta$'s do in Theorem 3.1. (For technical details see, for instance, R. G. Swan [1: p.39ff, especially, p.47, Note 3 and p.49, Theorem 6, as well as, p.51, Corollaries 1, 2 and p.57], or S. Lang [1:

p. 148ff, in particular, p.153, Corollary 7.2]). Furthermore, within the same context, one also concludes that;

(8.6) the (covariant) functor $h^*$, defined by (1.13) (cf. also (1.25)), and representing the *cohomology theory of $\mathcal{A}$-complexes,* can be viewed as a *(covariant) exact $\delta$-functor* between $\mathcal{A}$-$\mathcal{C}omp_X$, the (abelian) *category of $\mathcal{A}$-complexes* on $X$ and $\mathcal{A}$-$\mathcal{GM}od_X$, viz. *that one of graded $\mathcal{A}$-modules* on $X$ (still cf. (1.25)).

The assertion can be deduced from (1.25), Theorem 1.1 (*exactness* of $h^*$), while, as already noted above, the *naturality of $\delta$* in (1.36) (see also (1.38), viz. *the analogue* here *of (3.19) holds true,* as well. ∎

Now, before we proceed further, we still remark that,

(8.7) a *sheaf cohomology theory on a* given (arbitrary) *topological space $X$,* in the sense e.g. of (8.4), there *does exist* and *is,* in fact, *unique,* up to an *isomorphism.* Such a theory is, according to Theorem 3.1, the *derived functor cohomology,* or even *"the" cohomology theory of* the space $X$ concerned, *with coefficients in a* given *$\mathcal{A}$-module on $X$.*

Indeed, (3.24) yields a proof of the *existence of a (sheaf) cohomology theory of the space $X$* considered, in the sense of (8.5) (or, equivalently, by definition, of (8.4)), *based on any $\Gamma_X$-acyclic resolution of the $\mathcal{A}$-module $\mathcal{E}$* involved (*"sheaf of coefficients"*). Thus, such a resolution is always, for instance, the *canonical flabby* (or *Godement*) *resolution of $\mathcal{E}$* (see (3.34) and Lemma 3.1; in this regard, consult also (3.37) and (3.38)). Furthermore, the same Lemma 3.1, along with (3.24.2), prove the *uniqueness* of the cohomology theory, under discussion. (In this concern, see also, for instance, L. Kaup - B. Kaup [1: p.215, Theorem 50.2]. ∎ Now, concerning the previous notion of *uniqueness,* this amounts, in effect, as we shall presently see, to an *isomorphism of sheaf cohomology theories* on a given *topological space $X$,* as above, viz. to a *natural isomorphism of* the respective *(cohomology) functors, which* further *commutes with the $\delta$'s* (relative to *short exact $\mathcal{A}$-sequences,* in our case; see thus e.g. R. Kultze [1: p.85, Definition 19.2], or yet R. G. Swan [1: p.40, Definition; *"map of $\delta$-functors"*]).

Now, let us look, in particular, at the *Čech cohomology of a* given (arbitrary) *topological space $X$, with coefficients in an $\mathcal{A}$-module $\mathcal{E}$* on $X$ (we still recall that we are always concerned with a given fixed **C**-*algebraized space* $(X, \mathcal{A})$); viz. at the *Čech cohomology* $\mathcal{A}(X)$-*complex* (see (4.53′) and (4.15))

$$\check{H}^*(X, \mathcal{E}) \equiv \left\{\check{H}^n(X, \mathcal{E}), \delta \equiv (\delta^n)\right\}_{n \in \mathbf{Z}_+} . \tag{8.8}$$

However, as already noticed in the foregoing (cf., for instance, the introductory remarks to Section 4), the previous complex (8.8) *does not,* in general, *provide a sheaf cohomology theory,* in the sense, for instance, of (8.3) (or, equivalently, of (8.4) or (8.5)); indeed, by employing, for example, the language of $\delta$-functors, as above, one concludes that:

(8.9) *the Čech cohomology* $\mathcal{A}(X)$-*modules,* as given by (4.53), *do not, in general, yield a* $\delta$-*functor.*

See e.g. R. G. Swan [1: p.101]: So they lack, in general, *exactness* (of the cohomology functor they define), in the sense that *Theorem 5.3* (i.e., *cond. 3*) of *Theorem 3.1*) *does not hold, in general,* due to the fact that (the global section functor) $\Gamma_X$ *is not,* in general, *exact* (cf. Chapt. I; Example 12.1]). Yet, see R. Hartshorne [1: p.219, Caution 4.0.2], or even E. H. Spanier [1: p.326].

On the other hand, we have already seen that *this inconvenience of Čech (sheaf) cohomology disappears by employing paracompact Hausdorff spaces* (the same Theorem 5.3, as above). Thus, one actually proves that:

(8.10) *the properties which satisfy the (sheaf) cohomology functors* (3.12) in Theorem 3.1 *are also fulfilled by the Čech cohomology functors* $\check{H}^n(X, \cdot)$, $n \in \mathbf{Z}_+$ (cf. (8.8)) *in the case that* $X$ *is a paracompact Hausdorff space.*

Therefore, in that case too, *(sheaf) Čech cohomology provides a sheaf cohomology theory* on $X$, in the sense of (8.4) (or (8.5)).

As already noted, *"exactness"* of the said cohomology functors is supplied by Theorem 5.3. For the rest of the previous assertion cf., for instance, F. W. Warner [1: p.200, §5.33]. ∎ Furthermore, according to (8.7) and the comments after it, one still concludes, formally, the following basic fact:

*on any given topological space $X$ there exists, up to an isomorphism of the sheaf cohomology theories* (cf. (8.7)) *just one such;*
(8.11) *therefore* (cf. (8.10)), in the case of a *paracompact Hausdorff space $X$, this "is"*, for instance, (modulo, namely, the said isomorphism), *the Čech (sheaf) cohomology theory.*

In this context, see thus R. Kultze [1: p.86, Satz 19.3], or even F. W. Warner [1: p.181, Theorem 5.23 and p.184, Corollary]. ∎ So, as an immediate consequence of (8.11), one now arrives at the following fundamental fact, stated within the hitherto employed terminology. Namely, one has.

**Theorem 8.1.** *Let $(X, \mathcal{A})$ be a $\mathbf{C}$-algebraized space, with $X$ a paracompact Hausdorff space, and $\mathcal{E}$ an $\mathcal{A}$-module on $X$. Then, one obtains* (cf. (3.5) and (4.53))

(8.12) $$H^n(X, \mathcal{E}) = \check{H}^n(X, \mathcal{E}),\ n \in \mathbf{Z}_+,$$

*within an $\mathcal{A}(X)$-isomorphism of the $\mathcal{A}(X)$-modules concerned; i.e., sheaf and Čech cohomology on a paracompact* (Hausdorff) *space $X$ coincide* (within, in effect (cf. (8.11), a (natural) *isomorphism of* the respective *cohomology functors*). ∎

Furthermore, according to Theorem 6.2, *one extends (8.12) to* the case of *Čech cohomology with coefficients in a $\Gamma(\mathcal{A})$-presheaf* (whose sheafification is an $\mathcal{A}$-module $\mathcal{E}$; see (6.44)). Yet, one has an analogous conclusion to (8.12) for *de Rham cohomology* (on a $\mathcal{C}^\infty$-manifold, cf. Theorem 8.2 below), which we also consider for later use right below in the next Subsection 8.(a).

On the other hand, referring to the same relation (8.12), one obtains the following quite general result, as it concerns the space $X$, however, only in *dimensions* 0 (we already know it, for that matter!) *and* 1 (that we do not). So we have.

**Lemma 8.1.** *Let $(X, \mathcal{A})$ be a $\mathbf{C}$-algebraized space, with $X$ being any given topological space, and $\mathcal{E}$ an $\mathcal{A}$-module on $X$. Then, one has;*

(8.13) $$H^0(X, \mathcal{E}) = \check{H}^0(X, \mathcal{E}) = \Gamma(X, \mathcal{E}),$$

*as well as,*

(8.14) $$H^1(X, \mathcal{E}) = \check{H}^1(X, \mathcal{E}),$$

*within $\mathcal{A}(X)$-isomorphisms of the $\mathcal{A}(X)$-modules involved.*

**Proof.** Indeed, (8.13) is already a consequence of (3.14) (see thus (3.10)), in conjunction with (4.54). (In fact, we already anticipated (8.13), through (4.55)). On the other hand, concerning the proof of (8.14), consult e.g. R. Hartshorne [1: p.223, Ex. 4.4, c)]. (In this respect, see also Lemma 1.2 in the preceding). Yet, we can have a proof of (8.14) by employing the *spectral sequence, associated with the Čech cohomology* (of $\mathcal{E}$); the latter is derived from the Čech $\mathcal{A}$-bicomplex (cf. (7.30)) defined by the canonical (flabby) resolution of $\mathcal{E}$ (see (3.34)). A key-argument here is also (6.55). For details see thus e.g. R. Godement [1: p.227, Corollaire and Lemme], that is (6.55), as above. ∎

Within the *same context, as Lemma 8.1,* one still proves (ibid.) that

(8.15) $$\check{H}^2(X,\mathcal{E}) \overset{\subseteq}{\to} H^2(X,\mathcal{E}),$$

up to an $\mathcal{A}(X)$-*isomorphism (into)* of the $\mathcal{A}(X)$-modules concerned.

In this respect, we shall also apply in the sequel the so-called

(8.16) *cohomology with constant coefficients.*

Namely, by considering an arbitrary (discrete) *abelian group* $G$ (see also, however, Section 11 below), or any *module* $M$, or even the *(numerical) vector spaces* $\mathbf{R}$ and/or $\mathbf{C}$, one has, *by definition,* the following;

(8.16′)

given a *topological space* $X$ and a (discrete) *abelian group* $G$ one sets:

(8.16′.1) $$H^n(X,G) := H^n(X,G_X) \equiv H^n(X,G), \ n \in \mathbf{Z}.$$

So by the last relation one means the *n-th dimensional sheaf cohomology group* of $X$ *with coefficients* in the *constant sheaf*

(8.16′.2) $$G_X \equiv X \times G \equiv G.$$

(See Chapt. I; (4.1), (4.4), (4.4′), as well as, (3.5) above). Similarly, for *Čech cohomology with coefficients in $G$,* as above, one sets;

(8.16′.3) $$\check{H}^n(X,G) := \check{H}^n(X,G_X) \equiv \check{H}^n(X,G), \ n \in \mathbf{Z}_+.$$

(see (4.53)).

Thus, by taking now, more generally, the *constant presheaf* $G$, with $G$ as before (cf. Chapt. I; Subsection 10.(a)), and supposing also the given space $X$ to be *paracompact (Hausdorff)*, one obtains, according to (6.44), (8.16′.1) and (8.12) (see also Chapt. I; (10.8)), *for every* $n \in \mathbf{Z}_+$,

$$\begin{aligned}\check{H}^n(X, G) &= \check{H}^n(X, \mathbf{S}(G)) = \check{H}^n(X, G_X) \equiv \check{H}^n(X, G) \\ &= H^n(X, G_X) \equiv H^n(X, G),\end{aligned} \tag{8.17}$$

within the pertinent *isomorphisms*, as the case may be (regarding $G$, as above).

Of course, based on (6.12), one can define

$$\check{H}^n(X, G), \; n \in \mathbf{Z}_+, \tag{8.18}$$

for any given *topological space* $X$, with $G$ being viewed as a *constant presheaf.*

**Scholium 8.1.**- By further supplementing our previous information from (8.12), we still remark that: On a *paracompact (Hausdorff) space* $X$, apart from the *coincidence* (isomorphism) *of sheaf and Čech cohomology theories* (cf. Theorem 8.1), *the same holds true* with the classical *singular cohomology* theory, as well as, with the *Alexander - Spanier cohomology.* (Yet, this is still valid, as we shall presently see below, for the *de Rham cohomology,* if $X$ is, in particular, a $\mathcal{C}^\infty$*-manifold*). Similarly, under the same hypothesis for $X$, the *classical Čech* (or *singular*) *cohomology with constant coefficients* coincides with *that one,* we previously defined, *in* (8.16).

In this context, one usually has to *exhibit a* $\Gamma_X$*-acyclic resolution* of the particular *(constant) sheaf,* associated with the theory concerned (e.g. (pre)sheaf of *Alexander - Spanier cochains* or (pre)sheaf of *singular cochains,* see, for instance F. W. Warner [1: p.186, §5.26 and p.191, §5.31]), and then apply the *abstract de Rham theorem* (cf. (3.24.1)), proving further that the resulted thus graded objects (groups, modules etc) yield, indeed, a *cohomology theory,* hence, its *isomorphism* with the usual sheaf and/or Čech cohomology theories, as before. For details thereon, consult, for instance, the above quoted Ref.; yet, see also E. H. Spanier [1: p.334, Corollary 8], or even R. Hartshorne [1: p.211, Historical Note].

On the other hand, by still referring to Theorem 8.1, one gets an *extension of (8.12) to any topological space* $X$, as an application of *Leray's Theorem* (see e.g. Theorem 4.1). Namely, one has

**Theorem 8.2.** *Let* $(X, \mathcal{A})$ *be a* **C**-*algebraized space and* $\mathcal{E}$ *an* $\mathcal{A}$-*module on* $X$. *Moreover, suppose that* $\mathcal{U} = \{U_\alpha\}_{\alpha \in I}$ *is a Leray covering of* $X$ *relative to* $\mathcal{U}$; *viz.*, for any finite intersection $V \equiv U_{\alpha_0} \cap \cdots \cap U_{\alpha_p}$ of open sets of $\mathcal{U}$, *one has*

$$H^n(V, \mathcal{E}|_V) = 0, \; n \geqq 1 \tag{8.19}$$

(thus, $\mathcal{E}$ *"has no cohomology on any of the open sets"*). *Then one obtains*

$$H^n(X, \mathcal{E}) = \check{H}^n(X, \mathcal{E}), \; n \in \mathbf{Z}_+, \tag{8.20}$$

*up to* $\mathcal{A}(X)$-*isomorphisms of the* $\mathcal{A}(X)$-*modules concerned.*

**Proof.** See R. Hartshorne [1: p.225, Ex. 4.11]. ∎

Now, as a final remark here, we also note that (8.12) can still be derived by employing (see also the proof of Lemma 8.1 above) the so-called *spectral sequence, associated with the Čech cohomology;* cf. thus R. Godement [1: p.228, Théorème 5.10.1].

**8.(a). Fine resolutions. De Rham cohomology.**– As already mentioned, we are concerned below with the case where, by computing the (sheaf) cohomology of a space $X$ with coefficients in a given $\mathcal{A}$-module $\mathcal{E}$, one applies the so-called fine resolutions of $\mathcal{E}$. This type of resolutions will also be employed in later Sections of this book. So an important particular case is the classical *de Rham resolution,* that we also consider below for later use. We start thus with the necessary background material.

So assume, as usual; that we are given a **C**-*algebraized space* $(X, \mathcal{A})$ and let $\mathcal{E}$,$\mathcal{F}$ be two given $\mathcal{A}$-*modules* on $X$, along with an element

$$\phi \in \mathrm{Hom}_{\mathcal{A}}(\mathcal{E}, \mathcal{F}), \tag{8.21}$$

viz. an $\mathcal{A}$-*morphism* $\phi : \mathcal{E} \longrightarrow \mathcal{F}$ of the $\mathcal{A}$-modules concerned, the set in (8.21), that contains $\phi$, being, in particular, an $\mathcal{A}(X)$-*module* (cf. Chapt. II; (2.1) and (6.43)). In this context, one defines the *support of* $\phi$, denoted by $\mathrm{Supp}(\phi)$, as the closure in $X$ of those points $x \in X$, with $\phi_x \neq 0_x \in \mathrm{Hom}_{\mathcal{A}_x}(\mathcal{E}_x, \mathcal{F}_x)$ (see Chapt. II; (2.2)); that is, we set

$$\mathrm{Supp}(\phi) := \overline{\{x \in X : \phi_x \neq 0_x\}}. \tag{8.22}$$

Equivalently, Supp($\phi$) *is* thus *the least closed set* in $X$, *in the complement of which* $\phi$ *vanishes* (viz. equals the *zero-element* in $\mathrm{Hom}_{\mathcal{A}}(\mathcal{E},\mathcal{F})$; see also Chapt. I; (1.15)).

Now, suppose we have a *locally finite open covering* $\mathcal{U} = \{U_\alpha\}_{\alpha\in I}$ of $X$ (cf., for instance, the comments following (5.54) above). Then, by a *partition of unity* of a given $\mathcal{A}$-module $\mathcal{E}$ on $X$, *subordinate to the covering* $\mathcal{U}$, one means a family

(8.23) $$(\phi_\alpha)_{\alpha\in I} \subseteq End\,\mathcal{E}$$

(see Chapt. II; (6.30)), satisfying the following conditions:

(8.23′)
1) $\mathrm{Supp}\,(\phi_\alpha) \subseteq U_\alpha$, $\alpha \in I$.
2) $\sum_{\alpha\in I}\phi_\alpha = 1$ ($\equiv 1_{\mathcal{E}} \in \mathrm{End}\,\mathcal{E}$, *identity automorphism of* $\mathcal{E}$).

In this context, we still remark that, by the first of the above conditions for the family $(\phi_\alpha)$, one concludes that $\phi_\alpha(z) \neq 0_x$, $z \in \mathcal{E}_x$, *only if* $x \in \mathrm{Supp}(\phi_\alpha) \subseteq U_\alpha$ (cf. (8.22)); therefore, due to the hypothesis for $\mathcal{U}$, *cond. 2)*, as above, *has a meaning,* viz. *the sum* appeared therein *is,* in effect, *finite.*

**Remarks 8.1.**– i) The preceding notions can be formulated, of course, for any given *sheaf of* (abelian) *groups* on $X$. More generally, for any *presheaf* on $X$ with the pertinent algebraic structure (in the stalks; e.g. *presheaf of groups.* Cf., for instance, E. H. Spanier [1: p.330]). See also Chapt. I; (6.7), for that matter.

ii) *Cond. 1)* above is *equivalent* with the following one:

1′) $\phi_\alpha = 0|_V$, with $V$ an *open neighborhood of* $\complement U_\alpha$, $\alpha \in I$.

Thus, we come next to the following.

**Definition 8.1.** A given $\mathcal{A}$-*module* $\mathcal{E}$ on $X$ is said to be **fine,** whenever, for every locally finite open covering $\mathcal{U}$ of $X$, there exists a partition of unity of $\mathcal{E}$, subordinate to $\mathcal{U}$.

Now, a key-result for the ensuing discussion is the following:

(8.24) *on a paracompact (Hausdorff) space* $X$, *every fine sheaf* (of groups, in particular, $\mathcal{A}$-*module*) $\mathcal{E}$, *has trivial* (*sheaf,* hence, equivalently (cf. Theorem 8.1), *Čech*) *cohomology* (in positive dimensions, of course, cf. (8.13)); what amounts to the same thing (cf. (3.5), (2.20)), $\mathcal{E}$ *is* thus $\Gamma_X$-*acyclic.*

In this respect, we recall that a given $\mathcal{A}$-*module* $\mathcal{E}$ on $X$ (or any *sheaf* of sets, in general) is said to be *soft*, if for every closed set $B \subseteq X$, the canonical (restriction) map (see Chapt. I; (9.4))

$$\rho_B : \Gamma(X, \mathcal{E}) \longrightarrow \Gamma(B, X) \tag{8.25}$$

is surjective (viz., every continuous section of $\mathcal{E}$ over any closed subset of $X$ is extended to the whole $X$, as a (global) continuous section of $\mathcal{E}$).

Thus, concerning (8.24), what one really proves is the following stronger result;

(8.26) *on a paracompact (Hausdorff) space $X$, every soft sheaf (of (abelian) groups, in particular, $\mathcal{A}$-module) is $\Gamma_X$-acyclic, whereas every fine sheaf on such a space is soft,* as well (hence, (8.24)).

For the proof of (8.26) we refer, for instance, to G. E. Bredon [1: p.49, Theorem 9.8 and p.50, Theorem 9.12; but, see also p.109, Ex. (13)]. However, consult e.g. L. Kaup - B. Kaup [1: p.218, Lemma 50.8] for a direct proof of (8.24), independent of (8.26). ∎

So, as already promised, we come next to the *de Rham cohomology* of a given *(n-dimensional)* $\mathcal{C}^{\infty}$*-manifold* $X$:

In this context, one first defines the so-called *de Rham complex* of $X$; this is, in effect, an $\mathcal{A}$-*complex* on $X$, in the sense of (1.2), such that

$$\mathcal{A} := {}^{\mathbf{C}}\mathcal{C}^{\infty}_X \equiv \mathcal{C}^{\infty}_X \cong {}^{\mathbf{R}}\mathcal{C}^{\infty}_X \otimes_{\mathbf{R}} \mathbf{C}, \tag{8.27}$$

viz. the *sheaf of germs of* **C**-*valued* $\mathcal{C}^{\infty}$*-functions* on $X$ (see also Chapt. I; Subsection 4.(c)); i.e., one considers in the last term of (8.27) the complexification *of the sheaf* ${}^{\mathbf{R}}\mathcal{C}^{\infty}_X$, *sheaf of germs of* **R**-*valued* $\mathcal{C}^{\infty}$*-functions* on $X$, this being, in fact, an **R**-*algebra sheaf* on $X$, while **R** and **C** in "$\otimes_{\mathbf{R}}\mathbf{C}$" of (8.27) stand for the respective *constant (algebra) sheaves* on $X$ (cf. e.g. Chapt. II; (1.34)). So (8.27) is the sheaf on $X$, generated by the *(complete) presheaf*

$$U \longmapsto \mathcal{C}^{\infty}_{\mathbf{R}}(U) \otimes_{\mathbf{R}} \mathbf{C} \cong \mathcal{C}^{\infty}_{\mathbf{C}}(U) \equiv \mathcal{C}^{\infty}(U), \; U \text{ open in } X. \tag{8.28}$$

(Note also that up to this point *no supplementary hypothesis* for the topological space $X$ has been set, as we shall actually do, however, in the sequel). Thus, the pair

$$(X, \mathcal{C}^{\infty}_X) \tag{8.29}$$

is the corresponding to the case, under consideration, $\mathbf{C}$-*algebraized space.*

Now, we further define (briefly) the rest ingredients of the de Rham complex in question; thus, one sets

$$\Omega^p \cong {}^{\mathbf{R}}\Omega^p_X \otimes_{\mathbf{R}} \mathbf{C}, \; p \in \mathbf{Z}_+, \tag{8.30}$$

(see, however, (8.35′) below) such that one also defines

$$\Omega^0 := \mathcal{A} \equiv \mathcal{C}^\infty_X. \tag{8.30′}$$

So in (8.30) ${}^{\mathbf{R}}\Omega^p_X$, $p \geq 0$, stands for the *sheaf of germs of (real-valued)* $\mathcal{C}^\infty$*-forms* on $X$ of *degree* $p$, or for short (sheaf of) $p$*-forms* on $X$, this being an ${}^{\mathbf{R}}\mathcal{C}^\infty_X$*-module* on $X$. In turn, the *complexification* of the latter, as in (8.30), is (within a $\mathcal{C}^\infty_X$*-isomorphism*) $\Omega^p$, the *sheaf of germs of* $\mathbf{C}$*-valued* $p$*-forms* on $X$; thus, an object on $X$ depending on the *differential geometric structure* of $X$, defined through the given *structure sheaf* $\mathcal{A} \equiv \mathcal{C}^\infty_X$ of the $\mathbf{C}$-algebraized space (8.29), under discussion, or, by abusing language, of the $\mathcal{C}^\infty$-manifold $X$. Of course, one still concludes, by the very definitions, that

(8.31) $\Omega^p$, $p \in \mathbf{Z}_+$, *is an* $\mathcal{A}$*-module on* $X$*, with* $\mathcal{A} \equiv \mathcal{C}^\infty_X$.

Finally, one has to define the *"differentials"* (or *coboundary operators*) of the $\mathcal{A}$-complex in question (cf. (1.5)); indeed, a sequence of *differentials* (so here we mean it!, viz. no quotation marks), the latter being, in effect, the (standard) *exterior differentials* of forms on $X$: So one now defines;

$$d \equiv d^p : \Omega^p \longrightarrow \Omega^{p+1}, \; p \in \mathbf{Z}_+, \tag{8.32}$$

such that

$$d\omega \equiv d^p(\omega) := \sum_{i_1<\cdots<i_p} d\omega_{i_1\cdots i_p} \wedge dx_{i_1} \wedge \cdots \wedge dx_{i_p}, \tag{8.33}$$

where one sets

$$\omega \equiv \sum_{i_1<\cdots<i_p} \omega_{i_1\cdots i_p} dx_{i_1} \wedge \cdots \wedge dx_{i_p} \in \Omega^p(U), \tag{8.34}$$

for a *(local) section* of $\Omega^p$ over a *local chart* $U$ of $X$ at the point $x \in X$, with $\omega_{i_1\cdots i_p} \in \mathcal{C}^\infty(U)$ (see (8.28)) and $d\omega_{i_1\cdots i_p} \in \Omega^1(U)$ the usual (local) *differential of a* $\mathbf{C}$*-valued* $\mathcal{C}^\infty$*-function on* $U$*:* Thus, here, based on Chapt. I; Section 3 and/or Theorem 9.1, one may *identify* $\Omega^p$ *with* (the corresponding *sheaf* of) *its local sections, defining* thus the desired operators (8.32), *first, between the* respective (complete) *presheaves of sections,* through (8.33), hence, in turn, between the initially given sheaves (see also Chapt. I; Theorem 13.1, in particular, (13.7)). (*Caution!* In this context, we should also remark that by defining thus (8.33) *locally,* one has to assure *commutativity of the (differential) operators* involved *with* localizations of them, a fact requiring *paracompactness* of $X$, only *locally,* true, in general, due to the *local structure* of $X$. (However, see (8.38) below and the comments that follow). So one obtains the required commutativity relations (cf. Chapt. I; (6.5)) for *morphisms of* the respective *presheaves,* so that one can then apply Chapt. I; (13.7), as above. Moreover, one proves (chain rule) that the previous (local) *definition* is, of course, *independent of the local chart* of $X$ considered. In this regard, see also, for instance, A. Mallios [4: Chapt. X; (4.7)], or even W. D. Curtis - F. R. Miller [1: p.158, comments preceding Proposition 9.42]). So $\omega$ in (8.34) is, *equivalently,* a (local) *section* of the *p-th exterior power,* $\wedge^p \mathcal{T}^*(X)$, of the *cotangent bundle* $\mathcal{T}^*(X)$ of the given $\mathcal{C}^\infty$-manifold $X$.

Thus, one finally obtains the next (finite) sequence of *(exterior) differentials*

(8.35) $$d \equiv (d^p),\ p = 0, 1, \ldots, n = \dim X,$$

since, due to the hypothesis that $X$ is *finite-dimensional,* one has, of course,

(8.35′) $$\Omega^p = 0 \ \textit{and}\ d^p = 0,\ \textit{for every integer}\ p \geqq n+1.$$

(There are no higher (non-trivial) exterior powers of $\mathcal{T}^*(X)$). Yet, it is still true that

(8.36) $$d^2 \equiv d \circ d = 0$$

(see also (1.4), (1.6)). So one now obtains the following $\mathbf{C}_X$*-complex* on $X$ (viz. a *(cochain complex of* $\mathbf{C}$*-vector space sheaves* on $X$; cf. also the comments after (8.39′))

(8.37) $$\Omega^{\bullet}_{\mathrm{de\,R}} \equiv \left\{(\Omega^p, d \equiv \{d^p\})\right\}_{0 \leqq p \leqq n}.$$

We call it the *de Rham complex* of the given $\mathcal{C}^\infty$-manifold $X$; yet, by aplying the notation of (1.3) (see also (1.54), (8.30′) and (8.35′)) one has;

(8.37′) $$\Omega^{\bullet}_{\mathrm{de\,R}} : 0 \longrightarrow \Omega^0 \equiv \mathcal{C}^\infty_X \xrightarrow{d} \Omega^1 \xrightarrow{d} \Omega^2 \xrightarrow{d} \cdots \xrightarrow{d} \Omega^n \longrightarrow 0.$$

Now, as a prelude to the notion we are going to employ in a moment, consider next the image of (8.37′) in

$$\text{(8.38)} \qquad \mathbf{C}\text{-}\mathcal{L}inComp,$$

that is, in the *category* of **C**-*linear complexes,* or else *(cochain) complexes of* **C**-*vector spaces* (having as morphisms, **C**-*linear* morphisms (*maps*)), through the (*left exact* global) *section functor* $\Gamma_X$ (see Chapt. I; (9.3), (10.13.1), (11.11); ii), and (13.1), as well as, Chapt. II; (2.59)). So we have;

$$\text{(8.39)} \qquad \begin{aligned}\Gamma_X(\Omega^{\cdot}_{\mathrm{de\,R}}) : 0 \longrightarrow \Gamma_X(\Omega^0) \equiv \Gamma_X(\mathcal{C}^\infty_X) \xrightarrow{\Gamma_X(d^0)} \Gamma_X(\Omega^1) \xrightarrow{\Gamma_X(d^1)} \Gamma_X(\Omega^2)\\ \longrightarrow \cdots \xrightarrow{\Gamma_X(d^{n-1})} \Gamma_X(\Omega^n),\end{aligned}$$

or else (see Chapt. II; (6.41))

$$\text{(8.39}'\text{)} \quad \Gamma_X(\Omega^{\cdot}_{\mathrm{de\,R}}) : 0 \longrightarrow \mathcal{C}^\infty(X) \xrightarrow{d} \Omega^1(X) \xrightarrow{d} \Omega^2(X) \xrightarrow{d} \cdots \xrightarrow{d} \Omega^n(X),$$

where, for simplicity, we also set $\Gamma_X(d^p) \equiv d^p \equiv d$, $p \in [0,n] \subseteq \mathbf{Z}_+$ (see, however, (8.44) below, as well). As already said, (8.39′) is a **C**-*linear complex,* according to the very definition of (8.37).

Indeed, (8.37) or, *equivalently,* (8.37′) yields a (finite) sequence of $\mathcal{C}^\infty_X$-*modules,* so that, in turn, (8.39′) provides a similar one of $\mathcal{C}^\infty(X)$-*modules.* However, the *morphisms* involved in each case are, by the very definitions (see e.g. (8.33)) only **C**-*linear morphisms,* or **C**-*linear maps,* respectively, therefore, the preceding terminology. *Caution!* In this context, we should also note that *the* **C**-*linear complex (8.39),* or (8.39′), *is not exact(!),* due to the *non-exactness of (8.37′),* hence, the definition that follows, referring exactly to the *obstruction of being (8.39′) exact.* That is, we have.

**Definition 8.2.** Suppose we are given an $n$-dimensional Hausdorff $\mathcal{C}^\infty$-manifold $X$. Then, the **de Rham cohomology** of $X$ is the cohomology (cf. (1.13)) of the **C**-linear complex (8.39′), as above (viz. of the image, through the functor $\Gamma_X$, of the de Rham complex of $X$, (8.37′)). We denote it by

$$\text{(8.40)} \qquad H^*_{\mathrm{de\,R}}(X) \equiv \{H^p_{\mathrm{de\,R}}(X)\}_{0 \leqq p \leqq n}\,.$$

Hence, according to (1.13), one defines;

$$\text{(8.41)} \qquad H^*_{\mathrm{de\,R}}(X) := h^*(\Gamma_X(\Omega^{\cdot}_{\mathrm{de\,R}})) \equiv \{h^p(\Gamma_X(\Omega^{\cdot}_{\mathrm{de\,R}})\}_{0 \leqq p \leqq n}\,,$$

such that one has, in view of (8.39) (see also (1.9)),

$$H^p_{\mathrm{de\,R}}(X) := h^p(\Gamma_X(\Omega^{\cdot}_{\mathrm{de\,R}})) := \ker \Gamma_X(d^p)/\mathrm{im}\,\Gamma_X(d^{p-1}), \tag{8.42}$$

for every $p \in [0, n] \subseteq \mathbf{Z}_+$. We call (8.42) the $p$-**th dimensional de Rham cohomology C-vector space** of $X$, or even, by abusing terminology, the ($p$-th dimensional) **de Rham cohomology group** of $X$.

To arrive now, as promised, to de Rham's resolution (of the constant sheaf $\mathbf{C}$), thus, in turn, to the classical de Rham's theorem, we need a bit more terminology, which we supply right below:

So, by considering the *constant sheaf* (see also (8.27) above, as well as, Chapt. II; (1.34))

$$\mathbf{C}_X \equiv \mathbf{C} \underset{\varepsilon}{\overset{\subseteq}{\rightarrow}} \mathcal{A} \equiv \mathcal{C}^\infty_X, \tag{8.43}$$

one obtains the following *augmented* $\mathbf{C}_X$*-complex* (see also (8.37) for the termonilogy applied), with *augmentation map* (in effect, a $\mathbf{C}$*-linear isomorphism* (into) of the $\mathbf{C}$*-vector space sheaves* involved, cf. (8.47) below) $\varepsilon_X \equiv \varepsilon$

$$\mathbf{C}^{\cdot}_X : 0 \longrightarrow \mathbf{C} \overset{\varepsilon}{\longrightarrow} \mathcal{C}^\infty_X \equiv \Omega^0 \overset{d}{\longrightarrow} \Omega^1 \overset{d}{\longrightarrow} \cdots \overset{d}{\longrightarrow} \Omega^n \longrightarrow 0. \tag{8.44}$$

We call (8.44) the *augmented de Rham complex* of $X$, being regarded, as already said, a $\mathbf{C}_X$*-complex* on $X$ (see also the comments on (8.37), preceding Definition 8.2 above). Yet, concerning (8.43), one looks, via $\varepsilon$, at the *constant functions* (sections of $\mathbf{C}_X \equiv \mathbf{C}$) as $\mathcal{C}^\infty$*-functions* (sections of $\mathcal{C}^\infty_X$) on $X$; in this regard, see also the comments on Chapt. II; (1.34), as well as, Chapt. I; Theorem 11.1.

On the other hand, the first basic information one has on (8.44) is that;

(8.45) *the augmented de Rham complex (8.44) is exact,* so that it provides, in effect, a *(finite) resolution of (the constant sheaf)* $\mathbf{C}$. That is, one has (cf. also (8.37), as well as, (1.55′) in the foregoing);

$$0 \longrightarrow \mathbf{C} \overset{\varepsilon}{\longrightarrow} \Omega^{\cdot}_{\mathrm{de\,R}}, \text{ or simply } \mathbf{C} \overset{\varepsilon}{\longrightarrow} \Omega^{\cdot}_{\mathrm{de\,R}}. \tag{8.45.1}$$

This fundamental conclusion is, in effect, an immediate consequence of the following classical result.

**Lemma 8.2.** *(Poincaré's Lemma). Let $X$ be a given (Hausdorff, $n$-*

*dimensional)* $\mathcal{C}^\infty$*-manifold. Then, (locally)*

(8.46) *every closed form is exact* (and *conversely,* of course).

We have formulated here, in purpose, the above result in the *epigrammatic manner of (8.46),* whose technical meaning is actually the following: Suppose we are given a (local) $p$-form $\omega$ of $X$ (viz., cf. (8.34), $\omega \in \Omega^p(U)$, for some local chart $U$ of $X$), *such that* $d\omega = 0$ (we call then $\omega$ a *closed form,* in fact, a *closed p-form* of $X$). Thus, the assertion now, according to (8.46), is that, then *there exists a* $(p-1)$*-form* $\omega' \in \Omega^{p-1}(U)$ of $X$, *such that* $\omega = d\omega'$ (we call then $\omega$, an *exact form*). For the *proof of Lemma 8.2,* see, for instance, W. D. Curtis - F. R. Miller [1: p.161, Proposition 9.45]. ∎

On the other hand, we know already (cf. Lemma 1.1) that the *exactness of (8.44) is a "local" matter;* viz. it amounts to that one of the individual *stalks* of the sheaves involved (see (1.33)). But, then, according to what has been said just before, about the meaning of (8.46) and in view of Chapt. I; (9.8), (9.9), the latter is already straightforward from Lemma 8.2, which thus settles our claim in (8.45). ∎

Now, before we proceed further, we want to give a first *application of the exactness of (8.44), in connection with (8.42).* We still note that *no restriction* is made *on* $X$. So, based on the exactness of (8.44), one has the following *exact sequence of* **C**-*vector space sheaves* on $X$ (an extract, in fact, of (8.44))

$$0 \longrightarrow \mathbf{C}_X \xrightarrow{\varepsilon} \mathcal{C}^\infty_X \xrightarrow{d^0} \Omega^1; \tag{8.47}$$

hence, applying to (8.47) the *left exact functor* $\Gamma_X$, one gets the following *exact sequence of* **C**-*vector spaces* (cf. e.g. Chapt. II; (2.59), or even S. Lang [1: p.141, §6])

$$0 \longrightarrow \Gamma_X(\mathbf{C}_X) \xrightarrow{\Gamma_X(\varepsilon)} \mathcal{C}^\infty(X) \xrightarrow{\Gamma_X(d^0)} \Omega^1(X) \tag{8.48}$$

(see also (8.39), (8.39′) above for the notation applied). Therefore, based on (8.42) and the exactness of (8.48), one obtains

$$H^0_{\mathrm{deR}}(X) := \ker \Gamma_X(d^0) = \operatorname{im} \Gamma_X(\varepsilon) \cong \Gamma_X(\mathbf{C}_X) = H^0(X, \mathbf{C}_X) \equiv h^0(X, \mathbf{C}), \tag{8.49}$$

where the last item in (8.49) stands for the 0-*dimensional sheaf cohomology group* (in fact, **C**-*vector space*) of $X$, with *coefficients in* (the constant sheaf) **C** (see Theorem 3.1, or even (3.10), (3.11)); as usual, we have also set in (8.49)

$$\operatorname{im} \Gamma_X(d^{-1}) := (0). \tag{8.50}$$

Furthermore, in view also of Lemma 8.1 (obviously specialized for *cohomology coefficients* from **C**-$\mathcal{V}ect_X$, *category of* **C**-*vector space sheaves on* $X$, see also Chapt. II; (1.38)), one finally obtains the following supplement to (8.49);

$$H^0_{\text{de R}}(X) = H^0(X, \mathbf{C}) = \check{H}^0(X, \mathbf{C}) = \Gamma_X(\mathbf{C}_X) \equiv \Gamma_X(\mathbf{C}), \tag{8.51}$$

within **C**-*linear isomorphisms* of the **C**-vector spaces concerned.

Thus, (8.51) constitutes, in fact (and in full generality for $X$), a forrunner of what we are actually going to infer from de Rham's theorem (see Thorem 8.3 below) where, namely, (8.51) will be extended to *any dimension* $p \leq n = \dim X$ (including $p = 0$, of course), however, *at the cost of the generality of* the $\mathcal{C}^\infty$-manifold $X$ considered. (So the latter will be taken now *paracompact!*).

But, yet let us pause for a while at the last member of (8.51); thus, this stands (cf. also Chapt. I; Theorem 11.1) for the *set,* actually **C**-*vector space, of locally constant* (hence, ($\mathcal{C}^\infty$-) too) *functions* on $X$. Now, the same set admits obviously a *bijective* correspondence with that *power* of **C** (constant $\mathcal{C}^\infty$-functions on $X$), as the *cardinality* (viz. number of elements) of the *zero-homotopy set* of $X$, $\pi_0(X)$, or even the set of *(path-)components* of $X$. Thus, one obtains, by (8.51),

$$H^0_{\text{de R}}(X) = \mathbf{C}^{|\pi_0(X)|}, \tag{8.52}$$

up to a **C**-*linear isomorphism* of the **C**-vector spaces involved, where, of course, $|\pi_0(X)|$ denotes the number of elements (*cardinal*) of $\pi_0(X)$. Therefore, one concludes, in particular, that:

(8.53) in the case that the given $\mathcal{C}^\infty$-manifold $X$ is also *connected* (hence, *path-connected* too, cf. e.g. I. M. James [1: p.138, Proposition 5.4]), then one obtains

$$H^0_{\text{de R}}(X) = \mathbf{C}, \tag{8.53.1}$$

within an *isomorphism of* the **C**-*vector spaces* concerned.

So we finally come to the hitherto often quoted de Rham's theorem. To this end, however, we first make a supplementary assumption for $X$. Namely,

(8.54) henceforth, we assume that the given $n$-dimensional *Hausdorff* $\mathcal{C}^\infty$-manifold $X$ is also *paracompact.*

Now, to have (8.54), we can suppose, for example, that the (underlying) topological space (to the given $\mathcal{C}^\infty$-manifold) $X$ is *2nd countable;* thus, $X$ being also regular, as locally compact, it is *metrizable* (P. Uryson), therefore (A. H. Stone), paracompact too (cf. e.g. J. Dugundji [1: p.195, Corollary 9.2 and p.186, Theorem 5.3]). Now, the point here, concerning the previous supplementary assumption on $X$, is that;

(8.55) *every* (finite dimensional) *Hausdorff paracompact $\mathcal{C}^\infty$-manifold* $X$ admits a *$\mathcal{C}^\infty$-partition of unity.* This entails, in turn, that the respective *structure sheaf* $\mathcal{C}_X^\infty$ of the manifold in question admits a *partition of unity,* in the sense of (8.23), (8.23′) above. Therefore (see Definition 8.1), one concludes for the same manifold $X$, as before, that:

(8.55.1) $\mathcal{C}_X^\infty$ *is a fine sheaf.*

Now, concerning our first claim in (8.55), we recall that the terminology applied there (*$\mathcal{C}^\infty$-partition of unity*) means, by definition, a family of (real-valued) *$\mathcal{C}^\infty$-functions* on $X$, having analogous properties to those in (8.23′); on the other hand, its *existense* is assured from a *"$\mathcal{C}^\infty$-analogue of Uryson's Lemma"*, the latter being valid *for every paracompact (Hausdorff) $\mathcal{C}^\infty$-manifold* $X$. (Cf., for instance, F. W. Warner [1: p.11, Corollary], or even L. Conlon [1: p.85, Corollary 3.5.5]). Thus, having now such a collection of $\mathcal{C}^\infty$-functions, one further proves that *the (**C**-algebra) sheaf $\mathcal{C}_X^\infty$ admits a partition of unity,* in the sense of (8.23). In fact, one proves this, e.g., for the respective (complete) presheaf of sections of ${}^{\mathbf{R}}\mathcal{C}_X^\infty$ (cf., for instance, F. W. Warner [1: p.170, §5.10, (1)]), by also employing the pertinent *complexification* (cf. (8.28)); then, one can pass further to the corresponding *endomorphisms* of $\mathcal{C}_X^\infty$ (cf. (8.23)), as desired, according to Chapt. I; (13.9). ∎

On the other hand, one proves, in general, the following basic fact, when, in particular, the topological space $X$ considered is *paracompact (Hausdorff).* We shall also use it often in the sequel (cf. Part Two). So we have.

(8.56) Suppose we are given a **C**-*algebraized space* $(X, \mathcal{A})$, having $X$ a *paracompact (Hausdorff)* space, while the structure sheaf $\mathcal{A}$ is *fine*. Then, *every* $\mathcal{A}$-*module* $\mathcal{E}$ on $X$ *is fine,* as well. (Indeed, it would be enough here if $\mathcal{A}$ was just *soft;* cf. e.g. (8.25)).

See thus, for instance, G. E. Bredon [1: p.50, Theorem 9.12]. ∎ So taking now (8.26) and (8.45) into account, one infers that;

(8.57) *the augmented de Rham complex* (8.44) of any *paracompact (Hausdorff)* $\mathcal{C}^\infty$*-manifold* $X$ *is a* $\Gamma_X$*-acyclic resolution of* (the constant sheaf) **C**.

Accordingly (cf. (3.24.2)), one infers that:

(8.58) *the sheaf cohomology of* $X$ *with complex coefficients* (hence, in view of the hypothesis for $X$, *the Čech* one *too*) *is simply that one of the complex (8.39′), viz. the de Rham cohomology of* $X$.

In other words, one obtains the following relations (see also (8.41), (8.12), as well as, (1.13) and (3.5), for $\mathcal{E} \equiv \mathbf{C}_X \equiv \mathbf{C}$);

$$H^*(X, \mathbf{C}) = \check{H}^*(X, \mathbf{C}) = h^*(\Gamma_X(\Omega^{\cdot}_{\mathrm{de\,R}})) \equiv h^*_{\mathrm{de\,R}}(X), \tag{8.59}$$

within *isomorphisms of* the *graded* **C**-*vector spaces* concerned. ∎ But, this is actually the content of the celebrated *de Rham's theorem,* and we are done. However, we also consider, for convenience, the following less formal, viz. more standard, version of it. So we have.

**Theorem 8.3.** *(De Rham's Theorem). Let* $X$ *be an* $n$*-dimensional Hausdorff paracompact* $\mathcal{C}^\infty$*-manifold. Then, one obtains*

$$H^p(X, \mathbf{C}) = \check{H}^p(X, \mathbf{C}) = H^p_{\mathrm{de\,R}}(X), \ p \in [0, n] \subseteq \mathbf{Z}_+, \tag{8.60}$$

*within* **C**-*linear isomorphisms of the* **C**-*vector spaces concerned.*

**Proof.** As already said for (8.58), the assertion follows straightforwardly from (3.24.2), once we have ensured the pertinent conditions thereat, as already follows from the preceding discussion. Thus, one has (see also Theorem 8.1);

$$(8.61)\qquad \begin{aligned} \check{H}^p(X, \mathbf{C}) &= H^p(X, \mathbf{C}) = R^p\Gamma_X(\mathbf{C}) = h^p(\Gamma_X(\Omega^{\cdot}_{\mathrm{de\,R}})) \\ &= \ker \Gamma_X(d^p)/\mathrm{im}\,\Gamma_X(d^{p-1}) \equiv H^p_{\mathrm{de\,R}}(X), \end{aligned}$$

*for every* $p = 0, 1, \ldots, n$ (cf. (8.50)), up to $\mathbf{C}$-*linear isomorphisms* of the $\mathbf{C}$-vector spaces involved, as desired. ∎

Therefore, as a result of (8.60), one concludes that;

(8.62) *the de Rham cohomology of a paracompact (Hausdorff)* $\mathcal{C}^\infty$*-manifold* $X$, *defined in terms of the differentiable structure* (in particular, *via the* $\mathcal{C}^\infty$*-forms*) *of* $X$, *is, in effect, simply a topological* (actually, a *homotopy,* cf. e.g. (1.29)) *invariant* of $X$.

Analogously, for a (paracompact Hausdorff) complex (analytic) manifold $X$, which is also a *Stein manifold,* one has a

(8.63) *holomorphic version of de Rham's theorem.*

Here one employs *holomorphic* (differential) *forms* on $X$, getting thus analogous relations to (8.49), where now (8.37) is replaced by the corresponding *holomorphic de Rham complex* $\Omega^{\cdot}_{\partial}$ of $X$ ($\partial$*-cohomology*); see, for instance, L. Kaup - B. Kaup [1: p.273, 58.1 and p.274, 58.2]. ∎

Finally, within the same context, as above, consider a *complex manifold* $X$ (now not necessarily Stein), which we still suppose to be *finite-dimensional,* having the underlying topological space *Hausdorff* and *second countable* ("countable at infinity" or "countable basis" of the topology are synonymous of the latter term, also in use); therefore, $X$ is, in particular, *paracompact* (see the comments following (8.54) above). So here too,

(8.64) as an *application of (3.24.2),* within now the appropriate context (that is, $\bar{\partial}$*-cohomology*), one obtains the classical *Dolbeault's Theorem.*

For details, se e.g. K. Kodaira [1: p.140, Theorem 3.12], or even R. O. Wells [1: p.61, Theorem 3.17]. ∎ Now, if $X$ is a (finite-dimensional) *complex manifold,* in general, and $\mathcal{O}_X$ the *sheaf of* $\mathbf{C}$*-valued holomorphic functions on* $X$ (see also Chapt. I; (4.40), or even (10.14) and (11.11); iii)), the pair

$$(8.65)\qquad (X, \mathcal{O}_X)$$

yields in the case considered the corresponding **C**-*algebraized space,* given that $\mathcal{O}_X$ is, of course, a **C**-*algebra sheaf* on $X$ (cf. Chapt. I; (10.14) and (11.11; iii), as well as, Chapt. II; (1.67)). However, by contrast with what happens with (8.55.1),

(8.66) $\mathcal{O}_X$ *is not a fine sheaf.*

The assertion is, in effect, a consequence of the *principle of analytic continuation;* thus, see e.g. K. Kodaira [1: p.135]. ∎

In this context, as we shall see later on (cf. Part Two in the sequel), both (8.55.1) and (8.66) have important implications on the *differential geometric structure* of the manifolds concerned, pertaining, for instance, to the existence of $\mathcal{C}^{\infty}$-, resp. *holomorphic, connections of vector bundles* on the manifolds involved; indeed, these classical facts have here their abstract counterpart, stated within the general framework, adopted herewith, viz. that of the *geometry of vector sheaves,* as this will be exposed, in detail, in Part Two. In particular, one employs here the abstract analogue of the so-called *Atiyah's criterion,* referring to the *existence of connections,* the latter term taking also the appropriate meaning, within the aforesaid context; see thus Chapt. VI in the sequel.

# 9. The functors $\mathrm{Ext}_{\mathcal{A}}$ and $\mathcal{E}xt_{\mathcal{A}}$

The functors in the heading of this Section provide further examples of important *bifunctors,* apart from those already considered in the preceding (cf. Chapt. II; Section 6). Indeed, as we shall see in a moment, the present ones are the right derived functors of the former (cf. Section 2 above), being, in fact, among the "most important instances of derived functors" (cf. N. Jacobson [1: p.346, §6.7], the other case being the "torsion" functors not treated below). The same functors will be used in subsequent parts of this treatise.

As usual, we are always within a given **C**-*algebraized space*

(9.1) $$(X, \mathcal{A}),$$

while the categories, *domain* (of definition or *source*) and *target* of our functors will be the categories $\mathcal{A}$-$\mathcal{M}od_X$ and $\mathcal{A}(X)$-$\mathcal{M}od$, or in both cases $\mathcal{A}$-$\mathcal{M}od_X$, respectively, that we already met in the foregoing (cf. e.g. (1.1) and Chapt. II; (6.39)). So we begin with the first of the above functors.

**9.(a). The functor** $\mathrm{Ext}_{\mathcal{A}}$.– As is the case in the standard Homological Algebra, the (bi)functor in question is defined through the *bifunctor*

$$\mathrm{Hom}_{\mathcal{A}} \equiv \mathrm{Hom}_{\mathcal{A}}(\cdot\,,\cdot) : \mathcal{A}\text{-}\mathcal{M}od_X \times \mathcal{A}\text{-}\mathcal{M}od_X \longrightarrow \mathcal{A}(X)\text{-}\mathcal{M}od, \tag{9.2}$$

which we already considered in Chapt. II, Subsection 6.(a). So, for convenience, we recall that, for any two $\mathcal{A}$-*modules* $\mathcal{E}$,$\mathcal{F}$ on $X$, viz. objects of $\mathcal{A}$-$\mathcal{M}od_X$,

$$\mathrm{Hom}_{\mathcal{A}}(\mathcal{E},\mathcal{F}) \tag{9.3}$$

stands for the set of $\mathcal{A}$-*morphisms* of $\mathcal{E}$ into $\mathcal{F}$ (see Chapt. II; (2.1)). On the other hand,

(9.4) the set $\mathrm{Hom}_{\mathcal{A}}(\mathcal{E},\mathcal{F})$, as above, with fiber-wise defined operations, (cf. Chapt. II; (2.2)), becomes an $\mathcal{A}(X)$-*module* (see also (9.5) below), which thus *justifies* (9.2), as well. In particular, in view of the hypothesis for $\mathcal{A}$ (cf. (9.1) and Chapt. I; (1.34)), the same is a $\mathbf{C}$-*vector space.*

Precisely, based on Chapt. I; Proposition 2.1, we set, with $\phi \in \mathrm{Hom}_{\mathcal{A}}(\mathcal{E},\mathcal{F})$,

$$(\alpha\cdot\phi)\circ s := \alpha\cdot(\phi\circ s), \tag{9.5}$$

for any $\alpha \in \mathcal{A}(X)$ and $s \in \Gamma(U,\mathcal{E}) \equiv \mathcal{E}(U)$, a local section of $\mathcal{E}$ over an open $U \subseteq X$. That is, one has (see also Chapt. I; (1.14) and Chapt. II; (2.2))

$$((\alpha\cdot\phi)\circ s)(x) = \alpha(x)\cdot(\phi\circ s)(x) = \alpha(x)\cdot\phi(s(x)) = \alpha(x)\cdot\phi_x(s(x)), \tag{9.6}$$

for every point $x \in U$, as before, that also explains (9.5).

Now, given any fixed $\mathcal{A}$-*module* $\mathcal{E}$, one considers the following (particular) functor

$$\mathrm{Hom}_{\mathcal{A}}(\mathcal{E},\cdot) : \mathcal{A}\text{-}\mathcal{M}od_X \longrightarrow \mathcal{A}(X)\text{-}\mathcal{M}od, \tag{9.7}$$

defined, according to (9.2) and/or (9.3), for any object $\mathcal{F}$ in A-$\mathcal{M}od_X$. Thus, one concludes, in particular, that:

(9.8) for every *$\mathcal{A}$-module* $\mathcal{E}$, $\mathrm{Hom}_{\mathcal{A}}(\mathcal{E}, \cdot)$ is an *additive covariant left exact functor* between the categories $\mathcal{A}$-$\mathcal{M}od_X$ and $\mathcal{A}(X)$-$\mathcal{M}od$. Yet, as already observed (cf. Lemma 1.2), the *source* (and, in fact, the *target*, as well) of the functor in question is an *abelian category with enough injectives.* (In this regard, see also Chapt. I; (6.46) and (2.1) in the preceding, along with Scholium 2.1, as well as, e.g. Proof of the same Lemma 1.2, as before, concerning our claim for the target of (9.7)).

Therefore, according to the general pattern (cf., for instance, Definitions 2.1, 3.1 in the preceding), one can consider the *right-derived functors of* $\mathrm{Hom}_{\mathcal{A}}(\mathcal{E}, \cdot)$. So one sets now the following.

**Definition 9.1.** Given an $\mathcal{A}$-module $\mathcal{E}$ on $X$ (see also (9.1)) and an integer $n \geqq 0$, one defines the $n$**-th** $\mathrm{Ext}_{\mathcal{A}}$**-functor of** $\mathcal{E}$, denoted by $\mathrm{Ext}^n_{\mathcal{A}}(\mathcal{E}, \cdot)$, as the $n$-th right-derived functor of $\mathrm{Hom}_{\mathcal{A}}(\mathcal{E}, \cdot)$ (cf. (9.7)). That is, we set

$$\mathrm{Ext}^n_{\mathcal{A}}(\mathcal{E}, \cdot) := R^n\mathrm{Hom}_{\mathcal{A}}(\mathcal{E}, \cdot), \tag{9.9}$$

the latter yielding thus a functor between the same categories as in (9.7) (cf. also e.g. (2.6) in the preceding, as well as, (9.12), (9.13) below).

Thus, to compute (9.9) on a given *$\mathcal{A}$-module* $\mathcal{F}$ on $X$, suppose that

$$\mathcal{S}^{\cdot} : 0 \longrightarrow \mathcal{S}^0 \xrightarrow{d} \mathcal{S}^1 \xrightarrow{d} \cdots, \tag{9.10}$$

is an *injective $\mathcal{A}$-resolution of* $\mathcal{F}$ (cf. Definition 1.2 and Theorem 1.2), so that the corresponding *augmented $\mathcal{A}$-complex* $\mathcal{F} \xrightarrow{\varepsilon} \mathcal{S}^{\cdot}$, viz. the *$\mathcal{A}$-complex*

$$\widetilde{\mathcal{S}}^{\cdot} : 0 \longrightarrow \mathcal{F} \xrightarrow{\varepsilon} \mathcal{S}^0 \xrightarrow{d} \mathcal{S}^1 \xrightarrow{d} \cdots, \tag{9.11}$$

is *exact* and *acyclic* (cf. (1.55), (1.59)). Therefore, one obtains (see also e.g. (2.7), along with (9.9) and (2.10));

$$\begin{aligned} \mathrm{Ext}^n_{\mathcal{A}}(\mathcal{E}, \mathcal{F}) &:= [R^n\mathrm{Hom}_{\mathcal{A}}(\mathcal{E}, \cdot)](\mathcal{F}) \equiv R^n\mathrm{Hom}_{\mathcal{A}}(\mathcal{E}, \mathcal{F}) \\ &:= h^n(\mathrm{Hom}_{\mathcal{A}}(\mathcal{E}, \cdot)(\mathcal{S}^{\cdot})) \equiv h^n(\mathrm{Hom}_{\mathcal{A}}(\mathcal{E}, \mathcal{S}^{\cdot})). \end{aligned} \tag{9.12}$$

So one has, concerning (9.9);

$$\mathrm{Ext}^n_{\mathcal{A}}(\mathcal{E}, \cdot) : \mathcal{A}\text{-}\mathcal{M}od_X \longrightarrow \mathcal{A}(X)\text{-}\mathcal{M}od, \tag{9.13}$$

this being also an *additive* and *covariant functor* between the categories concerned. Yet, it is, in fact, *independent of the (injective) resolution* applied to define the second member of (9.9), for a given object in $\mathcal{A}$-$\mathcal{M}od_X$ (cf. (9.10), (9.12)), *up to a unique (natural) isomorphism of functors.* (See e.g. Theorem 2.1 above in the case of a similar functor $T$, or even S. Lang [1: p.142, Theorem 6.1]).

**Remark 9.1.**– By analogy with (9.13), for any given fixed $\mathcal{A}$-*module* $\mathcal{E}$ and every *integer* $n \geqq 0$, one defines

$$\text{(9.14)} \qquad \operatorname{Ext}^n_{\mathcal{A}}(\cdot\,,\mathcal{E}) : \mathcal{A}\text{-}\mathcal{M}od_X \longrightarrow \mathcal{A}(X)\text{-}\mathcal{M}od,$$

as an (additive) *contravariant functor,* being, by definition the $n$-*th right-derived functor of* (the additive *contravariant* functor) $\operatorname{Hom}_{\mathcal{A}}(\cdot, \mathcal{E})$ (see Chapt. II; (6.45)). In this context, one proves that

$$\text{(9.15)} \qquad \operatorname{Ext}^n_{\mathcal{A}}(\mathcal{E}, \cdot) = \operatorname{Ext}^n_{\mathcal{A}}(\cdot\,,\mathcal{E}),$$

*for any* $\mathcal{E}$ *and* $n$, as above, within a (natural) *isomorphism of functors.* The proof can be given through *spectral sequences;* see thus e.g. R. Godement [1: p.95, §5.3 and p.263, §7.3]. However, cf. also, for instance, N. Jacobson [1: p.351, Theorem 6.9].

On the other hand, the natural question on the *"sort of exactness"* of the previous functors is connected with the so-called *"functor dimension"* of the objects involved. Cf. e.g. S. Lang [1: p.160f], or even H. Cartan - S. Eilenberg [1: p.109, §2].

Now, before we proceed further, we want to notice that, concerning the definition of the functor $\operatorname{Ext}^n_{\mathcal{A}}(\mathcal{E}, \cdot)$ (see (9.9) and (9.12)), one can consider, *in place of an injective resolution,* as in (9.10), any other $\operatorname{Hom}_{\mathcal{A}}(\mathcal{E}, \cdot)$-*acyclic resolution* of the $\mathcal{A}$-*module* $\mathcal{F}$ involved, on which one is to apply the above functor. Indeed, this follows from our previous general principle, pertaining to *derived functors,* as stated in (3.38). For convenience, and due to its theoretical significance, as already remarked, we restate it in the following form:

(9.16) Suppose we are given an $\mathcal{A}(X)$-linear covariant left exact functor (see also (2.23), (2.24) for the terminology applied)

(9.16.1) $$T : \mathcal{A}\text{-}\mathcal{M}od_X \longrightarrow \mathcal{A}(X)\text{-}\mathcal{M}od.$$

Then, the *right-derived functors* of $T$ are given by the relations;

(9.16.2) $$(R^n T)(\mathcal{E}) = h^n(T(\mathcal{S}^{\cdot})),$$

within a natural *isomorphism of functors, for every* $n \in \mathbf{Z}_+$ and *any* $\mathcal{A}$*-module* $\mathcal{E}$ on $X$, where $\mathcal{E} \xrightarrow{\varepsilon} \mathcal{S}^{\cdot}$ stands for any $T$*-acyclic resolution of* $\mathcal{E}$ (cf. (1.58), (2.20)).

**Note.**– The categories in (9.16.1) can be replaced by any *abelian categories,* with *the first* one *having enough injectives* (cf. e.g. Lemma 1.2).

The significane of the above fundamental conclusion has been already evidenced in the preceding by its applications, for instance, in Section 8, even in the special case of (3.24) *("abstract de Rham theorem").* For a proof of (9.16), *without* employing *spectral sequences*(!), see S. Lang [1: p.145, Theorem 6.2]. Yet, cf. the comments after (3.24) in the foregoing. ∎

Now, in connection with our hypothesis in (9.16) that $\mathcal{S}^{\cdot}$ was a $T$*-acyclic resolution of* $\mathcal{E}$ and the analogous assumption in (9.12), we still remark that:

(9.17) *any injective* $\mathcal{A}$*-module* $\mathcal{S}$ on $X$ *is* $T$*-acyclic, with*

(9.17.1) $$T \equiv \mathrm{Hom}_{\mathcal{A}}(\mathcal{E}, \cdot),$$

where $\mathcal{E}$ stands for any $\mathcal{A}$*-module* $\mathcal{E}$ on $X$. Hence, *any injective resolution of* $\mathcal{E}$, as in (9.11), *is* $\mathrm{Hom}_{\mathcal{A}}(\mathcal{E}, \cdot)$*-acyclic.* (This further explains the meaning of (9.16), as well).

The assertion is straightforward by applying e.g. Theorem 2.1; 5), for the functor $T$, given by (9.17.1) and defined now as in (9.16.1) (viz., as usual, we can *replace* **C**-$\mathcal{V}ect$ *by the category* $\mathcal{A}(X)$-$\mathcal{M}od$, as above; thus, see also (2.21)). ∎ Therefore, in view of (9.9) and (2.20), one obtains, for any $\mathcal{A}$-*module* $\mathcal{E}$ on $X$;

(9.18) $$\mathrm{Ext}^n_{\mathcal{A}}(\mathcal{E}, \mathcal{S}) \equiv R^n\mathrm{Hom}_{\mathcal{A}}(\mathcal{E}, \mathcal{S}) = 0,$$

*for every integer* $n > 0$, and *any injective* $\mathcal{A}$*-module* $\mathcal{S}$ on $X$.

On the other hand, *for* $n = 0$, and *for every* $\mathcal{A}$*-module* $\mathcal{E}$ on $X$, one obtains the following (natural) *isomorphism of functors* (see (2.13))

$$\text{Ext}^0_{\mathcal{A}}(\mathcal{E},\cdot) \equiv R^0\text{Hom}_{\mathcal{A}}(\mathcal{E},\cdot) = \text{Hom}_{\mathcal{A}}(\mathcal{E},\cdot). \tag{9.19}$$

In particular, *for* $\mathcal{E} = \mathcal{A}$, we have;

$$\text{Ext}^0_{\mathcal{A}}(\mathcal{A},\cdot) \equiv R^0\text{Hom}_{\mathcal{A}}(\mathcal{A},\cdot) = \text{Hom}_{\mathcal{A}}(\mathcal{A},\cdot) = \Gamma_X(\cdot) = H^0(X,\cdot), \tag{9.20}$$

up to (natural) *isomorphisms of* the *functors* involved. Cf. thus Chapt. II; (6.47), or even (2.13) above, for $T$ given by (9.17.1), as well as, (3.14). ▮ Indeed, (9.20) *is valid, for every* $n \in \mathbf{Z}_+$; that is, one obtains

$$\text{Ext}^n_{\mathcal{A}}(\mathcal{A},\cdot) \equiv R^n\text{Hom}_{\mathcal{A}}(\mathcal{A},\cdot) = R^n\Gamma_X = H^n(X,\cdot),\ n \in \mathbf{Z}_+. \tag{9.21}$$

See Chapt. II; (6.47), along with (3.3) above, which thus proves (9.21) and, consequently (9.20) too. ▮ Therefore, one has, *for every* $\mathcal{A}$*-module* $\mathcal{E}$ on $X$,

$$\text{Ext}^0_{\mathcal{A}}(\mathcal{A},\mathcal{E}) = \Gamma(X,\mathcal{E}) \equiv \mathcal{E}(X) = H^0(X,\mathcal{E}), \tag{9.22}$$

while, *for* $n > 0$, one obtains

$$\text{Ext}^n_{\mathcal{A}}(\mathcal{A},\mathcal{E}) = H^n(X,\mathcal{E}), \tag{9.23}$$

up to $\mathcal{A}(X)$*-isomorphisms of* the $\mathcal{A}(X)$*-modules* involved in the above two last relations. In other words, *one thus obtains,* all in all;

$$\text{Ext}^n_{\mathcal{A}}(\mathcal{A},\mathcal{E}) = H^n(X,\mathcal{E}),\ n \in \mathbf{Z}_+, \tag{9.24}$$

within an $\mathcal{A}(X)$*-isomorphism of* the $\mathcal{A}(X)$*-modules* concerned, *for every* $\mathcal{A}$*-module* $\mathcal{E}$ on $X$. So we can also summarize (9.21) or, equivalently, (9.24) by simply writing, as usual;

$$\text{Ext}^*_{\mathcal{A}}(\mathcal{A},\cdot) = H^*(X,\cdot), \tag{9.25}$$

denoting thus a *(natural) isomorphism of* the respective *functors* (cf. also (3.12)). We have thus, through (9.25), another means of computing (sheaf) cohomology (and conversely, rather!). Some useful applications of (9.25) will be considered in the sequel (see Chapt. V; (7.6), (7.7) and Propositions 8.1, as well as, Chapt. VI; (13.7)).

Furthermore, based e.g. on Theorem 2.1, prop. 3) (and for $T$ given by (9.17.1)), one comes to an analogous conclusion, concerning the *second variable of* $\text{Ext}^*_{\mathcal{A}}$, for any given $\mathcal{A}$*-module* $\mathcal{E}$ on $X$; thus, for every *short exact* $\mathcal{A}$*-sequence,*

$$0 \longrightarrow \mathcal{S}'' \longrightarrow \mathcal{S} \longrightarrow \mathcal{S}' \longrightarrow 0, \tag{9.26}$$

one obtains the following *long exact sequence (of* $\mathcal{A}(X)$*-modules,* (cf. also (9.19))

$$\begin{aligned} &0 \longrightarrow \mathrm{Hom}_{\mathcal{A}}(\mathcal{E}, \mathcal{S}'') \longrightarrow \mathrm{Hom}_{\mathcal{A}}(\mathcal{E}, \mathcal{S}) \longrightarrow \mathrm{Hom}_{\mathcal{A}}(\mathcal{E}, \mathcal{S}') \xrightarrow{\delta} \mathrm{Ext}^1_{\mathcal{A}}(\mathcal{E}, \mathcal{S}'') \\ &\longrightarrow \mathrm{Ext}^1_{\mathcal{A}}(\mathcal{E}, \mathcal{S}) \longrightarrow \cdots \longrightarrow \mathrm{Ext}^n_{\mathcal{A}}(\mathcal{E}, \mathcal{S}') \xrightarrow{\delta} \mathrm{Ext}^{n+1}_{\mathcal{A}}(\mathcal{E}, \mathcal{S}'') \longrightarrow \cdots . \end{aligned} \tag{9.27}$$

In particular, *suppose that*

$$\mathrm{Ext}^1_{\mathcal{A}}(\mathcal{E}, \mathcal{F}) = 0, \tag{9.28}$$

*for every* $\mathcal{A}$*-module* $\mathcal{F}$ on $X$. *Then,* as a consequence of (9.27), *the* $\mathcal{A}(X)$*-sequence*

$$0 \longrightarrow \mathrm{Hom}_{\mathcal{A}}(\mathcal{E}, \mathcal{S}'') \longrightarrow \mathrm{Hom}_{\mathcal{A}}(\mathcal{E}, \mathcal{S}) \longrightarrow \mathrm{Hom}_{\mathcal{A}}(\mathcal{E}, \mathcal{S}') \longrightarrow 0 \tag{9.29}$$

*is exact* too, for any short exact $\mathcal{A}$-sequence (9.26). Accordingly, one comes to the conclusion that;

(9.30) for every $\mathcal{A}$*-module* $\mathcal{E}$ on $X$, *the functor*

$$\mathrm{Ext}^1_{\mathcal{A}}(\mathcal{E}, \cdot) \tag{9.30.1}$$

*measures the obstruction of being the* (left exact) *functor* $\mathrm{Hom}_{\mathcal{A}}(\mathcal{E}, \cdot)$ actually *exact* (viz. right exact too).

Applications of the previous functor $\mathrm{Ext}^*_{\mathcal{A}}$, in particular, of (9.30.1) will be considered in later chapters of this book, as well. (See, for instance, the next Chapter IV, or even Part Two of this treatise).

**Remark 9.2.**– The condition (9.28) *is fulfilled,* in the case, *for* instance, that $\mathcal{E}$ is *a projective* $\mathcal{A}$*-module* on $X$ (the "dual" notion to injective $\mathcal{A}$-module; cf. Definition 1.1). In fact, one then obtains

$$\mathrm{Ext}^n_{\mathcal{A}}(\mathcal{E}, \mathcal{F}) = 0, \ \textit{for } n > 0, \tag{9.31}$$

and *for every* $\mathcal{A}$*-module* $\mathcal{F}$ on $X$; indeed, (9.31) *is a characteristic property of* $\mathcal{E}$ *being projective* (see e.g. P. Griffits - J. Harris [1: p.686]). Of course, one has a *"dual situation"* to the preceding one, by considering the *functor* $\mathrm{Ext}^n_{\mathcal{A}}(\cdot, \mathcal{E})$ (cf. (9.14)), for a given $\mathcal{A}$*-module* $\mathcal{E}$ on $X$.

We come now, by the next Subsection, to the second of the functors considered in the title of the present Section.

**9.(b). The functor $\mathcal{E}xt_{\mathcal{A}}$.**– Suppose we are given an $\mathcal{A}$-module $\mathcal{E}$ on $X$ (cf. also (9.1)). Then, by looking at the functor

$$\mathcal{H}om_{\mathcal{A}}(\mathcal{E}, \cdot) : \mathcal{A}\text{-}\mathcal{M}od_X \longrightarrow \mathcal{A}\text{-}\mathcal{M}od_X, \tag{9.32}$$

one has a *covariant left exact functor* of the category $\mathcal{A}$-$\mathcal{M}od_X$ into itself (the same functor being also $\mathcal{A}(X)$-*linear.* Cf. Chapt. II; (6.11), along with (2.24) in the preceding). So one now sets the following (see also e.g. Definition 2.1).

**Definition 9.2.** Given an $\mathcal{A}$-module $\mathcal{E}$ on $X$ and an integer $n \geqq 0$, one defines the $n$**-th $\mathcal{E}xt_{\mathcal{A}}$-functor of $\mathcal{E}$,** denoted by $\mathcal{E}xt^n_{\mathcal{A}}(\mathcal{E}, \cdot)$, as the $n$-th right-derived functor of $\mathcal{H}om_{\mathcal{A}}(\mathcal{E}, \cdot)$ (see (9.32)). That is, one sets;

$$\mathcal{E}xt^n_{\mathcal{A}}(\mathcal{E}, \cdot) := R^n\mathcal{H}om_{\mathcal{A}}(\mathcal{E}, \cdot), \tag{9.33}$$

this being thus a functor (in fact, $\mathcal{A}(X)$-*linear*) of $\mathcal{A}$-$\mathcal{M}od_X$ into itself.

Now, we further look at certain immediate, yet useful, consequences of the above relation (9.33). So we first have the following.

**Lemma 9.1.** *Suppose we are given a* **C**-*algebraized space* $(X, \mathcal{A})$, *and let* $\mathcal{E}$ *be an* $\mathcal{A}$-*module on* $X$. *Then, one obtains* (cf. (9.33)):

$$\mathcal{E}xt^0_{\mathcal{A}}(\mathcal{E}, \cdot) = \mathcal{H}om_{\mathcal{A}}(\mathcal{E}, \cdot). \tag{9.34}$$

*In particular, one has*

$$\mathcal{E}xt^0_{\mathcal{A}}(\mathcal{A}, \cdot) = \mathrm{id} \equiv \mathrm{id}_{\mathcal{A}\text{-}\mathcal{M}od_X}, \tag{9.35}$$

*the second member of the last relation denoting the identity functor on* $\mathcal{A}$-$\mathcal{M}od_X$ (*viz.* $\mathrm{id}(\mathcal{E}) = \mathcal{E}$, *for any* $\mathcal{A}$-*module* $\mathcal{E}$ on $X$). *Finally, one still obtains*

$$\mathcal{E}xt^n_{\mathcal{A}}(\mathcal{A}, \cdot) = 0, \; \textit{for } n > 0. \tag{9.36}$$

(*the second member of* (9.36) *standing for the zero-functor on* $\mathcal{A}$-$\mathcal{M}od_X$). *Yet, all the previous three relations hold true up to (natural) isomorphisms of the functors concerned.*

**Proof.** First by applying (9.33) and Theorem 2.1, prop. 2) (viz. (2.13), for the case at hand, or else for $T \equiv \mathcal{H}om_{\mathcal{A}}(\mathcal{E}, \cdot)$, as in (9.32)), one obtains

$$\mathcal{E}xt^0_{\mathcal{A}}(\mathcal{E}, \cdot) := R^0\mathcal{H}om_{\mathcal{A}}(\mathcal{E}, \cdot) = \mathcal{H}om_{\mathcal{A}}(\mathcal{E}, \cdot), \tag{9.37}$$

up to (natural) *isomorphisms of* the respective *functors,* which *proves (9.34).* On the other hand, as a consequence of Chapt. II; (6.18), one has

$$\mathcal{H}om_{\mathcal{A}}(\mathcal{A}, \cdot) = \mathrm{id}, \tag{9.38}$$

(cf. the 2nd member of (9.35) for the meaning of "id"), within a (natural) *isomorphism of functors.* Therefore, applying now (9.33), for $\mathcal{E} = \mathcal{A}$, one obtains

$$\mathcal{E}xt^0_{\mathcal{A}}(\mathcal{A}, \cdot) := R^0\mathcal{H}om_{\mathcal{A}}(\mathcal{A}, \cdot) = \mathcal{H}om_{\mathcal{A}}(\mathcal{A}, \cdot) = \mathrm{id}, \tag{9.39}$$

which thus *proves (9.35).*

Now, (9.36) follows again straightforwardly from (9.33) and (9.38); that is, *one has,*

$$\mathcal{E}xt^n_{\mathcal{A}}(\mathcal{A}, \cdot) := R^n\mathcal{H}om_{\mathcal{A}}(\mathcal{A}, \cdot) = R^n\mathrm{id}, \ \textit{with } n \geqq 1. \tag{9.40}$$

However, one further obtains (see also e.g. (2.7))

$$(R^n\mathrm{id})(\mathcal{E}) = h^n(\mathrm{id}(\mathcal{S}^\cdot)) = h^n(\mathcal{S}^\cdot) = 0, \tag{9.41}$$

*for every integer* $n > 0$, where $\mathcal{E}$ is any $\mathcal{A}$-*module* on $X$ and $\mathcal{E} \xrightarrow{\varepsilon} \mathcal{S}^\cdot$ *any injective resolution of* $\mathcal{E}$. The last relation in (9.41) follows, for instance, from the *exactness of* $0 \longrightarrow \mathcal{E} \xrightarrow{\varepsilon} \mathcal{S}^\cdot$ (see e.g. (1.55)), by virtue of (1.9), *for* $n > 0$; in other words, one concludes that:

(9.42) *the* (strictly) *positive* (right-)*derived functors of the identity functor* (in effect, *of any exact functor,* as well) *vanish.*

Thus, this justifies (9.41), *proving* also (9.36), and the proof is complete. ∎

Furthermore, by analogy with (9.27), one gets a similar *long exact* $\mathcal{A}$-*sequence for* $\mathcal{E}xt_{\mathcal{A}}$, as an application of the pertinent result, for instance, in Theorem 2.1. Thus, for $T \equiv \mathcal{H}om(\mathcal{E}, \cdot)$, as in (9.32), and in view of Theorem 2.1, prop. 3), one concludes that; *for every short exact* $\mathcal{A}$-*sequence*

$$0 \longrightarrow \mathcal{S}'' \longrightarrow \mathcal{S} \longrightarrow \mathcal{S}' \longrightarrow 0, \tag{9.43}$$

one obtains the following *long exact* $\mathcal{A}$-*sequence* for $\mathcal{E}xt^*_{\mathcal{A}}(\mathcal{E}, \cdot)$, with $\mathcal{E}$ *any* $\mathcal{A}$-*module* on $X$ (see also (9.37)):

$$\begin{aligned}(9.44)\quad & 0 \longrightarrow \mathcal{H}om_{\mathcal{A}}(\mathcal{E}, \mathcal{S}'') \longrightarrow \mathcal{H}om_{\mathcal{A}}(\mathcal{E}, \mathcal{S}) \longrightarrow \mathcal{H}om_{\mathcal{A}}(\mathcal{E}, \mathcal{S}') \stackrel{\delta}{\longrightarrow} \mathcal{E}xt^1_{\mathcal{A}}(\mathcal{E}, \mathcal{S}'') \\ & \longrightarrow \mathcal{E}xt^1_{\mathcal{A}}(\mathcal{E}, \mathcal{S}) \longrightarrow \cdots \longrightarrow \mathcal{E}xt^n_{\mathcal{A}}(\mathcal{E}, \mathcal{S}') \stackrel{\delta}{\longrightarrow} \mathcal{E}xt^{n+1}_{\mathcal{A}}(\mathcal{E}, \mathcal{S}'') \longrightarrow \cdots .\end{aligned}$$

Now, an *interpretation of the functor* $\mathcal{E}xt^1_{\mathcal{A}}(\mathcal{E}, \cdot)$, *in connection with* $\mathcal{H}om_{\mathcal{A}}(\mathcal{E}, \cdot)$, *analogous to (9.30)*, for the functor (9.30.1), *can also be given* at this place, based on (9.44) (cf. (9.29)). Yet, concerning a similar relation, at this point, to (9.28), one still obtains the following, useful besides, result (in fact, a *generalization of (9.36)*).

**Lemma 9.2.** *Suppose we are given a* **C**-*algebraized space* $(X, \mathcal{A})$, *and let* $\mathcal{E}$ *be a vector sheaf on* $X$ (cf. Chapt. II; Definition 4.3). *Then, for every* $\mathcal{A}$-*module* $\mathcal{F}$ *on* $X$, *and for every integer* $n > 0$, *one obtains;*

$$\mathcal{E}xt^n_{\mathcal{A}}(\mathcal{E}, \mathcal{F}) = 0. \tag{9.45}$$

**Proof.** Based on (9.33) and (9.42), *it suffices to prove that;*

(9.46) *for every vector sheaf* $\mathcal{E}$ *on* $X$, *the functor* (see also (9.32))

$$\mathcal{H}om_{\mathcal{A}}(\mathcal{E}, \cdot) \tag{9.46.1}$$

*is exact.*

Indeed, we know already that *the functor (9.46.1) is left-exact* (see Chapt. II; (6.17)). So it remains to prove, when *in the particular case considered*, that *it is right-exact, as well:* Thus, for any given *exact* $\mathcal{A}$-*sequence*

$$0 \longrightarrow \mathcal{S}'' \longrightarrow \mathcal{S} \longrightarrow \mathcal{S}' \longrightarrow 0 \tag{9.47}$$

one has to prove that the following $\mathcal{A}$-*sequence*

$$0 \longrightarrow \mathcal{H}om_{\mathcal{A}}(\mathcal{E}, \mathcal{S}'') \longrightarrow \mathcal{H}om_{\mathcal{A}}(\mathcal{E}, \mathcal{S}) \longrightarrow \mathcal{H}om_{\mathcal{A}}(\mathcal{E}, \mathcal{S}') \longrightarrow 0 \tag{9.48}$$

*is exact* too. However, since we are dealing with $\mathcal{A}$-*modules, the question is "local"*; that is (cf. Lemma 1.1), one has to assure, *equivalently*, the *exactness of the* following $\mathcal{A}_x$-*sequence, for every* $x \in X$,

$$\mathcal{H}om_{\mathcal{A}}(\mathcal{E}, \mathcal{S}'')_x \longrightarrow \mathcal{H}om_{\mathcal{A}}(\mathcal{E}, \mathcal{S})_x \longrightarrow \mathcal{H}om_{\mathcal{A}}(\mathcal{E}, \mathcal{S}')_x \longrightarrow 0 \tag{9.49}$$

(but, see also Chapt. II; (6.10)!). Now, in view of Chapt. I; (9.8) and (9.9) along with Chapt. II; (6.8) and the fact that *direct* (viz. *inductive*) *limits preserve exactness* (cf. (4.103)), *exactness of (9.49) is assured by that one of the following sequence;*

(9.50) $\mathcal{H}om_{\mathcal{A}|_U}(\mathcal{E}|_U, \mathcal{S}''|_U) \longrightarrow \mathcal{H}om_{\mathcal{A}|_U}(\mathcal{E}|_U, \mathcal{S}|_U) \longrightarrow \mathcal{H}om_{\mathcal{A}|_U}(\mathcal{E}|_U, \mathcal{S}'|_U) \longrightarrow 0,$

where $U$ is an *open neighborhood of* $x \in X$ ( *"local expression"* of the exactness of (9.48)), that we can assume here to range over a *local basis* (of open neighborhoods) *at* $x$; namely, this is possible, since, by hypothesis for $\mathcal{E}$, *there exists an open covering of* $X$, say $\mathcal{U} = (U_\alpha)_{\alpha \in I}$, *such that*

(9.51) $$\mathcal{E}|_{U_\alpha} = \mathcal{A}^n|_{U_\alpha}, \ \alpha \in I,$$

within an $\mathcal{A}|_{U_\alpha}$*-isomorphism,* where $n = \mathrm{rk}\,\mathcal{E}$ (see Chapt. II; (4.2), (4.5)). But, then (cf. Chapt. II; (6.22), (6.23), along e.g. with (4.5)) *the exactness of (9.50) is reduced* to that of the following $\mathcal{A}|_U$*-sequence*

(9.52) $$(\mathcal{S}''|_U)^n \longrightarrow (\mathcal{S}|_U)^n \longrightarrow (\mathcal{S}'|_U)^n \longrightarrow 0.$$

This however, is obviously true, in view of the hypothesis for (9.47). (Here one also applies the fact that *direct sums* (of $\mathcal{A}$-modules) *preserve exactness,* as well as, that *the restriction functor,* being a *pull-back functor* ($U \overset{\subseteq}{\rightarrow} X$), *is exact*). ∎

In this context, we still note that by further employing the *bifunctor* $\mathcal{H}om_{\mathcal{A}}$ (cf. Chapt. II; (6.15)), we can state (9.46) in the following form too:

(9.53) given a **C**-algebraized space $(X, \mathcal{A})$, *the vector sheaves* on $X$ *are* $\mathcal{H}om_{\mathcal{A}}$*-exact,* in the sense that *the functor (9.46.1)* (see also (9.32)) *is exact, for every vector sheaf* $\mathcal{E}$ on $X$.

(See, for instance, S. Lang [1: p.157] for the terminology applied). An application of (9.46) (equivalently, of (9.53)) will be given right in the next Section 10. Yet, further relations connected with the previous two functors will be considered in subsequent paragraphs of the book. Thus, see also, for instance, the following Section for *another interpretation of* $\mathrm{Ext}^1_{\mathcal{A}}$ (see (9.9) above), justifying besides the terminology *"ext"* (for *extension*) of the above functors.

# 10. $\mathcal{A}$-extensions

We are concerned in the sequel with *"extensions"* of $\mathcal{A}$-modules, as this notion can be formulted in terms of the functors considered in the previous Section 9. Of course, one refers here to the counterpart in our case of the analogous classical theory of Homological Algebra (see, for instance, N. Jacobson [1: p.348ff]). The present discussion will find, of course, application in later chapters of the book. (So see e.g. Chapters V and VII in the sequel). Thus, we start with the following basic terminology.

**Definition 10.1.** Suppose we are given a **C**-algebraized spaces $(X, \mathcal{A})$ and two $\mathcal{A}$-modules $\mathcal{E}$,$\mathcal{F}$ on $X$. Then, by an **extension of $\mathcal{E}$ by $\mathcal{F}$,** is meant any short exact $\mathcal{A}$-sequence (see also Chapt. II; (2.18)) of the form;

$$0 \longrightarrow \mathcal{F} \longrightarrow \mathcal{S} \longrightarrow \mathcal{E} \longrightarrow 0. \tag{10.1}$$

(Of course, it concerns here an **$\mathcal{A}$-extension;** however, for brevity, the latter more precise terminology will usually be omitted in the sequel. Yet, for short, we well also employ for (10.1), simply, the term the "**extension $\mathcal{S}$**").

An obvious extension of the form (10.1), is, of course, the *Whitney sum* of $\mathcal{E}$ and $\mathcal{F}$, $\mathcal{E} \oplus \mathcal{F}$ (see Chapt. II; (3.1)), along with the obvious injection and projection (in fact, *$\mathcal{A}$-morphisms*), as in (10.1). We call it the *trivial extension* of $\mathcal{E}$ by $\mathcal{F}$ and will play a special role in the sequel; so we set for later use

$$\mathcal{E} \oplus \mathcal{F}, \ \textit{trivial extension} \text{ of } \mathcal{E} \text{ by } \mathcal{F}. \tag{10.2}$$

*However,* not every extension (10.1) is of the previous form! (Thus, any classical counter example can be used here too, by simply taking $X$ = pt. Indeed, we are going to find, by the ensuing discussion, less trivial examples. So one supplies such a (non-trivial) example, by taking, for instance, the *lack,* in general, *of holomorphic connections* into account; cf. Part Two).

Now, to confront with the previous question, namely, whether a given *"extension $\mathcal{S}$",* as in (10.1), is the trivial one, we first look at the following basic concept. That is, we have.

**Definition 10.2.** Given two extensions $\mathcal{S}$ and $\mathcal{S}'$ of $\mathcal{E}$ by $\mathcal{F}$, as in (10.1), we say that these are **equivalent**, whenever the following diagram is commutative (in terms of *$\mathcal{A}$-morphisms*)

(10.3)

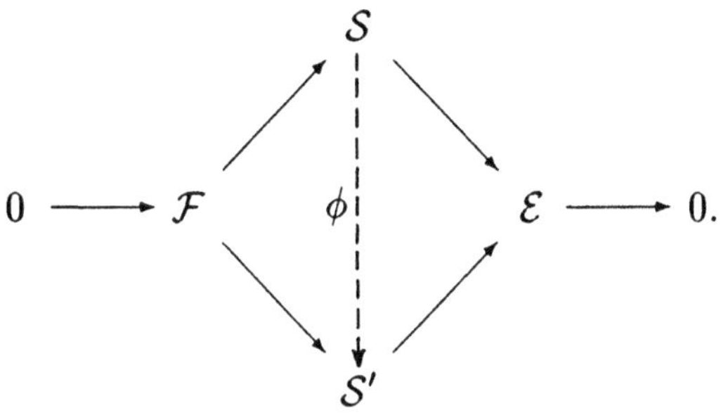

That is, one requires the *existence of an $\mathcal{A}$-morphism*

(10.4) $$\phi \in \mathrm{Hom}_{\mathcal{A}}(\mathcal{S}, \mathcal{S}'),$$

*making (10.3) commutative.*

In this context, and by analogy with the classical case, one also obtains the following.

**Lemma 10.1.** Suppose we are given the framework of the commutative diagram (10.3), as above. Then, *the $\mathcal{A}$-morphism* $\phi \in \mathrm{Hom}_{\mathcal{A}}(\mathcal{S}, \mathcal{S}')$ (cf. (10.4) *is,* in effect, *an $\mathcal{A}$-isomorphism* of the $\mathcal{A}$-modules concerned. That is, *one has, indeed,*

(10.5) $$\phi \in \mathrm{Isom}_{\mathcal{A}}(\mathcal{S}, \mathcal{S}') \subseteq \mathrm{Hom}_{\mathcal{A}}(\mathcal{S}, \mathcal{S}').$$

**Proof.** Based on Lemma 1.1, the commutative diagram (10.3), with $\phi$ given by (10.4), yields a similar one, *fiber-wise.* But, then the classical case is applicable, referred here to $\mathcal{A}_x$-modules (cf. e.g. (9.49)), $x \in X$, as a result of the so-called *"short five lemma"* (see, for instance, N. Jacobson [1: p.100, Ex. 4 and p.348], or even P. J. Hilton - U. Stammbach [1: p.14, Lemma 1.1]); thus, the given *$\mathcal{A}$-morphism $\phi$ is fiber-wise an isomorphism,* hence, *itself an isomorphism* too, by the very definitions (see Chapt. I; (1.14), (1.15) and (1.21)). ∎

As a result of the previous lemma, one concludes that;

(10.6) *equivalence of extensions is,* in effect, *an equivalence relation.*

See, for instance, J. J. Rotman [1: p.202]. ∎ Thus, given two *$\mathcal{A}$-modules* $\mathcal{E}$ and $\mathcal{F}$, as before, one defines

$$\text{(10.7)} \qquad \operatorname{ext}_{\mathcal{A}}(\mathcal{E}, \mathcal{F}),$$

as the *set of equivalence classes of extensions of* $\mathcal{E}$ *by* $\mathcal{F}$ (cf. Definitions 10.1, 10.2 above). Now, the latter set is *not empty;* thus, the *exact* $\mathcal{A}$*-sequence*

$$\text{(10.8)} \qquad 0 \longrightarrow \mathcal{F} \xrightarrow{i_{\mathcal{F}}} \mathcal{E} \oplus \mathcal{F} \xrightarrow{p_{\mathcal{E}}} \mathcal{E} \longrightarrow 0$$

defined by the *trivial extension of* $\mathcal{E}$ *by* $\mathcal{F}$ (cf. (10.2)), with the obvious meaning of the $\mathcal{A}$*-morphisms* $i_{\mathcal{F}}, p_{\mathcal{E}}$ involved, yields an element, at least, of (10.7).

In this context, we come now to the following important notion connected with (10.8); its geometrical meaning, within the appropriate framework, will be seen in Part Two of this discussion. So we first have the following.

**Proposition 10.1.** *Suppose we are given a* **C**-*algebraized space* $(X, \mathcal{A})$ *and let* $\mathcal{E}$,$\mathcal{F}$ *be* $\mathcal{A}$*-modules on* $X$. *Then, the following assertions are equivalent:*

*1) Given the short exact* $\mathcal{A}$*-sequence*

$$\text{(10.9)} \qquad 0 \longrightarrow \mathcal{F} \xrightarrow{\psi} \mathcal{S} \xrightarrow{\phi} \mathcal{E} \longrightarrow 0$$

(viz. an *extension of* $\mathcal{E}$ *by* $\mathcal{F}$, cf. Definition 10.1), *there exists an* $\mathcal{A}$*-morphism* $h \in \operatorname{Hom}_{\mathcal{A}}(\mathcal{E}, \mathcal{S})$, *such that*

$$\text{(10.10)} \qquad \phi \circ h = 1_{\mathcal{E}} \equiv \mathrm{id}_{\mathcal{E}} \in \operatorname{End} \mathcal{E}$$

(see also Chapt. II; (6.30). We call $h$, as before, a *section of* $\phi$ too).

*2) Given the exact sequence (10.9),* as before, *there exists an* $\mathcal{A}$*-morphism* $g \in \operatorname{Hom}_{\mathcal{A}}(\mathcal{S}, \mathcal{F})$ *such that*

$$\text{(10.11)} \qquad g \circ \psi = 1_{\mathcal{F}} \in \operatorname{End} \mathcal{F}.$$

(We call also the previous $g$ a *retraction* of $\mathcal{S}$, associated with $\psi$).

*3) The* $\mathcal{A}$*-module* $\mathcal{S}$, *as in (10.9) (extension of* $\mathcal{E}$ *by* $\mathcal{F}$*), is equivalent to the Whitney sum* $\mathcal{E} \oplus \mathcal{F}$ (viz. the *trivial extension of* $\mathcal{E}$ *by* $\mathcal{F}$, see (10.2) or (10.8)); *in other words* (see Definition 10.2), *the following diagram is commutative*

(10.12)

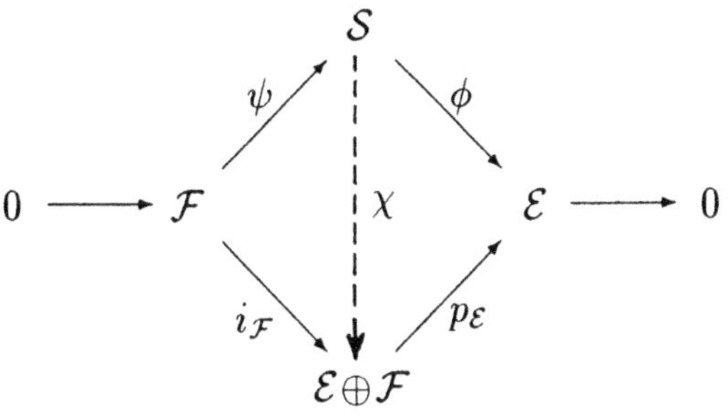


(see also (10.8)), *for some* $\mathcal{A}$-*morphism* $\chi \in \mathrm{Hom}_{\mathcal{A}}(\mathcal{S}, \mathcal{E} \oplus \mathcal{F})$ *(the latter being, in effect* (cf. Lemma 10.1), *an* $\mathcal{A}$-*isomorphism).*

**Proof.** Based on Lemma 10.1, it easy to prove, by the very definitions, that *either one of the first two conditions is equivalent with the third one,* so that *all* the stated *three conditions are equivalent;* however, for convenience, we do indicate briefly the *equivalence of 1) and 3):* So, *assuming 1),* one proves straightforwardly, in view of (10.10), that the $\mathcal{A}$-*morphism*

$$h + \psi : \mathcal{E} \oplus \mathcal{F} \longrightarrow \mathcal{S} \tag{10.13}$$

(see also Chapt. II; (6.21), (6.43) and (3.14)) yields a *morphism of* the $\mathcal{A}$-*extensions* concerned (in the sense that an analogous commutative diagram to (10.3) holds true, through (10.13)), so that (cf. Lemma 10.1) one sets

$$\chi := (h + \psi)^{-1}; \tag{10.14}$$

Hence, *1) implies 3).* Now for the *converse,* to find an $h \in \mathrm{Hom}_{\mathcal{A}}(\mathcal{E}, \mathcal{S})$, satisfying (10.10), one can apply the inverse of $\chi$ (still based on Lemma 10.1), composed with the canonical embedding $i_{\mathcal{E}} : \mathcal{E} \longrightarrow \mathcal{E} \oplus \mathcal{F}$, viz. one sets

$$h := \chi^{-1} \circ i_{\mathcal{E}}, \tag{10.15}$$

while (10.10) is obviously true, in view of the commutativity of (10.12). Namely, one obtains;

$$\phi \circ h = \phi \circ (\chi^{-1} \circ i_{\mathcal{E}}) = p_{\mathcal{E}} \circ i_{\mathcal{E}} = 1_{\mathcal{E}} \tag{10.16}$$

as claimed. A similar argument can also be applied for the other equivalence, and this completes the proof. ∎

Now, the previous proposition leads to the following.

**Definition 10.3.** Given two $\mathcal{A}$-modules $\mathcal{E}$,$\mathcal{F}$ on $X$, as above, an $\mathcal{A}$-extension of $\mathcal{E}$ by $\mathcal{F}$ (see Definition 10.1) is said to **split** (one still speaks of a **splitting extension** of $\mathcal{E}$ by $\mathcal{F}$), whenever any one of the three equivalent conditions of Proposition 10.1 holds true.

Thus, according to that result (cf. (10.12)), one has that;

(10.17) *any splitting extension of* $\mathcal{E}$ *by* $\mathcal{F}$ *is equivalent to the trivial one* (see (10.2)). Therefore, *any two splitting extensions of* $\mathcal{E}$ *by* $\mathcal{F}$ *are equivalent* (cf. also (10.6)).

On the other hand, the *equivalence class of the splitting extension of* $\mathcal{E}$ *by* $\mathcal{F}$, as before, viz. the corresponding *element of the set (10.7),* acquires an especially useful interpretation, when looking at the latter set through the functor $\mathrm{Ext}_{\mathcal{A}}$. We discuss this presently below in the following Subsection.

**10.(a). $\mathrm{Ext}^1_{\mathcal{A}}$ and $\mathcal{A}$-extensions.**– We start with the following fundamental result, which justifies also, as already mentioned in the preceding, the name "Ext" of the functor concerned. So we have.

**Theorem 10.1.** *Let* $(X, \mathcal{A})$ *be a* **C**-*algebraized space, and* $\mathcal{E}$,$\mathcal{F}$ *two given* $\mathcal{A}$-*modules on* $X$. *Then, one obtains* (see also (10.7) and (9.9))

$$\mathrm{ext}_{\mathcal{A}}(\mathcal{E}, \mathcal{F}) = \mathrm{Ext}^1_{\mathcal{A}}(\mathcal{E}, \mathcal{F}), \tag{10.18}$$

*up to a* (set-theoretic) *bijection of the respective sets.*

The construction of the correspondence, say,

$$\theta : \mathrm{ext}_{\mathcal{A}}(\mathcal{E}, \mathcal{F}) \longrightarrow \mathrm{Ext}^1_{\mathcal{A}}(\mathcal{E}, \mathcal{F}), \tag{10.19}$$

yielding the asserted bijection (10.18), is actually based on the so-called *"Fundamental Theorem of Homological Algebra"* (see e.g. S. Lang [1: p.139, Lemma 5.2]), converted here, of course, into a sheaf-theoretic language, in particular, within the category $\mathcal{A}$-$\mathcal{M}od_X$, our usual framework hitherto. For convenience, we do state it right below (without proof; in this regard see also, for instance R. G. Swan [1: p.45, Theorem 3], for an abstract categorical formulation). So we have.

**Lemma 10.2.** *Suppose we are given a* **C**-*algebraized space* $(X, \mathcal{A})$, *and consider the following diagram of (cochain* $\mathcal{A}$-*)complexes on* $X$ (see (1.2));

$$
\begin{array}{ccccccccccc}
0 & \longrightarrow & \mathcal{E} & \longrightarrow & \mathcal{E}^0 & \longrightarrow & \mathcal{E}^1 & \longrightarrow & \mathcal{E}^2 & \longrightarrow & \cdots \\
 & & \phi \big\downarrow & & \big\downarrow \phi^0 & & \big\downarrow \phi^1 & \cdots & \big\downarrow & & \\
0 & \longrightarrow & \mathcal{F} & \longrightarrow & \mathcal{J}^0 & \longrightarrow & \mathcal{J}^1 & \longrightarrow & \mathcal{J}^2 & \longrightarrow & \cdots .
\end{array}
\tag{10.20}
$$

*Furthermore, we assume that the top row in (10.20) is exact, while each* $\mathcal{J}^n$*, with* $n \geqq 0$*, is an injective* $\mathcal{A}$*-module. Finally, we are given* $\phi \in \mathrm{Hom}_{\mathcal{A}}(\mathcal{E}, \mathcal{F})$*. Then, there exists a morphism of complexes, say* $f \equiv \{f^n\}_{n \geqq 0}$ (cf. (1.15), (1.19)), *such that*

$$f_{-1} = \phi \tag{10.20$'$}$$

(we also say that $f$ is *over* $\phi$, or even that $\phi$ *is extended by* $f$). *Moreover, such a morphism* $f$ *is unique, up to homotopoy* (cf. (1.26), (1.28′)).

The proof of the above lemma can be patterned, verbatim, after the analogous one of the classical case; see thus, for instance, the previous quotation to S. Lang. ∎ So we come now to the following.

(Sketchy) **proof of Theorem 10.1.** Thus, consider the following diagram

$$
\begin{array}{ccccccccccc}
0 & \longrightarrow & \mathcal{F} & \xrightarrow{\ \psi\ } & \mathcal{S} & \xrightarrow{\ \phi\ } & \mathcal{E} & \longrightarrow & 0 & & \\
 & & \big\downarrow 1_{\mathcal{F}} & & \big\downarrow f^0 & & \big\downarrow \alpha & & & & \\
0 & \longrightarrow & \mathcal{F} & \xrightarrow{\ \varepsilon\ } & \mathcal{J}^0 & \xrightarrow{\ d^0\ } & \mathcal{J}^1 & \xrightarrow{\ d^1\ } & \mathcal{J}^2 & \longrightarrow & \cdots ,
\end{array}
\tag{10.21}
$$

where *the top row is exact,* corresponding to a given *extension of* $\mathcal{E}$ *by* $\mathcal{F}$ (cf. (10.1)), while the bottom row is an *injective resolution of the* $\mathcal{A}$*-module* $\mathcal{F}$ (cf. Lemma 1.2), and $1_{\mathcal{F}} \in \mathrm{End}\,\mathcal{F}$, the respective *identity endomorphism* of $\mathcal{F}$. So Lemma 10.2 can be applied, yielding for the given extension the $\mathcal{A}$*-morphism*

$$\alpha \in \mathrm{Hom}_{\mathcal{A}}(\mathcal{E}, \mathcal{J}^1), \tag{10.22}$$

as indicated in (10.21) (viz. $\alpha \equiv f^1$ in the notation of (10.20)). On the other hand, in view of (9.9) (see also (9.12)), one has (cf. (1.9))

$$\mathrm{Ext}^1_{\mathcal{A}}(\mathcal{E}, \mathcal{F}) := R^1\mathrm{Hom}_{\mathcal{A}}(\mathcal{E}, \mathcal{F}) = h^1(\mathrm{Hom}_{\mathcal{A}}(\mathcal{E}, \mathcal{J}^{\cdot})) = \ker d^1_* / \mathrm{im}\, d^0_*, \tag{10.23}$$

with $\mathcal{F} \longrightarrow \mathcal{J}^{\cdot}$ the given *injective resolution of* $\mathcal{F}$, as in (10.21), so that one obtains

$$(10.24) \quad 0 \longrightarrow \mathcal{H}om_{\mathcal{A}}(\mathcal{E}, \mathcal{J}^0) \xrightarrow{d^0_*} \mathcal{H}om_{\mathcal{A}}(\mathcal{E}, \mathcal{J}^1) \xrightarrow{d^1_*} \mathcal{H}om_{\mathcal{A}}(\mathcal{E}, \mathcal{J}^2 \longrightarrow \cdots,$$

which provides the last quotient ($\mathcal{A}(X)$-module) in (10.23) (cf. e.g. (2.8), (2.9) and (2.10)).

Now, by virtue of (10.21) and (10.22), one obtains

$$(10.25) \qquad d^1_*(\alpha) := d^1 \circ \alpha = 0,$$

that is, $\alpha$ is a 1-*cocycle of the* ($\mathcal{A}(X)$-)*complex* (10.24), where still we have set

$$(10.26) \qquad d^n_* \equiv \mathrm{Hom}_{\mathcal{A}}(\mathcal{E}, d^n), \ n \geqq 0.$$

Therefore, one can consider, in turn, its respective (cohomology) class in (10.23); that is, we set

$$(10.27) \qquad [\alpha] \equiv \alpha + \mathrm{im}\, d^0_* \in \mathrm{Ext}^1_{\mathcal{A}}(\mathcal{E}, \mathcal{F}).$$

Indeed, the above class defines now (10.19); viz. denoting by

$$(10.28) \qquad \xi : 0 \longrightarrow \mathcal{F} \xrightarrow{\psi} \mathcal{S} \xrightarrow{\phi} \mathcal{E} \longrightarrow 0$$

the given *extension of* $\mathcal{E}$ *by* $\mathcal{F}$, and $[\xi] \in \mathrm{ext}_{\mathcal{A}}(\mathcal{E}, \mathcal{F})$ the corresponding *equivalence class* in the set (10.7), one sets

$$(10.29) \qquad \theta([\xi]) \equiv \theta(\xi) := \alpha + \mathrm{im}\, d^0_* \equiv [\alpha].$$

In this context, one further proves that *the* $\mathcal{A}$-*morphism* $\alpha$, as above, *is uniquely defined, up to homotopy* (cf. also Lemma 10.1); so that *the given extension* $\xi$ (see (10.28)) *determines a unique element of* $\mathrm{Ext}^1_{\mathcal{A}}(\mathcal{E}, \mathcal{F})$, as in (10.27). Moreover, the latter is still *independent of the class of* $[\xi]$ *in* $\mathrm{Ext}_{\mathcal{A}}(\mathcal{E}, \mathcal{F})$, hence, *(10.29) is well defined,* therefore, (10.19) too. Now, the proof goes further, verbatim, as in the classical case, by "dualizing" the respective argument (given usually for "projective resolutions"). Thus, cf. e.g. J. J. Rotman [1: p.203ff; in particular, see the remarks at the end of p.206]; yet, cf. H. Cartan - S. Eilenberg [1: p.292, Remark], and this terminates our commentary on the proof of Theorem 10.1. ∎

On the other hand, based on the previous construction of (10.19) (see thus (10.29)), one concludes that;

(10.30) given two $\mathcal{A}$-*modules* $\mathcal{E}$,$\mathcal{F}$ on $X$, as above, *the class* in $\text{ext}_{\mathcal{A}}(\mathcal{E},\mathcal{F})$ *of the trivial extension of* $\mathcal{E}$ *by* $\mathcal{F}$ (cf. (10.18)), *therefore* (see (10.17)), *that one of any splitting extension* too (Definition 10.3), *corresponds* via (10.19) *to the zero-element of* (the $\mathcal{A}(X)$-module) $\text{Ext}^1_{\mathcal{A}}(\mathcal{E},\mathcal{F})$.

Indeed, the assertion is straightforward, by the very definition of (10.19), as given by (10.29); thus, due to the existence, in the case at hand, of a *right-inverse of* $\phi$ (viz. $\phi \circ h = 1_{\mathcal{E}}$, cf. (10.10)), one actually obtains for the $\mathcal{A}$-morphism $\alpha$, as in (10.22), that $\alpha \in \operatorname{im} d^0_*$, so that then $[\alpha] = 0$ (cf. (10.27)), as desired. Thus (cf. (10.21)), one obtains

$$d^0_*(f^0 \circ h) = d^0 \circ f^0 \circ h = \alpha, \tag{10.31}$$

which is just our claim. ∎ Now, another useful consequence of the same Theorem 10.1 is the following.

**Corollary 10.1.** *Assume that we have the framework of Theorem 10.1. Then, the following two assertions are equivalent:*

*1)* $\text{Ext}^1_{\mathcal{A}}(\mathcal{E},\mathcal{F}) = 0$.

*2) Any extension of* $\mathcal{E}$ *by* $\mathcal{F}$ *splits;* hence (see (10.17)), equivalently, *is* (equivalent to) the *trivial* one.

**Proof.** We first prove that *2) implies 1);* so if cond. 2) is valid, then the set $\text{ext}_{\mathcal{A}}(\mathcal{E},\mathcal{F})$ (cf. (10.7)) consists exactly of one element, viz. of the class of the trivial extension of $\mathcal{E}$ by $\mathcal{F}$ (see (10.8), (10.17)). Therefore, according to (10.18) and (10.30), cond. 1) holds true, as asserted.

*The converse* is more interesting(!) So let

$$0 \longrightarrow \mathcal{F} \xrightarrow{\psi} \mathcal{S} \xrightarrow{\phi} \mathcal{E} \longrightarrow 0 \tag{10.32}$$

be any given *extension of* $\mathcal{E}$ *by* $\mathcal{F}$. Thus, based on Theorem 2.1, prop. 3), one can further apply on the short exact $\mathcal{A}$-sequence (10.32) the (left-exact) functor $\text{Hom}_{\mathcal{A}}(\mathcal{E},\cdot)$, so that, in view of (9.9) (or else (2.13)) and our hypothesis (viz. *cond. 1) is valid*), one obtains (see also (9.27) and/or (2.16))

$$\begin{aligned} 0 \longrightarrow \text{Hom}_{\mathcal{A}}(\mathcal{E},\mathcal{F}) \longrightarrow \text{Hom}_{\mathcal{A}}(\mathcal{E},\mathcal{S}) \xrightarrow{\phi_*} \text{Hom}_{\mathcal{A}}(\mathcal{E},\mathcal{E}) \equiv \text{End}\,\mathcal{E} \\ \xrightarrow{\delta} \text{Ext}^1_{\mathcal{A}}(\mathcal{E},\mathcal{F}) = 0, \end{aligned} \tag{10.33}$$

where, of course, we set, as usual,

$$\text{Hom}_{\mathcal{A}}(\mathcal{E}, \cdot)(\phi) \equiv \text{Hom}_{\mathcal{A}}(\mathcal{E}, \phi) \equiv \phi_*, \tag{10.34}$$

with $\phi$, as in (10.32). That is, *the* $\mathcal{A}(X)$*-morphism* $\phi_*$, as in (10.33), *is surjective;* hence, in particular, the *identity endomorphism* of $\mathcal{E}$, $1_{\mathcal{E}} \in \text{End}\,\mathcal{E}$, is induced by an element of $\text{Hom}_{\mathcal{A}}(\mathcal{E}, \mathcal{S})$. That is, in other words, *there exists an* $\mathcal{A}$*-morphism* $h \in \text{Hom}_{\mathcal{A}}(\mathcal{E}, \mathcal{S})$, *such that*

$$\phi_*(h) = \phi \circ h = 1_{\mathcal{E}}. \tag{10.35}$$

But, then (Proposition 10.1, cf. also Definition 10.3), the given extension (10.32) *splits,* that was our claim; so *1) implies 2),* as well, and this completes the proof. ∎

Now, in view of Theorem 10.1, one can transfer, of course, the algebraic structure of the second member of (10.18) to the set in the first member of the same relation, so that the set $\text{ext}_{\mathcal{A}}(\mathcal{E}, \mathcal{F})$ *acquires through (10.18),* at least, *the* (additive) *structure of an abelian group.* However, this same structure on the latter set can also be given in an *intrinsic* manner, by defining thus in it the so-called *Baer sum* of (equivalence classes of) extensions. Furthermore, *this operation (sum)* in $\text{ext}_{\mathcal{A}}(\mathcal{E}, \mathcal{F})$ *is* proved to be *preserved by the* (set-theoretic) *isomorphism (10.18),* so that, in view of (10.17), one concludes that:

(10.36) *the set* $\text{ext}_{\mathcal{A}}(\mathcal{E}, \mathcal{F})$ (cf. (10.7)), *endowed with* the operation given by *the Baer sum becomes an abelian group, having as zero-element the trivial class* of $\text{ext}_{\mathcal{A}}(\mathcal{E}, \mathcal{F})$, *viz. the equivalence class of a splitting extension.* Moreover, *(10.18) is* thus *made into an isomorphism of* the *abelian groups* involved. We also call $\text{ext}_{\mathcal{A}}(\mathcal{E}, \mathcal{F})$ the *Baer group.*

Concerning the classical counterpart of the above, see e.g. J. J. Rotman [1: p.207f and p.209, Theorem 7.21] and also N. Jacobson [1: p.352, Ex. 4]. In this context, one can "dualize", of course, the classical arguments, which usually employ "projectives", as the cited Rfs. So see, for instance, the (sketchy) proof of Theorem 10.1 above. (Yet, see also e.g. P. J. Hilton - U. Stammbach [1: p.94, §3, as well as, p.143, §8], for "Ext *using injectives*"). ∎

On the other hand, *the above isomorphism (10.18) can also be established through the* so-called *characteristic class of an extension.* Thus, for the standard case of Homological Algebra, cf. e.g. H. Cartan - S. Eilenberg [1: p.290, Theorem 1.1 and Corollary 1.2, as well as, p.292, Remark]; yet, for the case at hand, see also, for instance, R. Hartshorne [1: p.237, Ex. 6.1], or even P. Griffiths - J. Harris [1: p.725, Lemma]. Now, since the aforementioned notion will also be of use in the sequel (see, for example, Part Two of this treatise), we do explain the corresponding situation, within the present framework.

Thus, assume that we are given the *$\mathcal{A}$-modules* $\mathcal{E}$ and $\mathcal{F}$ on $X$, as before, and let e.g.

$$0 \longrightarrow \mathcal{F} \xrightarrow{\psi} \mathcal{S} \xrightarrow{\phi} \mathcal{E} \longrightarrow 0 \tag{10.37}$$

be a given *extension of* $\mathcal{E}$ *by* $\mathcal{F}$ (cf. Definition 10.1). Thus, one can further consider the *long exact sequence of $\mathcal{A}(X)$-modules* associated with (10.37), after applying on it the *(left-exact) functor* $\mathrm{Hom}_{\mathcal{A}}(\mathcal{E}, \cdot)$, according e.g. to Theorem 2.1 (cf. (2.16)). So one obtains the following *long exact $\mathcal{A}(X)$-sequence*

$$\begin{aligned} 0 \longrightarrow \mathrm{Hom}_{\mathcal{A}}(\mathcal{E},\mathcal{F}) \longrightarrow \mathrm{Hom}_{\mathcal{A}}(\mathcal{E},\mathcal{S}) \xrightarrow{\phi_*} \mathrm{Hom}_{\mathcal{A}}(\mathcal{E},\mathcal{E}) \equiv \mathrm{End}\,\mathcal{E} \\ \xrightarrow{\delta} \mathrm{Ext}^1_{\mathcal{A}}(\mathcal{E},\mathcal{F}) \longrightarrow \mathrm{Ext}^1_{\mathcal{A}}(\mathcal{E},\mathcal{S}) \longrightarrow \cdots, \end{aligned} \tag{10.38}$$

so that one concludes the following.

(10.39) *a given extension of* $\mathcal{E}$ *by* $\mathcal{F}$ *splits* (cf. Definition 10.3) *if, and only if, one has*

$$\delta(1_{\mathcal{E}}) = 0, \tag{10.39.1}$$

where $1_{\mathcal{E}} \in \mathrm{End}\,\mathcal{E}$ denotes the *identity endomorphism of* $\mathcal{E}$. The element

$$\delta(1_{\mathcal{E}}) \in \mathrm{Ext}^1_{\mathcal{A}}(\mathcal{E},\mathcal{F}) \tag{10.39.2}$$

is called the *characteristic class* (or even the *obstruction* for the splitting) *of the extension* considered (see (10.37)).

Indeed, our claim is an immediate consequence of the relation

$$\phi_*(h) = \phi \circ h = 1_{\mathcal{E}}, \tag{10.40}$$

for some $h \in \mathrm{Hom}_{\mathcal{A}}(\mathcal{E},\mathcal{S})$ (cf. also (10.38)). So this *characterizes,* in fact, *the splitting of (10.37),* according to the very Definition 10.3 (cf. Proposition 10.1, (10.10)). Thus, *(10.40) is* actually *equivalent to (10.39.1),* in view of the *exactness* of (10.38), which proves the assertion. ∎

On the other hand, as an application of Lemma 9.2, we shall give later on (cf. Chapt. V; Section 8) and in case $\mathcal{E}$ is, in particular, a *vector sheaf* on $X$ (see Chapt. II; Definition 4.3), another useful interpretation of *the* above *characteristic class as* an appropriate 1-*dimensional cohomology class* of $X$.

# 11. The 1st cohomology set

We have considered above (cf. Sections 4, 5) *sheaf cohomology* of an arbitrary *topological space* $X$ with coefficients from an $\mathcal{A}$-*module,* where, as usual, $(X, \mathcal{A})$ stands for a given **C**-*algebraized space.* Yet, in the case of *Čech cohomology* an analogous procedure is valid (in fact, more generally, see thus e.g. (6.9)), when the coefficients are taken from a *presheaf* (we actually considered the case of a $\Gamma(\mathcal{A})$-*presheaf,* cf. Section 6). Of course, the above hold true, more generally, when the coefficient sheaf is a *sheaf of abelian groups* (certainly, every $\mathcal{A}$-module is such), or even a *presheaf of such* (an analogous remark, as before, for $\Gamma(\mathcal{A})$-presheaves). But, this is *not the case,* if the groups concerned are *not abelian!* (See, for instance, (11.11) in the sequel). Nevertheless, in *dimension one* we do define, presently below, a *"cohomology set"* endowed *with a distinguished element.* Yet, as we shall see, one has always an analogous set in *dimension zero,* as well. In the sequel we employ, in fact, *Čech cohomology* (cf. Section 4), having thus the standard general property of *uniqueness* for paracompact spaces (see Theorem 8.1, and (8.11) too).

On the other hand, for convenience, we do recall here, from Section 4 in the preceding, certain basic relevant notions, in the more general form, however, that we are actually going to employ them in the sequel. Thus, assume that we are given an arbitrary *topological space* $X$ and a *sheaf of* (not necessarily abelian) *groups* $\mathcal{S}$ on $X$. Furthermore, let

$$(11.1) \qquad \mathcal{U} = \{U_\alpha\}_{\alpha \in I}$$

be an *open covering* of $X$.

Now, by definition, a 1-*cochain of* $\mathcal{U}$ *with values* (or even *coefficients*) *in* $\mathcal{S}$ is a map

$$(11.2) \quad f : I \times I \longrightarrow \mathcal{S} : (\alpha, \beta) \longmapsto f(\alpha, \beta) \equiv f_{\alpha\beta} \in \mathcal{S}(U_\alpha \cap U_\beta) \equiv \Gamma(U_\alpha \cap U_\beta, \mathcal{S}).$$

So one further considers the *set of* 1*-cochains of* $\mathcal{U}$ *with values in* $\mathcal{S}$, denoted by

$$C^1(\mathcal{U},\mathcal{S}) := \prod_{(\alpha,\beta)\in I^2} \mathcal{S}(U_{\alpha\beta}), \tag{11.3}$$

where we also set, as usual, $U_{\alpha\beta} \equiv U_\alpha \cap U_\beta$ (the latter set being assumed *non-empty;* but, see also (4.13)); we denote an element of (11.3) by

$$f \equiv (f_{\alpha\beta}) \in C^1(\mathcal{U},\mathcal{S}). \tag{11.4}$$

In this regard, it is now clear that;

(11.5) the set (11.3) is actually a *group, not abelian, unless* $\mathcal{S}$ *is an abelian group sheaf.* (So consider e.g. the pairs $(\alpha,\alpha) \in I^2$). Of course, more generally, *(11.3) carries the same algebraic structure, as* $\mathcal{S}$ *does.*

Furthermore, one can consider, of course, $n$-cochains of $\mathcal{U}$ with coefficients in $\mathcal{S}$, as in (4.11), so that we can look, in particular, at the corresponding *coboundary operator*

$$\delta^1 \equiv \delta : C^1(\mathcal{U},\mathcal{S}) \longrightarrow C^2(\mathcal{U},\mathcal{S}), \tag{11.6}$$

according to (4.15). Thus, a given 1-cochain $(f_{\alpha\beta}) \in C^1(\mathcal{U},\mathcal{S})$ (cf. (11.4)) is called a 1*-cocycle,* in case one has

$$f \equiv (f_{\alpha\beta}) \in \ker \delta^1 \equiv Z^1(\mathcal{U},\mathcal{S}); \tag{11.7}$$

that is, one obtains

$$\delta(f) = \delta((f_{\alpha\beta})) := (\delta(f_{\alpha\beta})) \equiv (f_{\alpha\beta\gamma}), \tag{11.8}$$

in such a manner that (cf. (4.16))

$$(\delta f)_{\alpha\beta\gamma} \equiv f_{\alpha\beta\gamma} = f_{\beta\gamma} - f_{\alpha\gamma} + f_{\alpha\beta} = 0. \tag{11.9}$$

Equivalently, *a given* 1*-cochain,* as in (11.4), *is a* 1*-cocycle if (and only if,* by definition) *one has*

$$f_{\alpha\beta} + f_{\beta\gamma} = f_{\alpha\gamma}, \tag{11.9'}$$

for any $\alpha,\beta,\gamma$ in $I$, with $U_{\alpha\beta\gamma} \equiv U_\alpha \cap U_\beta \cap U_\gamma \neq \emptyset$ and $f_{\alpha\beta\gamma} \in \mathcal{S}(U_{\alpha\beta\gamma})$. (In this concern, see also (4.71) and (4.72)). Thus, following (4.21), we still set

(11.10) $$Z^1(\mathcal{U}, \mathcal{S}) := \ker \delta^1 \equiv \ker \delta,$$

for the *set of* 1-*cocycles of* $\mathcal{U}$ *with coefficients in* $\mathcal{S}$. However, in this context, it is to be remarked that:

(11.11) *the coboundary operator* $\delta^1 \equiv \delta$ (see (11.6), as well as, (4.16), for $n = 1$) *is not a group morphism,* to the extend that the given group sheaf $\mathcal{S}$ on $X$ is *not abelian,* so that *its kernel (11.10) is no more a group* (viz., otherwise, a *normal subgroup* of $C^1(\mathcal{U}, \mathcal{S})$).

Now, the above inconvenience can be reformed by bringing into play the 0-*dimensional cochain group of* $\mathcal{U}$ *with values in* $\mathcal{S}$,

(11.12) $$C^0(\mathcal{U}, \mathcal{S}) := \prod_{\alpha \in I} \mathcal{S}(U_\alpha)$$

(cf also Chapt. II; (1.67)). Indeed, one concludes that:

(11.13) *the* (not necessarily abelian) *group* $C^0(\mathcal{U}, \mathcal{S})$ *yields a (left) action on the set* (actually group, cf. (11.5)) *of* 1-*cochains* $C^1(\mathcal{U}, \mathcal{S})$. In other words, $C^1(\mathcal{U}, \mathcal{S})$ *is a* $C^0(\mathcal{U}, \mathcal{S})$*-set.*

That is, one sets;

(11.14) $$(f_\alpha) \cdot (f_{\alpha\beta}) := (f_\alpha + f_{\alpha\beta} - f_\beta),$$

for any $(f_\alpha) \in C^0(\mathcal{U}, \mathcal{S})$ and $(f_{\alpha\beta}) \in C^1(\mathcal{U}, \mathcal{S})$. (Here one takes care, of course, to restrict the $f_\alpha$'s to the sets $U_{\alpha\beta} \equiv U_\alpha \cap U_\beta$, as indicated in the second member of (11.14)). In this context, the crucial fact is the following assertion, which can be verified straightforwardly, based on the very definitions. So one concludes that:

(11.15) *the* above *group action,* as given by (11.14), *preserves the* 1-*cocycles;* that is, $Z^1(\mathcal{U}, \mathcal{S})$ (cf. (11.10)) *is a* $C^0(\mathcal{U}, \mathcal{S})$*-set,* or even, *the group* $C^0(\mathcal{U}, \mathcal{S})$ (see also (11.13)) *induces a (left) action on* $Z^1(\mathcal{U}, \mathcal{S})$.

Thus, the corresponding *set of orbits,* viz. the quotient set of $Z^1(\mathcal{U}, \mathcal{S})$, modulo the equivalence relation defined on it by (11.14), is denoted by

(11.16) $$H^1(\mathcal{U}, \mathcal{S}) := Z^1(\mathcal{U}, \mathcal{S}) / C^0(\mathcal{U}, \mathcal{S})$$

and called the *1st cohomology set of $\mathcal{U}$ with coefficients in $\mathcal{S}$.*

**Remark 11.1.**– In the case that the given group sheaf $\mathcal{S}$, as above, is a *sheaf of abelian groups,* then based on the same definitions, it is easy to see that *the coboundary operators* (see, for example, (4.16)) *are,* in effect, *group morphisms,* so that *the quotient set (11.16)* actually *coincides with the* usual *(abelian) cohomology group,* as given e.g. by (4.23).

Furthermore, as already said, the same set (11.16) is endowed with a *distinguished element,* namely, the (equivalence) *class of the trivial* 1-*cocycle* supplied by the *zeroes of $\mathcal{S}$;* so, if $0_x \in \mathcal{S}_x$, $x \in X$, stands for the *neutral element* of the particular stalk of $\mathcal{S}$ considered, the 1-*cocycle* in question, say, $f \equiv (f_{\alpha\beta}) \in Z^1(\mathcal{U}, \mathcal{S})$, is given by

$$f_{\alpha\beta}(x) := 0_x, \; x \in U_{\alpha\beta}, \tag{11.17}$$

*for any* $(\alpha, \beta) \in I^2$ *with* $U_{\alpha\beta} \equiv U_\alpha \cap U_\beta \neq \emptyset$ (cf. also (11.9′)). In other words, $f_{\alpha\beta}$, as defined by (11.17), is the *zero-section of* (the *group sheaf*) $\mathcal{S}$ *over* $U_{\alpha\beta}$*;* in this regard, see also Chapt. II; Lemma 1.1 and (1.8) too. The class of this cocycle in (11.16) is called the *neutral element* of the set $H^1(\mathcal{U}, \mathcal{S})$, still denoted by 0 (more precisely, by $0_{\mathcal{U}}$, in the case the individual groups, supplied by $\mathcal{S}$, are *additive,* otherwise, we denote the same class by $1_{\mathcal{U}}$, or just by 1. But, see also (11.27) in the sequel.

On the other hand, suppose now that we have a *refinement of the open* covering $\mathcal{U}$ of $X$, as above (see (11.1)), say

$$\mathcal{V} = \{V_\beta\}_{\beta \in J} \tag{11.18}$$

(cf. e.g. (4.2)), and let

$$\mu : J \longrightarrow I \tag{11.19}$$

be the corresponding *refinement map* (cf. (4.3), (4.4)). Therefore (see (4.34)), one can further consider the associated 1-*dimensional cochain map*

$$\mu^1 \equiv \mu : C^1(\mathcal{U}, \mathcal{S}) \longrightarrow C^1(\mathcal{V}, \mathcal{S}); \tag{11.20}$$

that is (cf. (4.34′)), for any given 1-*cochain* $f \equiv (f_{\alpha_0\alpha_1}) \in C^1(\mathcal{U}, \mathcal{S})$ and any pair $(\beta_0, \beta_1) \in J^2$, $\mu(f)_{\beta_0\beta_1}$ is the *restriction to* $V_{\beta_0\beta_1} \equiv V_{\beta_0} \cap V_{\beta_1} \subseteq U_{\mu(\beta_0)} \cap U_{\mu(\beta_1)} \equiv U_{\mu(\beta_0)\mu(\beta_1)}$ *of* (the *local section* of $\mathcal{S}$) $f_{\mu(\beta_0)\mu(\beta_1)} \in \mathcal{S}(U_{\mu(\beta_0)\mu(\beta_1)})$ (see

(11.2)). Furthermore, in view of (4.35), *the latter map commutes with the coboundary operator* $\delta^1 \equiv \delta$ (cf. (4.35.1) and (11.12)). Hence, by analogy with (4.38), one obtains a map between the respective cohomology sets

$$\mu^* : H^1(\mathcal{U}, \mathcal{S}) \longrightarrow H^1(\mathcal{V}, \mathcal{S}), \tag{11.21}$$

satisfying a similar relation to (4.39). Furthermore, one extends Lemma 4.2 to the present context, concluding thus that *(11.20) is,* in fact, *independent of the particular refining map* (cf. (4.40)). So one obtains the following map, corresponding to (4.49), viz.

$$\tau_{\mathcal{V},\mathcal{U}} : H^1(\mathcal{U}, \mathcal{S}) \longrightarrow H^1(\mathcal{V}, \mathcal{S}), \tag{11.22}$$

for any pair $(\mathcal{U}, \mathcal{V})$ of open coverings of $X$, with $\mathcal{V} \succeq \mathcal{U}$ (see (4.2)), while the *analogous relations to (4.50) and (4.51) are still in force.* Therefore, in other words,

(11.23) one obtains an *inductive system of sets*

$$(H^1(\mathcal{U}, \mathcal{S}), \tau_{\mathcal{V},\mathcal{U}}) \tag{11.23.1}$$

*with respect to the directed index set (4.9)* (cf. also (4.52)).

Thus, we now set the following.

**Definition 11.1.** Given a topological space $X$ and a sheaf of groups $\mathcal{S}$ on $X$, one defines as the **1st cohomology set of $X$ with coefficients in $\mathcal{S}$,** the inductive limit set of the inductive system (11.23.1); i.e., we set

$$H^1(X, \mathcal{S}) := \varinjlim_{\mathcal{U}} H^1(\mathcal{U}, \mathcal{S}). \tag{11.24}$$

Now, in this regard, we also remark that;

(11.25) *the analogous of Lemma 4.3 is* still *valid,* so that, *for any* $\mathcal{U}, V$, *as in (11.22), the* corresponding *map*

$$\mu^1 \equiv \tau_{\mathcal{V},\mathcal{U}} : H^1(\mathcal{U}, \mathcal{S}) \longrightarrow H^1(\mathcal{V}, \mathcal{S}) \tag{11.25.1}$$

*is one-to-one.*

Therefore, according to the general theory of *"inductive limits"* (cf., for instance, N. Bourbaki [1: p.92, Remarque 1)]), one concludes that:

(11.26) *the set (11.24) may be identified with the (disjoint) union of the individual sets (11.16); viz. one has*

(11.26.1) $$H^1(X,\mathcal{S}) = \bigcup_{\mathcal{U}} H^1(\mathcal{U},\mathcal{S}) = \sum_{\mathcal{U}} H^1(\mathcal{U},\mathcal{S}).$$

In particular, in the case that $\mathcal{S}$, as above, is a *sheaf of abelian groups* on $X$, then (11.26.1) coincides with the *1st Čech cohomology (abelian) group* of $X$ *with coefficients in* $\mathcal{S}$ (see Remark 11.1 and (6.12), for $n = 1$).

On the other hand, by looking at the *neutral element* $0_{\mathcal{U}} \equiv 0 \in H^1(\mathcal{U},\mathcal{S})$ (see (11.17)), one further remarks that;

(11.27) for every *pair* $(\mathcal{U},\mathcal{V})$ *of open coverings of* $X$, *with* $\mathcal{V} \succeq \mathcal{U}$, *the map (11.22) preserves the neutral elements of the sets concerned.*

The assertion follows straightforwardly by the very definitions (so see e.g. (4.34′) and (4.56)). ∎ Thus, by the same definition of the set (11.24) or, *equivalently,* (11.26.1) (see also, for instance, N. Bourbaki [1: p.89, (22)]), one infers that:

(11.28) *the 1st cohomology set* $H^1(X,\mathcal{S})$, as above, *is endowed with a distinguished element* too, viz. the image in it (see (11.24)) of the neutral element $0_{\mathcal{U}}$, as above. We call this particular element of the latter set still the *neutral element of* $H^1(X,\mathcal{S})$, denoting it again just by 0. (The notation 1 is also of use, in the case the groups involved are *multiplicative*).

Of course, if $\mathcal{S}$ is an *abelian group sheaf* on $X$, the preceding element coincides with the *usual neutral element* of the *abelian group* $\check{H}^1(X,\mathcal{S})$ (cf. also Remark 11.1).

Now, suppose that we are given a *morphism of group sheaves* over $X$, say

(11.29) $$\phi : \mathcal{S} \longrightarrow \mathcal{S}'$$

(see also Chapt. II; (2.1), (2.2), as well as, Remark 2.1). Thus, looking at (11.29) as a *morphism of* the respective (complete) *presheaves of sections* of the given sheaves (cf. Chapt. I; Theorem 13.1, in particular, (13.7)), one gets, based on the very definitions, a map

(11.30) $$\phi_{\mathcal{U}} : C^1(\mathcal{U},\mathcal{S}) \longrightarrow C^1(\mathcal{U},\mathcal{S}'),$$

for any given *open covering* $\mathcal{U}$ of $X$ (cf. e.g. (11.1), along with (11.3) and Chapt. I; (9.22)). So it is easy to see that *the latter map commutes with* the respective *coboundary operators* (cf. e.g. the first member of (11.9)), so that *it preserves cocycles;* yet, it is still readily verified that *the same map respects (11.14),* while *it also commutes with the* particular *maps* $\tau_{\mathcal{V},\mathcal{U}}$, as in (11.22). Therefore (see also, for instance, N. Bourbaki [1: p.92, Corollaire 1]), one obtains a *map between* the respective *1st cohomology sets*

$$\phi^* \equiv \varinjlim_{\mathcal{U}} \phi_{\mathcal{U}} : H^1(X,\mathcal{S}) \longrightarrow H^1(X,\mathcal{S}'). \tag{11.31}$$

Finally, it is still easily seen (viz. by a direct check of the commutativity of the pertinent maps) that *the map (11.31) preserves the* respective *neutral elements* of the sets concerned.

We close the present Section with some further relevant remarks to the preceding, referred, however, to dimension 0. So we set them out in the following Subsection.

**11.(a). 0-dimensional (non-abelian) cohomology.**- The discrepancies we met before in (Čech) cohomology of dimension one, stemming, as we saw, from the *non-abelianess of the coefficient group,* are no more present in *dimension zero:* Thus, according to the general definition (see e.g. (4.21)), a 0-*cochain*

$$f \equiv (f_\alpha) \in C^0(\mathcal{U},\mathcal{S}), \tag{11.32}$$

is said to be a 0-*cocycle,* whenever one has (we also set below $\delta^0 \equiv \delta$);

$$\delta(f) \equiv \delta((f_\alpha)) = (\delta(f_\alpha)) = (f_\beta - f_\alpha) \equiv (f_{\alpha\beta}) = 0, \tag{11.33}$$

for any $\alpha, \beta \in I$, with $U_{\alpha\beta} \equiv U_\alpha \cap U_\beta \neq \emptyset$ (see also (4.13)). That is, one obtains

$$f_\alpha = f_\beta|_{U_\alpha \cap U_\beta}, \tag{11.34}$$

for any $\alpha, \beta \in I$, as before, so that, in view of (11.12) and the hypothesis for $\mathcal{S}$ (see Chapt. I, Lemma 11.1), there exists a *global section* $f \in \mathcal{S}(X) \equiv \Gamma(X,\mathcal{S})$ *extending the given* $f_\alpha$*'s* (cf. also Chapt. I, Definition 11.1).

Now, denoting as usual by $Z^0(\mathcal{U},\mathcal{S})$ the *set of* 0-*cocycles of* $\mathcal{U}$ *with coefficients in* $\mathcal{S}$, and taking too (4.27) into account, one obtains

$$H^0(\mathcal{U},\mathcal{S}) = Z^0(\mathcal{U},\mathcal{S}), \tag{11.35}$$

the latter set being thus the 0-*dimensional cohomology set of* $\mathcal{U}$ *with coefficients in* $\mathcal{S}$.

Furthermore, it is easy to see (cf. Chapt. I; cond. (S1) in Definition 11.1) that *the correspondence*

$$H^0(\mathcal{U},\mathcal{S}) = Z^0(\mathcal{U},\mathcal{S}) \longrightarrow \Gamma(X,\mathcal{S}), \tag{11.36}$$

as given by our previous conclusion, connected with (11.34), *is* actually *a bijection,* its range being obviously a *group,* due to the hypothesis for $\mathcal{S}$ (it actually carries always the algebraic structure that $\mathcal{S}$ does; so see e.g. Chapt. II; (1.67)). Thus, one obtains (see also, for instance, N. Bourbaki [1: p.90, Example 2)]);

$$H^0(X,\mathcal{S}) := \varinjlim_{\mathcal{U}} H^0(\mathcal{U},\mathcal{S}) = \Gamma(X,\mathcal{S}), \tag{11.37}$$

where the second relation is valid up to a *bijection.* Therefore, one infers that:

(11.38) *the set* $H^0(X,\mathcal{S})$ *has a (natural) group structure* (that is, that one defined on it by $\Gamma(X,\mathcal{S})$, through the bijection (11.37)). We call the latter the 0-*dimensional cohomology group of* $X$ *with coefficients in* $\mathcal{S}$ (with $\mathcal{S}$ the given sheaf of groups on $X$).

The preceding extend, in effect, within the present framework (of considering, namely, *not necessarily abelian group sheaves*) the previous Lemma 4.1. On the other hand, applications of the above and, in particular, of the set (11.24), will be considered right in the next two chapters.

# CHAPTER IV

# Linear and Multilinear Algebra of Vector Sheaves

" ...sheaves are very useful in proving theorems ... the proofs given by the theory of sheaves are generally much simpler and clearer... and hold under much more general conditions..."

R. G. Swan in "The Theory of Sheaves" (The University of Chicago Press, 1964). p.1f.

The purpose of this Chapter is to present, within the general framework of the theory of $\mathcal{A}$-*modules* (see e.g. Chap. II), several fundamental notions and related results of the classical theory of *Linear and Multilinear Algebra,* to the extent that this will be necessary for the ensuing discussion (see, for instance, Chapt. VI in the sequel on the *Geometry of Vector Sheaves*). So we are actually concerned hereby with a sheaf-theoretic point of view of the standard *Theory of Modules* so far as this will be needed in the succeeding parts of this work. Now, as it concerns the *domain of coefficients* of the sheaves of modules involved, we make, as we actually did hitherto, the following.

(0.1) **Convention.**- In all that follows, and if no indication is made to the contrary, $\mathcal{A}$ will stand for a *sheaf of* **C**-*algebras, commutative* and *unital* on a given *topological space* $X$ (cf. also Chapt. II; (1.36), (1.34)). So applying the terminology adopted so far (thus, cf. Chapt. II; Scholium 1.2), we assume henceforth, even if it is not explicitly stated, that we are given a fixed **C**-*algebraized space*

(0.1.1) $$(X, \mathcal{A}),$$

with $\mathcal{A}$ as before and $X$ an arbitrary *topological space*. Yet, the base space $X$ will eventually be further specialized, as the case may be, but this will be explicitly indicated.

# 1. Classical group sheaves

We start with the discussion of certain basic "classical groups", as the latter are formulated in our case, that will be of a particular significance for the sequel. However, we first need the following preliminary material.

**1.(a). A (full) matrix algebra sheaf.**- So, as already noted (cf. (0.1) above), assume that we are given a **C**-*algebraized space* $(X, \mathcal{A})$.

Now, for any *natural number* $n \in \mathbf{N}$, consider the following correspondence

(1.1) $$U \longmapsto M_n(\mathcal{A}(U)),$$

where $U$ is varying over the *topology of* $X$; so the values of the previous map at a given *open* set $U \subseteq X$ is the *(full) matrix algebra* $M_n(\mathcal{A}(U))$, consisting of all $n \times n$ matrices with entries from the **C**-*algebra* (*unital* and *commutative,* cf. (0.1))

(1.2) $$\mathcal{A}(U) \equiv \Gamma(U, \mathcal{A})$$

(see also Chapt. I; (2.3)). In this context, one can now easily verify that;

(1.3) the above *correspondence* (1.1) yields a *presheaf of* **C**-*algebras* on $X$ (cf. Chapt. II; §1.(b)), in fact, a *presheaf of full matrix* **C**-*algebras* on $X$. Yet, it is readily seen (cf. e.g. Chapt. I; Definition 11.1) that the latter is a *complete presheaf* on $X$.

On the other hand, if

$$\Gamma(\mathcal{A}) \equiv (\mathcal{A}(U), \rho_V^U) \tag{1.4}$$

denotes, as usual, the *(complete) presheaf of sections* of $\mathcal{A}$ (see Chapt. I; (9.6), along with Lemma 11.1), then the presheaf considered by (1.3) is given by

$$(M_n(\mathcal{A}(U)), M_n(\rho_V^U)); \tag{1.5}$$

in this context, the **C**-*algebra morphism*

$$M_n(\rho_V^U) : M_n(\mathcal{A}(U)) \longrightarrow M_n(\mathcal{A}(V)) \tag{1.6}$$

where $U, V$ are *open* subsets of $X$, *with* $V \subseteq U$ (in effect, a *restriction morphism* of the presheaf concerned), is given by

$$M_n(\rho_V^U) := \rho_V^U \underbrace{\times \cdots \times}_{n^2-\text{times}} \rho_V^U. \tag{1.6'}$$

That is, one sets

$$M_n(\rho_V^U)(\alpha_{ij}^U) := (\rho_V^U(\alpha_{ij}^U) \equiv \alpha_{ij}^U|_V) \in M_n(\mathcal{A}(V)) \tag{1.7}$$

for any *section-matrix*

$$(\alpha_{ij}^U) \in M_n(\mathcal{A}(U)), \tag{1.7'}$$

*such that* $\alpha_{ij}^U \in \mathcal{A}(U) \equiv \Gamma(U, \mathcal{A})$, *with* $1 \leqq i, j \leqq n$ (see also Chapt. I; (5.1) and (9.5)).

In this context, we thus denote by

$$M_n(\mathcal{A}) \tag{1.8}$$

the *sheaf of* **C**-*algebras* on $X$, generated by the (complete) presheaf (1.5), and call it the *sheaf of full matrix algebras* on $X$, of *order* $n$, associated with (the given structure sheaf) $\mathcal{A}$.

So, with any given $n \in \mathbf{N}$, one obtains a *sheaf of unital* **C**-*algebras* on $X$, which are, in general, *non-commutative, unless* $n = 1$, according to our hypothesis for $\mathcal{A}$ (see (0.1)). Yet, the same sheaf (1.8), as above, will play a particularly useful role throughout the ensuing discussion.

On the other hand, based on the completeness of the presheaf (1.5) (cf. thus (1.3)), one still obtains

$$M_n(\mathcal{A})(U) \equiv \Gamma(U, M_n(\mathcal{A})) = M_n(\mathcal{A}(U)), \tag{1.9}$$

for any *open* set $U \subseteq X$, the last equality in (1.9) being valid, of course, within an *isomorphism of* **C**-*algebras* (see Chapt. I; Proposition 11.1). Thus, in view of (1.9), we still have for the *(complete) presheaf of sections of* $M_n(\mathcal{A})$ the relation (cf. also e.g. (1.4));

$$\Gamma(M_n(\mathcal{A})) = (M_n(\mathcal{A}(U)), M_n(\rho^U_V)), \tag{1.10}$$

which is thus to be compared with (1.5). Yet, due to the last relation and (1.4), we also write, for convenience,

$$\Gamma(M_n(\mathcal{A})) = M_n(\Gamma(\mathcal{A})), \tag{1.11}$$

in place of the *presheaf isomorphism* (1.9) (see also Chapt. I; (6.3), (13.3) and Chapt. II; (1.54)). Indeed, one has here a *functorial relation,* viz. one gets

$$\Gamma \circ M_n = M_n \circ \Gamma, \quad n \in \mathbf{N}, \tag{1.12}$$

*commutativity of* the *functors* concerned, where we also employed the *isomorphism of categories* supplied by the functor $\Gamma$ (see thus e.g. Chapt. I; (13.18). In this regard, the *(square matrix) functor* $M_n$, $n \in \mathbf{N}$, is defined in the obvious manner).

We come next to the following useful result, an application of which will be considered right in the next Subsection. So we now have.

**Lemma 1.1.** Let $\mathcal{A}$ be a *unital* **C**-*algebra sheaf* on a *topological space* $X$, and let

$$\mathcal{A}^{\cdot} \tag{1.13}$$

be the *sheaf on* $X$ *generated by the presheaf*

$$U \longmapsto \mathcal{A}(U)^{\cdot} \tag{1.14}$$

where $U$ varies over the topology of $X$ (the range of (1.14) being the *group of units,* viz. invertible elements, of the unital **C**-algebra $\mathcal{A}(U)$). Then, the correspondence (1.14) *yields a complete presheaf (of groups) on* $X$, such that one obtains

$$(1.15)\qquad \mathcal{A}^{\bullet}(U) = \mathcal{A}(U)^{\bullet},$$

up to a *group isomorphism,* for every open set $U \subseteq X$.

**Proof.** First, it is easy to see, according to the very definitions, that (1.14) yields a *presheaf of groups* on $X$ (cf. Chapt. II; (1.45)), the corresponding *restriction morphisms* being the restriction to $\mathcal{A}(U)^{\bullet} \subseteq \mathcal{A}(U)$, with $U$ *open* in $X$, of the respective maps of the (complete) presheaf (1.4), as above (see also e.g. (1.16) below, along with Chapt. I; (9.5)). Thus, we just prove that (1.14) defines, in fact, a *complete presheaf (of groups) on* $X$:

Indeed, let $U$ be an *open* subset of $X$ and $\mathcal{U} = \{U_\alpha\}_{\alpha \in I}$ an *open covering of* $\mathcal{U}$; moreover, let $s, t \in \mathcal{A}(U)^{\bullet} \subseteq \mathcal{A}(U)$ such that

$$(1.16)\qquad \rho^U_{U_\alpha}(s) \equiv s|_{U_\alpha} \equiv s_\alpha = t_\alpha \; (\equiv t|_{U_\alpha} \equiv \rho^U_{U_\alpha}(t)), \quad \alpha \in I$$

(see also e.g. Chapt. I; (1.14)). Accordingly (Chapt. I; Proposition 11.1), one gets $s = t$, given that $s, t$ are, by hypothesis, (local) sections over the open set $U \subseteq X$ of the sheaf $\mathcal{A}$, hence elements of the complete presheaf $\Gamma(\mathcal{A})$ (cf. (1.4), as well as, Chapt. I; Lemma 11.1). So the presheaf on $X$ given by (1.14) *satisfies cond. (S1)* in Chapt. I; Definition 11.1.

On the other hand, consider the element

$$(1.17)\qquad (s_\alpha) \in \prod_{\alpha \in I} \mathcal{A}(U_\alpha)^{\bullet} \subseteq \prod_{\alpha} \mathcal{A}(U_\alpha),$$

in such a manner that

$$(1.18)\qquad s_\alpha = s_\beta|_{U_\alpha \cap U_\beta},$$

for any $\alpha, \beta$ in $I$, with $U_\alpha \cap U_\beta \neq \emptyset$. Hence, by hypothesis for $(s_\alpha)$ (the same Chapt. I; Lemma 11.1 and Definition 11.1), there exists an element $s \in \mathcal{A}(U)$, such that one has

$$(1.19)\qquad s|_{U_\alpha} = s_\alpha, \quad \alpha \in I.$$

Yet, by virtue of (1.17), there exists an element $(t_\alpha) \in \prod_\alpha \mathcal{A}(U_\alpha)^{\bullet}$ such that (see also (1.18))

$$(1.20)\qquad s_\alpha \cdot t_\alpha = 1_\alpha = 1_\beta = s_\beta \cdot t_\beta = s_\alpha \cdot t_\beta,$$

for any $\alpha, \beta \in I$, with $U_\alpha \cap U_\beta \neq \emptyset$, that is, one obtains

$$t_\alpha = t_\beta|_{U_\alpha \cap U_\beta}, \tag{1.21}$$

with $\alpha, \beta$ as before. Therefore, by virtue of a similar argument as for (1.18), one concludes that there exists an element $t \in \mathcal{A}(U)$, extending the $t_\alpha$'s. So one has (cf. also e.g. (1.16), Chapt. II; (1.46), and (1.20) above);

$$(s \cdot t)_\alpha \equiv \rho^U_{U_\alpha}(s \cdot t) = \rho^U_{U_\alpha}(s) \cdot \rho^U_{U_\alpha}(t) \equiv s_\alpha \cdot t_\alpha = 1_\alpha, \quad \alpha \in I. \tag{1.22}$$

Hence (*cond.* (S1), as in the first part of the proof), one infers that $s \cdot t = 1 \in \mathcal{A}(U)$, so that $s \in \mathcal{A}(U)^\cdot$, satisfying already, by our hypothesis, (1.19). Thus, the same presheaf on $X$, as defined by (1.14), *fulfils cond. (S2), as well,* and the proof is complete. ∎

**Remark 1.1.**– The previous proof holds certainly true, for any *sheaf of rings with an identity element* on $X$ (see also Chapt. II; Definition 1.3′ for the terminology applied).

**Scholium 1.1.**– Before we proceed further, we also consider, by analogy with the standard situation, the case of *non-square matrices* that will be of use too, presently below: Thus, suppose we have a **C**-*algebraized space* $(X, \mathcal{A})$, as before, and let $(n, m)$ be a given pair of *natural numbers;* so we further look at the following correspondence,

$$U \longmapsto M_{n,m}(\mathcal{A}(U)), \tag{1.23}$$

where $U$ ranges over the *open* subsets of $X$, while the range of (1.23) is the set of all $n \times m$ *matrices with entries from* $\mathcal{A}(U)$, a unital commutative **C**-algebra, according to the hypothesis for $\mathcal{A}$ (cf. (0.1)). Now, it is easy to verify, by the same definitions (an analogous argument, in fact, to that one for (1.1)), that (1.23) is a *presheaf on* $X$, in effect, by what has been said, a *presheaf of* **C**-*vector spaces* on $X$; yet (the same analogy, as before, with (1.1)), it is still easily seen that the latter is a *complete* one.

So we denote by

$$M_{n,m}(\mathcal{A}) \tag{1.24}$$

the *sheaf of* **C**-*vector spaces* on $X$, generated by the previous presheaf (cf. also Chapt. II; (1.54)). Therefore (Chapt. I; Proposition 11.1), one obtains

$$M_{n,m}(\mathcal{A})(U) = M_{n,m}(\mathcal{A}(U)), \tag{1.25}$$

within an *isomorphism of* **C**-*vector spaces*, for every *open* set $U \subseteq X$.

In this context, we also note that, by analogy with the classical case, such sheaves (of **C**-vector spaces) on $X$ appear when dealing with *morphisms of free* $\mathcal{A}$-*modules of finite rank* (see thus Section 3 in the sequel).

**1.(b). The general linear group sheaf.**– We apply below our previous considerations on (1.13) to the particular case of the *algebra sheaf* (1.8). Thus, one defines;

$$\mathcal{GL}(n, \mathcal{A}) := M_n(\mathcal{A})^{\cdot}. \tag{1.26}$$

Therefore, given that (1.8) is, by the preceding, a unital **C**-algebra sheaf on $X$, one infers, according to Lemma 1.1, that:

(1.27) given the **C**-algebraized space $(X, \mathcal{A})$, $\mathcal{GL}(n, \mathcal{A})$, as defined by (1.26), is a *sheaf of groups on* $X$, *not necessarily abelian, unless* $n = 1$, in view of our hypothesis for $\mathcal{A}$ (cf. (0.1)). We call it the *general linear group sheaf* on $X$, of *order* $n \in \mathbf{N}$, associated with $(X, \mathcal{A})$.

The previously applied terminology can be justified by the following relations (cf. also (1.15) and (1.9)), as well,

$$\begin{aligned} \mathcal{GL}(n, \mathcal{A})(U) &:= M_n(\mathcal{A})^{\cdot}(U) = (M_n(\mathcal{A})(U))^{\cdot} \\ &= M_n(\mathcal{A}(U))^{\cdot} \equiv \mathrm{GL}(n, \mathcal{A}(U)), \end{aligned} \tag{1.28}$$

valid within *isomorphisms of* the *groups* involved, *for any open* $U \subseteq X$.

On the other hand, based on (1.28), one concludes that;

(1.29) *the group sheaf* $\mathcal{GL}(n, \mathcal{A})$, as given by (1.26), *is* (isomorphic to the group sheaf on $X$) *generated by the presheaf*

$$U \longmapsto \mathrm{GL}(n, \mathcal{A})(U)), \tag{1.29.1}$$

with $U$ varying over the *open* subsets of $X$. Moreover, the latter is a *complete presheaf* on $X$ too.

Indeed, it is readily seen that (1.29.1) *defines a presheaf of groups on* $X$ (*abelian*, if $n = 1$, cf. (0.1)), which is, moreover, *complete;* one can give here, directly, an analogous proof to that one of Lemma 1.1. Thus, the assertion in (1.29) is further justified by (1.28) and Chapt. I; (13.19). ∎

Yet, we also look for later use at the *constant general linear group sheaf* on $X$, of given *order* $n \in \mathbf{N}$; viz., by definition, the sheaf on $X$, *generated by the constant presheaf*

$$U \longmapsto \mathrm{GL}(n, \mathbf{C})(U) := \mathrm{GL}(n, \mathbf{C}), \tag{1.30}$$

with $U$ *open* in $X$ (see also Chapt. I; (10.2), (10.8), along with (11.15) and Theorem 11.1). As usual, we denote the above *constant sheaf* by

$$\mathrm{GL}(n, \mathbf{C})_X \equiv \mathrm{GL}(n, \mathbf{C}) \underset{\varepsilon}{\overset{\subset}{\rightarrow}} \mathcal{GL}(n, \mathcal{A}); \tag{1.31}$$

here the last inclusion, being, in effect, a *group sheaf isomorphism into*, is supplied by the *canonical injection* $\mathbf{C} \underset{\varepsilon_X}{\overset{\subset}{\rightarrow}} \mathcal{A}$ (see Chapt. II; (1.34)), so that one obtains, in particular,

$$\mathrm{GL}(n, \mathbf{C}) \overset{\subset}{\rightarrow} \mathrm{GL}(n, \mathcal{A}(U)) \cong \mathcal{GL}(n, \mathcal{A})(U), \tag{1.32}$$

for any *open* $U \subseteq X$ (see also (1.30), (1.28), as well as, (0.1)).

Now, apart from the above sheaf of groups (1.26), some other important classical groups as, for instance, the *orthogonal* and *unitary groups*, will also be considered in the sequel, within the present *sheaf-theoretic* framework; the latter sheaves of groups are going to be applied later on, as well. However, to define them, we shall need first, by still mimicking the classical counterpart, some more terminology, concerning the *structure sheaf* $\mathcal{A}$ of the $\mathbf{C}$-algebraized space $(X, \mathcal{A})$, under consideration. So one has to employ, for instance, the notion of a *sheaf-metric,* or else $\mathcal{A}$-*metric* (cf. Definition 8.3 below), on a given $\mathcal{A}$-module $\mathcal{E}$ on $X$; but, for this and other relevant material too, we have to refer to the pertinent places in the sequel. Thus, see, for example, Sections 8, 9 below.

## 2. An example

We consider below an example of a *general linear group sheaf,* as before, which is associated with a particular *algebra sheaf* $\mathcal{A}$ and will be of special importance later on (cf. Chapt. XI in the sequel). Thus, we also refer to that Chapter for more details on the terminology applied hereby, as well as, to A. Mallios [2] for the relevant material that we use below from *topological algebra theory* (the same example stems actually from that theory too).

So assume that we are given a *unital commutative topological* **C**-*algebra* **A**, as well as, an arbitrary *topological space* $X$. Yet, for any open set $U \subseteq X$, let

$$\mathcal{C}_c(U, \mathbf{A}) \tag{2.1}$$

be the **C**-*algebra* (unital and commutative too) *of* **A**-*valued continuous functions on* $U$, equipped with the *compact-open topology* in $U$ (denoted by "$c$"). So *(2.1) is a topological algebra,* as well, of the same sort as **A** and with algebraic operations defined pointwise (cf. also A. Mallios [1: Chapt. XI, Section 1]). Furthermore, consider the map

$$U \longmapsto \mathcal{C}_c(U, \mathbf{A}), \tag{2.2}$$

where $U$ ranges over the *open* subsets of $X$. Thus, it is now easy to see that;

(2.3) when $U$ is ranging, as above, and we also take the obvious restriction maps into account (cf. e.g. Chapt. I; (10.11), (10.12)), then the above correspondence (2.2) yields a *complete presheaf of* **C**-*algebras* on $X$ (see also Chapt. I; (11.10)). Indeed, the same provides actually a (complete) *presheaf of topological algebras* on $X$, in the same sense that (2.1) is a topological algebra, for every open $U \subseteq X$, the respective restriction maps being also continuous (**C**-algebra morphisms; yet, see Chapt. XI; (1.1)

Now, let us denote by

$$\mathcal{A} \equiv \mathcal{C}_X^{\mathbf{A}} \tag{2.4}$$

the *sheaf on* $X$, generated by (2.2). Thus, (2.4) is a *sheaf of topological algebras* (or else a *topological algebra sheaf*) on $X$, in the sense of Chapter XI; Subsection 1.(b) (see Vol. II). Yet, the same sheaf is (within an *isomorphism of sheaves*) the *sheaf of germs of* **A**-*valued continuous functions on* $X$, that also explains the notation in (2.4); in this regard, see also Chapt. I; Subsection 4.(b), in particular, (4.25). On the other hand, due to our hypothesis for the algebra **A**, the sheaf $\mathcal{A}$, as given by (2.4), is a unital commutative **C**-algebra sheaf on $X$ (cf. also e.g. Chapt. II; (1.54)), so that our argument in the previous Section 1 applies. Thus, based on (2.3) and Chapt. I; Proposition 11.1, we obtain

$$\mathcal{C}_X^{\mathbf{A}}(U) \equiv \mathcal{A}(U) = \mathcal{C}_c(U, \mathbf{A}), \tag{2.5}$$

up to a **C**-*algebra isomorphism,* for any *open* set $U \subseteq X$. Hence, by further considering the *full matrix algebra sheaf* on $X$, of *order* $n \in \mathbf{N}$, $M_n(\mathcal{A})$, associated with $\mathcal{A}$ (see (1.8)), one obtains, *for every open* $U \subseteq X$ (see also (1.3) and (2.5))

$$M_n(\mathcal{A})(U) = M_n(\mathcal{A}(U)) = M_n(\mathcal{C}_c(U, \mathbf{A})) = \mathcal{C}_c(U, M_n(\mathbf{A})). \tag{2.6}$$

The last equality in (2.6) is valid, of course, up to an *isomorphism of* **C**-*algebras, even of topological ones;* namely, one can consider the full matrix algebra $M_n(\mathbf{A})$ as a *topological algebra,* through the relation (in fact, **A**-*module isomorphism*)

$$M_n(\mathbf{A}) = \mathbf{A}^{n^2}. \tag{2.7}$$

Thus, the cartesian product topology in the second member of (2.7) makes the **C**-algebra $M_n(\mathbf{A})$ a *topological algebra of the same type as* **A**. (In this regard, cf. also A. Mallios [2: p.82], [1: p.458, (1.19)]).

Now, based on (1.28), (2.5) and (2.6), we further conclude that;

$$\mathcal{GL}(n, \mathcal{A})(U) = \mathrm{GL}(n, \mathcal{A}(U)) = \mathrm{GL}(n, \mathcal{C}_c(U, \mathbf{A})) = \mathcal{C}_c(U, \mathrm{GL}(n, \mathbf{A})), \tag{2.8}$$

up to *isomorphisms of* the *groups* concerned. Of course, one considers here $\mathrm{GL}(n, \mathbf{A})$ *with the relative topology from* $M_n(\mathbf{A})$ (cf. (2.7), as well as, the subsequent Scholium 2.1). Consequently, one infers that (see also (1.29));

(2.9) *the sheaf of groups* $\mathcal{GL}(n, \mathcal{A})$ *on* $X$, *with* $\mathcal{A} \equiv \mathcal{C}_X^{\mathbf{A}}$ (cf. (2.4)) *is* (up to a *group sheaf isomorphism,* that sheaf of groups on $X$, which is) *generated by the complete presheaf of groups on* $X$, given by the correspondence;

$$U \longmapsto \mathcal{C}_c(U, \mathrm{GL}(n, \mathbf{A})), \tag{2.9.1}$$

with $U$ varying over the *open* subsets of $X$.

Therefore, by analogy with (2.4), one further concludes that the (group) sheaf on $X$, generated by the presheaf (of groups) on $X$, given by (2.9.1), is, in effect (viz. up to a (group) sheaf isomorphism), the *sheaf of germs of* $\mathrm{GL}(n, \mathbf{A})$-*valued continuous functions on* $X$. So, denoting the latter sheaf (of groups) on $X$ by

$$\mathcal{C}_X^{\mathbf{GL}(n,\mathbf{A})}, \tag{2.10}$$

one finally obtains

$$\mathcal{GL}(n, \mathcal{C}_X^{\mathbf{A}}) = \mathcal{C}_X^{\mathbf{GL}(n,\mathbf{A})}, \tag{2.11}$$

up to a *group sheaf isomorphism,* for every (unital commutative) *topological algebra* $\mathbf{A}$, and any $n \in \mathbf{N}$.

**Scholium 2.1.**– The group $\mathrm{GL}(n, \mathbf{A}) \subseteq M_n(\mathbf{A})$, as above, endowed with the relative topology from $M_n(\mathbf{A})$ is certainly a *topological semigroup* (in the sense that the group multiplication is only separately continuous), due to the *separate continuity of the multiplication in* $\mathbf{A}$, in general (cf. also (2.7)). Now, suppose, in particular, that $\mathbf{A}$ is a *topological algebra with continuous multiplication and continuous inversion* (see A. Mallios [2: Chapt. II; §3]); then $\mathrm{GL}(n, \mathbf{A})$ topologized, as before, becomes a *topological group.* (Indeed, the assertion follows, for instance, from a mere application of the standard *Laplace Decomposition Theorem,* concerning the inverse of a given matrix in $\mathrm{GL}(n, \mathbf{A})$; cf. e.g. S. Lang [1: p.256, Proposition 4.16]). So, for convenience, we may assume, for example, that $\mathbf{A}$ is a (unital commutative) *locally convex* (topological ) *algebra with continuous multiplication and continuous inversion* (see A. Mallios [2]); thus, the following relation (cf. also (2.8))

$$\mathcal{C}_c(U, \mathrm{GL}(n, \mathbf{A})) = \mathrm{GL}(n, \mathcal{C}_c(U, \mathbf{A})), \tag{2.12}$$

with $U$ *open* in $X$, becomes an *isomorphism of topological groups.*

The above conditions on the topological algebra involved, occur, for instance, by restricting ourselves to a (unital commutative) *locally m-convex* (topological ) *algebra* $\mathbf{A}$; indeed, this concerns an important generalization of the standard notion of a *Banach algebra.* (cf. A Mallios [2: p.52, Lemma 3.1], as well as, Chapter XI in the sequel).

# 3. Free $\mathcal{A}$-modules (contn'd)

We have already considered in the preceding the notion of a *free $\mathcal{A}$-module* (of *finite rank,* say, $n \in \mathbf{N}$) $\mathcal{E}$ on $X$, where $(X, \mathcal{A})$ is a given $\mathbf{C}$-*algebraized space* (cf. Chapt. II; Definition 3.2, and also (0.1) above). Thus, the ensuing discussion supplements our previous considerations in Chapt. II; Section 3. Yet, for convenience of reference, we first recall certain basic facts from that account, that will also be of use below.

So consider, in particular,

$$\mathcal{A}^n, \tag{3.1}$$

viz. the standard *free $\mathcal{A}$-module of rank n on X* (prototype, by definition, of all the rest ones of the same rank), associated with the given **C**-algebraized space $(X, \mathcal{A})$. As already known, we get *n-many (canonical) global sections of $\mathcal{A}^n$*, i.e. (cf. also (3.3)),

$$\varepsilon_i := (\delta_{ij})_{1 \leqq j \leqq n} \in \mathcal{A}^n(X) = \mathcal{A}(X)^n, \tag{3.2}$$

with $\delta_{ij} \in \mathcal{A}(X)$, $1 \leqq i, j \leqq n$, the (canonical) *Kronecker sections of $\mathcal{A}$* over $X$ (cf. Chapt. II; (3.22)), while we still called the $\varepsilon_i$'s, as in (3.2), the (canonical) *Kronecker sections of $\mathcal{A}^n$*. We also speak of the *standard* (or *canonical*) *coordinate sections of $\mathcal{A}^n$*, while the above finite sequence

$$(\varepsilon_i)_{1 \leqq i \leqq n} \subseteq \mathcal{A}^n(X) = \mathcal{A}(X)^n \tag{3.3}$$

is called the *standard* (or *canonical*) *section frame* (or simply *frame* or even *basis*) *of $\mathcal{A}^n$*; yet, we still refer to (3.3) as the *standard* (or *canonical*) *gauge* of $\mathcal{A}^n$. Finally, the term *Kronecker gauge* of $\mathcal{A}^n$ will be also of use.

The above terminology is justified, of course, by the following standard fact, pertaining to the sections of $\mathcal{A}^n$, i.e., one obtains that

(3.4) *for any local section $\alpha \in \Gamma(U, \mathcal{A}^n)$*, with $U$ any *open* set in $X$, *one has* (see (3.3))

$$\alpha = \sum_{i=1}^{n} \alpha_i \cdot \varepsilon_i, \tag{3.4.1}$$

where *the sections $\alpha_i \in \Gamma(U, \mathcal{A}) \equiv \mathcal{A}(U)$ are uniquely determined by $\alpha$ (and determine the latter uniquely,* as well).

Indeed, the assertion is straightforward by (3.2) and Chapt. II; (3.7) and (3.14). In particular, the sections $\alpha_i \in \Gamma(U, \mathcal{A})$ are obtained by applying the obvious *projection $\mathcal{A}$-morphisms*

$$\mathrm{pr}_i : \mathcal{A}^n := \underbrace{\mathcal{A} \oplus \cdots \oplus \mathcal{A}}_{n-\text{times}} \longrightarrow \mathcal{A}, \quad 1 \leqq i \leqq n, \tag{3.5}$$

such that, concerning (3.4.1), one has

(3.6) $$\alpha_i := \mathrm{pr}_i \circ \alpha, \quad 1 \leqq i \leqq n. \blacksquare$$

On the other hand, suppose now we are given an arbitrary *free $\mathcal{A}$-module of rank $n$ on $X$*, say, $\mathcal{E}$, such that one has, by definition, the following *exact $\mathcal{A}$-sequence*

(3.7) $$0 \longrightarrow \mathcal{A}^n \xrightarrow{\phi} \mathcal{E} \longrightarrow 0.$$

Therefore, by applying to (3.7) the *section functor* $\Gamma$ (see Chapt. I; (13.1)), one has

(3.8) $$\Gamma(\mathcal{A}^n) \underset{\overline{\phi}=\phi}{=} \Gamma(\mathcal{E}),$$

within an *isomorphism of* the *(complete) presheaves* concerned (cf. also Chapt. I; (13.3) and (13.9)). Consequently (cf. also Chapt. I; (6.3), (6.10) and Chapt. II; (3.14)), *for every (local) section* $s \in \Gamma(U, \mathcal{E})$, with $U$ any open set in $X$, *there exists a (uniquely determined)* $\alpha \in \Gamma(U, \mathcal{A}^n)$ *such that* $s = \overline{\phi}(\alpha) \equiv \phi(\alpha)$. Therefore (see also (3.4.1) above), one obtains

(3.9) $$\begin{aligned} s = \phi(\alpha) = \phi\left(\sum_{i=1}^{n} \alpha_i \varepsilon_i\right) &= \phi_U\left(\sum_i \alpha_i \varepsilon_i\right) = \sum_i \phi_U(\alpha_i \varepsilon_i) \\ &= \sum_i \alpha_i \cdot \phi_U(\varepsilon_i) = \sum_{i=1}^{n} \alpha_i \cdot \phi(\varepsilon_i), \end{aligned}$$

that is, the rel. (3.25.1) in Chapt. II, if one sets

(3.10) $$e_i \equiv \phi(\varepsilon_i) \in \Gamma(U, \mathcal{E}), \quad 1 \leqq i \leqq n.$$

Thus, in other words, one infers that:

(3.11) *for any free $\mathcal{A}$-module $\mathcal{E}$ of rank $n$ on $X$, one obtains a canonical basis of (local) sections* (or else a *canonical gauge*) *of $\mathcal{E}$, depending on the particular identification* (viz. *$\mathcal{A}$-isomorphism*) $\phi$, as in (3.7); namely, *one has*

(3.11.1) $$e_i \equiv \phi(\varepsilon_i), \quad 1 \leqq i \leqq n.$$

*such that (3.9) holds true,* while the $\varepsilon_i$'s and $\phi(\equiv \overline{\phi})$ are given by (3.3) and (3.7), respectively. Thus, the map $\phi$ *in (3.10) is the same, for every open* $U \in X$; we say that *the gauge* in question, *viz. the (finite) sequence* $(e_i)$, *is global,* its restriction to an open $U \subseteq X$ entailing thus (cf. (3.9)) a basis (of sections) for $\mathcal{E}(U) \equiv \Gamma(U, \mathcal{E})$, the latter being a *free* $\mathcal{A}(U)$*-module of rank* $n$ (cf. Chapt. II; (3.19)).

In this context, a useful relevant matter, by direct analogy with the classical case of *Linear Algebra,* is to represent a given morphism of free $\mathcal{A}$-modules via appropriately defined section-matrices. We are dealt with this question right below in the next Subsection.

**3.(a). Matrix representations of $\mathcal{A}$-morphisms.**– According to the very definition of free $\mathcal{A}$-modules (of finite rank, see e.g. (3.7)), one can be confronted with the problem in the heading of this Subsection, when looking, equivalently, at the analogous question for *$\mathcal{A}$-morphisms of the standard free $\mathcal{A}$-modules* $\mathcal{A}^n$, $n \in \mathbf{N}$. So assume that

$$\phi : \mathcal{A}^n \longrightarrow \mathcal{A}^m \tag{3.12}$$

is a given *$\mathcal{A}$-morphism of* the *free $\mathcal{A}$-modules* concerned. Thus, by virtue of Chapt. I; (13.9), $\phi$ *is known whenever, equivalently,* $\overline{\phi} : \Gamma(\mathcal{A}^n) \longrightarrow \Gamma(\mathcal{A}^m)$ *is given;* therefore (cf. (3.4)), %m it suffices, for instance, *to know* $\overline{\phi}$ *on the Kronecker gauge of* $\mathcal{A}^n$ (see (3.3)). So one obtains

$$\overline{\phi}(\varepsilon_i) := \phi \circ \varepsilon_i = \sum_{j=1}^{m} \phi_{ij} \cdot \varepsilon_j, \quad 1 \leqq i \leqq n, \tag{3.13}$$

(we have still employed above the corresponding *Kronecker gauge of* $\mathcal{A}^m$, according to (3.2)), *such that* $\phi_{ij} \in \Gamma(X, \mathcal{A}) \equiv \mathcal{A}(X)$, with $1 \leqq i \leqq n$ and $1 \leqq j \leqq m$. In other words, one has (cf. also (1.25), for $U = X$)

$$(\phi_{ij}) \in M_{n,m}(\mathcal{A}(X)) = M_{n,m}(\mathcal{A})(X). \tag{3.14}$$

Consequently, we have thus established the following correspondence, say,

$$\begin{aligned} \eta : \mathrm{Hom}_{\mathcal{A}}(\mathcal{A}^n, \mathcal{A}^m) = \mathcal{H}om_{\mathcal{A}}(\mathcal{A}^n, \mathcal{A}^m)(X) &\longrightarrow M_{n,m}(\mathcal{A})(X) \\ &= M_{n,m}(\mathcal{A}(X)) \end{aligned} \tag{3.15}$$

(see also Chapt. II; (6.9)), given by

(3.16) $$\eta(\phi) := (\phi_{i,j}),$$

as above, for any given $\phi$, as in (3.12). Furthermore, based on the preceding, it is easy to see that;

(3.17) *the* above *map* $\eta$, as given by (3.16), *is an isomorphism of* $\mathcal{A}(X)$-*modules.* That is, *one obtains*

(3.17.1) $$\mathcal{H}om_{\mathcal{A}}(\mathcal{A}^n, \mathcal{A}^m)(X) = \mathrm{Hom}_{\mathcal{A}}(\mathcal{A}^n, \mathcal{A}^m) = M_{n,m}(\mathcal{A}(X)) = M_{n,m}(\mathcal{A})(X),$$

up to $\mathcal{A}(X)$-*isomorphisms,* as before.

Indeed, (3.17) is an immediate consequence of a more general result, given that (3.17.1) holds actually good, as we presently show, for the respective *sheaf of germs of $\mathcal{A}$-morphisms* (cf. Chapt. II; Definition 6.1). Namely, we prove that

(3.18) $$\mathcal{H}om_{\mathcal{A}}(\mathcal{A}^n, \mathcal{A}^m) = \mathcal{A}^{n\cdot m} = M_{n,m}(\mathcal{A}),$$

up to $\mathcal{A}$-*isomorphisms* of the $\mathcal{A}$-modules concerned: Thus, the first one of the latter isomorphisms (being also a *canonical* one) is already provided by Chapt. II; (6.22), applied to the $\mathcal{A}$-module $\mathcal{A}^m$, while the second one is derived as follows; viz., based on (1.25), one obtains, for every *open* $U \subseteq X$,

(3.19) $$M_{n,m}(\mathcal{A})(U) = M_{n,m}(\mathcal{A}(U)) = \mathcal{A}(U)^{n\cdot m} = \mathcal{A}^{n\cdot m}(U),$$

up to *isomorphisms of* the respective $\mathcal{A}(U)$-*modules* (cf. Chapt. II; (3.14), concerning the last one). In other words, *the* corresponding (complete) *presheaves of sections of the sheaves* $M_{n,m}(\mathcal{A})$ *and* $\mathcal{A}^{n\cdot m}$ *are isomorphic,* therefore (Chapt. II; (13.19)), the sheaves themselves, which thus proves our claim. ∎ Of course, (3.17.1) is now an immediate consequence of (3.18). Furthermore, as follows from (3.18),

(3.20) both $\mathcal{H}om_{\mathcal{A}}(\mathcal{A}^n, \mathcal{A}^m)$ *and* $M_{n,m}(\mathcal{A})$ *are* (isomorphic) *free* $\mathcal{A}$-*modules of rank* $n \cdot m$ on $X$.

Now, suppose we are given two *free* $\mathcal{A}$-*modules* $\mathcal{E}$ and $\mathcal{F}$ on $X$, such that $\operatorname{rk}\mathcal{E} = n$ and $\operatorname{rk}\mathcal{F} = m$. So, in view of Chapt. II; Definition 3.2 and of (3.18), as before, one obtains (cf. also Chapt. II; Definition 6.1 and (6.8))

(3.21) $$\mathcal{H}om_{\mathcal{A}}(\mathcal{E}, \mathcal{F}) = \mathcal{H}om_{\mathcal{A}}(\mathcal{A}^n, \mathcal{A}^m) = M_{n,m}(\mathcal{A}) = \mathcal{A}^{n\cdot m},$$

within $\mathcal{A}$-*isomorphisms* of the $\mathcal{A}$-modules involved. Consequently, one still obtains that

(3.22) for $\mathcal{E}$ and $\mathcal{F}$, as before, $\mathcal{H}om_{\mathcal{A}}(\mathcal{E}, \mathcal{F})$ *is a free* $\mathcal{A}$-*module on* $X$ too, *of rank* $n \cdot m$.

We have already seen a genuine generalization of (3.22) about locally free $\mathcal{A}$-modules (of finite rank, viz. *vector sheaves*) on $X$, in Chapt. II; Lemma 6.1.

On the other hand, for $\mathcal{E} = \mathcal{F} (\cong \mathcal{A}^n)$, one has (cf. also Chapt. II; (6.28))

$$\mathcal{H}om_{\mathcal{A}}(\mathcal{E}, \mathcal{E}) \equiv \mathcal{E}nd\, \mathcal{E} = M_n(\mathcal{A}) = \mathcal{A}^{n^2}, \tag{3.23}$$

up to $\mathcal{A}$-*isomorphisms* of the $\mathcal{A}$-modules concerned; yet, according to Chapt. II; (6.29), $\mathcal{E}nd\, \mathcal{E}$, hence (cf. also (3.18))

(3.24) $M_n(\mathcal{A}) = \mathcal{H}om_{\mathcal{A}}(\mathcal{A}^n, \mathcal{A}^n) \equiv \mathcal{E}nd_{\mathcal{A}}(\mathcal{A}^n)$ is an $\mathcal{A}$-*algebra sheaf*, as well,

supplementing thus our previous information about (1.8). In particular, one has (cf. (1.26));

$$\mathcal{A}ut(\mathcal{A}^n) \equiv \mathcal{A}ut_{\mathcal{A}}(\mathcal{A}^n) = M_n(\mathcal{A})^{\cdot} = \mathcal{GL}(n, \mathcal{A}). \tag{3.25}$$

# 4. The determinant morphism

By considering the *full matrix algebra sheaf* $M_n(\mathcal{A})$ (see (1.8)), one can define the following map

$$\partial\mathrm{et} : M_n(\mathcal{A}) \longrightarrow \mathcal{A}; \tag{4.1}$$

namely, by looking at the complete presheaves of sections of the sheaves concerned, one defines, equivalently (cf. Chapt. I; (13.3), (13.19)), the map

$$\overline{\partial\mathrm{et}} : \Gamma(M_n(\mathcal{A})) \longrightarrow \Gamma(\mathcal{A}). \tag{4.2}$$

Now (see Chapt. I; (6.3)), this means that we have a family of maps, indexed by the topology $\mathcal{T}$ of $X$, viz. one has

$$\overline{\partial\mathrm{et}} \equiv \left(\overline{\partial\mathrm{et}}_U\right)_{U\in\mathcal{T}} \tag{4.3}$$

such that one sets

$$\overline{\partial\mathrm{et}}_U : \Gamma(M_n(\mathcal{A}))(U) \longrightarrow \Gamma(\mathcal{A})(U); \tag{4.4}$$

on the other hand, one has, for every *open* set $U \subseteq X$ (cf. also Chapt. I; (9.3), as well as, (1.11) above and (1.9))

$$\Gamma(M_n(\mathcal{A}))(U) \equiv \Gamma(U, M_n(\mathcal{A})) \equiv M_n(\mathcal{A})(U) = M_n(\mathcal{A}(U)), \tag{4.5}$$

as well as,

$$\Gamma(\mathcal{A})(U) \equiv \Gamma(\cdot, \mathcal{A})(U) = \Gamma(U, \mathcal{A}) \equiv \mathcal{A}(U). \tag{4.6}$$

So one finally sets

$$\begin{aligned} \overline{\partial\mathrm{et}}_U : M_n(\mathcal{A}(U)) \longrightarrow \mathcal{A}(U) : \alpha \equiv (\alpha_{i,j}) \longmapsto \overline{\partial\mathrm{et}}_U(\alpha) := \\ \det{}_U(\alpha) \equiv \det{}_U(\alpha_{ij}), \end{aligned} \tag{4.7}$$

the last item of (4.7) being the usual $\mathcal{A}(U)$*-valued determinant map on* (the *full matrix* **C**-*algebra*) $M_n(\mathcal{A}(U))$; cf., for instance, S. Lang [1: p.451].

Of course, it is easy to see, by the very definitions, that the family (4.3), as defined by (4.7), yields actually a *morphism of presheaves,* as in (4.2) (see also Chapt. I; (6.5)), so that *the map (4.1) is* now uniquely defined, through (4.2), as *a morphism of sheaves,* as indicated, according to Chapt. I; (13.19).

We further consider the *restriction of the* above *sheaf morphism* $\partial\mathrm{et}$ *to* $\mathcal{GL}(n, \mathcal{A}) \subseteq M_n(\mathcal{A})$, retaining , as usual, the same symbol for the restricted map. (Concerning the last inclusion, see (1.26), as well as, e.g. Chapt. I; Lemma 12.1). So one actually obtains

$$\partial\mathrm{et}\,|_{\mathcal{GL}(n,\mathcal{A})} \equiv \partial\mathrm{et} : \mathcal{GL}(n, \mathcal{A}) \longrightarrow \mathcal{A}^{\cdot}. \tag{4.8}$$

Indeed, by looking at the element

$$\alpha \equiv (\alpha_{i,j}) \in \mathcal{GL}(n, \mathcal{A})(U) = \mathcal{GL}(n, \mathcal{A}(U)) \tag{4.9}$$

(cf. (1.28)), for some *open* $U \subseteq X$, *one obtains,* by virtue of (4.7),

$$\overline{\partial\mathrm{et}}_U(\alpha) = \det{}_U(\alpha_{ij}) \in \mathcal{A}(U)^{\cdot}. \tag{4.10}$$

Therefore, by referring to the $\mathcal{A}(U)$*-valued* $\det_U$, *the* above *last relation is a characteristic property for (4.9) to hold true* (see, for instance, S. Lang [1: p.256, Proposition 4.16]. Thus, the preceding argument proves now our claim about (4.8). ∎ On the other hand, one still infers that;

(4.11) *the map (4.8) is a morphism of* the *group sheaves* on $X$, under consideration.

Namely, based on Chapt. II; (1.67), it suffices to prove (4.11) for the map $\overline{\partial \mathrm{et}}$ (see (4.3)) restricted to $\Gamma(\mathcal{GL}(n,\mathcal{A}))$; viz. (cf. also (1.29.1), (1.15), as well as, (4.5)), for each one of the maps

$$\overline{\partial\mathrm{et}}_U : \Gamma(\mathcal{GL}(n,\mathcal{A}))(U) \equiv \Gamma(U, \mathcal{GL}(n,\mathcal{A})) \equiv \mathcal{GL}(n,\mathcal{A})(U) = \mathrm{GL}(n,\mathcal{A}(U)) \longrightarrow \mathcal{A}^{\cdot}(U) = \mathcal{A}(U)^{\cdot}, \tag{4.12}$$

with $U$ *open* in $X$. But, this is again standard (see, for instance, S. Lang [1: p.455, Proposition 4.15]), which thus proves our assertion in (4.11). ∎

On the other hand, by still extending to the present context the classical situation, we call *special linear group sheaf of order* $n$, associated with the unital commutative C-algebra sheaf $\mathcal{A}$ on $X$, as above, the *kernel of the morphism of group sheaves* (4.8) (see also Chapt. II; (2.26), (2.27)) and we denote it by

$$\mathcal{SL}(n,\mathcal{A}) := \ker(\partial\mathrm{et}\,|_{\mathcal{GL}(n,\mathcal{A})}). \tag{4.13}$$

In this context, one still obtains that;

(4.14) the above *sheaf of groups* on $X$, defined by (4.13), can be construed (modulo an isomorphism of group sheaves) as the *sheaf* (of groups) *on* $X$, generated by the *presheaf of groups* (see also (4.7))

$$U \longmapsto \mathrm{SL}(n,\mathcal{A}(U)) \equiv \ker(\overline{\partial\mathrm{et}}_U) \equiv \ker(\det{}_U), \tag{4.14.1}$$

for any *open* set $U \subseteq X$, along with the obvious restriction maps.

Indeed, (4.14.1) defines a *complete presheaf* (of groups) on $X$, so that one has;

$$\mathcal{SL}(n,\mathcal{A})(U) = \mathrm{SL}(n,\mathcal{A}(U)), \tag{4.14.2}$$

within a *group isomorphism,* for every *open* set $U \subseteq X$.

It is clear that (4.14.1) defines a presheaf of groups on $X$ and, in fact, a complete one, as this follows straightforwardly from the same definitions. Yet, in view of (4.13) and the fact that (see also (4.7))

$$(4.15) \qquad (\ker \partial\mathrm{et}\,)(U) = \ker \overline{\partial\mathrm{et}}_U = \ker \det{}_U = \mathrm{SL}(n, \mathcal{A}(U)),$$

up to a *group isomorphism,* for the first of the previous relations (cf. also the next Scholium 4.1), for every *open* $U \subseteq X$, one easily interprets (4.13), through (4.14.1), as asserted. ∎

**Scholium 4.1.**- Concerning our previous argument in (4.15), we have actually applied the following fact:

(4.16) given a *morphism of group sheaves*, say, $\phi : \mathcal{S} \longrightarrow \mathcal{S}'$, the sheaf of groups on $X$, $\ker\phi$ (cf. Chapt. II; (2.5)), *is isomorphic to the sheafification* (cf. Chapt. I; (7.23)) *of the (complete) presheaf of groups on* $X$, *defined by* $\ker\overline{\phi}_U$, $U$ *open* in $X$, along with the obvious restriction maps (see also Chapt. II; (13.3) and (6.3)).

(**Proof.** Straightforward). ∎ An analogous argument to (4.16) is certainly valid, *for any other algebraic structure.*

Now, in view of (4.13) and (4.8), one further obtains that;

(4.17) for every *open* set $U \subseteq X$, one has (cf. also (4.3))

$$(4.17.1) \qquad \alpha \in \mathcal{SL}(n, \mathcal{A})(U) \text{ if, and only if,}$$
$$\overline{\partial\mathrm{et}}\,(\alpha) = \overline{\partial\mathrm{et}}_U(\alpha) = 1.$$

Here we have set $1 \equiv 1_U \in \mathcal{A}^{\cdot}(U) = \mathcal{A}(U)^{\cdot}$ (see (1.15)) for the (constant) *section* 1 *of* $\mathcal{A}$, restricted to the open set $U \subseteq X$. (In this regard, cf. also Chapt. II; Definition 1.3′, along with (0.1) above).

Yet, as a result of the preceding, we still note for later use the following *exact sequence of group sheaves* on $X$;

$$(4.18) \qquad 1 \longrightarrow \mathcal{SL}(n, \mathcal{A}) \longrightarrow \mathcal{GL}(n, \mathcal{A}) \xrightarrow{\partial\mathrm{et}} \mathcal{A}^{\cdot}.$$

Apart from (4.13), other basic applications of the map $\partial\mathrm{et}$, as above, will be considered at the pertinent places in the sequel. (So see e.g. Section 7 below, as well as, the next Chapter V, Section 9).

**Remark 4.1.**- By analogy with the classical situation (cf., for instance, S. Mac Lane [1: p.16, §4]), we further note, in the sense of a general foundational comment, that *the map* $\partial\mathrm{et}$ as given by (4.8), *is,* indeed, *a natural transformation of* the *functors* involved, viz. of $\mathcal{GL}(n, \cdot)$ and $(\cdot)^{\bullet}$, the latter functors been defined between the obvious categories. (See also (1.26) and (1.13) above). Of course, the appearance of the notion (map) in question within the present abstract *sheaf-theoretic* context still witnesses its significance.

# 5. Dual $\mathcal{A}$-modules

We consider in the sequel the sheaf-theoretic version of the classical notion of a *"dual module"*, as this will be applied below. Thus, we start with the following.

**Definition 5.1.** Let $(X, \mathcal{A})$ be a given $\mathbf{C}$-algebraized space (cf. (0.1)) and let $\mathcal{E}$ be an $\mathcal{A}$-module on $X$. Then, the $\mathcal{A}$-module on $X$

$$\mathcal{E}^* := \mathcal{H}om_{\mathcal{A}}(\mathcal{E}, \mathcal{A}) \tag{5.1}$$

(see also Chapt. II; (6.7)) is called the *dual $\mathcal{A}$-module of $\mathcal{E}$*. (The notation $\mathcal{E}^{\vee}$ for (5.1) is still employed).

As a first consequence of (5.1), one remarks that:

(5.2) given a *free $\mathcal{A}$-module $\mathcal{E}$ of finite rank* on $X$, *its dual $\mathcal{A}$-module $\mathcal{E}^*$ is still a free $\mathcal{A}$-module* on $X$ *of the same (finite) rank as $\mathcal{E}$*. In particular, *one has*

$$\mathcal{E} = \mathcal{E}^*, \tag{5.2.1}$$

*within an $\mathcal{A}$-isomorphism.*

The assertion follows immediately from (3.21), as above, by setting, according to our hypothesis, $\mathcal{E} = \mathcal{A}^n$ (with $n = \operatorname{rk} \mathcal{E}$, cf. Chapt. II; Definition 3.2 and (4.2)) and $\mathcal{F} = \mathcal{A}$, so that one has

$$\mathcal{E}^* \equiv \mathcal{H}om_{\mathcal{A}}(\mathcal{E}, \mathcal{A}) = \mathcal{H}om_{\mathcal{A}}(\mathcal{A}^n, \mathcal{A}) = \mathcal{A}^n = \mathcal{E}, \tag{5.3}$$

within $\mathcal{A}$-*isomorphisms.* ▮ Thus, based on (5.3), we also conclude that

$$\mathcal{E}^{**} := (\mathcal{E}^*)^* = \mathcal{E}^* = \mathcal{E}, \tag{5.4}$$

*for any free $\mathcal{A}$-module* (of finite rank) on $X$, within $\mathcal{A}$-*isomorphisms.* In this context, we recall that, for a given $\mathcal{A}$-*module* $\mathcal{E}$ on $X$, the $\mathcal{A}$-module on $X$

$$\mathcal{E}^{**} := (\mathcal{E}^*)^* \tag{5.5}$$

is called the *bidual $\mathcal{A}$-module of $\mathcal{E}$*. Yet, if for an *$\mathcal{A}$-module $\mathcal{E}$* on $X$, as before, one has

$$\mathcal{E} = \mathcal{E}^{**}, \tag{5.6}$$

up to an *$\mathcal{A}$-isomorphism,* then the given $\mathcal{A}$-module $\mathcal{E}$ is said to be *reflexive.*

So, according to (5.4), one infers that:

(5.7) *every free $\mathcal{A}$-module* (of finite rank) on $X$ *is reflexive.*

Indeed, something much more is actually valid, for as we shall presently show (cf. Theorem 5.1 below), *(5.7) holds true for every vector sheaf* on $X$.

However, before we treat that case, we further remark that, as a consequence of Chapt. II; (6.20), (3.7), one obtains the following relation, for any given *$\mathcal{A}$-module $\mathcal{E}$* on $X$ (cf. also (5.1)), *and $n \in \mathbf{N}$*;

$$\mathcal{H}om_{\mathcal{A}}(\mathcal{E}, \mathcal{A}^n) = \mathcal{H}om(\mathcal{E}, \mathcal{A})^n \equiv (\mathcal{E}^*)^n, \tag{5.8}$$

within an *$\mathcal{A}$-isomorphism* (concerning the equality involved).

Now, turning to the promised extension of (5.7), we first define the following (canonical) *$\mathcal{A}$-morphism,* for any given *$\mathcal{A}$-module $\mathcal{E}$* on $X$. Namely, we have (see also (5.5));

$$\eta : \mathcal{E} \longrightarrow \mathcal{E}^{**} \tag{5.9}$$

such that one sets

$$\overline{\eta}(s)(u) \equiv \check{s}(u) := u(s), \tag{5.10}$$

*for any $s \in \mathcal{E}(V)$ and $u \in \mathcal{E}^*(V) \equiv \mathcal{H}om_{\mathcal{A}}(\mathcal{E}, \mathcal{A})(V)$* $= \mathrm{Hom}_{\mathcal{A}|_V}(\mathcal{E}|_V, \mathcal{A}|_V)$, where $V$ is any *open* set in $X$. (In this regard, see also Chapt. II; (6.1) and Chapt. I; (11.40). Yet, we have repeatedly employed in (5.10), as usual, in similar cases, hitherto, the technique of Chapt. I; (13.19) and (13.3)). So we now have the following useful result, that extends (5.7).

**Theorem 5.1.** *Let $\mathcal{E}$ be a vector sheaf* (viz. a locally free $\mathcal{A}$-module of finite rank) *on $X$*, base space of a given $\mathbf{C}$-algebraized space $(X, \mathcal{A})$. *Then $\mathcal{E}$ is reflexive; i.e. one has*

$$\mathcal{E} = \mathcal{E}^{**}, \tag{5.11}$$

*up to an $\mathcal{A}$-isomorphism* (given by (5.9), (5.10)).

**Proof.** We have to prove that *the* above (canonical) *$\mathcal{A}$-morphism (5.9) is,* in fact, *an $\mathcal{A}$-isomorphism (onto),* that is, one actually has, for the case considered, the following *exact $\mathcal{A}$-sequence;*

$$0 \longrightarrow \mathcal{E} \xrightarrow{\eta} \mathcal{E}^{**} \longrightarrow 0. \tag{5.12}$$

Therefore, based on the very definitions (see thus Chapt. I; (1.21)), one has to assure the following *exact $\mathcal{A}_x$-sequence*

$$0 \longrightarrow \mathcal{E}_x \xrightarrow{\eta_x} \mathcal{E}_x^{**} \longrightarrow 0, \tag{5.13}$$

*for every point $x \in X$* (cf. also Chapt. I; (1.14), as well as, Chapt. II; (6.10)). On the other hand, the *exactness of (5.13) at a* given *point $x \in X$ is equivalent with that one of the next sequence of $\mathcal{A}|_V$-modules*

$$0 \longrightarrow \mathcal{E}_V \xrightarrow{\eta_V} \mathcal{E}^{**}|_V \longrightarrow 0, \tag{5.14}$$

where $V$ stands for a (suitable) *open neighborhood of $x$*; see, for instance, Chapt. I; Theorem 12.1, along with (12.2), as well as, (13.19). Consequently, we would have proved our claim, as long as we had ensured the *existence of an open covering of $X$, on the* individual *members of which,* like $V$ above, (5.14) *hold good.* Indeed, such a covering of $X$ *there does exist,* due to the hypothesis for $\mathcal{E}$ (cf. Chapt. II; Definition 4.3, along with (4.8), as well as, (5.4) above), and this terminates the proof. ∎

**Remark.**– The sort of argument applied in the previous proof is, in effect, typical of *reducing a matter about a* given *sheaf morphism to a local one;* in this context, we still say that *the question* about the morphism, under consideration, *is of a local nature* (or even just *local*).

Now, by looking again at the standard free $\mathcal{A}$-module $\mathcal{A}^n$ on $X$, we further consider its *dual gauge,* viz. the (canonical) *Kronecker gauge of* $(\mathcal{A}^n)^*$, the *dual $\mathcal{A}$-module of $\mathcal{A}^n$*, this being too (cf. (5.2)) a free $\mathcal{A}$-module, of the same rank $n$, on $X$. We denote the latter gauge by

$$(\varepsilon_i^*) \subseteq (\mathcal{A}^n)^*(X), \tag{5.15}$$

$$\varepsilon_i^*(\varepsilon_j) = \delta_{i,j}, \quad 1 \leq i, j \leq n. \tag{5.16}$$

On the other hand, if $\mathcal{E}$ is an arbitrary *free $\mathcal{A}$-module* on $X$, with $\operatorname{rk}\mathcal{E} = n$, and $(e_i) \subseteq \mathcal{E}(X)$ its *canonical gauge* (cf. (3.11)), one further defines the *(canonical) dual gauge* of $\mathcal{E}$, viz. that one of its *dual $\mathcal{A}$-module* $\mathcal{E}^*$, denoted by

$$(e_i^*) \subseteq \mathcal{E}^*(X), \tag{5.17}$$

such that one has (see also Chapt. II; (3.22));

$$e_i^*(e_j) := \delta_{ij}, \quad 1 \leqq i, j \leqq n. \tag{5.18}$$

Applications of the preceding will be given in the sequel, even more generally, for *vector sheaves,* employing thus the previous situation *locally.* Yet, in this context, based on Chapt. II; Lemma 6.1, we also note for later use that;

(5.19) *the dual of a given vector sheaf $\mathcal{E}$ on $X$ is a vector sheaf on $X$, as well, of the same rank as $\mathcal{E}$. That is, one has* (cf. Chapt. II; (6.26))

$$\operatorname{rk}_{\mathcal{A}}\mathcal{E} = \operatorname{rk}_{\mathcal{A}}\mathcal{E}^*. \tag{5.19.1}$$

Finally, it is still appropriate to look at (5.1) as the value of a suitable *functor*, namely, of the following one

$$\mathcal{H}om_{\mathcal{A}}(\cdot, \mathcal{A}) := \mathcal{A}\text{-}\mathcal{M}od_X \longrightarrow \mathcal{A}\text{-}\mathcal{M}od_X, \tag{5.20}$$

called the *dual $\mathcal{A}$-module functor* between the categories concerned (see Chapt. II; (2.4)). Thus, we set

$$\mathcal{H}om_{\mathcal{A}}(\cdot, \mathcal{A})(\mathcal{E}) := \mathcal{H}om_{\mathcal{A}}(\mathcal{E}, \mathcal{A}) \equiv \mathcal{E}^*, \tag{5.21}$$

for every *$\mathcal{A}$-module* $\mathcal{E}$ on $X$. It is straightforward that (5.20) as given by (5.21) is a *contravariant functor* of the category $\mathcal{A}$-$\mathcal{M}od_X$ into itself.

So, for any given *$\mathcal{A}$-morphism* in $\mathcal{A}$-$\mathcal{M}od_X$, say

$$\phi : \mathcal{E} \longrightarrow \mathcal{F}, \tag{5.22}$$

one defines

$$\mathcal{H}om_{\mathcal{A}}(\cdot, \mathcal{A})(\phi) \equiv \mathcal{H}om(\phi, \mathcal{A}) \equiv {}^t\phi : \mathcal{F}^* \longrightarrow \mathcal{E}^*, \tag{5.23}$$

such that one has

$${}^t\phi(u) := u \circ \phi, \tag{5.24}$$

*for every* $u \in \mathcal{F}^*(V)$, with $V$ *open* in $X$.

In particular, *for every point* $x \in V$, with $V$ as before, *one obtains*

$$ {}^t\phi(u)(x) \equiv (u \circ \phi)(x) = u(x)(\phi(x)) \equiv u_x(\phi(x)), \tag{5.25}$$

such that

$$u(x) \equiv u_x : \mathcal{F}_x \longrightarrow \mathcal{A}_x \tag{5.26}$$

denotes the respective $\mathcal{A}_x$-*morphism,* provided by the given (local) *section*

$$u \in \mathcal{F}^*(V) \equiv \mathcal{H}om_{\mathcal{A}}(\mathcal{F}, \mathcal{A})(V) = \mathrm{Hom}_{\mathcal{A}|_V}(\mathcal{F}|_V, \mathcal{A}|_V) \tag{5.27}$$

(see also Chapt. II; (6.8), (6.3), along with Chapt. I; (12.52)). Applications of the above will also be considered by the ensuing discussion.

## 6. Certain basic formulae relating the functors $\mathcal{H}om_{\mathcal{A}}$ and $\otimes_{\mathcal{A}}$

We continue in this Section by further looking at some interrelations of the functors in the heading, which are instrumental in later applications (see, in particular, Part Two of this treatise). So, as our first relevant result, we have the following (cf. also Definition 5.1 above).

**Theorem 6.1.** *Let* $X, \mathcal{A})$ *be a* **C**-*algebraized space and* $\mathcal{E}$, $\mathcal{F}$ *two given* $\mathcal{A}$-*modules on* $X$, *such that either one of them is, in particular, a vector sheaf. Then, in the case e.g. that* $\mathcal{E}$ *is a vector sheaf on* $X$, *one obtains*

$$\mathcal{E}^* \otimes_{\mathcal{A}} \mathcal{F} \equiv \mathcal{H}om_{\mathcal{A}}(\mathcal{E}, \mathcal{A}) \otimes_{\mathcal{A}} \mathcal{F} = \mathcal{H}om_{\mathcal{A}}(\mathcal{E}, \mathcal{F}), \tag{6.1}$$

*while if* $\mathcal{F}$ *is a vector sheaf on* $X$, *one has*

$$\mathcal{E} \otimes_{\mathcal{A}} \mathcal{F}^* \equiv \mathcal{F}^* \otimes_{\mathcal{A}} \mathcal{E} = \mathcal{H}om_{\mathcal{A}}(\mathcal{F}, \mathcal{E}), \tag{6.2}$$

*both formulas being valid up to* $\mathcal{A}$-*isomorphisms, as indicated.*

**Proof.** We start by first looking at the *presheaves that generate the sheaves* (in fact, $\mathcal{A}$-*modules*) appeared *in* (6.1) (see Chapt. II; Definitions

5.1 and 6.1, in particular, (5.5) and (5.7), as well as, (6.4), (6.8)): So, *for every open* set $V \subseteq X$, one obtains the following *canonical morphism of* the *presheaves* involved

$$\mathcal{E}^*(V) \otimes_{\mathcal{A}(V)} \mathcal{F}(V) \xrightarrow{\phi_V} \mathrm{Hom}_{\mathcal{A}|_V}(\mathcal{E}|_V, \mathcal{F}|_V) \tag{6.3}$$

or even (see (5.1), (5.26) and Chapt. II; (6.8))

$$\mathrm{Hom}_{\mathcal{A}|_V}(\mathcal{E}|_V, \mathcal{A}|_V) \otimes_{\mathcal{A}(V)} \mathcal{F}(V) \xrightarrow{\phi_V} \mathrm{Hom}_{\mathcal{A}|_V}(\mathcal{E}|_V, \mathcal{F}|_V) \tag{6.4}$$

such that one has

$$(u \otimes t)(s) := u(s) \cdot t, \tag{6.5}$$

*for any* $u \in \mathcal{E}^*(V)$, $t \in \mathcal{F}(V)$ *and* $s \in \mathcal{E}(V)$. In fact, (6.5) entails that (6.4) is an $\mathcal{A}(X)$-*morphism* of the presheaves (of $\mathcal{A}(X)$-modules), under consideration. (Yet, we employed in (6.5), as usual, *"tensor products unadorned" of* the (local) *sections* involved, although they are always viewed with respect to the individual members of $\Gamma(\mathcal{A})$ (cf. (1.4)), as the case may be (here, in particular, relative to $\mathcal{A}(V)$). Moreover, one actually considers $u \otimes t$ in (6.5) as a *morphism between* $\Gamma(\mathcal{E}|_V)$ *and* $\Gamma(\mathcal{F}|_V)$ (see Chapt. I; (13.3)), so that one really applies $u \otimes t$ on a section $s \in (\mathcal{E}|_V)(U) = \mathcal{E}(U)$, for some *open* $U \subseteq V$; cf. also Chapt. I; (11.40)).

Now, suppose that $\mathcal{E}$ *is a vector sheaf* on $X$. Therefore, one may assume that $V$ *in (6.4) ranges over an open covering of* $X$, *on the individual members of which* $\mathcal{E}$ *reduces to a standard free* $\mathcal{A}$-*module of a given (constant) rank*, say $n = \mathrm{rk}\,\mathcal{E}$ (viz. $V$ is varying over a *local frame of* $\mathcal{E}$. See Chapt. II; Definitions 4.1, 4.2 and 4.3). Thus, *the source of* (6.4) becomes now (cf. also Chapt. II; (6.22) and (6.47));

$$\begin{aligned}\mathrm{Hom}_{\mathcal{A}|_V}(\mathcal{E}|_V, \mathcal{A}|_V) \otimes_{\mathcal{A}(V)} \mathcal{F}(V) &= \mathrm{Hom}_{\mathcal{A}|_V}(\mathcal{A}^n|_V, \mathcal{A}|_V) \otimes_{\mathcal{A}(V)} \mathcal{F}(V) \\ &= \mathcal{A}(V)^n \otimes_{\mathcal{A}(V)} \mathcal{F}(V) = \mathcal{F}(V)^n,\end{aligned} \tag{6.6}$$

while *its target* yields (still cf. Chapt. II; (6.47) and (3.14))

$$\mathrm{Hom}_{\mathcal{A}|_V}(\mathcal{E}|_V, \mathcal{F}|_V) = \mathrm{Hom}_{\mathcal{A}|_V}(\mathcal{A}^n|_V, \mathcal{F}|_V) = \mathcal{F}^n(V) = \mathcal{F}(V)^n, \tag{6.7}$$

both (6.6) and (6.7) being valid up to $\mathcal{A}(X)$-*isomorphisms* of the modules involved. Therefore, the map $\mathbf{S}(\phi) \equiv \tilde{\phi}$, obtained through the *sheafification of* (6.4) (cf. Chapt. I; (8.3)), has the property that *the* respective *fiber maps* $\tilde{\phi}_x$, $x \in X$, *are bijections* (see e.g. Chapt. I; Lemmas 12.1, 12.2), *hence* (Chapt. I; (1.21)), *the map* $\tilde{\phi}$ *itself.* That is, *the sheaves* appeared *in (6.1) are isomorphic.* Moreover, as already noted at the beginning of the proof, by looking e.g. at (6.5), *this isomorphism preserves the algebraic structures* involved (see also Chapt. II; (1.54)), which thus proves our claim about (6.1). Now, (6.2) follows, of course, straightforwardly, from (6.1), and the proof is complete. ∎

As an immediate application of the above, one now obtains the following.

**Corollary 6.1.** *Suppose we are given the* $\mathcal{A}$*-modules* $\mathcal{E}$, $\mathcal{F}$, $\mathcal{S}$ *on* $X$, *with* $(X, \mathcal{A})$ *as in Theorem 6.1, and let, in particular,* $\mathcal{E}$ *be a vector sheaf on* $X$. *Then, one has the following* $\mathcal{A}$*-isomorphism of the* $\mathcal{A}$*-modules concerned;*

$$\mathcal{H}om_{\mathcal{A}}(\mathcal{E}, \mathcal{F} \otimes_{\mathcal{A}} \mathcal{S}) = \mathcal{H}om_{\mathcal{A}}(\mathcal{E}, \mathcal{F}) \otimes_{\mathcal{A}} \mathcal{S}. \tag{6.8}$$

**Proof.** Indeed, if $\mathcal{E}$ is a *vector sheaf* on $X$, then a repeated application of (6.1) and the associativity law of tensor products (cf. Chapt. II; (5.16)), provide the following $\mathcal{A}$*-isomorphisms*

$$\begin{aligned} \mathcal{H}om_{\mathcal{A}}(\mathcal{E}, \mathcal{F} \otimes_{\mathcal{A}} \mathcal{S}) &= \mathcal{E}^* \otimes_{\mathcal{A}} (\mathcal{F} \otimes_{\mathcal{A}} \mathcal{S}) = (\mathcal{E}^* \otimes_{\mathcal{A}} \mathcal{F}) \otimes_{\mathcal{A}} \mathcal{S} \\ &= \mathcal{H}om_{\mathcal{A}}(\mathcal{E}, \mathcal{F}) \otimes_{\mathcal{A}} \mathcal{S}, \end{aligned} \tag{6.9}$$

which thus proves (6.8). ∎

Furthermore, as a consequence of (6.1), (5.2) and of Chapt. II; (6.22), one also obtains, *for every* $\mathcal{A}$*-module* $\mathcal{E}$ *on* $X$ *and* $n \in \mathbf{N}$, the following $\mathcal{A}$-*isomorphisms* of the $\mathcal{A}$-modules concerned;

$$\mathcal{H}om_{\mathcal{A}}(\mathcal{A}^n, \mathcal{A}) \otimes_{\mathcal{A}} \mathcal{E} = \mathcal{A}^n \otimes_{\mathcal{A}} \mathcal{E} = \mathcal{H}om_{\mathcal{A}}(\mathcal{A}^n, \mathcal{E}) = \mathcal{E}^n. \tag{6.10}$$

We come next to the *dual of a tensor product* (of $\mathcal{A}$-modules). However, we first have the following more general result.

**Theorem 6.2.** *Suppose we have a* **C***-algebraized space* $(X, \mathcal{A})$ *and* $\mathcal{E}$, $\mathcal{F}$, $\mathcal{S}$ *given* $\mathcal{A}$*-modules on* $X$. *In particular, assume that* $\mathcal{E}$ *is a vector sheaf on* $X$. *Then one obtains*

$$\mathcal{H}om_{\mathcal{A}}(\mathcal{E} \otimes_{\mathcal{A}} \mathcal{F}, \mathcal{S}) = \mathcal{H}om_{\mathcal{A}}(\mathcal{E}, \mathcal{H}om_{\mathcal{A}}(\mathcal{F}, \mathcal{S})), \tag{6.11}$$

*within an* $\mathcal{A}$*-isomorphism.*

**Proof.** By applying a similar argument to that one for the proof of Theorem 6.1, we first look at the *(complete) presheaves of sections of the sheaves ($\mathcal{A}$-modules)* appeared *in (6.11)*. Thus, for every *open set* $V \subseteq X$, one has the following (*V-component* of a) $\Gamma(\mathcal{A})$-*morphism* (see also Chapt. II; Definition 6.1 and (6.4), (6.23), as well as, Lemma 5.1);

$$\text{(6.12)} \quad \begin{gathered} \mathrm{Hom}_{\mathcal{A}|_V}(\mathcal{E}\otimes_{\mathcal{A}}\mathcal{F}|_V, \mathcal{S}|_V) = \mathrm{Hom}_{\mathcal{A}|_V}(\mathcal{E}|_V \otimes_{\mathcal{A}|_V} \mathcal{F}|_V, \mathcal{S}|_V) \xrightarrow{\phi|_V} \\ \mathrm{Hom}_{\mathcal{A}|_V}(\mathcal{E}|_V, \mathcal{H}om_{\mathcal{A}}(\mathcal{F},\mathcal{S})|_V) = \mathrm{Hom}_{\mathcal{A}|_V}(\mathcal{E}|_V, \mathcal{H}om_{\mathcal{A}|_V}(\mathcal{F}|_V, \mathcal{S}|_V)), \end{gathered}$$

such that one has (cf. also Chapt. I; (11.40))

$$\text{(6.13)} \qquad \tilde{u}(s)(t) := u(s\otimes t) \in (\mathcal{S}|_V)(U) = \mathcal{S}(U),$$

for some *open set* $U \subseteq X$, where we set $\phi_V(u) \equiv \tilde{u}$, for every $u$ in the source of (6.12), and any $s \in \mathcal{E}(U)$ and $t \in \mathcal{F}(U)$, with $U$, as before. (In this context, see also the comments in the proof of Theorem 6.1 about (6.5); yet, cf. Chapt. II; (6.24), concerning the (local) *sections* of the $\mathcal{A}|_V$-*module* "$\mathcal{H}om$" in the target of (6.12)).

Now, the proof proceeds further in exactly a similar manner, to the corresponding part of the proof of Theorem 6.1 in the preceding, so that we may leave out the details, which thus completes the proof. ∎

As already said, an immediate consequence of the previous theorem is the evaluation of the *dual of* $\mathcal{E}\otimes_{\mathcal{A}}\mathcal{F}$, with $\mathcal{E}$ and $\mathcal{F}$, as above, in terms of those of the factor sheaves, derived, of course (cf. (5.1)), from (6.11), for $\mathcal{S} = \mathcal{A}$. So we have the next.

**Corollary 6.2.** *Suppose we are given the $\mathcal{A}$-modules $\mathcal{E}$, $\mathcal{F}$ on $X$, as in Theorem 6.2, such that either one of them is a vector sheaf on $X$. So, if $\mathcal{E}$ is a vector sheaf on $X$, one obtains*

$$\text{(6.14)} \qquad (\mathcal{E}\otimes_{\mathcal{A}}\mathcal{F})^* = \mathcal{H}om_{\mathcal{A}}(\mathcal{E},\mathcal{F}^*) = \mathcal{E}^*\otimes_{\mathcal{A}}\mathcal{F}^*,$$

*while if $\mathcal{F}$ is a vector sheaf on $X$, one has*

$$\text{(6.15)} \qquad (\mathcal{E}\otimes_{\mathcal{A}}\mathcal{F})^* = \mathcal{H}om_{\mathcal{A}}(\mathcal{F},\mathcal{E}^*) = \mathcal{E}^*\otimes\mathcal{F}^*,$$

*up to $\mathcal{A}$-isomorphisms, in both of (6.14) and (6.15).*

*In particular, if both $\mathcal{E}$ and $\mathcal{F}$ are vector sheaves on $X$, then one has the following $\mathcal{A}$-isomorphisms;*

$$\text{(6.16)} \qquad \mathcal{H}om_{\mathcal{A}}(\mathcal{E},\mathcal{F}^*) = \mathcal{H}om_{\mathcal{A}}(\mathcal{F},\mathcal{E}^*) = \mathcal{E}^*\otimes_{\mathcal{A}}\mathcal{F}^* = (\mathcal{E}\otimes_{\mathcal{A}}\mathcal{F})^*.$$

**Proof.** Indeed, based on (5.1), (6.11) and (6.1), we obtain, in the case that is a vector sheaf on $X$;

$$(6.17)\qquad \begin{aligned}(\mathcal{E}\otimes_{\mathcal{A}}\mathcal{F})^* &\equiv \mathcal{H}om_{\mathcal{A}}(\mathcal{E}\otimes_{\mathcal{A}}\mathcal{F},\mathcal{A}) = \mathcal{H}om_{\mathcal{A}}(\mathcal{E},\mathcal{H}om_{\mathcal{A}}(\mathcal{F},\mathcal{A}))\\ &\equiv \mathcal{H}om_{\mathcal{A}}(\mathcal{E},\mathcal{F}^*) = \mathcal{E}^*\otimes_{\mathcal{A}}\mathcal{F}^*,\end{aligned}$$

within $\mathcal{A}$-*isomorphisms* of the $\mathcal{A}$-modules involved, which thus proves (6.14). Analogously, by just interchanging the roles of $\mathcal{E}$ and $\mathcal{F}$, one proves (6.15), while (6.16) is, of course, straightforward, in view of the hypothesis for $\mathcal{E}$, $\mathcal{F}$ and the rels. (6.14), (6.15). ∎

**Corollary 6.3.** *Let $(X,\mathcal{A})$ be a* **C**-*algebraized space and $\mathcal{E}$, $\mathcal{F}$ two given vector sheaves on $X$. Then, one has*

$$(6.18)\qquad \mathcal{H}om_{\mathcal{A}}(\mathcal{E},\mathcal{F})^* = \mathcal{H}om_{\mathcal{A}}(\mathcal{E}^*,\mathcal{F}^*) = \mathcal{H}om_{\mathcal{A}}(\mathcal{F},\mathcal{E}),$$

*up to $\mathcal{A}$-isomorphisms of the $\mathcal{A}$-modules involved.*

*In particular, for every vector sheaf $\mathcal{E}$ on $X$, one obtains*

$$(6.19)\qquad (\mathcal{E}nd\,\mathcal{E})^* = \mathcal{E}nd\,\mathcal{E},$$

*up to an $\mathcal{A}$-isomorphism.*

**Proof.** Based on (6.1), (6.14), Theorem 5.1 and 6.2, we get;

$$(6.20)\qquad \begin{aligned}\mathcal{H}om_{\mathcal{A}}(\mathcal{E},\mathcal{F})^* &= (\mathcal{E}^*\otimes_{\mathcal{A}}\mathcal{F})^* = \mathcal{E}^{**}\otimes_{\mathcal{A}}\mathcal{F}^* = \mathcal{H}om_{\mathcal{A}}(\mathcal{E}^*,\mathcal{F}^*)\\ &= \mathcal{E}\otimes_{\mathcal{A}}\mathcal{F}^* = \mathcal{H}om_{\mathcal{A}}(\mathcal{F},\mathcal{E}),\end{aligned}$$

within $\mathcal{A}$-*isomorphisms,* which proves (6.18). In particular, *setting* $\mathcal{E}=\mathcal{F}$ *in* (6.18), one obtains (6.19) (see also Chapt. II; (6.28)), and this terminates the proof. ∎

On the other hand, *for every $\mathcal{A}$-module $\mathcal{E}$ on $X$, and $n\in\mathbf{N}$*, one has the relation (cf. also Chapt. II; (3.7));

$$(6.21)\qquad (\mathcal{E}^n)^* = (\mathcal{E}^*)^n,$$

up to an $\mathcal{A}$-*isomorphism:* Indeed, one has, in view of (6.10), (6.14) and (5.2);

$$(6.22)\qquad \begin{aligned}(\mathcal{E}^n)^* &\equiv \mathcal{H}om_{\mathcal{A}}(\mathcal{E}^n,\mathcal{A}) = \mathcal{H}om_{\mathcal{A}}(\mathcal{A}^n\otimes_{\mathcal{A}}\mathcal{E},\mathcal{A}) \equiv (\mathcal{A}^n\otimes_{\mathcal{A}}\mathcal{E})^*\\ &= (\mathcal{A}^n)^*\otimes_{\mathcal{A}}\mathcal{E}^* = \mathcal{A}^n\otimes_{\mathcal{A}}\mathcal{E}^* = (\mathcal{E}^*)^n,\end{aligned}$$

which proves (6.21). ∎ Finally, based on the relation (5.8) in the preceding,

(6.21) is now supplemented, as follows;

$$\mathcal{H}om_{\mathcal{A}}(\mathcal{E},\mathcal{A}^n)=(\mathcal{E}^*)^n=(\mathcal{E}^n)^*, \tag{6.23}$$

up to *$\mathcal{A}$-isomorphisms,* valid *for every $\mathcal{A}$-module* $\mathcal{E}$ on $X$ *and* $n\in\mathbf{N}$.

# 7. Exterior algebra of $\mathcal{A}$-modules

As the title of this Section indicates, we are concerned below with the analogue, within the present *sheaf-theoretic* context, of the classical case, pertaining to the *exterior algebra* of a module, thus here to that one of a given *$\mathcal{A}$-module* $\mathcal{E}$ on a *topological space* $X$, where the pair $(X,\mathcal{A})$ has the meaning assigned to it hitherto (cf. (0.1)).

Thus, for every *integer* $n\geqq 2$, one defines the *$n$-th exterior power of* $\mathcal{E}$, denoted by

$$\bigwedge^n\mathcal{E}, \tag{7.1}$$

as the *sheafification of the* $\Gamma(\mathcal{A})$*-presheaf*

$$U\longmapsto \bigwedge^n\mathcal{E}(U)\equiv \bigwedge^n_{\mathcal{A}(U)}\mathcal{E}(U), \tag{7.2}$$

where $U$ ranges over the *open* subsets of $X$ (cf. also Chapt. I; (7.23)). So, according to (7.2), as well as, to Chapt. II; (1.54), one realizes that;

(7.3) $\bigwedge^n\mathcal{E}$ *is an $\mathcal{A}$-module on $X$ too, for any given $\mathcal{A}$-module $\mathcal{E}$ on $X$ and every integer $n\in\mathbf{N}$, with $n\geqq 2$.*

Namely, based on the classical theory of modules, one gets, for every *open* $U\subseteq X$, the following *$\mathcal{A}(U)$-module, nth exterior power of the $\mathcal{A}(U)$-module,* in view of the hypothesis for $\mathcal{E}$, $\mathcal{E}(U)$;

$$\bigwedge^n\mathcal{E}(U):=\bigotimes^n\mathcal{E}(U)/\mathcal{J}(U), \tag{7.4}$$

such that

$$\bigotimes^n\mathcal{E}(U):=\mathcal{E}(U)\underbrace{\otimes_{\mathcal{A}(U)}\cdots\otimes_{\mathcal{A}(U)}}_{n-\text{times}}\mathcal{E}(U), \tag{7.5}$$

hence, the "*index* $\mathcal{A}(U)$" in (7.2), while we have also set in (7.4)

$$(7.6)\qquad \mathcal{J}(U) := \left\langle \left\{ \bigotimes_{i=1}^{n} s_i \in \bigotimes^{n} \mathcal{E}(U) : s_i = s_j, \text{ for some } 1 \leqq i \neq j \leqq n \right\} \right\rangle$$

for the $\mathcal{A}(U)$*-submodule of (7.5), generated by the* (decomposable) *tensors of (7.5), as indicated* above. (Here our usual notice, concerning *"unadorned tensor products"*, as in (7.6), is still in force, even if not explicitly noted; so cf. (7.5)).

Furthermore, by still following the classical pattern, we extend the previous definition to $n = 0, 1$; namely, we first consider the *(complete)* $\Gamma(\mathcal{A})$*-presheaves*

$$(7.7)\qquad \bigwedge^0 \mathcal{E}(U) := \mathcal{A}(U), \quad \textit{as well as,} \quad \bigwedge^1 \mathcal{E}(U) := \mathcal{E}(U),$$

for every *open* set $U \subseteq X$. Therefore, by taking the respective *sheafifications* of the above presheaves, one obtains;

$$(7.8)\qquad \bigwedge^0 \mathcal{E} = \mathcal{A} \quad \textit{and} \quad \bigwedge^1 \mathcal{E} = \mathcal{E},$$

within $\mathcal{A}$*-isomorphisms* of the $\mathcal{A}$-modules concerned (cf. also Chapt. I; (9.17)).

Thus, by looking at the *(complete) presheaves* $\Gamma(\mathcal{A})$ and $\Gamma(\mathcal{E})$, as before, (viz. the *presheaves of sections* of the $\mathcal{A}$-modules involved, cf. Chapt. I; (9.7)), one can repeat, within the present *sheaf-theoretic* context, the standard *exterior algebra machinery* for a given module, by taking instead an $\mathcal{A}$*-module* $\mathcal{E}$ on $X$, as above. So we further consider, for later use, the *exterior algebra of a given* $\mathcal{A}$*-module* $\mathcal{E}$ on $X$, denoted by $\bigwedge \mathcal{E}$, and given by the relation (see also Chapt. II; (3.10), (3.12))

$$(7.9)\qquad \bigwedge \mathcal{E} := \bigoplus_{n=0}^{\infty} \bigwedge^n \mathcal{E}.$$

Indeed, one concludes that (see also Chapt. II; Definition 6.2);

(7.10) $\bigwedge \mathcal{E}$ is an $\mathcal{A}$*-algebra sheaf* on $X$, *for any* given $\mathcal{A}$*-module* $\mathcal{E}$ on $X$.

Namely, one still infers that:

the sheaf $\wedge\mathcal{E}$, as defined by (7.9), can also be construed as that one on $X$, generated by the *presheaf of $\mathcal{A}(U)$-algebras* (cf. also (7.2))

(7.11) (7.11.1)
$$U \longmapsto \bigoplus_{n=0}^{\infty} \wedge^n(\mathcal{E}(U)) \equiv \wedge\mathcal{E}(U),$$

with $U$ *open* in $X$, so that (cf. Chapt. II; (1.54)) $\wedge\mathcal{E}$ is, in fact, an *$\mathcal{A}$-algebra sheaf* on $X$, as claimed by (7.10).

Indeed, one obtains the following *$\mathcal{A}$-isomorphism*

$$\mathbf{S}\left(\bigoplus_{n=0}^{\infty} \wedge^n(\mathcal{E}(U))\right) \equiv \mathbf{S}(\wedge(\mathcal{E}(U))) = \wedge\mathcal{E} \left(:= \bigoplus_{n=0}^{\infty} \wedge^n\mathcal{E}\right). \tag{7.12}$$

That is, *for every open $U \subseteq X$*, one gets the following *(canonical) $\mathcal{A}(U)$-morphism of* the *$\mathcal{A}(U)$-modules* involved (cf. also Chapt. I; (11.5), along with Chapt. II; (3.11)),

$$\begin{aligned} \wedge\mathcal{E}(U) \equiv \bigoplus_{n=0}^{\infty} \wedge^n(\mathcal{E}(U)) &\xrightarrow{\phi_U} \Gamma(U, \wedge\mathcal{E}) \equiv (\wedge\mathcal{E})(U) \\ &:= \left(\bigoplus_{n=0}^{\infty} \wedge^n\mathcal{E}\right)(U) = \bigoplus_{n=0}^{\infty} (\wedge^n\mathcal{E})(U); \end{aligned} \tag{7.13}$$

therefore (cf. Chapt. I; (8.5) and (13.19)), one gets a *morphism,* say, $\tilde{\phi}$ *of* the *sheaves, generated by the $\Gamma(\mathcal{A})$-presheaves appeared in* the two members of (7.13). So (cf. Chapt. I; (1.21) and (8.11)), it suffices to prove that $\tilde{\phi}$ *is an $\mathcal{A}$-isomorphism fiber-wise;* however, this now is immediate, by virtue of Chapt. I; Corollary 9.1 (see thus, in particular, (9.16) there), and of the following relations, in effect, *$\mathcal{A}_x$-isomorphisms:*

$$\begin{aligned} \mathbf{S}\left(\bigoplus_{n=0}^{\infty} \wedge^n(\mathcal{E}(U))\right)_x &= \left(\bigoplus_{n} \wedge^n(\mathcal{E}(U))\right)_x = \varinjlim_U \left(\bigoplus_n \wedge^n(\mathcal{E}(U))\right) \\ &\equiv \varinjlim_U (\wedge(\mathcal{E}(U))) = \wedge\left(\varinjlim_U \mathcal{E}(U)\right) = (\mathcal{E}_x) \\ &= \text{by the 2nd term above and Chapt. II; (3.15))} \bigoplus_n (\wedge^n(\mathcal{E}(U)))_x \\ &= \bigoplus\left(\varinjlim_U \wedge^n(\mathcal{E}(U))\right) = \bigoplus_n (\wedge^n\mathcal{E})_x = \left(\bigoplus \wedge^n\mathcal{E}\right)_x \equiv (\wedge\mathcal{E})_x, \end{aligned} \tag{7.14}$$

*for every point* $x \in X$. (In this concern, see also, for instance, N. Bourbaki [2: Chap. 3; p.83, Proposition 9 and Chap. 2; p.91, Proposition 5]). ∎ In particular, the following extract from (7.14) will be useful right below; viz. we have

$$(7.14') \qquad (\bigwedge \mathcal{E})_x = \bigoplus \varinjlim_U \bigwedge^n \mathcal{E}(U) = \varinjlim_U \bigoplus_n \bigwedge^n \mathcal{E}(U) = \bigwedge \mathcal{E}_x, \ x \in X.$$

Therefore, defining the algebraic *operations in* $\bigwedge \mathcal{E}$, especially the *multiplication* in it, denoted by "$\wedge$", *section-wise* (cf. Chapt. I; (2.9) and Chapt. II; (1.67), yet (7.14′) above), for any given *(local) sections* $s, t$ of $\bigwedge \mathcal{E}$, with $s \in \bigwedge^n \mathcal{E}(U)$ and $t \in \bigwedge^m \mathcal{E}(U)$, for some *open* $U \subseteq X$, one obtains

$$(7.15) \qquad s \wedge t \in \bigwedge^{n+m} \mathcal{E}(U),$$

such that , according to the classical theory, one further gets

$$(7.16) \qquad s \wedge t = (-1)^{n \cdot m} t \wedge s,$$

the rest of *the standard properties of the exterior multiplication being* certainly *true*.

Now, in order to examine the *behaviour of vector sheaves relative to exterior products* (cf. thus Lemma 7.1 below), that will also be of use in the sequel, we have to look first at the case of *free* $\mathcal{A}$*-modules,* in particular, of the *standard* ones (see 3.1)): So, based on the classical theory (cf. e.g. S. Lang [1: p.590, Proposition 9.1]), one first remarks that, *for any open set* $U \subseteq X$ *and* $n \in \mathbf{Z}_+$, one obtains the following $\mathcal{A}(U)$*-isomorphism of* the $\mathcal{A}(U)$*-modules* concerned;

$$(7.17) \qquad \bigwedge^n \mathcal{A}^m(U) = \mathcal{A}^p(U), \ \ \textit{with} \ p \equiv \binom{m}{n}, \ \ n \leqq m,$$

(see also (3.3)). So one further obtains (cf. also (7.1), (7.2));

$$(7.18) \qquad \begin{aligned} \bigwedge^n \mathcal{A}^m := \mathbf{S}\left(\bigwedge^n \Gamma(\mathcal{A}^m)\right) = \mathbf{S}\left(\bigwedge^n \Gamma(\mathcal{A})^m\right) = \mathbf{S}(\Gamma(\mathcal{A})^p) \\ = \mathbf{S}(\Gamma(\mathcal{A}^p)) = \mathcal{A}^p, \end{aligned}$$

(cf. also Chapt. I (13.20)), up to $\mathcal{A}$*-isomorphisms of* the $\mathcal{A}$*-modules* involved, with $p$ as in (7.17). Thus, one concludes that;

$\bigwedge^n \mathcal{A}^m$ *is a free* $\mathcal{A}$*-module of rank* $p \equiv \binom{m}{n}$, *for every* $n \in \mathbf{Z}_+$, *with* $n \leqq m \equiv \mathrm{rk}_{\mathcal{A}}(\mathcal{A}^m)$ (see Chapt. II; (4.2)). *In particular, one obtains*

(7.19) (7.19.1) $$\bigwedge^n \mathcal{A}^m = 0,$$

*for every* $n > m$. Yet, it is clear (cf. Chapt. II; Definition 3.2) that *the* previous *conclusion is valid, for every free* $\mathcal{A}$*-module* $\mathcal{E}$ *on* $X$, *with* $\mathrm{rk}\,\mathcal{E} = m$.

So we come next to the following.

**Lemma 7.1.** *Let* $(X, \mathcal{A})$ *be a* $\mathbf{C}$*-algebraized space and* $\mathcal{E}$ *a given vector sheaf on* $X$, *of rank* $m \in \mathbf{N}$. *Then, for every* $n \in \mathbf{Z}_+$, $\bigwedge^n \mathcal{E}$ *is still a vector sheaf on* $X$, *whose rank is given by*

(7.20) $$\mathrm{rk}_{\mathcal{A}}(\bigwedge^n \mathcal{E}) \equiv \mathrm{rk} \bigwedge^n \mathcal{E} = \binom{m}{n} \equiv \binom{\mathrm{rk}\,\mathcal{E}}{n}.$$

**Proof.** Taking a *local frame of* $\mathcal{E}$, say, $\mathcal{U} = \{U\}$ (cf. Chapt. II; Definition 4.2), based further on (7.18) and Chapt. I; Lemma 9.2 and setting $p \equiv \binom{m}{n}$, as above, one obtains, *for every open set* $U \in \mathcal{U}$, the following $\mathcal{A}|_U$*-isomorphisms of* the $\mathcal{A}|_U$*-modules* concerned;

(7.21) $$\begin{aligned}(\bigwedge^n \mathcal{E})|_U &= \mathbf{S}(\bigwedge^n \Gamma(\mathcal{E}))|_U = \mathbf{S}((\bigwedge^n \Gamma(\mathcal{E}))|_U) = \mathbf{S}(\bigwedge^n (\Gamma(\mathcal{E})|_U)) \\ &= \mathbf{S}(\bigwedge^n (\Gamma(\mathcal{A}^m)|_U)) = \mathbf{S}((\bigwedge^n \Gamma(\mathcal{A}^m))|_U) = \mathbf{S}(\bigwedge^n \Gamma(\mathcal{A}^m))|_U = \mathcal{A}^p|_U,\end{aligned}$$

which thus proves our claim. ∎

In this context, we still note that in the last calculations above we have also employed the following relation

(7.22) $$(\bigwedge^n \Gamma(\mathcal{E}))|_U = \bigwedge^n (\Gamma(\mathcal{E})|_U),$$

*true,* by the very definitions (cf. (7.2) and Chapt. I; (5.11), (5.12)), *for every open* $U \subseteq X$, *and* $n \in \mathbf{Z}_+$.

Now, a useful application of the above is further considered by the following Subsection.

**7.(a). Determinant line sheaf.**– We examine below a particular *line sheaf,* which is *(canonically) associated with any* given *vector sheaf* on $X$, along with the counterpart in our case of the classical notion of *determinant*. So we start with the following.

**Definition 7.1.** Let $\mathcal{E}$ be a given vector sheaf on $X$. Then, we call the *determinant line sheaf of* $\mathcal{E}$, denoted by $\det(\mathcal{E})$, the *top* (or *maximal*) *exterior power of* $\mathcal{E}$; viz. we set

$$\det(\mathcal{E}) := \bigwedge^{\mathrm{rk}\mathcal{E}} \mathcal{E}. \tag{7.23}$$

Indeed, accordingto Lemma 71, $\det(\mathcal{E})$, as given by (7.23), is a vector sheaf on $X$, such that (cf. (7.20))

$$\mathrm{rk}\det(\mathcal{E}) = 1; \tag{7.24}$$

hence, $\det(\mathcal{E})$ *is* really *a line sheaf* on $X$ (cf. Chapt. II; Definition 4.3), which thus justifies, partly, the terminology for (7.23), while the rest justification has to do with the ensuing discussion, namely, the notion of the determinant map.

However, first, as a consequence of the last conclusion and of Chapt. II; Lemma 6.2, one further infers that, *for every vector sheaf* $\mathcal{E}$ on $X$, one has

$$\mathcal{E}nd(\det(\mathcal{E})) = \mathcal{A}, \tag{7.25}$$

up to an $\mathcal{A}$-*isomorphism of* the $\mathcal{A}$-*algebra sheaves* concerned (see also Chapt. II; (6.29)). On the other hand, *for any vector sheaf* $\mathcal{E}$ on $X$, *and* $m \in \mathbf{Z}_+$ (hence, for $n = \mathrm{rk}\,\mathcal{E}$, as well), one obtains

$$(\bigwedge^m \mathcal{E})_x = \bigwedge^m \mathcal{E}_x, \tag{7.26}$$

*for every* $x \in X$, up to an $\mathcal{A}_x$-*isomorphism of* the $\mathcal{A}_x$-*modules* involved: The assertion follows immediately from a similar argument to that in (7.21) (cf. also (7.1) and Chapt. II; (4.9)).

We come now, as already promised, to the notion of the *determinant map* (or even *morphism*), which is (canonically) associated with any given

$$u \in \mathrm{End}\,\mathcal{E} \equiv \mathrm{Hom}_{\mathcal{A}}(\mathcal{E}, \mathcal{E}) \tag{7.27}$$

(see Chapt. II; (6.30)); here, as before, $\mathcal{E}$ stands for a *vector sheaf* on $X$, *with* $\mathrm{rk}\,\mathcal{E} = n \in \mathbf{N}$. The map we are going to define generalizes also the determinant morphism, already considered in the preceding by (4.1), to the case of vector sheaves (cf. (7.48) below). Thus, for convenience, we shall employ for this extended notion too, the same notation.

Now, by virtue of Chapt. I; (13.19), for any given morphism $u$, as in (7.27), one obtains a (uniquely defined) *presheaf morphism* of $\Gamma(\mathcal{E})$ (viz. of the (complete) presheaf of sections of $\mathcal{E}$) into itself, that is, an element

$$\Gamma(u) \equiv \overline{u} = (\overline{u}_V)_{V \in \mathcal{T}_X} \in \operatorname{End} \Gamma(\mathcal{E}) \tag{7.28}$$

(cf., for instance, Chapt. I; (13.3) and (6.3)). Indeed, as easily follows from the same definitions (cf. Chapt. I; (12.49), (12.50)), the above map $\overline{u}$ provides a *morphism of* $\Gamma(\mathcal{A})$*-presheaves.* Thus, by considering the components of $\overline{u}$, as above, one obtains, *for every* $m \in \mathbf{Z}_+$, the corresponding *$m$-th exterior power* of each component *$\mathcal{A}(V)$-morphism,* for any *open* $V \subseteq X$, viz.

$$\bigwedge^m \overline{u}_V : \bigwedge^m \mathcal{E}(V) \longrightarrow \bigwedge^m \mathcal{E}(V). \tag{7.29}$$

Furthermore, based on the classical *functorial properties* of the "*operator* $\bigwedge^m$" (cf., for instance, L. Chambadal - J.L. Ovaert [1: p.214, 3. of Théorème 16.2]), one concludes that;

$$(\bigwedge^m \overline{u}_V) \equiv \bigwedge^m \overline{u} \in \operatorname{End} (\bigwedge^m \Gamma(\mathcal{E})) \tag{7.30}$$

(see also (7.2)). Hence (see Chapt. I; (13.19)), one finally obtains an *$\mathcal{A}$-morphism*

$$\widetilde{\bigwedge^m \overline{u}} \equiv \bigwedge^m u \in \operatorname{End} (\bigwedge^m \mathcal{E}) \equiv \operatorname{Hom}_{\mathcal{A}}(\bigwedge^m \mathcal{E}, \bigwedge^m \mathcal{E}). \tag{7.31}$$

The latter map is, in fact, an *$\mathcal{A}$-morphism of the vector sheaf $\bigwedge^m \mathcal{E}$ into itself* (cf. Lemma 7.1); we call it the *$m$-th exterior power of the* given *$\mathcal{A}$-morphism* $u \in \operatorname{End} \mathcal{E}$.

In particular, we call the *determinant of* $u \in \operatorname{End} \mathcal{E}$, denoted by $\partial\mathrm{et}\,(u)$, the *top exterior power of* $u$ (viz. the exterior power of $u$, corresponding to the top exterior power of $\mathcal{E}$); that is, we set

$$\partial\mathrm{et}\,(u) := \bigwedge^{\mathrm{rk}\mathcal{E}} u, \tag{7.32}$$

for any given $u \in \operatorname{End} \mathcal{E}$. Accordingly, one obtains

$$\partial\mathrm{et}\,(u) \equiv \bigwedge^{\mathrm{rk}\mathcal{E}} u \in \operatorname{End} (\det (\mathcal{E})) = \mathcal{A}(X) \tag{7.33}$$

(cf. (7.31), (7.23) and (7.25), along with Chapt. II; (6.33)), which thus justifies (however, still partly!, so see also (7.46) below) the terminology applied for $\partial\mathrm{et}\,(u)$. Yet, the last equality in (7.33) stands, in effect, for an *$\mathcal{A}(X)$-isomorphism of* the *$\mathcal{A}(X)$-algebras* involved (cf. Chapt. II; (6.31), (6.33)).

Now, to justify *completely*(!) the terminology applied so far for $\partial\mathrm{et}\,(u)$, one has to consider, according to the classical pattern, *free $\mathcal{A}$-modules,* yet more generally, concerning our present context, *locally free ones,* as was actually tha case above; so, in other words, one has to look at the preceding situation *locally.* Therefore, assume that

$$\mathcal{U} = \{V\} \tag{7.34}$$

is a *local frame* of a given *vector sheaf* $\mathcal{E}$ on $X$ (see Chapt. II; Definition 4.2) and let, for $V \in \mathcal{U}$,

$$(e_i^V)_{1 \leqq i \leqq n} \subseteq \mathcal{E}(V) = \mathcal{A}^n(V) = \mathcal{A}(V)^n, \quad \textit{with } n = \operatorname{rk} \mathcal{E}, \tag{7.35}$$

be the *canonical (Kronecker) local gauge* of the *free* $\mathcal{A}|_V$*-module* $\mathcal{E}|_V = \mathcal{A}^n|_V$ (cf. (3.11) and Chapt. II; (4.7)). Furthermore, consider the following (finite) sequence of *(local) sections of* $\mathcal{E}$ *over* $V \in \mathcal{U}$;

$$(s_i)_{1 \leqq i \leqq n} \subseteq \mathcal{E}(V), \quad \textit{viz.}\ (s_i) \in \mathcal{E}(V)^n = \mathcal{E}^n(V) \tag{7.36}$$

(cf also Chapt. II; (3.11)). *Equivalently* (cf. (7.55)), one has the following *section-matrix*

$$\alpha \equiv (\alpha_{ij}) \in M_n(\mathcal{A}(V)), \tag{7.37}$$

where one has

$$s_i = \sum_{j=1}^{n} \alpha_{ij} e_j^V \quad \textit{with} \quad \alpha_{ij} \in \mathcal{A}(V) \quad \textit{and} \quad 1 \leqq i, j \leqq n \tag{7.38}$$

(see (3.9), (3.10), or even Chapt. II; (3.25)). Thus, based on the classical argument, one further defines an $\mathcal{A}(V)$*-morphism*

$$u_V \in \operatorname{End} \mathcal{E}(V), \tag{7.39}$$

such that one sets (cf. (7.35))

$$u_V(e_i^V) := s_i, \quad 1 \leqq i \leqq n. \tag{7.40}$$

Thus, by still following the standard theory, one further obtains, by setting in (7.35), $e_V \equiv (e_i^V)$ (see also (7.39))

$$\begin{aligned} \det{}_{e_V}(s_1, \ldots, s_n) &\equiv \det{}_{e_V}(u_V(e_1^V), \ldots, u_V(e_n^V)) := \textstyle\bigwedge^n u_V \\ &\in \operatorname{End}(\textstyle\bigwedge^n \mathcal{E}(V)) \equiv \operatorname{End}(\det(\mathcal{E}(V))) = \mathcal{A}(V), \end{aligned} \tag{7.41}$$

such that, *by analogy with the classical situation, one still has*

$$(7.42)\quad (\bigwedge^n u_V)\left(\bigwedge_{i=1}^{n} e_i^V\right)=\bigwedge_{i=1}^{n} s_i=\det{}_{e_V}(s_1,\ldots,s_n)\cdot\bigwedge_{i=1}^{n} e_i^V\in\bigwedge^n\mathcal{E}(V)=\mathcal{A}(V),$$

the last equalities in (7.41) and (7.42) being, in fact, $\mathcal{A}(V)$-*isomorphisms,* due to the hypothesis for $\mathcal{E}$ and the open $V\subseteq X$ (cf. e.g. (7.35) and (7.34)). So we actually employed before a usual abuse of notation, by *identifying the* $\mathcal{A}(V)$-*endomorphism* $\bigwedge^n u_V$, as in (7.41), *with the corresponding section in* $\mathcal{A}(V)$. Furthermore, in view of (7.37), (7.41) and (7.42), one also obtains, by following again the classical pattern;

$$(7.43)\quad \det{}_V(\alpha)\equiv\det{}_V(\alpha_{ij})=\det{}_{e_V}(s_1,\ldots,s_n)=\det{}_V(u_V)\in\mathcal{A}(V).$$

On the other hand, starting now with an $\mathcal{A}$-*morphism* $u\in\operatorname{End}\mathcal{E}$, with $\mathcal{E}$, as before, and taking the $\mathcal{A}(V)$-*morphism* (cf. Chapt. II; (13.3))

$$(7.44)\quad u_V\equiv\overline{u}_V:=\Gamma(u)_V\in\operatorname{End}\mathcal{E}(V),$$

*with* $V\in\mathcal{U}$ (cf. (7.34)), one has

$$(7.45)\quad \overline{u}_V(e_i^V)\equiv s_i\in\mathcal{E}(V),\ \ 1\leqq i\leqq n,$$

such that one has (cf. (7.43), (7.37) and (7.33), along with (4.7))

$$(7.46)\quad \begin{aligned}\det{}_{e_V}(s_1,\ldots,s_n)&=\det{}_V(\overline{u}_V)=\bigwedge^n\overline{u}_V=\det{}_V(\alpha)\equiv\det{}_V(\alpha_{ij})\\&=\partial\mathrm{et}\,(u)_V\equiv\overline{\partial\mathrm{et}\,(u)}_V=\overline{\partial\mathrm{et}}_V(\alpha)\in\mathcal{A}(V),\end{aligned}$$

which thus *completely*(!) justifies the terminology for $\partial\mathrm{et}\,(u)$ applied hitherto. In this context, notice that we have also made use in (7.46) of the easy conclusion that (cf. also (7.26));

(7.47) (7.2) *is,* in effect, *a monopresheaf* (cf. Chapt. I; Remark 11.1, i) and (11.36′)).

Thus, by the preceding, one actually obtains the following map, which, according to (7.46), generalizes (4.1); viz. one has,

$$(7.48)\quad \partial\mathrm{et}\ :\operatorname{End}\mathcal{E}\longrightarrow\operatorname{End}(\bigwedge^{\mathrm{rk}\mathcal{E}}\mathcal{E})\equiv\operatorname{End}(\det(\mathcal{E}))=\mathcal{A}(X),$$

given by (7.33), for any *vector sheaf* $\mathcal{E}$ on $X$. We call (7.48) the *determinant map* (or *morphism*), associated with the given vector sheaf $\mathcal{E}$ on $X$, its values being thus, by definition, *global sections* of the *structure sheaf* $\mathcal{A}$.

Applications of the above will be given in the sequel, including Part II of this treatise. In particular, we shall be concerned below with the so-called *star* (or even *Hodge*) *operator,* as this classical notion appears, within the present *sheaf-theoretic* framework. However, to this end, we shall have to introduce first some more terminology, which thus we explain in the next Section.

# 8. Riemannian $\mathcal{A}$-modules

The terminology employed in the heading of this Section is, of course, *geometrically* inclined, this because our main motivation and application too, is from and to *Differential Geometry,* as the latter, to the extent we consider it hereby, is formulated, within the present *abstract sheaf-theoretic context* (see Part Two of this work). So the same $\mathcal{A}$-modules, we deal with in the sequel, might be called, equally well, *(pre-) Hilbert* $\mathcal{A}$*-modules* or even, in general, *inner* (or *semi-inner,* as the case may be) *product* $\mathcal{A}$*-modules.* On the other hand, *"sheaves endowed with metrics"* (in fact, *fiber-wise*) appear, of course, within some other context, already in various aspects of *Algebraic Geometry* (see, for instance, S. Lang [2]). Yet, the latter case was also the initial motivation to the ensuing discussion, by keeping, however, as already mentioned, a constant eye on *applications to Differential Geometry* (see thus Part Two).

So we start with the following basic notion for all that follows, as this is given by the next.

**Definition 8.1.** Let $(X, \mathcal{A})$ be an $\mathbf{R}$-*algebraized space* (viz. one now considers (cf. (0.1)), more generally, $\mathbf{R}$-*algebras* and, as usual, hitherto, always *unital* ones). Now, a given triple

$$(X, \mathcal{A}, \mathcal{P}), \tag{8.1}$$

*with* $\mathcal{P} \subseteq \mathcal{A}$, a *subsheaf* of $\mathcal{A}$ (cf. Chapt. I; (1.8′)), is said to be a *preordered* ($\mathbf{R}$-*algebraized*) *space,* whenever $\mathcal{P}$ (the *preorder* in $\mathcal{A}$) satisfies the following conditions:

$$\lambda \cdot \mathcal{P} \subseteq \mathcal{P}, \ \ \textit{for any } \lambda \in \mathbf{R}_+ \hookrightarrow \mathcal{A} \tag{8.2}$$

(here $\mathbf{R} \supset \mathbf{R}_+$ stands, of course, for the respective *constant* ($\mathbf{R}$-*algebra*) *sheaf* on $X$; see also e.g. Chapt. II; (1.34)),

$$\mathcal{P} + \mathcal{P} \subseteq \mathcal{P}, \tag{8.3}$$

as well as,

$$\mathcal{P} \cdot \mathcal{P} \subseteq \mathcal{P}. \tag{8.4}$$

(The indicated operations, as above, are meant, of course, *fiber-wise*). We call the previous set $\mathcal{P} \subseteq \mathcal{A}$ the *(pointed) multiplicative convex cone* of $\mathcal{A}$, or simply the *m-convex cone* of $\mathcal{A}$.

In particular, one speaks of a *(partially) ordered ($\mathbf{R}$-algebraized) space,* whenever the above set $\mathcal{P} \subseteq \mathcal{A}$ satisfies the relation;

$$\mathcal{P} \cap (-\mathcal{P}) = (0). \tag{8.5}$$

In this latter case we still say that $\mathcal{P}$ is a *(pointed) m-convex salient cone* of $\mathcal{A}$, or simply that $\mathcal{P}$ defines a *(partial) ordering,* or yet *order,* of $\mathcal{A}$.

Now, when we are given a $\mathbf{C}$-*algebraized space* $(X, \mathcal{A})$, as in the preceding (cf. (0.1)) and speak of the *ordering of $\mathcal{A}$,* we shall always mean that *the underlying $\mathbf{R}$-algebraized space has an ordering,* in the sense of the above Definition 8.1; that is, one thus assumes that there exists a *subsheaf* $\mathcal{P} \subseteq \mathcal{A}$, satisfying the previous conditions (8.2) – (8.5). In that case too, we just speak of an *ordered algebraized space* $(X, \mathcal{A})$.

In this context, the elements of $\mathcal{P} \subseteq \mathcal{A}$ are, by definition (in either case, as before), the *positive elements of $\mathcal{A}$*; so, denoting their set by $\mathcal{A}^+$, one has (alternatively)

$$\mathcal{A}^+ := \mathcal{P}. \tag{8.6}$$

In particular, a (local) *section of $\mathcal{A}$* is said to be *positive,* whenever its values are taken from $\mathcal{A}^+ \equiv \mathcal{P}$. Thus, *for every open set $U \subseteq X$,*

$$\Gamma(U, \mathcal{P}) \equiv \mathcal{P}(U) \equiv \mathcal{A}^+(U) \subseteq \mathcal{A}(U) \tag{8.7}$$

denotes the set of *positive sections of $\mathcal{A}$ over $U$.* Similarly, one considers $-\mathcal{P}$ (cf. (8.5)) as the set of *negative elements of $\mathcal{A}$,* denoting also their set by $\mathcal{A}^-$, as well, so that one still sets;

$$\mathcal{A}^- := -\mathcal{P} \subseteq \mathcal{A}. \tag{8.8}$$

Thus, by analogy with (8.7), one considers over any *open* $U \subseteq X$, the *negative sections of* $\mathcal{A}$, viz. the set $\Gamma(U, -\mathcal{P})$; indeed, by considering e.g. the *complete presheaf* $-\Gamma(\mathcal{P})$ (viz. $U \longmapsto -\mathcal{P}(U)$, $U \in \mathcal{T}_X$), it is easy to see (cf. also Chapt. I; (2.9) and Chapt. II; (1.67)) that $-\mathcal{P}$ *is a subsheaf of* $\mathcal{A}$, as well, so that one obtains

(8.9) $$\Gamma(U, -\mathcal{P}) = -\Gamma(U, \mathcal{P}),$$

within a *bijection, for every open* $U \subseteq X$.

On the other hand, it is still clear (see also Chapt. II; (1.44), (1.54) and (1.67)) that;

(8.10) *for every open* $U \subseteq X$, *the set of positive sections on* $U$, $\mathcal{P}(U)$, *defines a pointed m-convex salient cone of the* **R**-*algebra* $\mathcal{A}(U)$ (in the case, of course, that $\mathcal{P}$ has the corresponding properties, according to the above Definition 8.1). In other words, in that case too, $\mathcal{A}(U)$ *has a (partial) ordering,* supplied thus by $\mathcal{P}(U) \subseteq \mathcal{A}(U)$, *for every open* $U \subseteq X$.

So we come now to another fundamental notion for the sequel, as given by the following.

**Definition 8.2.** Suppose we are given an *ordered algebraized space* $(X, \mathcal{A})$ (cf. Definition 8.1, in conjuction with the comments right after it), and let $\mathcal{E}$ be an $\mathcal{A}$-*module* on $X$.

Now, by an $\mathcal{A}$-*valued inner product* on $\mathcal{E}$, one means a sheaf morphism (cee Chapt. I; (1.11) and Chapt. II; (3.3))

(8.11) $$\rho : \mathcal{E} \oplus \mathcal{E} \longrightarrow \mathcal{A},$$

satisfying the following conditions:

i) $\rho$ is an $\mathcal{A}$-*bilinear morphism,* concerning the $\mathcal{A}$-modules involved in (8.11) (see also Chapt. II; (3.6)).

ii) $\rho$ is *positive definite;* viz., for any (local) section $s \in \mathcal{E}(U)$ over an open $U \subseteq X$, one has

(8.12) $$\rho(s, s) \in \mathcal{P}(U), \quad \textit{such that} \quad \rho(s, s) = 0 \ \text{(if and)} \ \textit{only if} \quad s = 0.$$

(Here we have used, in fact, the *map on sections* of the $\mathcal{A}$-modules involved in (8.11), *provided by* $\rho$, retaining, for convenience, the same notation, as we always do below, as well; cf. also Chapt. II; (3.14) and Chapt. I; (12.49), as well as, (8.15) in the sequel).

iii) $\rho$ is *symmetric;* i.e., one has

$$\rho(s,t) = \rho(t,s), \tag{8.13}$$

for any two (local) sections $s, t \in \mathcal{E}(U)$, with $U$ open in $X$.

Thus, a pair

$$(\mathcal{E}, \rho), \tag{8.14}$$

with $\mathcal{E}, \rho$, as before, is called an *inner product $\mathcal{A}$-module* on $X$.

On the other hand, concerning the previous rels. (8.12) and (8.13), we have actually employed the *induced presheaf map by* $\rho$, applying, however, for simplicity, as already remarked, the same notation. That is, one sets

$$\Gamma(\rho) \equiv \rho : \Gamma(\mathcal{E} \oplus \mathcal{E}) = \Gamma(\mathcal{E}) \oplus \Gamma(\mathcal{E}) \longrightarrow \Gamma(\mathcal{A}), \tag{8.15}$$

where the equality involved stands for a $\Gamma(\mathcal{A})$*-isomorphism of* the $\Gamma(\mathcal{A})$-*presheaves* concerned; see Chapt. II; (3.14), as well as, Chapt. I; (12.49) and (6.10).

Now, given an *inner product $\mathcal{A}$-module* $(\mathcal{E}, \rho)$, as before (see (8.14)), one obtains the following (canonical) *$\mathcal{A}$-morphism of* the *$\mathcal{A}$-modules* concerned (see also Definition 5.1)

$$\tilde{\rho} : \mathcal{E} \longrightarrow \mathcal{E}^* \equiv \mathcal{H}om_{\mathcal{A}}(\mathcal{E}, \mathcal{A}), \tag{8.16}$$

such that one sets

$$\tilde{\rho}(s) \equiv \rho_s : t \longmapsto \rho_s(t) := \rho(s,t), \tag{8.17}$$

for any (local) sections $s, t \in \mathcal{E}(U)$, with $U$ open in $X$. (The assertion is justified by the properties of $\rho$; cf. Definition 8.2, i). Yet, as already noted before, one actually employs in (8.17) the *corresponding to $\tilde{\rho}$ section morphism* (see Chapt. I; (12.49) and also Chapt. II; (6.4)). On the other hand, the hypothesis that $\rho$ is positive - definite (cf. Definition 8.2, ii)) implies that *the* above *map* $\tilde{\rho}$, as defined by (8.17), *is*, in effect, *an $\mathcal{A}$-isomorphism;* viz. one obtains

$$\mathcal{E} \underset{\tilde{\rho}}{\overset{\subseteq}{\rightarrow}} \mathcal{E}^*, \tag{8.18}$$

within an *$\mathcal{A}$-isomorphism (into) of* the *$\mathcal{A}$-modules* concerned, given by (8.16).

Thus, an $\mathcal{A}$-valued inner product $\rho$ on a given $\mathcal{A}$-module $\mathcal{E}$, as above (cf. (8.11)), is said to be *strongly non-degenerate,* whenever the previous $\mathcal{A}$-isomorphism is *onto;* that is, one then obtains, by definition,

$$\mathcal{E} \underset{\tilde{\rho}}{\cong} \mathcal{E}^*, \tag{8.19}$$

up to an *$\mathcal{A}$-isomorphism of* the *$\mathcal{A}$-modules* involved, given by $\tilde{\rho}$.

So we finally come to the following basic notion for the ensuing discussion.

**Definition 8.3.** Suppose we are given the context of Definition 8.2. So by a *Riemannian $\mathcal{A}$-module $\mathcal{E}$* on $X$, one means an $\mathcal{A}$-module $\mathcal{E}$ on $X$, endowed with a strongly non-degenerate $\mathcal{A}$-valued inner product (cf. Definition 8.2 and (8.19)). Yet, we refer to the latter map (cf. (8.11)) as a *(Riemannian) $\mathcal{A}$-metric,* while we shall also employ the notation

$$(\mathcal{E}, \rho) \tag{8.20}$$

(cf. (8.14)) for a given Riemannian $\mathcal{A}$-module $\mathcal{E}$ on $X$. Moreover, the term *metrized $\mathcal{A}$-module* will be of use too, for a pair as in (8.20).

Now, consider, in particular, *a Riemannian vector sheaf* $(\mathcal{E}, \rho)$; viz. we are given a *vector sheaf* $\mathcal{E}$ on $X$ (cf. Chapt. II; Definition 4.3), endowed with an *$\mathcal{A}$-metric* $\rho$, as above. Yet, suppose that $\operatorname{rk} \mathcal{E} = n \in \mathbf{N}$, and let $\mathcal{U} = \{U\}$ be a given *local frame* of $\mathcal{E}$, so that one has;

$$\mathcal{E}|_U = \mathcal{A}^n|_U, \tag{8.21}$$

up to an *$\mathcal{A}|_U$-isomorphism of* the *$\mathcal{A}|_U$-modules* concerned, *for every* $u \in \mathcal{U}$ (see Chapt. II; Definition 4.2). So our purpose next is to *determine* $\rho$, as before, *in terms of the local frame* $\mathcal{U}$ (cf. Theorem 8.1 below). However, to this end, it is useful first to remark, in particular, that;

(8.22) *a given $\mathcal{A}$-metric $\rho$ on the* standard *free $\mathcal{A}$-module $\mathcal{A}^n$ is uniquely defined as*

$$\rho = (\alpha_{ij}) \in \mathcal{GL}(n, \mathcal{A})(X) = \mathrm{GL}(n, \mathcal{A}(X)), \tag{8.22.1}$$

such that *one sets*

$$\alpha_{ij} := \rho(\varepsilon_i, \varepsilon_j) \in \mathcal{A}(X), \quad 1 \leqq i, j \leqq n, \tag{8.22.2}$$

where $(\varepsilon_i) \subseteq \mathcal{A}^n(X) = \mathcal{A}(X)^n$ (cf. (3.3)) denotes the (standard) *Kronecker gauge of $\mathcal{A}^n$.*

Indeed, the assertion follows, first, from (8.22.2) and (3.4), along with the hypothesis that the map $\rho$ (cf. (8.11)), for $\mathcal{E} = \mathcal{A}^n$) *is an $\mathcal{A}$-bilinear morphism;* so, *for any* $\alpha, \beta \in \mathcal{A}^n(U)$, with $U$ *open* in $X$, one obtains

$$\begin{aligned}\rho(\alpha,\beta) = \rho\left(\sum_{i=1}^{n} \alpha_i \varepsilon_i^U, \sum_{j=1}^{n} \beta_j \varepsilon_j^U\right) &= \sum_{i,j} \alpha_i \beta_j \cdot \rho(\varepsilon_i^U, \varepsilon_j^U) \\ &\equiv \sum_{i,j} \alpha_i \beta_j (\alpha_{ij}|_U),\end{aligned} \tag{8.23}$$

where we set $\varepsilon_i^U \equiv \varepsilon_i|_U \in \mathcal{A}^n(U)$, $1 \leqq i \leqq n$, which thus proves the *uniqueness of the definition of* $\rho$, via (8.22.2). Furthermore, *the* resulting *section matrix*

$$\rho \equiv (\alpha_{ij}) \in M_n(\mathcal{A}(X)) = M_n(\mathcal{A})(X) \tag{8.24}$$

(see also (1.9)) *is invertible;* that is, one actually has

$$\rho \equiv (\alpha_{ij}) \in M_n(\mathcal{A}(X))^{\cdot} \equiv \mathrm{GL}(n, \mathcal{A}(X)) = \mathcal{GL}(n, \mathcal{A})(X) \tag{8.25}$$

(cf. (1.26) and (1.28)), due to the *positive-definiteness* (or even *non-degeneracy*) *of the map* $\rho$, as above (see also (8.18)), according to the corresponding classical argument (cf., for instance, S. Lang [1: p.465, Proposition 6.1]). ▮

Of course, one can consider, more generally, instead of $\mathcal{A}^n$, any *free $\mathcal{A}$-module* $\mathcal{E}$ on $X$, *of rank* $n \in \mathbf{N}$, such that ;

(8.26)

> *if* $(e_i)_{1 \leqq i \leqq n} \subseteq \mathcal{E}(X)$ *is a (global) gauge of* $\mathcal{E}$ (see (3.11)), *then any matrix* $\alpha \equiv (\alpha_{ij}) \in M_n(\mathcal{A}(X))$ *defines uniquely, via*
>
> $$\rho(e_i, e_j) := \alpha_{ij} \in \mathcal{A}(X), \ 1 \leqq i, j \leqq n, \tag{8.26.1}$$
>
> *an $\mathcal{A}$-bilinear morphism* (cf. also (8.23) and (3.9))
>
> $$\rho : \mathcal{E} \oplus \mathcal{E} \longrightarrow \mathcal{A}. \tag{8.26.2}$$

In particular, *the map* $\rho$, as above, *yields an* $\mathcal{A}$*-metric* if, in addition, the given section-matrix $\alpha \in M_n(\mathcal{A}(X))$ is:

i) *symmetric;* viz. one has

$$(8.26.3) \qquad {}^t\alpha = \alpha,$$

where we set ${}^t\alpha := (\alpha_{ji})$ (for the *transpose* of $\alpha$), and

ii) *invertible* in $M_n(\mathcal{A}(X))$, i.e., one has,

$$(8.26.4) \qquad \alpha \equiv (\alpha_{ij}) \in \mathrm{GL}(n, \mathcal{A}(X)).$$

*The converse* of the above *holds* obviously *true, for any given* $\mathcal{A}$*-metric* $\rho$ *on* $\mathcal{E}$.

(Same proposition of the classical theory, as before). ∎

Now, as already promised, we can *localize the preceding* by taking, more generally, *vector sheaves* on $X$. So one obtains the following.

**Theorem 8.1.** *Suppose we are given a Riemannian vector sheaf* $(\mathcal{E}, \rho)$ *and let* $\mathcal{U} = \{U_\alpha\}_{\alpha \in I}$ *be a local frame of* $\mathcal{E}$ (cf. (8.21)). *Then, the* $\mathcal{A}$*-metric* $\rho$ *restricted to* $\mathcal{U}$, we denote it by $\rho_{\mathcal{U}}$, *viz. the map*

$$(8.27) \qquad \rho_{\mathcal{U}} : \mathcal{E} \oplus \mathcal{E}|_{\mathcal{U}} \longrightarrow \mathcal{A}|_{\mathcal{U}},$$

pertaining to the respective (local) sections of the $\mathcal{A}$-modules involved (see also e.g. (8.15)), *is given by the relation;*

$$(8.28) \qquad \rho_{\mathcal{U}} = (\rho_\alpha) \in \prod_{\alpha \in I} M_n(\mathcal{A}(U_\alpha)).$$

*Indeed,* $\rho_{\mathcal{U}}$ *is characterized by the following* 0*-cocycle;*

$$(8.29) \qquad \begin{aligned} \rho_{\mathcal{U}} = (\rho_\alpha) \in Z^0(\mathcal{U}, \mathcal{GL}(n, \mathcal{A})) &= \Gamma(X, \mathcal{GL}(n, \mathcal{A})) \\ \equiv \mathcal{GL}(n, \mathcal{A})(X) &= \mathrm{GL}(n, \mathcal{A}(X)) \end{aligned}$$

(see also Chapt. III; Subsection 11. (a), for the notation applied).

**Proof.** To prove the assertion, we employ another useful *interpretation of the $\mathcal{A}$-metric $\rho$, in terms of tensor products,* following again the classical argument: Thus, by hypothesis for $\rho$ and (8.15), one obtains, *for every open* $U \subseteq X$, an *$\mathcal{A}(U)$-bilinear morphism*

$$\rho_U : \mathcal{E}(U) \oplus \mathcal{E}(U) \longrightarrow \mathcal{A}(U) \tag{8.30}$$

*of* the *$\mathcal{A}(U)$-modules* concerned, so that, according to the standard theory, one has an *$\mathcal{A}(U)$-linear morphism*

$$\overline{\rho}_U : \mathcal{E}(U) \otimes_{\mathcal{A}(U)} \mathcal{E}(U) \longrightarrow \mathcal{A}(U); \tag{8.31}$$

hence, we finally obtain a *presheaf morphism of* the *$\Gamma(\mathcal{A})$-presheaves* involved,

$$\overline{\rho} : \Gamma(\mathcal{E}) \otimes_{\mathcal{A}(U)} \Gamma(\mathcal{E}) \longrightarrow \Gamma(\mathcal{A}) \tag{8.32}$$

(cf. also Chapt. II; (5.5), concerning the notation applied). In particular, by taking the given *local frame* $\mathcal{U} = \{U_\alpha\}_{\alpha\in I}$ of $\mathcal{E}$ and setting for (8.31) $\overline{\rho}_{U_\alpha} \equiv \overline{\rho}_\alpha$, $\alpha \in I$, one has, in view of (8.32), *for every* $\alpha \in I$;

$$\begin{aligned}\overline{\rho}_\alpha &\in \mathrm{Hom}_{\mathcal{A}(U_\alpha)}(\mathcal{E}(U_\alpha) \otimes_{\mathcal{A}(U_\alpha)} \mathcal{E}(U_\alpha), \mathcal{A}(U_\alpha))\\ &= \mathrm{Hom}_{\mathcal{A}|_{U_\alpha}(U_\alpha)}(\mathcal{E}|_{U_\alpha}(U_\alpha) \otimes_{\mathcal{A}|_{U_\alpha}(U_\alpha)} \mathcal{E}|_{U_\alpha}(U_\alpha), \mathcal{A}|_{U_\alpha}(U_\alpha))\\ &= \mathrm{Hom}_{\mathcal{A}(U_\alpha)}(\mathcal{A}^n(U_\alpha) \otimes_{\mathcal{A}(U_\alpha)} \mathcal{A}^n(U_\alpha), \mathcal{A}(U_\alpha))\\ &= \mathrm{Hom}_{\mathcal{A}(U_\alpha)}(\mathcal{A}^{n^2}(U_\alpha), \mathcal{A}(U_\alpha)) = \mathcal{A}^{n^2}(U_\alpha) = M_n(\mathcal{A}(U_\alpha))\end{aligned} \tag{8.33}$$

(see also Chapt. I; (11.40), Chapt. II; (5.18) and (6.22), as well as, (3.17) above). Thus, one obtains

$$\overline{\rho} \equiv (\rho_{ij}^{(\alpha)}) \in M_n(\mathcal{A}(U_\alpha)), \quad \alpha \in I. \tag{8.34}$$

Furthermore, due to the *canonical $\mathcal{A}(U)$-isomorphism* between the respective *$\mathcal{A}(U)$-modules* of maps in (8.30) and (8.31), one may omit, for simplicity, the *"bar"* notation from the above, so that (8.34) implies, in effect, (8.28). Yet, by employing the *same abuse of notation,* one actually has

$$\rho_\alpha \equiv \rho_{U_\alpha} = \rho|_{U_\alpha}, \quad \alpha \in I, \tag{8.35}$$

where the last map means, in fact, the *map on sections derived from the given $\rho$,* as in (8.11), *restricted to $U_\alpha$.* Therefore, (8.28) *yields,* in effect, *a* 0-

*cocycle as in (8.29)* (cf. also (8.25)). Conversely, *any such* 0-*cocycle provides a global section of* $\mathcal{GL}(n, \mathcal{A})$, *viz. an element of* $\mathrm{GL}(n, \mathcal{A}(X))$, so that, in view of (8.26), *an* $\mathcal{A}$-*metric for* $\mathcal{E}|_{\mathcal{U}}$, as in (8.27) (that is, speaking in terms of sections, only for *those* (local) *sections of* $\mathcal{E}$ *carried by the members of* $\mathcal{U}$), and this terminates the proof. ∎

We call the previous 0-cocycle, as defined by (8.29), the *Gram cocycle of the* $\mathcal{A}$-*metric* $\rho$, *associated with the local frame* $\mathcal{U}$ of the given Riemannian vector sheaf $(\mathcal{E}, \rho)$.

Applications of the preceding will be given throughout the ensuing discussion. On the other hand, we further examine the *existence of an* $\mathcal{A}$-*metric on a given vector sheaf, in the case the structure sheaf* $\mathcal{A}$ *carries such a metric:* Of course, this is still reminiscent of the classical situation one has, for example, in the case of Riemannian manifolds and of the (smooth) vector bundles living on them. So, to start with, we first need the following easy remark; namely

(8.36) supposing that $(\mathcal{A}, \rho)$ *is a Riemannian* (free) $\mathcal{A}$-*module, then the same is true for every free* $\mathcal{A}$-*module (of finite rank) on* $X$, $\mathcal{E}$; viz. *there exists a* (uniquely defined Riemannian) $\mathcal{A}$-*metric* $\tilde{\rho}$ *on* $\mathcal{E}$ too.

For simplicity, we first consider, however, the *standard case* $\mathcal{E} = \mathcal{A}^n$, so that the claimed $\mathcal{A}$-*metric*, say $\overline{\rho} \equiv \rho^n$ on $\mathcal{A}^n$ is then given by

$$\rho^n := \underbrace{\rho \oplus \cdots \oplus \rho}_{n-\text{times}} \equiv \sum_{i=1}^{n} \rho_i \; (\equiv \rho); \tag{8.37}$$

that is, for any $s, t \in \mathcal{A}^n(U) = \mathcal{A}(U)^n$, such that $s = (s_1, \ldots, s_n)$ and $t = (t_1, \ldots, t_n)$, with $s_i, t_i \in \mathcal{A}(U)$, $1 \leqq i \leqq n$, for some open $U \subseteq X$, one sets;

$$\overline{\rho}(s, t) := \sum_{i=1}^{n} \rho(s_i, t_i) \in \mathcal{A}(U), \tag{8.37$'$}$$

which thus *proves our assertion for* $\mathcal{A}^n$.

Now, suppose that

$$0 \longrightarrow \mathcal{A}^n \stackrel{\phi}{\longrightarrow} \mathcal{E} \longrightarrow 0 \tag{8.38}$$

is the $\mathcal{A}$-*isomorphism,* defining a *free* $\mathcal{A}$-*module* $\mathcal{E}$ on $X$ *of rank* $n \in \mathbf{N}$ (see Chapt. II; Definition 3.2. Cf. also Section 3 in the preceding). So, within the context of (8.36), one obtains an *$\mathcal{A}$-metric on $\mathcal{E}$*, say $\tilde{\rho}$, by setting;

$$\tilde{\rho} := \rho^n \circ \left(\phi^{-1} \oplus \phi^{-1}\right) = \rho^n \circ (\phi \oplus \phi)^{-1}, \tag{8.39}$$

which thus proves our claim in (8.36). ∎

**Remark 8.1.**– The above relation (8.37′) suggests, *more generally,* the following fact:

(8.40) *any finite Whitney sum of Riemannian $\mathcal{A}$-modules is an $\mathcal{A}$-module of the same type.*

We come next, as already said, to the analogue, within the present framework, of a standard situation, occuring when one considers a pair $(X, \mathcal{A})$, as before, but having now the *base space $X$ paracompact* and the *structure sheaf $\mathcal{A}$ fine.* So, precisely speaking, one has the following.

**Theorem 8.2.** *Let $(X, \mathcal{A})$ be an ordered algebraized space, having $X$ paracompact (Hausdorff) and $\mathcal{A}$ a fine $\mathcal{A}$-module such that, in addition, $(\mathcal{A}, \rho)$ is a Riemannian $\mathcal{A}$-module. Then, every vector sheaf on $X$ is endowed with a symmetric $\mathcal{A}$-valued bilinear morphism* (cf. Definition 8.2; i) – iii)).

**Proof.** Let $\mathcal{E}$ be a *vector sheaf* on $X$, *of rank $n \in \mathbf{N}$*, and $\mathcal{V}$ a *local frame of $\mathcal{E}$*; hence, by hypothesis for $X$ (cf., for instance, J. Dugundji [1: p.162, Definition 2.1]), there exists a *locally finite (open) refinement of $\mathcal{V}$*, say,

$$\mathcal{U} = \{U_\alpha\}_{\alpha \in I}. \tag{8.41}$$

Now, $\mathcal{U}$ being a refinement of the local frame $\mathcal{V}$ of $\mathcal{E}$, it *is,* of course, *itself a local frame of $\mathcal{E}$* too; so, incidentally, we also conclude that:

(8.42) *for any vector sheaf $\mathcal{E}$ on a paracompact* (Hausdorff) *space $X$,* one can always take a *locally finite local frame* of $\mathcal{E}$. Yet, supplementing thus Chapt. II; (4.9), *such frames* still *constitute a cofinal subset of the set of* (proper) *open coverings of $X$.*

Accordingly, due to the hypothesis for $\mathcal{A}$ (cf. Chapt. II; Definition 8.1), one can further consider a *partition of unity of $\mathcal{A}$, subordinate to $\mathcal{U}$,* say,

$$(\phi_\alpha)_{\alpha \in I} \subseteq \operatorname{End} \mathcal{A} = \mathcal{A}(X) \tag{8.43}$$

(cf. also Chapt. III; (8.23), (8.23′) and Chapt. II; (6.30) and (6.48), for $\mathcal{E} = \mathcal{A}$).

Now, based on the hypothesis for $\mathcal{E}$ (cf. e.g. (8.21)), (8.36) and on (8.43), one obtains an *$\mathcal{A}$-metric*

$$\rho_\alpha : \mathcal{E} \times \mathcal{E}|_{U_\alpha} \longrightarrow \mathcal{A}|_{U_\alpha}, \tag{8.44}$$

*for every* $\alpha \in I$. So the *$\mathcal{A}$-bilinear morphism of $\mathcal{E}$,* we are looking for, is now given by the relation;

$$\tilde{\rho} = \sum_{\alpha \in I} \phi_\alpha \cdot \rho_\alpha. \tag{8.45}$$

The assertion follows, of course, straightforwardly, from the very definitions; thus, first, the sum in (8.45) *is well-defined,* due to the hypothesis for (8.43) and for the covering $\mathcal{U}$, by taking (8.45) *point-wise.* In the same way one also verifies the *continuity of* $\tilde{\rho}$ (cf. also Chapt. I; Lemma 1.1). On the other hand, $\tilde{\rho}$ is still *$\mathcal{A}$-bilinear* and *symmetric* (see (8.13)), by virtue of the same rels. (8.43) - (8.45), and this terminates the proof. ∎

The previous result comes, in effect, short of our initial aim, that is, of defining an *$\mathcal{A}$-metric* on a given *vector sheaf* $\mathcal{E}$, started from such a one on $\mathcal{A}$, this due, as the reader might suspect, already, to the question of whether the resulted (symmetric) *$\mathcal{A}$-bilinear form,* as in (8.45), is also *positive-definite,* the latter condition depending, as we shall presently see, on further assumptions for $\mathcal{A}$; so we discuss this matter straight away in the next Subsection.

**8.(a) Positive partitions of unity.**– We start by fixing the relevant terminology. Namely, we first set the following.

**Definition 8.4.** Assume that $(X, \mathcal{A})$ is an ordered algebraized space (cf. Definition 8.1) and let $\mathcal{U} = \{U_\alpha\}_{\alpha \in I}$ be a locally finite open covering of $X$. Moreover, consider the following family of $\mathcal{A}$-endomorphisms of $\mathcal{A}$,

$$(\phi_\alpha) \subseteq \operatorname{End} \mathcal{A} = \mathcal{A}(X), \tag{8.46}$$

satisfying the conditions;

1) $\operatorname{Supp}(\phi_\alpha) \subseteq U_\alpha$,

2) $\phi_\alpha|_{U_\alpha} \in \mathcal{P}(U_\alpha) \cap \mathcal{A}^{\cdot}(U_\alpha) = \mathcal{P}(U_\alpha) \cap \mathcal{A}(U_\alpha)^{\cdot}$

for every $\alpha \in I$, while we still assume that

3) $\sum_{\alpha \in I} \phi_\alpha = 1.$

We call a family, as above, a *strictly positive partition of unity of* $\mathcal{A}$, *subordinate to* $\mathcal{U}$. On the other hand, if, in place of cond. 2), one has

2′) $\phi_\alpha|_{U_\alpha} \in \mathcal{P}(U_\alpha)$, $\alpha \in I$,

one speaks of a *positive partition of unity of* $\mathcal{A}$, *subordinate to* $\mathcal{U}$.

Now, the previous definition is actually put in action through the following notion, as we shall see right below (cf. e.g. Theorem 8.3 in the sequel). So we further have the next.

**Definition 8.5.** Suppose that $(X, \mathcal{A})$ is an ordered algebraized space. Then, we shall say that $\mathcal{A}$ is a *strictly positive* (resp., *positively*) *fine sheaf* on $X$, provided that, for every locally finite open covering of $X$, there exists a strictly positive (resp., positive) partition of unity of $\mathcal{A}$, subordinate to the given covering (cf. Definition 8.4).

It is clear, of course (cf. thus Chapt. III; Definition 8.1, along with (8.23) and (8.23′) there), that

(8.47) *if* $\mathcal{A}$, as above, *is a strictly positive* (resp., *positively*) *fine sheaf*, then *it is* a `fortiori` *fine*.

**Examples 8.1**– The following classical $\mathbf{R}$-*algebra sheaves* admit, in effect, *strictly positive partitions of unity:*

1) $\mathcal{C}_X$; viz. the *sheaf of germs of* $\mathbf{R}$-*valued continuous functions on a paracompact (Hausdorff) space* $X$ (see also Chapt. I; (4.10)). The assertion is actually a consequence of the classical fact that $\mathcal{C}_X$ *is a fine sheaf*, if $X$ is *paracompact (Hausdorff)*, by further remarking that the resulted *partition of unity*, adapted to the corresponding complete presheaf $\Gamma(\mathcal{C}_X)$ *can be so chosen that cond. 2) of Definition 8.4 be satisfied:* Namely, suppose that $\mathcal{U} = \{U_\alpha\}_{\alpha \in I}$ is a *locally finite (open) covering of* $X$, so that, by hypothesis for $X$, there exists a *continuous partition of unity subordinate to* $\mathcal{U}$, say, $(\phi_\alpha)$. Therefore, *for every open* $U \subseteq X$, one obtains a $\mathcal{C}_X(U)$ $(= \mathcal{C}(U))$-*morphism*

$$h_U^\alpha : \mathcal{C}(U) \longrightarrow \mathcal{C}(U), \quad \textit{with} \quad \alpha \in I, \tag{8.48}$$

such that one sets;

$$h_U^\alpha(f) := \phi_\alpha \cdot f, \tag{8.49}$$

*for every* $f \in \mathcal{C}(U)$. It is now clear that *the family* $(h_U^\alpha)$, *with* $U$ *open in* $X$, and *for any given* $\alpha \in I$, *satisfies* the corresponding *compatibility conditions* (concerning *presheaf morphisms,* cf. Chapt. I; (6.5)). Hence, one obtains, *for every* $\alpha \in I$, a $\Gamma(\mathcal{C}_X)$*-endomorphism,*

$$h_\alpha \equiv (h_U^\alpha) \in \operatorname{End}\Gamma(\mathcal{C}_X), \tag{8.50}$$

therefore (see Chapt. I; (13.19)), a corresponding (uniquely defined) $\mathcal{C}_X$-*endomorphism of* $\mathcal{C}_X$,

$$\tilde{h}_\alpha \equiv h_\alpha \in \operatorname{End}\mathcal{C}_X = \mathcal{C}_X(X) = \mathcal{C}(X), \quad \alpha \in I; \tag{8.51}$$

for convenience, we retained the same notation. (Concerning the last relation, see also Chapt. II; (6.33), for $\mathcal{L} = \mathcal{A} = \mathcal{C}_X$, along with Chapt. I; (11.11), i) and (11.22)). Yet, by hypothesis for $(\phi_\alpha)$, one also concludes that

$$\operatorname{Supp}(h_\alpha) \subseteq U_\alpha, \quad \alpha \in I, \quad \textit{and} \quad \sum_{\alpha\in I} h_\alpha = 1 \tag{8.52}$$

(cf. also Chapt. III; (8.22), for the notation applied). On the other hand, by a standard argument (see, for instance, W.M. Boothby [1: p.192, Theorem 4.4]), one can assume that (cf. (8.51))

$$h_\alpha > 0, \quad \alpha \in I, \tag{8.53}$$

so that one finally obtains (cf. also (1.15))

$$h_\alpha \in \mathcal{C}(X)^{\cdot} = \mathcal{C}_X(X)^{\cdot} = \mathcal{C}_X^{\cdot}(X), \quad \alpha \in I, \tag{8.54}$$

which thus proves our claim. ∎

2) $\mathcal{C}_X^\infty$; ie., the *sheaf of germs of* **R**-*valued* $C^\infty$*-functions on a* (finite-dimensional) *paracompact Hausdorff* $C^\infty$*-manifold* $X$. (In this context, cf. also Chapt. I; (4.39)). Now, a quite analogous reasoning to that used in the preceding example, by employing, instead, $C^\infty$*-partitions of unity* (cf. also e.g. the last Ref. above), is here in force.

Concerning the previous examples, cf. also, for instance, R. Kultze [1: p.68, Beispiel 15.4 and 15.5]. On the other hand, further examples will be considered later on.

So we come now to our initial aim, which thus reads as follows.

**Theorem 8.3.** *Let* $(X, \mathcal{A})$ *be an ordered algebraized space, with* $X$ *a paracompact (Hausdorff) space and* $\mathcal{A}$ *a strictly positive fine* $\mathcal{A}$*-module. Moreover, suppose that* $(\mathcal{A}, \rho)$ *is a Riemannian* $\mathcal{A}$*-module. Then, every vector* *sheaf on $X$ admits a Riemannian $\mathcal{A}$-metric, as well.*

**Proof.** According to the hypothesis for $\mathcal{A}$, (8.47) and to Theorem 8.2, every *vector sheaf* $\mathcal{E}$ on $X$ is endowed with a *symmetric $\mathcal{A}$-bilinear form*

$$\tilde{\rho} : \mathcal{E} \oplus \mathcal{E} \longrightarrow \mathcal{A}, \tag{8.55}$$

*given by* (8.45) (cf. also (8.11)). So it remains to prove that, under the particular assumption for $\mathcal{A}$, $\tilde{\rho}$ *is,* in fact, *positive-definite* too. Thus, suppose that $s \in \mathcal{E}(U)$, with $U$ open in $X$, such that $\tilde{\rho}(s,s) = 0 \in \mathcal{A}(U)$ (cf. also (8.15)); hence, *for every* $x \in U$, one obtains (cf. (8.45))

$$\left(\sum_{\alpha\in I} \phi_\alpha \cdot \rho_\alpha(s,s)\right)(x) = \sum_{\alpha\in I} \phi_\alpha(x) \cdot \rho_\alpha(s(x), s(x))(x) = 0, \tag{8.56}$$

so that (cf. (8.5)) one has

$$\phi_\alpha(x) \cdot \rho_\alpha(s(x), s(x)) = 0 \quad \alpha \in I. \tag{8.57}$$

Now, given that, *for every* $\alpha \in I$, $\phi_\alpha|_{U_\alpha}$ *satisfies cond. 2) of Definition 8.4,* while we may also assume here that $x \in U_\alpha$, *for some* $\alpha \in I$, one concludes from (8.57) that;

$$\rho_\alpha(s(x), s(x)) = 0, \tag{8.58}$$

so that, by hypothesis for $\rho$ (cf. (8.44) and (8.45)), one has

$$s(x) = 0, \quad \textit{hence,} \quad s = 0. \tag{8.59}$$

That is, one gets *cond. (8.12) of Definition 8.2,* and this terminates the proof. ∎

**Scholium 8.1.**– Supposing that $\mathcal{A}$ is only a *positively fine sheaf* on the *paracompact Hausdorff* space $X$, with $(X, \mathcal{A})$ as in Definition 8.5, one gets from the previous proof that:

(8.60)

> $\tilde{\rho}$, as given by (8.45), is *positive-semidefinite;* viz., for any open $U \subseteq X$ and $s \in \mathcal{E}(U)$, one obtains (cf. also (8.7))
>
> (8.60.1) $\quad \tilde{\rho}(s,s) \in \mathcal{P}(U) \subseteq \mathcal{A}(U).$
>
> Of course, one still has that
>
> (8.60.2) $\quad \tilde{\rho}(s,s) = 0, \quad \textit{if} \quad s = 0.$

So the converse of the last relation is what can actually be assured, by imposing on $\mathcal{A}$ the more stringent *cond. 2) of Definition 8.4.* Accordingly,

(8.61) *Assuming that $\mathcal{A}$ is a positively fine sheaf,* while the rest of the conditions of Theorem 8.3 are satisfied, one concludes that *every vector sheaf on $X$ is endowed* with an *$\mathcal{A}$-bilinear morphism,* which is also *symmetric* (cf. also Theorem 8.2 and (8.47)) and *positive-semidefinite,* say $\tilde{\rho}$, as in (8.55).

On the other hand, it is clear by the preceding discussion that one can further examine, within the present *sheaf-theoretic* framework, several other classical notions and results, connected with the concept of a *"metric"*, as before. However, we have only considered so far those, which will also appear in the sequel, while we eventually complement later on the present account, by further results, as the case may demand. (So see, for instance, the following Sections 9, 10).

## 9. Hermitian $\mathcal{A}$-modules

Based on the classical counterpart, we consider next the case that the $\mathcal{A}$-modules involved are equipped with a *sesqui-linear* or *hermitian bilinear* ($\mathcal{A}$-valued) *morphism.* However, by contrast with the previous situation in Section 8, we consider a $\mathbf{C}$-*algebraized space* $(X, \mathcal{A})$ as we usually did in the preceding, while we also assume for $\mathcal{A}$ that it carries an *involutory automorphism* (or even *involution*), denoted by "$-$".

Thus, we assume that we are given a $\mathbf{Z}$-*additive morphism of sheaves* (of groups)

$$- : \mathcal{A} \longrightarrow \mathcal{A}, \tag{9.1}$$

such that the corresponding *presheaf morphism* (we retain the notation)

$$- : \Gamma(\mathcal{A}) \longrightarrow \Gamma(\mathcal{A}) \tag{9.1'}$$

(see Chapt. I; (13.19)) satisfies the following conditions:

$$\overline{s \cdot t} = \overline{s} \cdot \overline{t} \tag{9.2}$$

and

$$\bar{\bar{s}} = s, \tag{9.3}$$

*for any (local) sections* $s, t \in \mathcal{A}(U)$, with $U$ *open* in $X$. Furthermore, we assume that;

$$\overline{\lambda \cdot s} = \bar{\lambda} \cdot \bar{s}, \tag{9.4}$$

*for any* $\lambda \in \mathbf{C} \equiv \mathbf{C}_X(U)$ *and* $s \in \mathcal{A}(U)$, as before. Here, by an obvious *abuse of notation,* we set $\bar{\lambda}$ for the *complex conjugate* of $\lambda \in \mathbf{C}$.

**Scholium 9.1.**– Concerning the notation in (9.4), we still remark that the field of complexes $\mathbf{C}$ is taken hereby as the *constant* $\mathbf{C}$*-algebra sheaf* on $X$, $\mathbf{C}_X \equiv \mathbf{C}$ (cf. Chapt. I; comments prior to (1.32)), so that *one has* $\mathbf{C}_X(U) = \mathbf{C}$, *for any open* $U \subseteq X$ (cf. Chapt. I; (10.2), (10.8)). So one actually infers through (9.4) that the natural *involution in* $\mathbf{C}_X$, viz. the *complex conjugation,* coincides with the *induced involution on* $\mathbf{C} \equiv \mathbf{C}_X$ from the given one on $\mathcal{A}$ (see (9.1)), by virtue of the *canonical sheaf injection* $\mathbf{C} \equiv \mathbf{C}_X \overset{\subseteq}{\rightarrow} \mathcal{A}$ (cf. Chapt. II; (1.34, (1.34′)); thus, a justification of the notation applied in (9.1).

Now, a triple

$$(X, \mathcal{A}, -), \tag{9.5}$$

as above, with $(X, \mathcal{A})$ a given $\mathbf{C}$-algebraized space, such that the preceding conditions (9.1) - (9.4) are fulfilled by "$-$", is called an *involutive* $\mathbf{C}$*-algebraized space;* yet, equivalently, we say that the structure sheaf $\mathcal{A}$ is an *involutive* $\mathbf{C}$*-algebra sheaf* on $X$.

Thus, as already remarked (cf. Schilium 9.1), the triple $(X, \mathbf{C}_X \equiv \mathbf{C}, -)$ is a natural *involutive* $\mathbf{C}$*-algebraized space.* Now, further *standard examples* of such spaces are, of course, the $\mathbf{C}$*-algebraized spaces* $(X, {}^{\mathbf{C}}\mathcal{C}_X) \equiv (X, \mathcal{C}_X)$, $(X, {}^{\mathbf{C}}\mathcal{C}^{\infty}_X) \equiv (X, \mathcal{C}^{\infty}_X)$ and $(X, \mathcal{O}_X)$, considered in Chapt. I; (4.10), (4.39) and (4.40), with *involution* the usual *complex conjugation* of functions (sections; cf. (9.1′), in conjuction with Chap. I; (13.19)).

So we come now to a basic notion for the ensuing discussion, as this is given by the following.

**Definition 9.1.** Let $(X, \mathcal{A}, -)$ be a given involutive $\mathbf{C}$-algebraized space (cf. (9.5)), and $\mathcal{E}$ an $\mathcal{A}$-module on $X$. Thus, by an *$\mathcal{A}$-valued hermitian* (or *sesqui-linear*) *inner product* on $\mathcal{E}$, one means a sheaf morphism

$$\rho : \mathcal{E} \oplus \mathcal{E} \longrightarrow \mathcal{A}, \tag{9.6}$$

satisfying the following conditions:

i) $\rho$ is *biadditive,* or **Z**-*bilinear.*

ii) For any $s, t \in \mathcal{E}(U)$ and $\alpha \in \mathcal{A}(U)$, with $U$ open in $X$, one has;

$$\rho(\alpha s, t) = \alpha \cdot \rho(s,t) \ \ \textit{and} \ \ \rho(s, \alpha t) = \overline{\alpha} \cdot \rho(s,t). \tag{9.7}$$

iii) For any $s$ and $t$, as above, one has

$$\rho(s,t) = \overline{\rho(t,s)}. \tag{9.8}$$

Now, a given pair

$$(\mathcal{E}, \rho), \tag{9.9}$$

with $\mathcal{E}$ and $\rho$, as before, is called a *hermitian $\mathcal{A}$-module* on $X$, while we also say that $\rho$ is a *hermitian $\mathcal{A}$-metric* (or just a *hermitian metric*) on $\mathcal{E}$.

Following the classical pattern, one can now obtain analogous results within the present framework, provided we have established the appropriate *sheaf-theoretic context.* So, by analogy with our previous discussion in Section 8, we shall consider *hermitian vector sheaves* and get the analogue of Theorem 8.3. Thus, we first look, for the case at hand, at the corresponding situation to (8.2).

So assume that we have an *involutive* **C**-*algebraized space,* as in (9.5), such that $(\mathcal{A}, \rho)$ is a *hermitian $\mathcal{A}$-module* (cf. (9.9)). Thus, one concludes that;

(9.10) *every free $\mathcal{A}$-module $\mathcal{E}$ on $X$, of finite rank,* say $n \in \mathbf{N}$, *is endowed with a hermitian metric, as well.*

Indeed, the desired metric, say $\tilde{\rho}$, on $\mathcal{E}$ is given by the analogous relation, in the case at hand, to (8.39) (see also (8.37) and (8.37′)). ∎

On the other hand, by further employing a similar argument to (8.26), along with the corresponding notation, one infers that:

*the hermitian metric* $\tilde{\rho}$ *on* $\mathcal{E}$, as provided by (9.10), *is uniquely defined by a matrix* $\alpha \equiv (\alpha_{ij}) \in M_n(\mathcal{A}(X))$, *such that;*

$$\alpha_{i,j} := \tilde{\rho}(e_i, e_j), \quad 1 \leqq i, j \leqq n = \operatorname{rk} \mathcal{E}. \tag{9.11.1}$$

(9.11) *Conversely,* $\alpha$ *yields a hermitian metric on* $\mathcal{E}$, through the corresponding relation to (8.23), *if,* in addition, it is *hermitian;* viz. in the case one has (cf. also (9.1′)),

$$\alpha_{ji} = \overline{\alpha}_{ij}, \quad 1 \leqq i, j \leqq n. \tag{9.11.2}$$

More generally, one can treat the preceding situation *locally,* by looking, namely, at *vector sheaves* on $X$, as in (8.27) (see also (8.34)). So we finally come to the promised result, an analogous of Theorem 8.2. That is, one has the next.

**Theorem 9.1.** *Let* $(X, \mathcal{A}, -)$ *be an involutive* **C***-algebraized space, having* $X$ *a paracompact Hausdorff space and* $\mathcal{A}$ *a fine* $\mathcal{A}$*-module. Moreover, assume that* $(\mathcal{A}, \rho)$ *is a hermitian* $\mathcal{A}$*-module. Then, every vector sheaf on* $X$ *is equipped with a hermitian metric, as well.*

**Proof.** This goes verbatim, as the corresponding one of Theorem 8.2, such that the *analogous relation to (8.45) supplies* now *the desired hermitian metric.* ∎

**Scholium 9.2.**- By analogy with (8.16), given a *hermitian* $\mathcal{A}$*-module* $(\mathcal{E}, \rho)$ on $X$ (cf. (9.9)), one defines a *sheaf morphism* (see also (9.14) below)

$$\tilde{\rho} : \mathcal{E} \longrightarrow \overline{\mathcal{E}} := \overline{\mathcal{H}om}_{\mathcal{A}}(\mathcal{E}, \mathcal{A}) \tag{9.12}$$

according to the following relation (for sections, Chapt. II; (13.19))

$$[\tilde{\rho}(s)](t) \equiv \rho_s(t) := \rho(s, t), \tag{9.13}$$

*for any* $s, t \in \mathcal{E}(U)$ and $U$ *open* in $X$.

Now, due to the second relation in (9.7), the map $\rho_s \equiv \tilde{\rho}(s)$, with $s \in \mathcal{E}(U)$, as before, is an *additive (group sheaf) morphism,* such that one has

$$\rho_s(\alpha \cdot t) = \overline{\alpha} \cdot \rho_s(t), \tag{9.14}$$

*for every* $t \in \mathcal{E}(U)$, as above. We call $\tilde{\rho}(s) \equiv \rho_s$ an *anti-*$\mathcal{A}$*-morphism* of $\mathcal{E}$ into $\mathcal{A}$, hence, the "*bar*" notation in (9.12). Yet, we call

$$\overline{\mathcal{E}} \equiv \overline{\mathcal{H}om}_{\mathcal{A}}(\mathcal{E}, \mathcal{A}), \tag{9.15}$$

viz. the set of *anti-$\mathcal{A}$-morphisms of $\mathcal{E}$ into $\mathcal{A}$*, the *conjugate (dual) $\mathcal{A}$-module of $\mathcal{E}$*: indeed, arguing on the presheaf level, it is clear that (9.15) is a *sheaf of (abelian) groups* on $X$ (cf. Chapt. II; (6.4)), which, in effect, due to the *commutativity of $\mathcal{A}$* (cf. e.g. (0.1)), is still an *$\mathcal{A}$-module* on $X$.

Now, *if (9.12) is one-to-one,* we shall say that the given *hermitian metric* $\rho$ is *non-degenerate;* consequently, in that case one has

$$\mathcal{E} \overset{\subsetneq}{\to} \overline{\mathcal{E}}, \tag{9.16}$$

$\tilde{\rho}$ yielding an *$\mathcal{A}$-isomorphism (into) of* the *$\mathcal{A}$-modules* involved (cf. also the first relation in (9.7)).

On the other hand, by further employing *matrix notation* for the *non-degenerate hermitian metric* of a given *free hermitian $\mathcal{A}$-module* $(\mathcal{E}, \rho)$, one obtains (see also (9.9), (9.15) and (9.11));

$$\alpha = \overline{{}^t\alpha} = {}^t\overline{\alpha} \in M_n(\mathcal{A}(X))^{\boldsymbol{\cdot}} \equiv \mathrm{GL}(n, \mathcal{A}(X)) = \mathcal{GL}(n, \mathcal{A})(X) \tag{9.17}$$

(yet, cf. (8.25), or even (8.26.4). Concerning the classical theory, see also, for instance, L. Chambadal - J.L. Ovaert [1: p.356, Proposition 20.8]).

Finally, applying our previous notation in (7.43) to the given *free $\mathcal{A}$-module $\mathcal{E}$*, concerning the matrix of $\rho$, $\alpha \equiv (\alpha_{ij})$ (see (9.17)), one obtains (see also e.g. (7.46)):

$$\mathrm{Gram}_\rho(s_1, \ldots, s_n) := \det(\rho(s_i, s_j)) = |\det_e(s_1, \ldots, s_n)|^2 \cdot \det(\alpha_{ij}); \tag{9.18}$$

here we have set, *for any given $s \in \mathcal{A}(U)$*, with $U$ *open* in $X$,

$$|s|^2 := s \cdot \overline{s} \in \mathcal{A}(U), \tag{9.19}$$

where $s_i$, $1 \leq i \leq n = \mathrm{rk}\,\mathcal{E}$, stands for the $i$-th *row* of our matrix $\alpha \equiv (\alpha_{ij}) \in M_n(\mathcal{A}(X))$. (Concerning the notation "$\mathrm{Gram}_\rho$" in (9.18), as well as, the same relation in the classical case, see e.g. L. Chambadal - J.L. Ovaert [1: p.343, Definition 20.5, and p.256, Proposition 20.9]).

Further applications of the preceding will be given along the ensuing discussion. So we come now, by the following Section, to a useful extension, within the present context, of a classical device (see also the next Chapter V; Section 9).

# 10. Gram-Schmidt orthogonalization

To start with, suppose that we are given an *ordered algebraized space* $(X, \mathcal{A})$ (cf. Definition 8.1) such that

(10.1) $$\mathcal{P}^* := \mathcal{P} - \{0\} \subseteq \mathcal{A}^{\bullet}$$

(see also (8.2) and (1.13)). In other words and speaking in terms of sections, (10.1) means that:

(10.2) *every strictly positive section of $\mathcal{A}$ is invertible;* viz. for any $s \in \mathcal{P}(U)$, $U$ open in $X$, with $s(x) \neq 0$, $x \in U$, one has that $s \in \mathcal{A}(U)^{\bullet} = \mathcal{A}^{\bullet}(U)$.

Furthermore, we assume that:

(10.3) *every positive section of $\mathcal{A}$, has a square root;* viz. *for every* (local) *section $s \in \mathcal{P}(U) \subseteq \mathcal{A}(U)$* (cf. (8.7)), with $U$ open in $X$, *there exists a (unique) $t \in \mathcal{P}(U)$ such that*

(10.3.1) $$t^2 = s.$$

Yet, one defines

(10.3.2) $$\sqrt{s} := t,$$

such that (10.3.1) be satisfied, and call $t = \sqrt{s} \in \mathcal{P}(U)$ the *square root* of $s \in \mathcal{P}(U)$.

Now, it is clear that the correspondence $s \longmapsto \sqrt{s}$, $s \in \mathcal{P}(U)$, as above, entails a *morphism of* (complete) *presheaves*

(10.4) $$\sqrt{\ } : \Gamma(\mathcal{P}) \longrightarrow \Gamma(\mathcal{P})$$

(cf. also (8.7)), which we call, by an obvious abuse of language (cf. thus Chapt. I; (13.19)), the *square root* (morphism) *of $\mathcal{A}$*. Of course, we still assume that

(10.5) *(10.4) has the analogous properties of the corresponding classical function* (see e.g. Examples 10.1 below).

Now, to facilitate the terminology applied, we set the following.

**Definition 10.1.** We say that a given pair $(X, \mathcal{A})$ is an *enriched ordered algebraized space,* or even an *ordered algebraized space with square root,* we also denote it by

$$(X, \mathcal{A}, \surd), \tag{10.6}$$

whenever $(X, \mathcal{A})$ is an ordered algebraized space (see Definition 8.1), which further satisfies (10.2), (10.3) and (10.5).

**Examples 10.1.**– All the concrete *functional examples* of algebraized spaces, considered in Examples 8.1 above, can also be viewed as particular fundamental instances of *ordered algebraized spaces with square root,* where the latter morphism is given, of course, by the corresponding in each case *classical function, defining* now *the appropriate presheaf morphism,* as in (10.4).

Yet, the same remark, as before, is still valid for the *constant* **R**-*algebraized space* $(X, \mathbf{R}_X \equiv \mathbf{R})$, having the *standard ordered structure and square root.* Further examples will be considered in the sequel (see, for instance, Chapters X, XI).

Thus, given an *ordered algebraized space with square root* $(X, \mathcal{A})$ (cf. (10.6)), suppose that $(\mathcal{A}, \rho)$ is a *Riemannian* $\mathcal{A}$*-module* on $X$ (see Definition 8.3); hence, one can further consider the *extension of* $\rho$ *to* $\mathcal{A}^n$, viz. the $\mathcal{A}$*-metric*

$$\tilde{\rho} : \mathcal{A}^n \oplus \mathcal{A}^n \longrightarrow \mathcal{A}, \tag{10.7}$$

given by (8.37) (cf. also (8.37′)). So one now defines the following $\mathcal{A}$*-valued norm* (morphism) on $\mathcal{A}^n$;

$$|\cdot| : \mathcal{A}^n \longrightarrow \mathcal{P} \subseteq \mathcal{A}, \tag{10.8}$$

such that the corresponding *section morphism* between $\Gamma(\mathcal{A}^n) = \Gamma(\mathcal{A})^n$ and $\Gamma(\mathcal{P}) \subseteq \Gamma(\mathcal{A})$ (see Chapt. I; (13.19) and Chapt. II; (3.19)) is given by

$$|s| := \sqrt{\tilde{\rho}(s, s)} \in \mathcal{P}(U) \subseteq \mathcal{A}(U), \tag{10.9}$$

*for any* $s \in \mathcal{A}^n(U) = \mathcal{A}(U)^n$, with $U$ *open* in $X$ (cf. also (8.12) and (10.4)).

Thus, we are now ready to transcribe, within the present *sheaf-theoretic context,* the standard device of *transforming a given basis of a free module into an orthonormal one,* by the "inductive procedure of subtracting successive projections", that is, according to the classical terminology, by the *Gram-Schmidt orthogonalization process:*

So, for simplicity's sake, consider the standard *free $\mathcal{A}$-module $\mathcal{A}^n$*, with *$(X, \mathcal{A})$ and $(\mathcal{A}, \rho)$, as before;* moreover, let

$$(s_i)_{1 \leqq i \leqq n} \subseteq \mathcal{A}^n(X) = \mathcal{A}(X)^n \tag{10.10}$$

be a (global) *gauge of $\mathcal{A}^n$*. In other words, we are given a finite sequence of *global sections of $\mathcal{A}^n$, such that , for every $x \in X$,*

$$(s_i(x))_{1 \leqq i \leqq n} \subseteq \mathcal{A}_x^n \tag{10.11}$$

is a *basis of the free $\mathcal{A}_x$-module* $(\mathcal{A}_x)^n = (\mathcal{A}^n)_x \equiv \mathcal{A}_x^n$. Yet, it is *equivalent* to say that we are given the following *section-matrix* (cf. also (1.28))

$$(s_{ij}) \in \mathrm{GL}(n, \mathcal{A}(X)) = \mathcal{GL}(n, \mathcal{A})(X) \tag{10.12}$$

(concerning the classical case, see e.g. S. Lang [1: p.446, Proposition 3.1]).

Consequently, one has

$$s_i(x) \neq 0, \quad \text{for every } x \in X, \quad \text{with } 1 \leqq i \leqq n, \tag{10.13}$$

so that, by hypothesis for $\rho$ (cf. (8.12), (8.37) and (10.7)), one obtains

$$\overline{\rho}(s_i, s_i) > 0, \quad 1 \leqq i \leqq n; \tag{10.14}$$

that is, one has

$$\overline{\rho}(s_i, s_i) \in \mathcal{P}^*(X) = \mathcal{P}(X)^* := \mathcal{P}(X) - \{0\} \subseteq \mathcal{A}^{\boldsymbol{\cdot}}(X) = \mathcal{A}(X)^{\boldsymbol{\cdot}}, \tag{10.15}$$

*for every* $1 \leqq i \leqq n$ (cf. (10.1), (10.10)).

Now, according to (10.9) (see also (10.2), (10.15)), one defines

$$|s_i| := \sqrt{\overline{\rho}(s_i, s_i)} \in \mathcal{P}^*(X) \subseteq \mathcal{A}(X)^{\boldsymbol{\cdot}}, \quad 1 \leqq i \leqq n; \tag{10.16}$$

therefore, one can set

$$\tilde{s}_i := |s_i|^{-1} \cdot s_i \in \mathcal{A}(X), \quad 1 \leqq i \leqq n, \tag{10.17}$$

such that one obtains (cf. also Definition 8.2, cond. i), and (10.3))

$$|\tilde{s}_i| = 1 \in \mathcal{A}(X), \quad 1 \leqq i \leqq n. \tag{10.18}$$

Thus, one can now apply the aforementioned *"inductive procedure"*; namely, suppose that we are given the following (finite) sequence of (global) *sections of* $\mathcal{A}^n$

$$(10.19) \qquad (\tilde{s}_i)_{1 \leqq i \leqq m} \subseteq \mathcal{A}^n(X), \quad \textit{with} \;\; m \leqq n-1,$$

in such a manner that

$$(10.20) \qquad \overline{\rho}(\tilde{s}_i, \tilde{s}_j) = \delta_{ij}, \quad 1 \leqq i, j \leqq m.$$

In this context, by analogy with the classical terminology, a family sections, as in (10.19), which further satisfy the conditions (10.18) (for $n$ $m$) and (10.20), is called an *orthonormal* family of sections of $\mathcal{A}^n$.

Finally, we assume for (10.19) that

$$(10.21) \qquad [s_1, \ldots, s_m] = [\tilde{s}_1, \ldots, \tilde{s}_m],$$

where, of course, $(s_i) \subseteq \mathcal{A}^n(X)$ is given by (10.10). Here, by either o of the two members of (10.21), one means the *free* $\mathcal{A}(X)$*-module of ra* $m$, generated by the respective sections; in this regard, we still remark th *(10.20) forces (10.19) to consist of* $\mathcal{A}(X)$*-linearly independent sections* ( also cond. i) of Definition 8.2). Thus, (10.19) *yields,* in effect, *a (global) gau of* $\mathcal{A}^m$, such that the *free* $\mathcal{A}(X)$*-module* $\mathcal{A}^m(X) = \mathcal{A}(X)^m$ is given, with an $\mathcal{A}(X)$*-isomorphism,* by either one of the $\mathcal{A}(X)$-modules in (10.21). Th latter are still named (global) *flags* of the gauges concerned.

So we proceed, by further defining a *new (global) section of* $\mathcal{A}^n$, in term of the given ones in (10.10) and (10.19). That is, we set;

$$(10.22) \qquad t := s_{m+1} - \sum_{i=1}^{m} \overline{\rho}(s_{m+1}, \tilde{s}_i) \cdot \tilde{s}_i \in \mathcal{A}^n(X).$$

Now, we first remark that

$$(10.23) \qquad t \neq 0,$$

in the sense that *one has* $t(x) \neq 0$, *for every* $x \in X$. Indeed, otherwise, o would have, in view of (10.22);

$$(10.24) \qquad s_{m+1}(x) = \sum_{i=1}^{m} \overline{\rho}(s_{m+1}(x), \tilde{s}_i(x)) \cdot \tilde{s}_i(x) \in \mathcal{A}^n_x,$$

*for some* $x \in X$, that is, one obtains (cf. (10.21))

$$(10.25) \qquad s_{m+1}(x) \in [\tilde{s}_1, \ldots, \tilde{s}_m]_x = [s_1 \ldots, s_m]_x = \mathcal{A}^n_x,$$

which is a contradiction, according to the hypothesis for (10.10) (see al

(10.11)). ∎ Therefore, one has (cf. (8.12)) $\overline{\rho}(t,t) > 0$, so that one obtains (cf. (10.9), (10.14), (10.15) and (10.16));

$$|t| := \sqrt{\overline{\rho}(t,t)} \in \mathcal{P}^*(X) \subseteq \mathcal{A}^{\cdot}(X) = \mathcal{A}(X)^{\cdot}. \tag{10.26}$$

Thus, one further obtains a *new (global) section of $A$,* supplementing the given ones in (10.19), by setting;

$$\tilde{s}_{m+1} := |t|^{-1} \cdot t \in \mathcal{A}^n(X). \tag{10.27}$$

So one first concludes that (cf. e.g. (10.17), (10.18))

$$|\tilde{s}_{m+1}| = 1. \tag{10.28}$$

Furthermore, one proves that:

(10.29) *(10.20) is extended to all sections $\tilde{s}_i$, with $i = 1, 2, \ldots, m+1$;* viz., in addition to (10.20), one has

(10.29.1) $$\overline{\rho}(\tilde{s}_{m+1}, \tilde{s}_i) = 0, \quad 1 \leqq i \leqq m.$$

Indeed, *setting* for simplicity

$$\overline{\rho}(s,t) \equiv s \cdot t \in \mathcal{A}(X), \tag{10.30}$$

*for any $s, t \in \mathcal{A}^n(X)$*, one obtains, in view of the preceding;

$$\begin{aligned}\overline{\rho}(\tilde{s}_{m+1}, \tilde{s}_i) &\equiv \tilde{s}_{m+1} \cdot \tilde{s}_i = (|t|^{-1}t) \cdot \tilde{s}_i = |t|^{-1}(t \cdot \tilde{s}_i) \\ &= |t|^{-1}\left(s_{m+1} \cdot \tilde{s}_i - \sum_{j=1}^{m} (s_{m+1} \cdot \tilde{s}_j)(\tilde{s}_i \cdot \tilde{s}_j)\right) \\ &= |t|^{-1}(s_{m+1} \cdot \tilde{s}_i - s_{m+1} \cdot \tilde{s}_i)\end{aligned} \tag{10.31}$$

*for every* $1 \leqq i \leqq m$. ∎ On the other hand, *one* still *concludes that;*

$$[s_1, \ldots, s_m, s_{m+1}] = [\tilde{s}_1, \ldots, \tilde{s}_m, \tilde{s}_{m+1}]. \tag{10.32}$$

That is, in view of (10.21), and the hypothesis for (10.10), one first infers that $(\tilde{s}_1, \ldots, \tilde{s}_m, s_{m+1})$ constitutes a *(global) gauge of $\mathcal{A}^{m+1}$*, viz., *equivalently,* a *basis of the $\mathcal{A}(X)$-module $\mathcal{A}^{m+1}(X) = \mathcal{A}(X)^{m+1}$*. Therefore, by virtue of (10.22) and (10.27), one further obtains that;

$$t, \tilde{s}_{m+1} \in [s_1, \ldots, s_m, s_{m+1}]. \tag{10.33}$$

Furthermore, in view of (10.29.1), $\tilde{s}_{m+1}$ *is* $\mathcal{A}(X)$*-linearly independent of the* $\tilde{s}_i$'s, $1 \leqq i \leqq m$ (one can repeat here too the classical argument; cf. e.g. C.W. Curtis [1: p.123, Lemma 15.8]). This, in conjuction with (10.21) and (10.33), along with the previous conclusion about $(\tilde{s}_1, \ldots, \tilde{s}_m, s_{m+1})$, yield now the desired relation (10.32). ∎

The above constitute, in effect, the (proof of the) *sheaf-theoretic version of the* classical *Gram - Schmidt orthogonalization process.* Of course, it is still clear that (instead of considering the standard free $\mathcal{A}$-module $\mathcal{A}^n$),

(10.34) *the above procedure* of transforming, namely, a given (global) gauge to an orthonormal one, *holds true, for every free* $\mathcal{A}$*-module of finite rank* over $X$, as well.

Apply, for instance, (8.39) to our previous argument. ∎ Thus, the preceding supply now the proof of the following.

**Theorem 10.1.** *(Gram - Schmidt orthogonalization). Suppose that* $(X, \mathcal{A})$ *is an enriched ordered algebraized space* (cf. Definition 10.1), *such that* $(\mathcal{A}, \rho)$ *is a Riemannian* $\mathcal{A}$*-module. Moreover, let* $\mathcal{E}$ *be a free* $\mathcal{A}$*-module on* $X$*, of finite rank* $n \in \mathbf{N}$*, such that*

$$(s_1, \ldots, s_n) \subseteq \mathcal{E}(X)^n = \mathcal{E}^n(X) \tag{10.35}$$

*is a (global) gauge of* $\mathcal{E}$*. Then, there exists an orthonormal gauge of* $\mathcal{E}$*, associated with the given one, say,*

$$(\tilde{s}_1, \ldots, \tilde{s}_n) \subseteq \mathcal{E}(X)^n, \tag{10.36}$$

*in such a manner that one has* (cf. also (8.39))

$$\tilde{\rho}(\tilde{s}_i, \tilde{s}_j) = \delta_{ij}, \quad 1 \leqq i, j \leqq n, \tag{10.37}$$

as well as,

$$[\tilde{s}_1, \ldots, \tilde{s}_m] = [s_1, \ldots, s_m], \tag{10.38}$$

for every $1 \leqq m \leqq n$. ∎

**Remark 10.1.**– Concerning the hypothesis in the previous Theorem 10.1 that $(\mathcal{A}, \rho)$ *is a Riemannian* $\mathcal{A}$*-module,* one could, *equivalently,* assume that $(\mathcal{E}, \tilde{\rho})$ *is a Riemannian free* $\mathcal{A}$*-module* of finite rank over $X$, the corresponding argument being actually the same. Yet, notice that (8.39) would then supply $\mathcal{A}^n$ (*with* $n = \operatorname{rk}\mathcal{E}$), hence $\mathcal{A}$, as well, with a similar $\mathcal{A}$-*metric.*

Applications of the above will be supplied along the ensuing discussion. (See thus the following two Sections, as well as, the next Chapter V; Section 9).

# 11. Volume element

As an application of the preceding considerations, and following further the classical pattern, we discuss below the analogue in our case of the standard notion in the title of this Section. Here we mainly restrict ourselves to fixing the relevant terminology which, in fact, will be employed later on (cf. Part Two of this treatise). So, to start with, suppose that;

(11.1) $(X, \mathcal{A})$ stands for an *enriched ordered algebraized space* (cf. Definition 10.1), such that $(\mathcal{A}, \rho)$ is a *Riemannian* $\mathcal{A}$-*module* on $X$. Moreover, let $\overline{\rho} \equiv \rho^n$ be the *extension of* $\rho$ to the standard *free* $\mathcal{A}$-*module* $\mathcal{A}^n$ on $X$ (cf. (8.37), (8.37′)).

Now, consider further the *canonical coordinate* (global) *sections of* $\mathcal{A}^n$

$$\varepsilon \equiv (\varepsilon_i)_{1 \leqq i \leqq n} \subseteq \mathcal{A}^n(X) = \mathcal{A}(X)^n, \tag{11.2}$$

that is, the *Kronecker gauge of* $\mathcal{A}^n$ (see (3.3)). So motivated by the geometrical applications we shall be dealt with in the sequel, we next look at the following (global) section of $\mathcal{A}$; that is, we set (see also (7.23), (7.24))

$$\omega := \sqrt{|\overline{\rho}|}\, \varepsilon_1 \wedge \cdots \wedge \varepsilon_n \in (\wedge^n \mathcal{A}^n)(X) \equiv (\det \mathcal{A}^n)(X) = \mathcal{A}(X). \tag{11.3}$$

In this context, based on our hypothesis in (11.1), we further set

$$\sqrt{|\overline{\rho}|} := \sqrt{|\det(\overline{\rho}(\varepsilon_i, \varepsilon_j))|} \tag{11.4}$$

(see (10.4) and (10.8), for $n = 1$, as well as, (8.22), (4.7) and (4.10)).

Thus, by analogy with the standard terminology in (differential) geometry, we call the above function, in effect, global section of $\mathcal{A}$, as given by (11.3), the *volume element* of the given $\mathcal{A}$-metric $\overline{\rho}$.

Now, by the very definition, one gets

$$\omega(x) \neq 0, \quad \textit{for every } x \in X; \tag{11.5}$$

cf. (8.22.1), taking the hypothesis for (11.2) into account too. (Yet, see, for instance, L. Chambadal - J.L. Ovaert [1: p.229, Théorème 16.11]). Furthermore, based on the relation

$$(\wedge^n \mathcal{A}^n)^*(X) = (\wedge^n(\mathcal{A}^n)^*)(X) = (\det \mathcal{A}^n)^*(X) = \mathcal{A}(X), \tag{11.6}$$

*the volume element* $\omega$, as above, *can be viewed as an* $\mathcal{A}(X)$*-linear morphism on* $\det(\mathcal{A}^n)$; i.e., as such of $\mathcal{A}$ into itself, so that one has

$$\omega \in (\mathcal{E}nd\,\mathcal{A})(X) = \operatorname{End}\mathcal{A} = \mathcal{A}(X) \tag{11.7}$$

(cf. Chapt. II; (6.18) and (6.30), or even (6.33)).

On the other hand, suppose that

$$s \equiv (s_1, \ldots, s_n) \subseteq \mathcal{A}^n(X) = \mathcal{A}(X)^n \tag{11.8}$$

is a (global) *orthonormal gauge* of $\mathcal{A}^n$ (see (10.18), (10.20)). We note that one can always find such a gauge of $\mathcal{A}^n$, by a mere application, according to our hypothesis (cf. (11.1)), of the *Gram-Schmidt orthonormalization procedure* (see Theorem 10.1) on any given (global) gauge of $\mathcal{A}^n$. Consequently,

(11.9) by applying (11.3), one obtains

$$\omega = s_1 \wedge \cdots \wedge s_n \tag{11.9.1}$$

as the *volume element* of the $\mathcal{A}$-metric $\bar{\rho}$ of $\mathcal{A}^n$, *associated with the orthonormal gauge* of $\mathcal{A}^n$, given by (11.8).

Indeed, in view of (10.20), one has;

$$\operatorname{Gram}_{\bar{\rho}}(s_1, \ldots, s_n) := \det(\bar{\rho}(s_i, s_j)) = \det(\delta_{ij}) \equiv \det(\mathbf{1}_n) = 1, \tag{11.10}$$

where, of course, $\mathbf{1}_n \in M_n(\mathcal{A}(X))$ stands for the *identity* $n \times n$ (global) *section matrix* from $\mathcal{A}(X)$. So one then obtains

$$\sqrt{|\bar{\rho}|} = 1, \tag{11.11}$$

which, in conjuction with (11.3), yields thus (11.9.1). ∎

Now, by further employing the *Gram determinant of* $\bar{\rho}$, *with respect to* (11.8), one has (cf. also (11.10) and (11.2))

$$\begin{aligned} \text{Gram}_{\overline{\rho}}(s_1,\ldots,s_n) &= (\det{}_\varepsilon(s_1,\ldots,s_n))^2\cdot\det(\overline{\rho}(s_i,s_j)) \\ &= \det{}_\varepsilon(s_1,\ldots,s_n)^2\cdot\det(\alpha_{ij}) = 1. \end{aligned} \tag{11.12}$$

(Concerning the classical counterpart of (11.12), see also, for example, L. Chambadal - J.L. Ovaert [1: p.343, Proposition 20.3]). Therefore, one obtains

$$(\det{}_\varepsilon(s_1,\ldots,s_n))^2 = (\det(\alpha_{ij}))^{-1}, \tag{11.13}$$

so that by further assuming that (cf. (10.1))

$$\det{}_\varepsilon(s_1,\ldots,s_n) \in \mathcal{P}(X)^* \subseteq \mathcal{A}(X)^{\cdot}, \tag{11.14}$$

one obtains (cf. (10.3.1))

$$\det{}_\varepsilon(s_1,\ldots,s_n) = \left(\sqrt{\det(\alpha_{ij})}\right)^{-1}. \tag{11.15}$$

Thus, one finally obtains;

$$\omega = s_1\wedge\cdots\wedge s_n = \det{}_\varepsilon(s_1,\ldots,s_n)\cdot\varepsilon_1\wedge\cdots\wedge\varepsilon_n = \frac{1}{\sqrt{\det(\alpha_{ij})}}\cdot\varepsilon_1\wedge\cdots\wedge\varepsilon_n. \tag{11.16}$$

**Remark 11.1.**– It is clear that an *analogous statement to that one in Remark 10.1 is in force* here too. (Thus, apply again (8.39)).

Applications of the preceding will be given along the sequel. In particular, the notion of volume element, as given by (11.3) (see also (11.16)), will be employed right in the next Section.

# 12. The Hodge (or $*$-)operator

Assume that we have the *framework of (11.1)* (cf. also Remark 11.1). Moreover, suppose that we are given a *Riemannian free $\mathcal{A}$-module* (see Definition 8.3)

$$(\mathcal{E},\rho), \quad \textit{with} \ \ \text{rk}\,\mathcal{E} = n \in \mathbf{N}. \tag{12.1}$$

Thus, in view of (8.19) (see also (8.17)), one obtains the following $\mathcal{A}$-*isomorphism of* the $\mathcal{A}$-*modules* concerned (see Definition 5.1);

$$\mathcal{E} \underset{\tilde{\rho}}{=} \mathcal{E}^* \equiv \mathcal{H}om_{\mathcal{A}}(\mathcal{E}, \mathcal{A}). \tag{12.2}$$

Therefore, by further employing the terminology of the previous Section 11, one now obtains the following basic $\mathcal{A}$-*isomorphism of* $\mathcal{A}$-*modules*

$$* : \bigwedge^p \mathcal{E}^* \longrightarrow \bigwedge^{n-p} \mathcal{E}^*, \tag{12.3}$$

which at the *section* level is given by the following relation;

$$(*\alpha)(\beta) := \omega(\alpha\wedge\beta^{\#}) \equiv \langle \alpha\wedge\beta^{\#}, \omega\rangle \equiv \partial\mathrm{et}_{\omega}(\alpha\wedge\beta^{\#}) \in \mathcal{A}(X), \tag{12.4}$$

*for any* $\alpha \in \bigwedge^p \mathcal{E}^*(X) = \bigwedge^p \mathcal{E}(X)^*$ *and* $\beta \in \bigwedge^{n-p}\mathcal{E}(X)$, where we also set

$$\beta^{\#} := (\bigwedge^{n-p}\tilde{\rho})(\beta) \in \bigwedge^{n-p}\mathcal{E}^*(X) = (\bigwedge^{n-p}\mathcal{E}(X))^* \tag{12.5}$$

(cf. also (5.2), concerning the last relation above, in view of our hypothesis for $\mathcal{E}$). Accordingly, one has (see also (7.23), (7.24))

$$\alpha\wedge\beta^{\#} \in \bigwedge^n \mathcal{E}^*(X) = \bigwedge^n \mathcal{E}(X) \equiv \det(\mathcal{E}(X)) = \mathcal{A}(X). \tag{12.6}$$

Therefore, the notation applied in (12.4), where by the last terms one actually means

$$\langle \alpha\wedge\beta^{\#}, \omega\rangle := (\alpha\wedge\beta^{\#})\cdot\omega, \tag{12.7}$$

that is, *multiplication of* the respective *sections* in $\mathcal{A}(X)$ (see (12.6), along with (11.3), (11.6) and (5.2) too). Yet, it is clear from (12.4) (cf. also (12.7)) that

$$*1 = \omega, \tag{12.8}$$

where, of course, 1 stands for the *identity global section of* $\mathcal{A}$ (see also (7.8)).

On the other hand, (12.3) entails an $\mathcal{A}$-*isomorphism of the* $\mathcal{A}$-*module defined by the exterior algebra of* $\mathcal{E}^*$ *into itself,* viz. one has

$$* : \bigwedge\mathcal{E}^* \longrightarrow \bigwedge\mathcal{E}^*, \tag{12.9}$$

or even

$$* \in Aut_{\mathcal{A}}(\bigwedge\mathcal{E}^*), \tag{12.10}$$

where by the last item in (12.10) we mean, of course, the (group of) $\mathcal{A}$-*automorphisms of the* $\mathcal{A}$-*module* $\bigwedge\mathcal{E}^*$ (see also (7.9)).

Now, by still extending the classical terminology, we call the map (12.9) the *$*$-operator,* or even the *Hodge operator,* on (the free $\mathcal{A}$-module) $\mathcal{E}$, induced by the given $\mathcal{A}$-metric $\rho$ on $\mathcal{E}$.

The preceding indicate simply the way, as well as, the pertinent framework, within which one can treat the corresponding hereby standard notion of the above operator. Further relevant material will be considered in the sequel, as the case may be, where we also refer for details. Yet, this is made by still extending within the present context the classical model.

# CHAPTER V

# Vector Sheaves and Cohomology

" ...leading ideas in Mathematics have a variety of alternative formal expressions"

S. Mac Lane in "Mathematics Form and Function" (Springer - Verlag, 1986). p. 398.

Our aim in this Chapter is to give an existing connection between vector sheaves and basic notions of sheaf cohomology, as the latter were exhibited in Chapter III in the preceding. In fact, one can express vector sheaves in terms of sheaf cohomology, getting thus a *cohomological classification of vector sheaves* (cf. Section 2 below), the counterpart in our case of the analogous result in the geometry of vector bundles of the classical theory. As we shall see, the above point of view will be proved to be extremely useful, and, in effect, instrumental for the applications we shall be dealt with in Part Two of this discussion.

## 1. Vector sheaves and 1-cocycles

To begin with, suppose, as usual, that we have a $\mathbf{C}$-*algebraized space* $(X, \mathcal{A})$ and let $\mathcal{E}$ be a *vector sheaf* on $X$. Moreover, consider any *local frame* of $\mathcal{E}$, say,

$$\mathcal{U} = \{U_\alpha\}_{\alpha \in I} \tag{1.1}$$

(cf. Chapt. II; Definitions 4.2, 4.3). Thus, according to Chapt. II; (4.6), one has

$$\mathcal{E}_\alpha \equiv \mathcal{E}|_{U_\alpha} \underset{\phi_\alpha}{=} \mathcal{A}^n|_{U_\alpha} = (\mathcal{A}|_{U_\alpha})^n \equiv \mathcal{A}^n_\alpha, \tag{1.2}$$

up to an *$\mathcal{A}_\alpha$-isomorphism* (we thus set $\mathcal{A}_\alpha \equiv \mathcal{A}|_{U_\alpha}$, $\alpha \in I$) *of* the *$\mathcal{A}_\alpha$-modules* involved, *for every* $\alpha \in I$, where $n = \operatorname{rk} \mathcal{E}$ (see Chapt. II; (4.2)). Therefore, one infers that;

(1.3) *for any $\alpha, \beta$ in $I$, such that*

$$U_{\alpha\beta} \equiv U_\alpha \cap U_\beta \neq \emptyset, \tag{1.3.1}$$

one obtains the following *$\mathcal{A}|_{U_{\alpha\beta}}$-isomorphism of* the respective *$\mathcal{A}|_{U_{\alpha\beta}}$-modules*

$$\phi_{\alpha\beta} \equiv \phi_\alpha \circ \phi_\beta^{-1} : \mathcal{A}^n|_{U_{\alpha\beta}} \longrightarrow \mathcal{A}^n|_{U_{\alpha\beta}}. \tag{1.3.2}$$

That is, one has;

$$\begin{aligned} \phi_{\alpha\beta} &\in \operatorname{Aut}_{\mathcal{A}|_{U_{\alpha\beta}}}(\mathcal{A}^n|_{U_{\alpha\beta}}) = M_n(\mathcal{A}|U_{\alpha\beta})(U_{\alpha\beta})^{\cdot} \\ &= M_n((\mathcal{A}|_{U_{\alpha\beta}}(U_{\alpha\beta}))^{\cdot} = M_n(\mathcal{A}(U_{\alpha\beta}))^{\cdot} = M_n(\mathcal{A})(U_{\alpha\beta})^{\cdot} \\ &= M_n(\mathcal{A})^{\cdot}(U_{\alpha\beta}) = \mathcal{GL}(n, \mathcal{A})(U_{\alpha\beta}) = \mathrm{GL}(n, \mathcal{A}(U_{\alpha\beta})). \end{aligned} \tag{1.4}$$

In this context, see also Chapt. IV; (3.17), for $X = U_{\alpha\beta}$, as well as, (3.25), (1.9) and (1.28), along with Chapt. I; (11.40). Yet, we further remark that;

(1.5) *the relation*

$$\phi_{\alpha\beta} \in \mathcal{GL}(n, \mathcal{A})(U_{\alpha\beta}) = \mathrm{GL}(n, \mathcal{A}(U_{\alpha\beta})), \tag{1.5.1}$$

as supplied by (1.4), *yields,* in effect, *the change of a local gauge of $\mathcal{E}$* or, *equivalently, the transition from a local gauge of $\mathcal{E}$* (cf. (1.2)) *to another one.*

The above relation (1.5.1) concerns, in effect, a *fundamental relation,* for all that follows, provided by our hypothesis that $\mathcal{E}$ is a vector sheaf on $X$, while *another one* is still given by the following claim; namely,

(1.6) *for any* $\alpha, \beta, \gamma$ in $I$ (cf. (1.1)), *with*

(1.6.1) $$U_{\alpha\beta\gamma} \equiv U_\alpha \cap U_\beta \cap U_\gamma \neq \emptyset,$$

*one obtains* (see (1.32))

(1.6.2) $$\phi_{\alpha\gamma} = \phi_{\alpha\beta} \circ \phi_{\beta\gamma}.$$

Therefore, based on Chapt. III; (11.3), (11.7) and (11.9′), as well as, on (1.5) above, one obtains the following.

**Lemma 1.1.** *Let* $(X, \mathcal{A})$ *be a* **C**-*algebraized space and* $\mathcal{E}$ *a vector sheaf on* $X$. *Then, with any given local frame* $\mathcal{U} = \{U_\alpha\}_{\alpha \in I}$ *of* $\mathcal{E}$, *there is associated a 1-cocycle* $(\phi_{\alpha\beta})$ *of* $\mathcal{U}$ *with coefficients in* $\mathcal{GL}(n, \mathcal{A})$, *viz. the general linear group sheaf of* $\mathcal{A}$, *of order* $n = \operatorname{rk} \mathcal{E}$. *Thus, one has*

(1.7) $$(\phi_{\alpha\beta}) \in Z^1(\mathcal{U}, \mathcal{GL}(n, \mathcal{A})). \blacksquare$$

Concerning the terminology applied above, see also Chapt. II; (11.7) and Chapt. IV; (1.26), (1.27). Yet, to indicate the association with each other of $\mathcal{E}$, $\mathcal{U}$ and $(\phi_{\alpha\beta})$, as before, we shall also employ in the sequel, occasionally, the notation;

(1.8) $$\phi_{\mathcal{U}}(\mathcal{E}) \equiv (\phi_{\alpha\beta})_{\mathcal{U}} \equiv (\phi_{\alpha\beta}) \in Z^1(\mathcal{U}, \mathcal{GL}(n, \mathcal{A})).$$

Now, to fix further our terminology, we still set the following.

**Definition 1.1.** Suppose we have the framework of the above Lemma 1.1. Then, we call (1.7) the *coordinate 1-cocycle of* $\mathcal{E}$, associated with the given local frame $\mathcal{U}$ of $\mathcal{E}$, or even a *system of coordinate transformations of* $\mathcal{E}$, supplied by $\mathcal{U}$.

Therefore, based on the preceding and on Chapt. III; (11.16), one infers that, for any given *local frame* $\mathcal{U}$ *of* $\mathcal{E}$, as before (see (1.1), (1.2)), one obtains a corresponding *1-dimensional cohomology class of* $\mathcal{E}$, *with respect to* $\mathcal{U}$; namely, one has

(1.9) $$[(\phi_{\alpha\beta})]_{\mathcal{U}} \in H^1(\mathcal{U}, \mathcal{GL}(n, \mathcal{A})) \equiv Z^1(\mathcal{U}, \mathcal{GL}(n, \mathcal{A}))/C^0(\mathcal{U}, GL(n, \mathcal{A})).$$

Thus, one further concludes that:

(1.10) *two coordinate 1-cocycles of $\mathcal{E}$, associated with a given local frame $\mathcal{U}$ of $\mathcal{E}$* (see (1.8) and Definition 1.1), *differ by an element of the 0-dimensional cochain group of $\mathcal{U}$ with values in $\mathcal{GL}(n, \mathcal{A})$, viz. by an element of the group;*

(1.10.1)
$$C^0(\mathcal{U}, \mathcal{GL}(n, \mathcal{A})) \equiv \prod_{\alpha \in I} \mathrm{GL}(n, \mathcal{A}(U_\alpha)) \equiv \prod_\alpha \mathrm{GL}(n, \mathcal{A}(U_\alpha)).$$

See also Chapt. III; (11.12), (11.14). On the other hand, according to Chapt. II; (11.26.1), one has

(1.11)
$$H^1(X, \mathcal{GL}(n, \mathcal{A})) = \sum_{\mathcal{U}} H^1(\mathcal{U}, \mathcal{GL}(n, \mathcal{A})),$$

where $\mathcal{U}$ varies over the *set of proper open coverings of* $X$, (see Chapt. III; (4.9)), or for the case in question, *equivalently*, over *that one of local frames of* $\mathcal{E}$ (cf. Chapt. II; (4.9)). Therefore, by taking the *image of (1.9) in (1.11)*, one obtains

(1.12)
$$[(\phi_{\alpha\beta})] \in H^1(X, \mathcal{GL}(n, \mathcal{A})),$$

that is, the *1st cohomology class of (1.8)* (viz. of the coordinate 1-cocycle of $\mathcal{E}$) with coefficients in $\mathcal{GL}(n, \mathcal{A})$. As we shall presently see (cf. Section 2), (1.12) *completely determines the vector sheaf* $\mathcal{E}$, under consideration.

**Scholium 1.1.**– The following nomenclature concerning (1.12), has been also of use; namely, the *cohomology class in* (1.11), supplied by a given (coordinate) 1-cocycle of a vector sheaf $\mathcal{E}$ on $X$, as in (1.8), is also named a *coordinate $\mathcal{A}$-vector bundle,* or just *$\mathcal{A}$-bundle,* associated with the given locally free $\mathcal{A}$-module of finite rank (=vector sheaf) $\mathcal{E}$. See, for instance, R.C. Gunning [2: pp. 10,14]. However, we follow instead the descriptive (!) terminology, as adopted above by Definition 1.1.

Now, before we proceed further, we give by the following Subsection, a description, in terms of (local) sections, of being a vector sheaf, by definition, locally free. This affords also a better justification of the terminology adopted so far, referred to a *"local frame"* and/or *"local gauge"* of a given vector sheaf, while it will eventually be of use later on, as well (cf. thus Chapter VI in the sequel).

**1.(a). Section-wise definition of being a vector sheaf locally free.**– As already said, it is useful to give at this place, for eventual later use too, a *localization of* our previous argument in *Chapt. IV; (3.11):* Thus, combining the above $\mathcal{A}_\alpha$*-isomorphism* $\phi_\alpha$ (cf. (1.2)) with the *Kronecker gauge of* $\mathcal{A}^n$, *restricted to* $U_\alpha$ (cf. Chapt. IV; (3.3)),

(1.13) $$(\varepsilon_i^{(\alpha)})_{1 \leqq i \leqq n} \subseteq \mathcal{A}^n(U_\alpha) = \mathcal{A}(U_\alpha)^n,$$

*for any* $\alpha \in I$, as in (1.1), we obtain a corresponding local gauge of $\mathcal{E}$, by setting;

(1.14) $$e_i^{(\alpha)} := \phi_\alpha^{-1} \circ \varepsilon_i^{(\alpha)} \in (\mathcal{E}|_{U_\alpha})(U_\alpha) = \mathcal{E}(U_\alpha), \quad 1 \leqq i \leqq n = \operatorname{rk} \mathcal{E},$$

*for any* $\alpha \in I$. So the *local gauge of* $\mathcal{E}$, in question, is now given, *by definition* (see, instead, Chapt. II; Definition 4.2), by the following (finite) family of (local) sections of $\mathcal{E}$ over $U_\alpha$;

(1.15) $$e^{(\alpha)} \equiv \left\{ (e_i^{(\alpha)})_{1 \leqq i \leqq n} \right\} \subseteq \mathcal{E}(U_\alpha).$$

In this context, one further concludes the following fact, that also justifies completely the previously adopted terminology for (1.15). Namely,

(1.16) *$\mathcal{E}$ is a vector sheaf on $X$ if, and only if, there exists an open covering $\mathcal{U} = \{U_\alpha\}_{\alpha \in I}$ of $X$, providing a family $(e_i^{(\alpha)})_{\alpha \in I}$, as in (1.15);* viz. we assume that *the (local) sections $e_i^{(\alpha)} \in \mathcal{E}(U_\alpha)$, $1 \leqq i \leqq n$, yield a basis of the $\mathcal{A}(U_\alpha)$-module $\mathcal{E}(U_\alpha)$, for any $\alpha \in I$.*

The assertion is an immediate consequence of the very definitions; cf. also Chapt. II; comments following Definition 3.2. ∎ Now, the same comments therein provide also the proof of the following claim:

(1.17) the previous assertion *(1.16) is still equivalent with* the assumption that the given open covering $\mathcal{U} = \{U_\alpha\}_{\alpha \in I}$ of $X$ fulfils *the condition; the family*

(1.17.1) $$\{e_i^{(\alpha)}(x), 1 \leqq i \leqq n\} \subseteq \mathcal{E}_x$$

*yields a basis of the $\mathcal{A}_x$-module $\mathcal{E}_x$, for every $x \in U_\alpha$, and $\alpha \in I$;* here, $e^{(\alpha)} \subseteq \mathcal{E}(U_\alpha)$, $\alpha \in I$ (cf. e.g. (1.15)), stands for a given family of (local) sections in $\mathcal{E}(U_\alpha)$.

On the other hand, by looking now at the analogous situation to (1.5), taking thus (1.15) *as our definition of a local gauge of* $\mathcal{E}$, one may still consider an $n \times n$ matrix, like $\phi_{\alpha\beta} \in \mathrm{GL}(n, \mathcal{A}(U_{\alpha\beta}))$, as the corresponding *section-matrix of transforming,* say, $e^{(\alpha)}$ *in* $e^{(\beta)}$, *for any* $\alpha, \beta \in I$, *satisfying (1.3.1);* namely, one merely looks at the (natural) *identification of* $\mathcal{E}(U_\alpha)$ *with* $\mathcal{A}^n(U_\alpha) = \mathcal{A}(U_\alpha)^n$, *provided by* $e^{(\alpha)}$, $\alpha \in I$. Therefore, one gets the following *relation of transforming* the *local gauges* (viz. section-*bases*) in question;

$$e^{(\beta)} = \phi_{\alpha\beta}(e^{(\alpha)}) \equiv \phi_{\alpha\beta} \cdot e^{(\alpha)}, \tag{1.18}$$

where in the second member of (1.18) one means, of course, *multiplication of* the respective (section-)*matrices.*

Now, as already promised, we presently come, by the following Section, to a characterization of a given vector sheaf through the cohomology class, provided by any one of its local frames.

# 2. Cohomological classification of vector sheaves

In this Section we examine the fundamental relation which exists between vector sheaves and certain cohomology classes, that are naturally associated with them, as are the cohomology classes we already considered in the preceding Section. Indeed, these classes provide, as we shall see through the succeeding discussion, an important tool for the applications of vector sheaves (cf. Part Two of this treatise). So, as already said, our aim in the present Section is to identify the two previous notions, getting thus a *cohomological characterization of vector sheaves* (cf. Theorem 2.1 below).

To start with, suppose that we are given a $\mathbf{C}$-*algebraized space*

$$(X, \mathcal{A}), \tag{2.1}$$

as before (see e.g. Chapt. IV; (0.1)), and let $\mathcal{E}$ be a *vector sheaf* on $X$, along with a given *local frame* of $\mathcal{E}$, say

$$\mathcal{U} = \{U_\alpha\}_{\alpha \in I} \tag{2.2}$$

(see (1.1), (1.2) above). Furthermore, let

$$\mathcal{GL}(n, \mathcal{A}), \quad \textit{with} \quad n = \operatorname{rk} \mathcal{E}, \tag{2.3}$$

be the corresponding *general linear group sheaf of* $\mathcal{A}$, *of order* $n \in \mathbf{N}$ (cf. Chapt. IV; (1.26)), and

$$H^1(X, \mathcal{GL}(n, \mathcal{A})) \tag{2.4}$$

the associated *1st cohomology set,* as in (1.11). Now, it is our purpose through the succeeding discussion to show that;

(2.5) *the elements of (2.4) are exactly the vector sheaves on* $X$, such that *we do not distinguish between two vector sheaves differing by an* $\mathcal{A}$*-isomorphism.*

In this context, we recall (cf. Chapt. II; (4.15)) that we denote by

$$\Phi^n_{\mathcal{A}}(X) \tag{2.6}$$

the *set of isomorphism classes of vector sheaves on* $X$, *of rank* $n \in \mathbf{N}$. However, in practice, for simplicity's sake, we simply denote by $\mathcal{E}$ too, instead, say, of $[\mathcal{E}]$, an element of the set (2.6). Thus, we just write

$$\mathcal{E} \equiv [\mathcal{E}] \in \Phi^n_{\mathcal{A}}(X). \tag{2.7}$$

So our assertion in (2.5) amounts already to the proof of the following *(set-theoretic) bijection;*

$$\Phi^n_{\mathcal{A}}(X) = H^1(X, \mathcal{GL}(n, \mathcal{A})). \tag{2.8}$$

Now, to start finding the previous bijection, we first give a *(section)-matrix formulation of the equivalence relation in the set* $\Phi^n_{\mathcal{A}}(X)$, as above. So one has the following.

**Lemma 2.1.** *Suppose we have a* $\mathbf{C}$*-algebraized space* $(X, \mathcal{A})$ *and let* $\mathcal{E}, \mathcal{F}$ *be two given vector sheaves on* $X$, *of the same rank* $n \in \mathbf{N}$. *Then, the two following assertions are equivalent:*

*1) One has*

$$[\mathcal{E}] = [\mathcal{F}] \in \Phi^n_{\mathcal{A}}(X). \tag{2.9}$$

*2) There exists a common local frame, say,* $\mathcal{U} = \{U_\alpha\}_{\alpha \in I}$, *of* $\mathcal{E}$ *and* $\mathcal{F}$, *such that their coordinate 1-cocycles, with respect to* $\mathcal{U}$, *say,* $(\phi_{\alpha\beta})$ *and* $(\psi_{\alpha\beta})$, *respectively,* (cf. Definition 1.1), *provide similar (section-)matrices. That is, there exists a 0-cochain* (cf. (1.10.1))

(2.10) $$(\eta_\alpha) \in C^0(\mathcal{U}, \mathcal{GL}(n, \mathcal{A})),$$

*such that one has*

(2.11) $$\psi_{\alpha\beta} = \eta_\alpha \circ \phi_{\alpha\beta} \circ \eta_\beta^{-1},$$

*for any $\alpha, \beta$ in $I$, with $U_{\alpha\beta} \equiv U_\alpha \cap U_\beta \neq \emptyset$.*

**Proof.** Suppose that (2.9) is valid; viz. assume that there exists an $\mathcal{A}$-*isomorphism*

(2.12) $$h : \mathcal{E} \longrightarrow \mathcal{F}.$$

Now, based on Chapt. I; Theorem 12.1 (cf. also (12.1) therein) and on Chapt. II; (4.9), one can find an *open covering of* $X$

(2.13) $$\mathcal{U} = \{U_\alpha\}_{\alpha \in I},$$

being a *common local frame of* $\mathcal{E}$ *and* $\mathcal{F}$, as well, and such that one has the following $\mathcal{A}_\alpha$-*isomorphism* (we set $\mathcal{A}|_{U_\alpha} \equiv \mathcal{A}_\alpha$, $\alpha \in I$), derived from the given $\mathcal{A}$-isomorphism $h$, as in (2.12);

(2.14) $$h|_{U_\alpha} \equiv h_\alpha : \mathcal{E}_\alpha \equiv \mathcal{E}|_{U_\alpha} \longrightarrow \mathcal{F}|_{U_\alpha} \equiv \mathcal{F}_\alpha, \quad \alpha \in I.$$

Therefore, *for every* $\alpha \in I$, one defines the following *commutative diagram of* $\mathcal{A}_\alpha$-*isomorphisms;*

(2.15) $$\begin{array}{ccc} \mathcal{E}_\alpha & \xrightarrow{\ h_\alpha\ } & \mathcal{F}_\alpha \\ {\scriptstyle \phi_\alpha}\downarrow & & \downarrow{\scriptstyle \psi_\alpha} \\ \mathcal{A}_\alpha^n & \xrightarrow{\ \eta_\alpha\ } & \mathcal{A}_\alpha^n \end{array}$$

we again set above $\mathcal{A}^n|_{U_\alpha} \equiv \mathcal{A}_\alpha^n$, while $\phi_\alpha$ and $\psi_\alpha$, $\alpha \in I$, denote the corresponding $\mathcal{A}_\alpha$-*isomorphisms,* supplied by the previous local frame of $\mathcal{E}$ and $\mathcal{F}$ (cf. e.g. (1.2)). Namely, we have actually set in (2.15);

(2.16) $$\eta_\alpha := \psi_\alpha \circ h_\alpha \circ \phi_\alpha^{-1}, \quad \alpha \in I,$$

that is, one thus obtains a *0-cochain* (cf. also Scholium 2.1 below)

(2.17) $$(\eta_\alpha) \in C^0(\mathcal{U}, \mathcal{GL}(n, \mathcal{A})),$$

which, as we shall presently see, is the desired one, as in (2.10). Namely, due to the given $\mathcal{A}$-isomorphism (2.12), one has (see also (2.14) above)

$$h_\alpha = h_\beta = h|_{U_{\alpha\beta}}, \tag{2.18}$$

*for any* $\alpha, \beta \in I$, *with* $U_{\alpha\beta} \equiv U_\alpha \cap U_\beta \neq \emptyset$. Thus, *equivalently,* (cf. (1.3.2)), one obtains the following *commutative diagram;*

(2.19)

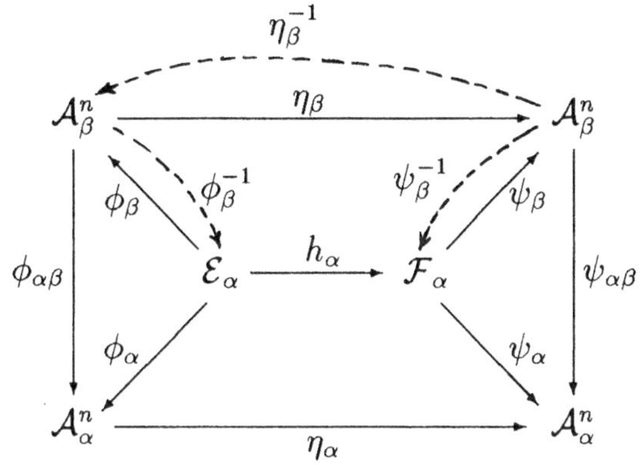


That is, one has

$$\psi_{\alpha\beta} = \eta_\alpha \circ \phi_{\alpha\beta} \circ \eta_\beta^{-1}, \tag{2.20}$$

for any $\alpha, \beta \in I$, as before, which thus along with (2.17), prove already that *1) implies 2).*

*Conversely,* suppose that (2.20) is valid, with $(\eta_\alpha)$ being a *0-cochain,* as in (2.17). Thus, defining

$$h_\alpha := \psi_\alpha^{-1} \circ \eta_\alpha \circ \phi_\alpha, \quad \alpha \in I, \tag{2.21}$$

with $\phi_\alpha, \psi_\alpha$, $\alpha \in I$, given by the corresponding relations to (1.2) (cf. e.g. (2.19)), one obtains a *0-cochain* (cf. also Scholium 2.1, ii) below)

$$(h_\alpha) \in C^0(\mathcal{U}, \mathcal{I}som_{\mathcal{A}}(\mathcal{E}, \mathcal{F})). \tag{2.22}$$

However, due to (2.20), and (2.21) (cf. also (1.3.2)), one has;

$$h_\alpha := \psi_\alpha^{-1} \circ \eta_\alpha \circ \phi_\alpha = \psi_\beta^{-1} \circ \eta_\beta \circ \phi_\beta \equiv h_\beta, \tag{2.23}$$

for any $\alpha, \beta \in I$, with $U_{\alpha\beta} \equiv U_\alpha \cap U_\beta \neq \emptyset$. That is, (2.22) is, in effect, a *0-cocycle,* hence, there exists an $\mathcal{A}$*-isomorphism* $h$, as in (2.12), such that;

(2.24) $h = (h_\alpha) \in Z^0(\mathcal{U}, \mathcal{I}som_{\mathcal{A}}(\mathcal{E}, \mathcal{F})) = \mathcal{I}som_{\mathcal{A}}(\mathcal{E}, \mathcal{F})(X) = \mathrm{Isom}_{\mathcal{A}}(\mathcal{E}, \mathcal{F})$

(see the next Scholium 2.1, ii)). So *2) implies 1)*, as well, and this terminates the proof. ∎

**Scholium 2.1.**– For completeness sake, it is profitable to give at this place *details of our previous argument, concerning (2.17) and (2.24):*

i) So looking at (2.17), one first remarks that, according to (2.16),

$$\eta_\alpha \in \mathrm{Aut}_{\mathcal{A}_\alpha}(\mathcal{A}_\alpha^n) \quad \alpha \in I. \tag{2.25}$$

(The last set stands for the *group of $\mathcal{A}_\alpha$-automorphisms of the $\mathcal{A}_\alpha$-module* $\mathcal{A}_\alpha^n$; cf. (2.15) for the notation applied). On the other hand, one also obtains (cf. Chapt. IV; (3.25));

$$\begin{aligned}\mathrm{Aut}_{\mathcal{A}_\alpha}(\mathcal{A}_\alpha^n) &= \mathcal{A}ut_{\mathcal{A}_\alpha}(\mathcal{A}_\alpha^n)(U_\alpha) = \mathcal{GL}(n,\mathcal{A}_\alpha)(U_\alpha) = \mathrm{GL}(n,\mathcal{A}_\alpha(U_\alpha)) \\ &= \mathrm{GL}(n,\mathcal{A}(U_\alpha)) = \mathcal{GL}(n,\mathcal{A})(U_\alpha), \quad \alpha \in I,\end{aligned} \tag{2.26}$$

given that one has (cf. Chapt. I; (11.40))

$$\mathcal{A}_\alpha(U_\alpha) \equiv (\mathcal{A}|_{U_\alpha})(U_\alpha) = \mathcal{A}(U_\alpha), \quad \alpha \in I. \tag{2.27}$$

Therefore, based now on (2.25) and (2.26), one obtains (see also Chapt. III; (11.12))

$$(\eta_\alpha) \in \prod_\alpha \mathrm{GL}(n,\mathcal{A}(U_\alpha)) = \prod_\alpha \mathcal{GL}(n,\mathcal{A})(U_\alpha) = C^0(\mathcal{U},\mathcal{GL}(n,\mathcal{A})), \tag{2.28}$$

*which* thus *justifies (2.17).* ∎

ii) An analogous argument to the previous one can actually be employed to get (2.24); in fact, one applies here the succeeding *technique, supplementary* to our previous considerations in *Chapt. II; Section 6:* So, more generally, suppose that we are given two *$\mathcal{A}$-modules* $\mathcal{E}$, $\mathcal{F}$ on $X$, and let us further consider the following *presheaf* (of sets) *on* $X$

$$U \longmapsto \mathrm{Isom}_{\mathcal{A}|_U}(\mathcal{E}|_U, \mathcal{F}|_U) \tag{2.29}$$

where $U$ varies over the open subsets of $X$, while the range of (2.29) stands for the set of *$\mathcal{A}|_U$-isomorphisms of* the *$\mathcal{A}|_U$-modules* concerned. We call the latter, the *set of local isomorphisms between* the ($\mathcal{A}$-modules) $\mathcal{E}$ *and* $\mathcal{F}$ *over* the open set $U \subseteq X$. Now, it is easy to prove (see e.g. Chapt. II; (6.4)) that;

(2.30) *(2.29) yields,* in effect, *a complete presheaf* of sets *on* $X$.

So we denote by

$$\mathcal{I}som_{\mathcal{A}}(\mathcal{E},\mathcal{F}) \tag{2.31}$$

the sheaf (of sets) on $X$, generated by the (complete) presheaf (2.29). We call it, the *sheaf of germs of $\mathcal{A}$-isomorphisms of $\mathcal{E}$ onto $\mathcal{F}$.*

Therefore, based on (2.30) and on Chapt. II; Proposition 11.1, we obtain

$$\mathcal{I}som_{\mathcal{A}}(\mathcal{E},\mathcal{F})(U) = \mathrm{Isom}_{\mathcal{A}|_U}(\mathcal{E}|_U,\mathcal{F}|_U), \tag{2.32}$$

within a *bijection of* the *sets* concerned, *for every open* $U \subseteq X$. In particular, one obtains;

$$\mathcal{I}som_{\mathcal{A}}(\mathcal{E},\mathcal{F})(X) \equiv \Gamma(X,\mathcal{I}som_{\mathcal{A}}(\mathcal{E},\mathcal{F})) = \mathrm{Isom}_{\mathcal{A}}(\mathcal{E},\mathcal{F}), \tag{2.33}$$

the last equality being valid up to a *bijection of* the *sets* concerned.

On the other hand, extending the respective terminology of Chapt. III; (11.12), one defines the *set of 0-cochains of* a given *open covering* $\mathcal{U} = \{U_\alpha\}_{\alpha\in I}$ *of* $X$ *with coefficients in* $\mathcal{I}som_{\mathcal{A}}(\mathcal{E},\mathcal{F})$, according to the relation;

$$C^0(\mathcal{U},\mathcal{I}som_{\mathcal{A}}(\mathcal{E},\mathcal{F})) := \prod_\alpha \mathcal{I}som_{\mathcal{A}}(\mathcal{E},\mathcal{F})(U_\alpha) = \prod_\alpha \mathrm{Isom}_{\mathcal{A}_\alpha}(\mathcal{E}_\alpha,\mathcal{F}_\alpha) \tag{2.34}$$

(see also (2.32), as well as, (2.14) for the notation applied). Now, by a similar extension of the terminology of Chapt. III; (11.35) (see also (11.36) therein), one can further consider the *set of 0-cocycles of* $\mathcal{U}$ *with coefficients in* $\mathcal{I}som_{\mathcal{A}}(\mathcal{E},\mathcal{F})$, denoted by

$$Z^0(\mathcal{U},\mathcal{I}som_{\mathcal{A}}(\mathcal{E},\mathcal{F})); \tag{2.35}$$

so, *by definition*, (2.35) is those elements $(h_\alpha)$ in (2.34), for which one has

$$h_\alpha = h_\beta|_{U_{\alpha\beta}}, \tag{2.36}$$

*for any* $\alpha,\beta \in I$, *with* $U_{\alpha\beta} \equiv U_\alpha \cap U_\beta \neq \emptyset$. So, based on (2.30), hence on (2.33), as well, and on *cond.* (S2) of Definition 11.1 in Chapt. I, we now conclude that

$$Z^0(\mathcal{U},\mathcal{I}som_{\mathcal{A}}(\mathcal{E},\mathcal{F}) = \mathrm{Isom}_{\mathcal{A}}(\mathcal{E},\mathcal{F}) = \mathcal{I}som_{\mathcal{A}}(\mathcal{E},\mathcal{F})(X), \tag{2.37}$$

thus, (2.34) together with (2.37), justify now completely (2.22) and (2.24), as asserted. ∎

Now, an immediate consequence of Lemma 2.1 is the following.

**Corollary 2.1.** *Suppose that we have the hypothesis of Lemma 2.1. Then, one has*

$$[\mathcal{E}] = [\mathcal{F}] \in \Phi^n_{\mathcal{A}}(X), \tag{2.38}$$

*if, and only if* (cf. (1.8)), *the following holds true* (cf. also (2.11))

$$\phi_{\mathcal{U}}(\mathcal{E}) \sim \phi_{\mathcal{U}}(\mathcal{F}) \ \text{ in } \ Z^1(\mathcal{U}, \mathcal{GL}(n, \mathcal{A})). \tag{2.39}$$

*Therefore* (see (1.9) - (1.12)), *(2.38) is valid, if, and only if, one has*

$$[(\phi_{\alpha\beta})] = [(\psi_{\alpha\beta})] \in H^1(X, \mathcal{GL}(n, \mathcal{A})). \ \blacksquare \tag{2.40}$$

Consequently, according to the previous corollary, one obtains

(2.41)

for every $n \in \mathbf{N}$, a *one-to-one map*

$$\rho : \Phi^n_{\mathcal{A}}(X) \longrightarrow H^1(X, \mathcal{GL}(n, \mathcal{A}) \tag{2.41.1}$$

*given by*

$$\rho([\mathcal{E}]) \equiv \rho(\mathcal{E}) := [(\phi_{\alpha\beta})], \tag{2.41.2}$$

with $(\phi_{\alpha\beta}) \in Z^1(\mathcal{U}, \mathcal{GL}(n, \mathcal{A}))$, viz. the *coordinate 1-cocycle of* $\mathcal{E}$, associated with a given *local frame* $\mathcal{U}$ of $\mathcal{E}$ (see Definition 1.1).

So the desired identification (2.8) is now supplied by the next.

**Theorem 2.1.** *Suppose we have a* **C**-*algebraized space* $(X, \mathcal{A})$. *Then, for any natural number* $n \in \mathbf{N}$, *one has the following set-theoretic bijection*

$$\Phi^n_{\mathcal{A}}(X) = H^1(X, \mathcal{GL}(n, \mathcal{A})), \tag{2.42}$$

*provided by the previous map* $\rho$ (cf. (2.41.2)).

**Proof.** As already said, (2.41.1) provides, via (2.41.2), a one-to-one map; so it remains to prove that the same map is *onto*, as well; namely, suppose that

$$(\phi_{\alpha\beta}) \in Z^1(\mathcal{U}, \mathcal{GL}(n, \mathcal{A})) \tag{2.43}$$

is a representative *1-cocycle* of a given *1-dimensional cohomology class* in the second member of (2.42) (see also (1.9), (1.11) and (1.12)). On the other hand, concerning the *open covering of* $X$ considered in (2.43), say,

(2.44) $$\mathcal{U} = \{U_\alpha\}_{\alpha \in I},$$

one may assume that;

(2.45) $\mathcal{U}$ *is a basis of the topology of* $X$.

Indeed, otherwise one can consider a suitable refinement of $\mathcal{U}$, that now satisfies (2.45).$\blacksquare$ Furthermore, by looking at the open covering $\mathcal{U}$ of $X$, as in (2.44), one defines the following correspondence (cf. also Chapt. I; (11.40))

(2.46) $$U_\alpha \longmapsto (\mathcal{A}^n|_{U_\alpha})(U_\alpha) = \mathcal{A}^n(U_\alpha), \quad \alpha \in I;$$

yet, *for any* $\alpha, \beta \in I$, *with* $U_\alpha \subseteq U_\beta$, one has a *restriction morphism*

(2.47) $$\rho_{\alpha\beta} : \mathcal{A}^n(U_\beta) \longrightarrow \mathcal{A}^n(U_\alpha),$$

according to the relation (cf. (1.5.1))

(2.48) $$\rho_{\alpha\beta} := \phi_{\alpha\beta} \circ \sigma_{\alpha\beta},$$

where

(2.49) $$\sigma_{\alpha\beta} \equiv \sigma^{U_\beta}_{U_\alpha} : \mathcal{A}^n(U_\beta) \longrightarrow \mathcal{A}^n(U_\alpha)$$

denotes the corresponding *restriction morphism of the* $\Gamma(\mathcal{A})$*-presheaf* (of sections) *associated with* $\mathcal{A}^n$ (cf. Chapt. II; (2.44)). Now, due to the *cocycle condition*

(2.50) $$\phi_{\alpha\gamma} = \phi_{\alpha\beta} \circ \phi_{\beta\gamma},$$

for any $U_\alpha \subseteq U_\beta \subseteq U_\gamma$ in $\mathcal{U}$ (cf. also (1.6.2)), one concludes that the family

(2.51) $$\{\mathcal{A}^n(U_\alpha), \rho_{\alpha\beta}; \alpha \in I\}$$

provides an *inductive system of* $\mathcal{A}(X)$*-modules.* Thus, let

(2.52) $$(\mathcal{E}, \pi, X)$$

be the *sheaf on* $X$*, associated with (2.51);* i.e., one sets

(2.53) $$\mathcal{E}_x := \varinjlim_{\alpha} \mathcal{A}^n(U_\alpha), \quad x \in X,$$

where *the inductive limit is taken over those sets in* $\mathcal{U}$ (cf. (2.44)), *which contain the point* $x \in X$ concerned, these sets providing a *local basis at* $x$ (cf.

also (2.45), along with Scholium 2.2 below). On the other hand, one proves that;

$$\text{(2.54)} \qquad \begin{array}{l} \mathcal{E} \textit{ is a vector sheaf on } X, \textit{ of rank } n, \textit{ such that} \\ \text{(2.54.1)} \qquad \phi_{\mathcal{U}}(\mathcal{E}) = (\phi_{\alpha\beta})_{\mathcal{U}} \equiv (\phi_{\alpha\beta}) \in Z^1(\mathcal{U}, \mathcal{GL}(n, \mathcal{A})), \\ \textit{as in (2.43)} \text{ (cf. also (1.8)).} \end{array}$$

Indeed, one proves that

$$\text{(2.55)} \qquad \mathcal{E}|_{U_\alpha} \underset{\phi'_\alpha}{=} \mathcal{A}^n|_{U_\alpha}, \quad \alpha \in I,$$

up to an $\mathcal{A}|_{U_\alpha}$*-isomorphism,* verifying it, at first, *fiber-wise,* in view of (2.53) (cf. also Chapt. I; (1.21)), and then as a *sheaf isomorphism,* according to Chapt. I; Theorem 3.1 (see also Scholium 2.2 below). Furthermore, by the same token, it is still easy to prove, in view of (2.55) (cf. also (2.47), (2.48)), that

$$\text{(2.56)} \qquad \phi'_{\alpha\beta} = \phi'_\alpha \circ \phi'^{-1}_\beta = \phi_{\alpha\beta},$$

*for any* $\alpha, \beta \in I$*, with* $U_{\alpha\beta} \equiv U_\alpha \cap U_\beta \neq \emptyset$*, which* thus *proves (2.54);* hence, *the map* $\rho$*,* as in (2.41.1), *is* an *onto* one, *as well,* and *this terminates the proof of Theorem 2.1.* ∎

**Scholium 2.2.**– Concerning the sheaf (2.52), one sets, of course (cf. (2.53), as well)

$$\text{(2.57)} \qquad \mathcal{E} := \sum_{x \in X} \mathcal{E}_x.$$

Moreover, consider the following family of $\mathcal{E}$*-valued local functions on* $X$

$$\text{(2.58)} \qquad \mathfrak{S} := \{\mathcal{A}^n(U_\alpha) : \alpha \in I\};$$

concerning our claim for the *range of the functions* (in fact, (local) *sections of* $\mathcal{A}^n$*) in* $\mathfrak{S}$, this follows easily from (2.53) and (2.44).

Yet, based on the very definitions (cf. also Chapt. I; (2.12)), it is easy to see that (2.58) can be viewed as a *defining family of sections of the sheaf* $\mathcal{E}$, as in (2.57), in the sense of Chapt. I; (3.17). (Indeed, based on (2.45), one easily verifies that *(2.58) can be construed as a defining family of sections of the sheaf* $\mathcal{A}^n$ *itself*). So this justifies now our claim in the preceding, concerning (2.52).

Now, taking in (2.42) $n = 1$, one obtains, in particular,

$$\Phi^1_{\mathcal{A}}(X) = H^1(X, \mathcal{A}^{\cdot}), \tag{2.59}$$

within a (set-theoretic) *bijection;* here the second member of (2.59) stands, according to the hypothesis for $\mathcal{A}$, for the *1st (sheaf) cohomology group of* $X$, *with coefficients in* $\mathcal{A}^{\cdot}$, the abelian group *sheaf of units of* $\mathcal{A}$ (cf. Chapt. IV; (1.13), as well as, Chapt. III; Remark 11.1). Hence,

(2.60) $H^1(X, \mathcal{A}^{\cdot})$ *classifies,* in terms of sheaf cohomology, the *line sheaves on* $X$.

More on (2.60) will be considered in Section 4 below.

**2.(a). Geometrical meaning of the 1st cohomology set** (à la Grothendieck).– We examine below a more general aspect about the identification (2.42), due to A. Grothendieck [1], which will be of use later on, as well; thus, cf., for instance, Theorem 8.1 in the sequel, as well as, the subsequent Chapt. VI.

So a given *sheaf* $\mathcal{E}$ on a *topological space* $X$ is called a *fiber space over* $X$ *of fiber* (or even *structure*) *type* $\mathcal{F}$, another *sheaf* on $X$, whenever $\mathcal{E}$ and $\mathcal{F}$ are *locally isomorphic;* viz.

(2.61) for every point $x \in X$, there exists an (open) neighborhood $U$ of $x$, such that one has

$$\mathcal{E}|_U = \mathcal{F}|_U, \tag{2.61.1}$$

within an *isomorphism of* the *sheaves* involved.

On the other hand, it is *equivalent* to assume that *(2.61.1) holds true with respect to a basis of the topology of* $X$ (see (2.45)).

Of course, the same terminology is reminiscent of Definition 4.3 in Chapter II, referring to *vector sheaves.* So, according to (2.61),

(2.62) a *vector sheaf* $\mathcal{E}$ on $X$ (the latter space being the base space of a given **C**-*algebraized space* $(X, \mathcal{A})$) is, in particular, a *fiber space over* $X$, *of fiber type some (finite) power* (equal to the rank of $\mathcal{E}$) *of the "sheaf of coefficients"* $\mathcal{A}$ (viz. of the *structure sheaf of* $(X, \mathcal{A})$; see Chapt. II; (4.1)).

In other words, *if* $\mathrm{rk}\,\mathcal{E} = n$, then $\mathcal{E}$ *is a fiber space over* $X$, *of fiber type* $\mathcal{A}^n$.

On the other hand, based on what has been said in the preceding, *a fiber space* $\mathcal{E}$, *as in (2.62) is cohomologically determined by a 1-cocycle ("coordinate 1-cocycle" of* $\mathcal{E}$*) with coefficients in the group sheaf* (of germs) *of automorphisms of the fiber type sheaf* (here, say, of $\mathcal{A}^n$). We call the latter group sheaf on $X$, *the structure sheaf of* $\mathcal{E}$. Thus, for the case at issue, this is actually that one appearing in Theorem 2.1; namely, what amounts to the same, one concludes that:

(2.63) *a vector sheaf* $\mathcal{E}$ *on* $X$, *with* $\mathrm{rk}\,\mathcal{E} = n$, *is a fiber space over* $X$, *of fiber type* $\mathcal{A}^n$ *and with structure sheaf*

(2.63.1) $$\mathcal{A}ut(\mathcal{A}^n) \equiv M_n(\mathcal{A})^{\cdot} = \mathcal{GL}(n, \mathcal{A})$$

(see Chapt. IV; (1.26) and (3.25)).

In this context, we also note that *the analogue of (2.42) is still in force, within the general context of (2.61)* (cf. thus e.g. (8.40) in the sequel). Thus, the preceding, within the general framework of *fiber spaces,* in conjunction with the aforementioned analogue of Theorem 2.1, constitute, in effect, the *"geometrical meaning"*, according to Grothendieck, *of the 1st cohomology set.* As already said, use of it, in the form of Theorem 2.1, in particular, will systematically be made throughout the sequel.

**2.(b). Examples.**– We consider below certain important special instances of the preceding, by taking as particular **C**-*algebraized spaces* $(X, \mathcal{A})$ the following ones, such that;

(2.64) $$\mathcal{A} = \mathcal{C}_X,\ \mathcal{C}^{\infty}_X \text{ and } \mathcal{O}_X, \text{ respectively.}$$

(cf., for instance, Chapt. I; (4.10), (4.39) and (4.40), respectively). Thus, the corresponding *1st cohomology sets:*

$$(2.65)\qquad \begin{aligned} &i)\ H^1(X, \mathcal{GL}(n, \mathcal{C}_X)),\\ &ii)\ H^1(X, \mathcal{GL}(n, \mathcal{C}_X^\infty)),\ \textit{and}\\ &iii)\ H^1(X, \mathcal{GL}(n, \mathcal{O}_X)), \end{aligned}$$

classify the *n-dimensional continuous*, or $\mathcal{C}^\infty$-, or *holomorphic* (**C**-)*vector bundles* on $X$ (as appropriate), respectively, through their *vector sheaves* (=locally free $\mathcal{A}$-modules, of rank $n$, on $X$ with $\mathcal{A}$ as in (2.64), respectively) *of* (germs of) *continuous*, $\mathcal{C}^\infty$-, or *holomorphic sections*, as the case may be.

Another useful example that will also be considered later on (cf. Chapter XI in the sequel) is the one, corresponding to $\mathcal{A} = \mathcal{C}_X^{\mathbf{A}}$, as in Chapt. IV; (2.4).

# 3. Behaviour of vector sheaves in cohomology

The purpose of the ensuing discussion is to sustain the following remark:

(3.1) Given a **C**-*algebraized space* $(X, \mathcal{A})$, *every locally free* $\mathcal{A}$-*module of finite rank,* (viz. a *vector sheaf*) *on* $X$ *behaves like a free one, of the same rank, when passing to* (Čech) *cohomology of dimension* $\geqq 1$.

Thus, suppose we are given a *vector sheaf* $\mathcal{E}$ on $X$, with $\operatorname{rk}\mathcal{E} = p \in \mathbf{N}$, while we also consider *Čech cohomology of* $X$ *with coefficients in* $\mathcal{E}$ (see Chapt. III; Section 4). So taking now the formula

$$(3.2)\qquad \check{H}^n(X, \mathcal{E}) := \varinjlim_{\mathcal{U}} \check{H}^n(\mathcal{U}, \mathcal{E}),\quad n \in \mathbf{N},$$

(ibid., (4.53)), we have, by definition, that $\mathcal{U}$ runs over the *directed set of the proper open coverings of* $X$ (ibid., (4.9)). However, *the local frames of* $\mathcal{E}$ *yield a cofinal subset of the previous set* (cf. Chapt. II; (4.9)), so that

(3.3) *one may assume, concerning (3.2), that* $\mathcal{U}$ *varies over the local frames of* $\mathcal{E}$.

(See also, for instance, N. Bourbaki [2: p.95, Proposition 8]). Therefore, taking now a particular *local frame of* $\mathcal{E}$, say,

$$(3.4)\qquad \mathcal{U} = \{U_\alpha\}_{\alpha \in I},$$

as well as, the corresponding $\mathcal{A}(X)$-*module of* $n$-*cochains of* $\mathcal{U}$ *with coefficients in* $\mathcal{E}$ (see Chapt. III; (4.14)), one obtains;

$$C^n(\mathcal{U},\mathcal{E}) = C^n(\mathcal{U},\mathcal{A}^p), \quad n \in \mathbf{N}, \tag{3.5}$$

within an $\mathcal{A}(X)$-*isomorphism of* the $\mathcal{A}(X)$-*modules* involved: Indeed, applying the notation of Chapt. III; (4.11), one has

$$C^n(\mathcal{U},\mathcal{E}) = \prod_{(\alpha_0,\ldots,\alpha_n)} \mathcal{E}(U_{\alpha_0,\ldots,\alpha_n}). \tag{3.6}$$

However, due to the hypothesis for $\mathcal{U}$ (see e.g. (1.2) in the preceding along with Chapt. I; (11.40)), *one obtains*

$$\mathcal{E}(U_{\alpha_0,\ldots,\alpha_n}) = \mathcal{A}^p(U_{\alpha_0,\ldots,\alpha_n}), \tag{3.7}$$

*for every* $(\alpha_0,\ldots,\alpha_n) \in I^{n+1}$, *with* $U_{\alpha_0,\ldots,\alpha_n} \equiv U_{\alpha_0} \cap \cdots \cap U_{\alpha_n} \neq \emptyset$, which thus justifies (3.5). ∎ Therefore, one concludes that;

(3.8) *the Čech cochain complex of* $\mathcal{U}$ *with coefficients in* $\mathcal{E}$ (cf. Chapt. III; (4.19)), $\mathcal{U}$ being a *local frame of* $\mathcal{E}$, *coincides* (up to an $\mathcal{A}(X)$-*isomorphism* of the individual issues) *with that one of* $\mathcal{U}$ *with coefficients in* $\mathcal{A}^p$ ($p = \operatorname{rk}\mathcal{E}$).

Accordingly (see also Chapt. III; (4.20)), one obtains

$$\check{H}^n(\mathcal{U},\mathcal{E}) = \check{H}^n(\mathcal{U},\mathcal{A}^p), \quad n \in \mathbf{N}, \tag{3.9}$$

up to an $\mathcal{A}(X)$-*isomorphism,* as before; hence, by virtue of (3.2), (3.3), *one* finally *obtains*

$$\check{H}^n(X,\mathcal{E}) = \check{H}^n(X,\mathcal{A}^p), \tag{3.10}$$

within an $\mathcal{A}(X)$-*isomorphism of* the $\mathcal{A}(X)$-*modules* involved, *for every* $n \geqq 1$, *where* $p = \operatorname{rk}\mathcal{E}$, which thus proves our claim in (3.1). ∎

Now, it is certainly clear that *(3.10) does not hold, in general, true for* $n = 0$; so see, for instance, Chapt. III; Lemma 4.1. The above formula (3.10) will be used in concrete cases later on (see Chapter VI).

## 4. Line sheaves. The Picard group

As already remarked by (2.59), *the abelian group*

$$(4.1) \qquad H^1(X, \mathcal{A}^\cdot)$$

*classifies the line sheaves* on the topological space $X$, base space of a given **C**-*algebraized space* $(X, \mathcal{A})$. Now, our aim by the following discussion is to show that the *set-theoretic bijection*

$$(4.2) \qquad \mathbf{\Phi}^1_{\mathcal{A}}(X) = H^1(X, \mathcal{A}^\cdot),$$

supplied by (2.59) in the preceding, is, in effect, an *isomorphism of abelian groups,* when the left-hand member of (4.2) is suitably endowed with an (abelian) *group structure:*

Thus, writing for simplicity $\mathcal{L} \equiv [\mathcal{L}] \in \mathbf{\Phi}^1_{\mathcal{A}}(X)$, as already done (see (2.7)), we first remark that, according to Chapt. II; (5.27),

(4.3) the *tensor product* $\otimes_{\mathcal{A}}$ defines in the set $\mathbf{\Phi}^1_{\mathcal{A}}(X)$ an *(abelian) group structure* (cf. Chapt. II; (5.14)), having as *neutral* (viz. *identity*) *element* the *line sheaf* (in fact, free $\mathcal{A}$-module of rank one) $\mathcal{A}$ (ibid., (5.15)) and *inverse of a* given *line sheaf* $\mathcal{L}$ on $X$, the *dual of* $\mathcal{L}$, $\mathcal{L}^* \equiv \mathcal{H}om_{\mathcal{A}}(\mathcal{L}, \mathcal{A})$, (see Chapt. IV; (5.1)).

Concerning the last assertion in (4.3), one has (see also Chapt. IV; (6.1) and Chapt. II; (6.32));

$$(4.4) \qquad \mathcal{L} \otimes_{\mathcal{A}} \mathcal{L}^* = \mathcal{H}om_{\mathcal{A}}(\mathcal{L}, \mathcal{L}) \equiv \mathcal{E}nd\, \mathcal{L} = \mathcal{A},$$

which thus proves our claim. ∎ (*Remark*.– Note that in (4.4) *we do not distinguish between $\mathcal{A}$-isomorphic line sheaves;* this was actually our practice in (4.3), as well, *by defining the group structure* in $\mathbf{\Phi}^1_{\mathcal{A}}(X)$, justified, of course, by the very nature of the elements in the latter set). On the other hand, by further extending the classical terminology (*vector bundle theory*), we now set the following.

**Definition 4.1.** Suppose we are given a **C**-algebraized space $(X, \mathcal{A})$. Then, the set $\mathbf{\Phi}^1_{\mathcal{A}}(X)$ of (isomorphism classes of) line sheaves on $X$, endowed with the (abelian) group structure provided by (4.3), is called the *Picard group* of $X$, with respect to the structure sheaf $\mathcal{A}$, and is denoted by

$$(4.5) \qquad \mathrm{Pic}_{\mathcal{A}}(X) \equiv \mathrm{Pic}\,(X).$$

In other words, one sets

$$(4.6) \qquad \mathrm{Pic}\,(X) := (\mathbf{\Phi}^1_{\mathcal{A}}(X), \otimes_{\mathcal{A}}) \equiv \mathbf{\Phi}^1_{\mathcal{A}}(X).$$

The elements of (4.5) are also named (in view of (4.4)) *invertible $\mathcal{A}$-modules* (or even *invertible sheaves*) on $X$.

We come next to establish the *group isomorphism,* inherent in (4.2), as already asserted, each member of the said relation having now its own (abelian) group structure:

So, more generally, let us take two *vector sheaves* $\mathcal{E}$ and $\mathcal{F}$ on $X$, *with* $\operatorname{rk}\mathcal{E} = n$ *and* $\operatorname{rk}\mathcal{F} = m$. Then (cf. Chapt. II; (5.27)), $\mathcal{E} \otimes_{\mathcal{A}} \mathcal{F}$ is still a *vector sheaf* on $X$, *such that* $\operatorname{rk}(\mathcal{E} \otimes_{\mathcal{A}} \mathcal{F}) = n \cdot m$. On the other hand, suppose that

$$(\phi_{\alpha\beta}) \in Z^1(\mathcal{U}, \mathcal{GL}(n, \mathcal{A})) \tag{4.7}$$

and

$$(\psi_{\alpha\beta}) \in Z^1(\mathcal{U}, \mathcal{GL}(m, \mathcal{A})) \tag{4.8}$$

are the *coordinate 1-cocycles* of $\mathcal{E}$ and $\mathcal{F}$, respectively, which are associated with a *common local frame $\mathcal{U}$ of $\mathcal{E}$ and $\mathcal{F}$* (cf., for instance, the beginning of the proof of Lemma 2.1 in the preceding, concerning the later assumption on $\mathcal{U}$). Therefore, one can further consider the *tensor* (or even *Kronecker*) *product of the (section-) matrices* involved in (4.7) and (4.8); namely, one has

$$\phi_{\alpha\beta} \otimes \psi_{\alpha\beta} \in \mathrm{GL}(n \cdot m, \mathcal{A}(U_\alpha \cap U_\beta)) \tag{4.9}$$

(see also (1.4), along with e.g. L. Chambadal - J.L. Ovaert [1: p.137f]). Thus, based further on standard properties of tensor product matrices (ibid.), it is easily seen that

$$(\phi_{\alpha\beta} \otimes \psi_{\alpha\beta}) \in Z^1(\mathcal{U}, \mathcal{GL}(n \cdot m, \mathcal{A})), \tag{4.10}$$

so that one has

$$(\phi_{\alpha\beta}) \otimes (\psi_{\alpha\beta}) = (\phi_{\alpha\beta} \otimes \psi_{\alpha\beta}), \tag{4.11}$$

*as 1-cocycles;* indeed, taking again a standard argument into account (cf., for instance, the above cited Ref.; p.86, Corollaire 7), one actually obtains the following (set-theoretic) *injection* (see also (4.9))

$$Z^1(\mathcal{U}, \mathcal{GL}(n, \mathcal{A})) \otimes Z^1(\mathcal{U}, \mathcal{GL}(m, \mathcal{A})) \stackrel{\subsetneq}{\rightarrow} Z^1(\mathcal{U}, \mathcal{GL}(n \cdot m, \mathcal{A})). \tag{4.12}$$

Furthermore, one still obtains (cf. the same Ref.; p.138, along with (1.3.2) in the preceding)

$$(4.13)\qquad \begin{aligned}\phi_{\alpha\beta}\otimes\psi_{\alpha\beta}\equiv\phi_\alpha\circ\phi_\beta^{-1}\otimes\psi_\alpha\circ\psi_\beta^{-1}&=(\phi_\alpha\otimes\psi_\alpha)\circ\left(\phi_\beta^{-1}\otimes\psi_\beta^{-1}\right)\\&=(\phi_\alpha\otimes\psi_\alpha)\circ(\phi_\beta\otimes\psi_\beta)^{-1}.\end{aligned}$$

So the foregoing, together with Chapt. II; Lemma 5.1 supply already the proof of the following.

**Lemma 4.1.** *Let* $(X,\mathcal{A})$ *be a* **C**-*algebraized space and* $\mathcal{E}$, $\mathcal{F}$ *two vector sheaves on* $X$, *of rank* $n$ *and* $m$, *respectively. Moreover, let* $\mathcal{U}$ *be a local frame of* $\mathcal{E}$ *and* $\mathcal{F}$ (cf. (2.13)), *such that let* $(\phi_{\alpha\beta})$ *and* $(\psi_{\alpha\beta})$ *be the coordinate 1-cocycles of* $\mathcal{E}$ *and* $\mathcal{F}$, *respectively, relative to* $\mathcal{U}$ (cf. (4.7), (4.8)). *Then,* $\mathcal{U}$ *is a local frame of the vector sheaf* $\mathcal{E}\otimes_{\mathcal{A}}\mathcal{F}$, *as well, such that*

$$(4.14)\qquad (\phi_{\alpha\beta}\otimes\psi_{\alpha\beta})\in Z^1(\mathcal{U},\mathcal{GL}(n\cdot m,\mathcal{A}))$$

*is a coordinate 1-cocycle of* $\mathcal{E}\otimes_{\mathcal{A}}\mathcal{F}$, *relative to* $\mathcal{U}$. ∎

Now, specializing in the case of line sheaves on $X$, we finally have the promised result, pertaining to Pic$(X)$, as above (see (4.6)). Namely, one obtains.

**Theorem 4.1.** *Let* $(X,\mathcal{A})$ *be a given* **C**-*algebraized space. Then, the Picard group of* $X$, *relative to* $\mathcal{A}$ (see Definition 4.1), *is isomorphic to the 1st cohomology group of* $X$ *with coefficients in* $\mathcal{A}^{\cdot}$, *viz. in the sheaf of units of* $\mathcal{A}$ (cf. Chapt. IV; (1.13)). *Thus, one has*

$$(4.15)\qquad \operatorname{Pic}(X)=H^1(X,\mathcal{A}^{\cdot}),$$

*up to an (abelian) group isomorphism.*

**Proof.** Let $\mathcal{L}$ and $\mathcal{L}'$ be two given *line sheaves* on $X$, having *coordinate 1-cocycles* $(\phi_{\alpha\beta})$ and $(\phi'_{\alpha\beta})$, respectively, relative to a *common local frame* $\mathcal{U}=\{U_\alpha\}_{\alpha\in I}$ *of* $\mathcal{L}$ *and* $\mathcal{L}'$ (cf. (4.7), (4.8)). Then, according to Lemma 4.1, $(\phi_{\alpha\beta}\otimes\phi'_{\alpha\beta})$ *is a coordinate 1-cocycle of the line sheaf* $\mathcal{L}\otimes_{\mathcal{A}}\mathcal{L}'$ (cf. Chapt. II; (5.27)), relative to the *same local frame* $\mathcal{U}$. Consequently, one obtains (cf. (4.11));

$$(4.16)\qquad (\phi_{\alpha\beta})\otimes(\phi'_{\alpha\beta})=(\phi_{\alpha\beta}\otimes\phi'_{\alpha\beta})=(\phi_{\alpha\beta}\cdot\phi'_{\alpha\beta})\in Z^1(\mathcal{U},\mathcal{A}^{\cdot}),$$

given that $\mathcal{GL}(1,\mathcal{A})\equiv\mathcal{A}^{\cdot}$ (see Chapt. IV; (1.26), for $n=1$), while, *one also has*

$$(4.17)\qquad \phi_{\alpha\beta}\otimes\phi'_{\alpha\beta}=\phi_{\alpha\beta}\cdot\phi'_{\alpha\beta}\in\mathcal{A}^{\cdot}(U_{\alpha\beta})=\mathcal{A}(U_{\alpha\beta})^{\cdot},$$

*for any* $\alpha, \beta \in I$, *with* $U_{\alpha\beta} \equiv U_\alpha \cap U_\beta \neq \emptyset$ (cf. also Chapt. IV; (1.15). Yet, in this context, see the Remark after (4.4) in the preceding). Therefore, *the set-theoretic bijection (2.41.1) preserves,* in the case under consideration, *the multiplication* (viz. the *group structure*) *in* $\mathbf{\Phi}^1_{\mathcal{A}}(X)$ *and* $H^1(X, \mathcal{A}^{\cdot})$, which is actually the assertion. ∎

**4.(a). The determinant morphism (contn'd).**– Our purpose here is to give an expression of the morphism in the title of this Subsection, in terms of cohomology classes (see thus (4.23) below), the same morphism being already considered in the preceding (cf. Chapt. IV; Section 4). In particular, we are interested in getting such an expression for the *determinant line sheaf* $\det(\mathcal{E})$ of a given *vector sheaf* $\mathcal{E}$ on $X$ (see Chapt. IV; Subsection 7.(a)).

So, according to Chapt. IV; (4.8), (4.10) and (4.11), one has the following *morphism of group sheaves*

$$\partial\mathrm{et} : \mathcal{GL}(n, \mathcal{A}) \longrightarrow \mathcal{A}^{\cdot}, \tag{4.18}$$

*for any* $n \in \mathbf{N}$, such that (see Chapt. IV; (4.12))

$$\overline{\partial\mathrm{et}}_U(\alpha) := \det{}_U(\alpha_{ij}) \in \mathcal{A}^{\cdot}(U) = \mathcal{A}(U)^{\cdot}, \tag{4.19}$$

*for every* $\alpha \equiv (\alpha_{ij}) \in \mathcal{GL}(n, \mathcal{A})(U) = \mathrm{GL}(n, \mathcal{A}(U))$, with $U$ *open* in $X$. Therefore (see Chapt. III; (11.29), (11.31)), one has, along with (4.18), a corresponding map between the respective *1st cohomology sets* (for simplicity, we retain the same notation for the second map too)

$$\partial\mathrm{et} : H^1(X, \mathcal{GL}(n, \mathcal{A})) \longrightarrow H^1(X, \mathcal{A}^{\cdot}). \tag{4.20}$$

Consequently, if $\mathcal{E}$ is a *vector sheaf* on $X$, *of rank* $n \in \mathbf{N}$, and

$$(\phi_{\alpha\beta}) \in Z^1(\mathcal{U}, \mathcal{GL}(n, \mathcal{A})) \tag{4.21}$$

stands for a *coordinate 1-cocycle of* $\mathcal{E}$, associated with a *local frame* $\mathcal{U}$ *of* $\mathcal{E}$ (cf. (1.8)), one obtains (see (4.19), together with Chapt. IV; (7.23));

$$\phi_{\mathcal{U}}(\det \mathcal{E}) = (\det{}_{U_{\alpha\beta}}(\phi_{\alpha\beta})) \equiv (\det(\phi_{\alpha\beta})) \in Z^1(\mathcal{U}, \mathcal{A}^{\cdot}). \tag{4.22}$$

Indeed, based on the standard *multiplication theorem* for the det *function* (cf., for example, S. Lang [1: p.455, Proposition 4.15]), it is easy to see that the *cocycle condition* for $(\phi_{\alpha\beta})$ (cf. (1.6.2) in the preceding) yields an analogous one for $(\det(\phi_{\alpha\beta}))$, which thus justifies the last relation in (4.22); now, the first issue in the same relation (4.22) is warranted by the fact that (cf. also (1.4))

(4.23) $\det(\phi_{\alpha\beta}) \in \mathrm{Aut}(\bigwedge^n \mathcal{A}^n|_{U_{\alpha\beta}}) = \mathrm{Aut}(\mathcal{A}|_{U_{\alpha\beta}}) = \mathcal{A}^{\cdot}(U_{\alpha\beta}) = \mathcal{A}(U_{\alpha\beta})^{\cdot}$,

*for any* $\alpha, \beta \in I$, *as in (1.3.1),* such that (see (1.3.2), together, for instance, with L. Chambadal - J.L. Ovaert [1: p.213, Théorème 16.2, prop. 3 and p.214, Corollaire]);

(4.24) $\det(\phi_{\alpha\beta}) = \bigwedge^n \phi_{\alpha\beta} = \bigwedge^n(\phi_\alpha \circ \phi_\beta^{-1}) = (\bigwedge^n \phi_\alpha) \circ (\bigwedge^n \phi_\beta)^{-1} = \phi_{\mathcal{U}}(\det \mathcal{E})$

(cf also Chapt. IV; (7.32), (4.3)). Thus, (4.22) is now completely justified. ∎

We call (4.22) the *determinant 1-cocycle* of (4.21); by what has been said before, it is thus the *coordinate 1-cocycle* of $\det \mathcal{E}$, the *determinant line sheaf* of the given *vector sheaf* $\mathcal{E}$ on $X$.

Thus, based on (4.22), one concludes that:

(4.25) if $[(\phi_{\alpha\beta})] \in H^1(X, \mathcal{GL}(n, \mathcal{A}))$ stands for the *1-dimensional cohomology class,* representing a given *vector sheaf* $\mathcal{E}$ on $X$ (see (1.12), (2.5)), then (see also (4.20), (4.15))

(4.25.1) $\partial\mathrm{et}([(\phi_{\alpha\beta})]) = [(\det(\phi_{\alpha\beta}))] \in H^1(X, \mathcal{A}^{\cdot}) = \mathrm{Pic}(X)$

yields the corresponding *1-dimensional cohomology class of the determinant line sheaf of* $\mathcal{E}$, $\det \mathcal{E}$.

# 5. Flat vector sheaves

We consider below a particular class of vector sheaves, that will be of importance later on (see the subsequent Chapters VI, VII). So suppose, as usual, that we are given a **C**-*algebraized space* $(X, \mathcal{A})$; moreover, consider the *constant general linear group sheaf* on $X$ (non-abelian, in general), *of order* $n \in \mathbf{N}$,

(5.1) $$\mathcal{GL}(n, \mathbf{C}) := X \times \mathrm{GL}(n, \mathbf{C}) \equiv \mathrm{GL}(n, \mathbf{C})_X \equiv \mathrm{GL}(n, \mathbf{C}).$$

(see also Chapt. I; (4.4), (4.4′) and Chapt. II; Example 1.1). Thus, for every *open* $U \subseteq X$, one obtains a *canonical injection*

(5.2) $$\mathrm{GL}(n, \mathbf{C}) \hookrightarrow \mathcal{GL}(n, \mathcal{A})(U) = \mathrm{GL}(n, \mathcal{A}(U))$$

(see Chapt. IV; (1.31), (1.32)). *Warning!* Here $\mathrm{GL}(n, \mathbf{C})$ stands for the corresponding *constant presheaf* (cf. Chapt. I; Subsection 10.(a)), not the constant sheaf, as in the last item of (5.1). Now, as a result of (5.2) (see also Chapt. I; (11.14), together with Lemma 12.1), one obtains a (canonical) *inclusion map* for the respective sheaves, in fact, a *group sheaf isomorphism* (into)

$$i_X : \mathrm{GL}(n, \mathbf{C})_X \equiv \mathrm{GL}(n, \mathbf{C}) \longrightarrow \mathcal{GL}(n, \mathcal{A}). \tag{5.3}$$

Consequently (cf. Chapt. II; (11.31)), one has, via (5.3), an induced map between the corresponding *1st cohomology sets;*

$$i_X^* : H^1(X, \mathrm{GL}(n, \mathbf{C})) \longrightarrow H^1(X, \mathcal{GL}(n, \mathcal{A})). \tag{5.4}$$

The preceding provide now the necessary framework for the following basic.

**Definition 5.1.** Suppose we have a $\mathbf{C}$-algebraized space $(X, \mathcal{A})$. Then, by a *flat* $\mathbf{C}$*-vector bundle, of rank* $n \in \mathbf{N}$, on $X$, one means an element of the set $H^1(X, \mathrm{GL}(n, \mathbf{C})) \equiv H^1(X, \mathcal{GL}(n, \mathbf{C}))$ (cf. (5.1)). On the other hand, we say that a given *vector sheaf* $\mathcal{E}$ on $X$ possesses a *flat representation,* whenever $\mathcal{E}$ belongs to the image of (5.4), as above, viz. in the case one has

$$\mathcal{E} \equiv [\mathcal{E}] \in \mathrm{im}\,(i_X^*) \subseteq H^1(X, \mathcal{GL}(n, \mathcal{A})) = \mathbf{\Phi}^n_{\mathcal{A}}(X) \tag{5.5}$$

(cf. (2.7), (2.8)). We also speak then of a *flat vector sheaf* $\mathcal{E}$ on $X$. Yet, we call

$$(i_X^*)^{-1}(\mathcal{E}) \subseteq H^1(X, \mathcal{GL}(n, \mathbf{C})) \equiv H^1(X, \mathrm{GL}(n, \mathbf{C})), \tag{5.6}$$

for any given $\mathcal{E} \in \mathbf{\Phi}^n_{\mathcal{A}}(X)$, as before, the *set of flat representatives of* $\mathcal{E}$.

Thus, according to the previous definition, *flatness of* a given $\mathcal{E} \in \mathbf{\Phi}^n_{\mathcal{A}}(X)$ *means,* in effect, *that*

$$(i_X^*)^{-1}(\mathcal{E}) \neq \emptyset; \tag{5.7}$$

*equivalently,* a given $\mathcal{E} \in \mathbf{\Phi}^n_{\mathcal{A}}(X)$ *possesses a flat representative* if one has

$$[\mathcal{E}] \equiv \mathcal{E} = i_X^*(\xi), \tag{5.8}$$

for some

$$\xi \equiv [\xi] \in H^1(X, \mathrm{GL}(n, \mathbf{C})). \tag{5.9}$$

**Scholium 5.1.**– We have already remarked in the foregoing that the *constant sheaf* $\mathbf{C}$ can be construed as a *(constant)* $\mathbf{C}$*-algebra sheaf* on $X$: See Chapt. II; Definition 1.4 and Scholium 1.1, together with (1.39) and (1.40), for $\mathbf{A} = \mathbf{C}$; *hence one has*

$$\mathcal{A} = \mathbf{C}_X \equiv \mathbf{C}. \tag{5.10}$$

Therefore, any element of the set

$$H^1(X, \mathrm{GL}(n, \mathbf{C})) \equiv H^1(X, \mathcal{GL}(n, \mathbf{C})) \tag{5.11}$$

can be viewed as a *locally free* $\mathbf{C}$*-module* of rank $n \in \mathbf{N}$ on $X$, that is, as a (particular) *vector sheaf* on $X$ (precisely, an isomorphism class of such. In this context, see also Theorem 2.1, for $\mathcal{A} = \mathbf{C}$, as in (5.10), along with Subsection 2.(a), especially, (2.63)). Thus, following the terminology of Definition 5.1, and based on the aforesaid result, for the particular case at issue, one may define

$$\mathbf{\Phi}^n_{\mathbf{C}}(X) = H^1(X, \mathrm{GL}(n, \mathbf{C})) \equiv H^1(X, \mathcal{GL}(n, \mathbf{C})), \tag{5.12}$$

as the *set of (isomorphism classes of) flat* $\mathbf{C}$*-vector bundles on* $X$.

Notice that (5.11), with $\mathrm{GL}(n, \mathbf{C})$ a *discrete group* classifies also *flat* $\mathbf{C}$*-vector bundles,* in the classical sense of (topological) *vector bundle theory;* thus, see e.g. R.C. Gunning [2: p.96ff], as well as, Chapt. III; (8.16′.1) in the preceding. Yet, cf. Chapters VII and XI in the sequel.

The hitherto applied (differential) geometric language will be justified in the sequel (see thus the subsequent Chapters VI, VII). Yet, the preceding nomenclature is further clarified by the notion of coordinate cocycles of vector sheaves, that we already applied so far, according to the following.

**Theorem 5.1.** *Suppose we are given a* $\mathbf{C}$*-algebraized space* $(X, \mathcal{A})$*, and let* $\mathcal{E}$ *be a vector sheaf on* $X$*. Then,* $\mathcal{E}$ *has a flat representative* (cf. (5.5) and/or (5.8)), *if, and only if, it has a coordinate 1-cocycle, consisting of* $\mathrm{GL}(n, \mathbf{C})$*-valued locally constant functions on* $X$*.*

**Proof.** Suppose that $\mathcal{E}$ is a given *vector sheaf* on $X$, for which (5.8) *is valid,* for some *flat* $\mathbf{C}$*-vector bundle* (cf. Definition 5.1)

$$\xi \equiv [\xi] \in H^1(X, \mathrm{GL}(n, \mathbf{C})) \equiv H^1(X, \mathcal{GL}(n, \mathbf{C})). \tag{5.13}$$

But (see Chapt. III; (11.26)), one has

(5.14) $$H^1(X, \mathrm{GL}(n, \mathbf{C})) = \sum_{\mathcal{U}} H^1(\mathcal{U}, \mathrm{GL}(n, \mathbf{C})),$$

where $\mathcal{U}$ *runs over the* (proper) *open coverings of* $X$, so that, in view of (5.13), one has (cf. Chapt. III; (11.16))

(5.15) $$\xi \in H^1(\mathcal{U}, \mathrm{GL}(n, \mathbf{C})) = Z^1(\mathcal{U}, \mathrm{GL}(n, \mathbf{C}))/C^0(\mathcal{U}, \mathrm{GL}(n, \mathbf{C})),$$

while we may further assume that

(5.16) $$\mathcal{U} = \{U_\alpha\}_{\alpha \in I}$$

in (5.15) is now a *local frame of* $\xi$ (cf. Scholium 5.1 above, together with Chapt. III; (11.23), (11.24) and Chapt. II; (4.9)). Thus, suppose that

(5.17) $$(\lambda_{\alpha\beta}) \in Z^1(\mathcal{U}, \mathrm{GL}(n, \mathbf{C})) \subseteq C^1(\mathcal{U}, \mathrm{GL}(n, \mathbf{C})) := \prod_{(\alpha,\beta)} \mathrm{GL}(n, \mathbf{C})(U_{\alpha\beta})$$

is a *coordinate 1-cocycle of* $\xi$, *associated with* $\mathcal{U}$ (see Definition 1.1 and Scholium 5.1 above, as well as, Chapt. III; (11.3), (11.6) and (11.7)). Therefore, due to (5.17), one concludes that;

(5.18) $$\lambda_{\alpha\beta} \in \mathrm{GL}(n, \mathbf{C})(U_{\alpha\beta}),$$

*for any* $\alpha, \beta \in I$, *with* $U_{\alpha\beta} \equiv U_\alpha \cap U_\beta \neq \emptyset$ (see (5.16)). In other words (see Chapt. I; Theorem 11.1), one infers that;

(5.19) $\lambda_{\alpha\beta}$ in (5.18) *is a* $\mathrm{GL}(n, \mathbf{C})$*-valued locally constant function on the open set* $U_{\alpha\beta} \equiv U_\alpha \cap U_\beta \subseteq X$, *for any* $\alpha, \beta \in I$, *with* $U_{\alpha\beta} \neq \emptyset$, as in (5.18). (We recall that $\mathrm{GL}(n, \mathbf{C})$ carries the *discrete topology*).

Furthermore, we remark that (see also Chapt. II; (1.34), (1.34′));

(5.20) $\lambda_{\alpha\beta} \in \mathrm{GL}(n, \mathbf{C})(U_{\alpha\beta})$, with $\alpha, \beta \in I$, as in (5.19), *is locally a constant multiple of the identity (section-)matrix* $1 \equiv 1_n \in \mathcal{GL}(n, \mathcal{A})(X) = \mathrm{GL}(n, \mathcal{A}(X))$. Of course, the (constant) section 1 has to be suitably restricted, in accordance with $\lambda_{\alpha\beta}$).

We are going to apply (5.20) right below.

Thus, suppose that $\mathcal{U}$, as in (5.16) is a *local frame of* $\mathcal{E}$, *as well* (cf. e.g. (4.7), (4.8)), so that let

(5.21) $$(\phi_{\alpha\beta}) \in Z^1(\mathcal{U}, \mathcal{GL}(n, \mathcal{A}))$$

be a *coordinate 1-cocycle of $\mathcal{E}$, relative to $\mathcal{U}$.* Therefore, by virtue of (5.8), (5.18) and (5.3), in conjunction with Theorem 2.1, one has

$$(\phi_{\alpha\beta}) = i_X^*((\lambda_{\alpha\beta})) \equiv i_X^*(\lambda_{\alpha\beta}) \in Z^1(\mathcal{U}, \mathcal{GL}(n, \mathcal{A})). \tag{5.22}$$

Namely, by virtue of (5.2) (cf. also Chapt. III; (11.3)), one still obtains the following (canonical) *injection* (cf. (5.20), as well)

$$C^1(\mathcal{U}, \mathrm{GL}(n, \mathbf{C})) \underset{i_X^*}{\overset{\subseteq}{\rightarrow}} C^1(\mathcal{U}, \mathcal{GL}(n, \mathcal{A})) \tag{5.23}$$

so that one further obtains a *corresponding restriction* (being) *one-to-one,* as well) *to 1-cocycles;* viz. one has

$$Z^1(\mathcal{U}, \mathrm{GL}(n, \mathbf{C})) \underset{i_X^*}{\overset{\subseteq}{\rightarrow}} Z^1(\mathcal{U}, \mathcal{GL}(n, \mathcal{A})). \tag{5.24}$$

(For simplicity, we retained above the notation $i_X^*$ for the *maps, induced by* (5.2), in all the previous correspondences (5.4), (5.23) and (5.24)). So one has (cf. (5.18), (5.22), (5.24) and (5.20));

$$\begin{aligned} \phi_{\alpha\beta} = i_X^*(\lambda_{\alpha\beta}) \in i_X^*(\mathrm{GL}(n, \mathbf{C})(U_{\alpha\beta})) &\underset{i_X^*}{\cong} \mathrm{GL}(n, \mathbf{C})(U_{\alpha\beta}) \\ &\overset{\subseteq}{\rightarrow} \mathrm{GL}(n, \mathcal{A}(U_{\alpha\beta})) = \mathrm{GL}(n, \mathcal{A})(U_{\alpha\beta}), \end{aligned} \tag{5.25}$$

*for any $\alpha, \beta \in I$, with $U_{\alpha\beta} \neq \emptyset$.* (We recall that *the section functor is left exact, for* any *sheaves of groups* too; cf. Chapt. II; (2.59). In this context, cf. also the next Scholium 5.2). This terminates the *"only if"* part of the assertion.

Now, suppose that we have (see also (1.8) and (5.24))

$$\phi_{\mathcal{U}}(\mathcal{E}) \equiv (\phi_{\alpha\beta}) = i_X^*(\lambda_{\alpha\beta}) \in Z^1(\mathcal{U}, \mathcal{GL}(n, \mathcal{A})), \tag{5.26}$$

*with $(\lambda_{\alpha\beta})$, as in (5.17).* Thus, based on the very definitions (yet, see Theorem 2.1), one concludes that $\mathcal{E}$ *has a flat representative,* and this completes the proof of the theorem. ∎

**Scholium 5.2.**– Concerning our previous argument on (5.25), we also have employed, in effect, the following general fact:

Namely, suppose we are given two *presheaves* (of groups) $E$ and $F$ on a *topological space $X$*, and let $\mathcal{E} \equiv \mathbf{S}(E)$, $\mathcal{F} \equiv \mathbf{S}(F)$ be their *sheafifications,* respectively (Chapt. I; Section 7). Then, one concludes that;

(5.27) *exactness of*

(5.27.1) $$0 \longrightarrow E \longrightarrow F$$

still *implies* the *exactness of*

(5.27.2) $$0 \longrightarrow \Gamma(\mathcal{E}) \longrightarrow \Gamma(\mathcal{F}).$$

(For the notation applied above, see e.g. Chapt. II; (2.31) and Chapt. I; (9.7)); now, concerning the truth of (5.27), one first remarks that exactness of (5.27.1) implies that one for the respective stalks of the presheaves involved (cf. e.g. Chapt. II; (2.34)), hence for those ones of the sheafifications of the latter (see Chapt. I; (7.6)). This, in turn, implies (by the very definition) the same for the corresponding sheaves (Chapt. I; (1.21)), which finally entails the desired exactness of (5.27.2), by virtue of the *left exactness of the section functor* (see Chapt. II; (2.59)). ∎

## 6. Line sheaves and Chern classes

We examine below the analogue in our case of the classical *Chern isomorphism* for line sheaves (viz., equivalently, for *line bundles,* when employing standard terminology). In this context, a crucial issue is certainly the *exponential function,* the existence of which in our abstract framework will be suitably postulated. Apart, of course, of the classical case (see Examples 6.1; i), ii) below), we do indicate the existence of the said function in concrete important instances, that will be treated in detail later on (cf. also the same Examples, iii) below). Now, to start with, we first set the following.

**Definition 6.1.** By an *exponential algebraized space,* one means a given **C**-algebraized space $(X, \mathcal{A})$ which is further endowed with a *(short exact) exponential sheaf sequence;* viz. with the following short exact sequence of sheaves of (abelian) groups

(6.1) $$0 \longrightarrow \mathbf{Z} \xrightarrow{\varepsilon} \mathcal{A} \xrightarrow{e} \mathcal{A}^{\cdot} \longrightarrow 1.$$

Concerning the above sequence, **Z** stands there for the *constant sheaf of integers;* viz. one has

(6.2) $$\mathbf{Z} \equiv \mathbf{Z}_X = X \times \mathbf{Z}$$

(cf. e.g. Chapt. I; (4.4). Yet, see Chapt. IV; (1.13) for $\mathcal{A}^{\cdot}$). Thus, equivalently, one infers from (6.1) that;

$$\ker(e) = \operatorname{im}(\varepsilon) = \mathbf{Z}, \tag{6.3}$$

the second equality above standing, in fact, for an (abelian) *group isomorphism.*

**Examples 6.1.**– We consider first certain important particular instances of the standard case, that realize (6.1). Besides, we refer to Chapter XI for other non-standard examples. So we have:

i) The *smooth* (viz. $\mathcal{C}^{\infty}$) or *holomorphic* case; so here $\mathcal{A}$ stands for the *structure sheaf* $\mathcal{C}_X^{\infty}$ (ie., the sheaf of germs of **C**-*valued* $\mathcal{C}^{\infty}$-*functions*) or for $\mathcal{O}_X$ (ie., the sheaf of germs of *holomorphic functions*), respectively. (In this regard, see also Chapt. I; (4.39), (4.40). Moreover, cf., for instance, R.O. Wells [1: p.45, Example 2.6, and p.105]). Thus, concerning the *holomorphic case,* one obtains the following *short exact sequence of* (abelian) *group sheaves;*

$$0 \longrightarrow \mathbf{Z} \longrightarrow \mathcal{O}_X \stackrel{e}{\longrightarrow} \mathcal{O}_X^{\cdot} \longrightarrow 1, \tag{6.4}$$

such that one sets

$$e(f) := \exp 2\pi i f, \tag{6.5}$$

*for every* $f \in \mathcal{O}_X(U) = \mathcal{O}(U)$ (see also Chapt. I; (11.10) and (11.11); iii), along with Proposition 11.1, concerning the last equality), and any *open* $U \subseteq X$, where $X$ stands here for a (Hausdorff) *connected complex* (analytic) *manifold.*

On the other hand, given a (Hausdorff) $\mathcal{C}^{\infty}$-*manifold* $X$, one obtains for the respective *structure sheaf* $\mathcal{C}_X^{\infty}$, as above, the *exact sequence*

$$0 \longrightarrow \mathbf{Z} \longrightarrow \mathcal{C}_X^{\infty} \stackrel{e}{\longrightarrow} (\mathcal{C}_X^{\infty})^{\cdot} \longrightarrow 1, \tag{6.6}$$

defining the *(group sheaf) morphism* $e$ involved, analogously to (6.5). (In this regard, see also e.g. F. Hirzebruch [1: p.24; (10)]).

ii) The *continuous case,* with $\mathcal{A} = \mathcal{C}_X$; viz. one considers the *sheaf of* germs of **C**-*valued continuous functions* on an arbitrary *topological space* $X$. See Chapt. I; (4.10) and (11.11); i), along e.g. with F. Hirzebruch [1: p.24; (9)]. Thus, one has the following *exact sequence;*

$$(6.7) \qquad 0 \longrightarrow \mathbf{Z} \longrightarrow \mathcal{C}_X \xrightarrow{e} \mathcal{C}^{\cdot}_X \longrightarrow 1,$$

setting again

$$(6.8) \qquad e(f) := \exp 2\pi i f,$$

*for every* $f \in \mathcal{C}_X(U) = \mathcal{C}(U)$, with $U$ *open* in $X$.

iii) As already mentioned, we are going to examine in Chapter XI certain particular *topological algebra sheaves* which have a special bearing on all the previous examples, and fulfil also the conditions of Definition 6.1 above.

Thus, assume that we are given an *exponential algebraized space* $(X, \mathcal{A})$ (cf. Definition 6.1). Therefore, the short exact sequence (6.1) entails the following *long exact sequence in* (sheaf) *cohomology* (see Chapt. III; Theorem 3.1, 3), along with Remark 6.1 below);

$$(6.9) \qquad \begin{aligned} &0 \longrightarrow \Gamma(X, \mathbf{Z}) \longrightarrow \Gamma(X, \mathcal{A}) \longrightarrow \Gamma(X, \mathcal{A}^{\cdot}) \xrightarrow{\delta} H^1(X, \mathbf{Z}) \\ &\longrightarrow H^1(X, \mathcal{A}) \longrightarrow H^1(X, \mathcal{A}^{\cdot}) \xrightarrow{\delta} H^2(X, \mathbf{Z}) \longrightarrow H^2(X, \mathcal{A}) \longrightarrow \cdots \end{aligned}$$

where $\delta$ stands for the corresponding *Bockstein operator* (see Chapt. III; (3.17)). In particular, the following extract from (6.9) will concern us often in the sequel; namely, consider from (6.9) the next *exact sequence of abelian groups*

$$(6.10) \qquad \cdots \longrightarrow H^1(X, \mathcal{A}) \longrightarrow H^1(X, \mathcal{A}^{\cdot}) \xrightarrow{\delta} H^2(X, \mathbf{Z}) \longrightarrow H^2(X, \mathcal{A}) \longrightarrow \cdots.$$

Now, before we proceed further, the following comments, as already mentioned, are fit in with (6.9). So they are embodied in the next.

**Remark 6.1.** By considering the long exact sequence of *sheaf cohomology* in (6.9), one actually applies the *analogue of Chapt. III; Theorem 3.1* in the case of the *abelian category* $\mathcal{A}b\mathcal{G}r\mathcal{S}h_X$, namely, that one of *abelian group sheaves* on $X$. Indeed, as already noted (see the comments before Theorem 2.1 of Chapt. III), *any abelian category* can also be employed, more generally, as the corresponding *source and target of the* respective *cohomology functors* (cf. Chapt. III; (2.5), (2.6) or even (3.2)).

Now, motivated further by, and also extending, the classical pattern, we set the following definition, which gives to the (categorical, in nature) Bockstein operator $\delta$ in (6.10) above a more (differential) *geometrical* significance.

Thus, the present material will be applied later on by examining, for instance, *characteristic classes of vector sheaves* (cf. Chapter IX in the sequel). So we now have the next.

**Definition 6.2.** Suppose we have an exponential algebraized space $(X, \mathcal{A})$ (cf. Definition 6.1). Then, the (connecting) cohomology (or else Bockstein) operator $\delta$, as in (6.10) above, is, in particular, called the *Chern morphism,* or even the *Chern operator* (on the line sheaves on $X$), which is thus associated with the given exponential sheaf sequence (6.1).

On the other hand, as already known (cf. Theorem 4.1), the source of the Chern morphism $\delta$ in (6.10) can be viewed as the Picard group of $X$ (see (4.6), (4.15)), viz. the set of (isomorphism classes of) line sheaves on $X$, therefore, the respective terminology employed in the previous Definition 6.2. In this context, the following nomenclature is also of use; that is, for every *line sheaf* on $X$,

(6.11) $$\mathcal{L} \equiv [\mathcal{L}] \in H^1(X, \mathcal{A}^{\cdot}) = \mathrm{Pic}\,(X) \equiv \mathbf{\Phi}^1_{\mathcal{A}}(X),$$

its image via $\delta$, viz.

(6.12) $$\delta(\mathcal{L}) \equiv \delta([\mathcal{L}]) \in H^2(X, \mathbf{Z}),$$

is called the *(integral) Chern class of* (the line sheaf) $\mathcal{L}$.

Yet, as we shall see in the sequel, under suitable (natural) supplementary restrictions for $(X, \mathcal{A})$ (cf. (6.33) below), the previous cohomology class $\delta(\mathcal{L})$ *uniquely determines* $\mathcal{L}$, in the sense, that then, $\delta$ *becomes an isomorphism* between the groups concerned (cf. (6.10)). However, before we come to that aspect, it is useful for the applications in the sequel, we have in mind, to look first at the following *complexification of the class* $\delta(\mathcal{L})$, for any given $\mathcal{L} \in \mathbf{\Phi}^1_{\mathcal{A}}(X)$, as above, that will be needed right below:

So, by condisering the (canonical) *injection* (indeed, *isomorphism into*) *of* the *(constant) group sheaves*

(6.13) $$\mathbf{Z} \underset{\varepsilon}{\hookrightarrow} \mathbf{C},$$

one obtains an associated to $\varepsilon$ *morphism of groups* between the respective *2-dimensional cohomology groups,*

(6.14) $$H^2(X, \mathbf{Z}) \xrightarrow{\varepsilon^* \equiv j} H^2(X, \mathbf{C})$$

(cf. Chapt. III; Definition 3.1, yet (3.12)). Thus, we further set the following.

**Definition 6.3.** Suppose we have an exponential algebraized space $(X, \mathcal{A})$. Then, the *complex Chern morphism*, associated with (6.1), is the composite map

$$c := j \circ \delta, \tag{6.15}$$

making thus, by definition, the ensuing diagram commutative;

$$\begin{array}{ccc} H^1(X, \mathcal{A}^{\cdot}) & \xrightarrow{\ \delta\ } & H^2(X, \mathbf{Z}) \\ & {\searrow}^{c} \quad {\swarrow}_{j} & \\ & H^2(X, \mathbf{C}). & \end{array} \tag{6.16}$$

Yet, for any line sheaf $\mathcal{L}$ on $X$, we call its image by $c$ (see also (6.11)), that is, the class

$$c(\mathcal{L}) := j(\delta(\mathcal{L})) \tag{6.17}$$

the *complex Chern class* of $\mathcal{L}$.

On the other hand, one always has the following standard *(complex) short exact exponential sequence* of the constant (abelian *group*) *sheaves* involved (cf. also (6.13));

$$0 \longrightarrow \mathbf{Z} \xrightarrow{\ \varepsilon\ } \mathbf{C} \xrightarrow{\ e\ } \mathbf{C}^{\cdot} \longrightarrow 0, \tag{6.18}$$

such that one sets

$$e(t) := \exp 2\pi i t, \quad t \in \mathbf{C}, \tag{6.19}$$

by looking, for instance, at (6.19) on the *section* level. Thus, by the same token, as e.g. in (6.10), one obtains the following *long exact sequence in cohomology* see also (6.14));

$$\cdots \longrightarrow H^1(X, \mathbf{C}^{\cdot}) \xrightarrow{\ \delta\ } H^2(X, \mathbf{Z}) \xrightarrow{\ j\ } H^2(X, \mathbf{C}) \longrightarrow \cdots. \tag{6.20}$$

The above two last sequences clarify also the way, that Definition 6.1 extends to our case the standard situation supplied by the classical *exponential map* (section), as in (6.19). Yet, we are going to employ (6.20) right below in the proof of the succeeding main theorem of this Section. However, for the same proof we still need the following (natural) *supplementary assumption* on the *map e in (6.1);* that is,

(6.21) suppose that *the restriction of* $e : \mathcal{A} \longrightarrow \mathcal{A}^{\cdot}$, as in (6.1) *to* (the constant subsheaf) $\mathbf{C} \underset{\varepsilon}{\overset{\subset}{\rightarrow}} \mathcal{A}$ *is the usual* (complex) *exponential map* $e$, as in (6.18). So *we assume the* validity of the *following relation* (cf. also the next diagram (6.21.2);

(6.21.1) $$\varepsilon \circ \exp = e \circ \epsilon.$$

Thus, *equivalently,* we accept the following *morphism of short exact* (exponential) *sheaf sequences;*

(6.21.2)
$$\begin{array}{ccccccccc}
0 & \longrightarrow & \mathbf{Z} & \longrightarrow & \mathbf{C} & \xrightarrow{\exp \equiv e} & \mathbf{C}^{\cdot} & \longrightarrow & 1 \\
 & & \| & & \downarrow{\scriptstyle \varepsilon} & & \downarrow{\scriptstyle \varepsilon \equiv i_X} & & \\
0 & \longrightarrow & \mathbf{Z} & \longrightarrow & \mathcal{A} & \xrightarrow{e} & \mathcal{A}^{\cdot} & \longrightarrow & 1 .
\end{array}$$

(See also, for instance, Chapt. III; (3.18)). Therefore (loc. cit.; (3.19)), one has the following *commutative diagram in cohomology*

(6.22)
$$\begin{array}{ccc}
H^1(X, \mathbf{C}^{\cdot}) & \xrightarrow{\delta} & H^2(X, \mathbf{Z}) \\
\downarrow{\scriptstyle i_X^*} & \searrow{\scriptstyle \delta} & \| \\
H^1(X, \mathcal{A}^{\cdot}) & \xrightarrow{\delta} & H^2(X, \mathbf{Z})
\end{array}$$

(concerning $i_X^*$, cf. (5.4), for $n = 1$). So, for convenience, we further set the next.

**Definition 6.4.** A *strictly exponential algebraized space* is a $\mathbf{C}$-algebraized space $(X, \mathcal{A})$, for which (the exact sequence) (6.1) and (the morphism of exact sequences) (6.21.2) are in force.

All the examples in the preceding, referred to the validity of (6.1) (cf. Examples 6.1 above), are still important particular instances, satisfying the conditions of the previous Definition 6.4.

Now, we come next to give the first *geometrical consequence* of the notion given by (6.17) above. As already mentioned, the geometric language, that is employed hereby, will be justified by concrete applications (of a geometric nature, abstract(!), of course) throughout the ensuing discussion. (So see, for instance, Chapters VII, or even XI, in the sequel). Thus, we now have the following.

**Theorem 6.1.** *Suppose we are given a strictly exponential algebraized space* $(X, \mathcal{A})$ (cf. Definition 6.4). *Then, the two following assertions are equivalent:*

$$c(\mathcal{L}) = 0 \tag{6.23}$$

*and* (see also (5.6))

$$(i_X^*)^{-1}(\mathcal{L}) \neq \emptyset. \tag{6.24}$$

*In other words, one concludes that;*

(6.25) *the complex Chern class of a given line sheaf* $\mathcal{L}$ *on* $X$ *vanishes if, and only if,* $\mathcal{L}$ *is a flat line sheaf on* $X$ (cf. Definition 5.1 above, along with (5.8)).

**Proof.** Assume that (cf. (6.17)) $c(\mathcal{L}) = j(\delta(\mathcal{L})) = 0$, viz. (see (6.20))

$$\delta(\mathcal{L}) \in \ker j = \operatorname{im}(H^1(X, \mathbf{C}^{\cdot}) \longrightarrow H^2(X, \mathbf{Z})). \tag{6.26}$$

Furthermore, suppose that

$$(g_{\alpha\beta}) \in Z^1(\mathcal{U}, \mathcal{A}^{\cdot}) \tag{6.27}$$

is a *coordinate 1-cocycle of* $\mathcal{L}$, representing it in $H^1(X, \mathcal{A}^{\cdot})$ (see (1.7), (1.12), for $n = 1$), hence, by virtue of the Chern morphism $\delta$ (cf. (6.10)) in $H^2(X, \mathbf{Z})$, as well. Therefore, in view of (6.26), one infers that $(g_{\alpha\beta})$ has to take values from $\mathbf{C}^{\cdot}$; that is, one has (cf. also the last perpendicular arrow in (6.21.2))

$$(g_{\alpha\beta}) \in Z^1(\mathcal{U}, \mathbf{C}^{\cdot}) \underset{\epsilon}{\overset{\subseteq}{\rightarrow}} Z^1(\mathcal{U}, \mathcal{A}^{\cdot}), \tag{6.28}$$

which actually proves (6.24) (cf. also e.g. (5.17), (5.22) and (5.24), for $n = 1$). *Conversely,* suppose that $\mathcal{L}$ *admits a flat representative,* viz. (Definition 5.1), suppose that (6.24) is valid; hence (Theorem 5.1), there exists a *1-cocycle*

$$(\lambda_{\alpha\beta}) \in Z^1(\mathcal{U}, \mathbf{C}^{\cdot}), \tag{6.29}$$

representing $\mathcal{L}$ in $H^1(X, \mathcal{A}^{\cdot})$. That is, one has

$$[(\lambda_{\alpha\beta})] \in \operatorname{im}(H^1(X, \mathbf{C}^{\cdot}) \longrightarrow H^1(X, \mathcal{A}^{\cdot})), \tag{6.30}$$

so that, in view of the commutative diagram (6.22) and the exactness of (6.20), one obtains

$$\delta(\mathcal{L}) \equiv \delta([(\lambda_{\alpha\beta})]) \in \operatorname{im}(H^1(X, \mathbf{C}^{\cdot}) \longrightarrow H^2(X, \mathbf{Z})) = \ker j. \tag{6.31}$$

Thus, one finally has

$$c(\mathcal{L}) = j(\delta(\mathcal{L})) = 0, \tag{6.32}$$

viz. (6.23), that was to be proved, and this terminates the proof. ∎

Now, as already mentioned, under suitable conditions on the structure sheaf $\mathcal{A}$ of $(X, \mathcal{A})$, one succeeds in having the *Chern morphism* $\delta$ as an *isomorphism* of the respective groups (cf. (6.10)). Such a natural demand on $\mathcal{A}$ is certainly the following relation;

$$H^1(X, \mathcal{A}) = H^2(X, \mathcal{A}) = 0. \tag{6.33}$$

So in that case one gets from (6.10) the exact sequence

$$0 \longrightarrow H(X, \mathcal{A}^{\cdot}) \xrightarrow{\delta} H^2(X, \mathbf{Z}) \longrightarrow 0, \tag{6.34}$$

that is, *equivalently,* one obtains;

$$H^1(X, \mathcal{A}^{\cdot}) = H^2(X, \mathbf{Z}), \tag{6.35}$$

within an *isomorphism of* the respective (abelian) *groups,* provided thus by the Chern morphism.

On the other hand, (6.33) holds true when, in particular, $\mathcal{A}$ is a *fine sheaf,* while $X$ is *paracompact* (Hausdorff), so that $\mathcal{A}$ is then $\Gamma_X$*-acyclic;* see Chapt. III; (3.20), (3.35), (3.36) and (8.24), (8.26).

Thus, concerning the Examples 6.1 in the preceding, *all except* $\mathcal{O}_X$ supply particular cases of a **C**-algebraized space $(X, \mathcal{A})$, with structure sheaf $\mathcal{A}$ fine (see also e.g. Chapt. III; (8.55.1) and (8.66). Yet, for the fact that $\mathcal{C}_X$ *is fine,* for $X$ *paracompact,* see, for instance, R. Godement [1:p.157, Example 3.7.1], or even R. Kultze [1: p.68, Beispiel 15.4]. On the other hand, as we shall see in the sequel, the ***topological algebra sheaves*** encountered in Chapter XI still share this property for $\mathcal{A}$.

Now, due to the preceding important particular cases that (6.33) holds true, we are tempted to call *fine exponential algebraized space,* a space as in Definition 6.1, which, in addition satisfies (6.33). Yet, we call just a *nice algebraized space,* a fine strictly exponential algebraized space $(X, \mathcal{A})$ (see also Definition 6.4 above).

## 7. The functors $\mathrm{Ext}_{\mathcal{A}}$ and $\mathcal{E}xt_{\mathcal{A}}$ (contn'd)

The functors in the title of this Section, together with applications to $\mathcal{A}$-extensions have already been considered in Chapt. III; Sections 9 and 10. So our purpose here is to relate further the notions at issue with appropriate cohomology groups (in fact, $\mathcal{A}(X)$*-modules;* see thus (7.5) below), a fact that will be of use in the sequel. So our first relevant result is the following lemma, while for the undefined terms hereby, we do refer to §§9,10 of Chapt. III. Thus, we have.

**Lemma 7.1.** *Let* $(X, \mathcal{A})$ *be a* $\mathbf{C}$*-algebraized space and* $\mathcal{E}$ *a given vector sheaf on* $X$*, while* $\mathcal{F}$ *and* $\mathcal{S}$ *are* $\mathcal{A}$*-modules on* $X$*. Then, for every* $n \in \mathbf{Z}_+$*, one has the following* $\mathcal{A}(X)$*-isomorphisms of the* $\mathcal{A}(X)$*-modules concerned;*

$$\text{(7.1)} \quad \mathrm{Ext}^n_{\mathcal{A}}(\mathcal{E} \otimes_{\mathcal{A}} \mathcal{F}, \mathcal{S}) = \mathrm{Ext}^n_{\mathcal{A}}(\mathcal{F}, \mathcal{E}^* \otimes_{\mathcal{A}} \mathcal{S}) = \mathrm{Ext}^n_{\mathcal{A}}(\mathcal{F}, \mathcal{H}om_{\mathcal{A}}(\mathcal{E}, \mathcal{S})).$$

*Furthermore, for every* $n \in \mathbf{Z}_+$*, one still obtains the next* $\mathcal{A}$*-isomorphisms of the* $\mathcal{A}$*-modules involved;*

$$\text{(7.2)} \quad \mathcal{E}xt^n_{\mathcal{A}}(\mathcal{E} \otimes_{\mathcal{A}} \mathcal{F}, \mathcal{S}) = \mathcal{E}xt^n_{\mathcal{A}}(\mathcal{F}, \mathcal{E}^* \otimes_{\mathcal{A}} \mathcal{S}) = \mathcal{E}xt^n_{\mathcal{A}}(\mathcal{F}, \mathcal{H}om_{\mathcal{A}}(\mathcal{E}, \mathcal{S})).$$

For the proof of the previous lemma we shall need the following *alternative of Theorem 6.2 of Chapt. IV;* namely, one has.

**Theorem 7.1.** *Suppose we have a* $\mathbf{C}$*-algebraized space* $(X, \mathcal{A})$ *and let* $\mathcal{E}$*,*$\mathcal{F}$*,*$\mathcal{S}$ *be given* $\mathcal{A}$*-modules on* $X$*, such that* $\mathcal{E}$ *is, in particular, a vector sheaf on* $X$*. Then, one obtains the next* $\mathcal{A}$*-isomorphisms of the* $\mathcal{A}$*-modules involved*

$$\text{(7.3)}\quad \mathcal{H}om_{\mathcal{A}}(\mathcal{E}\otimes_{\mathcal{A}}\mathcal{F},\mathcal{S})=\mathcal{H}om_{\mathcal{A}}(\mathcal{F},\mathcal{H}om_{\mathcal{A}}(\mathcal{E},\mathcal{S}))=\mathcal{H}om_{\mathcal{A}}(\mathcal{F},\mathcal{E}^{*}\otimes_{\mathcal{A}}\mathcal{S}).$$

**Proof.** We can follow here the proof of Theorem 6.2 of Chapter IV by obviously adapting (6.13) therein to the present context. Yet, concerning the last relation in (7.3) above, see also (6.1) of the same Chapter, as before. ∎

Furthermore, one has the following useful.

**Corollary 7.1.** *Assume that we have the conditions of the previous Theorem 7.1. Then, one obtains the following* $\mathcal{A}(X)$*-isomorphisms of the* $\mathcal{A}(X)$*-modules concerned;*

$$\text{(7.4)}\ \mathcal{H}om_{\mathcal{A}}(\mathcal{E}\otimes_{\mathcal{A}}\mathcal{F},\mathcal{S})(X)=\mathrm{Hom}_{\mathcal{A}}(\mathcal{F},\mathcal{H}om_{\mathcal{A}}(\mathcal{E},\mathcal{S}))=\mathrm{Hom}_{\mathcal{A}}(\mathcal{F},\mathcal{E}^{*}\otimes_{\mathcal{A}}\mathcal{S}).$$

**Proof.** Obvious, by virtue of (7.3) and Chapt. II; (6.9). ∎

So we come now to the

**Proof of Lemma 7.1.** The case $n = 0$, for both (7.1) and (7.2), follows from Chapt. III; (9.19) and (9.37), respectively, in conjunction with the previous formulae (7.3) and (7.4). Yet, for the general case, see e.g. R. Hartshorne [1:p.235, Proposition 6.7]. ∎

As a useful consequence of the preceding, we have now, as already promised, the following result, relating $\mathrm{Ext}_{\mathcal{A}}$ with sheaf cohomology modules. That is, one gets.

**Theorem 7.2.** *Let* $(X,\mathcal{A})$ *be a* **C***-algebraized space and* $\mathcal{E}$ *a vector sheaf on* $X$*, while* $\mathcal{F}$ *is any* $\mathcal{A}$*-module on* $X$*. Then, one has;*

$$\text{(7.5)}\qquad \mathrm{Ext}^{n}_{\mathcal{A}}(\mathcal{E},\mathcal{F})=H^{n}(X,\mathcal{H}om_{\mathcal{A}}(\mathcal{E},\mathcal{F}))=H^{n}(X,\mathcal{E}^{*}\otimes_{\mathcal{A}}\mathcal{F}),$$

*for every* $n\in\mathbf{Z}_{+}$*, up to an* $\mathcal{A}(X)$*-isomorphism of the* $\mathcal{A}(X)$*-modules involved.*

**Proof.** The assertion follows straightforwardly from (7.1) above and from the next $\mathcal{A}(X)$*-isomorphism*

$$\text{(7.6)}\qquad \mathrm{Ext}^{n}_{\mathcal{A}}(\mathcal{A},\mathcal{E})=H^{n}(X,\mathcal{E}),\quad n\geqq 0,$$

which is valid, *for any* $\mathcal{A}$*-module* $\mathcal{E}$ on $X$ (see Chapt. III; (9.24)). Namely, one then obtains the following $\mathcal{A}(X)$*-isomorphisms;*

$$\text{(7.7)}\qquad \begin{aligned}\operatorname{Ext}^n_{\mathcal{A}}(\mathcal{E},\mathcal{F}) &= \operatorname{Ext}^n_{\mathcal{A}}(\mathcal{A}\otimes_{\mathcal{A}}\mathcal{E},\mathcal{F}) = \operatorname{Ext}^n_{\mathcal{A}}(\mathcal{A},\mathcal{E}^*\otimes_{\mathcal{A}}\mathcal{F})\\ &= H^n(X,\mathcal{E}^*\otimes_{\mathcal{A}}\mathcal{F}) = H^n(X,\mathcal{H}om_{\mathcal{A}}(\mathcal{E},\mathcal{F})),\end{aligned}$$

(cf. also Chapt. II; (5.15)), which thus proves (7.5). ∎

It is still worthnoting here that another proof of the above basic relation (7.5) can be supplied in terms of *spectral sequences!* See, for instance, R. Godement [1:p.265, comments following Lemma 7.4.1]

# 8. Splittings of $\mathcal{A}$-extensions (contn'd)

The results of the preceding Section are applied below to splittings of $\mathcal{A}$-extensions, a notion that we have previously considered, as already mentioned, in Chapter III; Section 10, where we also refer for more details, concerning the ensuing terminology. Thus, as a prelude to this matter, consider first the set

$$\text{(8.1)}\qquad \operatorname{ext}_{\mathcal{A}}(\mathcal{F}^*, \mathcal{E}nd\,\mathcal{E});$$

viz. (cf. Chapt. III; (10.7)), the *set of isomorphism classes of extensions of* $\mathcal{F}^*$ (the *dual* of $\mathcal{F}$, cf. Chapt. IV; (5.1)) *by* $\mathcal{E}nd\,\mathcal{E}$, such that both $\mathcal{E}$ and $\mathcal{F}$ are given *vector sheaves* on a topological space $X$, base space of a **C**-*algebraized space* $(X, \mathcal{A})$. So a representative of an element of (8.1) is, by definition, a *short exact sequence* of the form

$$\text{(8.2)}\qquad 0 \longrightarrow \mathcal{E}nd\,\mathcal{E} \longrightarrow \mathcal{S} \longrightarrow \mathcal{F}^* \longrightarrow 0$$

(see Chapt. II; Definition 10.1). Furthermore, based on Theorem 10.1 of the same Chapter, one obtains the following (set-theoretic) *bijection;*

$$\text{(8.3)}\qquad \operatorname{ext}_{\mathcal{A}}(\mathcal{F}^*, \mathcal{E}nd\,\mathcal{E}) = \operatorname{Ext}^1_{\mathcal{A}}(\mathcal{F}^*, \mathcal{E}nd\,\mathcal{E}).$$

Therefore, in view of (7.5), one now obtains;

$$\text{(8.4)}\qquad \operatorname{ext}_{\mathcal{A}}(\mathcal{F}^*, \mathcal{E}nd\,\mathcal{E}) = \operatorname{Ext}^1_{\mathcal{A}}(\mathcal{F}^*, \mathcal{E}nd\,\mathcal{E}) = H^1(X, \mathcal{H}om_{\mathcal{A}}(\mathcal{F}^*, \mathcal{E}nd\,\mathcal{E})).$$

On the other hand, due to the hypothesis for $\mathcal{F}$ (cf. also Chapt. IV; Theorem 5.1 and 6.1), one has

$$\text{(8.5)}\qquad \mathcal{H}om_{\mathcal{A}}(\mathcal{F}^*, \mathcal{E}nd\,\mathcal{E}) = \mathcal{F}^{**}\otimes_{\mathcal{A}} \mathcal{E}nd\,\mathcal{E} = \mathcal{F}\otimes_{\mathcal{A}} \mathcal{E}nd\,\mathcal{E} \equiv \mathcal{F}(\mathcal{E}nd\,\mathcal{E}),$$

up to $\mathcal{A}$-*isomorphisms of* the $\mathcal{A}$-*modules* involved. Now, concerning the last term in (8.5), if $\mathcal{U}$ stands for a *local frame of* $\mathcal{E}$, one has, for every *open* $U \in \mathcal{U}$ (see also Chapt. I; (11.40) and Chapt. II; (5.25), (3.14));

$$\begin{aligned} &\mathcal{F}(\mathcal{E}nd\,\mathcal{E})(U) \equiv (\mathcal{F} \otimes_{\mathcal{A}} \mathcal{E}nd\,\mathcal{E})(U) = (\mathcal{E}nd\,\mathcal{E} \otimes_{\mathcal{A}} \mathcal{F})(U) = (\mathcal{E}nd\,\mathcal{E} \otimes_{\mathcal{A}} \mathcal{F})|_U(U) \\ (8.6) \quad &= ((\mathcal{E}nd\,\mathcal{E})|_U \otimes_{\mathcal{A}|_U} \mathcal{F}|_U)(U) = (\mathcal{A}^{n^2}|_U \otimes_{\mathcal{A}|_U} \mathcal{F}|_U)(U) = (\mathcal{A}^{n^2} \otimes_{\mathcal{A}} \mathcal{F})(U) \\ &\equiv (M_n(\mathcal{A}) \otimes_{\mathcal{A}} \mathcal{F})(U) \equiv M_n(\mathcal{F})(U) = \mathcal{F}^{n^2}(U) = \mathcal{F}(U)^{n^2} \equiv M_n(\mathcal{F}(U)). \end{aligned}$$

In this context, we also set above $n = \operatorname{rk} \mathcal{E}$, so that $\mathcal{E}nd\,\mathcal{E} \equiv \mathcal{H}om_{\mathcal{A}}(\mathcal{E}, \mathcal{E})$ being a *vector sheaf* on $X$ too, one has $\operatorname{rk}(\mathcal{E}nd\,\mathcal{E}) = n^2$ (see Chapt. II; Lemma 6.1), while we also have $\mathcal{A}^{n^2} = M_n(\mathcal{A})$ (cf. Chapt. IV; (3.23)). Yet, we also employed in the last terms of (8.6) the notation;

$$(8.7) \qquad M_n(\mathcal{E}) := M_n(\mathcal{A}) \otimes_{\mathcal{A}} \mathcal{E}, \quad n \in \mathbf{N},$$

for any $\mathcal{A}$-*module* $\mathcal{E}$ on $X$. Thus, by definition, *(8.7) entails just an $\mathcal{A}$-module on $X$*, as well; for convenience, we call it a *matrix $\mathcal{A}$-module.*

Therefore, continuing our calculations on (8.4) and, in particular, on the last term of it, one obtains (see also Chapt. III; (11.24), together with (11.3), as well as, the above relations (8.5) and (8.6));

$$\begin{aligned} &H^1(X, \mathcal{H}om_{\mathcal{A}}(\mathcal{F}^*, \mathcal{E}nd\,\mathcal{E})) = H^1(X, \mathcal{F}(\mathcal{E}nd\,\mathcal{E})) = \varinjlim_{\mathcal{U}} H^1(\mathcal{U}, \mathcal{F}(\mathcal{E}nd\,\mathcal{E})) \\ (8.8) \quad &\varinjlim_{\mathcal{U}} H^1(\mathcal{U}, M_n(\mathcal{F})) = H^1(X, M_n(\mathcal{F})). \end{aligned}$$

given that *the limits* in the above last relations *can be taken over the local frames of the vector sheaf* $\mathcal{F}(\mathcal{E}nd\,\mathcal{E})$ (see Chapt. II; (4.9), as well as, Chapt. III; (4.11)). Consequently, one finally obtains for (8.4);

$$\begin{aligned} (8.9) \quad \operatorname{ext}_{\mathcal{A}}(\mathcal{F}^*, \mathcal{E}nd\,\mathcal{E}) &= \operatorname{Ext}^1_{\mathcal{A}}(\mathcal{F}^*, \mathcal{E}nd\,\mathcal{E}) = H^1(X, \mathcal{H}om_{\mathcal{A}}(\mathcal{F}^*, \mathcal{E}nd\,\mathcal{E})) \\ &= H^1(X, \mathcal{F}(\mathcal{E}nd\,\mathcal{E})) = H^1(X, M_n(\mathcal{F})). \end{aligned}$$

Furthermore, we still note for use presently below that, for any *vector sheaf* $\mathcal{E}$ and $\mathcal{A}$-*module* $\mathcal{F}$ on $X$, one obtains

$$(8.10) \quad \operatorname{ext}_{\mathcal{A}}(\mathcal{E}, \mathcal{F}) = \operatorname{Ext}^1_{\mathcal{A}}(\mathcal{E}, \mathcal{F}) = H^1(X, \mathcal{H}om_{\mathcal{A}}(\mathcal{E}, \mathcal{F})) = H^1(X, \mathcal{E}^* \otimes_{\mathcal{A}} \mathcal{F});$$

see (7.5) above, for $n = 1$, in conjunction with Chapt. III; (10.18). ■ Thus, an $\mathcal{A}$-extension, in effect, the isomorphism class of it (cf. Chapt. III; (10.6)), belonging to (8.10), is a *short exact $\mathcal{A}$-sequence* of the form;

$$0 \longrightarrow \mathcal{F} \xrightarrow{\ i\ } \mathcal{S} \xrightarrow{\ \pi\ } \mathcal{E} \longrightarrow 0. \tag{8.11}$$

This, together with (8.2) will be systematically considered in the sequel to get *geometric informations,* when the relevant terminology will be at our disposal. So having such applications in mind (see e.g. the next Chapter VI), we are interested, in particular, in the *cohomological interpretation of splittings of the $\mathcal{A}$-extensions (8.2) and (8.11),* as above:

For convenience, we recall from Chapt. III; Corollary 10.1, that (8.11) *splits* (the same Chapter; Definition 10.3) *if* (and only if) *one has*

$$\mathrm{Ext}^1_{\mathcal{A}}(\mathcal{E},\mathcal{F}) = 0. \tag{8.12}$$

On the other hand, the *splitting of (8.11) amounts to*

$$\delta(1_{\mathcal{E}}) = 0, \tag{8.13}$$

that is, to the *vanishing of the characteristic class of the extension* in question (see Chapt. III; (10.38), (10.39)).

So, as already promised, we are now ready to examine the *cohomological interpretation of (8.13),* the latter condition expressing thus, *equivalently,* the *splitting of (8.11).* Now, first based on the above Theorem 7.2, for $n = 1$, (see also Chapt. III; (10.39.2)), we remark that;

$$\delta(1_{\mathcal{E}}) \in \mathrm{Ext}^1_{\mathcal{A}}(\mathcal{E},\mathcal{F}) = H^1(X, \mathcal{H}om_{\mathcal{A}}(\mathcal{E},\mathcal{F})). \tag{8.14}$$

Accordingly, one concludes that;

(8.15) *the condition*

(8.15.1) $$H^1(X, \mathcal{H}om_{\mathcal{A}}(\mathcal{E},\mathcal{F})) = 0$$

*entails* that $\delta(1_{\mathcal{E}}) = 0$, hence (cf. (8.13)), the *splitting* of (8.11).

In this context, see also Chapt. III; (10.40), together with the comments following it.

**Note.**- As we shall see later on, the above condition (8.15.1) is fulfilled in many concrete important examples, where the appropriate (natural) presuppositions for $X$ and $\mathcal{E}$ are satisfied, as well (as, for example, $X$ *paracompact* and $\mathcal{A}$ *soft,* concerning the pair $(X, \mathcal{A})$); see thus Chapters VI, XI in the sequel.

**Scholium 8.1**– Supplementing our previous considerations in Chapter III, concerning the *long exact sequence* (10.38), we further remark that, by applying to (8.11) above (cf. thus (10.37) therein) the *left exact contravariant functor* $\mathrm{Hom}_{\mathcal{A}}(\cdot, \mathcal{F})$ (see Chapt. II; (6.45), (6.46)), one obtains the following *long exact sequence* (cf. also, for intance, S. Lang [1:p.157, Remark and p.158, Lemma 8.3]);

$$(8.16)\qquad \begin{aligned} 0 \longrightarrow \mathrm{Hom}_{\mathcal{A}}(\mathcal{E},\mathcal{F}) \longrightarrow \mathrm{Hom}_{\mathcal{A}}(\mathcal{S},\mathcal{F}) &\xrightarrow{i_*} \mathrm{Hom}_{\mathcal{A}}(\mathcal{F},\mathcal{F}) \\ &\equiv \mathrm{End}\mathcal{F} \xrightarrow{\delta_*\equiv\delta} \mathrm{Ext}^1_{\mathcal{A}}(\mathcal{E},\mathcal{F}) \longrightarrow \cdots . \end{aligned}$$

Therefore, *the splitting of (8.11) is equivalent with the existence of* a global section, say

$$(8.17)\qquad \omega \in \mathcal{H}om_{\mathcal{A}}(\mathcal{S},\mathcal{F})(X) = \mathrm{Hom}_{\mathcal{A}}(\mathcal{S},\mathcal{F}),$$

*such that* one has (cf. (8.16), (8.11))

$$(8.18)\qquad i_*(\omega) := \omega \circ i = 1_{\mathcal{F}} \in \mathrm{End}\,\mathcal{F};$$

in this context, see also Chapt. III; Definition 10.3, together with Proposition 10.1, in particular, (10.11). ∎ Hence, due to the exactness of (8.16), one further infers, supplementing thus Chapt. III; (10.39), that:

(8.19) the short exact sequence (8.11) *splits if, and only if,* one has

(8.19.1) $$\delta_*(1_{\mathcal{F}}) \equiv \delta(1_{\mathcal{F}}) = 0 \in \mathrm{Ext}^1_{\mathcal{A}}(\mathcal{E},\mathcal{F}).$$

Now, to proceed further, we make hereafter the tacit *assumption* that;

(8.20) $X$, base space of our **C**-algebraized space $(X, \mathcal{A})$ involved, is a *paracompact (Hausdorff) space.*

Thus, one can employ henceforth (*equivalently,* see Chapt. III; (8.11)) *Čech cohomology* theory, so that one can use, for example, the relevant material, pertaining to *"lifting of cocycles"* (cf. Chapter III; Section 5). So, taking again the *short exact $\mathcal{A}$-sequence* (8.11), we can further apply the *left exact (covariant) functor* $\mathcal{H}om_{\mathcal{A}}(\mathcal{E}, \cdot)$ (see Chapt. II; (6.17)); however, due now to the hypothesis for $\mathcal{E}$, $\mathcal{H}om_{\mathcal{A}}(\mathcal{E}, \cdot)$ *is,* in effect, *an exact functor* on the category $\mathcal{A}$-$\mathcal{M}od_X$ (see Chapt. III; (9.46)). Accordingly, by applying it to (8.11), one obtains the next *short exact $\mathcal{A}$-sequence*

$$(8.21)\quad 0 \longrightarrow \mathcal{H}om_{\mathcal{A}}(\mathcal{E},\mathcal{F}) \stackrel{i}{\longrightarrow} \mathcal{H}om_{\mathcal{A}}(\mathcal{E},\mathcal{S}) \stackrel{\pi}{\longrightarrow} \mathcal{H}om_{\mathcal{A}}(\mathcal{E},\mathcal{E}) \equiv \mathcal{E}nd\,\mathcal{E} \longrightarrow 0.$$

(Concerning the *morphisms* in (8.21), we applied, for simplicity, an obvious *abuse of notation,* with respect to (8.11), by setting e.g. $i \equiv \mathcal{H}om_{\mathcal{A}}(\mathcal{E},\cdot)(i)$, etc). Now, (8.21) yields, in turn (cf. Chapt. III; Theorem 5.3), the following *long exact sequence in cohomology;*

$$(8.22)\quad \begin{aligned} 0 &\longrightarrow H^0(X,\mathcal{H}om_{\mathcal{A}}(\mathcal{E},\mathcal{F})) \longrightarrow H^0(X,\mathcal{H}om_{\mathcal{A}}(\mathcal{E},\mathcal{S})) \\ &\longrightarrow H^0(X,\mathcal{E}nd\,\mathcal{E}) \longrightarrow H^1(X,\mathcal{H}om_{\mathcal{A}}(\mathcal{E},\mathcal{F})) \longrightarrow \cdots \end{aligned}$$

or even (cf. Chapt. III; (4.55) and Chapt. II; (6.9))

$$(8.23)\quad \begin{aligned} 0 &\longrightarrow \mathrm{Hom}_{\mathcal{A}}(\mathcal{E},\mathcal{F})) \stackrel{i^*}{\longrightarrow} \mathrm{Hom}_{\mathcal{A}}(\mathcal{E},\mathcal{S}) \stackrel{\pi^*}{\longrightarrow} \mathrm{End}\mathcal{E} \\ &\stackrel{\delta^*\equiv\delta}{\longrightarrow} H^1(X,\mathcal{H}om_{\mathcal{A}}(\mathcal{E},\mathcal{F})) \longrightarrow \cdots. \end{aligned}$$

On the other hand, consider now the *0-cocycle*

$$(8.24)\quad 1_{\mathcal{E}} \in \mathrm{End}\mathcal{E} = (\mathcal{E}nd\,\mathcal{E})(X) \equiv \Gamma(X,\mathcal{E}nd\,\mathcal{E}) = \check{H}^0(\mathcal{U},\mathcal{E}nd\,\mathcal{E}) = \check{Z}^0(\mathcal{U},\mathcal{E}nd\,\mathcal{E})$$

(cf. Chapt. III; (4.55)). Thus, based on (8.21), in conjunction with *Chapt. III; Lemma 5.1, together with (5.52),* one concludes the existence of a *0-cochain*

$$(8.25)\quad h \equiv (h_\alpha) \in \check{C}^0(\mathcal{V},\mathcal{H}om_{\mathcal{A}}(\mathcal{E},\mathcal{S})),$$

*"lifting"* of (8.24), such that one has (cf. also (8.11))

$$(8.26)\quad \pi^*(h) := \pi \circ h = 1_{\mathcal{E}}.$$

(By what has been said in Chapt. III; Section 5, cf. the above quotation, $\mathcal{V}$ in (8.25) stands for an eventual *refinement of* $\mathcal{U}$, as in (8.24); yet, for simplicity, we do employ, occasionally, an obvious *abuse of notation,* for the respective *sets of indices* of the two coverings in question. Furthermore, concerning the second item in (8.26), see also Chapt. III; (4.11), for $n = 0$, together with Chapt. II; (6.8) and the above exact sequence (8.11)). Thus, $h$ in (8.25) *provides,* in fact, *a 0-cochain of (splitting) isomorphisms* (see (8.26)). Moreover (the same lemma of Chapter III, as before), the *1-cocycle associated with* (8.25) yields;

$$(8.27)\quad \delta(h) = (h_\beta - h_\alpha) \in Z^1(\mathcal{V},\ker\pi) = Z^1(\mathcal{V},\mathcal{H}om_{\mathcal{A}}(\mathcal{E},\mathcal{F})),$$

where, by virtue of (8.21), we have also set

$$(8.28)\quad \ker\pi = \mathrm{im}\, i = \mathcal{H}om_{\mathcal{A}}(\mathcal{E},\mathcal{F}),$$

the second equality, of course, for an *$\mathcal{A}$-isomorphism.* Therefore, in view of (8.14) and (8.23), in conjunction with Chapt. III; (5.21), one now obtains

$$(8.29) \quad \delta(1_{\mathcal{E}}) = [\delta(h)] = [(h_\beta - h_\alpha)] \in H^1(X, \mathcal{H}om_{\mathcal{A}}(\mathcal{E}, \mathcal{F})) = \mathrm{Ext}^1_{\mathcal{A}}(\mathcal{E}, \mathcal{F}).$$

Thus, by further applying the *splitting isomorphism* (8.25), one now defines;

$$(8.30) \qquad \phi_\alpha := h_\alpha \oplus i_\alpha, \quad \alpha \in I.$$

Let us first explain the notation applied in (8.30); so, once again, we repeat hereby our previous *abuse of notation, concerning the indices* "$\alpha$", assuming thus for simplicity, that we are referred henceforth to a given *open covering of* $X$, say

$$(8.31) \qquad \mathcal{U} = \{U_\alpha\}_{\alpha \in I}.$$

(See also the comments following (8.26) above). On the other hand, we have set in (8.30)

$$(8.32) \qquad i \equiv (i_\alpha)_{\alpha \in I},$$

concerning the *$\mathcal{A}$-isomorphism* in (8.21). (We recall here, see the comments after (8.21), that, for simplicity, we have actually set there too, $\mathcal{H}om_{\mathcal{A}}(\mathcal{E}, \cdot)(i) \equiv i$). Thus, (8.32) refers, in effect, to the *$\Gamma(\mathcal{A})$-isomorphism* (into) between the respective (complete) *$\Gamma(\mathcal{A})$-presheaves of sections* $\Gamma(\mathcal{F})$ and $\Gamma(\mathcal{S})$, derived from the exact sequence (8.11).

So, based now on (8.30), one obtains the following *$\mathcal{A}|_{U_\alpha}$-isomorphism of the $\mathcal{A}|_U$-modules* concerned (cf. also Chapt. II; (3.3));

$$(8.33) \qquad (\mathcal{E} \oplus \mathcal{F})|_{U_\alpha} \underset{\phi_\alpha}{=} \mathcal{S}|_{U_\alpha}, \quad \alpha \in I,$$

such that one sets

$$(8.34) \qquad \phi_\alpha(s \oplus t) = h_\alpha(s) + i_\alpha(t) \equiv h_\alpha(s) + t,$$

*for any* (local) *sections* $s \in \mathcal{E}(U_\alpha)$ *and* $t \in \mathcal{F}(U_\alpha)$, *with* $\alpha \in I$. (Note also that $\phi_\alpha$ is *onto,* as well, according to (8.11); viz. just apply $\pi$). In other words, by applying now the terminology of Subsection 2.(a) in the preceding, one thus concludes that:

(8.35) *the $\mathcal{A}$-modules $\mathcal{E} \oplus \mathcal{F}$ and $\mathcal{S}$ are locally isomorphic;* hence, the given *$\mathcal{A}$-extension $\mathcal{S}$ of $\mathcal{E}$ by $\mathcal{F}$* (cf. (8.11)) can be viewed as a *fiber space* over $X$ *of fiber type* $\mathcal{E} \oplus \mathcal{F}$ and with *structure sheaf* $\mathcal{A}ut(\mathcal{E} \oplus \mathcal{F})$ (see also (8.48) below).

Yet, as an easy byproduct of (8.35) (thus, of (8.33)) and of our hypothesis for $\mathcal{E}$ and $\mathcal{F}$, we still note (see Chapt. II; Definitions 4.1, 4.3) that:

(8.36) the $\mathcal{A}$-module $\mathcal{S}$, as in (8.11), is a *vector sheaf* on $X$, as well, in the case that $\mathcal{E}$ and $\mathcal{F}$ are, *such that;*

$$\operatorname{rk} \mathcal{S} = \operatorname{rk} \mathcal{E} + \operatorname{rk} \mathcal{F}.$$

Furthermore, as a consequence of the preceding, one infers that our *vector sheaf* $\mathcal{S}$ (cf. (8.36)) is *uniquely determined by the* following *coordinate transformations* (see (8.33), as well as, (1.4));

$$f_{\alpha\beta} := \phi_\alpha \circ \phi_\beta^{-1}, \tag{8.37}$$

*for any* $\alpha, \beta \in I$, *with* $U_\alpha \cap U_\beta \neq \emptyset$. Indeed, one essentially applies here our previous considerations in Section 2 and, in particular, the comments following (2.63); that is, the relevant technique, pertaining to the *cohomological classification of fiber spaces with structure sheaf* (à la Grothendieck), having, as already mentioned, a special bearing on the fundamental relation (2.42). (Of course, no *paracompactness* of $X$ is needed here, once we are given a family of *local isomorphisms,* as in (8.33)). ∎ So (8.37) entails, in effect, the following *1-cochain* (but, see finally (8.47))

$$(f_{\alpha\beta}) \in C^1(\mathcal{U}, \mathcal{A}ut(\mathcal{E} \oplus \mathcal{F})), \tag{8.38}$$

where, of course, we have set

$$\mathcal{A}ut(\mathcal{E} \oplus \mathcal{F}) \equiv \mathcal{A}ut_{\mathcal{A}}(\mathcal{E} \oplus \mathcal{F}) := \mathcal{E}nd\,(\mathcal{E} \oplus \mathcal{F})^{\cdot}; \tag{8.39}$$

see also Chapt. II; (6.28), as well as, the next.

**Scholium 8.2.**– Our purpose by the ensuing comments is to vindicate the above notation in (8.39):

So assume, as usual, that we are given a **C**-*algebraized space* $(X, \mathcal{A})$, and let $\mathcal{E}$ be an *$\mathcal{A}$-algebra sheaf* (or just an *$\mathcal{A}$-algebra*) on $X$ (see Chapt. II; Definition 6.2). Then, extending the situation in

Chapt. IV; Lemma 1.1, one can consider the *(group) sheaf of units* of $\mathcal{E}$, denoted by

$$\mathcal{E}^{\cdot}. \tag{8.40}$$

Thus, *by definition*, $\mathcal{E}^{\cdot}$ is the *sheaf of groups* (not necessarily abelian) on $X$, generated by the *complete presheaf of groups*, supplied by the correspondence;

$$U \longmapsto \mathcal{E}(U)^{\cdot}, \tag{8.41}$$

with $U$ *open* in X. (Note that, by definition of $\mathcal{E}$, $\mathcal{E}(U)$ is an $\mathcal{A}(U)$-*algebra*, so that one can take its *group of units* $\mathcal{E}(U)^{\cdot}$. Concerning the completeness of the presheaf at issue see e.g. Remark 1.1 in Chapt. IV). Thus, one obtains (see Chapt. I; Proposition 11.1)

$$\mathcal{E}^{\cdot}(U) = \mathcal{E}(U)^{\cdot}, \tag{8.42}$$

*for every open* $U \subseteq X$, *within a group isomorphism.*

Now, given an $\mathcal{A}$-*module* $\mathcal{S}$ on $X$, we know that $\mathcal{E}nd\,\mathcal{S}$ *is an* $\mathcal{A}$-*algebra sheaf* on $X$ (cf. Chapt. II; (6.29)). Therefore, according to the preceding, one now sets

$$\mathcal{A}ut_{\mathcal{A}}(\mathcal{S}) \equiv \mathcal{A}ut(\mathcal{S}) := (\mathcal{E}nd\,\mathcal{S})^{\cdot} \equiv \mathcal{E}nd\,\mathcal{S}^{\cdot}; \tag{8.43}$$

Thus, by virtue of (8.42) and (8.43) (cf. also Chapt. II; (6.30)), one further obtains

$$\text{Aut}(\mathcal{S}) = (\mathcal{A}ut\mathcal{S})(X) = (\mathcal{E}nd\,\mathcal{S})^{\cdot}(X) = (\mathcal{E}nd\,\mathcal{S})(X)^{\cdot} = (\text{End}\,\mathcal{S})^{\cdot}. \tag{8.44}$$

Of course, the preceding specialize, for instance, to Chapt. IV; (3.25), as well as, to the above formula (8.39).

Thus, after the previous useful intermezzo, we come again to (8.38), by further remarking that, in view of (8.37), (8.30), (8.34) and (8.27), one has;

$$f_{\alpha\beta}(s \oplus t) = s \oplus (t + (h_\beta - h_\alpha)(s)), \tag{8.45}$$

*for any s and t, as in* (8.34). Yet, as easily follows from (8.37), one infers that

(8.46) $(f_{\alpha\beta})$ *is,* in fact, *a 1-cocycle;* viz. *one has*

(8.46.1) $$(f_{\alpha\beta}) \in Z^1(\mathcal{U}, \mathcal{A}ut(\mathcal{E} \oplus \mathcal{F})).$$

However, in view of (8.45), it is certainly clear that

(8.47) $(f_{\alpha\beta})$ *is uniquely determined through the 1-cocycle*

(8.47.1) $$\delta(h) = (h_\beta - h_\alpha) \in Z^1(\mathcal{U}, \mathcal{H}om_{\mathcal{A}}(\mathcal{E}, \mathcal{F})).$$

(Notice again here the *abuse* with $\mathcal{U}, \mathcal{V}$!). Thus, as a consequence of (8.47), we may assert that:

(8.48) the *structure sheaf* of the fiber space $\mathcal{S}$, as in (8.35), is, in effect, the *vector sheaf* $\mathcal{H}om_{\mathcal{A}}(\mathcal{E}, \mathcal{F})$ (cf. Chapt. II; Lemma 6.1).

**Remark 8.1.**– It is clear from the preceding discussion that *(8.35) and (8.48) are still valid,* by *considering* just *$\mathcal{A}$-modules* on $X$, like $\mathcal{E}$,$\mathcal{F}$ *(not necessarily vector sheaves);* of course, one has to make in the previous argument the obvious adaptations, when vector sheaves are in sight. (We actually apply this in the next theorem).

Thus, the preceding supply now the proof of the following.

**Theorem 8.1.** *Suppose we are given a* **C***-algebraized space* $(X, \mathcal{A})$ *with* $X$ *a paracompact (Hausdorff) space. Moreover, let* $\mathcal{E}$ *and* $\mathcal{F}$ *be* $\mathcal{A}$*-modules on* $X$*, and*

(8.49) $$0 \longrightarrow \mathcal{F} \xrightarrow{\ i\ } \mathcal{S} \xrightarrow{\ \pi\ } \mathcal{E} \longrightarrow 0$$

*a given* $\mathcal{A}$*-extension of* $\mathcal{E}$ *by* $\mathcal{F}$*. Then,* $\mathcal{S}$ *can be construed as a fiber space over* $X$*, of fiber type* $\mathcal{E} \oplus \mathcal{F}$ (viz. the *trivial* $\mathcal{A}$*-extension of* $\mathcal{E}$ *by* $\mathcal{F}$ (see (8.33)), *whose structure sheaf is (uniquely determined by)* $\mathcal{H}om_{\mathcal{A}}(\mathcal{E}, \mathcal{F})$*; viz.*

(8.50) $\mathcal{S}$ *in (8.49) is classified by a 1-cocycle with coefficients from* $\mathcal{H}om_{\mathcal{A}}(\mathcal{E}, \mathcal{F})$.

*In particular,* $\mathcal{S}$ *is represented by the 1-cocycle (8.27), which corresponds* (cf. (8.29)) *to* $\delta(1_{\mathcal{E}})$, *the characteristic class of the* $\mathcal{A}$*-extension at issue. So, what amounts to the same,*

(8.51) $\mathcal{S}$ *is cohomologically classified by*

$$H^1(X, \mathcal{H}om_{\mathcal{A}}(\mathcal{E}, \mathcal{F})) \tag{8.51.1}$$

(viz. as an element of the above (abelian) group). ∎

Now, we further get the following general result, that will be applied in the sequel and, in particular, to questions concerning the *existence of* $\mathcal{A}$*-connections of vector sheaves* (see the next Chapter VI).

**Proposition 8.1.** *Suppose we have a* **C***-algebraized space* $(X, \mathcal{A})$ *and* $\mathcal{E}, \mathcal{F}$ *two vector sheaves on* $X$. *Then, the following* $\mathcal{A}$*-extensions*

$$0 \longrightarrow \mathcal{E} \otimes_{\mathcal{A}} \mathcal{F} \longrightarrow \mathcal{S} \longrightarrow \mathcal{E} \longrightarrow 0 \tag{8.52}$$

*and*

$$0 \longrightarrow \mathcal{E}nd\,\mathcal{E} \longrightarrow \mathcal{S}' \longrightarrow \mathcal{F}^* \longrightarrow 0 \tag{8.53}$$

*have the same* (in fact, *isomorphic*) *Baer groups; that is, one has the next isomorphism of (abelian) groups*

$$\mathrm{ext}_{\mathcal{A}}(\mathcal{E}, \mathcal{E} \otimes_{\mathcal{A}} \mathcal{F}) = \mathrm{ext}_{\mathcal{A}}(\mathcal{F}^*, \mathcal{E}nd\,\mathcal{E}) \tag{8.54}$$

(see also Chapt. III; (10.7), (10.36)).

**Proof.** Based on Chapt. III; Theorem 10.1, along with (10.36), and on Theorem 7.2 in the preceding, for $n = 1$, one obtains, for each one of the two members of (8.54), respectively;

$$\mathrm{ext}_{\mathcal{A}}(\mathcal{E}, \mathcal{E} \otimes_{\mathcal{A}} \mathcal{F}) = \mathrm{Ext}^1_{\mathcal{A}}(\mathcal{E}, \mathcal{E} \otimes_{\mathcal{A}} \mathcal{F}) = H^1(X, \mathcal{H}om_{\mathcal{A}}(\mathcal{E}, \mathcal{E} \otimes_{\mathcal{A}} \mathcal{F})), \tag{8.55}$$

as well as,

$$\mathrm{ext}_{\mathcal{A}}(\mathcal{F}^*, \mathcal{E}nd\,\mathcal{E}) = \mathrm{Ext}^1_{\mathcal{A}}(\mathcal{F}^*, \mathcal{E}nd\,\mathcal{E}) = H^1(X, \mathcal{H}om_{\mathcal{A}}(\mathcal{F}^*, \mathcal{E}nd\,\mathcal{E})), \tag{8.56}$$

within *isomorphisms of* the (abelian) *groups* concerned. On the other hand, by virtue of Chapt. IV; Corollary 6.1 and Theorem 5.1, one has;

$$\begin{aligned} &\mathcal{H}om_{\mathcal{A}}(\mathcal{E}, \mathcal{E} \otimes_{\mathcal{A}} \mathcal{F}) = \mathcal{H}om_{\mathcal{A}}(\mathcal{E}, \mathcal{E}) \otimes_{\mathcal{A}} \mathcal{F} \equiv \mathcal{E}nd\,\mathcal{E} \otimes_{\mathcal{A}} \mathcal{F} \equiv \mathcal{F}(\mathcal{E}nd\,\mathcal{E}) \\ &= \mathcal{F} \otimes_{\mathcal{A}} \mathcal{E}nd\,\mathcal{E} = \mathcal{F}^{**} \otimes_{\mathcal{A}} \mathcal{E}nd\,\mathcal{E} = \mathcal{H}om_{\mathcal{A}}(\mathcal{F}^*, \mathcal{E}nd\,\mathcal{E}), \end{aligned} \tag{8.57}$$

within *$\mathcal{A}$-isomorphisms of* the *$\mathcal{A}$-modules* (in effect, *vector sheaves*) involved. In other words, applying the terminology of (8.48), one concludes that

(8.58) *the two given $\mathcal{A}$-extensions (8.52) and (8.53) have the same* (in fact, *isomorphic*) *structure sheaves, so that the same holds true for the respective classifying 1-dimensional cohomology groups* (in effect, *$\mathcal{A}(X)$-modules*), *as well.*

Indeed, concerning the last part of (8.58), based on the *functorial character of "$H^1$"* (see Chapt. III; (3.2)), one infers from (8.57) that

$$H^1(X, \mathcal{H}om_{\mathcal{A}}(\mathcal{E}, \mathcal{E} \otimes_{\mathcal{A}} \mathcal{F})) = H^1(X, \mathcal{H}om_{\mathcal{A}}(\mathcal{F}^*, \mathcal{E}nd\, \mathcal{E})), \tag{8.59}$$

within an *isomorphism of groups* (in fact, *$\mathcal{A}(X)$-modules*), which thus vindicates the said part of (8.58); thus, (8.54) follows now from (8.59), in conjunction with (8.55) and (8.56). ∎

Furthemore, as a result of (8.54), one still infers that *the zero-elements* of the two groups in question *correspond* to each other in a *one-to-one* way, so that (cf. Chapt. III; (10.30), (10.36))

(8.60) *splitting of either one of the two $\mathcal{A}$-extensions (8.52) or (8.53) is equivalent to that one of the other.*

Indeed, in view of (8.55), (8.56) and (8.59), one also concludes that;

(8.61) *the above two $\mathcal{A}$-extensions (8.52) and (8.53) are classified by the same characteristic class* (see Chapt. III; (10.39.2), along with (8.14) above), or even they have *cohomologous coordinate 1-cocycles,* the latter entailing thus *the same 1-dimensional cohomology class.*

Of course, the previous statement implies thus, otherwise, again (8.60).

Yet, as a *complement to Chapt. III; Section 10,* we further examine through the next Subsection, another relevant notion to the preceding, which will also be of use in the ensuing discussion.

**8.(a). Dual $\mathcal{A}$-extensions.**– Suppose that we have again a **C**-*algebraized space* $(X, \mathcal{A})$, with $X$ now, as usual, an *arbitrary topological space.* Moreover, suppose that $\mathcal{E}$,$\mathcal{F}$ are two given *vector sheaves* on $X$, and

$$0 \longrightarrow \mathcal{F} \longrightarrow \mathcal{S} \longrightarrow \mathcal{E} \longrightarrow 0 \tag{8.62}$$

an *$\mathcal{A}$-extension of $\mathcal{E}$ by $\mathcal{F}$.*

So, by considering next the *dual $\mathcal{A}$-modules* $\mathcal{E}^*, \mathcal{F}^*$ of $\mathcal{E}$ and $\mathcal{F}$, respectively (cf. Chapt. IV; (5.1)), one defines the following *short exact $\mathcal{A}$-sequence*

$$0 \longrightarrow \mathcal{E}^* \longrightarrow \mathcal{S}' \longrightarrow \mathcal{F}^* \longrightarrow 0 \tag{8.63}$$

as the *dual $\mathcal{A}$-extension of $\mathcal{S}$.*

On the other hand, suppose that

$$\delta(1_{\mathcal{E}}) \equiv \xi \in H^1(X, \mathcal{H}om_{\mathcal{A}}(\mathcal{E}, \mathcal{F})) \tag{8.64}$$

and

$$\delta(1_{\mathcal{F}^*}) \equiv \eta \in H^1(X, \mathcal{H}om_{\mathcal{A}}(\mathcal{F}^*, \mathcal{E}^*)) \tag{8.65}$$

are the corresponding *characteristic classes* of $\mathcal{S}$ and $\mathcal{S}'$, respectively (see thus Chapt. IV; (10.39.2), as well as, (8.14) in the preceding).

However, in this context, one first obtains the following (canonical) *$\mathcal{A}$-isomorphism of* the *vector sheaves* involved;

$$\mathcal{H}om_{\mathcal{A}}(\mathcal{E}, \mathcal{F}) = \mathcal{H}om_{\mathcal{A}}(\mathcal{F}^*, \mathcal{E}^*). \tag{8.66}$$

Indeed, due to the hypothesis for $\mathcal{E}$ and $\mathcal{F}$, one obtains the next *$\mathcal{A}$-isomorphisms*

$$\mathcal{H}om_{\mathcal{A}}(\mathcal{E}, \mathcal{F}) = \mathcal{E}^* \otimes_{\mathcal{A}} \mathcal{F} = \mathcal{E}^* \otimes_{\mathcal{A}} \mathcal{F}^{**} = \mathcal{H}om_{\mathcal{A}}(\mathcal{F}^*, \mathcal{E}^*), \tag{8.67}$$

as desired. (In this regard, see thus Chapt. IV; (6.1), (5.11) and Chapt. II; (5.14)). ∎ Therefore, based on (8.66), one further concludes the following *isomorphism of groups* (in fact, of *$\mathcal{A}(X)$-modules*);

$$H^1(X, \mathcal{H}om_{\mathcal{A}}(\mathcal{E}, \mathcal{F})) = H^1(X, \mathcal{H}om_{\mathcal{A}}(\mathcal{F}^*, \mathcal{E}^*)). \tag{8.68}$$

In other words, as with (8.58),

(8.69) *the* given *$\mathcal{A}$-extension (8.62) and its dual (8.63) have the same* (viz. *isomorphic*) *structure sheaves, hence, isomorphic classifying cohomology groups* too.

In this context, see also our previous argument, concerning (8.59).

Now, denoting by $j, j^*$ the (canonical) isomorphisms (8.66) and (8.68), respectively, let

$$\xi^* := j^*(\xi) \in H^1(X, \mathcal{H}om_{\mathcal{A}}(\mathcal{F}^*, \mathcal{E}^*)); \tag{8.70}$$

that is, $\xi^*$ stands for the *image of* $\xi$ in (8.64) *under the* canonical *isomorphism* (8.68). Thus, our claim now is that;

$$\xi^* = -\eta, \tag{8.71}$$

under the additional assumption that $X$ is *paracompact (Hausdorff)*, so that Theorem 8.1 can be applied. Indeed, we employ (8.50) right away for the *proof of (8.71):*

Namely, based on (8.50), the given $\mathcal{A}$-extension $\mathcal{S}$ or, *equivalently*, $\xi$ (cf. (8.14), (8.64)), is determined by the *coordinate 1-cocycle*

$$f_{\alpha\beta} := \phi_\beta \circ \phi_\alpha^{-1} \in \mathcal{A}ut(\mathcal{E} \oplus \mathcal{F})(U_{\alpha\beta}), \tag{8.72}$$

*with* $\alpha, \beta \in I$ *such that* $U_{\alpha\beta} \equiv U_\alpha \cap U_\beta \neq \emptyset$ (see (8.37), (8.39), as well as, (8.44)). Thus, $\mathcal{S}'$ or, *equivalently* $\eta$ (see (8.65)), will be given by the following *1-cocycle* (cf. also Chapt. II; (6.21))

$$\begin{aligned} g_{\alpha\beta} := (\phi_\beta \circ \phi_\alpha^{-1})^* = (\phi_\alpha^{-1})^* \circ \phi_\beta^* \in \mathcal{A}ut(\mathcal{E} \oplus \mathcal{F})^*(U_{\alpha\beta}) \\ = \mathcal{A}ut(\mathcal{E}^* \oplus \mathcal{F})^*(U_{\alpha\beta}), \end{aligned} \tag{8.73}$$

*for any* $\alpha, \beta \in I$ *with* $U_{\alpha\beta} \neq \emptyset$. Here we have also set

$$\phi_\alpha^* \equiv {}^t\phi_\alpha : \mathcal{S}^*|_{U_\alpha} \longrightarrow (\mathcal{E} \oplus \mathcal{F})^* = \mathcal{E}^* \oplus \mathcal{F}^*|_{U_\alpha}, \tag{8.74}$$

*for any* $\alpha \in I$ (*transpose* $\mathcal{A}$*-morphism* of $\phi_\alpha$; cf. (8.33)). Therefore, if

$$\delta(h_\alpha) = (h_\beta - h_\alpha) \in Z^1(\mathcal{U}, \mathcal{H}om_{\mathcal{A}}(\mathcal{E}, \mathcal{F})) \tag{8.75}$$

is a *1-cocycle representing* $\xi$ (cf. (8.27), (8.29) and (8.64), see also Theorem 8.1), then

$$\delta(h_\alpha)^* = (h_\beta - h_\alpha)^* = (h_\alpha - h_\beta) \tag{8.76}$$

is the *1-cocycle representing* $\eta$ (cf. (8.73), (8.65)), which thus proves our assertion in (8.71). ∎

By the preceding we have actually proved the following useful result. (Concerning the classical counterpart of the above in vector bundle theory, see M.F. Atiyah [1:p.185,Proposition 3]). Thus, we now have.

**Proposition 8.2.** *Let* $(X, \mathcal{A})$ *be a given* **C***-algebraized space, having* $X$ *a paracompact (Hausdorff) space, and let* $\mathcal{E}, \mathcal{F}$ *be two vector sheaves on* $X$*. Furthermore, let*

$$(8.77) \qquad 0 \longrightarrow \mathcal{F} \longrightarrow \mathcal{S} \longrightarrow \mathcal{E} \longrightarrow 0$$

*be an $\mathcal{A}$-extension of $\mathcal{E}$ by $\mathcal{F}$and*

$$(8.78) \qquad 0 \longrightarrow \mathcal{E}^* \longrightarrow \mathcal{S}' \longrightarrow \mathcal{F}^* \longrightarrow 0$$

*the corresponding dual $\mathcal{A}$-extension of $\mathcal{F}^*$ by $\mathcal{E}^*$. Finally, suppose that $\xi, \eta$ are the respective 1-dimensional cohomology classes, defined by the $\mathcal{A}$-extensions in question, respectively* (cf. (8.64) and (8.65)). *Then, one has*

$$(8.79) \qquad \eta = -\xi^*,$$

*where $\xi^*$ stands for the "dual" cohomology class of $\xi$, determined by the (canonical) $\mathcal{A}$-isomorphism*

$$(8.80) \qquad \mathcal{H}om_{\mathcal{A}}(\mathcal{E}, \mathcal{F}) = \mathcal{H}om_{\mathcal{A}}(\mathcal{F}^*, \mathcal{E}^*),$$

*of the $\mathcal{A}$-modules* (in fact, *vector sheaves*) *concerned.* ∎

# 9. Classical group sheaves (contn'd)

We consider below certain of the corresponding to the present case *classical groups,* by analogy with our previous study in Chapter IV; Subsection 1.(b), concerning the general linear group sheaf (ibid.; (1.26), along with (4.18) therein). In particular, we are interested in connecting our previous cohomological characterization of vector sheaves (see Section 2 above) with the existence of suitable $\mathcal{A}$-*metrics,* in the sense of Chapt. IV; §§8,9. Thus, we first look at the following group sheaves.

**9.(a). The orthogonal group sheaf.**– Suppose, as usual, that we are given a **C**-*algebraized space* $(X, \mathcal{A})$, so that, for any *open* $U \subseteq X$, one defines (cf. Chapt. IV; (1.28));

$$(9.1) \qquad \mathcal{O}(n, \mathcal{A}(U)) := \{\alpha \in \mathcal{GL}(n, \mathcal{A})(U) = \mathrm{GL}(n, \mathcal{A}(U)) : \alpha \cdot {}^t\alpha = 1\}.$$

Now, it is easy to see that the above established correspondence

$$(9.2) \qquad U \longmapsto \mathcal{O}(n, \mathcal{A}(U)),$$

with $U$ running over the open subsets of $X$, along with the obvious restriction maps, yields a *complete presheaf of groups* on $X$. The corresponding *sheaf of groups* on $X$, *generated by (9.2)*, viz. the sheafification of (9.2) (cf., for intance, Chapt. II; (1.54) or even Lemma 1.3), is denoted by

$$\mathcal{O}(n, \mathcal{A}) \tag{9.3}$$

and called the *orthogonal group sheaf* on $X$, of *order* $n \in \mathbf{N}$, associated with $\mathcal{A}$. Thus, as a consequence of the completeness of (9.2), one has (cf. Chapt. I; Proposition 11.1);

$$\mathcal{O}(n, \mathcal{A})(U) = \mathcal{O}(n, \mathcal{A}(U)), \tag{9.4}$$

up to an *isomorphism of* the *groups* concerned, *for every open* $U \subseteq X$.

On the other hand, consider the (canonical) *Kronecker gauge of* $\mathcal{A}^n$,

$$(\varepsilon_i)_{1 \leqq i \leqq n} \subseteq \mathcal{A}^n(X) = \mathcal{A}(X)^n \tag{9.5}$$

(see Chapt. IV; (3.2), (3.3)), together with a (global) *section-matrix* of (9.3), say

$$\alpha \equiv (\alpha_{ij}) \in \mathcal{O}(n, \mathcal{A})(X) = \mathcal{O}(n, \mathcal{A}(X)). \tag{9.6}$$

Thus, by looking further at the $\mathcal{A}(X)$*-isomorphism of* $\mathcal{A}^n(X) = \mathcal{A}(X)^n$ *determined by* $\alpha$ (cf. (9.1), together with, for instance, Chapt. IV; (3.25) and (3.17), for $n = m$), one can prove, following standard argument (see, for example, I. Satake [1:p.143, Theorem 13]), that;

(9.7) *the* above $\mathcal{A}(X)$*-isomorphism of* $\mathcal{A}^n(X)$, *associated with* $\alpha$, *as in (9.6), is orthogonal, with respect to the* (canonical) *Riemannian* $\mathcal{A}$*-metric of* $\mathcal{A}^n$, *corresponding* to (9.5) (see (9.9) below).

Indeed, recalling that the latter metric is defined by

$$\overline{\rho}(\varepsilon_i, \varepsilon_j) = \delta_{ij}, \quad 1 \leqq i, j \leqq n, \tag{9.8}$$

(see (9.5) and Chapt. IV; (8.22.2)), one proves that

$$\overline{\rho}(\alpha \cdot s, \alpha \cdot t) = \overline{\rho}(s, t), \tag{9.9}$$

*for any* $s, t \in \mathcal{A}^n(X) = \mathcal{A}(X)^n$ (the latter sections being considered as column *section-matrices*).

Now, more generally, suppose that we are given an *enriched ordered algebraized space* $(X, \mathcal{A}, \surd)$ (cf. Chapt. IV; Definition 10.1), such that

$$(\mathcal{A}, \rho) \tag{9.10}$$

is a (free) *Riemannian $\mathcal{A}$-module;* moreover, let

$$(\mathcal{A}^n, \overline{\rho} \equiv \rho^n) \tag{9.11}$$

be the corresponding standard *free Riemannian $\mathcal{A}$-module,* of *rank n,* over $X$ (see Chapt. IV; (8.36), (8.37)). So assume further that $\mathcal{E}$ is a *vector sheaf* on $X$, of *rank n,* and let

$$\mathcal{U} = \{U_\alpha\}_{\alpha \in I} \tag{9.12}$$

be a *local frame* of $\mathcal{E}$. Thus, for any *local gauge* $U_\alpha \in \mathcal{U}$, $\alpha \in I$ (cf. Chapt. II; Definition 4.2), one can apply the corresponding *Gram-Schmidt orthogonalization* procedure (cf. Chapt. IV; Theorem 10.1). Hence, by considering the *coordinate 1-cocycle of $\mathcal{E}$, associated with $\mathcal{U}$,* say

$$(\phi_{\alpha\beta}) \in Z^1(\mathcal{U}, \mathcal{GL}(n, \mathcal{A})), \tag{9.13}$$

(cf. Lemma 1.1 in the preceding), *one can assume* that (see also (9.7), (9.4) above)

$$(\phi_{\alpha\beta}) \in \mathcal{O}(n, \mathcal{A}(U_{\alpha\beta})) = \mathcal{O}(n, \mathcal{A})(U_{\alpha\beta}) \subseteq \mathcal{GL}(n, \mathcal{A})(U_{\alpha\beta}), \tag{9.14}$$

*for any $\alpha, \beta \in I$ such that $U_\alpha \cap U_\beta \equiv U_{\alpha\beta} \neq \emptyset$.* Yet, one easily sees that (9.14) still supplies a *1-cocycle of $\mathcal{U}$ with coefficients* now *in $\mathcal{O}(n, \mathcal{A})$* (cf. also (1.3.2)); viz. one has

$$\phi_{\alpha\gamma} = \phi_{\alpha\beta} \circ \phi_{\beta\gamma}, \tag{9.15}$$

*for any $\alpha, \beta, \gamma \in I$ with $U_{\alpha\beta\gamma} \equiv U_\alpha \cap U_\beta \cap U_\gamma \neq \emptyset$.* Thus, the preceding provide already the proof of the following.

**Theorem 9.1.** *Suppose that we are given an enriched ordered algebraized space*

$$(X, \mathcal{A}, \surd) \tag{9.16}$$

(cf. Chapt. IV; (10.6)), *and let $\mathcal{E}$ be a vector sheaf on $X$. Then, one can assume that the coordinate 1-cocycle of $\mathcal{E}$, say, $(\phi_{\alpha\beta})$, relative to a given local frame $\mathcal{U}$ of $\mathcal{E}$, has coefficients from the orthogonal group sheaf $\mathcal{O}(n, \mathcal{A})$; that is, one obtains*

$$(\phi_{\alpha\beta}) \in Z^1(\mathcal{U}, \mathcal{O}(n, \mathcal{A})) \stackrel{\subset}{\rightarrow} Z^1(\mathcal{U}, \mathcal{GL}(n, \mathcal{A})). \blacksquare \tag{9.17}$$

On the other hand, applying a similar argument to that in Chapt. IV;

(4.14), one defines the *special orthogonal group sheaf* on $X$, *of order* $n \in \mathbf{N}$, denoted by

$$\mathcal{SO}(n, \mathcal{A}), \tag{9.18}$$

according to the relation (see also Chapt. IV; (4.1), (4.8), along with (9.4) above);

$$\mathcal{SO}(n, \mathcal{A}(U)) := \{\alpha \in \mathcal{O}(n, \mathcal{A}(U)) = \mathcal{O}(n, \mathcal{A})(U) : \partial\mathrm{et}\,(\alpha) = 1\}, \tag{9.19}$$

with $U$ *any open* set in $X$. Thus, with $U$ varying over the open subsets of $X$, one gets, in effect, through (9.19) (and the obvious restriction morphisms) a *complete presheaf of groups* on $X$, whose *sheafification* is, *by definition*, (9.18). Thus, one also infers that

$$\mathcal{SO}(n, \mathcal{A}(U)) = \mathcal{SO}(n, \mathcal{A})(U), \tag{9.20}$$

*for every open* $U \subseteq X$, within a *group isomorphism*.

Now, one further obtains the following *short exact sequence of group sheaves* on $X$;

$$\begin{array}{ccccccccc} 1 & \longrightarrow & \mathcal{SO}(n,\mathcal{A}) & \longrightarrow & \mathcal{O}(n,\mathcal{A}) & \xrightarrow{\partial\mathrm{et}} & \mathbf{Z}_2 & \longrightarrow & 1 \\ & & & & & \swarrow & & \searrow & \\ & & & & \mathbf{C}^{\cdot} & & \longrightarrow & & \mathcal{A}^{\cdot}. \end{array} \tag{9.21}$$

Here we denote by

$$\mathbf{Z}_2 \equiv \{\pm 1\} \hookrightarrow \mathbf{C}^{\cdot} \tag{9.22}$$

the respective *constant group sheaves* (canonically embedded in $\mathcal{A}^{\cdot}$, the (group) *sheaf of units of* $\mathcal{A}$; see e.g. Chapt. II; (1.34), (1.34′)). In this context, we still recall that *the* $\partial\mathrm{et}$ *morphism* (cf. Chapt. IV; (4.1), (4.7) and (4.8)) *satisfies the fundamental multiplicative property* of the classical counterpart (ibid.; (4.11)), while we further *assume the condition;*

$$\alpha^2 = 1 \ \textit{implies} \ \alpha = \pm 1, \tag{9.23}$$

*for any* $\alpha \in \mathcal{A}(U)$, with $U$ *open* in $X$. Finally, we also note that *from (9.18) onwards* the above material is certainly valid *for any* **C**-*algebraized space* $(X, \mathcal{A})$, in general (taking, of course, (9.23) into account, concerning (9.21)).

**9.(b). The unitary group sheaf.**- Suppose now that we are given an *involutive* **C**-*algebraized space*

$$(X, \mathcal{A}, -) \tag{9.24}$$

(cf. Chapt. IV; (9.5), (9.1)); thus, for every *open* $U \subseteq X$, one sets

$$\mathcal{U}(n, \mathcal{A}(U)) := \{\alpha \in \mathrm{GL}(n, \mathcal{A}(U)) : \alpha^{-1} = \overline{\alpha}\} \tag{9.25}$$

(see also ibid.; (9.1′), along with Scholium 9.1). Now, one defines the *unitary group sheaf* on $X$, of *order* $n \in \mathbf{N}$, denoted by

$$\mathcal{U}(n, \mathcal{A}), \tag{9.26}$$

as the *sheaf of groups* on $X$ (again, not necessarily, abelian, unless $n = 1$ in (9.25)), generated by the *complete* (as easily verified) *presheaf (of groups),* defined by (9.25) (along with the natural restriction morphisms), when $U$ varies over the open subsets of $X$.

Now, by further considering the *determinant morphism* (cf. Chapt. IV; (4.8), (4.11)),

$$\partial\mathrm{et} : \mathcal{GL}(n, \mathcal{A}) \longrightarrow \mathcal{A}^{\cdot}, \tag{9.27}$$

one obtains, in view of (9.25) (see also Chapt. IV; (9.19)),

$$\partial\mathrm{et}\,(\alpha \cdot \overline{\alpha}) = \partial\mathrm{et}\,(\alpha) \cdot \partial\mathrm{et}\,(\overline{\alpha}) = \partial\mathrm{et}\,(\alpha) \cdot \overline{\partial\mathrm{et}\,\alpha} = |\partial\mathrm{et}\,\alpha|^2 = 1, \tag{9.28}$$

so that one finally has

$$|\partial\mathrm{et}\,\alpha| = 1. \tag{9.29}$$

In this context, and in connection with the above two last assertions, we first remark that, by *complementing* here *our assumptions on Chapt. IV; (9.1),* we *assume that* the following holds true;

$$\partial\mathrm{et}\,\overline{\alpha} = \overline{\partial\mathrm{et}\,\alpha}, \tag{9.30}$$

*for every* $\alpha \in \mathcal{A}(U)$, with $U$ *open* in $X$. Furthermore, we also *assume that;*

(9.31) *for any* $U \subseteq X$, the *absolute value morphism*

(9.31.1) $|\cdot| : \mathcal{A}(U) \to \mathcal{S}(\mathcal{A}(U)^{\cdot}) := \{\alpha \in \mathcal{A}^{\cdot}(U) = \mathcal{A}(U)^{\cdot} : |\alpha| = 1\}$

*extends the* standard *absolute value morphism of the (constant* **C**-*algebra) sheaf* $\mathbf{C} \equiv \mathbf{C}_X \overset{\subseteq}{\rightarrow} \mathcal{A}$.

The above justify thus completely our previous argument in (9.28) and (9.29). ■ Yet, based on (9.31.1), one defines a *complete presheaf of groups*, whose *sheafification* is denoted by

$$\mathcal{SA}^{\cdot} \tag{9.32}$$

and called the *special sheaf of units of* $\mathcal{A}$, or even, the *unit circle sheaf of* $\mathcal{A}$. In particular, one thus obtains the next *group isomorphisms*

$$(\mathcal{SA}^{\cdot})(U) \equiv \mathcal{SA}^{\cdot}(U) = \mathrm{S}(\mathcal{A}(U)^{\cdot}) = \mathrm{S}(\mathcal{A}^{\cdot}(U)), \tag{9.33}$$

*for every open* $U \subseteq X$.

Thus, (9.27) becomes now (cf. (9.26), (9.29))

$$\partial\mathrm{et}\,|_{\mathcal{U}(n,\mathcal{A})} \equiv \partial\mathrm{et}\, : \mathcal{U}(n,\mathcal{A}) \longrightarrow \mathcal{SA}^{\cdot}, \tag{9.34}$$

so that one defines;

$$\mathcal{SU}(n,\mathcal{A}) := \ker(\partial\mathrm{et}\,|_{\mathcal{U}(n,\mathcal{A})}) \equiv \ker(\partial\mathrm{et}\,), \tag{9.35}$$

called also the *special unitary group sheaf*, of *order* $n \in \mathbf{N}$, on $X$ (which thus can be associated with an *appropriate* **C**-*algebraized space* $(X, \mathcal{A})$, as above, according to classical patterns). Of course, it is clear that (9.35) *can* also *be obtained* as the *sheafification* of the *(complete) presheaf of groups*, given by

$$\mathrm{SU}\,(n, \mathcal{A}(U)) := \{\alpha \in \mathcal{U}(n, \mathcal{A}(U)) : \partial\mathrm{et}\,\alpha = 1\}, \tag{9.36}$$

with $U$ varying over the *open* subsets of $X$, together with the obvious restriction morphisms. Hence, one has, in particular, the following *isomorphism of groups*

$$\mathcal{SU}(n,\mathcal{A})(U) = SU(n, \mathcal{A}(U)), \tag{9.37}$$

*for every open* $U \subseteq X$.

Thus, by considering a **C**-*algebraized space* $(X, \mathcal{A})$, *as before* (cf., for instance, (9.24), (9.30), (9.31)), one arrives at the following *short exact sequence of group sheaves* on $X$;

$$1 \longrightarrow \mathcal{SU}(n,\mathcal{A}) \longrightarrow \mathcal{U}(n,\mathcal{A}) \xrightarrow{\partial\mathrm{et}} \mathcal{SA}^{\cdot} \longrightarrow 1. \tag{9.38}$$

On the other hand, working within the previous context, one further obtains the following analogue of Theorem 9.1 above. Namely, we have.

**Theorem 9.2.** *Let* $(X, \mathcal{A})$ *be an involutive* **C**-*algebraized space, satisfy-*

*ing also (9.30), (9.31). Moreover, let $\mathcal{E}$ be a given vector sheaf on $X$, with* $\operatorname{rk}\mathcal{E} = n$. *Then, a coordinate 1-cocycle of $\mathcal{E}$ takes values from $\mathcal{SU}(n,\mathcal{A})$* (we also say then, that $\mathcal{E}$ admits an *$\mathcal{SU}(n,\mathcal{A})$-structure) if, and only if, the corresponding determinant line sheaf $\bigwedge^n\mathcal{E}$ of $\mathcal{E}$ is flat* (cf. Definition 5.1).

**Proof.** Suppose that

$$(g_{\alpha\beta}) \in Z^1(\mathcal{U}, \mathcal{GL}(n,\mathcal{A})) \tag{9.39}$$

is a *coordinate 1-cocycle of $\mathcal{E}$*, relative to some *local frame*

$$\mathcal{U} = \{U_\alpha\}_{\alpha\in I} \tag{9.40}$$

of $\mathcal{E}$ (cf. (1.7)), *such that* (cf. (9.36), (9.37))

$$g_{\alpha\beta} \in \mathcal{SU}(n,\mathcal{A})(U_{\alpha\beta}) = SU(n,\mathcal{A}(U)), \tag{9.41}$$

*for any $\alpha,\beta \in I$, with $U_\alpha \cap U_\beta \equiv U_{\alpha\beta} \neq \emptyset$;* thus (cf. (9.35) and Chapt. IV; (4.7)), one has

$$\overline{\partial\mathrm{et}}_{U_{\alpha\beta}}(g_{\alpha\beta}) \equiv \det(g_{\alpha\beta}) = 1. \tag{9.42}$$

Now, to proceed further, we first remark the following fact:

(9.43) if $(g_{\alpha\beta}) \in Z^1(\mathcal{U}, \mathcal{GL}(n,\mathcal{A}))$ is the *coordinate 1-cocycle* of a given *vector sheaf $\mathcal{E} \in \mathbf{\Phi}^n_{\mathcal{A}}(X)$, then*

$$\det(g_{\alpha\beta}) \in Z^1(\mathcal{U}, \mathcal{A}^{\cdot}) \tag{9.43.1}$$

*is that one of the* corresponding *determinant line sheaf of $\mathcal{E}$*, $\det(\mathcal{E}) \equiv \bigwedge^n\mathcal{E} \in \mathbf{\Phi}^1_{\mathcal{A}}(X)$ (cf. Chapt. IV; (7.23)).

Indeed, the above is, in effect, a reformulation of our previous conclusion, concerning (4.22) (see also (4.25)). ▮ Therefore, based now on (9.43.1) and (9.42), one already infers that

$$\det(g_{\alpha\beta}) \in i^*_X(\mathrm{GL}(n,\mathbf{C})(U_{\alpha\beta})) \equiv i^*_X(\mathrm{GL}(n,\mathbf{C})) \tag{9.44}$$

(see also (5.22), (5.20) and (5.25) in the preceding). So, in other words, one has

$$\det(\mathcal{E}) \in \operatorname{im}(i^*_X) \subseteq H^1(X, \mathcal{A}^{\cdot}), \tag{9.45}$$

that is, exactly *our claim for* $\det\mathcal{E}$ (see Definition 5.1).

Now, *conversely, suppose that (9.45) holds true,* so that let

$$(\lambda_{\alpha\beta}) \in Z^1(\mathcal{U}, \mathbf{C}^{\cdot}) \tag{9.46}$$

be a ***flat representative*** of $\det \mathcal{E}$ (cf. Theorem 5.1), such that

$$i_X^*(\lambda_{\alpha\beta}) = \det(g_{\alpha\beta}), \tag{9.47}$$

*with* $(g_{\alpha\beta})$, *as in* (9.39) (cf. also (5.22)). Thus, based now on (9.46), one defines a *coordinate 1-cocycle of* $\mathcal{E}$, by the relation;

$$g'_{\alpha\beta} := t_\alpha \cdot g_{\alpha\beta} \cdot t_\beta^{-1} \in \mathrm{GL}(n, \mathcal{A}(U_{\alpha\beta})), \tag{9.48}$$

such that

$$t_\alpha^n := s_\alpha \in \mathbf{C}^{\cdot}(U_\alpha) = \mathbf{C}^{\cdot}, \quad \textit{with} \ |s_\alpha| = 1, \ \alpha \in I. \tag{9.49}$$

In this context, we also note that *(9.48) yields,* in fact, *the same cohomology class for* $\mathcal{E}$ *in* $H^1(X, \mathcal{GL}(n, \mathcal{A}))$, *as (9.39) does;* see, for instance, Chapt. III; (11.14), (11.16)). However, one now obtains

$$\det(g'_{\alpha\beta}) = 1, \tag{9.50}$$

that is, the given vector sheaf $\mathcal{E}$ *admits an* $\mathcal{SU}(n, \mathcal{A})$*-structure* (see also the statement of the theorem), which just was to be proved, and this completes the proof of the theorem. ∎

# APPENDIX

# Category jargon

**1.** Our purpose by this Appendix is to make, for the reader's convenience, more explicit several basic notions from *Category Theory,* that we use throughout the text, in addition to the relevant comments that are eventually supplied at the pertinent places.

So it is standard for treatises of the present level, regarding the jargon in question, to accept the notion of a *category* as a "primitive" one, hence undefined, adopting, instead, as such, that one of a *"collection of objects"*, the latter being not necessarily a *set.* However, if the latter case happens to be in force, (viz. whenever the objects of the category, under consideration, constitute a *set,* in the sense of *Set Theory*), then one speaks of a *small category.* (Take e.g. the category of the open sets of a given topological space $X$; cf. p. 25).

Now, the second basic step, within this Category Theory context, is to assume (*axiom*) that;

> for every ordered pair $(A, B)$ of objects of a given category $\mathcal{A}$, there exists a set
>
> (1.1) $$\operatorname{Mor}_{\mathcal{A}}(A, B) \equiv \operatorname{Mor}(A, B)$$
>
> called the set of *morphisms* of $A$ into $B$.

Thus, it is important, any time we want to identify a given collection of objects (e.g. $\mathcal{A}$-*modules*) as a category, to define first the due "morphisms" between any given pair of the objects, we are interested in. We did it constantly (e.g. $\mathcal{A}$-*morphism*) throughout the previous text, whenever we had to deal with a particular assemblage of objects in the categorical point of view.

Of course, the above sets of morphisms, as in (1.1), for the various pairs of objects of $\mathcal{A}$, have to satisfy certain particular axioms; composition of morphisms, uniqueness of the set (1.1), associativity of the composition and (unique) existence of an *identity morphism,* say $1_A$, for any object $A$ of $\mathcal{A}$. (See, for instance, N. Jacobson [1: p.9, Definition 1.1].

Thus, a given morphism $\phi \in \mathrm{Mor}(A, B)$, as above, is said to be an *isomorphism* (in the given category $\mathcal{A}$), whenever there exists a (unique) morphism $\psi \in \mathrm{Mor}(B, A)$, such that one has;

$$
(1.2) \qquad \begin{aligned} \psi \circ \phi &= 1_A \in \mathrm{Mor}(A, A) \equiv \mathrm{End}\,(A) \quad \textit{and} \\ \phi \circ \psi &= 1_B \in \mathrm{Mor}(B, B) \equiv \mathrm{End}\,(B). \end{aligned}
$$

We set above $\mathrm{End}\,(A)$ for the set of *endomorphisms* of $A$, viz. morphisms of $A$ into itself. (We have actually employed the preceding definition in several places in our previous study; see e.g. Chapt. I; (1.20), for isomorphisms of sheaves). Furthermore, one defines a *monomorphism,* as a morphism $\phi \in \mathrm{Mor}(A, B)$ admitting a *left inverse,* i.e., a morphism (*contraction*) $\psi \in \mathrm{Mor}(B, A)$, such that the first of (1.2) hold true. The second relation in (1.2) characterizes a given morphism $\phi$, as above, as an *epimorphism,* in the sense, namely, that it admits a *right inverse* (or else *section*). Thus, an isomorphism is, by definition (cf. (1.2)), a *mono*- and *epi*- morphism. (In the preceding we have applied, however, for convenience, the *set-theoretic* version of the previous notions; see thus Chapt. I, (2.14) and (2.15)).

Now, a category $\mathcal{B}$ is said to be a *subcategory* of a given category $\mathcal{A}$, whenever every object of $\mathcal{B}$ is still an object of $\mathcal{A}$ too, while we also assume for the corresponding sets of morphisms that

$$
(1.3) \qquad \mathrm{Mor}_{\mathcal{B}}(A, B) \subseteq \mathrm{Mor}_{\mathcal{A}}(A, B),
$$

for any given objects $A, B$ of $\mathcal{B}$. (Yet, we further require in the above definition that identity morphisms and composition of morphisms are the same for the common objects of the previous two categories). In particular, if for the subcategory $\mathcal{B}$ of $\mathcal{A}$, as above, one has, instead of (1.3), the relation

$$
(1.4) \qquad \mathrm{Mor}_{\mathcal{B}}(A, B) = \mathrm{Mor}_{\mathcal{A}}(A, B),
$$

then one speaks of $\mathcal{B}$, as a *full subcategory* of $\mathcal{A}$. See, for example, Chapt. I; (11.63) and (11.64), concerning the categories $\mathcal{C}o\mathcal{PSh}_X$ and $\mathcal{PSh}_X$).

One of the important types of categories, we encountered in the preceding, was that of an *abelian category;* many particular instances of categories we considered above fall thus into this type, as already noted, for example, the category of (left) $\mathcal{A}$-*modules* on a topological space $X$, $\mathcal{A}$-$\mathcal{M}od_X$ (cf. Chapt. II; (2.4)), and its *full subcategory* of *vector sheaves* on $X$, $\mathcal{V}ect\mathcal{S}h_X$ (ibid., (4.10)). We also noted, at the pertinent places, the standard fact, for that matter, that an abelian category is, in effect, the natural setting for developing *Homological Algebra* and even *Sheaf Cohomology* (see A. Grothedieck [2]). For practical reasons, however, we have chosen in our treatment to deal, instead, with the above two particular abelian categories, in so far as this concerned the latter two disciplines, and to the extent, of course, that this was necessary for our study.

Among the various axioms that identify a given category $\mathcal{A}$ as an abelian one (see, for instance, R. Hartshorne [1: p.202, Definition]), we just note that such a category possesses a *zero object* $0_{\mathcal{A}} \equiv 0$, while each one of the respective morphism-sets, as in (1.1), is an (additive) *abelian group,* such that the corresponding composition law is biadditive. (We note that these properties define actually $\mathcal{A}$ already as a *preadditive category;* see, for instance, S. Mac Lane [1: pp. 28, 190].) Yet, another important feature of an abelian category, as the aforementioned ones, is the possibility of formulating in it the notion of an *exact sequence* and that of a *short exact sequence* (see e.g. P. J. Hilton – U. Stammbach [1: pp. 79, 80]). Again, for convenience, we employed these notions in the *concrete* (abelian) *category* $\mathcal{A}$-$\mathcal{M}od_X$, as above (cf. Chapt. I; p. 110, Definition 2.1).

**2.** By referring above to a "concrete" category, one actually applies the all important notion of a *functor.* Indeed, according to the creators of the theory, " *"category" theory has been defined in order to define "functor" and "functor" has been defined in order to define "natural transformation".* " (See S. Mac Lane [1: p.18]). So we come now to these notions that we often employed in our treatment. Thus, a *functor* $T$ is, in a sense, a kind of a map between two given categories $\mathcal{A}$ and $\mathcal{B}$, so we also write, for convenience

$$T : \mathcal{A} \longrightarrow \mathcal{B}, \tag{2.1}$$

in such a manner that;

i) to each object $A \in \mathcal{A}$ (forgiven abuse of notation here, concerning "$\in$"), there corresponds through $T$, an object $T(A) \in \mathcal{B}$, and

ii) for any two objects $A, A'$ in $\mathcal{A}$ and $f \in \mathrm{Mor}_{\mathcal{A}}(A, A')$, there corresponds a morphism

$$T(f) \in \mathrm{Mor}_{\mathcal{B}}(T(A), T(A')), \tag{2.2}$$

such that:

iii) for any $f \in \mathrm{Mor}_{\mathcal{A}}(A, A')$ and $g \in \mathrm{Mor}_{\mathcal{A}}(A', A'')$, one has

$$T(g \circ f) = T(g) \circ T(f). \tag{2.3}$$

What we have actually defined above is a so-called *covariant functor* $T$, as in (2.1), the specification *constravariant* for $T$ being in that, instead of (2.2) and (2.3) one has

$$T(f) \in \mathrm{Mor}_{\mathcal{B}}(T(A'), T(A)), \tag{2.2$'$}$$

and

$$T(g \circ f) = T(f) \circ T(g). \tag{2.3$'$}$$

Thus, as we have seen (cf. Chapt. I; (13.4) and (9.7)), for any given sheaf $\mathcal{S}$, the corresponding section functor, $\Gamma_{\mathcal{S}} \equiv \Gamma$, defines a *covariant functor*

$$\Gamma : \mathcal{S}h_X \longrightarrow \mathcal{C}o\mathcal{PS}h_X, \tag{2.4}$$

with $\Gamma(\mathcal{S}) = \Gamma_{\mathcal{S}}, \mathcal{S} \in \mathcal{S}h_X$, such that, for any *sheaf-morphism* $\phi \in \mathrm{Mor}(\mathcal{S}, \mathcal{S}')$, one defines a *presheaf morphism* $\Gamma(\phi) \equiv \overline{\phi} \in \mathrm{Mor}(\Gamma(\mathcal{S}), \Gamma(\mathcal{S}'))$ (ibid.). A similar functor was also the *sheafification functor* (ibid.; (8.1)). On the other hand, the same notion of a *presheaf* (of sets) is profitably defined, as we saw (ibid.; (5.10)), as a *contravariant functor.*

Indeed, by looking at the same example, as before, one has a nice particular instance of the all in all important notion (see the comments at the beginning of this section) of a *natural transformation of functors,* by taking as such a morphism between the two (contravariant) functors defined by two given presheaves. So by abstracting the corresponding situation therein (ibid., p.271), given two (covariant) functors $S$ and $T$ between the categories $\mathcal{A}$ and $\mathcal{B}$, a *natural transformation*

$$\alpha : S \longrightarrow T \tag{2.5}$$

is an assignment, such that for any objects $A, A' \in \mathcal{A}$ and a morphism

$f \in \mathrm{Mor}_{\mathcal{A}}(A, A')$, one has corresponding morphisms in $\mathcal{B}$, $\alpha(A) \equiv \alpha_A$ and $\alpha(A') \equiv \alpha_{A'}$, as depicted by the following diagram, which is also deemed to be commutative;

(2.6)
$$\begin{array}{ccc} S(A) & \xrightarrow{\quad S(f) \quad} & S(A') \\ {\scriptstyle \alpha_A}\downarrow & & \downarrow{\scriptstyle \alpha_{A'}} \\ T(A) & \xrightarrow[\quad T(f) \quad]{} & T(A') \end{array}$$
(diagram in the category $\mathcal{B}$)

(Of course, one has a complete analogy here with the analogous diagram in Chapt. I; (6.5)). Thus, *a morphism of presheaves is but a* natural transformation *of the respective (contravariant) functors, defined by the presheaves concerned.* (It is clear that in the case of *contravariant functors* $S$ and $T$ the "arrows" in (2.6), defined by $S$ and $T$ (only!) are reversed).

Another example of a natural transformation of functors that is connected with our previous considerations is that one provided by the *determinant morphism,* as defined in Chapt. IV; (4.1): Thus, by considering the (abelian) category

(2.7) $$\mathbf{C}\text{-}\mathcal{A}lg\mathcal{S}h_X$$

of $\mathbf{C}$-*algebra sheaves* on $X$ (cf. Chapt. II; Definition 1.4), one defines the *(covariant) functors*

(2.8) $$\mathcal{GL}(n, \cdot) \text{ and } (\ ,\ )^{\cdot} : \mathbf{C}\text{-}\mathcal{A}lg\mathcal{S}h_X \longrightarrow \mathcal{A}b\mathcal{G}r\mathcal{S}h_X$$

with target the *category of abelian group sheaves* on $X$, such that

(2.9) $$\mathcal{GL}(n, \cdot)(\mathcal{A}) := \mathcal{GL}(n, \mathcal{A})$$

and

(2.9′) $$(\ ,\ )^{\cdot}(\mathcal{A}) := \mathcal{A}^{\cdot}$$

with $\mathcal{A}$ any given $\mathbf{C}$-algebra sheaf on $X$ (cf. Chapt. IV; (1.26) and (1.13)). So for any given *morphism* $\phi \in \mathrm{Mor}(\mathcal{A}, \mathcal{A}')$ *of* $\mathbf{C}$-*algebra sheaves* on $X$, one has the next *commutative diagram*

$$(2.10)\qquad \begin{array}{ccc} \mathcal{GL}(n,\mathcal{A}) & \xrightarrow{\partial\mathrm{et}_{\mathcal{A}}} & \mathcal{A}^{\cdot} \\ {\scriptstyle \mathcal{GL}(n,\phi)}\downarrow & & \downarrow{\scriptstyle \phi^{\cdot}} \\ \mathcal{GL}(n,\mathcal{A}') & \xrightarrow[\partial\mathrm{et}_{\mathcal{A}'}]{} & (\mathcal{A}')^{\cdot}\,, \end{array}$$

as given by Chapt. IV; (4.8). Hence (for any $n \in \mathbf{N}$),

$$(2.11)\qquad \partial\mathrm{et} : \mathcal{GL}(n,\cdot) \longrightarrow (\ ,\ )^{\cdot}$$

yields a *natural transformation of* the *functors* concerned. (In this context, see also Chapt. IV; Remark 4.1).

Now, a given natural transformation of functors (called also a *morphism of functors*), as in (2.5), becomes a *natural equivalence* (or even *natural isomorphism*), whenever there exists a natural transformation

$$(2.12)\qquad \beta : T \longrightarrow S,$$

such that

$$(2.13)\qquad \alpha\beta = 1 \text{ and } \beta\alpha = 1,$$

in the sense that (cf. also (2.6))

$$(2.13')\qquad \alpha_A \circ \beta_A = 1_{T(A)} \text{ and } \beta_A \circ \alpha_A = 1_{S(A)},$$

for any $A \in \mathcal{A}$. (Thus, we assume that, for any object $A \in \mathcal{A}$, *the* corresponding *morphism* $\alpha_A \in \mathrm{Mor}(S(A), T(A))$ *is invertible* in $\mathcal{B}$; cf. also (1.2)). In that case we also write

$$(2.14)\qquad \beta = \alpha^{-1}.$$

We have encountered this situation in the preceding in the case of the functors $\Gamma$ and $\mathbf{S}$ ("*section*" and "*sheafification*" functors, respectively; cf. Chapt. I, Section 13). Thus, as we saw (ibid.; (13.20), (13.21)), the functors $\Gamma \circ \mathbf{S} \equiv \Gamma\mathbf{S}$ and $1_{\mathcal{C}o\mathcal{PS}h_X}$, as well as, $\mathbf{S} \circ \Gamma = \mathbf{S}\Gamma$ and $1_{\mathcal{S}h_X}$, are *naturally equivalent* (or *isomorphic*), respectively.

Indeed, one has here something more than a mere natural equivalence of functors, in that the functors at issue provide, in effect, an *equivalence of* the *categories* involved (ibid.; Theorem 13.1), which notion we futher explain presently below.

**3.** First, given a functor, say $T : \mathcal{A} \longrightarrow \mathcal{B}$, one gets, by the very definition of $T$, a map (still denoted by $T$)

$$T : \mathrm{Mor}_{\mathcal{A}}(A, A') \longrightarrow \mathrm{Mor}_{\mathcal{B}}(T(A), T(A')), \tag{3.1}$$

for any given pair $(A, A')$ of objects in $\mathcal{A}$. Thus, we say that $T$ is a *full functor,* in case the previous map (of sets) is onto, for any $A, A'$ in $\mathcal{A}$. (In other words, for any $A, A'$, as before, and any given $\psi \in \mathrm{Mor}(T(A), T(A'))$, there exists $\phi \in \mathrm{Mor}(A, A')$, with $\psi = T(\phi)$).

On the other hand, if the above map (3.1) is one-to-one, for any pair $(A, A')$, as before, then one speaks of $T$, as a *faithful functor.* A given functor $T : \mathcal{A} \longrightarrow \mathcal{B}$, which happens to be both full and faithful is simply called a *fully faithful functor.* (As we saw, cf. Chapt. I; (13.1), (13.5), the section functor $\Gamma$ is such a one, and, of course, something more. So, *by definition,*

(3.2) a given functor $T : \mathcal{A} \longrightarrow \mathcal{B}$ is said to be *fully faithful,* whenever one has

$$\mathrm{Mor}_{\mathcal{A}}(A, A') = \mathrm{Mor}_{\mathcal{B}}(T(A), T(A')), \tag{3.2.1}$$

within a *bijection,* for any objects $A, A'$ in A.

Yet, we can say that a *fully faithful functor* $T : \mathcal{A} \longrightarrow \mathcal{B}$ *describes* actually *its domain* (-category) $\mathcal{A}$ *as a full subcategory of* $\mathcal{B}$, in the sense that one has

$$\mathcal{A} \underset{\mathrm{T}}{\cong} T(\mathcal{A}) \subseteq \mathcal{B}. \tag{3.3}$$

More generally, if a given functor $T : \mathcal{A} \longrightarrow \mathcal{B}$ has the property that;

(3.4) for every object $B$ in $\mathcal{B}$, there exists an object $A$ in $\mathcal{A}$, such that $T(A) \cong B$ (viz. *isomorphism* in $\mathcal{B}$), then the given functor $T$ is said to be *essentially surjective.*

Thus, we come now to the important fact (*Theorem-Definition*) that a given functor $T : \mathcal{A} \longrightarrow \mathcal{B}$ defines a *category equivalence,* and we write

$$\mathcal{A} \underset{\mathrm{T}}{\cong} \mathcal{B}, \tag{3.5}$$

in the case that the given functor $T$ is *fully faithful and essentially surjective.* Indeed, the Theorem here is the fact (see thus e.g. S. Mac Lane [1: p.91,

Theorem 1]) that (3.4) *is equivalent with the existence of a functor* $S : \mathcal{B} \longrightarrow \mathcal{A}$, *such that one has;*

(3.6) $$TS \cong 1_{\mathcal{B}} \quad \text{and} \quad ST \cong 1_{\mathcal{A}},$$

as natural equivalences of the functors concerned. (So, that was actually the case with the *section functor* $\Gamma$ and its "adjoint" *sheafification functor* $\mathbf{S}$; Chapt. I; (13.5), along with (13.20) and (13.21)). Furthermore, the same relations (3.6) define $T$ as the *left adjoint* and also the *right adjoint of* $S$ (and analogously for the functor $S$, with respect to $T$. We also say in that case that either one of the functors $T$ or $S$ defines an *adjoint equivalence* of the categories concerned. See thus, for instance, Chapt. I; (13.22)). If, in particular, (3.6) are valid with equalities, one then speaks of *isomorphic categories* (instead of *equivalent* ones, as was the case above with (3.5) and (3.6)), or that $T$ defines then an *isomorphism of* the *categories* at issue.

Yet, if in place of (3.5) one has

(3.7) $$\mathcal{A} \underset{T}{\cong} \mathcal{B}^{\mathrm{op}},$$

with $\mathcal{B}^{\mathrm{op}}$ the *opposite category* of $\mathcal{B}$ (:same objects as $\mathcal{B}$ and "reversed arrows"), then one speaks of $T$, as a *category anti-equivalence.*

Finally, another categorical concept, that was used often in the preceding, is that of a *left exact* (covariant) *functor* $T : \mathcal{A} \longrightarrow \mathcal{B}$, as, for instance, in the case of $\Gamma_X$, the global section functor (see Chapt. II; (2.59)). Thus, by definition, the previous speciality for a given functor $T$, as above, means that;

(3.8) from any given, say, *exact* $\mathcal{A}$*-sequence,*

$$0 \longrightarrow \mathcal{E} \longrightarrow \mathcal{F} \longrightarrow \mathcal{S}$$

*(short left exact sequence),* one gets the exact sequence;

$$0 \longrightarrow T(\mathcal{E}) \longrightarrow T(\mathcal{F}) \longrightarrow T(\mathcal{S}).$$

In other words, *a left exact functor preserves short left exact sequences.*

Such a functor is usually assumed to be *additive* too, in the sense that the corresponding map (2.2) is a *group morphism* between the abelian additive groups, for the case at issue, as in (2.1); ii), so $\mathcal{A}$ and $\mathcal{B}$ are supposed to be, in particular, *additive categories.* (Cf. also the preceding comments in Section 1 on abelian categories).

So this is still the case for the *global section functor* $\Gamma_X$ (ibid.), as well as, for the (covariant) *functors* $\mathcal{H}om_{\mathcal{A}}(\mathcal{E}, \cdot)$ and $\mathrm{Hom}_{\mathcal{A}}(\mathcal{E}, \cdot)$ (ibid.; (6.17), (6.46)). Yet, concerning this property of $\Gamma_X$, it was the analogous situation to (3.8), that has been applied in Chapt. III; (3.9), (3.10). (Of course, "dual" arguments are valid for *right exact* functors).

**4.** Before we end off this brief annotation on the categorical notions that were employed throughout the preceding text, we still point out the importance of the so-called *injective resolutions* of a given object in an abelian category in order to define the *"derived functor cohomology"*. Thus, as already noted at the pertinent places in Chapter III, the cohomology at issue can be defined, in general, for functors, whose source is an *"abelian category with enough injectives"*. So this was actually the case with the category $\mathcal{A}$-$\mathcal{M}od_X$ and the global section functor $\Gamma_X$, through which one defines the *sheaf cohomology of a* given *topological space* $X$ (ibid.; p.156, Lemma 1.2 and p.164, Definition 3.1, along with (3.24) and (3.38)).

## BIBLIOGRAPHY

(The items of the bibliography are given separately, for each one of the two Volumes of this treatise; certain issues are thus repeated, for convenience)


**C. Andreian Cazacu**

1. *Theorie der Funktionen mehrerer komplexer Veränderlicher.* Birkhäuser, Basel, 1975.

**M. F. Atiyah**

1. *Complex analytic connections in fiber bundles.* Trans. Amer. Math. Soc. 85(1957), 181-207.

**W. M. Boothby**

1. *An Introduction to Differentiable Manifolds and Riemannian Geometry.* Academic Press, New York, 1975.

**R. Bott – L. W. Tu**

1. *Differential Forms in Algebraic Topology.* Springer - Verlag, New York, 1982.

**N. Bourbaki**

1. *Théorie des ensembles, Chap. 3 (2e éd.).* Hermann, Paris, 1967.
2. *Algèbre, Chap. 1–3.* Hermann, Paris, 1970.
3. *Topologie générale, Chap. 5–10.* Hermann, Paris, 1974.

**G. E. Bredon**

1. *Sheaf Theory.* McGraw-Hill, New York, 1967.

**J.-L. Brylinski**

1. *Loop Spaces, Characteristic Classes and Geometric Quantization.* Birkhäuser, Boston, 1993.

**N. P. Buchdahl**

1. *On the relative de Rham sequence.* Proc. Amer. Math. Soc. 87 (1983), 363-366.
2. *Analysis on analytic spaces and non-self-dual Yang-Mills fields.* Trans. Amer. Math. Soc. 288 (1985), 431-469.

**H. Cartan – S. Eilenberg**

1. *Homological Algebra.* Princeton Univ. Press, Princeton, New Jersey, 1973.

**L. Chambadal – J. L. Ovaert**

1. *Algèbre linéaire et algèbre tensorielle.* Dunod, Paris, 1968.

**L. Conlon**

1. *Differentiable Manifolds. A first Course.* Birkhäuser, Boston, 1993.

**C. W. Curtis**

1. *Linear Algebra. An Introductory Approach.* Springer-Verlag, New York, 1984.

**W. D. Curtis – F. R. Miller**

1. *Differential Manifolds and Theoretical Physics.* Academic Press, Orlando, Florida, 1985.

**C. H. Dowker**

1. *Lectures on Sheaf Theory.* Tata Institute, Bombay, 1956.

**J. Dugundji**

1. *Topology.* Allyn and Bacon, Boston, 1966.

**S. Eilenberg – S. MacLane**

1. *General theory of equivalences.* Trans. Amer. Math. Soc. 58 (1945), 231-294.

**S. Eilenberg – N. Steenrod**

1. *Foundations of Algebraic Topology.* Princeton Univ. Press, Princeton, New Jersey, 1952.

**O. Forster**

1. *Lectures on Riemann Surfaces.* Springer-Verlag, New York, 1981.

**J. Frenkel**

1. *Cohomologie non abélienne et spaces fibrés.* Bull. Soc. math. France 85 (1957), 135-220.

**R. Godement**

1. *Topologie algébrique et théorie des faisceaux (3ème éd.).* Hermann, Paris, 1973.

**H. Grauert – R. Remmert**

1. *Theory of Stein Spaces.* Springer-Verlag, Berlin, 1979.
2. *Coherent Analytic Sheaves.* Springer-Verlag, Berlin, 1984.

**P. Griffiths – J. Harris**

1 *Principles of Algebraic Geometry.* J. Wiley, New York, 1978.

**A. Grothendieck**

1 *A General Theory of Fiber Spaces with Structure Sheaf (2nd ed.).* Univ. of Kansas, 1958.

2. *Sur quelques points d' algèbre homologique.* Tôhoku Math. J. 9 (1957), 119-221.

**A. Grothendieck – J. Dieudonné**

1. *Eléments de Géométrie Algébrique I.* Springer - Verlag, Berlin, 1971.

**R. C. Gunning**

1. *Lectures on Riemann Surfaces.* Princeton Univ. Press, Princeton, New Jersey, 1966.

2. *Lectures on Vector Bundles over Riemann Surfaces,* Princeton Univ. Press, Princeton, New Jersey, 1967.

**R. C. Gunning – H. Rossi**

1. *Analytic Functions of Several Complex Variables.* Prentice - Hall, Englewood Cliffs, New Jersey, 1965.

**R. Hartshorne**

1. *Algebraic Geometry.* Springer - Verlag, New York, 1977.

**M. Heller**

1. *Algebraic foundations of the theory of differential spaces.* Demonstratio Math. 24 (1991), 349-364.

**P. J. Hilton – U. Stammbach**

1. *A Course in Homological Algebra.* Springer - Verlag, New York, 1971.

**F. Hirzebruch**

1. *Topological Methods in Algebraic Geometry.* Springer - Verlag, Berlin, 1978.

**L. Hörmander**

1. *An Introduction to Complex Analysis in Several Variables.* North - Holland, Amsterdam, 1973.

**S. Iitaka**

1. *Algebraic Geometry. An Introduction to Birational Geometry of Algebraic Varieties.* Springer - Verlag, New York, 1982.

**B. Iversen**

1. *Cohomology of Sheaves.* Springer - Verlag, Berlin, 1986.

**N. Jacobson**

1. *Basic Algebra II.* Freeman, San Fransisco, 1980.

**I. M. James**

1. *General Topology and Homotopy Theory.* Springer - Verlag, New York, 1984.

**L. Kaup – B. Kaup**

1. *Holomorphic Functions of Several Variables.* De Gruyter, Berlin, 1983.

**K. Kodaira**

1. *Complex Manifolds and Deformation of Complex Structures.* Springer - Verlag, New York, 1986.

**S. G. Krantz**

1. *Function Theory of Several Complex Variables.* J. Wiley, New York, 1982.

**R. Kultze**

1. *Garbentheorie.* Teubner, Stuttgart, 1970.

**A. Kyriazis**

1. *On topological algebra sheaves.* J. Austral. Math. Soc. 61 (1996), 14-29.
2. *On structure sheaf envelopes of vector-valued holomorphic functions* (manuscript).
3. *Theory of Topological* **A**-*Algebras. Tensor Products, Sheaves and Bundles.* [Greek]. Athens, 1993. pp. 235.

**S. Lang**

1. *Algebra (2nd Ed.).* Addison - Wesley, Reading, Massachusetts, 1984.
2. *Introduction to Arakelov Theory.* Springer - Verlag, New York, 1988.

**S. Mac Lane**

1. *Categories for the Working Mathematician.* Springer - Verlag, New York, 1971.

**A. Mallios**

1. *Vector bundles and K-theory over topological algebras.* J. Math. Anal. Appl. 92 (1983), 452-506.
2. *Topological Algebras. Selected Topics.* North - Holland, Amsterdam, 1986.
3. (i) *On the existence of $\mathcal{A}$-connections.* Abstracts Amer. Math. Soc. 9(6) (1988), 509. (ii) *$\mathcal{A}$-connections as splitting extensions.* Ibid. 10(2) (1989), 186. (iii) *On the abstract form of Weil's integrality theorem.* Ibid. 10(4) (1989), 312. (iv) *On the curvature of an $\mathcal{A}$-connection.* Ibid. 10(5) (1989), 410.
4. *Lectures on Differential Geometry. An Introduction: Theory of Differential Manifolds and of Lie Groups.* M. Kardamitsa Publ., Athens, 1992. [Greek]
5. *Topological Algebras. Localizations and Extensions.* (Book, in preparation).

**Yu. I. Manin**

1. *Gauge Field Theory and Complex Geometry.* Springer - Verlag, Berlin, 1988.

**M. A. Mostow**

1. *The differentiable space structure of Milnor classifying spaces, simplicial complexes, and geometric realizations.* J. Diff. Geom. 14 (1979), 255-293.

**J. R. Munkres**

1. *Elements of Algebraic Topology.* Addison - Wesley, Menlo Park, California, 1984.

**B. P. Palka**

1. *An Introduction to Complex Function Theory.* Springer - Verlag, New York, 1991.

**B. Pareigis**

1. *Categories and Functors.* Academic Press, New York, 1970.

**J. J. Rotman**

1. *An Introduction to Homological Algebra.* Academic Press, New York, 1979.

**I. Satake**

1. *Linear Algebra.* Marcel Dekker, New York, 1975.

**S. A. Selesnick**

1. *Watts cohomology for a class of Banach algebras and the duality of compact abelian groups.* Math. Z. 130 (1973), 313-323.
2. *Line bundles and harmonic analysis on compact groups.* Math. Z. 146 (1976), 53-67.
3. *Correspondence principle for the quantum net.* Intern. J. Theor. Phys. 30 (1991), 1273-1292.

**J.-P. Serre**

1. *Faisceaux algébriques cohérents.* Ann. Math. 61 (1955), 197-278.

**J. W. Smith**

1. *The de Rham theorem for general spaces.* Tôhoku Math. J. 18 (1966), 115-137.

**A. Solian**

1. *Theory of Modules. (An Introduction to the Theory of Module Categories).* J. Wiley, London, 1977.

**E. H. Spanier**

1 *Algebraic Topology.* Springer - Verlag, New York, 1966.

**J. R. Strooker**

1. *Introduction to Categories, Homological Algebra and Sheaf Cohomology.* Cambridge Univ. Press, Cambridge, 1978.

**R. G. Swan**

1. *The Theory of Sheaves.* Chicago Univ. Press, Chicago, 1964.

**R. M. Switzer**

1. *Algebraic Topology – Homotopy and Homology.* Springer - Verlag, Berlin, 1975.

**B. R. Tennison**

1. *Sheaf Theory.* Cambridge Univ. Press, Cambridge, 1975.

**R. S. Ward – R. O. Wells, Jr.**

1. *Twistor Geometry and Field Theory.* Cambridge Univ. Press, Cambridge, 1990.

**F. W. Warner**

1. *Foundations of Differentiable Manifolds and Lie Groups.* Springer - Verlag, New York, 1983.

**R. O. Wells**

1. *Differential Analysis on Complex Manifolds.* Springer - Verlag, New York, 1980.

# Notational Index

# Subject Index

Other *Mathematics and Its Applications* titles of interest:

---

B.A. Plamenevskii: *Algebras of Pseudodifferential Operators*. 1989, 304 pp. ISBN 0-7923-0231-1

Ya.I. Belopolskaya and Yu.L. Dalecky: *Stochastic Equations and Differential Geometry*. 1990, 288 pp. ISBN 90-277-2807-0

V. Goldshtein and Yu. Reshetnyak: *Quasiconformal Mappings and Sobolev Spaces*. 1990, 392 pp. ISBN 0-7923-0543-4

A.T. Fomenko: *Variational Principles in Topology. Multidimensional Minimal Surface Theory*. 1990, 388 pp. ISBN 0-7923-0230-3

S.P. Novikov and A.T. Fomenko: *Basic Elements of Differential Geometry and Topology*. 1990, 500 pp. ISBN 0-7923-1009-8

B.N. Apanasov: *The Geometry of Discrete Groups in Space and Uniformization Problems*. 1991, 500 pp. ISBN 0-7923-0216-8

C. Bartocci, U. Bruzzo and D. Hernandez-Ruiperez: *The Geometry of Supermanifolds*. 1991, 242 pp. ISBN 0-7923-1440-9

N.J. Vilenkin and A.U. Klimyk: *Representation of Lie Groups and Special Functions. Volume 1: Simplest Lie Groups, Special Functions, and Integral Transforms*. 1991, 608 pp. ISBN 0-7923-1466-2

A.V. Arkhangelskii: *Topological Function Spaces*. 1992, 206 pp. ISBN 0-7923-1531-6

Kichoon Yang: *Exterior Differential Systems and Equivalence Problems*. 1992, 196 pp. ISBN 0-7923-1593-6

M.A. Akivis and A.M. Shelekhov: *Geometry and Algebra of Multidimensional Three-Webs*. 1992, 358 pp. ISBN 0-7923-1684-3

A. Tempelman: *Ergodic Theorems for Group Actions*. 1992, 400 pp. ISBN 0-7923-1717-3

N.Ja. Vilenkin and A.U. Klimyk: *Representation of Lie Groups and Special Functions, Volume 3. Classical and Quantum Groups and Special Functions*. 1992, 630 pp. ISBN 0-7923-1493-X

N.Ja. Vilenkin and A.U. Klimyk: *Representation of Lie Groups and Special Functions, Volume 2. Class I Representations, Special Functions, and Integral Transforms*. 1993, 612 pp. ISBN 0-7923-1492-1

I.A. Faradzev, A.A. Ivanov, M.M. Klin and A.J. Woldar: *Investigations in Algebraic Theory of Combinatorial Objects*. 1993, 516 pp. ISBN 0-7923-1927-3

M. Puta: *Hamiltonian Mechanical Systems and Geometric Quantization*. 1993, 286 pp. ISBN 0-7923-2306-8

V.V. Trofimov: *Introduction to Geometry of Manifolds with Symmetry*. 1994, 326 pp. ISBN 0-7923-2561-3

Other *Mathematics and Its Applications* titles of interest:

---

J.-F. Pommaret: *Partial Differential Equations and Group Theory. New Perspectives for Applications*. 1994, 473 pp. ISBN 0-7923-2966-X

Kichoon Yang: *Complete Minimal Surfaces of Finite Total Curvature*. 1994, 157 pp. ISBN 0-7923-3012-9

N.N. Tarkhanov: *Complexes of Differential Operators*. 1995, 414 pp. ISBN 0-7923-3706-9

L. Tamássy and J. Szenthe (eds.): *New Developments in Differential Geometry*. 1996, 444 pp. ISBN 0-7923-3822-7

W.C. Holland (ed.): *Ordered Groups and Infinite Permutation Groups*. 1996, 255 pp. ISBN 0-7923-3853-7

K.L. Duggal and A. Bejancu: *Lightlike Submanifolds of Semi-Riemannian Manifolds and Applications*. 1996, 308 pp. ISBN 0-7923-3957-6

D.N. Kupeli: *Singular Semi-Riemannian Geometry*. 1996, 187 pp. ISBN 0-7923-3996-7

L.N. Shevrin and A.J. Ovsyannikov: *Semigroups and Their Subsemigroup Lattices*. 1996, 390 pp. ISBN 0-7923-4221-6

C.T.J. Dodson and P.E. Parker: *A User's Guide to Algebraic Topology*. 1997, 418 pp. ISBN 0-7923-4292-5

B. Rosenfeld: *Geometry of Lie Groups*. 1997, 412 pp. ISBN 0-7923-4390-5

A. Banyaga: *The Structure of Classical Diffeomorphism Groups*. 1997, 208 pp. ISBN 0-7923-4475-8

A.N. Sharkovsky, S.F. Kolyada, A.G. Sivak and V.V. Fedorenko: *Dynamics of One-Dimensional Maps*. 1997, 272 pp. ISBN 0-7923-4532-0

A.T. Fomenko and S.V. Matveev: *Algorithmic and Computer Methods for Three-Manifolds*. 1997, 346 pp. ISBN 0-7923-4770-6

A. Mallios: *Geometry of Vector Sheaves*. An Axiomatic Approach to Differential Geometry.
Volume I: Vector Sheaves. General Theory. 1998, 462 pp. ISBN 0-7923-5004-9
Volume II: Geometry. Examples and Applications. 1998, 462 pp. ISBN 0-7923-5005-7
(Set of 2 volumes: 0-7923-5006-5)

GPSR Compliance

*The European Union's (EU) General Product Safety Regulation (GPSR) is a set of rules that requires consumer products to be safe and our obligations to ensure this.*

*If you have any concerns about our products, you can contact us on ProductSafety@springernature.com*

In case Publisher is established outside the EU, the EU authorized representative is:

Springer Nature Customer Service Center GmbH
Europaplatz 3
69115 Heidelberg, Germany

**Batch number: 10823957**

Printed by Printforce, the Netherlands

Geometry of Vector Sheaves

# Mathematics and Its Applications



Volume 439

# Geometry of Vector Sheaves

## An Axiomatic Approach to Differential Geometry

### Volume II: Geometry. Examples and Applications

*by*

Anastasios Mallios

*Department of Mathematics,*
*University of Athens,*
*Athens, Greece*

KLUWER ACADEMIC PUBLISHERS
DORDRECHT / BOSTON / LONDON

A C.I.P. Catalogue record for this book is available from the Library of Congress.

Additional material to this book can be downloaded from http://extras.springer.com.

ISBN 0-7923-5004-9 (Vol. I)
ISBN 0-7923-5005-7 (Vol. II)
Set ISBN 0-7923-5006-5

---

Published by Kluwer Academic Publishers,
P.O. Box 17, 3300 AA Dordrecht, The Netherlands.

Sold and distributed in North, Central and South America
by Kluwer Academic Publishers,
101 Philip Drive, Norwell, MA 02061, U.S.A.

In all other countries, sold and distributed
by Kluwer Academic Publishers,
P.O. Box 322, 3300 AH Dordrecht, The Netherlands.

*Printed on acid-free paper*

## Contents



# Part Two. Geometry

## CHAPTER VI. Geometry of Vector Sheaves. $\mathcal{A}$-connections

**General Preface** (to both Volumes I, II)

Our aim by this study is to exhibit and also exploit a quite general technique, yet simple, by its very nature, which enables one to formulate, at least, a substantial amount (if not all!) of the fundamental notions of the classical differential geometry of $\mathcal{C}^{\infty}$-manifolds and obtain too several standard results thereof. More interestingly, this can be achieved (we might say here, surprisingly enough!), by no use at all (!) of any concept of tangent vectors or of differential forms, in the classical sense of these words; indeed, no calculus is employed altogether! This is, generally speaking, our program, through the present two-volume work, whose secondary title to the adopted main one might be, of course, *An Introduction to* (an) *Abstract* (treatise of) *Differential Geometry.* Such a perspective, even if subconciously evident, more or less, to the experts, has never been, however, quite explicit, to the extent and generality, at least, which is undertaken by this discussion, so that the popularization of this point of view might be of some interest, much more because, as we shall see, there do exist considerable possibilities for applications.

Treatments aiming at the exposition of several aspects of differential geometry, outside the classical framework of (finite-dimensional) $\mathcal{C}^{\infty}$-manifolds, range, of course, from the standard theory of infinite-dimensional $\mathcal{C}^{\infty}$-manifolds, with a vast already literature thereon, to the more sophisticated aspects of "differential geometry" on topological spaces, *more general than* $\mathcal{C}^{\infty}$*-manifolds.* However, this has always been done, through a suitable formulation of the basic notions of the standard case, as for instance, vector fields, differential forms and the like.

By contrast, our treatment is quite general and, so to speak, *axiomatic.* So the basic tools here are *sheaves of modules,* with respect to an appropriate *sheaf of* $\mathbf{C}$*-algebras* (alias, $\mathbf{C}$*-algebra sheaf*), our *"domain of coefficients"*, by analogy with the classical case. Yet, the sheaves involved are mostly over an arbitrary *topological space* $X$, that eventually is decreed to be *paracompact* (Hausdorff). The latter requirement for $X$, base space of the sheaves considered, is only entered into the discussion, to profit from it *cohomologically,* just as occasion serves; so this is, for instance, the case with the counterpart here of *Weil's integrality theorem,* pertaining, as we know, to the cohomol-

ogy class, determined by the curvature form of a connection of a given line bundle (sheaf).

Thus, *sheaf cohomology* is the second basic constituent of our armament, in order to carry out a rather ambitious(!) program, as is namely, the transfer, to the present abstract setting, of several fundamental results, of a global and/or local nature, of differential geometry of manifolds. Indeed, it is, at least, interesting (and also encouraging!, for an undertaking, like the one at hand) to realize how a considerable amount of the classical theory can be carried over to the present formalism, especially in what concerns an important differential geometric aspect, as is the theory of connections and its consequences, as e.g. curvature, characteristic classes (à la Chern-Weil) and so on.

Our study is, as already mentioned, quite general and essentially of an *algebraic* (topological) *nature,* based, as actually does, on *sheaf theory.* So it contains, as a very special case, to the extent, of course, that is accomplished herewith, the classical aspect of finite (and, partly, of infinite) dimensional differential geometry of $\mathcal{C}^{\infty}$-manifolds, as well as, several other generalizations thereon, as alluded to at the beginning of this Preface. The latter cases are also discussed, as particular non-standard examples, at the pertinent places in the sequel (see thus Volume II, Part Three).

In this context, another item worth pointing out here is the crucial rôle played, in a number of places, by the corresponding to our case *de Rham complex* of the standard theory. This refers, in particular, to the exactness of that complex (here deemed to be in force), in the classical case being, of course, a consequence of the famous *Poincarè Lemma* (viz. that every closed form is exact). Yet, in the aforementioned (non-standard) examples exactness of the previous complex is still valid, as a consequence again of the same lemma, which thus is proved to hold true in those extended cases, as well. On the other hand, in our abstract setting the situation can be remedied by demanding, for instance, the section algebras of the $\mathbf{C}$-algebra sheaf $\mathcal{A}$, our *"domain of coefficients",* to carry further a topological algebraic structure, being, thus, in effect, appropriate *topological algebras.* So our *"sheaf of coefficients"* becomes now, precisely speaking, a *topological* $\mathbf{C}$*-algebra sheaf* on $X$.

The insert at this place of *topological algebra theory* is both crucial and decisive. Of course, this can be construed as revealing, within the present general framework, the oblique, nontheless, significant rôle, played in the

classical case by the important (certainly, *non-normed!*) topological algebra $\mathcal{C}^\infty(X)$ of (say, **C**-valued) $\mathcal{C}^\infty$-functions on a given $\mathcal{C}^\infty$-manifold $X$. (We refer the reader to A. Mallios [2] for a general and detailed account on topological algebra theory, as well as, for the particular algebra at issue). Thus, by employing a suitable *topological algebra sheaf* $\mathcal{A}$, one is able to get an (exact) *resolution of* (the *constant sheaf*) **C**, of the form

$$(*) \qquad 0 \longrightarrow \mathbf{C} \xrightarrow{\varepsilon} \mathcal{A} \xrightarrow{\partial} \Omega^1 \xrightarrow{d^1} \Omega^2 \xrightarrow{d^2} \cdots,$$

where the $\Omega^i$'s stand for appropriately defined $\mathcal{A}$-*modules* (modules of *"differential forms"*) and the $d^i$'s, with $d^0 \equiv \partial$, for suitable also (**C**-linear) *morphisms* of the $\mathcal{A}$-modules concerned. Furthermore, a similar aspect (in fact, more feasible than $(*)$), which is still worth remarking here, is the exploitation of a *(short exact) exponential sheaf sequence* of the form

$$(**) \qquad 0 \longrightarrow \mathbf{Z} \longrightarrow \mathcal{A} \xrightarrow{e} \mathcal{A}^{\bullet} \longrightarrow 1,$$

where $\mathcal{A}^{\bullet}$ denotes the *sheaf of units* of $\mathcal{A}$, while $e$ represents the standard *exponential* (sheaf) *morphism* of the classical theory. Here again an exact sequence, like $(**)$, is further realizable in some particular cases of *topological algebra sheaves,* as before. Indeed, it was the relevant work of S.A. Selesnick [1; 2], in the context of *Banach(-Šilov) algebra theory,* aiming also at applications to harmonic analysis, which, by its intense geometrical flavour, kindled actually our first thoughts about this subject matter. (See thus the first announcement thereon in A. Mallios [3: (i) – (iv)]). Moreover, the same exponential sequence $(**)$ substitutes in many instances for the more stringent exact sequence $(*)$, more so, when the $\mathcal{A}$-modules involved are, in particular, the analogues, in our sheaf-theoretic approach, of the usual line bundles.

We mentioned already that the prominent features throughout this discussion are certain sheaves of $\mathcal{A}$-modules, in fact, *vector sheaves;* that is, locally free sheaves of $\mathcal{A}$-modules of finite rank, over an arbitrary, in general, topological space $X$, while $\mathcal{A}$ is our "sheaf of coefficients", as above. This type of $\mathcal{A}$-modules substitute here for the familiar sheaves (of germs) of sections (continuous or otherwise) of *vector bundles* (of finite rank) in the classical theory. If, in particular, the vector sheaf, under consideration, has rank 1, we then speak of a *line sheaf* on $X$. (See also the Preface to Volume I, concerning this particular terminology).

Within the same context, we further remark that the exploitation of *"structures more general than manifolds"*, pertaining to concrete problems, as for example, calculation of the cohomology of *quotient spaces,* or even of classifying spaces (à la Milnor) and of semi-simplicial manifolds (see e.g. M.A. Mostow [1] and J.W. Smith [1]), indicates a further potential concrete application of the point of view of this treatise. Thus, all the previous structures are indeed a very particular instance of those considered herein, so that the present technique exhibits a possible further extension, within a similar setting (in fact, much more general), not only of the topology, as is actually the case with the above structures, but also of the *geometry* (e.g. connections and curvature) *of* such *"generalized spaces"*.

Finally (but certainly, not least!), another fact that supports the framework and terminology adopted by this study, is the ever growing usage of a *sheaf-theoretic language,* compiled with that one of the standard differential geometry, in what actually concerns the *"geometrization of physical thought"*, as expressed, for instance, by the nowdays *"gauge theories"*. (See, for example, Yu.I. Manin [1] and/or S.A. Selesnick [3], or even R.S. Ward - R.O. Wells, Jr. [1]). All the preceding, along with an extremely positive reaction towards this language of several audiences, both at the University of Athens and elsewhere, has been a strong further inducement to the undertaking of the present more systematic account of the subject.

To make the exposition sufficiently self-contained, we prefix a pertinent introduction to *sheaf theory,* a theme in itself already classical, including the necessary for the present purpose material from *sheaf cohomology.* Yet, we adequately expose the indispensable elements for this study of what we may call *linear and multilinear algebra of vector sheaves,* or even of *sheaf modules,* in general, a subject for which a systematic account is still lacking. *Topological algebra sheaves* are also discussed, to the extent that we need it here, namely, in Part Two of this work. On the other hand, the individual Prefaces to each one of the two Volumes of this treatise contain more details and comments on the material of the corresponding Volume.

**Preface** (to Volume II)

As already mentioned hitherto (see the General Preface to both Volumes), the present Volume II of our study contains, in fact, the main material referred to differential geometric notions and results, as these are formulated and proved, within the abstract setting, which is advocated by this work. So this essentially concerns the Part Two of the present Volume, while in Part Three of the same we further discuss Applications and Examples of the technique developed so far, outside, of course, of the standard setting of (finite-dimensional) $\mathcal{C}^{\infty}$-manifolds and/or of holomorphic ones.

The individial chapters of the entire treatise have been consecutively numbered, through this two-volume work, the present Volume II containing thus Chapters VI – IX (Part Two) and X – XI (Part Three). Thus, in brief, these have, in particular, as follows: In Chapter VI we consider, in full generality, the fundamental notion, for the entire Volume, of an *$\mathcal{A}$-connection,* the latter been referred to a given *differential triad*

$$(i) \qquad (\mathcal{A}, \partial, \Omega)$$

on an arbitrary *topological space* $X$. Here $\mathcal{A}$ stands for a **C**-*algebra sheaf* on $X$, $\Omega$ for an *$\mathcal{A}$-module* (viz. for a sheaf of $\mathcal{A}$-modules on $X$, not necessarily locally free, or else a *vector sheaf*) and $\partial$ for a **C**-*linear morphism of $\mathcal{A}$ into* $\Omega$, satisfying the corresponding to our case *Leibniz rule* (concerning here the product of two local (continuous) sections of $\mathcal{A}$). So $\partial$, is in a sense, the given *flat $\mathcal{A}$-connection,* which is thus the counterpart here of the basic differential operator, deriving in the classical case from the coordinate-differentials of our (finite-dimensional) flat model (euclidean space). In turn, $\Omega$ substitutes thus for the (classical) space of *pfaffians* (or else smooth 1-forms); of course, one has to look at all these objects in a sheaf-theoretic disguise. So we next examine the fundamental concept of an *$\mathcal{A}$-connection* for an *$\mathcal{A}$-module* $\mathcal{E}$ on $X$, in general; this is, namely, a map

$$(ii) \qquad D : \mathcal{E} \longrightarrow \mathcal{E} \otimes_{\mathcal{A}} \Omega \equiv \Omega(\mathcal{E}),$$

which is, in particular, a **C**-*linear morphism* of the $\mathcal{A}$-modules involved, satisfying also the respective *Leibniz rule*

$$(iii) \qquad D(\alpha \cdot s) = \alpha \cdot Ds + s \otimes \partial\alpha,$$

for any given (continuous) *local sections* $\alpha \in \mathcal{A}(U)$ and $s \in \mathcal{E}(U)$. It is certain that such a map does not always exist in that generality, even if $\mathcal{E}$ is a vector sheaf on $X$. Indeed, it is the *vanishing of the* so-called *Atiyah class of* $\mathcal{E}$, which yields a *necessary and sufficient condition* in order a given *vector sheaf* on $X$ *to have an* $\mathcal{A}$*-connection;* concerning the (cohomology) class in question, one has, by definition,

$$(iv) \qquad \mathfrak{a}(\mathcal{E}) \in \check{H}^1(X, M_n(\Omega)),$$

where $n \in \mathbf{N}$ stands for the *rank* of the vector sheaf $\mathcal{E}$ on $X$, under consideration. Yet, the same class is always defined, when one is given a differential triad, as in (i) above. The situation is reminiscent of and generalizes, of course, the classical case, pertaining to the existence of holomorphic connections (à la Atiyah), which also was the motivation to the above point of view. So one gets a map (the *Atiyah map*)

$$(v) \qquad \mathfrak{a} : \mathbf{\Phi}^n_{\mathcal{A}}(X) \longrightarrow \check{H}^1(X, M_n(\Omega)),$$

whose domain of definition is the set of *isomorphism classes of vector sheaves on* $X$, of rank $n \in \mathbf{N}$, and whose *kernel*

$$(vi) \qquad \ker(\mathfrak{a}) \equiv \mathbf{\Phi}^n_{\mathcal{A}}(X)^{\nabla}$$

is thus those elements of the previous set, which admit an $\mathcal{A}$*-connection.* In particular, *for* $n = 1$, viz. for *line sheaves* on $X$, one gets a *cohomological description* of the corresponding set in (vi), in terms of *(Čech) hypercohomology,* the same set being now an abelian group, in such a manner that one has;

$$(vii) \qquad \mathbf{\Phi}^1_{\mathcal{A}}(X)^{\nabla} = \check{\mathbf{H}}^1(X, \mathcal{A}^{\bullet} \xrightarrow{\tilde{\partial}} \Omega)$$

(see Theorems 18.1, 18.2 of Chapt. VI). Indeed, the proof of (vii) requires results from the *local theory of* $\mathcal{A}$*-connections,* which thus constitutes the subject matter of the next Chapter VII. The *local aspects* of the classical theory of connections can be repeated here too, realizing thus, once more, the *essential absence of the calculus,* in the standard sense, *apart,* of course, *from its formal properties* (!) (viz. the formal local properties of standard derivations). So one can formulate, within the present context, the pertinent notions of a *Riemannian* $\mathcal{A}$*-metric,* as well as, that one of a *hermitian* $\mathcal{A}$*-metric;* indeed, one can prove several results from the standard theory as,

for instance, the *Fundamental Lemma of Riemannian Geometry,* pertaining here to the *existence of* an $\mathcal{A}$*-connection compatible with a* given *Riemannian* $\mathcal{A}$*-metric,* for every *Riemannian vector sheaf,* when the corresponding *differential triad,* as in (i) above, is *suitably adapted* to the present (riemannian) framework.

On the other hand, by contrast with the previous situation, concerning the existence of an $\mathcal{A}$-connection, one realizes that, *given an* $\mathcal{A}$*-connection, we can always associate with it* what we might consider here as the analogue of the classical notion of *curvature.* Thus, the study of this concept, together with the deduction of several standard results, within the present axiomatic framework, connected with the latter notion, constitute the subject matter of Chapter VIII. In this concern, we further note that, in order to formulate the same notion, one has to supplement the initially given differential triad, as in (i) above, with what we call a *curvature datum*

$$(viii) \qquad (\mathcal{A}, \partial, \Omega^1, d, \Omega^2).$$

Thus, in effect, by another *"differential operator"*

$$(ix) \qquad d \equiv d^1 : \Omega \equiv \Omega^1 \longrightarrow \Omega^2 \equiv \wedge^2 \Omega^1,$$

satisfying the appropriate *compatibility conditions,* with respect to $\partial$. We also speak then of a *curvature space* $X$. So the curvature of a given $\mathcal{A}$-connection $D$ is now defined by the formula

$$(x) \qquad R(D) \equiv D^1 \circ D,$$

where $D^1$ is the *1st prolongation of* $D$, by analogy with the classical case, the previous operator $d^1 \equiv d$ being thus the *1st prolongation of* $\partial$. So one of the compatibility conditions between $d$ and $\partial$, referred to above, is the assumption that

$$(xi) \qquad R(\partial) \equiv d^1 \circ \partial \equiv d \circ \partial = 0,$$

that is, the *flatness of the* given $\mathcal{A}$*-connection* $\partial$ of $\mathcal{A}$ (in the strict sense of the word, or else that the *curvature of* $\partial$ is *zero*). We also examine *flat* $\mathcal{A}$*-connections,* in general, in the sense that the corresponding curvature, as in (x) above, vanishes. The notion of *flatness* is further related with the analogues here of other classical concepts, as, for example, with those of *complete parallelism, integrability* and *triviality* of the corresponding $\mathcal{A}$-connections; yet, with the *flatness of* a given *vector sheaf* $E$, relative to the so-called *co-*

*ordinate 1-cocycle of* $E$, consisting in that case of *locally constant* **C**-*valued maps* on $X$. (We have examined that notion in Volume I; Section 5, as well as, in the present Volume II, Chapters VI; Subsection 11.(a) and VII; Section 6). Several classical *"identities"*, connected with the notion of curvature, as, for instance, *Cartan's structural equation, Bianchi's identities,* including *Ricci's Lemma,* are also extended to the present abstract framework. Furthermore, within the same flavour of ideas, one also obtains, in full generality, *Weil's integrality theorem,* referred, as known, to a *geometrical characterization of* certain *integral cohomology classes* of a given (paracompact Hausdorff) space $X$. For the case at hand, certain appropriate cohomological conditions on $X$ and $\mathcal{A}$ should also be imposed, being, however, fulfilled in all the important cases (take, for instance, $X$, as before, and $\mathcal{A}$, as in (i), *fine*). One further application of the same material, presented hitherto, is the examination, in the next Chapter IX, of *characteristic classes,* following *Chern-Weil theory.* The corresponding here *Fundamental Theorem* (Chern-Weil), as well as, an appropriate extension of the classical *Chern-Weil homomorphism* are also in force.

Finally, Part Three of this treatise (Chapters X, XI) is concerned with further examples of the preceding material: Thus, in Chapter X we consider various particular cases, apart, of course, of the classical one, which, as already said, fits well within the present framework, this being, in effect, a natural extension of the former. So the same examples constitute special instances of our abstract context, being also by themselves *concrete generalizations* of the standard case of $\mathcal{C}^\infty$-*manifolds.* Thus, the so-called *Satake manifolds,* or even some particular instances of *infinite-dimensional* ($\mathcal{C}^\infty$-) *manifolds,* fall into the previous framework; namely, within that one of an abstract *differential triad accompanied with* the pertinent sequence of *differential operators,* so that one has, in effect, an *exact de Rham complex,* or else a *resolution of the constant sheaf* **C** (see e.g. (∗) in the preceding General Preface to both Volumes I, II).

In the final Chapter XI we discuss certain particular examples, which are actually deduced from *topological algebra theory;* of course, it is standard that the same algebra ${}^{\mathbf{C}}\mathcal{C}^\infty(X) \equiv \mathcal{C}^\infty(X)$ of a $\mathcal{C}^\infty$-*manifold* $X$ is a very convenient *topological algebra,* which also *cannot be a Banach algebra*! Thus, for a suitable class of topological algebras, the so-called *geometric (topological) algebras,* one can have, through the corresponding *Gel'fand sheaf* (on the *spectrum* of such an algebra), a situation similar to that one of the classical

theory. This, in particular, in what essentially concerns the *structure sheaf* $\mathcal{C}_X^\infty$ of a $\mathcal{C}^\infty$*-manifold* $X$, as before, and the corresponding *de Rham complex.* We consider, instead, the *de Rham-Kähler complex* (and/or its *sheafification*) that corresponds to a given *topological algebra.* However, this *does not yield a resolution of* $\mathbf{C}$, in general. On the other hand, topological algebras, as before, do have a nice behaviour, relative to *sheaf exponential sequences* (see, for instance, the same Chapter, Section 7), a very useful device, when dealing, in particular, with *line sheaves.* Yet, a way out of the above inconvenience is to consider some other differential triad, namely, that one, which we call the *Watts-Selesnick differential triad;* this can be associated with a given topological algebra of the pertinent type, general enough, however, to include several important examples, ranging, for instance, from the standard case of the $\mathcal{C}^\infty$-manifolds to the *Fourier algebra* of a (not necessarily abelian) *compact (Lie) group.* Again the final lesson here is that, one can have, through the previous technique, a potential, eventually useful substitute for the *standard* $\mathbf{C}$*-algebra of* $\mathcal{C}^\infty$*-functions* by a *more perceptive* (!) *function algebra,* as the case may be.

Anastasios Mallios

Athens
June, 1997

## Acknowledgements

It is a special pleasure to record here my indebtedness to all those, who, in one way or another, contributed to giving me the indispensable impetus and countenance to bring this project to an end. First, to *S. A. Selesnick* for his continuing interest in my work, in particular, in what concerns its relation with topological algebra theory; thus, his own earlier work in harmonic analysis on compact groups, written, within the context of Banach algebra theory, with a strong geometrical flavour, was, in effect, my initial inspiration for the abstract framework that has finally been adopted by the present study. Second, to *E. E. Rosinger,* who, during my visits to the University of Pretoria (South Africa), gave me great encouragement to finish this project, by his extremely enthusiastic reaction to early versions of this material. The final boost, and while the manuscript was almost ready, came from *S. Negrepontis,* who, as Head of our Department in Athens, at that time, offered me kindly the first financial support for converting the manuscript into LaTeX form. Further financial help to finishing the project has been supplied by the University of Athens, thanks, especially, to the kind willingness of *Y. Dracopoulos,* Vice-Rector of the University of Athens. My special thanks are due here to *Dr. M. Gikas* for his excellent rendering of a quite difficult text in LaTeX form, as well as, for an extremely friendly and helpful cooperation. My singular thanks are also due to *my colleagues* in the *Math. Department* of the *University of Athens*, who allowed me an extended leave of absence, very helpful, indeed, to accomplish this undertaking.

Yet, a quite decisive influence to attempt the present venture, in spite of appearing circumspections, as to its applicability, due mainly to the entirely abstract (sheaf-theoretic) language employed, was the very positive final facing of the subject from *various colleagues in Greece and abroad.* For all those I want to keep my very best and thankful memories. For a similar interest I wished, especially, to mention and thank from this place too, *Prof. Ky Fan,* who, as member of the Editorial Board of JMAA, was very kind, as well as, enthusiastic in handling my earlier papers on the subject, connected, in particular, with topological algebra theory, while his response, at that time, was among the first strong impulses for me to continue working on the topic. For their kind help, for the LaTeX writing, during the final corrections of the camera-ready copies of this two-volume work, I want also to thank my (cousins and) colleagues, at the Math. Dept. in Athens, *M. Papatriantafillou*

and *E. Vassiliou*. (For any remaining errors, typing or otherwise, it is me, of course, who is to be blamed, not including, certainly, LaTeX itself!).

Last, but not least, I wish to express my sincere appreciation to *Prof. M. Hazewinkel* for accepting this book to his Series, as well as, to *Dr. P. Roos* for his extremely kind interest, enthusiasm and editorial endeavour in bringing this work to the light.

# Contents of Volume I

# PART TWO

## GEOMETRY

# CHAPTER VI

# Geometry of Vector Sheaves. $\mathcal{A}$-connections

*From now on we proceed to the main theme of this treatise*

"...motivation for the study of a new concept may, in great part, lie in the clarification and simplification of more familiar notions."

John L. Kelley in J. L. Kelley et al., *"Linear Topological Spaces"* (Springer - Verlag, 1963). p.v.

"There are many advantages to developing a theory in the most general context possible."

R. Hartshorne in *"Algebraic Geometry"* (Springer - Verlag, 1977). p.59.

As already remarked below the heading, the present chapter marks, in effect, the beginning of our *abstract* (viz. *axiomatic*) *approach to the standard Differential Geometry of smooth* (i.e. $\mathcal{C}^{\infty}$-)*manifolds.* A crucial aspect hereby is that *we do not employ* for this *any differential analysis;* that is, *no calculus* is involved at all(!), our tools being just *sheaf theory,* in particular, *vector sheaves and their algebra* (linear and multilinear) and/or *sheaf cohomology,* to the extent that these issues have been developed in the preceding chapters. Nontheless, as we shall realize through the succeeding discussion, several fundamental results of the classical theory, global, as well as, local ones, can still

be derived, within the present abstract framework. We start thus right away by displaying the necessary preliminaries, in order next to define (cf. Section 2 below) the important (differential geometric) notion of a *connection,* or even, in our terminology, of an *$\mathcal{A}$-connection,* which essentially lies in the core of all the subsequent material; this, with the exception of Chapter X, which, however, again is just an auxiliary one of the next Chapter XI, where anew *$\mathcal{A}$-connections of a particular type* are further examined.

# 1. C-derivations. Differential triads

A basic notion for the subsequent discussion is that of a *"derivation";* as we shall presently see, this essentially concerns, in our case, an appropriate *sheaf morphism* (see Chapt. I; Definition 1.2). Yet, we shall refer, as usual, to a general **C**-*algebraized space* $(X, \mathcal{A})$, unless otherwise specified (see Chapt. II; Scholium 1.2, or even Chapt. IV; (0.1)). So we begin with the following.

**Definition 1.1.** Suppose we have a **C**-algebraized space $(X, \mathcal{A})$, and let $\mathcal{E}$ be an $\mathcal{A}$-module on $X$. Thus, by a **C**-*derivation of $\mathcal{A}$ in $\mathcal{E}$,* one means a sheaf morphism

$$\partial : \mathcal{A} \longrightarrow \mathcal{E}, \tag{1.1}$$

such that the two following conditions are satisfied:

1) $\partial$ is a **C**-*linear morphism* (relative to the corresponding **C**-vector sheaves $\mathcal{A}$, $\mathcal{E}$ on $X$; cf. Chapt. II; (1.38)).

2) $\partial$ satisfies the following *Leibniz condition*

$$\partial(s \cdot t) = s \cdot \partial(t) + t \cdot \partial(s), \tag{1.2}$$

for any (local) sections $s, t$ in $\mathcal{A}(U)$, with $U$ any open set in $X$.

Of course, concerning the above rel. (1.2), one essentially employs here the corresponding to (1.1) *section-morphism* (cf. Chapt. I; (12.49), (12.50))

$$\overline{\partial} : \Gamma(\mathcal{A}) \longrightarrow \Gamma(\mathcal{E}); \tag{1.3}$$

however, for simplicity's sake, the latter is *still denoted by* $\partial$ (by an obvious *abuse of notation,* a practice that we shall often apply in the sequel, as well, without specific mention).

Now, a triplet

$$(\mathcal{A}, \partial, \mathcal{E}), \tag{1.4}$$

as before, is called a *differential triad* on $X$.

Thus, by analogy with the classical case, we first have the following easy, however, useful.

**Lemma 1.1.** *For any given differential triad* $(\mathcal{A}, \partial, \mathcal{E})$ (cf. (1.4)), *the restriction of* $\partial$ *to the constant (algebra) sheaf* $\mathbf{C} \equiv \mathbf{C}_X \overset{\subsetneq}{\rightarrow} \mathcal{A}$ (see also Chapt. II; (1.34)) *is zero. That is, one has*

$$\partial|_{\mathbf{C}} = 0 \tag{1.5}$$

(viz. the *zero* **C**-*linear morphism*).

**Proof.** The assertion follows from applying $\partial$, first, to the *identity section* $1 \equiv 1_{\mathcal{A}} \in \mathcal{A}(X)$ (see Chapt. II; (1.22)); indeed, one has (cf. also (1.3), (1.2))

$$\partial(1_{\mathcal{A}}) \equiv \partial(1) = \partial(1 \cdot 1) = 1 \cdot \partial(1) + 1 \cdot \partial(1) = \partial(1) + \partial(1), \tag{1.6}$$

such that one obtains

$$\partial(1_{\mathcal{A}}) \equiv \partial(1) = 0. \tag{1.7}$$

Therefore, by the **C**-*linearity of* $\partial$ (Definition 1.1), one has

$$\partial(\lambda) := \partial(\lambda \cdot 1_{\mathcal{A}}) \equiv \partial(\lambda \cdot 1) = \lambda \cdot \partial(1) = \lambda \cdot 0 = 0, \tag{1.8}$$

for every *"constant section"* $\lambda \in \mathbf{C} \overset{\subsetneq}{\rightarrow} \mathbf{C}_X(X) \overset{\subsetneq}{\rightarrow} \mathcal{A}(X)$ (see also Chapt. I; (10.2), (11.15)), and this completes the proof. ∎

Yet, as another consequence of (1.2), one also obtains the following useful analogue of the classical case.

**Lemma 1.2.** *("Quotient rule"). Suppose we have a differential triad* $(\mathcal{A}, \partial, \mathcal{E})$, *as before. Then, for any local sections* $s \in \mathcal{A}(U)$ *and* $t \in \mathcal{A}(U)^{\cdot} = \mathcal{A}^{\cdot}(U)$ (cf. Chapt. IV; (1.15)), *with* $U$ *open in* $X$, *one has;*

$$\partial(s \cdot t^{-1}) = t^{-2} \cdot (t \cdot \partial s - s \cdot \partial t). \tag{1.9}$$

**Proof.** Indeed, based on (1.2), one has;

$$(1.10)\qquad \partial(s\cdot t^{-1}) = s\cdot\partial t^{-1} + t^{-1}\cdot\partial s.$$

But, $t\cdot t^{-1} = 1 \equiv 1_{\mathcal{A}}|_U \in \mathcal{A}^{\bullet}(U)$, so that one obtains (cf. (1.7))

$$(1.11)\qquad \begin{aligned}\partial(t\cdot t^{-1}) &= t\cdot\partial(t^{-1}) + t^{-1}\cdot\partial(t) = \partial(1) \equiv \partial(1_{\mathcal{A}}|_U) \equiv \partial_U(1_{\mathcal{A}}|_U)\\ &\equiv \partial_U\left(\rho_U^X(1_{\mathcal{A}})\right) = \lambda_U^X(\partial_X(1_{\mathcal{A}})) \equiv \lambda_U^X(\partial(1_{\mathcal{A}})) \equiv \lambda_U^X(\partial(1)) = 0\end{aligned}$$

(see also e.g. Chapt. I; (6.5), (12.49) and Chapt. II; (1.46), (1.67), or even (2.21)). Therefore, *one* first *obtains;*

$$(1.12)\qquad \partial(t^{-1}) = -t^{-2}\cdot\partial(t),$$

*for any given* $t\in\mathcal{A}^{\bullet}(U) = \mathcal{A}(U)^{\bullet}$ and *open* $U\subseteq X$, as above. Hence, based on (1.10), (1.12), one has

$$(1.13)\qquad \left(s\cdot t^{-1}\right) = s\cdot\left(-t^{-2}\cdot\partial t\right) + t^{-1}\cdot\partial s,$$

which obviously yields the desired relation (1.9). ∎

Now, we next consider for later use a *square-matrix extension of the* **C**-*derivation* $\partial$, as above (cf. (1.4)):

That is, by looking at the $\mathcal{A}$-*algebra sheaf* $M_n(\mathcal{A})$ (cf. Chapt. IV; (3.24)) and the $M_n(\mathcal{A})$-*module* $M_n(\mathcal{E})$ (see also (1.21) below), one defines a **C**-*derivation,* as follows (cf. also Chapt. V; (8.7));

$$(1.14)\qquad \overline{\partial}\equiv\partial : M_n(\mathcal{A}) \longrightarrow M_n(\mathcal{E}) := M_n(\mathcal{A})\otimes_{\mathcal{A}}\mathcal{E}$$

such that one sets;

$$(1.15)\qquad \overline{\partial}(\alpha)\equiv\partial(\alpha) := (\partial(\alpha_{ij})) \in M_n(\mathcal{E}(U))$$

*for any* $\alpha\equiv(\alpha_{ij})\in M_n(\mathcal{A}(U)) = M_n(\mathcal{A})(U)$ and $U$ *open* in $X$. To explain the notation applied in the last relation, we first remark that one can look at $M_n(\mathcal{E})$ as an $M_n(\mathcal{A})$-*module* by defining it as the *sheafification of the complete* $\Gamma(\mathcal{A})$-*presheaf* $M_n(\Gamma(\mathcal{E}))$; that is, one defines

$$(1.16)\qquad M_n(\mathcal{E}) := \mathbf{S}(M_n(\Gamma(\mathcal{E}))),$$

such that one sets, for every *open* $U\subseteq X$ (see also Chapt. II; (3.11))

$$(1.17)\qquad M_n(\Gamma(\mathcal{E}))(U) := M_n(\mathcal{E}(U)) = \mathcal{E}(U)^{n^2} = \mathcal{E}^{n^2}(U) = M_n(\mathcal{E})(U),$$

the last three relations above standing for $\mathcal{A}(U)$-*isomorphisms of* the $\mathcal{A}(U)$-*modules* involved. Yet, the last item in (1.17) indicates also an *equivalent*

*definition* of the range of (1.14); namely, one can define $M_n(\mathcal{E})$, *equivalently,* as the *sheafification of the* $\Gamma(\mathcal{A})$*-presheaf* (cf. Chapt. II; (5.5))

(1.18) $$U \longmapsto M_n(\mathcal{E}(U)) := M_n(\mathcal{A}(U)) \otimes_{\mathcal{A}(U)} \mathcal{E}(U) \cong \mathcal{E}(U)^{n^2},$$

where $U$ *varies over the open subsets of* $X$, along with the appropriate *restriction morphisms.* (For simplicity, we employed an obvious *abuse of notation,* concerning the range of (1.18), in connection with (1.17)). Now, it is easy to see that

(1.19) $M_n(\Gamma(\mathcal{E}))$, as given by (1.17), together with the corresponding restriction morphisms, *yields a complete* $\Gamma(\mathcal{A})$*-presheaf.*

Furthermore, one defines

(1.20) $M_n(\mathcal{E})$ as an $M_n(\mathcal{A})$*-module,* according to the relation;

(1.20.1) $$\alpha \cdot s = (\alpha_{ij}) \cdot (s_{ij}) := \left(\sum_{k=1}^{n} \alpha_{ik} \cdot s_{kj}\right) \in M_n(\mathcal{E}(U)),$$

*for any* $\alpha \equiv (\alpha_{ij}) \in M_n(\mathcal{A}(U)) = M_n(\mathcal{A})(U)$ *and* $s \equiv (s_{ij}) \in M_n(\mathcal{E}(U))$, with $U$ *open* in $X$.

Now, based on the very definitions, it is readily seen that;

(1.21) the map (1.14) yields a **C**-*linear morphism* of the $\mathcal{A}$-modules (in fact, by the hypothesis for $\mathcal{A}$, **C**-*vector space sheaves*) on $X$ concerned, while *it* also *fulfils Leibniz condition (1.2).* Therefore, one obtains a **C**-*derivation of the* **C**-*algebra sheaf* $M_n(\mathcal{A})$ *into the* $M_n(\mathcal{A})$*-module* $M_n(\mathcal{E})$ (see also Chapt. II; (1.38), Chapt. IV; (3.24), along with (1.20) above).

The preceding justify now completely our claim about (1.14). ∎ Yet, the triplet, thus defined, *for any given* $n \in \mathbf{N}$,

(1.22) $$(M_n(\mathcal{A}), \bar{\partial} \equiv \partial, M_n(\mathcal{E})),$$

is called the *n-th square matrix extension of the* given *differential triad* $(\mathcal{A}, \partial, \mathcal{E})$ on $X$. Finally, to emphasize the *functorial nature* of $M_n(\cdot)$, with $n \in \mathbf{N}$, in (1.22), it is still appropriate to set

(1.23) $$\bar{\partial} \equiv M_n(\partial),$$

as the *n-th square matrix extension of* $\partial$.

**1.(a). Logarithmic derivation.**– We examine next another fundamental notion for the sequel, associated with the previously defined one of a **C**-derivation (see Definition 1.1); it will constantly appear throughout the ensuing discussion, together with its matrix extension, analogous to (1.14) above. Thus, we first set the following.

**Definition 1.2.** Suppose we have a differential triad $(\mathcal{A}, \partial, \mathcal{E})$ (cf. (1.4)), and let $\mathcal{A}^{\cdot}$ be the (group) sheaf of units of $\mathcal{A}$ (cf. Chapt. IV; Lemma 1.1). Then, we call *logarithmic derivation,* associated with the given **C**-derivation $\partial$ (see (1.1)), the following sheaf morphism

$$\tilde{\partial} : \mathcal{A}^{\cdot} \longrightarrow \mathcal{E}, \tag{1.24}$$

given (in terms of sections, cf. Chapt. I; (13.19)) by the relation;

$$\tilde{\partial}(\alpha) := \alpha^{-1} \cdot \partial(\alpha), \tag{1.25}$$

for any (local) section $\alpha \in \mathcal{A}^{\cdot}(U) = \mathcal{A}(U)^{\cdot}$ (cf. Chapt. IV; (1.15)) and $U$ open in $X$.

Now, as a first application of (1.25), we remark that

$$\varepsilon(\mathbf{C}^{\cdot}) \subseteq \ker \tilde{\partial} \tag{1.26}$$

(see also (1.5) and Chapt. II; (1.34)). On the other hand, as as immediate consequence of (1.25) and (1.2), one realizes that;

(1.27) $\tilde{\partial}$ *is,* in effect, *a morphism of the group sheaves involved;* that is, one has

$$\tilde{\partial}(s \cdot t) = \tilde{\partial}(s) + \tilde{\partial}(t), \tag{1.27.1}$$

*for any* $s, t \in \mathcal{A}^{\cdot}(U) = \mathcal{A}(U)^{\cdot}$ and $U$ *open* in $X$. In particular, one obtains (cf. also (1.26));

$$\tilde{\partial}\left(\alpha^{-1}\right) = -\tilde{\partial}(\alpha), \tag{1.27.2}$$

*for any* $\alpha \in \mathcal{A}^{\cdot}(U)$, with $U$ *open* in $X$.

Now, by analogy with (1.14), one also defines, *for any* $n \in \mathbf{N}$, the *n-th (square) matrix extension of* $\tilde{\partial}$ (by abusing notation, we continue to denote

it by $\tilde{\partial}$, as well), as the following *sheaf morphism*

$$\tilde{\partial} : \mathcal{GL}(n, \mathcal{A}) \longrightarrow M_n(\mathcal{E}), \tag{1.28}$$

given by the relation;

$$\tilde{\partial}(\alpha) := \alpha^{-1} \cdot \partial(\alpha), \tag{1.29}$$

*for any* $\alpha \in \mathcal{GL}(n, \mathcal{A})(U) = \mathrm{GL}(n, \mathcal{A}(U))$ and $U$ *open* in $X$ (cf. (1.20.1) and Chapt. IV; (1.26), (1.28)).

Thus, based on (1.29) and (1.5), (1.15), one still obtains that

$$\tilde{\partial}|_{\mathcal{GL}(n,\mathbf{C})} = 0 \tag{1.30}$$

(see also Chapt. IV; (1.31), or even Chapt. V; (1.5) for the notation applied). Yet, in view of (1.29), (1.21) and (1.30), one further infers, extending thus (1.27.2), that;

$$\tilde{\partial}\left(\alpha^{-1}\right) = -\tilde{\partial}(\alpha), \tag{1.31}$$

*for every* $\alpha \equiv (\alpha_{ij}) \in \mathcal{GL}(n, \mathcal{A})(U) = \mathrm{GL}(n, \mathcal{A}(U))$ and $U$ *open* in $X$.

In fact, something more is actually true; namely, based on the *Leibniz condition for (1.14)* (cf. (1.2), (1.21)), written now in the form (*equivalent to (1.2), for* $n = 1$, due to the commutativity of $\mathcal{A}$)

$$\partial(s \cdot t) = s \cdot \partial(t) + \partial(s) \cdot t, \tag{1.32}$$

*for any* $s, t \in M_n(\mathcal{A}(U)) = M_n(\mathcal{A})(U)$ and *open* $U \subseteq X$, *one obtains*

$$\tilde{\partial}(s \cdot t) = \mathrm{Ad}(t^{-1}) \cdot \tilde{\partial}(s) + \tilde{\partial}(t), \tag{1.33}$$

*for any* $s \equiv (s_{ij}), t \equiv (t_{ij})$ *in* $\mathcal{GL}(n, \mathcal{A})(U) = \mathrm{GL}(n, \mathcal{A}(U))$ and $U$ *open* in $X$: Indeed, one has (cf. (1.29), (1.32));

$$\begin{aligned} \tilde{\partial}(st) &= (st)^{-1} \cdot \partial(st) = t^{-1}s^{-1}(s \cdot \partial(t) + \partial(s) \cdot t) \\ &= t^{-1} \cdot \partial(t) + t^{-1}\left(s^{-1} \cdot \partial(s)\right)t = \tilde{\partial}(t) + t^{-1} \cdot \tilde{\partial}(s) \cdot t, \end{aligned} \tag{1.34}$$

*which* thus *proves (1.33),* when setting

$$t^{-1} \cdot \tilde{\partial}(s) \cdot t \equiv \mathrm{Ad}\left(t^{-1}\right) \cdot \tilde{\partial}(s). \tag{1.35}$$

Here the multiplication in the first member of (1.35) stands for the usual (section) *matrix multiplication,* as defined by (1.20.1) (see also (1.14), (1.15)). In this respect, we still note that, by analogy with (1.20.1), one infers that;

(1.36) $M_n(\mathcal{E})$ *is*, in fact, *an* $M_n(\mathcal{A})$*-bimodule* (for every $\mathcal{A}$*-module* $\mathcal{E}$ on $X$).

The notation in the second member of (1.35) is reminiscent, of course, of the standard *adjoint representation*

$$\mathrm{Ad} : \mathcal{GL}(n, \mathcal{A}) \longrightarrow (\mathcal{E}nd\, \mathcal{E}^n)^{\cdot}, \tag{1.37}$$

provided e.g. (1.36) hold true; see also, for intance, (1.17), as well as, Chapt. II; (6.29) and Chapt. V; Scholium 8.2, for the notation applied in the range of (1.37). Yet, (1.37) has certainly a meaning as a *matrix representation* in the case of a *free* $\mathcal{A}$*-module* $\mathcal{E}$ (of *finite rank* $n \in \mathbf{N}$) or, at least, of a *locally free one* (viz. a *vector sheaf*) on $X$. On the other hand, in the general case, one still defines an *adjoint representation* (cf. also (1.14))

$$\mathrm{Ad} : \mathcal{GL}(n, \mathcal{A}) \longrightarrow \mathcal{E}nd\,(M_n(\mathcal{E}) \equiv M_n(\mathcal{A}) \otimes_{\mathcal{A}} \mathcal{E})^{\cdot}, \tag{1.38}$$

by the relation;

$$\mathrm{Ad}(\alpha)(s \otimes t) := \alpha s \alpha^{-1} \otimes t, \tag{1.39}$$

*for any* $\alpha \in \mathcal{GL}(n, \mathcal{A})(U) = \mathrm{GL}(n, \mathcal{A}(U))$, $s \in M_n(\mathcal{A})(U) = M_n(\mathcal{A}(U))$ *and* $t \in \mathcal{E}(U)$, with $U$ *open* in $X$. So *one sets*

$$\mathrm{Ad} := i \otimes \mathrm{id}_{\mathcal{E}}, \tag{1.40}$$

*such that*

$$i(\alpha)(s) := \alpha s \alpha^{-1}, \tag{1.42}$$

*for any* $\alpha$ *and* $s$, *as above.*

**Note.**– When, in place of (1.29), we set

$$\tilde{\partial}(\alpha) := \partial(\alpha) \cdot \alpha^{-1}, \tag{1.43}$$

*for any* $\alpha$, *as in (1.29)* (cf. also (1.36) and (1.20.1)), one obtains, instead of (1.33), the relation;

$$\tilde{\partial}(s \cdot t) = \tilde{\partial}(s) + \mathrm{Ad}(s)\tilde{\partial}(t), \tag{1.43}$$

*for any* $s, t$, *as in (1.33).*

Use of the above will be made later on.

# 2. A standard example: The smooth triad $(\mathcal{C}^\infty_X, d, \Omega^1_X)$

We consider next a concrete example of the previous abstract setting, which, together with its ramifications, constitute, in fact, our fundamental motivation for all the ensuing discussion. Thus, the *concrete differential triad* $(\mathcal{A}, \partial, \mathcal{E})$, under consideration, as already indicated in the heading, has now as follows:

Consider any (finite dimensional Hausdorff) $\mathcal{C}^\infty$*-manifold* $X$, along with the pair

$$(X, \mathcal{C}^\infty_X) \tag{2.1}$$

as a **C**-*algebraized space;* namely, one thus sets

$$\mathcal{A} \equiv \mathcal{C}^\infty_X := {}^{\mathbf{C}}\mathcal{C}^\infty_X, \tag{2.2}$$

viz. the *sheaf of germs of* **C**-*valued* $\mathcal{C}^\infty$*-functions on* $X$ (cf. Chapt. I; (4.39)) is now taken as the *structure sheaf* of the concrete pair $(X, \mathcal{A})$ in question (cf. Chapt. II; Scholium 1.2), being certainly a **C**-*algebra sheaf* on $X$ (see Chapt. II; Definition 1.4, (1.67), as well as, Chapt. I; (11.11), ii) and (11.12)). In this respect, see also Chapt. III; (8.27), (8.29). Furthermore, one sets

$$\mathcal{E} \equiv \Omega^1_X, \tag{2.3}$$

that is, one further considers the $\mathcal{C}^\infty_X$*-module* (in fact, *vector sheaf*) on $X$, as this is given by the *sheaf of germs of* **C**-*valued smooth* (viz. $\mathcal{C}^\infty$-) *1-forms on* $X$; so one has, by definition,

$$\Omega^1_X := \mathbf{S}(\Gamma(\mathcal{T}^*_{\mathbf{C}}(X))). \tag{2.4}$$

Here $\mathcal{T}^*_{\mathbf{C}}(X)$ stands for the *complexified cotangent bundle of* $X$, such that $\Gamma(\mathcal{T}^*_{\mathbf{C}}(X))$ denotes, of course, the *(complete) presheaf of sections* of the (**C**-vector) bundle in question, being actually isomorphic to that one of the *sheaf (of germs) of sections of* $\mathcal{T}^*_{\mathbf{C}}(X)$. (In this concern, see also Chapt. III; (8.30), for $p = 1$, as well as, Chapt. XI in the sequel). Yet, we note, in passing, that (2.4) is a *locally free* $\mathcal{C}^\infty_X$*-module* on $X$, *of rank* $n = \dim X < +\infty$,

viz. a *vector sheaf* on $X$, a consequence, in effect, of the very definitions (see, for instance, Chapt. III; (8.34), still for $p = 1$; yet, cf. Section 4 in the sequel). Finally, as a **C**-*derivation* $\partial : \mathcal{A} \longrightarrow \mathcal{E}$, with $\mathcal{A}$ and $\mathcal{E}$ given by (2.2) and (2.3), respectively, one now sets

$$d : \mathcal{C}^{\infty}_{X} \longrightarrow \Omega^{1}_{X}, \tag{2.5}$$

where the above *sheaf morphism* is defined, *in terms of* the corresponding *sections* of the sheaves concerned (cf. Chapt. I; (13.19) and (2.17)); namely, to each **C**-valued (local) $\mathcal{C}^{\infty}$-function of $X$ one assigns its *differential,* being thus, by the same definitions, a (local $\mathcal{C}^{\infty}$-) 1-form on $X$. On the other hand, it is still a direct consequence of the very definitions that *(2.5) is* indeed *a* **C**-*derivation,* as indicated, in the sense of Definition 1.1 in the preceding.

Thus, one obtains the following triplet

$$(\mathcal{C}^{\infty}_{X}, d, \Omega^{1}_{X}), \tag{2.6}$$

called the *standard* $\mathcal{C}^{\infty}$*-triplet,* or even the *standard differential triad,* associated, as above, with the given $\mathcal{C}^{\infty}$-manifold $X$.

Another *concrete* (however, non standard!) example of a *differential triad* will be given in the sequel, taken from *topological algebra theory* (see Chapter XI) yet, general enough to specialize to several standard examples. In particular, the familiar *topological* **C**-*algebra* $\mathcal{C}^{\infty}_{\mathbf{C}}(X) = \Gamma(X, \mathcal{C}^{\infty}_{X})$ is here included. Furthermore, other examples, similar to (2.6), will also be given in the case of *generalized manifolds* (cf. Chapter X).

# 3. $\mathcal{A}$-connections. Basic definitions

We come now to consider one of the most important notions throughout the whole treatise. However, since a certain standard notation will be applied in the succeeding discussion, with obvious, for that matter, modifications in certain particular cases, we first set the following.

**Convention.**– Henceforth, we shall denote by

$$(\mathcal{A}, \partial, \Omega) \tag{3.1}$$

a fixed *differential triad* (cf. (1.4)), with respect to a fixed also **C**-algebraized space $(X, \mathcal{A})$, that was usually the case in the preceding, as well. Thus, by definition (see (1.1)), $\Omega$ stands here for a particular *$\mathcal{A}$-module* on $X$, range of the given **C**-*derivation*

$$\partial : \mathcal{A} \longrightarrow \Omega. \tag{3.2}$$

The notation hereby is, of course, reminiscent of the standard example (2.6), as above. Yet, further suitable alterations of this particular "$\Omega$" will be used in the sequel, as the case may be. (See thus, for instance, Section 4 below, as well as, in particular, the next Chapter VII).

In this context, we shall also employ throughout the sequel the following convenient abreviation, which, in a sense, has already been used in the preceding, as well (see thus e.g. (1.14)). So, for any given *$\mathcal{A}$-module* $\mathcal{E}$ on $X$, one sets;

$$\Omega(\mathcal{E}) := \mathcal{E} \otimes_{\mathcal{A}} \Omega = \Omega \otimes_{\mathcal{A}} \mathcal{E}, \tag{3.3}$$

the second member of (3.3) being, of course, by definition, an *$\mathcal{A}$-module* on $X$ too (cf. Chapt. II; Definition 5.1); yet, the equality thereby stands, in fact, for an *$\mathcal{A}$-isomorphism of* the *$\mathcal{A}$-modules* concerned (ibid.; (5.14)).

Thus, we come now to the definition of an important notion for all the succeeding discussion, as we have already mentioned in the foregoing. That is, one has the following.

**Definition 3.1.** Let $(\mathcal{A}, \partial, \Omega)$ be a given differential triad (cf. (3.1)) and $\mathcal{E}$ an arbitrary $\mathcal{A}$-module on $X$. Then, by an *$\mathcal{A}$-connection* of $\mathcal{E}$, one means a sheaf morphism (see also (3.3))

$$D : \mathcal{E} \longrightarrow \Omega(\mathcal{E}) \equiv \mathcal{E} \otimes_{\mathcal{A}} \Omega, \tag{3.4}$$

such that the two following conditions are fulfilled:

i) $D$ is a **C**-*linear morphism* of the **C**-vector space sheaves concerned; viz. one has

$$D(\lambda s + \mu t) = \lambda \cdot D(s) + \mu \cdot D(t), \tag{3.5}$$

for any (local) sections $s, t \in \mathcal{E}(U)$, with $U$ open in $X$, and $\lambda, \mu \in \mathbf{C} \equiv \mathbf{C}(U) \overset{\subseteq}{\rightarrow} \mathbf{C}_X(U)$.

ii) $D$ satisfies the following *Leibniz condition;*

$$D(\alpha \cdot s) = \alpha \cdot D(s) + s \otimes \partial(\alpha), \tag{3.6}$$

for any (local) sections $\alpha \in \mathcal{A}(U)$, and $s \in \mathcal{E}(U)$, with $U$ open in $X$,

We first comment a bit more on the previously applied notation: So the *properties (3.5) and (3.6)* of the given sheaf morphism ($\mathcal{A}$-connection) $D$, as in (3.4), *are* actually *referred to presheaves, generating* the sheaves and, in particular,, the $\mathcal{A}$-modules concerned. Thus, one has

$$\mathcal{E} = \mathbf{S}(\Gamma(\mathcal{E})), \quad \Omega(\mathcal{E}) = \mathbf{S}(\Gamma(\mathcal{E}) \otimes_{\Gamma(\mathcal{A})} \Gamma(\Omega)) \quad \textit{and} \quad \mathbf{C}_X = \mathbf{S}(\mathbf{C}) \tag{3.7}$$

(see Chapt. I; (9.17), (11.21) and Chapt. II; (5.7)). Therefore, the said properties are further extended to the sheaves in question, based on Chapt. I; (12.42), (8.11) and Chapt. II; (2.1) – (2.3) and (1.44). In this context, we also recall that, since $\mathcal{A}$ is, by assumption, a **C**-*algebra sheaf* on $X$ (cf. e.g. Chapt. IV; (0.1)), *each $\mathcal{A}$-module on $X$ is,* in particular, *a* **C**-*vector space sheaf* on $X$ (see Chapt. II; (1.38)).

Yet, concerning the *tensor product section* appeared in the second member of (3.6), one has, of course (cf. (3.7)), that;

$$s \otimes \partial(\alpha) \in \mathcal{E}(U) \otimes_{\mathcal{A}(U)} \Omega(U). \tag{3.8}$$

That is, we actually employ here, as usual, for simplicity, an obvious *abuse of notation,* by writing just $s \otimes \partial(\alpha)$, whereas the *(decomposable) tensor* at issue is always meant with respect to a certain **C**-*algebra of coefficients* (here $\mathcal{A}(U)$), different, in general, from **C** itself. However, to simplify things, as already said, we steadily follow this practice (see also e.g. Chapt. IV; (6.5) and the ensuing comments thereby). The above clarify now completely the notation applied in the above Definition 3.1.

On the other hand, regarding the employed term $\mathcal{A}$-*connection,* as above, we still remark that, since the latter notion is determined modulo the given differential triad, as in (3.1), the term $(\mathcal{A}, \partial, \Omega)$-*connection* of the $\mathcal{A}$-module $\mathcal{E}$ involved would be more appropriate, nevertheless somewhat awkward, as well. Thus, for simplicity, we shall retain the term, already used above, although the name just *connection* will also be employed, occasionally, in the sequel, the determination "$\mathcal{A}$" being, otherwise, easily understood from the context. Yet, the term $\partial$-*connection* might be applied for $D$ in (3.4), as well, by referring thus, of course, to the **C**-derivation, in effect, connection (cf. the next Section 4) $\partial$ of $\mathcal{A}$ involved; moreover, this would be also in agreement with another usage of the term. (See, for instance, M. Cornalba

- P. Griffiths [1]).

Now, before we proceed further to the study of $\mathcal{A}$-connections, along with the applications, we have in mind, within the present abstract setting, of this fundamental notion, we first consider right below certain particular instances of $\mathcal{A}$-connections. Further examples will be encountered throughout the succeeding discussion.

# 4. Examples of $\mathcal{A}$-connections

We examine below a few particular examples of $\mathcal{A}$-connections, which we have at our disposal from what has been already considered so far.

**4.(a). The standard (flat) $\mathcal{A}$-connection.**– As a prelude to more sophisticated instances of $\mathcal{A}$-connections, we first look at *the* given **C**-*derivation* $\partial : \mathcal{A} \longrightarrow \Omega$, *as an $\mathcal{A}$-connection* of $\mathcal{A}$ in $\Omega$ (see (3.1)):

Thus, we first note the following (canonical) *$\mathcal{A}$-isomorphism* (cf. also (3.3))

(4.1) $$\Omega(\mathcal{A}) \equiv \mathcal{A} \otimes_{\mathcal{A}} \Omega = \Omega$$

(cf. Chapt. II; (5.15)). Therefore, *the* given **C**-*derivation* (see (3.2))

(4.2) $$\partial : \mathcal{A} \longrightarrow \Omega = \Omega(\mathcal{A})$$

*provides,* in effect, *an $\mathcal{A}$-connection of* (the $\mathcal{A}$-module) $\mathcal{A}$ *in* $\Omega$, as this is easily verified by the same definitions. See thus Definition 1.1, while we also note hereby the following obvious relation in (1.2);

(4.3) $$t \cdot \partial(s) = \partial(s) \cdot t \equiv t \otimes \partial(s),$$

due to the (canonical) *$\mathcal{A}$-isomorphism* $\mathcal{A} \otimes_{\mathcal{A}} \mathcal{A} = \mathcal{A}$ (again Chapt. II; (5.15)), which thus justifies completely our claim for (4.2).

On the other hand, the geometrical term *flat,* which has been ascribed to $\partial$ will also be fully vindicated later on (see Chapter VIII below), referred, in fact, as we shall see (ibid.; (2.6)), to the *(zero) curvature* of the $\mathcal{A}$-connection $\partial$ in question.

**4.(b). $\mathcal{C}_X^\infty$-connections.**– Consider a finite-dimensional (Hausdorff) $\mathcal{C}^\infty$-*manifold* $X$, and let $\mathbf{E}$ be a $\mathcal{C}^\infty$-*complex vector bundle,* or *smooth* $\mathbf{C}$-*vector bundle,* of *rank* $n$ over $X$. Moreover, let $\mathcal{E}$ be the *sheaf of germs of* $\mathcal{C}^\infty$-*sections of* $\mathbf{E}$; that is, one has

$$\mathcal{E} := \mathbf{S}(\Gamma(\mathbf{E})), \tag{4.4}$$

such that

$$\Gamma^\infty(\mathbf{E}) \equiv \Gamma(\mathbf{E}) \tag{4.5}$$

stands for the *complete presheaf of sections of* $\mathbf{E}$, as given by the assignement

$$U \longmapsto \mathbf{E}(U) := \mathcal{C}^\infty(U, \mathbf{E}|_U)_s, \textit{ with } U \textit{ open in } X, \tag{4.6}$$

along with the obvious *restriction maps* (morphisms); the index "$s$" in the range of (4.6) is for *section* (see also (4.8) in the sequel). Thus, concerning the same set, we first remark that;

(4.7) by looking at the *indexed set of proper open coverings of* $X$ (see e.g. Chapt. III; (4.9)), one can always obtain a *cofinal subset* of it *such that*

$$\mathbf{E}|_U = U \times \mathbf{C}^n, \tag{4.7.1}$$

within a *diffeomorphism,* for every open set $U \subseteq X$ of the (open) covering of $X$, belonging to the particular cofinal set of coverings of $X$, as before. Something more, we may assume that the previous open sets, as in (4.7.1), belong to a (maximal) $\mathcal{C}^\infty$-*atlas of* $X$. We call such an open covering, a *coordinate open covering of* $X$.

In this context, we still note, in passing, that in view of (4.4), one has (see also Chapt. I; Proposition 11.1);

$$\Gamma(\mathbf{E})(U) = \mathbf{E}(U) \cong \mathcal{E}(U), \tag{4.7.2}$$

(within an *isomorphism* of the spaces involved), *for every open set* $U \subseteq X$.

Thus, for an open covering of X, as in (4.7), one obtains;

$$\begin{aligned}\mathcal{C}^{\infty}(U,\mathbf{E}|_U)_s &= \mathcal{C}^{\infty}(U, U\times \mathbf{C}^n)_s = \mathcal{C}^{\infty}(U,\mathbf{C}^n)\\ &= \mathcal{C}^{\infty}(U,\mathbf{C})^n \equiv \mathcal{C}^{\infty}_{\mathbf{C}}(U)^n \equiv \mathcal{C}^{\infty}(U)^n,\end{aligned} \tag{4.8}$$

the last two equalities denoting, in fact, *isomorphisms of* the **C**-*vector spaces* concerned, while the last item in (4.8) stands for the standard *free* $\mathcal{C}^{\infty}(U)$-*module* of *rank* $n$. Of course, *(4.8) is valid, for any open* $V \subseteq U$. Therefore, in view of (4.8), one concludes that

$$\Gamma(\mathbf{E})|_U = \Gamma(\mathbf{E}|_U) = \Gamma((\mathcal{C}^{\infty}_X)^n|_U) = \Gamma((\mathcal{C}^{\infty}_X)^n)|_U, \tag{4.9}$$

*for every open set* $U \subseteq X$, *as in (4.7.1)* (see also Chapt. I; (11.41) and Chapt. II; (3.7)). Hence, *one obtains* (cf. (4.4), as well as, Chapt. I; (9.23), (9.17));

$$\begin{aligned}\mathcal{E}|_U &= \mathbf{S}(\Gamma(\mathbf{E}))|_U = \mathbf{S}(\Gamma(\mathbf{E})|_U) = \mathbf{S}(\Gamma((\mathcal{C}^{\infty}_X)^n)|_U)\\ &= \mathbf{S}(\Gamma((\mathcal{C}^{\infty}_X)^n))|_U = (\mathcal{C}^{\infty}_X)^n|_U)\end{aligned} \tag{4.10}$$

up to a $\mathcal{C}^{\infty}_X|_U$-*isomorphism of* the $\mathcal{C}^{\infty}_X|_U$-*modules* involved, *for any open* $U \subseteq X$, *as above.* Thus, in other words, one infers that

(4.11) $\mathcal{E} \equiv \mathbf{S}(\Gamma(\mathcal{E}))$ *is a locally free* $\mathcal{C}^{\infty}_X$-*module on* $X$, *of rank* $n = \operatorname{rk}\mathbf{E} \in \mathbf{N}$; that is, a *vector sheaf* on $X$.

In this regard, see also Chapt. XI for an ample generalization of (4.11).

Now, for an *integer* $p \geqq 1$, let

$$\Omega^p(X,\mathbf{E}) \equiv \Omega^p(\mathbf{E}) \tag{4.12}$$

be the *sheaf of* **E**-*valued smooth* (viz. $\mathcal{C}^{\infty}$-) *p-forms* on $X$; thus, modulo a $\mathcal{C}^{\infty}_X$-*isomorphism,* one has

$$\Omega^p(\mathbf{E}) = \mathbf{S}(\Gamma(\Omega^p) \otimes_{\Gamma(\mathcal{C}^{\infty}_X)} \Gamma(\mathbf{E})), \tag{4.13}$$

such that one sets

$$\Omega^p \equiv \Omega^p_X := \bigwedge^p \Omega^1_X, \tag{4.14}$$

this standing for the *p-th exterior power of the* $\mathcal{C}^{\infty}_X$-*module* $\Omega^1_X$ (see (2.4) above, as well as, Chapt. IV; (7.1), (7.3)).

So, *for* $p = 1$, a $\mathcal{C}^{\infty}_X$-*connection,* for the case at hand, is given by the following *sheaf morphism* (cf. (4.4), (4.13))

$$D : \mathcal{E} \longrightarrow \Omega^1_X(\mathcal{E}) \equiv \Omega^1(\mathcal{E}) = \mathcal{E} \otimes_{\mathcal{C}^{\infty}_X} \Omega^1 = \Omega^1 \otimes_{\mathcal{C}^{\infty}_X} \mathcal{E}, \tag{4.15}$$

which further satisfies the conditions of Definition 3.1. Thus, (4.15) is, in fact, a $\mathcal{C}^{\infty}$*-connection,* or just a *connection, on the* given *smooth* **C**-*vector bundle* **E**, in the standard sense of Differential Geometry. In particular, suppose that **E**, as before, is the *complexified tangent bundle* of $X$,

$$\mathcal{T}_{\mathbf{C}}(X) := \mathcal{T}(X) \otimes_{\mathbf{R}} \mathbf{C}, \tag{4.16}$$

that is, the result of the *"complexification functor"*, applied on the *tangent bundle* of $X$, $\mathcal{T}(X)$ (the latter being, of course, a *smooth* **R**-*vector bundle* on $X$. See e.g. M. Karoubi [1: Example 4.8,e) p.21]). So in that case, a $\mathcal{C}_X^{\infty}$-connection on (4.16), by analogy with (4.15), is a $\mathcal{C}^{\infty}$*-connection,* or simply a *connection, on the* given $\mathcal{C}^{\infty}$*-manifold* $X$. (In this respect, see also, for instance, S. Kobayashi [1: p.1], or R.O. Wells [1: p. 70, Definition 1.5], or even J.W. Milnor - J.D. Stasheff [1: p. 289]).

The above example will often be used in the sequel too, as a link of several abstract results, within the present framework, with corresponding standard ones in Differential Geometry and/or Complex Manifold Theory. Further examples of $\mathcal{A}$-connections, along the lines of the classical theory, are considered right below.

# 5. Induced $\mathcal{A}$-connections

We examine below certain particular cases of $\mathcal{A}$-connections, which will be of special use throughout the succeeding discussion, corresponding, as already said, to *familiar instances of* the classical theory of $\mathcal{C}^{\infty}$*-connections,* according to the above Example 4.(b). Thus, we begin with the following standard, otherwise, notion.

**5.(a). Whitney sum of $\mathcal{A}$-connections.**– We consider, as usual, a basic *differential triad* $(\mathcal{A}, \partial, \Omega)$ over a given *topological space* $X$, common base space of all the sheaves, involved in our discussion, and let

$$(\mathcal{E}_i)_{1 \leqq i \leqq n} \tag{5.1}$$

be a finite sequence of $\mathcal{A}$*-modules* on $X$, along with a corresponding (finite) sequence of $\mathcal{A}$*-connections* on $\mathcal{E}_i$, $i = 1, \ldots, n$, respectively;

(5.2) $$(D_i)_{1 \leqq i \leqq n}.$$

Then, one concludes that;

(5.3) the *sheaf morphism*

(5.3.1) $$D := \bigoplus_{i=1}^n D_i : \bigoplus_{i=1}^n \mathcal{E}_i \equiv \mathcal{E} \longrightarrow \mathcal{E} \otimes_{\mathcal{A}} \Omega \equiv \Omega(\mathcal{E}) \equiv \Omega\left(\bigoplus_i \mathcal{E}_i\right) = \bigoplus_{i=1}^n (\mathcal{E}_i \otimes_{\mathcal{A}} \Omega) \equiv \bigoplus_i \Omega(\mathcal{E}_i),$$

which is given by

(5.3.2) $$D(s) \equiv D(s_1 \oplus \cdots \oplus s_n) := \bigoplus_{i=1}^n D_i(s_i) \in \bigoplus_i \Omega(\mathcal{E}_i) \equiv \Omega(\mathcal{E}),$$

*for any* given (local) *sections* $s_i \in \mathcal{E}(U)$, $1 \leq i \leq n$, with $U$ *open* in $X$ (so that $s = \bigoplus_{i=1}^n s_i \in (\bigoplus_i \mathcal{E}_i)(U) \equiv \mathcal{E}(U)$, as above), is an $\mathcal{A}$-*connection* on $\mathcal{E} \equiv \bigoplus_{i=1}^n \mathcal{E}_i$.

Concerning the notation applied above, see Chapt. II; (3.7), (3.14) and (5.17). Now, the assertion in (5.3) is an immediate consequence of (5.3.2) and Definition 3.1. ∎ Yet, the sheaf morphism $D$, as in (5.3.1), is also called the *Whitney sum of the $\mathcal{A}$-connections* $D_i$, $1 \leq i \leq n$.

On the other hand, an important *particular example of (5.3)*, which will also be applied in the sequel, is the $\mathcal{A}$-*connection*

(5.4) $$\partial^n := \underbrace{\partial \oplus \cdots \oplus \partial}_{n-\text{times}} : \mathcal{A}^n \longrightarrow \mathcal{A}^n \otimes_{\mathcal{A}} \Omega \equiv \Omega(\mathcal{A}^n) = \Omega^n \; (\equiv \Omega^{(n)})$$

(see also Chapt. II; (5.18)), *n-th Whitney sum of the* given *standard (flat)* $\mathcal{A}$-*connection* $\partial$ (cf. Example 4.(a); (4.2)), defined thus on the free $\mathcal{A}$-module $\mathcal{A}^n$, according to (5.3.2); that is, one sets

(5.5) $$\partial^n(s \equiv (s_i)) := \bigoplus_{i=1}^n \partial(s_i) \equiv (\partial(s_i))_{1 \leqq i \leqq n} \in \Omega(U)^n = \Omega^n(U),$$

*for any* $s \equiv (s_i) \in \mathcal{A}^n(U) = \mathcal{A}(U)^n$, and $U$ *open* in $X$. Yet, due to the particular form of $\partial^n$ (cf. (5.4)), the same is still named the *n-th power* (or even the *n-th dimensional extension*) *of* $\partial$.

So, for every *integer* $n \geq 2$, one considers the *n-th dimensional extension of the* given *differential triad* $(\mathcal{A}, \partial, \Omega)$, that is, the triple

(5.6) $$(\mathcal{A}^n, \partial^n, \Omega^n).$$

Of course, (5.6) has a special bearing on the *square-matrix extension of* $\partial$, already considered in the preceding by (1.22) (cf. (1.23) too); in this regard, see also (1.15), (1.18), together with Chapt. IV; (3.23) and Chapt. II; (5.18).

Now, more generally, by extending the situation in (5.4), one can look at the *n-th dimensional extension of an* $\mathcal{A}$*-connection* $D$ on a given $\mathcal{A}$*-module* $\mathcal{E}$ on $X$. So one has the triple

$$(\mathcal{E}^n, D^n, \Omega(\mathcal{E}^n)), \tag{5.7}$$

extending (5.6), such that one sets;

$$\begin{aligned} D^n(s \equiv (s_i)) &:= (D(s_i)) \equiv \bigoplus D(s_i) \in \bigoplus_{i=1}^n (\mathcal{E}_i \otimes_{\mathcal{A}} \Omega) \\ &= (\bigoplus_i \mathcal{E}_i) \otimes_{\mathcal{A}} \Omega = \mathcal{E}^n \otimes_{\mathcal{A}} \Omega \equiv \Omega(\mathcal{E}^n) = \Omega(\mathcal{E})^n, \end{aligned} \tag{5.8}$$

*for any* $s = (s_i) \in \mathcal{E}^n(U) = \mathcal{E}(U)^n$, with $U$ *open* in $X$ (see also Chapt. II; (3.7), (5.17) and (3.14)).

**Scholium 5.1.**– Based on (5.3.2), it is straightforward to consider *any family of summands (not necessarily finite)* in a *Whitney sum of* $\mathcal{A}$*-connections.*

The interest here lies, among other things, in the fact that, through e.g. such an *infinite Whitney sum of vector sheaves,* one has an example of an $\mathcal{A}$*-module (not necessarily a vector sheaf) endowed with an* $\mathcal{A}$*-connection.* Further examples of the same sort, as before, are also obtained by considering, for instance, *projective limits of vector sheaves with* $\mathcal{A}$*-connections* (E. Vassiliou), or *inductive limits* of such. We leave, however, the technical details hereby to the reader.

**5.(b). Tensor product of $\mathcal{A}$-connections.**– As usual, we tacitly assume, here too, the existence of a *differential triad* $(\mathcal{A}, \partial, \Omega)$ (see (3.1)) on a *topological space* $X$, while we further suppose that we are given two $\mathcal{A}$*-modules* $\mathcal{E}$ and $\mathcal{F}$ on $X$, endowed with $\mathcal{A}$*-connections* $D_{\mathcal{E}} \equiv D$ and $D_{\mathcal{F}} \equiv D'$. So our task is to define the pair

$$(\mathcal{E} \otimes_{\mathcal{A}} \mathcal{F}, D_{\mathcal{E} \otimes_{\mathcal{A}} \mathcal{F}}); \tag{5.9}$$

that is, to consider on the *tensor product* $\mathcal{A}$*-module* $\mathcal{E} \otimes_{\mathcal{A}} \mathcal{F}$ (cf. Chapt. II; (5.6)), an $\mathcal{A}$*-connection* $D_{\mathcal{E} \otimes_{\mathcal{A}} \mathcal{F}} \equiv \widetilde{D}$, which is actually given by the relation

$$\widetilde{D} \equiv D_{\mathcal{E} \otimes_{\mathcal{A}} \mathcal{F}} := (D_{\mathcal{E}} \otimes 1_{\mathcal{F}}) + (1_{\mathcal{E}} \otimes D_{\mathcal{F}}) \equiv D \otimes 1 + 1 \otimes D', \tag{5.10}$$

where $1_{\mathcal{E}}, 1_{\mathcal{F}}$ denote the *identity $\mathcal{A}$-automorphisms* of the $\mathcal{A}$-modules concerned. (In this regard, we still note that the notation "$\otimes$" in the second member of (5.10) stands, in effect, for an *"adorned tensor product"*, viz. for "$\otimes_{\mathcal{A}}$"; see also the comments about (3.8) in the preceding).

Now, our claim on (5.10) is, in fact, a straightforward implication of the same definitions (see thus Definition 3.1). So we check, for instance, the corresponding *Leibniz condition* for (5.10) (see (3.6)), by first applying *decomposable tensors (sections);* that is, one has

$$(5.11)\quad \begin{aligned} \widetilde{D}(\alpha\cdot(s\otimes t)) &= \widetilde{D}(s\otimes \alpha t) = [(D\otimes 1)+(1\otimes D')](s\otimes \alpha t) \\ &= (D\otimes 1)(s\otimes \alpha t)+(1\otimes D')(s\otimes \alpha t) = Ds\otimes \alpha t + s\otimes D'(\alpha t) \\ &= \alpha\cdot(Ds\otimes t)+s\otimes(\alpha D't + t\otimes \partial\alpha) = \alpha\cdot(Ds\otimes t + s\otimes D't)+s\otimes t\otimes \partial\alpha \\ &= \alpha\cdot \widetilde{D}(s\otimes t)+(s\otimes t)\otimes \partial\alpha, \end{aligned}$$

*for any* (local) *sections* $\alpha \in \mathcal{A}(U)$, $s \in \mathcal{E}(U)$ *and* $t \in \mathcal{F}(U)$, with $U$ *open* in $X$. The general case follows now *extending (5.11) by linearity,* which thus proves our assertion. ∎

**5.(c). $\mathcal{A}$-connections vis-à-vis to the functor $\mathcal{H}om_{\mathcal{A}}$.**– Suppose we are given the *same data, as in 5.(b)* above. Thus, our claim now is the following;

(5.12) under the previous hypothesis and the supplementary assumption that $\mathcal{E}$ is a *vector sheaf* on $X$, *one can define an $\mathcal{A}$-connection on the $\mathcal{A}$-module* $\mathcal{H}om_{\mathcal{A}}(\mathcal{E},\mathcal{F})$ on $X$, as follows:

That is, the sought $\mathcal{A}$-connection is supplied by the relation

$$(5.13)\qquad D_{\mathcal{H}om_{\mathcal{A}}(\mathcal{E},\mathcal{F})}(\phi) := D_{\mathcal{F}}\circ\phi - (\phi\otimes 1_{\Omega})\circ D_{\mathcal{E}},$$

where $\phi$ stands for any (local) section of $\mathcal{H}om_{\mathcal{A}}(\mathcal{E},\mathcal{F})$; viz. we assume that

$$(5.14)\qquad \phi \in \mathcal{H}om_{\mathcal{A}}(\mathcal{E},\mathcal{F})(U) = \mathrm{Hom}_{\mathcal{A}|_U}(\mathcal{E}|_U,\mathcal{F}|_U)$$

for some *open* $U \subseteq X$ (see Chapt. II; (6.8)), while $1_{\Omega}$ in (5.13) denotes the *identity $\mathcal{A}$-automorphism* of the given *$\mathcal{A}$-module* $\Omega$ on $X$. Yet, applying, for convenience, the following abbreviated notation,

$$(5.15)\qquad \widetilde{D} \equiv D_{\mathcal{H}om_{\mathcal{A}}(\mathcal{E},\mathcal{F})},\quad D \equiv D_{\mathcal{E}} \ \textit{and}\ D' \equiv D_{\mathcal{F}},$$

the defining relation (5.13) is written;

$$\widetilde{D}(\phi) := D' \circ \phi - (\phi \otimes 1_{\Omega}) \circ D \tag{5.16}$$

with $\phi$, *as in (5.14)*. In this regard, the following comments on (5.13) (or (5.16)) are still in order:

Namely, suppose that $s \in \mathcal{E}(U)$, with $U$ open in $X$; then setting, for simplicity,

$$\phi(s) \equiv \overline{\phi}(s) := \phi \circ s \tag{5.17}$$

(see Chapt. I; (12.49), (12.50)), one obtains

$$D_{\mathcal{F}}(\phi(s)) \in \mathcal{F}(U) \otimes_{\mathcal{A}(U)} \Omega(U) \tag{5.18}$$

(see (3.4) and the comments following Definition 3.1. Yet, concerning (5.14), see also Chapt. I; (11.40)). Analogously, since $s \in \mathcal{E}(U)$, one has

$$D_{\mathcal{E}}(s) \in \mathcal{E}(U) \otimes_{\mathcal{A}(U)} \Omega(U), \tag{5.19}$$

so that one further obtains;

$$(\phi \otimes 1_{\Omega})(D_{\mathcal{E}}(s)) \in \mathcal{F}(U) \otimes_{\mathcal{A}(U)} \Omega(U). \tag{5.20}$$

Thus, operating now on $\phi \equiv \overline{\phi}$, via $\widetilde{D}$ (see (5.14), (5.14)), and then on $s \in \mathcal{E}(U)$, that is, setting (see (5.16)),

$$\widetilde{D}(\phi)(s) = D'(\phi(s)) - (\phi \otimes 1_{\Omega})(D(s)), \tag{5.21}$$

one finally obtains a (local) *section over the open* $U \subseteq X$ of the following $\mathcal{A}$-*module* (cf. also (5.14), (5.19), (5.20)), together with Chapt. II; (5.25), (6.24))

$$\mathcal{H}om_{\mathcal{A}}(\mathcal{E}, \mathcal{F} \otimes_{\mathcal{A}} \Omega) = \mathcal{H}om_{\mathcal{A}}(\mathcal{E}, \mathcal{F}) \otimes_{\mathcal{A}} \Omega \equiv \Omega(\mathcal{H}om_{\mathcal{A}}(\mathcal{E}, \mathcal{F})), \tag{5.22}$$

as desired, for the case at issue, according to (3.4). In this regard, one should also notice that the equality (in fact, $\mathcal{A}$-*isomorphism*) in (5.22) *is* actually *based on the supplementary hypothesis for* $\mathcal{E}$(!) (cf. (5.12) and Chapt. IV; (6.8)). The preceding clarify now completely our notation in (5.13), in conjunction with (5.22) and (5.18), (5.20).

Finally, concerning our claim for (5.13), we check, for instance, that *the said sheaf morphism satisfies the Leibniz condition;* thus, *for any* $\phi, s$, as above (cf. e.g. (5.21)), *and* $\alpha \in \mathcal{A}(U)$, one has;

$$
\begin{aligned}
\widetilde{D}(\alpha\cdot\phi)(s) &= D'(\alpha\cdot\phi(s)) - (\alpha\phi\otimes 1_\Omega)(D(s))\\
&= \alpha\cdot D'(\phi(s)) + \phi(s)\otimes\partial\alpha - \alpha\cdot(\phi\otimes 1_\Omega)(D(s))\\
&= \alpha\cdot\widetilde{D}(\phi)(s) + \phi(s)\otimes\partial\alpha,
\end{aligned}
\tag{5.23}
$$

that was to be proved. (The $\mathbf{C}$-*linearity* of (5.13) is still straightforward from (5.21)). This justifies thus completely our assertion in (5.12). ∎

We consider next a particular instance of (5.13), namely, the case $\mathcal{E}=\mathcal{F}$; notice also that, according to our hypothesis in (5.12), $\mathcal{E}$ is a *vector sheaf* on $X$. Thus, for any given $\mathcal{A}$-*connection* $D$ of $\mathcal{E}$, one obtains from (5.13) (or (5.16)) the following $\mathcal{A}$-*connection* on the *vector sheaf* $\mathcal{H}om_{\mathcal{A}}(\mathcal{E},\mathcal{E})\equiv\mathcal{E}nd\,\mathcal{E}$ on $X$ (cf. Chapt. II; (6.28) and Lemma 6.1)

$$D_{\mathcal{E}nd\,\mathcal{E}}(\phi) = D\circ\phi - (\phi\otimes 1_\Omega)\circ D, \tag{5.24}$$

such that (ibid.; (6.30))

$$\phi\in(\mathcal{E}nd\,\mathcal{E})(U) = \mathcal{H}om_{\mathcal{A}|_U}(\mathcal{E}|_U,\mathcal{E}|_U)\equiv \mathrm{End}\,(\mathcal{E}|_U), \tag{5.25}$$

with $U$ *open* in $X$. The same relation (5.24) is still written, by an obvious *abuse of notation* (however, useful!), reminiscent of the *"Lie product"*, in the following form;

$$D_{\mathcal{E}nd\,\mathcal{E}}(\phi) = D\circ\phi - \phi\circ D\equiv [D,\phi]. \tag{5.26}$$

On the other hand, based on (5.13), and *taking* $\mathcal{E}=\mathcal{F}$, one obtains, for any two $\mathcal{A}$-connections $D, D'$

$$D_{\mathcal{E}nd\,\mathcal{E}}(\phi) = D'\circ\phi - (\phi\otimes 1_\Omega)\circ D, \tag{5.27}$$

with $\phi$, as in (5.25). *In particular, taking in (5.27)*

$$\phi\equiv 1_{\mathcal{E}}\in(\mathcal{E}nd\,\mathcal{E})(X) = \mathrm{Hom}_{\mathcal{A}}(\mathcal{E},\mathcal{E})\equiv\mathrm{End}\,\mathcal{E}, \tag{5.28}$$

that is, the *identity $\mathcal{A}$-automorphism* of $\mathcal{E}$ (cf. (3.7) and Chapt. II; (6.30)), one obtains the following useful relation;

$$D_{\mathcal{E}nd\,\mathcal{E}}(1) = D' - D\in\Omega(\mathcal{E}nd\,\mathcal{E})(X) = \mathrm{Hom}_{\mathcal{A}}(\mathcal{E},\mathcal{E}\otimes_{\mathcal{A}}\Omega). \tag{5.29}$$

Thus, as a consequence of (5.29), one realizes that *the difference of two $\mathcal{A}$-connections is* not merely a $\mathbf{C}$-linear morphism but, in particular, an *$\mathcal{A}$-morphism,* as well; this will be explained in Section 7 below (see thus Theorem 7.1).

Now, as a particular application of the preceding, we further examine in the next Subsection *the case* $\mathcal{F} = \mathcal{A}$.

**5.(d) Dual $\mathcal{A}$-connections.**– Suppose we are given an *$\mathcal{A}$-module* $\mathcal{E}$ on $X$, endowed with an *$\mathcal{A}$-connection* $D_{\mathcal{E}}$, with respect to a given *differential triad* $(\mathcal{A}, \partial, \Omega)$ (cf. (4.1)). Furthermore, let

$$\mathcal{E}^* \equiv \mathcal{H}om_{\mathcal{A}}(\mathcal{E}, \mathcal{A}) \tag{5.30}$$

be the *dual $\mathcal{A}$-module of* $\mathcal{E}$ on $X$ (see Chapt. IV; (5.1)). In particular, *due to (5.12)*, assume that $\mathcal{E}$ is a *vector sheaf* on $X$, hence, applying (5.13), for $\mathcal{F} = \mathcal{A}$ and $D_{\mathcal{F}} = D_{\mathcal{A}} = \partial$ (see (4.2)), one obtains an *$\mathcal{A}$-connection of* $\mathcal{E}^*$ by the relation;

$$D_{\mathcal{E}^*}(u) := \partial \circ u - (u \otimes 1_{\Omega}) \circ D_{\mathcal{E}}, \tag{5.31}$$

such that (cf. (5.30) and Chapt. II; (6.8))

$$u \in \mathcal{E}^*(U) = \mathrm{Hom}_{\mathcal{A}|_U}(\mathcal{E}|_U, \mathcal{A}|_U), \tag{5.32}$$

with $U$ *open* in $X$. Yet, cf. e.g. (5.21), one has

$$D_{\mathcal{E}^*}(u)(s) = \partial(u(s)) - (u \otimes 1_{\Omega})(Ds), \tag{5.33}$$

*for any $u$, as in (5.32), and $s \in \mathcal{E}(U)$.* The last relation is also written, in a more standard way, as;

$$\partial(u(s)) = u(Ds) + (D_{\mathcal{E}^*}(u))(s), \tag{5.34}$$

which is also taken as a *defining relation for* $D_{\mathcal{E}^*}$. We have still employed hereby an obvious *abuse of notation*, by simply writing

$$(u \otimes 1)(Ds) \equiv u(Ds). \tag{5.35}$$

Yet, setting

$$u(s) \equiv \langle s, u \rangle \quad \textit{and} \quad (u \otimes 1)(Ds) \equiv u(Ds) \equiv \langle Ds, u \rangle,$$

one still writes (5.34) in the form;

$$\partial(\langle s, u \rangle) = \langle Ds, u \rangle + \langle s, D^* u \rangle, \tag{5.35$'$}$$

taking it as a *defining relation* of $D^* \equiv D_{\mathcal{E}^*}$ (cf. (5.36) below). In this regard, notice also that, by definition, one has;

$$(5.36)\qquad D_{\mathcal{E}^*} : \mathcal{E}^* \longrightarrow \mathcal{E}^* \otimes_{\mathcal{A}} \Omega = \mathcal{H}om_{\mathcal{A}}(\mathcal{E}, \Omega)$$

(cf. Chapt. IV; Theorem 6.1), which thus justifies the notation applied in (5.33) or (5.34) (cf. also (5.35)).

The previously defined $\mathcal{A}$-connection $D_{\mathcal{E}^*}$ (cf. (5.31)) is called the *dual $\mathcal{A}$-connection* of the given $\mathcal{A}$-connection $D_{\mathcal{E}}$ of the *vector sheaf* $\mathcal{E}$ on $X$, under consideration; it is thus, by definition (cf. (5.36)), an *$\mathcal{A}$-connection of the dual vector sheaf $\mathcal{E}^*$ of* $\mathcal{E}$ on $X$ (cf. (5.12) and Chapt. IV; (5.19)).

On the other hand, as an application of (5.31), one has the following variant, concerning the definition of *$\mathcal{A}$-connections of $\mathcal{E}nd\,\mathcal{E}$*, where $\mathcal{E}$ is still a *vector sheaf* on $X$ (see (5.12)): So we first remark that;

$$(5.37)\qquad \mathcal{E}nd\,\mathcal{E} \equiv \mathcal{H}om_{\mathcal{A}}(\mathcal{E}, \mathcal{E}) = \mathcal{E}^* \otimes_{\mathcal{A}} \mathcal{E} = \mathcal{E} \otimes_{\mathcal{A}} \mathcal{E}^*,$$

within *$\mathcal{A}$-isomorphisms of* the *$\mathcal{A}$-modules* (in fact, *vector sheaves*) involved (see Chapt. II; (6.28), (5.14), Lemma 6.1 and Chapt. IV; Theorem 6.1, in conjunction with (5.12) above). Therefore, if $D_{\mathcal{E}}$ is a given *$\mathcal{A}$-connection of* $\mathcal{E}$, then, based on (5.10), one obtains (cf. also (5.36));

$$(5.38)\qquad D_{\mathcal{E}\otimes_{\mathcal{A}}\mathcal{E}^*} := (D_{\mathcal{E}} \otimes 1_{\mathcal{E}^*}) + (1_{\mathcal{E}} \otimes D_{\mathcal{E}^*}).$$

Now, the interesting fact here is that, *due to (5.37),*

(5.39) *the two $\mathcal{A}$-connections (5.38) and (5.27) are, in effect, identical, modulo the $\mathcal{A}$-isomorphism,* as given by (5.37), *say,*

$$(5.39.1)\qquad \theta : \mathcal{E} \otimes_{\mathcal{A}} \mathcal{E}^* \longrightarrow \mathcal{H}om_{\mathcal{A}}(\mathcal{E}, \mathcal{E}) \equiv \mathcal{E}nd\,\mathcal{E}.$$

Precisely, one has the following relation;

$$(5.40)\qquad D_{\mathcal{E}nd\,\mathcal{E}} = (\theta \otimes 1_{\Omega}) \circ D_{\mathcal{E}\otimes_{\mathcal{A}}\mathcal{E}^*} \circ \theta^{-1}.$$

The above relation (5.40) is explained through the following fundamental notion, whose particular instance is thus (5.40). The same general notion will be employed several times in the sequel; thus, anticipating hereby its use, in the general form, we set the following.

**Definition 5.1.** Suppose we have a differential triad $(\mathcal{A}, \partial, \Omega)$ and let $\mathcal{E}$,$\mathcal{F}$ be two given $\mathcal{A}$-modules on $X$, endowed with $\mathcal{A}$-connections $D_{\mathcal{E}}$ and $D_{\mathcal{F}}$, respectively. Moreover, let $\phi$ be a given $\mathcal{A}$-morphism of $\mathcal{E}$ in $\mathcal{F}$, viz. $\phi \in \mathrm{Hom}_{\mathcal{A}}(\mathcal{E}, \mathcal{F}) = \mathcal{H}om_{\mathcal{A}}(\mathcal{E}, \mathcal{F})(X)$ (cf. Chapt. II; (6.9)). Then, we say that

$D_{\mathcal{E}}$ and $D_{\mathcal{F}}$ are *$\phi$-related,* whenever the following diagram is commutative

(5.41)
$$\begin{array}{ccc} \mathcal{E} & \xrightarrow{D_{\mathcal{E}}} & \Omega(\mathcal{E}) \\ {\scriptstyle\phi}\big\downarrow & & \big\downarrow{\scriptstyle\phi\otimes 1_{\Omega}} \\ \mathcal{F} & \xrightarrow{D_{\mathcal{F}}} & \Omega(\mathcal{F}) \end{array}$$

that is, one has;

(5.42)
$$D_{\mathcal{F}} \circ \phi = (\phi \otimes 1_{\Omega}) \circ D_{\mathcal{E}}.$$

In particular, if $\phi$ in (5.42) is an *$\mathcal{A}$-isomorphism* of the $\mathcal{A}$-modules involved (cf. (5.41)), then one speaks of *gauge equivalent $\mathcal{A}$-connections;* thus, (5.42) can now be written in the form

(5.43)
$$D_{\mathcal{F}} = (\phi \otimes 1_{\Omega}) \circ D_{\mathcal{E}} \circ \phi^{-1}.$$

Thus, based now on (5.43) and the $\mathcal{A}$-isomorphism (5.39.1), our previous assertion in (5.39) and/or (5.40) says that;

(5.44) given the *vector sheaf* $\mathcal{E}$ on $X$ and an *$\mathcal{A}$-connection* $D_{\mathcal{E}}$ of it, *the* resulting *$\mathcal{A}$-connections* $D_{\mathcal{E}nd\mathcal{E}}$ (cf. (5.27)) *and* $D_{\mathcal{E}\otimes_{\mathcal{A}}\mathcal{E}^*}$ (see (5.38)) *are gauge equivalent,* with respect to the *gauge* (viz. *$\mathcal{A}$-isomorphism*), given by (5.39.1).

The proof of (5.40) is straightforward, by the very definitions; thus, cf. (5.38), (5.33) and (5.24) (or even (5.21) for the case in hand), in conjunction with our argument, throughout the preceding, connected e.g. with (3.7).

# 6. Induced $\mathcal{A}$-connections (contn'd). Pull-back of $\mathcal{A}$-connections

We continue in this Section our previous study of induced $\mathcal{A}$-connections, by considering the situation that results, when *one changes the base space* of

the sheaves involved.

Thus, suppose that we have a *continuous map*

$$f : X \longrightarrow Y \tag{6.1}$$

between the *topological spaces* $X$ and $Y$, while we also assume that we are given a *differential triad on* $Y$,

$$(\mathcal{A}_Y \equiv \mathcal{A}, \partial, \Omega). \tag{6.2}$$

So we first remark that:

(6.3) *(6.2) entails a differential triad on* $X$ *too,* such that one has;

$$f^*(\mathcal{A}_Y, \partial, \Omega) \equiv (f^*(\mathcal{A}), f^*(\partial), f^*(\Omega)) \tag{6.3.1}$$

(see the succeeding discussion for the notation applied). We call (6.3.1) the *pull-back of (6.2).*

Thus, *the pull-back of a differential triad, via a continuous map, is still a differential triad on the source of the given map.*

Now, concerning the *notation in (6.3.1),* we applied hereby our previous terminology in Chapt. I; Subsection 14.(b): So we know (cf. Chapt. II; (2.65) – (2.67)) that $f^*(\mathcal{A}) \equiv f^*(\mathcal{A}_Y)$ (see (6.2)) *is a* **C**-*algebra sheaf on* $X$ (of the *same type, as* $\mathcal{A}_Y$, concerning the **C**-algebras involved, cf., for instance, Chapt. IV; (0.1)), such that the pair

$$(X, \mathcal{A}_X \equiv f^*(\mathcal{A}_Y) \equiv f^*(\mathcal{A})) \tag{6.4}$$

is now our **C**-*algebraized space,* relative to $X$. Moreover, on the same space $X$, $f^*(\Omega)$ *is an* $f^*(\mathcal{A})$-*module* (cf. Chapt. II; (2.65)). On the other hand, by considering the given **C**-derivation $\partial : \mathcal{A}_Y \equiv \mathcal{A} \longrightarrow \Omega$, as in (6.2), one also obtains a *sheaf morphism,* as well,

$$f^*(\partial) : f^*(\mathcal{A}) \longrightarrow f^*(\Omega) \tag{6.5}$$

(cf. Chapt. I; (14.41)); indeed, one easily proves that

(6.6) $f^*(\partial)$ *is a* **C**-*derivation,* as indicated (cf. Definition 1.1).

Thus, one proves e.g. that (see also Chapt. I; (14.20))

$$f_V^* : \mathcal{A}(V) \longrightarrow f^*(\mathcal{A})\left(f^{-1}(V)\right) \tag{6.7}$$

is a **C**-*algebra morphism, for every open* $V \subseteq Y$ (apply Chapt. I; (14.23)), while by a similar argument one verifies the *Leibniz condition for (6.5)* (cf. (1.2)); This already justifies our claim in (6.3). ∎

Now suppose that

$$(\mathcal{E}, D) \tag{6.8}$$

is a given pair, consisting of an $\mathcal{A}$-*module* $\mathcal{E}$ on $Y$ (cf. (6.2)) and an $\mathcal{A}$-*connection* $D$ *of* $\mathcal{E}$. Thus, by considering $D$ as a *morphism of the appropriate* (**C**-*vector space*) *presheaves* (cf. Definition 3.1 and the relevant comments on (3.7)), one gets *a map* (we still retain, for convenience, the same symbol)

$$D : \Gamma(\mathcal{E}) \longrightarrow \Gamma(\mathcal{E}) \otimes_{\Gamma(\mathcal{A})} \Gamma(\Omega) \tag{6.9}$$

*satisfying (3.5), (3.6).* Therefore, one further obtains a corresponding *presheaf morphism;*

$$f^*(D) : \Gamma(f^*(\mathcal{E})) \longrightarrow \Gamma(f^*(\mathcal{E})) \otimes_{\Gamma(f^*(\mathcal{A}))} \Gamma(f^*(\Omega)), \tag{6.10}$$

such that one has

$$f^*(D)(f_V^*(t)) := f_V^*(D(t)), \tag{6.11}$$

*for every (local) section* $t \in \mathcal{E}(V)$, with $V$ *open* in $Y$. (In this context, see Chapt. I; (14.27), (14.20) and (14.40), (14.42). Yet, concerning (6.11), we also note that $D(t)$ thereby is, by definition, a (local) *section,* viz. element, *of the presheaf, range of (6.9),* hence the notation in (6.11), its second member being thus in the range of (6.10); in fact, *we abused notation* here, using, for simplicity, the same symbol for the *two* obvious *different uses of* "$f_V^*$", according to Chapt. I; (14.20)). So we are now in a position to set the following.

**Definition 6.1.** Suppose we have the situation described by (6.1), (6.2) and (6.8). Then, the sheaf morphism corresponding to (6.10) (see Chapt. I; (8.4), (8.5) and (8.10)) is called the *pull-back of the* given $\mathcal{A}$-*connection* $D$, still denoted by $f^*(D)$. That is, one has;

$$\begin{aligned} f^*(D) : f^*(\mathcal{E}) \longrightarrow f^*(\mathcal{E}) \otimes_{f^*(\mathcal{A})} f^*(\Omega) &\equiv f^*(\Omega)(f^*(\mathcal{E})) \\ &\equiv f^*(\Omega(\mathcal{E})). \end{aligned} \tag{6.12}$$

Now, it is, in fact, a routine matter to verify, by actually applying (6.10), that;

(6.13) $f^*(D)$, as given by (6.12), *is an* $f^*(\mathcal{A})$*-connection of* $f^*(\mathcal{E})$.

On the other hand, concerning the range of (6.12), one obtains, in effect, the following result, an extension indeed of Lemma 5.1 of Chapter II (see thus Subsection 6.(a) in the sequel).

**Lemma 6.1.** *Suppose we have a continuous map* $h : X \longrightarrow Y$, *with* $(Y, \mathcal{A})$ *a* **C**-*algebraized space, while* $\mathcal{E}, \mathcal{F}$ *are two given* $\mathcal{A}$*-modules on* $Y$. *Then, one obtains the following relation;*

$$h^*(\mathcal{E} \otimes_{\mathcal{A}} \mathcal{F}) = h^*(\mathcal{E}) \otimes_{h^*(\mathcal{A})} h^*(\mathcal{F}), \tag{6.14}$$

*valid, within an* $h^*(\mathcal{A})$*-isomorphism of the* $h^*(\mathcal{A})$*-modules on* $X$ *concerned.*

**Proof.** Applying Theorem 9.1 of Chapter I on the members of (6.14) and then based on the rels. (14.20), (14.23) and (14.27) of the same Chapter, one gets a *presheaf morphism*

$$\Gamma(h^*(\mathcal{E} \otimes_{\mathcal{A}} \mathcal{F})) \longrightarrow \Gamma(h^*(\mathcal{E}) \otimes_{\Gamma(h^*(\mathcal{A}))} \Gamma(h^*(\mathcal{F})); \tag{6.15}$$

now, this in turn yields (same Chapter I; (8.4), (8.5)) an associated to (6.15) *sheaf morphism* of the corresponding sheaves (in fact, $h^*(\mathcal{A})$-*modules* on $X$), as in (6.14). However, the latter is, in effect, a *sheaf isomorphism,* being such *fiber-wise* (cf. also Chapt. I; (1.21)); indeed, one has the following *(canonical) isomorphisms*

$$h^*(\mathcal{E} \otimes_{\mathcal{A}} \mathcal{F})_x = h^*(\mathcal{E})_x \otimes_{h^*(\mathcal{A})_x} h^*(\mathcal{F})_x = (h^*(\mathcal{E}) \otimes_{h^*(\mathcal{A})} h^*(\mathcal{F}))_x, \tag{6.16}$$

*for every point* $x \in X$ (cf. Chapt. I; (14.36) and Chapt. II; (5.9)). Yet, the above isomorphisms are, of course, *module structure preserving* (see also Chapt. II; (1.44)), which thus terminates the proof. ∎

Furthermore, referring still to (6.12), by analogy with the classical case (see, for instance, J.W. Milnor - J.D. Stasheff [1: p. 292, Lemma 3]), one has the following *commutative diagram* (cf. also (6.10))

$$
(6.17)\qquad
\begin{array}{ccc}
\Gamma(\mathcal{E}) & \xrightarrow{\quad D \quad} & \Gamma(\mathcal{E}) \otimes_{\Gamma(\mathcal{A})} \Gamma(\Omega) \equiv \Gamma(\Omega(\mathcal{E})) \\
\Gamma(f^*) \downarrow & & \downarrow \Gamma(f^*) \\
\Gamma(f^*(\mathcal{E})) & \xrightarrow{\quad f^*(D) \quad} & \Gamma(f^*(\mathcal{E}) \otimes_{\Gamma(f^*(\mathcal{A}))} \Gamma(f^*(\Omega)) = \Gamma(f^*(\Omega(\mathcal{E}))).
\end{array}
$$

Indeed, our claim can be proved by verifying it, for instance, *fiber-wise,* which thus follows from (6.16). (Yet, for simplicity, we employed in (6.17) an obvious *abuse of notation,* concerning the two *different* actually *morphisms* $\Gamma(f^*)$; see Chapt. I; (14.20)). ∎ Now, applying a similar argument, as for (6.15), one concludes the validity of an analogous *commutative diagram* for the sheaf morphisms, associated with (6.17). So extending, in an obvious way, the terminology of Definition 5.1 above, we still express (6.17) and/or its *sheaf analogue,* as before, by saying that;

(6.18) *$D$ and $f^*(D)$ are $f^*$-related.*

**6.(a). Restriction of $\mathcal{A}$-connections.**– The preceding have a special bearing, when one looks at an *open* subspace $U$ of a given *topological space* $X$, where the latter space is endowed with a *differential triad* $(\mathcal{A}, \partial, \Omega)$ (see (3.1)); yet, we assume that we are given a pair

$$(\mathcal{E}, D), \tag{6.19}$$

consisting now of an *$\mathcal{A}$-module $\mathcal{E}$ on $X$* and an *$\mathcal{A}$-connection $D$* of $\mathcal{E}$. Thus, as an application of our previous argument, our task now is to *transfer* the situation just described *on $U$, via the* canonical *inclusion map*

$$i_U \equiv i : U \hookrightarrow X. \tag{6.20}$$

So applying first (6.3.1), one concludes that;

(6.21) the *restriction on $U$ of the* given *differential triad* $(\mathcal{A}, \partial, \Omega)$ on $X$, thus, *by definition,* the triple

(6.21.1) $i^*(\mathcal{A}, \partial, \Omega) = (i^*(\mathcal{A}), i^*(\partial), i^*(\Omega)) \equiv (\mathcal{A}|_U, \partial|_U, \Omega|_U)$,

yelds a *differential triad on $U \subseteq X$.*

Therefore, by further employing (6.13), one infers that:

(6.22) $i^*(D) \equiv D|_U$, called the *restriction of $D$ on $U$*, (viz. the *pull-back of $D$ via (6.20)*), is an *$\mathcal{A}|_U$-connection* of the *$\mathcal{A}|_U$-module* $i^*(\mathcal{E}) \equiv \mathcal{E}|_U$, restriction of $\mathcal{E}$ to $U$.

Of course, one sets directly;

(6.23) $$(D|_U)(s|_U) := (Ds)|_U,$$

such that

(6.24) $$(D|_U)(s|_U)(x) = (Ds)(x),$$

*for any $s \in \mathcal{E}(U)$ and $x \in U$* (cf. Chapt. I; (14.23), (14.27), as well as, Chapt. II; (5.25)).

The preceding are going to be used right below (see thus Section 8). Yet, we can still examine the *local expression of an $\mathcal{A}$-connection, pull-back.* (Cf. also Chapt. VIII and Chapt. I; (14.23), (14.34)).

## 7. Affine space of $\mathcal{A}$-connections

Suppose that we are given a *differential triad* $(\mathcal{A}, \partial, \Omega)$ on a *topological space* $X$ (cf. e.g. (3.1)), and let $\mathcal{E}$ be an *$\mathcal{A}$-module* on $X$. Then, we denote by

(7.1) $$Conn_{\mathcal{A}}(\mathcal{E})$$

the *set of $\mathcal{A}$-connections of $\mathcal{E}$*; of course, the set (7.1) might be just the *empty one!*, even if $\mathcal{E}$ was a vector sheaf on $X$. (We shall discuss the relevant situation later on, giving also conditions guaranteeing the non-emptyness of the set (7.1). Thus, we make the *basic assumption,* throughout this Section, that;

(7.2) $$Conn_{\mathcal{A}}(\mathcal{E}) \neq \emptyset.$$

Now, suppose that we are given an *$\mathcal{A}$-connection $D$ of $\mathcal{E}$,* as before, viz. let

$$D \in Conn_{\mathcal{A}}(\mathcal{E}). \tag{7.3}$$

Then, according to Definition 3.1 (cf. (3.5)), one has

$$D \in \mathrm{Hom}_{\mathbf{C}}(\mathcal{E}, \Omega(\mathcal{E})) = \mathcal{H}om_{\mathbf{C}}(\mathcal{E}, \Omega(\mathcal{E})(X), \tag{7.4}$$

where one looks at $\mathbf{C}$ as the *constant* $\mathbf{C}$-*algebra sheaf* on $X$ (cf. Chapt. II; (1.27), (1.32), (1.34), along with (6.9), for $\mathcal{A} = \mathbf{C}_X \equiv \mathbf{C}$. Yet, see ibid.; (1.38)). Consequently, one obtains so far that;

$$Conn_{\mathcal{A}}(\mathcal{E}) \subseteq \mathrm{Hom}_{\mathbf{C}}(\mathcal{E}, \mathcal{E} \otimes_{\mathcal{A}} \Omega) \equiv \mathrm{Hom}_{\mathbf{C}}(\mathcal{E}, \Omega(\mathcal{E})), \tag{7.5}$$

while *the last set in (7.5) is* further *characterized by (3.6) (Leibniz condition)*. In this context, we next make the easy, however, very useful observation, that

(7.6) *given two* $\mathcal{A}$-*connections* $D$ *and* $D'$ *of* $\mathcal{E}$, as above, *one has;*

$$D' - D \in \mathrm{Hom}_{\mathcal{A}}(\mathcal{E}, \Omega(\mathcal{E})) \subseteq \mathrm{Hom}_{\mathbf{C}}(\mathcal{E}, \Omega(\mathcal{E})). \tag{7.6.1}$$

Indeed, we have already seen (7.6.1) in the preceding (cf. (5.29)), in the particular case that $\mathcal{E}$ is a vector sheaf on $X$. But, the same assertion is still valid, *for any* $\mathcal{A}$-*module* $\mathcal{E}$, as above, as an immediate consequence of (3.6) (viz. the *Leibniz condition*), applied to each one of the two summands in $D - D'$. ∎

Now, our next objective is to show that such differences constitute, in fact, the elements of the *"model"* (viz. the first set in (7.6.1)), which the previous set (7.1) can be built on, in a sense that we are going to specify right away (see thus (7.15) below). So our next result has now as follows.

**Lemma 7.1.** *Let* $(\mathcal{A}, \partial, \Omega)$ *be a differential triad on a topological space* $X$ *and* $\mathcal{E}$ *an* $\mathcal{A}$-*module on* $X$. *Then, for every* $\mathcal{A}$-*connection* $D$ *of* $\mathcal{E}$ (see also (7.2)), *one obtains a bijection*

$$\tau_D : \mathrm{Hom}_{\mathcal{A}}(\mathcal{E}, \Omega(\mathcal{E})) \longrightarrow Conn_{\mathcal{A}}(\mathcal{E}) \tag{7.7}$$

*given by the relation;*

$$\tau_D(u) := D + u. \tag{7.8}$$

*What amounts to the same,* given an $\mathcal{A}$-connection $D$ of $\mathcal{E}$, as above, *for every* $u \in \mathrm{Hom}_{\mathcal{A}}(\mathcal{E}, \Omega(\mathcal{E}))$, *there exists a uniquely defined* $\mathcal{A}$-*connection* $D'$ *of* $\mathcal{E}$, *such that*

$$u = D' - D; \tag{7.9}$$

*viz. one has*

$$D' := D + u. \tag{7.10}$$

**Proof.** We first remark that *the map (7.7)*, as given by (7.8), *is well-defined*, viz. *one has;*

$$D + u \in Conn_{\mathcal{A}}(\mathcal{E}), \tag{7.11}$$

*for any* $D \in Conn_{\mathcal{A}}(\mathcal{E})$ *and* $u \in \mathrm{Hom}_{\mathcal{A}}(\mathcal{E}, \Omega(\mathcal{E}))$. It is clear that it suffices only to check the *Leibniz condition for* $D + u \in \mathrm{Hom}_{\mathbf{C}}(\mathcal{E}, \Omega(\mathcal{E}))$ (cf. (3.6)). Thus, *one has;*

$$\begin{aligned}(D+u)(\alpha \cdot s) &= D(\alpha s) + u(\alpha s) = \alpha \cdot D(s) + s \otimes \partial\alpha + \alpha \cdot u(s) \\ &= \alpha \cdot (D+u)(s) + s \otimes \partial(\alpha),\end{aligned} \tag{7.12}$$

*for any* $\alpha \in \mathcal{A}(U)$ *and* $s \in \mathcal{E}(U)$, with $U$ *open* in $X$, as was to be proved. Moreover, as already noted, it is clear that;

$$D + u \in \mathrm{Hom}_{\mathbf{C}}(\mathcal{E}, \Omega(\mathcal{E})), \tag{7.13}$$

since this holds true for each one of the summands of $D + u$.

On the other hand, it is still obvious by (7.8), that *(7.7) is one-to-one*. Yet, for every $D' \in Conn_{\mathcal{A}}(\mathcal{E})$, one defines (see (7.6.1))

$$u := D' - D \in \mathrm{Hom}_{\mathcal{A}}(\mathcal{E}, \Omega(\mathcal{E})), \tag{7.14}$$

so that, in view of (7.8), the map (7.7) *is onto, as well,* which, in effect, terminates the proof. ∎

Now, as a result of the previous lemma, one obtains the following *basic relation;*

$$Conn_{\mathcal{A}}(\mathcal{E}) = D + \mathrm{Hom}_{\mathcal{A}}(\mathcal{E}, \Omega(\mathcal{E})), \tag{7.15}$$

within a *bijection,* defined through (7.8), *for any given* $D \in Conn_{\mathcal{A}}(\mathcal{E})$ (cf. (7.3)). Accordingly, it is instructive to look now at (7.15), in terms of the (geometric) language of *"affine spaces"*. So, for convenience, we first comment a bit on the relevant terminology through the following.

**Scholium 7.1.**- The succeeding terminological comments will facilitate the statement of our next result (see thus Theorem 7.1 below).

So assume that we are given a (commutative unital) *ring* $R$ and a (left) $R$*-module* $M$.

Thus, by an *affine space, modelled on the* $R$*-module* $M$, one means a non-empty set $E$ and a map

$$\phi : E \times E \longrightarrow M : (x, y) \longmapsto \phi(x, y) \equiv y - x \equiv \overrightarrow{xy} \tag{7.16}$$

such that the two following conditions are fulfilled:

i) *For every* $x \in E$, the partial map of $\phi$,

$$\phi_x : E \longmapsto M : y \longmapsto \phi_x(y) := \phi(x, y), \tag{7.17}$$

*is one-to-one.*

ii) *For any* $x, y, z \in E$, *one has* (cf. (7.18));

$$\overrightarrow{xy} + \overrightarrow{yz} + \overrightarrow{zx} = 0. \tag{7.18}$$

(We call (7.18) the *triangle axiom*, or even the *Chasles law*).

We come now to the statement of the following basic conclusion of the previous discussion. That is, one has.

**Theorem 7.1.** *Let* $(\mathcal{A}, \partial, \Omega)$ *be a given differential triad on a topological space* $X$ *and* $\mathcal{E}$ *an* $\mathcal{A}$*-module on* $X$. *Then, the set of* $\mathcal{A}$*-connections of* $\mathcal{E}$, $Conn_{\mathcal{A}}(\mathcal{E})$ (cf. also (7.2)), *is an affine space, modelled on the* $\mathcal{A}(X)$*-module* $\mathrm{Hom}_{\mathcal{A}}(\mathcal{E}, \Omega(\mathcal{E}))$.

*In particular, if* $\mathcal{E}$ *is a vector sheaf on* $X$, *then (7.15) becomes;*

$$Conn_{\mathcal{A}}(\mathcal{E}) = D + \Omega(\mathcal{E}nd\, \mathcal{E})(X), \tag{7.19}$$

*within a bijection* (see (7.7)), *for every* $D \in Conn_{\mathcal{A}}(\mathcal{E})$. *That is,* $Conn_{\mathcal{A}}(\mathcal{E})$ *is now an affine space, modelled on the* $\mathcal{A}(X)$*-module* $\Omega(\mathcal{E}nd\, \mathcal{E})(X)$.

**Proof.** Specializing the terminology of the previous Scholium 7.1 to the case, under consideration, $E$ (viz. the *affine space,* we look for) is *the set (7.1), assuming, of course, (7.2),* while $M$ (the *"model"*) the $\mathcal{A}(X)$*-module* $\mathrm{Hom}_{\mathcal{A}}(\mathcal{E}, \Omega(\mathcal{E}))$; the latter reduces to $\Omega(\mathcal{E}nd\, \mathcal{E})(X)$, whenever $\mathcal{E}$ is a *vector sheaf* on $X$ (see (7.22) below). Furthermore, the map corresponding to (7.16)

is that one, defined by the *difference of two* given *$\mathcal{A}$-connections* of $\mathcal{E}$, that is,

$$Conn_{\mathcal{A}}(\mathcal{E}) \times Conn_{\mathcal{A}}(\mathcal{E}) \longrightarrow \mathrm{Hom}_{\mathcal{A}}(\mathcal{E}, \Omega(\mathcal{E})), \tag{7.20}$$

such that (cf. (7.6.1))

$$(D, D') \longmapsto D' - D. \tag{7.21}$$

Thus, (7.18) is obvious, by the very definition of (7.21), while (7.17) is just (7.15); so this proves already our assertion, concerning the *affine space* $Conn_{\mathcal{A}}(\mathcal{E}) (\neq \emptyset)$.

Now, if $\mathcal{E}$ is a *vector sheaf* on $X$, one has (cf. Chapt. IV; Corollary 6.1)

$$\begin{aligned} \mathcal{H}om_{\mathcal{A}}(\mathcal{E}, \Omega(\mathcal{E})) &\equiv \mathcal{H}om_{\mathcal{A}}(\mathcal{E}, \mathcal{E} \otimes_{\mathcal{A}} \Omega) = \mathcal{H}om_{\mathcal{A}}(\mathcal{E}, \mathcal{E}) \otimes_{\mathcal{A}} \Omega \\ &\equiv (\mathcal{E}nd\, \mathcal{E}) \otimes_{\mathcal{A}} \Omega \equiv \Omega(\mathcal{E}nd\, \mathcal{E}), \end{aligned} \tag{7.22}$$

so that one obtains (see Chapt. II; (6.9));

$$\mathrm{Hom}_{\mathcal{A}}(\mathcal{E}, \Omega(\mathcal{E})) = \mathcal{H}om_{\mathcal{A}}(\mathcal{E}, \Omega(\mathcal{E})(X) = \Omega(\mathcal{E}nd\, \mathcal{E})(X), \tag{7.23}$$

which thus yields our assertion, concerning (7.19), and this completes the proof. ∎

**Corollary 7.1.** *Let $\mathcal{L}$ be a line sheaf on $X$, together with an $\mathcal{A}$-connection $D$ on $\mathcal{L}$. Then, one has*

$$Conn_{\mathcal{A}}(\mathcal{L}) = D + \Omega(X), \tag{7.24}$$

*within a bijection, given by (7.8). Thus, $Conn_{\mathcal{A}}(\mathcal{L})$ is an affine space, modelled on the $\mathcal{A}(X)$-module $\Omega(X)$.*

**Proof.** The assertion is, of course, a specialization of the second part of Theorem 7.1 to the case of a *line sheaf* $\mathcal{L}$ on $X$; thus, (7.24) follows from (7.19), in view of the *$\mathcal{A}$-isomorphism* (cf. Chapt. II; Lemma 6.2)

$$\mathcal{E}nd\, \mathcal{L} = \mathcal{A}, \tag{7.25}$$

so that one then obtains $\Omega(\mathcal{A}) \equiv \mathcal{A} \otimes_{\mathcal{A}} \Omega = \Omega$ (still up to an *$\mathcal{A}$-isomorphism*, ibid.; (5.15)), and this terminates the proof. ∎

Yet, as a further application of the classical *theory of affine spaces*, one obtains the following.

**Lemma 7.2.** *Let $(\mathcal{A}, \partial, \Omega)$ be a given differential triad on a topological space $X$ and $\mathcal{E}$ an $\mathcal{A}$-module on $X$. Then, for any two different and non-trivial $\mathcal{A}$-connections $D$ and $D'$ of $\mathcal{E}$, the corresponding $\mathcal{A}(X)$-line, they determine, that is, the set*

$$L_{D,D'}^{\mathcal{A}(X)} = \left\{\widetilde{D} \equiv \alpha \cdot D + (1-\alpha) \cdot D' : \alpha \in \mathcal{A}(X)\right\} \tag{7.26}$$

*is contained in $Conn_{\mathcal{A}}(\mathcal{E})$.*

**Proof.** First, one defines;

$$(\alpha \cdot D)(s) := \alpha \cdot D(s) \in \Omega(\mathcal{E})(U), \tag{7.27}$$

*for any $\mathcal{A}$-connection $D$ of $\mathcal{E}$, $\alpha \in \mathcal{A}(X)$ and $s \in \mathcal{E}(U)$*, with $U$ *open* in $X$. Hence, by hypothesis for $D$ and $D'$, the set (7.26) *is a subset of the* **C**-*vector space* $\mathrm{Hom}_{\mathbf{C}}(\mathcal{E}, \Omega(\mathcal{E}))$ (cf. (7.6)). Now, the assertion about (7.26) is straightforward by the very definitions. ∎

In particular, one concludes that;

(7.28) *for any two points $D \neq D'$ in $Conn_{\mathcal{A}}(\mathcal{E})$* (cf. also (7.2)), the corresponding **C**-*line*, which *they determine* in the **C**-*vector space* $\mathrm{Hom}_{\mathbf{C}}(\mathcal{E}, \Omega(\mathcal{E}))$, *lies actually in $Conn_{\mathcal{A}}(\mathcal{E})$*. That is, *one has;*

$$\lambda D + (1-\lambda) D' \in Conn_{\mathcal{A}}(\mathcal{E}), \tag{7.28.1}$$

*for any $\lambda \in \mathbf{C}$ and $D, D' \in Conn_{\mathcal{A}}(\mathcal{E})$, with $D \neq D'$.*

For the classical case of Differential Geometry, see, for instance, W.A. Poor [1: p.85, and p.86, Proposition 2.71], or even J.-L. Brylinski [1: p.71, Proposition 2.2.2].

So, according to the classical terminology, the $\mathcal{A}$-*morphism*

$$u = D' - D \in \mathrm{Hom}_{\mathcal{A}}(\mathcal{E}, \Omega(\mathcal{E})) \tag{7.29}$$

(see (7.14)), or in the case of a *vector sheaf* $\mathcal{E}$ on $X$, that one given by

$$u = D' - D \in \Omega(\mathcal{E}nd\,\mathcal{E})(X) \tag{7.30}$$

(see (7.22)), is, in effect, the so-called *connection difference (*End $\mathcal{E}$*-valued) 1-form* on $X$, a nomenclature justified by the *standard example* in Section 2 above (cf. also, for instance, (4.15)). In particular, this is really the case, viz. one has

$$(7.31) \qquad u = D' - D \in \Omega(X),$$

namely, a *"1-form"* on $X$ (the same example, as before, especially, (2.3), (2.4)), when we are given a *line sheaf* $\mathcal{L}$ on $X$ (cf. (7.24)).

On the other hand, the preceding rels. (7.15) and/or (7.19) permit one to employ, for convenience, even in the classical theory, an obvious *abuse of language,* by *identifying,* namely, *the two sets* appeared in either one of the relations in question (yet, in particular, of (7.19)). This happens, especially with (7.24), when one speaks usually of an *identification of $\mathcal{A}$-connections of* the given line sheaf $\mathcal{L}$ *with 1-forms on $X$.* (In this concern, see also, however, the next Chapter VII).

**Scholium 7.2.**- Before we leave this Section, we also note for later use the following more general aspect of the above Lemma 7.2:

Namely, suppose that we are given a *family*

$$(7.32) \qquad (\alpha_i)_{i\in I} \subseteq \mathcal{A}(X), \text{ with } \textit{finite support;}$$

viz. one has $\alpha_i = 0 \in \mathcal{A}(X)$, at the exception only of finite many indices $i \in I$. Yet, we assume that

$$(7.33) \qquad \sum_{i\in I} \alpha_i = 1 \in \mathcal{A}(X).$$

(The latter sum has, of course, a meaning, in view of (7.32)). Finally, consider any family of *$\mathcal{A}$-connections* of the given *$\mathcal{A}$-module* $\mathcal{E}$ on $X$

$$(7.34) \qquad (D_i)_{i\in I} \subseteq \mathcal{C}onn_{\mathcal{A}}(\mathcal{E})$$

(indexed by the same $I$, as in (7.32), but not necessarily of finite support). Thus, under the above circumstances, *one* now *concludes that;*

$$(7.35) \qquad \sum_{i\in I} \alpha_i D_i \in \mathcal{C}onn_{\mathcal{A}}(\mathcal{E}).$$

The previous sum is taken, in principle, in the *$\mathcal{A}(X)$-module* $\mathrm{Hom}_{\mathbf{C}}(\mathcal{E}, \Omega(\mathcal{E}))$ (see also (7.27)). Then, (7.33) contributes to prove that *(7.35) is,* in effect, *an $\mathcal{A}$-connection of $\mathcal{E}$*, as well. ∎

# 8. Localization of $\mathcal{A}$-connections

We come by the present Section to one of the main points of our discussion, as is the *existence of $\mathcal{A}$-connections for a given $\mathcal{A}$-module $\mathcal{E}$* on $X$. Thus, the suitable tool hereby, as we shall presently see, is *sheaf cohomology* (Chapter III), while we shall also restrict ourselves to *vector sheaves* on $X$. So, following the classical pattern (of *Differential Geometry*), our first task is to show that:

(8.1) every *vector sheaf* on $X$, can always be endowed with a *"connection 1-cocycle"*.

Now, the importance of the previous information lies in the fact that;

(8.2) *the vanishing of the* corresponding *cohomology class of the previous 1-cocycle guarantees the existence of an $\mathcal{A}$-connection for the vector sheaf* in question.

Indeed, we further note, in anticipation, that the aforementioned *vanishing* (of the *"connection cohomology class"*) yields, in fact, a *criterion* (the counterpart in our case of the standard *Atiyah's criterion*) *for the existence of an $\mathcal{A}$-connection.* However, this second information, as in (8.2), along with its consequences, will be our object of study in the ensuing few Sections. Thus, we start with fixing the relevant terminology, which we shall use throughout the sequel.

So assume, as usual, that we are given a *differential triad* $(\mathcal{A}, \partial, \Omega)$ on a *topological space* $X$, base space of a (fixed) **C**-*algebraized space* $(X, \mathcal{A})$.

On the other hand, we have already seen that:

i) we can always consider the *n-th dimensional extension of the* given *differential triad* $(\mathcal{A}, \partial, \Omega)$, as above, viz. the triple

$$(\mathcal{A}^n, \partial^n, \Omega^n), \tag{8.3}$$

*for every* $n \in \mathbf{N}$ (cf. (5.6)), getting thus a *free $\mathcal{A}$-module,* of *rank* $n$, equipped with an *$\mathcal{A}$-connection;* that is, one has

$$\partial^n : \mathcal{A}^n \longrightarrow \mathcal{A}^n \otimes_{\mathcal{A}} \Omega \equiv \Omega(\mathcal{A}^n) = \Omega(\mathcal{A})^n = \Omega^n \tag{8.4}$$

(see also (5.5) and (4.1)). Furthermore,

ii) one can always take the *localization,* or even *restriction,* of the previous pair $(\mathcal{A}^n, \partial^n)$, *to any open* set $U \subseteq X$, according to our argument in Section 6 in the preceding. That is, one has (cf. (6.21), (6.22));

$$(8.5) \qquad (\mathcal{A}^n, \partial^n)|_U = (\mathcal{A}^n|_U, \partial^n|_U) = ((\mathcal{A}|_U)^n, (\partial|_U)^n)$$

(see also e.g. Chapt. II; (4.5)), getting thus an $\mathcal{A}|_U$*-module* (in fact, a *free one,* of *rank* $n$), endowed with an $\mathcal{A}|_U$*-connection*

$$(8.6) \qquad \begin{aligned} \partial^n|_U &= (\partial|_U)^n : \mathcal{A}^n|_U = (\mathcal{A}|_U)^n \longrightarrow \mathcal{A}^n|_U \otimes_{\mathcal{A}|_U} \Omega|_U \\ &= (\mathcal{A}^n \otimes_{\mathcal{A}} \Omega)|_U = ((\mathcal{A} \otimes_{\mathcal{A}} \Omega)|_U)^n = \Omega^n|_U = (\Omega|_U)^n \end{aligned}$$

(*commutativity of Whitney sum and localization,* referred to a given *differential triad.* See also Chapt. II; Lemma 5.1 and (5.18)).

Now, it is clear that one can *transfer the previous data in ii), via any* given *local isomorphism of* the *sheaf modules* involved. This is, in effect, the *key idea* behind our next result, lying at the basis of all that follows, as well.

On the other hand, although the $\mathcal{A}$*-connection of* $\mathcal{A}^n$, which we shall use in the sequel is (8.4), *for the sake of generality* (but, no extra cost is paid for that, anyway!) we do consider right below an arbitrary $\mathcal{A}$*-connection* $d$ *of* $\mathcal{A}^n$. However, due to (7.19), one actually obtains;

$$(8.7) \qquad d = \partial^n + u,$$

for some *uniquely defined*

$$(8.8) \qquad \begin{aligned} u \in \Omega(\mathcal{E}nd\, \mathcal{A}^n)(X) &= \Omega(\mathcal{A}^{n^2})(X) = \Omega(\mathcal{A})^{n^2}(X) \\ &= \Omega^{n^2}(X) \equiv M_n(\Omega)(X) = M_n(\Omega(X)) \equiv \Omega(X)^{n^2} \end{aligned}$$

(see Chapt. IV; (3.23) and Chapt. II; (5.18), (3.14)). In this regard, we still note that the *existence of* $\partial$ *entails that one of* $\partial^n$, *for every* $n \in \mathbf{N}$ (cf. (5.4)), *hence, of* $d$ *too,* for some unique $u$, as in (8.8).

Thus, we are now in the position to start with our first basic information, pertaining to the *effect of the data in ii),* as above, *to a given vector sheaf on* $X$ (this, of course, *to a local extent!*). Indeed, as was already hinted at in the preceding, *the situation can be copied* (locally!), *through any* given *local frame of the vector sheaf* concerned. Precisely, one thus obtains the next.

**Lemma 8.1.** *Suppose we are given a differential triad* $(\mathcal{A}, \partial, \Omega)$ *on a topological space* $X$ *and a vector sheaf* $\mathcal{E}$ *on* $X$ *of rank* $n \in \mathbf{N}$. *Moreover,*

*assume that we are given an $\mathcal{A}$-connection of* (the free $\mathcal{A}$-module) $\mathcal{A}^n$,

$$d : \mathcal{A}^n \longrightarrow \mathcal{A}^n \otimes_{\mathcal{A}} \Omega \equiv \Omega(\mathcal{A}^n) = \Omega^n \tag{8.9}$$

(we can take, for instance, the $\mathcal{A}$-*connection* $\partial^n$, cf. (8.4), (8.7)), *while let*

$$\mathcal{U} = (U_\alpha)_{\alpha \in I} \tag{8.10}$$

*be a local frame of $\mathcal{E}$. Then, for every $\alpha \in I$, one defines the following commutative diagram;*

$$\begin{array}{ccc} \mathcal{E}|_{U_\alpha} & \overset{D_\alpha}{\dashrightarrow} & \mathcal{E} \otimes_{\mathcal{A}} \Omega|_{U_\alpha} \equiv \Omega(\mathcal{E})|_{U_\alpha} = \mathcal{E}|_{U_\alpha} \otimes_{\mathcal{A}|_{U_\alpha}} \Omega|_{U_\alpha} \\ \eta_\alpha \Big\downarrow & & \Big\uparrow \eta_\alpha^{-1} \otimes 1_{\Omega|_{U_\alpha}} \\ \mathcal{A}^n|_{U_\alpha} & \xrightarrow{d_\alpha \equiv d|_{U_\alpha}} & \mathcal{A}^n \otimes_{\mathcal{A}} \Omega|_{U_\alpha} = \Omega^n|_{U_\alpha} \end{array} \tag{8.11}$$

Here

$$\eta_\alpha : \mathcal{E}|_{U_\alpha} \longrightarrow \mathcal{A}^n|_{U_\alpha} = (\mathcal{A}|_{U_\alpha})^n, \quad \alpha \in I, \tag{8.12}$$

denotes the *(canonical) $\mathcal{A}|_{U_\alpha}$-isomorphism,* provided by our hypothesis for $\mathcal{E}$ and $\mathcal{U}$ (cf. Chapt. II; (4.7), (4.8)), *so that one then sets;*

$$D_\alpha := \left(\eta_\alpha^{-1} \otimes 1_{\Omega|_{U_\alpha}}\right) \circ d_\alpha \circ \eta_\alpha, \quad \alpha \in I. \tag{8.13}$$

*Then, for any $\alpha \in I$, $D_\alpha$ yields an $\mathcal{A}|_{U_\alpha}$-connection of the (free) $\mathcal{A}|_{U_\alpha}$-module $\mathcal{E}|_{U_\alpha}$.*

**Scholium 8.1.**– As already remarked in the foregoing, the idea of the previous lemma is that *any $\mathcal{A}$-connection of $\mathcal{A}^n$*, for which *one has, of course* (cf. (7.1)),

$$Conn_{\mathcal{A}}(\mathcal{A}^n) \neq \emptyset, \quad n \in \mathbf{N}, \tag{8.14}$$

*whenever we are given $\partial$* (cf. (5.4)), *can be transferred locally,* modulo a local $\mathcal{A}$-isomorphism, *to any other $\mathcal{A}$-module on $X$.* Therefore, naturally, to a given *vector sheaf* on $X$, through any *local frame* of it. So, roughly speaking, the lesson thus far is that;

(8.15) *given a differential triad* $(\mathcal{A}, \partial, \Omega)$ on a topological space $X$, *$\mathcal{A}$-connections always exist locally!*

(for *vector sheaves* on $X$), a fact, however, that, as we shall see, through the ensuing discussion, *does not hold, in general, true globally.*

Yet, another issue of (8.11), which should be clarified at this point, is that, *whenever one has an $\mathcal{A}$-isomorphism $\phi : \mathcal{E} \longrightarrow \mathcal{F}$ between* two given *$\mathcal{A}$-modules $\mathcal{E}$ and $\mathcal{F}$* on $X$, and an *$\mathcal{A}$-connection $D$ of $\mathcal{E}$*, then *the following diagram can be completed* (through the dashed map, becoming thus commutative), so as to give an *$\mathcal{A}$-connection, say $D'$, of $\mathcal{F}$*; i.e., one has

(8.16)
$$\begin{array}{ccc} \mathcal{E} & \xrightarrow{\ \ \phi\ \ } & \mathcal{F} \\ {\scriptstyle D}\downarrow & & \downarrow{\scriptstyle D'} \\ \Omega(\mathcal{E}) & \xrightarrow[\phi \otimes 1_{\Omega}]{} & \Omega(\mathcal{F}). \end{array}$$

The assertion, as already noted in the foregoing, is an immediate consequence of the very definitions and the hypothesis for $\phi \in \mathrm{Hom}_{\mathcal{A}}(\mathcal{E}, \mathcal{F})$. So one has, *by definition,*

(8.17) $$D' := (\phi \otimes 1_{\Omega}) \circ D \circ \phi^{-1};$$

that is, $D'$ is thus forced to be *gauge equivalent to $D$, with respect to* $\phi$ (see (5.43)). Said otherwise, one thus infers that;

(8.18) *elements of* $\mathrm{Isom}_{\mathcal{A}}(\mathcal{E}, \mathcal{F})$ (set of *$\mathcal{A}$-isomorphisms* between the *$\mathcal{A}$-modules* $\mathcal{E}$ and $\mathcal{F}$) *preserve $\mathcal{A}$-connections,* such that *the resulting $\mathcal{A}$-connections of $\mathcal{F}$*, through the element of $\mathrm{Isom}_{\mathcal{A}}(\mathcal{E}, \mathcal{F})$, under consideration, *are gauge equivalent to a given $\mathcal{A}$-connection of $\mathcal{E}$.*

Indeed, we are going to use (8.18), *suitably localized,* straight away. So we come next to the

**Proof of Lemma 8.1.** The only thing we have to check is that $D_{\alpha}$, $\alpha \in I$, as given by (8.13), is an *$\mathcal{A}|_{U_{\alpha}}$-connection of $\mathcal{E}|_{U_{\alpha}}$*: Now, based on the same

relation, it is clear that (see e.g. Chapt. II; (6.8))

$$D_\alpha \in \mathrm{Hom}_{\mathbf{C}}(\mathcal{E}|_{U_\alpha}, \Omega(\mathcal{E})|_{U_\alpha}) = \mathcal{H}om_{\mathbf{C}}(\mathcal{E}, \Omega(\mathcal{E}))(U_\alpha),\ \alpha \in I, \tag{8.19}$$

hence, it remains to prove the corresponding *Leibniz condition for* $D_\alpha$ (see Definition 3.1). However, taking into account that $\eta_\alpha$ (cf. (8.12)) is an $\mathcal{A}|_{U_\alpha}$-*isomorphism,* the desired property follows already from (8.18) *localized to* $U_\alpha \in \mathcal{U}$ (see (8.10), along with (8.17)) and the fact that

$$d_\alpha := d|_{U_\alpha},\quad \alpha \in I, \tag{8.20}$$

is an $\mathcal{A}|_{U_\alpha}$-*connection of* $\mathcal{A}^n|_{U_\alpha}$, according to the hypothesis for $d$ (cf. (8.9)) and to (6.22), which thus terminates the proof. ∎

Thus, following (8.17), we still conclude, in view of (8.13), that:

(8.21) the $\mathcal{A}|_{U_\alpha}$-*connections* $d_\alpha$ *and* $D_\alpha$ are *gauge equivalent,* through the $\mathcal{A}|_{U_\alpha}$-*isomorphism*

$$\eta_\alpha \in \mathrm{Isom}_{\mathcal{A}|_{U_\alpha}}(\mathcal{E}|_{U_\alpha}, \mathcal{A}^n|_{U_\alpha}) = \mathcal{I}som_{\mathcal{A}}(\mathcal{E}, \mathcal{A}^n)(U_\alpha) \tag{8.21.1}$$

(cf. (8.23) in the sequel), *for every* $\alpha \in I$.

In this context, it is still useful, for later use too, to comment a bit on the following *(physics oriented!)* terminology: Namely, every element of the set

$$\mathrm{Isom}_{\mathcal{A}}(\mathcal{E}, \mathcal{F}) = \mathcal{I}som_{\mathcal{A}}(\mathcal{E}, \mathcal{F})(X) \tag{8.22}$$

(see Chapt. V; (2.33) for the notation applied) is still a *gauge transformation* between the given $\mathcal{A}$-*modules* $\mathcal{E}$ and $\mathcal{F}$. (Of course, the terminology fits, in particular, well, when either one of $\mathcal{E}$,$\mathcal{F}$ is a *free* $\mathcal{A}$-*module,* of *finite rank*). Now, by a *local gauge transformation* between $\mathcal{E}$ and $\mathcal{F}$, one means any *local section of the sheaf* $\mathcal{I}som_{\mathcal{A}}(\mathcal{E}, \mathcal{F})$ (ibid.; (2.31)); viz. any element of the set

$$\mathcal{I}som_{\mathcal{A}}(\mathcal{E}, \mathcal{F})(U) = \mathrm{Isom}_{\mathcal{A}|_U}(\mathcal{E}|_U, \mathcal{F}|_U) \subseteq \mathrm{Hom}_{\mathcal{A}|_U}(\mathcal{E}|_U, \mathcal{F}|_U), \tag{8.23}$$

with $U$ any *open* set in $X$ (see e.g. (8.21.1) above).

Thus, the obvious, however, *fundamental fact* at this point is that (another formulation actually of (8.18));

(8.24) *(local) gauge transformations preserve* ($\mathcal{A}$-)*connections,* the *gauge transform of* a given $\mathcal{A}$-*connection* being supplied by (an analogous relation to) (8.17) (cf., for instance, (8.13)).

On the other hand, supplementing hereby our previous terminology in Section 5, we further remark that (5.42), in conjunction with (5.13), *can* also *be written in the form;*

(8.25) $$D_{\mathcal{H}om_{\mathcal{A}}(\mathcal{E},\mathcal{F})}(\phi) = 0,$$

*with* $\phi$, *as in (5.14).* We still say that $\phi$ is a *horizontal section* of $\mathcal{H}om_{\mathcal{A}}(\mathcal{E},\mathcal{F})$ (a terminology that will be justified later on).

Accordingly, based on (8.13), one concludes now that;

> *the local gauge transformation* (or even *"coordinate" local isomorphism*) $\eta_\alpha$ (see (8.12)) of the given *vector sheaf* $\mathcal{E}$ is always a *horizontal section* of the *vector sheaf* (actually *free* $\mathcal{A}|_{U_\alpha}$-*module;* see also Chapt. II; (6.23))

(8.26) (8.26.1) $$\mathcal{F}_\alpha \equiv \mathcal{H}om_{\mathcal{A}|_{U_\alpha}}(\mathcal{E}|_{U_\alpha}, \mathcal{A}^n|_{U_\alpha}) = \mathcal{H}om_{\mathcal{A}}(\mathcal{E}, \mathcal{A}^n)|_{U_\alpha}, \ \alpha \in I.$$

> *That is, one has*

(8.26.2) $$D_{\mathcal{F}_\alpha}(\eta_\alpha) = 0, \quad \alpha \in I.$$

**8.(a). Local Levi-Civita $\mathcal{A}$-connections.**– The connections in the title of this Subsection refer to the particular (however, *substantial!*) case, one takes in (8.13), when

(8.27) $$d = \partial^n : \mathcal{A}^n \longrightarrow \Omega^n = \Omega(\mathcal{A}^n)$$

(see also (8.9), (8.4), (5.5), (8.20), as well as, (8.7); viz. whenever one defines $D_\alpha$ in (8.13), by employing the *standard flat $\mathcal{A}$-connection $\partial$ of $\mathcal{A}$* (cf. Subsection 4.(a) in the preceding). Note, in passing, that the $\mathcal{A}$-connection $\partial^n$ *is* also *flat,* in a sense that will be specified later on (see thus Chapter VIII in the sequel).

Thus, setting now (see e.g. (8.6), along with (8.10)),

(8.28) $$\partial^n_\alpha \equiv \partial^n|_{U_\alpha} = (\partial|_{U_\alpha})^n, \quad \alpha \in I,$$

one obtains the following (local) $\mathcal{A}|_{U_\alpha}$*-connection of* $\mathcal{E}|_{U_\alpha}$ ($\alpha \in I$), which *there does exist!* (in view of our hypothesis for $\partial$; see also (8.15)). Namely, one now defines;

(8.29) $$D_\alpha := \left(\eta_\alpha^{-1} \otimes 1_{\Omega|_{U_\alpha}}\right) \circ \partial^n_\alpha \circ \eta_\alpha, \quad \alpha \in I.$$

Thus, stated otherwise, according to our previous terminology,

(8.30) *for every* $\alpha \in I$ (cf. (8.10)), one considers the *(local) gauge transform, through* $\eta_\alpha$*, of the standard flat (local)* $\mathcal{A}$*-connection of* $\mathcal{A}^n$, with $n = \operatorname{rk} \mathcal{E}$.

This also justifies already, formally, our previous claim in (8.15).

Now, extending the classical terminology, we call (8.29) the *local Levi-Civita* $\mathcal{A}|_{U_\alpha}$*-connection* of the given *vector sheaf* $\mathcal{E}$, the latter being restricted to the *local gauge* $U_\alpha$ of it, under consideration. Yet, for simplicity's sake, we also refer to (8.29), by abusing language, just as a *local Levi-Civita* $\mathcal{A}$*-connection of* $\mathcal{E}$ (always existing, given $\partial$, $\mathcal{E}$ and $U_\alpha$, as before). Finally, we call (8.11), in the particular case at issue (cf. (8.27)), the *Levi-Civita diagram*, associated with a given *vector sheaf* $\mathcal{E}$ on $X$ and a *local gauge* of it.

Therefore, according to (8.21), and taking (8.27) into account, one has that:

(8.31) *every local Levi-Civita* $\mathcal{A}$*-connection,* associated with a given vector sheaf $\mathcal{E}$ on $X$ and a local gauge of it, *is gauge equivalent to the standard flat (local)* $\mathcal{A}$*-connection of* $\mathcal{A}^n$ (cf. (8.28)), with $n = \operatorname{rk} \mathcal{E} \in \mathbf{N}$.

Now, as another consequence of the previous discussion, we already deduce that:

(8.32) given a *vector sheaf* $\mathcal{E}$ on $X$ and a *local frame* of it, $\mathcal{U} = (U_\alpha)_{\alpha \in I}$, one obtains, through (8.29) (see also e.g. (8.19)), the following *Levi-Civita 0-cochain*

(8.32.1) $$(D_\alpha) \in C^0(\mathcal{U}, \mathcal{H}om_{\mathbf{C}}(\mathcal{E}, \Omega(\mathcal{E})))$$

being *gauge equivalent* to the *standard flat* $\mathcal{A}$*-connection 0-cochain* (in fact, *0-cocycle,* cf. (8.33) below) *of* $\mathcal{A}^n$;

(8.32.2) $$\begin{aligned}(\partial_\alpha^n) &\in C^0(\mathcal{U}, \mathcal{H}om_{\mathbf{C}}(\mathcal{A}^n, \Omega(\mathcal{A}^n))) \\ &= C^0(\mathcal{U}, \mathcal{H}om_{\mathbf{C}}(\mathcal{A}^n, \Omega^n)).\end{aligned}$$

That is, *one has*

(8.32.3) $$(D_\alpha) \underset{\alpha}{\sim} (\partial_\alpha^n),$$

in the sense that *(8.29) holds true, for every* $\alpha \in I$.

Yet, as already said, *(8.32.2) is,* actually *a 0-cocycle,* a consequence, in fact, of (8.28). Thus, one obtains

$$\begin{aligned}\partial^n = (\partial^n_\alpha) \in Z^0(\mathcal{U}, \mathcal{H}om_{\mathbf{C}}(\mathcal{A}^n, \Omega^n)) &= \mathcal{H}om_{\mathbf{C}}(\mathcal{A}^n, \Omega^n)(X) \\ &= \mathrm{Hom}_{\mathbf{C}}(\mathcal{A}^n, \Omega^n) = \mathrm{Hom}_{\mathbf{C}}(\mathcal{A}^n, \Omega(\mathcal{A}^n))\end{aligned} \tag{8.33}$$

(cf. Chapt. III; Lemma 4.1, or even (3.11) therein).

On the other hand, taking, for instance, (7.22) and (7.30) into account, one concludes that (see also Chapt. III; (4.58));

(8.34) *the 1-cocycle, which corresponds to (8.32.1),* is given by

$$\begin{aligned}\delta(D_\alpha) = (D_\beta - D_\alpha) \in Z^1(\mathcal{U}, \mathcal{H}om_{\mathcal{A}}(\mathcal{E}, \Omega(\mathcal{E}))) \\ = Z^1(\mathcal{U}, \Omega(\mathcal{E}nd\,\mathcal{E})).\end{aligned} \tag{8.34.1}$$

We call it the *Levi-Civita ($\mathcal{A}$-connection) 1-cocycle* of the *vector sheaf $\mathcal{E}$*, which is thus associated with the given *local frame $\mathcal{U}$* of $\mathcal{E}$. We also call it the *standard $\mathcal{A}$-connection 1-cocycle of $\mathcal{E}$*, (canonically) *associated with $\mathcal{U}$*.

In this context, we still remark, in anticipation, that the $\mathcal{A}$*-module* on $X$, $\Omega(\mathcal{E}nd\,\mathcal{E})$, appeared in (8.34.1) is going to play a significant rôle throughout the sequel. (It will deliberately be called *"sheaf of $\mathcal{A}$-connection,* or just *connection, coefficients"* right below).

**Note 8.1.**– As a result of the preceding discussion one infers that:

(8.35) Given a *differential triad* $(\mathcal{A}, \partial, \Omega)$ on a topological space $X$, a *vector sheaf $\mathcal{E}$* on $X$ and a *local frame $\mathcal{U}$* of $\mathcal{E}$, one always gets the *standard $\mathcal{A}$-connection 1-cocycle of $\mathcal{E}$.*

Thus, its existence simply follows from the hypothesis that *$\partial$ is given* and that *$\mathcal{E}$ is locally free.* Yet, as we shall see (cf. Section 10 below), there is an *intimate (natural) relation between the above 1-cocycle and* the vector sheaf *$\mathcal{E}$ itself.*

However, the extent to which the same 1-cocycle provides an $\mathcal{A}$-connection of $\mathcal{E}$ (*in general, does not!*) depends, so to speak, on our *"arithmetic"*(!); viz. on the *cohomological behaviour,* as we shall see in the sequel, of the *sheaf of coefficients $\mathcal{A}$* and, in turn, of that one of *connection coefficients* $\Omega(\mathcal{E}nd\,\mathcal{E})$ (see the next Section 9, for the terminology applied). So the appeared ob-

struction to the classical case of Differential Geometry diversifies exactly the present abstract setting.

Yet, one *final remark*, of theoretical character only, is that the preceding machinery, through which one arrives at (8.35), *is*, of course, *always, in force, locally;* that is, once we are given just an $\mathcal{A}$*-module* $\mathcal{E}$ on $X$, for which there exists an *open set* $U \subseteq X$, such that, *for some* $n \in \mathbf{N}$, one has

(8.36) $$\mathcal{E}|_U = \mathcal{A}^n|_U,$$

within an $\mathcal{A}|_U$*-isomorphism of* the $\mathcal{A}|_U$*-modules* involved. (*Finite* $n$ here is still *not crucial*).

## 9. Sheaf of connection coefficients

We consider now the particular $\mathcal{A}$*-module* on $X$, appeared in (8.34.1), as the *"sheaf of coefficients"* of the relevant cochains (in fact, cocycles), which are naturally involved, when employing cohomology arguments. Indeed, such methods are going to be used systematically in the succeeding discussion. So our purpose here is to show that:

(9.1) by using *cohomology* $\mathcal{A}(X)$*-modules* (cf. Chapter III in the preceding), *the "sheaf of coefficients", that is finally involved, is actually independent of the vector sheaf* $\mathcal{E}$, under consideration, *depending only on the given differential triad* $(\mathcal{A}, \partial, \Omega)$*; precisely,* on the range of the standard flat $\mathcal{A}$-connection $\partial$ of $\mathcal{A}$, viz. *on the* $\mathcal{A}$*-module* $\Omega$.

To simplify notation, we also accept the following terminology (see also the comments at the end of the previous Section 8):

(9.2) Given a differential triad $(\mathcal{A}, \partial, \Omega)$, we say that *an* $\mathcal{A}$*-module* $\mathcal{E}$ on $X$ *has a local gauge,* whenever (8.36) *holds true, for some open* $U \subseteq X$ ($U$ being thus the *local gauge of* $\mathcal{E}$, in question, of *rank* $n \in \mathbf{N}$).

Of course, it is clear that any *vector sheaf* $\mathcal{E}$ on $X$ *has always*, by its very definition (see e.g. Chapt. II; (4.1.1)), a *local gauge* $U$, of *rank* $n = \mathrm{rk}_{\mathcal{A}}(\mathcal{E}) \in \mathbf{N}$ (ibid.; (4.2)), *for every point* $x \in X$, *with* $x \in U$. So our first relevant result is now the following.

**Lemma 9.1.** *Suppose we are given an* $\mathcal{A}$*-module* $\mathcal{E}$ *on* $X$, *having a local gauge* $U$, *of rank* $n \in \mathbf{N}$ (cf. (9.2) and (8.36)). *Then, one obtains* (see also Chapt. V; (8.7));

$$(9.3)\ \mathcal{H}om_{\mathcal{A}}(\mathcal{E}, \mathcal{E} \otimes_{\mathcal{A}} \Omega)(U) \equiv \mathcal{H}om_{\mathcal{A}}(\mathcal{E}, \Omega(\mathcal{E}))(U) = M_n(\Omega)(U) = M_n(\Omega(U)),$$

*within an* $\mathcal{A}(U)$*-isomorphism of the* $\mathcal{A}(U)$*-modules involved.*

*In particular, if* $\mathcal{E}$ *is a vector sheaf on* $X$, *of rank* $n \in \mathbf{N}$, *and* $U$ *a local gauge of it, then one has;*

$$(9.4) \qquad \Omega(\mathcal{E}nd\,\mathcal{E})(U) = M_n(\Omega)(U),$$

*up to an* $\mathcal{A}(U)$*-isomorphism.*

**Proof.** Indeed, one obtains;

$$\begin{aligned}
&\mathcal{H}om_{\mathcal{A}}(\mathcal{E}, \Omega(\mathcal{E}))(U) \equiv \mathcal{H}om_{\mathcal{A}}(\mathcal{E}, \mathcal{E} \otimes_{\mathcal{A}} \Omega)(U) = \mathrm{Hom}_{\mathcal{A}|_U}(\mathcal{E}|_U, (\mathcal{E} \otimes_{\mathcal{A}} \Omega)|_U) \\
&= \mathrm{Hom}_{\mathcal{A}|_U}(\mathcal{E}|_U, \mathcal{E}|_U \otimes_{\mathcal{A}|_U} \Omega|_U) = \mathrm{Hom}_{\mathcal{A}|_U}(\mathcal{A}^n|_U, \mathcal{A}^n|_U \otimes_{\mathcal{A}|_U} \Omega|_U) \\
(9.5)\ &= \mathcal{H}om_{\mathcal{A}}(\mathcal{A}^n, \mathcal{A}^n \otimes_{\mathcal{A}} \Omega)(U) = (\mathcal{H}om_{\mathcal{A}}(\mathcal{A}^n, \mathcal{A}^n) \otimes_{\mathcal{A}} \Omega)(U) \\
&\equiv (\mathcal{E}nd\,\mathcal{A}^n \otimes_{\mathcal{A}} \Omega)(U) \equiv \Omega(M_n(\mathcal{A}))(U) = M_n(\Omega)(U) \\
&\qquad \equiv \Omega^{n^2}(U) = \Omega(U)^{n^2} \equiv M_n(\Omega(U)),
\end{aligned}$$

which thus proves our claim, concerning (9.3). In this context, see also (8.36) above, together with Chapt. II; (6.8), (5.25), (3.14) and Chapt. IV; (6.8). On the other hand, (9.4) follows now from the preceding and (7.22), and this terminates the proof. ∎

Now, as a consequence of the above, which also motivates the terminology we apply in the sequel, one obtains the next.

**Corollary 9.1.** *Suppose we are given a differential triad* $(\mathcal{A}, \partial, \Omega)$ *on a topological space* $X$, $\mathcal{E}$ *a vector sheaf on* $X$, *of rank* $n \in \mathbf{N}$, *and* $\mathcal{U} = (U_\alpha)_{\alpha \in I}$ *a local frame of* $\mathcal{E}$. *Then, the Levi-Civita* ($\mathcal{A}$*-connection*) *1-cocycle of* $\mathcal{E}$, *associated with* $\mathcal{U}$ (see (8.34)),

(9.6) $$(D_\beta - D_\alpha) \in Z^1(\mathcal{U}, \Omega(\mathcal{E}nd\,\mathcal{E})),$$

*is such that* (cf. (9.4))

(9.7) $$D_\beta - D_\alpha \in \Omega(\mathcal{E}nd\,\mathcal{E})(U_{\alpha\beta}) = M_n(\Omega)(U_{\alpha\beta}) = M_n(\Omega(U_{\alpha\beta})),$$

*for any* $\alpha, \beta \in I$, *with* $U_{\alpha\beta} \equiv U_\alpha \cap U_\beta \neq \emptyset$. *Therefore, one has;*

(9.8) $$(D_\beta - D_\alpha) \in Z^1(\mathcal{U}, M_n(\Omega)).$$

**Proof.** The only thing we have to check, by applying (9.4) to derive (9.7), is that *the intersection of two local gauges of* $\mathcal{E}$ *is still such a one,* and this is certainly clear. The rest is obvious, by the very definitions (see e.g. Chapt. III; (4.11), (4.12), or even (11.3) therein). ∎

In addition to the above relation (9.8), where $M_n(\Omega)$ appears (in fact, just local sections of it (cf. (9.7)), as we shall see in the sequel (cf. the next Chapter VII), *any given* $\mathcal{A}$*-connection of a vector sheaf* $\mathcal{E}$ on $X$ *can be expressed locally through (local) sections of the* $\mathcal{A}$*-module* $M_n(\Omega)$; viz. via elements of the *complete presheaf*

(9.9) $$\Gamma(M_n(\Omega)) \equiv M_n(\Gamma(\Omega)).$$

So, as is also the case in the classical theory, this is a good reason to call the above $\mathcal{A}$*-module* on $X$,

(9.10) $$M_n(\Omega) \equiv M_n(\mathcal{A}) \otimes_{\mathcal{A}} \Omega \equiv \Omega(M_n(\mathcal{A})),$$

the *sheaf of* $\mathcal{A}$*-connection coefficients* of the vector sheaf $\mathcal{E}$, associated thus with the given differential triad $(\mathcal{A}, \partial, \Omega)$. Yet, for simplicity, we call (9.10) the *sheaf of connection coefficients of* $\mathcal{E}$.

The preceding establish thus completely, by now, our initial objective, as stated in (9.1). In this context, we still remark that the conclusion about (9.10) was actually emerged by employing local gauges, hence, *local frames,* of the vector sheaf involved. As a matter of fact, the point here is that:

(9.11) *by using,* for instance, *Čech cohomology* (see Chapt. III; Section 4), *one can always argue in terms of local frames of* $\mathcal{E}$, when considering (Čech) *cohomology groups* (in effect, $\mathcal{A}(X)$*-modules;* ibid.)

(In this regard, see also Chapt. V; Section 3, in particular, (3.10)). Now,

concerning our claim in (9.11), this is due, indeed, to the fact that (Chapt. III; (4.9), in conjunction with Chapt. II; (4.9)):

(9.12) *the local frames of a given vector sheaf $\mathcal{E}$ on $X$ constitute a cofinal subset of the set of all (proper) open coverings of $X$.*

So taking, for instance, the *first (Čech) cohomology group* (in fact, $\mathcal{A}(X)$-*module*) of $X$ with coefficients in the $\mathcal{A}$-*module* (9.10), as above, one obtains;

(9.13)
$$\check{H}^1(X, M_n(\Omega)) = \varinjlim_{\mathcal{U}} \check{H}^1(\mathcal{U}, M_n(\Omega)),$$

where $\mathcal{U}$ *ranges over the local frames of* $\mathcal{E}$. (Here (9.13) is valid, within an $\mathcal{A}(X)$-*isomorphism of* the $\mathcal{A}(X)$-*modules* involved; cf. Chapt. III; (4.53), as well as, e.g. N. Bourbaki [1: p.95, Proposition 8]).

Accordingly, as a consequence of (9.8) and (9.13), in conjunction with our previous argument in (8.34), we conclude now the following;

(9.14)

given a *vector sheaf* $\mathcal{E}$ on $X$ and a *local frame* $\mathcal{U}$ of it (cf. (8.10)), by using the *Levi-Civita 1-cocycle of* $\mathcal{E}$, associated with $\mathcal{U}$ (cf. (8.34.1)), one defines a *1-dimensional (Čech) cohomology class*

(9.14.1)
$$[(D_\beta - D_\alpha)] \in \check{H}^1(X, M_n(\Omega)).$$

Thus, we set

(9.14.2)
$$\mathfrak{a}(\mathcal{E}) := [(D_\beta - D_\alpha)].$$

The latter class is, of course, *uniquely determined by* $\mathcal{U}$ (see (9.8) and (9.13). We call $\mathfrak{a}(\mathcal{E})$ the *1-dimensional* $\mathcal{A}$-*connection (Čech) cohomology class of* $\mathcal{E}$; yet, by extending the classical terminology (cf., for instance, J.L. Koszul [2: p.119]), we simply call it too, the *Atiyah class of* $\mathcal{E}$.

Furthermore, it is still clear that *(9.14.1) depends only on the isomorphism class of a given vector sheaf* $\mathcal{E}$ on $X$ (cf. also Chapt. II; (4.15)). Therefore, the same definition (9.14.2) yields further, *for every* $n \in \mathbf{N}$, a map

(9.15)
$$\mathbf{\Phi}^n_{\mathcal{A}}(X) \longrightarrow \check{H}^1(X, M_n(\Omega)),$$

called the *Atiyah map* (or even *Atiyah operator*) of $X$; its definition is thus associated simply with the given *differential triad* $(\mathcal{A}, \partial, \Omega)$ on $X$.

Thus, one obtains an "exact sequence" of the form;

$$(9.16) \qquad 0 \longrightarrow \ker \mathfrak{a} \longrightarrow \Phi^n_{\mathcal{A}}(X) \xrightarrow{\mathfrak{a}} \check{H}^1(X, M_n(\Omega)).$$

As we shall see in the sequel (cf., for instance, Corollary 12.1), *the kernel of the Atiyah map,* as above,

$$(9.17) \qquad \ker \mathfrak{a} \equiv \Phi^n_{\mathcal{A}}(X)^{\nabla},$$

*is exactly the set of those vector sheaves* (of rank $n \in \mathbf{N}$) *on* $X$*, which admit an* $\mathcal{A}$*-connection* (*"Yang-Mills fields"* on $X$; of course, we do not distinguish here between *isomorphic vector sheaves* (see e.g. (18.6) below). In the particular *case* $n = 1$ (*line sheaves* on $X$), we further give below an appropriate *cohomological interpretation of* the previous set (9.17), in terms of (Čech) *hypercohomology* (see Theorem 18.2). Namely, as we are going to see right below, the above cohomology class, as given by (9.14.1), provides thus a means to examine the existence of an $\mathcal{A}$-connection for a given *vector sheaf* $\mathcal{E}$ on $X$. Precisely, by complete analogy with the classical case (cf. M.F. Atiyah [1]), one obtains through it a *criterion,* what, namely, is called in the sequel the *Atiyah criterion,* for the existence of an $\mathcal{A}$-connection of $\mathcal{E}$. This will actually occupy us in the succeeding few Sections. However, it is instructive first to consider (we shall use it, for that matter) an existing relationship between (8.34.1) and the *coordinate 1-cocycle of* $\mathcal{E}$, which, as we know (cf. Chapt. V; (1.8)), can be associated with any given *local frame* $\mathcal{U}$ of $\mathcal{E}$.

# 10. Levi-Civita 1-cocycles, as related to coordinate 1-cocycles of vector sheaves

As already said, we consider below a useful relation, that naturally emerges between the 1-cocycles in the heading of this Section. Yet, the latter cocycles are, as already known, canonically associated with any given local frame of a vector sheaf (cf. the following discussion).

So to fix the terminology applied, assume that we are given a *vector sheaf* $\mathcal{E}$ on a *topological space* $X$, the latter being, as usual, the base space of a

**C**-*algebraized space* $(X, \mathcal{A})$. Yet, we consider a *differential triad* $(\mathcal{A}, \partial, \Omega)$. Finally, suppose that

$$\mathcal{U} = (U_\alpha)_{\alpha \in I} \tag{10.1}$$

is a given *local frame of* $\mathcal{E}$.

Thus, the two following *1-cocycles of* $\mathcal{U}$ *with coefficients in* $M_n(\Omega)$ *and* $\mathcal{GL}(n, \mathcal{A})$, respectively, are canonically associated with the given vector sheaf $\mathcal{E}$, where *we set* $n = \mathrm{rk}\,(\mathcal{E}) \in \mathbf{N}$:

1) The *Levi-Civita 1-cocycle of* $\mathcal{E}$, canonically associated with $\mathcal{U}$, i.e., one has (see (8.34), (9.4) and (9.8))

$$(D_\beta - D_\alpha) \in Z^1(\mathcal{U}, \Omega(\mathcal{E}nd\,\mathcal{E})) = Z^1(\mathcal{U}, M_n(\Omega)). \tag{10.2}$$

2) The *coordinate 1-cocycle of* $\mathcal{E}$, as defined by $\mathcal{U}$; namely, one has (see Chapt. V; (1.7) and Definition 1.1),

$$(\eta_{\alpha\beta}) \in Z^1(\mathcal{U}, \mathcal{GL}(n, \mathcal{A})). \tag{10.3}$$

Here we have set;

$$\eta_{\alpha\beta} := \eta_\alpha \circ \eta_\beta^{-1}, \tag{10.4}$$

*for any* $\alpha, \beta \in I$, *with* $U_{\alpha\beta} \equiv U_\alpha \cap U_\beta \neq \emptyset$, while the $\mathcal{A}|_{U_\alpha}$*-isomorphism* $\eta_\alpha$, $\alpha \in I$, is given by (8.12) (see also Chapt. V; (1.3.2) and (1.5.1)).

Now, based on (8.29), *one obtains*, concerning (10.2);

$$D_\beta - D_\alpha = \left(\eta_\beta^{-1} \otimes 1\right) \partial_\beta^n \eta_\beta - \left(\eta_\alpha^{-1} \otimes 1\right) \partial_\alpha^n \eta_\alpha, \tag{10.5}$$

*for any* $\alpha, \beta \in I$, *as in (10.4)*. For convenience, *we* still *set* here

$$1_{\Omega|_{U_\alpha}} \equiv 1, \quad \alpha \in I, \tag{10.6}$$

while an *analogous notation will* also *be applied,* in the case one considers *any* $\alpha, \beta \in I$, *as in (10.4)*. Thus, in view of (10.4) and (10.5), one has;

$$(\eta_\alpha \otimes 1)(D_\beta - D_\alpha)\eta_\alpha^{-1} = (\eta_{\alpha\beta} \otimes 1)\partial_\beta^n \eta_{\alpha\beta}^{-1} - \partial_\alpha^n. \tag{10.7}$$

Now, based e.g. on our argument in (9.5), we understand that the *first member of (10.7)* is an *element of the* following *set;*

$$M_n(\Omega)(U_{\alpha\beta}) = \mathrm{Hom}_{\mathcal{A}|_{U_{\alpha\beta}}}(\mathcal{A}^n|_{U_{\alpha\beta}}, \Omega^n|_{U_{\alpha\beta}}) \tag{10.8}$$

(it concerns, in fact, an $\mathcal{A}(U_{\alpha\beta})$*-module*). Therefore (cf. Chapt. I; Proposition

2.1, yet (2.17)), the same element is known, *whenever we know its value at the "constant" section*

$$\mathbf{1} \in \mathcal{A}^n(X) = \mathcal{A}(X)^n, \tag{10.9}$$

suitably *localized,* as the case may be. Of course, we have set in (10.9)

$$\mathbf{1} \equiv \underbrace{1 \oplus \cdots \oplus 1}_{n-\text{times}}, \tag{10.10}$$

where $\mathbf{1} \equiv 1_{\mathcal{A}} \in \mathcal{A}(X)$ denotes the (global) *identity section of* (the $\mathbf{C}$*-algebra sheaf*) $\mathcal{A}$; see also Chapt. II; Definition 1.4 and (1.34), (1.34′). Thus, in other words, one may *replace the element in question, by its value at* (the *section*) 1, as before. (In this context, see also e.g. Chapt. I; (12.50)). On the other hand, for convenience, we still note that, if $u$ belongs, for instance, to (9.5), then, *for every* $\alpha \in \mathcal{A}^n(U)$, *one has* $u(\alpha) \equiv u(\alpha \cdot 1) = \alpha \cdot u(1)$, which thus vindicates completely our last assertion. ∎ Thus, in view now of (10.7), one obtains:

$$\begin{aligned}(\eta_\alpha \otimes 1)(D_\beta - D_\alpha)\eta_\alpha^{-1} &\equiv ((\eta_\alpha \otimes 1)(D_\beta - D_\alpha)\eta_\alpha^{-1})\,(\mathbf{1}) \\ &= \left((\eta_{\alpha\beta} \otimes 1) \circ \partial_\beta^n\right)\left(\eta_{\alpha\beta}^{-1}(\mathbf{1})\right) \equiv \left((\eta_{\alpha\beta} \otimes 1)\partial_\beta^n \eta_{\alpha\beta}^{-1}\right)(\mathbf{1}),\end{aligned} \tag{10.11}$$

given that, one has, of course,

$$\partial_\alpha^n(\mathbf{1}) = 0; \tag{10.12}$$

the last relation follows from (5.6) and (5.4) (cf. also (8.28)), in conjunction with (the obvious *extension to* $\mathcal{A}^n$ *of*) Lemma 1.1. Thus, here one considers, *fiber-wise,* $\mathcal{A}^n$ as an $\mathcal{A}$*-algebra sheaf* on $X$, with "$\mathbf{1}$" given by (10.10); see also Chapt. II; Definition 6.2). ∎

In a similar manner, one may consider, *equivalently,* $\eta_{\alpha\beta}(\mathbf{1})$, with $\mathbf{1}$ given by (10.10), *in place of*

$$\begin{aligned}\eta_{\alpha\beta} \in \mathcal{GL}(n, \mathcal{A})(U_{\alpha\beta}) &\subseteq \mathcal{H}om_{\mathcal{A}}(\mathcal{A}^n, \mathcal{A}^n)(U_{\alpha\beta}) \\ &= \operatorname{Hom}_{\mathcal{A}|_{U_{\alpha\beta}}}(\mathcal{A}^n|_{U_{\alpha\beta}}, \mathcal{A}^n|_{U_{\alpha\beta}}).\end{aligned} \tag{10.13}$$

So *we* further *set;*

$$g_{\alpha\beta} := \eta_{\alpha\beta}^{-1}(\mathbf{1}) \in \mathcal{A}^n(U_{\alpha\beta}) = \mathcal{A}(U_{\alpha\beta})^n, \tag{10.14}$$

*for any* $\alpha, \beta \in I$, *with* $U_{\alpha\beta} \equiv U_\alpha \cap U_\beta \neq \emptyset$. Yet, applying the same argument, as before (cf. (10.11)), and based on (10.14), one also obtains;

(10.15) $$\eta_{\alpha\beta} \otimes 1 \equiv \eta_{\alpha\beta}(\mathbf{1}) \cdot 1_{\Omega|_{U_{\alpha\beta}}} \equiv \eta_{\alpha\beta}(\mathbf{1}) = g_{\alpha\beta}^{-1}.$$

Therefore, one concludes from (10.11) that;

(10.16) $$(\eta_\alpha \otimes 1)(D_\beta - D_\alpha)\eta_\alpha^{-1} = g_{\alpha\beta}^{-1} \cdot \partial_\beta^n(g_{\alpha\beta}),$$

that is, in view of (1.29) (cf. also (1.28), for $\mathcal{E} = \Omega$, and (3.1)), along with (8.28), one finally obtains;

(10.17) $$(\eta_\alpha \otimes 1)(D_\beta - D_\alpha)\eta_\alpha^{-1} = \tilde{\partial}(g_{\alpha\beta}).$$

On the other hand, concerning the *1-cochain* of $\mathcal{U}$

(10.18) $$(g_{\alpha\beta}) \in C^1(\mathcal{U}, \mathcal{H}om_{\mathcal{A}}(\mathcal{A}^n, \mathcal{A}^n)),$$

as defined by (10.14), (10.13), one easily sees, according to the very definitions, that the same yields, in fact, a *1-cocycle*

(10.19) $$(g_{\alpha\beta}) \in Z^1(\mathcal{U}, \mathcal{GL}(n, \mathcal{A}));$$

moreover, the latter actually constitutes the *coordinate 1-cocycle of the vector sheaf* $\mathcal{E}$ at issue, which is *associated with the given local frame* $\mathcal{U}$ *of* $\mathcal{E}$ (see (10.1), and Chapt. V; (1.2), (1.3.2), together with the above rels. (8.12), (10.4) and (10.13), (10.14)).

So we can now summarize the preceding discussion into the form of the following.

**Theorem 10.1.** *Consider a differential triad* $(\mathcal{A}, \partial, \Omega)$ *on a topological space* $X$, $\mathcal{E}$ *a vector sheaf on* $X$, *of rank* $n \in \mathbf{N}$, *and* $\mathcal{U} = (U_\alpha)_{\alpha\in I}$ *a local frame of* $\mathcal{E}$. *Then, the Levi-Civita (*$\mathcal{A}$*-connection) 1-cocycle of* $\mathcal{E}$, *relative to* $\mathcal{U}$ (see (8.34), (9.8)),

(10.20) $$(D_\beta - D_\alpha) \in Z^1(\mathcal{U}, M_n(\Omega)),$$

*is gauge equivalent* (relative to the $\mathcal{A}|_{U_\alpha}$*-isomorphism* $\eta_\alpha$, $\alpha \in I$, cf. (8.12) or even (8.21.1)) *to the logarithmic derivative of the coordinate 1-cocycle of* $\mathcal{E}$, *associated with* $\mathcal{U}$, *say,*

(10.21) $$(g_{\alpha\beta}) \in Z^1(\mathcal{U}, \mathcal{GL}(n, \mathcal{A}))$$

(cf. (10.19)). *That is, one has* (see (10.17));

(10.22) $$(\eta_\alpha \otimes 1)(D_\beta - D_\alpha)\eta_\alpha^{-1} = \tilde{\partial}(g_{\alpha\beta}).$$

*for any* $\alpha, \beta \in I$, *with* $U_{\alpha\beta} \equiv U_\alpha \cap U_\beta \neq \emptyset$. *Yet, we write (10.22) in the following abbreviated form* (see also (8.32.3));

$$\delta(D_\alpha) \underset{\alpha}{\sim} \tilde{\partial}(g_{\alpha\beta}), \quad \alpha \in I. \blacksquare \tag{10.23}$$

Now, in view of (9.5) and (9.7) (see also e.g. (10.8)), one has;

$$\begin{aligned}(\eta_\alpha \otimes 1)(D_\beta - D_\alpha)\eta_\alpha^{-1} &\in Z^1(\mathcal{U}, \mathcal{H}om_{\mathcal{A}}(\mathcal{A}^n, \Omega(\mathcal{A}^n))) \\ &= Z^1(\mathcal{U}, M_n(\Omega)).\end{aligned} \tag{10.24}$$

Furthermore, concerning the *coordinate 1-cocycle of the vector sheaf* $\mathcal{E}$, under consideration (cf. (10.19)), *one* still *obtains* that;

$$\tilde{\partial}((g_{\alpha\beta})) \equiv \left(\tilde{\partial}(g_{\alpha\beta})\right) \in Z^1(\mathcal{U}, M_n(\Omega)). \tag{10.25}$$

Indeed, the last assertion follows from the relation;

$$\tilde{\partial}(g_{\alpha\gamma}) = \mathrm{Ad}\left(g_{\beta\gamma}^{-1}\right) \tilde{\partial}(g_{\alpha\beta}) + \tilde{\partial}(g_{\beta\gamma}), \tag{10.26}$$

*where* $\alpha, \beta, \gamma \in I$, *such that* $U_\alpha \cap U_\beta \cap U_\gamma \neq \emptyset$, which, in turn, in an immediate consequence of (10.21) and (1.33). ∎

Accordingly, as a result of (10.22), (10.24) and (10.25), *one obtains the following,* basic for the sequel, *relation* in (Čech) *cohomology;*

$$[(D_\beta - D_\alpha)] = \left[\left(\tilde{\partial}(g_{\alpha\beta})\right)\right] \in \check{H}^1(X, M_n(\Omega)). \tag{10.27}$$

Therefore, by virtue of (9.14), one concludes, by now, that;

(10.28) *the Atiyah class of a vector sheaf* $\mathcal{E}$ on $X$ *is given by*

(10.28.1) $\mathfrak{a}(\mathcal{E}) := [(D_\beta - D_\alpha)] = \left[\left(\tilde{\partial}(g_{\alpha\beta})\right)\right] \in \check{H}^1(X, M_n(\Omega)).$

Hence, by considering the Atiyah map (9.15), together with the (set-theoretic) *bijection*

$$\Phi_{\mathcal{A}}^n(X) = \check{H}^1(X, \mathcal{GL}(n, \mathcal{A})) \tag{10.29}$$

(see (9.14.2) and Chapt. V; Theorem 2.1), one obtains the following *commutative diagram;*

$$
\begin{array}{ccc}
\check{H}^1(X, \mathcal{GL}(n, \mathcal{A})) & \xrightarrow{\tilde{\partial}} & \check{H}^1(X, M_n(\Omega)) \\
\Vert & \nearrow_{\mathfrak{a}} & \\
\Phi^n_{\mathcal{A}}(X) & &
\end{array}
\tag{10.30}
$$

The maps appeared in the previous diagram are, of course, the relevant *extensions to cohomology* (due to the *functorial behaviour*(!) of the cohomology functors, see e.g. Chapt. III; Theorem 3.1 and Lemma 8.1, along with (8.3)) *of the maps (1.28)* (for $\mathcal{E} = \Omega$, cf. (3.1) and (9.15)). Yet, the *commutativity of (10.30)* holds certainly true, by virtue of (10.28.1), so that one actually obtains the following relation;

$$
\mathfrak{a} = \tilde{\partial}, \tag{10.31}
$$

a fact that will be of use right in the next Section 11.

**Note 10.1.**– Concerning the above calculations about the *basic relation (10.22),* we still note, from a historical perspective, that the motive thereby was initially a seminal (for our whole subject matter(!), in effect) paper of S.A. Selesnick [2], which we shall also use later on. The same was actually referred to certain relevant questions, within the framework of *Banach algebra theory*! Yet, we shall have the opportunity to look at it, more closely, in terms also of *topological algebra theory,* subsequently (see thus Chapter XI below).

## 11. Existence of $\mathcal{A}$-connections

As already promised, we start with this Section to seek for conditions guaranteeing the *existence of $\mathcal{A}$-connections of a* given *vector sheaf* on $X$, an inquiry that deviates from our standard experience, pertaining e.g. to smooth (viz. $\mathcal{C}^\infty$-) $\mathbf{C}$-vector bundles. The relevant reason, classically exhibited already by M.F. Atiyah [1], attains also, within the present abstract

framework, an analogous formulation; so this will be our objective in this and the two following Sections.

Thus, to begin with, suppose that we are given a vector sheaf $\mathcal{E}$ on $X$, and let

$$\mathfrak{a}(\mathcal{E}) \equiv [(D_\beta - D_\alpha)] \in \check{H}^1(X, M_n(\Omega)) \tag{11.1}$$

be the *Atiyah class of* $\mathcal{E}$, determined (uniquely) through a given local frame of $\mathcal{E}$ (cf. (9.14.2)). So our program now is to show that:

(11.2) *a vector sheaf $\mathcal{E}$ has an $\mathcal{A}$-connection if, and only if, one has;*

$$\mathfrak{a}(\mathcal{E}) = 0. \tag{11.2.1}$$

Thus, we start right away, with considering first the *"if"* part (in fact, the easy one) of our assertion, while the rest (viz. the *"only if"* part) will be treated in the sequel (see Sections 12, 13 below). So we now have the following.

**Theorem 11.1.** *Let $\mathcal{E}$ be a given vector sheaf on $X$, such that the Atiyah class of it, associated with a local frame of $\mathcal{E}$, vanishes. That is, we assume that*

$$\mathfrak{a}(\mathcal{E}) = 0. \tag{11.3}$$

*Then, $\mathcal{E}$ admits an $\mathcal{A}$-connection.*

**Proof.** Suppose that

$$\mathfrak{a}(\mathcal{E}) \equiv [(D_\beta - D_\alpha)] = 0 \in \check{H}^1(X, M_n(\Omega)) \tag{11.4}$$

is the Atiyah class of the given vector sheaf $\mathcal{E}$ on $X$, with $n = \operatorname{rk}\mathcal{E}$, corresponding to a *local frame of* $\mathcal{E}$,

$$\mathcal{U} = (U_\alpha)_{\alpha\in I} \tag{11.5}$$

(see (9.14.2)). Thus, in view of (11.4) (see also, for instance, Chapt. II; (11.26.1), (11.16), along with Remark 11.1 and Lemma 8.1), one has

$$(D_\beta - D_\alpha) = \delta(\theta_\alpha) \in \delta^0(\check{C}^0(\mathcal{U}, M_n(\Omega))), \tag{11.6}$$

for some *0-cochain*

$$\theta \equiv (\theta_\alpha) \in \check{C}^0(\mathcal{U}, M_n(\Omega)). \tag{11.7}$$

Therefore, one obtains (cf. Chapt. III; (4.16))

$$\delta(D_\alpha) = (D_\beta - D_\alpha) = \delta(\theta_\alpha), \tag{11.8}$$

so that one has

$$\delta(D_\alpha) - \delta(\theta_\alpha) = \delta(D_\alpha - \theta_\alpha) = 0. \tag{11.9}$$

Furthermore, one has

$$(D_\alpha - \theta_\alpha) \in \check{C}^0(\mathcal{U}, \mathcal{H}om_{\mathbf{C}}(\mathcal{E}, \Omega(\mathcal{E}))), \tag{11.10}$$

which follows from (8.32.1) and (11.7), in conjunction with (9.3). Yet, we still note, in passing, that from the same rel. (9.3), *one obtains;*

$$\check{C}^0(\mathcal{U}, M_n(\Omega)) = \check{C}^0(\mathcal{U}, \mathcal{H}om_{\mathcal{A}}(\mathcal{E}, \Omega(\mathcal{E}))). \tag{11.11}$$

Now, by virtue of (11.9), *the 0-cochain (11.10) is,* in effect, *a 0-cocycle,* so that one has;

$$\begin{aligned} (D_\alpha) - (\theta_\alpha) = (D_\alpha - \theta_\alpha) &\in \check{Z}^0(\mathcal{U}, \mathcal{H}om_{\mathbf{C}}(\mathcal{E}, \Omega(\mathcal{E}))) \\ &= \mathcal{H}om_{\mathbf{C}}(\mathcal{E}, \Omega(\mathcal{E}))(X) = \mathrm{Hom}_{\mathbf{C}}(\mathcal{E}, \Omega(\mathcal{E})) \end{aligned} \tag{11.12}$$

(see Chapt. III; Lemma 8.1). Therefore, let

$$\widetilde{D} := (D_\alpha - \theta_\alpha) \tag{11.13}$$

be the corresponding $\mathbf{C}$*-linear morphism, as given by (11.12).* On the other hand, based on (11.13), (11.7) and (9.4), as well as, on (7.19), *one* still *obtains;*

$$\begin{aligned} \widetilde{D}_\alpha := \widetilde{D}|_{U_\alpha} = D_\alpha - \theta_\alpha &\in D_\alpha + \Omega(\mathcal{E}nd\,\mathcal{E})(U_\alpha) \\ &= Conn_{\mathcal{A}|_{U_\alpha}}(\mathcal{E}|_{U_\alpha}) \equiv Conn_{\mathcal{A}}(\mathcal{E})|_{U_\alpha}, \end{aligned} \tag{11.14}$$

*for every* $\alpha \in I$. Thus, $\widetilde{D}_\alpha$ *is an* $\mathcal{A}|_{U_\alpha}$*-connection of* $\mathcal{E}|_{U_\alpha}$, *for every* $U_\alpha \in \mathcal{U}$ (cf. (11.5)); accordingly, in view of (11.13) and (11.14),

$$\widetilde{D} = \left(\widetilde{D}_\alpha\right) = (D_\alpha - \theta_\alpha) \tag{11.15}$$

*is an* $\mathcal{A}$*-connection of* $\mathcal{E}$ too, since the remaining yet (see thus (11.12)) *Leibniz condition for* $\widetilde{D}$ is now fulfilled, this holding true *locally,* according to (11.14), which thus terminates the proof. ∎

Now, by analogy with the terminology, already employed for (8.29) (see the comments following (8.30)), we set the next.

**Definition 11.1.** Let $\mathcal{E}$ be a vector sheaf on $X$. Then, the $\mathcal{A}$-connection of $\mathcal{E}$, which is provided by (11.15), under the cond. (11.3), is called a *Levi-Civita $\mathcal{A}$-connection of $\mathcal{E}$*. Yet, for simplicity, the term *Levi-Civita connection of $\mathcal{E}$* will occasionally be employed for the same notion, as before.

In this context, we still note, in anticipation (see also (11.2)), that;

(11.16) given a *vector sheaf* $\mathcal{E}$ on $X$, *the condition* (cf. (7.1))

(11.16.1) $$Conn_{\mathcal{A}}(\mathcal{E}) \neq \emptyset$$

*is necessary and sufficient in order $\mathcal{E}$ to have a Levi-Civita $\mathcal{A}$-connection.* In other words, *a vector sheaf $\mathcal{E}$ on $X$, having an $\mathcal{A}$-connection, has also a Levi-Civita connection* (*and conversely,* of course).

Full justification of the previous claim will be given in the sequel (see thus Theorem 12.1, as well as, Corollary 12.1 below along with (12.31)).

**11.(a). Integrable $\mathcal{A}$-connections.**– The notion in the title of this Subsection will be fully clarified later on (see Chapter VIII in the sequel). However, we do look at it hereby, in anticipation, giving its formal definition, while, as we shall see, right away, the same notion is naturally connected with the previous one of a Levi-Civita connection. So we start with the following.

**Definition 11.2.** Suppose we have a vector sheaf $\mathcal{E}$ on $X$. Then, by an *integrable $\mathcal{A}$-connection* of $\mathcal{E}$, one means an $\mathcal{A}$-connection $\widetilde{D}$ of $\mathcal{E}$, for which there exists a local frame $\mathcal{U} = (U_\alpha)_{\alpha \in I}$ of $\mathcal{E}$, such that one has

(11.17) $$\widetilde{D}|_{U_\alpha} \equiv \widetilde{D}_\alpha = D_\alpha, \quad \alpha \in I,$$

where $D_\alpha$, $\alpha \in I$, is given by (8.13).

As already said, $\mathcal{A}$-connections of the previous type will be amply considered in the sequel, namely, when the notion of *curvature* of a given $\mathcal{A}$-connection will be available. Nonetheless, we do examine in the meantime the above kind of $\mathcal{A}$-connections in conjunction with our argument in Section 8, a fact which also explains the reason for the succeeding discussion.

Thus, suppose that the *Levi-Civita 0-cochain*

(11.18) $$(D_\alpha) \in \check{C}^0(\mathcal{U}, \mathcal{H}om_{\mathbf{C}}(\mathcal{E}, \Omega(\mathcal{E}))),$$

which can be defined for any given *local frame* $\mathcal{U} = (U_\alpha)_{\alpha \in I}$ of a *vector sheaf* $\mathcal{E}$ on $X$, is, in particular, a *0-cocycle* of $\mathcal{U}$; that is, *assume that:*

(11.19) $$(D_\alpha) \in \check{Z}^0(\mathcal{U}, \mathcal{H}om_{\mathbf{C}}(\mathcal{E}, \Omega(\mathcal{E}))) = \mathrm{Hom}_{\mathbf{C}}(\mathcal{E}, \Omega(\mathcal{E}))$$

(see also, for instance, (11.12)). Hence, in view of our hypothesis for the individual sections $D_\alpha$, $\alpha \in I$, in (11.19) (cf. Lemma 8.1, in particular, (8.13)), one concludes that;

(11.20) $$D := (D_\alpha),$$

as in (11.19), *is an $\mathcal{A}$-connection of $\mathcal{E}$*, such that, by definition, one has

(11.21) $$D|_{U_\alpha} = D_\alpha, \quad \alpha \in I.$$

Consequently,

(11.22) *(11.20) is an integrable $\mathcal{A}$-connection* of $\mathcal{E}$.

Yet, in this context, we further remark that, according to the very definitions (see (11.15) and (11.17)),

(11.23) *every integrable $\mathcal{A}$-connection* of a given *vector sheaf $\mathcal{E}$ is,* in particular, *a Levi-Civita $\mathcal{A}$-connection* of $\mathcal{E}$, *in such a manner that*

(11.23.1) $$(\theta_\alpha) = 0 \in C^0(\mathcal{U}, M_n(\Omega)).$$

Concerning (11.23.1), see also (11.7), (11.14) and (11.17). ∎ The same relation (11.23.1), for an integrable $\mathcal{A}$-connection, will be of use later on (cf. Chapter VIII in the sequel). Finally, based on the preceding, one concludes that:

(11.24) *the Levi-Civita 0-cochain (11.18) is a 0-cocycle,* yielding *thus a (Levi-Civita) $\mathcal{A}$-connection* (cf. (11.20)), *if, and only if, the latter is an integrable $\mathcal{A}$-connection.*

**11.(b). Levi-Civita $\mathcal{A}$-connections of $\mathcal{A}^n$, $n \in \mathbf{N}$.**– Our purpose by the following discussion is to show that;

(11.25) *the existence of an $\mathcal{A}$-connection of a* given *vector sheaf* $\mathcal{E}$ on $X$ *implies that one too of a Levi-Civita $\mathcal{A}$-connection of $\mathcal{A}^n$, where $n = \operatorname{rk}\mathcal{E}$.*

The *converse* of the above claim is, of course, *not always true, unless (11.3) is fulfilled,* as this is easily seen by the standard *$\mathcal{A}$-connection $\partial^n$ of $\mathcal{A}^n$*; see (8.4), (11.2) and (11.16), along with the succeeding comments to the two last quotations. Yet, cf. also Lemma 8.1.

Thus, suppose we are given a *vector sheaf* $\mathcal{E}$ on $X$, of *rank* $n \in \mathbf{N}$, admitting an *$\mathcal{A}$-connection* $\widetilde{D}$; moreover, let $\mathcal{U}$ be a *local frame of* $\mathcal{E}$, as in (11.5). Then, by setting

$$\widetilde{D}_\alpha := \widetilde{D}|_{U_\alpha}, \quad \alpha \in I, \tag{11.26}$$

one defines, *for every* $\alpha \in I$, by analogy with (8.11), a *commutative diagram,* as follows;

$$\begin{array}{ccc} \mathcal{E}|_{U_\alpha} & \xrightarrow{\widetilde{D}_\alpha} & \mathcal{E} \otimes_{\mathcal{A}} \Omega|_{U_\alpha} \\ \Big\downarrow{\scriptstyle \eta_\alpha^{-1}} & & \Big\downarrow{\scriptstyle \eta_\alpha \otimes 1_{\Omega|_{U_\alpha}}} \\ \mathcal{A}^n|_{U_\alpha} & \overset{d_\alpha}{\dashrightarrow} & \mathcal{A}^n \otimes_{\mathcal{A}} \Omega|_{U_\alpha} \end{array} \tag{11.27}$$

That is, we set

$$d_\alpha := \left(\eta_\alpha \otimes 1_{\Omega|_{U_\alpha}}\right) \circ \widetilde{D} \circ \eta_\alpha^{-1}, \quad \alpha \in I. \tag{11.28}$$

As a result of the preceding, one thus infers, so far, that;

*for any $\mathcal{A}$-connection $\widetilde{D}$ of the vector sheaf $\mathcal{E}$ and any given local frame $\mathcal{U} = (U_\alpha)_{\alpha \in I} \in \mathcal{E}$, one obtains a 1-cocycle*

(11.29.1) $$\delta(d_\alpha) = (d_\beta - d_\alpha) \in Z^1(\mathcal{U}, M_n(\Omega))$$

(11.29) (see also e.g. (10.8)). We still refer to (11.29.1) as the *Levi-Civita 1-cocycle of $\mathcal{A}^n$, associated with $\mathcal{U}$,* while the *0-cochain*

(11.29.2) $$\begin{aligned}(d_\alpha) &\in C^0(\mathcal{U}, \mathcal{H}om_{\mathbf{C}}(\mathcal{A}^n, \Omega(\mathcal{A}^n))) \\ &= C^0(\mathcal{U}, \mathcal{H}om_{\mathbf{C}}(\mathcal{A}^n, \Omega^n)),\end{aligned}$$

as defined by (11.28), is still named the *Levi-Civita 0-cochain of $\mathcal{A}^n$, corresponding to* the given $\mathcal{U}$, as before.

Now, the question *whether the* above *0-cochain (11.29.2) yields an $\mathcal{A}$-connection of $\mathcal{A}^n$*, is treated by the following.

**Theorem 11.2.** *Let $\mathcal{E}$ be a vector sheaf on $X$, of rank $n \in \mathbf{N}$, $\mathcal{U} = (U_\alpha)_{\alpha \in I}$ a local frame of $\mathcal{E}$ and $\widetilde{D}$ an $\mathcal{A}$-connection of $\mathcal{E}$. Then, the Levi-Civita 0-cochain of $\mathcal{A}^n$, as derived from $\widetilde{D}$ and $\mathcal{U}$* (see (11.29.2)), *yields an $\mathcal{A}$-connection of $\mathcal{A}^n$, if, and only if, the respective Levi-Civita 1-cocycle of $\mathcal{A}^n$* (cf. (11.29.1)) *is a coboundary; i.e., equivalently, if, and only if, one has*

(11.30) $$[(D_\beta - D_\alpha)] = 0 \in \check{H}^1(X, M_n(\Omega)).$$

*Furthermore, in that case too, one obtains* (see (9.14))

(11.31) $$\mathfrak{a}(\mathcal{E}) = [(D_\beta - D_\alpha)] = 0 \in \check{H}^1(X, M_n(\Omega));$$

*hence* (Theorem 11.1, along with Definition 11.1), *the given vector sheaf $\mathcal{E}$ admits a Levi-Civita $\mathcal{A}$-connection, as well.*

**Proof.** Assuming (11.30), one obtains (see also e.g. (11.6))

(11.32) $$\delta(d_\alpha) = \delta(\psi_\alpha),$$

for some *0-cochain*

(11.33) $$\begin{aligned}\psi \equiv (\psi_\alpha) \in C^0(\mathcal{U}, \mathcal{H}om_{\mathcal{A}}(\mathcal{A}^n, \mathcal{A}^n \otimes_{\mathcal{A}} \Omega)) &= C^0(\mathcal{U}, \mathcal{H}om_{\mathcal{A}}(\mathcal{A}^n, \Omega^n)) \\ &= C^0(\mathcal{U}, M_n(\Omega))\end{aligned}$$

(cf. also, for instance, (10.8) or even the argument in (9.5)). Hence, one obtains from (11.32),

$$\delta(d_\alpha) - \delta(\psi_\alpha) = \delta(d_\alpha - \psi_\alpha) = 0, \tag{11.34}$$

(see Chapt. III; (4.15)), that is, we thus obtain a *0-cocycle,*

$$\begin{aligned}(d_\alpha - \psi_\alpha) \in Z^0(\mathcal{U}, \mathcal{H}om_{\mathbf{C}}(\mathcal{A}^n, \mathcal{A}^n \otimes_{\mathcal{A}} \Omega)) &= \mathcal{H}om_{\mathbf{C}}(\mathcal{A}^n, \mathcal{A}^n \otimes \Omega)(X) \\ = \mathrm{Hom}_{\mathbf{C}}(\mathcal{A}^n, \mathcal{A}^n \otimes_{\mathcal{A}} \Omega) &= \mathrm{Hom}_{\mathbf{C}}(\mathcal{A}^n, \Omega^n).\end{aligned} \tag{11.35}$$

Therefore, setting

$$\tilde{d} := (d_\alpha - \psi_\alpha), \tag{11.36}$$

or, *equivalently,*

$$\tilde{d}|_{U_\alpha} \equiv \tilde{d}_\alpha := d_\alpha - \psi_\alpha, \quad \alpha \in I, \tag{11.37}$$

one concludes, by (11.28) and (8.24), that

(11.38) $\tilde{d}$, as given by (11.36), *yields an $\mathcal{A}$-connection of $\mathcal{A}^n$.*

*Conversely,* suppose that the 0-cochain $(d_\alpha)$, as in (11.29.2), provides an $\mathcal{A}$-connection of $\mathcal{A}^n$; so, in particular, one has that

$$(d_\alpha) \in Z^0(\mathcal{U}, \mathcal{H}om_{\mathbf{C}}(\mathcal{A}^n, \Omega^n)) = \mathrm{Hom}_{\mathbf{C}}(\mathcal{A}^n, \Omega^n), \tag{11.39}$$

so that one obtains that $\delta(d_\alpha) = (d_\beta - d_\alpha) = 0$, which obviously implies (11.30). So *this completes the proof of the first part of the assertion.*

On the other hand, *to prove (11.31),* we first remark that, by virtue of (11.30) and since $\partial^n$ *is an $\mathcal{A}$-connection of $\mathcal{A}^n$* (cf. (8.4)), one concludes for $\tilde{d}$ (cf. (11.36), (11.38)), in view of Theorem 7.1, that:

$$\tilde{d} = \partial^n + \phi, \tag{11.40}$$

for some *uniquely defined*

$$\begin{aligned}\phi \equiv (\phi_\alpha) \in \Omega(\mathcal{E}nd\,\mathcal{A}^n)(X) &\equiv M_n(\Omega)(X) = M_n(\Omega(X)) \\ &= Z^0(\mathcal{U}, M_n(\Omega)) \subseteq C^0(\mathcal{U}, M_n(\Omega))\end{aligned} \tag{11.41}$$

(see also Chapt. IV; (3.24) and Chapt. V; (8.7), for the notation applied). Therefore, by (11.37) and (11.40), we obtain;

$$d_\alpha = \tilde{d}_\alpha + \psi_\alpha = \partial^n_\alpha + (\phi_\alpha + \psi_\alpha), \quad \alpha \in I, \tag{11.42}$$

such that, by (11.33) and (11.41), we have

$$\chi \equiv (\chi_\alpha) := \phi + \psi \equiv (\phi_\alpha + \psi_\alpha) \in C^0(\mathcal{U}, M_n(\Omega)). \tag{11.43}$$

Furthermore, based now on our hypothesis for $\widetilde{D}$ (cf. also (11.26)), on (11.28) and on (10.4), (10.6) and (10.7), along with (11.42) and (11.43), we obtain;

$$\begin{aligned} 0 &= (\eta_\alpha \otimes 1)(\widetilde{D}_\beta - \widetilde{D}_\alpha)\eta_\alpha^{-1} = (\eta_\alpha \otimes 1)\left(\eta_\beta^{-1} \otimes 1\right) d_\beta \eta_\beta \eta_\alpha^{-1} - d_\alpha \\ &= (\eta_{\alpha\beta} \otimes 1)\partial_\beta^n \eta_{\alpha\beta}^{-1} - \partial_\alpha^n + (\eta_{\alpha\beta} \otimes 1)\chi_\beta \eta_{\alpha\beta}^{-1} - \chi_\alpha \\ &= (\eta_\alpha \otimes 1)(D_\beta - D_\alpha)\eta_\alpha^{-1} + (\eta_{\alpha\beta} \otimes 1)\chi_\beta \eta_{\alpha\beta}^{-1} - \chi_\alpha, \end{aligned} \tag{11.44}$$

*for any* $\alpha, \beta \in I$, *with* $U_\alpha \cap U_\beta \equiv U_{\alpha\beta} \neq \emptyset$. Thus, one has (still cf. (10.4))

$$D_\beta - D_\alpha = \left(\eta_\beta^{-1} \otimes 1\right) \chi_\beta \eta_\beta - \left(\eta_\alpha^{-1} \otimes 1\right) \chi_\alpha \eta_\alpha, \tag{11.45}$$

*for any* $\alpha, \beta \in I$, *as above.* Therefore, setting now

$$\psi'_\alpha := \left(\eta_\alpha^{-1} \otimes 1\right) \circ \chi_\alpha \circ \eta_\alpha, \quad \alpha \in I, \tag{11.46}$$

such that (see also (9.4))

$$(\psi'_\alpha) \in C^0(\mathcal{U}, \Omega(\mathcal{E}nd\, \mathcal{E})) = C^0(\mathcal{U}, M_n(\Omega)), \tag{11.47}$$

one thus obtains from (11.45);

$$\delta(D_\alpha) = \delta(\psi'_\alpha). \tag{11.48}$$

In other words, one infers that (cf. e.g. Chapt. III; (4.22))

$$\begin{aligned} \delta(D_\alpha) = (D_\beta - D_\alpha) \in \operatorname{im}\delta^0 &= B^1(\mathcal{U}, \Omega(\mathcal{E}nd\, \mathcal{E})) \\ &= B^1(\mathcal{U}, M_n(\Omega)), \end{aligned} \tag{11.49}$$

so that (ibid.; (4.23)) one finally concludes that;

$$[(D_\beta - D_\alpha)] = 0 \in \check{H}^1(X, M_n(\Omega)), \tag{11.50}$$

*that was* actually *to be proved* (cf. (11.31)), and this, in effect, terminates the proof of the theorem. ∎

**Note.**– The question that was formulated before the statement of Theorem 11.2 is actually treated by the same theorem in a *round-about way.* So it would be interesting, at least, to have a *direct attack* to the same question, *in terms,* namely, *of the* $\mathcal{A}$*-connections* $\widetilde{D}$ *and* $\tilde{d}$ (cf. (11.36)) *themselves.*

We come next to handle, as already said, by the following Section 12, the *"only if"* part of (11.2).

## 12. $\mathcal{A}$-connections as splittings of jet-line sheaves

As a result of the succeeding discussion and in conjunction with the above Theorem 11.1, we are going actually to see that:

(12.1) *the Atiyah class* of a given *vector sheaf* $\mathcal{E}$, which is associated with a *local frame* $\mathcal{U} = (U_\alpha)_{\alpha \in I}$ of $\mathcal{E}$,

(12.1.1) $$\mathfrak{a}(\mathcal{E}) \equiv [(D_\beta - D_\alpha)] \in \check{H}^1(X, M_n(\Omega))$$

(see (9.14)), *constitutes,* in effect, *the obstruction to the existence of an $\mathcal{A}$-connection of $\mathcal{E}$.*

Indeed, see Theorem 12.1 and Corollary 12.1 below. ∎ So we start with fixing again the terminology we are going to apply.

Thus, suppose, as usual, that we are given a *differential triad* $(\mathcal{A}, \partial, \Omega)$, with respect to a fixed **C**-*algebraized space* $(X, \mathcal{A})$, and let $\mathcal{E}$ be an $\mathcal{A}$-*module* on $X$. Moreover, consider the following **C**-*vector space sheaf* on $X$

(12.2) $$\mathcal{J}^1_{\mathcal{A}}(\mathcal{E}) := \mathcal{E} \oplus (\mathcal{E} \otimes_{\mathcal{A}} \Omega) \equiv \mathcal{E} \oplus \Omega(\mathcal{E})$$

(cf. also Chapt. II; (1.38)). Hereby we have, of course, employed the fact that $\mathcal{J}^1_{\mathcal{A}}(\mathcal{E})$ *is,* according to our hypothesis for $\mathcal{E}$ and $\Omega$, *an $\mathcal{A}$-module* on $X$ (see Chapt. II; (3.6)). However, for reasons that will become apparent right below, we do define *another $\mathcal{A}$-module structure* on the above **C**-vector space sheaf $\mathcal{J}^1_{\mathcal{A}}(\mathcal{E})$, as follows;

(12.3) $$\alpha \cdot (s \oplus t) := \alpha \cdot s \oplus (\alpha \cdot t + s \otimes \partial(\alpha)),$$

*for any (local) sections $\alpha \in \mathcal{A}(U)$, $s \in \mathcal{E}(U)$ and $t \in \Omega(\mathcal{E})(U)$,* with $U$ *open* in $X$.

Now, the $\mathcal{A}$-*module*

$$\mathcal{J}^1_{\mathcal{A}}(\mathcal{E}) \equiv \mathcal{J}^1(\mathcal{E}), \tag{12.4}$$

as defined by (12.3), is called the *jet-line sheaf* (or even *jet sheaf of order 1*) of the *$\mathcal{A}$-module $\mathcal{E}$*, relative to the given differential triad $(\mathcal{A}, \partial, \Omega)$. The terminology employed extends, of course, the classical one of *vector bundle* (differential) *geometry;* see, for instance, M. Karoubi - G. Leruste [1: p. 338, §B7].

Thus, our purpose now, by the ensuing discussion, is to relate the above *$\mathcal{A}$-module* $\mathcal{J}^1(\mathcal{E})$ with certain *$\mathcal{A}$-extensions of $\mathcal{E}$* (cf. Chapt. III; Section 10 and Chapt. V; Sections 7, 8), in such a manner that *splittings of the latter assure the existence of $\mathcal{A}$-connections of $\mathcal{E}$, and conversely:*

Indeed, by considering $\mathcal{J}^1(\mathcal{E})$ *as an $\mathcal{A}$-module* on $X$, *according to (12.3),* one obtains the following *short exact $\mathcal{A}$-sequence*

$$0 \longrightarrow \Omega(\mathcal{E}) \stackrel{i}{\longrightarrow} \mathcal{J}^1(\mathcal{E}) \stackrel{\pi}{\longrightarrow} \mathcal{E} \longrightarrow 0, \tag{12.5}$$

where $\pi$ denotes the obvious *projection $\mathcal{A}$-morphism* (with the *$\mathcal{A}$-module structure of $\mathcal{I}^1(\mathcal{E})$, as before!*); yet, $i$ stands again for the natural *injection $\mathcal{A}$-morphism* of $\Omega(\mathcal{E})$ in the *Whitney sum of* the **C**-*vector space sheaves* concerned, as in (12.2) (see also (12.3), in conjunction with Lemma 1.1). Therefore (cf. Chapt. III; Definition 10.1),

(12.6) given a *differential triad* $(\mathcal{A}, \partial, \Omega)$ and an *$\mathcal{A}$-module* $\mathcal{E}$ on $X$, *the* corresponding *jet-line sheaf of* $\mathcal{E}$, $\mathcal{J}^1(\mathcal{E})$ (cf. (12.4)), *provides an $\mathcal{A}$-extension of $\mathcal{E}$ by $\Omega(\mathcal{E}) \equiv \mathcal{E} \otimes_{\mathcal{A}} \Omega$.*

(*Warning!* not the trivial one; cf. (12.3) and Chapt. III; (10.2)). In other words, *(12.5) defines an element of the* respective (abelian) *Baer group*

$$\mathrm{ext}_{\mathcal{A}}(\mathcal{E}, \Omega(\mathcal{E})) = \mathrm{Ext}^1_{\mathcal{A}}(\mathcal{E}, \Omega(\mathcal{E})) \tag{12.7}$$

(see Chapt. III; Theorem 10.1 and (10.36)).

On the other hand, for use right below, we further note that, in the case that $\mathcal{E}$ is, in particular, a *vector sheaf* on $X$, *with* $\mathrm{rk}\,\mathcal{E} = n \in \mathbf{N}$, one obtains the following (abelian) *group isomorphisms;*

$$H^1(X, \mathcal{H}om_{\mathcal{A}}(\mathcal{E}, \Omega(\mathcal{E}))) = H^1(X, \Omega(\mathcal{E}nd\,\mathcal{E})) = H^1(X, M_n(\Omega)). \tag{12.8}$$

Indeed, based on Chapt. III; Lemma 8.1 and (4.53), as well as, on our previous remark in (9.12), along with (9.13), (9.4) and (7.22) in the preceding, one infers that;

$$
\begin{aligned}
&H^1(X, \mathcal{H}om_{\mathcal{A}}(\mathcal{E}, \Omega(\mathcal{E}))) = H^1(X, \Omega(\mathcal{E}nd\,\mathcal{E})) = \check{H}^1(X, \Omega(\mathcal{E}nd\,\mathcal{E})) \\
(12.9) \quad &= \varinjlim_{\mathcal{U}} \check{H}^1(\mathcal{U}, \Omega(\mathcal{E}nd\,\mathcal{E})) = \varinjlim_{\mathcal{U}} \check{H}^1(\mathcal{U}, M_n(\Omega)) = \check{H}^1(X, M_n(\Omega)) \\
&= H^1(X, M_n(\Omega)),
\end{aligned}
$$

as was actually to be proved, according to (12.8). ∎ Thus, based now on Chapt. V; Theorem 7.2, one concludes that;

$$
\begin{aligned}
(12.10) \quad \operatorname{ext}_{\mathcal{A}}(\mathcal{E}, \Omega(\mathcal{E})) &= \operatorname{Ext}^1_{\mathcal{A}}(\mathcal{E}, \Omega(\mathcal{E})) = H^1(X, \mathcal{H}om_{\mathcal{A}}(\mathcal{E}, \Omega(\mathcal{E}))) \\
&= H^1(X, \Omega(\mathcal{E}nd\,\mathcal{E})) = H^1(X, M_n(\Omega)),
\end{aligned}
$$

up to (abelian) *group isomorphisms,* as indicated, for every *vector sheaf* $\mathcal{E}$ on $X$, *with* $\operatorname{rk}\mathcal{E} = n \in \mathbf{N}$, and a given *differential triad* $(\mathcal{A}, \partial, \Omega)$.

The previous rels. (12.10) will be applied presently below in the proof of (12.1). But, first we have to state the following fundamental lemma, on which the same proof is based too. So we have.

**Lemma 12.1.** *Let* $(\mathcal{A}, \partial, \Omega)$ *be a given differential triad on a topological space* $X$, $\mathcal{E}$ *an* $\mathcal{A}$*-module on* $X$ *and* $\mathcal{J}^1(\mathcal{E})$ *the corresponding jet-line sheaf of* $\mathcal{E}$ (cf. (12.4)). *Moreover, let*

$$(12.11) \qquad \pi : \mathcal{J}^1(\mathcal{E}) \longrightarrow \mathcal{E}$$

*be the (canonical) projection* $\mathcal{A}$*-morphism* (see (12.5)). *Then,* $\pi$ *admits a section, viz. an* $\mathcal{A}$*-morphism*

$$(12.12) \qquad D' : \mathcal{E} \longrightarrow \mathcal{J}^1(\mathcal{E}),$$

*satisfying the relation*

$$(12.13) \qquad \pi \circ D' = 1_{\mathcal{E}},$$

*or, equivalently* (cf. Chapt. III; Proposition 10.1 and Definition 10.3), *one has a splitting of (12.5), if, and only if, there exists a uniquely defined* $\mathcal{A}$*-connection* $D$ *of* $\mathcal{E}$, *such that one has* (cf. (12.2));

$$(12.14) \qquad D' = 1_{\mathcal{E}} \oplus D.$$

**Proof.** Indeed, suppose that

$$(12.15) \qquad D : \mathcal{E} \longrightarrow \mathcal{E} \otimes_{\mathcal{A}} \Omega \equiv \Omega(\mathcal{E})$$

is a given $\mathcal{A}$*-connection* of the $\mathcal{A}$-module $\mathcal{E}$. Then, the relation

$$D' := 1_{\mathcal{E}} \oplus D \equiv (1_{\mathcal{E}}, D) : \mathcal{E} \longrightarrow \mathcal{E} \oplus \Omega(\mathcal{E}) \equiv \mathcal{J}^1(\mathcal{E}) \tag{12.16}$$

provides a *splitting of (12.5);* that is, *equivalently* (cf. the statement of the lemma), *a section of* $\pi$ (cf. (12.11)). Furthermore,

(12.17) *$D'$ is an $\mathcal{A}$-morphism* of the $\mathcal{A}$-modules involved, as in (12.16) (cf. also (12.3)), *satisfying (12.13).*

Thus, *for any* (local) *section* $s \in \mathcal{E}(U)$, with $U$ *open* in $X$, one has, by (12.16),

$$D'(s) = s \oplus D(s) \equiv (s, D(s)), \tag{12.18}$$

so that one obtains

$$(\pi \circ D')(s) = \pi(D'(s)) \equiv \pi_{\mathcal{E}}(s, Ds) = s, \tag{12.19}$$

*as desired.* (We set here $\pi \equiv \pi_{\mathcal{E}}$, as in (12.5)). Yet, for any $\alpha \in \mathcal{A}(U)$ and $s \in \mathcal{E}(U)$, as before, one has, *by virtue of (12.3);*

$$\begin{aligned} D'(\alpha \cdot s) &= \alpha s \oplus D(\alpha s) = \alpha s \oplus (\alpha \cdot Ds + s \otimes \partial\alpha) \\ &= \alpha \cdot (s \oplus Ds) = \alpha \cdot D'(s), \end{aligned} \tag{12.20}$$

given that $D(s) \in \Omega(\mathcal{E})(U)$, by (12.15). (We still note here the *significance of the definition (12.3)*). So $D'$ *is an $\mathcal{A}$-morphism,* which thus completely justifies our claim about it in (12.17).

*Conversely,* we shall prove now that;

(12.21) *any splitting $D' : \mathcal{E} \longrightarrow \mathcal{J}^1(\mathcal{E})$ of (12.5)* (in other words, any *section of* $\pi$, such that $\pi \circ D' = 1_{\mathcal{E}}$) *is of the form (12.16), for some uniquely defined $\mathcal{A}$-connection $D$ of $\mathcal{E}$.*

Namely, by first considering the following diagram

(12.22)

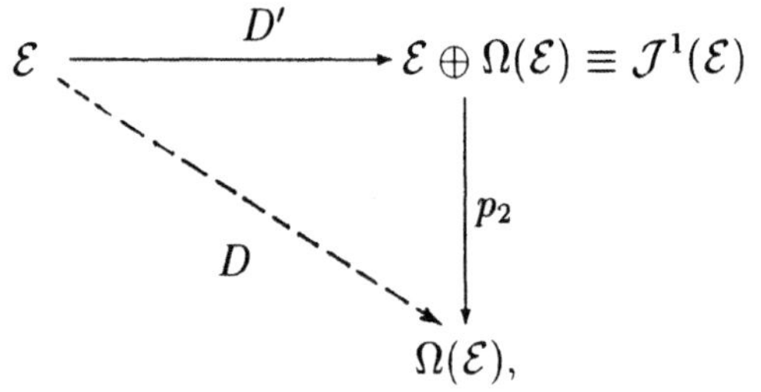


one makes it *commutative,* by defining,

$$D := p_2 \circ D', \tag{12.23}$$

where $p_2$ stands for the (canonical) *projection of* $\mathcal{J}^1(\mathcal{E}) \equiv \mathcal{E} \oplus \Omega(\mathcal{E})$ *onto the second factor;* as already remarked in the preceding, the latter sheaf is the *Whitney sum of the* **C**-*vector space sheaves involved,* so that $p_2$ *is* certainly *a* **C**-*linear morphism.* Hence, *the same holds true for* $D$, by virtue of (12.23) and of our hypothesis for $D'$ (cf. (12.21), as well as, Chapt. III; (10.10), along with Definition 10.3 and Chapt. II; (2.3)).

Thus, we further prove that $D$, as given by (12.23), *satisfies the Leibniz condition (3.6):* First, by hypothesis for $D'$ (cf. (12.21)), along with (12.13) and (12.23), one has

$$D' = \pi \circ D' \oplus p_2 \circ D' = 1_{\mathcal{E}} \oplus D. \tag{12.24}$$

Therefore, *for any* given $\alpha \in \mathcal{A}(U)$ *and* $s \in \mathcal{E}(U)$, with $U$ *open* in $X$, one obtains (cf. also (12.3));

$$\begin{aligned} D(\alpha \cdot s) &= (p_2 \circ D')(\alpha s) = p_2(D'(\alpha s)) = p_2(\alpha \cdot D'(s)) \\ &= p_2(\alpha \cdot (s \oplus Ds)) = p_2(\alpha s \oplus (\alpha \cdot Ds + s \otimes \partial\alpha)) = \alpha \cdot Ds + s \otimes \partial\alpha, \end{aligned} \tag{12.25}$$

that was actually to be proved. Finally, the same relation (12.24) *proves,* of course, *the uniqueness of* $D$, and this terminates the proof of the lemma. ∎

Now, as an application of the previous lemma, we come next, as already promised, to the *vindication of (12.1).* However, we do state first the following more convenient (although less precise) formulation of the same lemma. That is, one has.

**Theorem 12.1.** *Let* $(\mathcal{A}, \partial, \Omega)$ *be a given differential triad on a topological space* $X$, $\mathcal{E}$ *an* $\mathcal{A}$-*module on* $X$ *and* $\mathcal{J}^1(\mathcal{E})$ *the respective jet-line sheaf of* $\mathcal{E}$ (cf. (12.2)). *Moreover, consider the following short exact* $\mathcal{A}$-*sequence*

$$0 \longrightarrow \Omega(\mathcal{E}) \xrightarrow{i} \mathcal{J}^1(\mathcal{E}) \xrightarrow{\pi} \mathcal{E} \longrightarrow 0; \tag{12.26}$$

*viz. an* $\mathcal{A}$-*extension of* $\mathcal{E}$ *by* $\Omega(\mathcal{E})$, *as defined by* $\mathcal{J}^1(\mathcal{E})$ (cf. (12.5)).

*Then, the given* $\mathcal{A}$-*module* $\mathcal{E}$ *admits an* $\mathcal{A}$-*connection if, and only if, (12.26) splits.*

**Proof.** The assertion is, in effect, an immediate consequence of the above Lemma 12.1. (We do not actually refer here to the *uniqueness of the connection of* $\mathcal{E}$ supplied by the splitting of (12.26)). ∎

Thus, as a result of the preceding, we now have the following reformulation of our claim in (12.1).

**Corollary 12.1.** *Suppose we are given a differential triad* $(\mathcal{A}, \partial, \Omega)$ *on a topological space* $X$*, and let* $\mathcal{E}$ *be a vector sheaf on* $X$*, of rank* $n \in \mathbf{N}$*. Then, the two following assertions are equivalent:*

*i) The given vector sheaf* $\mathcal{E}$ *admits an* $\mathcal{A}$*-connection.*

*ii) The Atiyah class of* $\mathcal{E}$*, relative to a given local frame*

$$\mathcal{U} = (U_\alpha)_{\alpha \in I} \tag{12.27}$$

*of* $\mathcal{E}$ *vanishes; viz. one has* (cf. (9.14))

$$\mathfrak{a}(\mathcal{E}) = [(D_\beta - D_\alpha)] = 0 \in \check{H}^1(X, M_n(\Omega)). \tag{12.28}$$

**Proof.** We already know that *ii) implies i)* (cf Theorem 11.1). Thus, *conversely,* assume that $\mathcal{E}$ *admits an* $\mathcal{A}$*-connection,* so that, according to Theorem 12.1, *(12.26) splits.* Now, this actually means that the equivalence class, defined by $\mathcal{J}^1(\mathcal{E})$, in the Baer group $\mathrm{ext}_{\mathcal{A}}(\mathcal{E}, \Omega(\mathcal{E}))$ is the trivial one; that is, the zero element of the latter group or, equivalently (cf. also (12.7)), of the (abelian) group (in fact, $\mathcal{A}(X)$-module) $\mathrm{Ext}^1_{\mathcal{A}}(\mathcal{E}, \Omega(\mathcal{E}))$ (cf. Chapt. III; (10.30), (10.36)). Hence, based on (12.10), the same class corresponds to the zero element of (the $\mathcal{A}(X)$-module) $H^1(X, M_n(\Omega))$. However, due to the same rel. (12.10), *the* $\mathcal{A}$*-extension at issue* (viz. of $\mathcal{E}$ by $\Omega(\mathcal{E})$, as given by $\mathcal{J}^1(\mathcal{E})$) *defines uniquely an element* of $H^1(X, M_n(\Omega))$, while *the same element is uniquely determined, as well, by the Atiyah class of* $\mathcal{E}$, as the latter is defined by any given local frame of $\mathcal{E}$ (cf. (12.27)), according to the relation (see also (9.14), (10.28), along with (10.29), (10.31));

$$\mathfrak{a}(\mathcal{E}) = [(D_\beta - D_\alpha)] = [\tilde{\partial}(g_{\alpha\beta})] \in \check{H}^1(X, M_n(\Omega)). \tag{12.29}$$

Therefore, by what has been said before, one concludes that $\mathfrak{a}(\mathcal{E}) = 0$, that was actually to be proved, so that *i) implies ii), as well,* and this terminates the proof. ∎

**Scholium 12.1.**– As a clarification of our argument in the last part of the preceding proof, we still set out the following comments:

Given a *vector sheaf* $\mathcal{E}$ on $X$, as before, the *Levi-Civita* ($\mathcal{A}$*-connection) 1-cocycle of* $\mathcal{E}$, associated with any given *local frame of* $\mathcal{E}$ (cf. (12.27) and (8.34.1)), *yields* further *an* $\mathcal{A}$*-extension of* $\mathcal{E}$ *by* $\Omega(\mathcal{E})$, as determined e.g.

by the *jet-line sheaf of* $\mathcal{E}$, $\mathcal{J}^1(\mathcal{E})$ (see (12.5)), according to the following (abelian) *group isomorphisms;*

$$\begin{aligned} &\mathrm{ext}_{\mathcal{A}}(\mathcal{E},\Omega(\mathcal{E})) = \mathrm{Ext}^1_{\mathcal{A}}(\mathcal{E},\Omega(\mathcal{E})) = \mathrm{Ext}^1_{\mathcal{A}}(\mathcal{E}\otimes_{\mathcal{A}}\mathcal{A},\Omega(\mathcal{E})) \\ (12.30)\qquad &= \mathrm{Ext}^1_{\mathcal{A}}(\mathcal{A},\mathcal{E}^*\otimes_{\mathcal{A}}\Omega(\mathcal{E})) = \mathrm{Ext}^1_{\mathcal{A}}(\mathcal{A},\mathcal{H}om_{\mathcal{A}}(\mathcal{E},\Omega(\mathcal{E}))) \\ &= \mathrm{Ext}^1_{\mathcal{A}}(\mathcal{A},\Omega(\mathcal{E}nd\,\mathcal{E})) = H^1(X,\Omega(\mathcal{E}nd\,\mathcal{E})) = H^1(X,M_n(\Omega)) \end{aligned}$$

(see (12.7), (12.8), (12.10) and Chapt. V; (7.7)).

Thus (cf. Theorem 12.1), *the aforementioned 1-cocycle yields,* in fact, *an $\mathcal{A}$-connection of $\mathcal{E}$ if, and only if, the $\mathcal{A}$-extension at issue* (cf. (12.5)) *splits. Equivalently* (cf. Chapt. III; (10.30), (10.36)), *if, and only if, the* same *1-cocycle is a coboundary,* or yet (cf. (10.28.1)) *if, and only if, the Atiyah class of $\mathcal{E}$* (defined by the given local frame of $\mathcal{E}$, as above) *vanishes.*

In this context, see also Note 8.1 in the foregoing, along with Theorem 14.1 and (14.10) below. Yet, the above picture is supplemented by further *specializing,* for instance, *on* $\Omega$, by requiring this to be a *vector sheaf* on $X$, as well (cf. the next Section 13), or even *on the* base *space* $X$, assuming it now *(Hausdorff) paracompact* (see Section 14 below).

Finally, as another consequence of the preceding, we still infer that;

(12.31) a given *vector sheaf* $\mathcal{E}$ on $X$ *admits an $\mathcal{A}$-connection if, and only if, it also admits a Levi-Civita $\mathcal{A}$-connection.*

Indeed, if $\mathcal{E}$ has an $\mathcal{A}$-connection, then (Corollary 12.1) $\mathfrak{a}(\mathcal{E}) = 0$, so that (cf. Theorem 11.1 and Definition 11.1) $\mathcal{E}$ admits a Levi-Civita $\mathcal{A}$-connection, as well, the rest of (12.31) being clear. ∎ This explains also completely (11.16) in the preceding.

## 13. $\mathcal{A}$-connections as splitting extensions (contn'd)

We examine in this Section another short exact $\mathcal{A}$-sequence, save (12.5) above, that is, another $\mathcal{A}$-*extension,* (yet, naturally) connected with a given

*vector sheaf* $\mathcal{E}$ on $X$ (we always refer, of course, to a given fixed, as usual, differential triad $(\mathcal{A}, \partial, \Omega)$) and whose splitting is still equivalent with the existence of an $\mathcal{A}$-*connection* of $\mathcal{E}$. However, now, under an extra hypothesis for the $\mathcal{A}$-module $\Omega$; yet, a further restriction on the topological space $X$ considered, will involve, as we shall see, but in the next Section 14, the so-called *"coordinate 1-cocycle"* of the extension concerned.

So assume that we are given a *differential triad* $(\mathcal{A}, \partial, \Omega)$ on an arbitrary *topological space* $X$ (cf. (3.1)), such that

(13.1) $\Omega$ is a *vector sheaf* on $X$, as well.

Therefore, by considering the *dual $\mathcal{A}$-module* of $\Omega$,

$$\Omega^* := \mathcal{H}om_{\mathcal{A}}(\Omega, \mathcal{A}), \tag{13.2}$$

one gets another *vector sheaf* on $X$ *with* $\operatorname{rk}\Omega^* = \operatorname{rk}\Omega$ (cf. Chapt. IV; Definition 5.1, and (5.19)). Furthermore, due to (13.1), $\Omega$ *is reflexive;* that is, one has

$$\Omega^{**} = \Omega, \tag{13.3}$$

up to an $\mathcal{A}$-*isomorphism of* the $\mathcal{A}$-*modules* (in fact, *vector sheaves*) involved (cf. Chapt. IV; Theorem 5.1). In this context, we also note that one has, of course, a similar situation to (13.1) in the classical case of differential geometry (see thus Section 2 in the preceding, as well as, Section 16 below).

On the other hand, we still recall that, given a *vector sheaf* $\mathcal{E}$ on $X$, *with* $\operatorname{rk}\mathcal{E} = n \in \mathbf{N}$, the set

$$\mathcal{H}om_{\mathcal{A}}(\mathcal{E}, \mathcal{E}) = \mathcal{E}nd\,\mathcal{E} \tag{13.4}$$

is a *vector sheaf* on $X$ too, *with* $\operatorname{rk}(\mathcal{E}nd\,\mathcal{E}) = n^2$ (cf. Chapt. II; Lemma 6.1).

Thus, under the above circumstances, consider further the *short exact $\mathcal{A}$-sequence,*

$$0 \longrightarrow \mathcal{E}nd\,\mathcal{E} \longrightarrow \mathcal{S} \longrightarrow \Omega^* \longrightarrow 0; \tag{13.5}$$

that is, an $\mathcal{A}$-*extension of* $\Omega^*$ *by* $\mathcal{E}nd\,\mathcal{E}$ (cf. Chapt. III; Definition 10.1), or even when looking at the corresponding *isomorphism class of (13.5),* an element of the *Baer group*

$$\operatorname{ext}_{\mathcal{A}}(\Omega^*, \mathcal{E}nd\,\mathcal{E}) = \operatorname{Ext}^1_{\mathcal{A}}(\Omega^*, \mathcal{E}nd\,\mathcal{E}), \tag{13.6}$$

while the latter group is also suitably expressed through the functor $\mathrm{Ext}^1_{\mathcal{A}}$, as above (cf. Chapt. III; Theorem 10.1). Therefore, based on Chapt. V; Proposition 8.1, we conclude that;

(13.7) under the assumption (13.1), *the* previous *$\mathcal{A}$-extensions (12.5) and (13.5) have the same* (in fact *isomorphic*) *Baer groups.* That is, one has the following (abelian) *group isomorphisms;*

$$\text{(13.7.1)} \quad \mathrm{ext}_{\mathcal{A}}(\mathcal{E}, \Omega(\mathcal{E})) = \mathrm{ext}_{\mathcal{A}}(\Omega^*, \mathcal{E}nd\,\mathcal{E}) = H^1(X, M_n(\Omega)).$$

However, for convenience, we do give the relevant calculations, concerning (13.7.1); see thus (12.8), (12.20) *and* (13.3) above, along with Chapt. IV; (6.1). So we have;

$$\begin{aligned} &\mathrm{ext}_{\mathcal{A}}(\mathcal{E}, \Omega(\mathcal{E})) = \mathrm{Ext}^1_{\mathcal{A}}(\mathcal{E}, \Omega(\mathcal{E})) = H^1(X, \mathcal{H}om_{\mathcal{A}}(\mathcal{E}, \Omega(\mathcal{E}))) \\ (13.8) \quad &= H^1(X, \Omega(\mathcal{E}nd\,\mathcal{E})) = H^1(X, M_n(\Omega)) = H^1(X, \Omega^{**}(\mathcal{E}nd\,\mathcal{E})) \\ &= H^1(X, \mathcal{H}om_{\mathcal{A}}(\Omega^*, \mathcal{E}nd\,\mathcal{E})) = \mathrm{Ext}^1_{\mathcal{A}}(\Omega^*, \mathcal{E}nd\,\mathcal{E}) = \mathrm{ext}_{\mathcal{A}}(\Omega^*, \mathcal{E}nd\,\mathcal{E}), \end{aligned}$$

which thus justifies (13.7). ■ Accordingly (see also Chapt. V; (8.60)), one further concludes that;

(13.9) under the hypothesis (13.1), *the splitting of either one of the two $\mathcal{A}$-extensions (12.5) and (13.5) is tantamount to the existence of an $\mathcal{A}$-connection of $\mathcal{E}$.*

Indeed, the two extensions, under consideration, have the same splittings, corresponding to the zero elements of the isomorphic Baer groups in (13.7.1), respectively (see Chapt. III; (10.30) and (10.36)). Thus, our claim in (13.9) follows already from Theorem 12.1. ■

For convenience, we summarize the information, we gathered hitherto, into the following.

**Theorem 13.1.** *Suppose we have a differential triad $(\mathcal{A}, \partial, \Omega)$ on a topological space $X$, with $\Omega$ being, in particular, a vector sheaf on $X$. Moreover, let also $\mathcal{E}$ be another given vector sheaf on $X$. Then, all the next four assertions are equivalent:*

*1) The given vector sheaf $\mathcal{E}$ admits an $\mathcal{A}$-connection.*

*2)* $\mathfrak{a}(\mathcal{E}) = [(D_\beta - D_\alpha)] = 0$; viz. *the Atiyah class of $\mathcal{E}$*, relative to a given local frame $\mathcal{U} = (U_\alpha)_{\alpha \in I}$ of $\mathcal{E}$ (cf. (9.14)), *vanishes.*

*3) The following $\mathcal{A}$-extension splits;*

$$0 \longrightarrow \Omega(\mathcal{E}) \longrightarrow \mathcal{J}^1(\mathcal{E}) \longrightarrow \mathcal{E} \longrightarrow 0 \tag{13.10}$$

(see also (12.5)), *and*

*4) the next $\mathcal{A}$-extension splits*

$$0 \longrightarrow \mathcal{E}nd\,\mathcal{E} \longrightarrow \mathcal{S} \longrightarrow \Omega^* \longrightarrow 0. \tag{13.11}$$

**Proof.** First, according to (13.9), the assertions *1), 3) and 4) are* all *equivalent* to one another. On the other hand, *1) and 2) are* already *equivalent,* by Corollary 12.1 (even without any restriction on $\Omega$), and this completes the proof. ∎

We come now, by the succeeding Section 14, to relate the above *$\mathcal{A}$-extensions,* in effect, their respective *coordinate 1-cocycles,* with another familiar such cocycle, connected with a given *vector sheaf* $\mathcal{E}$ on $X$ and a *local frame* of it, viz. the *Levi-Civita $\mathcal{A}$-connection 1-cocycle of $\mathcal{E}$.* However, this under the further assumption of *paracompactness* of $X$.

# 14. Coordinate 1-cocycles of $\mathcal{A}$-extensions, as related to $\mathcal{A}$-connections

To begin with, suppose that we are given a *differential triad* $(\mathcal{A}, \partial, \Omega)$ on a topological space $X$, such that the two following supplementary conditions are fulfilled:

(14.1)
- i) $X$ is a *paracompact (Hausdorff)* space.
- ii) $\Omega$ is a *vector sheaf* on $X$.

Furthermore, suppose that $\mathcal{E}$ is a *vector sheaf* on $X$, *with* $\operatorname{rk}\mathcal{E} = n \in \mathbf{N}$, and consider again the *$\mathcal{A}$-extension,* given by (13.5) above,

$$0 \longrightarrow \mathcal{E}nd\,\mathcal{E} \longrightarrow \mathcal{S} \longrightarrow \Omega^* \longrightarrow 0. \tag{14.2}$$

Thus, by looking now at the *characteristic class of (14.2)* (see Chapt. III; (10.39)), one obtains ((14.1); ii) is needed here, *only!*);

$$\begin{aligned}\delta(1_{\Omega^*}) \in \mathrm{Ext}^1_{\mathcal{A}}(\Omega^*, \mathcal{E}nd\,\mathcal{E}) &= \mathrm{Ext}^1_{\mathcal{A}}(\mathcal{A}, \Omega^{**} \otimes_{\mathcal{A}} \mathcal{E}nd\,\mathcal{E}) \\ &= \mathrm{Ext}^1_{\mathcal{A}}(\mathcal{A}, \Omega \otimes_{\mathcal{A}} \mathcal{E}nd\,\mathcal{E}) = \mathrm{Ext}^1_{\mathcal{A}}(\mathcal{A}, \Omega(\mathcal{E}nd\,\mathcal{E})) \\ &= H^1(X, \Omega(\mathcal{E}nd\,\mathcal{E})) = H^1(X, M_n(\Omega)).\end{aligned} \tag{14.3}$$

Concerning the above calculations, cf. Chapt. V; (7.1), or even (7.7), for $n = 1$, along with (13.8) above. Yet, as already known (Chapt. III; (10.39)), $\delta(1_{\Omega^*})$ *measures the obstruction to the splitting of (14.2).* Therefore (cf. Theorem 13.1), one further concludes that;

(14.4) by assuming (14.1); ii) *only* (hence, for any space $X$), one infers that $\delta(1_{\Omega^*})$ *measures the obstruction to the existence of an $\mathcal{A}$-connection of $\mathcal{E}$.*

In this context, see also (14.12) below.

By the same token, one can actually get a better *criterion for the existence of an $\mathcal{A}$-connection* of the *vector sheaf* $\mathcal{E}$ involved, patterned after Theorem 12.1, that is, by considering the *$\mathcal{A}$-extension* of $\mathcal{E}$ by $\Omega(\mathcal{E})$, as given by (12.5);

$$0 \longrightarrow \Omega(\mathcal{E}) \longrightarrow \mathcal{J}^1(\mathcal{E}) \longrightarrow \mathcal{E} \longrightarrow 0. \tag{14.5}$$

Thus, one now concludes that;

(14.6) *without* any *restriction on $\Omega$ or $X$,* one infers that *the characteristic class $\delta(1_{\mathcal{E}})$ of (14.5) measures the obstruction to the existence of an $\mathcal{A}$-connection of the vector sheaf $\mathcal{E}$.*

It is clear, of course, in view of Chapt. III; (10.39), that (14.6) is, in effect, another formulation of the previous Theorem 12.1.

On the other hand, *assuming* now (14.1) and based on Chapt. V; (8.61), we further conclude that:

(14.7) *the $\mathcal{A}$-extensions (14.2) and (14.5) are* (cohomologically) *classified by the same characteristic class, or even the same coordinate 1-cocycle;* viz. one has

$$\delta(1_{\mathcal{E}}) = \delta(1_{\Omega^*}) = [(h_\beta - h_\alpha)] \in H^1(X, M_n(\Omega)) \tag{14.7.1}$$

(see also Chapt. V; (8.27), (8.29)).

Now, the last conclusion with (12.29), in concert, provide already the proof of the following.

**Theorem 14.1.** *Let $(\mathcal{A}, \partial, \Omega)$ be a given differential triad on a paracompact (Hausdorff) space $X$, with $\Omega$ being also a vector sheaf on $X$. Moreover, let $\mathcal{E}$ be a vector sheaf on $X$, of rank $n \in \mathbf{N}$, together with the following 1-cocycles of a local frame $\mathcal{U} = (U_\alpha)_{\alpha \in I}$ of $\mathcal{E}$:*

*i)* A *coordinate 1-cocycle of $\mathcal{E}$*, associated with $\mathcal{U}$, say,

$$(g_{\alpha\beta}) \in Z^1(\mathcal{U}, \mathcal{GL}(n, \mathcal{A})) \tag{14.8}$$

(see e.g. (10.19)), along with *its image via the logarithmic derivation $\tilde{\partial}$* (cf. (10.25)); i.e.,

$$\tilde{\partial}(g_{\alpha\beta}) \in Z^1(\mathcal{U}, M_n(\Omega)). \tag{14.9}$$

*ii)* The *Levi-Civita ($\mathcal{A}$-connection) 1-cocycle of $\mathcal{E}$* (cf. (9.8));

$$(D_\beta - D_\alpha) \in Z^1(\mathcal{U}, M_n(\Omega)). \tag{14.10}$$

*iii)* The *coordinate 1-cocycle of an $\mathcal{A}$-extension of $\Omega^*$* (dual vector sheaf of $\Omega$) *by $\mathcal{E}nd\,\mathcal{E}$*;

$$\delta(h_\alpha) = (h_\beta - h_\alpha) \in Z^1(\mathcal{U}, M_n(\Omega)) \tag{14.11}$$

(see Chapt. V; (8.27), for $\mathcal{E} = \Omega^*$ and $\mathcal{F} = \mathcal{E}nd\,\mathcal{E}$, along with Chapt. IV; (6.1) and (5.11)). *The latter coincides* (through a (canonical) *group isomorphism;* see e.g. (13.8)) *with that one of (14.5)* (cf. also (14.7.1)).

*Then, the 1-dimensional cohomology classes of $X$, which correspond to all the preceding $M_n(\Omega)$-valued 1-cocycles of $\mathcal{U}$* (viz. to (14.9), (14.10) and (14.11)) *are all equal; that is, one has* (see also (14.7.1), concerning the first of the succeeding equalities)

$$\begin{aligned} \delta(1_{\mathcal{E}}) = \delta(1_{\Omega^*}) &= [(h_\beta - h_\alpha)] = [(D_\beta - D_\alpha)] \\ &= [\tilde{\partial}(g_{\alpha\beta})] \in H^1(X, M_n(\Omega)). \end{aligned} \tag{14.12}$$

(In this regard, cf. also Chapt. V; (8.54), *with $\mathcal{E}$ and $\mathcal{F}$, as in iii) above,* along with (8.59), (8.61) therein).

Consequently, once more (see also (14.6)), the *triviality of any one of the previous 1-cocycles, as* appeared *in (14.12),* viz. if any one of them is cohomologous to zero, then this *is a necessary and sufficient condition for the existence of an $\mathcal{A}$-connection of $\mathcal{E}$.* ∎

The above beautiful and all embracing result constitutes, in effect, the *echo,* within the present abstract framework, of a classical one due to M.F.

Atiyah [1] and referred to (holomorphic) *vector bundle geometry.*

Now, by further applying the terminology of Chapt. V; (8.48), and based also on (8.59) (cf. (8.51.1) too) of the same Chapter, as well as, for instance, on (9.3), (9.4) in the preceding, we can say that:

(14.13) the *sheaf of $\mathcal{A}$-connection coefficients*

(14.13.1) $$M_n(\Omega) := M_n(\mathcal{A}) \otimes_{\mathcal{A}} \Omega \equiv \Omega(M_n(\mathcal{A}))$$

(cf. (9.10)), in fact, another *vector sheaf* on $X$ too, in view of (14.1) (cf. also Chapt. II; (5.27)), *is* still (cohomologically) *the structure sheaf of any $\mathcal{A}$-extension of the form (14.2), or even of (14.5).* In particular, one concludes that;

(14.13.2) *every element of* $H^1(X, M_n(\Omega))$, viz. *any 1-cocycle* with coefficients from the sheaf in question, *determines uniquely* (cf. (13.7.1)) *an $\mathcal{A}$-extension of any one of the previous two equivalent forms.*

# 15. Examples (contn'd)

We consider below certain particular important instances from the classical theory, on which our previous considerations have a special bearing, while the same classical examples often motivated the present abstract context. Now, to put things in perspective, we first describe the relevant abstract setting, in accordance with the hitherto general framework.

Thus, suppose that we have a *differential triad* $(\mathcal{A}, \partial, \Omega)$ on a *topological space* $X$, as usual, for which, however, we further assume the following cohomological condition;

(15.1) $$H^1(X, M_n(\Omega)) = 0, \quad n \in \mathbf{N}.$$

An important *special case,* for which (15.1) is fulfilled is, for example, that one which occurs when,

(15.2) concerning a given $\mathbf{C}$-algebraized space $(X, \mathcal{A})$, the structure sheaf $\mathcal{A}$ is *soft* (cf. Chapt. III; (8.25)), while the base space $X$ is *paracompact* (Hausdorff).

Of course, the important thing here, in connection with (15.4) below, is the fact that;

(15.3) *if $\mathcal{A}$ is soft over a paracompact* (Hausdorff) *space $X$,* then *every $\mathcal{A}$-module $\mathcal{E}$* on *$X$ is fine* (cf. Chapt. III; Definition 8.1), while, *every fine sheaf* on such a space *is soft* too.

Concerning (15.3), see, for instance, G.E. Bredon [1: p.50, Theorem 9.12]. ∎ Furthermore, one obtains that:

(15.4) *every soft sheaf over a paracompact* (Hausdorff) *space is acyclic;* that is, *it has trivial cohomology.*

(ibid.; p.49, Theorem 9.8). ∎ Thus, by virtue of (15.3) and (15.4), it is now clear that;

(15.5) *if (15.2) holds true,* then, *every $\mathcal{A}$-module* on $X$, therefore $M_n(\Omega)$ itself, *with $n \in \mathbf{N}$* (cf. (14.3.1) and Chapt. V; (8.7)) *is fine,* hence, soft and *thus acyclic.* So, *(15.1) is in force.*

Now, the fundamental fact that interests us here is, of course, the following.

(15.6) *Assume that (15.2) holds true, so* that, in particular, *(15.1) is valid.* Then, *every vector sheaf* on $X$ *admits an $\mathcal{A}$-connection.*

This assertion is certainly clear, in view of (15.1) and Theorem 11.1 (cf. (11.4)). ∎ On the other hand, as we shall presently see, by the ensuing Subsections, (15.6) represents the situation, one actually meets in the classical theory:

**15.(a). $\mathcal{C}^\infty$-manifolds.**– In Section 2 in the preceding we have already considered the *standard differential triad*

(15.7)
$$(\mathcal{C}^\infty_X, d, \Omega^1_X),$$

which is naturally associated with any given (finite dimensional) $\mathcal{C}^\infty$ (or *smooth*)*-manifold* $X$ (see (2.6)). In particular, assuming that $X$ is *paracompact* (Hausdorff) (cf. Chapt. III; (8.54), along with the subsequent comments there), one concludes that the respective *structure sheaf*

(15.8)
$$\mathcal{A} \equiv \mathcal{C}^\infty_X$$

*is fine* (see (2.2), as well as, Chapt. III; (8.27) and (8.55.1)). Therefore, due to the hypothesis for $X$ and (15.3), the sheaf $\mathcal{C}_X^\infty$ is soft, as well, so that (yet, (15.3) and (15.4)) one concludes that;

(15.9) *every $\mathcal{C}_X^\infty$-module on $X$ is* a fine sheaf on $X$ and, therefore, *acyclic.*

In particular, *the last conclusion is valid for the sheaf of connection coefficients $M_n(\Omega_X^1)$, $n \in \mathbf{N}$,* where now $\Omega_X^1$ is given by (2.4) (cf. also (14.13.1) and Chapt. III; (8.30), for $p = 1$). Thus, (15.1) *is in force,* concerning now the triple (15.7).

On the other hand, as already mentioned (cf. (4.11) above), for any given (finite dimensional) *smooth* (viz. $\mathcal{C}^\infty$-) **C**-*vector bundle* **E** on $X$, the corresponding *sheaf $\mathcal{E}$ of germs of smooth sections of* **E** is a *vector sheaf* on $X$, with respect to (15.8), and whose *rank* equals *the dimension of* **E**. Furthermore, a $\mathcal{C}_X^\infty$*-connection of* $\mathcal{E}$, as above, in the sense of Definition 3.1, is (still, by definition, cf. (4.15)) a standard (linear) *connection* (viz. *law of derivation* of smooth sections (again!)) *of* **E**, in the usual sense of differential geometry. Therefore, as a consequence of (15.6) and the preceding, one thus arrives at the classical result, according to which:

(15.10) *every smooth* **C**-*vector bundle* **E** *over a paracompact $\mathcal{C}^\infty$-manifold $X$ admits a connection.*
In particular, *this holds true for $\mathcal{T}(X)$, the* (complexified) *tangent bundle of $X$* (cf. (4.16)); in that case we also say that *the manifold $X$ itself has a* (smooth) *connection* (i.e., a *law of derivation* of the *smooth vector fields* on $X$, the latter being the *smooth sections of $\mathcal{T}(X)$*).

In this context, see e.g. W.A. Poor [1: p.61, Theorem 2.35], or even J.W. Milnor - J.D. Stasheff [1: p.291, Lemma 2]. Yet, see also the next Section 16 for another proof of the same result. ∎

Now, it is still standard (Atiyah) that *(15.10) does not hold, in general, true* in the case of *holomorphic connections* (cf. Definition 15.1 below). Thus, we do discuss it briefly right away in the following Subsection.

**15.(b). Complex manifolds. Holomorphic connections.**– Suppose that we have a *complex manifold $X$*, such that let

(15.11) $$(X, \mathcal{O})$$

be the corresponding **C**-*algebraized space,* where $\mathcal{O}$ stands for the *sheaf of germs of* **C**-*valued holomorphic functions on* $X$ (cf. also Chapt. III; (8.65)).

Now, as already noted (ibid.; (8.66)), $\mathcal{O}$ *is not a fine sheaf* on $X$, the latter property being violated, by the principle of analytic continuation. Therefore, (15.4) *cannot be applied:* Precisely, supposing further that our (complex) manifold $X$ is Hausdorff and 2nd countable, hence, *paracompact,* we remark that *the* (*unital* **C**-*algebra,* cf. Chapt. III; comments following (8.65)) *sheaf* $\mathcal{O}$, being an $\mathcal{O}$-module over itself, *is fine if, and only if, it is soft* (cf. (15.3). See also G.E. Bredon [1: p.267; (j)]). ∎

However, thanks to the classical *Cartan's Theorem B* (see (15.16) below), the previous situation, about the non validity of (15.4), can be remedied, when restricting ourselves to a *Stein manifold* $X$:

In this context, we first recall, for convenience (see, for instance, L. Hörmander [1: p. 109, Definition 5.1.3]), that by the last term one means a (Hausdorff, 2nd countable, hence, *paracompact*) *complex manifold* $X$, satisfying the following properties;

i) $\mathcal{O}(X) = \Gamma(X, \mathcal{O})$ (viz. the algebra of **C**-*valued holomorphic functions* on $X$) *separates points* in $X$ and *supplies global-local coordinates at every point of* $X$.

ii) $X$ is *holomorphically convex,* viz. the holomophically convex hull of any compact subset of $X$ is (relatively) compact, as well.

> **Scholium 15.1.**– Note that the condition of *separation of points* in $X$ by $\mathcal{O}(X)$ in the above definition of a Stein manifold *is* actually *redundant;* cf. H. Grauert [1], as well as, S.G. Krantz [1: p. 276; ex. 14,(iv)].
>
> There are, of course, other equivalent, more sophisticated (nevertheless useful!) definitions of this important class of complex manifolds, through e.g. functional analysis methods (cf., for instance, A. Mallios [4: p. 230; Theorem 3.1]) or cohomological ones (see L. Kaup - B. Kaup [1: p. 93, Theorem 63.2]).

Now, given a *complex manifold* $X$, one considers the $\mathcal{O}$-*module,* in effect, *vector sheaf,* on $X$ (with respect to the structure sheaf $\mathcal{O}$, cf. (15.11))

(15.12) $$\Omega^1 := \Omega^1_X$$

of (germs of) **C**-*valued holomorphic 1-forms* on $X$, *with*

(15.13) $$\mathrm{rk}_{\mathcal{O}}(\Omega^1) = n = \dim_{\mathbf{C}} X.$$

Thus, we have that;

(15.14) $\Omega^1$ is a *coherent* (analytic) *sheaf* on $X$.

See, for instance, H. Grauert - R. Remmert [1: p.67, Corollary]. ∎ Hence,

(15.15) *the same holds true for the vector sheaf* (cf. (15.31))

(15.15.1) $$M_n(\Omega^1) := M_n(\mathcal{O}) \otimes_{\mathcal{O}} \Omega^1 = (\Omega^1)^{n^2} \equiv \Omega^{(n^2)},$$

*with* $n \in \mathbf{N}$.

(See also Chapt. II; (5.27)). Here, as in (15.14) too, one actually applies the fact that, since $\mathcal{O}$ *is a coherent sheaf (Oka's Coherence Theorem), every vector sheaf* on $X$ *is coherent,* as well. (See e.g. H. Grauert - R. Remmert [2: pp. 12, 16]).

Thus, according to *Cartan's Theorem B,* one now concludes that

(15.16) *every coherent* (analytic) *sheaf* $\mathcal{S}$ *on a Stein manifold* $X$ *is acyclic; viz. one has*

(15.16.1) $$H^p(X, \mathcal{S}) = 0,$$

Cf. e.g. L. Hörmander [1: p.182, Theorem 7.4.3]. ∎ Consequently, based on (15.15) and (15.16), one infers that;

(15.17) *for every Stein manifold* $X$, *one has*

(15.17.1) $$H^1(X, M_n(\Omega^1)) = 0, \quad \textit{with} \quad n \in \mathbf{N}.$$

Therefore, in view of (15.1) and the last part of Theorem 14.1, one infers that (see also (15.32) in the sequel);

(15.18) *Let* $X$ *be a Stein manifold. Then, every* (principal holomorphic fiber bundle, hence, any) *holomorphic vector bundle on* $X$ *admits a holomorphic connection.*

However, it is instructive to clarify at this place, a bit more, the term *holo-*

*morphic connection,* which has been used before. So, according to our general notion, as given by Definition 3.1 in the foregoing, the term is, in principle, referred to a (*differential* or, rather, for the case at hand, to a) *holomorphic triad;* in particular, this concerns here the triple

$$(\mathcal{O}, d, \Omega^1) \tag{15.19}$$

(cf. (15.11), (15.12)), where, of course,

$$d : \mathcal{O} \longrightarrow \Omega^1 \tag{15.20}$$

stands for the usual *differential of holomorphic functions* (viz. *sections of* $\mathcal{O}$; see also Chapt. I; (10.14), (11.11); iii), along with Proposition 2.1 and (12.49), (12.50) therein). In this context, see also the next Chapter VII, concerning a *local description of any given* $\mathcal{A}$*-connection,* hence, of a *holomorphic connection* too. However, following M.F. Atiyah [1], we do display right away a *holomorphic connection of an* $\mathcal{O}$*-extension,* relative to the particular **C**-*algebraized space* $(X, \mathcal{O})$, under consideration, according to the general pattern, as already presented in the foregoing:

Thus, suppose that we are given a *principal holomorphic fiber bundle* $P$ over a *complex manifold* $X$ (the latter being, as usual, Hausdorff and 2nd countable, hence *paracompact*), having as *structure group* a (connected) *complex Lie group* $G$. Furthermore, let $T(X)$ be the *(holomorphic) tangent bundle of* $X$; hence, denoting by $\mathcal{T}(X)$ the *sheaf of* (germs of) *holomorphic vector fields* on $X$ (viz. the respective *sheaf of germs of (holomorphic) sections of* $T(X)$, cf. e.g. (4.4)), one has

$$\mathcal{T}(X) = (\Omega^1)^* \equiv \mathcal{H}om_{\mathcal{O}}(\Omega^1, \mathcal{O}), \tag{15.21}$$

that is, the *dual* (vector) *sheaf of* $\Omega^1$, as above (cf. also Chapt. IV; (5.1)). Yet, let $L(P)$ be the *holomorphic vector bundle* over $X$, associated to $P$, via the *adjoint representation of* $G$ on its Lie algebra; so $L(P)$ is, in fact, an *algebra bundle* over $X$, its fiber being isomorphic to $\mathcal{G}$, the *Lie algebra of* $G$. On the other hand, one has a natural *action of* $G$ on $T(P)$, the (holomorphic) *tangent bundle of* $P$, induced on it by the given action of $G$ on $P$; hence, one can further consider the *quotient vector bundle*

$$Q := T(P)/G. \tag{15.22}$$

(In this concern, see e.g. S. Kobayashi - K. Nomizu [1: p.50ff], [2: Chapt. IX], along with M.F. Atiyah [2: p.10]). So the first relevant result here is the

following *short exact sequence of the holomorphic vector bundles* concerned (*Atiyah exact sequence*);

$$0 \longrightarrow L(P) \longrightarrow Q \longrightarrow T(X) \longrightarrow 0 \tag{15.23}$$

(see M.F. Atiyah [1: p.187, Theorem 1]).

Therefore, by considering the respective sheaves of germs of (holomorphic) sections of the vector bundles in (15.23), one gets the following *short exact $\mathcal{O}$-sequence of vector sheaves* on $X$;

$$0 \longrightarrow \mathcal{L}(P) \longrightarrow \mathcal{Q} \longrightarrow \mathcal{T}(X) \longrightarrow 0 \tag{15.24}$$

(see M.F. Atiyah [1: p.183, Proposition 1]. Yet, cf. (4.11) and (4.10) in the foregoing). So, by applying the terminology of $\mathcal{A}$-extensions (Chapt. III; Section 10), we first conclude that;

(15.25) *every principal holomorphic fiber bundle* $P$ over a complex manifold $X$ *entails an $\mathcal{O}$-extension of* $\mathcal{T}(X)$ *by* $\mathcal{L}(P)$ (see (15.24)).

Consequently, *the same $\mathcal{O}$-extension is classified by the* respective *characteristic class;*

$$\delta(1_{\mathcal{T}(X)}) \equiv \mathfrak{a}(P) \in H^1(X, \mathcal{H}om_{\mathcal{O}}(\mathcal{T}(X), \mathcal{L}(P)) \tag{15.26}$$

(cf. Chapt. III; (10.39.2) and Chapt. V; (8.4). See also (10.28) and (14.12) in the preceding, concerning the above applied notation). Thus, by still following M.F. Atiyah [1: p.188, Definition], one sets the following.

**Definition 15.1.** Let $P$ be a holomorphic principal fiber bundle on a complex manifold $X$. Then, we call a *holomorphic connection* of $P$, any splitting of (15.24).

Now, as we presently show, the previous terminology can also be embodied in the hitherto applied framework: Indeed, one first obtains (see also (15.21));

$$\begin{aligned}\mathcal{H}om_{\mathcal{O}}(\mathcal{T}(X), \mathcal{L}(P)) &= \mathcal{T}(X)^* \otimes_{\mathcal{O}} \mathcal{L}(P) = \mathcal{L}(P) \otimes_{\mathcal{O}} \mathcal{T}(X)^* \\ &= \mathcal{L}(P) \otimes_{\mathcal{O}} (\Omega^1)^{**} = \mathcal{L}(P) \otimes_{\mathcal{O}} \Omega^1(\mathcal{L}(P)).\end{aligned} \tag{15.27}$$

(Concerning the previous calculations, cf. Chapt. IV; (6.1), (5.11), Chapt. II; (5.14), along with (15.12) above). On the other hand, by employing any *local gauge* $U$ of the *vector sheaf* $\mathcal{L}(P)$, one obtains;

$$
\begin{aligned}
&\Omega^1(\mathcal{L}(P))(U) = (\Omega^1(\mathcal{L}(P))|_U)(U) \equiv (\mathcal{L}(P)\otimes_{\mathcal{O}}\Omega^1|_U)(U)\\
(15.28)\qquad &= (\mathcal{L}(P)|_U \otimes_{\mathcal{O}|_U} \Omega^1|_U)(U) = (\mathcal{O}^m|_U \otimes_{\mathcal{O}|_U} \Omega^1|_U)\\
&= (\mathcal{O}^m \otimes_{\mathcal{O}} \Omega^1)(U) = (\Omega^1)^{(m)}(U) \equiv \Omega^{(m)}(U),
\end{aligned}
$$

*such that*

$$
(15.29)\qquad m = \operatorname{rk}\mathcal{L}(P) = \operatorname{rk}L(P) = \dim_{\mathbf{C}}\mathcal{G} = \dim_{\mathbf{C}}G
$$

(see thus (4.10), (4.11) above, in conjunction with Chapt. I; (11.40) and Chapt. II; Lemma 5.1, for the calculations in (15.28)). Therefore, one has (see also, for instance, the argument used for (12.8) above: (12.9), in conjunction with Lemma 9.1)

$$
(15.30)\quad H^1(X, \mathcal{H}om_{\mathcal{O}}(\mathcal{T}(X), \mathcal{L}(P))) = H^1(X, \Omega^1(\mathcal{L}(P))) = H^1(X, \Omega^{(m)}).
$$

Now, given that the *vector sheaf* (Whitney sum)

$$
(15.31)\qquad \Omega^{(m)} \equiv (\Omega^1)^m := \underbrace{\Omega^1 \oplus \cdots \oplus \Omega^1}_{m-\text{times}} \quad (m \in \mathbf{N}),
$$

being, by its definition (cf. (15.29), as well as, Chapt. II; (3.7)), another $\mathcal{O}$*-module, is coherent* (see the comments after (15.15) above), one concludes that (see also (15.16));

(15.32) for every *Stein manifold* $X$, one has

$$
(15.32.1)\qquad H^1(X, \Omega^{(m)}) = 0, \quad m \in \mathbf{N}.
$$

Therefore (cf. (15.26) and Definition 15.1), *every holomorphic principal fiber bundle* $P$ *over* $X$ *admits a holomorphic connection.*

The above contributes to the vindication of (15.18) (see also (15.36) below).

On the other hand, given a *holomorphic vector bundle* $\mathbf{E}$ on a *complex manifold* $X$, one can further consider the corresponding *principal holomorphic fiber bundle* $P$ on $X$ of the *complex linear frames of* $\mathbf{E}$, having *structure group* $G = \mathrm{GL}(n, \mathbf{C})$, *where* $n = \operatorname{rk}\mathbf{E}$. Hence, by considering the respective *vector sheaf of* (germs of holomorphic) *sections of* $\mathbf{E}$, $\mathcal{E}$ (see, for instance, (4.4)), one obtains, according to the very definitions;

$$
(15.33)\qquad \mathcal{L}(P) = \mathcal{E}nd\,\mathcal{E}.
$$

Therefore, (15.24) takes now the form

(15.34) $$0 \longrightarrow \mathcal{E}nd\,\mathcal{E} \longrightarrow \mathcal{Q} \longrightarrow (\Omega^1)^* \longrightarrow 0,$$

whose *sructure sheaf* (cf. Chapt. V; Theorem 8.1, in particular, (8.50)), is given by

(15.35) $$\mathcal{H}om_{\mathcal{O}}((\Omega^1)^*, \mathcal{E}nd\,\mathcal{E}) = \Omega^1(\mathcal{E}nd\,\mathcal{E})$$

(cf. (15.27), (15.21) and (15.33)). Accordingly, one infers that;

(15.36) *a holomorphic vector bundle* **E** *on a complex manifold* $X$ *has a holomorphic connection* (viz., by definition, the respective *vector sheaf* $\mathcal{E}$, as above, *has* an $\mathcal{O}$*-connection,* with respect to (15.19)) *if, and only if, this happens for the* corresponding *principal holomorphic fiber bundle of (complex) linear frames of* **E**.

In particular, *this is always the case, if* $X$ is a *Stein manifold.*

The above, in conjunction with (15.32), justifies now completely our assertion in (15.18).

Finally, concerning the $\mathcal{O}(X)$*-module* ("cohomology group"), *which classifies* (15.34) (cf. Chapt. V; Theorem 8.1, in particular, (8.51)), one has

(15.37) $$\begin{aligned} H^1(X, \Omega^1(\mathcal{L}(P))) = H^1(X, \Omega^1(\mathcal{E}nd\,\mathcal{E})) &= H^1(X, M_n(\Omega^1)) \\ &= H^1(X, \Omega^{(k)}), \end{aligned}$$

*such that*

(15.38) $$k = n^2 = \operatorname{rk}_{\mathcal{O}}(\mathcal{E}nd\,\mathcal{E}), \quad \textit{with} \quad n = \operatorname{rk}\mathcal{E} = \operatorname{rk}\mathbf{E}$$

(see (15.15) and Chapt. II; Lemma 6.1).

## 16. Fine vector sheaves

The sheaves in the title of this Section provide another important instance of *sheaves admitting* $\mathcal{A}$*-connections,* once the base space $X$ is *paracompact* and, of course, a standard *(law of) derivation,* viz. a *differential triad* is available.

Now, *fine sheaves,* in fact, *fine $\mathcal{A}$-modules,* have already been considered in the preceding (see Chapt. III; Definition 8.1); thus, we still refer to Chapt. III; Subsection 8.(a) for eventual unexplained terminology, we shall use below.

So assume that we are given a *differential triad* $(\mathcal{A}, \partial, \Omega)$ on a *topological space* $X$, and let $\mathcal{E}$ be an *$\mathcal{A}$-module* on $X$. Moreover, let

$$\mathcal{U} = (U_\alpha)_{\alpha \in I} \tag{16.1}$$

be an *open covering* of $X$, and

$$(D_\alpha)_{\alpha \in I} \tag{16.2}$$

a family of *local $\mathcal{A}$-connections of $\mathcal{E}$*. That is, each $D_\alpha$, $\alpha \in I$, is an $\mathcal{A}|_{U_\alpha}$-*connection of $\mathcal{E}|_{U_\alpha}$*, so that one has

$$\begin{aligned}(D_\alpha)_{\alpha \in I} \in \prod_\alpha \mathrm{Hom}_{\mathbf{C}}(\mathcal{E}|_{U_\alpha}, \mathcal{E} \otimes_{\mathcal{A}} \Omega|_{U_\alpha}) &= C^0(\mathcal{U}, \mathcal{H}om_{\mathbf{C}}(\mathcal{E}, \mathcal{E} \otimes_{\mathcal{A}} \Omega)) \\ &\equiv C^0(\mathcal{U}, \mathcal{H}om_{\mathbf{C}}(\mathcal{E}, \Omega(\mathcal{E})))\end{aligned} \tag{16.3}$$

(see, for example, Chapt. III; (11.12)); thus, (16.2) is, by definition, an *$\mathcal{O}$-cochain of $\mathcal{U}$* with coefficients in the **C**-*vector space sheaf* $\mathcal{H}om_{\mathbf{C}}(\mathcal{E}, \Omega(\mathcal{E}))$, satisfying (coordinate-wise) the corresponding *Leibniz condition* (cf. (3.6)).

Thus, one has, for instance, the previous situation, by supposing that $\mathcal{E}$ is, in particular, a *vector sheaf* on $X$, while (16.1) stands for a *local frame of $\mathcal{E}$*. Then, one gets a family, like (16.2), by considering the *Levi-Civita $\mathcal{A}$-connection 0-cochain, associated with $\mathcal{U}$* (cf. (8.32.1)).

In this context, we first have the following general result, which will be of use presently below; yet, we refer to Chapt. III; (8.23), (8.23′) for the terminology applied.

**Lemma 16.1.** *Let $(\mathcal{A}, \partial, \Omega)$ be a given differential triad on a topological space $X$ and $\mathcal{E}$ an $\mathcal{A}$-module on $X$. Moreover, suppose that*

$$\mathcal{U} = (U_\alpha)_{\alpha \in I} \tag{16.4}$$

*is a locally finite open covering of $X$, while*

$$(\phi_\alpha)_{\alpha \in I} \subseteq \mathrm{End}\,\mathcal{E} \tag{16.5}$$

*is a partition of $1 \equiv 1_{\mathcal{E}}$ ($\in \mathrm{Aut}\,\mathcal{E}$, identity automorphism of $\mathcal{E}$) and*

$$(D_\alpha)_{\alpha \in I} \in C^0(\mathcal{U}, \mathcal{H}om_{\mathbf{C}}(\mathcal{E}, \Omega(\mathcal{E}))) \tag{16.6}$$

*a given family of local $\mathcal{A}$-connections of $\mathcal{E}$* (cf. (16.2), (16.3)). *Then, the relation;*

$$\widetilde{D} := \sum_{\alpha \in I} (\phi_\alpha \otimes 1) \circ D_\alpha \tag{16.7}$$

*provides an $\mathcal{A}$-connection of $\mathcal{E}$.*

**Scholium 16.1.**– As we shall see right away (see the proof of Theorem 16.1 below), one has the situation of the previous lemma by taking a *fine vector sheaf $\mathcal{E}$* over a *paracompact (Hausdorff)* space $X$, along with a *locally finite local frame $\mathcal{U}$ of $\mathcal{E}$.*

**Proof of Lemma 16.1.** Of course, for convenience, we have employed in (16.7) the obvious abuse of notation;

$$1 \equiv 1_{\Omega|_{U_\alpha}}, \quad \alpha \in I. \tag{16.8}$$

Thus, each one of the **C**-*linear morphisms*

$$\widetilde{D}_\alpha := (\phi_\alpha \otimes 1) \circ D_\alpha : \mathcal{E}|_{U_\alpha} \longrightarrow \Omega(\mathcal{E})|_{U_\alpha}, \quad \alpha \in I \tag{16.9}$$

(cf. (16.3), (16.6)), may be viewed as a *local $\mathcal{A}$-connection of $\mathcal{E}$*, when the latter (sheaf) is *restricted to $U_\alpha \subseteq X$* (along, of course, with a similar restriction of the given differential triad $(\mathcal{A}, \partial, \Omega)$; cf. (6.21)). However, what one really needs here is just the fact that

$$\widetilde{D}_\alpha \in \mathrm{Hom}_{\mathbf{C}}(\mathcal{E}|_{U_\alpha}, \Omega(\mathcal{E})|_{U_\alpha}), \quad \alpha \in I. \tag{16.10}$$

Yet, based on the hypothesis for the family (16.5) (cf. Chapt. III; (8.22), (8.23′)), one still obtains that

$$\mathrm{Supp}\,(\phi_\alpha \otimes 1) = \mathrm{Supp}\,(\phi_\alpha) \subseteq U_\alpha, \quad \alpha \in I; \tag{16.11}$$

Therefore, in view of (16.9), and *extending $\widetilde{D}_\alpha$, $\alpha \in I$, in $\complement$* Supp $(\phi_\alpha \otimes 1)$ *by zero*, one may further assume that

$$\widetilde{D}_\alpha \in \mathrm{Hom}_{\mathbf{C}}(\mathcal{E}, \Omega(\mathcal{E})), \quad \alpha \in I, \tag{16.12}$$

which, of course, implies, in turn, that;

$$\widetilde{D} := \sum_{\alpha \in I} \widetilde{D}_\alpha \equiv \sum_\alpha (\phi_\alpha \otimes 1) \circ D_\alpha \in \mathrm{Hom}_{\mathbf{C}}(\mathcal{E}, \Omega(\mathcal{E})). \tag{16.13}$$

(Here the *local finiteness* of $\mathcal{U}$ guarantees, as usual, the meaning of the "infinite sum" in (16.13)).

Now, based on (16.11)) and the hypothesis for (16.5) (see Chapt. III; (8.23′), 2)), one concludes that *the family*

$$\phi_\alpha \otimes 1_\Omega \equiv \phi_\alpha \otimes 1 \in \mathrm{End}(\Omega(\mathcal{E})), \quad \alpha \in I, \tag{16.14}$$

yields a *partition of* $1_{\Omega(\mathcal{E})} \equiv 1$, so that, in addition to (16.11), one obtains;

$$\sum_{\alpha \in I} (\phi_\alpha \otimes 1) = 1. \tag{16.15}$$

Thus, the last relation contributes now, substantially (see, for example, Scholium 7.2 in the preceding), in proving that *the* **C**-*linear morphism* $\widetilde{D}$, as defined by (16.13), *fulfils the* corresponding *Leibniz condition,* as well, which thus completes the proof. ∎

As already noted, we come now to clarify, by the next more concrete result, the framework adopted by the previous Lemma 16.1. So we have.

**Theorem 16.1.** *Let* $(\mathcal{A}, \partial, \Omega)$ *be a given differential triad on a (Hausdorff) paracompact space* $X$. *Then, every fine vector sheaf* $\mathcal{E}$ *on* $X$ *admits an* $\mathcal{A}$-*connection.*

**Proof.** We first remark that, due to the hypothesis for $X$, for any given *local frame* of $\mathcal{E}$, one can consider a *locally finite open refinement* of it, which thus will provide a *locally finite local frame of* $\mathcal{E}$ too. However, for any *(locally finite) local frame* $\mathcal{U}$ of $\mathcal{E}$, one can further consider the *Levi-Civita* $\mathcal{A}$-*connection 0-cochain, associated with* $\mathcal{U}$, as in (8.32.1) in the preceding. Thus, the proof of our claim is now an immediate consequence of the previous Lemma 16.1, due again to the hypothesis for $X$ and now for $\mathcal{E}$, as well (cf. thus Chapt. III; Definition 8.1). ∎

**Scholium 16.2.**– In connection with our argument in the previous proof, we still remark here, in passing, that:

(16.16) given a *vector sheaf* $\mathcal{E}$ on a *paracompact (Hausdorff)* space $X$, *the locally finite local frames of* $\mathcal{E}$ *constitute a cofinal subset of the set of all* (proper) *open coverings of* $X$, the latter set being upwards directed by refinement.

See also Chapt. II; (4.9) and Chapt. III; (4.9).

As already remarked in the preceding (cf. the comments following (15.10) above), the situation described by Theorem 16.1, is what one actually applies

in the classical case of $\mathcal{C}^{\infty}$*-manifolds*. Thus, due to its significance in that important context, we restate Theorem 16.1, for convenience, less formally, as follows:

(16.17) *every fine vector sheaf $\mathcal{E}$ over a paracompact (Hausdorff) space $X$ admits an $\mathcal{A}$-connection, once we are given the standard* (flat) *$\mathcal{A}$-connection $\partial : \mathcal{A} \longrightarrow \Omega$* (viz. a *differential triad* $(\mathcal{A}, \partial, \Omega)$).

So in the case of the usual *differential geometry of smooth* (viz. $\mathcal{C}^{\infty}$-)*manifolds,* the standard result, as expressed by (15.10), has also the following, more classical justification, as an application now of (16.17):

Namely, as already remarked in (15.8), when taking the base $\mathcal{C}^{\infty}$-manifold $X$ *paracompact* (Hausdorff), *the structure sheaf $\mathcal{C}^{\infty}_X$ is fine,* so that (cf. (15.9)) *this is the case for every $\mathcal{C}^{\infty}_X$-module* too; in particular, the same holds true for the *vector sheaves* appeared as *sheaves of* (germs of) $\mathcal{C}^{\infty}$*-sections of smooth* $\mathbf{C}$*-vector bundles* on $X$ (see, for instance, (4.11)). As a consequence of the above discussion, one thus arrives at the following conclusion, as desired. That is,

(16.18) *every vector sheaf, as above, admits a $\mathcal{C}^{\infty}_X$-connection; hence* (by definition), *any* given *smooth vector bundle,* as well.

Yet, concerning the situation in (16.18), one can further employ, of course, the existing one-to-one (and onto) correspondence between (isomorphism classes of) the two objects, under consideration. Indeed, we actually have the following *bijections;*

$$\Phi^n_{\mathcal{A}}(X) = H^1(X, \mathcal{GL}(n, \mathcal{A})) = H^1\left(X, \mathcal{A}^{\mathrm{GL}(n,\mathbf{C})}\right), \tag{16.19}$$

*for every $n \in \mathbf{N}$, with $\mathcal{A} \equiv \mathcal{C}^{\infty}_X$*. (See Chapt. V; Theorem 2.1 and (2.65), ii); yet, Chapt. IV; (2.11), for $\mathbf{A} = \mathbf{C}$ and $\mathcal{C}^{\infty}_X$ insted of $\mathcal{C}_X$. Finally, cf. also Chapter XI in the sequel, for a more general point of view than in (16.19)).

## 17. Moduli space of $\mathcal{A}$-connections

Our aim in this Section is to show the way one can group together *gauge*

*equivalent connections* of a given $\mathcal{A}$-*module* $\mathcal{E}$, a procedure, which, along with its consequences, is of a particular importance in the classical context *(gauge field theories)*, the respective technique amounting thus to the title of the heading. The relevant basic idea has already been tackled in the preceding, in its general form (cf. Definition 5.1), while here we restrict ourselves to the case of one and the same $\mathcal{A}$-module or, in particular, vector sheaf $\mathcal{E}$.

So to start with, suppose, as usual, that we are given a *differential triad*

$$(\mathcal{A}, \partial, \Omega) \tag{17.1}$$

on an arbitrary *topological space* $X$, while let $\mathcal{E}$ be an $\mathcal{A}$-*module* on $X$, for which we further assume that

$$Conn_{\mathcal{A}}(\mathcal{E}) \neq \emptyset, \tag{17.2}$$

viz. we still assume that $\mathcal{E}$ *admits an* $\mathcal{A}$-*connection* (cf. (7.2), (7.15)).

On the other hand, let

$$Aut_{\mathcal{A}}(\mathcal{E}) \equiv Aut\,(\mathcal{E}) := (\mathcal{E}nd\,\mathcal{E})^{\cdot} \tag{17.3}$$

be the *group sheaf of* $\mathcal{A}$-*automorphisms* (or just *automorphisms*) *of* $\mathcal{E}$ (i.e., $\mathcal{A}$-*isomorphisms of* $\mathcal{E}$ *onto itself;* cf. also Chapt. II; (6.29) and Chapt. V; (2.31) together with Scholium 8.1). Of course, (17.3) is a *non-abelian group sheaf,* in general, unless $\mathcal{E}$ is a *line sheaf* on $X$ (see Chapt. II; Lemma 6.2 and Chapt. IV; Lemma 1.1).

Thus, extending the classical terminology, we call (17.3) the *gauge group of* $\mathcal{E}$. (Concerning the standard theory, see e.g. T. Petrie - J. Randall [1: p.23ff]). Yet, it is still usual to employ (by abusing language again!) the previous term, *gauge group of* $\mathcal{E}$, for

$$\Gamma(Aut\,\mathcal{E}), \tag{17.4}$$

viz. for the *(complete) presheaf of* (local) *sections of* $Aut\,\mathcal{E}$ (see also Chapt. V; Scholium 8.2). Finally, the same term, as above, is ascribed to the *group of global sections of* $Aut\,\mathcal{E}$,

$$\mathrm{Aut}\,\mathcal{E} := (Aut\,\mathcal{E})(X) = (\mathcal{E}nd\,\mathcal{E})^{\cdot}(X) = (\mathcal{E}nd\,\mathcal{E})(X)^{\cdot} = (\mathrm{End}\,\mathcal{E})^{\cdot} \tag{17.5}$$

(see (17.3) and Chapt. V; (8.44)).

So, as an application of Definition 5.1 in the foregoing, our next objective is now to show, under the assumption (17.2), that;

(17.6) *the gauge group Aut* $\mathcal{E}$ (in fact, *the group* Aut $\mathcal{E}$, cf. (17.5)) *acts on the (affine) space of* $\mathcal{A}$*-connections of* $\mathcal{E}$, $Conn_{\mathcal{A}}(\mathcal{E})$,

in a sense that will be made precise. Thus, we first set the following definition, a particular case, in effect (for $\mathcal{F} = \mathcal{E}$), of our previous Definition 5.1. So one has.

**Definition 17.1.** Assuming the previous data (cf. (17.1)), two given $\mathcal{A}$-connections $D$ and $D'$ of an $\mathcal{A}$-module $\mathcal{E}$ are said to be *gauge equivalent,* if there exists an element $\phi \in \mathrm{Aut}\,\mathcal{E}$ (see (17.5)), making the following diagram commutative;

$$\begin{array}{ccc} \mathcal{E} & \xrightarrow{D} & \Omega(\mathcal{E}) \\ {\scriptstyle \phi}\downarrow & & \downarrow{\scriptstyle \phi\otimes 1_{\Omega} \equiv \phi \otimes 1} \\ \mathcal{E} & \xrightarrow{D'} & \Omega(\mathcal{E}) \end{array} \tag{17.7}$$

That is, one has

$$D' \circ \phi = (\phi \otimes 1) \circ D, \tag{17.8}$$

or, equivalently,

$$D' = (\phi \otimes 1) \circ D \circ \phi^{-1}. \tag{17.9}$$

**Scholium 17.1.**- Concerning the above relation (17.9), it is still convenient to employ here a simpler (however, *abused*) notation, by just *writing, instead of (17.9);*

$$D' = \phi \circ D \circ \phi^{-1} \equiv \phi D \phi^{-1} \equiv \mathrm{Ad}(\phi)D. \tag{17.10}$$

In this context, it is clear now that;

(17.11) *(17.9) defines an equivalence relation on the set of $\mathcal{A}$-connections of the given $\mathcal{A}$-module $\mathcal{E}$, $Conn_{\mathcal{A}}(\mathcal{E})$* ($\neq \emptyset$, cf. (17.2)). We still denote it by

(17.11.1) $$D \underset{\phi}{\sim} D', \quad \textit{where } \phi \in \mathrm{Aut}\,\mathcal{E}.$$

On the other hand, one further concludes that;

(17.12) *the* above *equivalence relation (17.11.1) in the set $Conn_{\mathcal{A}}(\mathcal{E})$ is,* in effect, *that one defined by the action of the group* Aut $\mathcal{E}$ *on* $Conn_{\mathcal{A}}(\mathcal{E})$.

Indeed, one has the following correspondence

(17.13) $$\tau : \mathrm{Aut}\,\mathcal{E} \times Conn_{\mathcal{A}}(\mathcal{E}) \longrightarrow Conn_{\mathcal{A}}(\mathcal{E}),$$

defined by (cf. (17.10))

(17.14) $$(\phi, D) \longmapsto \tau(\phi, D) \equiv \phi \cdot D \equiv \phi(D) := \phi D \phi^{-1} \equiv \mathrm{Ad}(\phi) D.$$

That is, we first note that *(17.13) is well-defined,* according to (17.14): Thus, by hypothesis *for $\phi$ and $D$, as in (17.14),* we first have, of course, that (see also (17.9), (17.10));

(17.15) $$\phi D \equiv \phi D \phi^{-1} \equiv (\phi \otimes 1) \circ D \circ \phi^{-1} \in \mathrm{Hom}_{\mathbf{C}}(\mathcal{E}, \Omega(\mathcal{E})).$$

Furthermore, *for any* (local) *sections $\alpha \in \mathcal{A}(U)$ and $s \in \mathcal{E}(U)$,* with $U$ *open* in $X$, one obtains;

(17.16) $$\begin{aligned}(\phi D)(\alpha \cdot s) &\equiv (\phi D \phi^{-1})(\alpha \cdot s) = (\phi D)(\phi^{-1}(\alpha s)) = (\phi D)(\alpha \cdot \phi^{-1}(s)) \\ &\equiv (\phi D)(\alpha \cdot t) = \phi(D(\alpha \cdot t)) = \phi(\alpha \cdot Dt + t \otimes \partial \alpha) \\ &= \phi(\alpha \cdot Dt) + \phi(t \otimes \partial\alpha) \equiv \alpha \cdot \phi D(\phi^{-1}(s)) + (\phi \otimes 1)(t \otimes \partial \alpha) \\ &= \alpha (\phi D \phi^{-1})(s) + \phi(t) \otimes \partial\alpha \equiv \alpha \cdot (\phi D)(s) + s \otimes \partial \alpha,\end{aligned}$$

that is, $D' \equiv \phi \cdot D$ (cf. (17.14)) satisfies also the respective *Leibniz condition,* which thus, along with (17.15), justifies now our claim about (the *range of*) the map (17.13). ∎ Now, it is easy to see that;

(17.17) *the map (17.13), as given by (17.14),* yields an *action of the group* $\operatorname{Aut}\mathcal{E}$ *on the set* $Conn_{\mathcal{A}}(\mathcal{E})$. *That is, one has:*

(17.17.1) $$1_{\mathcal{E}} \cdot D = D, \quad D \in Conn_{\mathcal{A}}(\mathcal{E}),$$

as well as,

(17.17.2) $$\phi(\psi D) = (\phi\psi)(D),$$

*for any* $\phi, \psi \in \operatorname{Aut}\mathcal{E}$ *and* $D \in Conn_{\mathcal{A}}(\mathcal{E})$.

Thus, *for every* $D \in Conn_{\mathcal{A}}(\mathcal{E})$, the corresponding *orbit of* $D$ is given by the set (see also (17.10), (17.11) and (17.14));

(17.18) $$\mathcal{O}_D := \{\phi \cdot D \in Conn_{\mathcal{A}}(\mathcal{E}) : \phi \in \operatorname{Aut}\mathcal{E}\} = \Big\{D' \in Conn_{\mathcal{A}}(\mathcal{E}) : D' \underset{\phi}{\sim} D, \ \text{for some } \phi \in \operatorname{Aut}\mathcal{E}\Big\}.$$

In other words,

(17.19) the *orbit* of an element $D \in Conn_{\mathcal{A}}(\mathcal{E})$, under the action of (the *gauge group*) $\operatorname{Aut}\mathcal{E}$ on the latter set, consists of *those* $\mathcal{A}$*-connections of* $\mathcal{E}$, which are *gauge equivalent to* $D$.

The above yields now a complete justification of our previous claim, concerning (17.11.1). ∎

Now, the corresponding *orbit space* of the action considered, i.e., the set;

(17.20) $$M(\mathcal{E}) \equiv Conn_{\mathcal{A}}(\mathcal{E})/\operatorname{Aut}\mathcal{E} := \bigcup_{D \in Conn_{\mathcal{A}}(\mathcal{E})} \mathcal{O}_D = \sum_D \mathcal{O}_D,$$

is called the (global) *moduli space of the* $\mathcal{A}$*-connections* of $\mathcal{E}$.

On the other hand, the specification *"global"* in the last definition indicates the possibility of considering the *local aspect of (17.20)* by restricting ourselves to an *open* $U \subseteq X$: So consider the *restriction to* $U$ of the preceding data; thus, we first have that one of (17.1), viz. the *differential triad* on $U$ (see (6.21))

(17.21) $$(\mathcal{A}, \partial, \Omega)|_U = (\mathcal{A}|_U, \partial|_U, \Omega|_U),$$

yet the $\mathcal{A}|_U$*-module* $\mathcal{E}|_U$, restriction to $U$ of the given $\mathcal{A}$*-module* $\mathcal{E}$, as well as, the restriction to $U$ of the set $Conn_{\mathcal{A}}(\mathcal{E})$ (*non-void,* by hypothesis, cf. (17.2)), so that one has

$$Conn_{\mathcal{A}}(\mathcal{E})|_U \subseteq Conn_{\mathcal{A}|_U}(\mathcal{E}|_U) \tag{17.22}$$

(see (6.22) – (6.24)). Finally, we also consider a similar restriction of the *group sheaf* $\mathcal{A}ut\,\mathcal{E}$ and then (guided by (17.5)) take sections over $U$ of the same restriction; thus, we have (cf. Chapt. I; (11.40))

$$\begin{aligned}(\mathcal{A}ut\,\mathcal{E})|_U(U) &= (\mathcal{A}ut\,\mathcal{E})(U) = \mathrm{Isom}_{\mathcal{A}|_U}(\mathcal{E}|_U, \mathcal{E}|_U) \\ &= \mathcal{I}som_{\mathcal{A}|_U}(\mathcal{E}|_U), \mathcal{E}|_U)(U) \equiv \mathcal{A}ut\,(\mathcal{E}|_U)(U) = \mathrm{Aut}\,(\mathcal{E}|_U)\end{aligned} \tag{17.23}$$

(cf. also e.g. Chapt. V; (2.32)). Thus, one finally obtains

$$(\mathcal{A}ut\,\mathcal{E})(U) = \mathrm{Aut}\,(\mathcal{E}|_U), \tag{17.24}$$

for every open set $U \subseteq X$ (up to an *isomorphism of* the *groups* concerned). So one has an *action of the last group on the set(s) in (17.22)*); viz. one gets the following map

$$\mathrm{Aut}\,(\mathcal{E}|_U) \times Conn_{\mathcal{A}|_U}(\mathcal{E}|_U) \longrightarrow Conn_{\mathcal{A}|_U}(\mathcal{E}|_U), \tag{17.25}$$

defined *analogously to (17.14)*.

In this context, it is not difficult to see that one actually has here a *presheaf of orbit spaces,* by analogy with (17.20), when $U$ ranges over the open subsets of $X$, together with the obvious restriction maps. Accordingly, one can further speak of the *sheaf (of germs) of moduli spaces of $\mathcal{A}$-connections* of the given $\mathcal{A}$-*module* $\mathcal{E}$, denoted by

$$\mathcal{M}(\mathcal{E}) = Conn_{\mathcal{A}}(\mathcal{E})/\mathcal{A}ut\,(\mathcal{E}). \tag{17.26}$$

However, we are not going to use (17.26) in the sequel, formally (at least!). Nevertheless, the following special case, that mostly concerns the applications, as well, has an interest per se:

Thus, assume that $\mathcal{E}$ is, in particular, a *vector sheaf* on $X$, *with*

$$\mathrm{rk}\,\mathcal{E} = n \in \mathbf{N}. \tag{17.27}$$

Moreover, consider an *open* $U \subseteq X$, which is a *local gauge of* $\mathcal{E}$, viz. one has

$$\mathcal{E}|_U = \mathcal{A}^n|_U, \tag{17.28}$$

up to an $\mathcal{A}|_U$-*isomorphism of* the $\mathcal{A}|_U$-*modules* involved. (Of course, we could consider here, more generally, just an $\mathcal{A}$-*module* $\mathcal{E}$ on $X$, together with a *local gauge* of it *of rank* $n \in \mathbf{N}$; cf., for instance, the comments for (8.36) in the foregoing). Thus, one now obtains

$$
\begin{aligned}
(Aut\,\mathcal{E})(U) &= \mathrm{Aut}\,(\mathcal{E}|_U) = \mathrm{Aut}\,(\mathcal{A}^n|_U) = (Aut\mathcal{A}^n)(U) \\
&= M_n(\mathcal{A})^{\cdot}(U) \equiv \mathcal{GL}(n,\mathcal{A})(U) = \mathrm{GL}(n,\mathcal{A}(U))
\end{aligned}
\tag{17.29}
$$

(cf. (17.24), (17.28), along with Chapt. IV; (1.28)). Consequently, one thus concludes that;

(17.30) *any local automorphism of a* given *vector sheaf* $\mathcal{E}$ *over a local gauge* $U$ of $\mathcal{E}$ *is* (given, in effect, by) *a local automorphism of* $\mathcal{A}^n$, *where* $n = \mathrm{rk}\,\mathcal{E}$, *viz. an element of*

$$
\mathrm{GL}(n,\mathcal{A}(U)) = \mathcal{GL}(n,\mathcal{A})(U) \equiv \mathrm{GL}(n,\mathcal{A}|_U).
\tag{17.30.1}
$$

Thus, in the special case of a *vector sheaf* $\mathcal{E}$, as mostly happens, of course, in the applications, the respective *gauge group* $Aut\,\mathcal{E}$, or $\mathrm{Aut}\,\mathcal{E}$ (see (17.5)) is actually reduced *locally* (precisely, in terms of a *local frame* $\mathcal{U}$ of $\mathcal{E}$) to that one of $\mathcal{A}^n$ ($n = \mathrm{rk}\,\mathcal{E}$), viz. to the *group sheaf* $\mathcal{GL}(n,\mathcal{A})$, or its *(complete) presheaf of sections* $\Gamma(\mathcal{GL}(n,\mathcal{A}))$ (cf. Chapt. IV; (1.29)). Therefore, the terminology one usually employs hereby, referring rather to the latter as the respective *gauge groups.* This also in conjunction with the particular rôle, the *local frames* of a given *vector sheaf* play, with respect to the open coverings of $X$. (See e.g. Chapt. II; (4.9) and Chapt. V; (3.1), as well as, Scholium 16.2 in the preceding).

Applications of the above will be considered by the ensuing discussion. See thus the succeeding Chapters VII and VIII.

## 18. Cohomological classification of line sheaves, admitting $\mathcal{A}$-connections

Given an arbitrary $\mathbf{C}$-*algebraized space* $(X,\mathcal{A})$ (cf., for instance, Chapt. IV; (0.1)), we already know that the *line sheaves* on $X$ are classified, within $\mathcal{A}$-*isomorphisms,* by what we have called the Picard group of $X$; thus, one first has

$$
\Phi^1_{\mathcal{A}}(X) = H^1(X,\mathcal{A}^{\cdot}),
\tag{18.1}
$$

within a (set-theoretic) *bijection* (cf. Chapt. V; (2.59), (2.60)). Now, the previous identification becomes an *abelian group isomorphism,* when the set of isomorphism classes of line sheaves on $X$, $\mathbf{\Phi}^1_{\mathcal{A}}(X)$, is further endowed with the *abelian group structure,* defined on it by the *tensor product of line sheaves* (see Chapt. V; (4.3)). The resulting abelian group, viz. the *Picard group of* $X$, denoted by $\mathrm{Pic}_{\mathcal{A}}(X) \equiv \mathrm{Pic}(X)$ (ibid.; Definition 4.1), is now isomorphic to the *1st cohomology group* in (18.1); i.e., one has

$$(18.2) \qquad (\mathbf{\Phi}^1_{\mathcal{A}}(X), \otimes_{\mathcal{A}}) \equiv \mathrm{Pic}(X) = H^1(X, \mathcal{A}^{\cdot}) = \check{H}^1(X, \mathcal{A}),$$

up to *isomorphisms of* the (abelian) *groups* involved (ibid.; Theorem 4.1, along with Chapt. III; Lemma 8.1, regarding the last equality).

On the other hand, suppose now that we are given on $X$ a *differential triad*

$$(18.3) \qquad (\mathcal{A}, \partial, \Omega)$$

(cf. (3.1)). So let further

$$(18.4) \qquad \mathbf{\Phi}^1_{\mathcal{A}}(X)^{\nabla}$$

be the *set of equivalence classes of pairs*

$$(18.5) \qquad (\mathcal{L}, D),$$

consisting of a *line sheaf* $\mathcal{L}$ on $X$ and an $\mathcal{A}$*-connection* $D$ *of* $\mathcal{L}$, where we set

$$(18.6) \qquad (\mathcal{L}, D) \sim (\mathcal{L}', D')$$

for two such pairs, whenever there exists an $\mathcal{A}$*-isomorphism*

$$(18.7) \qquad \phi \in \mathrm{Isom}_{\mathcal{A}}(\mathcal{L}, \mathcal{L}'),$$

making the $\mathcal{A}$-connections $D$ and $D'$ of $\mathcal{L}$ and $\mathcal{L}'$, respectively, *gauge equivalent;* viz. one has

$$(18.8) \qquad D' = (\phi \otimes 1_{\Omega}) \circ D \circ \phi^{-1} \equiv (\phi \otimes 1) D \phi^{-1} \equiv \phi D \phi^{-1}$$

(see (5.43), as well as, e.g. (17.9), (17.10), for the notation applied).

Thus, in view of (18.6), it is easy to see that every element

$$(18.9) \qquad [(\mathcal{L}, D)] \in \mathbf{\Phi}^1_{\mathcal{A}}(X)^{\nabla}$$

constitutes, indeed, an *equivalence class* of pairs $(\mathcal{L}, D)$, as in (18.5). In this

context, we can further say that:

(18.10) *(18.4) is that subset* (possibly empty) *of* $\mathbf{\Phi}^1_{\mathcal{A}}(X)$, *the elements of which have a representative admitting an* $\mathcal{A}$*-connection.*

Indeed, if $[\mathcal{L}] \in \mathbf{\Phi}^1_{\mathcal{A}}(X)$ and $\mathcal{L}$ admits an $\mathcal{A}$-connection $D$, then *every* $\mathcal{L}' \in [\mathcal{L}]$, $\phi$*-isomorphic* to $\mathcal{L}$, with $\phi$ given by (18.7), *has an* $\mathcal{A}$*-connection, as well,* say $D'$ defined by (18.8), so that $(\mathcal{L}, D) \sim (\mathcal{L}', D')$, in the sense of (18.6) (see also (8.18)). ∎

Now, for reasons that will become clear later on (see thus Chapter X in the sequel), one employs for (18.5) a physics originated nomenclature; thus, a pair

$$(\mathcal{L}, D), \tag{18.11}$$

as above, consisting of a line sheaf $\mathcal{L}$ on $X$ and an $\mathcal{A}$-connection $D$ of $\mathcal{L}$, is called a *Maxwell field* on $X$. Yet, the same term is employed, occasionally, to the respective equivalence class,

$$[(\mathcal{L}, D)] \in \mathbf{\Phi}^1_{\mathcal{A}}(X)^{\nabla}, \tag{18.12}$$

as well. Thus, we also speak of (18.4), as the *set of Maxwell fields* on $X$.

In this regard, as an application of our considerations in previous Sections, our first result says that *the above set inherits from the Picard group of* $X$ *the respective abelian group structure, becoming,* in fact, a *subgroup* of the latter. That is, one has the next.

**Theorem 18.1.** *Let* $(\mathcal{A}, \partial, \Omega)$ *be a differential triad on a topological space* $X$*. Then, the set of Maxwell fields on* $X$*,*

$$\mathbf{\Phi}^1_{\mathcal{A}}(X)^{\nabla} \overset{\subset}{\rightarrow} \operatorname{Pic}(X) \equiv (\mathbf{\Phi}^1_{\mathcal{A}}(X), \otimes_{\mathcal{A}}) \tag{18.13}$$

(cf. also (18.10, (18.2)), *endowed with the relative (abelian) group structure from the Picard group of* $X$*, becomes a subgroup of the latter, thus an abelian group itself, as well.*

**Proof.** First, supplementing our argument in (18.10), by justifying thus too the (set-theoretic) *injection* appeared in (18.13), we remark that the map in question is, in effect, the correspondence,

$$[(\mathcal{L}, D)] \longmapsto [\mathcal{L}] : \mathbf{\Phi}^1_{\mathcal{A}}(X)^{\nabla} \longrightarrow \mathbf{\Phi}^1_{\mathcal{A}}(X), \tag{18.14}$$

as already follows from the comments after (18.10); yet, the same map is certainly *one-to-one.*

Now, given the elements $(\mathcal{L}, D)$ and $(\mathcal{L}', D')$ of (18.4) (in fact, two *representatives,* respectively, of the elements in question), we further consider the pair

$$(\mathcal{L} \otimes_{\mathcal{A}} \mathcal{L}', D \otimes D'), \tag{18.15}$$

which is again a *representative of an element of the same set:* Indeed, this follows from Chapt. V; (4.3), by further setting

$$D \otimes D' \equiv D_{\mathcal{L} \otimes_{\mathcal{A}} \mathcal{L}'} := (D \otimes 1_{\mathcal{L}'}) + (1_{\mathcal{L}} \otimes D') \equiv D \otimes 1 + 1 \otimes D', \tag{18.16}$$

according to (5.10) in the preceding.

On the other hand, based further on Chapt. V; (4.3) and on Subsection 5.(d) of the present Chapter, one defines the *inverse of an element* $[(\mathcal{L}, D)] \in \mathbf{\Phi}^1_{\mathcal{A}}(X)^{\nabla}$ (in fact, *of a representative* of the latter) by the relation

$$(\mathcal{L}, D)^{-1} := (\mathcal{L}^*, D^*), \tag{18.17}$$

where, of course, $\mathcal{L}^*$ stands for the *dual line sheaf* of $\mathcal{L}$ and $D^*$ for the respective *dual $\mathcal{A}$-connection of the* given *$\mathcal{A}$-connection $D$ of $\mathcal{L}$*, being thus an *$\mathcal{A}$-connection of the line sheaf $\mathcal{L}^*$*; so one has (cf. (5.31))

$$D^*(u) := \partial \circ u - (u \otimes 1_{\Omega}) \circ D \equiv \partial \circ u - (u \otimes 1) \circ D, \tag{18.18}$$

*for every*

$$u \in \mathcal{L}^*(U) \equiv \mathcal{H}om_{\mathcal{A}}(\mathcal{L}, \mathcal{A})(U) = \mathrm{Hom}_{\mathcal{A}|_U}(\mathcal{L}|_U, \mathcal{A}|_U) \tag{18.19}$$

and any *open* $U \subseteq X$. Therefore, *for every (local) section* $s \in \mathcal{L}(U)$, one has, in view of (18.18);

$$D^*(u)(s) = \partial(u(s)) - (u \otimes 1)(Ds). \tag{18.20}$$

So one obtains (cf. also (18.17))

$$[(\mathcal{L}, D)]^{-1} := \left[(\mathcal{L}, D)^{-1}\right] \equiv [(\mathcal{L}^*, D^*)] \in \mathbf{\Phi}^1_{\mathcal{A}}(X)^{\nabla}, \tag{18.21}$$

which thus, together with (18.15), completes now the proof. ∎

Thus, based on Theorem 18.1, one may speak of the *abelian group of Maxwell fields* (in fact, equivalence classes of such) on $X$, *a subgroup,* in effect (viz. modulo the injection in (18.13), see (18.14)), *of the corresponding Picard group of* $X$. So our next objective is now to identify the former group

with an appropriate cohomology group, by analogy with the already familiar situation in (18.1). Indeed, one has the following.

**Theorem 18.2.** *Suppose that we have a differential triad* $(\mathcal{A}, \partial, \Omega)$ *on topological space* $X$ *and let*

$$\tilde{\partial} : \mathcal{A}^{\cdot} \longrightarrow \Omega \tag{18.22}$$

*be the logarithmic derivation, induced by* $\partial$ *on* $\mathcal{A}^{\cdot}$, *the group sheaf of units of* $\mathcal{A}$. *Then, the group of Maxwell fields on* $X$ *coincides, within an (abelian) group isomorphism, with the 1st (Čech) hypercohomology group of* $X$ *with coefficients in the 2-term* $\mathcal{A}$*-complex on* $X$, *defined by (18.22). That is, one has* (cf. Chapt. III; (7.43))

$$\mathbf{\Phi}^1_{\mathcal{A}}(X)^{\nabla} = \check{\mathbf{H}}^1(X, \mathcal{A}^{\cdot} \xrightarrow{\tilde{\partial}} \Omega), \tag{18.23}$$

*up to an isomorphism of the abelian groups concerned.*

The proof of the above theorem is actually based on material related with the *"local theory" of* $\mathcal{A}$*-connections,* a subject that we are dealt with right in the next Chapter (see thus ibid.; Section 11). Notwithstanding, the previous statement was aiming, in effect, to give a particular application of Theorem 18.1, the latter result belonging, for that matter, to the global theory of $\mathcal{A}$-connections, as this has already been exposed in the foregoing.

Concerning the classical counterpart of the previous result, see, for instance, J.-L. Brylinski [1: p.76, Theorem 2.2.11, (2)]. In this context, we also note that, by analogy with the classical pattern (ibid.), one can further obtain the corresponding result to the previous theorem, by employing *Deligne cohomology* (see, for instance, ibid.; p.77, Theorem 2.2.12). However, we have not treated the cohomology in question in the preceding, hence we omit here the relevant discussion, concerning the group (18.4) (cf. Theorem 18.1) in that context.

# CHAPTER VII

# $\mathcal{A}$-connections. Local Theory

" ...to understand what's what ... is a vital aspect of Mathematics."

S. Mac Lane in "Mathematics Form and Function" (Springer - Verlag, 1986). p. 288.

Our aim in this Chapter is to study the *local aspects* of certain fundamental concepts, which we were dealt with in the previous Chapter VI, where we mainly examined the *global point of view* of the notion of an $\mathcal{A}$-connection. As we shall presently see, several standard results of the classical *(local) differential geometry of $C^\infty$-manifolds* are still in force, within the present abstract (axiomatic) framework. This means, at least, that *even in the classical setting no calculus* (notwithstanding, its *formal properties,* yes!) *is actually needed* for the results in question.

Furthermore, various applications of this local point of view of $\mathcal{A}$-connections are also considered, according to the classical patterns, in this and the following Chapter VIII; yet, as another application of this same aspect, we can still mention our treatment of *characteristic classes* of vector sheaves *(Chern-Weil theory),* as this is given in Chapter IX in the sequel.

# 1. Local form of $\mathcal{A}$-connections

We start with looking at the form, which is taken by an $\mathcal{A}$-connection $D$ of an $\mathcal{A}$-module $\mathcal{E}$, when we restrict ourselves to a given local gauge of $\mathcal{E}$, as this latter notion is specified in the sequel (see thus (1.15) below; but, cf. also, for instance, Chapt. VI; (8.36), (9.2)). Of course, our argument applies when, in particular, one looks at a given vector sheaf and a local frame of it.

Thus, to begin with, suppose, as usual, that we are given a *differential triad*

$$(\mathcal{A}, \partial, \Omega) \tag{1.1}$$

on a *topological space* $X$ (see Chapt. VI; (3.1)). Moreover, let $\mathcal{E}$ be an *$\mathcal{A}$-module* on $X$, along with an *open* $U \subseteq X$, being a *local gauge of* $\mathcal{E}$, of *rank* $n \in \mathbf{N}$; viz. one has, by definition,

$$\mathcal{E}|_U = \mathcal{A}^n|_U, \tag{1.2}$$

within an *$\mathcal{A}|_U$-isomorphism of* the *$\mathcal{A}|_U$-modules* involved.

Now, assume further that

$$(e_1, \ldots, e_n) \equiv (e_i)_{1 \leqq i \leqq n} \subseteq \mathcal{E}(U) \tag{1.3}$$

is a (finite) sequence of *local sections* of $\mathcal{E}$ over the *open* $U \subseteq X$, as above, providing a *basis* of the *free* $\mathcal{A}(U)$*-module* $\mathcal{E}(U)$; indeed, one obtains

$$\mathcal{E}(U) = (\mathcal{E}|_U)(U) = (\mathcal{A}^n|_U)(U) = \mathcal{A}^n(U) = \mathcal{A}(U)^n \tag{1.4}$$

(see (1.2), and Chapt. I; (11.40). Of course, all of the equalities in (1.4) stand for *$\mathcal{A}(U)$-isomorphisms of* the *$\mathcal{A}(U)$-modules* involved).

**Scholium 1.1.**– Concerning the *$\mathcal{A}(U)$-basis* (1.3) of $\mathcal{E}(U)$, as above, we remark that this is, in effect, a *consequence* of the *$\mathcal{A}|_U$-isomorphism* (1.2). Indeed, one gets such a basis by composing, for instance, the *inverse* of the previous isomorphism with the (canonical) *Kronecker sections* of $\mathcal{A}^n$, restricted to $U$; see Chapt. IV; (3.3) and (3.11.1).

On the other hand, for convenience of reference, we adopt, heretofore, the following succinct terminology, concerning the previous situation: Namely, we shall speak of a *local gauge* of an *$\mathcal{A}$-module $\mathcal{E}$* over an *open* $U \subseteq X$, of *rank* $n \in \mathbf{N}$, and write

$$e^U \equiv \{U; (e_1, \ldots, e_n)\}, \tag{1.5}$$

whenever *one has (1.2), together with a given basis (1.3),* as before. Yet, the term *local coordinate system* (or simply *local coordinates*) *of $\mathcal{E}$ over $U$* are applied, occasionally, for (1.5).

Thus, based on the preceding, one now concludes that (cf. also Chapt. IV; (3.11)):

(1.6) given a *local gauge of the $\mathcal{A}$-module $\mathcal{E}$*, as in (1.5), any *local section* $s \in \mathcal{E}(U)$ is written in the form;

$$s = \sum_{i=1}^{n} s_i e_i, \tag{1.6.1}$$

such that (cf. (1.4))

$$(s_i)_{1 \leqq i \leqq n} \in \mathcal{A}(U)^n = \mathcal{A}(U)^n. \tag{1.6.2}$$

We call the (local) sections $s_i \in \mathcal{A}(U)$, $1 \leqq i \leqq n$, the *coordinates of* the given $s \in \mathcal{E}(U)$, with respect to $e^U$, as in (1.5). Thus, when (1.5) is given, they are, of course, in view of (1.4), *uniquely defined, for any* $s \in \mathcal{E}(U)$.

Now, assume further that we are given an *$\mathcal{A}$-connection $D$* of the *$\mathcal{A}$-module* $\mathcal{E}$, under consideration (cf. Chapt. VI; Definition 3.1). Thus, by applying $D$ to (1.6.1) above, one obtains;

$$D(s) = D\left(\sum_{i=1}^{n} s_i e_i\right) = \sum_i D(s_i \cdot e_i) = \sum_i (s_i De_i + e_i \otimes \partial s_i). \tag{1.7}$$

So this is the *local form of the* given *$\mathcal{A}$-connection $D$ on $U$*, precisely speaking on $\mathcal{E}|U$, when $U$ is a local gauge of $\mathcal{E}$, as in (1.5).

**Remark 1.1.**– In fact, one considers in (1.7) the *restriction of the $\mathcal{A}$-connection $D$ to the open $U \subseteq X$*, in the sense of Chapt. VI; (6.23).

Accordingly, by virtue of (1.7), one concludes the following:

(1.8) Suppose we are given a *local gauge* $e^U$ *of an* $\mathcal{A}$*-module* $\mathcal{E}$, as in (1.5), and an $\mathcal{A}$*-connection* $D$ *of* $\mathcal{E}$. Then, *the* given $\mathcal{A}$*-connection* $D$ *is known on* $U$ (cf. Remark 1.1), *when we know its values at the basis (1.3); viz. once we know*

$$(1.8.1)\quad D(e_i) \in (\mathcal{E} \otimes_{\mathcal{A}} \Omega)(U) \equiv \Omega(\mathcal{E})(U) = \mathcal{E}(U) \otimes_{\mathcal{A}(U)} \Omega(U),$$

*with* $1 \leqq i \leqq n$. (See also (1.10) below).

Now, the following argument justifies (1.8.1); thus, by (1.2) and Chapt. II; Lemma 5.1, along with (5.18), one first has the next $\mathcal{A}|_U$*-isomorphisms,*

$$(1.9)\qquad \begin{aligned} \Omega(\mathcal{E})|_U \equiv (\mathcal{E} \otimes_{\mathcal{A}} \Omega)|_U &= \mathcal{E}|_U \otimes_{\mathcal{A}|_U} \Omega|_U = \mathcal{A}^n|_U \otimes_{\mathcal{A}|_U} \Omega|_U \\ &= (\mathcal{A}^n \otimes_{\mathcal{A}} \Omega)|_U = \Omega^n|_U = (\Omega|_U)^n. \end{aligned}$$

Therefore, one obtains (cf. also (1.4), Chapt. I; (11.40) and Chapt. II; (5.25));

$$(1.10)\qquad \begin{aligned} \Omega(\mathcal{E})(U) &\equiv (\mathcal{E} \otimes_{\mathcal{A}} \Omega)(U) = (\mathcal{E} \otimes_{\mathcal{A}} \Omega)|_U(U) \\ &= (\Omega^n|_U)(U) = \Omega^n(U) = \Omega(U)^n = \mathcal{A}(U)^n \otimes_{\mathcal{A}(U)} \Omega(U) \\ &= \mathcal{E}(U) \otimes_{\mathcal{A}(U)} \Omega(U), \end{aligned}$$

which thus completely explains (1.8.1). ▮

On the other hand, in view of (1.4), one has that ($\mathcal{E}(U)$ *is a free* $\mathcal{A}(U)$*-module of rank* $n \in \mathbf{N}$. Accordingly, *the values of the* given $\mathcal{A}$-connection $D$ of $\mathcal{E}$ at the elements of the basis (1.3), being, by virtue of (1.8.1), *elements of the* $\mathcal{A}(U)$*-module of the last term in (1.10),* are now determined by the following relations (cf. also, for instance, W. Greub [1: p.7, Lemma 1.5.2]);

$$(1.11)\qquad D(e_j) = \sum_{i=1}^{n} e_i \otimes \omega_{ij}, \quad 1 \leqq j \leqq n,$$

for some *uniquely defined local sections*

$$(1.12)\qquad \omega_{ij} \in \Omega(U), \quad 1 \leqq i, j \leqq n.$$

Therefore, *one has*

$$(1.13)\qquad \omega \equiv (\omega_{ij}) \in M_n(\Omega(U)) = M_n(\Omega)(U)$$

(cf. also Chapt. V; (8.7) and Chapt. VI; (9.3)).

We call (1.13) the *local* $\mathcal{A}$*-connection matrix of* $D$, which is thus associated with the given *local gauge* $e^U$ of the $\mathcal{A}$*-module* $\mathcal{E}$ (cf. (1.5)).

Thus, as a consequence of the preceding discussion, we now have the following basic result.

**Theorem 1.1.** *(Local form of an $\mathcal{A}$-connection). Suppose we are given a differential triad $(\mathcal{A}, \partial, \Omega)$ on a topological space $X$, an $\mathcal{A}$-module $\mathcal{E}$ on $X$ and an $\mathcal{A}$-connection $D$ of $\mathcal{E}$. Moreover, let*

$$e^U \equiv \{U; (e_1, \ldots, e_n)\} \tag{1.14}$$

*be a local gauge of $\mathcal{E}$ over an open set $U \subseteq X$, of rank $n \in \mathbf{N}$* (cf.(1.5)). *Then, $D$* (restricted to $U$) *is given by the relation*

$$D(s) = \sum_{i=1}^{n} e_i \otimes \left(\partial s_i + \sum_{j=1}^{n} s_j \omega_{ij}\right), \tag{1.15}$$

*for any (local) section of $\mathcal{E}$ over $U$,*

$$s = \sum_{i=1}^{n} s_i e_i \in \mathcal{E}(U) \tag{1.16}$$

(cf. (1.6.1)), *while*

$$\omega \equiv (\omega_{ij}) \in M_n(\Omega(U)) = M_n(\Omega)(U) \tag{1.17}$$

*is a uniquely determined* (given $D$ and $e^U$) *$n \times n$ matrix with entries in $\Omega(U)$* (alias *"1-forms"* on $U$).

*Conversely, given (1.16), by further assuming (1.13), the relation (1.15) determines uniquely an $\mathcal{A}$-connection of $\mathcal{E}$ over $U$, viz. precisely speaking, an $\mathcal{A}|_U$-connection of $\mathcal{E}|_U$.*

**Proof.** First, we prove (1.15): Thus, for any local section $s \in \mathcal{E}(U)$, as in (1.16), one obtains, *in view of (1.11);*

$$\begin{aligned} D(s) &= D\left(\textstyle\sum_{j=1}^{n} s_j e_j\right) = \textstyle\sum_j D(s_j e_j) = \textstyle\sum_j (s_j De_j + e_j \otimes \partial s_j) \\ &\textstyle\sum_j (s_j (\sum_{i=1}^{n} e_i \otimes \omega_{ij}) + e_j \otimes \partial s_j) = \textstyle\sum_j (s_j (\sum_i e_i \otimes \omega_{ij})) + \textstyle\sum_j e_j \otimes \partial s_j \\ &= \textstyle\sum_j (\sum_i e_i \otimes s_j \omega_{ij}) + \textstyle\sum_j e_j \otimes \partial s_j = \textstyle\sum_i (e_i \otimes \sum_j s_j \omega_{ij}) + \textstyle\sum_i e_i \otimes \partial s_i \\ &= \textstyle\sum_{i=1}^{n} e_i \otimes (\partial s_i + \sum_{j=1}^{n} s_j \omega_{ij}), \end{aligned} \tag{1.18}$$

as was actually to be proved. Now, this, in conjunction also with our previous conclusion about the *uniqueness of (1.12)*, already establishes the first part of the assertion.

*Conversely,* we prove now that, *given the* $n \times n$ *matrix* (1.17), the relation (1.15) *yields,* in fact, *an* $\mathcal{A}|_U$*-connection of the* $\mathcal{A}|_U$ *module* $\mathcal{E}|_U$ (cf. e.g. Chapt. VI; (6.21), (6.22)): First, the **C**-*linearity* of $D$ (in fact, of $D|_U$), with respect to $s \in \mathcal{E}(U)$, as the latter is given by (1.16), is certainly clear by the same form of (1.15). So, for convenience, we check the corresponding *Leibniz condition* (see Chapt. VI; (3.6)); thus, *for any (local) sections* $s \in \mathcal{E}(U)$, as given by (1.16) (see also (1.6.1)), *and* $\alpha \in \mathcal{A}(U)$, one obtains, by virtue of (1.15) (cf. also Chapt. VI; (1.2)),

$$\begin{aligned} D(\alpha \cdot s) &= \textstyle\sum_{i=1}^n e_i \otimes (\partial(\alpha s_i) + \sum_{j=1}^n (\alpha s_j)\omega_{ij}) \\ (1.19) \quad &= \textstyle\sum_i e_i \otimes (\alpha \cdot \partial s_i + s_i \cdot \partial\alpha + \alpha \cdot \sum_j s_j \omega_{ij}) = \alpha \cdot (\sum_i \otimes (\partial s_i + \sum_j s_j \omega_{ij})) \\ &+ \textstyle\sum_i e_i \otimes s_i \partial\alpha = \alpha \cdot D(s) + (\sum_i s_i e_i) \otimes \partial\alpha = \alpha \cdot Ds + s \otimes \partial\alpha, \end{aligned}$$

as desired. Furthermore, it is clear that *assuming (1.11),* the relation (1.15) *defines* $D|_U$ *uniquely* (cf. also (1.7)), which thus completes the proof. ∎

Now, by analogy with (1.11), it is still useful to put (1.15) into the following form;

$$(1.20) \qquad D(s) = \sum_{i=1}^n e_i \otimes (Ds)_i,$$

where, of course, we have set

$$(1.21) \qquad (Ds)_i := \partial(s_i) + \sum_{j=1}^n s_j \omega_{ij} \in \Omega(U), \quad 1 \leqq i \leqq n,$$

*for any* $s \in \mathcal{E}(U)$, *given by (1.6.1),* in view always of our hypothesis for $e^U$. As a consequence of the above, one can further look at (1.20), as given by the relation (cf. also (1.10))

$$(1.22) \qquad D(s) = ((Ds)_i)_{1 \leqq i \leqq n} \in \Omega(U)^n = \Omega(\mathcal{E})(U).$$

This also explains the obvious abuse of language, one usually applies to this point, by referring to the elements (1.21) of $\Omega(U)$, as the *(local) coordinates of*

$$(1.23) \qquad D(s) \in \Omega(\mathcal{E})(U) = \mathcal{E}(U) \otimes_{\mathcal{A}(U)} \Omega(U)$$

(cf. (1.10)), *with respect to the given local gauge* $e^U$ *of* $\mathcal{E}$ (see (1.5)). They are, of course, *uniquely defined, for any* given $s \in \mathcal{E}(U)$, as before, once $D$ *and* $e^U$ are kept *fixed* (cf. e.g. (1.11), (1.12)).

Furthermore, the same relation (1.22), in conjunction with (1.21), permits also another, *very convenient expression of* $D$, in fact, of $D|_U$, *in matrix form;* that is, we have

$$D = \partial + \omega \ (\textit{locally on } U), \tag{1.24}$$

such that one has (cf. also (1.29) below)

$$D(s) = \partial(s) + \omega \cdot s \equiv \partial(s) + \omega(s), \tag{1.25}$$

*for every (local) section*

$$s \equiv (s_i)_{1 \leqq i \leqq n} \in \mathcal{A}(U)^n = \mathcal{E}(U) \tag{1.26}$$

(cf. (1.4)). So concerning (1.25), we have also set

$$\partial(s) \equiv (\partial(s_i))_{1 \leqq i \leqq n} \in \Omega(U)^n; \tag{1.27}$$

that is, we abused notation here, for convenience, by setting $\partial$, in place actually of its *$n$-dimensional extension, restricted to* $U$, viz.

$$(\partial \equiv)\, \partial^n|_U : \mathcal{A}^n|_U \longrightarrow \Omega^n|_U, \tag{1.28}$$

given exactly by (1.27) (cf. Chapt. VI; (5.4), (5.5) and (8.6)). Yet, concerning the last item in (1.25), we have also set (cf. (1.21))

$$\left(\sum_{j=1}^{n} s_j \cdot \omega_{ij}\right) = \left(\sum_{j=1}^{n} \omega_{ij} \cdot s_j\right) = \omega \cdot s \equiv \omega(s), \tag{1.29}$$

*for any* $s \in \mathcal{E}(U)$, *as given by (1.26).*

On the other hand, by hypothesis for $D$ and (1.2), (1.9), one obtains

$$D|_U \in \mathrm{Hom}_{\mathbf{C}}(\mathcal{E}|_U, \Omega(\mathcal{E})|_U) = \mathrm{Hom}_{\mathbf{C}}(\mathcal{A}^n|_U, \Omega^n|_U), \tag{1.30}$$

such that *one has,* in effect, *that*

$$D|_U \in Conn_{\mathcal{A}|_U}(\mathcal{A}^n|_U). \tag{1.31}$$

Now, this along with (1.28), serve to *justify* (1.24): Namely, the resulting *$\mathcal{A}|_U$-connection difference*

$$D - \partial \,(= \omega) \tag{1.32}$$

*is exactly an element,* indeed, an $n \times n$ *section matrix,* as follows;

$$\begin{aligned}\omega \in \mathrm{Hom}_{\mathcal{A}|_U}(\mathcal{A}^n|_U, \Omega^n|_U) &= \mathcal{H}om_{\mathcal{A}}(\mathcal{A}^n, \Omega^n)(U)\\ &= \mathcal{H}om_{\mathcal{A}}(\mathcal{A}^n, \mathcal{A}^n \otimes_{\mathcal{A}} \Omega)(U) = (\mathcal{H}om_{\mathcal{A}}(\mathcal{A}^n, \mathcal{A}^n) \otimes_{\mathcal{A}} \Omega)(U)\\ &\equiv (M_n(\mathcal{A}) \otimes_{\mathcal{A}} \Omega)(U) \equiv M_n(\Omega)(U) = M_n(\Omega)(U)),\end{aligned} \tag{1.33}$$

which is actually the case with (1.24), *according to (1.17)*. In this context, cf. also Chapt. VI; (7.14), (7.30), as well as, e.g. (8.7), (8.8)). So this, in conjunction with (1.21), (1.22) and (1.28), (1.29), *explains* now *completely (1.24).* ▮

**Scholium 1.2.**– The above *(local) matrix form of* $D$, as given by (1.24), leads often to a convenient *abuse of language* (even in the standard theory of differential geometry), by saying that:

(1.34) *the* $\mathcal{A}$*-connection* $D$ *is (uniquely) determined by* $\omega$, a fact *holding true,* of course, *only locally*(!) (see thus Theorem 1.1 above), viz. for every *local gauge* $e^U$ of the $\mathcal{A}$*-module* $\mathcal{E}$, under consideration.

Thus, by further looking at the things *locally,* as before, we still realize that:

(1.35) a given $\mathcal{A}$*-connection* $D$ of an $\mathcal{A}$*-module* $\mathcal{E}$ on $X$, *restricted to a local gauge* $e^U$ *of* $\mathcal{E}$ (see (1.5)), becomes, in fact, an $\mathcal{A}|_U$*-connection of* $\mathcal{A}^n|_U$ (cf. (1.31)), *determining* thus *a change (perturbation) of the standard (flat)* $\mathcal{A}|_U$*-connection of* $\mathcal{A}^n|_U$, $\partial^n|_U \equiv \partial$ (the latter being already supplied by the given *differential triad* $(\mathcal{A}, \partial, \Omega)$ on $X$; cf. Chapt. VI; (8.6)). Yet, *this perturbation is expressed* (*locally,* always!) *by* (the *potential*) $\omega \in M_n(\Omega(U))$, according to (1.24) (see also (1.31)).

So, putting it differently, we can say that;

(1.36) $\omega$ in (1.24) measures the *deviation from "differentiating flatly"*, viz. in terms of $\partial^n|_U \equiv \partial$ (again (1.31)).

A word of *terminology,* before we proceed: So we recall that we have already named in the preceding $\omega$, the *local* $\mathcal{A}$*-connection matrix* of $D$. Yet, we call $\omega$, as in (1.24) (cf. also (1.13)), the *(local) connection 1-form* of $D$, or just *(local) connection form.* Moreover, the terms *potential form* (*of the connection* involved), or simply *potential* (as applied at the end of (1.35)

above), are in use, as well, especially when referring to applications in *physics.*

So, by terminating this discussion, we can still say that, as a result of (1.35), one can repeat, here too (along with the classical theory) that, roughly speaking, *(locally)*

(1.37) *a connection is a potential.*

Finally, as we shall see later on (cf. Chapter X in the sequel), the same *(local)* relation (1.24) admits a nice *physical interpretation,* as well.

On the other hand, by still referring to the terminology of the last Scholium, we also note, in anticipation, that a means of estimating the deviation at issue (cf. (1.36)), or even the way one realizes the above *potential* (see (1.37)), is the respective *"field strength"* of the potential, under consideration, in other words, the corresponding *curvature* of the connection involved. Indeed, this fundamental notion is going to be treated in the subsequent Chapter VIII.

Now, the preceding have, of course, a natural application when one is concerned with *vector sheaves* on $X$, a fact that will also be considered in the sequel. In particular, the previous formulae about (1.24) become extremely simple in the case one is dealt with a *Maxwell field* $(\mathcal{L}, D)$ on $X$ (see Chapt. VI; (18.11), (18.12), for the relevant terminology). Thus, by looking at a *local gauge* of (the *line sheaf*) $\mathcal{L}$, say

(1.38) $$e^U \equiv (U; e)$$

(see (1.5), for $n = 1$), one has

(1.39) $$e \in \mathcal{L}(U) = \mathcal{A}(U)$$

(see (1.4), for $n = 1$, the equality in (1.39) standing, of course, for an $\mathcal{A}(U)$-*isomorphism*).

Accordingly, for the case in question, one obtains for the (local) *potential of the $\mathcal{A}$-connection $D$ of $\mathcal{L}$*,

(1.40) $$\omega \in \Omega(U),$$

viz. a *"1-form"* on $U$; yet, concerning (1.25), *one has*

(1.41) $$(D|_U)(s) \equiv D(s) = \partial(s) + \omega \cdot s = \partial(s) + s \cdot \omega,$$

*for any local section*

$$s \in \mathcal{L}(U) = \mathcal{A}(U) \tag{1.42}$$

(see (1.39); yet, concerning (1.41), we recall that, according to our hypothesis for $\mathcal{A}$, $\mathcal{A}(U)$ is a *commutative* **C**-algebra. The same argument has been applied in (1.29)).

Now, we are interested next in the way the various local $\mathcal{A}$-connection matrices, as above, vary when (the local gauge) $U$ *is running over a local frame* $\mathcal{U}$ *of a* given *vector sheaf* on $X$. Indeed, this point of view is dealt with straight off in the following Section.

## 2. Change of local gauges

We examine in this Section the manner that the local connection forms of a given $\mathcal{A}$-connection of a vector sheaf are transformed when a change of a local gauge involved is on foot (see thus Theorem 2.1 below).

So to start with, suppose first that we are given an $\mathcal{A}$-*module* $\mathcal{E}$ on $X$, as in the preceding Section 1, along with the *local gauges* of it

$$e^U \equiv \{U; (e_1, \dots, e_n)\} \text{ and } f^V \equiv \{V; (f_1, \dots, f_n)\} \tag{2.1}$$

over the *open* sets $U, V$ of $X$, of the *same dimension* (or *rank*) $n \in \mathbf{N}$ (see (1.5)). Thus, by further assuming that

$$U \cap V \neq \emptyset, \tag{2.2}$$

one obtains, by *restriction of* the (local) gauges in (2.1) *to* $U \cap V$, *two* $n$-*dimensional local gauges of* $\mathcal{E}$, *with common domain of definition the open set* $U \cap V \subseteq X$.

Therefore, by first applying (1.6.1), one gets

$$f_j = \sum_{i=1}^{n} g_{ij} \cdot e_i, \quad 1 \leqq j \leqq n, \tag{2.3}$$

(of course, every thing here is *restricted to* $U \cap V$; cf. the comments above, following (2.2)). So one actually obtains

(2.4) $$g \equiv (g_{ij}) \in \mathrm{GL}(n, \mathcal{A}(U \cap V)) = \mathcal{GL}(n, \mathcal{A})(U \cap V)$$

(cf. also (1.6.2) and Chapt. IV; (1.28)). Accordingly,

(2.5) *the* $n \times n$ *matrix* $g$ in (2.4), as given by (2.3), is the *matrix of change between the local gauges* $e^U$ *and* $f^V$. Thereby, we call $g$ a *local gauge transformation;* yet, in addition to the comments, following (17.30) in the previous Chapter VI, the form of $g$ in (2.4) explains also the reason that we still apply, occasionally, for $\mathcal{GL}(n, \mathcal{A})$ the mame *gauge group (sheaf).*

Now, our next objective is to determine the *form, that is taken by a local $\mathcal{A}$-connection matrix* $\omega$ (see (1.13)), *under the action of a local gauge transformation* $g$, as in (2.4) (cf. also (2.5)):

First, for such a transformed $\omega$, we employ thc notation;

(2.6) $$g^*(\omega) \equiv \omega' \equiv (\omega'_{ij}) \in M_n(\Omega(U \cap V)) = M_n(\Omega)(U \cap V).$$

So, *by definition,* $\omega'$ is the *local $\mathcal{A}$-connection matrix* of $D$ (the latter being a given $\mathcal{A}$-connection of the $\mathcal{A}$-module $\mathcal{E}$ involved), which is associated with the *local gauge of $\mathcal{E}$ over $U \cap V$*, given by

(2.7) $$f^{U\cap V} \equiv f^V|_{U\cap V} \equiv \{U \cap V; (f_1, \ldots, f_n)|_{U\cap V}\}$$

(viz. the *restriction to $U \cap V$ of the* given *local gauge* $f^V$ of $\mathcal{E}$, as in (2.1)).

Therefore, according to the general pattern, as given by (1.11), we further consider the corresponding *defining relation for* $\omega'$, with respect to the local gauge (2.7); that is, we have

(2.8) $$D(f_j) = \sum_{i=1}^{n} f_i \otimes \omega'_{ij}, \quad 1 \leqq j \leqq n,$$

such that one has

(2.9) $$\omega'_{ij} \in \Omega(U \cap V), \quad 1 \leqq i, j \leqq n.$$

Moreover, the latter sections are *uniquely defined,* given $D$ and the local gauge $f^{U\cap V}$ of $\mathcal{E}$, as in (2.7) (see the comments about (1.12) above). So this also explains the notation adopted by (2.6).

Thus, in view of (2.3) and (2.8), one now obtains

$$\begin{aligned}D(f_j) &= \textstyle\sum_i f_i \otimes \omega'_{ij} = \sum_i (\sum_k g_{ki} \cdot e_k) \otimes \omega'_{ij} \\ &= \textstyle\sum_k e_k \otimes (\sum_i g_{ki} \cdot \omega'_{ij}).\end{aligned} \tag{2.10}$$

On the other hand, according to the same relation (2.3) and the very properties of the given $\mathcal{A}$-connection $D$ of $\mathcal{E}$ (cf. thus Chapt. VI; (3.5), (3.6)), one further obtains;

$$\begin{aligned}D(f_j) &= D(\textstyle\sum_i g_{ij} e_i) = \sum_i D(g_{ij} e_i) = \sum_i (g_{ij} De_i + e_i \otimes \partial g_{ij}) \\ &= \textstyle\sum_i ((\sum_k e_k \otimes \omega_{ki}) g_{ij} + e_i \otimes \partial g_{ij}) = \sum_i (\sum_k e_k \otimes \omega_{ki}) g_{ij} + \sum_i e_i \otimes \partial g_{ij} \\ &= \textstyle\sum_k e_k \otimes (\sum_i \omega_{ki} g_{ij} + \partial g_{kj}).\end{aligned} \tag{2.11}$$

Yet, concerning both of the previous relations (2.10) and (2.11), one should also remark that, in addition to (2.7), we do consider here, in effect, an analogous restriction to $U \cap V$ of the given local gauge $e^U$ of $\mathcal{E}$, as in (2.1); that is, one has the following *local gauge of $\mathcal{E}$ over $U \cap V$*

$$e^{U \cap V} \equiv e^U|_{U \cap V} \equiv \{U \cap V; (e_1, \ldots, e_n)|_{U \cap V}\}, \tag{2.12}$$

which thus, along with (2.7), appeared, in fact, in the above two relations.

> **Remark 2.1.**- In this context, by rectifying our previous argument in (2.5), we should say, precisely speaking, that the (section) matrix $g \equiv (g_{ij})$ in (2.4) is actually the *matrix of change of the* given *local gauges $e^U$ and $f^V$* of $\mathcal{E}$, as in (2.1), when both are *restricted to $U \cap V$*, as in (2.7) and (2.12), respectively. Thus, one essentially has the *change of (local) gauges*
>
> $$e^{U \cap V} \equiv (e_i)|_{U \cap V} \xrightarrow[g \equiv (g_{ij})]{} (f_j)|_{U \cap V} \equiv f^{U \cap V}, \tag{2.13}$$
>
> as expressed by (2.3) above.

So, based on the same relations (2.10) and (2.11), we finally conclude that;

$$\sum_{i=1}^{n} g_{ki} \omega'_{ij} = \sum_{i=1}^{n} \omega_{ki} g_{ij} + \partial g_{kj} \in \Omega(U \cap V). \tag{2.14}$$

(In this regard, see also, for instance, W. Greub [1: p. 6, Lemma 1.5.1], or even our previous argument on (1.11), concerning the uniqueness of the $\omega_{ij}$'s). *Equivalently,* one can write (2.14) in the following *matrix form* (cf. also (2.4) and (2.6));

$$g \cdot \omega' = \omega \cdot g + \partial(g). \tag{2.15}$$

We still *abused notation* above, given that $\partial$ stands here, in fact, for the respective *extension to* $M_n(\mathcal{A})$ (cf. Chapt. VI; (1.22), (1.23)). Thus, from (2.15) and (2.4) one now obtains

$$\omega' = g^{-1}\omega g + g^{-1}\partial g, \tag{2.16}$$

or by employing the notation of Chapt. VI; (1.29), (1.35) and (1.41), one finally has the relation;

$$\begin{aligned} \omega' = g^{-1}\omega g + \tilde{\partial}(g) \equiv \mathrm{Ad}(g^{-1})\omega + \tilde{\partial}(g) &\in M_n(\Omega(U \cap V)) \\ &= M_n(\Omega)(U \cap V). \end{aligned} \tag{2.17}$$

We can now summarize the preceding into the form of the following.

**Lemma 2.1.** *Suppose we are given a differential triad* $(\mathcal{A}, \partial, \Omega)$ *on a topological space* $X$, *an* $\mathcal{A}$*-module* $\mathcal{E}$ *on* $X$, *as well as, two local gauges of* $\mathcal{E}$,

$$e^U \equiv \{U; (e_1, \ldots, e_n)\} \text{ and } f^V \equiv \{V; (f_1, \ldots, f_n)\} \tag{2.18}$$

*over the open sets* $U$ *and* $V$ *of* $X$, *respectively, such that*

$$U \cap V \neq \emptyset, \tag{2.19}$$

*both being of the same rank* $n \in \mathbf{N}$. *Moreover, let*

$$g \equiv (g_{ij}) \in \mathcal{GL}(n, \mathcal{A})(U \cap V) = \mathrm{GL}(n, \mathcal{A}(U \cap V)) \tag{2.20}$$

*be the respective local gauge transformation over* $U \cap V$ (cf. (2.19)) *between the above two local gauges of* $\mathcal{E}$ (see (2.4), (2.5) and (2.13)). *Finally, let* $D$ *be a given* $\mathcal{A}$*-connection of* $\mathcal{E}$, *whose local* $\mathcal{A}$*-connection matrices over* $U, V$ *are*

$$\omega \equiv (\omega_{ij}) \in M_n(\Omega(U)) = M_n(\Omega)(U) \tag{2.21}$$

*and*

$$\omega' \equiv (\omega'_{ij}) \in M_n(\Omega(V)) = M_n(\Omega)(V), \tag{2.22}$$

*respectively* (cf. (1.13)).

*Then,* $\omega$ *and* $\omega'$ *are gauge equivalent through* $g$; *viz. their restrictions to* $U \cap V$ (we keep the same notation) *are related by*

$$\omega' = \mathrm{Ad}(g^{-1})\omega + \tilde{\partial}(g). \tag{2.23}$$ ∎

Thus, as a consequence of the preceding, one further realizes that;

(2.24) the above relation (2.23) *is a necessary condition for being two local matrices* $\omega$ *and* $\omega'$, as before (see (2.21), (2.22)), *the* (local) $\mathcal{A}$*-connection matrices* of an $\mathcal{A}$-connection $D$ of the $\mathcal{A}$-module $\mathcal{E}$, *resulted from a change of local gauges* of $\mathcal{E}$, as this is expressed by a section matrix $g$, *as in (2.4)* (cf. also (2.13)).

Yet, as we shall see in the succeeding Section 3, the same relation (2.23) is still a *sufficient condition for the existence of an* $\mathcal{A}$*-connection* $D$ of a given *vector sheaf* $\mathcal{E}$ (cf. Theorem 3.1 in the sequel).

**Scholium 2.1..**– By analogy with the *"global case"* (cf. (2.25) below, together with (17.10) in Chapt. VI; ), two *local (matrix) forms* $\omega$ *and* $\omega'$, as above, are said to be *gauge equivalent,* whenever a relation like (2.23) is in force, while $g$ is then called a *local gauge transformation* (see also (2.5) above); yet, we say that $\omega'$ is the *gauge transform* of $\omega$. In this concern, it is still easy to see that;

(2.25) *(2.23) defines, in fact, an equivalence relation among the various local* (matrix) *forms with common domain of definition.*

Thus, denoting by

$$\omega \underset{g}{\sim} \omega' \tag{2.26}$$

the *condition that (2.23) hold true,* we check, for instance, that (2.26) *is transitive:* That is, *if* $\omega \underset{g}{\sim} \omega'$ *and* $\omega' \underset{f}{\sim} \omega''$, *then* $\omega \underset{gf}{\sim} \omega''$, such that

$$gf := g \circ f \in \mathrm{GL}(n, \mathcal{A}(U \cap V \cap W)), \tag{2.27}$$

the notation applied having the obvious meaning (cf. (2.20)); indeed, one has, by hypothesis (see also Chapt. VI; (1.33)),

$$\begin{aligned} \omega'' &= \mathrm{Ad}(f^{-1})\omega' + \tilde{\partial}(f) = \mathrm{Ad}(f^{-1})(\mathrm{Ad}(g^{-1})\omega + \tilde{\partial}(g)) + \tilde{\partial}(f) \\ &= \mathrm{Ad}((gh)^{-1})\omega + \mathrm{Ad}(f^{-1})\tilde{\partial}(g) + \tilde{\partial}(f) = \mathrm{Ad}((gf)^{-1})\omega + \tilde{\partial}(gf), \end{aligned} \tag{2.28}$$

as desired. (For the *symmetry* of (2.26), cf. also Chapt. VI; (1.31)). ∎

In this context, another relevant terminology in connection with (2.23), is to call the latter relation the *gauge transformation law,* or even the *transformation law of potentials,* a nomenclature of obvious origin, of course (see

thus Scholium 1.2 in the preceding, as well as, Theorem 3.1 in the sequel).

Now, in conclusion, we also remark that *the above constitute,* in fact, a *localization of our previous considerations in Chapt. VI; Section 17,* when, in the particular case considered, $\mathcal{E}$ is a *vector sheaf* on $X$ (see especially Chapt. VI; (17.30)). Thus,

(2.29) the *effect of a (global) automorphism* $\phi \in \operatorname{Aut}\mathcal{E}$ *on a* given $\mathcal{A}$*-connection* $D$ *of* $\mathcal{E}$ (see e.g. Chapt. VI; (17.10)) is locally expressed by an *analogous transformation of* the respective *local* $\mathcal{A}$*-connection matrix.* Thus, $g$ in (2.4) can be conceived as but the *(local) action of* $\phi$ *on the* particular *local gauge of* $\mathcal{E}$ involved (cf. Chapt. VI; (17.29)).

**2.(a). Vector sheaves.**– The foregoing will be applied in particular, in the special case one considers a *vector sheaf* $\mathcal{E}$ on $X$, say, of *rank* $n \in \mathbf{N}$ (cf. thus also Section 3 in the sequel). Now, assume further that

$$\mathcal{U} = \{U_\alpha\}_{\alpha \in I} \tag{2.30}$$

is a *local frame* of $\mathcal{E}$. So it is clear that anything, has been said above holds, in particular, true for any *local gauge* $U_\alpha \in \mathcal{U}$ of $\mathcal{E}$; thus, precisely speaking (see e.g. (1.5)), the latter is given by

$$e^{(\alpha)} \equiv \left\{U_\alpha; (e_1^{(\alpha)}, \ldots, e_n^{(\alpha)})\right\} \tag{2.31}$$

such that

$$(e_i^{(\alpha)})_{1 \leqq i \leqq n = \operatorname{rk}\mathcal{E}} \in \mathcal{E}(U_\alpha)^n = \mathcal{A}(U_\alpha)^{n^2} \tag{2.32}$$

stands for the respective *basis (of sections)* of the *free* $\mathcal{A}(U_\alpha)$*-module* (see also (1.3), (1.4));

$$\mathcal{E}(U_\alpha) = (\mathcal{E}|_{U_\alpha})(U_\alpha) = (\mathcal{A}^n|_{U_\alpha})(U_\alpha) = \mathcal{A}^n(U_\alpha) = \mathcal{A}(U_\alpha)^n. \tag{2.33}$$

Now, suppose further that we are given an $\mathcal{A}$*-connection* of $\mathcal{E}$

$$D : \mathcal{E} \longrightarrow \Omega(\mathcal{E}) \equiv \mathcal{E} \otimes_{\mathcal{A}} \Omega; \tag{2.34}$$

so, based on the preceding (cf. (1.17), (1.35)), one defines, *for every* $\alpha \in I$, the respective *local* $\mathcal{A}$*-connection matrix* (or even the *potential*) *of* $D$ *over* $U_\alpha \subseteq X$,

(2.35) $$\omega^{(\alpha)} \equiv (\omega_{ij}^{(\alpha)}) \in M_n(\Omega(U_\alpha)) = M_n(\Omega)(U_\alpha),$$

according to the relation,

(2.36) $$\omega^{(\alpha)} = D_\alpha - \partial_\alpha^n.$$

Here we have set, of course,

(2.37) $$D_\alpha := D|_{U_\alpha},$$

so that one has

(2.38) $$D_\alpha \equiv D|_{U_\alpha} \in Conn_{\mathcal{A}|_{U_\alpha}}(\mathcal{E}|_{U_\alpha}) = Conn_{\mathcal{A}|_{U_\alpha}}(\mathcal{A}^n|_{U_\alpha});$$

in this context, see (1.31) - (1.33), (1.28) and (1.30) above, as well as, Chapt. VI; (8.28). Yet, cf. (2.47) below. In conclusion, one thus infers that:

(2.39) given an *$\mathcal{A}$-connection* $D$ of a *vector sheaf* $\mathcal{E}$ on $X$, then, *for every local frame* $\mathcal{U} = (U_\alpha)_{\alpha \in I}$ *of* $\mathcal{E}$, one obtains a *0-cochain of matrix-valued "1-forms"*,

(2.39.1) $$(\omega^{(\alpha)}) \in C^0(\mathcal{U}, M_n(\Omega)).$$

We call it the *$\mathcal{A}$-connection 0-cochain matrix form,* associated with the given local frame $\mathcal{U}$ of $\mathcal{E}$.

On the other hand, suppose now that

(2.40) $$g \equiv (g_{\alpha\beta}) \in Z^1(\mathcal{U}, \mathcal{GL}(n, \mathcal{A}))$$

is a *coordinate 1-cocycle* of the *vector sheaf* $\mathcal{E}$, associated with the given *local frame* $\mathcal{U}$ *of* $\mathcal{E}$, as in (2.30) (cf. Chapt. V; (1.8), along with Definition 1.1). Thus, *for any* $\alpha, \beta \in I$, *with*

(2.41) $$U_{\alpha\beta} \equiv U_\alpha \cap U_\beta \neq \emptyset,$$

the respective *section-matrix*

(2.42) $$g_{\alpha\beta} \in \mathcal{GL}(n, \mathcal{A})(U_{\alpha\beta}) = \mathrm{GL}(n, \mathcal{A}(U_{\alpha\beta}))$$

entails, in effect, the *(local) gauge transformation between the local gauges* $U_\alpha$ *and* $U_\beta$ *of* $\mathcal{E}$ (cf. Chapt. V; (1.5), (1.18), in conjunction with (2.13) above).

Consequently, by considering now the respective *transformation law of potentials,* as the latter are defined by (2.35), one obtains, by virtue of (2.23) (see also (2.40)), the following *basic relation;*

$$\omega^{(\beta)} = \mathrm{Ad}(g_{\alpha\beta}^{-1})\omega^{(\alpha)} + \tilde{\partial}(g_{\alpha\beta}). \tag{2.43}$$

In this regard, it is instructive (yet, in anticipation for later applications) to give here a bit further *clarification of the notation applied in (2.43):*

Thus, looking at the *local gauge* (2.31) of $\mathcal{E}$, one has

$$e_i^{(\alpha)} := \eta_\alpha^{-1} \circ \varepsilon_i^{(\alpha)}, \quad 1 \leqq i \leqq n = \mathrm{rk}\,\mathcal{E}, \quad \alpha \in I, \tag{2.44}$$

with $(\varepsilon_i^{(\alpha)}) \subseteq \mathcal{A}^n(U_\alpha)$ the *Kronecker gauge of* $\mathcal{A}^n$ *over* $U_\alpha$; cf. Chapt. IV; (3.2), (3.3) and (3.11.1), along with Chapt. VI; (8.12). Yet, see Chapt. V; (1.13), (1.15). Moreover, we also set (cf. Chapt. VI; (10.4))

$$g_{\alpha\beta} := \eta_\alpha \circ \eta_\beta^{-1}, \text{ for any } \alpha, \beta \in I, \text{ satisfying (2.41)}. \tag{2.45}$$

So, based now on (1.11), one obtains (cf. also (2.35), (2.37))

$$D_\alpha(e_j^{(\alpha)}) \equiv (D|_{U_\alpha})(e_j^{(\alpha)}) \equiv D(e_j^{(\alpha)}) = \sum_{i=1}^{n} e_i^{(\alpha)} \otimes \omega_{ij}^{(\alpha)}, \tag{2.46}$$

*with* $1 \leqq j \leqq n$ *and* $\alpha \in I$. From this and from (1.15), (1.21) and (1.32), one now obtains;

$$D_\alpha(e_j^{(\alpha)})_i = \omega_{ij}^{(\alpha)} = (D_\alpha - \partial_\alpha^n)(e_j^{(\alpha)}), \quad 1 \leqq i, j \leqq n \ \text{ and } \ \alpha \in I, \tag{2.47}$$

which also *justifies* now *completely* (2.36) in the foregoing.

On the other hand, one still obtains (see also (2.42), (2.45))

$$e_j^{(\beta)} = \sum_{i=1}^{n} g_{ij}^{(\alpha\beta)} e_i^{(\alpha)}, \quad 1 \leqq j \leqq n, \tag{2.48}$$

where we set

$$g_{\alpha\beta} \equiv (g_{ij}^{(\alpha\beta)}) \in \mathrm{GL}(n, \mathcal{A}(U_{\alpha\beta}) = \mathcal{GL}(n, \mathcal{A})(U_{\alpha\beta}), \tag{2.49}$$

*for any* $\alpha, \beta \in I$, *as in (2.41).*

The preceding can be summarized now into the following.

**Theorem 2.1.** *Suppose that we are given a differential triad* $(\mathcal{A}, \partial, \Omega)$ *on a topological space* $X$. *Moreover, let* $\mathcal{E}$ *be a vector sheaf on* $X$, *of rank* $n \in \mathbf{N}$, $D$ *an* $\mathcal{A}$*-connection of* $\mathcal{E}$ *and* $(g_{\alpha\beta}) \in Z^1(\mathcal{U}, \mathcal{GL}(n, \mathcal{A}))$ *a coordinate 1-cocycle of* $\mathcal{E}$, *with respect to a given local frame* $\mathcal{U} = (U_\alpha)_{\alpha \in I}$ *of* $\mathcal{E}$.

*Then, the local $\mathcal{A}$-connection matrices of $D$, which are associated with $\mathcal{U}$* (cf. (2.35), along with (2.39))

$$\omega^{(\alpha)} \in M_n(\Omega)(U_\alpha), \quad \alpha \in I, \tag{2.50}$$

*are pairwise gauge equivalent, through* $g_{\alpha\beta} \in \mathrm{GL}(n, \mathcal{A}(U_{\alpha\beta}))$ (cf. (2.49)); *that is, one has the following relation (transformation law of potentials), valid, for any $\alpha, \beta \in I$, satisfying (2.41)* (cf. (2.43):

$$\omega^{(\beta)} = \mathrm{Ad}(g_{\alpha\beta}^{-1})\omega^{(\alpha)} + \tilde{\partial}(g_{\alpha\beta}). \tag{2.51}$$

*In particular, for every point $x \in U_{\alpha\beta} \equiv U_\alpha \cap U_\beta$, one obtains;*

$$\omega^{(\beta)}(x) = \mathrm{Ad}(g_{\alpha\beta}^{-1}(x))\omega^{(\alpha)}(x) + \tilde{\partial}(g_{\alpha\beta})(x). \blacksquare \tag{2.52}$$

The converse of the previous statement is also true, as we shall presently see in the next Section 3 (see thus Theorem 3.1 below). Indeed, one can realize it straightaway, at least, for *line sheaves*.

So it is still worth looking at the *transformation law of potentials*, as given above by (2.51), in the particular case of a *line sheaf* $\mathcal{L}$ on $X$, endowed with an *$\mathcal{A}$-connection* $D$, viz. for a *Maxwell field* $(\mathcal{L}, D)$ on $X$ (see Chapt. VI; (18.11)):

Thus, by taking again a *local frame* $\mathcal{U}$ of $\mathcal{L}$ (cf. (2.30)), one now obtains for the corresponding *$\mathcal{A}$-connection 0-cochain matrix form*, the relation (see (2.39.1), *for* $n = 1$);

$$(\omega_\alpha) \in C^0(\mathcal{U}, \Omega) = \prod_{\alpha \in I} \Omega(U_\alpha) \tag{2.53}$$

(cf. also Chapt. III; (11.12)). So, for any *coordinate 1-cocycle of* $\mathcal{L}$,

$$(g_{\alpha\beta}) \in Z^1\mathcal{U}, \mathcal{A}^{\cdot}), \tag{2.54}$$

so that, for any $\alpha, \beta \in I$, with (2.41) in force, one has

$$g_{\alpha\beta} \in \mathcal{A}^{\cdot}(U_{\alpha\beta}) = \mathcal{A}(U_{\alpha\beta})^{\cdot} \tag{2.55}$$

(see (2.4), (2.49), *for* $n = 1$, *one obtains for (2.51);*

$$\omega_\beta = g_{\alpha\beta}^{-1}\omega_\alpha g_{\alpha\beta} + \tilde{\partial}(g_{\alpha\beta}) \tag{2.56}$$

(cf. also e.g. (2.16)). However, due to (2.55), along with the *commutativity of the* **C**-*algebra* $\mathcal{A}(U)$, for any *open* $U \subseteq X$, according to our hypothesis for $\mathcal{A}$, one has from (2.56);

$$\omega_\beta = \omega_\alpha + \tilde{\partial}(g_{\alpha\beta}). \tag{2.57}$$

Therefore, by further applying the *Bockstein operator* $\delta$ (Chapt. III; (4.16), (4.29)), so that one has

$$\omega_\beta - \omega_\alpha = \delta(\omega_\alpha) \in B^1(\mathcal{U}, \Omega) \subseteq Z^1(\mathcal{U}, \Omega), \tag{2.58}$$

*one* finally *obtains* from (2.57) *the relation;*

$$\delta(\omega_\alpha) = \tilde{\partial}(g_{\alpha\beta}). \tag{2.59}$$

In other words, one thus infers that:

(2.60) *the 1-cocycle*

(2.60.1) $$\tilde{\partial}(g_{\alpha\beta}) \in Z^1(\mathcal{U}, \Omega)$$

(cf. Chapt. VI; (10.25), *for* $n = 1$) *is,* by virtue of (2.59), *a (1-)coboundary,* so that one has;

(2.60.2) $$\left[\tilde{\partial}(g_{\alpha\beta})\right] = 0 \in H^1(X, \Omega).$$

On the other hand, we already know (Chapt. VI; (10.28.1) and Corollary 12.1) that (2.60.2) *is a necessary and sufficient condition for a line sheaf to admit an $\mathcal{A}$-connection (Atiyah's criterion).* So one concludes here the aforementioned *converse of Theorem 2.1,* concerning, namely, the *existence of an $\mathcal{A}$-connection,* in the particular case of a *line sheaf.* In fact, as already said, the analogous conclusion holds true, *for every vector sheaf,* in general.

**Scholium 2.2.**- Some final comments, in conjunction with our previous remarks in (2.25), pertaining to the notion of an $\mathcal{A}$-connection, are still here in order: Thus, as follows from the preceding, given a *vector sheaf* $\mathcal{E}$ and a *local frame* $\mathcal{U} = (U_\alpha)_{\alpha \in I}$ *of* $\mathcal{E}$, an $\mathcal{A}$*-connection* $D$ of $\mathcal{E}$ gives rise to an *equivalence class* (cf. (2.26))

$$[\omega^{(\alpha)}], \tag{2.61}$$

where $\omega^{(\alpha)} \in M_n(\Omega)(U_\alpha)$, $\alpha \in I$, is given by (2.35) (cf. also (2.39)). Indeed, *such an equivalence class entails an $\mathcal{A}$-connection* too, as we shall see in the sequel; of course, *we have seen it* already *in the case of a line sheaf* (see (2.60) above, along with the comments following it).

Thus, according to the preceding and in anticipation also of Theorem 3.1

in the sequel, we can say, *roughly speaking,* that:

(2.62) an $\mathcal{A}$-*connection* of a *vector sheaf* is an *equivalence class of local* $\mathcal{A}$-*connection matrices* (viz. *matrix-valued "1-forms"*), which is *supplied by any* given *local frame of the vector sheaf* involved.

## 3. Transformation law of potentials. Existence of $\mathcal{A}$-connections

As already mentioned, our purpose by the ensuing discussion is to examine whether (2.51) yields a *sufficient condition for the existence of an* $\mathcal{A}$-*connection.* Indeed, our main objective hereby is to prove the following *basic result,* justifying thus completely our previous comments.

**Theorem 3.1.** *Let* $(\mathcal{A}, \partial, \Omega)$ *be a given differential triad on a topological space* $X$, $\mathcal{E}$ *a vector sheaf on* $X$, *with* $\operatorname{rk}\mathcal{E} = n \in \mathbf{N}$, *and*

$$(g_{\alpha\beta}) \in Z^1(\mathcal{U}, \mathcal{GL}(n, \mathcal{A})) \tag{3.1}$$

*a coordinate 1-cocycle of* $\mathcal{E}$, *with respect to a given local frame* $\mathcal{U} = (U_\alpha)_{\alpha \in I}$ *of* $\mathcal{E}$. *Moreover, let*

$$\omega \equiv (\omega^{(\alpha)}) \in C^0(\mathcal{U}, M_n(\Omega)) \tag{3.2}$$

*be a 0-cochain of* $\mathcal{U}$ *with values in the sheaf of* $\mathcal{A}$-*connection coefficients of* $\mathcal{E}$, *such that the following "transformation law of potentials" holds true;*

$$\omega^{(\beta)} = \operatorname{Ad}(g_{\alpha\beta}^{-1})\omega^{(\alpha)} + \tilde{\partial}(g_{\alpha\beta}), \tag{3.3}$$

*for any* $\alpha, \beta \in I$, *such that*

$$U_{\alpha\beta} \equiv U_\alpha \cap U_\beta \neq \emptyset. \tag{3.4}$$

*Then, there exists a uniquely defined* $\mathcal{A}$-*connection* $D$ *of* $\mathcal{E}$, *whose local* $\mathcal{A}$-*connection matrices, relative to* $\mathcal{U}$, *are given by* (the components of) *(3.2).*

**Proof.** Based on Theorem 1.1, one can define, by employing tha $\alpha$-*th component of* the 0-cochain $\omega$ in (3.2), i.e. the *section-matrix*

(3.5) $$\omega^{(\alpha)} \equiv (\omega_{ij}^{(\alpha)}) \in M_n(\Omega)(U_\alpha) = M_n(\Omega(U_\alpha))$$

(see also (2.35)), an $\mathcal{A}|_{U_\alpha}$*-connection of* $\mathcal{E}|_{U_\alpha}$, *for every* $\alpha \in I$; namely, *the* (local) *connection,* we are looking for, *is* simply *given by the corresponding relation to (1.15).* Therefore, if

(3.6) $$e^{(\alpha)} \equiv \{U_\alpha; (e_i^{(\alpha)})_{1 \leqq i \leqq n}\}$$

is the respective *local gauge of* $\mathcal{E}$ *over* $U_\alpha$ (cf. (2.31), (2.44)), then the values of the latter connection, denoting it by $D_\alpha$, on the *section-basis* $(e_i^{(\alpha)})_{1 \leqq i \leqq n}$ *of* $\mathcal{E}(U_\alpha)$ (see (2.32), (2.33)) are given by the relation;

(3.7) $$D_\alpha(e_j^{(\alpha)}) = \sum_{i=1}^{n} e_i^{(\alpha)} \otimes \omega_{ij}^{(\alpha)} \in \Omega(\mathcal{E})(U_\alpha), \ 1 \leqq j \leqq n$$

(cf. also (1.10)). So *our claim* now *is that;*

(3.8) $$D_\alpha(e_j^{(\beta)}) = D_\beta(e_j^{(\beta)}), \ 1 \leqq j \leqq n,$$

*for every* $\alpha, \beta \in I$, *satisfying (3.4);* of course, an obvious *restriction to* $U_{\alpha\beta}$ for both *of* $D_\alpha$ *and* $D_\beta$, along with a *similar one of* $e^{(\alpha)}$, are here in sight (see also e.g. (2.12)).

Thus, under the validity of (3.8), the two **C**-*linear morphisms* $D_\alpha, D_\beta$, in effect, *their restrictions,* as before, being equal on the (similarly restricted) *section-basis* $(e_i^{(\beta)})|_{U_{\alpha\beta}}$ *of* $\mathcal{E}(U_{\alpha\beta})$, *would be the same, for any* $\alpha, \beta \in I$, *as in (3.4).* This, in turn, would imply now that the *0-cochain*

(3.9) $$(D_\alpha) \in C^0(\mathcal{U}, \mathcal{H}om_{\mathbf{C}}(\mathcal{E}, \Omega(\mathcal{E}))),$$

as defined by the preceding (cf. (1.15), (3.5)), would be actually a *0-cocycle;* viz. we would have then (cf. also e.g. Chapt. II; (6.4))

(3.10) $$(D_\alpha) \in Z^0(\mathcal{U}, \mathcal{H}om_{\mathbf{C}}(\mathcal{E}, \Omega(\mathcal{E}))) = \mathcal{H}om_{\mathbf{C}}(\mathcal{E}, \Omega(\mathcal{E}))(X) = \mathrm{Hom}_{\mathbf{C}}(\mathcal{E}, \Omega(\mathcal{E})).$$

That is, one thus obtains a **C**-*linear morphism*

(3.11) $$D \equiv (D_\alpha) \in \mathrm{Hom}_{\mathbf{C}}(\mathcal{E}, \Omega(\mathcal{E})),$$

as above, satisfying also the corresponding *Leibniz condition,* this being actually true, according to the preceding, for each one of its components $D_\alpha$, $\alpha \in I$. In other words, (3.11) *would give* us then *an* $\mathcal{A}$*-connection of* $\mathcal{E}$, which, in addition, would be *uniquely defined,* as well, *given the 0-cochain*

(3.2). Finally, the corresponding *local $\mathcal{A}$-connection matrices of $D$*, when the latter is given by (3.11), will be, by virtue of (1.11) and (3.7), those provided by (3.5). Thus, to complete the proof of the theorem, all that we still have to do, is just to give the *proof of the relation (3.8):*

Indeed, based on (3.7) and (2.48), one first obtains;

$$\begin{aligned} D_\beta(e_j^{(\beta)}) &= \textstyle\sum_{i=1}^n e_i^{(\beta)} \otimes \omega_{ij}^{(\beta)} = \sum_i (\sum_{k=1}^n g_{ki}^{(\alpha\beta)} e_k^{(\alpha)}) \otimes \omega_{ij}^{(\beta)} \\ &= \textstyle\sum_k e_k^{(\alpha)} \otimes (\sum_i g_{ki}^{(\alpha\beta)} \omega_{ij}^{(\beta)}), \end{aligned} \tag{3.12}$$

*for any* $1 \leqq j \leqq n$ *and* $\beta \in I$. On the other hand, by employing the notation of (1.20), one gets from (3.12);

$$D_\beta(e_j^{(\beta)})_k := \sum_{i=1}^n g_{ki}^{(\alpha\beta)} \cdot \omega_{ij}^{(\beta)} \equiv (g_{\alpha\beta} \cdot \omega^{(\beta)})_{kj}, \quad 1 \leqq k, j \leqq n, \tag{3.13}$$

where the *matrix-notation* in the last item above is, of course, clear (see, for instance, (2.35) and (2.49)). Now, the same matrix-notation facilitates, in effect, the application here, according to our hypothesis, of the *transformation law of potentials* in the form of (2.15). Thus, for the case at hand, one now obtains;

$$g_{\alpha\beta} \cdot \omega^{(\beta)} = \omega^{(\alpha)} \cdot g_{\alpha\beta} + \partial(g_{\alpha\beta}), \tag{3.14}$$

true, *for any* $\alpha, \beta \in I$, *satisfying (3.4)*. Indeed, we have here an *equivalent formulation,* of course, of (3.3). Yet, we should also recall at this place that in the last relation $\partial$ stands, in effect, for its *extension to* $M_n(\mathcal{A})$; (cf. e.g. (1.28)). Consequently, one now obtains, by (3.14),

$$\begin{aligned} (g_{\alpha\beta} \cdot \omega^{(\beta)})_{kj} &= (\omega^{(\alpha)} \cdot g_{\alpha\beta} + \partial(g_{\alpha\beta}))_{kj} = \partial(g_{\alpha\beta})_{kj} + (\omega^{(\alpha)} \cdot g_{\alpha\beta})_{kj} \\ &= \partial(g_{kj}^{(\alpha\beta)}) + \textstyle\sum_{i=1}^n \omega_{ki}^{(\alpha)} \cdot g_{ij}^{(\alpha\beta)} = D_\alpha(\sum_{i=1}^n g_{ij}^{(\alpha\beta)} e_i^{(\alpha)})_k = D_\alpha(e_j^{(\beta)})_k, \end{aligned} \tag{3.15}$$

*for any* $1 \leqq k, j \leqq n$. (In this regard, cf. also (2.48), (1.15), as well as, for instance, (1.29), or even (2.11)). Thus, in view of the *uniqueness of an expression, like (1.11), in terms of the* $\omega_{ij}$'s, the above relations (3.13) and (3.15) *prove* already (3.8), which, thus, according to the preceding, terminates the proof of the theorem. ∎

Now, in view of its significance, we do reformulate our information, gathered so far by the previous two Theorems 2.1 and 3.1, into the form of the following.

**Theorem 3.2.** Suppose that we are given a *differential triad* $(\mathcal{A}, \partial, \Omega)$ on a *topological space* $X$ and $\mathcal{E}$ a *vector sheaf* on $X$, of *rank* $n \in \mathbf{N}$. Moreover, let $\mathcal{U} = (U_\alpha)_{\alpha \in I}$ be a *local frame of* $\mathcal{E}$, along with the following *0-cochain* of $\mathcal{U}$;

$$\omega \equiv (\omega^{(\alpha)}) \in C^0(\mathcal{U}, M_n(\Omega)). \tag{3.16}$$

Then,

(3.17) *the 0-cochain (3.16) yields an $\mathcal{A}$-connection of the vector sheaf* $\mathcal{E}$ (cf. (3.11), (3.7)) *if, and only if, the* corresponding *transformation law of potentials holds true,*

the latter law (cf. (3.3), or even (3.14)) being associated with $\omega$, as in (3.16), and with a given *coordinate 1-cocycle of* $\mathcal{E}$, relative to $\mathcal{U}$ (cf. (3.1)). ∎

Concerning the classical case of differential geometry, see, for instance, E. Binz - J. Śniatycki - H. Fischer [1: p. 172, Remarks, (i)], or even M. Nakahara [1: p. 335, (10.9)]; yet, see M. Göckeler - T. Schüker [1: p. 50, (4.42)], as well as, Y. Choquet-Bruhat, C. DeWitt-Morette and M. Dillard-Bleick [1: p. 364, Theorem].

## 4. Dual $\mathcal{A}$-connections. Local form

We have already considered in the foregoing (Chapt. VI; Subsection 5.(d)) *dual $\mathcal{A}$-connections* for *vector sheaves.* As the heading indicates, we further examine in this Section the local aspect of the same notion, yet, for *$\mathcal{A}$-modules,* in general, by employing the notion of a *local gauge* (cf., for example, (1.2)). We shall find applications of the present material later on (see thus the next Chapter VIII).

Now, to start with, suppose, as usual, that we are given a *differential triad* $(\mathcal{A}, \partial, \Omega)$ on a *topological space* $X$ (see Chapt. VI; (3.1)), and let $\mathcal{E}$ be an *$\mathcal{A}$-module* on $X$. Moreover, assume that

$$e^U \equiv \{U; (e_i)_{1 \leqq i \leqq n}\} \tag{4.1}$$

is a *local gauge of* $\mathcal{E}$ over an *open* set $U \subseteq X$, of *rank* $n \in \mathbf{N}$ (cf. (1.2) - (1.4)). On the other hand, consider the *dual* $\mathcal{A}$*-module of* $\mathcal{E}$,

$$\mathcal{E}^* := \mathcal{H}om_{\mathcal{A}}(\mathcal{E}, \mathcal{A}) \tag{4.2}$$

(see Chapt. IV; Definition 5.1), and let

$$(e^U)^* \equiv \{U; (e_i^*)_{1 \leqq i \leqq n}\} \tag{4.3}$$

the *dual local gauge of* $e^U$, being thus a *local gauge of* $\mathcal{E}^*$, of the *same sort, as* $e^U$: That is, first, due to our hypothesis for $e^U$ (cf. (1.2)), one obtains (see Chapt. II; (6.23), (6.22));

$$\begin{aligned} \mathcal{E}^*|_U = \mathcal{H}om_{\mathcal{A}}(\mathcal{E}, \mathcal{A})|_U &= \mathcal{H}om_{\mathcal{A}|_U}(\mathcal{E}|_U, \mathcal{A}|_U) = \mathcal{H}om_{\mathcal{A}|_U}(\mathcal{A}^n|_U, \mathcal{A}|_U) \\ &= \mathcal{H}om_{\mathcal{A}}(\mathcal{A}^n, \mathcal{A})|_U = \mathcal{A}^n|_U, \end{aligned} \tag{4.4}$$

within $\mathcal{A}|_U$*-isomorphisms.* Therefore, one also obtains

$$\mathcal{E}^*(U) = \mathcal{A}^n(U) = \mathcal{A}(U)^n, \tag{4.5}$$

up to $\mathcal{A}(U)$*-isomorphisms* (cf. (1.4)); viz. again a *free* $\mathcal{A}(U)$*-module,* of *rank* $n \in \mathbf{N}$. Furthermore, one has, by definition (see e.g. Chapt. IV; (5.18)),

$$e_i^*(e_j) := \delta_{ij} \in \mathcal{A}(U), \quad 1 \leqq i, j \leqq n \tag{4.6}$$

*(Kronecker's sections),* so that

$$(e_i^*)_{1 \leqq i \leqq n} \subseteq \mathcal{E}^*(U) \tag{4.7}$$

becomes the *dual* (section) *basis* of the given one of $\mathcal{E}(U)$, as in (4.1). The above explain also the meaning of (4.3).

Now, suppose further that we are given an $\mathcal{A}$*-connection* $D$ *of* $\mathcal{E}$. Thus, our purpose now is to define an $\mathcal{A}|_U$*-connection of* $\mathcal{E}^*|_U$, which, by *abusing notation,* we denote by $D^*|_U$, such that one has;

$$D^*|_U : \mathcal{E}^*|_U \longrightarrow \mathcal{E}^*|_U \otimes_{\mathcal{A}|_U} \Omega|_U \tag{4.8}$$

(see Scholium 4.1 below for an explanation too of the notation applied in (4.8); cf. thus (4.15)). Yet, concerning the range of the last map, one has, by virtue of (4.4) (see also Chapt. II; (5.25) and (5.18)),

$$\mathcal{E}^*|_U \otimes_{\mathcal{A}|_U} \Omega|_U = \mathcal{A}^n|_U \otimes_{\mathcal{A}|_U} \Omega|_U = \mathcal{A}^n \otimes_{\mathcal{A}} \Omega|_U = \Omega^n|_U, \tag{4.9}$$

so that one finally obtains for (4.8);

$$D^*|_U : \mathcal{E}^*|_U = \mathcal{A}^n|_U \longrightarrow \Omega(\mathcal{E}^*)|_U = \Omega^n|_U. \tag{4.10}$$

Thus, consider *any (local) section* (cf. also (1.2))

$$\begin{aligned} u \in \mathcal{E}^*(U) = \mathcal{H}om_{\mathcal{A}}(\mathcal{E}, \mathcal{A})(U) = \mathrm{Hom}_{\mathcal{A}|_U}(\mathcal{E}|_U, \mathcal{A}|_U) \\ \mathrm{Hom}_{\mathcal{A}|_U}(\mathcal{A}^n|_U, \mathcal{A}|_U) = \mathcal{H}om_{\mathcal{A}}(\mathcal{A}^n, \mathcal{A})(U) = \mathcal{A}^n(U) = \mathcal{A}(U)^n. \end{aligned} \tag{4.11}$$

We thus arrived at (4.5) again; however, the first row above is of importance for our next definition in (4.12). Thus, *for* $u \in \mathcal{E}^*(U)$ *and* $s \in \mathcal{E}(U)$, *one has* $u(s) \in \mathcal{A}(U)$, so that, *for every* $s \in \mathcal{E}(U)$, one now defines;

$$(D^*|_U)(u)(s) \equiv D^*(u)(s) := \partial(u(s)) - (u \otimes 1)(D(s)), \tag{4.12}$$

where we set, as usual, $1 \equiv 1_{\Omega|_U} \in \mathrm{Aut}(\Omega|_U) = (\mathcal{A}ut\Omega)(U)$. (Of course, similar *restrictions to* $U$ *of* the operators $\partial$ *and* $D$, as appeared in (4.2), are, in fact, implicitly understood, as well, cf. also Chapt. VI; (6.21)). Thus, we still set;

$$(D^*|_U)(u) \equiv D^*(u) := \partial \circ u - (u \otimes 1) \circ D, \tag{4.13}$$

*for every* $u \in \mathcal{E}^*(U)$ (cf. (4.11)).

Now, our claim is that (4.12), or even (4.13), provides an $\mathcal{A}|_U$*-connection of* $\mathcal{E}^*|_U$: Indeed, this follows easily from Chapt. VI; (5.13), when we *localize things* on the given by (4.1) *local gauge* $e^U$ of the $\mathcal{A}$*-module* $\mathcal{E}$ considered (cf. e.g. (4.9), (4.10) above). ∎ In this context, we further remark that (see also Chapt. II; (6.23));

$$\mathcal{E}^*|_U = \mathcal{H}om_{\mathcal{A}}(\mathcal{E}, \mathcal{A})|_U = \mathcal{H}om_{\mathcal{A}|_U}(\mathcal{E}|_U, \mathcal{A}|_U) \equiv (\mathcal{E}|_U)^*, \tag{4.14}$$

*for any* given $\mathcal{A}$*-module* $\mathcal{E}$ on $X$, and *open* $U \subseteq X$.

**Scholium 4.1.**– What we have actually defined above, by (4.13), is $(D|_U)^*$; that is, the *dual* $\mathcal{A}|_U$*-connection of* $D|_U$, which is *defined on* $(\mathcal{E}|_U)^* = \mathcal{E}^*|_U$ (see (4.14)). Of course, when $D^* \equiv D_{\mathcal{E}^*}$, viz. the *dual* $\mathcal{A}$*-connection of* $D$, there does exist (take, for instance, $\mathcal{E}$ a vector sheaf on $X$, as in Chapt. VI; (5.31)), one obtains

$$(D|_U)^* = D^*|_U. \tag{4.15}$$

Thus, this also clarifies the *abused notation,* as already remarked, that was employed in (4.8) (and in (4.12) too), however, simply, for

convenience. Yet, it also justifies better the heading of the running Section.

On the other hand, concerning (4.12), one also obtains (cf. (1.10), as well as, for instance, (4.18) below)

$$D(s)\in\Omega(\mathcal{E})(U)\equiv(\mathcal{E}\otimes_{\mathcal{A}}\Omega)(U)=\mathcal{E}(U)\otimes_{\mathcal{A}(U)}\Omega(U)=\Omega^n(U)=\Omega(U)^n, \tag{4.16}$$

*for every* $s\in\mathcal{E}(U)$ and $U\subseteq X$, as in (4.1). Moreover, one has (cf. (4.13), (4.10) and (4.5))

$$\begin{aligned} D^*(u)\in\Omega(\mathcal{E}^*)(U)&\equiv(\mathcal{E}^*\otimes_{\mathcal{A}}\Omega)(U)=\mathcal{E}^*(U)\otimes_{\mathcal{A}(U)}\Omega(U)\\ &=\Omega^n(U)=\Omega(U)^n,\end{aligned} \tag{4.17}$$

*for every* $u\in\mathcal{E}^*(U)$ (see (4.11), the *open* $U\subseteq X$ being, by hypothesis, as before, (the domain of) a *local gauge* of the $\mathcal{A}$*-module* $\mathcal{E}$ involved.

Accordingly, by considering now the *local gauges (4.1) and (4.3)* of the $\mathcal{A}$*-module* $\mathcal{E}$ and its *dual* $\mathcal{A}$*-module* $\mathcal{E}^*$, respectively, one obtains from (4.12) (cf. also (1.11) and (4.6), as well as, Chapt. VI; (1.7));

$$\begin{aligned} D^*(e_j^*)(e_i)&=\partial(e_j^*(e_i))-(e_j^*\otimes 1)(D(e_i))\\ &=\partial(\delta_{ij})-(e_j^*\otimes 1)(\textstyle\sum_{k=1}^n e_k\otimes\omega_{ki})=-\sum_k e_j^*(e_k)\otimes\omega_{ki}\\ &\equiv-\textstyle\sum_k e_j^*(e_k)\cdot\omega_{ki}=-\omega_{ji},\end{aligned} \tag{4.18}$$

*for any* $1\leqq i,j\leqq n$. Consequently, one now sets;

$$\omega_{ij}^*:=D^*(e_j^*)(e_i)=-\omega_{ji},\quad 1\leqq i,j\leqq n. \tag{4.19}$$

Thus, by extension of the previous terminology, connected with (1.11), we can say that the (local) $\mathcal{A}|_U$*-connection matrix* of the $\mathcal{A}|_U$*-connection* $(D|_U)^*$ of $\mathcal{E}^*|_U$ is given by (4.19). So *one* finally *sets;*

$$\omega^*\equiv(\omega_{ij}^*)=-{}^t\omega\equiv(-\omega_{ji})\in M_n(\Omega(U))=M_n(\Omega)(U), \tag{4.20}$$

*with* $U$ *given by (4.1).*

In particular, one then concludes that;

(4.21) given a *vector sheaf* $\mathcal{E}$ on $X$, *with* $\operatorname{rk}\mathcal{E}=n\in\mathbf{N}$, $e^U$ a *local gauge* of it, as in (4.1), and $D$ an $\mathcal{A}$*-connection of* $\mathcal{E}$, then (4.15) *holds true,* so that $\omega^*$, *as given by (4.20), is,* according to (1.11), *the local* $\mathcal{A}$*-connection matrix of* $D^*\equiv D_{\mathcal{E}^*}$, the *dual* $\mathcal{A}$*-connection of* $D$, with respect to the given local gauge $e^U$ of $\mathcal{E}$.

Applications of the preceding will appear throughout the sequel; see thus the proof of Theorem 5.1 in the next Section 5, as well as, the subsequent Chapter VIII, where (4.21) will be employed in deriving the fundamental *Bianchi's identity,* within the present abstract context.

## 5. Christoffel sections

As an application of our considerations in Section 1 above, we reproduce here, within our abstract framework, the familiar situation from the classical theory of the *Christoffel functions ("symbols").* So, as in the standard differential geometry of smooth manifolds, one obtains another way of identifying *locally* (but, see also the next Section 6) a given $\mathcal{A}$-connection *("law of (covariant) differentiation").* Furthermore, we shall have the opportunity of looking at an $\mathcal{A}$-connection, as a type of the familiar, from the classical theory, "*$\nabla$-operator*" (yet, *locally*! See thus (5.34) below).

So assume now that we have our hitherto standard framework, viz. a *differential triad*

$$(\mathcal{A}, \partial, \Omega) \tag{5.1}$$

on a *topological space* $X$. Yet, for the first time, thus far, we are going to specialize the given $\mathcal{A}$-module $\Omega$, right away (see thus (5.3) below), by further imitating the classical case (cf. Chapt. VI; (2.3), (2.4)).

On the other hand, suppose we have a *vector sheaf* $\mathcal{E}$ on $X$, with

$$\operatorname{rk} \mathcal{E} = n \in \mathbf{N}, \tag{5.2}$$

which is also endowed with an *$\mathcal{A}$-connection $D$.* Furthermore, as already said, we assume, in particular, that

$$\Omega = \mathcal{E}^* \equiv \mathcal{H}om_{\mathcal{A}}(\mathcal{E}, \mathcal{A}); \tag{5.3}$$

hence, $\Omega$ *is a vector sheaf* on $X$ too, of the *same rank, as* $\mathcal{E}$, being by (5.3), the *dual* ($\mathcal{A}$-module) *sheaf* of the given vector sheaf $\mathcal{E}$ on $X$ (cf. Chapt. IV; Definition 5.1 and (5.19), along with Chapt. II; Lemma 6.1). That is, we have (cf. also (5.2));

(5.4) $$\operatorname{rk}\Omega = \operatorname{rk}\mathcal{E}^* = \operatorname{rk}\mathcal{E} = n \in \mathbf{N}.$$

**Scholium 5.1.**– Given that *any vector sheaf* $\mathcal{E}$ on $X$ *is reflexive* (cf. Chapt. IV; Theorem 5.1), viz. one has

(5.5) $$\mathcal{E} = \mathcal{E}^{**},$$

*we further obtain* from (5.3), *concerning the given vector sheaf* $\mathcal{E}$, the relation;

(5.6) $$\mathcal{E} = \Omega^*$$

(within, in effect, an $\mathcal{A}$-*isomorphism of* the $\mathcal{A}$-*modules* concerned). Moreover, given the $\mathcal{A}$-*connection* $D$ *of* $\mathcal{E}$, one also defines, according to the general pattern (see Chapt. VI; Subsection 5.(d)), its *dual* $\mathcal{A}$-*connection* $D_{\mathcal{E}^*} \equiv D^*$ on the dual $\mathcal{A}$-module (in fact, vector sheaf) of $\mathcal{E}$, $\mathcal{E}^* = \Omega$ (cf. (5.3)), so that (the vector sheaf now) $\Omega$ *admits an* $\mathcal{A}$-*connection,* as well, *the dual of* $D$.

Consequently, it would be *equivalent to* say, concerning *the previous data,* that;

(5.7) we are given a *differential triad* $(\mathcal{A}, \partial, \Omega)$ on a *topological space* $X$, such that $\Omega$ is, in particular, a *vector sheaf* on $X$, of *rank* $n \in \mathbf{N}$, admitting also an $\mathcal{A}$-*connection* $D^*$.

Thus, the succeeding discussion refers, in effect, to the *dual (vector) sheaf* of $\Omega$, $\Omega^* \equiv \mathcal{E}$ (cf. (5.6)). So we still obtain that;

(5.8) $$\operatorname{rk}\mathcal{E} \equiv \operatorname{rk}\Omega^* = \operatorname{rk}\Omega = n \in \mathbf{N},$$

while (the *vector sheaf*) $\mathcal{E} \equiv \Omega^*$ admits an $\mathcal{A}$-*connection,* the *dual of* $D^*$, viz. we have

(5.9) $$(D^*)^* \equiv D^{**} = D.$$

We wrote here (5.9) in this form, in purpose; that is, as if $D^*$ was the dual of an $\mathcal{A}$-connection $D$. So we call then $D^{**} \equiv (D^*)^*$, the *bidual* $\mathcal{A}$-*connection* of $D$. Indeed, what we really have here, is that;

(5.10) *(5.9) holds true, for every $\mathcal{A}$-connection $D$ of a given vector sheaf $\mathcal{E}$ on $X$.*

The assertion follows from Theorem 1.1 in the preceding, in conjunction with (4.19), or even (4.20) above, and Theorem 5.1 in the sequel. ∎ The latter also supplements our previous considerations in Section 1 above.

**Theorem 5.1.** *Suppose that we are given a differential triad $(\mathcal{A}, \partial, \Omega)$ on a topological space $X$, and a vector sheaf $\mathcal{E}$ on $X$, of rank $n \in \mathbf{N}$. Moreover, let $D$ and $\widetilde{D}$ be two given $\mathcal{A}$-connections of $\mathcal{E}$. Then, one obtains*

$$D = \widetilde{D}, \tag{5.11}$$

*if, and only if, one has;*

$$\omega^{(\alpha)} = \tilde{\omega}^{(\alpha)} \in M_n(\Omega(U_\alpha)), \quad \alpha \in I, \tag{5.12}$$

*for the respective local $\mathcal{A}$-connection matrices of $D$ and $\widetilde{D}$, relative to some local frame*

$$\mathcal{U} = (U_\alpha)_{\alpha \in I} \tag{5.13}$$

*of $\mathcal{E}$* (cf. (1.13), along with (2.35)).

**Proof.** In view of what has been said above for (1.11), it is enough, of course, to prove the *"if"* part of the assertion: Thus, *assuming (5.12)*, we shall prove that

$$D(z) = \widetilde{D}(z), \tag{5.14}$$

*for every point $z \in \mathcal{E}$.* So if

$$\pi : \mathcal{E} \longrightarrow X \tag{5.15}$$

stands for the given (vector) *sheaf* $\mathcal{E}$ on $X$ (Chapt. I; (1.2)), such that $\pi(z) = x \in X$, for some given $z \in \mathcal{E}$, let $x \in U_\alpha \in \mathcal{U}$, *for some* $\alpha \in I$, according to the hypothesis for $\mathcal{U}$. Therefore (cf. Chapt. I; (2.9)), one may assume that there exists a *(local) section $s \in \mathcal{E}(U)$*, for a particular *open $U \subseteq U_\alpha$, such that*

$$z = s(x). \tag{5.16}$$

On the other hand, by employing the *local gauge*

$$e^{(\alpha)} \equiv \left\{U_\alpha; (e_i^{(\alpha)})_{1 \leqq i \leqq n}\right\} \tag{5.17}$$

*of* $\mathcal{E}$, which is associated with $U_\alpha \in \mathcal{U}$, as above, we shall have (cf., for instance, (2.31) along with (1.6.1))

$$s = \sum_{i=1}^{n} s_i \cdot e_i^{(\alpha)}. \tag{5.18}$$

(Note that one actually employs here a *restriction to* $U \subseteq U_\alpha$ *of* the local gauge $e^{(\alpha)}$, as in (5.17), getting thus a *new local gauge of* $\mathcal{E}$ *over* $U$). So one now obtains, by virtue of (5.18), (1.15) and (5.12);

$$D(s) = \sum_i e_i^{(\alpha)} \otimes \left(\partial s_i + \sum_j s_j \omega_{ij}^{(\alpha)}\right) = \sum_i e_i^{(\alpha)} \otimes \left(\partial s_i + \sum_j s_j \tilde{\omega}_{ij}^{(\alpha)}\right) = \widetilde{D}(s). \tag{5.19}$$

Consequently, in view of (5.16), one finally has;

$$D(z) = D(s(x)) = \widetilde{D}(s(x)) = \widetilde{D}(z) \tag{5.20}$$

(see also the succeeding Scholium 5.2 for the notation applied above), so we arrived at *the desired relation (5.14)*, which thus finishes the proof. ∎

**Scholium 5.2.**– Concerning the notation that was employed in (5.20) above, we should also make the following explanatory comments: Namely, when we consider a given $\mathcal{A}$-*connection*

$$D : \mathcal{E} \longrightarrow \Omega(\mathcal{E}) \tag{5.21}$$

of an $\mathcal{A}$-*module* $\mathcal{E}$ as a ($\mathbf{C}$-linear) *morphism of the sheaves* (of $\mathcal{A}$-modules) involved, then (cf. Chapt. I; (12.49), (12.50)), we also obtain a similar *morphism of* the corresponding *(complete) presheaves of sections,* viz.

$$\overline{D} : \Gamma(\mathcal{E}) \longrightarrow \Gamma(\Omega(\mathcal{E})). \tag{5.22}$$

It is still *worth noting* on this occasion that *it is the last (presheaf) morphism,* which is *actually employed, hitherto,* in all the preceding calculations, *when* $D$ *is applied on sections;* thus, (5.14) is, in fact, the first genuine evaluation of $D$, so far(!), *viewed,* by its very definition (cf. (5.21)), *as a sheaf morphism.* (Cf. also e.g. the relevant comments

on $\partial$ (cf. (5.1)) in Chapt. VI; (1.3)).

Thus (the same Chapt. I; (12.50)), for every *(local) section* $s \in \mathcal{E}(U)$, and any *open* $U \subseteq X$, one sets;

$$\overline{D}(s) := D \circ s \equiv D(s), \tag{5.23}$$

employing thus, for simplicity, an obvious *abuse of notation,* as in the last term above. (In this context, see also Chapt. I; Proposition 2.1). Hence, evaluating now (5.23) at *any point* $x \in U \equiv \mathrm{Dom}(s)$, as above, *one has;*

$$\overline{D}(s)(x) := (D \circ s)(x) \equiv D(s(x)) \equiv D(s)(x), \tag{5.24}$$

where we still *abuse notation by the last term* of (5.24).

Therefore, more precisely, by applying the notation of (5.22) to (5.19), one essentially obtains

$$\overline{D}(s) = \overline{\overline{\widetilde{D}}}(s), \tag{5.25}$$

so that, *for every* $x \in \mathrm{Dom}(s) \equiv U$, one finally has, by virtue of (5.16), (5.24) and (5.25);

$$\begin{aligned} D(z) = D(s(x)) = (D \circ s)(x) = \overline{D}(s)(x) = \overline{\overline{\widetilde{D}}}(s)(x) \\ = (\widetilde{D} \circ s)(x) = \widetilde{D}(s(x)) = \widetilde{D}(z). \end{aligned} \tag{5.26}$$

So this unambiguously explains now (5.20).

Thus, by returning now to the given $\mathcal{A}$-*connection* $D$ of the *vector sheaf* $\mathcal{E}$ condidered and looking further at the *range of* $D$, one has, by our hypothesis for $\Omega$ (cf. (5.3));

$$D : \mathcal{E} \longrightarrow \Omega(\mathcal{E}) \equiv \mathcal{E} \otimes_{\mathcal{A}} \Omega = \mathcal{E} \otimes_{\mathcal{A}} \mathcal{E}^* = \mathcal{H}om_{\mathcal{A}}(\mathcal{E}, \mathcal{E}) \equiv \mathcal{E}nd\, \mathcal{E} \tag{5.27}$$

(see also Chapt. IV; (6.1) and Chapt. II; (6.28)). In other words, one obtains a **C**-*linear morphism*

$$D \in \mathrm{Hom}_{\mathbf{C}}(\mathcal{E}, \mathcal{E}nd\, \mathcal{E}), \tag{5.28}$$

which further satisfies the *Leibniz condition,* as well (cf. Chapt. VI; (3.6)).

Therefore, according to the general theory (see Chapt. I; (13.19)), $D$, as in (5.28), entails a corresponding *morphism of* (complete) *presheaves,* still denoted by $D$ (cf. (5.22), (5.23) above); that is (cf. Chapt. I; (6.7)), $D$ provides, in effect, a *family of local* (presheaf) *morphisms,* indexed by the open subsets of $X$; viz. one has

$$D = (D^U \equiv D|_U)_{U \subseteq X,\, open}, \tag{5.29}$$

such that

$$D^U \in \mathrm{Hom}_{\mathbf{C}}(\mathcal{E}(U), (\mathcal{E}nd\,\mathcal{E})(U)), \tag{5.30}$$

*for every open* set $U \subseteq X$. Consequently, *for every (local) section* $s \in \mathcal{E}(U)$, one gets, by (5.30),

$$\begin{aligned}(D|_U)(s) \equiv D^U(s) \equiv D(s) \in (\mathcal{E}nd\,\mathcal{E})(U) &= \mathrm{Hom}_{\mathcal{A}|_U}(\mathcal{E}|_U, \mathcal{E}|_U) \\ &\equiv \mathrm{End}(\mathcal{E}|_U).\end{aligned} \tag{5.31}$$

Thus, by further considering *any other (local) section*

$$t \in \mathcal{E}(U) = (\mathcal{E}|_U)(U) \tag{5.32}$$

(cf. Chapt. I; (11.40)), one has, by (5.31);

$$D(s)(t) \in \mathcal{E}(U). \tag{5.33}$$

We summarize the preceding by the following statement;

> under the *assumption (5.3),* given an *$\mathcal{A}$-connection $D$ of a vector sheaf* $\mathcal{E}$ on $X$ and an *open* $U \subseteq X$, one obtains a map
>
> $$\nabla : \mathcal{E}(U) \times \mathcal{E}(U) \longrightarrow \mathcal{E}(U), \tag{5.34}$$
>
> defined by the relation (see (5.31), (5.33))
>
> $$\nabla(t, s) \equiv \nabla_t(s) := [D(s)](t), \tag{5.35}$$
>
> *for any* $s, t \in \mathcal{E}(U)$.

In this context, one further concludes that:

*the operator (5.34),* as given by (5.35), *displays the properties of a* classical *"linear connection"*; viz., *for any* $\alpha \in \mathcal{A}(U)$ *and* $s, t \in \mathcal{E}(U)$, with $U$ *open* in $X$, *one has*

$$\nabla(\alpha \cdot t, s) = \alpha \cdot \nabla(t, s), \tag{5.36}$$

*as well as (Leibniz condition)*

$$\nabla(t, \alpha \cdot s) = \alpha \cdot \nabla(t, s) + (\partial\alpha)(t) \cdot s. \tag{5.37}$$

Yet, as a consequence of (5.36) and (5.37), one also infers that;

(5.38) the operator (5.34) *is bilinear.*

(see Chapt. VI; (1.5), along with Chapt. II; (1.38), (1.67)).

**Proof of (5.36) and (5.37).** First, concerning (5.36), one obtains, by (5.35);

$$\nabla(\alpha \cdot t, s) = D(s)(\alpha \cdot t) = \alpha \cdot D(s)(t) = \alpha \cdot \nabla(t, s), \tag{5.39}$$

*which* thus *proves* (5.36).

On the other hand, one obtains,

$$\begin{aligned} \nabla(t, \alpha s) = D(\alpha s)(t) &= (\alpha \cdot Ds + s \otimes \partial\alpha)(t) = \alpha \cdot D(s)(t) + (s \otimes \partial\alpha)(t) \\ &\equiv \alpha \cdot D(s)(t) + (\partial\alpha)(t) \cdot s = \alpha \cdot \nabla(t, s) + (\partial\alpha)(t) \cdot s, \end{aligned} \tag{5.40}$$

*which,* of course, *proves (5.37), proviso* the following relation, that was actually employed in the last calculations in (5.40); thus, *we have set*

$$(s \otimes \partial\alpha)(t) = (\partial\alpha)(t) \cdot s, \tag{5.41}$$

*for any* $\alpha \in \mathcal{A}(U)$ *and* $s, t \in \mathcal{E}(U)$, as in (5.37): Indeed, one first obtains the next *canonical* $\mathcal{A}(U)$*-morphism,* by virtue of (5.3),

$$\begin{aligned} \mathcal{E}(U) \otimes_{\mathcal{A}(U)} \Omega(U) = \mathcal{E}(U) \otimes_{\mathcal{A}(U)} \mathcal{E}^*(U) &= \mathcal{E}(U) \otimes_{\mathcal{A}(U)} \mathrm{Hom}_{\mathcal{A}|_U}(\mathcal{E}|_U, \mathcal{A}|_U) \\ &\longrightarrow \mathrm{Hom}_{\mathcal{A}|_U}(\mathcal{E}|_U, \mathcal{E}|_U) = (\mathcal{E}nd\, \mathcal{E})(U), \end{aligned} \tag{5.42}$$

so that *one* actually *defines;*

$$(s \otimes u)(t) := u(t) \cdot s, \tag{5.43}$$

*for any* $s, t \in \mathcal{E}(U)$ *and* $u \in \mathcal{E}^*(U) \equiv \Omega(U)$, with $U$ *open* in $X$ (see also Chapt. IV; (6.4), (6.5)). Thus, (5.41) is just a particular instance of (5.43), given that, in view of (5.3), *one has;*

$$(5.44) \qquad \partial(\alpha) \in \Omega(U) = \mathcal{E}^*(U) = \mathcal{H}om_{\mathcal{A}}(\mathcal{E}, \mathcal{A})(U) = \mathrm{Hom}_{\mathcal{A}|_U}(\mathcal{E}|_U, \mathcal{A}|_U),$$

*for every* $\alpha \in \mathcal{A}(U)$. Hence, *for every* $t \in \mathcal{E}(U)$ (cf. also (5.32)), one obtains

$$(5.45) \qquad [\partial(\alpha)](t) \equiv (\partial\alpha)(t) \equiv \overline{\partial(\alpha)}_U(t) \in \mathcal{A}(U),$$

since one has (cf. e.g. (5.29), (5.30))

$$(5.46) \qquad \overline{\partial(\alpha)}_U(t) \in \mathrm{Hom}_{\mathcal{A}(U)}(\mathcal{E}(U), \mathcal{A}(U)) \equiv \mathcal{E}(U)^*,$$

by referring, in particular, to the given *open* $U \subseteq X$, when $\partial(\alpha)$ is viewed as an element of the last set in (5.44); thus, an analogous relation to (5.46) is certainly valid, *for every open* $V \subseteq U$. The preceding supply thus already the *full justification of (5.37),* and we are done. ▮

We are thus ready now to accomplish our program, enunciated at the beginning of this Section; we turn to this task straight away, through the succeeding discussion.

So consider first a *local gauge,* say

$$(5.47) \qquad e^U \equiv \{U; (e_1, \ldots, e_n)\},$$

of the given *vector sheaf* $\mathcal{E}$ over an *open* set $U \subseteq X$. (Notice that, due to the hypothesis for $\mathcal{E}$, such objects are come across in abundance; in fact, every point $x \in U$ is contained in a set $U$, as above. See, for instance, (2.30) – (2.33) in the preceding and (5.2)). Thus, by virtue of our hypothesis (5.3), the *dual local gauge of* $e^U$

$$(5.48) \qquad (e^U)^* \equiv \{U; (e_1^*, \ldots, e_n^*)\},$$

is a *local gauge of the vector sheaf* $\Omega = \mathcal{E}^*$ (cf. (4.3), (4.6)).

Thus, we come next to define, through (5.35) and (5.47), the following *(local) sections of* (the *structure sheaf*) $\mathcal{A}$ *over* $U$; viz. we set (see also (5.33), along with e.g. (2.33)):

$$(5.49) \qquad \begin{aligned} \Gamma_{ij}^k &:= D(e_j)(e_i)_k \equiv \nabla(e_i, e_j)_k \equiv \nabla_{e_i}(e_j)_k \\ &= e_k^*(D(e_j)(e_i)) \in \mathcal{A}(U), \end{aligned}$$

*for any* $1 \leq i, j, k \leq n = \mathrm{rk}\,\mathcal{E}$. We call (5.49) the *Christoffel sections* of the

$\mathcal{A}$-connection $D$ of $\mathcal{E}$, which are associated with the given local gauge $e^U$ of the vector sheaf $\mathcal{E}$, as in (5.47).

Therefore, one obtains from (5.49);

$$\nabla_{e_i}(e_j) = \nabla(e_i, e_j) = \sum_{k=1}^{n} \Gamma_{ij}^k \cdot e_k, \quad 1 \leqq i, j \leqq n. \tag{5.50}$$

As a result, our next observation is now that;

(5.51) *the* given *$\mathcal{A}$-connection $D$ of the vector sheaf $\mathcal{E}$*, when *restricted to $U$* (domain of definition of a *local gauge $e^U$ of $\mathcal{E}$*, cf. (5.47)), *is uniquely determined through the* corresponding to $U$, $n^3$ ($n = \operatorname{rk}\mathcal{E}$) *Christoffel sections* of $D$ (cf. (5.49)). Of course, this always, *under the* particular *assumption (5.3).*

Indeed, *one proves that* (cf. (5.35)):

$$\nabla(t, s) = \sum_{k=1}^{n} \left( \partial(s_k)(t) + \sum_{i,j} t_i s_j \Gamma_{ij}^k \right) e_k, \tag{5.52}$$

*for any (local) sections*

$$t = \sum_{i=1}^{n} t_i e_i \ \textit{and} \ s = \sum_{j=1}^{n} s_j e_j \ \textit{in} \ \mathcal{E}(U) = \mathcal{A}(U)^n \tag{5.53}$$

(see also e.g. (1.6.1) and (1.4)). Therefore, based on (5.52), the *coordinates of* $\nabla(t, s) \in \mathcal{E}(U) = \mathcal{A}(U)^n$, *with respect to* $e^U$, as in (5.47), are given by the relation (cf. also (4.6));

$$\nabla(t, s)_k = e_k^*(\nabla(t, s)) = \partial(s_k)(t) + \sum_{i,j} t_i s_j \Gamma_{ij}^k, \ 1 \leqq k \leqq n. \tag{5.54}$$

**Proof of (5.52):** Our claim on (5.52) is a routine consequence of the properties (5.36), (5.37) and (5.38) of the operator $\nabla$, applied to the elements of (5.53); thus, one has

$$\begin{aligned} \nabla(t, s) &= \nabla(\textstyle\sum_i t_i e_i, \sum_j s_j e_j) = \sum_{i,j} t_i \nabla(e_i, s_j e_j) \\ &= \textstyle\sum_{i,j} t_i (s_j \nabla(e_i, e_j) + \partial(s_j)(e_i) e_j) \\ &= \textstyle\sum_{i,j} t_i s_j (\sum_k \Gamma_{ij}^k e_k) + \sum_{i,j} t_i \partial(s_j)(e_i) e_j \\ &= \textstyle\sum_k (\sum_{i,j} t_i s_j \Gamma_{ij}^k) e_k + \sum_j \partial(s_j)(\sum_i t_i e_i) e_j \\ &= \textstyle\sum_k (\partial(s_k)(t) + \sum_{i,j} t_i s_j \Gamma_{ij}^k) e_k, \end{aligned} \tag{5.55}$$

that is, *the desired relation (5.52).* (In the last calculations above we have also made use of the analogous facts to (5.44) – (5.46), as before, for the particular case at hand). ∎

Now, for convenience, we gather the preceding together into the following.

**Theorem 5.2.** *Suppose that we are given a differential triad* $(\mathcal{A}, \partial, \Omega)$ *on a topological space* $X$ *and a vector sheaf* $\mathcal{E}$ *on* $X$, *of rank* $n \in \mathbf{N}$, *such that one has;*

$$\Omega = \mathcal{E}^*. \tag{5.56}$$

*Moreover, assume that* $\mathcal{E}$ *admits an* $\mathcal{A}$*-connection, while let also*

$$e^U \equiv \{U; (e_1, \ldots, e_n)\} \tag{5.57}$$

*be a local gauge of* $\mathcal{E}$.

*Then,* $D$ *restricted to* $U$ *is uniquely defined through the Christoffel sections of* $D$, *associated with (5.57)* (see (5.49)).

**Proof.** The assertion follows already, as explained before, from (5.52), in conjunction with (5.35). (See also (5.31), (5.32)). ∎

On the other hand, the standard familiar situation, pertaining to the *Christoffel* functions (*sections* in our case, cf. (5.49)), is further extended, within the present framework, by realizing here too that one can actually obtain the converse of the information supplied by the previous Theorem 5.2; namely, *the aforementioned sections define* (*locally,* always(!), but, see also the next Section 6) *an* $\mathcal{A}$*-connection of* $\Omega^* \equiv \mathcal{E}$ (see (5.3), (5.6)). Thus, precisely speaking, one has the following.

**Theorem 5.3.** *Let* $(\mathcal{A}, \partial, \Omega)$ *be a given differential triad on a topological space* $X$, *with* $\Omega$ *being a vector sheaf on* $X$, *of rank* $n \in \mathbf{N}$, *and let*

$$e^U \equiv \{U; (e_1, \ldots, e_n)\} \tag{5.58}$$

*be a local gauge of* $\Omega^* \equiv \mathcal{E}$ (the *dual of* $\Omega$). *Moreover, suppose that we are given the following* $n^3$ *(local) sections of* $\mathcal{A}$ *over the open* $U \subseteq X$;

$$\Gamma^k_{ij} \in \mathcal{A}(U), \quad 1 \leqq i, j, k \leqq n = \operatorname{rk} \Omega. \tag{5.59}$$

*Then, there exists a uniquely defined* $\mathcal{A}|_U$*-connection of* $\Omega^*|_U \equiv \mathcal{E}|_U$, *having the given sections (5.59) as the corresponding Christoffel sections, which are associated with (5.58).*

**Proof.** By considering the given *(local) sections* (5.59) and the *local gauge* $e^U$ (cf. (5.58)) of the *dual of* $\Omega$,

$$\mathcal{E} \equiv \Omega^*, \tag{5.60}$$

for any *local sections* of $\mathcal{E}$ over the *open* set $U \subseteq X$,

$$t = \sum_{i=1}^{n} t_i e_i \ \textit{and} \ s = \sum_{j=1}^{n} s_j e_j, \tag{5.61}$$

*where* $n = \operatorname{rk}\mathcal{E} \equiv \operatorname{rk}\Omega^* = \operatorname{rk}\Omega$ (cf., for instance, (5.4) or even (5.8)), one sets;

$$\nabla(t,s) := \sum_{k=1}^{n} \alpha_k \cdot e_k, \tag{5.62}$$

*such that*

$$\alpha_k := \partial(s_k)(t) + \sum_{i=1}^{n}\sum_{j=1}^{n} t_i s_j \Gamma_{ij}^k \in \mathcal{A}(U),\ 1 \leq k \leq n. \tag{5.63}$$

Thus, one actually obtains a map

$$\nabla^U \equiv \nabla : \mathcal{E}(U) \times \mathcal{E}(U) \longrightarrow \mathcal{E}(U), \tag{5.64}$$

having the formal properties of a *linear* $\mathcal{A}(U)$*-connection on* $\mathcal{E}(U)$; namely, this is:

i) *bilinear,* relative to the **C**-vector spaces concerned (cf. Chapt. II; (1.38), (1.51), or even (1.67)), as easily follows from (5.62) and (5.63).

ii) $\mathcal{A}(U)$*-linear,* with respect to the *first variable;* this is still obvious by (5.61), (5.63), along with the fact that, *for any* $\alpha \in \mathcal{A}(U)$, *one obtains* (cf. also (5.44) - (5.46), as well as, (5.5) and (5.60));

$$\partial(\alpha) \in \Omega(U) = \mathcal{E}^*(U) = \mathcal{H}om_{\mathcal{A}}(\mathcal{E},\mathcal{A})(U) = \operatorname{Hom}_{\mathcal{A}|_U}(\mathcal{E}|_U, \mathcal{A}|_U). \tag{5.65}$$

Finally, one concludes that;

iii) *the same map* $\nabla^U \equiv \nabla$, as given by (5.64) and (5.62), *satisfies the Leibniz condition* (5.37): Indeed, for any $s, t \in \mathcal{E}(U)$, as in (5.61), and $\alpha \in \mathcal{A}(U)$, one obtains (see also (5.63), (5.41), along with Chapt. VI; (1.2));

$$
\begin{aligned}
\nabla(t,\alpha s) &= \textstyle\sum_k \left(\partial(\alpha s_k)(t) + \sum_{i,j} t_i(\alpha s_j)\Gamma_{ij}^k\right) e_k \\
&= \textstyle\sum_k \left(\alpha\cdot\partial(s_k)(t) + (s_k \otimes \partial\alpha)(t) + \alpha\cdot\sum_{i,j} t_i s_j \Gamma_{ij}^k\right) e_k \\
&= \alpha\cdot\textstyle\sum_k(\partial(s_k)(t) + \sum_{i,j} t_i s_j \Gamma_{ij}^k)e_k + \partial(\alpha)(t)\cdot\sum_k s_k e_k \\
&= \alpha\cdot\nabla(t,s) + \partial(\alpha)(t)\cdot s,
\end{aligned}
\tag{5.66}
$$

that is, the assertion.

Now, by setting

$$D(s)(t) := \nabla(t,s), \tag{5.67}$$

*for any* $s, t \in \mathcal{E}(U)$ (note that, by virtue of our hypothesis for $e^U$, the latter sections can always be put in the form (5.61)), one realizes that

$$D(s) \in (\mathcal{E}nd\,\mathcal{E})(U) = \mathrm{End}(\mathcal{E}|_U) \equiv \mathrm{Hom}_{\mathcal{A}|_U}(\mathcal{E}|_U, \mathcal{E}|_U), \tag{5.68}$$

*for every* $s \in \mathcal{E}(U)$, by the very properties of (5.64), as proved above; in particular, by still employing here an obvious *abuse of notation* for (5.68) (applying thus the same notation too for its *"U-component"*, see e.g. (5.29)), *one actually has*

$$D(s) \in \mathrm{Hom}_{\mathcal{A}(U)}(\mathcal{E}(U), \mathcal{E}(U)), \tag{5.69}$$

*for every* $s \in \mathcal{E}(U)$. Furthermore, *one obtains* (caution! See also e.g. (5.30))

$$D(\alpha\cdot s) = \alpha\cdot D(s) + s \otimes \partial(\alpha), \tag{5.70}$$

*for any* $\alpha \in \mathcal{A}(U)$ *and* $d \in \mathcal{E}(U)$: Indeed, one has (cf. also (5.67), in conjunction with (5.66) and (5.41));

$$
\begin{aligned}
D(\alpha s)(t) &= \nabla(t,\alpha s) = \alpha\cdot\nabla(t,s) + (\partial\alpha)(t)\cdot s \\
&= \alpha\cdot D(s)(t) + (s\otimes\partial\alpha)(t) = (\alpha\cdot D(s) + s\otimes\partial\alpha)(t),
\end{aligned}
\tag{5.71}
$$

*for every* $t \in \mathcal{E}(U)$. Therefore, one gets (5.70), given that, both summands in the second member of (5.70), viz. $\alpha\cdot D(s)$ *and*, by virtue of (5.42), (5.43) (see also (5.44)), $s\otimes\partial(\alpha)$ *belong to the same* $\mathcal{A}(U)$*-module, as* $D(s)$ *does* (cf. (5.69)).

Consequently, *the map*

$$D^U \equiv D : \mathcal{E}|_U \longrightarrow \mathcal{E}\otimes_{\mathcal{A}}\Omega|_U = \mathcal{E}\otimes_{\mathcal{A}}\mathcal{E}^*|_U = (\mathcal{E}nd\,\mathcal{E})|_U \tag{5.72}$$

(cf also (5.60), (5.27)), which is defined, *section-wise*, by (5.67), *yields an* $\mathcal{A}|_U$*-connection of* $\mathcal{E}|_U$ ($\equiv \Omega^*|_U$). Furthermore, *the same connection has*

*(5.59) as Christoffel sections,* associated with the given local gauge $e^U$, as this follows from the relation;

$$D(e_j)(e_i)_k = \nabla(e_i, e_j)_k = \Gamma^k_{ij}, \quad 1 \leqq i, j, k \leqq n \tag{5.73}$$

(see also (5.47), (5.49), (5.67) and (5.63), along with Chapt. VI; (1.5)). Finally, such an $\mathcal{A}|_U$-connection on $\mathcal{E}|_U$ (viz. with given Christoffel sections on $U$), is *uniquely defined,* as this follows straightforwardly from a relevant localization to $U$ of Theorem 5.2 above, and this terminates the proof. ∎

On the other hand, by pursuing further the classical pattern, one also obtains *another expression for the Christoffel sections,* as in (5.49), of the given $\mathcal{A}$-connection $D$, *by means of the local $\mathcal{A}$-connection matrices* of $D$, as they were defined in the preceding by (1.13). (See thus (5.77) below).

Thus, suppose, in particular, that *we have the hypotheses of Theorem 5.2* above. Moreover, let

$$\omega \equiv (\omega_{ij}) \in M_n(\Omega(U)) = M_n(\Omega)(U) \tag{5.74}$$

be the *local $\mathcal{A}$-connection matrix* of the given *$\mathcal{A}$-connection $D$ of $\mathcal{E}$,* which can be associated with a *local gauge $e^U$ of $\mathcal{E}$* (cf. (5.57) and (1.13)), such that one has (cf. (5.56))

$$\omega_{ij} \in \Omega(U) = \mathcal{E}^*(U), \quad 1 \leqq i, j \leqq n = \operatorname{rk} \mathcal{E}. \tag{5.75}$$

Accordingly, using the *dual local gauge* $(e^U)^*$ of $e^U$ (cf. (5.48)), as well as, the corresponding to these data *Christoffel sections* (see (5.49)), *one can define* the previous sections (5.75), hence, *the matrix* $\omega$, through the following relation;

$$\omega_{ij} := \sum_{k=1}^{n} \Gamma^k_{ij} \cdot e^*_k \in \mathcal{E}^*(U) = \Omega(U), \quad 1 \leqq i, j \leqq n. \tag{5.76}$$

Thus, *equivalently,* one obtains that (see (4.6));

$$\omega_{ij}(e_k) = \Gamma^k_{ij}, \quad 1 \leqq i, j, k \leqq n. \tag{5.77}$$

Now, as a result of the preceding discussion (cf. (5.76) above and Theorem 5.2), one concludes that:

(5.78) under the *hypothesis of Theorem 5.2,* an *$\mathcal{A}$-connection $D$* of *$\mathcal{E}$ is, locally, uniquely defined either through the local $\mathcal{A}$-connection matrix $\omega$, or the* corresponding *Christoffel sections $\{\Gamma^k_{ij}\}$ of $D$,* where both objects are *referred to a* given *local gauge $e^U$ of $\mathcal{E}$.*

In this context, we also note that, by further following the classical pattern, one can give a *relation connecting the Christoffel sections, when a gauge transformation is under consideration.* We do not give it, however, due to its somewhat elaborate (routine, anyway!) technical nature. Yet, we shall be concerned with the Christoffel sections of a given $\mathcal{A}$-connection, several times in the sequel.

## 6. Complete parallelism

The concept in the title of this Section belongs, in effect, to the next Chapter VIII, that is dealt with the *"curvature"* of a given $\mathcal{A}$-connection. However, we have already anticipated in the preceding that notion, by considering *flat vector sheaves* (Chapt. V; Section 5), as well as, *integrable $\mathcal{A}$-connections* (Chapt. VI; Subsection 11.(a)), to which the present notion is intimately related; in a sense, all these concepts are (in concrete instances, including the classical case, of course), equivalent, as we are going to see (partly, the whole story will be told in Chapter VIII), straight away, by the ensuing discussion.

Thus, suppose that we are given a *differential triad* $(\mathcal{A}, \partial, \Omega)$ on a *topological space* $X$, along with an *$\mathcal{A}$-module $\mathcal{E}$* on $X$ and an *$\mathcal{A}$-connection $D$ of $\mathcal{E}$*. Moreover, let

$$e^U \equiv \{U; (e_1, \ldots, e_n)\} \tag{6.1}$$

be a *local gauge of* $\mathcal{E}$, of *rank* $n \in \mathbf{N}$, over the *open* set $U \subseteq X$ (see (1.5)).

Now, we say that the $\mathcal{A}$-connection $D$, as before, defines a *complete parallelism of $\mathcal{E}$ over $U$*, whenever one has

$$D(e_i) = 0, \quad 1 \leqq i \leqq n. \tag{6.2}$$

In this context, we also write, for short, instead of (6.2), just

$$D(e^U) = 0. \tag{6.3}$$

On the other hand, a given *(local) section* $s \in \mathcal{E}(V)$ over the *open* set $V \subseteq X$ is said to be *parallel* (or even *horizontal*), with respect to the $\mathcal{A}$-connection $D$ of $\mathcal{E}$, as above, if one has

$$D(s) = 0. \tag{6.4}$$

Thus, when (6.2) is satisfied (cf. also (1.3)), one also speaks for $e^U$, as a *parallel* (or even *horizontal*) *local gauge* of $\mathcal{E}$.

The preceding nomenclature is further specialized by the following.

**Definition 6.1.** Suppose we have a vector sheaf $\mathcal{E}$ on $X$, as before, of rank $n \in \mathbf{N}$, which is also endowed with an $\mathcal{A}$-connection $D$. Then, we say that $\mathcal{E}$ has a *complete parallelism,* with respect to $D$ (or even that $D$ *provides a complete parallelism* of $\mathcal{E}$), whenever $\mathcal{E}$ has a *parallel* (or *horizontal*) *local frame,* with respect to $D$.

Namely, we assume that there exists a *local frame* of $\mathcal{E}$,

$$\mathcal{U} = (U_\alpha)_{\alpha \in I}, \tag{6.5}$$

such that *the* corresponding *local gauges* (see (2.31), (2.32))

$$e^{(\alpha)} \equiv \{U_\alpha; (e_i^{(\alpha)})_{1 \leqq i \leqq n}\} \tag{6.6}$$

*are parallel;* that is, one has (cf. (6.2), (6.3))

$$D(e^{(\alpha)}) = 0, \quad \alpha \in I. \tag{6.7}$$

Now, in order also to respond to the comments at the beginning of this Section, it is instructive to consider at this place a *generalization* of the notion *of flat vector sheaf* (cf. thus (6.12) in the sequel). The same extended notion of the latter will be of importance later on too. So we further set the following.

**Definition 6.2.** Suppose that we are given a differential triad $(\mathcal{A}, \partial, \Omega)$ on a topological space $X$, and let $\mathcal{E}$ be a vector sheaf on $X$, with $\operatorname{rk} \mathcal{E} = n \in \mathbf{N}$. Now, we say that $\mathcal{E}$ is $\partial$*-flat,* whenever there exists a coordinate 1-cocycle of $\mathcal{E}$,

$$(g_{\alpha\beta}) \in Z^1(\mathcal{U}, \mathcal{GL}(n, \mathcal{A})), \tag{6.8}$$

with respect to a local frame $\mathcal{U}$ of $\mathcal{E}$ (cf. Chapt. V; (1.8)), such that

$$\partial(g_{\alpha\beta}) = 0. \tag{6.9}$$

In the last relation $\partial$ stands, in effect, for its *extension to* $M_n(\mathcal{A})$ (*Whitney sum,* cf. Chapt. VI; (1.15)). So by extending the terminology of (6.4) (to the *$\mathcal{A}$-connection $\partial$ of $\mathcal{A}$*, in fact, to its *extension to* $M_n(\mathcal{A})$), we can say that;

(6.10) a given *vector sheaf* $\mathcal{E}$ on $X$ is *$\partial$-flat,* whenever it possesses a coordinate 1-cocycle, consisting of *$\partial$-parallel,* or even *$\partial$-horizontal, (coordinate) matrices.*

Notice that, by definition (see (6.8)), one has;

$$g_{\alpha\beta} \in \mathcal{GL}(n, \mathcal{A})(U_{\alpha\beta}) = \mathrm{GL}(n, \mathcal{A}(U_{\alpha\beta})), \tag{6.11}$$

*for any $\alpha, \beta \in I$, with $U_{\alpha\beta} \equiv U_\alpha \cap U_\beta \neq \emptyset$.*

So, as already mentioned, one concludes, in fact, according to the very definitions, that;

(6.12) *every flat vector sheaf* on $X$ *is $\partial$-flat.* (See also Chapt. V; Definition 5.1).

Thus, see Chapt. V; (5.7), Theorem 5.1; in particular, (5.18) and (5.24), along with Chapt. VI; Lemma 1.1. ∎

We are ready now to come to our first result, pertaining to the preceding notions. So, as we shall presently see, *every $\partial$-flat vector sheaf possesses always* a certain special type of *$\mathcal{A}$-connection,* we already met before (viz. an *integrable $\mathcal{A}$-connection*). Thus, more precisely, one has the following.

**Theorem 6.1.** *Suppose that we have a differential triad $(\mathcal{A}, \partial, \Omega)$ on a topological space $X$, and let $\mathcal{E}$ be a vector sheaf on $X$, of rank $n \in \mathbf{N}$. Then, the three following assertions are equivalent:*

*1) $\mathcal{E}$ is $\partial$-flat.*

*2) $\mathcal{E}$ has an integrable $\mathcal{A}$-connection* (cf. Chapt. VI; Definition 11.2).

*3) $\mathcal{E}$ admits an $\mathcal{A}$-connection $D$, providing a complete parallelism.*

**Proof.** We prove first that *1) implies 2):* So let

$$\mathcal{U} = (U_\alpha)_{\alpha \in I} \tag{6.13}$$

be a *local frame of* $\mathcal{E}$, with respect to which we have a *coordinate 1-cocycle* (6.8), for which (6.9) *holds true.* Thus, by considering the corresponding to (6.13) *Levi-Civita ($\mathcal{A}$-connection) 1-cocycle* of $\mathcal{E}$, one obtains in view of (6.9) (see also Chapt. VI; (8.34) and Theorem 10.1, in particular (10.22), (10.23)),

$$\delta(D_\alpha) = (D_\beta - D_\alpha) \underset{\alpha}{\sim} \tilde{\partial}(g_{\alpha\beta}) = g_{\alpha\beta}^{-1}\partial(g_{\alpha\beta}) = 0. \tag{6.14}$$

That is, *the Levi-Civita 0-cochain* $(D_\alpha)$, as above (cf. also Chapt. VI; (8.32.1)), *is,* in fact, *a 0-cocycle,* so that it defines an $\mathcal{A}$-connection of $\mathcal{E}$, which is, of course, integrable (see Chapt. VI; (11.24)), viz. the assertion. *Convesely,* assume that $\mathcal{E}$ admits an *integrable $\mathcal{A}$-connection* $\widetilde{D}$, with respect to (6.13) (cf. Chapt. VI; Definition 11.2). Therefore (see Chapt. VI; (6.23)), one has

$$0 = \delta(\widetilde{D}_\alpha) = \delta(D_\alpha) \underset{\alpha}{\sim} \tilde{\partial}(g_{\alpha\beta}) = g_{\alpha\beta}^{-1}\partial(g_{\alpha\beta}), \tag{6.15}$$

hence, $\partial(g_{\alpha\beta}) = 0$, so that *2) implies 1),* as well.

Furthermore, we prove that *1) and 3) are equivalent:* Indeed, *if 1) is valid,* and we are given (6.13), one can further define a *0-cochain of local $\mathcal{A}$-connection matrices,* associated with it, viz.

$$\omega \equiv (\omega^{(\alpha)}) \in C^0(\mathcal{U}, M_n(\Omega)), \tag{6.16}$$

by simply setting

$$\omega^{(\alpha)} \equiv (\omega_{ij}^{(\alpha)}) := 0 \in M_n(\Omega)(U_\alpha) = M_n(\Omega(U_\alpha)), \tag{6.17}$$

*for every* $\alpha \in I$. However, *by virtue of (6.9), the same 0-cochain satisfies* now *the transformation law of potentials* (cf. (3.3)), so that (Theorem 3.1) $\mathcal{E}$ *admits (uniquely) an $\mathcal{A}$-connection $D$,* having the above $\omega^{(\alpha)}$'s. In particular, *1) implies 3),* since now, by (6.17) and (2.46), one has;

$$D(e_j^{(\alpha)}) = \sum_i e_i^{(\alpha)} \otimes \omega_{ij}^{(\alpha)} = 0, \tag{6.18}$$

*for any* $1 \leqq i, j \leqq n$, *and* $\alpha \in I$, that is, (6.7) *is fulfilled.* Conversely, *if (6.7) is valid,* then by the same relation (6.18), one concludes that

$$\omega^{(\alpha)} = 0, \quad \alpha \in I, \tag{6.19}$$

so that by still employing the transformation law of potentials, one further obtains that

$$\tilde{\partial}(g_{\alpha\beta}) = g_{\alpha\beta}^{-1}\partial(g_{\alpha\beta}) = 0, \tag{6.20}$$

hence, $\partial(g_{\alpha\beta}) = 0$. Thus, *3) implies 1), as well,* and this completes the proof of the theorem. ∎

**Scholium 6.1.**- The aim of the following lines is to offer *another proof of prop. 1) implies 2) of* the previous *Theorem 6.1.* (Although the present proof is rather longer, than the one given above, it has, however, an interest per se; cf. thus (6.27), (6.28) and (6.29) below). So, following the argument in the first step of the proof given before, one defines an *$\mathcal{A}$-connection* $\widetilde{D}$ of $\mathcal{E}$, *whose local $\mathcal{A}$-connection matrices were defined by (6.17).* We prove here that *this same* $\widetilde{D}$ (not only the Levi-Civita one, as above) *is an integrable $\mathcal{A}$-connection of $\mathcal{E}$ too;* viz. one has

$$\widetilde{D}_\alpha \equiv \widetilde{D}|_{U_\alpha} = D_\alpha, \quad \alpha \in I, \tag{6.21}$$

where the last item of (6.21), an $\mathcal{A}|_{U_\alpha}$*-connection of* $\mathcal{E}|_{U_\alpha}$, is given by Chapt. VI; (8.29): Indeed, *for any*

$$s = \sum_{i=1}^{n} s_i e_i^{(\alpha)} \in \mathcal{E}(U_\alpha), \quad \alpha \in I, \tag{6.22}$$

(see (1.6)), *one obtains* (cf. (1.15))

$$\widetilde{D}_\alpha(s) = \sum_{i=1}^{n} e_i^{(\alpha)} \otimes \left(\partial(s_i) + \sum_{j=1}^{n} s_j \omega_{ij}^{(\alpha)}\right), \tag{6.23}$$

so that, in view of (6.17), *one has;*

$$\widetilde{D}_\alpha(s) = \sum_{i=1}^{n} e_i^{(\alpha)} \otimes \partial(s_i). \tag{6.24}$$

On the other hand, based on Chapt. VI; (8.29), one sets

$$D_\alpha := (\eta_\alpha^{-1} \otimes 1) \circ \partial_\alpha \circ \eta_\alpha \quad \alpha \in I. \tag{6.25}$$

Therefore, since, by definition (cf. Scholium 1.1), one has

$$e_i^{(\alpha)} := \eta_\alpha^{-1} \circ \varepsilon_i^{(\alpha)} \quad 1 \leqq i \leqq n, \tag{6.26}$$

*for every $\alpha \in I$, with* $\varepsilon_i^{(\alpha)} = (\delta_{ij}) \in \mathcal{A}^n(U_\alpha) = \mathcal{A}(U_\alpha)^n$ (canonical *Kronecker section-basis of* $\mathcal{A}^n$ over $U_\alpha$), *one obtains,* by virtue of (6.25) (see also (5.5), (1.7));

(6.27) $$D_\alpha(e_i^{(\alpha)}) = 0, \quad 1 \leqq i \leqq n, \quad \textit{and} \quad \alpha \in I.$$

(As a *byproduct,* one thus concludes that *when the Levi-Civita $\mathcal{A}$-connection $D = (D_\alpha)$ of $\mathcal{E}$ exists,* then, according to (6.27), one concludes (directly!) that *the latter* still *supplies a complete parallelism;* a consequence, otherwise, of Chapt. VI; (11.24), in conjunction with Theorem 6.1 above). Hence (see, for instance, (6.18)), if $\tilde{\omega}^{(\alpha)} \in M_n(\Omega)(U_\alpha)$, $\alpha \in I$, denote the local $\mathcal{A}$-connection matrices of $D$, one has, by (6.27),

(6.28) $$\tilde{\omega}^{(\alpha)} \equiv (\omega_{ij}^{(\alpha)}) = 0,$$

so that one still concludes from (1.15) that

(6.29) $$D_\alpha(s) = \sum_{i=1}^{n} e_i^{(\alpha)} \otimes \partial(s_i),$$

*for every $s \in \mathcal{E}(U_\alpha)$, and $\alpha \in I$,* which thus proves (6.21), in view of (6.24). ∎

On the other hand, by supplementing our argument in Chapt. VI; (11.24), we further remark that;

(6.31) *if $\widetilde{D}$ is an integrable $\mathcal{A}$-connection of* (a vector sheaf) $\mathcal{E}$, then *the Levi-Civita 0-cochain*

(6.31.1) $$(D_\alpha) \in C^0(\mathcal{U}, M_n(\Omega))$$

(see Chapt. VI; (11.18)) *is,* in fact, *a 0-cocycle, yielding thus the integrable* (always! whenever it exists) *Levi-Civita $\mathcal{A}$-connection $D = (D_\alpha)$ of $\mathcal{E}$.* (Cf. also Chapt. VI; Definition 11.1).

Furthermore, as a consequence of (6.23) and (6.17), in conjunction with Theorem 6.1, we also conclude that;

(6.32) *a given $\mathcal{A}$-connection $\widetilde{D}$ of* (a vector sheaf) $\mathcal{E}$ *is integrable, if, and only if, one has*

(6.32.1) $$\widetilde{D}_\alpha(s) = \sum_{i=1}^{n} e_i^{(\alpha)} \otimes \partial(s_i),$$

*for any (local) section $s = \sum_{i=1}^{n} s_i e_i^{(\alpha)} \in \mathcal{E}(U_\alpha)$, and for every $\alpha \in I$* (cf. (6.24) for the notation applied).

For the *"if"* part of our claim apply (6.32.1) to $e_i^{(\alpha)}$ to get (6.7) (cf. also

(6.26), (6.27), so that the assertion follows then from Theorem 6.1. The rest is already provided by the preceding. ∎

Finally, it is worth noticing at this place that, as a result of (6.21), along with (6.24) and (6.29), one also obtains for the corresponding (local) *components* of the *$\mathcal{A}$-connections $\widetilde{D} = (\widetilde{D}_\alpha)$ and $D = (D_\alpha)$ of $\mathcal{E}$*, as above, the relations;

$$\widetilde{D}_\alpha(s)_i = D_\alpha(s)_i = \partial(s_i), \quad 1 \leqq i \leqq n, \tag{6.33}$$

*for any $s \in \mathcal{E}(U_\alpha)$, as in (6.22), and $\alpha \in I$.* (This should also be compared with our previous remark in Chapt. VI; (8.32.3)). Thus, stated otherwise, (6.33) means that:

(6.34) *an integrable $\mathcal{A}$-connection* of a given vector sheaf $\mathcal{E}$, *hence also* (cf. Theorem 6.1), *an $\mathcal{A}$-connection of $\mathcal{E}$, supplying a complete parallelism, are locally the* given *standard flat $\mathcal{A}$-connection $\partial$ of $\mathcal{A}$* (indeed, *its extension to $M_n(\mathcal{A})$, with $n = \text{rk}\,\mathcal{E} \in \mathbf{N}$*).

**Scholium 6.2.**- The previous notion of *$\partial$-flatness* is proved to be, in effect, the *appropriate generalization,* within the present abstract setting of the classical notion *of flatness* (viz. *zero curvature,* cf. the next Chapter VIII), which in standard terms reduces to the fact that

$$\ker \partial = \mathbf{C} \tag{6.35}$$

(here, of course, its *extension to $M_n(\mathcal{A})$*, see (6.9), (6.11) above, in conjunction with Chapt. VI; (1.22), (1.23), and Chapt. V; (5.25)).

Now, the last relation, as above, *is likely not to hold true, in general.* Yet, as we shall see in the subsequent Chapter *$\partial$-flatness implies curvature zero* for the respective (integrable) $\mathcal{A}$-connection (Theorem 6.1) and conversely, as well, while the same conditions are implied, in turn, by a *flat vector sheaf*, in the sense of (6.12) above. However, to all these things we shall return, with more details, at the pertinent places in the sequel.

In this context, by further looking at the local expressions of a given $\mathcal{A}$-connection $D$, in terms of the "$\omega$'s", as above, as well as, of the corresponding Christoffel sections, one concludes the following:

Namely, assume that $\mathcal{E}$ is a *vector sheaf* on $X$, of *rank* $n \in \mathbf{N}$, endowed with an *$\mathcal{A}$-connection* $D$, and let $\mathcal{U} = (U_\alpha)_{\alpha \in I}$ be a *local frame* of $\mathcal{E}$ (cf. (6.5), (6.6)). Moreover, let

$$\omega \equiv (\omega^{(\alpha)}) \in C^0(\mathcal{U}, M_n(\Omega)) \tag{6.36}$$

be the corresponding *0-cochain of local ($\mathcal{A}$-connection) matrices* of $D$, which are associated with $\mathcal{U}$ (cf., for instance, (2.39.1) in the foregoing). Then

(6.37) *$D$ yields a complete parallelism* (cf. Definition 6.1), *if, and only if, one has*

$$\omega^{(\alpha)} = 0, \tag{6.37.1}$$

*for every* $\alpha \in I$.

On the other hand, if, in addition, (5.3) is satisfied, then one concludes that;

*(6.37) is still equivalent with the condition*

$$^{(\alpha)}\Gamma^k_{ij} = 0, \tag{6.38}$$

*for any* $1 \leqq i, j, k \leqq n = \operatorname{rk} \mathcal{E}$, *and* $\alpha \in I$.

(The notation applied in (6.38) indicates obviously the dependence of the respective Christoffel sections on the various indices $\alpha$ in $I$). So (6.38) is, in fact, an immediate consequence of (6.7) and (2.46), while (6.38) follows from (6.7), in conjunction with (5.77) (cf. also (5.76)). ∎ Thus, putting now (6.37) and (6.38) together, we certainly infer that:

(6.39) by *assuming (5.3), the following conditions are mutually equivalent:*
i) $D(e^{(\alpha)}) = 0$, ii) $\omega^{(\alpha)} = 0$, *for every* $\alpha \in I$, *and* iii) $^{(\alpha)}\Gamma^k_{ij} = 0$, *for* any $1 \leqq i, j \leqq n$, *and* $\alpha \in I$.
So, *if any one of the above three equivalent conditions is satisfied,* then *the* given *$\mathcal{A}$-connection* $D$, as before, *provides a complete parallelism* and conversely, of course.

The intimate relation of all the preceding notions with that one of a *"flat $\mathcal{A}$-connection"* (cf. Scholium 6.2 above) will be examined, in detail, as already mentioned, in the next Chapter VIII.

# 7. $\partial$-flat line sheaves. Generalized Selesnick - Chern class of a line sheaf

We consider by the ensuing discussion, a certain particular 2-dimensional cohomology class, which as we shall presently see (cf. Theorem 7.1 below), by analogy with its complex counterpart, the (complex) Chern class of a line sheaf (cf. Chapt. V; Theorem 6.1), characterizes $\partial$-flatness; in this context, we have already seen (cf. (6.12), along with Scholium 6.2) that the latter notion generalizes the classical one of flatness. Yet, to keep track of the evolution of the idea around the aforementioned cohomology class, we still note that this has first been applied by S.S. Selesnick [2] to the particular case of a *line bundle* over the *spectrum* of a suitable *Banach algebra*. (In this regard, see also Chapt. VI; Note 10.1, as well as, Chapter XI in the sequel).

Throughout the succeeding discussion we shall make repeated use of our previous terminology and results in Chapt. V; Section 6. Thus, to start with, suppose that we are given an *exponential algebraized space* $(X, \mathcal{A})$ (cf. Chapt. V; Definition 6.1), which we further assume to be endowed with a *differential triad* $(\mathcal{A}, \partial, \Omega)$, in such a manner that the following diagram is *commutative;*

$$\begin{array}{ccccccccc} 0 & \longrightarrow & \mathbf{Z} & \longrightarrow & \mathcal{A} & \xrightarrow{\;e\;} & \mathcal{A}^{\cdot} & \longrightarrow & 1 \\ & & & & & \partial\searrow & \swarrow \frac{1}{2\pi i}\tilde{\partial} & & \\ & & & & & \partial\mathcal{A}. & & & \end{array} \tag{7.1}$$

We say that (7.1) defines an *exponential sheaf diagram.* Yet, for brevity's sake we shall refer to the previous data, as a *topological space $X$ with an exponential sheaf diagram.* The even shorter term, that, namely, $X$ is a *sheaf exponential space,* will often be of use below.

Now, as a result of the *short exact sequence,* that appears in (7.1) and of Chapt. VI; Lemma 1.1, one has the following *morphism of short exact sequences;*

$$
\begin{array}{ccccccccc}
0 & \longrightarrow & \mathbf{Z} & \longrightarrow & \mathcal{A} & \xrightarrow{e} & \mathcal{A}^{\cdot} & \longrightarrow & 1 \\
 & & \Big\downarrow \zeta & & \Big\| & & \Big\downarrow \frac{1}{2\pi i}\tilde{\partial} & & \\
0 & \longrightarrow & \ker \partial & \longrightarrow & \mathcal{A} & \longrightarrow & \partial\mathcal{A} & \longrightarrow & 0 .
\end{array}
\tag{7.2}
$$

Therefore (see e.g. Chapt. III; Theorem 3.1, prop. 4)), one gets the following *commutative diagram in (sheaf) cohomology;*

$$
\begin{array}{ccc}
H^1(X, \mathcal{A}^{\cdot}) & \xrightarrow{\quad\delta\quad} & H^2(X, \mathbf{Z}) \\
\Big\downarrow \frac{1}{2\pi i}\tilde{\partial} & c_{\mathcal{A}} & \Big\downarrow \zeta \\
H^1(X, \partial\mathcal{A}) & \xrightarrow{\quad\gamma\quad} & H^2(X, \ker \partial).
\end{array}
\tag{7.3}
$$

Thus we can now set the following.

**Definition 7.1.** Suppose we have a sheaf exponential space $X$ (see (7.1)), and let $\mathcal{L}$ be a given line sheaf on $X$. Moreover, consider the Picard group of $X$ (cf. Chapt. V; Theorem 4.1), so that one has;

$$\mathcal{L} \equiv [\mathcal{L}] \in \mathrm{Pic}(X) = H^1(X, \mathcal{A}^{\cdot}). \tag{7.4}$$

Thus, one defines the *Chern class of* $\mathcal{L}$*, relative to* $\mathcal{A}$, or just the *Selesnick-Chern class of* $\mathcal{L}$, as the 2-dimensional cohomology class $c_{\mathcal{A}}(\mathcal{L})$, which is given by the relation (cf. (7.3));

$$c_{\mathcal{A}}(\mathcal{L}) := \gamma\left(\frac{1}{2\pi i}\tilde{\partial}([(g_{\alpha\beta})])\right) \in H^2(X, \ker \partial), \tag{7.5}$$

where $(g_{\alpha\beta}) \in Z^1(\mathcal{U}, \mathcal{A}^{\cdot})$ stands for a coordinate 1-cocycle of $\mathcal{L}$ (cf. Chapt. V; (1.8), for $n = 1$).

Now, as already mentioned at the beginning of this Section, a basic application of the previous cohomology class is that its vanishing yields a criterion for the $\partial$*-flatness* of the given line sheaf $\mathcal{L}$ (which, in effect, means (Theorem 6.1) that $\mathcal{L}$ admits an integrable $\mathcal{A}$-connection); so cf. Theorem 7.1, or even Corollary 7.1 below. However, before we come to the cited results, it is con-

structive first to give the following key-lemmas for the proof of the former. (In this context, see also S.A. Selesnick [2: p. 66, Theorem 3.6]). So we first have the following result, referring actually to the *kernel* and *image of* the *logarithmic derivation* $\tilde{\partial}$, with respect to those of $\partial$; viz. one obtains.

**Lemma 7.1.** *Suppose we have a differential triad* $(\mathcal{A}, \partial, \Omega)$ *on a topological space* $X$. *Then,* applying the terminology of Chapt. VI; Definition 1.2, *one has;*

$$\ker \tilde{\partial} = (\ker \partial) \cap \mathcal{A}^{\cdot}. \tag{7.6}$$

*In particular, suppose that we are given, in addition, the following commutative triangle of morphisms of sheaves* (of abelian) groups, see also (7.1), for the notation applied),

$$\begin{array}{ccc} \mathcal{A} & \xrightarrow{\ e\ } & \mathcal{A}^{\cdot} \\ & {\scriptstyle\partial}\searrow \quad \swarrow{\scriptstyle \frac{1}{2\pi i}\tilde{\partial}} & \\ & \Omega, & \end{array} \tag{7.7}$$

*with the morphism* $e$ *being also surjective. Then, one has;*

$$\tilde{\partial}(\mathcal{A}^{\cdot}) = \partial(\mathcal{A}) = 2\pi i \cdot \partial(\mathcal{A}). \tag{7.8}$$

**Proof.** Our claim for (7.6) is, of course, an immediate consequence of the very definitions (cf. thus Chapt. VI; (1.24), (1.25)).

Now, concerning (7.8), based on our supplementary assumption for (7.1) and the map $e$, as well as, on the **C**-*linearity of* $\partial$, one obtains;

$$\tilde{\partial}(\mathcal{A}^{\cdot}) = \tilde{\partial}(e(\mathcal{A})) = 2\pi i \cdot \partial(\mathcal{A}) = \partial(2\pi i \cdot \mathcal{A}) = \partial(\mathcal{A}), \tag{7.9}$$

which thus proves (7.8), and this completes the proof. ∎

Now, as an application of the previous two elementary, however, useful formulae (7.6) and (7.8), one further obtains the following key-result for our subsequent considerations. Thus, we have the next.

**Lemma 7.2.** *(S.A. Selesnick). Assume that we are given a differential triad* $(\mathcal{A}, \partial, \Omega)$ *on a topological space* $X$, *along with a commutative triangle, as in (7.7), the morphism* $e$ *being surjective. Moreover, let*

$$(g_{\alpha\beta}) \in Z^1(\mathcal{U}, \mathcal{A}^{\cdot}) \tag{7.10}$$

*be an $\mathcal{A}^{\cdot}$-valued 1-cocycle of an open covering $\mathcal{U} = (U_\alpha)_{\alpha\in I}$ of $X$. Then, the two following assertions are equivalent:*

i) $\partial(g_{\alpha\beta}) = 0$.

ii) $[\tilde{\partial}(g_{\alpha\beta})] = 0 \in H^1(X, \Omega)$.

**Proof.** By definition (see Chapt. VI; (1.24), (1.25)), one has for the given 1-cocycle (7.10);

$$\tilde{\partial}(g_{\alpha\beta}) = g_{\alpha\beta}^{-1}\partial(g_{\alpha\beta}). \tag{7.11}$$

Therefore, it suffices only to prove that *ii) implies i), as well:* Thus, consider the following *short exact sequence of* (abelian) *group sheaves;*

$$1 \longrightarrow \ker\tilde{\partial} \longrightarrow \mathcal{A}^{\cdot} \longrightarrow \tilde{\partial}(\mathcal{A}^{\cdot}) \longrightarrow 0. \tag{7.12}$$

However, according to our hypothesis, the above *formulae (7.6) and (7.8) hold true,* so that (7.12) yields now the *short exact sequence;*

$$1 \longrightarrow (\ker\partial) \cap \mathcal{A}^{\cdot} \longrightarrow \mathcal{A}^{\cdot} \longrightarrow \partial(\mathcal{A}) \longrightarrow 0. \tag{7.13}$$

Hence, by looking now at the corresponding *long exact sequence in cohomology* (see e.g. Chapt. III; Theorem 2.1, prop. 3)), one has

$$\cdots \longrightarrow H^1(X, \ker\partial \cap \mathcal{A}^{\cdot}) \longrightarrow H^1(X, \mathcal{A}^{\cdot}) \xrightarrow{(1/2\pi i)\tilde{\partial}} H^1(X, \partial\mathcal{A}) \longrightarrow \cdots \tag{7.14}$$

Therefore, *if cond. ii) is valid,* then one has;

$$\left(\frac{1}{2\pi i}\tilde{\partial}\right)([(g_{\alpha\beta})]) = \frac{1}{2\pi i}\left[\tilde{\partial}(g_{\alpha\beta})\right] = 0, \tag{7.15}$$

that is, one obtains, in view of (7.14),

$$\begin{aligned}[(g_{\alpha\beta})] &\in \ker(H^1(X, \mathcal{A}^{\cdot}) \xrightarrow{(1/2\pi i)\tilde{\partial}} H^1(X, \partial\mathcal{A})) \\ &= \operatorname{im}(H^1(X, \ker\partial \cap \mathcal{A}^{\cdot}) \longrightarrow H^1(X, \mathcal{A}^{\cdot})).\end{aligned} \tag{7.16}$$

In other words, *the* given *1-cocycle (7.10) should take values from* $\ker\partial \cap \mathcal{A}^{\cdot}$; viz. one has

$$(g_{\alpha\beta}) \in Z^1(\mathcal{U}, \ker\partial \cap \mathcal{A}^{\cdot}) \subsetneq Z^1(\mathcal{U}, \ker\partial), \tag{7.17}$$

so that $\partial(g_{\alpha\beta}) = 0$, as desired, and this terminates the proof. ∎

Stated otherwise, Selesnick (ibid.) says that;

(7.17′) *the logarithmic derivative of a coordinate 1-cocycle of a line sheaf* (on a space $X$, as in Lemma 7.2) *can be a coboundary* (if, and) *only if, the line sheaf in question is $\partial$-flat.*

We now come to the promised result, *relating $c_{\mathcal{A}}(\mathcal{L})$ with $\partial$-flatness.* That is, one has the following.

**Theorem 7.1.** *Suppose that we have a sheaf exponential space $X$* (cf. (7.1)), *and let $\mathcal{L}$ be a given line sheaf on $X$. Moreover, consider the following assertions:*

*1) $\mathcal{L}$ is $\partial$-flat; i.e., one has*

$$\partial(g_{\alpha\beta}) = 0, \tag{7.18}$$

*for some coordinate 1-cocycle of $\mathcal{L}$, $(g_{\alpha\beta}) \in Z^1(\mathcal{U}, \mathcal{A}^{\cdot})$.*

*2) One has;*

$$c_{\mathcal{A}}(\mathcal{L}) = 0. \tag{7.19}$$

*Then, 1) implies 2).*

*In particular, assume, in addition, that the following condition holds true;*

$$H^1(X, \mathcal{A}) = H^2(X, \mathcal{A}) = 0. \tag{7.20}$$

*Then, 2) implies 1), as well, so that the two above conditions are then equivalent.*

**Note 7.1.**– The above condition (7.20) has already been considered in the preceding (see Chapt. V; (6.33)), to assure, for instance, within the present abstract framework, the so-called *Chern isomorphism $\delta$* (ibid.; (6.34), (6.35)). Thus, we refer to the relevant comments there, for particular instances, where the said condition is in force.

**Proof of Theorem 7.1.** The first part of the assertion is certainly clear, in view of the very definition of $c_{\mathcal{A}}(\mathcal{L})$ (see (7.5)) and of the above Lemma 7.2 (i) implies ii); note that $\gamma$ in (7.3) is a **C**-*linear map* between the **C**-*vector spaces* concerned). That is, one has

$$\begin{aligned} c_{\mathcal{A}}(\mathcal{L}) &:= \gamma\left(\tfrac{1}{2\pi i}\tilde{\partial}([(g_{\alpha\beta})])\right) = \tfrac{1}{2\pi i}\gamma\left(\tilde{\partial}([(g_{\alpha\beta})])\right) \\ &= \tfrac{1}{2\pi i}\gamma([\tilde{\partial}(g_{\alpha\beta})]) \in H^2(X, \ker\partial), \end{aligned} \tag{7.21}$$

which thus justifies our argument.

On the other hand, based on the *short exact sequence* at the bottom row of (7.2), one has the following *long exact sequence in (sheaf) cohomology* (cf. e.g. Chapt. III; Theorem 3.1, prop. 3))

$$(7.22)\quad \cdots \longrightarrow H^1(X,\mathcal{A}) \longrightarrow H^1(X,\partial\mathcal{A}) \longrightarrow H^2(X,\ker\partial) \longrightarrow H^2(X,\mathcal{A}) \longrightarrow \cdots$$

Consequently, *if* the supplementary *cond. (7.20) is valid,* then one gets from (7.22) the next *exact sequence* (see also (7.3));

$$(7.23)\qquad 0 \longrightarrow H^1(X,\partial\mathcal{A}) \xrightarrow{\gamma} H^2(X,\ker\partial) \longrightarrow 0.$$

Thus, the $\mathbf{C}$-linear map $\gamma$ *becomes,* in fact, *a* $\mathbf{C}$*-vector space isomorphism;* therefore, *if (7.19) is valid,* then *one has* (cf. also (7.21))

$$(7.24)\qquad c_{\mathcal{A}}(\mathcal{L}) = \frac{1}{2\pi i}\gamma\left(\left[\tilde{\partial}(g_{\alpha\beta})\right]\right) = 0,$$

so that, in view of the *isomorphism* $\gamma$, one obtains

$$(7.25)\qquad \left[\tilde{\partial}(g_{\alpha\beta})\right] = 0.$$

Hence (Lemma 7.2), *one has* $\partial(g_{\alpha\beta}) = 0$ *too,* viz. *2) implies 1), as well* (under (7.20)!), and the proof is complete. ∎

Now, due to its significance for the sequel, we reset the second part of the previous theorem into the form of the following corollary. (We come to it again in the next Chapter VIII). So we have.

**Corollary 7.1.** *Let $X$ be a given sheaf exponential space, in such a manner that the following condition holds true;*

$$(7.26)\qquad H^1(X,\mathcal{A}) = H^2(X,\mathcal{A}) = 0.$$

*Moreover, let $\mathcal{L}$ be a line sheaf on $X$. Then, $\mathcal{L}$ is $\partial$-flat if, and only if, one has $c_{\mathcal{A}}(\mathcal{L}) = 0$.* ∎

The previous data will often be applied in the sequel too. So, for convenience, we still set the following.

**Definition 7.2.** A topological space satisfying the hypothesis of the previous Corollary 7.1, is called a *strict sheaf exponential space.*

Yet, as we shall see later on, under further natural assumptions for the space $X$, the preceding conditions, as given e.g. by Corollary 7.1, are still equivalent with the line sheaf $\mathcal{L}$ having a *flat $\mathcal{A}$-connection* (see thus the next Chapter VIII).

**7.(a). Complex Chern class of a line sheaf (contn'd).**– By anticipating later applications, it is still appropriate to give here the relation of the above cohomology class $c_{\mathcal{A}}(\mathcal{L})$, that is associated with any given line sheaf $\mathcal{L}$ on $X$, with the usual *complex Chern class of* $\mathcal{L}$; the latter class has been already considered in the preceding (cf. Chapt. V; (6.17)).

Thus, suppose that we are given a *line sheaf*

$$\mathcal{L} \equiv [\mathcal{L}] \equiv [(g_{\alpha\beta})] \in H^1(X, \mathcal{A}^{\cdot}) = \Phi^1_{\mathcal{A}}(X), \tag{7.27}$$

as before, such that

$$(g_{\alpha\beta}) \in Z^1(\mathcal{U}, \mathcal{A}^{\cdot}) \tag{7.28}$$

is a *coordinate 1-cocycle of* $\mathcal{L}$ (cf. Chapt. V; (6.11), (2.59)). Furthermore, consider the following *(canonical) commutative triangle;*

$$\begin{array}{ccc} \mathbf{Z} & \xrightarrow{\;j\;} & \mathbf{C} \\ & {\scriptstyle\zeta}\searrow \quad \swarrow{\scriptstyle\varepsilon} & \\ & \ker \partial, & \end{array} \tag{7.29}$$

where $\mathbf{Z}$ and $\mathbf{C}$ are viewed, of course, as the corresponding *constant sheaves* (of *abelian groups*) on $X$, the latter space being endowed, as usual, with a *differential triad* $(\mathcal{A}, \partial, \Omega)$. Notice that (7.29) is a consequence of the fact that

$$\mathbf{C} \underset{\varepsilon}{\overset{\subseteq}{\rightarrow}} \ker \partial \subseteq \mathcal{A} \tag{7.30}$$

(cf. Chapt. VI; Lemma 1.1).

In particular, suppose now that $X$ is a given *sheaf exponential space* (see (7.1), so that the commutative diagram (7.3) *is in force.* Therefore, based further on Chapt. III; (3.22), (3.23), along with (7.29) above, one obtains the following *commutative diagram,* supplementing (7.3);

(7.31)
$$
\begin{array}{ccccc}
 & & \overset{c}{\dashrightarrow} & & H^2(X,\mathbf{C}) \\
 & & & \nearrow j & \\
H^1(X,\mathcal{A}^{\cdot}) & \xrightarrow{\quad\delta\quad} & H^2(X,\mathbf{Z}) & & \\
\Big\downarrow \frac{1}{2\pi i}\tilde{\partial} & \overset{c_{\mathcal{A}}}{\searrow} & \Big\downarrow \zeta & \swarrow \varepsilon & \\
H^1(X,\partial\mathcal{A}) & \xrightarrow{\quad\gamma\quad} & H^2(X,\ker\partial). & &
\end{array}
$$

Thus, one now defines the *complex Chern class* of the given line sheaf $\mathcal{L}$ (see also (7.27)), by the relation (cf. (7.31));

(7.32) $$c(\mathcal{L}) := j(\delta(\mathcal{L})).$$

That is, one thus considers the *"complexification" of the integral Chern class of $\mathcal{L}$*, $\delta(\mathcal{L})$ (cf. (7.31) and Chapt. V; (6.12)). So we still call, occasionally,

(7.33) $$z(\mathcal{L}) := \delta(\mathcal{L}) \in H^2(X,\mathbf{Z}),$$

the *(integral) Chern class of $\mathcal{L}$*. (In this regard, see also Chapter IX in the sequel). Yet, based on (7.31) and (7.5), *one obtains;*

(7.34) $$c_{\mathcal{A}}(\mathcal{L}) = \varepsilon(c(\mathcal{L})) = \frac{1}{2\pi i}\gamma([\tilde{\partial}(g_{\alpha\beta})]).$$

In particular, *if (7.20) holds true*, then (7.34) yields;

(7.35) $$2\pi i \cdot c_{\mathcal{A}}(\mathcal{L}) = [\tilde{\partial}(g_{\alpha\beta})].$$

In this context, our main conclusion is displayed, however, by the following.

**Theorem 7.2.** *Suppose that $X$ is a sheaf exponential space, and let $\mathcal{L}$ be a given line sheaf on it. Then, the condition*

(7.36) $$c(\mathcal{L}) = 0$$

*implies that $\mathcal{L}$ is a flat line sheaf on $X$, so that $\mathcal{L}$ is ∂-flat, as well* (see also (6.12)).

**Proof.** Consider the standard *short exact exponential sequence*

$$0 \longrightarrow \mathbf{Z} \xrightarrow{j} \mathbf{C} \xrightarrow{e} \mathbf{C}^{\cdot} \longrightarrow 1, \tag{7.37}$$

such that

$$e(t) := \exp 2\pi i t, \quad t \in \mathbf{C} \tag{7.38}$$

(see also Chapt. V; (6.18), (6.19)). Hence, by considering further the corresponding to (7.37) *long exact sequence in cohomology,* one obtains

$$\cdots \longrightarrow H^1(X, \mathbf{C}^{\cdot}) \longrightarrow H^2(X, \mathbf{Z}) \xrightarrow{j} H^2(X, \mathbf{C}) \longrightarrow \cdots \tag{7.39}$$

Therefore, supposing that $c(\mathcal{L}) = 0$, one also gets (cf. (7.32))

$$c(\mathcal{L}) = j(\delta(\mathcal{L})) = 0, \tag{7.40}$$

that is (see also (7.39)),

$$\begin{aligned} \delta(\mathcal{L}) &\in \ker(H^2(X, \mathbf{Z}) \xrightarrow{j} H^2(X, \mathbf{C})) \\ &= \operatorname{im}(H^1(X, \mathbf{C}^{\cdot}) \longrightarrow H^2(X, \mathbf{Z})). \end{aligned} \tag{7.41}$$

Consequently, by considering a *coordinate 1-cocycle of $\mathcal{L}$, as in (7.28),* representing $\mathcal{L}$ in $H^1(X, \mathcal{A}^{\cdot})$ (see (7.27)), hence, in turn, in $H^2(X, \mathbf{Z})$, via the *Chern morphism* $\delta$ (cf. (7.31)), one infers from (7.41) that *this 1-cocycle should take values from* $\mathbf{C}^{\cdot}$; that is, one actually has that

$$(g_{\alpha\beta}) \in Z^1(\mathcal{U}, \mathbf{C}^{\cdot}) \underset{\varepsilon}{\hookrightarrow} Z^1(\mathcal{U}, \mathcal{A}^{\cdot}). \tag{7.42}$$

However, *this proves,* in effect, *our assertion,* according to the very definition of a *flat line sheaf* (cf. Chapt. V; (5.24), for $n = 1$), and the proof is complete. ∎

Therefore, based on (6.12) and on Theorem 7.1 (viz. 1) implies 2)), or even just (7.34), one now concludes that;

$$c(\mathcal{L}) = 0 \ \textit{implies} \ c_{\mathcal{A}}(\mathcal{L}) = 0. \tag{7.43}$$

On the other hand, based on the very definitions, it is not hard to see that;

(7.44) *in the* particular *case* that the *cond. (6.35) holds true, all the preceding notions coincide.*

Of course, this happens already in the classical theory; however, see also Chapters X and XI in the sequel. Yet, on all these we shall return, when the

notion of *"curvature"* of a given $\mathcal{A}$-connection will be available. So we still defer the relevant discussion for the next Chapter VIII.

## 8. Symmetric $\mathcal{A}$-connections

We apply below our earlier considerations in Chapt. IV; Section 8, pertaining to *Riemannian $\mathcal{A}$-modules.* Of course, as is the case with our classical model, viz. the *differential geometry of $C^\infty$-manifolds,* the preponderant notion hereby is that one of a *"compatible $\mathcal{A}$-connection"*, in a sense, that will be made precise presently below.

Thus, suppose that we are given an *ordered algebraized space* $(X, \mathcal{A})$ (cf. Chapt. IV; Definition 8.1, together with the comments after it). Moreover, we assume that

$$(\mathcal{A}, \rho) \tag{8.1}$$

is a *Riemannian $\mathcal{A}$-module;* we recall (ibid.; Definition 8.3) that this means that

$$\rho : \mathcal{A} \oplus \mathcal{A} \longrightarrow \mathcal{A} \tag{8.2}$$

is an *$\mathcal{A}$-valued strongly non-degenerate inner product* on $\mathcal{A}$, that is, a *Riemannian $\mathcal{A}$-metric on $\mathcal{A}$*, which thus implies a *similar one on $\mathcal{A}^n$*, denoted by $\overline{\rho}$ (viz. *Whitney sum of $\rho$, $n$-times:* ibid.; (8.37), (8.37′), or even (8.40)).

On the other hand, we know that, under further suitable restrictions on $X$ and $\mathcal{A}$, one can prove the existence of a Riemannian $\mathcal{A}$-metric on every vector sheaf $\mathcal{E}$ on $X$, provided we are given (8.1), as above (ibid.; Theorem 8.3), something that we are going to apply in the sequel. However, for generality's, as well as, for simplicity's sake, we just assume now that

$$(\mathcal{E}, \rho) \tag{8.3}$$

is a given *Riemannian vector sheaf* on $X$ (ibid.; Definition 8.3). Then, as already known (ibid.; Theorem 8.1), given a *local frame of $\mathcal{E}$*,

$$\mathcal{U} = (U_\alpha)_{\alpha \in I}, \tag{8.4}$$

the *(Riemannian) $\mathcal{A}$-metric* $\rho$ on $\mathcal{E}$ is (uniquely) determined by a *0-cocycle*

$$\rho_{\mathcal{U}} \equiv \rho \equiv (\rho_\alpha) \in \prod_\alpha M_n(\mathcal{A})(U_\alpha) = \prod_\alpha M_n(\mathcal{A}(U_\alpha)), \tag{8.5}$$

such that one actually has

$$\rho \equiv (\rho_\alpha) \in Z^0(\mathcal{U}, \mathcal{GL}(n, \mathcal{A})) = \mathrm{GL}(n, \mathcal{A}(X)), \tag{8.6}$$

*with* $n = \operatorname{rk} \mathcal{E} \in \mathbf{N}$; viz., as we say, one considers the *Gram cocycle of the $\mathcal{A}$-metric* $\rho$, associated with $\mathcal{U}$ (Chapt. IV; (8.29), along with the comments, following the end of the proof of Theorem 8.1).

Furthermore, suppose now that we are also given on the *topological space* $X$, as a bove, a *differential triad* $(\mathcal{A}, \partial, \Omega)$, while we still assume the relation

$$\mathcal{E}^* = \Omega \tag{8.7}$$

(see, for instance, (5.3) together with Scholium 5.1 in the preceding for a similar situation). Thus, $\Omega$ ia s *vector sheaf* on $X$ too, of the *same rank, as* $\mathcal{E}$ (cf. (5.4)).

Furthermore, by considering a *local gauge of* $\mathcal{E}$,

$$e^U \equiv \{U; (e_1, \ldots, e_n)\} \tag{8.8}$$

(cf., for example (4.1) in the preceding), one defines the corresponding *"nabla" operator,*

$$\nabla : \mathcal{E}(U) \times \mathcal{E}(U) \longrightarrow \mathcal{E}(U), \tag{8.9}$$

according to (5.35), given the *$\mathcal{A}$-connection $D$ of $\mathcal{E}$*, along with the *validity of (8.7)*. Now, *for any local sections*

$$s = \sum_{i=1}^n s_i e_i \quad and \quad t = \sum_{j=1}^n t_j e_j \quad in\ \mathcal{E}(U), \tag{8.10}$$

*with $U$, as in (8.8),* one defines the following *Lie product of (local) sections;*

$$[s, t] := \sum_{i=1}^n \left( \sum_{j=1}^n (t_j \partial(s_i) - s_j \partial(t_i))(e_j) \right) e_i \in \mathcal{E}(U). \tag{8.11}$$

Thus, *equivalently,* by looking at the *$i$-th component of* $[s, t] \in \mathcal{E}(U)$, *relative to* $e^U$ (see (8.8)), *one sets;*

$$[s,t]_i := \sum_{j=1}^{n}(t_j\partial(s_i) - s_j\partial(t_i))(e_j), \quad 1 \leqq i \leqq n = \operatorname{rk}\mathcal{E}. \tag{8.12}$$

So, as a result of (8.12), by further taking the corresponding *(local) section basis of* $\mathcal{E}(U)$ (cf. (1.3), (1.4)),

$$(e_i)_{1\leqq i\leqq n} \subseteq \mathcal{E}(U) \subseteq \mathcal{A}(U)^n, \tag{8.13}$$

one obtains (cf. also Chapt. VI; (1.6), as well as, (6.26) above);

$$[e_i, e_j] = 0, \quad 1 \leqq i, j \leqq n = \operatorname{rk}\mathcal{E}. \tag{8.14}$$

Thus, we come now to the following basic notion for the ensuing discussion, as given by the next.

**Definition 8.1.** Suppose we have a Riemannian vector sheaf $(\mathcal{E}, \rho)$ on a topological space $X$, the latter being endowed with a differential triad $(\mathcal{A}, \partial, \Omega)$ such that (8.7) be satisfied. Then, a given *$\mathcal{A}$-connection $D$ of $\mathcal{E}$* is said to be *compatible with the Riemannian $\mathcal{A}$-metric $\rho$ of $\mathcal{E}$*, or even, for short, a *Riemannian $\mathcal{A}$-connection* of $\mathcal{E}$, whenever the two following conditions are fulfilled:

1) For any (local) sections $s, t \in \mathcal{E}(U)$, with $U$, as in (8.8), one has (*symmetry of $D$*);

$$\nabla(s,t) - \nabla(t,s) = [s,t]. \tag{8.15}$$

2) For any $s, t, z \in \mathcal{E}(U)$, as before, the following relation holds true (*Ricci identity*);

$$\partial(\rho(s,t))(z) = \rho(\nabla(z,s),t) + \rho(s,\nabla(z,t)). \tag{8.16}$$

Now, by applying the local section basis (8.13) of $\mathcal{E}(U)$, one first obtains, in view of (8.15) and (8.14), the following *symmetry condition of a Riemannian $\mathcal{A}$-connection $D$* (*equivalent*, of course, *to (8.14)*);

$$\nabla(e_i, e_j) = \nabla(e_j, e_i), \quad 1 \leqq i, j \leqq n. \tag{8.17}$$

Furthermore, one has the following *equivalent formulation of Ricci's identity;*

$$\partial(\rho_{ij})(e_k) = \rho(\nabla(e_k, e_i), e_j) + \rho(e_i, \nabla(e_k, e_j)), \tag{8.18}$$

*with* $1 \leqq i, j, k \leqq n$. Here

(8.19) $$\rho_U \equiv (\rho_{ij}) \in M_n(\mathcal{A})(U) = M_n(\mathcal{A}(U))$$

stands for the local form of (the restriction of) $\rho$ on $U \subseteq X$, such that

(8.20) $$\rho_{ij} := \rho(e_i, e_j) \in \mathcal{A}(U), \quad 1 \leqq i, j \leqq n,$$

(see also Chapt. IV; (8.34), (8.35), for $U \equiv U_\alpha$ $\alpha \in I$, as in (8.4)). Yet, in view of (8.6) (cf. also e.g. Chapt. IV; (8.26.4)), one has

(8.21) $$\rho_U \equiv (\rho_{ij}) \in \mathrm{GL}(n, \mathcal{A}(U)) = \mathcal{GL}(n, \mathcal{A})(U).$$

Now, our assertion about (8.18) follows from the very properties of $\nabla$ (see (5.34) - (5.38)), from those of $\partial$ (cf. Chapt. VI; Definition 1.1) and the *$\mathcal{A}$-bilinearity of $\rho$* (Chapt. IV; Definition 8.2), as well as, from our hypothesis for (8.13) (see (1.6.1)). ∎

On the other hand, as already seen in Section 5 in the preceding, the previous data supply also the appropriate framework for the definition of the *(local) Christoffel sections* of the given *$\mathcal{A}$-connection* $D$ of $\mathcal{E}$, which are associated with the *local gauge* $e^U$ *of* $\mathcal{E}$, as in (8.8). Thus, for convenience, we recall that one can define the following local sections of $\mathcal{A}$ over $U$ (cf. (5.49));

(8.22) $$\Gamma^k_{ij} := \nabla(e_i, e_j)_k \in \mathcal{A}(U), \quad 1 \leqq i, j, k \leqq n,$$

that is, *equivalently,* one has (see also (1.6))

(8.23) $$\nabla(e_i, e_j) = \sum_{k=1}^{n} \Gamma^k_{ij} e_k, \quad 1 \leqq i, j \leqq n.$$

Therefore, by virtue of (8.17), one has, here too, the classical *symmetry condition* of a Riemannian $\mathcal{A}$-connection, *in terms of the* corresponding *Christoffel sections;* viz. one obtains

(8.24) $$\Gamma^k_{ij} = \Gamma^k_{ji}, \quad 1 \leqq i, j, k \leqq n,$$

which thus, in view of (8.22), is an *equivalent expression of (8.17).* Furthermore, as a consequence of (8.23) (see also (8.20)), one first obtains;

(8.25) $$\begin{aligned}\rho(\nabla(e_i, e_j), e_k) = \rho(\textstyle\sum_{l=1}^{n} \Gamma^l_{ij} e_l, e_k) &= \textstyle\sum_l \Gamma^l_{ij} \rho(e_l, e_k) \\ &= \textstyle\sum_l \Gamma^l_{ij} \rho_{lk}.\end{aligned}$$

Thus, based further on (8.18), one can now compute the *Christoffel sections, in terms of the Riemannian $\mathcal{A}$-metric $\rho$*:

Namely, by symmetry, one first obtains the two other analogous relations to (8.18), so that by further applying the standard procedure, and due also to (8.17) and the *symmetry of* $\rho$ (cf. Chapt. IV; (8.13) and (8.22.2)), one finally gets, for the case under consideration, the *1st Christoffel identity;*

$$(8.26) \qquad \rho(\nabla(e_i, e_j), e_k) = \frac{1}{2}(\partial(\rho_{jk})(e_i) + \partial(\rho_{ik})(e_j) - \partial(\rho_{ij})(e_k)).$$

Consequently, based on (8.25) and (8.26), one now obtains

$$(8.27) \qquad \sum_{l=1}^{n} \Gamma^l_{ij}\rho_{lk} = \frac{1}{2}A,$$

with the obvious meaning of "$A$", so that, by still applying the standard procedure, one finds, within the present framework the *2nd Christoffel identity,* as actually promised at the beginning; that is, one has

$$(8.28) \qquad \Gamma^l_{ij} = \frac{1}{2}\sum_{k=1}^{n}(\partial(\rho_{jk})(e_i) + \partial(\rho_{ik})(e_j) - \partial(\rho_{ij})(e_k))\rho^{kl},$$

*for any* $1 \leqq i, j, l \leqq n$, where, as usual (from the classical theory!), we set

$$(8.29) \qquad (\rho^{kl}) := (\rho_{lk})^{-1} \in \mathrm{GL}(n, \mathcal{A}(U))$$

(see e.g. (8.21)).

Now, as a consequence of the previous discussion, we thus arrived at the following basic conclusion, as given by the next.

**Lemma 8.1.** *Consider a topological space* $X$, *endowed with a differential triad* $(\mathcal{A}, \partial, \Omega)$, *such that* $(X, \mathcal{A})$ *is an ordered algebraized space. Moreover, let* $(\mathcal{E}, \rho)$ *be a Riemannian vector sheaf on* $X$, *with* $\operatorname{rk}\mathcal{E} = n \in \mathbf{N}$, *and* $e^U \equiv \{U; (e_1, \ldots, e_n)\}$ *a local gauge of* $\mathcal{E}$ *over an open set* $U \subseteq X$, *while we also assume that*

$$(8.30) \qquad \mathcal{E}^* = \Omega.$$

*Then, the* $n^3$ *(local) sections of* $\mathcal{A}$ *over* $U$,

$$(8.31) \qquad \Gamma^l_{ij} \in \mathcal{A}(U), \quad 1 \leqq i, j, l \leqq n,$$

*as they are given by (8.28), may be considered as the (local) Christoffel sections of a unique Riemannian* $\mathcal{A}|_U$*-connection of* $\mathcal{E}|_U$.

**Proof.** Indeed, according to Theorem 5.3 (see also (5.78)), *such* $n^3$ *local*

*sections of $\mathcal{A}$ over $U$ provide a uniquely defined $\mathcal{A}|_U$-connection of $\mathcal{E}|_U$.* Thus, we further prove that *the latter is, in effect, a Riemannian $\mathcal{A}|_U$-connection.*

So, concerning the *symmetry of the $\mathcal{A}|_U$-connection* at issue, we remark that, since the given Riemannian $\mathcal{A}$-metric $\rho$ of $\mathcal{E}$ is, by hypothesis, symmetric, one has, by definition (cf. (8.20))

$$\rho_{ij} = \rho_{ji}, \quad 1 \leqq i, j \leqq n = \operatorname{rk} \mathcal{E}, \tag{8.32}$$

or even in *matrix form* (cf. (8.21)) the relation

$$\rho = {}^t\rho \tag{8.33}$$

(see Chapt. IV; (8.26.1), (8.26.3)). Therefore, due to (8.28), one further obtains

$$\Gamma^l_{ij} = \Gamma^l_{ji}, \quad 1 \leqq i, j, l \leqq n. \tag{8.34}$$

However, the last relations are, in fact, *equivalent with the symmetry of the* (local) *$\mathcal{A}|_U$-connection of $\mathcal{E}|_U$*, defined by the given Christoffel sections (8.31), as this has already been explained above, in the case of the relation (8.24).

Thus, one has still to prove the *Ricci identity,* in its equivalent formulation, for instance, *via (8.18);* indeed, this will be the result of the subsequent complementary discussion:

So to start with, assume that we are given, of course, the *framework of Lemma 8.1,* and consider the *Ricci identity* (cf. (8.16)), in its *equivalent version, as given by (8.18).* Thus, by applying now (8.23) to (8.18), and by taking the *$\mathcal{A}$-bilinearity of the $\mathcal{A}$-metric $\rho$* into account (see Chapt. IV; (8.26.2)), one gets

$$\begin{aligned} \partial(\rho_{ij})(e_k) &= \rho(\nabla(e_k, e_i), e_j) + \rho(e_i, \nabla(e_k, e_j)) \\ &= \rho(\textstyle\sum_l \Gamma^l_{ki} e_l, e_j) + \rho(e_i, \textstyle\sum_l \Gamma^l_{kj} e_l) \\ &= \textstyle\sum_l (\Gamma^l_{ki} \rho(e_l, e_j) + \Gamma^l_{kj} \rho(e_i, e_l)). \end{aligned} \tag{8.35}$$

Therefore, *one* finally *obtains;*

$$\partial(\rho_{ij})(e_k) = \sum_{l=1}^{n} (\Gamma^l_{ki} \rho_{lj} + \Gamma^l_{kj} \rho_{li}), \quad 1 \leqq i, j, k \leqq n. \tag{8.36}$$

(In the last term of (8.36) we made use of (8.32)). Now, it is easy to check that *the above argument can* actually *be reversed,* so that one concludes, in

fact, that:

(8.37) *the Ricci identity,* as given by (8.18), *is equivalent with the rel. (8.36).*

Yet, in view of (8.28) and (8.36), one further obtains, as in the classical case, that;

(8.37′) $\partial(\rho_{ij}) = 0$, *if, and only if,* $\Gamma_{ij}^k = 0$,

with $1 \leqq k \leqq n$, and $1 \leqq i, j \leqq n$.

Thus, based further on the last conclusion, we can now proceed, as already promised, to the

**End of the proof of Lemma 8.1.** Based on (8.28) and (8.29), and by making the obvious substitutions in (8.36), one can prove, following thus the analogous to our case classical argument, that *the 2nd Christoffel identity, viz. (8.28), entails (8.36).* However, according to (8.37), the rel. (8.36) is equivalent to Ricci's identity, as the latter is given, equivalently, by (8.18), which thus terminates the proof of the lemma. (The last argument, in conjunction with our considerations in the foregoing, leading to (8.28), prove also, incidentally, that, within the context e.g. of Definition 8.1, the rels. *(8.28) and (8.18) are* still *equivalent*). ∎

We proceed now to give further *equivalent expressions of the Ricci identity;* in particular, we wish to have (8.18), in terms of the *local $\mathcal{A}$-connection matrix-valued forms* of a given *Riemannian $\mathcal{A}$-connection,* as before. As we shall see these alternatives of (8.18) will be of much use right in the sequel of our discussion:

Thus, assuming again that we have the *context of Lemma 8.1,* we first recall, for convenience, that, by virtue of the given *Christoffel sections* (8.28), one can further define the following (local) sections of $\Omega = \mathcal{E}^*$ (see (8.30)) over $U$, by the relations;

(8.38) $$\omega_{ij}(e_k) := \Gamma_{ij}^k, \quad 1 \leqq i, j, k \leqq n.$$

That is, *equivalently,* one has

(8.39) $$\omega_{ij} = \sum_{k=1}^{n} \Gamma_{ij}^k e_k^* \in \Omega(U), \quad 1 \leqq i, j \leqq n,$$

such that

$$(e^U)^* \equiv \{U; (e_i^*)_{1 \leqq i \leqq n}\} \tag{8.40}$$

stands for the *dual local gauge of* $e^U$, hence, a *local gauge of* (the *vector sheaf*) $\mathcal{E}^* = \Omega$ over the *open* $U \subseteq X$. (In this context, cf. (5.75) and (5.76) in the foregoing).

Accordingly, based now on (8.34) and (8.38), *one obtains;*

$$\omega_{ij} = \omega_{ji} \in \Omega(U), \quad 1 \leqq i, j \leqq n, \tag{8.41}$$

or in *matrix notation,*

$$\omega = {}^t\omega \in M_n(\Omega(U)) = M_n(\Omega)(U). \tag{8.42}$$

On the other hand, as already noticed (cf. (8.37)), *the Ricci identity* (cf. (8.16)), as (equivalently) given by (8.18), *is* further *equivalent with* (8.36). Thus, applying now (8.38) to (8.36), one obtains (cf. also (8.32));

$$\begin{aligned} \partial(\rho_{ij})(e_k) &= \textstyle\sum_l(\Gamma^l_{ki}\rho_{lj} + \Gamma^l_{kj}\rho_{li}) = \sum_l(\omega_{ki}(e_l)\rho_{lj} + \omega_{kj}(e_l)\rho_{li}) \\ &= \textstyle\sum_l(\rho_{il}\omega_{kj} + \rho_{jl}\omega_{ki})(e_l), \end{aligned} \tag{8.43}$$

*with* $1 \leqq k \leqq n$, so that we have

$$\partial(\rho_{ij}) = \sum_{k=1}^{n}(\rho_{ik}\omega_{kj} + \rho_{jk}\omega_{ki}), \quad 1 \leqq i, j \leqq n. \tag{8.44}$$

Therefore, by employing *matrix notation,* and in view also of (8.32) and (8.42), one has;

$$\partial(\rho) = \rho \cdot \omega + {}^t\rho \cdot {}^t\omega = \rho\omega + \rho\omega = (\rho + \rho) \cdot \omega = 2\rho\omega, \tag{8.45}$$

so that *one* finally *obtains* (cf. also (8.21))

$$\omega = \frac{1}{2}\rho^{-1}\partial(\rho) \equiv \frac{1}{2}\tilde{\partial}(\rho). \tag{8.46}$$

By extending the classical terminology, we call (8.46) the *Levi-Civita identity.*

In this context, it is now easy to see that the previous argument, that led to (8.46) *can,* in fact, *be reversed,* under the further assumption, of course, that (8.41), or *equivalently, (8.42), holds true.* Therefore, one thus concludes that;

(8.47) supposing that (8.41), or equivalently (8.42), holds true, *the Ricci and the Levi-Civita identities,* (8.18) and (8.46), respectively, *are,* in effect, *equivalent.* Indeed, under the said assumption (but, see also (8.48) below), *the rels. (8.18), (8.44) and (8.46) are equivalent.*

The above conclusion will be applied just below, however, in a strengthened form. Thus, we still remark that:

(8.48) as already noted by (8.47), *the Levi-Civita identity (8.46) implies the Ricci identity,* under the assumption, however, of (8.41). Now, the point here is that, as easily seen, *this* hypothesis about (8.41), *is* actually *redundant,* due to (8.46) itself.

As we said, it is in the form of (8.48) that we are going to apply the preceding. Thus, although a more general case will be treated right in the next Section 9, we do give here an application of the above discussion, pertaining to the *Fundamental Lemma* of Riemannian geometry, as this classical result appears within the present abstract setting. So one has the next.

**Theorem 8.1.** (Fundamental Lemma of Riemannian $\mathcal{A}$-connections). *Suppose we are given an ordered algebraized space* $(X, \mathcal{A})$, *with* $X$ *a paracompact (Hausdorff) space, such that* $(\mathcal{A}, \overline{\rho})$ *is a strictly positive fine* $\mathcal{A}$*-module, with* $\overline{\rho}$ *a Riemannian* $\mathcal{A}$*-metric. Furthermore, assume that* $X$ *carries a differential triad* $(\mathcal{A}, \partial, \Omega)$, *while we still adopt that* $\Omega$ *is a vector sheaf on* $X$, *with* $\mathrm{rk}\,\Omega = n \in \mathbf{N}$. *Then, the dual (vector sheaf) of* $\Omega$, *say,*

$$\mathcal{E} \equiv \Omega^* := \mathcal{H}om_{\mathcal{A}}(\Omega, \mathcal{A}), \tag{8.49}$$

*is (canonically) endowed with a unique Riemannian* $\mathcal{A}$*-connection.*

**Proof.** Based on our hypothesis, we first remark (cf. Chapt. IV; Theorem 8.3) that the *vector sheaf* $\mathcal{E} = \Omega^*$ (see (8.49)) can be endowed with a *Riemannian* $\mathcal{A}$*-metric* $\rho$, so that the pair

$$(\mathcal{E}, \rho) \tag{8.50}$$

constitutes a *Riemannian vector sheaf* on $X$, such that

$$\mathcal{E}^* = \Omega \tag{8.51}$$

(cf. (8.49) above and Chapt. IV; Theorem 5.1), that is, (8.30) *is* thus *fulfilled.*

Now, suppose that

$$\mathcal{U} = (U_\alpha)_{\alpha \in I} \tag{8.52}$$

is a *local frame* of our vector sheaf $\mathcal{E}$, as above; therefore, for each *local gauge* $e^{(\alpha)}$ *of* $\mathcal{E}$ *over* $U_\alpha \in \mathcal{U}$, $\alpha \in I$ (see, for instance, (2.31), (2.32)), one considers the respective *Riemannian* $\mathcal{A}|_{U_\alpha}$*-metric of* $\mathcal{E}|_{U_\alpha}$, *restriction of* $\rho$ in (8.50) *to* $U_\alpha$,

$$\rho_\alpha := \rho|_{U_\alpha} \equiv (\rho_{ij}^{(\alpha)}) \in \mathrm{GL}(n, \mathcal{A}(U_\alpha)) = \mathcal{GL}(n, \mathcal{A})(U_\alpha) \tag{8.53}$$

(cf. (8.6), (8.20) and (8.21)), hence, the corresponding *Christoffel sections in* $\mathcal{A}(U_\alpha)$, as they are given now, *for each index* $\alpha \in I$, by an analogous formula to (8.28). Consequently, in view of Lemma 8.1, one infers that *each one of* the vector sheaves $\mathcal{E}|_{U_\alpha}$ over $U_\alpha$, $\alpha \in I$, *is endowed with a uniquely defined Riemannian* $\mathcal{A}|_{U_\alpha}$*-connection,* say $\overline{D}_\alpha$, whose *local* $\mathcal{A}|_{U_\alpha}$*-connection matrix* is given (cf. (8.46)) by the relation

$$\overline{\omega}^{(\alpha)} = \frac{1}{2}\tilde{\partial}(\rho_\alpha), \quad \alpha \in I. \tag{8.54}$$

On the other hand, by hypothesis, $\mathcal{A}$ is a *fine* $\mathbf{C}$*-algebra sheaf* over the *paracompact* (Hausdorff) *space* $X$, so that the given $\mathcal{A}$*-module* $\mathcal{E} = \Omega^*$ *is a fine sheaf* on $X$ too (cf., for example, Chapt. III; (8.56)). Therefore, the standard *Levi-Civita 0-cochain,* which is (canonically) associated with the given local frame $\mathcal{U}$ of our vector sheaf $\mathcal{E}$, as before (cf. (8.52) and Chapt. VI; (8.32.1)),

$$(D_\alpha) \in C^0(\mathcal{U}, \mathcal{H}om_{\mathbf{C}}(\mathcal{E}, \Omega(\mathcal{E}))), \tag{8.55}$$

yields now an $\mathcal{A}$*-connection* $D$ *of* $\mathcal{E}$ (see Chapt. VI; Lemma 16.1. In this context, it is certainly clear that (8.52) may be taken to be a *locally finite local frame of* $\mathcal{E}$, in view of our hypothesis for $X$; see e.g. Chapt. VI; (16.16)). Thus (see also Chapt. VI; Theorem 7.1), one can define now, *for every* $\alpha \in I$, a (local) $\mathcal{A}|_{U_\alpha}$*-connection of* $\mathcal{E}|_{U_\alpha}$, by the relation,

$$\widetilde{D}_\alpha := D_\alpha + \overline{\omega}^{(\alpha)}; \tag{8.56}$$

that is, *equivalently,* one sets (cf. also (8.54))

$$\widetilde{D} := D + \overline{\omega} = D + \frac{1}{2}\tilde{\partial}(\rho), \tag{8.57}$$

such that *one actually has*

(8.58) $$\overline{\omega} \equiv \frac{1}{2}\tilde{\partial}(\rho) \in \Omega(\mathcal{E}nd\,\mathcal{E})(X),$$

which thus, in view of the cited result of Chapter VI, yields that $\widetilde{D}$, as given by (8.57), is a well-defined *$\mathcal{A}$-connection of $\mathcal{E}$*. Now, concerning our last claim for $\overline{\omega}$, one obtains (see also (8.6) above, together with Chapt. III; (4.55), (4.28) and Chapt. VI; (9.4));

(8.59) $$\begin{aligned}\tilde{\partial}(\rho) &\equiv \tilde{\partial}((\rho_{ij})) := (\tilde{\partial}(\rho_{ij})) \in M_n(\Omega(X)) = M_n(\Omega)(X)\\ &= Z^0(\mathcal{U}, M_n(\Omega)) \subseteq \textstyle\prod_\alpha M_n(\Omega)(U_\alpha) = \prod_\alpha \Omega(\mathcal{E}nd\,\mathcal{E})(U_\alpha),\end{aligned}$$

so that, in particular, *one has*

(8.60) $$\tilde{\partial}(\rho) \in Z^0(\mathcal{U}, \Omega(\mathcal{E}nd\,\mathcal{E})) = \Omega(\mathcal{E}nd\,\mathcal{E})(X),$$

which thus justifies (8.58). ▮ Now, we prove further that $\widetilde{D}$, as defined by (8.57), *is*, in fact, *a Riemannian $\mathcal{A}$-connection of $\mathcal{E}$*:

Indeed, as already known, *the Levi-Civita $\mathcal{A}$-connection* $D$, as defined above, *is integrable* (see Chapt. VI; (11.24)); hence, one still obtains that (cf. (6.28));

(8.61) $$\omega^{(\alpha)} = 0, \quad \alpha \in I.$$

Accordingly, in view of (8.56) and (8.54), one has

(8.62) $$\tilde{\omega}^{(\alpha)} = \overline{\omega}^{(\alpha)} = \frac{1}{2}\tilde{\partial}(\rho_\alpha), \quad \alpha \in I,$$

the first item of (8.62) standing, of course, for the *local $\mathcal{A}$-connection matrix of the $\mathcal{A}$-connection* $\widetilde{D}$, as the latter is given by (8.57). However, as already remarked (cf. (8.48) above), the previous rel. (8.62) *provides the symmetry of* $\widetilde{D}$, *as well as, the Ricci identity* (cf. (8.47)). Yet, by virtue of (8.51), the corresponding *Christoffel sections of* $\widetilde{D}$ are given (locally) by the analogous relations to (8.28) above (see also (8.38)), yielding thus the *uniqueness of* $\widetilde{D}$ (cf. also Lemma 8.1), which now completes the proof of the theorem. ▮

**Scholium 8.1.**- It is profitable, in anticipation of a later application, to notice at this point the use, we made above, of the *extra hypothesis* (8.49) or, *equivalently,* (8.51) (see Scholium 5.1 in the foregoing); so the latter condition has been employed above, in order *to get* the form (8.28) of the Christoffel sections, hence (Lemma 8.1), *the uniqueness of the* (Riemannian) *$\mathcal{A}$-connection* $\widetilde{D}$, as in (8.57). Indeed,

as we shall see in the next Section 9, the same proof, as above, works for the general case too, viz. if *no extra assumption, like (8.51), is made* (cf. Theorem 9.1 in the sequel). However, this *at the cost of the uniqueness of* $\widetilde{D}$, as before, while we also have to, appropriately, reformulate Definition 8.1 (see thus Definition 9.1 below).

Now, before we end this Section, we *still* consider *another equivalent formulation of the Levi-Civita identity,* viz. of (8.46):

Thus, suppose that we are given, as usual, a *differential triad*

$$(\mathcal{A}, \partial, \Omega) \tag{8.63}$$

on a *topological space* $X$, in such a manner that

$$(X, \mathcal{A}) \tag{8.64}$$

is, in particular, an *ordered algebraized space* (Chapt. IV; Definition 8.1). Furthermore, assume that

$$(\mathcal{E}, \rho) \tag{8.65}$$

is a *Riemannian vector sheaf* on $X$ (see Chapt. IV; Definition 8.3). Thus (ibid.; (8.19), (8.17)), the given *Riemannian $\mathcal{A}$-metric* $\rho$ on $\mathcal{E}$, as in (8.65), induces, by definition, an *$\mathcal{A}$-isomorphism* of $\mathcal{E}$ onto its *dual* (vector sheaf) $\mathcal{E}^* := \mathcal{H}om_{\mathcal{A}}(\mathcal{E}, \mathcal{A})$,

$$\mathcal{E} \underset{\tilde{\rho}}{\cong} \mathcal{E}^*, \tag{8.66}$$

such that one sets;

$$\tilde{\rho}(s)(t) := \rho(s, t), \tag{8.67}$$

*for any* (local) sections $s, t$ *in* $\mathcal{E}(U)$, and *open* $U \subseteq X$. So our next objective is to show now the following.

**Theorem 8.2..** *Suppose we are given an ordered algebraized space* $(X, \mathcal{A})$, *endowed with a differential triad* $(\mathcal{A}, \partial, \Omega)$. *Moreover, let*

$$(\mathcal{E}, \rho) \tag{8.68}$$

*be a Riemannian vector sheaf on* $X$, *with* $\operatorname{rk}\mathcal{E} = n \in \mathbf{N}$, *admitting also an $\mathcal{A}$-connection* $D$. *Then,* $D$ *satisfies the Levi-Civita identity* (cf. (8.46)) *if, and only if, it is gauge equivalent to* $D^*$, *the dual $\mathcal{A}$-connection of* $D$ (cf. Chapt. VI; (5.31)), *via the $\mathcal{A}$-isomorphism*

(8.69) $$\tilde{\rho} : \mathcal{E} \longrightarrow \mathcal{E}^* \equiv \mathcal{H}om_{\mathcal{A}}(\mathcal{E}, \mathcal{A}),$$

*provided by* $\rho$ (see (8.66), (8.67)).

*Equivalently, (8.46) holds true, if, and only if,* $\tilde{\rho}$ *is horizontal, with respect to the* $\mathcal{A}$*-connection of the vector sheaf* $\mathcal{H}om_{\mathcal{A}}(\mathcal{E}, \mathcal{E}^*) = (\mathcal{E} \otimes_{\mathcal{A}} \mathcal{E})^*$ (cf. Chapt. IV; (6.16), along with Chapt. VI; (5.13) and (5.25)); *viz. if, and only if, one has*

(8.70) $$D_{\mathcal{H}om_{\mathcal{A}}(\mathcal{E},\mathcal{E}^*)}(\tilde{\rho}) = 0.$$

**Proof.** First, we remark that *(8.70) is,* indeed, by the vey definitions, *tantamount to* the $\mathcal{A}$-connection $D$ *being gauge equivalent with* $D^*$, *via the* $\mathcal{A}$*-isomorphism* $\tilde{\rho}$; viz. to the relation

(8.71) $$D^* = (\tilde{\rho} \otimes 1) \circ D \circ \tilde{\rho}^{-1},$$

or even to the *commutativity* of the following *diagram*,

(8.72) $$\begin{array}{ccc} \mathcal{E} & \xrightarrow{\quad D \quad} & \Omega(\mathcal{E}) \equiv \mathcal{E} \otimes_{\mathcal{A}} \Omega \\ \tilde{\rho} \downarrow \; \uparrow \tilde{\rho}^{-1} & & \downarrow \tilde{\rho} \otimes 1 \\ \mathcal{E}^* & \xrightarrow{\quad D^* \quad} & \Omega(\mathcal{E}^*) \equiv \mathcal{E}^* \otimes_{\mathcal{A}} \Omega \end{array}$$

(see thus Chapt. VI; (5.41), (5.43) and (5.13), along with (8.25)).

Thus, *assume* first *that (8.70), or,* as already explained, *equivalently, (8.71) holds true;* yet, *suppose,* namely, *that we have*

(8.73) $$D^* \circ \tilde{\rho} = (\tilde{\rho} \otimes 1) \circ D.$$

*Then,* we assert that *(8.46) is in force:* In this context, we recall first that $\omega$ (cf. also (8.58)) is determined, *locally,* by a relation, like (1.11). Hence, assuming (8.73), we are going to verify (8.46), locally, in terms of a *local gauge of* $\mathcal{E}$, say, $e^U$ (see e.g. (8.8)), and the corresponding *(local) section basis of* $\mathcal{E}(U)$, as in (8.13). So, applying now (8.73) to the same basis, one obtains;

(8.74) $$D^*(\tilde{\rho}(e_i))(e_j) = (\tilde{\rho} \otimes 1)(D(e_i))(e_j), \quad 1 \leqq i, j \leqq n = \operatorname{rk} \mathcal{E}.$$

Accordingly, one has from the 1st member of the last relation (cf. Chapt. VI; (5.33), together with (8.67), (8.20) and (1.11) in the preceding);

$$\begin{aligned} D^*(\tilde{\rho}(e_i))(e_j) &= \partial(\tilde{\rho}(e_i)(e_j)) - (\tilde{\rho}(e_i)\otimes 1)(D(e_j)) \\ &= \partial(\rho(e_i,e_j)) - (\tilde{\rho}(e_i)\otimes 1)(\textstyle\sum_{k=1}^{n} e_k \otimes \omega_{kj}) \\ &= \partial(\rho_{ij}) - \textstyle\sum_k \tilde{\rho}(e_i)(e_k)\otimes\omega_{kj} = \partial(\rho_{ij}) - \textstyle\sum_k \rho(e_i,e_k)\cdot\omega_{kj} \\ &= \partial(\rho_{ij}) - \textstyle\sum_k \rho_{ik}\omega_{kj}. \end{aligned} \tag{8.75}$$

On the other hand, the 2nd member of (8.74) yields (cf. also Chapt. IV; (6.5)),

$$\begin{aligned} (\tilde{\rho}\otimes 1)(D(e_i))(e_j) &= (\tilde{\rho}\otimes 1)(\textstyle\sum_k e_k\otimes\omega_{ki})(e_j) \\ &= (\textstyle\sum_k \tilde{\rho}(e_k)\otimes\omega_{ki})(e_j) = \textstyle\sum_k \tilde{\rho}(e_k)(e_j)\otimes\omega_{ki} \\ &= \textstyle\sum_k \rho(e_k,e_j)\cdot\omega_{ki} = \textstyle\sum_k \rho_{kj}\omega_{ki}. \end{aligned} \tag{8.76}$$

Consequently, in view of (8.74), (8.75) and (8.76), one finally obtains;

$$\partial(\rho_{ij}) - \sum_k \rho_{ik}\omega_{kj} = \sum_k \rho_{kj}\omega_{ki} = \sum_k \rho_{jk}\omega_{ki} \tag{8.77}$$

(the last equality in (8.77) is due, of course, to (8.32). Now, (8.77) is written in *matrix form;*

$$\partial(\rho) - \rho\omega = \rho\omega. \tag{8.78}$$

Therefore, one has

$$\partial(\rho) = \rho\omega + \rho\omega = (\rho+\rho)\cdot\omega = 2\rho\omega, \tag{8.79}$$

so that one finally obtains the *Levi-Civita identity* (cf. (8.46));

$$\omega = \frac{1}{2}\tilde{\partial}(\rho) \in M_n(\Omega(U)). \tag{8.80}$$

Of course, we have arrived at (8.80), by employing a *local argument,* with respect, namely, to a *local gauge* $e^U$ of $\mathcal{E}$, as e.g. in (8.8). However, by further considering a *local frame of* $\mathcal{E}$ (see, for instance, (8.52)), and by still following a similar procedure to that one for (8.58) (see also (8.59), (8.60)), one can think of (8.80) *globally;* viz. one has

$$\omega = \frac{1}{2}\tilde{\partial}(\rho) \in \Omega(\mathcal{E}nd\,\mathcal{E})(X). \tag{8.81}$$

On the other hand, it is quite obvious that *the above argument can be*

*reversed;* thus, (8.80) entails (8.87), so that, in turn, (8.74) or, equivalently, (8.73), hence, (8.70) too, which thus terminates the proof of the theorem. ∎

Now, as already remarked (see (8.48)), *the Levi-Civita identity (8.46) implies the symmetry of* $\omega$, hence, in the case that the Christoffel sections can be defined (cf. (8.51) and (8.22)), *cond. (8.24) too.* Accordingly, taking now (8.47) into account, one concludes, on the basis of the previous Theorem 8.2, that:

(8.82)

Given an $\mathcal{A}$*-connection* $D$ of a *Riemannian vector sheaf* $(\mathcal{E}, \rho)$, as before, *the following* assertions *are equivalent:*

1) *The Levi-Civita identity is valid* (see (8.80) or, *equivalently,* (8.81)).

2) $D$ *and its dual* $\mathcal{A}$*-connection* $D^*$ *are gauge equivalent, through* $\tilde{\rho}$ (cf. (8.69)).

3) The *cond. (8.70) holds true; viz.* $\tilde{\rho}$, as in 2), *is horizontal, with respect to the* $\mathcal{A}$*-connection of* (the vector sheaf) $\mathcal{H}om_{\mathcal{A}}(\mathcal{E}, \mathcal{E}^*) = \mathcal{E}^* \otimes_{\mathcal{A}} \mathcal{E}^* = (\mathcal{E} \otimes_{\mathcal{A}} \mathcal{E})^*$ (cf. also Chapt. IV; (6.16)).

4) The *cond. (8.42) and the Ricci identity,* as in (8.44), *are in force.* ∎

# 9. Riemannian vector sheaves. Fundamental lemma (general case)

As the heading indicates, we give in this Section the analogue in our case of the *"Fundamental lemma of Riemannian geometry"* of the standard theory. So, under suitable further restrictions on $X$ and $\mathcal{A}$ of a given pair $(X, \mathcal{A})$, as in the preceding Section 8, one succeeds, by analogy with the classical situation, to glue together local informations about Riemannian $\mathcal{A}$-connections, to get such a one globally. However, before we embark on the statement of our main result, we have first to appropriately modify, as already mentioned, our previous Definition 8.1, to fit the general case, that

is to be considered in the sequel. In this context, we further remark that *the two definitions,* thus resulted, *are,* in effect, *equivalent* in the special case one has *cond. (8.7) fulfilled,* as follows from (8.47) and (8.48), in conjunction also with Theorem 8.2 above. Thus, we now have the next.

**Definition 9.1.** Suppose we have an ordered algebraized space $(X, \mathcal{A})$, such that $(\mathcal{A}, \partial, \Omega)$ is a differential triad on $X$. Moreover, let $(\mathcal{E}, \rho)$ be a Riemannian vector sheaf on $X$. Then, a given *$\mathcal{A}$-connection $D$ of $\mathcal{E}$* is said to be *compatible with the Riemannian $\mathcal{A}$-metric $\rho$*, whenever one has;

$$D_{\mathcal{H}om_{\mathcal{A}}(\mathcal{E},\mathcal{E}^{*})}(\tilde{\rho}) = 0 \tag{9.1}$$

(see (8.70) for the notation applied above).

Now, we continue by formulating straight away our main object of the succeeding discussion. For details, concerning the terminology applied, we still refer to Chapt. IV; Section 8. Thus, one has the following basic result.

**Theorem 9.1. (Fundamental lemma of Riemannian vector sheaves).** *Suppose we are given an ordered algebraized space $(X, \mathcal{A})$, with $X$ (Hausdorff) paracompact and $\mathcal{A}$ a strictly positive fine $\mathcal{A}$-module, endowed with a Riemannian $\mathcal{A}$-metric $\overline{\rho}$. Moreover, assume that $X$ is equipped with a differential triad $(\mathcal{A}, \partial, \Omega)$. Then, every vector sheaf $\mathcal{E}$ on $X$ becomes a Riemannian vector sheaf $(\mathcal{E}, \rho)$, admitting further an $\mathcal{A}$-connection compatible with the Riemannian $\mathcal{A}$-metric $\rho$.*

**Proof.** First, based on our hypothesis for $X$ and $\mathcal{A}$, we know that *the* given *vector sheaf $\mathcal{E}$ admits a Riemannian $\mathcal{A}$-metric $\rho$* (see Chapt. IV; Theorem 8.3), so that

$$(\mathcal{E}, \rho) \tag{9.2}$$

is a *Riemannian vector sheaf* on $X$ (ibid.; Definition 8.3).

Now, consider a *local frame of $\mathcal{E}$*,

$$\mathcal{U} = (U_\alpha)_{\alpha \in I}, \tag{9.3}$$

which further might be taken to be (as an open covering of $X$) *locally finite* (see e.g. Chapt. IV; (8.42)). Moreover, since by hypothesis $\mathcal{A}$ is a *fine* **C**-*algebra sheaf* over the *paracompact* (Hausdorff) space $X$, *the* given *$\mathcal{A}$-module* (in fact, vector sheaf) *$\mathcal{E}$ is a fine sheaf* too (cf. Chapt. III; (8.56)),

so that the above *locally finite local frame of* $\mathcal{E}$ has a *subordinate partition of unity* for $\mathcal{E}$ (see Chapt. III; Definition 8.1). Therefore, by further considering the *Levi-Civita 0-cochain,* which is associated eith $\mathcal{U}$,

$$(D_\alpha) \in C^0(\mathcal{U}, \mathcal{H}om_{\mathbf{C}}(\mathcal{E}, \Omega(\mathcal{E}))) \tag{9.4}$$

(cf. Chapt. VI; (8.32.1)), one gets an *$\mathcal{A}$-connection $D$ of $\mathcal{E}$,* which is given by the relation;

$$D := \sum_\alpha (\phi_\alpha \otimes 1) \circ D_\alpha \tag{9.5}$$

with

$$(\phi_\alpha)_{\alpha \in I} \subseteq \operatorname{End} \mathcal{E} \tag{9.6}$$

the *partition of unity* for $\mathcal{E}$, subordinate to $\mathcal{U}$, as above (see Chapt. VI; (16.7)). In this context, it is still useful to note for use right below (anyway, we have done it again in the preceding, in the proof of Theorem 8.1) that the *Levi-Civita $\mathcal{A}$-connection $D$,* as given by (9.5), is integrable, so that it has *trivial local $\mathcal{A}$-connection matrices,* still associated with $\mathcal{U}$; viz. we have

$$\omega^{(\alpha)} \equiv (\omega_{ij}^{(\alpha)}) = 0 \in M_n(\Omega(U_\alpha)), \quad \alpha \in I \tag{9.7}$$

(see Chapt. VI; (11.22), along with (6.28) above). So consider now the following *$\mathcal{A}$-connection of $\mathcal{E}$;*

$$\widetilde{D} := D + \frac{1}{2}\tilde{\partial}(\rho). \tag{9.8}$$

Now, as already remarked by (8.58), one has

$$\frac{1}{2}\tilde{\partial}(\rho) \in \Omega(\mathcal{E}nd\,\mathcal{E})(X), \tag{9.9}$$

so that the above *rel. (9.8)* certainly *provides an $\mathcal{A}$-connection of $\mathcal{E}$* (cf. Chapt. VI; Theorem 7.1). On the other hand, as already pointed out, by the proof of Theorem 8.1 (see thus (8.57)), the previous *$\mathcal{A}$-connection (9.8) satisfies prop. 4) of (8.82).* Accordingly, still by (8.82), *the $\mathcal{A}$-connection at issue fulfills (8.70), as well,* which thus terminates the proof (see also Definition 9.1). ∎

**Remarks 9.1.**– i) Looking more closely at the previous proof, one ascertains that what we really need is our *vector sheaf* $\mathcal{E}$ to satisfy the following

condition:

(9.10) $\mathcal{E}$ has a *locally finite local frame* $\mathcal{U}$, such that there exists a *strictly positive partition of unity subordinate to* $\mathcal{U}$. We then call $\mathcal{E}$ a *numerable vector sheaf* on $X$.

So the form, which the above Theorem 9.1 can take, is actually the following one:

(9.11) We have an *ordered algebraized space* $(X, \mathcal{A})$, with $X$ any *topological space,* which is further endowed with a *differential triad* $(\mathcal{A}, \partial, \Omega)$. Then, *every numerable Riemannian vector sheaf* $(\mathcal{E}, \rho)$ on $X$, *admits an* $\mathcal{A}$*-connection, compatible with the* given *Riemannian* $\mathcal{A}$*-metric* $\rho$. $\blacksquare$

Furthermore, we can suppose *instead,* that $(\mathcal{A}, \overline{\rho})$ is a *Riemannian* $\mathcal{A}$*-module* and $\mathcal{E}$ just a *numerable vector sheaf* on $X$ (cf. (9.10)). Then, we have the *same conclusion, as in (9.11).* $\blacksquare$ Thus, guided by the preceding, we might also consider, in the spirit of (9.10) above, an *analogous strengthening of* various similar results, as, for instance, of *Chapt. IV; Theorem 8.3.*

ii) Concerning the same notion above, as given by (9.10), we still note the following:

(9.12) Suppose that we are given a *continuous map* $f : X \longrightarrow Y$ and a *numerable vector sheaf* $\mathcal{E}$ *on* $Y$. Then, the *pull-back of* $\mathcal{E}$*, via* $f$*, over* $X$*,* $f^*(\mathcal{E})$ (see Chapt. I; (14.15)), *is a numerable vector sheaf on* $X$, as well.

First, if $(Y, \mathcal{B})$ is an *ordered algebraized space* over $Y$, *its pull-back, via* $f$*,* $(X, f^*(\mathcal{B}) \equiv \mathcal{A})$*, is an ordered algebraized space* on $X$, as one easily verifies, based on the very definitions (cf. Chapt. I; (14.23) and Chapt. IV; Definition 8.1). Thus, if now $\mathcal{V} = (V_\alpha)_{\alpha \in I}$ is a *locally finite local frame of* $\mathcal{E}$, then the *open covering of* $X$, $\mathcal{U} \equiv f^{-1}(\mathcal{V}) := (f^{-1}(V_\alpha))_{\alpha \in I}$ is a *similar one for* $f^*(\mathcal{E})$. Yet, if $(\phi_\alpha) \subseteq \operatorname{End} \mathcal{E}$ is a *strictly positive partition of unity for* $\mathcal{E}$*, subordinate to* $\mathcal{V}$, then

$$(f^*(\phi_\alpha))_{\alpha \in I} \subseteq \operatorname{End} f^*(\mathcal{E}) \tag{9.13}$$

is an *analogous family for* $f^*(\mathcal{E})$*, subordinate to* $\mathcal{U}$, as above (see Chapt. I; (14.43) and (14.26), in conjunction with Chapt. IV; Definition 8.4).

The previous considerations suggest now, of course, the case that, by

suitably choosing $Y$ and $\mathcal{E}$, as before, one can look at a *numerable vector sheaf,* like $f^*(\mathcal{E})$, as considered here, however, *over a not necessarily paracompact* (Hausdorff) *space* $X$. (For the classical case of *vector bundles,* see e.g. D. Husemoller [1: p.48, Proposition 9.3]).

# 10. Hermitian $\mathcal{A}$-connections

We have already considered in the foregoing *hermitian $\mathcal{A}$-modules* over a topological space $X$, base space of a given *involutive* **C**-*algebraized space*

(10.1) $$(X, \mathcal{A}, -)$$

(see Chapt. IV; (9.5)); here we still assume, tacitly, that

(10.2) $$(X, \mathcal{A})$$

is, in particular, an *ordered algebraized space* (ibid.; Definition 8.1), when one has to consider *"positivity"* questions. Furthermore, suppose that $X$ is also endowed with a *differential triad*

(10.3) $$(\mathcal{A}, \partial, \Omega),$$

and let

(10.4) $$(\mathcal{E}, \rho)$$

be a *hermitian $\mathcal{A}$-module* on $X$ (cf. Chapt. IV; Definition 9.1). So one can first obtain an *extension of* $\rho$ (for simplicity, we retain the previous symbol for the extended $\rho$ too),

(10.5) $$\rho : \Omega(\mathcal{E}) \oplus \mathcal{E} \longrightarrow \Omega,$$

such that, by taking the *decomposable tensors* of $\Omega(\mathcal{E})_{\mathbf{C}} \equiv \mathcal{E} \otimes \Omega$, one sets;

(10.6) $$\rho(s \otimes t, s') := \rho(s, s') \cdot t,$$

*for any* (local) sections $s, s'$ *in* $\mathcal{E}(U)$ *and* $t \in \Omega(U)$, with $U$ *open* in $X$.

Now, since the given *hermitian $\mathcal{A}$-metric* $\rho$ on $\mathcal{E}$, as in (10.4), is *$\mathcal{A}$-linear* in the *first variable* (see Chapt. IV; (9.7)), the previous relation (10.6) can be

*extended by $\mathcal{A}$-linearity to* all of $\Omega(\mathcal{E}) \equiv \mathcal{E} \otimes_{\mathcal{A}} \Omega$, so that (10.5) *is well-defined.* Yet, based on (10.6) and on Chapt. IV; (9.8) *one* still *sets;*

$$\rho(s, s' \otimes t) := \overline{\rho(s' \otimes t, s)} = \overline{\rho(s', s) \cdot t} = \overline{\rho(s', s)} \cdot \bar{t} = \rho(s, s') \cdot \bar{t}, \tag{10.7}$$

*for any $s, s' \in \mathcal{E}(U)$ and $t \in \Omega(U)$.* Here the "$-$" over $t \in \Omega(U)$ refers, *by definition,* to the *(coordinates) components of $t$,* relative to a given *local gauge of $\Omega$* over the *open $U \subseteq X$, the former being* thus (local) sections of $\mathcal{A}$ over $U$, viz. *elements of $\mathcal{A}(U)$* (cf., for instance, (1.6.1)). So we can set now the following.

**Definition 10.1.** Suppose we have the framework of (10.1), (10.3) and let $(\mathcal{E}, \rho)$ be a hermitian $\mathcal{A}$-module on $X$. Then, a given *$\mathcal{A}$-connection $D$ of $\mathcal{E}$* is said to be *compatible with the hermitian $\mathcal{A}$-metric,* or just a *hermitian $\mathcal{A}$-connection,* whenever the following relation holds true (cf. also (10.5), (10.6), (10.7));

$$\partial(\rho(s, t)) = \rho(Ds, t) + \rho(s, Dt), \tag{10.8}$$

*for any $s, t \in \mathcal{E}(U)$,* with *$U$ open* in $X$.

Thus, suppose now that

$$e^U \equiv \{U; (e_1, \ldots, e_n)\} \tag{10.9}$$

is a *local gauge,* of *rank $n \in \mathbf{N}$,* over the *open* set $U \subseteq X$ of the given *$\mathcal{A}$-module* $\mathcal{E}$, with

$$(e_i)_{1 \leqq i \leqq n} \subseteq \mathcal{E}(U) \tag{10.10}$$

the corresponding *section-basis of $\mathcal{E}(U)$* (see e.g. (1.2) – (1.5) in the preceding).

So, by considering again a *hermitian $\mathcal{A}$-connection $D$* of $\mathcal{E}$ and *translating* further (10.8), *in terms of the* above *section-basis* of $\mathcal{E}(U)$, one has

$$\partial(\rho(e_i, e_j)) = \rho(De_i, e_j) + \rho(e_i, De_j), \quad 1 \leqq i, j \leqq n. \tag{10.11}$$

Furthermore, setting, as before,

$$\rho_{ij} := \rho(e_i, e_j) \in \mathcal{A}(U), \quad 1 \leqq i, j \leqq n, \tag{10.12}$$

and using also the *local $\mathcal{A}$-connection matrix of $D$,* associated with the given

local gauge $e^U$,

$$\omega \equiv (\omega_{ij}) \in M_n(\Omega(U)) = M_n(\Omega)(U), \tag{10.13}$$

(see (1.13)), one has from (10.11) (cf. also (1.11));

$$\begin{aligned} \partial(\rho_{ij}) &= \rho(\textstyle\sum_k e_k \otimes \omega_{ki}, e_j) + \rho(e_i, \textstyle\sum_k e_k \otimes \omega_{kj}) \\ &= \textstyle\sum_k \rho(e_k \otimes \omega_{ki}, e_j) + \textstyle\sum_k \rho(e_i, e_k \otimes \omega_{kj}). \end{aligned} \tag{10.14}$$

Thus, by further applying (10.6) and (10.7), along with Chapt. IV; (9.8), one obtains, by (10.14),

$$\partial(\rho_{ij}) = \sum_k \rho_{kj} \cdot \omega_{ki} + \sum_k \rho_{ik} \cdot \overline{\omega}_{kj}; \tag{10.15}$$

hence, in view again of the *hermitian symmetry of* $\rho$ (see Chapt. IV; (9.8)), one gets, in *matrix form*

$$\partial(\rho) = \rho \cdot \omega + \rho \cdot {}^t\overline{\omega} = \rho(\omega + {}^t\overline{\omega}), \tag{10.16}$$

so that *one* finally *obtains*

$$\omega + {}^t\overline{\omega} = \tilde{\partial}(\rho). \tag{10.17}$$

So we arrived at a *necessary condition for being the* given $\mathcal{A}$*-connection* $D$ *of* $\mathcal{E}$ *hermitian.*

On the other hand, using the expression (1.6.1), for any $s \in \mathcal{E}(U)$, as before, and following further the classical technique (cf., for instance, R.O. Wells [1: p. 76]), one easily sees that *the above condition (10.17) is a sufficient one, as well.* Thus, we can now summarize the preceding, as follows:

(10.18) Suppose we have the *framework of (10.1), (10.3)* and let $(\mathcal{E}, \rho)$ ba a *hermitian* $\mathcal{A}$*-module* on $X$. Then, a given $\mathcal{A}$*-connection* $D$ *of* $\mathcal{E}$ *is hermitian if, and only if,* one has the relation;

$$\tilde{\partial}(\rho) = \omega + {}^t\overline{\omega}, \tag{10.18.1}$$

with respect to a *local gauge* $e^U$ *of* $\mathcal{E}$ (cf. (10.9)).

Now, another fact that will be of use here, is that;

(10.19) by imitating the *Gram-Schmidt process,* given in Chapt. IV; Section 10, in the case of a Riemannian $\mathcal{A}$-module, we can repeat the relevant argument for a *hermitian $\mathcal{A}$-module*, as well, obtaining thus *orthonormal local frames* of a given *hermitian vector sheaf.* (See also, in particular, Chapt. IV; Theorem 10.1).

So, based on the preceding and by further mimiking the classical pattern, we have in the case, under consideration, the following analogue of our previous Theorem 9.1.

**Theorem 10.1.** *Consider an enriched ordered and involutive algebraized space* $(X, \mathcal{A})$ (cf. also Chapt. IV; Definition 10.1), *with* $X$ *a (Hausdorff) paracompact space and* $\mathcal{A}$ *a fine* $\mathcal{A}$*-module such that* $(\mathcal{A}, \rho)$ *is a hermitian* $\mathcal{A}$*-module. Moreover, suppose that* $X$ *is further endowed with a differential triad* $(\mathcal{A}, \partial, \Omega)$*, with* $\Omega$ *being a vector sheaf on* $X$*, while* $\mathcal{E}$ *is also a given vector sheaf on* $X$*. Then,* $\mathcal{E}$ *admits a hermitian* $\mathcal{A}$*-connection. Thus, in particular, this holds true for (the vector sheaf)* $\Omega$*, as well.* ∎

Concerning the standard case, see, for instance, R.O. Wells [1: p. 76, Proposition 1.11]. Finally, analogous comments to those in Remark 9.1; i) above, are here also in force, as it regards our previous theorem.

## 11. Cohomological classification of Maxwell fields (contn'd)

Our aim by the following discussion is to prove Theorem 18.2 of Chapter VI in the foregoing. It refers to a *cohomological characterization of line sheaves, having $\mathcal{A}$-connections,* what actually we have called (Chapt. VI; (18.11)) *Maxwell fields* on $X$.

So, to begin with, suppose, as usual, that we are given a *differential triad* $(\mathcal{A}, \partial, \Omega)$ on a *topological space* $X$. Moreover, let

$$(\mathcal{L}, D) \tag{11.1}$$

be a pair, consisting of a *line sheaf* $\mathcal{L}$ on $X$, along with an *$\mathcal{A}$-connection $D$ of $\mathcal{L}$,* that is, as already said, a *Maxwell field* on $X$.

Thus, we further denote by

$$\mathbf{\Phi}^1_{\mathcal{A}}(X)^{\nabla} \tag{11.2}$$

the *set of* (isomorphism classes of) *Maxwell fields on* $X$ (see Chapt. VI; (18.12)). Now, the latter set is, in effect, an *abelian group,* under "$\otimes_{\mathcal{A}}$" (ibid.; Theorem 18.1). So we come next, as promised, to the

**Proof of Theorem 18.2 of Chapt. VI.** Indeed, what we are actually going to prove, is the following *isomorphism of abelian groups;*

$$\mathbf{\Phi}^1_{\mathcal{A}}(X)^{\nabla} = \check{\mathbf{H}}^1(X, \mathcal{A}^{\cdot} \xrightarrow{\tilde{\partial}} \Omega). \tag{11.3}$$

Now, the second member of (11.3) stands for the *1st hypercohomology group of* $X$*, with coefficients in the 2-term $\mathcal{A}$-complex* of $X$,

$$\mathcal{E}^{\cdot} : \mathcal{A}^{\cdot} \xrightarrow{\tilde{\partial}} \Omega \longrightarrow 0 \cdots, \tag{11.4}$$

which is defined by the *logarithmic derivation* of $\partial$,

$$\tilde{\partial} : \mathcal{A}^{\cdot} \xrightarrow{\tilde{\partial}} \Omega \tag{11.5}$$

(see Chapt. VI; (1.24), (1.25), for $\mathcal{E} \equiv \Omega$, along with (3.1)). Thus, we can further employ our previous argument in Chapt. III; Example 7.1; indeed, the latter example has been manufactured with the present application in mind:

So, by considering a *Maxwell field* $(\mathcal{L}, D)$ on $X$, as in (11.1), let

$$g \equiv (g_{\alpha\beta}) \in Z^1(\mathcal{U}, \mathcal{A}^{\cdot}) \tag{11.6}$$

be a *coordinate 1-cocycle of* $\mathcal{L}$, relative to a given *local frame* of $\mathcal{L}$,

$$\mathcal{U} = (U_{\alpha})_{\alpha \in I}. \tag{11.7}$$

Furthemore, suppose that

$$\theta \equiv (\theta_{\alpha}) \in C^0(\mathcal{U}, \Omega) \equiv \prod_{\alpha \in I} \Omega(U_{\alpha}) \tag{11.8}$$

is the corresponding to $\mathcal{U}$, *0-cochain of local $\mathcal{A}$-connection "forms"* defining the given $\mathcal{A}$-connection $D$ of $\mathcal{L}$ (cf. (2.39.1), for $n = 1$). Thus, as already known (cf. (2.59)), one obtains;

$$(11.9) \qquad \delta(\theta_\alpha) = \tilde{\partial}(g_{\alpha\beta}).$$

However, according to Chapt. III; (7.55), the pair

$$(11.10) \qquad z \equiv (g, \theta) \equiv ((g_{\alpha\beta}), (\theta_\alpha)),$$

as resulted by the previous data, determines a *1-dimensional (hyper)cohomology class,* as follows,

$$(11.11) \qquad [z] \in \check{\mathbf{H}}^1(X, \mathcal{E}^\cdot).$$

Thus, we have already established a map

$$(11.12) \qquad \chi : \mathbf{\Phi}^1_{\mathcal{A}}(X)^\nabla \longrightarrow \check{\mathbf{H}}^1(X, \mathcal{E}^\cdot),$$

given by (11.10) and (11.11): Indeed, suppose we have;

$$(11.13) \qquad (\mathcal{L}, D) \sim (\mathcal{L}', D') \ \ in \ \ \mathbf{\Phi}^1_{\mathcal{A}}(X)^\nabla.$$

Thus (see Chapt. VI; (18.6) and Chapt. V; (2.24)), one concludes that; i) there exists an *$\mathcal{A}$-isomorphism*

$$(11.14) \qquad h \in \mathrm{Isom}_{\mathcal{A}}(\mathcal{L}, \mathcal{L}') = Z^0(\mathcal{U}, \mathcal{I}som_{\mathcal{A}}(\mathcal{L}, \mathcal{L}')),$$

such that one has

$$(11.15) \qquad \mathcal{L} \underset{h}{\sim} \mathcal{L}' \ \ in \ \ \mathbf{\Phi}^1_{\mathcal{A}}(X).$$

That is (cf. Chapt. V; Lemma 2.1), there exists a *0-cochain*

$$(11.16) \qquad (s_\alpha) \in \check{C}^0(\mathcal{U}, \mathcal{A}^\cdot) = \prod_\alpha \mathcal{A}^\cdot(U_\alpha) = \prod_\alpha \mathcal{A}(U_\alpha)^\cdot,$$

such that *the following diagram commutes;*

$$(11.17) \qquad \begin{array}{ccc} \mathcal{L}|_{U_\alpha} & \xrightarrow{\ h_\alpha\ } & \mathcal{L}'|_{U_\alpha} \\ \phi_\alpha \big\downarrow & & \big\downarrow \psi_\alpha \\ \mathcal{A}|_{U_\alpha} & \xrightarrow{\ s_\alpha\ } & \mathcal{A}|_{U_\alpha}, \end{array}$$

where one still has, *by definition,*

$$(11.18) \qquad g_{\alpha\beta} = \phi_\alpha \circ \phi_\beta^{-1} \ \text{ and } \ g'_{\alpha\beta} = \psi_\alpha \circ \psi_\beta^{-1},$$

for the *coordinate 1-cocycles of* the given line sheaves $\mathcal{L}$ *and* $\mathcal{L}'$, *respectively,* that are associated with $\mathcal{U}$, as in (11.7), the latter open covering of $X$, being now considered as a *common local frame of* $\mathcal{L}$ *and* $\mathcal{L}'$ (cf. also Chapt. V; (1.3.2)). Moreover, one gets, *equivalently to (11.15),* the relation;

$$(11.19)\qquad g'_{\alpha\beta} = s_\alpha \cdot g_{\alpha\beta} \cdot s_\beta^{-1} = (s_\beta^{-1} s_\alpha) \cdot g_{\alpha\beta} = \delta(s_\alpha^{-1}) \cdot g_{\alpha\beta}.$$

*Therefore, one has;*

$$(11.20)\ (g'_{\alpha\beta})\cdot(g_{\alpha\beta})^{-1} \equiv g'\cdot g^{-1} = \delta(s_\alpha^{-1}) \in \delta(\check{C}^0(\mathcal{U},\mathcal{A}^\cdot)) \equiv \mathrm{im}\delta \subseteq \check{C}^1(\mathcal{U},\mathcal{A}^\cdot).$$

the multiplication in the first member of (11.20) being considered in the *abelian group* $Z^1(\mathcal{U},\mathcal{A}^\cdot)$ (again Chapt. V; (2.11), for $n = 1$).

ii) *As a result of (11.13),* one further obtains, for the $\mathcal{A}$*-connections* $D$ *and* $D'$ involved, the relation (see Chapt. VI; (18.8));

$$(11.21)\qquad D' = (h \otimes 1) \circ D \circ h^{-1}.$$

Now, the above global relation is *equivalently* expressed, in view of (11.7), through the *local* $\mathcal{A}$*-connection "forms"* of $D$ and $D'$ (see e.g. (11.8)) and the corresponding to the given $h$, as in (11.14) *"gauge transformation"*

$$(11.22)\qquad (s_\alpha^{-1}) \in \check{C}^0(\mathcal{U},\mathcal{A}^\cdot),$$

by the following relation (see also, for instance, (2.15) or even (1.41));

$$(11.23)\qquad s_\alpha^{-1} \cdot \theta'_\alpha = \theta_\alpha \cdot s_\alpha^{-1} + \partial(s_\alpha^{-1}),\ \ \alpha \in I,$$

so that *one has*

$$(11.24)\qquad \theta'_\alpha = s_\alpha \cdot \theta_\alpha \cdot s_\alpha^{-1} + s_\alpha \cdot \partial(s_\alpha^{-1}) = \theta_\alpha + \tilde{\partial}(s_\alpha^{-1}),\ \ \alpha \in I.$$

*Therefore, one obtains* (cf. also (11.22))

$$(11.25)\qquad \begin{aligned} \theta' - \theta \equiv (\theta'_\alpha) - (\theta_\alpha) = (\theta'_\alpha - \theta_\alpha) = \tilde{\partial}(s_\alpha^{-1}) \in \tilde{\partial}(\check{C}^0(\mathcal{U},\mathcal{A}^\cdot)) \\ \equiv \mathrm{im}\tilde{\partial} \subseteq \check{C}^0(\mathcal{U},\Omega). \end{aligned}$$

Thus, *in view of (11.20) and (11.25), one has;*

$$(11.26)\qquad \begin{aligned} &(\mathcal{L}',D') - (\mathcal{L},D) \equiv (g',\theta') - (g,\theta) = (g'\cdot g^{-1}, \theta' - \theta) \\ &= (\delta(s_\alpha^{-1}), \tilde{\partial}(s_\alpha^{-1})) = (\delta,\tilde{\partial})(s_\alpha^{-1}) \equiv (\delta \oplus \tilde{\partial})(s_\alpha^{-1}) \\ &\in \mathrm{im}\,(\delta + \tilde{\partial}) \equiv \mathrm{im}\, D^0 \subseteq \check{C}^1(\mathcal{U},\mathcal{A}^\cdot) \oplus \check{C}^0(\mathcal{U},\Omega) \equiv \mathcal{S}^1. \end{aligned}$$

Here we have also set (cf. (11.4) and Chapt. III; (7.13), (7.16))

$$D^0 = \delta \oplus \tilde{\partial} : \check{C}^0(\mathcal{U}, \mathcal{A}^{\cdot}) \equiv \mathcal{S}^0 \longrightarrow \mathcal{S}^1 \equiv \check{C}^1(\mathcal{U}, \mathcal{A}^{\cdot}) \oplus \check{C}^0(\mathcal{U}, \Omega). \tag{11.27}$$

Similarly, one obtains (ibid., and (11.27))

$$D^1 : \mathcal{S}^1 \longrightarrow \mathcal{S}^2 \equiv \check{C}^2(\mathcal{U}, \mathcal{A}^{\cdot}) \oplus \check{C}^1(\mathcal{U}, \Omega), \tag{11.28}$$

such that one has

$$D^1 = (\delta - \tilde{\partial}) + \delta. \tag{11.29}$$

(*Warning!* For convenience, we abused notation here, by setting, $\delta^{0,1} \equiv \delta$, as well as, $\delta^{1,0} \equiv \delta$; (cf. Chapt. III; (7.16) and the following diagram (11.32)). Therefore, in view of (11.9) and (11.6), *one obtains*

$$(g, \theta) \in \ker D^1 = \ker(\delta - \tilde{\partial}) \oplus \ker \delta, \tag{11.30}$$

*for any* $(\mathcal{L}, D) \equiv (g, \theta)$, as above. Accordingly, *by virtue of (11.26) and (11.30), one* now *obtains* (cf. also (11.4) and Chapt. III; (7.45));

$$[(\mathcal{L}, D)] = [(\mathcal{L}', D')] \in \check{\mathbf{H}}^1(X, \mathcal{E}^{\cdot}) := \ker D^1 / \operatorname{im} D^0. \tag{11.31}$$

That is, *(11.13) entails (11.31),* which thus proves that *the map (11.12),* as given by (11.10) and (11.11) (cf. also (11.30)), *is well-defined,* as asserted.

For convenience, we still depict the previous situation, concerning the various *hypercohomology* spaces and operators, through the *following diagram (11.32),* which will also be of use presently below;

(11.32)
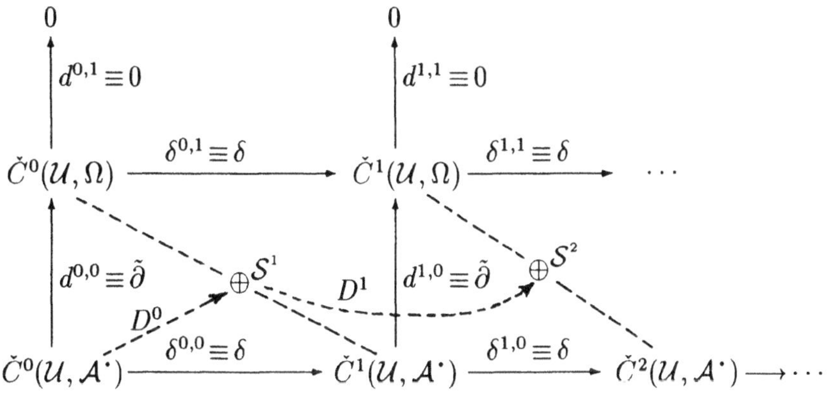


Now, we further prove that *the map (11.12) is one-to-one:* Thus, suppose

that

$$[(\mathcal{L}, D)] = 0 \in \check{\mathbf{H}}^1(X, \mathcal{E}^{\cdot}), \tag{11.33}$$

for some $(\mathcal{L}, D) \equiv (g, \theta)$, as above (cf. (11.10)). Therefore, one concludes that (see (11.31) and (11.27))

$$(\mathcal{L}, D) \equiv (g, \theta) \in \operatorname{im} D^0 \equiv \operatorname{im}(\delta + \tilde{\partial}); \tag{11.34}$$

in other words, there exists a *0-cochain*

$$(s_\alpha) \in \check{C}^0(\mathcal{U}, \mathcal{A}^{\cdot}), \tag{11.35}$$

in such a manner that one has

$$(\mathcal{L}, D) \equiv (g, \theta) = (\delta + \tilde{\partial})(s_\alpha) \in \operatorname{im} D^0. \tag{11.36}$$

Consequently, one obtains (cf. also (11.6) and (11.8));

$$g \equiv (g_{\alpha\beta}) = \delta(s_\alpha) \equiv (\delta(s_\alpha)) = (s_\beta s_\alpha^{-1}), \tag{11.37}$$

as well as,

$$\theta \equiv (\theta_\alpha) = \tilde{\partial}(s_\alpha) \equiv (\tilde{\partial}(s_\alpha)) = (s_\alpha^{-1}\partial(s_\alpha)). \tag{11.38}$$

In this context, we still note, in passing, that, in view of (11.37) and (11.38), in conjunction with Chapt. VI; (1.27), one gets, of course,

$$\delta(\theta_\alpha) = \tilde{\partial}(g_{\alpha\beta}), \tag{11.39}$$

a fact, that was to be expected, for that matter, by virtue of (11.33) and (11.9).

Accordingly, setting now

$$g'_{\alpha\beta} := s_\alpha \cdot g_{\alpha\beta} \cdot s_\beta^{-1}, \tag{11.40}$$

one obtains, in view of (11.37), that

$$g'_{\alpha\beta} = 1 \ (\equiv 1_{\alpha\beta}), \tag{11.41}$$

while we still have, by (11.40), (11.35), in conjunction with Chapt. V; Lemma 2.1, that;

$$\mathcal{L} \equiv (g_{\alpha\beta}) \sim (g'_{\alpha\beta}) \equiv \mathcal{L}' \tag{11.42}$$

*in* $\mathbf{\Phi}^1_{\mathcal{A}}(X)$. That is (cf. (11.41) and Chapt. V; (4.3)),

(11.43) *the line sheaf* $\mathcal{L}$, as in (11.33), given further by (11.37), *is* $\mathcal{A}$*-isomorphic* to the *"trivial" line sheaf* (viz. the *free* $\mathcal{A}$*-module* of *rank* 1) $\mathcal{A}$. Thus, to the *neutral element* of the *multiplicative* (under $\otimes_{\mathcal{A}}$) *abelian group* $\mathbf{\Phi}^1_{\mathcal{A}}(X)$ (cf. Chapt. V; Theorem 4.1).

On the other hand, by further setting

$$\theta'_\alpha = \theta_\alpha + \tilde{\partial}(s_\alpha^{-1}) \in \Omega(U_\alpha), \quad \alpha \in I \tag{11.44}$$

(see also (11.35), or even (11.22)), one gets a *0-cochain*

$$(\theta'_\alpha) \in \check{C}^0(\mathcal{U}, \Omega) = \prod_\alpha \Omega(U_\alpha), \tag{11.45}$$

*which,* in effect, *supplies an* $\mathcal{A}$*-connection* $D'$ *of* $\mathcal{L}'$; in other words, one verifies that *(11.45) satisfies the* corresponding *transformation law of potentials* (cf. Theorem 3.2, along with (2.59)): Indeed, one obtains;

$$\begin{aligned} \delta(\theta'_\alpha) &= \delta(\theta_\alpha) + \delta\tilde{\partial}(s_\alpha^{-1}) = \tilde{\partial}(g_{\alpha\beta}) + \tilde{\partial}(s_\beta^{-1}) - \tilde{\partial}(s_\alpha^{-1}) \\ &= \tilde{\partial}(s_\beta s_\alpha^{-1}) - \tilde{\partial}(s_\beta) + \tilde{\partial}(s_\alpha) = \tilde{\partial}(s_\beta) - \tilde{\partial}(s_\alpha) - \tilde{\partial}(s_\beta) + \tilde{\partial}(s_\alpha) \\ &= 0 = \tilde{\partial}(1) = \tilde{\partial}(g'_{\alpha\beta}). \end{aligned} \tag{11.46}$$

(In this regard, cf. Chapt. VI; (1.24) and (1.7), along with the previous rels. (11.44), (11.39), (11.37) and (11.41)). Thus, (11.46) justifies now our assertion. Furthermore, *the* two $\mathcal{A}$*-connections* $D$ *and* $D'$ *of* $\mathcal{L}'$, as given by (11.38) and (11.45), respectively, *satisfy the corresponding relation to (11.24),* by the very definition of $(\theta'_\alpha)$ (see (11.44)). Thus, one actually obtains that;

$$(\mathcal{L}, D) \sim (\mathcal{L}', D') \equiv (1, D') \tag{11.47}$$

in $\mathbf{\Phi}^1_{\mathcal{A}}(X)^\nabla$ (cf. Chapt. VI; (18.6)). So we have proved that *(11.33) implies (11.47), that was,* in fact (cf. also (11.43)), *to be proved.*

Furthermore, we prove that *the map (11.12) is onto:* Thus, suppose we have an element

$$[z] \in \check{\mathbf{H}}(X, \mathcal{E}^\cdot). \tag{11.48}$$

Therefore (see e.g. (11.31), (11.30)), one has

$$z \in \ker D^1 = \ker \delta \oplus \ker(\delta - \tilde{\partial}), \tag{11.49}$$

so that one finally concludes that (cf. also, for instance, (11.32), or even

(11.27), (11.28));

(11.50) $$z \equiv (g, \theta) \equiv ((g_{\alpha\beta}), (\theta_\alpha)) \in \check{C}^1(\mathcal{U}, \mathcal{A}^{\cdot}) \oplus \check{C}^0(\mathcal{U}, \Omega)$$

(for any *representative* $z$ of the cohomology class considered in (11.48)). Consequently, in view of (11.49) and (11.50), one *first* obtains

(11.51) $$\delta(g) \equiv \delta(g_{\alpha\beta}) = 0,$$

so that one thus concludes that

(11.52) $$(g_{\alpha\beta}) \in Z^1(\mathcal{U}, \mathcal{A}^{\cdot});$$

in other words, one thus obtains a *line sheaf* $\mathcal{L}$ on $X$, which is determined through its *coordinate 1-cocycle* (11.52) (see Chapt. V; Theorem 4.1, or even (2.60) therein). *Secondly,* according to the same rels. (11.49) and (11.50), one further obtains

(11.53) $$\delta(\theta_\alpha) = \tilde{\partial}(g_{\alpha\beta}) \in Z^1(\mathcal{U}, \Omega)$$

(cf. also Chapt. VI; (10.25) above, for $n = 1$). Thus, *the 1-cocycle* $\tilde{\partial}(g_{\alpha\beta})$, as above, *is* actually *a coboundary,* so that one has

(11.54) $$[\tilde{\partial}(g_{\alpha\beta})] = 0 \in \check{H}^1(X, \Omega).$$

However, the last condition implies already (*Atiyah's criterion,* cf. Chapt. VI; Theorem 11.1, along with (10.28.1)) that *our line sheaf* $\mathcal{L}$, as given by (11.52), *admits an* $\mathcal{A}$*-connection* $D$, whose *local* $\mathcal{A}$*-connection forms* are given by the *0-cochain*

(11.55) $$(\theta_\alpha) \in \check{C}^0(\mathcal{U}, \Omega)$$

as appeared in (11.53) (see Chapt. VI; (11.7), (11.8) and (10.23); yet, see (1.24) in the preceding). Thus, we arrived at an element

(11.56) $$[(\mathcal{L}, D)] \in \mathbf{\Phi}^1_{\mathcal{A}}(X)^{\nabla},$$

whose image, through the map (11.12), is the given element $z$, as in (11.48), according to (11.50) and our previous interpretation of (11.52) and (11.55); i.e., *we have*

(11.57) $$\chi([(\mathcal{L}, D)]) = z,$$

*as asserted.*

So we have established, so far, through the map (11.12), a *bijection between* the two *(abelian) groups* involved. Now, our final claim is that *the same map is,* in effect, *a group morphism:* Namely, suppose that we have *two elements of (11.2)*

$$(\mathcal{L}, D) \equiv (g, \theta) \quad and \quad (\mathcal{L}', D') \equiv (g', \theta') \tag{11.58}$$

(see e.g. (11.10). As usual so far, we do not distinguish, for convenience, the latter elements from their corresponding (isomorphism) classes in $\mathbf{\Phi}^1_{\mathcal{A}}(X)^{\nabla}$). Thus, one obtains (cf. Chapt. VI; (18.2), (18.15), (18.16), (18.17) and (18.21), along with Chapt. V; (4.3));

$$\begin{aligned}(\mathcal{L}', D') \otimes (\mathcal{L}, D)^{-1} &= (\mathcal{L}' \otimes_{\mathcal{A}} \mathcal{L}^{-1}, D' \otimes D^{-1}) \\ &= (\mathcal{L}' \otimes_{\mathcal{A}} \mathcal{L}^{*}, D' \otimes D^{*}) \equiv (h, \omega),\end{aligned} \tag{11.59}$$

such that one has(see also Chapt. V; (4.16), (4.17))

$$h = g' \cdot g^{-1} \in Z^1(\mathcal{U}, \mathcal{A}^{\cdot}), \tag{11.60}$$

while we still obtain that

$$\omega = \theta' - \theta \in \check{C}^0(\mathcal{U}, \Omega), \tag{11.61}$$

as an *$\mathcal{A}$-connection 0-cochain for the $\mathcal{A}$-connection $D' \otimes D^*$*. The assertion here follows from Chapt. VI; (5.10), in conjunction with (1.11) and (4.20) in the preceding. On the other hand, by looking at the pairs (11.58), as above, and based on (11.39), one further obtains from (11.61) and (11.60);

$$\begin{aligned}\delta(\omega) = \delta(\theta' - \theta) = \delta(\theta') - \delta(\theta) = \tilde{\partial}(g') - \tilde{\partial}(g) &= \tilde{\partial}(g') + \tilde{\partial}(g^{-1}) \\ &= \tilde{\partial}(g'g^{-1}) = \tilde{\partial}(h).\end{aligned} \tag{11.62}$$

That is, one has here again *the analogous relation to (11.39), concerning the pair* $(h, \omega)$, as in (11.59); hence, *the pair* in question,

$$(h, \omega) = (g'g^{-1}, \theta' - \theta) \tag{11.63}$$

*determines* thus, in view of (11.62), *an element of* $\check{\mathbf{H}}^1(X, \mathcal{E}^{\cdot})$. Indeed, as such, viz. *the image of (11.63)* (in fact, of the respective *isomorphism class*), *through the map (11.12), has as representative in the latter (additive) group the element*

$$(\mathcal{L}', D') - (\mathcal{L}, D) \equiv (g', \theta') - (g, \theta) = (g'g^{-1}, \theta' - \theta), \tag{11.64}$$

as this follows from the very *group structure* of

$$\operatorname{dom} D^1 = \mathcal{S}^1 \equiv \check{C}^1(\mathcal{U}, \mathcal{A}^\cdot) \oplus \check{C}^0(\mathcal{U}, \Omega) \tag{11.65}$$

(see also (11.27), or even (11.32)). So this proves already our last claim, concerning the map (11.12), and, in turn, *the proof* of Theorem 18.2 of Chapter VI, *is* thus *completely established,* as desired. ∎

Yet, an appropriate *physical interpretation* of the content of the same theorem, as above, will be given later on; see thus Chapter X, Section 5 in the sequel.

# CHAPTER VIII

# Curvature

"Matter tells space how to curve."

C. W. Milner – K. S. Thorne – J. A. Wheeler in "Gravitation" (Freeman, 1973). p. 5.

"...Mathematics aims to understand, to manipulate, to develop, and to apply those aspects of the universe which are formal."

S. Mac Lane in "Mathematics Form and Function" (Springer – Verlag, 1986). p. 456.

We consider in this Chapter the notion of *"curvature"*, which, under supplementary suitable assumptions for a given differential triad, can be associated, in a natural way, with any given $\mathcal{A}$-connection. Thus, in complete contrast with the situation one encounters with the case of $\mathcal{A}$-connections, which, as we know already (cf. Chapter VI in the preceding), do not always exist, we realize here that an $\mathcal{A}$-connection in the case there does exist is, instead, always accompanied with a curvature. This fact, as we shall see later on (cf. thus Chapter X in the sequel), can acquire an appropriate physical interpretation, as well. Yet, as was the case hitherto, one still witnesses here that several fundamental notions and results of the classical theory can also be recasted within the present abstract (axiomatic) framework, that has been advocated so far by this study.

So we start with establishing straight away the necessary basic terminology.

# 1. Preliminaries. Basic definitions

Before we proceed, we first comment a bit on a change of the notation applied hitherto, that will be employed hereafter:

So the notation for a *differential triad* used heretofore (cf. Chapt. VI; (3.1)), is now changed to

$$(\mathcal{A}, \partial, \Omega^1). \tag{1.1}$$

Again $(X, \mathcal{A})$ stands here for a given **C**-*algebraized space* on an arbitrary *topological space* $X$, as before, while we now set;

$$\Omega \equiv \Omega^1 \equiv \wedge^1 \Omega, \tag{1.2}$$

for the given $\mathcal{A}$-*module* $\Omega$, range of the **C**-*derivation* $\partial$, as above. Thus, $\Omega^1$ stands for the *1st exterior power of* $\Omega$, the given $\mathcal{A}$-module on $X$, as was actually the case hitherto. Indeed, the reason for (1.2) is that we are going to consider right below *higher exterior powers of* $\Omega \equiv \Omega^1$ (see e.g. (1.6) in the sequel).

On the other hand, to distinguish the present situation from the *Whitney sum* notation, occasionally employed so far, we retain, henceforth, for the latter the (usual, for that matter) symbolism;

$$\Omega^{(n)} := \underbrace{\Omega \oplus \cdots \oplus \Omega}_{n\text{-times}}, \quad n \in \mathbf{N}. \tag{1.3}$$

Thus, to start with, consider a *differential triad*

$$(\mathcal{A}, \partial, \Omega^1) \tag{1.4}$$

(cf. (1.1)) on a given **C**-*algebraized space*

$$(X, \mathcal{A}). \tag{1.5}$$

Furthermore, consider the *2nd exterior power of* $\Omega^1$;

$$\Omega^2 := \Omega^1 \wedge \Omega^1 \equiv \bigwedge^2 \Omega \equiv \bigwedge^2 \Omega^1 \tag{1.6}$$

(cf. (1.2), along with Chapt. IV; (7.1), (7.3) and (7.8). As already said the above is an instance justifying the previous change of notation, as in (1.1)).

On the other hand, apart from $\partial$, as in (1.4), we further assume that we are given a second **C**-*linear morphism*

$$d^1 \equiv d : \Omega^1 \longrightarrow \Omega^2 \tag{1.7}$$

between the **C**-*vector space sheaves* concerned, satisfying the two following conditions:

i) *The "differential operators"* $\partial$ *and* $d^1 \equiv d$, as above, *satisfy the relation;*

$$d \circ \partial = 0. \tag{1.8}$$

ii) *For any (local) sections* $\alpha \in \mathcal{A}(U)$ *and* $s \in \Omega^1(U)$, with $U$ *open* in $X$ *one has;*

$$d(\alpha \cdot s) = \alpha \cdot ds - s \wedge \partial\alpha = \alpha \cdot ds + \partial(\alpha) \wedge s. \tag{1.9}$$

Now, by extending the standard terminology, we call the previous map $d^1 \equiv d$, the *1st exterior derivation* (or even *1st exterior derivative operator*) of (1.4). Yet, by still mimicking the classical case, we also call occasionally the elements of the $\mathcal{A}$-*module* on $X$,

$$\Omega^n := \underbrace{\Omega^1 \wedge \cdots \wedge \Omega^1}_{n\text{-times}} \equiv \bigwedge^n \Omega \equiv \bigwedge^n \Omega^1, \quad n \in \mathbf{N}. \tag{1.10}$$

(cf. also (1.2)), "*n-forms*" on $X$. (Indeed, the terminology refers actually to *sections* of the latter $\mathcal{A}$-module).

Of course, (1.8) *is equivalent with* the relation;

$$\operatorname{im} \partial \subseteq \ker d. \tag{1.11}$$

Yet, as a first application of (1.8), (1.9), one also concludes, by the very definitions, that

$$d \circ \tilde{\partial} = 0, \tag{1.12}$$

as well, that is, *equivalently,*

$$\operatorname{im} \tilde{\partial} \subseteq \ker d. \tag{1.13}$$

Indeed, one has, *for any* $\alpha \in \mathcal{A}^{\cdot}(U) = \mathcal{A}(U)^{\cdot}$ and *open* $U \subseteq X$,

$$(1.14)\qquad \begin{aligned} d(\tilde{\partial}(\alpha)) &= d(\alpha^{-1}\partial\alpha) = \alpha^{-1}d(\partial(\alpha)) + \partial(\alpha^{-1})\wedge\partial(\alpha) \\ &= -\alpha^{-2}\partial(\alpha)\wedge\partial(\alpha) = -\alpha^{-2}(\partial\alpha\wedge\partial\alpha) = 0, \end{aligned}$$

which proves (1.12). (See also Chapt. VI; (1.12)). ∎

So we come now to the following basic notion for the sequel, as exhibited by the next.

**Definition 1.1.** Suppose we have a **C**-algebraized space $(X, \mathcal{A})$, endowed with a differential triad $(\mathcal{A}, \partial, \Omega^1)$ (cf. (1.1)), along with a **C**-linear morphism

$$(1.15)\qquad d : \Omega^1 \longrightarrow \Omega^2 \equiv \Omega^1 \wedge \Omega^1,$$

satisfying (1.8) and (1.9). Then, the quintuple

$$(1.16)\qquad (\mathcal{A}, \partial, \Omega^1, d, \Omega^2)$$

is called a *curvature datum* on $X$. Yet, for short, we also speak in that case of a *curvature space* $X$.

As follows from the above definition, we have a curvature datum, once we are given, of course, the *pair*

$$(1.17)\qquad (\partial, d)$$

*of "differential operators"*, as before, *such that (1.8) and (1.9) be satisfied.*

Furthermore, based on (1.9), and for reasons that will be made apparent presently below, we may also call the previously given operator $d^1 \equiv d$, as in (1.15), the *1st prolongation of* $\partial$.

So our next objective is to define an analogous *"prolongation"* of a given $\mathcal{A}$*-connection* $D$. Indeed, one has the following.

**Lemma 1.1.** Suppose we are given a *curvature space* $X$ (see Definition 1.1), and let $\mathcal{E}$ be an $\mathcal{A}$*-module* on $X$, endowed with an $\mathcal{A}$*-connection* $D$. Then, one defines the *1st prolongation of* $D$, as the map

$$(1.18)\qquad D^1 : \Omega^1(\mathcal{E}) \equiv \mathcal{E} \otimes_{\mathcal{A}} \Omega^1 \longrightarrow \mathcal{E} \otimes_{\mathcal{A}} \Omega^2 \equiv \Omega^2(\mathcal{E}),$$

such that

$$(1.19)\qquad D^1(s \otimes t) := s \otimes dt - t \wedge Ds = s \otimes dt + Ds \wedge t,$$

*for any* $s \in \mathcal{E}(U)$ *and* $t \in \Omega^1(U)$, with $U$ *open* in $X$.

Moreover, $D^1$ is *the only* **C**-*linear morphism,* as in (1.18), *satisfying* (1.19).

**Proof.** At first, by looking at the corresponding *presheaves,* generating the $\mathcal{A}$-*modules* appeared in (1.18) (see Chapt. II; (5.5) and Chapt. IV; (7.1), (7.2)), one defines, *for every open* $U \subseteq X$, a **C**-*bilinear map*

$$\begin{aligned}\rho_U : \mathcal{E}(U) \times \Omega^1(U) &\longrightarrow \mathcal{E}(U) \otimes_{\mathbf{C}} \Omega^2(U) \hookrightarrow \mathcal{E}(U) \otimes_{\mathcal{A}(U)} \Omega^2(U) \\ &= (\mathcal{E}(U) \otimes_{\mathcal{A}(U)} \Omega^1(U)) \wedge \Omega^1(U) : (s,t) \longmapsto \rho_U(s,t) := s \otimes dt + Ds \wedge t,\end{aligned} \tag{1.20}$$

with $s \in \mathcal{E}(U)$ and $t \in \Omega^1(U)$. (Concerning the range of $\rho_U$, see, for instance, N. Bourbaki [3: Chap. 2; p.53, Proposition 2]). Fuerthermore, *the same map (1.20) is* $\mathcal{A}(U)$-*balanced.* That is, for every $\alpha \in \mathcal{A}(U)$ and $s, t$, as before, one has the relation;

$$\rho_U(\alpha s, t) = \rho_U(s, \alpha t), \tag{1.21}$$

as one easily sees by (1.20). Indeed, one has (cf. also (1.9) and Chapt. VI; (3.6));

$$\begin{aligned}\rho_U(\alpha s, t) &= (\alpha s) \otimes dt + D(\alpha s) \wedge t = \alpha s \otimes dt + (\alpha\, Ds + s \otimes \partial\alpha) \wedge t \\ &= \alpha s \otimes dt + (s \otimes \partial\alpha) \wedge t + \alpha\, Ds \wedge t = s \otimes (\alpha\, dt + \partial\alpha \wedge t) + \alpha\, Ds \wedge t \\ &= s \otimes d(\alpha t) + Ds \wedge (\alpha t) = \rho_U(s, \alpha t),\end{aligned} \tag{1.22}$$

as just was to be proved. Accordingly,

(1.23) *the map* $\rho_U$, as given by (1.20), *can be extended,* by virtue of (1.21), *uniquely, to a* **C**-*linear map*

$$\bar{\rho} : \mathcal{E}(U) \otimes_{\mathcal{A}(U)} \Omega^1(U) \longrightarrow \mathcal{E}(U) \otimes_{\mathcal{A}(U)} \Omega^2(U). \tag{1.23.1}$$

Now, *these various* **C**-*linear maps* $\rho_U$, when $U$ runs over the open subsets of $X$, *satisfy the* corresponding *compatibility conditions* for a *presheaf morphism* (see Chapt. I; (6.5), (6.7)); in turn, this entails, still *uniquely,* a corresponding *sheaf morphism* between the sheaves generated by the presheaves involved (see e.g. (1.23.1), along with Chapt. I; (8.5) and Lemma 8.1). However, due to (1.23), the latter map is, in effect, a **C**-*linear morphism* of the **C**-*vector space sheaves* involved, which, *by definition,* is now the **C**-*linear morphism* $D^1$, as in (1.18), we were looking for; the same is, of course, *uniquely defined,* as well, as follows from (1.23) and the previous argument, and this terminates

the proof. ∎

As a first application of the above, one obtains the following *two properties of the 1st prolongation $D^1$ of an $\mathcal{A}$-connection $D$*, which will be applied presently below. Thus, we have that:

*For any (local) sections $\alpha \in \mathcal{A}(U)$, $s \in \mathcal{E}(U)$ and $t \in \Omega^1(U)$*, with $U$ *open* in $X$, *one obtains;*

$$(1.24)\qquad \begin{aligned} D^1(\alpha\cdot(s\otimes t)) &= \alpha\cdot D^1(s\otimes t) - (s\otimes t)\wedge\partial\alpha \\ &= \alpha\cdot D^1(s\otimes t) + \partial\alpha\wedge(s\otimes t), \end{aligned}$$

*as well as,*

$$(1.25)\qquad D^1(\alpha\cdot Ds) = \alpha\cdot D^1(Ds) - Ds\wedge\partial\alpha = \alpha\cdot D^1(Ds) + \partial\alpha\wedge Ds.$$

Indeed, one has (cf. also (1.21), (1.23), together with (1.19) and (1.9));

$$(1.26)\qquad \begin{aligned} &D^1(\alpha\cdot(s\otimes t)) = D^1(\alpha s\otimes t) = D^1(s\otimes\alpha t) = s\otimes d(\alpha t) - \alpha t\wedge Ds \\ &= s\otimes(\alpha\, dt - t\wedge\partial\alpha) - \alpha t\wedge Ds = s\otimes(\alpha\, dt) - (s\otimes t)\wedge\partial\alpha - \alpha t\wedge Ds \\ &= \alpha(s\otimes dt - t\wedge Ds) - (s\otimes t)\wedge\partial\alpha = \alpha\cdot D^1(s\otimes t) - (s\otimes t)\wedge\partial\alpha, \end{aligned}$$

That is, *the rel. (1.24), as desired.* Now, to prove (1.25), we first remark that *(1.24) is* already *one aspect of (1.25), in the case of decomposable tensors* $s\otimes t \in \mathcal{E}(U)\otimes_{\mathcal{A}(U)}\Omega^1(U)$. Hence, the general case of an element $D(s) \in \mathcal{E}(U)\otimes_{\mathcal{A}(U)}\Omega^1(U)$, with $s \in \mathcal{E}(U)$, that is, the rel. (1.25), *follows* now *from (1.24) and the* **C**-*linearity of* $D^1$ (Lemma 1.1). ∎

Now, for convenience, as well as, as a particular instance of a more general situation, which we deal with later on (see thus Section 7 in the sequel), we still write (1.19), the defining relation of the *1st prolongation* $D^1$ *of the* given *$\mathcal{A}$-connection* $D$, in the form;

$$(1.27)\qquad D^1 := 1_{\mathcal{E}}\otimes d^1 - 1_{\Omega^1}\wedge D \equiv 1\otimes d - 1\wedge D = 1\otimes d + D\wedge 1$$

where $1_{\mathcal{E}}$ and $1_{\Omega^1}$ have the obvious meaning.

## 2. Curvature of an $\mathcal{A}$-connection

Based on the material of the preceding Section, we are now in the position to give the following fundamental definition, concerning the concept in the heading of this Section. So we have.

**Definition 2.1.** Suppose we are given a curvature space $X$ (cf. Definition 1.1), as well as, an $\mathcal{A}$-module $\mathcal{E}$ on $X$, endowed with an $\mathcal{A}$-connection $D$. Then, the *curvature* of the given $\mathcal{A}$-connection $D$, denoted by

$$R(D) \equiv R, \tag{2.1}$$

is defined by the relation;

$$R(D) := D^1 \circ D, \tag{2.2}$$

where $D^1$ stands for the 1st prolongation of $D$ (cf. Lemma 1.1). In other words, one has the following defining triangle for $R$ (see (1.18));

$$\begin{array}{ccc} \mathcal{E} & \xrightarrow{D} & \Omega^1(\mathcal{E}) \\ & {\scriptstyle R \equiv D^1 \circ D} \searrow \quad \swarrow {\scriptstyle D^1} & \\ & \Omega^2(\mathcal{E}). & \end{array} \tag{2.3}$$

**Note 2.1.**– In complete contrast with what happens in the case of *an $\mathcal{A}$-connection,* which, as we know (cf. Chapt. VI; (11.2)), *does not exist, in general,* we remark here that:

(2.4) *one can always define the curvature of a given $\mathcal{A}$-connection,*

*provided,* of course, *one has* the appropriate framework, to formulate this notion; hence, according to the preceding, *a curvature space* (cf. (1.16)).

Having now at our disposal the classical notion of curvature, we say that a given $\mathcal{A}$-connection $D$ is *flat,* whenever one has;

$$R(D) = 0 \tag{2.5}$$

(see also the previous Note 2.1, concerning the relevant context for the formulation of (2.5)).

So, by looking at what we have called hitherto the *standard flat $\mathcal{A}$-connection* $\partial$ (cf. Chapt. VI; Subsection 4.(a), in conjunction with (1.1), (1.2) above), we see that, according to our hypothesis (1.8), one has;

$$R(\partial) := d^1 \circ \partial = 0. \tag{2.6}$$

That is, $\partial$ *is* really *a flat $\mathcal{A}$-connection,* in the sense of (2.5). Of course, we have used in (2.6) the fact that (the given exterior derivative operator) $d^1 \equiv d$ (cf. (1.7)) may be considered as the *1st prolongation of* $\partial$: Indeed, one has, *by definition* (cf. also Chapt. II; (5.15)),

$$d^1 : \Omega^1 = \Omega^1(\mathcal{A}) \longrightarrow \Omega^2(\mathcal{A}) = \Omega^2, \tag{2.7}$$

such that (1.9) *be valid;* however, the same rel. (1.9) corresponds to (1.19), when $\mathcal{E} \equiv \mathcal{A}$. That is, one can have here for (the $\mathcal{A}$-connection) $\partial$ an analogue of Lemma 1.1, pertaining, of course, for the case at hand to the *uniqueness of* (the given!) $d^1 \equiv d$ *only.* ∎

**Remark 2.1.**– Alternatively, concerning (2.6), we further note that one still obtains the relation;

$$D^1|_{\Omega^1} = d^1 \equiv d, \tag{2.8}$$

in the sense that one has then $\mathcal{E} \equiv \mathcal{A}$, as above. That is, we have by (1.19) (see also Chapt. VI; (1.7));

$$D^1(1 \otimes s) = 1 \otimes ds - s \wedge \partial(1) = 1 \otimes ds - 0 = ds, \tag{2.9}$$

*for any* $s \in \Omega^1(U)$ and *open* $U \subseteq X$, which thus again leads to (2.6), in view of (2.2). ∎

Now, our next basic information, concerning the curvature of a given $\mathcal{A}$-connection, is included in the following.

**Lemma 2.1.** *Suppose we have a curvature space $X$, an $\mathcal{A}$-module $\mathcal{E}$ on $X$, and an $\mathcal{A}$-connection $D$ of $\mathcal{E}$. Then, the curvature of $D$, $R(D) \equiv R$* (cf. (2.1), (2.2)), *is an $\mathcal{A}$-morphism between the $\mathcal{A}$-modules $\mathcal{E}$ and $\Omega^2(\mathcal{E})$ on $X$; viz. one has* (see also Chapt. II; (6.9))

$$R \in \mathrm{Hom}_{\mathcal{A}}(\mathcal{E}, \Omega^2(\mathcal{E})) = \mathcal{H}om_{\mathcal{A}}(\mathcal{E}, \Omega^2(\mathcal{E}))(X). \tag{2.10}$$

*In particular, in the case of a vector sheaf $\mathcal{E}$ on $X$, one has*

$$R \in \Omega^2(\mathcal{E}nd\,\mathcal{E})(X). \tag{2.11}$$

**Proof.** To prove (2.10), we just apply the definition of $R$, according to (2.2), based also on (1.25), (1.19) and (1.8), along with Chapt. VI; (3.6). So, *for any* given $\alpha \in \mathcal{A}(U)$ *and* $s \in \mathcal{E}(U)$, with $U$ *open* in $X$, one obtains;

$$\begin{aligned} & R(\alpha \cdot s) = (D^1 \circ D)(\alpha \cdot s) = D^1(D(\alpha s)) = D^1(\alpha \cdot Ds + s \otimes \partial\alpha) \\ (2.12) \quad & = D^1(\alpha \cdot Ds) + D^1(s \otimes \partial\alpha) = \alpha \cdot D^1(Ds) - Ds \wedge \partial\alpha + s \otimes d(\partial\alpha) - \partial\alpha \wedge Ds \\ & = \alpha \cdot D^1(Ds) = \alpha \cdot R(s), \end{aligned}$$

as desired. On the other hand, if $\mathcal{E}$ is a *vector sheaf* on $X$, then, by virtue of (2.10) (see also Chapt. IV; (6.8) and Chapt. II; (6.28)), one has;

$$(2.13) \quad \begin{aligned} & R \in \mathrm{Hom}_{\mathcal{A}}(\mathcal{E}, \Omega^2(\mathcal{E})) = \mathcal{H}om_{\mathcal{A}}(\mathcal{E}, \Omega^2(\mathcal{E}))(X) \equiv \mathcal{H}om_{\mathcal{A}}(\mathcal{E}, \mathcal{E} \otimes_{\mathcal{A}} \Omega^2)(X) \\ & = (\mathcal{H}om_{\mathcal{A}}(\mathcal{E}, \mathcal{E}) \otimes_{\mathcal{A}} \Omega^2)(X) \equiv ((\mathcal{E}nd\, \mathcal{E}) \otimes_{\mathcal{A}} \Omega^2)(X) \equiv \Omega^2(\mathcal{E}nd\, \mathcal{E})(X), \end{aligned}$$

as was to be proved, which also terminates the proof. ▮

Now, assume again that $\mathcal{E}$ is a *vector sheaf* on $X$, a *curvature spave,* as before, such that

$$(2.14) \qquad \mathrm{rk}\, \mathcal{E} = n \in \mathbf{N},$$

and let

$$(2.15) \qquad \mathcal{U} = (U_\alpha)_{\alpha \in I}$$

be a *local frame of* $\mathcal{E}$. Yet, suppose that $\mathcal{E}$ is endowed with an $\mathcal{A}$-*connection* $D$. Then, in view of (2.13) (see also Chapt. III; (4.25), (4.28) and (8.13)), one has;

$$(2.16) \qquad R \in \Omega^2(\mathcal{E}nd\, \mathcal{E})(X) = Z^0(\mathcal{U}, \Omega^2(\mathcal{E}nd\, \mathcal{E})).$$

On the other hand, by looking at a *local gauge* $U_\alpha \in \mathcal{U}$ *of* $\mathcal{E}$, as in (2.15), one obtains;

$$(2.17) \qquad \Omega^2(\mathcal{E}nd\, \mathcal{E})(U_\alpha) = M_n(\Omega^2)(U_\alpha) = M_n(\Omega^2(U_\alpha))$$

(see (2.14) above, in conjunction with Chapt. VI; (9.4), (9.3) and Chapt. V; (8.7), (8.6)). Thus, by virtue of (2.16) and (2.17), one now obtains;

$$(2.18) \quad \begin{aligned} & R \in \Omega^2(\mathcal{E}nd\, \mathcal{E})(X) = Z^0(\mathcal{U}, \Omega^2(\mathcal{E}nd\, \mathcal{E})) \\ & \subseteq C^0(\mathcal{U}, \Omega^2(\mathcal{E}nd\, \mathcal{E})) = \prod_\alpha \Omega^2(\mathcal{E}nd\, \mathcal{E})(U_\alpha) = \prod_\alpha M_n(\Omega^2(U_\alpha)), \end{aligned}$$

such that one has

(2.19) $$R = \left(R^{(\alpha)}\right)_{\alpha \in I} \equiv \left(\left(\omega_{ij}^{(\alpha)}\right)\right),$$

where we have set

(2.20) $$R^{(\alpha)} \equiv \left(\omega_{ij}^{(\alpha)}\right) \in M_n(\Omega^2)(U_\alpha) = M_n(\Omega^2(U_\alpha)),$$

*for any* $\alpha \in I$, as in (2.15). So one concludes that:

(2.21) the *curvature* $R(D) \equiv R$ of a given $\mathcal{A}$*-connection* $D$ of a *vector sheaf* $\mathcal{E}$ (on a *curvature space* $X$), *with* $\operatorname{rk} \mathcal{E} = n \in \mathbf{N}$, is a *0-cocycle of (local)* $n \times n$ *matrices with entries from (local sections of)* $\Omega^2 \equiv \Omega^1 \wedge \Omega^1$ (viz. *(local) "2-forms"* on $X$).

A detailed *local treatment of* the *curvature* $R(D)$, as above, along with applications thereon, will further be considered straight away in the subsequent sections.

# 3. Local form of the curvature. Cartan's structural equation

At first, suppose, as also was done hitherto, that we are given a *curvature space* $X$; that is (see Definition 1.1), we assume that we have the following quintuple

(3.1) $$(\mathcal{A}, \partial, \Omega^1, d^1 \equiv d, \Omega^2),$$

the *topological space* $X$ being the common *base* of the $\mathcal{A}$*-modules* involved in (3.1).

Furthermore, consider an $\mathcal{A}$*-module* $\mathcal{E}$ on $X$, endowed with an $\mathcal{A}$*-connection* $D$, and let

(3.2) $$e^U \equiv \{U; (e_1, \ldots, e_n)\}$$

be a *local gauge* of $\mathcal{E}$ over an *open* set $U \subseteq X$ of *rank* $n \in \mathbf{N}$ (see, for instance, Chapt. VII; (1.5)). Thus, *for any (local) section* $s \in \mathcal{E}(U)$, one has a *unique expression* of the form (ibid. (1.6.1));

(3.3) $$s = \sum_{i=1}^{n} s_i e_i,$$

*such that* $s_i \in \mathcal{A}(U)$, *with* $1 \leqq i \leqq n$.

On the other hand, we already know (cf. Chapt. VII; (1.8.1), (1.11) and (1.13) that *the* given $\mathcal{A}$-*connection* $D$, as above, *is locally expressed through the* corresponding to (3.2) *local* $\mathcal{A}$-*connection matrix*

(3.4) $$\omega^U \equiv \omega = (\omega_{ij}) \in M_n(\Omega^1(U)) = M_n(\Omega^1)(U),$$

such that one has;

(3.5) $$D(e_j) = \sum_{i=1}^{n} e_i \otimes \omega_{ij}, \quad 1 \leqq j \leqq n.$$

Thus, by further applying the curvature of $D$ to the given $\mathcal{A}(U)$-*basis of* $\mathcal{E}(U)$,

(3.6) $$(e_i)_{1 \leqq i \leqq n} \subseteq \mathcal{E}(U),$$

as in (3.2), one obtains (see also (3.5) and (1.19), or even (1.27));

(3.7) $$\begin{aligned} R(e_j) &= (D^1 \circ D)(e_j) = D^1(D(e_j)) = D^1(\textstyle\sum_i e_i \otimes \omega_{ij}) \\ &= \textstyle\sum_i D^1(e_i \otimes \omega_{ij}) = \sum_i (e_i \otimes d\omega_{ij} - \omega_{ij} \wedge D(e_i)) \\ &= \textstyle\sum_i e_i \otimes d\omega_{ij} + \sum_i D(e_i) \wedge \omega_{ij}. \end{aligned}$$

Furthermore, one obtains (cf. (3.5))

(3.8) $$\begin{aligned} \textstyle\sum_i D(e_i) \wedge \omega_{ij} &= \textstyle\sum_i (\sum_k e_k \otimes \omega_{ki}) \wedge \omega_{ij} \\ &= \textstyle\sum_k e_k \otimes (\sum_i \omega_{ki} \wedge \omega_{ij}). \end{aligned}$$

Thus, by (3.7) and (3.8), and by also changing, appropriately, indices in (3.8), *one gets;*

(3.9) $$\begin{aligned} R(e_j) &= \textstyle\sum_i e_i \otimes d\omega_{ij} + \sum_i e_i \otimes (\sum_k \omega_{ik} \wedge \omega_{kj}) \\ &= \textstyle\sum_{i=1}^{n} e_i \otimes (d\omega_{ij} + \sum_k \omega_{ik} \wedge \omega_{kj}) \in \mathcal{E}(U) \otimes_{\mathcal{A}(U)} \Omega^2(U), \end{aligned}$$

*such that* $1 \leqq j \leqq n$. Accordingly, *we* now *set;*

(3.10) $$R(e_j)_i \equiv R_{ij} := d\omega_{ij} + \sum_k \omega_{ik} \wedge \omega_{kj} \in \Omega^2(U),$$

*for any* $1 \leqq i, j \leqq n$, such that one has

$$R \equiv (R_{ij}) \in M_n(\Omega^2(U)) = M_n(\Omega^2)(U), \tag{3.11}$$

a fact, already noticed by (2.20) above in the particular case of a *local gauge* of a *vector sheaf* on $X$ (cf. also (2.21)). Thus, by applying *matrix notation,* one obtains from (3.10),

$$R = d\omega + \omega \wedge \omega, \tag{3.12}$$

as an *element* (local section) *of the* $\mathcal{A}(U)$*-module,* appeared *in* (3.11). The last relation yields in our case the so-called *("second") Cartan's structural equation.*

Now, by specializing, for convenience, to the case of a vector sheaf on $X$, we summarize the preceding into the form of the following.

**Theorem 3.1.** *(Cartan's structural equation). Suppose we are given a curvature space $X$ and a vector sheaf $\mathcal{E}$ on $X$, of rank $n \in \mathbf{N}$, endowed with an $\mathcal{A}$-connection $D$. Moreover, let*

$$\mathcal{U} = (U_\alpha)_{\alpha \in I} \tag{3.13}$$

*be a local frame of $\mathcal{E}$, and*

$$\omega \equiv \left(\omega^{(\alpha)}\right) = \left(\left(\omega_{ij}^{(\alpha)}\right)\right) \in C^0(\mathcal{U}, M_n(\Omega^1)) \tag{3.14}$$

*the corresponding to (3.13) 0-cochain of local $\mathcal{A}$-connection matrices of $D$* (see Chapt. VII; (2.39.1)). *Then, the curvature of $D$ is given by the following 0-cocycle of $\mathcal{U}$; viz. by a global section of $\Omega^2(\mathcal{E}nd\,\mathcal{E})$, as follows* (cf. also (2.18)):

$$\begin{aligned} R(D) \equiv R = \left(R^{(\alpha)}\right) = \left(\left(R_{ij}^{(\alpha)}\right)\right) &\in Z^0(\mathcal{U}, \Omega^2(\mathcal{E}nd\,\mathcal{E})) \\ &= \Omega^2(\mathcal{E}nd\,\mathcal{E})(X) \hookrightarrow \textstyle\prod_\alpha M_n(\Omega^2(U_\alpha)), \end{aligned} \tag{3.15}$$

*such that one has;*

$$R^{(\alpha)} \equiv \left(R_{ij}^{(\alpha)}\right) = d\omega^{(\alpha)} + \omega^{(\alpha)} \wedge \omega^{(\alpha)} \in M_n(\Omega^2(U_\alpha)), \tag{3.16}$$

*for every* $\alpha \in I$ (cf. (3.13), as well as, e.g. (3.10)). ∎

Occasionally, it is still useful to write (3.12) in the following form: *Maurer-Cartan formula;*

$$R = d\omega + \frac{1}{2}[\omega, \omega]. \tag{3.17}$$

That is, one has, where $\omega$ is given by (3.4);

$$2(\omega \wedge \omega) = \omega \wedge \omega + \omega \wedge \omega = \omega \wedge \omega - \omega \wedge \omega \equiv [\omega, \omega], \tag{3.18}$$

which thus yields (3.17), by virtue of (3.12). ∎

On the other hand, assume again that we have the *general framework, referred to (3.12),* and let us further apply the curvature $R$ to an arbitrary (local) section $s \in \mathcal{E}(U)$, as the latter is given by (3.3); thus, since $R$ *is an* $\mathcal{A}$*-morphism* (cf. (2.10)), one has (see also (3.9))

$$\begin{aligned} R(s) &= R(\textstyle\sum_{i=1}^{n} s_i e_i) = \sum_i R(s_i \cdot e_i) = \sum_i s_i \cdot R(e_i) \\ &= \textstyle\sum_i s_i (\sum_k e_k \otimes (d\omega_{ki} + \sum_l \omega_{kl} \wedge \omega_{li})) \\ &= \textstyle\sum_k e_k \otimes (\sum_i s_i (d\omega_{ki} + \sum_l \omega_{kl} \wedge \omega_{li})) \in \mathcal{E}(U) \otimes_{\mathcal{A}(U)} \Omega^2(U). \end{aligned} \tag{3.19}$$

Thereore, one obtains (cf. (3.9));

$$R_{ij}(s) = \sum_{j=1}^{n} s_j \left( d\omega_{ij} + \sum_k \omega_{ik} \wedge \omega_{kj} \right) \in \Omega^2(U), \quad 1 \leqq i \leqq n, \tag{3.20}$$

or even in *matrix notation* one has (cf. (3.12))

$$R(s) \equiv s \cdot R = s \cdot (d\omega + \omega \wedge \omega), \tag{3.21}$$

that is, a *row-matrix,* whose individual terms are given by (3.20). (The abuse of notation, applied for convenience in (3.19) and (3.21), is certainly clear).

Now, in the case of a *line sheaf* on $X$, one obtains, in particular, the following.

**Corollary 3.1.** *Suppose we have a curvature space $X$ and a line sheaf $\mathcal{L}$ on $X$, endowed with an $\mathcal{A}$-connection $D$. Moreover, assume that*

$$\mathcal{U} = (U_\alpha)_{\alpha \in I} \tag{3.22}$$

*is a local frame of $\mathcal{L}$ and let*

$$\omega \equiv (\omega_\alpha) \in C^0(\mathcal{U}, \Omega^1) = \prod_\alpha \Omega^1(U_\alpha) \tag{3.23}$$

*be the 0-cochain of local $\mathcal{A}$-connection "1-forms" of $D$. Then, the curvature of $D$ is given by the 0-cocycle;*

$$\begin{aligned} R(D) \equiv R = (d\omega_\alpha) \in Z^0(\mathcal{U}, d\Omega^1) &\hookrightarrow Z^0(\mathcal{U}, \Omega^2) \\ &= \Omega^2(X) \subseteq \textstyle\prod_\alpha \Omega^2(U_\alpha). \end{aligned} \tag{3.24}$$

**Proof.** The assertion follows immediately from (3.15), (3.16), by just remarking that $\mathcal{E}nd\,\mathcal{L} = \mathcal{A}$ (up to an *$\mathcal{A}$-isomorphism.* See Chapt. II; (6.32), as well as, (5.15)). ∎

# 4. Change of local gauge

As the heading indicates, we examine in this Section the way the local matrices of the curvature (see e.g. (3.16)) alter, when one changes the local gauges that produce these matrices. However, before we embark on the relevant discussion, we first have to comment a bit more on the notion of an *"adjoint representation"* (cf. thus Chapt. VI; (1.37)), which will be of use in our subsequent calculations.

**4.(a). Adjoint representation of a sheaf of groups in a given $\mathcal{A}$-module.**– Suppose we have a **C**-*algebraized space* $(X, \mathcal{A})$, with $X$ an arbitrary *topological space, $\mathcal{G}$ a sheaf of groups* (not necessarily abelian), as well as, an *$\mathcal{A}$-module $\mathcal{E}$* on $X$. In particular, we consider presently below the case of the *general linear group sheaf,* of *order* $n \in \mathbf{N}$, on $X$, $\mathcal{GL}(n, \mathcal{A})$ (see Chapt. IV; (1.26), (1.29)).

Thus, by the *adjoint representation of $\mathcal{GL}(n, \mathcal{A})$ in $\mathcal{E}$,* one means the following group sheaf morphism (see, for instance, Chapt. V; (8.43))

$$\text{Ad} : \mathcal{GL}(n, \mathcal{A}) \longrightarrow (\mathcal{E}nd\, M_n(\mathcal{E}))^{\cdot}, \tag{4.1}$$

such that one has;

$$[\text{Ad}(\alpha)](s) := \alpha s \alpha^{-1}, \tag{4.2}$$

for any $\alpha \in \mathcal{GL}(n, \mathcal{A})(U) = \text{GL}(n, \mathcal{A}(U))$ and $s \in M_n(\mathcal{E})(U) = M_n(\mathcal{E}(U))$, with $U$ open in $X$. (In the second member of (4.2) one considers the usual *matrix multiplication* (of sections), given that *$M_n(\mathcal{E})$ is an $M_n(\mathcal{A})$-bimodule,* according to the hypothesis for $\mathcal{A}$; see Chapt. VI; (1.36)). Of course, it is easy to see that *the map (4.1),* as given by (4.2), *has the following properties,* which, in effect, have been alluded to already, by its definition; viz. *one has*

$$\text{Ad}(\alpha \cdot \beta) = \text{Ad}(\alpha) \cdot \text{Ad}(\beta), \tag{4.3}$$

*for any* $\alpha, \beta \in \mathcal{GL}(n, \mathcal{A})(U)$, as above, so that *one still obtains*

$$\mathrm{Ad}(\alpha^{-1}) = \mathrm{Ad}(\alpha)^{-1}, \tag{4.4}$$

*for any* $\alpha \in \mathrm{GL}(n, \mathcal{A}(U))$.

In particular, suppose now that we are given a *curvature space* $X$ (cf. Definition 1.1), so that let us further consider the $\mathcal{A}$*-modules* $\Omega^1$ and $\Omega^2$, together with the given *exterior derivative operator*

$$d^1 \equiv d : \Omega^1 \longrightarrow \Omega^2 \tag{4.5}$$

(see (1.7), along with (1.16)). Now, it is clear that *the operator (4.5) can be extended* ("coordinate-wise") *to a similar one between the corresponding matrix spaces for* $\Omega^1$ *and* $\Omega^2$, as above (for simplicity, we retain the same notation for the extended map), so that *one obtains the* following *commutative diagram;*

$$\begin{array}{ccc} M_n(\Omega^1) & \xrightarrow{Ad(\alpha)} & M_n(\Omega^1) \\ {\scriptstyle d}\downarrow & & \downarrow{\scriptstyle d} \\ M_n(\Omega^2) & \xrightarrow{Ad(\alpha)} & M_n(\Omega^2), \end{array} \tag{4.6}$$

*for any* $\alpha \in \mathcal{GL}(n, \mathcal{A})(U)$, and $U$ *open* in $X$. *Equivalently,* one has

$$d \circ \mathrm{Ad}(\alpha) = \mathrm{Ad}(\alpha) \circ d, \tag{4.7}$$

*for every* $\alpha \in \mathcal{GL}(n, \mathcal{A})(U) = \mathrm{GL}(n, \mathcal{A}(U))$, as before: Indeed, *for every* $s \in M_n(\Omega^1)(U) = M_n(\Omega^1(U))$, one has (cf. also (4.2), (1.9) and Chapt. VI; (1.27.2));

$$\begin{aligned} (d \circ \mathrm{Ad}(\alpha))(s) &= d(\mathrm{Ad}(\alpha)(s)) = d(\alpha s \alpha^{-1}) \\ &= \alpha\, d(s\alpha^{-1}) - (s\alpha^{-1}) \wedge \partial\alpha = \alpha((ds)\alpha^{-1} - s \wedge \partial(\alpha^{-1})) + \alpha^{-1}\partial\alpha \wedge s \\ &= \alpha(ds)\alpha^{-1} + (\alpha\partial(\alpha^{-1}) + \alpha^{-1}\partial\alpha) \wedge s \\ &= [\mathrm{Ad}(\alpha)](ds) + (\tilde{\partial}(\alpha^{-1}) + \tilde{\partial}(\alpha)) \wedge s = (\mathrm{Ad}(\alpha) \circ d)(s), \end{aligned} \tag{4.8}$$

that was actually to be proved. ∎

We come now to our main objective of this discussion, as this is exposed by the ensuing Subsection.

**4.(b). Change of curvature under a gauge transformation.**– Suppose that we have a *curvature space* $X$ and an $\mathcal{A}$*-module* $\mathcal{E}$ on $X$, equipped with an $\mathcal{A}$*-connection* $D$. Moreover, assume that we are given two *local gauges of* $\mathcal{E}$, say,

$$e^U \equiv \{U; (e_i, \ldots, e_n)\} \tag{4.9}$$

*and*

$$f^V \equiv \{V; (f_i, \ldots, f_n)\} \tag{4.10}$$

over the *open sets* $U$ *and* $V$ of $X$, *such that*

$$U \cap V \neq \emptyset, \tag{4.11}$$

and of the *same (finite) rank* $n \in \mathbf{N}$ see also Chapt. VII; (2.1)). Thus, suppose further that

$$g \equiv (g_{ij}) \in \mathrm{GL}(n, \mathcal{A}(U \cap V)) = \mathcal{GL}(n, \mathcal{A})(U \cap V) \tag{4.12}$$

is the *local gauge transformation*, which realizes the change from $e^U$ to $f^V$, such that one has;

$$f_j = \sum_{i=1}^{n} g_{ij} e_i, \quad 1 \leqq j \leqq n \tag{4.13}$$

(see Chapt. VII; (2.3), (2.5), along with Remark 2.1).

Now, denoting by $\omega$ and $\omega'$ the *local* $\mathcal{A}$*-connection matrices of* $D$, which are associated with the given local gauges $e^U$ and $f^V$, respectively (see e.g. (3.4)), one has the relation;

$$\omega' \equiv g^*(\omega) = \mathrm{Ad}(g^{-1})\omega + \tilde{\partial}(g) \tag{4.14}$$

(*transformation law of potentials*. Cf. Chapt. VII; (2.23)).

Thus, our next objective is now to find the corresponding relation to (4.14), pertaining to the *curvatures* $R$ *and* $R'$, which are defined by $\omega$ and $\omega'$, respectively (cf., for instance, (3.12)), when one takes the previous gauge transformation $g$ into account. So, by *"differentiating"* (4.14), viz. by applying to it the *1st exterior derivation* $d^1 \equiv d$ (see (1.7) and (3.1)), one obtains (cf. also (1.12), (4.7) and (3.12), concerning the calculations that follow);

$$
\text{(4.15)}\qquad
\begin{aligned}
d\omega' &= d(\mathrm{Ad}(g^{-1})\omega + \tilde{\partial}(g)) = d(\mathrm{Ad}(g^{-1})\omega) + d(\tilde{\partial}(g))\\
&= d(\mathrm{Ad}(g^{-1})\omega) = \mathrm{Ad}(g^{-1})(d\omega) = \mathrm{Ad}(g^{-1})(R - \omega\wedge\omega)\\
&= \mathrm{Ad}(g^{-1})R - \mathrm{Ad}(g^{-1})(\omega\wedge\omega).
\end{aligned}
$$

On the other hand, one still obtains (see (3.12) and (4.14));

$$
\text{(4.16)}\qquad
\begin{aligned}
d\omega' &= R' - \omega'\wedge\omega' = R' - (\mathrm{Ad}(g^{-1})\omega + \tilde{\partial}(g))\wedge(\mathrm{Ad}(g^{-1})\omega + \tilde{\partial}(g))\\
&= R' - \mathrm{Ad}(g^{-1})(\omega\wedge\omega) - \mathrm{Ad}(g^{-1})\omega\wedge\tilde{\partial}(g) - \tilde{\partial}(g)\wedge\mathrm{Ad}(g^{-1})\omega\\
&\quad + \tilde{\partial}(g)\wedge\tilde{\partial}(g) = R' - \mathrm{Ad}(g^{-1})(\omega\wedge\omega).
\end{aligned}
$$

(In this context, we have also applied above the relations (cf. (4.2) and (4.12)): $\mathrm{Ad}(g^{-1})\omega\wedge\mathrm{Ad}(g^{-1})\omega \equiv g^{-1}\omega g\wedge g^{-1}\omega g = g^{-1}(\omega g\wedge g^{-1}\omega)g \equiv g^{-1}(\omega\wedge g\wedge g^{-1}\wedge\omega)g = g^{-1}(\omega\wedge(gg^{-1})\wedge\omega)g = g^{-1}(\omega\wedge\omega)g \equiv \mathrm{Ad}(g^{-1})(\omega\wedge\omega)$). Thus, by (4.15) and (4.16), one finally has;

$$
\text{(4.17)}\qquad g^*(R) \equiv R' = \mathrm{Ad}(g^{-1})R \equiv g^{-1}Rg.
$$

So the last relation represents the *local behaviour of the curvature* $R$ of a given $\mathcal{A}$-connection, *under a gauge transformation.*

Thus, an immediate consequence of (4.17) is now the fact that:

(4.18) *any gauge transformation preserves* (the eventual) *flatness* of a given $\mathcal{A}$-connection (cf. (2.5)).

Now, suppose, in particular, that $\mathcal{E}$ is a *vector sheaf,* of *rank* $n \in \mathbf{N}$, on a *curvature space* $X$, as before, such that $\mathcal{E}$ is also endowed with an $\mathcal{A}$*-connection* $D$. Moreover, let

$$
\text{(4.19)}\qquad \mathcal{U} = (U_\alpha)_{\alpha\in I}
$$

be a *local frame* of $\mathcal{E}$, and

$$
\text{(4.20)}\qquad (g_{\alpha\beta}) \in Z^1(\mathcal{U}, \mathcal{GL}(n,\mathcal{A}))
$$

a *coordinate 1-cocycle of* $\mathcal{E}$, relative to $\mathcal{U}$ (see Chapt. VII; Lemma 1.1, or even Theorem 2.1). Thus, for any *(local) gauge transformation* of $\mathcal{E}$,

$$
\text{(4.21)}\qquad g_{\alpha\beta} \in \mathcal{GL}(n,\mathcal{A})(U_{\alpha\beta}) = \mathrm{GL}(n, \mathcal{A}(U_{\alpha\beta})),
$$

*such that*

$$
\text{(4.22)}\qquad U_{\alpha\beta} := U_\alpha \cap U_\beta \neq \emptyset,
$$

with $\alpha, \beta \in I$, as in (4.19) (see also Chapt. V; (1.3) – (1.5)), one can consider, according to (4.17), the corresponding *local expression of changing the curvature* $R(D) \equiv R$ of the given $\mathcal{A}$-connection $D$; viz. one has

$$g^{*}_{\alpha\beta}(R) \equiv R' = \mathrm{Ad}(g^{-1}_{\alpha\beta})R \equiv g^{-1}_{\alpha\beta} R g_{\alpha\beta}. \tag{4.23}$$

In this context, we also note (as was actually already the case with (4.17), cf. (4.12), as well) that $R$ above *stands, in effect, for*

$$R^{(\alpha\beta)} \equiv \left(R^{(\alpha\beta)}_{ij}\right) \in M_n(\Omega^2(U_{\alpha\beta})) \tag{4.24}$$

(similar observation for $R'$; cf. also (3.16), while we further notice that $M_n(\Omega^2)$ *is an* $M_n(\mathcal{A})$*-bimodule* on $X$, a fact actually *true, for every* $M_n(\Omega^p)$, *with* $p \in \mathbf{Z}_+$).

In particular, consider a *line sheaf* $\mathcal{L}$ on $X$, which is further endowed with an $\mathcal{A}$*-connection* $D$; viz. we are given a *Maxwell field*

$$(\mathcal{L}, D) \tag{4.25}$$

on $X$ (cf. thus Chapt. VI; (18.11)). Therefore, if

$$(g_{\alpha\beta}) \in Z^1(\mathcal{U}, \mathcal{A}^{\cdot}) \tag{4.26}$$

is a *coordinate 1-cocycle of* $\mathcal{L}$, associated with a given *local frame* $\mathcal{U}$ *of* $\mathcal{L}$ (cf., for instance, (4.19), as well as, (4.20) for $n = 1$, along with Chapt. V; (2.59), (2.60)), one gets (cf. (4.23))

$$R' = g^{-1}_{\alpha\beta} R g_{\alpha\beta}, \tag{4.27}$$

such that (see (3.24))

$$R \in \Omega^2(X), \tag{4.28}$$

while one still has (cf. (4.22) and (4.26))

$$g_{\alpha\beta} \in \mathcal{A}^{\cdot}(U_{\alpha\beta}) = \mathcal{A}(U_{\alpha\beta})^{\cdot}; \tag{4.29}$$

of course, by analogy with (4.24) (viz., for $n = 1$), one actually has for $R$ (and analogously for $R'$) in (4.27):

$$R^{(\alpha\beta)} \equiv R|_{U_{\alpha\beta}}. \tag{4.30}$$

Therefore, in view of (4.27), (4.29) and of the *commutativity of* (the $\mathbf{C}$-algebra sheaf) $\mathcal{A}$, one finally obtains;

(4.31) $$R' = R.$$

That is,

(4.32) in the case of a *Maxwell field* $(\mathcal{L}, D)$ on a *curvature space* $X$, the *curvature of $D$ does not change* (form) *under* (the action of) *a gauge transformation*. So, said otherwise, the *curvature* created by (or corresponding to) a Maxwell field *is space invariant.*

In this regard, see also Chapter X; Sections 4, 5 in the sequel for an appropriate *physical interpretation* of the last conclusion.

## 5. Flat $\mathcal{A}$-connections (contn'd)

The purpose of this Section is to relate the notion of a *flat $\mathcal{A}$-connection* $D$, as given above, that is, when one has

(5.1) $$R(D) \equiv R = 0$$

(see (2.5)), with the various notions of *"flatness"*, pertaining to a given *$\mathcal{A}$-module,* or, in particular, to a *vector sheaf,* which have already been considered in the preceding (cf. Chapt. V; Section 5, Chapt. VI; Subsection 11.(a) and Chapt. VII; Section 6). As we shall see, *there are important* particular *instances,* including, of course, that one of the standard differential geometry, where *all notions of flatness,* as alluded to before, *are,* in fact, *equivalent.*

So, by supplementing our previous information, as in the places just quoted above, we first state the following.

**Theorem 5.1.** *Suppose we are given a differential triad* $(\mathcal{A}, \partial, \Omega^1)$ *on a topological space* $X$*, along with a vector sheaf* $\mathcal{E}$ *on* $X$*, of rank* $n \in \mathbf{N}$*. Then, the following three assertions are equivalent* (see Chapt. VII; Theorem 6.1):

i) $\mathcal{E}$ is *$\partial$-flat.*

ii) $\mathcal{E}$ has an *integrable $\mathcal{A}$-connection,* and

iii) $\mathcal{E}$ has an $\mathcal{A}$-connection, supplying a *complete parallelism.*

In particular, if $X$, as above, is a *curvature space* (cf. Definition 1.1), then *any one of the* previous *three* equivalent assertions *implies* that;

iv) $\mathcal{E}$ has a *flat $\mathcal{A}$-connection* (cf. (5.1)).

**Proof.** Assume, for example, that the given vector sheaf $\mathcal{E}$ *has an $\mathcal{A}$-connection $D$, providing a complete parallelism.* Therefore (cf. Chapt. VII; Definition 6.1), there exists a *(parallel) local frame of $\mathcal{E}$*, say,

$$\mathcal{U} = (U_\alpha)_{\alpha \in I}, \tag{5.2}$$

such that the 0-cochain of local $\mathcal{A}$-connection matrices of $D$, which is associated with $\mathcal{U}$,

$$\omega \equiv \left(\omega^{(\alpha)}\right) \in C^0(\mathcal{U}, M_n(\Omega^1)), \tag{5.3}$$

is trivial; viz. one has

$$\omega^{(\alpha)} \equiv \left(\omega_{ij}^{(\alpha)}\right) = 0 \in M_n(\Omega^1(U_\alpha)), \tag{5.4}$$

*for every* $\alpha \in I$ (cf. Chapt. VII; (6.7), (6.18) and (6.19)). Accordingly, by virtue of *Cartan's structural equation* (cf. (3.16)), one obtains

$$R^{(\alpha)} = 0 \in M_n(\Omega^2(U_\alpha)), \quad \textit{with } \alpha \in I, \tag{5.5}$$

so that *one gets* $R = 0$, as well (cf. also (2.19), (2.20) and (5.4) above); thus, the above proves actually that:

(5.6) *any $\mathcal{A}$-connection, providing a complete parallelism is flat.*

This, of course, yields now the assertion of the theorem, and the proof is complete. ∎

**Scholium 5.1.**– We have already seen (cf. Chapt. VII; (6.12)) that *every flat vector sheaf is $\partial$-flat* (see Chapt. V; Definition 5.1 and Chapt. VII; Definition 6.2). Consequently, as follows from the previous Theorem 5.1, one further obtains that:

(5.7) *every flat vector sheaf* (on a given curvature space) *admits a flat $\mathcal{A}$-connection.*

This, of course, justifies also the terminology applied so far. True, *in the classical case the two notions that appear in (5.7) are,* in effect, *equivalent.* However, *this is not necessarily the case here,* a fact that will become more apparent through the ensuing discussion. On the other hand, as we shall presently see, something better is actually valid

in the particular case of a *line sheaf* over a suitable curvature space (see thus Theorem 5.2 in the sequel).

Now, the question of *whether flatness of a* given $\mathcal{A}$-*connection implies any one of the three equivalent notions,* which *appear in* the previous *Theorem 5.1,* is encountered by the next Subsection.

**5.(a). Frobenius integrability condition.**– We start with the following useful result, providing also half of the answer to the above question. So we have the next.

**Lemma 5.1.** *Suppose we are given a curvature space* $X$ *and let* $\mathcal{E}$ *be an* $\mathcal{A}$-*module on* $X$, *endowed with an* $\mathcal{A}$-*connection* $D$. *Moreover, let*

$$e^U \equiv \{U; (e_i, \ldots, e_n)\} \tag{5.8}$$

*be a local gauge of* $\mathcal{E}$, *of rank* $n \in \mathbf{N}$, *over an open set* $U \subseteq X$ (see, for instance, (4.9) in the preceding, or even Chapt. VII; (2.1), *and*

$$g \in \mathrm{GL}(n, \mathcal{A}(U \cap V)) = \mathcal{GL}(n, \mathcal{A})(U \cap V) \tag{5.9}$$

*a gauge transformation, sending (5.8) to another local gauge of* $\mathcal{E}$, (of the same rank, as (5.8), over an open $V \subseteq X$, with $U \cap V \neq \emptyset$), *this being also a parallel one, with respect to the given* $\mathcal{A}$-*connection* $D$ (see Chapt. VII; (6.3), along with (4.10), (4.12) and (4.13) in the preceding).

*Then,* $D$ *restricted to* $U \cap V$ (cf. Chapt. VI; (6.22)) *is flat; that is, one has:*

$$R(D|_{U\cap V}) = R(D)|_{U\cap V} \equiv R|_{U\cap V} = 0 \tag{5.10}$$

(cf. also (2.11) and (3.11), as well as, (9.28) in the sequel).

**Proof.** Suppose that

$$f^V \equiv \{V; (f_j)_{1 \leqq j \leqq n}\} \tag{5.11}$$

is a *parallel local gauge of* $\mathcal{E}$, with respect to $D$ (as in the statement of the theorem), arising from a *gauge transformation* $g$, like (5.9), acting on (5.8). Thus, by further considering the *local* $\mathcal{A}$-*connection matrices of* $D$, that are associated with (5.8) and (5.11), say,

$$\omega_U \equiv \omega \in M_n(\Omega^1(U)) \;\; \textit{and} \;\; \omega_V \equiv \omega' \in M_n(\Omega^1(V)), \tag{5.12}$$

respectively, one has, according to the *transformation law of potentials* (cf. Chapt. VII; Lemma 2.1), the relation;

$$\omega' = \mathrm{Ad}(g^{-1})\omega + \tilde{\partial}(g) \tag{5.13}$$

(of course, all the *section-matrices* in (5.13) are defined *over* $U \cap V \neq \emptyset$). However, in view of the *hypothesis for (5.11),* one has (cf. Chapt. VII; (6.3))

$$\omega' = 0, \tag{5.14}$$

so that one now obtains, by (5.13),

$$\mathrm{Ad}(g^{-1})\omega + \tilde{\partial}(g) = 0. \tag{5.15}$$

In other words, one has

$$g^{-1}\omega g + g^{-1}\partial(g) = 0, \tag{5.16}$$

or even

$$\omega g + \partial(g) = 0. \tag{5.17}$$

Thus, one further obtains;

$$\partial(g) = -\omega \cdot g. \tag{5.18}$$

Now, we next apply the operator $d^1 \equiv d$ (see e.g. (3.1)) to the above rel. (5.17), alias, we "differentiate" (5.17). Thus, we obtain (cf. also (1.8), (1.9), (5.18) and (3.12));

$$\begin{aligned} 0 &= d(\omega g + \partial g) = d(\omega g) + d(\partial g) = d(\omega g) \\ &= (d\omega)g - \omega \wedge \partial g = (d\omega)g - \omega \wedge (-\omega g) = (d\omega)g + (\omega \wedge \omega)g \\ &= (d\omega + \omega \wedge \omega) \cdot g = R \cdot g. \end{aligned} \tag{5.19}$$

That is, one finally has

$$R = 0, \tag{5.20}$$

as actually was to be proved. ∎

**Scholium 5.2.**- Concerning (5.18), we further note, for use presently below, that:

the above rel. (5.18) *is* still *equivalent with the relation*

(5.21) $$\tilde{\partial}(g^{-1}) = \omega.$$

(See also Chapt. VI; (1.27.2), as well as, (1.42) therein, concerning the notation applied in (5.21)). Thus, *one* finally *concludes that:*

(5.22) $$\omega \in \operatorname{im}\tilde{\partial}.$$

Now, the last relation characterizes, in fact, what we are going to call, right below, *trivial $\mathcal{A}$-connections.* See thus (5.47) and (5.49) in the sequel, along with (5.53) for *another justification of (5.10), based on (5.22).*

Thus, by summarizing the preceding, we are led to the following comments:

(5.23) Suppose we are given an *$\mathcal{A}$-module $\mathcal{E}$* on a *curvature space $X$*, endowed with an *$\mathcal{A}$-connection $D$.* Then, (5.10), viz. the *vanishing of the curvature,* is a *necessary condition for the existence of a* (local) *gauge transformation* (like (5.9)), *sending a given local gauge of $\mathcal{E}$ to a parallel one,* with respect to $D$.

On the other hand, the *sufficiency of (5.10),* to get the above situation, as described by (5.23) (or even by the *hypothesis of Lemma 5.1*), is actually, as it concerns the classical context, the *Frobenius integrability condition;* in other words,

(5.24) (5.10), viz. the *vanishing of the curvature, is a sufficient condition* too, *in order the equation* (cf. (5.17))

$$\omega g + \partial(g) = 0,$$

or, *equivalently* (see (5.21)), *the equation*

$$\tilde{\partial}(g^{-1}) = \omega,$$

*to have a solution, with respect to $g$,* as in (5.9), *when $\omega$* (the *"potential" of $D$*) *is known.*

In this context, we are able to give a positive answer to the above question,

by further imposing suitable *topological and cohomological conditions on* $\mathcal{A}$, as we shall see later on (cf. thus Chapter XI in the sequel. Yet, concerning the classical case, see, for instance, M. Crampin - F.A.E. Pirani [1: p. 146, and p. 282f], or even S. Kobayashi [1: p. 5]).

However, by restricting ourselves to *line sheaves,* we can have a complete answer to our previous question, through other means, that we already have at our disposal; yet, this by considering a suitable *restriction on the curvature space* $X$ involved, as above. So one has the following.

**Theorem 5.2.** *Suppose we have a curvature space* $X$, *being also a sheaf exponential* (cf. Chapt. VII; (7.1)) *and paracompact (Hausdorff) space. Moreover, let* $\mathcal{L}$ *be a line sheaf on* $X$. *Then, the following assertions are equivalent:*

1) $\mathcal{L}$ is $\partial$-*flat.*
2) $\mathcal{L}$ has an *integrable* $\mathcal{A}$-*connection.*
3) $\mathcal{L}$ admits an $\mathcal{A}$-*connection, yielding a complete parallelism,* and
4) $\mathcal{L}$ has a *flat* $\mathcal{A}$-*connection.*

**Proof.** We already know, by Theorem 5.1, that *the first three assertions are equivalent, while any one of them implies the fourth.* So we prove now, for instance, that *4) implies 1),* as well; viz. *if* $\mathcal{L}$ *has a flat* $\mathcal{A}$-*connection* $\widetilde{D}$, then, $\mathcal{L}$ *is* $\partial$-*flat* (cf. Chapt. VII; Definition 6.2):

Thus, assume that

$$\mathcal{U} = (U_\alpha)_{\alpha \in I} \tag{5.25}$$

is a *local frame of* $\mathcal{L}$, and let

$$(g_{\alpha\beta}) \in Z^1(\mathcal{U}, \mathcal{A}^{\cdot}) \tag{5.26}$$

be a *coordinate 1-cocycle of* $\mathcal{L}$, relative to $\mathcal{U}$. Furthermore, suppose that

$$\theta \equiv (\theta_\alpha) \in C^0(\mathcal{U}, \Omega^1) \tag{5.27}$$

is a *0-cochain of local* $\mathcal{A}$-*connection "1-forms"* of the given $\mathcal{A}$-connection $\widetilde{D}$ of $\mathcal{L}$, with respect to $\mathcal{U}$, while

$$\delta(D_\alpha) \in Z^1(\mathcal{U}, \Omega^1) \tag{5.28}$$

is the corresponding *Levi-Civita* $\mathcal{A}$-*connection 1-cocycle of* $\mathcal{L}$, associated with $\mathcal{U}$. So one has

$$\delta(D_\alpha) = \delta(\theta_\alpha) = \tilde{\partial}(g_{\alpha\beta}) \tag{5.29}$$

(see Chapt. VI; (11.8), in conjunction with Theorem 10.1. Yet, see also, for example, Chapt. VII; (2.59)).

On the other hand, according to our hypothesis for $X$ (cf. Chapt. VII; (7.1), and Chapt. V; (6.1)), one has the following *short exact exponential sheaf sequence;*

$$0 \longrightarrow \mathbf{Z} \longrightarrow \mathcal{A} \xrightarrow{e} \mathcal{A}^{\cdot} \longrightarrow 1. \tag{5.30}$$

Thus, since, by hypothesis, $X$ is a *paracompact (Hausdorff)* space too, given the 1-cocycle (5.26), *there exists a 1-cochain*

$$(f_{\alpha\beta}) \in C^1(\mathcal{V}, \mathcal{A}) \tag{5.31}$$

*such that* one has

$$g_{\alpha\beta} = e(f_{\alpha\beta}) \tag{5.32}$$

(*"liftable cohomology"*; cf. Chapt. III; Lemma 5.1). Moreover, one obtains (loc. cit.), in view also of (5.30);

$$\delta(f_{\alpha\beta}) \in Z^2(\mathcal{V}, \ker e = \mathbf{Z}). \tag{5.33}$$

Furthermore, according to the *exponential sheaf diagram,* that is associated with (5.30), by hypothesis for $X$ (cf. Chapt. VII; (7.1)), one has the following *commutative triangle,* supplementing (5.30);

$$\begin{array}{ccc} \mathcal{A} & \xrightarrow{e} & \mathcal{A}^{\cdot} \\ & \partial \searrow \quad \swarrow \frac{1}{2\pi i}\tilde{\partial} & \\ & \Omega^1. & \end{array} \tag{5.34}$$

Therefore, one has, in view of (5.32),

$$\tilde{\partial}(e(f_{\alpha\beta})) = \tilde{\partial}(g_{\alpha\beta}) = 2\pi i \cdot \partial(f_{\alpha\beta}). \tag{5.35}$$

Moreover, one still obtains;

$$\partial(f_{\alpha\beta}) \in Z^1(\mathcal{V}, \partial\mathcal{A}) \hookrightarrow Z^1(\mathcal{V}, \Omega^1). \tag{5.36}$$

Indeed, one has

$$\delta(\partial(f_{\alpha\beta})) = \partial(\delta(f_{\alpha\beta})) = 0, \tag{5.37}$$

as immediately follows from (5.33) and Chapt. VI; (1.5), which thus justifies (5.36). Therefore, one gets a cohomology class

$$[\partial(f_{\alpha\beta})] \in H^1(X, \Omega^1), \tag{5.38}$$

that will be of use right below.

Namely, consider now the following *short exact sequence* (of $\mathbf{C}$-*vector space sheaves*),

$$0 \longrightarrow \ker d \longrightarrow \Omega^1 \longrightarrow d\Omega^1 \longrightarrow 0, \tag{5.39}$$

along with the corresponding *long exact sequence in (Čhech) cohomology,*

$$\cdots \longrightarrow \Gamma(X, d\Omega^1) \longrightarrow H^1(X, \ker d) \longrightarrow H^1(X, \Omega^1) \longrightarrow \cdots \tag{5.40}$$

(cf. Chapt. III; Theorem 5.3. See also Theorem 8.1 therein). Thus, the exactness of the last sequence and (1.8) yield now for (5.38) the relation;

$$\begin{aligned} [\partial(f_{\alpha\beta})] &\in \ker(H^1(X, \ker d) \longrightarrow H^1(X, \Omega^1)) \\ &= \operatorname{im}(\Gamma(X, d\Omega^1) \longrightarrow H^1(X, \ker d)). \end{aligned} \tag{5.41}$$

On the other hand, based on the hypothesis for (5.27) and on (3.24), one obtains;

$$R = (d\theta_\alpha) \in Z^0(\mathcal{U}, d\Omega^1) = \Gamma(X, d\Omega^1) \tag{5.42}$$

(cf. also Chapt. III; (4.55) and (8.13)). At the same time, one infers from (5.41) that (cf. Chapt. III; (5.21), for the notation applied)

$$\delta^*(R) = [\partial(f_{\alpha\beta})] \in H^1(X, \ker d). \tag{5.43}$$

Furthermore, by applying again Lemma 5.1 of Chapter III to (5.42), in conjunction with (5.39), one concludes the existence of a *0-cochain*

$$(\theta'_\alpha) \in C^0(\mathcal{V}, \Omega^1), \tag{5.44}$$

such that one has;

$$R = (d\theta'_\alpha) \text{ and } \delta^*(R) = [(\delta\theta'_\alpha)] \in H^1(X, \ker d). \tag{5.45}$$

Therefore, based on (5.45), (5.41) (or even (5.43)), along with (5.29) and (5.35), we conclude that

(5.46) $$\delta(\theta'_\alpha) = \partial(f_{\alpha\beta}) = \frac{1}{2\pi i}\delta(\theta_\alpha) = \frac{1}{2\pi i}\tilde{\partial}(g_{\alpha\beta}).$$

Therefore, since *by* our *hypothesis,* $R = 0$ (and $\delta^*$, as in (5.43) or (5.45), is a **C**-*linear map,* cf. Chapt. III; (5.20)), one has from (5.46)

(5.47) $$[\tilde{\partial}(g_{\alpha\beta})] = 0 \in H^1(X, \Omega^1).$$

However, the last relation implies already that

(5.48) $$\partial(g_{\alpha\beta}) = 0$$

(*Selesnick's Lemma;* see Chapt. VII; Lemma 7.2). That is, the given line sheaf $\mathcal{L}$ *is* $\partial$-*flat,* that actually was our claim, and this terminates the proof. ∎

In this context, a byproduct of the previous proof, which is of a particular significance for the sequel, as well (see thus Theorem 11.1 below), is now the following.

**Lemma 5.2.** *Suppose we have a sheaf exponential (Hausdorff) paracompact curvature space* $X$. *Moreover, let* $\mathcal{L}$ *be a line sheaf on* $X$, *endowed with an* $\mathcal{A}$-*connection* $D$. *Then, the curvature of* $D$, $R(D) \equiv R$, *supplies a 2-dimensional integral cohomology class of* $X$. *That is, denoting the latter class by* $c(R)$, *one has;*

(5.49) $$c(R) \in H^2(X, \mathbf{Z}).$$

**Scholium 5.3.**– By looking at the *canonical embedding*

(5.50) $$\mathbf{Z} \overset{\subseteq}{\rightarrow} \mathbf{C}$$

*of* the *constant sheaves* concerned, along with the corresponding map in cohomology,

(5.51) $$H^p(X, \mathbf{Z}) \longrightarrow H^p(X, \mathbf{C}), \quad p \in \mathbf{Z}_+,$$

one calls *(p-dimensional) integral cohomology class,* any element in the image of (5.51).

So this explains now the terminology applied before for (5.49). Yet, since by our hypothesis in the previous Lemma 5.2, $X$ is a (Hausdorff) *paracompact* space, one may consider, of course, in the same relation *Čech cohomology* (cf. Chapt. III; (8.11), or even Theorem 8.1).

**Proof of Lemma 5.2.** Having, by hypothesis, the framework of the above Theorem 5.2, we first remark that (see (5.42))

$$R = (d\theta_\alpha) \in \Gamma(X, d\Omega^1) = Z^0(\mathcal{U}, d\Omega^1). \tag{5.52}$$

Furthermore, as a consequence of (5.52) and the exactness of (5.39), one also obtains (see (5.45))

$$R = (d\theta'_\alpha) = (d\theta_\alpha), \tag{5.53}$$

such that one has (cf. (5.46));

$$\delta(\theta'_\alpha) = \partial(f_{\alpha\beta}) = \frac{1}{2\pi i}\tilde{\partial}(g_{\alpha\beta}), \tag{5.54}$$

while, we still have (see (5.33)):

$$\delta(f_{\alpha\beta}) \equiv (\lambda_{\alpha\beta\gamma}) \in Z^2(\mathcal{V}, \mathbf{Z}). \tag{5.55}$$

Thus, based now on the last relation, we further obtain

$$c(R) = \left[\frac{1}{2\pi i}R\right] = \frac{1}{2\pi i}[R] := [(\lambda_{\alpha\beta\gamma})] \in H^2(\mathcal{V}, \mathbf{Z}), \tag{5.56}$$

which is just our claim. (See also the next Scholium 5.4, concerning the last definition in (5.56); yet, cf. Chapt. IX; (5.15) in the sequel). ▮

**Scholium 5.4.**- Concerning the last defining relation in (5.56), this is justified as follows:

First, one has (see (5.54))

$$\begin{aligned}\delta(\theta'_\alpha) = \partial(f_{\alpha\beta}) \in (\partial\mathcal{A})(U_{\alpha\beta}) &\subseteq (\ker d)(U_{\alpha\beta}) \\ &\subseteq \Omega^1(U_{\alpha\beta})\end{aligned} \tag{5.57}$$

(see also e.g. (4.22) for the notation applied above). Hence, in particular, one obtains;

$$\partial(f_{\alpha\beta}) \in Z^1(\mathcal{U}, \partial\mathcal{A}) \overset{\subseteq}{\rightarrow} Z^1(\mathcal{U}, \Omega^1). \tag{5.58}$$

Yet, since (cf. (5.53))

$$R = (d\theta'_\alpha) = (d\theta_\alpha), \tag{5.59}$$

one now defines the *2-dimensional cohomology class of* $R$ (a *closed "2-form"*, as we shall see right below), by the relation;

$$(5.60) \qquad [R] := [2\pi i \cdot \delta(f_{\alpha\beta})] = 2\pi i \cdot [\delta(f_{\alpha\beta})],$$

such that, in view of (5.33), one obtains

$$(5.61) \qquad [\delta(f_{\alpha\beta})] \equiv [(\lambda_{\alpha\beta\gamma})] \in H^2(X, \mathbf{Z}).$$

So this clarifies the situation in (5.56). (In this regard, see also Chapt. IX; Section 5. Yet, the appearance of the *"coefficient* $2\pi i$*"* in (5.60), or (5.56), is essentially due to the classical definition of the "exp" *function,* in our case, *section;* cf. also Chapter XI in the sequel).

Of course, connections of the above with what we have called in the preceding *Selesnick-Chern class of* $\mathcal{L}$, $c_{\mathcal{A}}(\mathcal{L})$, with $\mathcal{L}$ a given *line sheaf,* are obviously in force; so see, for instance, Chapt. VII; Theorem 7.1 and its Corollary 7.1.

Now, although not directly connected with the material of this particular Subsection, we do give, for convenience of reference, the following result, being also a further application of the aforementioned *Selesnick's lemma,* in conjunction with *Atiyah's criterion* about the existence of an $\mathcal{A}$-connection (cf. Chapt. VI; Theorem 11.1, or even Corollary 12.1). Thus, we have.

**Proposition 5.1.** *Suppose we are given a curvature space* $X$*, which is further endowed with a commutative triangle, like (5.34), where, in particular, the (exponential) morphism* $e$ *is also surjective* (take, for instance, a *sheaf exponential curvature space* $X$). *Moreover, let* $\mathcal{L}$ *be a line sheaf on* $X$*, admitting an* $\mathcal{A}$*-connection. Then,* $\mathcal{L}$ *has a flat* $\mathcal{A}$*-connection too.*

**Proof.** Since, $\mathcal{L}$ admits, by hypothesis, an $\mathcal{A}$-connection, the rel. (5.47) is satisfied (*Atiyah's criterion;* loc. cit.). Hence (*Selesnick's lemma,* cf. Chapt. VII; Lemma 7.2), (5.48) is still in force; in other words, (see Chapt. VII; Definition 6.2), *the* given *line sheaf* $\mathcal{L}$ *is* $\partial$*-flat,* so that, according to the previous Theorem 5.1, it also admits a flat $\mathcal{A}$-connection. ∎

# 6. Trivial $\mathcal{A}$-connections

We consider below another useful application of the notion of *flatness;* for the classical counterpart see, for instance, B.A. Dubrovin - A.T. Fomenko - S.P. Novikov [1: p. 444, Theorem 41.2.2].

Thus, suppose that we are given a *curvature space* $X$ and let $\mathcal{E}$ be a *vector sheaf* on $X$, such that

$$\operatorname{rk}\mathcal{E} = n \in \mathbf{N}. \tag{6.1}$$

(We could also take, instead, more generally, an $\mathcal{A}$-*module,* as, for instance, in Section 3; cf. e.g. (3.2)). Furthermore, consider a *local frame*

$$\mathcal{U} = (U_\alpha)_{\alpha\in I} \tag{6.2}$$

of $\mathcal{E}$, as well as, an $\mathcal{A}$-*connection* $D$ of $\mathcal{E}$, with

$$\omega \equiv \left(\omega^{(\alpha)}\right) \equiv \left(\left(\omega_{ij}^{(\alpha)}\right)\right) \in C^0(\mathcal{U}, M_n(\Omega^1)) \tag{6.3}$$

the *0-cochain of local* $\mathcal{A}$-*connection matrices of* $D$, associated with $\mathcal{U}$ (cf. Chapt. VII; (1.13), or even (2.39)).

Now, we say that $D$ is a *trivial* $\mathcal{A}$-*connection,* whenever, concerning (6.3), as above, one has, *first;*

$$\begin{aligned}\omega \equiv \left(\omega^{(\alpha)}\right) \in \tilde{\partial}(C^0(\mathcal{U}, \mathcal{GL}(n,\mathcal{A}))) = C^0(\mathcal{U}, \tilde{\partial}(\mathcal{GL}(n,\mathcal{A})))\\ \subseteq C^0(\mathcal{U}, M_n(\Omega^1)).\end{aligned} \tag{6.4}$$

That is, there exists a *0-cochain,*

$$g \equiv \left(g^{(\alpha)}\right) \in C^0(\mathcal{U}, \mathcal{GL}(n,\mathcal{A})), \tag{6.5}$$

such that one further sets;

$$\omega := \tilde{\partial}(g^{-1}), \tag{6.6}$$

thus, an *equivalent expression of (6.4).*

Now, by considering further the $n \times n$ *(section-)matrix*

$$g^{(\alpha)} \equiv \left(g_{ij}^{(\alpha)}\right) \in \mathcal{GL}(n,\mathcal{A})(U_\alpha) = \mathrm{GL}(n, \mathcal{A}(U_\alpha)), \tag{6.7}$$

as in (6.5), one actually obtains an element of

$$(\mathcal{A}ut\mathcal{E})(U_\alpha) = \mathrm{Aut}(\mathcal{E}|_{U_\alpha}) \tag{6.8}$$

(see Chapt. VI; (17.24)); thus, if $U_{\alpha\beta} \equiv U_\alpha \cap U_\beta \neq \emptyset$, with $\alpha, \beta \in I$, as in (6.2), and one sets (cf. (6.7))

$$g_{\alpha\beta} := g^{(\alpha)}|_{U_{\alpha\beta}} \equiv \left(g_{ij}^{(\alpha)}|_{U_{\alpha\beta}}\right) \in \mathrm{GL}(n, \mathcal{A}(U_{\alpha\beta})), \tag{6.9}$$

then, *second,* we also assume that the *1-cochain of $\mathcal{U}$*, as given by (6.9), yields, in effect, a *gauge transformation of $\mathcal{E}$* (loc. cit., comments following (17.30)), such that, in particular,

$$\tilde{g} \equiv (g_{\alpha\beta}) \in Z^1(\mathcal{U}, \mathcal{GL}(n, \mathcal{A})) \tag{6.10}$$

is a *coordinate 1-cocycle of $\mathcal{E}$.* (See, for example, the proof of Theorem 6.1 below, in particular, (6.38)).

Thus, by looking now at the corresponding to $\tilde{g}$, as in (6.10), *transformation law of potentials* (cf. Chapt. VII; (2.43), or even (2.51)), one has the following. (See also (6.6), (6.9) above, together with Chapt. VI; (1.33), (1.30) and (1.7)):

$$\begin{aligned} \tilde{g}^*\left(\omega^{(\alpha)}\right) &\equiv \omega^{(\beta)} = \mathrm{Ad}(g_{\alpha\beta}^{-1})\omega^{(\alpha)} + \tilde{\partial}(g_{\alpha\beta}) \\ &= \mathrm{Ad}(g_{\alpha\beta}^{-1})\tilde{\partial}(g_{\alpha\beta}^{-1}) + \tilde{\partial}(g_{\alpha\beta}) = \tilde{\partial}(g_{\alpha\beta}^{-1} \cdot g_{\alpha\beta}) = \tilde{\partial}(1) = 0. \end{aligned} \tag{6.11}$$

Accordingly, the corresponding local form of the *curvature* of the given $\mathcal{A}$-connection $D$ (cf. (6.3)) has now as follows;

$$\begin{aligned} \tilde{g}^*\left(R^{(\alpha)}\right) &= R^{(\beta)} = d\omega^{(\beta)} + \omega^{(\beta)} \wedge \omega^{(\beta)} = 0 \\ &= g_{\alpha\beta}^{-1} R^{(\alpha)} g_{\alpha\beta} \end{aligned} \tag{6.12}$$

(see also (3.16) and (4.23)). Hence, one obtains

$$R^{(\alpha)} = 0, \quad \text{for every } \alpha \in I, \tag{6.13}$$

that is (cf. (2.19), (2.20), or even (3.15)), finally, one has;

$$R = 0. \tag{6.14}$$

In other words (cf. (2.5)), one thus concludes that:

(6.15) *every trivial $\mathcal{A}$-connection is flat.* (It concerns here an *$\mathcal{A}$-connection* of a *vector sheaf $\mathcal{E}$* on a given *curvature space $X$*).

**Scholium 6.1.**– The previous conclusion (6.15) furnishes also another way of inferring that the *"canonical flat $\mathcal{A}$-connection"*

$$\partial : \mathcal{A} \longrightarrow \Omega^1 \tag{6.16}$$

(cf. Chapt. VI; (4.2)) *is indeed flat,* by showing that $\partial$ *is trivial,* in the sense of (6.6) (while (6.10) is trivially satisfied): Thus, by taking the *identity section of* $\mathcal{A}$ (alias, canonical *section-basis* of the *line sheaf* $\mathcal{A}$), one has

$$\tilde{\partial}(1) = \partial(1) = 0. \tag{6.17}$$

(See Chapt. VI; (1.7), as well as, e.g. Chapt. VII; (1.11)). So (6.17) proves our claim, hence, by virtue of (6.15), $\partial$ *is a flat* $\mathcal{A}$*-connection of* $\mathcal{A}$, in the sense that,

$$R(\partial) = 0. \tag{6.18}$$

(Of course, we already know (6.18), according to (2.6), while one comes to the *same conclusion,* by virtue of (6.17) and (3.12)).

On the other hand, the *converse of (6.15)* is still related with the aforementioned *Frobenius integrability condition;* see thus (5.24) in the foregoing, as well as, Chapter XI in the sequel. It seems, namely, that some *extra conditions have to be imposed on* $\mathcal{A}$*, for the converse of (6.15).* Indeed, this is demonstrated here in the particular case of a *line sheaf,* where an additional *cohomological condition* is put *on* $\mathcal{A}$ (cf. (6.19) below), the latter being otherwise just a $\mathbf{C}$*-algebra sheaf,* as was the case hitherto. Precisely, one thus obtains the following.

**Theorem 6.1.** *Suppose we have a curvature space* $X$ (see (1.16)), *such that* (cf. (1.8), (1.11))

$$\operatorname{im}\partial = \ker d; \tag{6.19}$$

*viz., equivalently, we assume the following short exact sequence*

$$0 \longrightarrow \partial\mathcal{A} \longrightarrow \Omega^1 \xrightarrow{d^1 \equiv d} d\Omega^1 \longrightarrow 0. \tag{6.20}$$

*Furthermore, suppose we are given the following commutative triangle;*

$$\begin{array}{ccc} \mathcal{A} & \xrightarrow{e} & \mathcal{A}^{\cdot} \\ & \partial \searrow \quad \swarrow \frac{1}{2\pi i}\tilde{\partial} & \\ & \partial\mathcal{A}. & \end{array} \tag{6.21}$$

*Finally, let $\mathcal{L}$ be a line sheaf on $X$. Then, a given $\mathcal{A}$-connection $D$ of $\mathcal{L}$ is trivial if, and only if, $D$ is flat.*

**Proof.** Based on (6.15), it suffices only to prove the *"if"* part of the assertion; that is, assume that *the* given *$\mathcal{A}$-connection $D$* of *$\mathcal{L}$ is flat,* so that one obtains (cf. (3.24), or even (5.42))

$$R = (d\omega_\alpha) = 0 \in \Gamma(X, d\Omega^1) = Z^0(\mathcal{U}, d\Omega^1). \tag{6.22}$$

Therefore, concerning the *local $\mathcal{A}$-connection "1-form" of $D$*, as in (6.22),

$$\omega \equiv (\omega_\alpha) \in C^0(\mathcal{U}, \Omega^1), \tag{6.23}$$

where

$$\mathcal{U} = (U_\alpha)_{\alpha \in I} \tag{6.24}$$

is a given *local frame* of $\mathcal{L}$, one has, by virtue of (6.22), that;

$$d\omega \equiv d(\omega_\alpha) := (d\omega_\alpha) = 0, \tag{6.25}$$

so that, *equivalently,* one obtains

$$\omega \equiv (\omega_\alpha) \in \ker d, \tag{6.26}$$

or even (cf. (6.19))

$$\omega_\alpha \in \ker d = \operatorname{im} \partial, \quad \alpha \in I. \tag{6.27}$$

Now, since the *section functor* $\Gamma$ is *left exact* (cf. Chapt. II; (2.59)), one concludes from the exactness of (6.20), an analogous *exactness of the sequence*

$$0 \longrightarrow \Gamma(\partial\mathcal{A}) \xrightarrow{\partial} \Gamma(\Omega^1) \xrightarrow{d} \Gamma(d\Omega^1) \tag{6.28}$$

(loc. cit.; (2.57), (2.58)), where, of course, we still have set, for convenience,

$$\Gamma(\partial) \equiv \partial \ \ \textit{and} \ \ \Gamma(d) \equiv d \tag{6.29}$$

(explaining thus the notation in (6.27), as well; see also Chapt. I; (13.3)). Hence, one obtains from (6.27)

(6.30) $$\omega_\alpha \in (\ker d)(U_\alpha) = (\operatorname{im} \partial)(U_\alpha) \equiv (\partial \mathcal{A})(U_\alpha),$$

*for every* $\alpha \in I$. Accordingly, since *a sheaf morphism,* like $\partial$, *is fiber-preserving* (Chapt. I; Definition 1.2), one finds a *(local) section*

(6.31) $$f_\alpha \in \mathcal{A}(V_\alpha), \text{ for some } open \ V_\alpha \subseteq U_\alpha,$$

in such a manner that

(6.32) $$\omega_\alpha = \partial(f_\alpha)|_{V_\alpha}, \quad \alpha \in I,$$

(see also Chapt. I; (2.9), (2.12)).

Therefore, by further applying (6.21), one first sets;

(6.33) $$g_\alpha := e\left(\frac{1}{2\pi i} f_\alpha\right)^{-1} \in \mathcal{A}^{\cdot}(V_\alpha) = \mathcal{A}(V_\alpha)^{\cdot}, \quad \alpha \in I,$$

so that, by virtue of the hypothesis for (6.21), one has

(6.34) $$\begin{aligned}\tilde{\partial}(g_\alpha^{-1}) &= \tilde{\partial}\left(e\left(\tfrac{1}{2\pi i} f_\alpha\right)\right) = 2\pi i \cdot \partial\left(\tfrac{1}{2\pi i} f_\alpha\right)\\ &= 2\pi i \cdot \tfrac{1}{2\pi i}\partial(f_\alpha) = \partial(f_\alpha) \in \partial(\mathcal{A}(V_\alpha)) \subseteq (\partial\mathcal{A})(V_\alpha)\end{aligned}$$

(see also Chapt. I; Proposition 2.1, concerning the last relation in (6.34)). Hence, in view of (6.32), one has

(6.35) $$\omega_\alpha = \tilde{\partial}(g_\alpha^{-1})|_{V_\alpha}, \quad \alpha \in I.$$

In other words, one thus arrives at a *0-cochain*

(6.36) $$g \equiv (g_\alpha) \in C^0(\mathcal{V}, \mathcal{A}^{\cdot})$$

(see (6.33), such that one has, by (6.35),

(6.37) $$\omega = \tilde{\partial}(g^{-1}),$$

that is, the desired relation (6.6) for the *triviality of* $D$. (Yet, concerning (6.36), $\mathcal{V}$ stands there for the *local frame of* $\mathcal{L}$, defined by the $V_\alpha$'s, as in (6.31), in fact, a *refinement* of the given $\mathcal{U}$, as in (6.24); in this regard, see e.g. Chapt. II, proof of (4.9)). On the other hand, one further verifies (6.10), by just setting;

(6.38) $$\tilde{g} \equiv (g_{\alpha\beta}) := \delta(g^{-1}) \equiv (g_\alpha g_\beta^{-1}) \in Z^1(\mathcal{V}, \mathcal{A}^{\cdot}),$$

which thus, together with (6.37), prove that the given $\mathcal{A}$-connection $D$ *is trivial,* as was our claim, and the proof is complete. ∎

As an immediate consequence of the preceding, one has now the following.

**Corollary 6.1.** *Suppose we have the framework of the above Theorem 5.2, while we also assume* (cf. (6.19)) *that;*

(6.39) $$\ker d = \operatorname{im} \partial.$$

*Furthermore, suppose that $\mathcal{L}$ is a line sheaf on $X$, equipped with an $\mathcal{A}$-connection $D$.*

*Then, triviality of $D$* (cf. (6.6), (6.10)) *is a fifth further equivalent condition to the rest four ones for $D$, as in Theorem 5.2.* ∎

**Note 6.1.**– It is worth remarking here, in anticipation, that:

(6.40) *the context of the previous Corollary 6.1 is what one actually needs to prove "Weil's integrality theorem".*

(See thus Section 11 in the sequel). That is, one considers hereby a

(6.41) *paracompact (Hausdorff) sheaf exponential curvature space, such that (6.39) be satisfied.*

In this context, due to the aforementioned Weil's theorem (see Theorem 11.1 in the sequel), we are thus tempted to call a topological space $X$, satisfying (6.41), a *Weil space.* (In fact, this is in disagreement, somewhat, with our teminology previously applied in A. Mallios [6: p. 192, Definition 7.1], the present term being thus weakened, hence, improved).

# 7. Bianchi's identity

We consider below the counterpart in our case of the classical identity referred to in the title of this Section. Yet, by analogy with the standard theory, we examine in the next Chapter IX an important application of the same identity, by defining *characteristic classes* of vector sheaves.

So assume that we have a *curvature space* $X$ and an $\mathcal{A}$-*module* $\mathcal{E}$ on $X$, endowed with an $\mathcal{A}$-*connection* $D$. Moreover, suppose that we are given a *local gauge*

$$e^U \equiv \{U; (e_1, \dots, e_n)\} \tag{7.1}$$

of $\mathcal{E}$ over an *open* set $U \subseteq X$, of *rank* $n \in \mathbf{N}$ (cf. Chapt. VII; (1.5)). Now, we already know that the given $\mathcal{A}$-connection $D$ and its *curvature* $R(D) \equiv R$ have *local expressions* on $U$, through the corresponding *local matrix-valued "forms"* (see Chapt. VII; (1.13), along with (3.11) above).

Thus, by considering the *curvature of* $D$, *restricted to* $U$, as above, as an *element of* $M_n(\Omega^2(U)) = M_n(\Omega^2)(U)$, we further look at the form, which is attained to by $R$, if one assumes, in addition, the action of a *second exterior derivation* (cf. also (1.10))

$$d^2 : \Omega^2 \longrightarrow \Omega^3. \tag{7.2}$$

Namely, suppose further that we are given a $\mathbf{C}$-*linear morphism*

$$d^2 \equiv d, \tag{7.3}$$

as in (7.2) above, between the $\mathbf{C}$-*vector space sheaves* concerned, in such a manner that one actually sets, by definition,

$$d^2(s \wedge t) := d^1 s \wedge t - s \wedge d^1 t \equiv ds \wedge t - s \wedge dt, \tag{7.4}$$

*for any* $s, t \in \Omega^1(U)$, while we still *assume that* (cf. also (8.9) below);

$$d^2 \circ d^1 \equiv d \circ d = 0. \tag{7.5}$$

**Note 7.1.**- Concerning the defining relation (7.4), as above, one easily proves that the map (7.2), as given by (7.4), is indeed a $\mathbf{C}$-*linear morphism of* $\Omega^1$ *into* $\Omega^2$. Yet, an analogous argument to that one used for the map (1.20) is here still in force. (See also the relevant comments about the map (1.7) in the preceding). Thus, *the extra condition* here, against a curvature datum (cf. Definition 1.1), *is,* in effect, *the rel. (7.5).*

Thus, for convenience of reference, we further set the following.

**Definition 7.1.** By a *Bianchi datum* on a topological space $X$, one means a curvature datum (see Definition 1.1), such that, in addition, we are

given a **C**-linear morphism (7.2), defined by (7.4), in such a manner that (7.5) be satisfied.

We denote the above data by

$$\{(\mathcal{A}, \partial, \Omega^1, d^1 \equiv d, \Omega^2);\ d^2 \equiv d\}. \tag{7.6}$$

Moreover, for simplicity's sake, we also apply for the same data the following notation;

$$(\mathcal{A}, \partial, \Omega^1, d^1 \equiv d, \Omega^2, d^2 \equiv d) \equiv (\partial, d^1, d^2). \tag{7.7}$$

Finally, we say that $X$, as before, is a *Bianchi space*.

Thus, to be clear, a *Bianchi space* is a curvature space $X$ (Definition 1.1), for which we further consider an exterior derivative operator $d^2 \equiv d$, as before, satisfying (7.5) (cf. also the previous Note 7.1).

Now, this same operator is "coordinate-wise" *extended to* a similar operator (we retain for simplicity the same symbol for the extended map too) between *the* following $\mathcal{A}$-*modules*

$$d^2 \equiv d : M_n(\Omega^2) \longrightarrow M_n(\Omega^3) \tag{7.8}$$

(cf. also Chapt. V; (8.7), for the notation applied). Thus, we can further apply the last *extended operator* to the curvature $R$ (restricted to $U$, as in (7.1)), as the latter is given by *Cartan's structural equation* (cf. (3.12)). So one obtains (see (7.4), (7.5));

$$\begin{aligned} d^2 R \equiv dR &= d(d\omega + \omega \wedge \omega) = d(d\omega) + d(\omega \wedge \omega) \\ &= d(\omega \wedge \omega) = d\omega \wedge \omega - \omega \wedge d\omega \\ &= (R - \omega \wedge \omega) \wedge \omega - \omega \wedge (R - \omega \wedge \omega) \\ &= R \wedge \omega - \omega \wedge \omega \wedge \omega - \omega \wedge R + \omega \wedge \omega \wedge \omega, \end{aligned} \tag{7.9}$$

so that one finally has;

$$dR = R \wedge \omega - \omega \wedge R \equiv [R, \omega]. \tag{7.10}$$

We call the last relation the *("second") Bianchi's identity*.

Thus, still for convenience of reference, we collect now the preceding, formally, into the following.

**Proposition 7.1.** *Suppose we are given a Bianchi space $X$ and an $\mathcal{A}$-module $\mathcal{E}$ on $X$, endowed with an $\mathcal{A}$-connection $D$. Moreover, let $e^U$ be a given*

*local gauge of* $\mathcal{E}$ (cf. (7.1)) *such that*

$$\omega \in M_n(\Omega^1(U)) \text{ and } R \in M_n(\Omega^2(U)) \tag{7.11}$$

*are the local (section) matrix-valued "forms" of $D$ and $R$ on $U$, respectively* (see Chapt. VII; (1.13), as well as, (3.11) above). *Then, one obtains (second Bianchi's identity);*

$$dR = R \wedge \omega - \omega \wedge R \equiv [R, \omega]. \blacksquare \tag{7.12}$$

Now, suppose that we have a *Bianchi space* $X$, as above, and let

$$(\mathcal{E}, D) \tag{7.13}$$

be a given *pair, as in* the previous *Proposition 7.1.* Thus, one can further define another **C**-*linear morphism*

$$D^2 : \Omega^2(\mathcal{E}) \longrightarrow \Omega^3(\mathcal{E}), \tag{7.14}$$

according to the relation (cf. also (8.17) below, for $p = 2$);

$$\begin{aligned} D^2(s \otimes t) &:= s \otimes d^2 t + t \wedge Ds \equiv s \otimes dt + t \wedge Ds \\ &= s \otimes dt - Ds \wedge t, \end{aligned} \tag{7.15}$$

*for any* $s \in \mathcal{E}(U)$ and $t \in \Omega^2(U)$, with $U$ *open* in $X$. (In this regard, see also the comments on (1.20) in the preceding).

We call the above map $D^2$, as defined by (7.14), (7.15), the *2nd prolongation of the* given $\mathcal{A}$-*connection* $D$; yet, by extending the classical terminology, we call $D^2$ the *2nd exterior covariant derivative operator* (or just *derivation*).

Thus, our next objective is to *evaluate* $D^2 R$ (cf. (7.14) and (2.3)) in the case of a given pair $(\mathcal{E}, D)$, as in (7.13), when, in particular, $\mathcal{E}$ is a *vector sheaf* on our *Bianchi space* $X$.

Thus, supposing again that $e^U$ is a *local gauge of* $\mathcal{E}$ (cf., for instance, (7.1)), assume that

$$(e^U)^* \equiv \{U; (e_i^*)_{1 \leqq i \leqq n}\}, \tag{7.16}$$

with $n = \operatorname{rk} \mathcal{E} \in \mathbf{N}$, is the *dual local gauge of* $e^U$, viz. a corresponding to $e^U$ *local gauge over the* same *open* set $U \subseteq X$ of the *dual vector sheaf* of $\mathcal{E}$,

$$\mathcal{E}^* \equiv \mathcal{H}om_{\mathcal{A}}(\mathcal{E}, \mathcal{A}) \tag{7.17}$$

(see Chapt. VII; (4.2) - (4.7)). On the other hand, due to the hypothesis for

$\mathcal{E}$, one has (cf. Chapt. II; (6.28) and Chapt. IV; (6.1))

$$\mathcal{E}nd\,\mathcal{E} \equiv \mathcal{H}om_{\mathcal{A}}(\mathcal{E},\mathcal{E}) = \mathcal{E} \otimes_{\mathcal{A}} \mathcal{E}^*. \tag{7.18}$$

Therefore, by considering the curvature $R$, restricted to $U$, one obtains in view also of the hypothesis for $U$ (cf., for example, Chapt. V; (8.6));

$$\begin{aligned} R|_U \equiv R \in \Omega^2(\mathcal{E}nd\,\mathcal{E})(U) &\equiv ((\mathcal{E}nd\,\mathcal{E}) \otimes_{\mathcal{A}} \Omega^2) \\ &= ((\mathcal{E} \otimes_{\mathcal{A}} \mathcal{E}^*) \otimes_{\mathcal{A}} \Omega^2)(U) = (\mathcal{E} \otimes_{\mathcal{A}} \mathcal{E}^*)(U) \otimes_{\mathcal{A}(U)} \Omega^2(U) \\ &(\mathcal{E}(U) \otimes_{\mathcal{A}(U)} \mathcal{E}^*(U)) \otimes_{\mathcal{A}(U)} \Omega^2(U). \end{aligned} \tag{7.19}$$

On the other hand, by hypothesis for (7.1) and (7.16), the double sequence

$$(e_i \otimes e_j^*)_{1 \leqq i,j \leqq n} \tag{7.20}$$

yields a *local gauge of the vector sheaf* $\mathcal{E} \otimes_{\mathcal{A}} \mathcal{E}^*$ over the *open* $U \subseteq X$, of *rank* $n^2$ (see Chapt. IV; (5.19) and Chapt. II; (5.27)). Accordingly (see also e.g. W. Greub [1: p.7, Lemma 1.5.2]), one obtains, by (7.19),

$$R|_U \equiv R = \sum_{i,j} (e_i \otimes e_j^*) \otimes R_{ij}, \tag{7.21}$$

for some *uniquely defined* (sections)

$$R_{ij} \in \Omega^2(U), \quad 1 \leqq i,j \leqq n = \operatorname{rk} \mathcal{E}, \tag{7.22}$$

for given $R$, as in (7.19), along with (7.20), so that one actually gets;

$$(R_{ij}) \in M_n(\Omega^2(U)) = M_n(\Omega^2)(U). \tag{7.23}$$

Now, based on (7.14) and (7.15), we can further consider the **C**-*linear morphism*

$$D^2 \equiv D : \Omega^2(\mathcal{E} \otimes_{\mathcal{A}} \mathcal{E}^*) \longrightarrow \Omega^3(\mathcal{E} \otimes_{\mathcal{A}} \mathcal{E}^*), \tag{7.24}$$

which, by analogy with (7.15), is defined by the relation;

$$D^2 \equiv D := 1_{\mathcal{E} \otimes_{\mathcal{A}} \mathcal{E}^*} \otimes d^2 + 1_{\Omega^2} \wedge D_{\mathcal{E} \otimes_{\mathcal{A}} \mathcal{E}^*}, \tag{7.25}$$

where, of course,

$$D_{\mathcal{E} \otimes_{\mathcal{A}} \mathcal{E}^*} : \mathcal{E} \otimes_{\mathcal{A}} \mathcal{E}^* \longrightarrow \Omega^1(\mathcal{E} \otimes_{\mathcal{A}} \mathcal{E}^*) \tag{7.26}$$

stands for the $\mathcal{A}$-*connection of* (the *vector sheaf*) $\mathcal{E} \otimes_{\mathcal{A}} \mathcal{E}^*$. The latter corresponds to the given $\mathcal{A}$-*connection* $D$ of $\mathcal{E}$ and its *dual* $\mathcal{A}$-*connection*

(7.27) $$D_{\mathcal{E}^*} \equiv D^* : \mathcal{E}^* \longrightarrow \Omega^1(\mathcal{E}^*),$$

as this is given by Chapt. VI; (5.31); thus, one has (loc. cit.; (5.38))

(7.28) $$D_{\mathcal{E}\otimes_{\mathcal{A}}\mathcal{E}^*} := D \otimes 1_{\mathcal{E}^*} + 1_{\mathcal{E}} \otimes D^*.$$

**Note 7.2.**- Concerning (7.24), we further remark, in anticipation (see thus the next Section 9), that the operator at issue is, in effect, the *2nd prolongation of the $\mathcal{A}$-connection (7.26) of* $\mathcal{E} \otimes_{\mathcal{A}} \mathcal{E}^*$; in this regard, see also (8.17), for $p = 2$, in the sequel.

In this context, we still recall, for convenience, that if

(7.29) $$\omega \equiv (\omega_{ij}) \in M_n(\Omega^1(U)) = M_n(\Omega^1)(U)$$

is the *local $\mathcal{A}$-connection matrix of* $D$, associated with (7.1) (see Chapt. VII; (1.13)), then the corresponding one for $D^*$, as in (7.27), is given by

(7.30) $$\omega^* \equiv (\omega^*_{ij}) = -{}^t\omega \equiv -(\omega_{ji}) \in M_n(\Omega^1(U))$$

(cf. (7.29), together with Chapt. VII; (4.20)), such that one has (see Chapt. VII; (4.19), as well as, e.g. Chapt. VI; (5.36));

(7.31) $$D^*(e^*_j)(e_i) = \omega^*_{ij} = -\omega_{ji}, \quad 1 \leqq i, j \leqq n,$$

or even, *equivalently*,

(7.32) $$D^*(e^*_j) = \sum_{i=1}^{n} e^*_i \otimes \omega^*_{ij} = -\sum_i e^*_i \otimes \omega_{ji}, \quad 1 \leqq j \leqq n.$$

So we are now in the position to formulate our main result in this Section. That is, we have.

**Theorem 7.1.** *Let $X$ be a Bianchi space* (cf. Definition 7.1), *$\mathcal{E}$ a vector sheaf on $X$, with* $\operatorname{rk}\mathcal{E} = n \in \mathbf{N}$, *and $U$ an open set in $X$, domain of definition of a local gauge of $\mathcal{E}$, while we also assume that $\mathcal{E}$ has an $\mathcal{A}$-connection $D$. Then, the two following assertions about the curvature of $D$, $R(D) \equiv R$, are equivalent:*

*1) One has (Bianchi's identity;* see (7.12));

(7.33) $$dR = R \wedge \omega - \omega \wedge R \equiv [R, \omega].$$

*2) The following relation holds true ("differential" Bianchi's identity);*

$$DR = 0 \ \ (\text{i.e., } D^2_{\mathcal{E}nd\,\mathcal{E}}(R) = 0) \tag{7.34}$$

(cf. also (7.18), (7.24) and Note 7.2).

**Proof.** First, by applying the operator (7.24) to (7.21), one gets;

$$\begin{aligned} D^2(R|_U) \equiv DR &= D(\textstyle\sum_{i,j}(e_i \otimes e_j^*) \otimes R_{ij}) \\ &= \textstyle\sum_{i,j} D((e_i \otimes e_j^*) \otimes R_{ij}). \end{aligned} \tag{7.35}$$

On the other hand, based on (7.25), one has (cf. also (7.3) and (7.22));

$$D((e_i \otimes e_j^*) \otimes R_{ij}) = (e_i \otimes e_j^*) \otimes dR_{ij} + R_{ij} \wedge D_{\mathcal{E}\otimes_{\mathcal{A}}\mathcal{E}^*}(e_i \otimes e_j^*). \tag{7.36}$$

However, by virtue of (7.28) and (7.32), in conjunction with Chapt. VII; (1.11), one further obtains;

$$\begin{aligned} D_{\mathcal{E}\otimes_{\mathcal{A}}\mathcal{E}^*}(e_i \otimes e_j^*) &= D(e_i) \otimes e_j^* + e_i \otimes D^*(e_j^*) \\ &= (\textstyle\sum_k e_k \otimes \omega_{ik}) \otimes e_j^* + e_i \otimes \sum_k e_k^* \otimes \omega_{jk}^* \\ &\approx \textstyle\sum_k (e_k \otimes e_j^*) \otimes \omega_{ik} - \sum_k (e_i \otimes e_k^*) \otimes \omega_{kj}. \end{aligned} \tag{7.37}$$

(The notation "$\approx$" indicates here the appropriate change, that can be made *on the* relevant *(sheaf) spaces,* according to Chapt. II; (5.14), (5.16)). Therefore, based now on (7.37), one obtains;

$$\begin{aligned} &\textstyle\sum_{i,j} R_{ij} \wedge D_{\mathcal{E}\otimes_{\mathcal{A}}\mathcal{E}^*}(e_i \otimes e_j^*) \\ &= \textstyle\sum_{i,j} R_{ij} \wedge (\sum_k (e_k \otimes e_j^*) \otimes \omega_{ik}) - \sum_{i,j} R_{ij} \wedge (\sum_k (e_i \otimes e_k^*) \otimes \omega_{kj}) \\ &= \textstyle\sum_{i,j} (e_i \otimes e_j^*) \otimes (\sum_k \omega_{ik} \wedge R_{kj} - \sum_k R_{ik} \wedge \omega_{kj}). \end{aligned} \tag{7.38}$$

Consequently, one now obtains, in view also of (7.35) and (7.36),

$$DR = \sum_{ij} (e_i \otimes e_j^*) \otimes \left( dR_{ij} + \sum_k \omega_{ik} \wedge R_{kj} - \sum_k R_{ik} \wedge \omega_{kj} \right), \tag{7.39}$$

so that one has

$$(DR)_{ij} = dR_{ij} + \sum_k \omega_{ik} \wedge R_{kj} - \sum_k R_{ik} \wedge \omega_{kj} \tag{7.40}$$

*for any* $1 \leqq i, j \leqq n$. Thus, by employing *matrix notation* to (7.40), one finally obtains;

$$DR = dR + \omega \wedge R - R \wedge \omega \equiv dR + [\omega, R], \tag{7.41}$$

which thus proves our assertion. ∎

# 8. Higher exterior derivations and prolongations

By analogy with the classical theory, we further consider below, within our abstract (viz. axiomatic) framework, the case of *exterior derivations of higher order,* along with the resulted thereby *prolongations.*

Thus, suppose that we are given a *differential triad*

$$(\mathcal{A}, \partial, \Omega) \tag{8.1}$$

on a *topological space* $X$. (We knowingly adopt here anew the notation of Chapt. VI; (3.1), concerning (8.1), which, however, will be justified straight away). So we call an $(\Omega)$ $\mathcal{A}$-*sequence* on $X$, the following *(cochain) complex of* $\mathbf{C}$-*vector space sheaves* (under the *constant* ($\mathbf{C}$-vector space) *sheaf* $\mathbf{C}$);

$$\begin{aligned} (\Omega) : 0 \longrightarrow \mathbf{C} \xrightarrow{\varepsilon} \Omega^0 \equiv \mathcal{A} \xrightarrow{d^0 \equiv \partial} \Omega^1 \equiv \Omega \\ \xrightarrow{d^1} \Omega^2 \xrightarrow{d^2} \Omega^3 \longrightarrow \cdots \longrightarrow \Omega^p \\ \xrightarrow{d^p} \Omega^{p+1} \longrightarrow \cdots, \end{aligned} \tag{8.2}$$

where we have set (see also (1.10))

$$\Omega^p := \underbrace{\Omega \wedge \cdots \wedge \Omega}_{p-\text{times}} \equiv \wedge^p \Omega \equiv \wedge^p \Omega^1, \quad p \in \mathbf{N}, \tag{8.3}$$

while we also assumed that (cf. (8.2));

$$\Omega^0 \equiv \mathcal{A} \ \textit{and} \ \Omega \equiv \Omega^1. \tag{8.4}$$

Thus, the individual terms in the given sequence (8.2) are, apart from the *constant sheaf* $\mathbf{C}$, in effect, $\mathcal{A}$-*modules* (by their very definition; cf. (8.1), (8.3) and Chapt. IV; (7.3), (7.8)). Hence, in particular, $\mathbf{C}$-*vector space sheaves,* where, as already noted, the *constant one* $\mathbf{C}$ is also included (cf. Chapt. II; (1.38)); yet, the connecting maps appeared in (8.2) are, by definition, $\mathbf{C}$-*linear morphisms.* Furthermore, we first remark that, by hypothesis for (8.1), one obtains (cf. Chapt. VI; Lemma 1.1)

$$\partial \circ \varepsilon = 0; \tag{8.5}$$

on the other hand, *we* also *assume,* by virtue of (8.2), *the existence of* the $\mathbf{C}$-linear morphism $d^1$, *such that*

$$d^1 \circ \partial = 0 \tag{8.6}$$

(cf. (1.7), (1.8)). Moreover, we define the operators $d^p$ in (8.2), section-wise, by the relation;

$$d^{p+q}(s \wedge t) := d^p s \wedge t + (-1)^p s \wedge d^q t, \tag{8.7}$$

*for any* $p, q \in \mathbf{N}$, *with* $s \in \Omega^p(U)$, $t \in \Omega^q(U)$, and every *open* $U \subseteq X$.

**Remark.**– Since the *exterior powers* in (8.3) are viewed *with respect to* $\mathcal{A}$ (cf. Chapt. IV; (7.4), (7.5)), we actually use in (8.7) an *analogous argument* to that one, pertaining *to (1.20)* above.

Thus, *we can prove* now *that;*

$$d^{p+1} \circ d^p = 0, \quad \textit{for every} \;\; p \geqq 2, \tag{8.8}$$

while *we* finally *assume* that

$$d^2 \circ d^1 = 0, \tag{8.9}$$

*as well.* So the preceding rels. (8.6), (8.8) and (8.9) justify now our previous terminology for (8.2). In other words, we thus have

$$d^{p+1} \circ d^p = 0, \quad \textit{for every} \;\; p \in \mathbf{Z}_+. \tag{8.10}$$

Yet, for convenience, we just put (8.10) into the form;

$$d \circ d \equiv d^2 = 0. \tag{8.11}$$

**Note.**– In this context, we should also remark that: *We could have (1.9)* by suitably *extending (8.7) to the pair* of indices $(1, 0) \in \mathbf{Z}_+$; that is, one then obtains

$$\begin{aligned} d^1(\alpha \cdot s) &\equiv d^{0+1}(\alpha \wedge s) = d^0 \alpha \wedge s + \alpha \wedge d^1 s \\ &\equiv \partial \alpha \wedge s + \alpha \cdot d^1 s = \alpha \cdot ds - s \wedge \partial \alpha, \end{aligned} \tag{8.12}$$

*for any* $\alpha \in \mathcal{A}(U)$ *and* $s \in \Omega(U) \equiv \Omega^1(U)$, with $U$ *open* in $X$, which is just our claim. ∎

Thus, the previous complex ($\Omega$), as given by (8.2), yields, in particular, a *curvature datum*

$$(\mathcal{A}, \partial, \Omega^1, d, \Omega^2) \tag{8.13}$$

on $X$, by further setting $d^1 \equiv d$ (see (1.7)); viz. ($\Omega$) furnishes already what we have called in the preceding a *curvature space* $X$ (cf. Definition 1.1). So assume now that we are also given a pair

$$(\mathcal{E}, D), \tag{8.14}$$

consisting of an *$\mathcal{A}$-module $\mathcal{E}$* on $X$ and an *$\mathcal{A}$-connection $D$ of $\mathcal{E}$*. Thus, consider now the following **C**-*sequence* on $X$ (cf. also (8.4));

$$\begin{gathered}\mathcal{E} \xrightarrow{D^0 \equiv D} \Omega^1(\mathcal{E}) \xrightarrow{D^1} \Omega^2(\mathcal{E}) \xrightarrow{D^2} \Omega^3(\mathcal{E}) \longrightarrow \cdots \\ \cdots \longrightarrow \Omega^p(\mathcal{E}) \xrightarrow{D^p} \Omega^{p+1}(\mathcal{E}) \longrightarrow \cdots,\end{gathered} \tag{8.15}$$

where, as usual, we have set

$$\Omega^p(\mathcal{E}) := \mathcal{E} \otimes_{\mathcal{A}} \Omega^p, \quad p \in \mathbf{Z}_+ \tag{8.16}$$

(see also (8.4), along with Chapt. II; (5.15)). Yet, one also defines

$$D^p := 1_{\mathcal{E}} \otimes d^p + (-1)^p 1_{\Omega^p} \wedge D^0 \tag{8.17}$$

(cf. (8.2)), while we have already set (see (8.15))

$$D^0 := D. \tag{8.18}$$

Thus, writing (8.17) in a section-wise manner, one has, *by definition* (see also (8.16));

$$D^p(s \otimes t) := s \otimes d^p t + (-1)^p t \wedge Ds, \tag{8.19}$$

*for any $s \in \mathcal{E}(U)$ and $t \in \Omega^p(U)$*, with $U$ *open* in $X$.

> **Scholium 8.1.**- In practice, the *open* set $U \subseteq X$, as used above, will be, in effect, (the domain of definition of) a local gauge of the $\mathcal{A}$-module (in fact, *vector sheaf*) $\mathcal{E}$ involved, so that one can then apply, concerning the definition of $D^p$, as in (8.19), an analogous argument to Chapt. VII; (1.10) (see thus e.g. (7.19) in the preceding).
> On the other hand, by considering a *section of $\Omega^p$ over $U$*, as above, one actually looks at a *similar section of the* corresponding *presheaf*

*generating* $\Omega^p$ (see Chapt. IV; (7.2), along with (7.47) therein). So it is, in effect, this point of view that one actually applies, when we have written, hitherto, by usually *abusing notation,* $\Omega^p(U)$. Yet, concerning the same definition (8.19), as above, an analogous argument to that one applied to the previous map (1.23), in conjunction with (1.19), is here still in force.

Now, we call the above *operator* $D^p$, $p \in \mathbf{N}$, (cf. (8.15)), as given by (8.17), or even (8.19), the *p-th prolongation of the $\mathcal{A}$-connection* $D$ ($\equiv D^0$; cf. (8.18)).

Thus, (8.17) proves exactly that $D^1$ *is,* indeed, the *1st prolongation of the given $\mathcal{A}$-connection* $D$, as the latter has already been defined in the preceding, by (1.18) or (1.19). On the other hand, it is worth remarking here that:

(8.20) *the* **C**-*sequence (8.15) does not, in general, yield a complex,* in the sense that *we do not always have;*

(8.20.1) $$D^p \circ D^{p-1} = 0, \quad p \in \mathbf{N}.$$

Indeed, we have already set in the preceding, by defining the *curvature of* $D$ (cf. (2.2));

$$R(D) \equiv R := D^1 \circ D \equiv D^1 \circ D^0 \tag{8.21}$$

(see (8.18)). Hence, even the relation $D^1 \circ D = 0$ *is not valid, in general, unless* the given $\mathcal{A}$-connection $D$ *is flat* (cf. (2.5)). In this context, *one proves,* indeed, *that;*

$$(D^p \circ D^{p-1})(s \otimes t) = t \wedge R(s), \tag{8.22}$$

*for any* $p \in \mathbf{N}$, *and* $s \in \mathcal{E}(U)$, $t \in \Omega^{p-1}(U)$, with $U$ *open* in $X$. Therefore, one concludes that;

(8.23) *the curvature* $R$ *of a given $\mathcal{A}$-connection* $D$ ($\equiv D^0$, see (8.18)) *measures the obstruction of being the above* **C**-*sequence (8.15) a complex* (of **C**-vector space sheaves and **C**-linear morphisms, such that (8.20.1) be satisfied).

Now, *to prove (8.22),* one employs a *local gauge of* (the $\mathcal{A}$-*module*) $\mathcal{E}$, as e.g. in (3.2), in conjunction with the corresponding *local forms of* $R$ *and* $D$. (See thus (3.19) above, along with Chapt. VI; (1.11) and (1.15). ∎ (In this regard,

we also observe that one can still recapture (8.21) from (8.22), by virtue of (8.4) and (2.11)).

**8.(a). Trace of curvature.**– Suppose we are given the previous *complex* $(\Omega)$, as in (8.2), on a *topological space* $X$ (cf. also (8.6), (8.8) and (8.9)). Moreover, assume that $\mathcal{E}$ is an $\mathcal{A}$*-module* on $X$, which is further endowed with an $\mathcal{A}$*-connection* $D$ (cf., for instance, (8.13) above for the relevant context). Finally, suppose that

$$e^U \equiv \{U; (e_i)_{1 \leqq i \leqq n}\} \tag{8.24}$$

is a *local gauge of* $\mathcal{E}$ (cf. (3.2)). Thus, our objective by the following discussion, as already indicated by the title of this Subsection, is to consider the *trace of the curvature* (operator) $R$ of $D$, when restricted to the *open* $U \subseteq X$, as above.

Namely, given that one has

$$R|_U \equiv R = (R_{ij}) \in M_n(\Omega^2(U)) = M_n(\Omega^2)(U), \tag{8.25}$$

one further sets;

$$\operatorname{tr} R \equiv \operatorname{tr}(R|_U) := \operatorname{tr}(R_{ij}) = \sum_{i=1}^{n} R_{ii} \in \Omega^2(U). \tag{8.26}$$

On the other hand, let now

$$R^2 \equiv R \wedge R \in M_n(\Omega^4(U)), \tag{8.27}$$

where $R$ is still given by (8.25). In particular, one obtains;

$$d^4(\operatorname{tr} R^2) \equiv d(\operatorname{tr} R^2) = \operatorname{tr}(dR^2) \tag{8.28}$$

(as easily follows from (8.26) and the **C**-*linearity of* $d \equiv d^4$). However, one further obtains, in view of (8.27) and (8.7) (for $p = q = 2$);

$$dR^2 \equiv d(R \wedge R) = dR \wedge R + R \wedge dR = 0. \tag{8.29}$$

Now, in this context, one further obtains (cf. (8.27), (8.26));

$$\begin{aligned} \operatorname{tr}(R^2) &\equiv \operatorname{tr}((R^2_{ij})) \equiv \operatorname{tr}((R \wedge R)_{ij}) \\ &= \operatorname{tr}((\textstyle\sum_{k=1}^n R_{ik} \wedge R_{kj})) = \sum_{i=1}^n (\sum_{k=1}^n R_{ik} \wedge R_{ki}) \equiv \sum_{i,k} R_{ik} \wedge R_{ki}, \end{aligned} \tag{8.30}$$

so that one has

$$(8.31)\qquad \begin{aligned} d(\operatorname{tr} R^2) &\equiv d^4(\operatorname{tr} R^2) = d(\textstyle\sum_{i,k} R_{ik}\wedge R_{ki}) \\ &= \textstyle\sum_{i,k} d(R_{ik}\wedge R_{ki}) = \sum_{i,k}(dR_{ik}\wedge R_{ki} + R_{ik}\wedge dR_{ki}) = 0. \end{aligned}$$

Accordingly, as a result of (8.31), or even of (8.29), one concludes for the *curvature* $R$ of the given $\mathcal{A}$-connection $D$, as above, that:

(8.32) *the trace of* $R^2 \equiv R\wedge R$ (cf. (8.27)) *is a "closed" 4-form; viz.* one has

(8.32.1) $$d(\operatorname{tr} R^2) = 0.$$

The above is reminiscent, of course, of the analogous classical fact, pertaining to the so-called *"topological charge"*. Yet, as we shall see in the next Chapter IX, the same *"4-form"*

$$(8.33)\qquad \operatorname{tr}(R^2) \equiv \operatorname{tr}(R\wedge R) \in \Omega^4(U)$$

(see (8.27), as well as, Scholium 8.1 in the preceding) is an example of what we shall call in the sequel a *characteristic class of* (the *vector sheaf*) $\mathcal{E}$; thus, as already seen (cf. (8.32.1)), it is a *"closed 4-form"* (on the *open* set $U \subseteq X$, so that one actually needs a *0-cocycle of such,* as (8.33) yields, for that matter, in the case of a *vector sheaf* $\mathcal{E}$, the local gauge (8.24), as above, being then just a member of a given local frame of $\mathcal{E}$; cf. Chapter IX). Furthermore, it is also *"gauge invariant"*, in a sense to be made precise (loc. cit.). In this regard, concerning the classical case see, for instance, B.A. Dubrovin - A.T. Fomenko - S.P. Novikov [1: p.452, Definition 42.2].

# 9. Curvatures of induced $\mathcal{A}$-connections

We examine below curvatures of $\mathcal{A}$-connections, which are induced by given ones (see Chapter VI), in relation to the curvatures of the initially given $\mathcal{A}$-connections. So we start with the

**Curvature of a Whitney sum of $\mathcal{A}$-connections** (cf. Chapt. VI; Subsection 5.(a)): For simplicity's sake, we just consider two *$\mathcal{A}$-modules* $\mathcal{E}$ and $\mathcal{F}$, endowed with *$\mathcal{A}$-connections* $D_{\mathcal{E}}$ and $D_{\mathcal{F}}$, respectively.

Thus, by taking the *Whitney sum* $\mathcal{E} \oplus \mathcal{F}$ of the previous $\mathcal{A}$-modules, this is further equipped with an *$\mathcal{A}$-connection* too, given by

$$D_{\mathcal{E}\oplus\mathcal{F}} = D_{\mathcal{E}} \oplus D_{\mathcal{F}} \tag{9.1}$$

(see Chapt. VI; (5.3)). Therefore, based now on (3.4) and (3.12), one easily sees, by virtue of (9.1), that the *curvature* of the latter $\mathcal{A}$-connection is given by;

$$R(D_{\mathcal{E}\oplus\mathcal{F}}) = R(D_{\mathcal{E}}) \oplus R(D_{\mathcal{F}}). \tag{9.2}$$

Yet, as alluded to already at the beginning of this Subsection, one further remarks that;

(9.3) the above argument, concerning (9.2), *is valid,* of course, *for any finite,* or even *infinite, many summands.* (Cf. also Chapt. VI; Scholium 5.1).

Now, by looking again at a *differential triad* $(\mathcal{A}, \partial, \Omega)$ on a *topological space* $X$, as also was the case before, then, as an immediate conclusion from (9.2) (cf. (9.3) too), together with Chapt. VI; (5.4), (5.5) and (2.6) in the preceding, one obtains that;

$$R(\partial^n) = R(\partial \oplus \cdots \oplus \partial) = R(\partial) \underbrace{\oplus \cdots \oplus}_{n-\text{times}} R(\partial) = 0, \tag{9.4}$$

*for every* $n \in \mathbf{N}$. In other words, one thus concludes that:

(9.5) $\partial^n$, $n \in \mathbf{N}$, *is a flat $\mathcal{A}$-connection of* $\mathcal{A}^n$ (see also Chapt. VI; (5.6)).

More generally, one actually infers from (9.2) (in fact, from its extension to *any finite number of summands,* see (9.3)) that;

(9.6) *the Whitney sum of flat $\mathcal{A}$-connections is* still *flat.*

On the other hand, we next examine the

**Curvature of a tensor product of $\mathcal{A}$-connections** (see Chapt. VI; Subsection 5.(b)): Let us consider again the same framework, as before; then, by analogy with (9.2), one can still apply *Cartan's structural equation* (cf. (3.12) in the preceding) to compute the *curvature of the $\mathcal{A}$-connection of*

$\mathcal{E} \otimes_{\mathcal{A}} \mathcal{F}$, *viz.* of

$$D_{\mathcal{E}\otimes_{\mathcal{A}}\mathcal{F}} = D_{\mathcal{E}} \otimes 1_{\mathcal{F}} + 1_{\mathcal{E}} \otimes D_{\mathcal{F}} \tag{9.7}$$

(see Chapt. VI; (5.10)). Thus, first, based on (9.7), one easily sees that the corresponding *local (matrix-)form of the $\mathcal{A}$-connection* (9.7) is given by the relation

$$\omega(D_{\mathcal{E}\otimes_{\mathcal{A}}\mathcal{F}}) \equiv \tilde{\omega} = \omega \otimes 1 + 1 \otimes \omega' \equiv \omega + \omega', \tag{9.8}$$

where, of course, we have set $1_{\mathcal{E}} \equiv 1$ and $1_{\mathcal{F}} \equiv 1$, along with $\omega \equiv \omega(D_{\mathcal{E}})$ and $\omega' \equiv (D_{\mathcal{F}})$ (see also e.g. Chapt. VII; (1.13)). Thus, in view of (3.12) and (9.8), one now obtains;

$$\begin{aligned} R(D_{\mathcal{E}\otimes_{\mathcal{A}}\mathcal{F}}) &= d\tilde{\omega} + \tilde{\omega}\wedge\tilde{\omega} \\ &= d(\omega \otimes 1 + 1 \otimes \omega') + (\omega \otimes 1 + 1 \otimes \omega')\wedge(\omega \otimes 1 + 1 \otimes \omega') \\ &= d\omega \otimes 1 + (\omega\wedge\omega) \otimes 1 + 1 \otimes d\omega' + 1 \otimes (\omega'\wedge\omega') + (\omega \otimes 1)\wedge(1 \otimes \omega') \\ &\qquad + (1 \otimes \omega')\wedge(\omega \otimes 1) \\ &= (d\omega + \omega\wedge\omega) \otimes 1 + 1 \otimes (d\omega' + \omega'\wedge\omega'). \end{aligned} \tag{9.9}$$

In this context, we should also notice that, for convenience, we have taken above that *both $\omega$ and $\omega'$ are elements of the same $\mathcal{A}(U)$-module, $M_n(\Omega(U)) =$ $M_n(\Omega)(U)$* (cf. also Chapt. VII; (1.17)).

Accordingly, *one* finally *obtains the relation* (see the same (3.12));

$$R(D_{\mathcal{E}\otimes_{\mathcal{A}}\mathcal{F}}) = R(D_{\mathcal{E}}) \otimes 1 + 1 \otimes R(D_{\mathcal{F}}) \equiv R(D_{\mathcal{E}}) + R(D_{\mathcal{F}}). \tag{9.10}$$

Hence, as an immediate consequence of (9.10), one infers that;

(9.11) *the tensor product of flat $\mathcal{A}$-connections is a flat $\mathcal{A}$-connection,* as well.

Yet, by recapitulating (9.6) and (9.11), for convenience, one thus concludes that:

(9.12) the *Whitney sum and* the *tensor product preserve flatness of $\mathcal{A}$-connections.*

Now, a similar conclusion can be deduced, concerning the *dual* of a given (flat) *$\mathcal{A}$-connection:* Thus, by taking a pair $(\mathcal{E}, D)$, as before, along with a *local gauge $e^U$ of $\mathcal{E}$* (cf. e.g. (8.24)), let us further look at the *dual $\mathcal{A}$-mo-*

*dule of* $\mathcal{E}$ and the corresponding *dual* $\mathcal{A}$*-connection of* $D$ (see Chapt. VI; Subsection 5.(d)). Then, the *local* $\mathcal{A}$*-connection matrix (-"form") of the latter,* with respect to $e^U$, is given by the relation (cf. Chapt. VII; (4.20))

$$\omega(D^*) \equiv \omega^* = -{}^t\omega, \tag{9.13}$$

with $\omega \equiv \omega(D)$, while we have also set, as usual, $D_{\mathcal{E}^*} \equiv D^*$ (see thus Chapt. VI; (5.31), (5.36)). Therefore (cf. (3.12)), one has;

$$\begin{aligned} R(D^*) &= d\omega^* + \omega^* \wedge \omega^* = d(-{}^t\omega) + (-{}^t\omega)\wedge(-{}^t\omega) \\ &= -{}^t d\omega + {}^t\omega \wedge {}^t\omega, \end{aligned} \tag{9.14}$$

or even, by further setting, $R(D^*) \equiv R^*$, one finally obtains for the *curvature of* the *dual* $\mathcal{A}$*-connection,*

$$R^* = -{}^t d\omega + {}^t(\omega \wedge \omega). \tag{9.15}$$

In particular, by considering a *line sheaf* $\mathcal{L}$ on $X$, equipped with an $\mathcal{A}$-connection $D$, together with the *dual line sheaf* $\mathcal{L}^*$ and the *dual* $\mathcal{A}$*-connection* $D^* \equiv D_{\mathcal{L}^*}$, one obtains for the *curvature of* $D^*$ (see (9.15));

$$R(D^*) \equiv R^* = d\omega^* = d(-\omega) = -d\omega. \tag{9.16}$$

In other words, one thus obtains;

$$R^* = -R. \tag{9.17}$$

On the other hand, as already alluded to at the beginning, it is easy to prove that:

(9.18) *the dual of a flat* $\mathcal{A}$*-connection is* still *flat.*

Indeed, assuming that $R(D) = d\omega + \omega \wedge \omega = 0$ (cf. (3.12), (2.5)), one has;

$$d\omega = -\omega \wedge \omega = \omega \wedge \omega \tag{9.19}$$

(see also, for instance, (3.18) in the foregoing for the meaning of the last equality in (9.19)). So, based now on (9.13) and (9.19), one obtains;

$$\begin{aligned} d\omega^* &= d(-{}^t\omega) = -d({}^t\omega) = -{}^t(d\omega) = -{}^t(\omega \wedge \omega) \\ &= {}^t\omega \wedge {}^t\omega = -{}^t\omega \wedge (-{}^t\omega) = \omega^* \wedge \omega^*, \end{aligned} \tag{9.20}$$

so that, by applying a similar argument, as in (9.19), one gets

(9.21) $$R^* \equiv R(D^*) = 0,$$

that was our claim. ∎

Finally, we further consider the situation which arises, when one looks at the *pull-back* of a given *$\mathcal{A}$-connection* (see Chapt. VI; Section 6), as it concerns the corresponding *curvature:*

Thus, suppose that we are given a *differential triad* $(\mathcal{A}, \partial, \Omega^1)$ on a *topological space* $Y$, as well as, a *continuous map*

(9.22) $$f : X \longrightarrow Y,$$

while, we still assume that we have a pair $(\mathcal{E}, D)$ *on* $Y$, as e.g. in (8.14). Therefore, by further considering the *pull-back of the* given *differential triad* on $Y$, one gets a *similar one on* $X$ (cf. Chapt. VI; (6.3)). Consequently, the pair on $X$,

(9.23) $$f^*(\mathcal{E}, D) \equiv (f^*(\mathcal{E}), f^*(D))$$

has now an analogous meaning, as the given one on $Y$ (loc. cit.; (6.13)). Now, based on the very definitions, one can prove, first, that:

(9.24) *the pull-back of a curvature space,* via a continuous map, like (9.22), *is a space of the same sort;* viz.

(9.24.1) $$(f^*(\mathcal{A}), f^*(\partial), f^*(\Omega^1), f^*(d), f^*(\Omega^2))$$

is a *curvature space on* $X$.

(Concerning the $f^*(\mathcal{A})$*-module* on $X$, $f^*(\Omega^2)$, cf. also Chapt. IV; (7.4), (7.6), as well as, Chapt. VI; Lemma 6.1). Thus, based further on the same definitions and on Cartan's structural equation (cf. (3.12)), one can prove now that;

(9.25) $$R(f^*(D)) = f^*(R(D)),$$

that is,

(9.26) the *curvature is preserved by the pull-back.*

In particular, one thus infers that:

(9.27) the *pull-back preserves flatness;* precisely, *the pull-back of a flat $\mathcal{A}$-connection is a flat $f^*(\mathcal{A})$-connection* (cf. (9.22), (9.24.1)).

On the other hand, as a particular instance of the above, one can consider the *restriction of a* given *curvature space* $X$ (cf. (1.16)) to an *open* subset $U$ of $X$ (see Chapt. VI; (6.21)). So (9.25) yields then the following relation (*restriction of the curvature* $R$), which we already used, in anticipation, in the preceding (see thus (5.10));

(9.28) $$R(D|_U) = R(D)|_U.$$

# 10. Torsion of an $\mathcal{A}$-connection. Local form

Assume, as usual, that we are given a *differential triad*

(10.1) $$(\mathcal{A}, \partial, \Omega^1)$$

on a *topological space* $X$, while we still suppose that

(10.2) $$\theta^U \equiv \{U; (\theta_1, \ldots, \theta_n)\}$$

is a *local gauge* of the given $\mathcal{A}$-*module* $\Omega^1$ on $X$, of *rank* $n \in \mathbf{N}$, over the *open* $U \subseteq X$, such that

(10.3) $$(\theta_i)_{1 \leqq i \leqq n} \subseteq \Omega^1(U)$$

is a *section-basis* of the *free* $\mathcal{A}(U)$-*module*

(10.4) $$\Omega^1(U) = \mathcal{A}^n(U) = \mathcal{A}(U)^n$$

(see, for instance, Chapt. VII; (1.3), (1.4), (1.5)).

On the other hand, consider the *dual* $\mathcal{A}$-*module of* $\Omega^1$; that is, we set

(10.5) $$\mathcal{E} \equiv (\Omega^1)^* := \mathcal{H}om_{\mathcal{A}}(\Omega^1, \mathcal{A}).$$

Thus,

(10.6) $$(\theta^U)^* \equiv e^U \equiv \{U; (e_1, \ldots, e_n)\}$$

is a *local gauge of* $\mathcal{E}$ *(dual of (10.2)) over* $U$, of the *same rank* $n$, as (9.2), such that

$$(10.7) \qquad e_i := \theta_i^*, \quad 1 \leqq i \leqq n,$$

with

$$(10,8) \qquad e_i(\theta_j) := \delta_{ij}, \quad 1 \leqq i, j \leqq n$$

(see Chapt. VII; (4.3), (4.6). In practice (cf. e.g. the classical theory), $\mathcal{E}$ is a given *vector sheaf* on $X$, whose *dual* (vector sheaf, then, too) *is* assumed to be $\Omega^1$; cf., for instance, Chapt. VII; (5.3) and Scholium 5.1). Thus, we are led to the following.

**Definition 10.1.** Suppose we have a curvature space $X$, and let $\mathcal{E}$ be an $\mathcal{A}$-module on $X$, with $(\Omega^1)^* = \mathcal{E}$. Moreover, let $D$ be an $\mathcal{A}$-connection of $\mathcal{E}$, while $\theta^U$ and $e^U \equiv (\theta^U)^*$ are local gauges of the $\mathcal{A}$-modules concerned, as in (10.2) and (10.6), respectively. Then, the *local torsion of $D$ over $U$* is given by the relation;

$$(10.9) \qquad \Theta|_U \equiv \Theta := d\theta + \omega \wedge \theta,$$

where $\omega \in M_n(\Omega^1(U))$ stands for the local $\mathcal{A}$-connection matrix of $D$, which is associated with the given local gauge $e^U$ of $\mathcal{E}$ (cf. Chapt. VII; (1.13)).

**Note.**- Concerning the notation in (10.9), the term $\theta$ stands for the *section-basis* of the given *local gauge* $\theta^U$ of $\Omega^1$, as in (10.3); viz. we have set

$$(10.10) \qquad \theta \equiv (\theta_1, \ldots, \theta_n) \in \Omega^1(U)^n,$$

considering it as a *"column vector"* of $\Omega^1(U)^n$.

The same relation (10.9) constitutes, in fact, the counterpart in our case of the so-called *Cartan's first structural equation* of the classical theory. (See, for instance, M. Crampin - F.A.E. Pirani [1: p. 278]).

On the other hand, suppose now that, in addition to the previously given curvature space $X$, we have, in particular, the framework of a *Bianchi space* $X$, in the sense of Definition 7.1 in the preceding (see also the comments following the same definition). Thus (loc. cit.), assume that we are given further the *second exterior derivation*

$$(10.11) \qquad d^2 \equiv d : \Omega^2 \longrightarrow \Omega^3,$$

as this is given by (7.4), in such a manner that (7.5) *hold true.* So we are led now to the following classical result (see e.g. M. Crampin - F.A.E. Pirani [1: p. 279; *Bianchi identities,* or even p.281, Ex. 35]).

**Theorem 10.1.** *(Ricci's Lemma). Suppose we are given a Bianchi space* $X$ (cf. Definition 7.1) *and an* $\mathcal{A}$*-module* $\mathcal{E}$ *on* $X$*, such that* $\mathcal{E} = (\Omega^1)^*$*, which is further endowed with an* $\mathcal{A}$*-connection* $D$*, while we also assume the existence of a local gauge of* $\Omega^1$*, hence, of* $\mathcal{E}$ *too, as in (10.2) and (10.6) above, respectively. Then, concerning the (local) torsion of* $D$ *(over* $U$, cf. (10.9)), *one has the relation;*

$$d\Theta = R\wedge\theta - \omega\wedge\Theta \tag{10.12}$$

*(first Bianchi's identity),* with $\omega$ and $\theta$, as in (10.9) (see also (10.10)).

**Proof.** Based on (7.5), (7.4) and (3.12), and by applying $d^2 \equiv d$, as in (10.11), one obtains, in view of the same relation;

$$\begin{aligned} d\Theta &= d(d\theta + \omega\wedge\theta) = d(d\theta) + d(\omega\wedge\theta) = d(\omega\wedge\theta) \\ &= d\omega\wedge\theta - \omega\wedge d\theta = (R - \omega\wedge\omega)\wedge\theta - \omega\wedge(\Theta - \omega\wedge\theta) \\ &= R\wedge\theta - \omega\wedge\omega\wedge\theta - \omega\wedge\Theta + \omega\wedge\omega\wedge\theta = R\wedge\theta - \omega\wedge\Theta, \end{aligned} \tag{10.13}$$

*as desired.* ∎

## 11. Weil's integrality theorem

We finally come, by this Section, to the full proof of the theorem in the heading. Indeed, one half of it, namely, the *"only if"* part (cf., for instance, Theorem 11.1′ below) has already been established by Lemma 5.2 in the preceding. So we give right away the statement of the theorem in question.

**Theorem 11.1.** *Suppose that we are given a sheaf exponential (Hausdorff) paracompact curvature space* $X$*. Moreover, let* $\mathcal{L}$ *be a line sheaf on* $X$*, endowed with an* $\mathcal{A}$*-connection* $D$*. Then, the curvature of* $D$*,* $R(D) \equiv R$*, yields a 2-dimensional integral cohomology class of* $R$*,* $c(R)$*; viz. one has* (see (5.49), (5.51))

$$c(R) \in H^2(X, \mathbf{Z}). \tag{11.1}$$

*Conversely, assume that our space $X$, as before, satisfies the condition*

$$H^1(X, \mathcal{A}) = H^2(X, \mathcal{A}) = 0, \tag{11.2}$$

*while we still suppose that;*

(11.3) *every line sheaf on $X$, admits an $\mathcal{A}$-connection*

(take, for instance, $H^1(X, \Omega^1) = 0$, or even, concerning both (11.2) and (11.3), assume that $\mathcal{A}$ is a fine sheaf. See e.g. Chapt. III; (8.24)). *Then, given a 2-dimensional integral cohomology class of $X$, there exists a (uniquely defined) line sheaf on $X$, admitting an $\mathcal{A}$-connection, whose curvature yields the given (integral) cohomology class.*

**Proof.** As already noted, the first part of the statement is exactly Lemma 5.2 in the foregoing. So consider now a *2-dimensional integral cohomology class of $X$*, say,

$$z \in H^2(X, \mathbf{Z}) \tag{11.4}$$

(cf. also the above Scholium 5.3, regarding the meaning of (11.4)). Now, by hypothesis for $X$, one has the following *short exact exponential sheaf sequence*

$$0 \longrightarrow \mathbf{Z} \longrightarrow \mathcal{A} \xrightarrow{e} \mathcal{A}^{\cdot} \longrightarrow 1 \tag{11.5}$$

(see Chapt. VII; (7.1) and Chapt. V; (6.1)). Thus, by further considering the corresponding long exact sequence in cohomology one obtains (cf., for instance, Chapt. III; Theorem 5.3);

$$\cdots \longrightarrow H^1(X, \mathcal{A}) \longrightarrow H^1(X, \mathcal{A}^{\cdot}) \xrightarrow{\delta} H^2(X, \mathbf{Z}) \longrightarrow H^2(X, \mathcal{A}) \longrightarrow \cdots. \tag{11.6}$$

Therefore, taking now (11.2) into account, one gets the exact sequence

$$0 \longrightarrow H^1(X, \mathcal{A}^{\cdot}) \xrightarrow{\delta} H^2(X, \mathbf{Z}) \longrightarrow 0, \tag{11.7}$$

so that one finally obtains;

$$H^1(X, \mathcal{A}^{\cdot}) = H^2(X, \mathbf{Z}), \tag{11.8}$$

within an *isomorphism of* (the abelian) *groups* concerned. (The *Bockstein operator* $\delta$, as in (11.6), is called, in this context, the *Chern isomorphism*, as

well; see also e.g. Chapt. V; (6.33), (6.35)). Thus, one now sets (cf. (11.4))

(11.9) $$\mathcal{L} := \delta^{-1}(z) \in H^1(X, \mathcal{A}^{\cdot}),$$

this being, as we shall presently see, the *line sheaf* on $X$, that we are looking for (cf. Chapt. V; (2.59), (2.60)): Indeed, suppose that

(11.10) $$(g_{\alpha\beta}) \in Z^1(\mathcal{U}, \mathcal{A}^{\cdot})$$

is a representative *(coordinate) 1-cocycle of* $\mathcal{L}$ (see loc. sit.; (2.43), for $n = 1$), such that

(11.11) $$\mathcal{U} = (U_\alpha)_{\alpha \in I}$$

is a *local frame* of $\mathcal{L}$. On the other hand, based on (11.3), one has for the respective *Atiyah class of* $\mathcal{L}$ the relation;

(11.12) $$\mathfrak{a}(\mathcal{L}) \equiv \left[\tilde{\partial}(g_{\alpha\beta})\right] = 0 \in H^1(X, \Omega^1)$$

(see Chapt. VI; (10.28.1), for $n = 1$, along with Corollary 12.1). Therefore, one obtains

(11.13) $$\tilde{\partial}(g_{\alpha\beta}) = \delta(\theta_\alpha),$$

for some *0-cochain* of *"1-forms"*

(11.14) $$(\theta_\alpha) \in C^0(\mathcal{U}, \Omega^1).$$

Furthermore, in view of the short exact sequence (11.5) and the hypothesis for $X$, the 1-cocycle (11.10) provides a *1-cochain*

(11.15) $$(f_{\alpha\beta}) \in C^1(\mathcal{V}, \mathcal{A}),$$

such that one has

(11.16) $$g_{\alpha\beta} = e(f_{\alpha\beta}),$$

where $\mathcal{V}$ in (11.15) stands for an eventual refinement of $\mathcal{U}$ (cf. Chapt. III; Lemmas 5.1, 5.2). Therefore, based now on the commutative triangle (5.34), according to our hypothesis for $X$ (see Chapt. VII; (7.1)), one obtains, by (11.16),

(11.17) $$\tilde{\partial}(g_{\alpha\beta}) = \tilde{\partial}(e(f_{\alpha\beta})) = 2\pi i \cdot \partial(f_{\alpha\beta}) = \partial(2\pi i \cdot f_{\alpha\beta}),$$

so that, in view of (11.13), one has;

$$\delta(\theta_\alpha) = \tilde{\partial}(g_{\alpha\beta}) = 2\pi i \cdot \partial(f_{\alpha\beta}). \tag{11.18}$$

(Of course, the first equality in the last relation expresses in our case the corresponding *transformation law of potentials;* cf. Chapt. VII; (2.59)).

Now, suppose that $D$ is an $\mathcal{A}$-*connection* of $\mathcal{L}$ (cf. (11.3)), such that let

$$D|_{U_\alpha} = D_\alpha - \theta_\alpha, \quad \alpha \in I, \tag{11.19}$$

be the *local form of* $D$, with respect to the local frame $\mathcal{U}$ of $\mathcal{L}$, as in (11.11) (see Chapt. VI; (8.13), (11.14), for $\widetilde{D} \equiv D$). On the other hand, the set of $\mathcal{A}$-connections of $\mathcal{L}$ being an *affine space*, modelled on $\Omega^1(X)$ (cf. Chapt. VI; Corollary 7.1), one may assume that the 0-cochain of "1-forms", $(\theta_\alpha) \in C^0(\mathcal{U}, \Omega^1)$, as appeared in (11.19), is that one defined by (11.13) and (11.14). Hence, the *curvature* of $D$, $R(D) \equiv R$, is given by the relation (cf. Corollary 3.1)

$$R = (d\theta_\alpha). \tag{11.20}$$

So the last relation yields now a *closed 2-form on* $X$ (loc. cit.), which can further be construed as that one supplying the given cohomology class $z \in H^2(X, \mathbf{Z})$, as in (11.4): Indeed, one has, by (11.18),

$$\partial(f_{\alpha\beta}) = \frac{1}{2\pi i}\delta(\theta_\alpha) = \frac{1}{2\pi i}\tilde{\partial}(g_{\alpha\beta}) \tag{11.21}$$

while one further obtains, by virtue of (11.5), (11.15) and (11.16), that

$$\delta(f_{\alpha\beta}) \equiv (\lambda_{\alpha\beta\gamma}) \in Z^2(\mathcal{V}, \mathbf{Z} = \ker e), \tag{11.22}$$

which thus explains our claim about $z$ and $R$, as the latter is given by (11.20) (see also Scholium 5.4 in the preceding), and this terminates the proof. ∎

Due to its significance (see e.g. Chapter X in the sequel), we restate the above result in the following more compact form. That is, we have.

**Theorem** 11.1′. *Let $X$ be a Hausdorff paracompact sheaf exponential curvature space, such that (11.2) and (11.3) hold true. Then, the only 2-dimensional integral cohomology classes of $X$ are* (those associated with) *field strengths of Maxwell fields* (viz. curvatures of $\mathcal{A}$-connections of line sheaves) *on $X$.* ∎

# CHAPTER IX

# Characteristic classes

"...an axiomatic approach ... emphasizes the mathematical structures involved..."

Sir Michael F. Atiyah in "The Geometry and Physics of Knots" (Cambridge Univ. Press, 1990). p. 12.

We consider below (Chern) characteristic classes of vector sheaves, the analogue in our case of the same classes of vector bundles (over a $\mathcal{C}^\infty$-manifold). In particular, we apply the counterpart here of the (differential-geometric) *Chern-Weil description* of these classes, *in terms of the curvature* (tensor). So, as happens in the standard case, the *Bianchi identity* (see the previous Chapter, Theorem 7.1) becomes here a key-result. On the other hand, one has to assume further, within this abstract treatment, the existence of an *appropriate analogue of de Rham's complex,* to ensure, by analogy with the standard case, a cohomology class, which is thus (canonically) associated with any given *"closed form"*. So we start with the necessary preliminary material.

# 1. Preliminaries

Suppose that we are given, as usual, a **C**-*algebraized space*

$$(X, \mathcal{A}) \tag{1.1}$$

on an arbitrary *topological space* $X$ (cf. Chapt. II; Scholium 1.2). Furthermore, we assume that $X$ is endowed with a *differential triad*

$$(\mathcal{A}, \partial, \Omega^1) \tag{1.2}$$

(see Chapt. VI; (3.1) and Chapt. VIII; (1.1), (1.2)).

On the other hand, consider a pair

$$(\mathcal{E}, D), \tag{1.3}$$

consisting of a *vector sheaf* $\mathcal{E}$ on $X$, of *rank* $n \in \mathbf{N}$, and a given $\mathcal{A}$-*connection* $D$ *of* $\mathcal{E}$.

Now, if the *rank* of $\mathcal{E}$, as before, is *at least* 2, viz. in the case one has

$$\operatorname{rk}_{\mathcal{A}}(\mathcal{E}) \equiv \operatorname{rk} \mathcal{E} = n \geqq 2, \tag{1.4}$$

then, by extending to the present abstract setting, the classical terminology, connected with elementary particle physics, a pair as in (1.3) above, will also be called a *Yang-Mills field* on $X$. In a similar manner, we have named in the preceding an analogous pair $(\mathcal{L}, D)$, where now $\mathcal{L}$ is a *line sheaf* on $X$, viz. when $\operatorname{rk} \mathcal{L} = 1$, a *Maxwell field* on $X$ (cf. Chapt. VI; (18.11). See also Chapter X in the sequel).

Now, suppose that

$$\mathcal{U} = (U_\alpha)_{\alpha \in I} \tag{1.5}$$

is a *local frame of* our *vector sheaf* $\mathcal{E}$. Then, as already known, *the* given $\mathcal{A}$-*connection* $D$ *of* $\mathcal{E}$ *is determined* by a *0-cochain of local* $\mathcal{A}$-*connection matrices,* associated with $\mathcal{U}$,

$$\omega \equiv \left(\omega^{(\alpha)}\right) = \left(\left(\omega_{ij}^{(\alpha)}\right)\right) \in C^0(\mathcal{U}, M_n(\Omega^1)) \tag{1.6}$$

such that

$$\omega^{(\alpha)} \equiv \left(\omega_{ij}^{(\alpha)}\right) \in M_n(\Omega^1(U_\alpha)) = M_n(\Omega^1)(U_\alpha), \tag{1.7}$$

*for every* $\alpha \in I$, as in (1.5) (cf. Chapt. VII; (2.39.1) and (2.21)). Yet, (1.6) has to satisfy the corresponding *transformation law of potentials;*

$$\omega^{(\beta)} = \mathrm{Ad}(g_{\alpha\beta}^{-1})\omega^{(\alpha)} + \tilde{\partial}(g_{\alpha\beta}) \tag{1.8}$$

(ibid.; Theorem 2.1. See also Theorem 3.2 therein). In this context,

$$(g_{\alpha\beta}) \in Z^1(\mathcal{U}, \mathcal{GL}(n, \mathcal{A})) \tag{1.9}$$

stands here for a *coordinate 1-cocycle of* $\mathcal{E}$, associated with $\mathcal{U}$, as in (1.5), and yielding the *gauge transformation* in (1.8).

On the other hand, let us assume now that $X$ is, in particular, a *curvature space,* viz. (cf. Chapt. VIII; Definition 1.1) suppose further that we are given a *curvature datum*

$$(\mathcal{A}, \partial, \Omega^1, d^1, \Omega^2). \tag{1.10}$$

Thus, by considering a pair $(\mathcal{E}, D)$, as in (1.3), one has naturally associated with the given $\mathcal{A}$-connection $D$ its *curvature* $R(D) \equiv R$, the latter being a *global section* of the *vector sheaf* $\Omega^2(\mathcal{E}nd\,\mathcal{E})(X)$. That is, we have

$$\begin{aligned} R(D) \equiv R \equiv \left(R^{(\alpha)}\right) \in \Omega^2(\mathcal{E}nd\,\mathcal{E})(X) &= Z^0(\mathcal{U}, \Omega^2(\mathcal{E}nd\,\mathcal{E})) \\ &= Z^0(\mathcal{U}, M_n(\Omega^2)), \end{aligned} \tag{1.11}$$

so that, in particular, one obtains;

$$R^{(\alpha)} \equiv \left(R_{ij}^{(\alpha)}\right) \in M_n(\Omega^2)(U_\alpha) = M_n(\Omega^2(U_\alpha)), \tag{1.12}$$

*for every* $\alpha \in I$ (see Chapt. VIII; (2.16) – (2.20), along with (2.21)). Yet, one has, by virtue of (1.12);

$$R_{ij}^{(\alpha)} \in \Omega^2(U_\alpha), \tag{1.13}$$

*for every* $\alpha \in I$, *and* $1 \leqq i, j \leqq n = \mathrm{rk}\,\mathcal{E}$, (In this regard, cf. also, for instance, Scholium 8.1 in the previous Chapter VIII). Furthermore, one has the analogue of (1.8), pertaining now to the corresponding *transformation law of curvature* (cf. also Chapt. VIII; (4.23)),

$$R^{(\beta)} = \mathrm{Ad}(g_{\alpha\beta}^{-1})R^{(\alpha)}, \tag{1.14}$$

which is thus associated with the given local frame $\mathcal{U}$ of $\mathcal{E}$ and the corresponding coordinate 1-cocycle $(g_{\alpha\beta})$ of it, as in (1.5), (1.9).

On the other hand, consider now the following $\mathcal{A}$-*module* on $X$

(1.15) $$\Omega^{\cdot}_{\text{even}} \equiv \Omega^{\cdot}_{\text{ev}} := \bigoplus_{m=0}^{\infty} \wedge^{2m}\Omega^1 \equiv \bigoplus_{m \geqq 0} \Omega^{2m}$$

(see also e.g. Chapt. VIII; (8.3)); viz. one takes the *Whitney sum of the even exterior powers of the* given *$\mathcal{A}$-module* $\Omega^1$ (cf. Chapt. IV; (7.9), (7.11.1)). Yet, we still set in (1.15);

(1.16) $$\wedge^0\Omega^1 \equiv \Omega^0 := \mathcal{A}$$

(see, for example, Chapt. VIII; (8.4)). Thus, one now concludes that;

(1.17) *the* above *$\mathcal{A}$-module* on $X$, $\Omega^{\cdot}_{\text{ev}}$, as given by (1.15) and (1.16), *when* further *endowed with the* corresponding *exterior multiplication* (of *sections,* cf. e.g. Chapt. II; (1.67)), *provides a commutative unital* **C**-*algebra sheaf* on $X$.

Indeed, $\Omega^{\cdot}_{\text{ev}}$ is an *$\mathcal{A}$-algebra sheaf* on $X$. Cf. Chapt. II; Definition 1.4, 1.6 and Chapt. IV; (7.10).

In this context, it is still useful to remark that, in view of the relation,

(1.18) $$\Omega^n_x \equiv (\wedge^n\Omega^1)_x = \wedge^n(\Omega^1_x), \quad n \in \mathbf{N},$$

which is valid, up to an *$\mathcal{A}_x$-isomorphism, for every* $x \in X$, one can work with the corresponding *presheaf of sections, generating (1.15)* (cf. Chapt. IV; (7.11) and (7.14), (7.14′), along with Chapt. VIII; Scholium 8.1). Indeed, this is what we essentially employ in the sequel. Thus, as a direct consequence of the last remarks, we further note that, by virtue of (1.12) and (1.15) (see also e.g. Chapt. IV; (3.23) and Chapt. II; (5.18), (3.14)), one now obtains;

(1.19) $$\begin{aligned} R^{(\alpha)} \equiv \left(R^{(\alpha)}_{ij}\right) &\in M_n(\Omega^2(U_\alpha)) = M_n(\Omega^2)(U_\alpha) \\ &= (\Omega^2)^{n^2}(U_\alpha) = \Omega^{2n^2}(U_\alpha) \subsetneq \left(\bigoplus_{m=0}^{\infty}\Omega^{2m}\right)(U_\alpha) \\ &= \bigoplus_{m \geqq 0}\Omega^{2m}(U_\alpha) \equiv \mathcal{A}(U_\alpha) \oplus \Omega^2(U_\alpha) \oplus \cdots \\ &= \Omega^{\cdot}_{\text{ev}}(U_\alpha) \subsetneq (\Omega^{\cdot}_{\text{ev}})(U_\alpha), \end{aligned}$$

*for any* $\alpha \in I$ (cf. (1.5)), and *any given* $n \in \mathbf{N}$. (Concerning the last inclusion in (1.19), see also Chapt. IV; (7.47), together with Chapt. I; (11.36′)). *Warning!* In this regard, it is still to be noticed that, in general, one has

(1.20) $$(\Omega^{\cdot}_{\text{ev}})(U) \neq \Omega^{\cdot}_{\text{ev}}(U),$$

for a given *open* set $U$ in $X$. (However, see also (1.18) above and the relevant comments after it, as well as, Chapt. VIII; Scholium 8.1).

Thus, we are now ready to embark on our main theme in this chapter, as indicated by its title. However, for convenience, we still recall in the next Section 2 certain basic facts, pertaining to *"invariant polynomials"*, that will be of essential use throughout the sequel.

## 2. Invariant polynomials

The polynomials in the title of this Section, that are going to be considered by the ensuing discussion are, in fact, *complex polynomials,* whose "indeterminates" will be elements of the *unital commutative* **C**-*algebra*

$$\Omega^{\cdot}_{\mathrm{ev}}(U_\alpha), \tag{2.1}$$

as above; see thus (1.5), (1.17), as well as, (1.19) and (1.20) of the previous Section. Notice that, in view of (1.19) (see also Chapt. II; (1.34)), *one* always *has;*

$$\mathbf{C} \underset{\varepsilon}{\hookrightarrow} \mathcal{A}(U_\alpha) \hookrightarrow \Omega^{\cdot}_{\mathrm{ev}}(U_\alpha), \tag{2.2}$$

*for every* $\alpha \in I$, as in (1.5).

So consider next the *polynomial algebra,*

$$\mathbf{C}[(\lambda_{ij})], \quad 1 \leqq i, j \leqq n \in \mathbf{N}. \tag{2.3}$$

That is, one thus takes a *commutative unital* **C**-*algebra,* consisting of all the *complex polynomials* in the $n^2$ many *indeterminates* $(\lambda_{ij})$, $1 \leqq i, j \leqq n$, so that let us consider an element

$$p \in \mathbf{C}[(\lambda_{ij})]. \tag{2.4}$$

Hence, one can further look at the respective *polynomial function,* relative to (2.1) (cf. (1.19), as well);

$$\tilde{p} : (\Omega^{\cdot}_{\mathrm{ev}}(U_\alpha))^{n^2} = M_n(\Omega^{\cdot}_{\mathrm{ev}}(U_\alpha)) \longrightarrow \Omega^{\cdot}_{\mathrm{ev}}(U_\alpha) \hookrightarrow (\Omega^{\cdot}_{\mathrm{ev}})(U_\alpha), \tag{2.5}$$

*for every* $\alpha \in I$, as in (1.5), *and* $n \in \mathbf{N}$ (cf. also Chapt. V; (8.7) and Chapt.

VI; (9.3)). Yet, due to our previous *warning*, about (1.20), we also have, in general,

$$M_n(\Omega^{\cdot}_{\mathrm{ev}})(U_\alpha) \neq M_n(\Omega^{\cdot}_{\mathrm{ev}}(U_\alpha)). \tag{2.6}$$

But, see also (2.36) below; however, the above inconvenience *does not affect*, actually, our argument in the sequel.

On the other hand, the *correspondence*

$$p \longmapsto \tilde{p}, \tag{2.7}$$

as defined by (2.4) and (2.5), *is*, in effect, *one-to-one*, so that we further set, for convenience,

$$\tilde{p} \equiv p \tag{2.8}$$

(see, for instance, N. Bourbaki [4: p.28, Remarque 1)]).

In particular, suppose that we are given a *curvature space* $X$, along with a pair

$$(\mathcal{E}, D), \tag{2.9}$$

consisting of a *vector sheaf* $\mathcal{E}$ on $X$, *with* $\mathrm{rk}\,\mathcal{E} = n \in \mathbf{N}$, and an $\mathcal{A}$*-connection* $D$ *of* $\mathcal{E}$. Furthermore, let

$$\mathcal{U} = (U_\alpha)_{\alpha \in I} \tag{2.10}$$

be a *local frame of* $\mathcal{E}$; then, associated with $\mathcal{U}$, there is a *0-cochain*, defined by the *curvature of* $D$ (cf. (1.11) and (1.19)),

$$\begin{aligned} R \equiv \left(R^{(\alpha)}\right) \equiv \left(\left(R^{(\alpha)}_{ij}\right)\right) &\in C^0(\mathcal{U}, M_n(\Omega^2)) \\ &= C^0(\mathcal{U}, (\Omega^2)^{n^2}) \subsetneqq C^0(\mathcal{U}, \Omega^{\cdot}_{\mathrm{ev}}). \end{aligned} \tag{2.11}$$

Accordingly, *for every polynomial*

$$p \in \mathbf{C}[(\lambda_{ij})], \tag{2.12}$$

as in (2.4), one gets a *0-cochain*

$$\tilde{p}(R) \equiv p(R) \equiv \left(p\left(R^{(\alpha)}\right)\right) \in C^0(\mathcal{U}, \Omega^{\cdot}_{\mathrm{ev}}) \tag{2.13}$$

(cf. (2.5), (2.8), (2.11) and (1.17)).

Now, in this context, we still recall that a given (complex) polynomial $p$, as in (2.12), is said to be *invariant*, whenever one has

$$p(\alpha\lambda\alpha^{-1}) = p(\text{ad}(\alpha)\lambda) = p(\lambda), \tag{2.14}$$

for every $\lambda \equiv (\lambda_{ij}) \in M_n(\mathbf{C})$ and $\alpha \in \text{GL}(n, \mathbf{C})$. (Of course, the multiplication, appeared in the first member of (2.14) is matrix-multiplication, within $M_n(\mathbf{C})$).

**Example 2.1.**- The following is classical, being also of an essential use in the following: Thus, given a *complex* $n \times n$ *matrix*

$$\lambda \equiv (\lambda_{ij}) \in M_n(\mathbf{C}), \tag{2.15}$$

the corresponding *determinant* of $\lambda$ can be construed as a (complex) *polynomial* with *indeterminates the entries of* $\lambda$, $\lambda_{ij} \in \mathbf{C}$, $1 \leqq i, j \leqq n$; namely, one has (cf. (2.3))

$$\det(\lambda) \in \mathbf{C}[(\lambda_{ij})]. \tag{2.16}$$

Thus, it is now a direct consequence of the same properties of det that;

(2.17) *the determinant function*

$$\det : M_n(\mathbf{C}) \longrightarrow \mathbf{C} \tag{2.17.1}$$

*is an invariant polynomial* (cf. also (2.8), concerning the map (2.17.1)).

That is, (2.14) is fulfilled (cf. e.g. C.W. Curtis [1: p.149]). Indeed, the above situation will be amply applied in the sequel. (See thus, for instance, Section 5 below).

Now, our first relevant result pertains to the above 0-cochain (2.13) which *is, in fact, a 0-cocycle, whenever* our polynomial $p$ *is invariant,* in the previous sense. Precisely, one thus obtains the following.

**Lemma 2.1.** *Suppose we are given a curvature space* $X$*, a pair* $(\mathcal{E}, D)$*, as in (2.9), and an invariant polynomial* $p \in \mathbf{C}[(\lambda_{ij})]$*, such that* $1 \leqq i, j \leqq n = \text{rk}\,\mathcal{E}$*. Then, the curvature 0-cochain* $p(R)$ (cf. (2.13))*, which is associated with a given local frame* $\mathcal{U}$ *of* $\mathcal{E}$ (cf. (2.10))*, is actually a 0-cocycle, viz. one has*

$$p(R) \equiv \left(p\left(R^{(\alpha)}\right)\right) \in Z^0(\mathcal{U}, \Omega^{\cdot}_{\text{ev}}) = \Omega^{\cdot}_{\text{ev}}(X). \tag{2.18}$$

*In particular, if*

$$\deg p = m, \tag{2.19}$$

*then, one obtains;*

$$p(R) \in \wedge^{2m}\Omega^1(X) \hookrightarrow (\wedge^{2m}\Omega^1)(X) \equiv \Omega^{2m}(X) \hookrightarrow \Omega^{\bullet}_{\mathrm{ev}}(X). \tag{2.20}$$

**Proof.** Suppose that

$$p \in \mathbf{C}[(\lambda_{ij})] \tag{2.21}$$

is an *invariant polynomial.* Then, by considering a *coordinate 1-cocycle of* $\mathcal{E}$, relative to the given local frame $\mathcal{U}$ of $\mathcal{E}$, say,

$$(g_{\alpha\beta}) \in Z^1(\mathcal{U}, \mathcal{GL}(n, \mathcal{A})), \tag{2.22}$$

one obtains (cf. (1.14), (2.14));

$$p\left(R^{(\beta)}\right) = p\left(\mathrm{Ad}(g_{\alpha\beta}^{-1})R^{(\alpha)}\right) = p\left(R^{(\alpha)}\right), \tag{2.23}$$

for any $\alpha, \beta \in I$, with $U_{\alpha\beta} \equiv U_\alpha \cap U_\beta \neq \emptyset$ (cf. (2.10)). That is, *the 0-cochain* $\left(p\left(R^{(\alpha)}\right)\right)$ *is a 0-cocycle,* as was to be proved. On the other hand, (2.20) is certainly clear, by virtue of (2.19) and (1.19). ∎

To proceed further, we first recall some standard facts from the classical *theory of polynomials* pertaining, in particular, to *formal derivations:*

So let us take again a complex *polynomial*

$$p \in \mathbf{C}[(\lambda_{ij})], \quad 1 \leqq i, j \leqq n \tag{2.24}$$

(of $n^2$ *indeterminates*). Furthermore, consider the *matrix of the (formal) first partial derivatives of* $p$,

$$\left(\frac{\partial p}{\partial \lambda_{ij}}\right) \in M_n(\mathbf{C}[(\lambda_{ij})]) \tag{2.25}$$

(see e.g. N. Bourbaki [3: Chap. 4; p. 38, Définition 2]). Hence, by taking now the *derivative* (or even *differential*) *of* $p$, one has;

$$dp = \sum_{i,j} \frac{\partial p}{\partial \lambda_{ij}} d\lambda_{ij} \tag{2.26}$$

(loc. cit.; p.38, Définition 1, as well as, rels. (2), (3)). On the other hand, setting

$$(2.27) \qquad p'(\boldsymbol{\lambda}) := {}^{t}\left(\frac{\partial p}{\partial \lambda_{ij}}\right) \in M_n(\mathbf{C}[(\lambda_{ij})]),$$

viz. the *transpose* (matrix) *of (2.25),* one obtains, in view of (2.26),

$$(2.28) \qquad dp = \sum_{i,j} \frac{\partial p}{\partial \lambda_{ij}} d\lambda_{ij} = \operatorname{tr}(p'(\boldsymbol{\lambda})\, d\boldsymbol{\lambda}),$$

where, of course, $\boldsymbol{\lambda} \equiv (\lambda_{ij}) \in M_n(\mathbf{C})$, while "tr" stands for the *trace* of the (product) matrix concerned. (In this regard, see also e.g. L. Smith [1: p.249, *"trace scalar product"*]; yet, cf. J.W. Milnor - J.D. Stasheff [1: p.297, proof of *"Fundamental Lemma"*]).

In particular, by taking an *invariant polynomial* (cf. (2.14)),

$$(2.29) \qquad p \equiv p(\boldsymbol{\lambda}) \equiv p((\lambda_{ij})) \in \mathbf{C}[(\lambda_{ij})], \quad 1 \leqq i, j \leqq n,$$

such that

$$(2.30) \qquad \boldsymbol{\lambda} \equiv (\lambda_{ij}) \in M_n(\mathbf{C}),$$

one concludes that (cf. also (2.27));

$$(2.31) \qquad p'(\boldsymbol{\lambda}) \cdot \boldsymbol{\lambda} = \boldsymbol{\lambda} \cdot p'(\boldsymbol{\lambda}).$$

See the last Ref. above for the proof of (2.31). ∎

So we are now in the position to give the proof of the following basic result for what follows. It is thus the analogue in our case of the aforementioned "Fundamental Lemma" (loc. cit.). So the point is that, *for every invariant polynomial* $p$, as in (2.29), *the global "form"* $p(R)$, as given by (2.20), *is, moreover, a "closed"* one (see (2.36) below). Precisely speaking, one thus obtains the next.

**Theorem 2.1.** *(Fundamental Lemma). Suppose we are given a Bianchi space* $X$ (cf. Chapt. VIII; Definition 7.1) *and a vector sheaf* $\mathcal{E}$ *on* $X$, *with* $\operatorname{rk}\mathcal{E} = n \in \mathbf{N}$, *endowed also with an* $\mathcal{A}$*-connection* $D$. *Moreover, let*

$$(2.32) \qquad \mathcal{U} = (U_\alpha)_{\alpha \in I}$$

*be a local frame of* $\mathcal{E}$ *and*

$$(2.33) \qquad p \equiv p(\boldsymbol{\lambda}) \equiv p((\lambda_{ij})) \in \mathbf{C}[(\lambda_{ij})]$$

*an invariant polynomial, such that*

$$\lambda \equiv (\lambda_{ij}) \in M_n(\mathbf{C}). \tag{2.34}$$

*Then, the curvature 0-cocycle, associated with $\mathcal{U}$ and $p$* (Lemma 2.1),

$$p(R) = p\left(\left(R^{(\alpha)}\right)\right) \in \Omega^{\cdot}_{\mathrm{ev}}(X), \tag{2.35}$$

*yields a "closed form"; viz. one has*

$$dp(R) = 0. \tag{2.36}$$

**Proof.** First, based on (1.12) and (1.19), one obtains;

$$\begin{aligned} R^{(\alpha)} \equiv \left(R^{(\alpha)}_{ij}\right) \in M_n(\Omega^2(U_\alpha)) &\stackrel{\subsetneq}{\to} M_n(\Omega^{\cdot}_{\mathrm{ev}}(U_\alpha)) \\ &= M_n(\Omega^{\cdot}_{\mathrm{ev}})(U_\alpha), \end{aligned} \tag{2.37}$$

*for every* $\alpha \in I$, as in (2.32), *where* also $1 \leqq i, j \leqq n = \operatorname{rk}\mathcal{E}$. Thus, by considering now the given (invariant) polynomial $p$, as in (2.33), one has (cf. (2.5), (2.8) and (2.18));

$$p\left(R^{(\alpha)}\right) \equiv p\left(\left(R^{(\alpha)}_{ij}\right)\right) \equiv p(R) \in \Omega^{\cdot}_{\mathrm{ev}}(U_\alpha), \tag{2.38}$$

*for every* $\alpha \in I$, as before. (As already said, multiplication involved by (the form of) $p$ is *exterior multiplication*. Yet, we should notice here the *abuse of notation* employed for $p(R)$ in (2.38) and (2.35) above, the obvious difference being, however, clear from the context). Thus, our previous rel. (2.31) yields now;

$$p'(R) \wedge R = R \wedge p'(R). \tag{2.39}$$

In this context, we still recall here that one always understands in (2.39) and in the sequel too, $R \equiv \left(R^{(\alpha)}_{ij}\right)$, as in (2.37).

Hence, in view of (2.28), one further obtains,

$$dp(R) = \operatorname{tr}(p'(R) \wedge dR). \tag{2.40}$$

On the other hand, *Bianchi's identity* (cf. Chapt. VIII; (7.12)) yields

$$dR = [R, \omega] \equiv R \wedge \omega - \omega \wedge R, \tag{2.41}$$

so that, by virtue of (2.40) and (2.39), one now obtains;

$$
\begin{aligned}
dp(R) &= \operatorname{tr}(p'(R)\wedge dR) = \operatorname{tr}(p'(R)\wedge(R\wedge\omega - \omega\wedge R)) \\
&= \operatorname{tr}(p'(R)\wedge R\wedge\omega - p'(R)\wedge\omega\wedge R) \\
&= \operatorname{tr}((p'(R)\wedge R - R\wedge p'(R))\wedge\omega) = 0.
\end{aligned} \tag{2.42}
$$

Now, the last conclusion is valid, in fact, *for every* $\alpha \in I$, which thus, in view of (2.18), proves our assertion, viz. (2.36). ∎

It is still useful for the sequel to have the statement of the previous Theorem 2.1 less formally, combined too with the above Lemma 2.1 (cf. (2.20)). Thus, we have the next.

**Corollary 2.1.** *Suppose we have the framework of the above Theorem 2.1. Then, for every invariant polynomial*

$$p \in \mathbf{C}[(\lambda_{ij})] \tag{2.43}$$

(see (2.33)), *such that*

$$\deg p = m, \tag{2.44}$$

$p(R)$ (cf. (2.20)) *is a "closed 2m-form"; that is, one has the two following conditions fulfilled:*

$$p(R) \in \wedge^{2m}\Omega^1(X) \subsetneq (\wedge^{2m}\Omega^1)(X) \equiv \Omega^{2m}(X) \subsetneq \Omega^{\cdot}_{\mathrm{ev}}(X), \tag{2.45}$$

*as well as,*

$$dp(R) = 0. \quad ∎ \tag{2.46}$$

**Remark 2.1.**– The *derivation* in (2.46) (the *same* holds true, of course, *for* (2.36)) is meant, in effect, for the operator

$$d^{2m} : \Omega^{2m} \longrightarrow \Omega^{2m+1}, \tag{2.47}$$

so that, as usual, *we have set;*

$$d^{2m} \equiv d. \tag{2.48}$$

Yet, we note that *the* same *operator does exist,* in view of the hypothesis for $X$ (cf. Chapt. VIII; (8.7)).

# 3. Generalized de Rham spaces. Fundamental theorem

Our objective by the following discussion is to define, via (the *"closed form"*) $p(R)$, as above (see (2.45), (2.46)), an appropriate cohomology class of the base space $X$ involved. In the classical case of (paracompact Hausdorff) $\mathcal{C}^{\infty}$*-manifolds,* this is accomplished through the corresponding *de Rham complex* of the manifold concerned (our base space $X$). Now, although, as we shall see in the sequel (cf. the succeeding Chapters X, XI), we do have particular examples, exhibiting analogous complexes, this is not valid, in general. Thus, keeping further the axiomatic point of view, that we have employed hitherto, we do give straight a way the following.

**Definition 3.1.** By a *generalized de Rham space of order* $m \in \mathbf{Z}_{+}$, or simply a *generalized de Rham* $m$*-space* (with $m$ a positive integer), one means a paracompact (Hausdorff) space $X$, endowed with the following exact sequence of $\mathbf{C}$-vector space sheaves on $X$;

$$\begin{aligned} &0 \longrightarrow \ker\partial \longrightarrow \mathcal{A} \xrightarrow{d^0\equiv\partial} \Omega^1 \xrightarrow{d^1\equiv d} \Omega^2 \xrightarrow{d^2\equiv d} \cdots \\ &\qquad \cdots \longrightarrow \Omega^m \xrightarrow{d^m\equiv d} d\Omega^m \longrightarrow 0 \end{aligned} \tag{3.1}$$

(see also Scholium 3.1 below, concerning the notation applied in (3.1)).

In particular, if in the previous exact sequence one has

$$\ker\partial = \mathbf{C}, \tag{3.2}$$

then one speaks of a *de Rham* $m$*-space.*

On the other hand, in the case that (3.1) is valid, for every $m \in \mathbf{Z}_{+}$ (but, see also (3.5) below), then one speaks of a *generalized de Rham space,* or (if (3.2) is fulfilled) of a *de Rham space,* respectively.

**Scholium 3.1.**– By definition, a *generalized de Rham* $m$*-space* is a Hausdorff (paracompact) topological space $X$, endowed, first, with a *differential triad*

$$(\mathcal{A}, \partial, \Omega^1), \tag{3.3}$$

in the usual sense, applied so far (cf., for instance, Chapt. VIII; (1.1)). In particular, $X$ is a *curvature space*

$$(\mathcal{A}, \partial, \Omega^1, d^1 \equiv d, \Omega^2) \tag{3.4}$$

(loc. cit.; (1.16)); hence, the respective **C**-*linear morphisms* $\partial$ and $d^1 \equiv d$ have the pertinent properties, as before (see the same Chapter VIII; Sections 1, 7, 8). Furthermore, the $\mathcal{A}$-*modules* $\Omega^i$, $1 \leqq i \leqq m$, are defined as in Chapt. VIII; (8.3), together with the corresponding *higher derivative operators,* as the case may be, depending on the given $m \in \mathbf{Z}_+$, as in (3.1) (see also Chapt. VIII; (8.7)). Therefore, *if* $m \geqq 2$, our space $X$ is a *Bianchi space* too (loc. cit.; (7.6)).

In this context, we should also remark that

$$\text{(3.1) is already exact at } \mathcal{A}, \tag{3.5}$$

by the very definitions; *equivalently,* the following

$$0 \longrightarrow \ker \partial \longrightarrow \mathcal{A} \xrightarrow{\partial} \partial\mathcal{A} \longrightarrow 0 \tag{3.6}$$

is a *short exact sequence* (of **C**-vector space sheaves on $X$).

Thus, *exactness of (3.1) refers,* in effect, to subsequent places of the sequence at issue, that is, *to* $m \geqq 1$.

Finally, the relation

$$\ker \partial = \mathbf{C} \tag{3.7}$$

is a crucial restriction on $\mathcal{A}$ (nevertheless, important!, for its consequences), which might be called *Poincaré condition,* having to do in the standard case (again $\mathcal{C}^\infty$-*manifolds*) with the classical *Poincaré Lemma* (cf. Chapt. III; Lemma 8.2, together with (8.47)). Indeed, as we shall see later on (see the next Chapters X and XI), there do exist several particular *important examples* (apart, of course, from the aforementioned classical case), for which (3.7) *is satisfied.*

Thus, our first relevant result, which, in fact, was also the motive for the previous Definition 3.1, is now the following lemma, being also the counterpart in our case of the analogous standard situation. That is, we have.

**Lemma 3.1.** *Suppose we are given a generalized de Rham m-space X, as well as, a "closed m-form" ω on X* (with $m \in \mathbf{N}$); *that is, one has* (see (3.1))

$$\omega \in \Omega^m(X) \ \ \textit{and} \ \ d^m(\omega) \equiv d\omega = 0. \tag{3.8}$$

*Then, one can associate with ω an m-dimensional (Čech) cohomology class of X with coefficients in* (the **C**-*vector space sheaf* on $X$) $\ker \partial$, *denoted by* $c(\omega)$; *viz. one has*

$$c(\omega) \in H^m(X, \ker \partial) = \check{H}^m(X, \ker \partial). \tag{3.9}$$

**Scholium 3.2.**- For the proof of the above lemma we shall systematically employ Lemma 5.1 of Chapter III in the preceding, referring to the so-called *"liftable (Čech) cohomology"*, hence the appearance of the *paracompact* (Hausdorff) *space X*. Indeed, what we really apply is a strengthening of the aforementioned result, as given by Chapt. III; (5.52) (see thus loc. sit., Lemma 5.2, as well as, (5.86). In this context, one has to consider an eventual *refinement of the* initial *covering* $\mathcal{U}$ *of* $X$ involved; for simplicity's sake, however, we do not mark it in the succeeding proof of Lemma 3.1, employing thus an obvious *abuse of notation* (see Chapt. III; (5.54)).

On the other hand, we further notice that, due to the *paracompactness* of $X$, *Čech and sheaf cohomology theories are* essentially *the same* (ibid.; Theorem 8.1 and (8.1)).

Thus, we are now ready to embark on the next.

**Proof of Lemma 3.1.** By hypothesis of the lemma, one assumes the *exactness of (3.1)*, so that let us, first, consider the following extract of the same sequence; viz. we take the next *short exact sequence,*

$$0 \longrightarrow \ker d^m \longrightarrow \Omega^m \xrightarrow{d^m \equiv d} d\Omega^m \longrightarrow 0, \tag{3.10}$$

along with the corresponding *long exact sequence in (Čech) cohomology* (cf.

also Chapt. III; Theorem 2.1, prop. 3), the comments before it, and Theorem 5.3),

$$0 \longrightarrow \Gamma(X, \ker d^m) \longrightarrow \Gamma(X, \Omega^m) \xrightarrow{\Gamma(d^m)} \Gamma(X, d\Omega^m) \longrightarrow \cdots . \tag{3.11}$$

Thus, in view of (3.6), one has

$$\omega \in \Omega^m(X) \equiv \Gamma(X, \Omega^m), \tag{3.12}$$

in such a manner, that one further has;

$$d^m(\omega) = \Gamma_X(d^m)(\omega) = \Gamma(d^m)(\omega) \equiv d\omega = 0. \tag{3.13}$$

Hence, by (3.13) and the exactness of (3.11) and (3.1), one now obtains;

$$\begin{aligned} \omega &\in \ker(\Gamma(X, \Omega^m) \longrightarrow \Gamma(X, d\Omega^m)) \\ &= \operatorname{im}(\Gamma(X, \ker d^m) \longrightarrow \Gamma(X, \Omega^m)) \\ &= \Gamma(X, \ker d^m) = \Gamma(X, \operatorname{im} d^{m-1}) = Z^0(\mathcal{U}, \operatorname{im} d^{m-1}) \\ &= Z^0(\mathcal{U}, d^{m-1}(\Omega^{m-1})) \equiv Z^0(\mathcal{U}, d\Omega^{m-1}), \end{aligned} \tag{3.14}$$

where $\mathcal{U}$ stands for an *open covering of* $X$. In conclusion, one thus obtains, according to (3.14), that;

$$\omega \in Z^0(\mathcal{U}, d\Omega^{m-1}) = \Gamma(X, d\Omega^{m-1}) \equiv (d\Omega^{m-1})(X) \subseteq \Omega^m(X), \tag{3.15}$$

a better information, of course, than the one supplied already, by the first of (3.8). (Concerning the above calculations in (3.14) and (3.15), cf. also Chapt. III; (4.55). In this context, we should still note that one could get (3.15) *directly from (3.8) and the exactness of (3.1)*, according to the very definitions; so cf. also Chapt. I; (13.19) and Theorem 12.1, along with (2.9) and (2.12)).

Now, consider further the following *short exact sequence of* $\mathbf{C}$*-vector space sheaves* on $X$;

$$0 \longrightarrow \ker d^{m-1} \longrightarrow \Omega^{m-1} \longrightarrow d^{m-1}\Omega^{m-1} \longrightarrow 0. \tag{3.16}$$

Accordingly, based now on (3.15) and Chapt. III; (5.86) (cf. also Scholium 3.2 above), one thus concludes, by employing *"liftable cohomology"*, that there exists a *0-cochain*

$$(\theta_\alpha) \in C^0(\mathcal{U}, \Omega^{m-1}), \tag{3.17}$$

such that one has, concerning the given $\omega$, a *0-cocycle,* as in (3.15);

(3.18) $$\omega = d^{m-1}((\theta_\alpha)) := (d^{m-1}(\theta_\alpha)) \equiv (d\theta_\alpha).$$

(Of course, one has in (3.17) an eventual *refinement* $\mathcal{V}$ of $\mathcal{U}$, the latter covering already appeared in (3.14), or even (3.15); however, for simplicity, as already mentioned above (see Scholium 3.2), we do not keep track of it here, as well as, in the sequel). On the other hand, (3.17) satisfies the relation (cf. Chapt. III; (5.86));

(3.19) $$\delta(\theta_\alpha) \in Z^1(\mathcal{U}, \ker d^{m-1} = \operatorname{im} d^{m-2}).$$

Therefore, by the same token, one is finally led, inductively to an $m$-*cocycle*

(3.20) $$z(\omega) \in Z^m(\mathcal{U}, \ker \partial)$$

(we recall that one has eventually here $\mathcal{U} = \mathcal{U}^{(m)}$, as explained above). Thus, the corresponding cohomology class to $z(\omega)$ will now be, *by definition*, the desired one, as in (3.9), supplied by the given closed $m$-form $\omega$; that is, we set

(3.21) $$c(\omega) := [z(\omega)] \in H^m(X, \ker \partial),$$

which thus terminates the proof. ∎

As an application of the preceding, we are now in the position to state the following *fundamental result.*

**Theorem 3.1.** *(Chern-Weil). Suppose we are given a generalized de Rham $m$-space $X$, such that*

(3.22) $$m = 2k \in \mathbf{Z}_+.$$

*Moreover, let $\mathcal{E}$ be a vector sheaf on $X$, with* $\operatorname{rk}\mathcal{E} = n \in \mathbf{N}$, *endowed with an $\mathcal{A}$-connection $D$, and consider further an invariant polynomial*

(3.23) $$p \in \mathbf{C}[(\lambda_{ij})], \quad \textit{with} \quad 1 \leqq i, j \leqq n,$$

*and such that*

(3.24) $$\deg p = k.$$

*Finally, let $R(D) \equiv R$ be the curvature of the $\mathcal{A}$-connection $D$ of $\mathcal{E}$.*

*Then* (see also (3.22) and (2.45) above),

$$(3.25) \qquad p(R) \in \wedge^m \Omega^1(X) \hookrightarrow (\wedge^m \Omega^1)(X) \equiv \Omega^m(X)$$

*is a closed m-form on X* (cf. (3.8)), *providing an m-dimensional cohomology class of X* (see (3.9)),

$$(3.26) \qquad c(p(R)) \equiv c(R) \in \check{H}^m(X, \ker \partial),$$

*which, in addition, is independent of the given $\mathcal{A}$-connection D.*

**Proof.** First, according to Corollary 2.1 (see (2.45), (2.46)), in conjunction with (3.22) and (3.24) (cf. also Scholium 3.1, concerning the relevant framework), one concludes that

$$(3.27) \qquad p(R) \in \wedge^m \Omega^1(X),$$

that is, *(3.25) is fulfilled,* while *the same "m-form" on X is closed,* as well. Hence (Lemma 3.1), $p(R)$, as above, supplies an *m-dimensional* (*Čech*, same Scholium 3.1) *cohomology class* $c(R)$, *as in (3.26)* (see (3.9)). So it only remains to prove that the latter cohomology class $c(R)$ *is,* in fact, *independent of the $\mathcal{A}$-connection D* involved:

Namely, consider the *unit interval*

$$(3.28) \qquad I \equiv [0,1] \subseteq \mathbf{R},$$

such that, *for every* $t \in I$, let us further consider the *continuous map*

$$(3.29) \qquad \phi_t : X \longrightarrow X \times I,$$

given by the relation;

$$(3.30) \qquad \phi_t(x) := (x, t).$$

Furthermore, consider the (canonical) *1st projection* of the range of $\phi_t$,

$$(3.31) \qquad \psi \equiv \mathrm{pr}_1 : X \times I \longrightarrow X,$$

such that one has

$$(3.32) \qquad \psi(x,t) := x, \quad (x,t) \in X \times I.$$

Now, suppose that we are given two *$\mathcal{A}$-connections* of our vector sheaf $\mathcal{E}$ on $X$, say $D_0$ *and* $D_1$. Thus, by looking further at the *pull-back* of the given *differential triad* (3.3) on $X$, *via the* above (continuous) *map* $\psi$, let

$$(3.33) \qquad \mathcal{A}' \equiv \psi^*(\mathcal{A})$$

(cf. Chapt. VI; (6.3.1)). Furthermore, let

$$D'_i := \psi^*(D_i), \quad (i = 0, 1) \tag{3.34}$$

be the *induced $\mathcal{A}'$-connections of* (the *vector sheaf*) $\psi^*(\mathcal{E})$ on $X \times I$, provided by the previously two given ones of $\mathcal{E}$. (See thus loc. cit., Definition 6.1, along with Chapt. II; (4.14)). So let

$$\widetilde{D} := D'_0 + t(D'_1 - D'_0), \quad t \in [0,1] \equiv I, \tag{3.35}$$

be the *interpolation of* the above two *$\mathcal{A}'$-connections* of $\psi^*(\mathcal{E})$ (cf. (3.34)), yielding thus, for every $t \in [0,1] \subseteq \mathbf{R}$, *another $\mathcal{A}'$-connection* of the same vector sheaf $\psi^*(\mathcal{E})$ (cf. Chapt. VI; (7.28.1)).

On the other hand, by looking at the continuous maps $\phi_t$ and $\psi$, as above (see (3.29) and (3.31), respectively), one has, concerning the corresponding *pull-back (functors;* cf. Chapt. I; (14.45))

$$\phi_t^* \circ \psi^* = (\psi \circ \phi_t)^* = \mathrm{id}_X^*. \tag{3.36}$$

Therefore, based on (3.34), (3.35) and (3.36), one now obtains;

$$\phi_t^*(\widetilde{D}) = D_0 + t(D_1 - D_0) \equiv D_t, \quad t \in I, \tag{3.37}$$

that is, *another $\mathcal{A}$-connection of $\mathcal{E}$*, due to the hypothesis for $D_0, D_1$ and again to Chapt. VI; (7.28.1), as before. Thus, setting now

$$R(\widetilde{D}) \equiv \widetilde{R} \ \ and \ \ R(D_t) \equiv R_t, \tag{3.38}$$

for any $t \in I$, as in (3.37), one has the following relations (see (3.37), (3.38) and Chapt. VIII; (9.25). Yet, cf., for example, R. Godement [1: p.199, §4.16]);

$$\begin{aligned} \phi_t^*(c(p(\widetilde{R}))) &= c(p(\phi_t^*(\widetilde{R}))) = c(p(\phi_t^*(R(\widetilde{D})))) \\ &= c(p(R(\phi_t^*(\widetilde{D})))) = c(p(R(D_t))) = c(p(R_t)). \end{aligned} \tag{3.39}$$

In conclusion, *one* thus *obtains*

$$c(p(R_t)) = \phi_t^*(c(p(\widetilde{R}))), \tag{3.40}$$

*for any $t \in [0,1]$.*

However, *the continuous maps $\phi_0$ and $\phi_1$* (cf. (3.28) – (3.30)) *are homotopic,* through the *identity map on $X \times I$* (see, for instance, J. Dugundji [1: p.315, Definition 1.1]). Therefore, *they define the same functors in cohomo-*

*logy* (*"Homotopy axiom"*; see, for example, E.H. Spanier [1: p.240, 12], or even J.J. Rotman [1: p.379, Theorem 12.4]). Accordingly, one obtains, by virtue of (3.40) (for $t = 0, 1$, respectively);

$$c(p(R_0)) = \phi_0^*(c(p(\tilde{R}))) = \phi_1^*(c(p(\tilde{R}))) = c(p(R_1)). \tag{3.41}$$

In other words, one thus concludes that;

(3.42) $c(R) \equiv c(p(R)) \in H^m(X, \ker \partial)$ (cf. (3.26)), *is*, in fact, *independent of the $\mathcal{A}$-connection $D$*, considered,

once we are given $R \equiv R(D)$ (for some $\mathcal{A}$-connection $D$ of $\mathcal{E}$), which actually was our claim, and this also completes the proof of the theorem. ∎

**Scholium 3.3.**– Concerning the first equality in (3.39), we have actually employed there the fact that for every *continuous map*

$$f : X \longrightarrow Y, \tag{3.43}$$

*the* corresponding *pull-back functor* (cf. Chapt. I; (14.44.1))

$$f^* : \mathcal{S}h_Y \longrightarrow \mathcal{S}h_X \tag{3.44}$$

*commutes with* (sheaf) *cohomology:* Namely, by considering, for instance, *cohomology as in (3.9)*, one obtains, via (3.44), a (canonical) *morphism in cohomology*

$$f^* : H^*(Y, \ker \partial) \longrightarrow H^*(X, \ker f^*(\partial)) \tag{3.45}$$

(for simplicity we retain the same notation with (3.44)), in such a manner, that *one has;*

$$f^*([z]) = [f^*(z)], \tag{3.46}$$

*where* $z \in Z^m(\mathcal{U}_Y, \ker \partial)$, *with* $m \in \mathbf{Z}_+$. (Again, see also R. Godement [1: p.199; §4.16]).

**Scholium 3.4.**– We comment below a bit more on the *cohomology* (**C**-*vector) spaces,* appeared e.g. *in (3.9):* For simplicity's sake, we also look at a

*generalized de Rham space* $X$, so that, by hypothesis (cf. Definition 3.1), one has the following *exact sequence;*

$$(3.47)\qquad \widetilde{\Omega}^{\cdot}_{\mathrm{gdeR}} : 0 \longrightarrow \ker\partial \longrightarrow \Omega^0 \equiv \mathcal{A} \xrightarrow{d^0\equiv\partial} \Omega^1 \xrightarrow{d^1\equiv d} \Omega^2 \longrightarrow \cdots \longrightarrow \Omega^m \xrightarrow{d^m\equiv d} \Omega^{m+1} \longrightarrow \cdots.$$

Thus, by further employing here our previous notation in Chapter III (ibid.; (1.55′)), we still write (3.47) in the form;

$$(3.48)\qquad 0 \longrightarrow \ker\partial \longrightarrow \Omega^{\cdot}_{\mathrm{gdeR}},$$

where we set

$$(3.49)\qquad \Omega^{\cdot}_{\mathrm{gdeR}} : 0 \longrightarrow \mathcal{A} \equiv \Omega^0 \xrightarrow{\partial} \Omega^1 \xrightarrow{d} \cdots \longrightarrow \Omega^m \xrightarrow{d} \cdots,$$

this being thus the *(generalized) de Rham complex* of $X$ (having the same *(C-linear) morphisms* with (3.47). In this regard, see also Chapt. III; (8.37′) for the analogous notation in the standard case of a $\mathcal{C}^\infty$*-manifold, viz. when* $\mathcal{A} \equiv \mathcal{C}^\infty_X$). Accordingly, (3.47), or even (3.48), is a *resolution of* $\ker\partial$, the latter object being, as already noted above, another **C**-*vector space sheaf* on $X$.

Now, given that, by hypothesis, our (base) space $X$ is *paracompact (Hausdorff),* let us further assume that the coefficient **C**-*algebra sheaf* $\mathcal{A}$ is, in particular, *soft* (cf. Chapt. III; (8.25)). Thus, (3.49) *is* then $\Gamma_X$*-acyclic,* alias *the resolution (3.48) of* $\ker\partial$ *is* $\Gamma_X$-acyclic (cf. Chapt. III; (2.20), (8.26) and (8.56). Apart from the standard case, mentioned above, the previous hypothesis for $\mathcal{A}$ will be fulfilled in some particular concrete examples, we are going to consider later on (see Chapters X, XI in the sequel). Something more, $\mathcal{A}$ will be, in fact, a *fine* sheaf on a paracompact (Hausdorff) space $X$, as above). So, as a consequence of the *abstract de Rham theorem* (cf. Chapt. III; (3.24), (3.25)), one now infers that;

(3.50) in the case that $\mathcal{A}$ is a *soft* sheaf on (the *paracompact* (Hausdorff) space) $X$, *the sheaf cohomology of* $X$ *with coefficients in* $\ker\partial$ *is* simply *the cohomology of* $\Gamma_X(\Omega^{\cdot}_{\mathrm{gdeR}})$, as above.

That is, *one has;*

$$(3.51)\qquad H^*(X, \ker\partial) = \check{H}^*(X, \ker\partial) = h^*(\Gamma_X(\Omega^{\cdot}_{\mathrm{gdeR}})).$$

In particular, *one* thus *obtains*

$$(3.52)\qquad \begin{aligned}(R^m\Gamma_X)(\ker\partial) &\equiv H^m(X,\ker\partial) := h^m(\Gamma_X(\Omega^{\cdot}_{\mathrm{gdeR}}))\\ &:= \ker\Gamma_X(d^m)/\mathrm{im}\,\Gamma_X(d^{m-1}),\end{aligned}$$

for every *positive integer* $m$ (loc. cit.; (3.5)).

# 4. Symmetric polynomials

For convenience of reference, we collect in this Section several standard facts, pertaining to the polynomials in the title, which will be of use right in the sequel. For simplicity, we also consider *complex polynomials* (the only ones, for that matter, that were and are going to be employed below, as well), although, concerning the material of this Section, any commutative ring with identity could be considered instead. (Cf., for example, S. Lang [1: p. 204]).

Thus, consider the *(complex) polynomial algebra*

$$(4.1)\qquad \mathbf{C}[t_1,\ldots,t_n],$$

with respect to the $n$ *indeterminates* $t_1,\ldots,t_n$ ($n\in\mathbf{N}$). Furthermore, setting

$$(4.2)\qquad A\equiv\mathbf{C}[t_1,\ldots,t_n]$$

(a *unital commutative* $\mathbf{C}$*-algebra*), consider now a *polynomial* $p$ of one *indeterminate* $x$, with coefficients from $A$; viz. let

$$(4.3)\qquad p\in A[x],$$

such that, in particular, we set

$$(4.4)\qquad \begin{aligned}p(x) &:= \textstyle\prod_{i=1}^n(x-t_i)\equiv(x-t_1)(x-t_2)\cdots(x-t_n)\\ &= x^n - s_1x^{n-1}+\cdots+(-1)^n s_n,\end{aligned}$$

where, of course, we have

$$(4.5)\qquad s_i\equiv s_i(t_1,\ldots,t_n)\in\mathbf{C}[t_1,\ldots,t_n],\quad 1\leqq i\leqq n.$$

The above polynomials (4.5), as they are given by (4.4), are called the *elementary symmetric polynomials,* with respect to (the indeterminates) $t_1, \ldots, t_n$.

Therefore, by virtue of (4.4), it is easy to see that one has the following *form of the* $s_i$'s, as above;

$$
(4.6) \qquad \begin{aligned}
s_1 &= t_1 + t_2 + \cdots + t_n \\
s_2 &= \textstyle\sum_{1 \leqq i<j \leqq n} t_i t_j \\
&\cdots\cdots\cdots \\
s_k &= \textstyle\sum_{1 \leqq i_1 < i_2 < \cdots < i_k \leqq n} t_{i_1} t_{i_2} \cdots t_{i_k} \\
&\cdots\cdots\cdots \\
s_n &= t_1 \cdot t_2 \cdots t_n .
\end{aligned}
$$

Now, a given polynomial $p \in \mathbf{C}[t_1, \ldots, t_n]$ is said to be *symmetric,* whenever it remains unchanged under any permutation of the independent variables (indeterminates); precisely speaking, in the case one has

$$
(4.7) \qquad p^{\pi}(t_1, \ldots, t_n) := p(t_{\pi(1)}, \ldots, t_{\pi(n)}) = p(t_1, \ldots, t_n),
$$

or just,

$$
(4.8) \qquad p^{\pi} = p,
$$

for every *permutation* $\pi \in \mathfrak{S}(n)$ (viz. the *symmetric group of order* $n$ of all permutations of $[1, n] \subseteq \mathbf{N}$).

So it is an easy consequence of the preceding definition of the $s_i$'s, $1 \leqq i \leqq n$, namely, as coefficients of the polynomial (4.4) (see thus (4.6)), that the "elementary symmetric functions", as above, are indeed *symmetric* in the previous sense.

Thus, the *fumdamental theorem* on symmetric polynomials has now as follows:

(4.9) *every symmetric polynomial in* $\mathbf{C}[t_1, \ldots, t_n]$ (cf. (4.8)) *is, in a unique way, a polynomial of the elementary symmetric polynomials* (cf. (4.6)).

That is, *for every symmetric* $p \in \mathbf{C}[t_1, \ldots, t_n]$, *there exists a unique* $f \in \mathbf{C}[y_1, \ldots, y_n]$, *such that one has;*

$$
(4.9.1) \qquad p(t_1, \ldots, t_n) = f(s_1(t_1, \ldots, t_n), \ldots, s_n(t_1, \ldots, t_n)).
$$

For the proof of (4.9.1) see, for instance S. Lang [1: p.204, Theorem 9.1], or even A.I. Kostrikin [1: p.281, Theorem 1]. ∎ (Here one needs to consider, concerning the general case (cf. the comments at the introduction to this Section), an *integral domain*, where the polynomials involved have their coefficients).

Now, let us consider again the *polynomial algebra*

(4.10) $$\mathbf{C}[(\lambda_{ij})], \quad 1 \leqq i, j \leqq n,$$

as in Section 2 in the preceding, as well as, the *invariant polynomial* (cf. (2.14))

(4.11) $$\det(\mathbf{1} + t\boldsymbol{\lambda}),$$

such that $\boldsymbol{\lambda} \equiv (\lambda_{ij}) \in M_n(\mathbf{C})$, $\mathbf{1}$ stands for the respective $n \times n$ *identity matrix,* while $t \in \mathbf{C}$. (See thus, for instance, Example 2.1 above, in particular, cf. (2.17)). Thus, one has

(4.12) $$\det(\mathbf{1} + t\boldsymbol{\lambda}) = \mathbf{1} + \sum_{i=1}^{n} t^i p_i(\boldsymbol{\lambda}),$$

in such a manner that;

(4.13) *each one of the* $p_i$*'s,* $1 \leqq i \leqq n$*, is a (homogeneous) invariant polynomial* (with respect to $\boldsymbol{\lambda} \equiv (\lambda_{ij})$, cf. (4.10)), *such that* $\deg p_i = i$, $1 \leqq i \leqq n$.

In this context, one actually proves that;

(4.14) *every invariant polynomial in* $\mathbf{C}[(\lambda_{ij})]$ *can be put (uniquely) in the form of a polynomial of the elementary symmetric polynomials in* $\mathbf{C}[t_1, \ldots, t_n]$ (see (4.6)).

For the proof of (4.14), consult e.g. R. Bott [1: p.77, Appendix A]. ∎

The previous result is based on the fact that *the restriction of an invariant polynomial of the algebra (4.10) to the diagonal elements of the same algebra yields,* in effect, *a symmetric polynomial of the polynomial algebra (4.1)* (loc. cit.); in turn, this is an immediate consequence of the fact that *every automorphism of* $M_n(\mathbf{C})$*, hence* a permutation of the diagonal entries of a matrix, as before, too, *is inner* (viz. it *comes from an inner automorphism,* defined *by an element of* $\mathrm{GL}(n, \mathbf{C})$. (Cf. also (2.14) and (4.7); yet, see e.g.

L. Chambadal - J.L. Ovaert [1: p.302, Remarque 2, and p.511, Ex. 158 A; I.4]).

Thus, by applying the above remarks to the invariant polynomial (4.11), one obtains (setting also $t = 1$);

$$\begin{aligned}
&\det(\mathbf{1}+\boldsymbol{\lambda}) = \det(\mathrm{diag}(1+t_1,\ldots,1+t_n)) = \prod_{i=1}^{n}(1+t_i)\\
&\equiv (1+t_1)(1+t_2)\cdots(1+t_n)\\
&= 1+(t_1+t_2+\cdots+t_n)+(t_1t_2+t_1t_3+\cdots t_{n-1}t_n)+\cdots\\
&\qquad\cdots+\textstyle\sum_{1\leqq i_1<i_2<\cdots<i_k\leqq n} t_{i_1}t_{i_2}\cdots t_{i_k}+\cdots+t_1\cdot t_2\cdots t_n\\
&= 1+s_1+s_2+\cdots+s_k+\cdots+s_n\\
&= 1+\mathrm{tr}\,\boldsymbol{\lambda}+\tfrac{1}{2}((\mathrm{tr}\,\boldsymbol{\lambda})^2-\mathrm{tr}\,\boldsymbol{\lambda}^2)+\cdots+\det\boldsymbol{\lambda}
\end{aligned} \tag{4.15}$$

(cf. also (4.6) above). Yet, concerning the last equality in (4.15), one employs here the so-called *Newton's relations,* connecting the elementary symmetric polynomials and the respective to (4.1) *power sums,* being also symmetric polynomials of the same algebra. (See, for instance, R. Kochendörffer [1: p.171 and p.176]).

Applications of the preceding will be made right below in the succeeding Section.

# 5. Generalized Chern classes

We have already at our disposal, as a result of the preceding discussion, the necessary material to define the particular (cohomology) classes, as in the title of this Section.

So, to fix the terminology applied, suppose that we are given a *curvature space* $X$ (cf. Chapt. VIII; Definition 1.1), and let $\mathcal{E}$ be a *vector sheaf* on $X$, with

$$\mathrm{rk}\,\mathcal{E} = n \in \mathbf{N}, \tag{5.1}$$

while we further assume that $\mathcal{E}$ has an $\mathcal{A}$-*connection* $D$, whose *curvature* is $R(D) \equiv R$ (ibid.; (2.1)). Furthermore, suppose that

(5.2) $$\mathcal{U} = (U_\alpha)_{\alpha \in I}$$

is a *local frame of* $\mathcal{E}$, so that one has, concerning the curvature $R$, as before (see (2.11) in the foregoing);

(5.3) $$R \equiv \left(R^{(\alpha)}\right) \equiv \left(\left(R_{ij}^{(\alpha)}\right)\right) \in C^0(\mathcal{U}, M_n(\Omega^2)) \subsetneq C^0(\mathcal{U}, \Omega^{\cdot}_{\mathrm{ev}}).$$

Thus, by further considering the *invariant polynomial* (4.11), as above, one now defines;

(5.4) $$c(\mathcal{E}) := \det\left(1 - \frac{1}{2\pi i} R\right).$$

We call (5.4) the *total Chern class* of the given *vector sheaf* $\mathcal{E}$. (The applied terminology will be fully justified right below).

In this context, we first remark that, in view of (4.11) and the previous Lemma 2.1, we actually have

(5.5) $$c(\mathcal{E}) \in \Omega^{\cdot}_{\mathrm{ev}}(X)$$

(see (2.18)). Yet, given that

(5.6) $$\deg \det(\lambda) = n, \quad \textit{for every } \lambda \equiv (\lambda_{ij}) \in M_n(\mathbf{C}),$$

(cf. also (2.16)), one infers, in particular, that;

(5.7) $$c(\mathcal{E}) \in \Omega^{2n}(X) \subsetneq \Omega^{\cdot}_{\mathrm{ev}}(X).$$

Accordingly (see also (1.15), (1.17)), the above polynomial (5.7) can be put in the form;

(5.8) $$c(\mathcal{E}) = 1 + c_1(\mathcal{E}) + c_2(\mathcal{E}) + \cdots,$$

such that one has

(5.9) $$c_i(\mathcal{E}) \in \wedge^{2i}\Omega^1(X) \subsetneq \Omega^{2i}(X), \quad i \in \mathbf{N}.$$

We call (5.9) the *i-th Chern class* of $\mathcal{E}$ (see (5.28) below for the justification of the terminology applied). Furthermore, we extend the latter definition, by still setting for the *0-th Chern class* of $\mathcal{E}$;

(5.10) $$c_0(\mathcal{E}) := 1 \in \mathcal{A}(X) \equiv \Omega^0(X),$$

viz. the *identity* (global) *section of* $\mathcal{A}$ (cf. also (1.16), as well as, Chapt. II; (1.22)).

In this regard, we should remark, of course, that *the sum in (5.8)* (finite, anyway, according to (5.7) and (1.15)) *is terminated at*

$$c_n(\mathcal{E}) = \det\left(-\frac{1}{2\pi i}R\right), \tag{5.11}$$

where $n = \operatorname{rk}\mathcal{E}$ (see (5.1), (5.6), as well as, (5.22) below). That is, one has;

$$c_i(\mathcal{E}) = 0, \quad \textit{for} \ \ i > n = \operatorname{rk}\mathcal{E} = \deg\left(\det\left(-\frac{1}{2\pi i}R\right)\right). \tag{5.12}$$

Now, we can have at this point the following byproduct, when looking at a *line sheaf* $\mathcal{L}$ on $X$. Namely, in view of (5.12), one concludes that;

(5.13) in the particular case of a *line sheaf* $\mathcal{L}$ on $X$, *the only non-zero Chern class of* $\mathcal{L}$ (apart from the *0-th one* (see (5.10))) *is*

$$c_1(\mathcal{L}) \in \Omega^2(X). \tag{5.13.1}$$

Therefore, based on (5.11), one further obtains;

$$c_1(\mathcal{L}) = -\frac{1}{2\pi i}R \in \Omega^2(X) \tag{5.14}$$

(see also, for instance, Chapt. VIII; (3.24)). Hence, the corresponding *total Chern class* of the *line sheaf* $\mathcal{L}$ is given by

$$c(\mathcal{L}) = 1 + c_1(\mathcal{L}) = 1 - \frac{1}{2\pi i}R \tag{5.15}$$

(see (5.4), or even (5.8), (5.13) and (5.14)).

In this context, we have still to remark that in the case the given $\mathcal{A}$-module $\Omega^1$, as above, is, in particular, a *vector sheaf* on $X$ too (this happens, of course, in the classical theory, cf. Chapt. VI; Section 2), then one gets a result similar to (5.12). Thus, if, for example, in that case one has,

$$\operatorname{rk}\Omega^1 = m \in \mathbf{N}, \tag{5.16}$$

then, in view of (5.9) (cf. also Chapt. IV; (7.20)), one gets again, concerning (5.8);

$$c_i(\mathcal{E}) = 0, \quad \textit{for any} \ \ 2i > m = \operatorname{rk}\Omega^1. \tag{5.17}$$

Apart from the classical case of $\mathcal{C}^\infty$*-manifolds,* as already noted above, we shall also see later on *further examples, for which (5.16) holds true* (cf. thus Chapters X, XI in the sequel).

Now, by direct application of our previous argument on (4.15) to the present context, one obtains;

$$g^{-1}\left(-\frac{1}{2\pi i}R\right)g \equiv \mathrm{Ad}(g^{-1})\left(-\frac{1}{2\pi i}R\right) = \mathrm{diag}(t_1, \ldots, t_n) \equiv \boldsymbol{\lambda}, \tag{5.18}$$

where one has, of course (see also (5.2)),

$$g \equiv \left(g_{ij}^{(\alpha)}\right) \in \mathcal{GL}(n, \mathcal{A})(U_\alpha) = \mathrm{GL}(n, \mathcal{A}(U_\alpha)) \tag{5.19}$$

for some *coordinate 1-cocycle of* $\mathcal{E}$,

$$(g_{\alpha\beta}) \in Z^1(\mathcal{U}, \mathcal{GL}(n, \mathcal{A})). \tag{5.20}$$

Accordingly, one has (cf. (5.4), (5.8) and (5.12))

$$c(\mathcal{E}) \equiv \det(1 + \boldsymbol{\lambda}) = \det\left(1 - \frac{1}{2\pi i}R\right) = 1 + c_1(\mathcal{E}) + \cdots + c_n(\mathcal{E}), \tag{5.21}$$

so that, by virtue of the same rel. (4.15), as above, in conjunction with (5.18), one now obtains;

$$\begin{aligned} &c_0(\mathcal{E}) = 1 \\ &c_1(\mathcal{E}) = \mathrm{tr}\,\boldsymbol{\lambda} = \mathrm{tr}\left(g^{-1}\left(-\tfrac{1}{2\pi i}R\right)g\right) = \mathrm{tr}\left(-\tfrac{1}{2\pi i}R\right) = -\tfrac{1}{2\pi i}\mathrm{tr}\,R \\ &\quad\vdots \qquad\qquad\qquad \cdots\cdots\cdots \\ &c_n(\mathcal{E}) = \det\boldsymbol{\lambda} = \det\left(g^{-1}\left(-\tfrac{1}{2\pi i}R\right)g\right) = \det\left(-\tfrac{1}{2\pi i}R\right) = \left(-\tfrac{1}{2\pi i}\right)^n \det R. \end{aligned} \tag{5.22}$$

**Scholium 5.1.**- Concerning the above calculations, one can construed the *curvature* $R$, as a *local* $n \times n$ *matrix* (see e.g. (5.18)); viz. one has

$$\begin{aligned} R \equiv \left(R^{(\alpha)}\right) &\in M_n(\Omega^2(U_\alpha)) = M_n(\Omega^2)(U_\alpha) \\ &\overset{\subsetneq}{\to} M_n(\Omega^{\cdot}_{\mathrm{ev}}(U_\alpha)) = M_n(\Omega^{\cdot}_{\mathrm{ev}})(U_\alpha), \end{aligned} \tag{5.23}$$

*for every* $\alpha \in I$ (cf. (5.2), as well as, e.g. (2.37)). Therefore (see also e.g. (2.17.1)), *one has*

(5.24) $$\det R \equiv \det\left(R^{(\alpha)}\right) \in \Omega^{\boldsymbol{\cdot}}_{\mathrm{ev}}(U_\alpha), \quad \alpha \in I.$$

However, due to the fact that det is an *invariant polynomial* (cf. (2.17)), one actually concludes that (see Lemma 2.1);

(5.25) $$\det R \equiv \left(\det R^{(\alpha)}\right) \in Z^0(\mathcal{U}, \Omega^{\boldsymbol{\cdot}}_{\mathrm{ev}}) = \Omega^{\boldsymbol{\cdot}}_{\mathrm{ev}}(X).$$

In this context, we further recall the standard fact that, within the context of (2.3), (2.15),

*the trace function*

(5.26) $$\mathrm{tr} : M_n(\mathbf{C}) \longrightarrow \mathbf{C}$$

*is an invariant polynomial.*

See (2.14), as well as, e.g. S. Lang [1: p.443]. ∎

As a consequence of (5.26) one gets, of course, an analogous situation to (5.25) for (the *global section,* same Lemma 2.1) $\mathrm{tr}\, R$.

Thus, according to the preceding, we conclude now the following basic fact:

(5.27) *each one of the Chern classes* of the given *vector sheaf* $\mathcal{E}$, as they were given by (5.22), *yields a* (homogeneous) *invariant polynomial,* with respect to the *curvature matrix* $R$ (cf. (5.23)), *whose degree is at most* $n = \mathrm{rk}\,\mathcal{E}$ (cf. (5.12), (4.13)).

Therefore, in the case our curvature space $X$, as given above, is in particular, a *(generalized) de Rham* $2n$*-space* (cf. Definition 3.1, along with (5.1)), then in view of the *Chern-Weil Theorem* (see Theorem 3.1), one further infers that:

(5.28) *each one of the above Chern classes* (see (5.22)) *supplies a* $2\deg p_i$*-dimensional,* that is (cf. (4.13)), *a* $2i$*-dimensional (Čech) cohomology class of* $X$, $1 \leqq i \leqq n$, *with coefficients in* $\ker\partial$ (see (3.26)). Namely, *one has;*

(5.28.1) $$c_i(\mathcal{E}) \equiv [c_i(\mathcal{E})] \in H^{2i}(X, \ker\partial),$$

*with* $1 \leqq i \leqq n = \mathrm{rk}\,\mathcal{E}$.

The previous rels. (5.28.1) provide also the aforementioned vindication of the nomenclature applied to the terms defined by (5.9).

In the classical theory the corresponding classes take their values from $\mathbf{R}$ ($\stackrel{\subseteq}{\to}$ $\mathbf{C}$) and, indeed, from $\mathbf{Z}$; that is, as we say, they are *integral,* in the sense that *each one lies in*

$$\text{(5.29)} \qquad \operatorname{im}(\check{H}^{2i}(X, \mathbf{Z}) \longrightarrow H^{2i}(X, \mathbf{R})),$$

with $i = 1, 2, \ldots, n$, *respectively,* as the latter map is defined by the (canonical) *injection* $\mathbf{Z} \stackrel{\subseteq}{\to} \mathbf{R}$. In fact, this is a consequence of the so-called *splitting principle* (see e.g. R. Shanahan [1: p.191ff]).

In our case we do have a counterpart of the above for the particular instance of a *line sheaf,* through the *Weil integrality theorem* (see Chapt. VIII; Theorem 11.1, or even (5.49), (5.51) and (5.56)). On the other hand, concerning the previous principle, one could also keep track of the classical procedure in the general case of *vector sheaves,* as well, on the basis of the hitherto presented material. However, we do not carry it out here, leaving this matter, instead, to the interested reader.

## 6. Chern-Weil homomorphism

We consider below the analogous to our case *"functorial properties" of* the previously defined *(generalized) Chern classes* (see (5.28.1)) for a given *vector sheaf,* by analogy with the standard theory, pertaining, in effect, to the so-called *Weil homomorphism* (cf. (6.15) below) and its *"naturality".*

So to start with, suppose that we are given a *curvature space* $X$, together with a *vector sheaf* $\mathcal{E}$ on $X$, of *rank* $n \in \mathbf{N}$, which is further endowed with an $\mathcal{A}$*-connection* $D$, whose *curvature* is $R(D) \equiv R$.

On the other hand, consider the following *matrix algebra,* consisting of the *global sections* of the *(matrix)* $\mathbf{C}$*-algebra sheaf* involved (see also (1.15) and (2.1));

$$\text{(6.1)} \qquad M_n(\Omega^{\cdot}_{\text{ev}})(X) = Z^0(\mathcal{U}, M_n(\Omega^{\cdot}_{\text{ev}})),$$

where, of course, as usual,

$$\mathcal{U} = (U_\alpha)_{\alpha\in I} \tag{6.2}$$

stands for a given *local frame* of $\mathcal{E}$. Thus, we denote by

$$\mathcal{I}^*(\mathcal{GL}(n, \mathcal{A})) \tag{6.3}$$

the *(graded)* **C***-algebra of invariant polynomials,* in effect, *polynomial functions* on (6.1) (see, for example, (2.5), (2.8) and (2.14)), the *multiplication in (6.3) being,* by definition, the *exterior product.* Namely, one sets;

$$\mathcal{I}^*(\mathcal{GL}(n, \mathcal{A})) = \bigoplus_{k=0}^{\infty} \mathcal{I}_k(\mathcal{GL}(n, \mathcal{A})), \tag{6.4}$$

where the summands in the second member of (6.4) correspond to the various *degrees of* the (invariant) *polynomials* concerned.

Now, let us consider further the *curvature matrix* of $D$, which is *associated with the* given *local frame* $\mathcal{U}$ of $\mathcal{E}$; viz. one has (see (1.11) and (1.12));

$$\begin{aligned} R \equiv \left(R^{(\alpha)}\right) \equiv \left(\left(R^{(\alpha)}_{ij}\right)\right) &\in Z^0(\mathcal{U}, M_n(\Omega^2)) \\ &\subseteq \textstyle\prod_\alpha M_n(\Omega^2)(U_\alpha) = \prod_\alpha M_n(\Omega^2(U_\alpha)). \end{aligned} \tag{6.5}$$

In particular, one obtains;

$$\left(R^{(\alpha)}_{ij}\right) \in M_n(\Omega^2)(U_\alpha) = M_n(\Omega^2(U_\alpha)), \quad \alpha \in I, \tag{6.6}$$

such that (cf. (1.13))

$$R^{(\alpha)}_{ij} \in \Omega^2(U_\alpha), \quad \textit{with} \quad 1 \leqq i, j \leqq n = \operatorname{rk} \mathcal{E}, \quad \textit{and} \quad \alpha \in I. \tag{6.7}$$

Therefore, by looking now at an element

$$p \in \mathcal{I}_k(\mathcal{GL}(n, \mathcal{A})) \overset{\subseteq}{\rightarrow} \mathcal{I}^*(\mathcal{GL}(n, \mathcal{A})) \tag{6.8}$$

(see (6.4)), that is, at an *invariant polynomial* of *degree* $k \in \mathbf{N}$, one obtains that (cf. also (2.20))

$$\begin{aligned} p(R) \in \wedge^{2k}\Omega^1(X) &\overset{\subseteq}{\rightarrow} (\wedge^{2k}\Omega^1)(X) \equiv \Omega^{2k}(X) \overset{\subseteq}{\rightarrow} \Omega^{\cdot}_{\mathrm{ev}}(X) \\ &\overset{\subseteq}{\rightarrow} M_n(\Omega^{\cdot}_{\mathrm{ev}}(X)) = M_n(\Omega^{\cdot}_{\mathrm{ev}})(X). \end{aligned} \tag{6.9}$$

On the other hand, suppose now that our space $X$ is, in particular, a *Bianchi space.* Thus (cf. Theorem 2.1), one then obtains;

$$dp(R) = 0, \tag{6.10}$$

viz. $p(R)$, as in (6.9), is a *closed* $2k$*-form on* $X$, with $k = \deg p \in \mathbf{N}$.

Furthermore, let us assume now that the same Bianchi space $X$, as before, is, in addition, a *(generalized) de Rham space* (no restriction on order; see Definition 3.1, together with Scholium 3.1. However, cf. (6.13) in the sequel, as well). Thus, in view of (6.10) and based further on Lemma 3.1 (cf. (3.25), (3.26)), one can associate with $p(R)$, as in (6.9), a $2k$*-dimensional (Čech) cohomology class of* $X$*, with coefficients in the* **C***-vector space sheaf* $\ker \partial$; viz. we have

$$c(p(R)) \equiv c(R) \equiv [p(R)] \in \check{H}^{2k}(X, \ker \partial) \hookrightarrow \bigoplus_{n=0}^{\infty} \check{H}^{2n}(X, \ker \partial) \equiv \check{H}^{**}(X, \ker \partial). \tag{6.11}$$

The last item in (6.11) denotes the *(graded)* **C***-algebra of even dimensional (Čech) cohomology* **C***-vector spaces* of $X$, with *coefficients in* $\ker \partial$. In this concern, the *multiplication in the latter algebra* is still defined *by exterior multiplication* on the representatives (cocycles) of the cohomology classes involved (cf. (3.1) too).

**Scholium 6.1.**- Concerning the *order of the (generalized) de Rham space* $X$, as above (cf. Definition 3.1), it is to be noticed that *this cannot exceed* $\mathrm{rk}\,\Omega^1$, in the case the given $\mathcal{A}$*-module* $\Omega^1$ *is,* in particular, a *vector sheaf* on $X$ *too.* Thus, in tha latter case it is, in effect, enough to consider a *(generalized) de Rham* $m$*-space, with*

$$m = \mathrm{rk}\,\Omega^1 \in \mathbf{N}. \tag{6.12}$$

Furthermore, based now on (6.9), one obtains;

$$p(R) = 0, \quad \textit{if} \quad 2\deg p > m = \mathrm{rk}\,\Omega^1. \tag{6.13}$$

On the other hand, since

$$\check{H}^{**}(X, \ker \partial) \hookrightarrow \check{H}^{*}(X, \ker \partial) := \bigoplus_{k \in \mathbf{Z}_+} \check{H}^{k}(X, \ker \partial) \tag{6.14}$$

(see also (6.11)), the multiplication in the second member of (6.14) being defined in an analogous manner to that one in (6.11), *one* finally *obtains the following map;*

(6.15) $$\chi_{\mathcal{E}} : \mathcal{I}^*(\mathcal{GL}(n, \mathcal{A})) \longrightarrow \check{H}^*(X, \ker \partial),$$

*such that* (see (6.11))

(6.16) $$\chi_{\mathcal{E}}(p) := c(p(R)) \equiv [p(R)] \in \check{H}^{\deg p}(X, \ker \partial),$$

the same map being also *well-defined,* by virtue of Theorem 3.1, *for any* given *pair* $(\mathcal{E}, D)$, as above.

The previous map (6.15), as given by (6.16), is named the *Chern-Weil homomorphism,* being thus associated with any vector sheaf having an $\mathcal{A}$-connection, when, of course, the appropriate context is provided, as well (cf. the following theorem).

Now, the above terminology for the map (6.15) is further justified according to the next.

**Theorem 6.1.** *Suppose we are given a (generalized) de Rham space $X$, and let $\mathcal{E}$ be a vector sheaf on $X$, endowed with an $\mathcal{A}$-connection $D$. Then, the corresponding Chern-Weil homomorphism $\chi_{\mathcal{E}}$* (cf. (6.15), (6.16)) *is a morphism of the respective (graded)* **C***-algebras* (see (6.4) and (6.14)).

*On the other hand, for any given continuous map*

(6.17) $$f : Y \longrightarrow X,$$

*one obtains ("naturality of $\chi_{\mathcal{E}}$");*

(6.18) $$\chi_{f^*(\mathcal{E})} = f^* \circ \chi_{\mathcal{E}},$$

*where $f^*(\mathcal{E})$ denotes the vector sheaf on $Y$, pull-back, via $f$, of the given vector sheaf $\mathcal{E}$ on $X$.*

**Proof.** The first assertion about $\chi_{\mathcal{E}}$ is an immediate consequence of the very definition of the multiplications in the **C**-algebras involved, as already explained at the pertinent places above (see the comments after (6.11) and (6.14), respectively, in the preceding).

Now, suppose that

(6.19) $$(g_{\alpha\beta}) \in Z^1(\mathcal{U}, \mathcal{GL}(n, \mathcal{A})), \quad \textit{with} \quad n = \operatorname{rk} \mathcal{E} \in \mathbf{N},$$

is a *coordinate 1-cocycle* of the *vector sheaf* $\mathcal{E}$, with respect to a *local frame of* $\mathcal{E}$,

(6.20) $$\mathcal{U} = (U_\alpha)_{\alpha \in I}.$$

Thus, it is an easy consequence of the very definitions that

$$f^*(\mathcal{U}) := (f^{-1}(U_\alpha))_{\alpha\in I} \tag{6.21}$$

is a *local frame of* $f^*(\mathcal{E})$, the latter sheaf being a *vector sheaf on* $Y$, *relative to* $f^*(\mathcal{A}) \equiv \mathcal{A}_Y$, such that

$$\mathrm{rk}_{f^*(\mathcal{A})} f^*(\mathcal{E}) = \mathrm{rk}_{\mathcal{A}}\mathcal{E} = n \tag{6.22}$$

(see Chapt. II; (4.14), (2.68) and (2.69), together with Chapt. I; (14.26)).

On the other hand, based on the *exactness of the pull-back functor* $f^*$ (Chapt. II; (2.69)), one has transported to $Y$, through $f$, the corresponding framework on $X$, namely, that one of a *(generalized) de Rham space* and its paraphernalia, which, by hypothesis, is already given on $X$ (see, for instance, Chapt. VIII; (9.24), in conjunction with Definition 3.1 in the preceding). Finally, *for every invariant polynomial* $p \in \mathcal{I}^*(\mathcal{GL}(n, \mathcal{A}))$ (see (6.3), (6.4)), *one has*

$$f^* \circ p = p \circ f^* \in \mathcal{I}^*(\mathcal{GL}(n, f^*(\mathcal{A}))), \tag{6.23}$$

so that one still obtains;

$$p(f^*(R)) = f^*(p(R)). \tag{6.24}$$

In this concern, we also recall that $f^*(R)$ is the *curvature of the* $f^*(\mathcal{A})$-*connection of* $f^*(\mathcal{E})$ (see Chapt. VIII; (9.25)). Therefore, *one has* (cf. (6.16));

$$f^*(\chi_{\mathcal{E}}(p)) = \chi_{f^*(\mathcal{E})}(p), \tag{6.25}$$

*for every* $p \in \mathcal{I}^*(\mathcal{GL}(n, \mathcal{A}))$, as desired, and this terminates the proof. ∎

# PART THREE

# EXAMPLES AND APPLICATIONS

# CHAPTER X

# Classical theory

"Progress in any aspect is a movement through changes in terminology."

W. Stevens in "Adagia" (included in "Opus Posthumous", ed. by Vintage Books, 1990). p.184.

We consider in this Chapter several particular instances, that have to do, essentially, with the classical theory of $\mathcal{C}^{\infty}$*-manifolds* (in particular, *differential geometry* of such), for which the framework, exhibited hitherto in the preceding Chapters, constitutes their abstract (axiomatic) description. Of course, this happens in the *finite-dimensional* case, while certain *infinite-dimensional examples* also are fitted in well, within the same context, as in the foregoing. A particularly interesting case is that one of the so-called *generalized manifolds,* or even *orbifolds,* which we are dealt with in Section 2. The case of *complex manifolds* is also discussed (Section 3). Finally, we also examine a potential application of the previous framework to the description of certain basic concepts of *theoretical physics,* including the case of *supermanifolds.* Further applications, as well as, examples of the preceding discussion, within the framework of *topological algebra theory,* will be given

in the next Chapter XI.

## 1. Differential geometry of $\mathcal{C}^\infty$-manifolds. (Finite and infinite-dimensional cases)

We have already pointed out in the preceding several places, where the abstract framework employed so far specialized to the standard *differential geometry of finite-dimensional smooth manifolds:*

So, to avoid recurences, we simply indicate by the ensuing discussion concrete places in the foregoing, where one finds relevant references to the *standard case,* the latter being, for that matter, the *life-motif of our whole exposition.* In this context, we also note that just the headings too of the particular Sections in the second part of this treatise (viz. of *Part Two, Geometry,* in particular, Chapters VI - IX) indicate, of course, clearly the relation between the present material and the classical one; yet, the two aspects are linked up through the fundamental concept of a *differential triad* (Chapt. VI; (1.4), (3.1) and Chapt. VIII; (1.1)). In particular, here, for the case at issue, by the *standard differential triad*

$$(\mathcal{C}_X^\infty, d, \Omega_X^1), \tag{1.1}$$

*with* $X$ a *(finite-dimensional)* $\mathcal{C}^\infty$*-manifold;* see thus Chapt. VI; Section 2.

On the other hand, we have already noted (same Chapt. VI; Subsection 4.(b)), how the present (axiomatic) point of view specializes to the standard *theory of connections* (viz. *covariant differentiation*) *of* $\mathcal{C}^\infty$*-(complex) vector bundles* (see also Note 1.1 below, concerning the more general case of *principal (fiber) bundles).* In this regard, we have also fully explained the fundamental fact that, in the particular case when our base $\mathcal{C}^\infty$-manifold $X$ is *paracompact (Hausdorff),* then

(1.2) $\mathcal{C}^\infty$*-connections always exist*

(see Chapt. VI; (15.10), (16.18)), by contrast with what happens, in general, within our abstract framework; see thus, for instance, Chapt. III; Corollary

12.1, where one realizes that the existence of an *$\mathcal{A}$-connection* for a given *vector sheaf* $\mathcal{E}$ is characterized by the *vanishing of* what we have called (generalized) *Atiyah class of* $\mathcal{E}$.

Now, as already indicated in the pertinent places of the preceding discussion, the important conclusion (1.2), for *(finite-dimensional paracompact) $\mathcal{C}^\infty$-manifolds,* is actually due to the basic fact that, for the case at issue, the *sheaf of coefficients,* or even the *structure sheaf,* viz. the $\mathbf{C}$-*algebra sheaf*

$$ {}^{\mathbf{C}}\mathcal{C}_X^\infty \equiv \mathcal{C}_X^\infty \tag{1.3} $$

(cf. Chapt. I; (4.39)) on a *smooth manifold* $X$, as before, is *fine* (Chapt. III; (8.55.1)). However, see also the next Chapter XI for a similar conclusion in certain particular instances in topological algebra theory, that also include the previous case.

On the other hand, *the local aspect of the classical theory has also been exposed, within the axiomatic framework* of the previous discussion (see thus Chapters VII and VIII). In this context, we should further remark that the *(partial) differential operator*

$$ \frac{\partial}{\partial x} \tag{1.4} $$

of the *standard theory* is to be replaced here by an expression of the form;

$$ \partial(s)(t) \tag{1.5} $$

(see, for example, Chapt. VII; (5.41) for the notation applied above, or even (5.63) therein). Analogously, one can consider *higher derivations,* by setting, for instance,

$$ \partial(s')(\partial(s)(t)), \tag{1.6} $$

as a substitute of the classical operator

$$ \frac{\partial^2}{\partial x\, \partial y}. \tag{1.7} $$

Here *one cannot expect, in general,* the usual *"symmetry"* of the second (partial) derivative, as in the classical theory, pertaining to the order of the (partial) differentiation considered. Of course, one still understands in (1.6) the relevant *context of Chapt. VII; Section 5.*

On the other hand, *metric ($\mathcal{A}$-)connections* have been considered in

Chapt. VII; Sections 8 and 9, extending thus, within the present abstract context, the classical subject of *Riemannian and hermitian connections* on a $\mathcal{C}^\infty$-manifold, as above. (Concerning the classical case, see e.g. R.O. Wells [1: p.76, Proposition 1.11]).

Now, another aspect that is also of interest, within the preceding framework, is the analogue in our case of the standard *de Rham complex;* thus, applying the notation of Chapt. IX; Definition 3.1, we consider the following *generalized de Rham complex*

$$(1.8)\qquad \begin{aligned} &0 \longrightarrow \mathcal{A} \equiv \Omega^0 \xrightarrow{d^0 \equiv \partial} \Omega^1 \xrightarrow{d^1 \equiv d} \Omega^2 \longrightarrow \cdots \\ &\cdots \longrightarrow \Omega^n \xrightarrow{d^n \equiv d} \Omega^{n+1} \longrightarrow \cdots, \end{aligned}$$

being, *by hypothesis, a complex of* **C**-*vector space sheaves* on $X$ (see also Chapt. VIII; (8.2) - (8.4), as well as, Chapt. IX; Scholium 3.1). In particular, *we assume* that the following is an *exact sequence,* as before;

$$(1.9)\qquad \begin{aligned} &0 \longrightarrow \ker \partial \longrightarrow \mathcal{A} \equiv \Omega^0 \xrightarrow{p} \Omega^1 \xrightarrow{d^1} \Omega^2 \longrightarrow \cdots \\ &\cdots \longrightarrow \Omega^n \xrightarrow{d^n} \Omega^{n+1} \longrightarrow \cdots. \end{aligned}$$

We have already seen in the previous Chapter IX the significance of (1.9), in order to get in our case the analogous procedure of *associating a cohomology class to a closed form* on $X$ (ibid.; Lemma 3.1), and its impact, in turn, in proving the corresponding here *Chern-Weil Theorem,* the basic tool for extending to the present abstract framework the theory of *Chern (characteristic) classes* (loc. cit., Theorem 3.1. See also Scholium 3.4 therein, along with Section 5).

Furthermore, we can still consider the following *augmented generalized (strict) de Rham complex,* being also, by hypothesis, exact;

$$(1.10)\qquad 0 \longrightarrow \mathbf{C} \xrightarrow{\varepsilon} \mathcal{A} \xrightarrow{\partial} \Omega^1 \xrightarrow{d} \cdots \xrightarrow{d} \Omega^n \xrightarrow{d} \cdots,$$

the rest being, as in (1.8). Of course, when one looks at a $\mathcal{C}^\infty$*-manifold* $X$, the last complex (1.10) is the usual *(augmented) de Rham complex of* $X$ of the classical theory, where one further employs **C**-*valued functions* (cf. (1.3). See also Chapt. III; (8.37′) and (8.44)). In this context, we have already noted (loc. cit. Lemma 8.2) the significant rôle played here by *Poincaré's Lemma. In the general case,* however, as the one considered herewith, the

aforementioned result, is *unattainable,* hence, the *exactness of (1.10),* as well. Nevertheless, as we shall see later on (cf. Chapter XI), we do have *exactness of (1.9) at the first place,* at least, in the following form;

$$(1.11) \qquad 0 \longrightarrow \mathbf{A} \longrightarrow \mathcal{A} \xrightarrow{\partial} \Omega^1 \longrightarrow \cdots,$$

were $\mathbf{A}$ stands for a suitable *topological algebra;* this is of a particular significance for *differential geometric applications* to the so-called $\mathbf{A}$*-vector bundles,* pertaining, for instance, to their *flatness* (loc. cit.).

On the other hand, apart from the standard $\mathcal{C}^\infty$-manifolds, as above, we are going to consider right below *certain particular examples,* even *infinite dimensional,* where the *exactness of (1.10) holds true.* We thus exhibit in the sequel a number of such examples (in fact, *infinite-dimensional* ones), the rest being deferred for Section 3. But, before this, we still comment on another basic point of view, according to the classical theory, which, however, we do not examine by this treatise.

**Note 1.1.**- We have treated heretofore, while this will also be the case in the sequel, *vector sheaves,* viz. *locally free* $\mathcal{A}$*-modules* (of *finite rank*), with respect to a given fixed $\mathbf{C}$*-algebra sheaf* $\mathcal{A}$. As already explained (cf. Chapt. VI; (4.11)), these correspond to $\mathbf{C}$*-vector bundles* of the classical theory.

On the other hand, the counterpart in our case of the standard notion of a *principal fiber bundle* can also be examined within the abstract (axiomatic) framework, as above. Indeed, such a study has already been started by the recent work of E. Vassiliou [1] - [4].

**1.(a). Infinite-dimensional examples.**- We consider below examples of *infinite-dimensional* $\mathcal{C}^\infty$*-manifolds,* for which the *augmented de Rham complex* (1.10) yields an *exact sequence:*

The discussion that follows is based on the work of M.E. Verona [1]. Thus, consider a *projective limit topological space*

$$(1.12) \qquad X = \varprojlim_{\alpha} X_\alpha$$

(see, for instance, N. Bourbaki [1: Chap. 1; p.28ff]), such that each one of the factor spaces $X_\alpha$, $\alpha \in I$ (the latter being an upwards directed set) is a usual (finite dimensional) $\mathcal{C}^\infty$*-manifold, Hausdorff* and *2nd countable,* hence,

*metrizable* and a fortiori *paracompact* (see e.g. J. Dugundji [1: p.195, Corollary 9.2 and p.186, Theorem 5.3]). Moreover, we assume that the connecting maps

$$\phi_{\alpha\beta} : X_\beta \longrightarrow X_\alpha, \quad \alpha \leqq \beta, \tag{1.13}$$

are *proper submersions;* thus, $X$ is *locally compact* and $\sigma$*-compact,* hence, *paracompact* (see N. Bourbaki [1; p.77, Proposition 7] and J. Dugundji [1: p.241, Theorem 6.5]).

Now, one defines *"$\mathcal{C}^\infty$-functions"* on $X$, as follows: Taking an *open* $U \subseteq X$, a **C**-*valued map*

$$f : U \longrightarrow \mathbf{C} \tag{1.14}$$

is said to be a $\mathcal{C}^\infty$*-map on* $U$, whenever, for every $x \in U$, there exists an *open* $U_x \subseteq X$, with $x \in U_x$, and some $g \in \mathcal{C}^\infty(X_\alpha) \equiv {}^{\mathbf{C}}\mathcal{C}^\infty(X_\alpha)$ (that is, **C**-valued $\mathcal{C}^\infty$-functions on the smooth manifold $X_\alpha$), where $\alpha \in I$, as above, such that;

$$f = g \circ \pi_\alpha|_{U_x}, \tag{1.15}$$

where $\pi_\alpha : X \longrightarrow X_\alpha$, $\alpha \in I$, stands for the corresponding canonical map. We denote the set of such maps on the *open* $U \subseteq X$ by

$$\mathcal{C}^\infty(U). \tag{1.16}$$

It is a *unital commutative* **C**-*algebra;* yet, one can consider, following standard reasoning, the **C**-*algebra sheaf* on X

$$\mathcal{A} \equiv \mathcal{C}^\infty_X, \tag{1.17}$$

viz. the *sheaf of germs of* **C**-*valued* $\mathcal{C}^\infty$*-functions on* $X$. (See also, for instance, Chapt. I; Subsection 4.(b)). It is proved that (1.17) *is* actually a *fine sheaf* on $X$ (cf. the aforementioned work of M.E. Verona).

On the other hand, one further defines the $\mathcal{A}$*-module* on $X$ of *"1-forms"*,

$$\Omega^1_X \equiv \Omega^1, \tag{1.18}$$

as follows; for every $x \in X$, one first considers the **C**-vector space of *point derivations at* $x$ *of* $\mathcal{C}^\infty(X)$ (cf. (1.16), for $U = X$). Then, the *dual of the* latter *space* is the **C**-*vector space of 1-forms of* $X$ *at* $x \in X$, denoted by $\Omega^1_x(X)$. (It is further canonically associated with the *classical quotient* of the ideal $I_x$ of $\mathcal{C}^\infty(X)$, consisting of those elements of $\mathcal{C}^\infty(X)$, which vanish

at $x$; viz. one has $I_x/I_x^2 \cong \Omega_x^1(X)$, $x \in X$).

Now, for any open $U \subseteq X$, one defines the **C**-*vector space* (in fact, a $\mathcal{C}^\infty(U)$-*module,* cf. (1.16)) *of 1-forms on* $U$, as those maps

$$\omega : U \longrightarrow \bigcup_{x \in U} \Omega_x^1(X), \tag{1.19}$$

which are *locally* (with respect to $U$) *of the form;*

$$\omega = \sum_{i=1}^{n} g_i (df_i)_x |_{V \subseteq U, \text{ open}} \tag{1.20}$$

with $x \in V$, and $g_i, f_i \in \mathcal{C}^\infty(U)$, $1 \leqq i \leqq n$, such that $(df)_x := [f - f(x)] \in I_x/I_x^2 \equiv \Omega_x^1$, for any $f \in \mathcal{C}^\infty(U)$. (More precisely, one can actually define $I_x$, as those **C**-*valued* maps only *locally defined* and *smooth at* $x \in X$, see e.g. (1.16), *vanishing* also *at the same point;* thus, (1.20) acquires a more direct meaning). So we can further define $\Omega^1$ (cf. (1.18)), as the *sheaf on* $X$, in effect, a $\mathcal{C}_X^\infty$-*module, generated by the 1-forms on the* various *open sets of* $X$, as before.

On the other hand, we still define (see also, for instance, Chapt. VIII; (8.3), (8.4))

$$\Omega^p := \wedge^p \Omega^1, \quad p \in \mathbf{N}, \tag{1.21}$$

while we also set (cf. (1.17));

$$\Omega^0 := \mathcal{A} \equiv \mathcal{C}_X^\infty. \tag{1.22}$$

Thus, one now concludes that

$$0 \longrightarrow \mathbf{C} \longrightarrow \Omega^0 \equiv \mathcal{A} \xrightarrow{d^0 \equiv d} \Omega^1 \xrightarrow{d^1} \Omega^2 \longrightarrow \cdots \tag{1.23}$$

is a *fine resolution of* **C**; that is, (1.23) is an *exact sequence of* **C**-*vector space sheaves,* such that *each one of the* $\mathcal{A}$-*modules* $\Omega^i$, $i \in \mathbf{Z}_+$, *is a fine sheaf* on $X$ (cf. (1.17), along with Chapt. III; (8.57)). Yet, the *"differential operators"* in (1.23), apart from $d^0 \equiv d$, already defined above (cf. (1.20)), can still be given on the basis of the same relation (1.20), following the classical pattern (see, for instance, Chapt. III; (8.33), (8.34)).

So, as a result of the preceding, in conjunction too with our discussion in the previous chapters, one concludes that;

(1.24) *the machinery developed so far, by the preceding exposition, is still available, within the particular case of the "generalized manifold" $X$, as in (1.12).*

But, there is, however, a *proviso*! Namely, the validity of the *exponential framework,* as the latter is given for instance, by Definition 6.1 of Chapter V. Indeed, as we shall see in the next Chapter XI, this is further assured by considering as our *structure sheaf* $\mathcal{A}$, an appropriate *topological algebra sheaf;* again, apart from the classical case (see (1.3)), this too specializes to certain particular examples (loc. cit.).

On the other hand, by considering a weakened version of a *"generalized manifold",* as above, to the extent, namely, that $X$ *is not a locally compact* space, but, employing, however, *Banach manifolds* in place of the $X_\alpha$'s, as in (1.12), $X$ can then be equipped with an appropriate *infinite dimensional manifold structure;* in particular, it can be modeled on a *Fréchet* (locally convex) *space.* Indeed, such a point of view is recently examined by the thesis of G. Galanis [1]. (In this context, still consult the recent work by M. Papatriantafillou [1]). Furthermore, concerning (1.12), see also, for instance, G.E. Bredon [1: p.69, §14].

Now, our *second* relevant *example,* where the framework of the preceding chapters can be realized, being still outside of the classical context, is supplied by the work of H.R. Fischer [1]: Thus, our $\mathcal{C}^\infty$*-manifold* $X$ is here a given *locally convex space* $E$, on which one further considers $\mathcal{C}^\infty$*-functions* with values from another *complete locally convex space* $F$. (We thus have, so to speak, an *"affine" infinite dimensional* $\mathcal{C}^\infty$*-manifold;* the *"localization"* of this line of thoughts, according to the classical pattern, would be worthwhile. See, for instance, our next quotation right below). Thus, within the same context of ideas, one further defines a *de Rham complex*

$$\Omega^{\cdot}(E,F) \equiv (\Omega^p(E,F), d^p)_{p\in\mathbf{Z}_+} \tag{1.25}$$

of $\mathcal{C}^\infty$*-differentiable p-forms* ($p \in \mathbf{Z}_+$) *on* $E$, with values in $F$; in fact, *sheaves of germs* of such. Indeed, one has here again a *resolution of* $\mathbf{C}$, that is, an *exact sequence of (*$\mathbf{C}$*-vector space) sheaves*

$$0 \longrightarrow \mathbf{C} \longrightarrow \Omega^0 \longrightarrow \Omega^1 \longrightarrow \Omega^2 \longrightarrow \cdots . \tag{1.26}$$

Accordingly, one has here too a *conclusion similar to (1.24)* above.

Another *example* of the same nature, as before, is supplied within the

context recently considered by J.-L. Brilynski [1]: Here by a suitably chosen $E$, as above, (in fact, an appropriate *projective limit of Hilbert spaces* is employed), one has that the corresponding *structure sheaf* $\mathcal{A}$ is *fine,* so that one finally obtains a *generalized de Rham theorem,* in the sense of Chapt. III; (3.24) (cf. also the above cited Ref.; p.35, Lemma 1.4.1, and p.39, Theorem 1.4.15. Thus, again

(1.27) a *conclusion analogous to that in (1.24) is here too in force.*

Yet, another example, within the same vein of ideas, as before, is the one appeared in the recent work of J. Bernstein - V. Lunts [1]; here one considers a *paracompact* (Hausdorff) *space*

$$X \cong \varinjlim X_n \tag{1.28}$$

homeomorphic with the *inductive limit space* of a suitable *increasing sequence of* standard $\mathcal{C}^\infty$-(paracompact) *manifolds.* Thus, by further taking the *de Rham complex* of each one of the individual manifolds $X_n$, $n \in \mathbf{N}$, as in (1.28), say $\Omega^{\cdot}_{X_n}$ (cf. Chapt. III; (8.44)), one defines the *de Rham complex of* $X$, by the relation,

$$\Omega^{\cdot}_X := \varprojlim \Omega^{\cdot}_{X_n}, \tag{1.29}$$

while the latter is proved to be a *(soft) resolution of the constant sheaf* $\mathbf{C}_X \equiv \mathbf{C}$ (ibid.; p.96, Proposition 12.2.3). Consequently, *a similar argument to (1.24) is still valid* in the case under consideration.

# 2. Holomorphy

We have already explained in the pertinent places, throughout our discussion in the preceding Chapters, the way the present point of view is related with the standard theory of *(finite-dimensional) complex manifolds;* thus, the whole treatment, concerning the *existence of $\mathcal{A}$-connections,* has been motivated, as already noted in the foregoing, by the relevant work of M.F. Atiyah

[1] in the case of *"holomorphic connections"*. (See, for example, Chapt. VI; §§10-13, where the so-called *Atiyah class* of a given *vector sheaf* plays a significant rôle; loc. cit. (10.28)). Precisely, we refer to Chapt. VI; Subsection 15.(b) for a concrete presentation of the relationship between our abstract framework, as exposed in the preceding, and that one of the classical theory. See thus, for instance, loc. cit. Definition 15.1, or even (15.36).

Furthermore, the connection of the above with the *exponential framework,* as the latter has been presented herein, was discussed in Chapt. V; Example 6.1, i) (cf., in particular, (6.4) therein). On the other hand, as already mentioned above, another realization of the same connection, as before, will be given in the next Chapter XI, through appropriate *topological algebra sheaves,* by still abstracting the corresponding *structure sheaf* $\mathcal{O}_X$ of a given *complex manifold* $X$. (See also, for example, Chapt. I; (4.40), Chapt. III; (8.65), (8.66) and Chapt. VI; (15.11)).

## 3. Generalized manifolds (contn'd)

We further consider in this Section certain particular instances of our axiomatic framework, as exposed in the preceding, which still correspond to suitable *generalizations of* the classical notion of *a (finite-dimensional) smooth* (viz. $\mathcal{C}^\infty$-)*manifold.* Indeed, as we shall see in the sequel, this extended point of view of the standard theory, which also fits well, within the preceding context, specializes to concrete cases and applications of the usual differential geometry of smooth manifolds (e.g. *Satake manifolds, orbifolds* and the like. See thus Subsection 3.(a) below).

The succeeding discussion corresponds to material based on earlier relevant work of J.W. Smith [1], as well as, on the more recent one by M.A. Mostow [1]; yet, the latter is also related with previous similar considerations in a series of papers by K.-T. Chen (see e.g. the said author [1]). So for details we also refer to the cited works; however, we deviate occasionally from the notation of the aforementioned authors, by employing instead a more contiguous notation to that one of our preceding exposition.

So we start with an arbitrary *topological space* $X$, along with a certain $\mathbf{C}$-*algebra sheaf* $\mathcal{A}$ on $X$ (cf. Chapt. II; Definition 1.4) such that one has

$$\mathcal{A} \subseteq \mathcal{C}_X \tag{3.1}$$

(see also Chapt. I; (4.10)); that is, $\mathcal{A}$ is a *subsheaf of the sheaf* of germs *of continuous* $\mathbf{C}$-*valued maps on* $X$, which is further characterized by the following *"closure" condition:*

(3.2) *for any finite many global sections of* $\mathcal{A}$ *("smooth functions"),* $\alpha_1, \ldots, \alpha_n$ *and any* $\mathbf{C}$*-valued* $\mathcal{C}^\infty$*-function* $f$ *on* $\mathbf{R}^n$*, one has;*

$$f(\alpha_1, \ldots, \alpha_n) \in \mathcal{A}(X) \equiv \Gamma(X, \mathcal{A}). \tag{3.2.1}$$

On the other hand, one further defines the $\mathcal{A}$-*module*

$$\Omega_X^1 \equiv \Omega^1 \tag{3.3}$$

on $X$, viz. the *sheaf of* (germs of) *"1-forms" on* $X$, as follows: First, one considers certain *local* $X$*-valued maps* on open subsets of some euclidean space, say $\mathbf{R}^n$ (*"plots"*), that are *"stable"*, with respect to *"*$\mathcal{C}^\infty$*-functions"* on their domain and range (cf. (3.2)); so these *plots* are then used as *"building blocks"* for the definition of the *"1-forms"* on $X$ (or on any open subset of $X$. For details see M.A. Mostow [1: p.261], or even K.-T. Chen [2: p. 219]). Thus, the previous sheaf (3.3) is, by definition, that one, generated by the *presheaf of "1-forms"*, defined over the various open subsets of $X$. Furthermore, one can repeat here the classical argument, concerning the definition of *"p-forms"* on $X$, *for any* $p \in \mathbf{N}$. (In this regard, see also the relevant exposition by J.W. Smith [1: p. 119ff]).

On the other hand, one also defines an *exterior derivative operator*

$$d^p \equiv d : \Omega^p \longrightarrow \Omega^{p+1}, \quad p \in \mathbf{N}, \tag{3.4}$$

as well as, the corresponding one *for* $p = 0$, that is,

$$d^0 \equiv \partial : \Omega^0 \equiv \mathcal{A} \longrightarrow \Omega^1, \tag{3.5}$$

such that one finally gets a *de Rham complex* for $X$, yielding the *exact sequence;*

$$
\begin{aligned}
(3.6) \qquad & 0 \longrightarrow \mathbf{C} \longrightarrow \mathcal{A} \equiv \Omega^0 \xrightarrow{\partial} \Omega^1 \xrightarrow{d} \Omega^2 \longrightarrow \cdots \\
& \cdots \longrightarrow \Omega^p \xrightarrow{d} \Omega^{p+1} \longrightarrow \cdots .
\end{aligned}
$$

Precisely, one has (3.6) for a suitable *"differentiable space"*

$$(3.7) \qquad (X, \mathcal{A}),$$

as above; indeed, $X$ is further supposed to be a (Hausdorff) *paracompact locally smoothly contractible* space and $\mathcal{A}$ a *fine sheaf* on $X$ (see thus M.A. Mostow [1: p.268, Theorem 5.2]).

Now, within the same vein of ideas, on which our abstract context of the preceding Chapters is further specialized, one should also mention the earlier works by R. Sikorski [1], [2] and C.D. Marshall [1]; a comparison of these and other relevant studies (including the aforementioned ones by J.W. Smith and K.-T. Chen) has also been given by M.A. Mostow (loc. cit.). Finally, an early announcement by A. Verona [1] about *abstract prestratifications,* in conjunction with a *de Rham* type *theorem,* still belongs to the previous class of spaces. (An extension of the latter work was further given by M. Saralegi - G. Hector [1]; see also J.P. Brasselet - G. Hector - M. Saralegi [1], or even G. Hector - M. Saralegi [1]).

**3.(a). Orbifolds (or Satake manifolds).**– The term *V-manifold* is still applied for the type of (generalized) manifold we are dealt with below. This sort of manifolds is also contained in the class of spaces considered above (see, for example, J.W. Smith [1: p.134]), so that it further supplies another particular instance of the abstract (axiomatic) treatment, as this has been exhibited in the preceding.

Roughly speaking, a space of the above type is a *quotient* of a usual (finite dimensional $\mathcal{C}^\infty$-)*manifold* modulo a finite group of diffeomorphisms of $X$, that acts effectively on $X$.

Furthermore, *regular G-varieties,* in the sense of J.L. Koszul [1], fall also into the case of a generalized manifold, as above; again we have here the *quotient* space of a (finite dimensional) $\mathcal{C}^\infty$*-manifold* $X$, modulo a *compact Lie group* of transformations (viz. diffeomorphisms) of $X$. On the other hand, *orbit spaces* resulting from the action of *non-compact groups* still supply similar examples of generalized manifolds, as before. (In this context, see also J.W. Smith [1: p.131, comments following Theorem II]).

In this regard, it is further to be noticed that in all the preceding instances

the base space $X$ is *paracompact* (Hausdorff), while the *structure sheaf* $\mathcal{A}$ is *fine;* hence, all the respective $\mathcal{A}$*-modules* $\Omega^p$, $p \in \mathbf{N}$, are also *fine* (cf. Chapt. III; (8.56)). As a result, one thus concludes that:

(3.8) *every vector sheaf* on such a space $X$ *has always an* $\mathcal{A}$*-connection.* Hence, the same holds true, for every $\mathcal{C}^\infty$*-vector bundle* on $X$, while the corresponding theory of *Chern characteristic classes is* still *in force.*

See the preceding Chapter IX, as well as, for instance, T. Kawasaki [1: p.78].

# 4. Electromagnetism

It is a well known aspect that *quantization* arguments in elementary particle physics (e.g. electromagnetism) lead, via their mathematical formulation, to *cohomology* theory; it concerns here, of course, the so-called *geometric quantization* (see, for instance, J. Śniatycki [1], [2]).

So it is a standard procedure to describe *electromagnetism in terms of differential forms* on the *space-time manifold* (cf., for example, M. Göckeler - T. Schücker [1: p.43ff]). Indeed, a crucial item in the so-called *prequantization* of the *symplectic manifold,* representing the *phase space* of an *elementary (relativistic) particle,* for the case at issue, of an *electron,* is the classical *Weil's integrality theorem,* whose abstract formulation, within the present axiomatic framework, as exhibited in the preceding, has been given already by Chapt. VIII; Theorem 11.1 (see also ibid., Theorem 11.1′).

Thus, as we shall see right below (cf. the next Section 5), given that the *photon* (viz. the *"quantum"*, or even the *carrier* of the *electromagnetic field*) obeys *Bose-Einstein statistics,* equivalently, has *integral spin,* its *states* can be conceived as sections of a *line sheaf* (or in the standard terminology, equivalently, of a *line bundle*). Now, due to the *type of* $X$ *and* $\mathcal{A}$, as above, the previous line sheaf always admits a *connection,* whose *curvature* supplies the required (for the prequantization) *2-dimensional integral cohomology class* (loc. cit.).

In this context, we further mention, in anticipation (see thus (5.13) below) that;

(4.1) *every* (free) *elementary particle can be conceived in a prequantized form.*

The preceding should also be compared with *Śniatycki's criterion* on the (pre)*quantization of (relativistic) charged* (elementary) *particles* (cf. e.g. J. Śniatycki [1]). In this respect, cf. also a recent account by M. Puta [1], or even the one by E. Prugovečki [2].

## 5. Elementary particles

We give in this Section a brief exposition of a potential application of our abstract framework to a description of the *elementary particles* and the accompanied *(pre)quantization,* in terms of *vector sheaves* and the associated *connections.* Indeed, the idea is based on an earlier study by S.A. Selesnick [3], to exhibit the so-called *second quantization* (alias *quantum field theory*), in terms of *projective finitely generated modules,* with respect to the algebra of *classical observables.* All that follows is more or less an adaptation, within the preceding abstract context, of the relevant account in the above cited work, along with the pertinent comments thereon.

Thus, one first considers *relativistic elementary particles,* for which the *space-time manifold* (or else *configuration space* of the physical system at issue) is a (Hausdorff) compact connected 4-dimensional *Lorentz ($\mathcal{C}^\infty$-)manifold $X$*, whose *lorentzian metric* represents the *gravitational field.* (We further assume that the various *differential geometric items,* which will be involved in the sequel, are *complexified;* cf., for instance, Chapt. VI; (4.16)). So the aforementioned *algebra of classical observables* of a physical system, as above, is defined to be

$$\mathbf{A} := \mathcal{C}^\infty_{\mathbf{C}}(X) \equiv \mathcal{C}^\infty(X); \tag{5.1}$$

viz. the algebra (with point-wise defined operations) of $\mathbf{C}$*-valued* $\mathcal{C}^\infty$*-functions* on $X$. (As we shall see in the next Chapter XI, the previous algebra supplies

an important example of a *topological algebra*, which is *not Banachable*!). Furthermore, the same algebra can be construed as the *algebra of global sections* of the *structure sheaf*

$$\mathcal{A} := {}^{\mathbf{C}}\mathcal{C}^{\infty}_{X} \tag{5.2}$$

(see also Chapt. I; (10.13), along with (11.11) and (11.12)). Yet, (5.2) provides an important instance of a *topological algebra sheaf*, in the sense of the next Chapter XI (cf. Theorem 1.1 therein).

> **Remark 5.1.**– Indeed, what we actually consider is the so-called *phase space* (physics terminology) of the physical system, under discussion, viz. the *cotangent bundle* $\mathcal{T}^{*}(X)$ of the manifold $X$, as above. So it is, in effect, the algebra
>
> $$\mathcal{C}^{\infty}(\mathcal{T}^{*}(X)), \tag{5.3}$$
>
> which one considers as the *algebra of classical observables* of the system. (We can still assume *complexifications*, as before).

Now, according to an *axiomatic presentation of elementary particles*, due to S.A. Selesnick (loc. cit.), one concludes that:

(5.4) *bare* (viz. free) *particles have states that correspond to sections of suitable vector sheaves;* precisely, (bare) *bosons* are associated with *line sheaves*, while (bare) *fermions* with *vector sheaves*, viz. those of *rank greater that* 1.

As a matter of fact, by still following S.A. Selesnick [3: p.39], one realizes that:

(5.5) $\Omega^1$ (viz. the sheaf of germs of sections of the complexified cotangent bundle of $X$, $\mathcal{T}^{*}(X)$) *represents* (by its *sections*, as before, *states* of) *fermions*, viz. *spin*-1/2 *particles.*

Now, by our hypothesis $X$ is a *paracompact* (Hausdorff) space, while the *structure sheaf* $\mathcal{A}$ (see (5.2)) *is fine* (cf., for instance, (1.3)), so that *the same holds true for the* $\mathcal{A}$-*module* (in fact, *vector* sheaf) $\Omega^1$ (cf. Chapt. III; (8.26) and Chapt. VI; (15.3), (15.4)).

Therefore, a *line sheaf* $\mathcal{L}$, as in (5.4), always *admits a* ($\mathcal{C}^{\infty} \equiv \mathcal{A}$-)*connection* $D$ (cf. (1.2)); consequently, one infers that

*the* (free) *boson* involved, as in (5.4), *may be construed, as yielding a pair*

(5.6) $$(\mathcal{L}, D)$$

*on* $X$, viz. a *Maxwell field.*

The above, also explains our previous terminology, concerning a pair, as in (5.6). On the other hand, as we shall presently see, the previous assertion also entails that;

(5.7) *the free* (elementary) *particle considered* (in our case, a *bare boson) has already, from the outset, a "prequantizing line bundle"* (equivalently, *line sheaf*), *associated with it.*

*We have* thus *circumvented,* through Selesnick's axiomatic (5.4), the need to resort to the standard *Hamiltonian formalism,* in order, namely, to find first the appropriate *quantum-mechanical* representative of the physical system at issue; that is, one thus avoids the *"correspondence principle"* (or even the *first quantization mechanism.* In this regard, see also e.g. D.J. Simms - N.M.J. Woodhouse [1]). Indeed, based on *Weil's integrality theorem* (see Chapt. VIII; Theorem 11.1′, or even Lemma 5.2 there), one concludes that *the curvature* of the $\mathcal{A}$-connection $D$, involved, $R(D) \equiv R$, *yields a* (2-dimensional) *integral cohomology class of* $X$, in the sense that one has (cf. Chapt. VIII; (5.56))

(5.8) $$\frac{1}{2\pi i}[R] \in H^2(X, \mathbf{Z}).$$

However, this is exactly the *prequantization condition* (or even *integrality condition*), viz. a criterion in order the *physical system,* under consideration, to be *susceptible of (pre)quantization.* (See the previously quoted authors, or even, for instance, N. Woodhouse [1: p.116ff]). In conclusion, we thus infer (see (4.1) in the preceding);

(5.9) *every* (free) *boson is prequantizable.*

Next, we are going to infer a *similar statement, for every (free) fermion,* as well: Thus, as already noted before (cf. (5.5)), such an elementary particle is represented (through its states) by (sections of) $\Omega^1$. Hence, by further letting

a boson act on $\Omega^1$, one has a *new particle,* which *locally* is still $\Omega^1$; indeed, one has (see Chapt. II; Lemma 5.1 and (5.15), along with (5.4) above)

$$\begin{aligned}\Omega^1(\mathcal{L})|_U \equiv \mathcal{L} \otimes_{\mathcal{A}} \Omega^1|_U = \mathcal{L}|_U \otimes_{\mathcal{A}|_U} \Omega^1|_U = \mathcal{A}|_U \otimes_{\mathcal{A}|_U} \Omega^1|_U \\ = (\mathcal{A} \otimes_{\mathcal{A}} \Omega^1)|_U = \Omega^1|_U.\end{aligned} \tag{5.10}$$

Here $U$ stands, of course, for a *common local gauge* of $\Omega^1$ and $\mathcal{L}$ (loc. cit.). In this context, the previous argument, which can also be construed as a kind of *"mediation of forces through the exchange of bosons"*, furnishes $\Omega^1$ with the *prequantizable framework,* that we are looking for. Indeed, one has the following *commutative diagram;*

$$\begin{array}{ccc} \rho^*(\mathcal{L}) & \longrightarrow & \mathcal{L} \\ \downarrow & & \downarrow \\ \Omega^1 & \xrightarrow{\ \rho\ } & X, \end{array} \tag{5.11}$$

where $\rho$ denotes the respective (canonical) *projection* (of the *sheaf* $\Omega^1$ onto the *base space* $X$) and $\rho^*(\mathcal{L})$ the corresponding *pull-back of* $\mathcal{L}$, *via* $\rho$ (cf. Chapt. I; (14.14)). Therefore, one obtains;

$$\frac{1}{2\pi i}[\rho^*(R)] = \rho^*\left(\frac{1}{2\pi i}[R]\right) \in H^2(X, \mathbf{Z}) \tag{5.12}$$

(see also Chapt. VIII; (9.25)). Accordingly, by the same token, as for (5.9), one arrives at the *analogous conclusion for spin-1/2 particles* too, so that, due to the standard *spin-classification* of elementary particles, one infers that;

(5.13) *every (free) elementary particle is prequantizable.* What amounts to the same thing, every such physical system is eo ipso (canonically) associated with a *prequantizing line sheaf.*

Indeed, the framework of the previous considerations is actually the classical differential geometry of $C^\infty$-manifolds. Thus, their formulation here, in terms of the preceding terminology, is simply a further justification of our previous *sheaf-theoretic point of view* of treating the relevant notions. In this context, we also remark that the latter aspect seems to correspond to an

*esoteric need* for more recent accounts *of quantum physics.* (See thus, for instance, Yu. Manin [1], or even S.A. Selesnick [5]). The classical *field of coefficients,* for example, that is, the complexes **C**, is now replaced by suitable **C**-*algebra sheaves,* such as, e.g. (5.2) above. Furthermore, as we shall see in the subsequent Chapter XI, the extra hypothesis of carrying the individual *section algebras* a *compatible topology* might be of interest, as well. On the other hand, this still makes explicit, for that matter, as the case may be, an analogous structure already existing in algebra sheaves of the form (5.2) (viz. *function algebra sheaves*).

Finally, within the same sheaf-theoretic context of a *differential triad,* seems also to fit well another application, that still refers to a mathematical formulation of *quantum field theory,* namely, that one derived from the so-called theory of *"supermanifolds"*. We discuss it, in brief, right away by the next Section.

## 6. Supermanifolds

By the ensuing short discussion, we just want to point out that the framework of the theory, concerning the spaces in the heading is very akin to that one of an abstract *differential triad*

$$(\mathcal{A}, \partial, \Omega^1), \tag{6.1}$$

as this was the case in the preceding (see e.g. Chapt. VI; (1.4), (2.6) and (3.1)). Roughly speaking, every thing here *is* (appropriately) *graded* (in effect, $\mathbf{Z}_2$-*grading* is considered). In this cursory exposition we rather follow closely the recent account on the subject, as given by C. Bartocci, U. Bruzzo, D. Hernández-Ruipérez [1]:

Thus, the *base space* $X$ is again a *topological space,* being in particular, a *graded space* over **R**. (However, one can always take *complexifications,* as we actually assume here; see also, for instance, loc. cit., p.164, Section 2). Precisely, the underlying space $X$, as above, is actually a *smooth* (viz. ($\mathcal{C}^\infty$-)*manifold,* in the usual sense, while (here comes the gradation), the structure sheaf $\mathcal{A}$ *is a sheaf of* (*graded,* as above) **C**-*algebras;* so, as this was

the case hitherto in the preceding, one has at the basis what we have named in the foregoing (see, for instance, Chapt. II; (1.36)), a **C**-*algebraized space*

$$(X, \mathcal{A}). \tag{6.2}$$

However, as already mentioned, the deviation here from our previous context lies in that $\mathcal{A}$ is no more just *"commutative"*, so to speak, as was the case before, but now a *(unital) graded-commutative* **C**-*algebra sheaf* on $X$ (in the sense, of course, that the corresponding local section algebras of $\mathcal{A}$ are of the said type). Specifically here, due to this hypothesis for $\mathcal{A}$, the above pair (6.2) is named a *graded space,* $X$ being, otherwise, a smooth manifold, having thus too *its own structure sheaf* $\mathcal{C}^{\infty}_X \equiv \mathcal{S}$, which we shall also use right away. Indeed, one assumes (loc. cit., p.57) the following *short exact sequence of* (**C**-*algebra) sheaves;*

$$0 \longrightarrow \mathfrak{N} \longrightarrow \mathcal{A} \longrightarrow \mathcal{C}^{\infty}_X \longrightarrow 0, \tag{6.3}$$

where $\mathfrak{N}$ stands for the (2-sided) (graded) *ideal*) *sheaf of nilpotents of* $\mathcal{A}$. Moreover, we suppose that

$$\mathcal{E} \equiv \mathfrak{N}/\mathfrak{N}^2 \tag{6.4}$$

is a *vector sheaf,* with respect to $\mathcal{S}$ (viz. a *locally free* $\mathcal{S}$-*module of finite rank,* say $n$), in such a way that (loc. cit., p.58), for every point $x \in X$, there exists an open neighborhood $U$ of $x$, such that

$$\mathcal{A}(U) = \Gamma(U, \wedge\mathcal{E}), \tag{6.5}$$

within an *isomorphism of graded-commutative* **C**-*algebras;* here $\wedge\mathcal{E}$ stands for the *exterior* (**C**-*)algebra sheaf of* $\mathcal{E}$, the previous vector sheaf, as in (6.4) (see also e.g. Chapt. IV; (7.9) in the preceding). The above constitute the issues in the definition of a *graded manifold* (loc. cit., p.57, Definition 1.2).

In this context, we still note, in passing, that one actually proves that (*Batchelor's theorem*)

$$\mathcal{A} = \wedge(\mathfrak{N}/\mathfrak{N}^2), \tag{6.6}$$

within an *isomorphism of graded-commutative* **C**-*algebra sheaves* (ibid., p.71, Corollary 1.9). The assertion is based on the fact that in the case of a *graded manifold* $X$, as above, *the structure sheaf* $\mathcal{A}$ *is fine* and hence acyclic (see e.g. Chapt. III; (8.26). Of course, we may still assume that the given smooth manifold $X$ is, in particular, (Hausdorff) *paracompact*).

On the other hand, the situation changes entirely when one considers, more generally, *G-supermanifolds* (or even *graded-supermanifolds;* ibid., p.82ff). We shall not enter here into the technical details of the respective definition, our aim being, as already mentioned, just to point out the *relevance of the present discussion with our technique,* as exhibited in the preceding. Thus, roughly speaking, a crucial item here is in that the *constants* **C** (or **R**) are now replaced by the *exterior* (alias *Grassmann*) **C**-*algebra of* $\mathbf{R}^k$, $k \in \mathbf{N}$, *(complexified);*

$$C_k := (\wedge \mathbf{R}^k) \otimes_{\mathbf{R}} \mathbf{C}. \tag{6.7}$$

(As already said, we almost follow the notation and terminology of the previously quoted work). Thus, the corresponding pair $(X, \mathcal{A})$, as in (6.2), is now locally isomorphic to a pair based on $C_k$, having as structure sheaf an appropriate *graded-commutative* $C_k$*-algebra sheaf* on $C_k$ (cf. ibid., p.85, Definition 4.1. See also, for instance, Chapt. II; (1.40) in the foregoing). Yet, we also assume that the base space $X$ is a *(Hausdorff) paracompact* topological space. Moreover, $\mathcal{A}$ is suitably considered as a *subsheaf of the sheaf of* $C_k$*-valued continuous functions* on $X$. But, now $\mathcal{A}$ *is not, in general, fine* (cf. loc. cit., p.147, in particular, the comments at the end of p.148).

Accordingly, we do not have always what we have called in the preceding $\mathcal{A}$*-connections* (cf. Chapt. VI; Definition 3.1), a notion which has an analogous formulation for the case at hand (see the above cited work, p.160, Definition 1.1). Of course, one has first to define the $\mathcal{A}$*-module* $\Omega^1$ of *(graded) differential 1-forms* on $X$. In fact, one sets

$$\Omega^1 := \mathcal{D}er^* \mathcal{A}, \tag{6.8}$$

viz. the *dual sheaf* of the sheaf of *(graded) derivations* of $\mathcal{A}$ (cf. also, for instance, Chapt. IV; (5.1), for the terminology applied). Yet, one considers the *differential operator*

$$d^0 \equiv \partial : \mathcal{A} \longrightarrow \Omega^1, \tag{6.9}$$

in effect, a *graded linear operator,* satisfying the corresponding *Leibniz condition.* Precisely, one sets;

$$\partial(s)(u) := (-1)^{|s|\cdot|u|} u(s), \tag{6.10}$$

for any *homogeneous (local) sections* $s \in \mathcal{A}(U)$ and $u \in \mathcal{D}er(\mathcal{A}(U))$, with $U$ *open* in $X$. (Here $|\cdot|$ means the *degree* (of homogenuity) of the element

(section) at issue, being, according to our assumption, always an *element of* $\mathbf{Z}_2 = \{0, 1\}$; see also loc. cit., p.127).

Furthermore, one defines the *differential operators*

$$d^n : \Omega^n \longrightarrow \Omega^{n+1}, \quad n \in \mathbf{N}, \tag{6.11}$$

by analogy with the classical case (cf., for instance, Chapt. III; (8.32), (8.33)). Of course, one always considers here the appropriate $\mathbf{Z}_2$*-grading,* as before (see the above cited book; p.115). Thus, one finally obtains a *(graded) de Rham complex,* which, in turn, yields a *resolution of the constant (*$\mathbf{C}$*-algebra) sheaf* $C_k$ (cf. (6.7), along with Chapt. III; Definition 1.2); viz. one has the following *exact sequence of* $C_k$*-modules:*

$$0 \longrightarrow C_k \longrightarrow \mathcal{A} \xrightarrow{\partial} \Omega^1 \xrightarrow{d} \Omega^2 \longrightarrow \cdots \tag{6.12}$$

(cf. also the aforementioned work, p.145, Proposition 2.1).

On the other hand, we have already noted before that *the structure sheaf* $\mathcal{A}$, in the case *of a G-supermanifold* $X$, *is not, in general, acyclic.* Nevertheless, the situation can be remedied by taking the base space $X$ to be a so-called *de Witt G-supemanifold* (loc. cit., p.122), a supermanifold very similar to a graded manifold, as considered at the beginning of this discussion. So in that case *the* corresponding *structure sheaf* $\mathcal{A}$ *is acyclic* (ibid., p.151, Proposition 3.2). In sum, one thus infers the following;

(6.13) the methodology of *the previous considerations,* in what concerns the *differential geometric point of view, are very akin to that one of the preceding* chapters, *modulo,* of course, the $\mathbf{Z}_2$-*grading* that is involved here. Indeed, the *graded-commutativity of the structure sheaf* $\mathcal{A}$, as above, must *replace* our assumption in the foregoing that the structure sheaf $\mathcal{A}$ of our algebraized space $(X, \mathcal{A})$, involved therein, had the corresponding (local) section algebras $\mathcal{A}(U)$, with $U$ *open* in $X$, *commutative* (*"commutativity" of* the structure sheaf $\mathcal{A}$), in the usual sense, the same being, otherwise, unital $\mathbf{C}$-algebras, as happens here too. (A replacement, not particularly hard, for that matter; one can consider, for instance, in lieu of $\mathcal{A}$, a *non-necessarily commutative* $\mathcal{A}$).

Now, another feature, that is worth pointing out here, in connection with

our considerations in the next Chapter XI, is that, for crucial reasons in the theory of supermanifolds (see, for example, loc. cit., as above, p.90, Proposition 4.5, together with p.93 and p.132, Axiom 6), one takes the structure sheaf $\mathcal{A}$, as before, to be, in effect, a *topological algebra sheaf,* in the sense of Chapter XI below (see Theorem 11.1 there); indeed, the individual (local) section algebras of $\mathcal{A}$ are, in fact *(Fréchet) locally m-convex (topological) algebras.* (See the above quoted book, p.62, Proposition 1.2, as well as, A. Mallios [4: p. 131; (4.17), along with the relevant comments therein]). So, as we shall see in the sequel, this, among other things, permits one to employ *the exponential function,* together with relevant desired properties of the latter: In particular, questions pertaining to *exponential sheaf sequences and applications* thereon, as was the case in the foregoing (see e.g. Chapters V, VII and VIII), can thus be dealt with in the framework of the previous discussion, as well. (See, for instance, the same citation, as above, on supermanifolds; p.165ff).

## 7. Some more applications

Further recent justifications of the point of view, that is advocated by the present treatise, come from the work of M. Heller and the rest researchers of the Polish group (see, for instance, M. Heller [1] - [5], or W. Sasin [1] - [3]), where there is applied an appropriate version of the notion of *"differential space"* à la Sikorski (see R. Sikorski [1], [2]). The above considerations fall also within the framework of our treatment. Yet, the same can be said about a recent study of S. Kobayashi [2]; in this context, see also our considerations in the next Chapter XI, where one finds, as already mentioned, a further application of the preceding in *topological algebra theory,* and vice-versa(!), as well as, the current work of M. Papatriantafillou [1], [2]. Potential applications of the present framework are also very likely to be found in the theory of PDOs, as this results, for instance, from the relevant work of S.I. Andersson [1] and A. Asada [1], [2]. Finally, *Lipschitz manifolds* and *"generalized functions"* (J.-F. Colombeau et al.) are still eventual areas of applications. (I am indebted to I. Colojoară and E.E. Rosinger, respectively, for drawing my attention, recently, to the latter subjects).

# CHAPTER XI

# Sheaves and presheaves with topological algebraic structures

"...[that] the utmost abstractions are the true weapons with which to control our thoughts of concrete fact."

A.N. Whitehead

"Reality is the motif."

W. Stevens in "Adagia" (included in "Opus Posthumous", ed. by Vintage Books, 1990). p.195

The purpose of the subsequent discussion is to show that, apart from the classical framework of differential ($\mathcal{C}^\infty$-)manifolds (generalized or not), as exposed in the preceding chapter, certain types of *topological algebras* too furnish further non-trivial examples, where one can apply the point of view of the present treatment, getting thus a considerable part of the standard machinery of the classical differential geometry. In this context, we still notice that, here too, the key-idea is to replace the standard sheaf of germs of ($\mathbf{C}$-valued) smooth functions, as appears in the classical theory (see e.g. Chapt. X; (1.3)), through the sheaf of germs of (continuous) sections of a suitable *topological algebra sheaf,* the aforementioned one of the classical theory constituting, in effect, simply a very special (important, nevertheless!) example of the latter type of sheaf; yet, one further seeks for a *de Rham complex* (generalized or not) analogous to the one already considered in the preceding (see, for instance, (1.8) or even (1.10) of the previous Chapter).

# 1. Preliminaries. Topological algebra sheaves

The notions of a *topological vector space presheaf,* respectively *sheaf,* are already standard, while that one of a *topological algebra presheaf,* or *sheaf,* is still standard, althouh less applied so far in the literature. Thus, we first recall the latter notions, as they will be exploited in the succeeding discussion. In this context, concerning the general theory of *topological algebras,* that will be used presently below, we shall steadily refer to A. Mallios [4], which, *for convenience, we quote henceforth* simply *by* [TA].

Now, as was the case hitherto, we always consider linear associative algebras over the complexes **C**, while we briefly refer to such objects as **C**-*algebras;* such an algebra will be further endowed with a *topology compatible with the* given *algebraic structure* in it. So one arrives at the notion of a *topological algebra.* (For the sake of generality, we also suppose that the *multiplication* in such an algebra is *separately continuous,* unless stated otherwise; ibid.). Thus, we start with the following basic notion.

**1.(a). Topological algebra presheaves.**– Now, following the categorical definition of a presheaf of sets (cf. Chapt. I; (5.10)), one has, by definition, that;

(1.1) by a *topological algebra presheaf* on a topological space $X$, one means a contravariant functor from the category of the open subsets of $X$ to that one of topological algebras.

So in ordinary language (cf. Chapt. I; Definition 5.1) this means that;

(1.2) a *topological algebra presheaf*

(1.2.1) $$E \equiv \left(E(U), \rho_V^U\right)_{U \subseteq X,\text{ open}}$$

on a topological space $X$, is a presheaf of sets on $X$, as above, such that;

i) $E(U)$ is a *topological algebra,* for every open $U \subseteq X$, and

ii) each one of the (restriction) maps

$$\rho_V^U : E(U) \longrightarrow E(V) \tag{1.2.2}$$

is a *continuous morphism of the topological algebras* concerned ($U, V$ open sets in $X$, with $V \subseteq U$).

Examples of such mathematical objects will appear in abundance through the succeeding discussion. We just mention here that *all the concrete particular instances of (functional) presheaves of sets,* considered in the preceding, furnish such examples, when suitably topologizing the individual terms (algebras) of the presheaves involved (see, for instance, Chapt. I; (11.11) and Example 12.1).

**1.(b). Topological algebra sheaves.**– We come now to our main objective of this Section, namely, the sheaves in the heading of this Subsection. Thus, the following theorem supplies three equivalent definitions of the notion under consideration.

**Theorem 1.1.** *(Definition). Let $(\mathcal{E}, \pi, X)$ be a sheaf (of sets) over a topological space $X$. Then, the following assertions are equivalent:*

1) $\mathcal{E}$ is a *topological algebra sheaf;* viz. the set of local (continuous) sections of $\mathcal{E}$ over any *open* set $U \subseteq X$,

$$\Gamma(U, \mathcal{E}) \equiv \mathcal{E}(U), \tag{1.3}$$

is a *topological algebra,* while the respective *restriction ($\mathbf{C}$-algebra) morphisms* (cf. Chapt. I; (9.4), along with Chapt. II; (1.67) and (1.71)) are also *continuous.*

2) *$\mathcal{E}$ is (isomorphic to a sheaf) generated by a topological algebra presheaf.* (See (1.2) above, as well as, Chapt. I; (7.23)).

3) *Each local section set $\mathcal{E}(U)$ of $\mathcal{E}$,* associated with an *open* set $U \subseteq X$, is a $\mathbf{C}$-*algebra* such that *the corresponding stalks $\mathcal{E}_x$, $x \in X$, of $\mathcal{E}$ are topological algebras;* in other words, $\mathcal{E}$ is a $\mathbf{C}$-*algebra sheaf on $X$, whose stalks are topological algebras.*

**Proof.** By considering the (complete) *presheaf of* (continuous) *sections of $\mathcal{E}$,*

$$\Gamma(\mathcal{E}) \equiv \left(\Gamma(U, \mathcal{E}), \sigma_V^U\right) \tag{1.4}$$

(see Chapt. I; (9.6), along with Lemma 11.1), one has the conditions of the previous definition (1.2) fulfilled, in the case we assume that 1) is valid. That is, (1.4) *is* then *a topological algebra presheaf,* while $\mathcal{E}$ is, of course, generated by it, in the sense that $\mathcal{E}$ *is* (isomorphic, as a sheaf, to that one generated by $\Gamma(\mathcal{E})$, or else to) *the sheafification of* $\Gamma(\mathcal{E})$. Namely, one has;

$$\mathcal{E} = \mathbf{S}(\Gamma(\mathcal{E})) \tag{1.5}$$

(up to a *sheaf isomorphism.* Cf. Chapt. I; Theorem 9.1). Thus, *1) implies 2).*

Now, suppose that 2) is valid; viz. we assume that *a* given *topological algebra presheaf,* as, for instance, (1.2.1) above, has *its sheafification,* say $\mathcal{E}'$, (sheaf) *isomorphic to* $\mathcal{E}$. Thus, consider next the sets;

$$\mathcal{E}'_x := \varinjlim_{U \in \mathcal{V}(x)} E(U), \quad x \in X, \tag{1.6}$$

stalks of the sheaf $\mathcal{E}'$, as before, where $\mathcal{V}(x)$ stands for an open basis of neighborhoods of $x \in X$ (cf. Chapt. I; (7.2), (7.6)). Therefore, $\mathcal{E}'_x$, $x \in X$, endowed with the corresponding *inductive limit* (vector space) *topology,* is made into a *topological algebra* (see [TA]; p.115, Lemma 2.2]; in this regard, we still remark that the type of the topology on $\mathcal{E}'_x$, as above, depends on that one of the particular (topological) algebras $E(U)$, $U \subseteq X$, open (loc. cit., p.115ff). Furthermore, we observe that $\mathcal{E}'_x$ *is a* $\mathbf{C}$*-algebra* (cf. Chapt. II; (1.55)), *so that,* by hypothesis (see Chapt. I; (1.21)), $\mathcal{E}_x$, *the stalk of the* given *sheaf* $\mathcal{E}$ *at* $x \in X$, *as well.* Thus, the same holds true, for each one of the local section sets of $\mathcal{E}$, $\Gamma(U, \mathcal{E})$, with $U$ open in $X$ (see also, for instance, Chapt. II; Remark 1.1). Accordingly, the given sheaf $\mathcal{E}$ itself *is a* $\mathbf{C}$*-algebra sheaf* on $X$ (cf. also e.g. Chapt. II; (1.76), together with the comments following it). Thus, *2) implies 3),* as well.

Finally, we prove that *3) implies 1):* Indeed, if $U$ is any *open set in* $X$, one obtains the following *bijection;*

$$\begin{aligned}\Gamma(U, \mathcal{E}) \cong \Big\{ (s_x)_{x \in U} \in \prod_{x \in U} \mathcal{E}_x : \forall\, x \in U \ \exists\, V \subseteq U, \text{ open, with} \\ x \in V, \text{ and } s \in \mathcal{E}(V), \text{ with } s(x) = s_x,\ x \in V \Big\} \subseteq \prod_{x \in U} \mathcal{E}_x.\end{aligned} \tag{1.7}$$

Now, the above is conceived on the basis of Chapt. I; Lemma 9.1, see, in particular (9.11) there. So the map

$$\tilde{s} \equiv (s_x)_{x \in U} : U \longrightarrow \mathcal{E} : x \longmapsto \tilde{s}(x) := s_x \in \mathcal{E}_x \subseteq \mathcal{E} \tag{1.8}$$

defines an element of $\mathcal{E}(U) \equiv \Gamma(U, \mathcal{E})$, and conversely (cf. also loc. cit.; (2.13)). Thus, the same rel. (1.7) proves now, in view of our hypothesis, that $\Gamma(U, \mathcal{E})$ *is a topological algebra* (see [TA: p.81; (1.6) and subsequent comments]). Moreover, for any *open sets* $U, V$ in $X$, *with* $V \subseteq U$, *the* corresponding *algebra morphism*

$$\sigma_V^U : \mathcal{E}(U) \longrightarrow \mathcal{E}(V) \tag{1.9}$$

(cf. (1.4)) *is continuous,* as easily verified, by the same definition of the topology of $\mathcal{E}(U)$, as in (1.7). This proves already our assertion, and the proof of the theorem is complete. ∎

We specialize next to a particular topological algebra sheaf, that can be associated with any given topological algebra, being thus of a special importance for our subsequent discussion.

## 2. The Gel'fand sheaf of a topological algebra

Suppose that $E$ is a given *topological algebra,* whose *spectrum* is $\mathfrak{M}(E)$; the latter is, by definition, the set of all *non-zero continuous morphisms of* $E$ *into* $\mathbf{C}$, topologized by the *"weak* (or else *Gel'fand*) *topology"*, defined on it by the following $\mathbf{C}$-valued maps;

$$\hat{x} : \mathfrak{M}(E) \longrightarrow \mathbf{C} : f \longmapsto \hat{x}(f) := f(x), \tag{2.1}$$

*with* $x \in E$ (cf. [TA: Chapt. V]). Thus, associated with $E$ is the so-called *Gel'fand transform algebra of* $E$, viz. the algebra (of the *Gel'fand transforms of* the *elements of* $E$, as in (2.1))

$$E^{\wedge} \equiv \{\hat{x} : x \in E\} \subseteq C_c(\mathfrak{M}(E)). \tag{2.2}$$

We topologize $E^{\wedge}$ with the relative topology from $C_c(\mathfrak{M}(E))$; the latter denotes the *algebra of* $\mathbf{C}$*-valued continuous functions on* $\mathfrak{M}(E)$, endowed with the *compact-open topology* "c" in $\mathfrak{M}(E)$. Therefore, both sets appeared in (2.2) are (commutative) *locally m-convex* (topological) *algebras* (loc. cit.).

As already said, our purpose here is to associate with $E$ a suitable $\mathbf{C}$-*algebra sheaf on* $\mathfrak{M}(E)$ (cf. Chapt. II; Definition 1.4), which at the end will appear to be, in effect, a *topological algebra sheaf* over $\mathfrak{M}(E)$, in the sense of the previous Section (cf. Theorem 1.1). Thus, we start with first defining a *(topological) algebra presheaf over* $\mathfrak{M}(E)$, as follows:

Given an *open* set $U \subseteq \mathfrak{M}(E)$, consider the following $\mathbf{C}$-*algebra;*

$$E\hat{}|_U := \{\hat{x}|_U : x \in E\}, \tag{2.3}$$

that is, the *restriction on* $U$ *of the Gel'fand transform algebra of* $E$, as the latter is given by (2.1), (2.2). However, due to the definition of the topology of $\mathfrak{M}(E)$, each one of the maps $\hat{x}$, $x \in E$, is continuous (see e.g. (2.2)), so that the same holds true, of course, for their restrictions on the open set $U \subseteq \mathfrak{M}(E)$. Therefore, one actually gets;

$$E\hat{}|_U \subseteq \mathcal{C}(U), \tag{2.4}$$

*for every open set* $U \subseteq \mathfrak{M}(E)$, where $\mathcal{C}(U)$ stands, as usual hitherto, for the $\mathbf{C}$-*algebra of* $\mathbf{C}$-*valued continuous functions on* $U$. The latter algebra is endowed with the *compact-open topology in* $U$, we denote it by $\mathcal{C}_c(U)$, becoming thus a (commutative) *locally m-convex algebra;* therefore (see also [TA: Chapts. I, III]);

(2.5) *for every open* $U \subseteq \mathfrak{M}(E)$,

(2.5.1) $$E\hat{}|_U$$

(see (2.3)) *is a* (commutative) *locally m-convex algebra.*

Furthermore, *for any open sets* $U, V$ *in* $\mathfrak{M}(E)$, *with* $V \subseteq U$, *the* corresponding *restriction morphism*

$$\rho_V^U : E\hat{}|_U \longrightarrow E\hat{}|_V \tag{2.6}$$

*is continuous,* as easily follows from (2.4). Consequently, based now on (1.2) one infers that;

(2.7) the family

(2.7.1) $$\mathbf{E} \equiv \left(E\hat{}|_U, \rho_V^U\right)$$

provides a *presheaf of topological algebras on* $\mathfrak{M}(E)$.

Indeed, what we actually need for the next notion is the fact that (2.7.1) yields just a presheaf of **C**-algebras on $\mathfrak{M}(E)$. Thus, we come now to the following.

**Definition 2.1.** Let $E$ be a given topological algebra, whose spectrum is $\mathfrak{M}(E)$. Then, the *Gel'fand sheaf of* $E$ is the (**C**-algebra) sheaf on $\mathfrak{M}(E)$, which is generated by the presheaf (2.7.1). We denote it by

$$(\mathcal{E}, \pi, \mathfrak{M}(E)), \tag{2.8}$$

or, for short, just by $\mathcal{E}$ (see Chapt. I; (7.7), as well as, (7.23)).

Now, by applying our previous terminology in Chapt. I; Section 3, we conclude that;

(2.9)

a *defining family of (continuous) sections for the Gel'fand sheaf* $\mathcal{E}$ *of* $E$, as above, is given by the maps

$$\tilde{x} : U \longrightarrow \mathcal{E}, \tag{2.9.1}$$

*for any* $x \in E$ *and open* $U \subseteq \mathfrak{M}(E)$, such that

$$\tilde{x}(f) := [\hat{x}]_f \in \mathcal{E}_f, \tag{2.9.2}$$

where we set

$$\mathcal{E}_f := \varinjlim_{U \in \mathcal{V}(f)} (E^{\hat{}}|_U), \tag{2.9.3}$$

*stalk of* $\mathcal{E}$ *at* $f \in \mathfrak{M}(E)$, with $\mathcal{V}(f)$ a basis of open neighborhoods of $f$ in $\mathfrak{M}(E)$. (See (2.7), together with Definition 2.1 above and Chapt. I; (7.2) and (9.16)).

**Note.**– Since $\hat{x}|_{\mathfrak{M}(E)} = \hat{x}$, for every $x \in E$ (see (2.1)), the *germ of* $\hat{x}|_U \in E^{\hat{}}|_U$ *in* $\mathcal{E}_f$, with $U$ open in $\mathfrak{M}(E)$, is given by

$$[\hat{x}|_U]_f = [\hat{x}]_f, \tag{2.10}$$

which thus justifies the previous notation in (2.9.2).

In other words, our claim in (2.9) means that *the family of the (local) maps*

$$\{\tilde{x}\}_{x \in E}, \tag{2.11}$$

as given by (2.9.1) and (2.9.2), *satisfies the* respective *conditions of the analogous family in* Chapt. I; (3.2), concerning the set (sheaf) here $\mathcal{E}$ and its fibers $\mathcal{E}_f$, $f \in \mathfrak{M}(E)$, as above, so that one has;

(2.12) $$\mathcal{E} = \sum_{f \in \mathfrak{M}(E)} \mathcal{E}_f .$$

The assertion follows straightforwardly from the very definitions (see e.g. (2.9.3) above and loc. cit. (3.1), (3.4)). Thus, our claim about the *continuity of the* previous *maps* (2.9.1) follows now from Chapt. I; Theorem 3.1 (cf. also Lemma 3.1 therein).

In this context, one further infers that;

(2.13) $$E\hat{}|_U \underset{\rho_U}{\hookrightarrow} \Gamma(U, \mathcal{E}) \equiv \mathcal{E}(U),$$

for any *open* $U \subseteq \mathfrak{M}(E)$, where *one sets* (cf. (2.9.1), (2.9.2))

(2.14) $$\rho_U(\hat{x}|_U) := \tilde{x},$$

*for every* $x \in E$. Precisely, one has that

(2.15) *the map* $\rho_U$, as given by (2.14), being also, in fact, a $\mathbf{C}$-*algebra morphism, is one-to-one.*

Indeed, suppose that

(2.16) $$\rho_U(\hat{x}|_U) \equiv \tilde{x} = 0,$$

so that, by virtue of (2.9.2) and (2.10), one obtains

(2.17) $$\tilde{x}(f) = [\hat{x}]_f = [\hat{x}|_U]_f = 0,$$

*for every* $f \in U \subseteq \mathfrak{M}(E)$. Therefore, *one has* (cf. (2.9.3))

(2.18) $$\hat{x}|_V = 0,$$

*for some open* $V \subseteq U$, *with* $f \in V$; that is, one has $(\hat{x}|_V)(f) = \hat{x}(f) = 0$, concerning the particular, otherwise arbitrary (cf. (2.17)), element $f \in U$. Hence, one obtains $\hat{x}|_U = 0$, which thus proves our assertion in (2.15). ∎

Thus, as a consequence of (2.15) (in fact, is an equivalent assertion), one concludes that (see also Chapt. I; (11.14), p. 50);

(2.19) (2.7.1), the *Gel'fand presheaf* of $E$, *is a monopresheaf.*

(See also Chapt. I; Remark 11.1, i), together with (11.36′)). In particular, setting in (2.13) $U = \mathfrak{M}(E)$, *one obtains;*

$$E^{\wedge} \underset{\rho}{\hookrightarrow} \Gamma(\mathfrak{M}(E), \mathcal{E}) \equiv \mathcal{E}(\mathfrak{M}(E)). \tag{2.19}$$

In other words, one thus concludes hitherto that;

(2.20) *the Gel'fand transform algebra of any given topological algebra is represented* (within an *isomorphism of* **C**-*algebras*) *as a subalgebra of the global section algebra* (viz. of the **C**-algebra of the global (continuous) sections) *of the respective Gel'fand sheaf of the given algebra.*

Specifically, suppose that the corresponding *Gel'fand map* of the given topological algebra $E$,

$$\mathcal{G} : E \longrightarrow E^{\wedge} : x \longmapsto \mathcal{G}(x) := \hat{x} \tag{2.21}$$

(cf. (2.1)) is *one-to-one;* viz. we assume that we are given a so-called *semisimple* (topological) algebra (see [TA: p.266, Definition 3.2]). Thus, one has, by definition,

$$E \underset{\mathcal{G}}{=} E^{\wedge}, \tag{2.22}$$

within an *isomorphism of* **C**-*algebras,* established by $\mathcal{G}$. So (2.19) specializes now to the following relation;

$$E \underset{\mathcal{G}}{=} E^{\wedge} \underset{\rho}{\hookrightarrow} \Gamma(\mathfrak{M}(E), \mathcal{E}) \equiv \mathcal{E}(\mathfrak{M}(E)). \tag{2.23}$$

In other words, one thus infers by the last relation that;

(2.24) *any given semisimple topological algebra admits,* via its Gel'fand sheaf, *a faithful sectional representation.*

Of course, our final objective here is to find, as already said, suitable conditions for the topological algebras involved, so that the above *faithful representation,* as in (2.23), be also *full;* thus, $E^{\wedge}$ will then be identified with the global section algebra of $\mathcal{E}$, $\mathcal{E}(\mathfrak{M}(E))$. However, to this end, we first look at *another* useful *version of* $\mathcal{E}$, viz. of the Gel'fand sheaf of the given topological algebra, that we presently explain. So one considers

(2.25) *the Gel'fand sheaf* $\mathcal{E}$ of a given topological algebra $E$ *as the sheaf of germs of* $\mathbf{C}$*-valued maps, locally belonging to* $E^{\wedge}$ (viz. to the *Gel'fand transform algebra* of $E$, see (2.2)).

Indeed, as we shall see right away (cf. (2.44) below), the above asserted identification is accomplished *modulo a sheaf isomorphism* between $\mathcal{E}$ and the sheaf descibed in (2.25); precisely, one means by the latter terminology the following: Take an *open* set $U \subseteq \mathfrak{M}(E)$, along with the set;

$$\tilde{E}_U = \{\alpha : U \longrightarrow \mathbf{C} : \forall f \in U \ \exists \text{ open } V \subseteq \mathfrak{M}(E), \text{ with } f \in V \subseteq U, \text{ together with } x \in E, \text{ such that } \alpha = \hat{x}|_V\}. \tag{2.26}$$

So $\tilde{E}_U$ is, by definition, the *set of* $\mathbf{C}$*-valued maps on* $U$*, which locally belong to* $E^{\wedge}$. It is easy to see that the latter set is, in effect, *a* $\mathbf{C}$*-algebra, consisting of* ($\mathbf{C}$*-valued*) *continuous functions on* $U$. That is, one has;

$$\tilde{E}_U \subseteq \mathcal{C}(U) \tag{2.27}$$

(see also (2.4)). Thus, one actually obtains a *presheaf of* $\mathbf{C}$*-algebras on* $\mathfrak{M}(E)$,

$$\left(\tilde{E}_U, \tau_V^U\right), \tag{2.28}$$

by letting $U$ varying over the open subsets of $\mathfrak{M}(E)$ and taking the corresponding restriction ($\mathbf{C}$-algebra) morphisms. So we denote by

$$(\tilde{\mathcal{E}}, \pi', \mathfrak{M}(E)) \tag{2.29}$$

the $\mathbf{C}$*-algebra sheaf* on $\mathfrak{M}(E)$, generated by the ($\mathbf{C}$-algebra) presheaf (2.28) (see also Chapt. II; (1.54)). Now, our aim is to show that one has;

$$\mathcal{E} = \tilde{\mathcal{E}}, \tag{2.30}$$

within a *sheaf isomorphism*.

Indeed, what we actually prove is that *the* above *presheaf (2.28) is isomorphic to* $\Gamma(\mathcal{E})$, the (complete) presheaf of (continuous) sections of $\mathcal{E}$ (see (1.4)). So one has the following. (But see also (4.4) in the sequel).

**Lemma 2.1.** *Let* $E$ *be a given topological algebra with spectrum* $\mathfrak{M}(E)$ *and* $\mathcal{E}$ *the corresponding Gel'fand sheaf of* $E$ *over* $\mathfrak{M}(E)$. *Then, for every open set* $U \subseteq \mathfrak{M}(E)$*, one obtains;*

$$\widetilde{E}_U = \Gamma(U, \mathcal{E}) \equiv \mathcal{E}(U), \tag{2.31}$$

*within a* **C**-*algebra isomorphism.*

**Proof.** For every open $U \subseteq \mathfrak{M}(E)$ and *for every map* $\alpha : U \longrightarrow \mathbf{C}$, *locally in* $E\hat{}$, so that (cf. (2.26)) $\alpha \in \widetilde{E}_U$, one obtains a *continuous section of* $\mathcal{E}$ *over* $U$, that is, *an element* $\tilde{\alpha} \in \mathcal{E}(U)$, as follows;

$$\tilde{\alpha} : U \longrightarrow \mathcal{E} : f \longmapsto \tilde{\alpha}(f) := [\hat{x}]_f \in \mathcal{E}_f \subseteq \mathcal{E} \tag{2.32}$$

(see also e.g. (2.12)), where the element $x \in E$ is determined, according to our hypothesis for $\alpha$ and the same condition (2.26). Moreover, the previous section $\tilde{\alpha}$ *is,* as asserted, *continuous,* since, due again to (2.26), one has (see also (2.9.2)),

$$\tilde{\alpha} = \tilde{x}|_V, \tag{2.33}$$

for some open $V \subseteq \mathfrak{M}(E)$, with $f \in V \subseteq U$; but, we have seen already that $\tilde{x}|_V \in \mathcal{E}(V)$ (see (2.9)). (At this point, we should also remark that the map $\tilde{\alpha}$ in (2.32) is well-defined, concerning the element $x \in E$, due to the same definition of $x \in E$, as in (2.26), together with (2.9.2) and (2.9.3). An analogous argument holds certainly true for the last relation (2.33)).

Accordingly, *for every open* $U \subseteq \mathfrak{M}(E)$, one obtains a map;

$$h_U : \widetilde{E}_U \longrightarrow \Gamma(U, \mathcal{E}) \equiv \mathcal{E}(U) : \alpha \longmapsto h_U(\alpha) := \tilde{\alpha} \tag{2.34}$$

(see (2.32)). Now, it is easy to see, according to the very definitions, that the previous map (2.34) *yields,* in fact, *a* **C**-*algebra morphism,* which also is *one-to-one:* Indeed, if $\tilde{\alpha} = 0$, for some $\alpha \in \widetilde{E}_U$, and $f \in U$, then (see (2.32)), $\tilde{\alpha}(f) = [\hat{x}]_f = 0$; therefore (cf. (2.33), along with (2.9.1), (2.9.2) and (2.26)), one concludes that $\alpha(f) = 0$, which thus proves our claim. Finally, we prove that *(2.34) is an onto map;* thus, if $s \in \mathcal{E}(U)$, consider the map

$$\bar{s} : U \longrightarrow \mathbf{C} : f \longmapsto \bar{s}(f) := \hat{x}(f) \equiv [\hat{x}]_f(f), \tag{2.35}$$

such that

$$s(f) = [\hat{x}]_f = [\hat{x}|_V]_f = \rho^f_V(\hat{x}|_V) = \tilde{x}(f), \tag{2.36}$$

(see (2.9.2), (2.9.3), (2.10) and (2.14), as well as, Chapt. I; (7.9), (7.10)). So one obtains (see also (2.14) and Chapt. I; (2.13))

$$s = \tilde{x}|_W, \tag{2.37}$$

with $W$ an *open neighborhood of* $f$ in $V$, hence, *in* $\mathfrak{M}(E)$ too. Therefore, in view of (2.35), (2.36) and (2.37), one obtains;

$$\overline{s} = \hat{x}|_W, \tag{2.38}$$

that is (cf. (2.26)), one finally has that

$$\overline{s} \in \widetilde{E}_U. \tag{2.39}$$

Thus, based further on (2.32), (2.33) and (2.37), one obtains (see also (2.34));

$$h_U(\overline{s}) = \tilde{\overline{s}} = s, \tag{2.40}$$

which thus proves our assertion, and this completes the proof of the lemma. ∎

Now, as a result of (2.31) (see also Chapt. I; (6.10), Lemma 11.1 and (11.44)), one further obtains that:

(2.41) (2.28) yields a *complete presheaf of* **C**-*algebras* on $\mathfrak{M}(E)$, such that one actually has (loc. cit., Proposition 11.1)

(2.41.1) $$\widetilde{E}_U = \Gamma(U, \widetilde{\mathcal{E}}) = \Gamma(U, \mathcal{E}) \equiv \mathcal{E}(U),$$

within an *isomorphism of* **C**-*algebras, for every open* set $U \subseteq \mathfrak{M}(E)$ (cf. also (2.29)).

In particular, *for* $U = \mathfrak{M}(E)$, one gets;

$$\widetilde{E}_{\mathfrak{M}(E)} = \Gamma(\mathfrak{M}(E), \widetilde{\mathcal{E}}) = \Gamma(\mathfrak{M}(E), \mathcal{E}) \equiv \mathcal{E}(\mathfrak{M}(E)), \tag{2.42}$$

up to an *isomorphism of* the **C**-*algebras* concerned. The first term in (2.42) is, by definition (cf. (2.26)), the *algebra of* (*continuous*, cf. (2.27)) **C**-*valued maps on* $\mathfrak{M}(E)$, *which locally belong to* $E^{\wedge}$.

On the other hand, the same relation (2.41.1) yields now, as already promised (cf. (2.30)), that;

$$\mathcal{E} = \widetilde{\mathcal{E}}, \tag{2.43}$$

within a *sheaf isomorphism* (see also Chapt. I; (13.7)). Furthermore, due to the fact that (2.7.1) is a presheaf of topological algebras, while the same holds true for (2.28) (see, for instance, (2.27)), one concludes, in view of Theorem 1.1, that:

(2.44) *the Gel'fand sheaf* $\mathcal{E}$ of a given topological algebra $E$, *as well as, the sheaf* $\widetilde{\mathcal{E}}$, *are,* in effect, *topological algebra sheaves on* $\mathfrak{M}(E)$, the spectrum of the initial algebra $E$.

For the moment, we shall not use the last conclusion, until Section 4 below. Instead, as already said, our purpose now is to provide suitable conditions for *the topological algebra E involved,* as above, so that the latter can be expressed *as the algebra of global sections of* $\mathcal{E}$ (viz. of the *Gel'fand sheaf* of $E$), hence, of $\widetilde{\mathcal{E}}$ too (cf. (2.43)), a fact, that will be of use presently below. So we undertake this task in the next Section.

# 3. Geometric topological algebras

The algebras in the title of this Section are those which will mostly be of concern to us throughout the sequel, while they also comprise the standard ones of the classical theory (viz. smooth and holomorphic functions). Moreover, they provide the type of topological algebras, alluded to at the end of the previous Section; it is, in effect, this latter property, which motivated their name, according to the classical terminology (see e.g. P. de Bartolomeis [1]). So we start with the following basic.

**Definition 3.1.** By a *geometric topological algebra,* one means a topological algebra $E$, which is ($\mathbf{C}$-algebra isomorphic with) the algebra of global continuous sections of a $\mathbf{C}$-algebra sheaf over (a topological space $X$, homeomorphic with) the spectrum $\mathfrak{M}(E)$ of the initial algebra.

Thus, our purpose by the ensuing discussion is to exhibit *sufficient conditions,* which enable a given topological algebra to be a geometric one, in the previous sense. In this context, the key-result is the so-called *Local Theorem,* a standard conclusion already from the classical Banach algebra theory (see, for instance, I.M. Gelfand - D. Raikov - G. Shilov [1: p.201, Theorem 1]). Indeed, one has the following basic result.

**Lemma 3.1.** *(Local Theorem). Let E be a strongly hereditarily complete regular topological* $\mathbf{C}$*-algebra, having an identity element and an equiconti-*

*nuous spectrum* $\mathfrak{M}(E)$. *Then, every* **C**-*valued map on* $\mathfrak{M}(E)$, *which locally belongs to* $E^{\wedge}$ (viz. to the Gel'fand transform algebra of $E$), *belongs globally to it, as well.*

For convenience, we first have to comment a bit on the terminology applied in the statement of the previous lemma:

Thus, a given topological algebra $E$ is said to be *strongly hereditarily complete,* whenever, for every closed (2-sided) ideal $I$ of $E$, the respective quotient algebra

$$E/I, \tag{3.1}$$

endowed with the quotient topology, is a *complete* topological algebra, while the same holds true, for every such quotient topological algebra, as in (3.1).

It is clear that every *Banach algebra* and, more generally, every *Fréchet,* or even, every *Pták (topological) algebra* is an algebra of the previous type. (See [TA] for the terminology applied herewith). Furthermore, it is an immediate consequence of the above definition, that *a topological algebra of the above type is itself complete.*

Now, the *proof of* the above *Lemma 3.1* can actually be patterned after that one of its classical (Banach algebra) analogue (loc. cit.). Indeed, a version of it for (Pták) *locally m-convex* (topological) *algebras* has been given already in [TA: p. 348, Theorem 5.1], while its present form appears in A. Mallios [7: p. 307, Lemma 2.1]. In this context, a crucial point is the fact that;

(3.2) *a topological algebra E* having the properties of that one, *as in the statement* of Lemma 3.1 *is,* in fact, *normal.* (Thus, by definition, $E^{\wedge}$, the Gel'fand transform algebra of $E$ (cf. (2.2)), *"separates closed sets" in* $\mathfrak{M}(E)$, the spectrum of $E$; so, not only "closed sets and points" (*regular*). See also [TA] for the relevant terminology).

As already mentioned, the above property that acquires $E$ is a decisive ingredient of the proof in the classical case (loc. cit., p.200, Corollary, and p.201, proof of the Lemma). For an idea of the proof of (3.2), cf. also [TA: p.345, Corollary 4.5], given there still for *Pták* (locally $m$-convex) *algebras,* while the same result is an extension, within the present context (viz. *no convexity* assumptions for the topological algebras concerned), of an earlier conclusion

of M. Rosenfeld [1] for *Fréchet-Šilov* (locally $m$-convex) *algebras.* So, what one actually needs here, is the *type of completeness of E, as above* (cf. (3.1) and subsequent comments there). ∎

Therefore, as a consequence of the preceding, one now infers that:

(3.3) given a *topological algebra E, satisfying the conditions of Lemma 3.1,* one then obtains (see also (2.42));

(3.3.1) $$E^{\wedge} = \widetilde{E}_{\mathfrak{M}(E)},$$

where $\mathfrak{M}(E)$ is the *spectrum of E.*

Thus, by further specializing to *semisimple* (topological) *algebras* (see (2.22)), one gets the awaited so far result, according to the following.

**Theorem 3.1.** *Let E be a unital topological* **C**-*algebra, which is strongly hereditarily complete regular and semisimple, having also an equicontinuous spectrum* $\mathfrak{M}(E)$. *Moreover, let* $\mathcal{E}$ *be the Gel'fand sheaf of E over* $\mathfrak{M}(E)$ (cf. (2.8)). *Then, one has*

(3.4) $$E = \Gamma(\mathfrak{M}(E), \mathcal{E}) \equiv \mathcal{E}(\mathfrak{M}(E)),$$

*within a* **C**-*algebra isomorphism.*

**Proof.** Obvious, according to our hypothesis for $E$, in conjunction with (2.22), (3.3.1) and (2.42). ∎

To fix the terminology employed so far, we further set the following definition. (The relevant motivation will appear later on; cf. thus Theorem 4.1, or even (4.18) below). So we have.

**Definition 3.2.** By a *soft algebra,* we mean a unital topological **C**-algebra $E$ with an equicontinuous spectrum $\mathfrak{M}(E)$, such that $E$ is also strongly hereditarily complete regular and semisimple. (In other words, any algebra satisfying the conditions of Theorem 3.1).

Thus, in view of the previous Definitions 3.1, 3.2 and of Thorem 3.1, one concludes that;

(3.5) *every soft algebra is,* relative to its Gel'fand sheaf, *a geometric topological algebra.*

**Note 3.1.**- If $E$ is a *soft* (topological) *algebra*, then (Definition 3.2), its *spectrum* $\mathfrak{M}(E)$ is *equicontinuous*, so that *the* corresponding *Gel'fand map of* $E$ (see (2.21)) *is continuous* (cf. [TA: p. 184, Corollary 1.4]). Thus, one further obtains

$$\mathfrak{M}(E) = \mathfrak{M}(E^\wedge), \tag{3.6}$$

within a *homeomorphism* of the topological spaces concerned (see R.I. Hadjigeorgiou [1: Chapt. II; Theorem 2.1], or even [2: Theorem 3.1]). Of course, here $E^\wedge$ is a *normed algebra*, when endowed with the relative topology from $\mathcal{C}_u(\mathfrak{M}(E))$ ($\mathfrak{M}(E)$ being also *compact* (Hausdorff)); yet, $\mathfrak{M}(E^\wedge)$ *is equicontinuous*, $E^\wedge$ being unital too, by hypothesis for $E$ (see e.g. [TA: p. 146, Lemma 2.2]). Finally, $E^\wedge$ *is* (always) *semisimple*, as well as, *regular*, according to the hypothesis for $E$ and to (3.6). Nevertheless, *no information* is available, in general, concerning the *completeness* of $E^\wedge$. However, in view of (3.6), (3.3.1), (2.42) and (2.43), *one has;*

$$E^\wedge = \widetilde{E}_{\mathfrak{M}(E)} = \Gamma(\mathfrak{M}(E^\wedge), \mathcal{E}). \tag{3.7}$$

Therefore,

(3.8) *for any given soft algebra* $E$, hence (cf. (3.5)), a geometric one, its Gel'fand transform algebra $E^\wedge$ *is* also *a geometric topological* (in fact, *normed*) *algebra.*

In this context, we still remark that, due to the semisimplicity of $E^\wedge$, $\mathcal{E}$ (viz. the Gel'fand sheaf of $E$) may also be viewed as the Gel'fand sheaf of $E^\wedge$, by an obvious adaptation to the present case of the respective terminology for the presheaf (2.7.1) above. Anyway, our claim in (3.8) is justified, according to Definition 3.1 and to the $\mathbf{C}$-*algebra isomorphism*

$$E^\wedge = \Gamma(\mathfrak{M}(E), \mathcal{E}), \tag{3.9}$$

in view of (3.6).

On the other hand, by specializing to *locally m-convex algebras,* one proves that:

(3.10) every *unital Fréchet regular and semisimple* (for short, *Fréchet-Šilov*) *locally m-convex algebra is a geometric topological algebra,* relative to its Gel'fand sheaf.

The preceding is based on an earlier result of R.M. Brooks [1], pertaining to a corresponding version of *the Local Theorem* (see Lemma 3.1 above), *in terms of the topological algebras, as in (3.10)* (see loc. cit.; p.271, Theorem 2.6). In this context, we also remark that the *spectrum* $\mathfrak{M}(E)$ of the topological algebra involved here *is no longer assumed* to be *equicontinuous, hence* (Alaoglou - Bourbaki), *compact*. However, *it is* actually a *paracompact* (Hausdorff) space (see the above quoted work, p.266), a fact that will be of use in the sequel. ∎

Notice that some of the *important examples of geometric topological algebras* in the classical theory *are of the last type, as* namely *in (3.10);* for instance, the (topological) algebras $\mathcal{C}^\infty(X)$ (with $X$ *not compact*) and $\mathcal{O}(X)$ (cf. e.g. Chapt. II; (10.13) and (10.14)). Concerning this kind of examples, we refer to A. Mallios [7: p. 315, Section 5]; we can find there too another example of a *Pták-Šilov locally m-convex Q-algebra,* hence, of a *geometric* one (cf. (3.5), along with [TA: p. 75, Proposition 7.1]), through the holomorphic functions on a *compact Stein set* of a *Stein manifold* $X$. (See thus A. Mallios [7: p. 318, Theorem 5.1], as well as, (11.40) in the sequel).

Now, looking at the above relation (3.4), as the defining condition that $E$ be a *geometric* topological *algebra,* we can also express, roughly speaking, our previous information in (3.5) or (3.10), by saying that;

(3.11) *the Gel'fand sheaf of a geometric topological algebra* (of either form (3.5) or (3.10)) *can be reconstructed* (equivalently, *be determined*) *through the* **C**-*algebra of its global (continuous) sections* (the latter algebra being actually isomorphic to the given geometric topological algebra).

Yet, for convenience of reference, we also gather here our previous information (see comments following (3.10)) about the *spectra of geometric* topological *algebras,* as above. So, as already said, we have that:

(3.12) *the spectrum of every soft algebra* (cf. Definition 3.2) *is a compact* (Hausdorff) space, while *that one of a Fréchet-Šilov locally m-convex algebra* (cf. (3.10)) a *paracompact* (Hausdorff) space.

On the other hand, before we end this discussion, we still point out a further property of soft algebras, or that ones in (3.10), in connection with being the algebras in question geometric, in the sense of (3.4). Yet, the terminology

involved will be of use right in the next Section 4. First, for convenience, we recall a relevant notion, which we shall need presently below.

Thus, suppose we are given a *topological algebra* $E$, whose *spectrum* is $\mathfrak{M}(E)$. Then, we say that an *ideal* $I$ in $E$ is a *kernel* of $E$, in the case one has;

$$I = k(h(I)) \equiv \bigcap_{f \in \mathfrak{M}(E),\, \ker f \supseteq I} \ker f. \tag{3.13}$$

It is clear that $I$ is then a *closed 2-sided ideal of* $E$ (being thus, by definition the *kernel of its hull;* see [TA: Chapter IX]). So we set now the following.

**Definition 3.3.** A given topological algebra $E$ is called a *hereditarily geometric algebra,* whenever the given algebra $E$ itself, as well as, every quotient algebra of it, modulo a kernel ideal of $E$ is (in the quotient topology) a geometric topological algebra, with respect to the corresponding Gel'fand sheaf of the topological algebra, under consideration.

On the other hand, the significance of the preceding notion, as formulated by the previous definition, will become apparent after a while through the next Section 4; however, in the mean-time we do remark the following, that will also be of use below.

(3.14) *every soft algebra* (cf. Definition 3.2), as well as, *every Fréchet-Šilov locally m-convex algebra* (see (3.10)) *is a hereditarily geometric algebra.*

What one actually proves here is the fact that;

(3.15) *both types of topological algebras in (3.14) yield, modulo a kernel ideal, quotient algebras of the same type, as the initial ones; viz. the said algebras are "hereditarily" such.*

But, the latter are geometric algebras, according to (3.5) and (3.10). On the other hand, our claim in (3.15) is provided by standard results of the general topological algebra theory. (See [TA: p.339, Theorem 4.1, Corollary 4.1 and p.340, Corollaries 4.2 and 4.3]). ∎

We prove now, right in the next Section, the important property that the above algebras share, in having the Gel'fand sheaves *soft* and, in effect, *fine.*

# 4. Softness of the Gel'fand sheaf

Our objective by the following discussion has already been mentioned in the preceding, being, for that matter, clearly indicated by the heading of this Section.

Thus, to start with, suppose we are given a *topological algebra* $E$, whose *spectrum* is $\mathfrak{M}(E)$. Now, we say that a set $B \subseteq \mathfrak{M}(E)$ is a *hull* (of $\mathfrak{M}(E)$), in the case one has;

$$B = h(k(B)) \equiv \left\{ f \in \mathfrak{M}(E) : \ker f \supseteq k(B) \equiv \bigcup_{g \in B} \ker g \right\} \tag{4.1}$$

(see also [TA: Chapter IX]). Furthermore, by extending the preceding notation in (2.26), one sets, *for any subset,* say $B$, *of* $\mathfrak{M}(E)$;

$$\begin{aligned} \tilde{E}_B := \Big\{ \alpha : B \longrightarrow \mathbf{C} : \forall f \in B \ \exists \ \textit{open } V \subseteq \mathfrak{M}(E), \ \textit{with } f \in V, \\ \textit{and } x \in E, \ \textit{such that } \alpha = \hat{x}|_{V \cap B} \Big\}. \end{aligned} \tag{4.2}$$

That is, again one considers *the set* (in fact, $\mathbf{C}$-*algebra*) *of* $\mathbf{C}$-*valued maps on* $B$, *that locally belong to* $E^{\wedge}$; as the analogous relation to (2.27), it is also clear that;

$$\tilde{E}_B \subseteq \mathcal{C}(B). \tag{4.3}$$

On the other hand, if $\mathcal{E}$ is the *Gel'fand sheaf of* $E$ over $\mathfrak{M}(E)$ (see Definition 2.1), then by applying a similar argument to that one for (2.31), *one* still *obtains that;*

$$\tilde{E}_B = \Gamma(B, \mathcal{E}) \equiv \mathcal{E}(B), \tag{4.4}$$

*for every* $B \subseteq \mathfrak{M}(E)$, within an *isomorphism of* the $\mathbf{C}$-*algebras* involved.

Thus, one gets through (4.4) a genuine *extension of* our previous *Lemma 2.1 to an arbitrary* $B \subseteq \mathfrak{M}(E)$, which we shall need presently below. Now, by further considering some $B$, as before, let

$$I = k(B) := \bigcap_{f \in B} \ker f \tag{4.5}$$

be its *kernel,* thus a *closed 2-sided ideal* in $E$ (loc. cit.). So we can further take the corresponding *quotient topological algebra*

$$F = E/I, \tag{4.6}$$

whose *spectrum* is given by (see also (4.1), (4.5))

$$\mathfrak{M}(F) \equiv \mathfrak{M}(E/I) \underset{\rho}{=} h(I) \equiv h(k(B)) \subseteq \mathfrak{M}(E); \tag{4.7}$$

here $\rho$ stands for a *homeomorphism* (cf. [TA: p.339, Theorem 4.1]). Precisely, if $\phi : E \longrightarrow E/I \equiv F$ denotes the (canonical) quotient map, the map $\rho$ is given by

$$\rho \equiv {}^t\phi : \mathfrak{M}(E/I) \longrightarrow \mathfrak{M}(E), \tag{4.8}$$

its image in $\mathfrak{M}(E)$ being $h(I)$, viz. the *hull of* $I$.

In particular, if the set $B \subseteq \mathfrak{M}(E)$ involved hitherto is a *hull,* as in (4.1), then, in view of (4.7), one obtains

$$\mathfrak{M}(E/I) \equiv \mathfrak{M}(F) = B, \tag{4.9}$$

up to a *homeomorphism,* established by the map $\rho$, as above.

Specifically, by considering the (quotient) *topological algebra* $F$, as in (4.6), and by still assuming $B$ *a hull,* one concludes that the $\mathbf{C}$*-algebra* $\widetilde{F}_{\mathfrak{M}(F)}$ (cf., for instance, (2.26) or even (2.42)) consists of those maps (cf. (4.9))

(4.10) $\alpha : \mathfrak{M}(F) = B \longrightarrow \mathbf{C}$, *such that, for every* $g \in \mathfrak{M}(F)$, *there exists an open neighborhood* $V$ *of* $\rho(g) \in B \subseteq \mathfrak{M}(E)$, *along with an element* $y \in F$, *such that one has;*

$$\alpha = \hat{y}|_{\rho^{-1}(V)}. \tag{4.10.1}$$

Thus, as an application of the above, one is led to the following conclusion, having a particular interest **per se** (cf. Scholium 4.1 below), being also the crucial ingredient in the proof of our main result in the sequel. So we have.

**Lemma 4.1.** *Let* $E$ *be a topological algebra with spectrum* $\mathfrak{M}(E)$ *and* $B \subseteq \mathfrak{M}(E)$ *a hull, with kernel* $I = k(B)$ (cf. (4.1), (4.5)), *while* $F \equiv E/I$ *is the corresponding quotient topological algebra. Finally, let* $\mathcal{E}$ *and* $\mathcal{F}$ *be the Gel'fand sheaves of the topological algebras* $E$ *and* $F$, *respectively. Then, one has;*

(4.11) $$\Gamma(B, \mathcal{E}) = \Gamma(B, \mathcal{F}),$$

*within a* **C**-*algebra isomorphism.*

**Proof.** Based on (4.4) and (2.42), it is enough to prove the following *isomorphism of* **C**-*algebras;*

(4.12) $$\tilde{E}_B = \tilde{F}_B = \tilde{F}_{\mathfrak{M}(F)} = \Gamma(B, \mathcal{F}).$$

Here one looks at the given hull $B \subseteq \mathfrak{M}(E)$, as the *spectrum of* $F$, via the *homeomorphism* $\rho$ (see (4.8), (4.9)).

Thus, our assertion in (4.12) is now a consequence of (4.2) and (4.10), in conjunction with the relation

(4.13) $$\alpha = \widehat{\phi(x)} = \hat{x} \circ \rho|_{\rho^{-1}(V)},$$

which holds true, *for any map* $\alpha$, *as in (4.10), with* $y = \phi(x) \in E/I \equiv F$, and this terminates the proof of the lemma. ∎

**Scholium 4.1.**– The algebra $\tilde{E}_B$, with $B \subseteq \mathfrak{M}(E)$, may also be considered, as the *localization of* $E$, *relative to* $B$. (We refer to A. Mallios [9] for a general account thereon). Thus, (4.12) or even the **C**-*algebra isomorphism*

(4.14) $$\tilde{E}_B = (\widetilde{E/I})_B,$$

with $I = k(B)$ (see (4.5)), can now be expressed, as follows:

(4.15) by looking at the *localization* of a given topological algebra $E$, with respect to a *hull* of its spectrum, *the* (Gel'fand transforms of the) *elements of* $E$ *that vanish on this hull* (hence, those in its kernel) *do not count.*

We come now to the promised assertion, concerning the *softness of a* given *Gel'fand sheaf.* Indeed, one has the following result (see also Definition 3.3 above).

**Theorem 4.1.** *The Gel'fand sheaf of a regular hereditarily geometric* (topological) *algebra is soft.*

**Proof.** Let $E$ be a given topological algebra with spectrum $\mathfrak{M}(E)$ and Gel'fand sheaf $\mathcal{E}$, satisfying the conditions of the theorem (cf. Definition

3.3), while let $B$ ba a *closed* subset of $\mathfrak{M}(E)$. So one has to prove that *the* corresponding *restriction map* (in fact, $\mathbf{C}$-*algebra morphism*)

$$\rho_B : \Gamma(\mathfrak{M}(E), \mathcal{E}) \longrightarrow \Gamma(B, \mathcal{E}) \tag{4.16}$$

*is surjective* (see also e.g. Chapt. III; (8.25)): Indeed, since, by hypothesis, $E$ is regular, *the closed set* $B \subseteq \mathfrak{M}(E)$ is still closed, with respect to the *hull-kernel topology* in $\mathfrak{M}(E)$, therefore, *a hull in* $\mathfrak{M}(E)$ (cf. [TA: p.332, Theorem 2.1, along with p.330, Lemma 1.1 and p.331, Definition 1.2]). Consequently, if $I = k(B)$ denotes the *kernel of* $B$ (see (4.5)), one has (cf. (4.1)) $B = h(k(B)) \equiv h(I)$. Therefore, based now on our hypothesis for $E$ (see Definition 3.3), and on (4.11), one obtains;

$$\begin{aligned} \Gamma(\mathfrak{M}(E), \mathcal{E}) = E \xrightarrow{\phi} E/I \equiv F = \Gamma(\mathfrak{M}(F), \mathcal{F}) \\ = \Gamma(B, \mathcal{F}) = \Gamma(B, \mathcal{E}), \end{aligned} \tag{4.17}$$

that is, exactly our assertion for (4.16), and this completes the proof. ∎

On the other hand, we have already remarked in the previous Section (see (3.14)) that soft, as well as, Fréchet-Šilov (locally $m$-convex) algebras are herediterily geometric ones; moreover, both type of the latter algebras are, by their definition, *regular* (cf. Definition 3.2 and (3.10)). Thus, the last remark, together with the previous Theorem 4.1, provide now the promised assertion, alluded to by the heading of this Section. Namely, one concludes that;

(4.18) *the Gel'fand sheaf of any soft algebra, or* that one *of any Fréchet-Šilov* (locally $m$-convex) *algebra, is soft.*

The above also justifies the terminology adopted thus, in anticipation, in the preceding by Definition 3.2. In this context, something more than (4.18) is actually true; namely, due to the topological properties of the spectra of the topological algebras appeared in (4.18) (cf. (3.12)), as well as, to the topological-algebraic (viz. *"geometric"*) structure of the same algebras (see (3.2) above, and S.A. Selesnick [1: p.319, Proposition 2.4, (iii)]), one concludes that;

(4.19) *the Gel'fand sheaf of either type of* the *topological algebras* appeared *in (4.18) is fine.* (See e.g. Chapt. III; Definition 8.1). Hence (loc. cit., (8.26)), *the same sheaf is soft* too.

The significance of the previous remark is, of course, evident, concerning *cohomological questions* (cf. Chapt. III; Section 8), connected with the Gel'fand sheaf, as we shall see in the sequel.

## 5. De Rham-Kähler complex of a topological algebra

Our purpose by the following discussion is to associate a (de Rham-) *Kähler complex of differentials* with a given topological algebra, the latter being, as we shall presently see, of a fairly general type. Precisely, we thus assume that;

(5.1) the *topological algebras* involved in the sequel are *locally convex* ones with (jointly) *continuous multiplication* [TA], being also *unital* and *commutative* **C**-*algebras*. Furthermore, if $E$ is such an algebra, we always assume, as usual, that $\mathfrak{M}(E)$ stands for its *spectrum* (loc. cit.).

Of course, we are based here upon the standard theory of *Kähler differentials,* which is actually valid for any unital commutative ring (see, for instance, N. Bourbaki [2: Chap. 3; p.134], or even E. Kunz [1]). However, we consider instead, as already mentioned (cf. (5.1)), *unital commutative algebras over the complexes,* for short, **C**-*algebras,* which are also *suitably topologized.* The interference here of a toplology, compatible with the algebraic structure involved, is due, as we shall see right below, to our desire to cope with certain *analytical aspects,* as, for instance, with the existence of the *exponential map;* thus, the latter is actually supplied by a particular type of topological algebras, general enough, however, to cover the relevant applications (see Section 7 in the sequel). On the other hand, as we have already seen in our abstract treatment in the preceding (see e.g. Chapter VIII), the map in question has important *geometrical applications;* so we have employed it, for instance, in our treatment of *line sheaves, in connection with Chern classes,* the counterpart here of the classical *Chern isomorphism* (for *line bundles,* see Chapt.

V; Section 6), or even in the proof of *Weil's integrality theorem* (cf. Chapt. VIII; Theorem 11.1. See also (7.19) below).

Thus, to start with, suppose that we are given a *topological algebra* $E$, *as in (5.1)*. Furthermore, we consider the *locally convex* (tensor product) *algebra*

$$E \underset{\pi}{\otimes} E, \tag{5.2}$$

endowed with the *projective tensorial topology* $\pi$ [TA: Chapter X] and the *multiplication morphism*

$$\mu : E \underset{\pi}{\otimes} E : x \otimes y \longmapsto \mu(x \otimes y) := xy, \tag{5.3}$$

being, in fact, a *continuous algebra morphism* of the topological algebras concerned. Thus, the *kernel of* $\mu$ is a *closed 2-sided ideal* of $E \underset{\pi}{\otimes} E$, so that by further setting

$$I \equiv \ker \mu \subseteq E \underset{\pi}{\otimes} E, \tag{5.4}$$

one has the following *continuous exact sequence*

$$0 \longrightarrow \ker \mu \longrightarrow E \underset{\pi}{\otimes} E \xrightarrow{\mu} E \longrightarrow 0. \tag{5.5}$$

As a result, one obtains;

$$E \underset{\pi}{\otimes} E = E \oplus \ker \mu, \tag{5.6}$$

a *topological direct sum decomposition:* The assertion on (5.6) is based on the *exactness of (5.5)* and the *continuity of the projection* $i \circ \mu$ of $E \underset{\pi}{\otimes} E$, corresponding to (5.5), where one also employs the *continuous* (canonical) *injection* (in fact, a *section* of $\mu$);

$$i : E \longrightarrow E \underset{\pi}{\otimes} E : x \longmapsto i(x) := x \otimes 1. \tag{5.7}$$

(In this context, see also e.g. J. Horváth [1: p.121, comments following Definition 2.1]. Yet, we still remark here that one can take in (5.6) the *complete algebra* $E \widehat{\underset{\pi}{\otimes}} E$ [TA], having again a similar decomposition, by assuming now $E$ to be *complete;* see e.g. A. Mallios [8: p. 153; (4.5)]).

On the other hand, consider now the vector subspace of $E \otimes E$, generated by $I \cdot I$ (see (5.4)), denoted by $I^2 \equiv [I \cdot I]$, taking, in effect, an *ideal* of $E \otimes E$, such that $I^2 \subseteq I$; hence, by further looking at the quotient space $I/I^2$, endowed with the respective *quotient topology,* one obtains a *complete locally convex $E$-module,* by setting

$$\Omega_E^1 := \widehat{I/I^2}. \tag{5.8}$$

(For technical reasons, having to do with *"analytical questions"* (see, for example, the comments following (5.1) above), we consider in (5.8) the corresponding *completed* space). Now, following classical terminology (see, for instance, R. Hartshorne [1: p.172], or even E. Kunz [1: p.39]),

(5.9) we call $\Omega_E^1$, as given by (5.8), the *Kähler module of the 1st-order differentials of $E$*, or even simply the (*1st-order*) *Kähler module of $E$*.

Furthermore, setting

$$\wedge^p \Omega_E^1 := \widehat{\wedge^p (I/I^2)} \equiv \Omega_E^p, \tag{5.10}$$

*for any $p \in \mathbf{Z}_+$*, where we also set

$$\wedge^0 \Omega_E^1 \equiv \Omega_E^0 := E \quad and \quad \wedge^1 \Omega_E^1 := \Omega_E^1, \tag{5.11}$$

one finally obtains a *sequence of complete locally convex $E$-modules* (we assume, of course, due to the first of (5.11), that $E$ is *complete* too). Thus, we now define (see (5.10));

$$\Omega_E^* \equiv \wedge \Omega_E^1 := \bigoplus_{p=0}^{\infty} \widehat{\wedge^p (I/I^2)} \equiv \bigoplus_{p \geqq 0} \wedge^p \Omega_E^1 \equiv \bigoplus_{p \geqq 0} \Omega_E^p, \tag{5.12}$$

viz. the *exterior* (or even *Grassmann*) *algebra of the* (complete convex) *$E$-module $\Omega_E^1$*, as the latter is given by (5.8).

On the other hand, one can further consider on the above algebra $\Omega_E^*$ the *(locally convex) direct sum topology;* in other words, the finest locally convex (vector space) topology on it, making the canonical inclusion (**C**-linear) maps (cf. (5.12))

$$\Omega_E^p \longrightarrow \Omega_E^*, \quad p \in \mathbf{Z}_+, \tag{5.13}$$

continuous. Thus, by virtue of (5.10), (5.11) and of our hypothesis for $E$, one concludes that;

(5.14) *$\Omega_E^*$ is a complete locally convex algebra with continuous multiplication and an identity element* (viz. that one of $E$, cf. (5.11)).

In this concern, see also, for instance, G. Köthe [1: p.212, (3)], along with A. Colojoară [1], [2]. ∎

Furthermore, by still following the classical algebraic counterpart (see e.g. N. Bourbaki [2: Chap. 3; p. 132], or even E. Kunz [1]), one defines a *continuous derivation of $E$ into* (the $E$-module) $\Omega_E^1$, as follows;

$$(5.15)\quad d : E \longrightarrow \Omega_E^1 : x \longmapsto d(x) := (1 \otimes x - x \otimes 1) + I^2 \equiv [(1 \otimes x - x \otimes 1)];$$

viz. $d$, as above, is a $\mathbf{C}$-*linear map,* which still satisfies the corresponding *Leibniz condition* (i.e., one has $d(x, y) = x \cdot dy + y \cdot dx$, for any $x, y$ in $E$). Yet, setting

$$(5.16)\qquad \tilde{d} : E \longrightarrow E \otimes E : x \longmapsto \tilde{d}(x) := 1 \otimes x - x \otimes 1 \in \ker \mu \equiv I,$$

one infers that;

$$(5.17)\qquad I = \overline{[\operatorname{im} \tilde{d}]} \subseteq E \mathop{\widehat{\otimes}}_{\pi} E.$$

In other words, one thus obtains that:

(5.18) $\tilde{d}(E) \subseteq I$ is a *set of topological generators of $I$,* the latter being considered as an *$E$-module;* thus, equivalently, every element of $I$ can be approximated by elements of the form,

$$(5.18.1)\qquad \sum_i x_i \tilde{d}(y_i),$$

such that $\sum_i x_i \otimes y_i \in E \otimes E$.

In this regard, see also e.g. N. Bourbaki [2: Chap. 3; p.132, Lemme 1]. ∎

We call the $\mathbf{C}$-linear map (5.15), as above, the *Kähler derivation* (precisely, the *1st-order Kähler derivation*) *of $E$.*

Thus, our next objective is to find now a *(continuous) exterior derivation of $\Omega_E^*$ into itself,* satisfying the appropriate conditions, according to the classical pattern, provided by the *de Rham algebra of a smooth manifold:*

Indeed, given a *topological algebra $E$, as in (5.1),* one defines a *continuous* $\mathbf{C}$-*linear map*

$$(5.19)\qquad \delta : \Omega_E^1 \longrightarrow \Omega_E^2,$$

as follows; thus, one first sets (cf. (5.15), together with (5.10)),

$$(5.20) \qquad \delta\left(\sum_i x_i \otimes y_i\right) := \sum_i dx_i \wedge dy_i,$$

*for any* $\sum_i x_i \otimes y_i \in E \otimes E$, so that *one* actually *obtains* $I^2 \subseteq \ker \delta$. Therefore, due to the *continuity of the* **C**-*linear map, as given by (5.20),* one finally gets the desired map, as in (5.19), (we still denote the latter map by $\delta$); viz. one has (cf. (5.10)),

$$(5.21) \qquad \delta : \widehat{I/I^2} \equiv \Omega_E^1 \longrightarrow \Omega_E^2 \equiv \wedge^2 \Omega_E^1 \equiv \widehat{\wedge^2(I/I^2)}.$$

(In this regard, see also A. Mallios [8: p. 155, Lemma 4.1]). ∎ So the relevant situation is now exhibited by the following result, patterned after the standard algebraic case (see also loc. cit.; p.156, Theorem 4.1).

**Theorem 5.1.** *Let $E$ be a unital commutative locally convex algebra with continuous multiplication. Then, there exists a continuous exterior derivation of degree one* (cf. (5.12))

$$(5.22) \qquad D : \Omega_E^* \longrightarrow \Omega_E^*,$$

*viz. a continuous* **C**-*linear map, as above, such that the following conditions hold true;*

$$(5.23) \qquad D(\Omega_E^p) \subseteq \Omega_E^{p+1}, \quad p \in \mathbf{Z}_+,$$

*and, for any $\alpha, \beta$ in $\Omega_E^*$, with $\alpha \in \Omega_E^p$, one has,*

$$(5.24) \qquad D(\alpha \wedge \beta) = D(\alpha) \wedge \beta + (-1)^p \alpha \wedge D(\beta).$$

*Moreover, the following relations are still valid* (cf. (5.11), (5.12), (5.15));

$$(5.25) \qquad D|_E = d,$$

*which thus implies the uniqueness of $D$* (cf. also (5.19), (5.20)), *as well as,*

$$(5.26) \qquad D^2 = 0. \ \blacksquare$$

Thus, given a *topological algebra $E$*, as in the above Theorem 5.1, we are now justified, by virtue of the same theorem, and by following standard terminology, to call the pair (see (5.12), (5.22))

$$(5.27) \qquad (\Omega_E^*, D) \equiv (\wedge \Omega_E^1, D)$$

the *de Rham-Kähler complex of* $E$.

We are going now to look at the preceding, right in the next Section, within the appropriate sheaf-theoretic framework, that will be also of use in the sequel.

# 6. Sheafification of the de Rham-Kähler complex. Logarithmic derivation

Suppose we are given a *topological algebra sheaf* $\mathcal{E}$ on a *topological space* $X$ (cf. Theorem 1.1) in such a manner that:

(6.1) for any *open* set $U \subseteq X$, the corresponding *(local) section algebra of* $\mathcal{E}$ *over* $U$,

(6.1.1) $$\Gamma(U, \mathcal{E}) \equiv \mathcal{E}(U)$$

is a *topological algebra, as in (5.1)* above.

Thus, by further employing the notation of (5.8) in the preceding, one can consider the following *presheaf on* $X$,

$$U \longmapsto \Omega^1_{\Gamma(U,\mathcal{E})} := \widehat{I_U/I_U^2}, \tag{6.2}$$

where $U$ is varying over the open sets in $X$, while one also takes the obvious restriction morphisms. In other words, with any open $U \subseteq X$, one associates the corresponding *Kähler module of* $\Gamma(U, \mathcal{E})$. So we call the previous presheaf on $X$ the *Kähler presheaf of* $\mathcal{E}$, while the respective sheaf on $X$, generated by the previous presheaf, is named the *Kähler sheaf* (or even the *sheaf of 1st-order Kähler differentials*) *of* $\mathcal{E}$ and is denoted by

$$\Omega^1_{\mathcal{E}}. \tag{6.3}$$

Consequently, given that *each one of the sets (6.2)*, as above, when $U$ varies over the open sets in $X$, *is a* $\Gamma(U, \mathcal{E})$*-module* (the corresponding *restriction morphisms* being also such *of the pertinent type*), one obtains that (see also Chapt. II; (1.54));

(6.4) *the sheaf* $\Omega^1_{\mathcal{E}}$, as defined above, *is,* in fact, *an* $\mathcal{E}$*-module over* $X$.

(See Chapt. II; Definition 1.5, along with (1.66), or even Proposition 1.1 therein, taking also our hypothesis for $\mathcal{E}$ into account).

Now, one further defines the *Kähler derivation of* $\mathcal{E}$, as the following *morphism of sheaves,*

$$\partial \equiv (\partial_U) : \mathcal{E} \longrightarrow \Omega^1_{\mathcal{E}}, \tag{6.5}$$

such that one sets;

$$\partial_U(s) := (1 \otimes s - s \otimes 1) + I_U^2 \equiv [\tilde{d}_U(s)], \tag{6.6}$$

*for any* (local) *section* $s \in \mathcal{E}(U) \equiv \Gamma(U, \mathcal{E})$, *and* an arbitrary *open* $U \subseteq X$. (Here $\tilde{d}_U$ is defined analogously with (5.16); yet, concerning the same rel. (6.6), *one actually has* $\partial_U = d_U$ (cf. (5.15)). In this context, cf. also Chapt. I; Definition 1.2, along with (6.3), (8.5) and Lemma 8.1). Thus, by the very definitions, one infers that:

(6.7) $\partial$ *is a* $\mathbf{C}$*-linear morphism* of the $\mathbf{C}$*-vector space sheaves* involved in (6.5) (see Chapt. II; (1.38)), such that the following *Leibniz condition* holds true, as well;

$$\partial(s \cdot t) = (\partial s) \cdot t + s \cdot \partial t, \tag{6.7.1}$$

*for any* (continuous local) *sections* $s, t$ *in* $\mathcal{E}(U)$, *and any open* $U \subseteq X$ (see also e.g. Chapt. VI; (1.2)).

The above thus justifies too the previously applied terminology, concerning the map (6.5) (cf. Chapt. VI; Definition 1.1). So, in other words, the triple

$$(\mathcal{E}, \partial, \Omega^1_{\mathcal{E}}), \tag{6.8}$$

as given above, supplies here, in effect, a particular example of a *differential triad,* in the sense of Chapt. VI; (1.4) (see also (3.1) there). We call it the *Kähler differential triad* of the given *topological algebra sheaf* $\mathcal{E}$, as above (cf. (6.1)). In fact, we are going right away to specialize the previous triple, as in (6.8), to the case of a *topological algebra* $E$, by employing the corresponding *Gel'fand sheaf* of $E$ (cf. (2.8)).

Thus, by virtue of (2.41.1) and (2.27) (see also (2.44)), one concludes, in particular, that:

given a *topological algebra* $E$, *unital* (not necessarily commutative), whose *spectrum* is

(6.9.1) $$\mathfrak{M}(E) \equiv X,$$

the corresponding *Gel'fand sheaf* $\mathcal{E}$ of $E$ (see (2.8)) yields a *topological algebra sheaf on* $X$, such that one has;

(6.9) *for any open* $U \subseteq X$, the respective *(local) section algebra of* $\mathcal{E}$ *over* $U$,

(6.9.2) $$\Gamma(U, \mathcal{E}) \equiv \mathcal{E}(U)$$

*is a unital commutative locally m-convex (topological* $\mathbf{C}$-*)algebra.*

In other words (cf. Theorem 1.1), *the Gel'fand sheaf* $\mathcal{E}$ *is a unital commutative locally m-convex topological* $\mathbf{C}$-*algebra sheaf on* $X$, the *spectrum* of the given topological algebra $E$, as above (cf. (6.9.1)).

Consequently, our previous *cond. (5.1) is fulfilled* for the topological algebras involved. (Of course, in view of the relation $\widetilde{E}_U \subseteq \mathcal{C}_c(U)$ (see (2.72)), one considers on $\widetilde{E}_U$ the *relative topology from* $\mathcal{C}_c(U)$; cf., for instance, (2.4) and the ensuing comments therein, or even (2.5)). *Warning!* At this place, we should also remark that the given topological algebra $E$ *is not necessarily a geometric* (topological) *algebra* (Definition 3.1), so that *we do not have, in general, (3.4) fulfilled,* in either sense (viz. either algebraically or even topologically).

On the other hand, within the same vein of ideas, and based on the respective definitions in (5.19), (5.20), along with Theorem 5.1, one further defines *exterior differential operators*

(6.10) $$d^p \equiv d : \Omega^p_{\mathcal{E}} \longrightarrow \Omega^{p+1}_{\mathcal{E}}, \quad p \geqq 0,$$

where we have set (see also (5.10), together with Chapt. IV; (7.1), (7.3))

(6.11) $$\Omega^p_{\mathcal{E}} := \wedge^p \Omega^1_{\mathcal{E}}$$

(cf. (6.3) above), while we still set (see also (5.11));

(6.12) $$\wedge^0 \Omega^1_{\mathcal{E}} := \mathcal{E} \text{ and } \wedge^1 \Omega^1_{\mathcal{E}} := \Omega^1_{\mathcal{E}}.$$

Thus, one has, for instance, that (see (6.5) and (6.10) for $p = 1$);

$$d^1 \circ \partial \equiv d \circ \partial = 0 \tag{6.13}$$

(note that $d(1) = 0$; cf. (5.15). See also (5.26) in the preceding; so we have found again the relation that guarantees the *flatness of the* $\mathcal{E}$*-connection* $\partial$ *of* $\mathcal{E}$. See Chapt. VIII; (1.8), (2.6)). Of course, one can further formulate here the corresponding *sheaf-theoretic version of* the same *Theorem 5.1,* as above. Thus, one obtains the (sheaf-theoretic) *de Rham-Kähler complex of the* given *topological algebra sheaf* $\mathcal{E}$, as before; viz.

$$(\Omega^*_{\mathcal{E}}, d), \tag{6.14}$$

with the obvious meaning of the notation involved here; see also (5.27), with $D \equiv d$, along with (5.12), (6.10), (6.11) and (6.12).

**6.(a). Logarithmic derivation.**– The notion in the heading of this Subsection has been already examined in the preceding, within the appropriate abstract axiomatic context (cf. Chapt. VI; Definition 1.2). So here we assume that we are given a *topological algebra sheaf* $\mathcal{E}$, as above (see (6.1)), and let

$$\mathcal{E}^{\cdot} \tag{6.15}$$

be the corresponding *sheaf of units of* $\mathcal{E}$, an *abelian group sheaf* on $X$; thus, for every *open* $U \subseteq X$, we denote by $\mathcal{E}(U)^{\cdot}$ the *group of units* of the given algebra $\mathcal{E}(U) \equiv \Gamma(U, \mathcal{E})$ (cf. (6.1.1) and (5.1)), while one still obtains,

$$\mathcal{E}^{\cdot}(U) \equiv \Gamma(U, \mathcal{E}^{\cdot}) = \Gamma(U, \mathcal{E})^{\cdot} \equiv \mathcal{E}(U)^{\cdot}, \tag{6.16}$$

within a *group isomorphism* (see Chapt. IV; Lemma 1.1, in particular, (1.15)).

So we further define the following *sheaf morphism* (cf. (6.3))

$$\tilde{\partial} : \mathcal{E}^{\cdot} \longrightarrow \Omega^1_{\mathcal{E}}, \tag{6.17}$$

such that one sets;

$$\tilde{\partial}(s) := s^{-1} \cdot \partial(s), \tag{6.18}$$

*for any* $s \in \mathcal{E}^{\cdot}(U) = \mathcal{E}(U)^{\cdot}$, and $U$ *open* in $X$.

Now, according to the general theory (see Chapt. VI; (1.27)), one thus concludes that the map (6.17) is a *morphism of sheaves of (abelian) groups;* viz. one has

(6.19) $$\tilde{\partial}(s \cdot t) = \tilde{\partial}(s) + \tilde{\partial}(t),$$

for any $s, t$ in $\mathcal{E}^{\cdot}(U)$ and $U$ open in $X$. Moreover, one gets, straightforwardly from (6.18), that;

(6.20) $$\ker \tilde{\partial} = (\ker \partial) \cap \mathcal{E}^{\cdot}.$$

In particular, we are going to apply the above in the sequel (see thus the next Section 7 below).

Finally, as was the case already in Chapter VI, due to (6.18), we call the map (6.17), for the case under consideration here, the *(kählerian) logarithmic derivation of* $\mathcal{E}$, while the element $\tilde{\partial}(s) \in \Omega^1_{\mathcal{E}}(U)$, as in (6.18), the *(kählerian) logarithmic derivative of* (the local section) $s \in \mathcal{E}^{\cdot}(U) = \mathcal{E}(U)^{\cdot}$ (see (6.16)).

## 7. The sheaf exponential sequence

We consider in this Section conditions for a given topological algebra, so that its Gel'fand sheaf (cf. (2.8)) yield an exact exponential sheaf sequence, in the sense already examined within our abstract context in the preceding (see Chapt. V; Definition 6.1).

Indeed, one first obtains the following preliminary result, valid under fairly general conditions for the topological algebras involved. Thus, for convenience, we recall, however, at this place that:

(7.1) given a *unital locally m-convex* (topological) **C**-*algebra* $E$, which is also $\sigma$-*complete* (i.e., every *Cauchy sequence* in $E$ converges), then the *exponential function acts* (or *operates*) *on* $E$, viz. *one has*

(7.1) $$\exp x := \sum_{n=0}^{\infty} \frac{1}{n!} x^n \in E,$$

*for every element* $x \in E$.

(In this context, one just considers the corresponding numerical series $\sum_n \frac{1}{n!} p(x)^n$, for every (continuous) *submultiplicative seminorm* $p$ on $E$, taken

from a family of such, defining the topology of $E$; cf. [TA: p.18, Proposition 3.2]. See also A. Mallios [3: p. 492; (5.1)], or even [6: p. 196; (8.5)]). ∎

Thus, we come now right away to the promised result. That is, one has the following.

**Lemma 7.1.** *Let $E$ be a unital $\sigma$-complete locally m-convex algebra with Gel'fand map continuous* (cf. (2.21), (2.2)). *Moreover, let $\mathcal{E}$ be the corresponding Gel'fand sheaf of $E$ over $\mathfrak{M}(E)$, the spectrum of the given topological algebra $E$* (see Definition 2.1), *and $\mathcal{E}^{\cdot}$ the respective sheaf of units of $\mathcal{E}$* (cf. (6.15)). *Then, there exists the exponential sheaf morphism*

$$e : \mathcal{E} \longrightarrow \mathcal{E}^{\cdot}, \tag{7.2}$$

*such that the following sequence (of sheaves of groups,* cf. Chapt. V; (6.1), (6.2)) *is exact;*

$$0 \longrightarrow \mathbf{Z} \longrightarrow \mathcal{E} \xrightarrow{e} \mathcal{E}^{\cdot}. \tag{7.3}$$

**Proof.** Consider an *open* set $U \subseteq \mathfrak{M}(E)$, along with the following *(local) exponential function* (cf. also (2.3))

$$\begin{aligned} e_U : E^{\hat{}}|_U \longrightarrow (E^{\hat{}}|_U)^{\cdot} : \hat{x}|_U \longmapsto e_U(\hat{x}|_U) \\ := \exp 2\pi i(\hat{x}|_U) = \widehat{\exp 2\pi i x}|_U. \end{aligned} \tag{7.4}$$

Indeed, according to our hypothesis for $E$, *one has*

$$\widehat{\exp 2\pi i x} \in E^{\hat{}}, \tag{7.5}$$

*for any element $x \in E$* (cf. (7.1.1)). But, in view of the same definition (7.1.1) and the *continuity of the Gel'fand map* (in fact, a *continuous* **C**-*algebra morphism*), one further obtains (see also (2.21));

$$\begin{aligned} \widehat{\exp 2\pi i x} &\equiv \mathcal{G}(\exp 2\pi i x) = \mathcal{G}\left(\lim_n \textstyle\sum_{k=0}^{n} \frac{1}{k!}(2\pi i x)^k\right) \\ &= \lim_n \mathcal{G}\left(\textstyle\sum_{k=0}^{n} \frac{1}{k!}(2\pi i x)^k\right) = \lim_n \textstyle\sum_{k=0}^{n} \frac{1}{k!}\, \widehat{2\pi i x}^{\,k} \\ &= \exp \widehat{2\pi i x} = \exp 2\pi i \hat{x} \in E^{\hat{}}. \end{aligned} \tag{7.6}$$

(The last relation proves, in effect, that *the Gel'fand transform algebra $E^{\hat{}}$ of*

$E$ *is* still *"closed" with respect to the exponential function,* or even *the latter function acts* also *on* $E^{\wedge}$). Therefore, due, in turn, to the *continuity of the restriction* (**C**-*algebra) morphism*

$$\rho_U : E^{\wedge} \longrightarrow E^{\wedge}|_U, \tag{7.7}$$

*for any open* $U \subseteq \mathfrak{M}(E)$ (cf. also (2.2), (2.6)), one finally obtains;

$$\widehat{\exp 2\pi i x|_U} \equiv \exp(2\widehat{\pi i x}|_U) = \exp 2\pi i(\hat{x}|_U) \in E^{\wedge}|_U, \tag{7.8}$$

which thus completely justifies now (7.4).

On the other hand, by considering the *Gel'fand presheaf of* $E$ *over* $\mathfrak{M}(E)$, as given in the preceding by (2.7.1), in conjunction with the above exponential functions, as in (7.4), one gets the following *commutative diagram;*

$$\begin{array}{ccc} E^{\wedge}|_U & \xrightarrow{e_U} & (E^{\wedge}|_U)^{\cdot} \\ \rho^U_V \downarrow & & \downarrow \rho^U_V \\ E^{\wedge}|_V & \xrightarrow{e_V} & (E^{\wedge}|_V)^{\cdot}, \end{array} \tag{7.9}$$

*for any open* $U, V$ *in* $\mathfrak{M}(E)$, *with* $V \subseteq U$. Therefore, one obtains the following *morphism of* the *presheaves* concerned;

$$e_U : E^{\wedge}|_U \longrightarrow (E^{\wedge}|_U)^{\cdot} \stackrel{\subseteq}{\rightarrow} \Gamma(U, \mathcal{E})^{\cdot} = \Gamma(U, \mathcal{E}^{\cdot}), \tag{7.10}$$

*for any open* $U \subseteq \mathfrak{M}(E)$ (see (2.15) and (6.16)). Thus, one finally gets the desired sheaf morphism (7.2) (cf. Chapt. I; (8.4), (8.5)). Furthermore, it is clear, in view of (7.4), that;

$$\ker(e_U) \cong \mathbf{Z} \text{ with } U \text{ open in } \mathfrak{M}(E). \tag{7.11}$$

(We assume, of course, here that *the* given *topology of* $E$ *is Hausdorff*). Hence, *one has*

$$\ker(e) \cong \mathbf{Z}, \tag{7.12}$$

as desired, $\mathbf{Z}$ being considered here as a *constant sheaf* (of abelian groups) *over* $\mathfrak{M}(E)$ (cf. Chapt. I; (4.1), Scholium 4.1 and (10.2), (10.8)). Thus, (7.12) stands, in fact, for an *isomorphism of* the *(abelian group) sheaves* concerned,

and this terminates the proof. ∎

Now, to have the previous sequence (7.3) in the form of a *short exact sequence* (a fact, that is now our objective), one has to supplement appropriately our hypothesis for the topological algebra $E$ involved in the above Lemma 7.1. Thus, the pertinent condition here seems to be the assumption that $E$ is a *soft algebra,* being further a *self-adjoint* one.

Thus, for convenience, we recall at this point (see also, for instance, [TA: p.482, (1.4)]) that a given *topological algebra* $E$, which is further endowed with a *(continuous) involution* "$*$" (we then refer to a *topological* $^*$*-algebra*) is said to be *self-adjoint,* if the corresponding Gel'fand map of $E$ (see (2.21)) is *involution* (or, as we also say, $^*$-)*preserving;* viz. one has

$$\mathcal{G}(x^*) \equiv \widehat{x^*} = \bar{\hat{x}} \equiv \overline{\mathcal{G}(x)}, \tag{7.13}$$

*for any* $x \in E$.

Now, coming back to our abovementioned objective, viz. to get a *short exact (exponential sheaf) sequence* through (7.5), we note that a crucial step here is Lemma 7.2, which follows; its Banach algebra analogue has been employed already, within the appropriate framework, by S.A. Selesnick [2: p.58, Proposition 2.5], while the same is based further on a classical argument, still within the Banach algebra theory, as applied to abstract harmonic analysis. (See thus, Y. Katznelson [1: p.236, Lemma]). So we now have the next.

**Lemma 7.2.** *Suppose we have a soft self-adjoint locally* $m$*-convex* $^*$*-algebra* $E$ (cf. Definition 3.2 and (7.13)). *Moreover, let* $f \in \mathfrak{M}(E)$ (the *spectrum* of $E$) *and* $U$ *an open neighborhood of* $f$ *in* $\mathfrak{M}(E)$. *Then, there exists an element* $x \in E$ *and an open neighborhood* $V$ *of* $f$ *in* $\mathfrak{M}(E)$, *such that one has;*

$$\begin{aligned} &\text{i) } 0 \leqq \hat{x} \leqq 1|_{\mathfrak{M}(E)}, \\ &\text{ii) } \operatorname{Supp} \hat{x} \subseteq U, \text{ and} \\ &\text{iii) } \hat{x} = 1_V. \end{aligned} \tag{7.14}$$

**Proof.** As a consequence of our hypothesis (see Definition 3.2, (3.2) and [TA: p.334, Definition 2.2]), $E$ is a *normal* (locally $m$-convex) *algebra.* Now, this latter property of $E$ is what one actually needs from the classical argument in Banach algebra theory (loc. cit. before the statement of the lemma). So the desired element $x \in E$ is given by

$$x := \sin^2 \frac{\pi}{2} yy^* \in E, \tag{7.15}$$

for some suitably chosen $y \in E$, according to our hypothesis, for $f$ and $U$, by applying the *normality* of $E$ (ibid.). In this context, we also note that *(7.15) defines,* of course, *an element of* $E$, as this follows from the convergence in $E$ of the *power series* concerned. (At this place, we further note that, as follows from the hypothesis, $E$ *is complete;* see thus our comments on (3.1) in the preceding. However, $\sigma$*-completeness* of $E$ would also suffice for the existence of $x \in E$, as given by (7.15); see, for example, the analogous situation in (7.1.1) above).

On the other hand, $\mathfrak{M}(E)$ being, by hypothesis (Definition 3.2), *equicontinuous,* one concludes (see [TA: p.184, Theorem 1.2, or even Corollary 1.4]) that *the Gel'fand map of* $E$ *is continuous.* Therefore, in view of (7.13) (see also e.g. (7.6)), one finally obtains that;

$$\hat{x} = \sin^2 \frac{\pi}{2} \hat{y}\bar{\hat{y}} \in E\hat{}\,. \tag{7.16}$$

Now, it is easy to see (cf. also the classical argument for Banach algebras, as quoted at the beginning) that *the* above *element* $\hat{x} \in E\hat{}$, *as given by (7.16), satisfies* all *the* three *conds. in (7.14),* and this terminates the proof of the lemma. ∎

Now, consider the *Gel'fand sheaf* $\mathcal{E}$ of a *topological algebra* $E$, *as in* the previous *Lemma 7.2,* and let $f \in \mathfrak{M}(E)$. So, by looking at the corresponding *fiber map*

$$e_f : \mathcal{E}_f \longrightarrow \mathcal{E}^{\boldsymbol{\cdot}}_f, \tag{7.17}$$

as this is defined by (7.4) (see also (2.9.3)), one proves that $e_f$ *is surjective:* Indeed, based on Lemma 7.2, one can repeat here the respective argument of S.A. Selesnick [2: p.58, Proposition 2.1], given there for the appropriate class of Banach algebras (viz. for *"self-adjoint (Banach-)Šilov algebras"*, ibid.). Therefore (see Chapt. I; (1.21)), one infers that *the (exponential) sheaf morphism*

$$e : \mathcal{E} \longrightarrow \mathcal{E}^{\boldsymbol{\cdot}} \tag{7.18}$$

(Lemma 7.1) *is surjective,* a fact that we were looking for, of course. In conclusion, we can thus state the desired result, as follows, while its proof is already provided by Lemma 7.1 and by what has been said before. So we

have now the next.

**Theorem 7.1.** *Let $E$ be a soft self-adjoint locally m-convex $^*$-algebra, $\mathcal{E}$ its Gel'fand sheaf, and $\mathcal{E}^{\cdot}$ the sheaf of units of $\mathcal{E}$. Then, one has the following short exact sequence of sheaves* (of abelian groups; *exact exponential sheaf sequence of $E$*)

$$0 \longrightarrow \mathbf{Z} \longrightarrow \mathcal{E} \xrightarrow{e} \mathcal{E}^{\cdot} \longrightarrow 1. \tag{7.19}$$

**Scholium 7.1.**– As already said, the class of topological algebras, appeared in the above Theorem 7.1, specializes to that one of *self-adjoint (Banach-)Šilov algebras,* already employed by S.A. Selesnick (loc. cit.). Yet, we have used in A. Mallios [6] *unital commutative Pták-Šilov self-adjoint (locally m-convex $^*$-)algebras with equicontinuous spectra,* or else *Selesnick algebras,* as they have been named there (due to the abovementioned work of S.A. Selesnick in the case of Banach algebras), to which class the algebras involved in Theorem 7.1 have also a special bearing.

In this context, according to what has been already said in (3.12), we also remark that all the above classes of (topological) algebras have their *spectra compact (Hausdorff) spaces,* as follows from their very definitions. (We still recall here too that *equicontinuous spectra* (of topological algebras [TA]) *are always (relatively) compact Hausdorff spaces* (Alaoglou-Bourbaki)). On the other hand, we consider right away, by the next Theorem 7.2, another class of topological algebras, to which a similar argument to that one employed for (7.19), can still be applied and whose *spectra* are no more compact; however, the same are indeed *paracompact (Hausdorff) spaces* (see the same remark (3.12) in the preceding).

The last conclusions above, about the spectra of the (topological) algebras involved, are also of a particular importance, in conjunction with our previous remarks in (4.18) and (4.19), in what concerns certain *cohomological applications,* we are going to consider in the sequel. So we come first to employ the other class of topological algebras, than those in Theorem 7.1, that we alluded to above. That is, we have.

**Theorem 7.2.** *Let $E$ be a unital commutative Fréchet regular and semi-simple self-adjoint locally $m$-convex $^*$-algebra;* thus, alias, a *self-adjoint Fréchet-Šilov (locally $m$-convex) algebra. Then,* by considering the Gel'fand sheaf $\mathcal{E}$ of $E$, *one obtains an exact exponential sheaf sequence of the form (7.19).*

**Proof.** By hypothesis $E$ is a Fréchet (locally $m$-convex topological) algebra, hence, it is a *barrelled* (and `a fortiori` a *spectrally barrelled*) one, having thus the respective *Gel'fand map continuous* (see [TA: p.264, Scheme 2.1 and p.183, Corollary 1.1]). Therefore, Lemma 7.1 can be applied (the given topological algebra $E$ being also *complete,* as by definition Fréchet). On the other hand, *any Fréchet-Šilov* (locally $m$-convex) *algebra is normal* (see the comments following (3.2) in the preceding); accordingly, one can further apply the argument in the proof of Lemma 7.2, hence, also that one concerning the *surjectivity of (7.18),* as above, a fact that now completes the proof. ∎

We come now, and we shall continue in the next Section 8, to a number of *applications of the* above *exact exponential sheaf sequence,* as given by (7.19):

Thus, we first have a variant, within the previous context, of the classical *Arens-Royden theorem,* initially formulated for (unital commutative) Banach algebras (see, for instance, E.L. Stout [1: p.99, Theorem 10.1]). The same result has been extended by R.M. Brooks [2] to *unital commutative Fréchet (locally $m$-convex) algebras* with *spectra $k$-spaces* (loc. cit., p.332, Corollary 3.5). To fix our terminology, we first set the following.

**Definition 7.1.** Let $E$ be a unital commutative $\sigma$-complete locally $m$-convex algebra, whose spectrum is $\mathfrak{M}(E)$. Then, $E$ is called an *Arens-Royden algebra,* whenever one has;

$$H^1(\mathfrak{M}(E), \mathbf{Z}) = E^{\cdot}/e(E), \tag{7.20}$$

within an *isomorphism of* the *abelian groups* concerned.

Here we still denote by

$$e : E \longrightarrow E^{\cdot} \tag{7.21}$$

the *exponential function* on $E$ (with $E^{\cdot}$ the *group of units of $E$*), as given by (7.1.1); in particular, we set

$$e(x) := \exp 2\pi i x, \quad x \in E, \tag{7.22}$$

(see also e.g. (7.4). So we actually apply here an obvious *abuse of notation,* by comparison with our previous one in (7.2)). Thus, our clain now is that;

(7.23) *both types of* (the locally $m$-convex) *algebras, appeared in* the above *Theorems 7.1 and 7.2, are Arens-Royden algebras* (Definition 7.1).

An analogous remark was made by S.A. Selesnick (loc. cit., as above, p.60), for the particular class of Banach algebras employed by the said author. (See, for instance, our previous Scholium 7.1 about that type of algebras).

Now, as already remarked in Scholium 7.1, the *spectra of the* (topological) *algebras, involved in Theorems 7.1 and 7.2,* are *compact,* respectively, *paracompact (Hausdorff) spaces,* so that (see, for instance, Chapt. III; Theorem 8.1) one may consider in the first member of (7.20) *Čech cohomology.* Yet, by virtue of the said theorems, both types of algebras yield a *short exact* (sheaf) *sequence of the form* (7.19), so that by looking at the corresponding *long exact sequence in (Čech) cohomology* (cf. Chapt. III; Theorem 5.3), one obtains;

$$\begin{aligned} &0 \longrightarrow \Gamma(X, \mathbf{Z}) \longrightarrow \Gamma(X, \mathcal{E}) \xrightarrow{e} \Gamma(X, \mathcal{E}^{\cdot}) \\ &\xrightarrow{\delta} H^1(X, \mathbf{Z}) \longrightarrow H^1(X, \mathcal{E}) \longrightarrow \cdots . \end{aligned} \tag{7.24}$$

For convenience, we have set here

$$\mathfrak{M}(E) \equiv X, \tag{7.25}$$

concerning the *spectrum* of a topological algebra $E$ of either one of the types in question. However, referring to the same sort of topological algebras, as before, we have already remarked (cf. (4.18)) that *their* respective *Gel'fand sheaves are soft, hence* (see Chapt. III; (8.26)), *acyclic,* so that one has;

$$H^1(X, \mathcal{E}) = 0. \tag{7.26}$$

Therefore, in view of (7.24) and (7.26), one now obtains the following *exact sequence;*

$$\cdots \longrightarrow \Gamma(X, \mathcal{E}) \xrightarrow{e} \Gamma(X, \mathcal{E}^{\cdot}) \longrightarrow H^1(X, \mathbf{Z}) \longrightarrow 0, \tag{7.27}$$

hence, the relation

$$(7.28)\qquad H^1(X, \mathbf{Z}) = \Gamma(X, \mathcal{E}^{\cdot})/e(\Gamma(X, \mathcal{E})),$$

up to an *isomorphism of* the (abelian) *groups* concerned.

On the other hand, the same types of (topological) algebras, as above, are geometric ones (see (3.14), in conjunction with Definition 3.3), so that one has (cf. Definition 3.1), along with Theorem 3.1);

$$(7.29)\qquad E = \Gamma(X, \mathcal{E}) \equiv \mathcal{E}(X),$$

within a $\mathbf{C}$-*algebra isomorphism*. Furthermore, by virtue of (6.16), by taking $U = \mathfrak{M}(E) \equiv X$, and of (7.29), one obtains

$$(7.30)\qquad \Gamma(X, \mathcal{E}^{\cdot}) = \Gamma(X, \mathcal{E})^{\cdot} \equiv \mathcal{E}(X)^{\cdot} = E^{\cdot}$$

(isomorphisms of the (abelian) groups involved). Thus, in view of (7.28), (7.30) and (7.29), one finally obtains;

$$(7.31)\qquad H^1(X, \mathbf{Z}) = E^{\cdot}/e(E),$$

as desired (cf. (7.20)). ∎

Now, as already mentioned, we continue presently below by the next Section our discussion on some further *applications of the exactness of (7.19).*

## 8. Line sheaves over the spectrum of a geometric topological algebra

We consider below certain further *implications of the short exact exponential sheaf sequence in (7.19)*. Indeed, what we are actually going to do is to derive two particular *morphisms of short exact sequences* (one of which will be, of course, that one in (7.19)), which further will be of use right away in the ensuing discussion.

Thus, suppose that we are given the *framework of either one of* the above two *Theorems 7.1 or 7.2,* while we also adopt the *notation of (7.25).* So assume now, as usual (see Chapt. I; Subsection 4.(b)), that

$$(8.1)\qquad \mathcal{C}_X$$

stands for the *sheaf* (of germs) *of* **C**-*valued continuous functions* on $X \equiv \mathfrak{M}(E)$, where the latter space is, as already said (cf. (7.25)), the *spectrum* of the given *topological algebra* $E$, *which satisfies the hypotheses of Theorem 7.1 or those of Theorem 7.2.* Thus, one first obtains the following *commutative diagram,* or even a *morphism of short exact sequences* (cf. also e.g. Chapt. III; (2.17), concerning the applied terminology);

$$\begin{array}{ccccccccc} 0 & \longrightarrow & \mathbf{Z} & \longrightarrow & \mathcal{E} & \xrightarrow{e} & \mathcal{E}^{\cdot} & \longrightarrow & 1 \\ & & \| & & \downarrow & & \downarrow & & \\ 0 & \longrightarrow & \mathbf{Z} & \longrightarrow & \mathcal{C}_X & \xrightarrow{e} & \mathcal{C}^{\cdot}_X & \longrightarrow & 1. \end{array} \tag{8.2}$$

Indeed, the assertion follows straightforwardly from the very definitions of the sheaves $\mathcal{E}$ (Definition 2.1) and $\mathcal{C}_X$ (cf. Chapt. I; (4.25.1). Yet, see loc. cit., (10.9) and (11.12.1), as well as, (8.4) and (8.5), along with (2.4) and (2.7.1) in the preceding).

On the other hand, we already know (cf. (4.19)) that the Gel'fand sheaf $\mathcal{E}$ of a topological algebra $E$, as above, is *fine* over the *paracompact* (or even compact) *Hausdorff space* $X \equiv \mathfrak{M}(E)$ (see the above Scholium 7.1). Furthermore, based on *Urysohn's Lemma* (see e.g. J. Dugundji [1: p.146, Theorem 4.1]) and on its use in obtaining (continuous) *partitions of unity* (loc. cit., p.173; (2), or even F. Hirzebruch [1: p.30, Theorem 2.8.4]), one also concludes that;

(8.3) *the sheaf* $\mathcal{C}_X$ *of germs of continuous* **C**-*valued functions on a paracompact (Hausdorff) space* $X$ *is fine.*

See, for instance, F. Hirzebruch [1: p.35, Theorem 2.11.1]. ▮ However, as we know (cf. Chapt. III; (8.24)), *fine sheaves over paracompact spaces are* ($\Gamma_X$-) *acyclic* (viz. they yield *trivial (Čech) cohomology*). Thus, this fact applied to the sheaves $\mathcal{E}$ and $\mathcal{C}_X$, as above, in conjunction with the *long exact cohomology sequences,* which are associated with each one of the two short exact (sheaf) sequences in (8.2), respectively, (see, for instance, (7.24) above, concerning the sequence in question for $\mathcal{E}$), yield now the following *commutative diagram* (of exact sequences of abelian groups);

$$\begin{array}{ccccccc} & & & & & & 0\\ & & & & & \nearrow & \\ 0 & \longrightarrow & H^1(X,\mathcal{E}^{\cdot}) & \xrightarrow{\;\delta\;} & H^2(X,\mathbf{Z}) & \longrightarrow & 0\\ & & \Big\downarrow & \nearrow_{\delta} & & & \\ & & H^1(X,\mathcal{C}^{\cdot}_X) & & & & \\ & \nearrow & & & & & \\ 0 & & & & & & \end{array} \tag{8.4}$$

In other words, one has the next *isomorphisms of* the *abelian groups* concerned;

$$H^1(X,\mathcal{C}^{\cdot}_X) = H^2(X,\mathbf{Z}) = H^1(X,\mathcal{E}^{\cdot}). \tag{8.5}$$

Before we proceed to the *geometric* interpretation of (8.5), we sum up the preceding discussion by stating that;

(8.6) given a *topological algebra* $E$ *of* the *type appeared in* either one of the *Theorems 7.1 or 7.2,* the respective *Gel'fand sheaf* $\mathcal{E}$ *of* $E$ (cf. (2.8)) *over the spectrum* $\mathfrak{M}(E) \equiv X$ *of the* given *algebra* yields the corresponding *sheaf of units* $\mathcal{E}^{\cdot}$ (see (6.15)) *such that* (8.5) (or, *equivalently,* (8.4)) *holds true.*

We come now straight away, as already said, to *explain (8.5) in geometric terms* (see thus Theorem 8.1 below):

Thus, first of all, the (abelian) group on the left-hand side of (8.5) classifies (cohomologically) *continuous* **C**-*line bundles over* $X \equiv \mathfrak{M}(E)$*; equivalently,* since a (continuous) **C**-*line bundle is (uniquely) determined by the* respective *sheaf (of germs) of continuous sections* (see also, for instance, Chapt. IV; 2.11), with $\mathbf{A} = \mathbf{C}$, and Chapt. VI; (4.11), taking $\mathcal{C}_X$ in place of $\mathcal{C}^{\infty}_X$), the latter being a $\mathcal{C}_X$*-line sheaf,* in our terminology (loc. cit. and Chapt. II; Definition 4.3), *the* same *group*

$$H^1(X,\mathcal{C}^{\cdot}_X) \tag{8.7}$$

*classifies* $\mathcal{C}_X$*-line sheaves on* $X$ (cf. also Chapt. V; Theorem 2.1 and (2.64), (2.65)). Of course, this is a classical result; see e.g. F. Hirzebruch [1: p.41,

Theorem 3.2.1], where one sets $G \equiv \mathbf{C}^{\cdot} := \mathbf{C} - \{0\}$. Namely, one has

$$H^1(X, \mathbf{C}^{\cdot}) = H^1(X, \mathcal{C}^{\cdot}_X), \tag{8.8}$$

within an *isomorphism of the abelian groups* concerned; see also (8.11) below, for $n = 1$ and $\mathbf{A} = \mathbf{C}$, along with (8.13), as well as, Chapt. V; (4.15).

Now, the above situation has been amply extended in A. Mallios [1] to *continuous* **A**-*vector bundles over* $X$; viz. to locally trivial fiber bundles, in the standard sense, over an *arbitrary topological space* $X$, whose fibers are *finitely-generated projective* **A**-*modules,* with **A** a *unital* (commutative) *locally* $m$-*convex* (topological) *algebra.* So suppose that

$$\mathcal{A} := \mathcal{C}^{\mathbf{A}}_X, \tag{8.9}$$

that is, consider the *sheaf of germs of continuous* **A**-*valued maps on* $X$ (see Chapter IV, Section 2), indeed, a **C**-*algebra sheaf on* $X$ (cf. Chapt. II; Definition 1.4), so that

$$\mathcal{GL}(n, \mathcal{A}) \tag{8.10}$$

is the associated with (8.9) *sheaf of germs of continuous* $\mathrm{GL}(n, \mathbf{A})$-*valued maps* on $X$, thus a *sheaf of groups* on $X$ (*non-abelian,* unless $n = 1$, with **A** *commutative*); cf. Chapt. IV; (2.11). (In this context, see also A. Mallios [1: p. 460ff] for a systematic account on **A**-*vector bundles,* as well as, A. Mallios [5: p. 419, Section 5], concerning the possibility of applying *more general topological algebras* **A**, than the ones considered above). Thus, by looking now at

$$H^1(X, \mathrm{GL}(n, \mathbf{A})), \tag{8.11}$$

viz. at the *set of isomorphism classes of (continuous)* **A**-*vector bundles over* $X$, of *rank* $n \in \mathbf{N}$, one obtains;

$$H^1(X, \mathrm{GL}(n, \mathbf{A})) = H^1(X, \mathcal{GL}(n, \mathcal{A})), \tag{8.12}$$

within a *(set-theoretic) bijection,* where $\mathcal{A}$ is given by (8.9) (see loc. cit. p.406, Theorem 1.1, as well as, concerning the second member of (8.12), Chapt. V; (1.11), (2.5) and Theorem 2.1). (*Warning!* In (8.11) $\mathrm{GL}(n, \mathbf{A})$ is considered as a *topological group,* in the relative topology from $M_n(\mathbf{A})$ (see Chapt. IV; Scholium 2.1), thus *not* necessarily *in the discrete topology;* however, see Chapt. V; Section 5, in particular, Scholium 5.1, concerning *flat* **C**-*vector bundles,* or even A. Mallios [5; §2, p. 410ff] for *flat* **A**-*vector bundles,* in

general, $\mathrm{GL}(n, \mathbf{A})$ carrying now the discrete topologogy.

Thus, setting now in (8.12) $n = 1$, one obtains the following *isomorphism of* (abelian, when $\mathbf{A}$ commutative) *groups;*

$$H^1(X, \mathbf{A}^{\cdot}) = H^1(X, \mathcal{A}^{\cdot}), \tag{8.13}$$

such that one has, in view of (8.9) (cf. also the above *warning*),

$$\mathcal{A}^{\cdot} \equiv \left(\mathcal{C}_X^{\mathbf{A}}\right)^{\cdot} = \mathcal{C}_X^{\mathbf{A}^{\cdot}} \tag{8.14}$$

(in fact, a *group sheaf isomorphism*). Indeed, (8.14) follows from the fact that;

$$\begin{aligned} \mathcal{A}^{\cdot}(U) = \mathcal{A}(U)^{\cdot} \equiv \left(\mathcal{C}_X^{\mathbf{A}}\right)(U)^{\cdot} &= \mathcal{C}(U, \mathbf{A})^{\cdot} \\ = \mathcal{C}(U, \mathbf{A}^{\cdot}) &= \left(\mathcal{C}_X^{\mathbf{A}^{\cdot}}\right)(U), \end{aligned} \tag{8.15}$$

*for every open set* $U \subseteq X$ (see also Chapt. IV; Lemma 1.1 and Scholium 2.1, along with Chapt. I; (11.10) and (9.17)). ∎ However, we already know (Chapt. V; (2.60)) that *the second member of (8.13) classifies* $\mathcal{A}$*-line sheaves* on $X$ (see also loc. cit., Theorem 4.1).

Now, after the above short deviation, pertaining to the *more general situation, supplied by (8.9)* (instead, namely, just of (8.1)), let us *look again at (8.5)*. Thus, since $X \equiv \mathfrak{M}(E)$, according to our hypothesis for $E$, is *paracompact Hausdorff* (cf. Scholium 7.1 above), one still obtains;

$$H^1(X, \mathcal{E}^{\cdot}) = H^1(X, \mathbf{E}^{\cdot}), \tag{8.16}$$

within an *isomorphism of* the *abelian groups* involved. Here $\mathbf{E}^{\cdot}$ stands for the *unit presheaf* (or even *presheaf of units*) of the corresponding *Gel'fand* ($\mathbf{C}$*-algebra) presheaf* $\mathbf{E}$, that is associated with the given topological algebra $E$, as above (see (2.7.1)); so one sets

$$U \longmapsto \mathbf{E}^{\cdot}(U) := (E^{\wedge}|_U)^{\cdot}, \tag{8.17}$$

*for every open* $U \subseteq X$, which thus, along with the obvious *restriction morphisms,* defines the *presheaf* $\mathbf{E}^{\cdot}$ at issue. Concerning (8.16), see also Chapt. III; (6.35), in particular, Theorem 6.2, in conjunction with (8.11) there. Therefore, based now on (8.5) and (8.16), one infers that;

*any (continuous)* **C**-*line bundle on* $X$, viz. *every element* $\xi \in H^1(X, \mathcal{C}^{\boldsymbol{\cdot}}_X)$ *corresponds (bijectively) to a 1-cocycle from* $\mathbf{E}^{\boldsymbol{\cdot}}$.

(8.18) That is, *one has;*

$$\xi \longleftrightarrow (g_{\alpha\beta}) \in Z^1(\mathcal{U}, \mathbf{E}^{\boldsymbol{\cdot}}). \tag{8.18.1}$$

In the last relation above (see also e.g. Chapt. III; (6.7)), one considers a *local frame of* $\xi$ *over* $X$, say, $\mathcal{U} = (U_\alpha)_{\alpha \in I}$ (cf. Chapt. II; Definition 4.2, in conjunction with (8.8) above), such that, *for any* $U_\alpha, U_\beta$ *in* $\mathcal{U}$, *with* $U_{\alpha\beta} \equiv U_\alpha \cap U_\beta \neq \emptyset$, one has

$$g_{\alpha\beta} \in \mathbf{E}^{\boldsymbol{\cdot}}(U_\alpha \cap U_\beta) = (E^{\wedge}|_{U_{\alpha\beta}})^{\boldsymbol{\cdot}} \subseteq \mathcal{C}(U_{\alpha\beta})^{\boldsymbol{\cdot}} = \mathcal{C}^{\boldsymbol{\cdot}}_X(U_{\alpha\beta}) \tag{8.19}$$

(see also (8.17) and (2.4), as well as, e.g. (6.16), concerning the last relation in (8.19)).

In other words, *any constituent* $g_{\alpha\beta}$ *of the 1-cocycle* $(g_{\alpha\beta})$, as in (8.18.1), *is an* (invertible) *element of the Gel'fand transform algebra* $E^{\wedge}$ of $E$, *restricted to* $U_{\alpha\beta}$, hence, according to our hypothesis for $E$ (we assumed $E$ *semisimple*), *of the algebra* $E$ *itself.* Furthermore, based on the last issues of the same relation (8.19), as above, one infers that, *every such element of* $E$ may be construed as an *invertible* (hence, *nowhere vanishing*) *continuous* **C**-*valued map on* $U_{\alpha\beta} \subseteq X$, something that we knew, for that matter, due to the hypothesis for $\xi$ and to (8.8); thus, our new information comes actually from the first items in (8.19).

The preceding supply now the proof of the next result. The corresponding statement, within the Banach algebra theory, has been given by S.A. Selesnick [2: p.60, Corollary 2.3]. So one has the following.

**Theorem 8.1.** *Suppose that* $E$ *is either a self-adjoint soft, or a self-adjoint Fréchet-Šilov, locally m-convex algebra* (see e.g. Theorems 7.1, 7.2 above), *whose spectrum is* $\mathfrak{M}(E)$. *Then, every continuous* **C**-*line bundle over a topological space* $X$, *homotopic to* $\mathfrak{M}(E)$, *can be expressed locally by elements of* $E$, viz. *the line bundle in question can be represented by a 1-cocycle, whose elements belong locally to* $E$. ∎

**Warning!** Concerning the previously applied terminology, the phrasing *"belong locally to* $E$*"* means, in fact, here a relation, like (8.19), taken also into account that $E^{\wedge} = E$ (within a **C**-*algebra isomorphism,*

provided by the Gel'fand map of $E$ (cf. (2.21)), due to the *semisimplicity* of $E$). This terminology should thus not be mixed up with the analogous one, already employed in (2.35), (2.26), or even in (2.42).

On the other hand, we can give to *the previous Theorem 8.1* a more general form, valid *for any (continuous)* **C**-*vector bundle,* not necessarily of rank 1, as was the case before, by restricting ourselves, however, to *soft* (locally $m$-convex) *algebras,* which thus, as follows from their definition (cf. Scholium 7.1), have their *spectra compact (Hausdorff)* spaces. Our argument has been motivated by some similar considerations in the realm rather of *(Real) Analytic* and/or of *Algebraic Geometry* (cf., for instance, K. Lønsted [1], or even R. Bkouche [1]; the same result has been given already in A. Mallios [1: p.488, Theorem 6.1], within the context of *Pták* (locally $m$-convex) *algebras,* where we still refer for relevant details. Thus, we now have the following.

**Theorem 8.2.** *Let $E$ be a soft locally $m$-convex* (topological) *algebra* (see Definition 3.2), *having spectrum $\mathfrak{M}(E) \equiv X$ and Gel'fand sheaf $\mathcal{E}$. Then, for every positive integer $n \in \mathbf{N}$, the following sets are (canonically) equivalent:*

*(i) The set of isomorphism classes of vector sheaves, with respect to $\mathcal{E}$, of rank $n \in \mathbf{N}$, over $X$, denoted by $\mathbf{\Phi}^n_{\mathcal{E}}(X)$* (cf. Chapt. II; (4.15)).

*(ii) The set of isomorphism classes of $n$-generated projective $E$-modules, denoted by $\mathcal{P}_n(E)$.*

*(iii) The set of isomorphism classes of $n$-generated projective $\mathcal{C}(X)$-modules, denoted by $\mathcal{P}_n(\mathcal{C}(X))$.*

*(iv) The set of isomorphism classes of continuous (complex) $n$-plane bundles over $X$.*

*In other words, one has the following (canonical) set-theoretic bijections:*

$$\begin{aligned} H^1(X, \mathrm{GL}(n,E)) &= H^1(X, \mathcal{GL}(n,\mathcal{E})) = \mathbf{\Phi}^n_{\mathcal{E}}(X) = \mathcal{P}_n(E) = \mathcal{P}_n(\mathcal{C}(X)) \\ &= H^1(X, \mathcal{GL}(n, \mathcal{C}_X)) = H^1(X, \mathrm{GL}(n, \mathbf{C})). \end{aligned} \tag{8.20}$$

**Proof.** First, it is standard that one has

$$H^1(X, \mathcal{GL}(n, \mathcal{C}_X)) = \mathcal{P}_n(\mathcal{C}(X)), \tag{8.21}$$

*for any $n \in \mathbf{N}$,* and *compact (Hausdorff) space $X$,* within a (set-theoretic)

*bijection* (*Serre-Swan Theorem;* see e.g. M. Karoubi [1: p.32, Theorem 6.18]). Thus, (8.21) identifies, in fact, the last three sets in the statement, the fifth one being identified with the 1st cohomology set, as in (8.21) (see, for instance, Chapt. V; Theorem 2.1, along with Chapt. VI; (4.11), applied to $\mathcal{C}_X$, in place of $\mathcal{C}_X^\infty$; cf. also (8.12) above, for $\mathbf{A} = \mathbf{C}$).

On the other hand, according to our hypothesis for $E$ (see Definition 3.2), the algebra in question is *unital commutative complete with an equicontinuous spectrum* $\mathfrak{M}(E) \equiv X$; therefore (S. Warner [1: p.7, Theorem 6]), a *Q-algebra,* as well (viz. *the group of units of* $E$ *is* an *open* subset of the latter algebra; see also e.g. [TA: p.187, Lemma 1.3 and p.184, Theorem 1.2]). Now, the following result has been obtained in A. Mallios [1: p.485, (5.6)] (an extension to the sort of topological algebras considered, as before, of an earlier result for *commutative unital Banach algebras,* due to O. Forster [1: p. 10, Satz 6]);

(8.22) *continuous (complex) n-plane bundles over a compact (Hausdorff) space* $X$ *are "algebraic"* (cf. (8.23) below), *relative to any commutative unital complete locally m-convex Q-algebra* $E$ (viz. to a *Waelbroeck algebra* [TA]), *whose spectrum* $\mathfrak{M}(E)$ *is homotopic to* $X$.

Thus, one obtains the following *bijections;*

$$\mathcal{P}_n(E) = \mathcal{P}_n(\mathcal{C}(\mathfrak{M}(E))) = \mathcal{P}_n(\mathcal{C}(X)), \tag{8.23}$$

*for every* $n \in \mathbf{N}$. (See also A. Mallios [2: p.297, (1.1), and p.298, Theorem 1.1, as well as, p.305, Appendix]). So our assertion follows now from (8.21) and (8.23), in conjunction with the fact that, for any topological algebra $E$, as in the statement, one further obtains;

$$\mathcal{P}_n(E) = H^1(X, \mathcal{GL}(n, \mathcal{E})), \tag{8.24}$$

within a *bijection* (see A. Mallios [1: p.488, Theorem 6.1, in particular, (6.2)]. In this context, we still remark that it is actually enough to consider here just *soft algebras,* as above, in place of *Pták-Šilov* ones as was the case therein). Finally, we also notice that the second equality (in fact, set-theoretic bijection) in (8.20) is already a consequence of Chapt. V; Theorem 2.1 (being valid, in effect, for any **C**-*algebraized space* whatsoever), and this also terminates the proof. (See also (8.25) below, concerning the 1st member of (8.20)). ∎

Now, taking in the previous Theorem 8.2, $n = 1$, *we have* the preceding *Theorem 8.1, more generally, for any soft locally m-convex algebra.* (We need *no involution* here, in particular, self-adjointness of the algebra involved; however, the *compact spectrum* of the topological algebra considered is a crucial issue here. So the previous argument does not apply to the type of topological algebras employed by Theorem 7.2 in the foregoing (see also Theorem 8.1 above). Thus, applying now a similar argument to that one in (8.16), one obtains (see Chapt. III; Theorem 6.2);

$$\mathcal{P}_n(E) = \mathcal{P}_n(\mathcal{C}(X)) = H^1(X, \mathcal{GL}(n, \mathcal{C}(X)) = H^1(X, \mathcal{GL}(n, \mathcal{E})) \\ = H^1(X, \mathrm{GL}(n, \mathbf{E})). \tag{8.25}$$

Here $\mathrm{GL}(n, \mathbf{E})$ stands for the *presheaf of groups* on $X$, defined by (cf. (2.3));

$$U \longmapsto \mathrm{GL}(n, \mathbf{E})(U) := \mathrm{GL}(n, E^{\wedge}|_U), \tag{8.26}$$

for any *open* $U \subseteq X \equiv \mathfrak{M}(E)$ (cf. also Theorem 8.2), along with the obvious *restriction morphisms.* Thus, taking first (2.19) into account (cf. also e.g. Chapt. I; (11.14)), it is easy to see that;

(8.27) *the above presheaf* $\mathrm{GL}(n, \mathbf{E})$, as given by (8.26), *is a mono-presheaf* (of groups) on $X$.

Furthermore, it is not hard to see, based on the very definitions, that;

$$\mathbf{S}(\mathrm{GL}(n, \mathbf{E})) = \mathbf{S}(\Gamma(\mathcal{GL}(n, \mathcal{E}))) = \mathcal{GL}(n, \mathcal{E}) \tag{8.28}$$

(see also Chapt. I; Theorem 9.1). In other words, one concludes that;

(8.29) *the sheafification of* $\mathrm{GL}(n, \mathbf{E})$ *is,* in effect, $\mathcal{GL}(n, \mathcal{E})$.

See thus Chapt. I; (7.23), (7.6) and (7.10), along with Chapt. IV; (1.28). Indeed, one proves that;

$$\mathrm{GL}(n, \mathbf{E})_f = \mathcal{GL}(n, \mathcal{E})_f, \tag{8.30}$$

*for every* $f \in \mathfrak{M}(E) \equiv X$ (loc. cit., along with (8.27) and (2.9.3) above). Yet, see Chapt. I; Theorem 3.1 and (2.14) in the foregoing, which thus justifies (8.28). ∎ So one finally concludes that;

(8.31) given a *soft locally m-convex algebra* $E$ (Definition 3.2), *every continuous (complex) n-plane bundle* $\xi$ *on* (a topological space *homotopic to*) $\mathfrak{M}(E) \equiv X$, the *spectrum of* $E$, can be *(uniquely) represented by a 1-cocycle, whose elements belong (locally) to*

(8.31.1) $$\mathrm{GL}(n, \mathbf{E})$$

(cf. (8.26)). That is, one has (cf. e.g. (8.18.1) and (8.19), concerning the notation applied);

(8.31.2) $$\xi \longleftrightarrow (g_{\alpha\beta}) \in Z^1(\mathcal{U}, \mathrm{GL}(n, \mathbf{E})).$$

In particular, by looking at (8.31.2), one has;

(8.32) $$\begin{aligned} g_{\alpha\beta} \in \mathrm{GL}(n, \mathbf{E})(U_{\alpha\beta}) &= \mathrm{GL}(n, E^{\hat{}}|_{U_{\alpha\beta}}) \cong \mathrm{GL}(n, E|_{U_{\alpha\beta}}) \\ &\subseteq \mathrm{GL}(n, \mathcal{C}(U_{\alpha\beta})) = \mathcal{GL}(n, \mathcal{C}_X)(U_{\alpha\beta}), \end{aligned}$$

*for any* $U_\alpha, U_\beta$ *in* $\mathcal{U}$ (a given *local frame of* $\xi$), *with* $U_\alpha \cap U_\beta \equiv U_{\alpha\beta} \neq \emptyset$. Thus, one infers that;

(8.33) *the 1-cocycle* $(g_{\alpha\beta})$, as in (8.32.1), *representing* $\xi$, *consists (locally) of invertible* $n \times n$ *matrices, with entries from the* given (topological) *algebra* $E$ ($\cong E^{\hat{}}$, cf. (8.31)).

The previous conclusion is, in effect, an *equivalent way of expressing* the fact *that;*

(8.34) *continuous (complex) n-plane bundles over a compact (Hausdorff) space* $X$ *are "algebraic"* (see (8.22), (8.23)).

**Scholium 8.1.**– In the case of *commutative unital Banach algebras* O. Forster (loc. cit., p.10) refers to the analogous result (cf. (8.22) above) as the *"Banach algebra analogue of Grauert's theorem"* (the latter classical result pertaining to a similar situation in the case of *holomorphic and continuous* **C**-*vector bundles on a Stein manifold;* cf. H. Grauert [1: p. , Satz ]). Thus, following O. Forster (loc. cit.), one has, according to (8.22) (see also (8.23)),

(8.35) *a Waelbroeck (locally m-convex) algebra analogue of Grauert's theorem.*

Indeed, (8.35) might also be construed as the analogue, within the present context, of the classical *"Oka's principle"* (see, for instance, [TA: p.300]). Yet, see also J. Wagner [1: p.166, Proposition], as well as, M.E. Novodvorskii [1: p.490, Theorem 2] for another, respectively, independent, point of view, concerning the above mentioned Forster's result.

On the other hand, as a byproduct of the previous discussion and, in particular, of (8.23), for $n = 1$, *one obtains;*

$$\mathcal{P}_1(E) = \mathcal{P}_1(\mathcal{C}(X)), \tag{8.36}$$

with $X \equiv \mathfrak{M}(E)$ the *spectrum* of the given *topological algebra* $E$, *as in* (8.22). Thus, employing the notion of the *Picard group of* $E$, (8.36) is written, by definition, equivalently, in the form;

$$\mathrm{Pic}(E) = \mathrm{Pic}(\mathcal{C}(X)) \tag{8.37}$$

(see also e.g. Chapt. V; Definition 4.1, along with (4.15), concerning the previously applied terminology. Yet, cf., for instance, N. Bourbaki [4: Chapt. II, p. 115, Proposition 7, as well as, Remark (2)], pertaining to the concept of the *Picard group of a* commutative unital ring, in general). Therefore, taking now (8.5) and (8.21) (for $n = 1$) into account, one obtains

$$\mathrm{Pic}(E) = \mathrm{Pic}(\mathcal{C}(X)) = H^1(X, \mathcal{C}^{\bullet}_X) = H^2(X, \mathbf{Z}), \tag{8.38}$$

within *isomorphisms of* the *abelian groups* involved. Accordingly, one now arrives at the following result, an extension to our case of an earlier conclusion of O. Forster (loc. cit., p.15, Satz 8), for *unital commutative Banach algebras.* Thus, we have the next.

**Theorem 8.3.** *Let $E$ be a Waelbroeck algebra* (viz. a *commutative unital complete locally m-convex Q-algebra*), *having spectrum $\mathfrak{M}(E)$. Then, one has;*

$$\mathrm{Pic}(E) = H^2(X, \mathbf{Z}), \tag{8.39}$$

*up to an isomorphism of abelian groups, where $X$ stands for any paracompact (Hausdorff) space, homotopic to $\mathfrak{M}(E)$.* ∎

**8.(a). Morphisms of short exact exponential sheaf sequences**

**(contn'd).**– We continue below, as already promised, with our second morphism of short exact sequences, that can be derived from (7.19), having the latter sequence as one of its constituents (see e.g. (8. ) in the sequel): Thus, suppose again, as in Theorem 7.1 above, that $E$ is a *soft self-adjoint locally m-convex *-algebra*, whose *Gel'fand sheaf* is $\mathcal{E}$, and let $\mathcal{E}^{\cdot}$ be the corresponding *sheaf of units of* $\mathcal{E}$ (cf. (2.8) and (6.15) in the foregoing). Moreover, let $\mathfrak{M}(E) \equiv X$ be the *equicontinuous spectrum* of $E$ (cf. Definition 3.2), thus a *compact (Hausdorff) space* (see Scholium 7.1).

Furthermore, consider now the *exponential sheaf morphism*

$$e : \mathcal{E} \longrightarrow \mathcal{E}^{\cdot} \tag{8.40}$$

(see Lemma 7.1), where thus $e$ *is the sheafification of the presheaf maps*

$$e_U : E^{\wedge}|_U \longrightarrow (E^{\wedge}|_U)^{\cdot} \tag{8.41}$$

with $U$ *open* in $X$ (see (7.10)), as they are defined by (7.4) (cf. also (7.8); in this context, see Chapt. I; Section 8, in particular, (8.4) and (8.10)). Thus, *for any (local) section* $s \in \Gamma(U, \mathcal{E}) \equiv \mathcal{E}(U)$, and $U$ *open* in $X$, one sets;

$$e(s) := \exp 2\pi i s \in \mathcal{E}^{\cdot}(U) = \mathcal{E}(U)^{\cdot} \tag{8.42}$$

(see (6.16)), *such that one has*

$$e(s)(f) := \exp 2\pi i s(f) \in \mathcal{E}^{\cdot}_f = \mathbf{E}^{\cdot}_f, \tag{8.43}$$

*for every* $f \in U \subseteq X \equiv \mathfrak{M}(E)$; of course, here $\mathbf{E}^{\cdot}_f$ stands for the *stalk at* $f \in U \subseteq \mathfrak{M}(E)$ of the *unit presheaf* over $\mathfrak{M}(E)$ *of* $\mathbf{E}$, the *Gel'fand presheaf of the* given (topological) *algebra* $E$ (see (2.7.1) and (8.17), along with Chapt. I; (9.16), concerning the last relation in (8.43)). Thus, by further considering the *"formal derivative"* of (8.43), one obtains (cf. also (7.1.1));

$$(\exp 2\pi i s(f))' = 2\pi i \cdot \exp 2\pi i s(f) \in \mathcal{E}^{\cdot}_f, \quad f \in U. \tag{8.44}$$

Therefore, one concludes that;

$$(\exp 2\pi i s)' = 2\pi i \cdot \exp 2\pi i s \in \mathcal{E}^{\cdot}(U), \tag{8.45}$$

*for every* $s \in \mathcal{E}(U)$, as in (8.42), the latter relation (8.45) *yielding* thus *the "formal derivative" of (8.43)*. Accordingly (see also A. Mallios [8: p. 157, Lemma 5.1]), by applying now to (8.42) the corresponding *Kähler derivation of* $\mathcal{E}$ (cf. (6.5)), one obtains

$$\partial(\exp 2\pi i s) = (\exp 2\pi i s)' \cdot \partial s = 2\pi i \cdot (\exp 2\pi i s) \cdot \partial s, \tag{8.46}$$

for any $s \in \mathcal{E}(U)$, as before. Hence, by considering, in turn, the *(kählerian) logarithmic derivation of* $\mathcal{E}$ (see (6.18)), one concludes, by virtue of (8.46) and (8.42), that;

$$\tilde{\partial} \circ e = 2\pi i \cdot \partial. \tag{8.47}$$

Now, the previous relation supplies, in effect, the desired *commutative diagram*, or even *morphism of short exact (sheaf) sequences*, as follows;

$$\begin{array}{ccccccccc} 0 & \longrightarrow & \mathbf{Z} & \longrightarrow & \mathcal{E} & \xrightarrow{\ e\ } & \mathcal{E}^{\cdot} & \longrightarrow & 1 \\ & & {\scriptstyle \varepsilon}\downarrow & & \| & & \downarrow{\scriptstyle \tilde{\partial}} & & \\ 0 & \longrightarrow & \ker \partial & \longrightarrow & \mathcal{E} & \xrightarrow{\ 2\pi i \cdot \partial\ } & \partial\mathcal{E} & \longrightarrow & 0 \end{array} \tag{8.48}$$

(see also (6.7), (6.8), as well as, Chapt. VI; Lemma 1.1 and Chapt. V; (6.13)). In conclusion, one thus infers that;

(8.49) *(8.48) holds true* for any *soft self-adjoint locally m-convex *-algebra,* or even for every *self-adjoint Fréchet-Šilov* (locally *m*-convex *-)*algebra.*

Thus, the types of topological algebras appeared in (8.49) provide, through their Gel'fand sheaves and the corresponding Kähler differential triads (see (6.8)), concrete *examples* of *exponential algebraized spaces* (see Chapt. V: Definition 6.1). It is also an immediate consequence of the very definitions that the same too are, in effect, *examples* of *strictly exponential algebraized spaces* (loc. cit., Definition 6.4), in the sense that one has the following *morphism of short exact (sheaf) sequences;*

$$\begin{array}{ccccccccc} 0 & \longrightarrow & \mathbf{Z} & \longrightarrow & \mathbf{C} & \xrightarrow{\ e \equiv \exp 2\pi i\ } & \mathbf{C}^{\cdot} & \longrightarrow & 1 \\ & & \| & & {\scriptstyle \varepsilon}\downarrow & & \downarrow & & \\ 0 & \longrightarrow & \mathbf{Z} & \longrightarrow & \mathcal{E} & \xrightarrow{\ e\ } & \mathcal{E}^{\cdot} & \longrightarrow & 1. \end{array} \tag{8.50}$$

Therefore, all in all, taking (8.48) and (8.50) together, one finally obtains the following *commutative diagram;*

(8.51)
$$\begin{array}{ccccccccc}
0 & \longrightarrow & \mathbf{Z} & \longrightarrow & \mathbf{C} & \xrightarrow{e\equiv\exp 2\pi i} & \mathbf{C}^{\cdot} & \longrightarrow & 1 \\
 & & \| & & \downarrow & & \downarrow{\scriptstyle \varepsilon} & & \\
0 & \longrightarrow & \mathbf{Z} & \longrightarrow & \mathcal{E} & \xrightarrow{e} & \mathcal{E}^{\cdot} & \longrightarrow & 1 \\
 & & {\scriptstyle \varepsilon}\downarrow & & \| & & \downarrow{\scriptstyle \tilde{\partial}} & & \\
0 & \longrightarrow & \ker\partial & \longrightarrow & \mathcal{E} & \xrightarrow{2\pi i\cdot\partial} & \partial\mathcal{E} & \longrightarrow & 0.
\end{array}$$

Furthermore, supplementing our previous information in (6.20), and based on (8.48), in conjunction with Chapt. VII; Lemma 7.1, one still obtains

(8.52)
$$\tilde{\partial}(\mathcal{E}^{\cdot}) = \partial(\mathcal{E}) = 2\pi i\cdot\partial(\mathcal{E}).$$

On the other hand, we already know (see (4.18), or even (4.19), in the preceding) that *the Gel'fand sheaf* $\mathcal{E}$ of a topological algebra $E$, as in (8.49) (in fact, something less is needed, loc. cit.) *is fine on the paracompact* (if not *compact,* when $E$ is soft) *Hausdorff space* $X \equiv \mathfrak{M}(E)$, so that (cf. Chapt. III; (8.24)) $\mathcal{E}$ *is acyclic;* therefore, one has

(8.53)
$$H^1(X,\mathcal{E}) = H^2(X,\mathcal{E}) = 0.$$

So we have here again, via the topological algebras considered in (8.49), another *concrete example* of a *strict sheaf exponential space* (see Chapt. VII; Definition 7.2). Thus, by looking further at the appropriate context of that chapter, one can conclude, for instance, that

(8.54) given a *topological algebra* $E$, *as in* (8.49), having *spectrum* $\mathfrak{M}(E) \equiv X$ and *Gel'fand sheaf* $\mathcal{E}$, *a line sheaf* $\mathcal{L}$ *on* $X$ *is* $\partial$-*flat if, and only if, one has*

(8.54.1)
$$c_{\mathcal{E}}(\mathcal{L}) = 0$$

(*Selesnick - Chern class of* $\mathcal{L}$; loc. cit.).

See Chapt. VII; Corollary 7.1, along with Definition 7.1. In this context, we also recall (ibid., (7.5), as well as, (7.20) and (7.23)) that the above class of $\mathcal{L}$ is given by the relation;

$$c_{\mathcal{E}}(\mathcal{L}) = \frac{1}{2\pi i}\tilde{\partial}(g_{\alpha\beta}), \tag{8.55}$$

where $(g_{\alpha\beta}) \in Z^1(\mathcal{U}, \mathcal{E}^{\cdot})$ is a *coordinate 1-cocycle of* $\mathcal{L}$ (see Chapt. V; (1.8), for $n = 1$. We still recall that we are here applying the *Kähler differential triad* (cf. (6.8) above), that is (canonically) associated with the given topological algebra $E$, as before). Furthermore, according to our previous Theorem 8.1, one may consider *the above cocycle* entirely as *being expressed (locally) through elements of the* (topological) *algebra* $E$ at issue (cf. (8.19), as well as, e.g. (8.32), for $n = 1$).

Concerning the same conclusion (8.54), as above, we also refer to Chapt. VII; Theorem 6.1, or even *Chapt. VIII;* Theorem 5.1 and, in particular, *Theorem 5.2* for further relevant *geometric consequences of* (8.54) (in fact, for *equivalent interpretations* of the same assertion). In this respect, one also remarks that;

(8.56) by considering the *Kähler differential triad,* defined by a *topological algebra* $E$, *as in* (8.49) (see (6.8)), one realizes that *a Bianchi datum* (cf. Chapt. VIII: Definition 7.1) *is* actually *obtained.*

See thus (6.13), by employing, at the same time, a corresponding *sheafification of Theorem 5.1* above, in the spirit of Section 6 of this Chapter. ∎

Thus, the aforementioned *Theorem 5.2 of Chapter VIII* acquires a meaning, within the present *concrete framework,* which is derived, according to the preceding, from the particular *topological algebra* $E$ at issue, *as in* (8.49). On the other hand, an *extension,* under appropriate presuppositions, *of the situation in* (8.54) above *to vector sheaves,* in general (viz. *not* necessarily *of rank* 1, as was the case before), is presently examined in the next Section 9.

# 9. Frobenius integrability condition (contn'd)

As already promised, we consider in the sequel the possibility of *extending,* under appropriate presuppositions, *Theorem 5.2 of Chapter VIII to* general *vector sheaves,* a fact that in the classical case (of complex $C^{\infty}$-vector bundles, for instance) is assured by employing the integrability condition of the heading of this Section (cf., for example, S. Kobayashi [1: p. 5]). In the particular case of *line sheaves,* as was the case so far (see e.g. the proof of the aforesaid theorem, in particular, (5.30) there, or even (8.54) above), the corresponding result was attained, due mainly to the *existence of the short exact exponential (sheaf) sequence,* as e.g. in (7.19), the latter permitting thus one to circumvent the said condition. However, in the general case of vector sheaves this seems to be possible only (so far!) under some *extra assumptions* on the part of the differential triad involved (see (9.3) below), this also including the consideration of the respective *structure sheaf,* as a suitable *topological algebra sheaf* on our *base space* $X$ (see (9.2) in the sequel).

Thus, suppose that we have, as in the preceding chapters, a *differential triad* (cf. Chapt. VI; (3.1), or even Chapt. VIII; (1.1))

$$(\mathcal{A}, \partial, \Omega^1) \tag{9.1}$$

on a *topological space* $X$, such that;

(9.2) $\mathcal{A}$ is a *topological algebra sheaf* (cf. Theorem 1.1), in such a manner that *the local section algebras of* $\mathcal{A}$, $\mathcal{A}(U) \equiv \Gamma(U, \mathcal{A})$, for every *open* $U \subseteq X$, *are unital* commutative $\sigma$*-complete locally m-convex* (topological) $\mathbf{C}$*-algebras.*

On the other hand, suppose also that $X$ is a *curvature space* (cf. Chapt. VIII: Definition 1.1), while we still assume that

$$0 \longrightarrow \partial\mathcal{A} \longrightarrow \Omega^1 \xrightarrow{d^1 \equiv d} d\Omega^1 \longrightarrow 0 \tag{9.3}$$

is an *exact sequence of* ($\mathbf{C}$-vector space) *sheaves.* Now, one obtains, *for every* $n \in \mathbf{N}$, a *matrix analogue of (9.3)* in the form of the following *exact sequence*

$$(9.4)\qquad \begin{aligned} 0 \longrightarrow \partial M_n(\mathcal{A}) \equiv M_n(\partial\mathcal{A}) \longrightarrow M_n(\Omega^1) \longrightarrow dM_n(\Omega^1) \\ \equiv M_n(d\Omega^1) \longrightarrow 0. \end{aligned}$$

Thus, assuming that (9.3) is exact, we shall prove that the same holds true for (9.4), for every $n \in \mathbf{N}$; so one actually infers that:

(9.5) *exactness of (9.3) is equivalent to that one of (9.4), for every* $n \in \mathbf{N}$.

Indeed, one can prove the assertion *fiber-wise,* our claim being then a consequence of the hypothesis for (9.3) and a standard argument, pertaining to the direct sum decomposition of the individual spaces *(stalks),* corresponding to (9.4), for any given $n \in \mathbf{N}$. (See also Chapt. I; Lemmas 12.2 and 12.3, as well as, e.g. N. Bourbaki [2: Chap. 2; p. 13, Proposition 7]).

On the other hand, due to our hypothesis for $\mathcal{A}$ (see (9.2), in conjunction with (7.1) in the preceding), one concludes that;

(9.6) *the exponential function,* in fact, *sheaf morphism e acts on* $\mathcal{A}$ (viz., in effect, *on the local (continuous) sections* of the latter sheaf).

That is, one sets;

$$(9.7)\qquad e(s) := \exp 2\pi i s,$$

*for any (local) section* $s \in \mathcal{A}(U)$ and $U$ *open* in $X$. Indeed, one can obtain (9.6) *for every matrix algebra sheaf* $M_n(\mathcal{A})$, $n \in \mathbf{N}$, *with* $\mathcal{A}$ *as in* (9.2), by first remarking that;

(9.8) $M_n(\mathcal{A})$ is a *topological algebra sheaf on* $X$ *of the same sort as* $\mathcal{A}$ (cf. (9.2)), apart, of course, of the commutativity, unless $n = 1$.

The assertion follows from the relation;

$$(9.9)\qquad M_n(\mathcal{A})(U) = M_n(\mathcal{A}(U)),$$

valid, *for any open* $U \subseteq X$, *with* $n \in \mathbf{N}$ (cf. Chapt. IV; (1.9)), in conjunction with the fact that *the second member of (9.9) is* still *a topological algebra of the same type as* $\mathcal{A}(U)$, when on the same algebra one considers the corresponding *cartesian product topology* (cf. also A. Mallios [1: p.458], as well as, the same author [3: comments at the beginning of p.465]) ∎ Therefore,

one obtains the following *exact (sheaf) sequence;*

$$0 \longrightarrow \mathbf{Z} \longrightarrow M_n(\mathcal{A}) \xrightarrow{e} M_n(\mathcal{A})^{\cdot} \equiv \mathcal{GL}(n, \mathcal{A}) \tag{9.10}$$

(see also Chapt. IV: (1.26)). Furthermore, based on (8.47), one still concludes that;

$$\tilde{\partial} \circ e = 2\pi i \cdot \partial, \tag{9.11}$$

with the obvious extension ("*section-wise*") to $M_n(\mathcal{A})$ and $M_n(\mathcal{A})^{\cdot}$ of the sheaf morphisms involved in (9.11). (In this concern, see also A. Mallios [8: p. 157, Lemma 5.1]).

So after the preceding preliminary discussion, we are now in the position to formulate the desired *general form of Theorem 5.2 of Chapter VIII.* That is, one has the following.

**Theorem 9.1.** *Suppose we are given a curvature space $X$ with corresponding curvature datum*

$$(\mathcal{A}, \partial, \Omega^1, d, \Omega^2) \tag{9.12}$$

(cf. Chapt. VIII; Definition 1.1), *whose structure sheaf $\mathcal{A}$ satisfies (9.2), while we also assume the following short exact sequence of* $\mathbf{C}$*-vector space sheaves on $X$;*

$$0 \longrightarrow \partial\mathcal{A} \longrightarrow \Omega^1 \longrightarrow d\Omega^1 \longrightarrow 0. \tag{9.13}$$

*Moreover, let $\mathcal{E}$ be a vector sheaf on $X$, with* $\operatorname{rk}\mathcal{E} = n \in \mathbf{N}$. *Then, the following are equivalent:*

*i) $\mathcal{E}$ is $\partial$-flat.*

*ii) $\mathcal{E}$ admits a flat $\mathcal{A}$-connection.*

**Proof.** We have already seen in Chapt. VIII; Theorem 5.1 that *prop. i)* of the statement *implies that $\mathcal{E}$ has a flat $\mathcal{A}$-connection,* under the much more general condition that $X$ is just a *curvature space* (thus, without any supplementary assumption for the structure sheaf $\mathcal{A}$ and the exactness of the sequence in (9.13)). So we prove now that *ii) implies i),* as well.

Namely, suppose that

$$\omega \equiv (\omega_U) \in C^0(\mathcal{U}, M_n(\Omega^1)), \tag{9.14}$$

is an *$\mathcal{A}$-connection 0-cochain of 1-forms* on $X$, defining the given $\mathcal{A}$-*connecti-*

*on,* say $D$, of $\mathcal{E}$ (cf. Chapt. VII; (2.39)). Thus, our objective now is to find a *gauge transformation*

$$h \in C^0(\mathcal{U}, \mathcal{GL}(n, \mathcal{A})) \tag{9.15}$$

in such a manner that one has (see also Chapt. VI; (1.28), (1.29));

$$\tilde{\partial}(h^{-1}) = \omega, \tag{9.16}$$

where $\omega$ is given by (9.14). (Again, what our supplementary assumption assures here is that (9.16) *has a solution, given that,* by hypothesis, $R = 0$ (cf. Chapt. VIII; (2.5)), the latter relation being a *necessary condition, in order (9.16) to have a solution;* see thus Chapt. VIII; (5.23), (5.24)). Now, by applying $d$ to (9.16) (or even, as we also say, by *"differentiating"* (9.16)), one obtains (cf. Chapt. VIII; (1.12))

$$d\omega = 0, \tag{9.17}$$

so that, in view of the exactness of (9.13), *one has*

$$\omega \in \ker d = \operatorname{im} \partial. \tag{9.18}$$

Therefore, *one obtains (locally)*

$$\omega = \partial(g), \tag{9.19}$$

*for some* $g \in M_n(\mathcal{A})(V) = M_n(\mathcal{A}(V))$, *with* $V$ *open* in $X$ (see also, for example, (9.9)). In this context, we still note that (9.19) *can be obtained by applying (9.18) fiber-wise.* Thus, one finally gets an *open covering of* $X$, for the individual members of which one has an analogous relation to (9.19), for any given $\omega$, as in (9.14); of course, one can choose too a common *open covering of* $X$, *corresponding to both* the relations (9.14) *and* (9.19).

Accordingly, in view now of (9.11) and (9.19), one obtains

$$\tilde{\partial}\left(e\left(\frac{1}{2\pi i} g\right)\right) = 2\pi i \cdot \partial\left(\frac{1}{2\pi i} g\right) = \partial(g) = \omega; \tag{9.20}$$

hence, by further setting

$$h := e\left(\frac{1}{2\pi i} g\right)^{-1}\Big|_V \in \mathcal{GL}(n, \mathcal{A})(V) = \mathrm{GL}(n, \mathcal{A}(V)) \tag{9.21}$$

(see (9.15), along with our previous comments on (9.19), as well as, Chapt. IV; (1.28)), one finally obtains

$$\tilde{\partial}(h^{-1}) = \omega, \tag{9.22}$$

which was actually looked for (cf. (9.16)). On the other hand, the last relation is equivalent, in effect, with the following one;

$$\mathrm{Ad}(h^{-1})\omega + \tilde{\partial}(h) = 0 \tag{9.23}$$

(cf. also, for instance, Chapt. VII; (2.17)). Therefore, $h$ (see (9.15)) *can now be construed as that gauge transformation, which sends* $\omega$, as the latter is given by (9.14), *to the 0-cochain*

$$\omega' \equiv (\omega'_V) \in C^0(\mathcal{U}, M_n(\Omega^1)), \tag{9.24}$$

*such that* one has;

$$\omega'_V = 0, \quad \textit{for every } V \in \mathcal{U} \tag{9.25}$$

(see also the comments following (9.19) above). Our previous argument is still justified by the corresponding *transformation law of potentials,* as the latter is given here by (9.23) (cf. also Chapt. VII; (2.24)). Accordingly, $\omega'$ as defined by (9.24) and (9.25), provides now a *complete parallelism for the* given *vector sheaf* $\mathcal{E}$ (cf. Chapt. VII; Definition 6.1 and (6.37)), which *equivalently,* means that $\mathcal{E}$ *is* $\partial$*-flat* (see Chapt. VIII; Theorem 5.1), that actually was our claim, and this also completes the proof of the theorem. ∎

Now, by the same token we have also proved, in effect, that;

(9.26) a *vector sheaf* $\mathcal{E}$ on a *topological space* $X$, *satisfying the conditions of Theorem 9.1, admits a flat* $\mathcal{A}$*-connection if, and only if,* $\mathcal{E}$ *has a trivial* $\mathcal{A}$*-connection.*

See thus Chapt. VIII; (6.4), or even (6.6), along with (9.22) above. (In this respect, cf. also (9.27) right in the sequel). ∎ Therefore, under the *supplementary topological assumptions for* $\mathcal{A}$ (cf. (9.2)), the preceding provide also the already promised *extension of Theorem 6.1 in Chapter VIII to vector sheaves,* in general. Thus, in sum, one concludes that;

(9.27) under the *hypothesis of Theorem 9.1,* one obtains, all in all, *five equivalent expressions of flatness for* a given $\mathcal{A}$*-connection of a vector sheaf.*

See the same Theorem 5.1 of Chapter VIII, along with (6.4) there; yet, see also Corollary 6.1 of that chapter for the analogous situation in the particular

case of *line sheaves* (no extra hypothesis for $\mathcal{A}$ is needed here). ∎

**Note.**- By looking more closely at the proof of the previous Theorem 9.1, one realizes that;

(9.28) the conclusion of *Theorem 9.1 holds true, under the assumption of (9.11), in place of (9.2).*

(See also, for instance, Chapt. VIII; (6.21)). Indeed, it was in effect (9.11), which we derived from the above discussion, through our extra hypothesis on $\mathcal{A}$, as in (9.2). (In this respect, see thus, in particular, (9.20) above).

**Scholium 9.1.**- Our previous remark in (9.28) can still be applied to the case of a *topological algebra sheaf,* as defined by the *Gel'fand sheaf* of a given *topological algebra* (cf. (2.44)), the latter being, in particular, of the type appeared, for instance, in (8.49), or even, more generally, in Lemma 7.1 above: Thus, according to that lemma, one has then the *exponential sheaf morphism, as in* (7.2) (cf. also (7.3)), so that *we* can *further obtain* (8.47). (Notice here that, concerning the last relation, we do not actually need the exponential morphism $e$, as before, to be *surjective,* therefore, the full hypothesis of Theorem 7.1, that is, *self-adjointness* of the topological algebras involved).

The above framework remedy thus the disadvantage that we might have, in the general case by taking a *Gel'fand sheaf,* as before, whose *local section algebras* are, namely, *not necessarily $\sigma$-complete* (as was, instead, the case with (9.2); see also (2.4) and (2.7.1)). Of course, we could start, in place of the Gel'fand presheaf, as in (2.7), with a *topological algebra presheaf* of the form;

$$U \longmapsto \overline{E^{\wedge}|_U} \subseteq \mathcal{C}_c(U), \tag{9.29}$$

along with the obvious *restriction* (algebra) *morphisms.* Here $U$ varies over the *open* subsets of the *spectrum* $\mathfrak{M}(E) \equiv X$ of a given *topological algebra* $E$, while we also assume for $X$ that it is, in particular, a *hereditarily $k$-space* (viz. we suppose that every open subset of $X$ is a *$k$-space;* take, for instance, a locally compact, or a 1st countable space. See also e.g. J.L. Kelley [1: p.231, Theorems 12, 13], as well as, concerning (9.29), A. Mallios [8: p. 146, (1.12)]).

As a result of our discussion in the previous Scholium 9.1, one thus infers that;

(9.30) *(9.28) remains valid for a Gel'fand sheaf $\mathcal{E}$*, as before, (that is, corresponding to a *topological algebra E, as in Lemma 7.1, or* more specifically, *in Theorem 7.1*), *provided that* one has the relation

(9.30.1) $$\ker d = \operatorname{im} \partial,$$

that is, *equivalently, that (9.13) be exact.* (Of course, one applies here the *Kähler differential triad* (cf. (6.8)), which can be associated with $E$, as above).

Now, it is still our purpose to give in the subsequent Section 11 a further *example of topological algebras, where (9.30.1) holds true,* by suitably changing the (differential) complex, under consideration.

# 10. Flat A-vector bundles vis-à-vis to ∂-flatness of the associated vector sheaves

We have seen already in the preceding (cf. (8.12)) the relation between **A**-*vector bundles* and *vector sheaves* on a given *topological space* $X$; namely, an object of the first type is thus (canonically) associated, through its local (continuous) sections, to a vector sheaf on $X$. Indeed, *the two notions coincide, under the aforesaid correspondence, for every unital* (commutative) *locally m-convex* (topological **C**-)*algebra,* at least. (See also Chapt. IV; Scholium 2.1, along with A. Mallios [5: p. 419, Section 5, especially the comments at the beginning of p. 420, and rel. (5.1) there]).

In particular, *flat* **A***-vector bundles are classified through the 1st cohomology set*

(10.1) $$H^1(X, \mathrm{GL}(n, \mathbf{A})),$$

where one considers

(10.2) $$\mathrm{GL}(n, \mathbf{A}) \equiv \mathrm{GL}(n, \mathbf{A})_X \hookrightarrow \mathcal{GL}(n, \mathcal{A})$$

as a *constant sheaf* (of groups) *on* $X$ with fiber the *discrete group* $\mathrm{GL}(n, \mathbf{A}) := M_n(\mathbf{A})^{\cdot}$, viz. the *group of units* of the full *matrix-algebra* $M_n(\mathbf{A})$, while the $\mathbf{C}$-*algebra sheaf* $\mathcal{A}$ is given by (8.9). (Yet, as usual, $n \in \mathbf{N}$ stands here for the *rank* of the vector bundles, respectively sheaves considered). In this concern, see also, for instance, Chapt. V; (5.1), (5.3), regarding the above relation (10.2). Furthermore, by complete analogy with the particular case of (continuous) *flat* $\mathbf{C}$-*vector bundles* on $X$ (take thus $\mathbf{A} = \mathbf{C}$; loc. cit.), one obtains that:

> *every flat* $\mathbf{A}$-*vector bundle on* $X$ *can be determined by a 1-cocycle,* say $(g_{\alpha\beta})$, *consisting of locally constant* $\mathrm{GL}(n, \mathbf{A})$-*valued maps; viz. one has*
>
> (10.3) (10.3.1)) $$(g_{\alpha\beta}) \in Z^1(\mathcal{U}, \mathrm{GL}(n, \mathbf{A})),$$
>
> where $\mathcal{U} = (U_\alpha)_{\alpha \in I}$ stands for a *"trivializing" open covering of* $X$ (the analogue here of a *local frame* for the corresponding vector sheaf).

(See loc. cit. (5.19), (5.20)). In this respect, we also note that $\mathrm{GL}(n, \mathbf{A})$ in (10.3.1) still stands for the corresponding *constant sheaf* on $X$, as in (10.2). On the other hand, it is a direct consequence of the very definitions that:

> (10.4) by considering the above group $\mathrm{GL}(n, \mathbf{A})$ as an ordinary *topological group* with the canonical topology it inherits from $M_n(\mathbf{A}) \cong \mathbf{A}^{n^2}$ (*cartesian product topology,* cf. Chapt. IV; (2.7), along with Scholium 2.1), the *sheaf of germs of locally constant* $\mathrm{GL}(n, \mathbf{A})$-*valued maps on* $X$ (hence, continuous) *is the same, as the constant sheaf* (10.2), as above.

In this respect, see also Chapt. I; Theorem 11.1. (Of course, an analogous conclusion to the previous one in (10.4) is valid too, quite generally, *if in place of the topological group* $\mathrm{GL}(n, \mathbf{A})$, as in (10.4), *one considers any topological space* $Y$, concerning the *constant sheaf,* determined by the corresponding *discrete space* $Y$). ∎

Now, suppose that the *topological space* $X$ considered is, in particular, *path-connected,* while $\mathbf{A}$ is still a *unital locally m-convex* (topological) *algebra.* Then, by analogy with the classical case (where one has $\mathbf{A} = \mathbf{C}$, see e.g. R.C.

Gunning [2: p.123ff]), one infers that;

(10.5) *for every flat* **A**-*vector bundle*

(10.5.1) $$\xi \in H^1(X, \mathrm{GL}(n, \mathbf{A}))$$

(cf. (10.1)), *there is associated an element*

(10.5.2) $$\tilde{\xi} \in \mathrm{Hom}(\pi_1(X), \mathrm{GL}(n, \mathbf{A})).$$

We call $\tilde{\xi}$, as above, the *characteristic representation of* $\xi$ (see also (10.6) below).

Thus, "Hom" in (10.5.2) above stands for the *set of* the *group morphisms* concerned, where $\pi_1(X)$ denotes, of course, the *fundamental* (or even *Poincaré*) *group of* $X$. So the previous element $\tilde{\xi}$, as in (10.5.2), yields by its definition, a *representation of* $\pi_1(X)$ *in the* (free locally convex) **A**-*module* $\mathbf{A}^n$, in such a manner that $\tilde{\xi}$ *characterizes,* in fact, $\xi$, *within an inner automorphism of* $\mathrm{GL}(n, \mathbf{A})$. Speaking formally, one thus obtains the following *bijection;*

(10.6) $$H^1(X, \mathrm{GL}(n, \mathbf{A})) = \mathrm{Hom}(\pi_1(X), \mathrm{GL}(n, \mathbf{A}))^{\mathrm{GL}(n, \mathbf{A})},$$

the second member of (10.6) denoting the *set of isomorphism classes of representations of* $\pi_1(X)$ *in* $\mathbf{A}^n$, *modulo inner automorphisms of* $\mathrm{GL}(n, \mathbf{A})$. (In this concern, see also A. Mallios [5: p. 412f] for further relevant details, as well as, consequences of (10.6). Yet, cf. e.g. R.C. Gunning [1: p. 186, Lemma 27], where the corresponding result to (10.6), for *principal G-bundles,* is proved *for any group G,* whatsoever).

Accordingly, by further applying the above rel. (10.6) to the particular case of **A**-*line bundles* (viz., equivalently (cf. (8.12)), to *line sheaves,* with respect to the **C**-algebra sheaf $\mathcal{A}$, as the latter is defined by (8.9)) on (the *path-connected* space) $X$ and for a *commutative* **A**, *as before,* one obtains the following (set-theoretic) *bijection*

(10.7) $$H^1(X, \mathbf{A}^\cdot) = \mathrm{Hom}(\pi_1(X), \mathbf{A}^\cdot);$$

the last relation *characterizes* thus *flat* **A**-*line bundles on* $X$, with **A** and $X$, as above.

**10.(a). $\partial$-flatness of A-vector bundles.**– We have seen already, according to (10.1) and (8.12), that

(10.8) $$H^1(X, \mathrm{GL}(n, \mathbf{A})) \hookrightarrow H^1(X, \mathcal{GL}(n, \mathcal{A})),$$

the second member of the last relation classifying $\mathbf{A}$-vector bundles, where $\mathcal{A}$ is given by (8.9) (cf. (8.12)), while the first one identifies flat $\mathbf{A}$-vector bundles on $X$. (Of course, (10.8) indicates just the resulting *canonical map between the cohomology sets* concerned, *in view of the* corresponding *map in* (10.2); see thus, for instance, Chapt. III; (11.31)).

On the other hand, suppose now that we are given a *differential triad* (cf. Chapt. VIII; (1.1))

(10.9) $$(\mathcal{A}, \partial, \Omega^1)$$

on a *topological space* $X$, where $\mathcal{A}$ is still given by (8.9). (See, for instance, a variant of the situation at issue, as supplied by the ensuing Example 10.1). Furthermore, *we still assume* here (a crucial hypothesis, for that matter, pertaining to our main conclusion in (10.11) below) *that one has;*

(10.10) $$\mathbf{A} \subseteq \ker \partial$$

(a natural assumption, otherwise, for the case at hand, corresponding to Chapt. VI; Lemma 1.1. See also e.g. (10.13) below).

Thus, based already on (10.3), one concludes now that;

(10.11) *every flat* $\mathbf{A}$*-vector bundle on* $X$ *has the corresponding vector sheaf* (of its local sections, cf. (8.12) or even (10.8)) *on* $X$ $\partial$*-flat.* Of course, *the converse holds true,* as well, *in the particular case* that one has (*generalized Poincar/'e condition,* see e.g. Chapt. IX; (3.7))

(10.11.1) $$\mathbf{A} = \ker \partial$$

(see the subsequent example).

**Example 10.1.**– Consider a (Hausdorff *connected*) $\mathcal{C}^\infty$*-manifold* $X$ and the $\mathbf{C}$*-algebra of* $\mathbf{A}$*-valued* $\mathcal{C}^\infty$*-funtions* on $X$, where $\mathbf{A}$ is a *unital commutative* $\sigma$*-complete locally* $m$*-convex* $\mathbf{C}$*-algebra.* Thus, the corresponding *structure sheaf* here is

(10.12) $$\mathcal{A} \equiv {}^{\mathbf{A}}\mathcal{C}^\infty_X = \mathcal{C}^\infty_X \varepsilon \mathbf{A},$$

where "$\varepsilon$" denotes the respective "$\varepsilon$-(topological) *tensor product*" (see A.

Mallios [1: p. 489] and [4: p. 392ff]). So one can prove now that

$$\ker \partial = \mathbf{A}; \tag{10.13}$$

indeed, one proves that *an* $\mathbf{A}$*-valued smooth funtion* on $X$ *is locally constant if, and only if,* (the respective germ of) *it belongs to* $\ker \partial$. (See, for instance, L. Nachbin [2: p. 124, Proposition 19.1 and its proof]).

Therefore (cf. (10.11)), a given (smooth) $\mathbf{A}$*-vector bundle* on $X$ *is flat if, and only if, it is* $\partial$*-flat;* hence (Chapt. VIII; Theorem 5.1) *if, and only if, it has a flat* $\mathcal{A}$*-connection.* In this context, we further remark that by assuming, in particular, the $\mathcal{C}^\infty$-manifold $X$ to be *paracompact* (Hausdorff), one concludes that;

(10.14) *every* (smooth) $\mathbf{A}$*-vector bundle* on $X$ *admits an* $\mathcal{A}$*-connection, where* $\mathcal{A}$ *is given by (10.12).*

Indeed, $\mathcal{A}$ *is a fine sheaf* on $X$, this being already the case for $\mathcal{C}^\infty_X$, so that *the same holds true, for every vector sheaf on* $X$ (see Chapt. III; (8.56)). Therefore, (10.14) is already a consequence of Chapt. VI; Theorem 16.1. ∎ Finally, we still note that *every continuous flat* $\mathbf{A}$*-vector bundle on* $X$, as above, *can be construed as a smooth one,* as well, by virtue of (10.3). (See also A. Mallios [5: p. 412, Remark 2.1], for relevant details thereon).

## 11. Watts-Selesnick complex

We consider below another complex, different from the de Rham-Kähler one (cf. (5.27) and/or (6.14) in the preceding), that can be associated with a *topological algebra* of an appropriate type (see e.g. (11.34) or even (11.37) in the sequel). In its algebraic formulation the notion, we are going to be dealt with, is the standard *Watts complex,* introduced by C.E. Watts [1] and pertaining to the cohomology of algebras (essentially à la Amitsur) with an application to the ring (algebra) of $\mathbf{R}$-valued continuous functions on a compact space $X$. (In this respect, see also, for instance, A.T. Kitchen [1] and A.A. Korobov [1]). Now, as already said, the study of Watts is quite *algebraic,* involving, in particular, tensor products of the algebra concerned. On the

other hand, independently of that work, an analogous complex has also been considered by S.A. Selesnick [1], who has employed, instead, (topological) *completed tensor products of Banach*(-Šilov) *algebras,* while was still looking at the *cohomology of the spectrum* of such algebras, viz. of the *compact space* $X$ in the particular case of Watts' ring $\mathcal{C}(X)$ (cf. also [TA: p.233, Theorem 1.2]). We call in the sequel the complex thus obtained the *Watts-Selesnick complex* of the topological algebra involved (see (11.21) below).

More generally, we shall finally consider in the sequel (see thus (11.34) below) *soft* (topological) *algebras* (cf. Definition 3.2), although we do start straight away by employing *as general topological algebras, as the case* at issue *permits.* Yet, for convenience, we recall, in brief, the construction of the aforementioned complex, within the present more general context, referring, on the other hand, to the above cited work of Selesnick for further relevant details, that can also be adapted straightforwardly to the present framework.

Thus, suppose that $E$ is a *unital locally convex* (topological) *algebra* (with *continuous multiplication* [TA]), so that one further defines the following sequence of *locally convex* $E$*-modules* and *continuous* $\mathbf{C}$*-linear morphisms;*

$$
(11.1)\qquad
\begin{aligned}
&E^0 := E\\
&E^1 := E^0 \underset{\pi}{\otimes} E = E \underset{\pi}{\otimes} E\\
&\cdots\cdots\cdots\cdots\cdots\cdots\\
&E^p := E^{p-1} \underset{\pi}{\otimes} E = E \underbrace{\otimes \cdots \otimes}_{p+1-\text{times}} E\\
&\cdots\cdots\cdots\cdots\cdots\cdots,
\end{aligned}
$$

*with* $p \in \mathbf{N}$, where $\pi$ denotes the corresponding *projective tensor product topology* (cf. [TA: p.376, Lemma 3.1]). So each one of the previous spaces $E^p$, $p \in \mathbf{Z}_+$, is made, in the obvious way, into an $E$*-module,* indeed, into a *locally convex* $E$*-module* (see, for instance, loc. cit.; p.365, (2.5)).

On the other hand, one also defines the following $\mathbf{C}$*-linear operators* (on decomposable tensors, extending then by $\mathbf{C}$-linearity);

$$
(11.2)\qquad d^p : E^p \longrightarrow E^{p+1}, \quad p \in \mathbf{Z}_+,
$$

such that one sets

$$
(11.3)\qquad d^p := \sum_{i=0}^{p+1} (-1)^i d_i^p,
$$

where one further defines

$$d_i^p : E^p \longrightarrow E^{p+1}, \quad i = 0, \ldots, p+1, \quad \textit{and} \quad p \in \mathbf{Z}_+, \tag{11.4}$$

according to the following relations (cf. also (11.1));

$$d_i^p(x_0 \otimes \cdots \otimes x_p) := x_0 \otimes \cdots \otimes x_{i-1} \otimes 1 \otimes x_i \otimes \cdots \otimes x_p, \tag{11.5}$$

*for any* $0 < i \leqq p$, *with* $p \in \mathbf{N}$. (Of course, $1 \equiv 1_E$ stands here, and in the sequel too, for the *identity element* of the given algebra $E$). Thus, we still set;

$$d_0^p(x_0 \otimes \cdots \otimes x_p) := 1 \otimes x_0 \otimes \cdots \otimes x_p, \tag{11.6}$$

as well as,

$$d_{p+1}^p(x_0 \otimes \cdots \otimes x_p) := x_0 \otimes \cdots \otimes x_p \otimes 1, \tag{11.7}$$

*for any* $p \in \mathbf{N}$ (extending thus (11.5) *for* $i = 0$ *and* $p+1$, respectively). Yet, one finally sets;

$$d_0^0(x) := 1 \otimes x, \quad \textit{as well as}, \quad d_1^0(x) := x \otimes 1, \tag{11.8}$$

*for every* $x \in E$, extending thus the preceding *for* $p = 0$ (see also (11.2) and (11.3)).

Accordingly, we further note at this place, in anticipation, that by virtue of (11.3), one has, in particular;

$$d^0 := d_0^0 - d_1^0, \tag{11.9}$$

so that, in view of (11.8), one obtains

$$d^0(x) = 1 \otimes x - x \otimes 1, \tag{11.10}$$

*for every* $x \in E$.

Now, it is an easy consequence of the very definitions, along that one of the tensor product topology $\pi$ (loc. cit.), that *all the previously defined* **C**-*linear maps are continuous;* hence, they can be extended by continuity to the respective *completed spaces.* Thus, one obtains the following extended *continuous* **C**-*linear maps:*

$$\widetilde{d^p} : \widetilde{E^p} \longrightarrow \widetilde{E^{p+1}}, \quad p \in \mathbf{Z}_+. \tag{11.11}$$

On the other hand, suppose that

$$\mathfrak{M}(E) \equiv X \tag{11.12}$$

is the *spectrum* of the given topological algebra $E$ (see beginning of Section 2 in the preceding). Thus, for every *continuous character* $f \in \mathfrak{M}(E)$, one can consider the map;

$$f^p := f \underbrace{\otimes \cdots \otimes}_{p+1-\text{times}} f : \widetilde{E^p} \longrightarrow \mathbf{C}, \quad p \in \mathbf{Z}_+, \tag{11.13}$$

given by the relation

$$(f \otimes \cdots \otimes f)(x_0 \otimes \cdots \otimes x_p) := f(x_0) \cdots f(x_p) \in \mathbf{C}, \tag{11.14}$$

*with* $x_i \in E$, $0 \leqq i \leqq p$ (for decomposable tensors), and then extending by linearity and continuity to the whole of $\widetilde{E^p}$, proving, in effect, that $f^p$ is also a *continuous character of the complete locally convex* $\mathbf{C}$*-algebra* $\widetilde{E^p}$. That is, one has (see (11.1))

$$\begin{aligned} f^p \in \mathfrak{M}(\widetilde{E^p}) = \mathfrak{M}(E^p) &= \mathfrak{M}(E) \underbrace{\times \cdots \times}_{p+1-\text{times}} \mathfrak{M}(E) \\ &\equiv \mathfrak{M}(E) \otimes \cdots \otimes \mathfrak{M}(E). \end{aligned} \tag{11.15}$$

In this context, the first equality above is referred to a *set-theoretic bijection,* which is also *continuous,* while the second one is a *homeomorphism,* with respect to the natural *(Gel'fand) topologies* of the spaces involved (see [TA: p.145, Lemma 2.1 and p.407, Lemma 1.1]; yet, see e.g. (2.1) in the preceding, concerning the aforesaid topologies). In particular, the same first equality in (11.15) above becomes, in fact, a *homeomorphism,* if $\mathfrak{M}(E)$ is *(locally) equicontinuous* (loc. cit., p.411, Theorem 1.2). Indeed, this will be, for instance, the case in the sequel when employing *soft algebras* (cf. Definition 3.2, as well as, (11.34) below). On the other hand, one has, of course (see (11.15)), that

$$\ker(f^p) \subseteq \widetilde{E^p} \tag{11.16}$$

is a *closed maximal* (2-sided) *ideal* in $\widetilde{E^p}$, so that, by still following S.A. Selesnick (loc. cit., p.317), we further denote by

$$\lambda_f^p : \widetilde{E^p} \longrightarrow (\widetilde{E^p})_{\ker(f^p)} \equiv (\widetilde{E^p})_{f^p} \tag{11.17}$$

the corresponding (canonical) *localizing map of* (the topological algebra) $\widetilde{E^p}$ at $f^p \in \mathfrak{M}(\widetilde{E^p})$, given by the relation

$$\lambda_f^p(z) := z/\otimes^{p+1} 1 \equiv [z] \in (\widetilde{E^p})_{\ker(f^p)}, \tag{11.18}$$

*for every* $z \in \widetilde{E^p}$. (In this concern, see also, for instance, A. Mallios [9] for a detailed account on *"localizations of topological algebras"*). Thus, we further set;

$$N^p := \bigcap_{f \in \mathfrak{M}(E)} \ker(f^p) \subseteq \widetilde{E^p}, \quad p \in \mathbf{Z}_+, \tag{11.19}$$

getting a *closed* (2-sided) *ideal of* $\widetilde{E^p}$, such that one has (cf. (11.11))

$$\widetilde{d^p}(N^p) \subseteq N^{p+1}, \tag{11.20}$$

*for every* $p \in \mathbf{Z}_+$. (See also S.A. Selesnick [1: p.318, Proposition 2.4]). Hence, one can now consider the following (quotient) complex

$$W_S(E) \equiv (\widetilde{E^p}/N^p, \widetilde{d^p})_{p \in \mathbf{Z}_+}, \tag{11.21}$$

which we call the *Watts-Selesnick complex of* $E$, that can thus be associated, according to the preceding, with every $E$, a *unital locally convex* **C**-*algebra* with *continuous multiplication.*

On the other hand, following further Selesnick (loc. cit.) and based on a suitable *sheafification of the* previous *complex* (11.21), one can compute the (Čech) *cohomology of the spectrum* of an appropriate topological algebra, via the aforementioned sheafified complex, hence, entirely, *in terms of the* given *algebra itself.* Yet, the same sheafified complex, as before, yields also *another example of an* $\mathcal{E}$-*complex* (with $\mathcal{E}$ a **C**-*algebra sheaf*), providing an *exact sequence of de Rham type* (see (11.44) below). This becomes also of importance, in view of the potential (differential-) geometric applications we were considering in the preceding chapters. We discuss these topics straight away in the ensuing Subsection.

**11.(a). Sheafification of the Watts-Selesnick complex.**– Suppose we are given an *algebra* $E$, satisfying the following conditions:

(i) $E$ is a *locally convex* (topological) **C**-*algebra* with *continuous multiplication* and an *identity element.*

(ii) The *spectrum* of $E$, $\mathfrak{M}(E) \equiv X$, is *locally equicontinuous* (viz. every $f \in \mathfrak{M}(E)$ has an *equicontinuous* neighborhood).

Therefore, by considering the *(topological tensor product) algebras* $E^p$, $p \in \mathbf{N}$, as in (11.1), one concludes that;

(11.22) *the completed algebras* $\widetilde{E^p}$, $p \in \mathbf{Z}_+$, are *(complete) locally convex algebras, which also satisfy the previous conds. (i) and (ii).*

See [TA: p.410, Lemma 1.2 and p.146, Lemma 2.2]. ∎ As a result, one thus obtains, concerning the *spectrum of* $\widetilde{E^p}$ (cf. also (11.12))

$$\mathfrak{M}(\widetilde{E^p}) = \underbrace{X \times \cdots \times X}_{p+1-\text{times}} \equiv X^{p+1}, \tag{11.23}$$

within a *homeomorphism* (of the *Gel'fand spaces* concerned; ibid., p.411, Theorem 1.2).

Thus, for each one of the preceding (complete) topological algebras $\widetilde{E^p}$, $p \in \mathbf{Z}_+$, one can consider the following *presheaf on* $X$; namely, for every *open* $U \subseteq X$, one defines (see also e.g. (2.26))

$$\begin{aligned} \mathbf{E}^p(U) := \left\{\alpha : U^{p+1} \longrightarrow \mathbf{C} : \alpha = \hat{z}|_{U^{p+1}}, \ \text{for some } z \in \widetilde{E^p}\right\} \\ \subseteq \mathcal{C}(U^{p+1}), \end{aligned} \tag{11.24}$$

where, of course, we set (cf. (11.23))

$$U^{p+1} := \underbrace{U \times \cdots \times U}_{p+1-\text{times}} \subseteq X^{p+1} \equiv \mathfrak{M}(\widetilde{E^p}), \tag{11.25}$$

for every open set $U \subseteq X$, as above. So the above correspondence, as given by (11.24), together with the obvious *restriction morphisms*, supplies already a *presheaf of (unital commutative)* **C**-*algebras on* $X \equiv \mathfrak{M}(E)$. (The same can still be construed as a *presheaf of topological algebras on* $X$ (in effect, *locally m-convex* ones), in view of the last issue in (11.24), by applying a similar argument here, as that one in (2.7) in the preceding; see also e.g. (2.27), (2.28). Hence (cf. Theorem 1.1), *an analogous remark is* actually *valid for the sheaf* (11.26), we are going to define presently below). Thus, we denote by

$$\mathcal{E}^p, \quad p \in \mathbf{Z}_+, \tag{11.26}$$

the *sheaf of* **C**-*algebras on* $X$, generated by the previous presheaf on $X$, as defined by (11.24). In particular, *for* $p = 0$, one obtains

$$\mathcal{E}^0 = \mathcal{E}, \tag{11.27}$$

where the second member of the last relation stands for the *Gel'fand sheaf of*

*the* given *topological algebra* $E$ *over* $\mathfrak{M}(E) \equiv X$ (cf. (2.3), (2.7.1), Definition 2.1 and Lemma 2.1).

Now, our next objective is to define the appropriate *"differentials"* (viz. **C**-*linear morphisms*) between the previous sheaves, as in (11.26) above, the same sheaves being, in fact, according to their definition (see e.g. (11.24)), $\mathcal{E}$-*modules on* $X \equiv \mathfrak{M}(E)$, with $\mathcal{E}$ the *Gel'fand sheaf* of $E$ (cf. (11.27)): Thus, we start by first defining the differentials, we are seeking for, between the elements of the *presheaf of* **C**-*algebras* on $X$, as the latter is given by (11.24). That is, *for every open set* $U \subseteq X$, we set

$$(11.28) \qquad \delta^p : \mathbf{E}^p(U) \longrightarrow \mathbf{E}^{p+1}(U), \quad p \in \mathbf{Z}_+,$$

such that one defines (cf. (11.24), (11.25));

$$(11.29) \quad \delta^p(\alpha|_{U^{p+1}})(f_0, \ldots, f_{p+1}) := \sum_{i=0}^{p+1} (-1)^i (\alpha|_{U^{p+1}})(f_0, \ldots, f_{i-1}, f_{i+1}, \ldots, f_{p+1}).$$

Thus, one obtains, based on the very definitions (cf. also (11.11));

$$(11.30) \qquad \delta^p(\hat{z}|_{U^{p+1}}) = \widehat{d^p(z)}|_{U^{p+2}}, \quad p \in \mathbf{Z}_+,$$

*for every* $z \in \widetilde{E^p}$ (see also loc. cit., p.319, Proposition 2.5, (i)), while we still define (cf. (2.3) and (2.7.1))

$$(11.31) \qquad \varepsilon : \mathbf{C} \longrightarrow \mathbf{E}^0(U) = \mathbf{E}(U) = E^{\hat{}}|_U : \lambda \longmapsto \varepsilon(\lambda) := \lambda \cdot 1_{E^{\hat{}}}|_U .$$

(Here **C** stands for the respective *constant presheaf* on $X$, defined by the (discrete) *set* **C**; cf. Chapt. I; (10.2)). Thus, it is clear, according to the same definitions, that (11.28) *defines a* **C**-*linear map,* indeed, *a* **C**-*linear morphism of the* (**C**-*vector space*) *presheaves concerned* (see (11.29), in conjunction with Chapt. I; (6.3), (6.5)). Therefore (Chapt. I; (8.4), (8.5) and Lemma 8.1), one thus obtains **C**-*linear morphisms of the* respective *sheafifications* of the same presheaves, viz. the maps

$$(11.32) \qquad \delta^p : \mathcal{E}^p \longrightarrow \mathcal{E}^{p+1}, \quad p \in \mathbf{Z}_+.$$

(We applied above an obvious *abuse of notation,* retaining the same symbol for the sheafifications of the maps in (11.28)). Of course, an analogous conclusion holds true for the map (11.31), referring now to the *constant sheaf* $\mathbf{C} \equiv \mathbf{C}_X$ *on* $X$ (see also Chapt. I; (4.1), (4.4') and (11.21)). Thus, one also obtains the following **C**-*linear morphism* (cf. (11.27), along with the above

remark on the notation applied);

$$\varepsilon : \mathbf{C} \longrightarrow \mathcal{E}^0 = \mathcal{E}. \tag{11.33}$$

Now, to get further a suitable *exact sequence of the sheaves,* in fact, $\mathcal{E}$*-modules, defined by* (11.26), (11.27) (in particular, they are here considered as **C**-*vector space sheaves* on $X$), along with the previous **C**-*linear morphisms, as in* (11.32), (11.33), we shall restrict ourselves to an appropriate class of topological algebras. So *hereafter we assume that;*

(11.34) $E$ is a *soft* topological *algebra* (cf. Definition 3.2), whose *spectrum* is $\mathfrak{M}(E) \equiv X$ (a *compact Hausdorff space* (cf. (3.12)).

As an application of (11.34), one obtains, for instance, the following useful relation (see also (11.36) below), pertaining to the *"local behaviour"* of $E$;

$$\mathcal{E}_f = E_{\ker f}, \tag{11.35}$$

*for any* $f \in \mathfrak{M}(E)$, the second member of (11.35) denoting the *localization of* $E$ *at* $f \in \mathfrak{M}(E)$. (See A. Mallios [9], for the case considered by (11.34), or even S.A. Selesnick [1: p.314, Proposition 1.1 (ii)] for Banach algebras). On the other hand, we have already noticed that

(11.36) *the Gel'fand sheaf of* $E$, $\mathcal{E} = \mathcal{E}^0$ (cf. (11.27)), *is a fine sheaf on* $X$.

See (4.18) and (4.19). ∎ Indeed, *the previous conclusion is* actually *valid for all the sheaves* $\mathcal{E}^p$, $p \in \mathbf{N}$, as in (11.26), as we shall presently see; however, *under* certain *additional assumptions for* $E$. Thus, *we* next *suppose that:*

(11.37) $E$ is a *soft locally convex* (topological) *algebra* (cf. (11.34)), *closed under* the formation of *completed projective* (topological) *tensor products.*

**Scholium 11.1.**– We have already considered from the beginning of this Subsection *locally convex algebras, satisfying the conds. (i) and (ii),* so that the respective (topological) tensor product algebras, as in (11.1), along with their completions, be of the same type (a fact, that we applied above; see thus (11.22), as well as, for instance, (11.23)). On the other hand, for technical

reasons, having to do, for example, with the application of Lemma 3.1 (*Local Theorem*), we also want the same (topological) tensor product algebras, as before (viz. the *completions* of the algebras appeared in (11.1), *for* $p > 0$), to *retain the softness* of the initial algebra: Thus, we refer to A. Mallios [4: Chapt. XIII] for details on this matter, within the context of (locally convex) topological algebras, in general; for example, *semisimplicity* is thus preserved, under *additional assumptions for the algebras* involved. Such an extra hypothesis is, for instance, the so-called *approximation property* (loc. cit., p.446), that has to be shared by the locally convex algebra, under consideration. Now, a familiar class of such (topological) algebras is, of course, the *nuclear* (locally convex) *algebras* (ibid., p.302f). Accordingly, a class of algebras satisfying (11.37) is that one of

(11.38) *nuclear Fréchet-Šilov* (locally convex) *Q-algebras.*

See also (3.10) in the preceding, concerning the terminology applied. An *example* of the above type of (topological) algebras is certainly supplied by the (*non-normed!*) algebra $C^\infty(X)$, where $X$ is a (connected Hausdorff) *2nd countable compact* $C^\infty$*-manifold.* In this respect, we still note, in passing, that:

(11.39) a (finite-dimensional connected Hausdorff) *2nd countable* $C^\infty$*-manifold* $X$ *is compact if, and only if, the algebra* $C^\infty(X)$ (endowed with the *Schwartz topology*) *is a* $Q$*-algebra.* Moreover, *the same topology is the only one making* $C^\infty(X)$ *an* $FQ$*-algebra.*

(See also loc. cit.; p. 227, Theorem 2.1, as well as, p. 187, Lemma 1.3, p. 183, Corollary 1.1 and p. 143, Proposition 1.1. Yet, concerning the last part of the assertion, cf. M. Fragoulopoulou [1: p. 967, Proposition 3.3, or even p. 966, Corollary 2.8, (iii)]). ∎ On the other hand, there do exist *Banach(-Šilov) algebras having the approximation property,* so that the previous cond. (11.37) is still in force, for the particular case at hand. The latter algebras have been employed, for instance, by S.A. Selesnick [1]. Moreover, *Fourier algebras* of the type considered by the next Section 12 are also Banach algebras of the previous sort. However, (topological) *algebras satisfying (11.38) cannot be Banach, unless they are finite-dimensional.* (See, for instance, A. Grothendieck [1: Chap. II; p.38, Corollary 2]).

Finally, we also note, supplementing thus our previous discussion in Section 3 in the preceding (see the comments following (3.10)), that;

(11.40) the algebra

$$\mathcal{O}(K) := \varinjlim_{U \supseteq K} \mathcal{O}(U) \tag{11.40.1}$$

of *"locally holomorphic"* **C**-*valued functions* on a *Stein compact* set $K$ of a (2nd countable) *Stein manifold* $X$ (when endowed with the respective inductive limit *van Hove* topology) is a

(11.40.2) *nuclear Pták-Šilov (locally m-convex) Q-algebra.*

In this context, see also A. Mallios [7: p. 318, Theorem 5.1], as well as, A. Grothendieck [1: Chap. II; p.56, Corollaire and p.48, Corollaire 1]. Yet, it is worth noting here that *the van Hove topology on* $\mathcal{O}(K)$, as above, *is the only one making it an* $\mathcal{L}FQ$*-algebra.* See M. Fragoulopoulou [1: p. 966, Proposition 3.1, or even Corollary 2.8, (i)]. ∎ Thus, one obtains, through (11.40.1), another *example* of a topological algebra *satisfying,* by virtue of (11.40.2), the *cond.* (11.37). On the other hand, *the class of* (topological) *algebras of the type (11.40.2) is,* of course, *bigger than the one defined by* (11.38). (Namely, *every Fréchet* (locally convex) *space is Pták (Krein-Šmulian Theorem);* see, for instance, H.H. Schaefer [1: p.162]).

Thus, assuming now that the given topological algebra $E$ *satisfies* (11.37), one first obtains an *extension of (11.35), for every* $p > 0$; that is, one has (cf. also (11.26) and (11.17))

$$\mathcal{E}^p_f = (\widetilde{E^p})_{f^p}, \quad p \in \mathbf{N}, \tag{11.41}$$

*for every* $f \in \mathfrak{M}(E) \equiv X$, within a **C**-*algebra isomorphism.* (See also S.A. Selesnick [1: p.320,Proposition 2.6]). On the other hand, as already said, one further concludes the following, supplementing thus (11.36);

(11.42) *each one of the sheaves* $\mathcal{E}^p$, $p \in \mathbf{N}$ (cf. (11.26)), is a *fine sheaf on* $X$.

See also loc. cit., p.319, Proposition 2.5. ∎ Furthermore (ibid.), one now proves that, *for every open* $U \subseteq X$, *the following sequence is exact* (see also (11.29), (11.31));

$$(11.43) \qquad 0 \longrightarrow \mathbf{C} \xrightarrow{\varepsilon} \mathbf{E}(U) \xrightarrow{\delta^0} \mathbf{E}^1(U) \xrightarrow{\delta^1} \mathbf{E}^2(U) \longrightarrow \cdots .$$

(In this context, we still observe that the relevant argument in loc. cit., p.319, proof of Proposition 2.5, (iii), can be based on [TA: p.447, Lemma 5.1], avoiding thus the use of *Fréchet algebras;* see also loc. cit., p.449, Scholium 5.1, as well as, the comments at the end of the above Scholium 11.1). Thus, one finally infers that *the next sequence of sheaves on* $X$ *is exact* too;

$$(11.44) \qquad \mathcal{E}^* : 0 \longrightarrow \mathbf{C} \xrightarrow{\varepsilon} \mathcal{E} \xrightarrow{\delta^0} \mathcal{E}^1 \xrightarrow{\delta^1} \mathcal{E}^2 \longrightarrow \cdots .$$

In other words, based thus on (11.36) and (11.42), one now concludes that;

(11.45) (11.44) *provides a fine* (torsionless) *resolution of (the constant sheaf)* $\mathbf{C}$.

Therefore, in view of the *compactness of* $X$ (cf. (11.37)), one obtains a $\Gamma_X$-*acyclic resolution of* $\mathbf{C}$ (cf. Chapt. III; (8.24)). Accordingly, based on the *abstract de Rham Theorem* (ibid.; (3.24) or even (8.25)), one now comes to the following basic conclusion, extension to our case of the analogous result of S.A. Selesnick [1: p.317, Theorem 2.1], for *Banach-Šilov algebras* having the property that *completed projective tensor products* of such algebras *are* still *of the same type* (viz. the corresponding condition here to (11.37) above; thus, one may consider, for instance (see also the previous Scholium 11.1), *Banach-Šilov algebras with the* (Banach) *approximation property*. Cf. the next Section 12). Thus, one has the following.

**Theorem 11.1.** *Let* $E$ *be a soft (locally convex) algebra, closed under the formation of completed projective* (topological) *tensor products* (see e.g. (11.38), (11.39), (11.40)), *with spectrum* $\mathfrak{M}(E) \equiv X$. *Moreover, let* $\mathcal{E}^*$ *be the fine resolution of* $\mathbf{C}$ *on* $X$, *as defined by (11.44), hence a* $\Gamma_X$-*acyclic resolution of* $\mathbf{C}$. *Then, one has;*

$$(11.46) \qquad \check{H}^*(X, \mathbf{C}) = H^*(X, \mathbf{C}) = H^*(\Gamma_X(\mathcal{E}^*)) \equiv W_S^*(E),$$

*within isomorphisms of the graded* $\mathbf{C}$-*algebras concerned.*

Indeed, by considering the *constant* ($\mathbf{C}$-*vector space) sheaf* $\mathbf{C}$ and its *fine resolution* (11.44) (cf. (11.45)), our claim about (11.46) is now just a consequence, as already said, of Chapt. III; (3.24.2). Thus, according to that relation, one has;

$$(11.47)\qquad (R^n\Gamma_X)(\mathbf{C}) = H^n(X,\mathbf{C}) \cong h^n(\Gamma_X(\mathcal{E}^*)) = \check{H}^n(X,\mathbf{C}),$$

*for every* $n \in \mathbf{Z}_+$, within *isomorphisms of* the $\mathbf{C}$-*vector spaces* concerned, the last relation being a consequence of Theorem 8.1 of that Chapter, which thus proves (11.46). ∎

We call the last item in (11.46) the *Watts-Selesnick cohomology of* $E$, the latter algebra being, as in (11.37). Specifically, one has

$$(11.48)\qquad W_S^*(E) \equiv H^*(\Gamma_X(\mathcal{E}^*)) := \{h^n(\Gamma_X(\mathcal{E}^*))\}_{n\in\mathbf{Z}_+}$$

(see also Chapt. III; (3.38.2), or even (1.13)).

On the other hand, concerning the *complex* $\Gamma_X(\mathcal{E}^*)$, one has (cf. Chapt. III; (3.8), along with (11.27) above)

$$(11.49)\qquad \begin{aligned} &0 \longrightarrow \Gamma(X,\mathcal{E}) \xrightarrow{\delta^0} \Gamma(X,\mathcal{E}^1) \xrightarrow{\delta^1} \Gamma(X,\mathcal{E}^2) \longrightarrow \cdots \\ &\cdots \longrightarrow \Gamma(X,\mathcal{E}^p) \xrightarrow{\delta^p} \Gamma(X,\mathcal{E}^{p+1}) \longrightarrow \cdots . \end{aligned}$$

Therefore, one obtains (same Chapter; (3.5). See also (11.47), (11.48) above);

$$(11.50)\qquad (R^n\Gamma_X)(\mathbf{C}) = h^n(\Gamma_X(\mathcal{E}^*)) := \ker\Gamma_X(\delta^n)/\operatorname{im}\Gamma_X(\delta^{n-1}),$$

*for every* $n \in \mathbf{Z}_+$. However, one really concludes that;

(11.51) *the cohomology of the complex* $\Gamma_X(\mathcal{E}^*)$ in (11.49), as this is given by (11.50), *is actually that one of the Watts-Selesnick complex*, as the latter is given by (11.21).

The above also justifies the terminology applied in (11.48). *Indeed, one has*

$$(11.52)\qquad \Gamma(X,\mathcal{E}^p) = \widetilde{E^p}/N^p, \quad p \in \mathbf{Z}_+,$$

within an *isomorphism of* $\mathbf{C}$-*algebras* (cf. also (11.19)); furthermore, one concludes here that

$$(11.53)\qquad N^p = \ker\tau^p,$$

where

$$(11.54)\qquad \tau^p : \mathbf{E}^p(X) \cong \widetilde{E^p} \longrightarrow \Gamma(X,\mathcal{E}^p)$$

is an *onto* $\mathbf{C}$-*algebra morphism,* given by the relation

$$(11.55)\qquad \tau^p(z) := [z]_{f^p} \in (\widetilde{E^p})_{f^p} = \mathcal{E}^p_f,$$

*for every* $z \in \widetilde{E^p} (\cong \mathbf{E}^p(X))$ *and* $p \in \mathbf{Z}_+$. (See (11.15), (11.24), (11.25), as well as, (11.35), (11.41) and (11.17); yet, cf. also S.A. Selesnick [1: p.321, proof of Theorem 2.1 and p.320, Proposition 2.7]). ∎ In particular, one obtains

(11.56) $$\Gamma(X, \mathcal{E}^0) = \Gamma(X, \mathcal{E}) = \tilde{E} = E,$$

based on (11.27) and the fact that the given *topological algebra* $E$, *as in* (11.37), is, by definition, *complete,* as well as, a *geometric* one, being, by our hypothesis, a *soft algebra* (see (3.14) and (3.4)).

Thus, recapitulating the preceding and taking also (11.51) into account, one can restate the previous Theorem 11.1 into the form;

(11.57) *the Watts-Selesnick cohomology of a topological algebra, satisfying* (11.37) (see (11.48)), *equals* (viz. is *isomorphic to*) *the Čech cohomology of its spectrum with complex coefficients.*

Finally, in this context and for other purposes, as well (see, for instance, A. Mallios [9]) we further remark that, in view of (11.35) (cf. also S.A. Selesnick [1: p.316, Corollary 1.6]), one concludes that;

(11.58) *the Gel'fand sheaf of a soft locally m-convex algebra* (see, for example, (11.40.1)) *is strictly henselian* (in the sense that its *stalks* are *strictly henselian local* **C**-*algebras*).

Now, another point that is worth pointing out at this place, in connection with our considerations in the preceding chapters, is the following fact. Namely,

(11.59) *the zero-th dimensional operator*

(11.59.1) $$\delta^0 \equiv \partial : \mathcal{E} \longrightarrow \mathcal{E}^1$$

(see (11.32)) *supplies,* in effect, *a* **C**-*derivation* (viz. a *flat* $\mathcal{E}$-*connection*) *of the* **C**-*algebra sheaf* $\mathcal{E}$ *into the* $\mathcal{E}$-*module* $\mathcal{E}^1$.

Indeed, by virtue of (11.10), one obtains

(11.60) $$\partial(s \cdot t) = \partial(t) \cdot s + t \cdot \partial(s),$$

*for any (local) sections* $s, t$ *in* $\mathcal{E}(U)$, where $U \subseteq X$ is *open;* thus, one actually applies (11.10) to the corresponding individual element of the *presheaf*

(11.61) $$\left\{\mathbf{E}^0(U) = E^\frown|_U; \sigma_V^U\right\}$$

(see (11.24) and (2.3)), verifying thus (11.60) for the case at hand, which in turn is obtained by *sheafification* (see also (11.35)). An analogous argument can be applied, of course, to prove that $\partial$ is a $\mathbf{C}$-*linear morphism of* the $\mathbf{C}$-*vector space sheaves* involved (see also Chapt. VI; Definition 1.1). ∎

Therefore, one concludes that;

(11.62) the triple

(11.62.1) $$(\mathcal{E}, \partial, \mathcal{E}^1)$$

yields a *differential triad,* in the sense of Chapt. VI; (1.4), *for which the* corresponding *sequence (11.44) is exact.*

We call (11.62.1) the *Watts-Selesnick differential triad,* which thus can be associated with any given *topological algebra, satisfying (11.37).*

Of course, the above statement makes it possible to apply, within the framework of (11.62.1), several of our considerations in the preceding chapters of a *differential-geometric nature* (see thus, for instance, Chapt. VI; (16.17)). In particular, by considering *soft topological algebras, as in* (11.37), which are, moreover, *self-adjoint locally m-convex *-algebras,* the relevant considerations of the foregoing, concerning *exponential sheaf spaces* and the like, are here too applicable (see also (7.13), or even Theorems 7.1 and 7.2 in the foregoing).

## 12. Further applications

We consider in this Section two more particular instances, where the technique developed in the foregoing could be applied. So we begin with the following example, pertaining to topological groups, with an eventual special *differential-geometric application to* the case of (compact) *Lie groups.* The same example supplies also, as already mentioned in the preceding, an important particular instance of a *Banach-Šilov algebra, having the approximation*

*property;* as we have seen in the previous Section 11, the latter class of topological algebras, together with its pertinent generalizations, constituted the appropriate framework for that discussion. Yet, as already said, this class of (Banach) algebras, along with the example we discuss below, was also the case that has been considered earlier by S.A. Selesnick [1: p.321, §3].

**12.(a). Fourier algebras.**– For convenience, we first recall, in brief, the relevant basic terminology: So, to start with, suppose that we are given a *group* $G$ (not necessarily abelian); thus, a **C**-valued function $\phi$ on $G$ is said to be *positive-definite* (or even a *function of positive type*), whenever one has;

$$\sum_{i,j} \lambda_i \bar{\lambda}_j \phi(x_i x_j^{-1}) \geqq 0, \tag{12.1}$$

for any (finite-many) elements $x_1, \ldots, x_n$ in $G$ and $\lambda_1, \ldots, \lambda_n$ in **C**. (Yet, we may restrict ourselves to *distinct elements* of $G$, as before).

In the sequel we denote by $P_+(G)$ the *set of* **C**-*valued positive-definite functions* on a given group $G$, while we also set

$$P(G) = [P_+(G)] \tag{12.2}$$

for the **C**-*vector space spanned by the set* $P_+(G) \subseteq \mathbf{C}^G$. (The latter item here stands, of course, for the **C**-*vector space of* **C**-*valued maps on* $G$, with *point-wise defined operations*).

On the other hand, it is not hard to see that;

(12.3) in the case of a *topological group* $G$, $\mathcal{P}(G)$, viz. the **C**-*linear span of* the set of *all continuous positive-definite functions on* $G$, $\mathcal{P}_+(G)$, with point-wise defined operations (cf. (12.2)), is a *unital commutative self-adjoint* (relative to complex conjugation) *subalgebra of* $\mathcal{C}(G)$ (algebra of **C**-valued continuous functions on $G$).

See, for instance, E. Hewitt-K.A. Ross [1: p.260, Theorem 32.10] ▮

On the other hand, a fact that also interests us here, is the interpretation of *the algebra* $\mathcal{P}(G)$, as in (12.3) above, *as* the space of *absolutely convergent Fourier series on* the given topological group $G$. Now, as already said at the beginning of this Subsection, $G$ is an arbitrary *group* viz. *not necessarily abelian;* on the other hand, *we assume henceforth that;*

(12.4) $G$ is a given *compact* (Hausdorff topological) *group.*

Thus, the familiar from the abelian case character group (i.e., *1-dimensional representations*) of $G$ has to be replaced here by *finite-dimensional,* in general, *representations of* $G$; indeed, we know that *continuous irreducible (unitary) representations of a compact group are finite-dimensional* (see, for instance, M.A. Naĭmark [1: p. 442, Theorem 4]). So we denote by

$$\hat{G} \equiv K \tag{12.5}$$

the *set of equivalence classes* of such representations of $G$. (The name *dual object,* or just *dual,* of $G$ is also of use for $K$). On the other hand, there exists a *one-to-one and onto correspondence* (modulo, of course, the appropriate identifications) *between continuous positive-definite functions on a locally compact group $G$ and continuous unitary representations of $G$* (ibid.; p.400, Theorem 1), connecting thus the two notions in (12.3) and (12.5), as above, concerning, in particular, the given compact group $G$.

Thus, if $\alpha \in K$, as in (12.5), we denote by $\pi_\alpha$ any *representative of the class $\alpha$* at issue, hence, a *continuous irreducible (unitary) representation* of the *compact group $G$* in, say, an *$n_\alpha$-dimensional Hilbert space $H_\alpha$, with* $n_\alpha \in \mathbf{N}$. Yet, we further denote by $dx$ the (unique) *Haar measure* on $G$, which we choose thus *normalized,* viz. we assume that

$$\int_G dx = 1. \tag{12.6}$$

Now, for every $f \in L^1(G)$ (viz. the *group algebra* of $G$), one further defines the *Fourier series* of $f$

$$\sum_{\alpha \in K} n_\alpha \operatorname{tr}(T_\alpha \pi_\alpha) \tag{12.7}$$

(here "tr" stands for the *trace* of the corresponding operator (matrix) in $L(H_\alpha) \equiv \mathcal{L}(H_\alpha)$, cf. also (12.9) below), such that the *Fourier coefficient* is the following (continuous) endomorphism of $H_\alpha$

$$T_\alpha = \int_G \pi_\alpha(x^{-1}) f(x)\, dx \in \mathcal{L}(H_\alpha), \tag{12.8}$$

where one sets;

$$T_\alpha(z) = \int_G \pi_\alpha(x^{-1})(z) f(x)\, dx \in H_\alpha, \tag{12.9}$$

*for every* $z \in H_\alpha$. (Of course, one has here a typical case of an *"operator-valued" integration;* see e.g. loc. cit., p.149, §19). Thus, we say that the *Fourier series* in (12.7) is *absolutely convergent,* whenever one has;

(12.10) $$\sum_{\alpha \in K} n_\alpha \mathrm{tr}\, |T_\alpha| < +\infty.$$

Here we have set (cf. also (12.8))

(12.11) $$|T_\alpha| := \sqrt{T_\alpha T_\alpha^*}$$

in the (finite-dimensional) $C^*$*-algebra* $\mathcal{L}(H_\alpha)$ of the *(finite-dimensional) Hilbert space* $H_\alpha$ (viz. the *representation space of* $\pi_\alpha$, as above). So in that case *one obtains;*

(12.12) $$f(x) = \sum_{\alpha \in K} n_\alpha \mathrm{tr}\, (T_\alpha \pi_\alpha(x)),$$

*for every* $x \in G$. (in this respect, we further note that here, as well as, in (12.10), *the set of those* $\alpha \in K$, *with* $|T_\alpha| > 0$, *is denumerable;* see E. Hewitt - K.A. Ross [1: p.331, Theorem 34.5, or even p.81, Corollary 28.41]). Thus, the fundamental result that interests us here, is the following fact:

(12.13) the set of *those* $f \in L^1(G)$ which possess *absolutely convergent Fourier series* on $G$ (cf. (12.10)), constitutes a *unital commutative* **C**-*algebra,* being actually (in the *norm defined by* (12.10), see also (12.12)) a *Banach algebra;* the same algebra is also *self-adjoint regular* and *semisimple.* We denote it by

(12.13.1) $$A(G),$$

while its *spectrum* is given by

(12.13.2) $$\mathfrak{M}(A(G)) = G,$$

within a *homeomorphism.*

We call $A(G)$ the *Fourier algebra of the compact group* $G$, under consideration.

Thus, in other words, one concludes that;

(12.14) *the Fourier algebra* $A(G)$ *of a compact* (Hausdorff) *group* $G$ *is a* (complex) unital commutative *self-adjoint Banach-Šilov algebra, whose spectrum is* (homeomorphic to) $G$.

Furthermore, connecting the preceding with (12.3), one also proves that

$$(12.15) \qquad A(G) = \mathcal{P}(G),$$

*for any given compact group* $G$, so that the *operations in* $A(G)$ can still be defined *point-wise*. (Details on all the foregoing and relevant proofs can be found in E. Hewitt - K.A. Ross [1: p.336, Theorem 34.13], concerning (12.15), while for (12.13) see ibid., p.502, Theorem 39.31, p.340, Theorem 34.20, along with the relevant comments at the end of its proof, as well as, p.260, Theorem 32.10). ∎ On the other hand, one finally concludes that;

(12.16) *the Fourier algebra* $A(G)$, as in (12.13.1), *satisfies the* (Banach) *approximation property*.

See (12.13), as well as, A. Grothendieck [1: Chap. I; p.187, Corollaire 2], in conjunction with the cited work of Hewitt and Ross; p.260, Theorem 32.10, p.341, Remark 34.22 and p.346, Theorem 34.26. ∎

**Scholium 12.1.** In the case that $G$ is, in particular, a *compact abelian group,* then the above argument is reduced to the standard *Gel'fand description* of the corresponding situation: Namely, one can then consider the (complex) *unital commutative Banach algebra*

$$(12.17) \qquad l^1(\widehat{G}),$$

(viz. the respective *group algebra* of $\widehat{G}$), where now $\widehat{G}$ stands for the *(discrete) character group* of $G$. Thus, by further considering the respective *Gel'fand map of* $l^1(\widehat{G})$ (cf. (2.21) in the preceding, along with (12.19) below) one now defines the *Fourier algebra* of $G$, $A(G)$ (we retain the same notation for the latter algebra, as before), according to the following relation (see also, for instance, (2.1) and (2.2));

$$(12.18) \qquad A(G) := l^1(\widehat{G})^{\widehat{}} \subseteq \mathcal{C}(G).$$

Concerning the last item of the previous relation, one has, in effect, regarding the *spectrum of* $l^1(\widehat{G})$, the following *homeomorphisms;*

$$(12.19) \qquad \mathfrak{M}(l^1(\widehat{G})) = \widehat{\widehat{G}} = G,$$

where the second one is due to *Pontryagin's duality.* (In this respect, see also, for instance, [TA: p. 237, Theorem 4.1], as well as, M.A. Naĭmark [1: p. 415ff,

in particular, p. 425, Theorem 5]).

On the other hand, it is a standard result that *the group algebra of any locally compact* (Hausdorff) *abelian group,* like (12.17) above, *is regular and semisimple.* (See, for example, the above cited book of Naĭmark; p. 427, Theorem 7 and p.380, VII; in this respect, see also loc. cit., p. 376, concerning the *unital* group algebra (12.17), as before). Therefore, one infers that (cf. also (12.19));

(12.20) $l^1(\widehat{G})$, as in (12.17), is a *unital commutative Banach-Šilov algebra,* whose *spectrum* is (homeomorphic to the given compact group) $G$. Furthermore, *the same algebra satisfies the* (Banach) *approximation property.*

Concerning our last assertion in (12.20), see also e.g. A. Grothendieck [1: Chap. I; p. 185, Proposition 41]. ∎

Now, the *Fourier algebra of* $G$, $A(G)$, as defined by (12.18), *is* further *topologized by the relation*

$$l^1(\widehat{G}) \underset{\mathcal{G}}{\cong} l^1(\widehat{G})^{\widehat{}}, \tag{12.21}$$

where $\mathcal{G}$ stands for the respective *Fourier(-Gel'fand) transform of* (the *semisimple* Banach algebra) $l^1(\widehat{G})$ (cf. (2.21), (2.22) amd (12.20) above). Thus, in view of (12.21), $A(G)$ becomes a *Banach algebra isomorphic* (as such) *to* $l^1(\widehat{G})$. Now,

(12.22) it is actually *under the previous topologization,* through (12.21) (in conjunction with (12.18)), that *one understands the Fourier algebra* $A(G)$ *of a* given *compact abelian group* $G$, *as a Banach algebra.*

(In this respect, see also e.g. H. Reiter [1: p.104, (4)]). Therefore, one infers that;

(12.23) *the Fourier algebra* $A(G)$, as before, *being* (by its very definition, as above) *Banach algebra-isomorphic to* $l^1(\widehat{G})$, *shares* the same properties with the latter algebra, hence, (12.20) *too.*

Of course, our previous discussion, throughout the present Scholium, has only the meaning of a more direct access to $A(G)$, than in (12.13.1), due now, according to our hypothesis, to the *commutativity* of the given compact

group $G$.

Accordingly, based now on the foregoing, we conclude that;

(12.24) given a *compact group* $G$ (abelian or not), its *Fourier algebra* $A(G)$ (cf. (12.18), (12.22), or (12.13.1)) is a *(complex) unital commutative self-adjoint Banach-Šilov algebra, having the* (Banach) *approximation* property, and whose *spectrum* is (homeomorphic to) $G$.

Thus, as an application of the preceding considerations, one realizes that, by looking at the *Fourier algebra* $A(G)$ of a *compact group* $G$, one obtains a very convenient indeed *"geometric"* algebra (even in the technical sense of the term, as in Section 3 above), in such a manner that the *differential geometric applications,* as exhibited in the previous chapters of this treatise, *are* here too *available.* Thus, suppose, in particular, that;

(12.25) $G$ is a given *compact* (Hausdorff) *Lie group* (abelian or not).

Consequently, as an *application of the previous discussion,*

(12.26) one can formulate several *differential geometric notions and* obtain related *results, in terms of the,* eventually more convenient, *Fourier algebra of* $G$, $A(G)$, *than of the algebra* $\mathcal{C}^\infty(G)$ (of $\mathbf{C}$-*valued* $\mathcal{C}^\infty$*-functions on* $G$), that traditionally is associated with $G$ in the case, under consideration.

What amounts to the same thing, one thus concludes that;

(12.27) one can do *differential geometry on a* given *compact Lie group* (at least, to the extent we consider it in the preceding Chapters), *by means of absolutely convergent Fourier series,* rather, *than of* $\mathcal{C}^\infty$*-functions* (as usual) on the group.

**12.(b). Formal power series algebras.**- We terminate our discussion, by exhibiting one more *example* of a *topological algebra,* which indeed provides a *resolution of* $\mathbf{C}$, through an appropriate *"de Rham complex",* that is associated with such an algebra, by constructing the pertinent to the algebra at issue *modules of "differentials".*

Now, as the heading of this subsection indicates, the example we are going to consider below, contains as a particular instance the *algebra of,* say, *complex polynomials in n* (independent) *variables,*

(12.28) $$E \equiv \mathbf{C}[t_1, \ldots, t_n],$$

that also exhibits an analogous situation, as that one mentioned above: Namely, by taking, for example, the corresponding to $E$ *de Rham-Kähler complex,* as in (5.27), one obtains a *resolution of* $\mathbf{C}$, in the sense that *the following sequence of* $\mathbf{C}$*-vector spaces is exact;*

(12.29) $$0 \longrightarrow \mathbf{C} \longrightarrow E \longrightarrow \Omega_E^1 \longrightarrow \cdots \longrightarrow \Omega_E^n \longrightarrow 0.$$

In this context, one considers on $E$, as given by (12.28), the *discrete topology,* so that $E$ is just a $\mathbf{C}$*-algebra.* (We do not make actually here any use of the aforementioned topology; see also S. Lang [1: p.612, Theorem 12.3]). ∎

On the other hand, one can consider on the same algebra $E$, as in (12.28), a canonical topology, making it into a *locally m-convex* $\mathbf{C}$*-algebra,* via the relation;

(12.30) $$E \equiv \mathbf{C}[t_1, \ldots, t_n] \overset{\subsetneq}{\rightarrow} \mathcal{C}_c(\mathbf{C}^n),$$

such that, concerning the *spectrum* of $E$, one has

(12.31) $$\mathfrak{M}(E) \equiv \mathfrak{M}(\mathbf{C}[t_1, \ldots, t_n]) = \mathbf{C}^n,$$

within a *homeomorphism.* In this context, *we* still *remark, in passing, that;*

(12.32) *every maximal ideal of* (the unital commutative locally $m$-convex $\mathbf{C}$-algebra) $E \equiv \mathbf{C}[t_1, \ldots, t_n]$, as the latter algebra is topologized by (12.30), *is closed,* given by a point of $\mathbf{C}^n$ (*Nullstellensatz*).
On the other hand, $E$ *cannot be a Q-algebra.*

The last assertion is true, since, otherwise, $E$ would have a *compact* (viz. closed and equicontinuous) *spectrum* [TA: p.187, Lemma 1.3]), a contradiction, of course, to (12.31). ∎

Now, the same (polynomial) *topological algebra* $E$, as in (12.30), is also *semisimple,* a direct consequence of the very definition, in conjunction with (12.30), (12.31) (the corresponding *Gel'fand map* of $E$ being also *continuous*). Yet, $E$ is *metrizable* (as a topological algebra, $\mathbf{C}^n$ being a *hemicompact* space) and *self-adjoint* (with respect to complex-conjugation); therefore, its

*completion,* being, in effect, the whole algebra $\mathcal{C}_c(\mathbf{C}^n)$ (Stone-Weierstrass, cf. e.g. L. Nachbin [1: p.48, Corollary 2]) is a *self-adjoint Fréchet-Šilov (locally m-convex) algebra,* whose *spectrum* is again *(homeomorphic to)* $\mathbf{C}^n$ (cf. [TA: p.223, Theorem 1.2]). Therefore, one can further apply here our previous considerations in Section 7 (see, for instance, Theorem 7.2), concerning, for example, the *existence of a sheaf exponential sequence* (cf. (7.19)).

Thus, after the above prelude, we come back again to a description, in brief, of the situation one has in the more general case of a *formal power series* **C**-*algebra* (in *finite-many variables*). That is, of an algebra, say, of the form;

$$C_n \equiv \mathbf{C}[[t_1, \ldots, t_n]]. \tag{12.33}$$

Now, concerning the purely algebraic part of our exposition, see e.g. T.W. Hungerford [1: p.154ff], while for the rest we mainly follow the relevant account in the work of H. Grauert - R. Remmert [1]. So (12.33) is, by its very definition, first, a *unital commutative* **C**-*algebra,* being also a *local* **C**-*algebra.* Thus, the *unique maximal ideal of* $C_n$ consists of those *formal power series without* (viz. having zero) *constant term,* the rest elements of $C_n$ yielding its *group of units* (cf. e.g. Hungerford, loc. cit.; p.155, Corollary 5.10).

Indeed, we restrict ourselves below to a certain subalgebra of $C_n$, consisting of the so-called *convergent formal power series:* That is, following, as we have said already, the terminology of H. Grauert - R. Remmert [1: p.15ff], we set

$$E \equiv E_n := \mathbf{C}\langle t_1, \ldots, t_n\rangle = \bigcup_{\lambda \in \mathbf{R}^n_+} B_\lambda\langle t_1, \ldots, t_n\rangle, \tag{12.34}$$

such that one further defines (cf. (12.33))

$$B_\lambda\langle t_1, \ldots, t_n\rangle := \{f \in C_n : \|f\|_\lambda < \infty\}, \tag{12.35}$$

for any $\lambda \equiv (\lambda_1, \ldots, \lambda_n) \in \mathbf{R}^n_+$, where we also set

$$\|f\|_\lambda := \sum_{\nu \in \mathbf{N}^n} |\alpha_\nu| \lambda^\nu, \tag{12.36}$$

with $\nu \equiv (\nu_1, \ldots, \nu_n) \in \mathbf{N}^n$ and $\lambda^\nu := \lambda_1^{\nu_1} \cdots \lambda_n^{\nu_n}$, for every *formal power series,*

$$f = \sum_{\nu \in \mathbf{N}^n} \alpha_{\nu_1 \cdots \nu_n} t_1^{\nu_1} \cdots t_n^{\nu_n} \in C_n. \tag{12.37}$$

Thus,

(12.38) we call $E$, as given above by (12.34), the *algebra of convergent formal power series* (in $n$ variables, $n \in \mathbf{N}$), being again a *unital commutative* $\mathbf{C}$*-algebra,* a subalgebra in fact of $C_n$, as in (12.33).

Furthermore, each one of the sets $B_\lambda \subseteq C_n$, $\lambda \in \mathbf{R}^n_+$, as in (12.34) (cf. also (12.35)), is, in effect, a (unital commutative) *Banach* $\mathbf{C}$*-algebra,* in the norm given by (12.36). The same algebras contain the polynomial algebra (12.30) as a *dense subalgebra* (see loc. cit.; p.16, Satz 1, *infra*). Furthermore,

(12.39) $E$ is a *local* $\mathbf{C}$*-algebra, its only maximal ideal* $\mathfrak{m}$ being the set of those elements of $E$, which vanish at the *zero-element,* viz., equivalently, those elements in $E$, *whose constant term is zero.* Yet, one obtains

(12.39.1) $$E/\mathfrak{m} \cong \mathbf{C},$$

within an *isomorphism of* $\mathbf{C}$*-algebras.*

(See loc. cit.; p.27, Satz 1, *infra,* and p.210, §2). ▮

In this context, a crucial ingredient for what concerns us in the sequel, is the existence of *partial derivatives in $E$, with respect to the variables* (indeterminates) $t_i$, $1 \leq i \leq n$; all these are *formally defined,* while one further obtains, yet *formally,* an *indefinite integral,* for every $f \in E$ (ibid., p.9ff). Furthermore, given the $\mathbf{C}$*-algebra* $E$, as in (12.34), one still defines the *$E$-module of 1st order differentials* (or *1-forms,* or even *pfaffian forms,* yet just *pfaffians*) of $E$, via the aforementioned partial derivatives and the corresponding differentials of the (independent) variables, by analogy with the classical case (of $\mathcal{C}^\infty$-functions). We denote the $E$-module, under discussion, by

(12.40) $$\Omega(E) \equiv \Omega^1,$$

while one further defines;

(12.41) $$\Omega^p := \bigwedge_1^p \Omega^1, \quad \textit{with} \;\; p \geq 2,$$

along with the pertinent *(exterior) differentials,* by following the standard case. Thus, one obtains the next basic result;

given the $\mathbf{C}$-*algebra* $E$, as defined by (12.34) above, one has the following *exact sequence of* $\mathbf{C}$-*vector spaces*

(12.42) (12.42.1)
$$0 \longrightarrow \mathbf{C} \longrightarrow E \xrightarrow{d} \Omega^1 \longrightarrow \Omega^2 \longrightarrow \cdots$$
$$\cdots \longrightarrow \Omega^p \longrightarrow \Omega^{p+1} \longrightarrow \cdots.$$

(See loc. cit.; p.176, Satz 10). ∎ Here one also applies the fact that *the partial derivatives and* the respective *integration* are *continuous* ($\mathbf{C}$-*linear*) *maps,* with respect to an appropriate *topology on* $E$*,* the so-called *sequential topology* (or even *sequence topology; "Folgentopologie"*. See loc. cit. p.58, and p.59, Satz 6). Yet, the latter topology makes $E$ into a *Fréchet locally convex* (topological) $\mathbf{C}$-*algebra* with a *unique closed maximal ideal* (given as in (12.39). See also loc. cit. p.66, Satz 7, p.74, Satz 7, and p.82, Satz 1, along with p.85, Satz 6, the Korollar and Satz 7). Thus, $E$ is, in particular, a *Fréchet (locally convex) Q-algebra* [TA].

We call (12.42.1) the *Poincaré-de Rham (exact) sequence,* which thus is associated with the given algebra $E$, as in (12.34).

Thus, as a consequence of the previous discussion, we conclude that one can consider, for the case at hand, the triple

(12.43)
$$(E, d, \Omega^1),$$

as defined above, as a *differential triad* (over the *one-point space*(!) $\{\mathfrak{m}\}$, as in (12.39). Hence, one has here a *trivial sheafification over the spectrum of* $E$). So in conclusion;

(12.44) we can have here, once more, *with respect to the above triple* $(E, d, \Omega^1)$, as in (12.43), we call the latter, the *Grauert-Remmert differential triad,* the *differential geometric apparatus,* that has been developed so far in the preceding.

Thus, here again, one has the possibility of *doing differential geometry* (or employing, at least, its language) *not by* means of $C^\infty$-*functions* (hence, when disposing a usual $C^\infty$-manifold context), *but through another appropriate* (eventually more profitable or, at least, less expensive!) *algebra of functions*!

# BIBLIOGRAPHY

(The items of the bibliography are given, for each one of the two Volumes of this treatise, separately; thus, for convenience, certain issues of the following list are repeated from the references of Volume I)

**S. I. Andersson**

1. *Vector bundle connections and liftings of partial differential operators.* In "Differential Geometric Methods in Mathematical Physics". Lecture Notes in Mathematics, No 905, pp. 119-132. Springer-Verlag, Berlin, 1982.

**A. Asada**

1. *Algebraic cohomology of loop spaces.* J. Fac. Sci. Shinshu Univ. 4 (1969), 1-21.
2. *Connections on differential operators.* J. Fac. Sci. Shinshu Univ. 13 (1978), 87-102.

**M. F. Atiyah**

1. *Complex analytic connections in fiber bundles.* Trans. Amer. Math. Soc. 85(1957), 181-207.
2. *K-Theory.* Benjamin, New York, 1967.
3. *Elliptic Operators in Compact Groups.* Lecture Notes in Mathematics, No 401. Springer-Verlag, Berlin, 1974.
4. *Geometry of Yang–Mills Fields.* Accad. Naz. Lincei (Scuola Normale Superiore), Pisa, 1979.

**C. Bartocci, U. Bruzzo, D. Hernández-Ruipérez**

1. *The Geometry of Supermanifolds.* Kluwer Academic Publs. Dordrecht, 1991.

**J. Bernstein – V. Lunts**

1. *Equivalent Sheaves and Functors.* Lecture Notes in Mathematics, No 1578. Springer-Verlag, Berlin, 1994.

**E. Binz – J.Śniatycki – H. Fischer**

1. *Geometry of Classical Fields.* North-Holland, Amsterdam, 1988.

**R. Bkouche**

1. *Couples spectraux et faisceaux associés. Applications aux anneaux de fonctions.* Bull. soc. math. France 98(1970), 253-295.

**R. Bott**

1. *Lectures on characteristic classes and foliations.* In "Lectures on Algebraic and Differential Topology". Lecture Notes in Mathematics, No 279, pp. 1-80. Springer-Verlag, Berlin, 1972.

**N. Bourbaki**

1. *Théorie des ensembles, Chap. 3 (2e éd.).* Herman, Paris, 1967.
2. *Algèbre, Chap. 1–3.* Hermann, Paris, 1970.
3. *Algèbre, Chap. 4–5 (2e éd.)* Hermann, Paris, 1959.
4. *Commutative Algebra.* Addison-Wesley, Reading, Massachusetts, 1972.

**J.P. Brasselet – G. Hector – M. Saralegi**

1. *Théorème de de Rham pour les variétés stratifiées.* Ann. Global Annal. Geom. 9 (1991), 211-243.

**G. E. Bredon**

1. *Sheaf Theory.* McGraw-Hill, New York, 1967.

**R. M. Brooks**

1. *Partitions of unity in F-algebras.* Math. Ann. 177(1968), 265-272.
2. *On Arens' "Weierstrass products" paper.* Math. Ann. 190(1971), 329-337.

**J.-L. Brylinski**

1. *Loop Spaces, Characteristic Classes and Geometric Quantization.* Birkhäuser, Boston, 1993.

**L. Chambadal – J. L. Ovaert**

1. *Algèbre linéaire et algèbre tensorielle.* Dunod, Paris, 1968.

**K.-T. Chen**

1. *Algebras of iterated path integrals and fundamental groups.* Trans. Amer. Math. Soc. 156(1971), 359-379.
2. *Iterated integrals of differential forms and loop space homology.* Ann. Math. 97(1973), 217-246.
3. *Iterated integrals, fundamental groups and covering spaces.* Trans. Amer. Math. Soc. 206(1975), 83-98.
4. *Iterated path integrals.* Bull. Amer. Math. Soc. 8(1977), 115-137.

**Y. Choquet-Bruhat C. De Witt-Morette M. Dillard-Bleick**
1. *Analysis, Manifolds and Physics (Revised Ed.).* North-Holland, Amsterdam, 1982.

**E. Cioranu-Verona**
1. *Funcţii şi forme pe varietăţi generalizate.* St. Cerc. Mat. 26(1974), 133-143. [Romanian]

**A. Colojoăra**
1. *Algèbre symmetrique du dual d' un espace nucléaire – DF.* Rev. roum. math. pures appl. 23(1978), 1317-1339.
2. *Sur le théorème de Poincaré – Birkhoff – Witt.* Rev. roum. math. pures appl. 25(1980), 1327-1347.

**M. Cornalba – P. Griffiths**
1. *Analytic cycles and vector bundles on noncompact algebraic varieties.* Invent. Math. 26(1975), 402-406.

**M. Crampin – F. A. E. Pirani**
1. *Applicable Differential Geometry.* Cambridge Univ. Press, Cambridge, 1986.

**C. W. Curtis**
1. *Linear Algebra. An Introductory Approach.* Springer-Verlag, New York, 1984.

**P. de Bartolomeis**
1. *Algebre di Stein nel case reale.* Rend. Accad. Naz. dei Lincei 40 [Quaranta] 5 (1) (2) (1975) (1976); (1977), 105-144.

**P. Deligne**
1. *Équations différentielles à points singuliers réguliers.* Lecture Notes in Mathematics, No 163. Springer-Verlag, Berlin, 1970.

**R. Dijkgraaf – E. Witten**
1. *Topological gauge theories and group cohomology.* Comm. Math. Phys. (in press).

**B. A. Dubrovin – A. T. Fomenko – S. P. Novikov**
1. *Modern Geometry – Methods and Applications. Part I. The Geometry of Surfaces, Transformation Groups, and Fields.* Springer-Verlag, New York, 1984.

**J. Dugundji**
1. *Topology.* Allyn and Bacon, Boston, 1966.

**T. Eguchi – P.B. Gilkey – A.J. Hanson**

1. *Gravitation, Gauge Theories and Differential Geometry.* Phys. Rep. 66 (1980), 213-393.

**D. Eisenbud**

1 *Commutative Algebra with a View Toward Algebraic Geometry.* Springer-Verlag, New York, 1995.

**C. Farsi**

1. *Index Theory and Positive Scalar Curvature for Orbifolds.* Ph. D. Thesis, Univ. of Maryland, College Park, MD, 1989.
2. *K-theoretical index theorems for orbifolds.* J. Math. Oxford 43 (1992), 183-200.

**H.R. Fischer**

1. *Differentialkalkül für nicht-metrische Strukturen II: Differentialformen.* Arch. Math. 8 (1957), 428-443.

**F.J. Flaherty**

1. *The de Rham theorem for function spaces.* J. Math. Mech. 19 (1969), 317-320.
2. *A differentiable structure for* Diff$(X)$ (manuscript)

**M. Fragoulopoulou**

1. *Uniqueness of topology for semisimple LFQ-algebras.* Proc. Amer. Math. Soc. 117 (1993), 963-969.

**O. Forster**

1. *Funktionentheoretische Hilfsmittel in der Theorie der kommutativen Banach-Algebren.* Jber. Deutsch. Math. -Verein. 76 (1974), 1-17.

**G. Galanis**

1. *Differential Geometry of Projective Limits of Fiber Bundles.* Doctoral Dissertation, Univ. of Athens, 1995. [Greek]

**I.M. Gelfand – D. Raikov – G. Shilov**

1. *Commutative Normed Rings.* Chelsea Publ. Co. Bronx, New York, 1964.

**M. Göckeler – T. Schücker**

1. *Differential Geometry, Gauge Theories, and Gravity.* Cambridge Univ. Press, Cambridge, 1987.

**H. Grauert**

1. *Charakterisierung der holomorph-vollständigen Räumen.* Math. Ann. 129 (1955), 233-259.
2. *Holomorphe Funktionen mit Werten in komplexen Lieschen Gruppen.* Math. Ann. 133 (1957), 450-472.

**H. Grauert – R. Remmert**

1. *Analytische Stellenalgebren.* Springer-Verlag, Berlin, 1971.
2. *Theory of Stein Spaces.* Springer - Verlag, Berlin, 1979.

**W. Greub**

1. *Multilinear Algebra (2nd Ed.).* Springer-Verlag, New York, 1978.

**P. Griffiths – J. Adams**

1 *Topics in Algebraic and Analytic Geometry.* Princeton Univ. Press, Princeton, New Jersey, 1974.

**A. Grothendieck**

1 *Produits tensoriels topologiques et espaces nucléaires.* Memoirs Amer. Math. Soc., no 16 (1955).

**R. C. Gunning**

1. *Lectures on Riemann Surfaces.* Princeton Univ. Press, Princeton, New Jersey, 1966.
2. *Lectures on Vector Bundles over Riemann Surfaces,* Princeton Univ. Press, Princeton, New Jersey, 1967.

**R. I. Hadjigeorgiou**

1. *Spectral Geometry of Topological Algebras.* Doctoral Dissertation, Univ. of Athens, 1995. [Greek]
2. *Choquet boundaries in topological algebras.* Comment. Math. 36(1996), 131-148.

**R. Hartshorne**

1. *Algebraic Geometry.* Springer - Verlag, New York, 1977.

**G. Hector – M. Saralegi**

1. *Intersection cohomology of $S^1$-actions.* Trans. Amer. Math. Soc. 338 (1993), 263-288.

**M. Heller**

1. *Algebraic foundations of the theory of differential spaces.* Demonstr. Math. 24 (1991), 349-364.
2. *Einstein algebras and general relativity.* Intern. J. Theor. Phys. 31 (1992), 277-288.
3. *Generalized Friedman's equation and its singularities.* Acta Cosmol. 19 (1993), 23-33.
4. *Structured spaces and their application to relativistic physics.* J. Math. Phys. 36 (1995), 3644-3662.
5. *Sheaves of Einstein algebras.* Intern. J. Theor. Phys. 34 (1995), 387-398.

**E. Hewitt – K. A. Ross**

1. *Abstract Harmonic Analysis II.* Springer-Verlag, Berlin, 1970.

**F. Hirzebruch**

1. *Topological Methods in Algebraic Geometry.* Springer-Verlag, Berlin, 1978.

**L. Hörmander**

1. *An Introduction to Complex Analysis in Several Variables.* North-Holland, Amsterdam, 1973.

**T. W. Hungerford**

1. *Algebra.* Springer-Verlag, Berlin, 1974.

**D. Husemoller**

1. *Fiber Bundles (2nd Ed.).* Springer-Verlag, New York, 1975.

**M. Karoubi**

1. *K-Theory.* Springer-Verlag, Berlin, 1978.

**M. Karoubi – G. Leruste**

1. *Algebraic Topology via Differential Geometry.* Cambridge Univ. Press, Cambridge, 1987.

**Y. Katznelson**

1. *An Introduction to Harmonic Analysis.* Wiley, New York, 1968.

**L. Kaup – B. Kaup**

1. *Holomorphic Functions of Several Variables.* W. de Gruyter, Berlin, 1983.

**T. Kawasaki**

1. *The signature theorem for V-manifolds.* Topology 17 (1978), 75-83.
2. *The Riemann – Roch theorem for complex V-manifolds.* Osaka J. Math 16 (1979), 151-159.
3. *The index of elliptic operators over V-manifolds.* Nagoya Math. J. 84 (1981), 135-157.

**J. L. Kelley**

1. *Topology.* Springer-Verlag, New York, 1975.

**A. T. Kitchen**

1. *Algebraic Alexander-Spanier Cohomology.* Ph. D. Thesis, Univ. of Rochester, Rochester, New York, 1971.

**S. Kobayashi**

1. *Differential Geometry of Complex Vector Bundles.* Princeton Univ. Press, Princeton, New Jersey, 1987.
2. *Manifolds over function algebras and mapping spaces.* Tôhoku Math. J. 41 (1989), 263-282.

**S. Kobayashi – K. Nomizu**

1. *Foundations of Differential Geometry, Vol. I.* Interscience Publ., New York, 1963.
2. *Foundations of Differential Geometry, Vol. II.* Interscience Publ., New York, 1969.

**R. Kochendörffer**

1. *Introduction to Algebra.* Wolters-Noordhoff, Groningen, 1972.

**A. A. Korobov**

1. *Cohomology of algebras with unit.* Soviet Math., Doklady 10 (1969), 17-20.

**B. Kostant**

1. *Quantization and unitary representations.* In "Lectures in Modern Analysis and Applications, II". Lecture Notes in Mathematics, No 170, pp. 87-208. Springer-Verlag, Berlin, 1970.

**A. I. Kostrikin**

1. *Introduction to Algebra.* Springer-Verlag, New York, 1982.

**J. L. Koszul**
1. *Sur certaines groupes de transformations de Lie.* Coll. Geom. Diff., Strasbourg, 1953.
2. *Lectures on Fiber Bundles and Differential Geometry.* Tata Institute, Bombay, 1960.

**G. Köthe**
1. *Topological Vector Spaces I.* Springer-Verlag, Berlin, 1969.

**S. G. Krantz**
1. *Function Theory of Several Complex Variables.* J. Wiley, New York, 1982.

**E. Kunz**
1. *Kähler Differentials.* Vieweg, Braunschweig, 1986.

**S. Lang**
1. *Algebra (2nd Ed.).* Addison - Wesley, Reading, Massachusetts, 1984.

**K. Lønsted**
1. *An algebraization of vector bundles on compact manifolds.* J. Pure Appl. Algebra 2 (1972), 193-207.
2. *Vector bundles over finite CW-complexes are algebraic.* Proc. Amer. Math. Soc. 38 (1973), 27-31.

**A. Mallios**
1. *Vector bundles and K-theory over topological algebras.* J. Math. Anal. Appl. 92 (1983), 452-506.
2. *Homotopy invariants of the spectrum of a topological algebra.* J. Math. Anal. Appl. 101 (1984), 297-307.
3. *Hermitian K-theory over topological ∗-algebras.* J. Math. Anal. Appl. 106 (1985), 454-539.
4. *Topological Algebras. Selected Topics.* North - Holland, Amsterdam, 1986.
5. *Continuous vector bundles over topological algebras, II.* J. Math. Anal. Appl. 132 (1988), 401-423.
6. *On an abstract form of Weil's integrality theorem.* Note Mat. 12 (1992), 167-202.
7. *On geometric topological algebras.* J. Math. Anal. Appl. 173 (1993), 301-322.
8. *The de Rham-Kähler complex of the Gel'fand sheaf of a topological algebra.* J. Math. Anal. Appl. 175 (1993), 143-168.
9. *Topological Algebras. Localizations and Extensions.* (Book, in preparation).

**Yu. I. Manin**

1. *Gauge Field Theory and Complex Geometry.* Springer - Verlag, Berlin, 1988.

**C. D. Marshall**

1. *The de Rham cohomology of subcartesian spaces.* J. Diff. Geom. 10 (1975), 575-588.

**J. W. Milnor – J. D. Stasheff**

1. *Characteristic classes.* Princeton Univ. Press, Princeton, New Jersey, 1974.

**M. A. Mostow**

1. *The differentiable space structures of Milnor classifying spaces, simplicial complexes, and geometric realizations.* J. Diff. Geom. 14 (1979), 255-293.

**L. Nachbin**

1. *Elements of Approximation Theory.* D. van Nostrand, Princeton, New Jersey, 1967.
2. *Introduction to Functional Analysis: Banach Spaces and Differential Calculus.* Marcel Dekker, New York, 1981.

**M. Nakahara**

1. *Geometry, Topology and Physics.* Adam Hilger, Bristol, 1990.

**M. A. Naĭmark**

1. *Normed Algebras.* Wolters-Noordhoff, Groningen, 1972.

**M. E. Novodvorskiĭ**

1. *On certain homotopic invariants of the space of maximal ideals.* Mat. Zametki 1 (1967), 487-494. [Russian]

**M. Papatriantafillou**

1. *On the de Rham's complex in abstract spaces* (manuscript).
2. *The de Rham's complex in function spaces* (to appear).

**T. Petrie – J. Randall**

1. *Connections, Definite forms, and Fourmanifols.* Oxford Univ. Press, Oxford, 1990.

**W. A. Poor**

1. *Differential Geometric Structures.* McGraw-Hill, New York, 1981.

**E. Prugovečki**

1. *Quantum Mechanics in Hilbert Space (2nd Ed.).* Academic Press, New York, 1981.
2. *Quantum Geometry. A Framework for Quantum General Relativity.* Kluwer, Dordrecht, 1992.

**M. Puta**

1. *Hamiltonian Mechanical Systems and Geometric Quantization.* Kluwer, Dordrecht, 1993.

**H. Reiter**

1. *Classical Harmonic Analysis and Locally Compact Groups.* Oxford Univ. Press, Oxford, 1968.

**M. Rosenfeld**

1. *Commutative F-algebras.* Pacific J. Math. 16 (1966), 159-166.

**J.J. Rotman**

1. *An Introduction to Algebraic Topology.* Springer-Verlag, New York, 1988.

**M. Saralegi – G. Hector**

1. *Formes différentielles d' intersection: un théorème de de Rham pour l' homologie d' intersection des préstratifications abstraites.* C. R. Acad. Sci. Paris 308 (1989), 25-28.

**E. Sasin**

1. *On some exterior algebra of the differential forms over a differential space.* Demonstr. Math. 19 (1986), 1063-1075.
2. *The de Rham cohomology of differential spaces.* Demonstr. Math. 22 (1989), 249-270.
3. *Differential spaces and singularities in differential space-times.* Demonstr. Math. 24 (1991), 601-634.

**W. Sasin – K. Spallek**

1. *Gluing of differentiable spaces and applications.* Math. Ann. 292 (1992), 85-102.

**I. Satake**

1. *On a generalization of the notion of manifold.* Proc. Nat. Acad. Sci. 42 (1956), 359-363.

**H.H. Schaefer**

1. *Topological Vector Spaces.* Springer-Verlag, New York, 1980.

**J. Schwartz**

1. *De Rham's theorem for arbitrary spaces.* Amer. J. Math. 77 (1995), 29-44.

**S. A. Selesnick**

1. *Watts cohomology for a class of Banach algebras and the duality of compact abelian groups.* Math. Z. 130 (1973), 313-323.
2. *Line bundles and harmonic analysis on compact groups.* Math. Z. 146 (1976), 53-67.
3. *Second quantization, projective modules, and local gauge invariance.* Intern. J. Theor. Phys. 22 (1983), 29-53.
4. *Geometrical symmetry breaking.* Nuovo Cimento 83A (1984), 175-190.
5. *Correspondence principle for the quantum net.* Intern. J. Theor. Phys. 30 (1991), 1273-1292.

**R. Sikorski**

1. *Abstract covariant derivative.* Colloq. Math. 18 (1967), 251-271.
2. *Differential modules.* Colloq. Math. 24 (1971), 45-70.

**D. J. Simms – N. M. J. Woodhouse**

1. *Lectures on Geometric Quantization.* Lecture Notes in Physics, No 53. Springer-Verlag, Berlin, 1976.

**P. Shanahan**

1. *The Atiyah-Singer Index Theorem. An Introduction.* Lecture Notes in Mathematics, No 638. Springer-Verlag, Berlin, 1978.

**L. Smith**

1. *Linear Algebra (2nd Ed.).* Springer-Verlag, New York, 1984.

**J. W. Smith**

1. *The de Rham theorem for general spaces.* Tôhoku Math. J. 18 (1966), 115-137.

**J. Śniatycki**

1. *Prequantization of charge.* J. Math. Phys. 15 (1974), 619-620.
2. *Geometric Quantization and Quantum Mechanics.* Springer-Verlag, New York, 1980.

**J.-M. Souriau**

1. *Structure de systèmes dynamiques.* Dunod, Paris, 1970.

**K. Spallek**

1. *Differezierbare Raüme.* Math. Ann. 180 (1969), 269-296.

**E. H. Spanier**

1 *Algebraic Topology.* Springer - Verlag, New York, 1966.

**E. L. Stout**

1. *The theory of Uniform Algebras.* Bogden & Quigley, Inc., Tarrytown-on-Hudson, New York, 1971.

**J. N. Tavares**

1. *Chen integrals, generalized loops and loop calculus* (preprint) (1993).

**E. Vassiliou**

1. *Connections on principal sheaves* (manuscript).
2. *On Mallios $\mathcal{A}$-connections as connections on principal sheaves.* Note Mat. (in press).
3. *Transformations of sheaf connections* (to appear).
4. *From principal connections to connections on principal sheaves.* Anal. Stiint. Univ. "Al. I. Cusa" 42(1996), 149-160.

**A. Verona**

1. *Le théorème de de Rham pour les préstratifications abstraites.* C. R. Acad. Sci. Paris 273 (1971), 886-889.

**M. E. Verona**

1. *A de Rham theorem for generalized manifolds.* Proc. Edinburg Math. Soc. 22 (1979), 127-135.

**J. Wagner**

1. *Faisceau structural associé à une algèbre de Banach.* Sém. P. Lelong (1968/69), Lecture Notes in Mathematics, No 116. pp. 164-167. Springer-Verlag, Berlin, 1970.

**S. Warner**

1. *Polynomial completeness in locally multiplicatively-convex algebras.* Duke Math. J. 23 (1956), 1-11.

**C. E. Watts**

1. *Alexander-Spanier cohomology and rings of continuous functions.* Proc. Nat. Acad. Sci. USA 54 (1965), 1027-1028.

**A. Weil**

1. *Introduction à l' étude des variétés kählériennes (nouv. éd.).* Hermann, Paris, 1971.

**R. O. Wells**

1. *Differential Analysis on Complex Manifolds.* Springer - Verlag, New York, 1980.

**K. Whyburn**

1. *Differentiable cohomology on locally compact groups.* Proc. Nat. Acad. Sci. USA 67 (1970), 45-51.

**N. Woodhouse**

1. *Geometric Quantization.* Oxford Univ. Press, Oxford, 1980.

# Notational Index

# Subject Index

(Notions already appeared in Volume I, but still applied in the present volume, are not included in the following list)

Other *Mathematics and Its Applications* titles of interest:

---

V.I. Istratescu: *Fixed Point Theory. An Introduction.* 1981, 488 pp.
*out of print,* ISBN 90-277-1224-7

A. Wawrynczyk: *Group Representations and Special Functions.* 1984, 704 pp.
ISBN 90-277-2294-3 (pb), ISBN 90-277-1269-7 (hb)

R.A. Askey, T.H. Koornwinder and W. Schempp (eds.): *Special Functions: Group Theoretical Aspects and Applications.* 1984, 352 pp. ISBN 90-277-1822-9

A.V. Arkhangelskii and V.I. Ponomarev: *Fundamentals of General Topology. Problems and Exercises.* 1984, 432 pp. ISBN 90-277-1355-3

J.D. Louck and N. Metropolis: *Symbolic Dynamics of Trapezoidal Maps.* 1986, 320 pp. ISBN 90-277-2197-1

A. Bejancu: *Geometry of CR-Submanifolds.* 1986, 184 pp.
ISBN 90-277-2194-7

R.P. Holzapfel: *Geometry and Arithmetic Around Euler Partial Differential Equations.* 1986, 184 pp. ISBN 90-277-1827-X

P. Libermann and Ch.M. Marle: *Sympletic Geometry and Analytical Mechanics.* 1987, 544 pp. ISBN 90-277-2438-5 (hb), ISBN 90-277-2439-3 (pb)

D. Krupka and A. Svec (eds.): *Differential Geometry and its Applications.* 1987, 400 pp. ISBN 90-277-2487-3

Shang-Ching Chou: *Mechanical Geometry Theorem Proving.* 1987, 376 pp.
ISBN 90-277-2650-7

G. Preuss: *Theory of Topological Structures. An Approach to Categorical Topology.* 1987, 318 pp. ISBN 90-277-2627-2

V.V. Goldberg: *Theory of Multicodimensional (n+1)-Webs.* 1988, 488 pp.
ISBN 90-277-2756-2

C.T.J. Dodson: *Categories, Bundles and Spacetime Topology.* 1988, 264 pp.
ISBN 90-277-2771-6

A.T. Fomenko: *Integrability and Nonintegrability in Geometry and Mechanics.* 1988, 360 pp. ISBN 90-277-2818-6

L.A. Cordero, C.T.J. Dodson and M. de Leon: *Differential Geometry of Frame Bundles.* 1988, 244 pp. *out of print,* ISBN 0-7923-0012-2

E. Kratzel: *Lattice Points.* 1989, 322 pp. ISBN 90-277-2733-3

E.M. Chirka: *Complex Analytic Sets.* 1989, 396 pp. ISBN 0-7923-0234-6

Kichoon Yang: *Complete and Compact Minimal Surfaces.* 1989, 192 pp.
ISBN 0-7923-0399-7

A.D. Alexandrov and Yu.G. Reshetnyak: *General Theory of Irregular Curves.* 1989, 300 pp. ISBN 90-277-2811-9

Other *Mathematics and Its Applications* titles of interest:

---

J.-F. Pommaret: *Partial Differential Equations and Group Theory. New Perspectives for Applications*. 1994, 473 pp. ISBN 0-7923-2966-X

Kichoon Yang: *Complete Minimal Surfaces of Finite Total Curvature*. 1994, 157 pp. ISBN 0-7923-3012-9

N.N. Tarkhanov: *Complexes of Differential Operators*. 1995, 414 pp. ISBN 0-7923-3706-9

L. Tamássy and J. Szenthe (eds.): *New Developments in Differential Geometry*. 1996, 444 pp. ISBN 0-7923-3822-7

W.C. Holland (ed.): *Ordered Groups and Infinite Permutation Groups*. 1996, 255 pp. ISBN 0-7923-3853-7

K.L. Duggal and A. Bejancu: *Lightlike Submanifolds of Semi-Riemannian Manifolds and Applications*. 1996, 308 pp. ISBN 0-7923-3957-6

D.N. Kupeli: *Singular Semi-Riemannian Geometry*. 1996, 187 pp. ISBN 0-7923-3996-7

L.N. Shevrin and A.J. Ovsyannikov: *Semigroups and Their Subsemigroup Lattices*. 1996, 390 pp. ISBN 0-7923-4221-6

C.T.J. Dodson and P.E. Parker: *A User's Guide to Algebraic Topology*. 1997, 418 pp. ISBN 0-7923-4292-5

B. Rosenfeld: *Geometry of Lie Groups*. 1997, 412 pp. ISBN 0-7923-4390-5

A. Banyaga: *The Structure of Classical Diffeomorphism Groups*. 1997, 208 pp. ISBN 0-7923-4475-8

A.N. Sharkovsky, S.F. Kolyada, A.G. Sivak and V.V. Fedorenko: *Dynamics of One-Dimensional Maps*. 1997, 272 pp. ISBN 0-7923-4532-0

A.T. Fomenko and S.V. Matveev: *Algorithmic and Computer Methods for Three-Manifolds*. 1997, 346 pp. ISBN 0-7923-4770-6

A. Mallios: *Geometry of Vector Sheaves*. An Axiomatic Approach to Differential Geometry.
Volume I: Vector Sheaves. General Theory. 1998, 462 pp. ISBN 0-7923-5004-9
Volume II: Geometry. Examples and Applications. 1998, 462 pp. ISBN 0-7923-5005-7
(Set of 2 volumes: 0-7923-5006-5)

GPSR Compliance

*The European Union's (EU) General Product Safety Regulation (GPSR) is a set of rules that requires consumer products to be safe and our obligations to ensure this.*

*If you have any concerns about our products, you can contact us on ProductSafety@springernature.com*

In case Publisher is established outside the EU, the EU authorized representative is:

Springer Nature Customer Service Center GmbH
Europaplatz 3
69115 Heidelberg, Germany

**Batch number: 10823957**

Printed by Printforce, the Netherlands